2006 5th International Power Electronics and Motion Control Conference

Shanghai, China
13-16 August 2006

Volume 1 of 4

IEEE Catalog Number: 06EX1405
ISBN: 1-4244-0448-7

Copyright © 2006 by The Institute of Electrical and Electronics Engineers, Inc.
All Rights Reserved

Copyright and Reprint Permissions: Abstracting is permitted with credit to the source. Libraries are permitted to photocopy beyond the limit of U.S. copyright law for private use of patrons those articles in this volume that carry a code at the bottom of the first page, provided the per-copy fee indicated in the code is paid through Copyright Clearance Center, 222 Rosewood Drive, Danvers, MA 01923.

For other copying, reprint or republications permission, write to IEEE Copyrights Manager, IEEE Operations Center, 445 Hoes Lane, Piscataway, New Jersey USA 08854. All rights reserved.

IEEE Catalog Number: 06EX1405

ISBN: 1-4244-0448-7

Library of Congress: 2006925601

Additional Copies of This Publication Are Available from:

IEEE Service Center
445 Hoes Lane
Piscataway, NJ 08854
IEEE Service Center
445 Hoes Lane
Piscataway, NJ 08854
Phone: (800) 678-IEEE
        (732) 981-1393
Fax: (732) 981-9667
E-mail: customer-service@ieee.org

2006 5th International Power Electronics and Motion Control Conference

Shanghai, China
13-16 August 2006

IEEE Catalog Number: CFP06792-POD
ISBN: 978-1-42440-448-3

Table of Contents

Design Challenges For Distributed Power Systems 1
Fred C. Lee, Ming Xu, Shuo Wang, Bing Lu

A Smarter Grid for Improving System Reliability and Asset Utilization 16
D. Divan, H. Johal

Medium-Voltage Power Conversion Systems in the Next Generation 23
Hirofumi Akagi, Shigenori Inoue

Modern Electrical Drives: Design and Future Trends 31
R. W. De Doncker

Power Semiconductors development trends 39
L. Lorenz

Power Electronics in Wind Turbine Systems 46
F. Blaabjerg, Z. Chen, R. Teodorescu, F. Iov

Sustainable Energy and Mobility, and Challenges to Power Electronics 57
C.C.Chan

Wind farms with increased transient stability margin provided by a STATCOM 63
Marta Molinas, Jon Are Suul, Tore Undeland

A New Super Junction LDMOS with N+-Floating Layer 70
Baoxing Duan, Bo Zhang, Zhaoji Li

Unified Power Flow Controller: Comparison of Two Advanced Control Schemes and Performance Analysis for Power Flow Control 74
Liu Liming, Zhu Pengcheng, Kang Yong, Chen Jian

A New Analytical Model for the Surface Electrical Field Distribution of Double RESURF LDMOS 79
Qi Li, Zhaoji Li

A Novel Centralized HID Ballast System with Power-Bus 83
Xiaodong Lu, Bo Yang, Jiande Wu, Xiangning He

Research on a Novel Structure of SiGeC/Si Heterojunction Power Diodes 88
Liu Jing, Gao Yong, Ma Li

Gate driving of high power IGBT by wireless transmission 92
Stéphane Bréhaut, François Costa

The Characteristics of Thyristor Controlled Reactance Series Compensation by Adjustable Coupling 97
Guo-rong Zhu, Min-zu Li, Yong Kang

Dual-Side Cooled Novel IPM and Improved Capability of Inverter for Elevated-Temperature Operations 102
Jie (Jay) Chang, Changming Liao

An Improved Current-Doubler with Coupled Inductors 108
T.-F. Wu, C.-T. Tsai, W.-C. Lin, Y.-M. Chen

Monolithic Integration of Trench Power JFET with Schottky Diode 113
Yang Gao, Jie Chen, Alex. Q. Huang

Sequential Color LED Back-Light Driving System for LCD Panels 117
C.-C. Chen, C.-Y. Wu, P.-C. Lu, Y.-M. Chen, T.-F. Wu

Development of Large Capacity Programmable Harmonic Current Generator Based on Three-phase-four-wire Configuration 122
LIU Tao, ZHUO Fang, CHEN Bo, ZHAI Xi, WANG Zhao-an

A Universal Digital Platform and Software Library for Power Electronic Systems Integration 127
Haibing HU, Tianjun Jin, Wenxi YAO, Zhengyu LU, Zhaoming Qian

Table of Contents

Unipolar SiC Devices - Latest Achievements on the Way to a New Generation of High Voltage Power Semiconductors 132
Peter Friedrichs

Implementation of GA-trained GRNN for Intelligent Fast Charger for Ni-Cd Batteries 137
Panom Petchjatuporn, Noppadol Khaehintung, Khamron Sunat, Phaophak Sirisuk, Wiwat Kiranon

Modelling and Analysis of a Novel Transformer with Ability to Suppress Conducted interference 142
Zongxiang Chen, Pengsheng Ye, Junmin Pan

An Observer-Based Three-Phase Current Reconstruction using DC Link Measurement in PMAC Motors 147
Li Ying, Nesimi Ertugrul

Experiment Research of Chaotic PWM Suppressing EMI in Converter 152
R. Yang, B. Zhang , F.Li, J.J. Jiang

Emitter Size Effect in 4H-SiC BJT 157
Yan Gao, Alex Q. Huang, Sumi Krishnaswami, Anant K. Agarwal, Charles Scozzie

PSIM and SIMULINK Co-simulation for Threelevel Adjustable Speed Drive Systems 161
Zhang Yongchang, Zhao Zhengming, Baihua, Yuan Liqiang, Zhang Haitao

Three-Phase Z-Source AC-AC Converter for Motor Drives 166
Xu-Peng Fang

Construction and Application of Macro Model for ZVS Resonant Mode Controller MC34067 171
Wei Chen, Yilei Gu, Zhengyu Lu, Zhaoming Qian

Optimum Design of Hollow Conductor in Stator Winding for Large Evaporative Hydro-generator 175
Z. Wen , L. Ruan, G. Gu

Rotor Suspension Principle and Decoupling Control for Self-bearing Induction Motors 179
Tengchao Zhang, Huangqiu Zhu, Yuxin Sun

Field Oriented Control of Linear Induction Motor Considering Attraction Force & End-Effects 184
Jianqiang Liu, Fei Lin, Zhongping Yang, Trillion Q. Zheng

Series Resonant High Frequency Link Sine-wave Inverter System Modeling using Sampled Data 189
Jin Xiaoyi, Dong Wei, Sun Xiaofeng, Wu Weiyang

Maximal Power Point Tracking under Speed-Mode Control for Wind Energy Generation System with Doubly Fed Introduction Generator 194
Y. Zhao, X. D. Zou, Y. N. Xu, Y. Kang, J. Chen

Effective Mobility in Nano-Scaled n-MOSFETs 199
Yue-Hua Dai, Jun-Ning Chen, Dao-Ming Ke, Jia-E Sun

Investigation on the Factors Affecting Inrush Current of Transformers Based on Finite Element Modeling 204
M. Reza Feyzi, M. B. B. Sharifian

An Improved Support Vector Machine Method for Harmonic and Inter-harmonic Detecting 209
Ma Li, Liu Kaipei, Lei Xiao

A Common Mode and Differential Mode Integrated EMI Filter 214
Liu Nan, Yang Yugang

Electromagnetism Model and Characteristic Simulation of Novel Claw Pole Generator with Permanent Magnet Outer Rotor 219
Fengge Zhang, Haijun Bai, Shifu Zhang, Hans Pert Gruenberger, Eugen Nolle

An Improved Adaptive Filter for Voltage and Current Reference Extraction 224
A. Abedini, A. Nasiri

Simulation Analysis on Current SVM Algorithm of Matrix Rectifier 229
Xi-jun Yang, Peng-sheng Ye, Xiang Liu, Xing-hua Yang, Jian-quan Wang, Luan-guo Zhang

Table of Contents

Study of Measurement Approach of Loop Gain of Converter 236
Weiping Zhang, Yunpeng Chen, Yuanchao Liu, Dongyan Zhang, Zheng Meng

A Stand-Alone Hybrid Generation System Combining Solar Photovoltaic and Wind Turbine with Simple Maximum Power Point Tracking Control 242
Nabil A. Ahmed, Masafumi Miyatake

Design Optimization of Industrial Motor Drive Power Stage Using Genetic Algorithms 249
F. Wang, W. Shen, D. Boroyevich, S. Ragon, V. Stefanovic, M. Arpilliere

FEM Based Simulation of a Permanent Magnet Synchronous Motor Performance Characteristics 254
L. Petkovska, G. Cvetkovski

Analytical Modeling of Semiconductor Losses in Matrix Converters 259
Bingsen Wang, Giri Venkataramanan

Nonlinear Robust Sliding Mode Control for PM Linear Synchronous Motors 267
Xi Zhang, Junmin Pan

Dynamic Analysis of PWM Switching DC-DC Converters 272
Liu Jian, Wang Yuanbin

A Novel LLC Resonant Converter Topology: Voltage Stresses of All Components in Secondary Side Being Half of Output Voltage 276
Yilei Gu, Zhengyu Lu, Zhaoming Qian

On the hybrid automaton models and control synthesis of a single inductor, double output boost converter 281
Sreekumar C, Vivek Agarwal

Complex Intermittency in Voltage-Mode Controlled Buck Converter 286
Zheng-Ping Li, Yu-Fei Zhou, Jun-Ning Chen

Dual Mode Control Multiphase DC/DC Converter for CPU Power 291
Li-Wei Lin, Chung-Hsing Chang, Huang-Jen Chiu, Shann-Chyi Mou

An Analog Implementation of Pulse-Width-Modulation Based Sliding Mode Controller for DC-DC Boost Converters 296
Siew-Chong Tan, Y. M. Lai, Chi K. Tse

Low Cost Electronic Ballast with Buck Converter as PFC Stage 301
Li Xiangrong, Xu Dianguo, Zhang Xiangjun

A New Converter Architecture for Future Generations of Microprocessors 306
Dodi Garinto

A Combined ZVS Converter with Naturally Sharing Input-Current and High Voltage Gain 311
Linbing Wang, Bo Yang

Matrix Coefficient Polynomial Description Model of DC-DC Converters Based on Switched Linear Systems 316
Yongping Zhang, Bo Zhang, Zongbo Hu, Dongyuan Qiu, Guiping Du

Development of DC-DC Multiple Converter based on Push-pull Forward Topology accomplished 321
Weihao Hu, Yunqing Pei, Zhaoan Wang

Voltage Fed and Current Fed Full Bridge Converter for the Use in Three Phase Grid Connected Fuel Cell Systems 325
M. Mohr, F.-W. Fuchs

Small-Signal Modeling of Asymmetrical Half Bridge Flyback Converter 332
Tso-Min Chen, Chern-Lin Chen

A DSP Based Controller for High Power Dual-Phase DC-DC Converters 337
Xin Guo, Xuhui Wen, Ermin Qiao

v

Table of Contents

Effective Load Resistance; A New Method to Evaluate DC/DC converters Efficiency 342
Alan Elbanhawy

Calculation of Power Loss in Output Diode of a Flyback Switching DC-DC Converter 346
Jiaxin Chen, Jianguo Zhu, Youguang Guo

A Multiple Output Forward Converter Adopting Weighted Time-Sharing Control and Switch-Linear Hybrid Scheme 351
Xiaodong Liu, Songqin Hu, Sizhou Sun

A Novel Soft-Switching PWM Full-Bridge DC/DC Converter with DC Busline Series Switch-Parallel Capacitor Edge Resonant Snubber Assisted by High-Frequency Transformer Leakage Inductor 356
Khairy Fathy, Toshimitsu Doi, Keiki Morimoto, Hyun Woo Lee, Mutsuo Nakaoka

High-Efficiency Cascode Forward Converter of Low Power PEMFC System 361
Jiann-Fuh Chen, Wei-Shih Liu, Ray-Lee Lin, Tsorng-Juu Liang, Ching-Hsiung Liu

Control of Bifurcation by Fuzzy Logic Controller for Current-mode Boost Converters 368
Noppadol Khaehintung, Phaophak Sirisuk, Anantawat Kunakorn

An Improved Three-Level Soft-Switching DC/DC Converter 373
Z. L. Lou, Z. S. Wang

A Novel Soft Switching Bidirectional DC/DC Converter and Design Consideration 378
Ma Gang, Qu Wenlong, Liu Yuanyuan

State-Variable Description and Analysis of a DC-Rail ZVT Inverter Feeding a Permanent Magnet Synchronous Motor 382
Ming Zhengfeng, Zhong Yanru

Analysis, Simulations and Experiments Of A Novel ZVS -ZCS Inverter With Pulse Current Feedback Transformer Auxiliary Commutation 386
Yaogang, Mahamnad Mansoor Khan, Chenchen

A Novel Eddy-Current Based Far-Infrared Rays Radiant Planner Heater using High-Frequency ZVT-PWM Inverter 392
Hisayuki Sugimura, Bishwajit Saha, Hideki Omori, Hyun Woo Lee, Mutsuo Nakaoka

3 Phases-3 Devices AC Voltage Regulator With Quasi-Zero Switching 397
Qianzhi Zhou, Wenhua Hu, Bin Wu

Study on Power Decoupling Control of Three Phase Voltage Source PWM Rectifiers 401
Wang Jiuhe, Yin Hongren, Zhang Jinlong, Li Huade

A Fully Digital Controlled 3KW, Single-Stage Power Factor Correction Converter Based on Full-Bridge Topology 406
HANG Li-jun, YANG Yue-feng, SU Bin, LU Zheng-yu, QIAN Zhao-ming

A New ZVT Power Factor Corrected Three-Phase AC-AC Converter with Single-Phase HF Link 411
T. H. Abdelhamid, A. Sabzali

Simple Bridge-Type AC/DC Converters with Natural Input-Current-Shaper 417
Hsing-Fu Liu, Chih-Yu Wu, Chin Sun, Lon-Kou Chang

Rough Controlling TSC for Reactive Current Compensation in Traction Substations 423
Hongsheng Su, Qunzhan Li

A Digitally Controlled 4-kW Single-Phase Bridgeless PFC Circuit for Air Conditioner Motor Drive Applications 428
Yong Li, Toshio Takahashi

Optimized Electrical Design for Single Phase PFC Active IPEM 433
Qiaoliang Chen, Xu Yang, Zhao-an Wang

Table of Contents

A Novel Topology of APFC with On-Line Half-Bridge UPS Controlled by DSP 438
Xuejun Ma, Xuezhi Hu, Hongxia Wu, XuWu Chen

Nonlinear Current Control of Single-Phase PFC Suitable for Mixed-Signal IC Implementation 442
Min Chen, Anu Mathew, Jian Sun

A Novel Detection Method for Three-Phase Reactive Current 449
Zong Ming, Wang Fengxiang, Hua Funian, Sun Yidan

Selective Harmonic Controlling for Three-Level High Power Active Front End Converter with Low Switching Frequency 453
Hui Zhang, Kaipei Liu,

A Unity Power Factor Three-Phase Buck Type SVPWM Rectifier Based on Direct Phase Control Scheme 458
LI Yabin, Li Heming, Peng Yonglong

3-Phase Current-Source SMES-UPS Based on TFSC and its Control Strategies Control Strategies 463
WANG Fu-sheng, LI Hong-mei

A novel control scheme of 230kA DC power source using thyristor, Phase-shifting rectifier transformer and On-load tap changer 468
Qiao Shutong, Jiang Jianguo, Zuo Dongsheng, Wu Xiaojie

Research on Control Method of Double-Mode Inverter with Grid-Connection and Stand-Alone 473
Herong Gu, Zilong Yang, Deyu Wang, Weiyang Wu

Power and Energy Management of a Dual- Energy Source Electric Vehicle - Policy Implementation Issues 478
P.C.K. Luk, L.C. Rosario

Study on Non Contact Automatic on-Load Voltage Regulating Distributing Transformer Based on Solid State Relay 484
Zhao-Yulin, Dong-Shoutian, Li-Jiahui, Yao-Xin, Zheng-Na, Liu-Xueli

The Principle of a Novel Arc-suppression Coil and its Implementation 489
Cheng Lu, Chen Qiaofu, Zhang Yu, Zhang Changzheng

Grid Connection to Stand Alone Transitions of Slip Ring Induction Generator During Grid Faults 494
G. Iwanski, W. Koczara

System Control of Power Electronics Interfaced Distribution Generation Units 499
D. Feng, Z. Chen

Test Loadability of Power Systems using A Networked Power Electronic Devices Control and Measurement System 505
Sheng Yang, Venkataramana Ajjarapu, Bo Zhang

Test-Bed of Doubly Fed Induction Generator for Variable-Speed Constant-Frequency Wind Power Generation 510
S. Y. Yang, X. Zhang, C. W. Zhang, R. X. Cao

Control strategy of Hybrid sources for Transport applications using supercapacitors and batteries 515
M.B. Camara, H. Gualous, F. Gustin, A. Berthon

Wind Generator Stabilization With Doubly-Fed Asynchronous Machine 520
Li Wu, Zhixin Wang

Design Consideration of a Novel Digital Bidirectional Constant Current Source Used in Hybrid Electric Vehicle 526
Qingbo Hu, Zhengyu Lü

A Single-Phase Grid-Connected Inverter System With Zero Steady-State Error 532
Guo Xiaoqiang, Zhao Qinglin, Wu Weiyang

DC Transformer with Line Frequency Ripple Cancellation 537
Sen Dou, Wilson Wu, Annabelle Pratt, Pavan Kumar

vii

Table of Contents

A Novel PWM Method for Stacked Flying Capacitor Inverter 542
Gangui Yan, Gang Mu, Yafeng Huang, Wenhua Liu

Study on a New Method of Voltage-Source Induction Heating Load-Matched 549
Li Jin-gang, Zhong Yan-ru, Zhao Miao

An Alternating-master-salve Parallel Control Research for Single Phase Paralleled Inverters Based on CAN Bus 554
Zhang Chunjiang, Chen Guitao, Guo Zhongnan, Wu Weiyang

Analysis and Design of a Novel Dual Secondary Winding and Dual Power Bridge High Frequency Link Inverter 559
Zhang Zhe, Zhang Chunjiang, Wu Weiyang, Gu Herong, Shen Hong

Reduction of Common Mode EMI in a Full-Bridge Converter through Automatic Tuning of Gating Signals 564
Kai Zhang, Yunbin Zhou, Yonggao Zhang, Yong Kang

Phase Multilevel Inverter Fault Diagnosis and Tolerant Control Technique 569
Wang Baocheng, Wang Jie, Sun Xiaofeng Wu Junjuan, Wu Weiyang

Microcontroller-Based Single Phase Inverter Using a New Switching Strategy 574
K. Meghriche, O. Mansouri, A. Cherifi

Study of Stability Regions in Parallel Connected Boost Converters 580
Yuehui Huang, Chi. K. Tse

A Novel Analysis and Design Method for Integrated Magnetics 585
Zheng Feng, Weihao Hu, Pei Yun-qing

Investigation on the Space Vector PWM for Large Power Three-Level DC-Link Voltage Source Inverter Equipped with IGCTs 589
Wang Chengsheng, Li Chongjian, Li Yaohua, Zhao Xiaotan

Status and Opportunities of Photovoltaic Inverters in Grid-Tied and Micro-Grid Systems 593
Xiaoming Yuan, Yingqi Zhang

Adaptive Neuro-Fuzzy Control with Fuzzy Supervisory Learning Algorithm for Speed Regulation of 4-Switch Inverter Brushless DC Machines 597
A. Halvaei Niasar, H. Moghbelli, A. Vahedi

Combined Modulation and Harmonic Suppression 602
Cheng Weibin, Zhong Yanru, Jin Shun

Application Research of Maximum Wind-energy Tracing Controller Based Adaptive Control Strategy in WECS 607
Changhong Shao, Xiangjun Chen, Zhonghua Liang

Research on Synchrodrive Control Technology for Wind Turbine Adjustable-Pitch System Based on Adaptive decoupling Control 612
Hongche Guo, Qingding Guo

Limit-Trajectory Single- and Two-Mode Overmodulation Technology 617
Shun Jin, Yan-ru Zhong

Multiphase Permanent Magnet Motor Drive System Based on A Novel Multiphase SVPWM 622
Shan Xue, Xuhui Wen, Zhao Feng

Novel Random-Harmonic Elimination PWM Technique for Single-Switch Three-Phase AC-DC Buck Converter 627
Guang-Hui Tan, Wenchuan Ma, Yanchao Ji, Hongxiang Yu, Wancai Xu

viii

Table of Contents

FPGA Based Multichannel PWM Pulse Generator for Multi-modular Converters or Multilevel Converters 632
Liqiao Wang, Weiyang Wu

Cascaded Multilevel Converters with Non-Integer or Dynamically Changing DC Voltage Ratios 637
Shuai Lu, Keith A. Corzine

Practical Thermal Design Considerations for IPEM-based Converter 642
Qiaoliang Chen, Xu Yang, Zhao-an Wang

Realization of an FPGA-Based Space-Vector PWM Controller 647
Zhou Yuan, Xu Fei-peng, Zhou Zhao-yong

Chaotifying Control of Permanent Magnet Synchronous Motor 652
Hai Peng Ren, Chong Zhao Han

Analysis of PMLSM Direct Thrust Control System Based on Sliding Mode Variable Structure 657
Junyou Yang, Guofeng He, Jiefan Cui

Carrier-based Pulse Width Modulation for Three-Level Inverters: Neutral Point Potential and Output Voltage Distortion 662
Jang-Hwan Kim, Seung-Ki Sul

AC Current Sensorless Control of Three-Phase Three-Wire PWM rectifiers under the Unbalanced Source Voltage 669
Jia-peng Xu, Yu-peng Tang

Waveform Library Control of Converter 674
Xiaofeng Sun, Bin Wang, Meng Lingjie, Weiyang Wu

d-model Adaptive Algorithm Based on Plant-Parameterization 679
Zhao Feng, Liu Weiguo

Dynamics and Control of Electronic Cascaded Systems 684
Wen Wei, Xu Haiping, Wen Xuhui, Shi Wenqing

The Controlling Strategy for Electronic Ballast of HID Lamps 688
Weiping Zhang, Xiaohan Guan, Xusen Zhao, Hongtao Li, Zhengang Liu

Voltage Spectra of Three-Level Inverters with Three-Phase Modulation 693
S. Halász, I. Varjasi

Design of Motion Control System Used for Filter Rod Production Machine 699
Yang Qingyu, Ge Sibo, Ye Kesong, Shi Ren

Magnetic Pole Identification for PMSM at Zero Speed Based on Space Vector PWM 703
Jiangang Hu, Longya Xu, Jingbo Liu

Study on Stagewise Control of Connecting DFIG to the Grid 708
Xueguang Zhang, Dianguo Xu, Yongqiang Lang, Hongfei Ma

Generalized Control Approach for Active Power Filters 713
Xiaoyu Wang, Jinjun Liu, Chang Yuan, Zhaoan Wang

Novel Circuit Configuration for Hybrid Reactive Power Compensator 718
H.L Jou, J.C Wu, J.J. Yang, W.P. Hsu

Shunt Active Power Filter with Sample Time Staggered Space Vector Modulation Based Cascade Multilevel Converters 724
Liqiao Wang, Weiyang Wu

Shunt Active Power Filter Synthesizing Resistive Loads by Means of Adaptive Inverse Control 729
Wu Yanfeng, Wu Zhengguo, Li Hua, Li Hui

ix

Table of Contents

Single Neutral Element Self-Adaptive PID Controller Used In SVC..**734**
Zeng Guang, Ke Min-qian, Su Yan-min, Fu Qi-gang

A Novel Shunt Single-Phase Active Power Filter for High Voltage Application **739**
Zhang Changzheng, Chen Qiaofu, Zhao Youbin, Chen Yuda, Cheng Lu

Three-phase Active Power Filter Based on Space Vector and One-cycle Control....................................**744**
Wang Yong, Shen Songhua, Guan Miao

Implementation of a Shunt-Series Compensator for Nonlinear and Voltage Sensitive Load....................**748**
Bor-Ren Lin, Chien-Lan Huang

Three-Phase Active Filter using a Single-Phase STATCOM Structure with Asymmetrical Dead-band Control..**753**
Seyyed Hossein Hosseini, Mehran Sabahi

Mitigation of Voltage Sag Using Adaptive Neural Network with Dynamic Voltage Restorer.....................**759**
M. R. Banaei, S. H. Hosseini, M. Darkalee Khajee

Mitigation of Current Harmonic Using Adaptive Neural Network with Active Power Line Conditioner **764**
M. R. Banaei, S. H. Hosseini

A direct control strategy for UPQC in three-phase four-wire system...**769**
Tan Zhili, Li Xun, Chen Jian, Kang Yong, Duan Shanxu

Three-Phase Harmonic Selective Active Filter Using Multiple Adaptive Feed Forward Cancellation Method..**774**
Lewei Qian, David Cartes, Qiang Zhang

Reactive Power Compensation in Distribution Networks with STATCOM by Fuzzy Logic Theory Application **779**
Seyyed Hossein Hosseini, Reza Rahnavard, Yousef Ebrahimi

A Distributed Fuel Cell Based Generation and Compensation System to Improve Power Quality............**784**
Haimin Tao, Jorge L. Duarte, Marcel A. M. Hendrix

Parallel Control of Three-Phase Three-Wire Shunt Active Power Filters **789**
Xueliang Wei, Ke Dai, Xin Fang, Pan Geng, Fang Luo, Yong Kang

Study and Design of Noninductive Bus bar for high power switching converter **794**
Zhiling Qiu, Hongyan Zhang, Guozhu Chen

A New Minimum Torque-ripple and Sensorless Control Scheme of BLDC Motors Based on RBF Networks...**798**
Juan Wang, Hongwei Liu, Yuran Zhu, Bo Cui, Huijuan Duan

Improved Modelling and Calculation on Electromagnetic Transient of Power Transformer......................**802**
Chen Zhe, Wen Yuanfang, Lu Guojun

The Simulation and the Experimental Research of the Stator Bars' Evaporative Cooling System in the Three Gorges' Hydrogenerator **808**
Ruan Lin, Gu Guobiao, Tian Xindong, Yuan JiaYi

An Investigation of Multi-phase Transverse Flux Permanent Magnet Machine **813**
G.Q. Bao, J.K.Wang, D.Zhang, J.Z. Jiang

Suspension Principle and Digital Control for Bearingless Permanent Magnet Slice Motors **817**
Huangqiu Zhu, Liang Fang

The effect of parameter variations on the performance of indirect vector controlled induction motor drive **821**
A. Shiri, A. Vahedi, A. Shoulaie

Magnetic Field Analysis and Performance Calculation for New Type of Claw Pole Motor with Permanent Magnet Outer Rotor **826**
Fengge Zhang, Shifu Zhang, Haijun Bai, Eugen Nolle, Hans Pert Gruenberger

x

Table of Contents

Performance Analysis of a PM Claw Pole SMC Motor with Brushless DC Control Scheme.........................831
Youguang Guo, Jianguo Zhu, Jiaxin Chen, Jianxun Jin

Solving Induction Motor Equivalent Circuit using Numerical Methods for an In-Service and Nonintrusive Motor Efficiency Estimation Method.........................836
Bin Lu, Wei Qiao, Thomas G. Habetler, Ronald G. Harley

Fault Investigation of X-by-wire Permanent Magnet Synchronous Machine.........................842
L. Feng, A. Binder, A. Rentschler, A. Paweletz, D. Guenther

PLC-Based Speed Control of DC Motor.........................847
Ashraf Salah El Din Zein El Din

H8 Control of Adjustable-Pitch Wind Turbine Adjustable-Pitch System.........................853
Hongche Guo, Qingding Guo

The Motion Control Algorithm based on Quaternion Rotation for a Permanent Magnet Spherical Stepper Motor.........................857
Qun-jing Wang, Kun Xia

Research on Restraining Thrust Force Ripple for Permanent Magnet Linear Synchronous Motor.........................862
Cui Jiefan, Wu Hui, Sun Qing, Zhang Yi, Zhao Lijun

Using Recurrent Fuzzy Wavelet Neural Network to Control AC Servo System.........................866
Yan Tang, Wei Sun, Yaonan Wang, Xiaohua Zhai

new topology of multi - level - converter for harmonic reduction.........................870
Frank Grundmann, Jian Xie

PWM Based Sensing and Control of Magnetic Bearings.........................875
Zhuliang Yeic, Flalph Vansencc

Position Sensorless Direct Torque Control of Synchronous Reluctance with Permanent Magnet Motor.........................880
Jiang Dong, Zhao Zhengming, Duan Yao, Guo Wei

Counter-Rotating Permanent Magnet Brushless DC Motor for Underwater Propulsion.........................885
Jianqi Qiu, Cenwei Shi, Mengjia Jin, Ruiguang Lin

A Special Flux-weakening Control Scheme of PMSM - Incorporating and Adaptive to Wide-Range Speed Regulation.........................890
Song Chi, Longya Xu

Model-based Disturbance Attenuation for Linear Motor Servo System.........................896
Guiqiu Liu, Qingding Guo

A Fuzzy-Wavelet-Network-Based Position Control for PMSM.........................899
Wang Jun, Peng Hong, Xia Ling

Stability Analysis of Magnetic Bearing with Resonance Circuit.........................903
Zong Ming, Wang Fengxiang, Sun Yidan, Wang Jiqiang

Flux-Weakening Characteristics of Trapezoidal Back-EMF Machines in Brushless DC and AC Modes.........................908
Z.Q. Zhu, J.X. Shen, D. Howe

A Cost Effective Sensorless Control Method for Permanent Magnet Synchronous Motors Based on Average Terminal Voltage.........................913
Cheng-Hu Chen, Wei-Chih Tai, Ming-Yang Cheng

DSP-based Discrete-Time Reaching Law Control of Switched Reluctance Motor.........................918
Ge Baoming, Zhao Nan

Digital Control System on Bearingless Permanent Magnet-type Synchronous Motors.........................923
Jianming Deng, Huangqiu Zhu, Yang Zhou

xi

Table of Contents

Practical Issues in Sensorless Control of PM Brushless Machines Using Third-Harmonic Back-EMF **928**
J.X. Shen, Z.Q. Zhu, D. Howe

Switched Reluctance Motors Drive for the Electrical Traction in Shearer **933**
H. Chen

Research on Three-level Inverter of Six-phase Synchronous Motor **937**
Yao Wenxi, Hu Haibing, Lu Zhengyu, Xu Haijie

Doubly-Salient Permanent-Magnet Machine with Skewed Rotor and Six-State Commutating Mode **942**
Yongbin Li, Chris Mi

Sensorless Control and PMSM Drive System for Compressor Applications **947**
Dongsheng Li, Takahiro Suzuki, Kiyoshi Sakamoto, Yasuo Notohara, Tsunehiro Endo, Chikara Tanaka, Tatsuo Ando

Analysis and Experimental Study of Slot Effect in Synchronous Reluctance Permanent Magnet Motors **952**
Wei Guo, Zhengming Zhao, Yingchao Zhang

A New BLDC Motor Drives Method Based on BUCK Converter for Torque Ripple Reduction **958**
Zhang Xiaofeng, Lu Zhengyu

Performance Investigation of a Fault-Tolerant Brushless Permanent Magnet AC Motor Drive **962**
Jingwei Zhu, Nesimi Ertugrul, Wen Liang Soong

Current sensorless integral variable structure controller of synchronous reluctance motor **967**
Huann-Keng Chiang, Chien-An Chen, Bor-Ren Lin, Kai-Sheng Hsu

An Improved Sliding Mode Observer for Speed Sensorless Vector Control Drive of PMSM **972**
K. Paponpen, M. Konghirun

Analysis of an AC fed direct converter for a switched reluctance machine in aerospace applications **977**
S. J. Forrest, J. Wang, G. W. Jewell, C. M. Johnson, S.D. Calverley

Direct Torque Control of an Interior Permanent Magnet Synchronous Machine fed by a Direct AC-AC Converter **983**
D. Xiao, M. F. Rahman

A Novel Modular Permanent Magnet Drive System Design **989**
Wen Ouyang, Nicholas Lemberg, Ruoping Yao, T.A.Lipo

Research on Digital Control Systems for Large Power AC-DC-AC Converters with Synchronous Motor Load **995**
Xiaotan Zhao, Chongjian Li, Weihui Sheng, Yaohua Li

About the Prediction of Undesired Higher Current and Torque Harmonics of Inverter Driven Motors with Numerical Methods **999**
C. Grabner

A Method of Stator Voltage Error Compensation in MRAS Sensorless Vector Control of Induction Motor **1006**
Wen Xuhui, Chen Guilan, Han Li

Systematic Design of Fuzzy Logic Based Hybrid On-Line Minimum Input Power Search Control Strategy for Efficiency Optimization of IM **1012**
Zhang Liwei, Liu Jun, Wen Xuhui, Trillion Q. Zheng

Research on an AC Variable-frequency Power Dynamometer Based on PWM Rectifier and Fuzzy Direct Torque Control **1017**
Jia-qiang Yang, Jin Huang

Characteristic Research of Bearing Currents in Inverter-Motor Drive Systems **1023**
Xing Shancheng, Wu Zhengguo

Research on a New Motor Drive Control System for Electric Transit Bus **1027**
SHAO Gui-xin, ZHANG Cheng-ning

xii

Table of Contents

New Micro-Drive Series For Induction Motors & Survey of Market Trends 1032
Henrik Rosendal Andersen, Ruimin Tan, Zhang Hui

Robust Backstepping Control of Induction Motor Drives Using Artificial Neural Networks 1038
J. Soltani, R. Yazdanpanah

Robust Nonlinear Control of Linear Induction Motor taking into account the Primary End Effects 1043
J. Soltani, M.A. Abbasian

A Novel Adaptive Scheme for Stator Resistance Estimation in Sensorless Induction Motor Drives 1049
Han Li, Wen Xuhui, Chen Guilan

Ripple-Free Sampling of Current Signals in Drives with Carrier-based PWM Patterns 1054
Haihui Lu, Qiang Yin, Russel J. Kerkman, Thomas A. Nondahl

**Study of Speed Sensorless Control Methodology for Single Inverter Parallel Connected Dual Induction
Motors Based on the Dynamic Model** 1061
Shi Wei, Wang Ruxi, Wang Yue, He Yanhui, Wang Zhaoan, Liu Jinjun

ADC architecture with direct binary output for digital controllers of high-frequency SMPS 1066
Tao Zhou, Jianping Xu

Analysis and Evaluation of a High-Voltage AC Amplifier for Electrostatic Suspension 1071
F. T. Han, Q. P. Wu, K. Liu, Z. Y. Gao

Design and Development of a 50kW Z-Source Inverter for Fuel Cell Vehicles 1076
Miaosen Shen, Alan Joseph, Yi Huang, Fang Z. Peng, Zhaoming Qian

Identification and improvement of stray coupling effect in an L-C-L common mode EMI filter 1081
Junping He, Wei Chen, Jianguo Jiang

**High Step-up Converter Associated with Soft-Switching Circuit with Partial Energy Processing for
Livestock Stunning Applications** 1086
S. -Y. Tseng, S.-H. Tseng, J. -Z. Shiang

**A Computationally Intelligent Methodologies and Sliding Mode Control Based Traction control System
for in-wheel driven EV** 1091
Ming Zhengfeng, NI Guangzheng

A Low-Cost Gate Driver Design Using Bootstrap Capacitors for Multilevel MOSFET Inverters 1096
J. J. Graczkowski, K. L. Neff, X. Kou

An Effective Method to Suppress Resonance in Input LC Filter of a PWM Current-Source Rectifier 1101
Y.W. Li, B. Wu, N. Zargari, J. Wiseman, D. Xu

Topological and Modulation Design of Three-Level Z-Source Inverters 1107
P. C. Loh, F. Gao F. Blaabjerg

Investigation of Power Supplies for a Piezoelectric Brake Actuator in Aircrafts 1112
Rongyuan Li, Norbert Fröhleke, Hermann Wetzel, Joachim Böcker

A Line Power-Supply for LED Lighting using Piezoelectric Transformers in Class-E Topology 1117
F.E. Bisogno, S. Nittayarumphong, M. Radecker, A. V. Carazo, R. N. do Prado

Integrating Large Wind Farms into Weak Power Grids with Long Transmission Lines 1122
Richard Piwko, Nicholas Miller, Juan Sanchez-Gasca, Xiaoming Yuan, Renchang Dai, James Lyons

Turn-on Condition and Characteristics of Highpower Semiconductor Switch RSD 1129
Y. M. Zhou, Y. H. Yu, H. G. Chen, L. Liang

The analysis and simulation of power circuits for high voltage converter 1133
S. I. Volskiy, Y. Y. Skorokhod, V. V. Shergin

A novel IGCT-based Half-controlled Bridge Type Fault Current Limiter 1138
Wanmin Fei, Yanli Zhang

xiii

Table of Contents

Influence of Proton Irradiation dose on the Performance of Local Lifetime Controlled Power Diode with Proximity Gettering of Platinum 1143
B.D. Han, D.Q. Hu, S.S. Xie, Y.P. Jia, B.W. Kang

IMPLEMENTATION OF A HIGHER QUALITY DC POWER CONVERTER 1148
Barsoum, N.N., YII, M.L.

Design of a Digital Programmable Control IC for Single-Phase Controlled Rectifiers 1154
Ming-Fa Tsai, Fu-Jing Ke, Ying-De Lin, Jui-Kum Wang

Feasibility Study of AlGaN/GaN HEMT for Multimegahertz DC/DC Converter Applications 1159
Yang Gao, Alex Q. Huang

The Mechanism Analysis of IGBT Module Invalidation 1162
Xu Aide, Fan Yinhai, Wang Xinxin, Liu Yuanyuan

A New Injection Efficiency Controlled GTO 1167
Wang Cailin, Gao Yong, Zhang Ruliang

Implementation and Analysis of 3-phase Voltage Sourced Regenerative Rectifier 1171
Rui Chen, Qiongxuan Ge, Shijie Li

Design and Implementation of Electronic Ballast for Fluorescent Lamps with Low Lighting Flicker 1178
Yang-Sheng Lin, Chun-An Cheng, Jiann-Fuh Chen, Tsorng-Juu Liang, Wei-Shih Liu

A Floating-point Coprocessor Configured by a FPGA in a Digital Platform Based on Fixed-point DSP for Power Electronics 1183
Haibing HU, Tianjun Jin, Xianmiao Zhang, Zhengyu LU, Zhaoming Qian

An Analytical Model for 4H-SiC Super-Junction Devices 1188
L.C. Yu, K. Sheng

Architecture Implementation of Class-D Amplifiers Using Digital-Controlled Multiphase-Interleaved PWM Technique 1192
Yu-Tzung Lin, Chi-Yang Lee, Ying-Yu Tzou,

Integrated IC-like Thyristor–based Switching Structure for Pulse Current Generation to Electronic Ignition 1198
C. L. Zhang, K. S. Jeon, C. H. Ahn, J. D. Park, E. D. Kim, Na Zhi, Yong Gao

A Wide Bandwidth Current Probe Based on Rogowski Coil and Hall Sensor 1202
Dong Li, Guiyou Chen

Voltage Dip Detection Based on an Efficient Least Squares Algorithm for D-STATCOM Application 1207
Thip Manmek, Chathura P. Mudannayake, Colin Grantham

Optimal Design and Analysis on Bearingless Permanent Magnet-type Synchronous Motors Using Finite Element Method 1213
Chang Jiang , Huangqiu Zhu, Zhenyue Huang

The Restrain of Harmonic Circulating Currents between Parallel Inverters 1218
Yu Zhang, Shanxu Duan, Yong Kang, Jian Chen

Simulation of Permanent Magnet Synchronous Motor with Dual Closed Loop by Time-Stepping Finite Element Model 1223
Xinhua Liu, Jianzhong Jiang, Yu Gong, Ye Ding

Online Dynamic Parameter Estimation of Transformer Equivalent Circuit 1228
M. Reza Feyzi, Mehran Sabahi

Worst-Case Tolerance Analysis for a Power Electronic System by Modified Genetic Algorithms 1233
Toshiji Kato, Kaoru Inoue, Kazuya Nishimae

The Reduction of Force Ripples of PMLSM Using Field Oriented Control Method 1238
Yu-wu Zhu, Kun-seok Jung, Yun-hyun Cho

xiv

Table of Contents

Analysis and Design of Signal Stage AC/DC Converter with Resonant Model PFC 1243
Weiping Zhang, Liangrui Lin, Dongyan Zhang, Xusen Zhao

Low Frequency Model for the Metal Halide Lamp 1248
Weiping Zhang, Yuanchao Liu, Xiaoqiang Zhang, Hongtao Li, Wenji Liu

H8 Robust Controller Based on Local Feedback Recurrent Neural Network for Permanent Magnet Linear Synchronous Motor 1253
Junyou Yang, Naiguang Fa, Ruijuan Chen

Parameter Estimate Modeling of Electronic Transformer 1258
Jiaju Wu, Hidehiko Sugimoto, Changkun Wang

Analysis and Design of Boost DC-DC Converters for Intrinsic Safety 1267
Shu-Lin Liu, Jian Liu, Hong Mao

Modeling and Fuzzy Logic with Integrator Control for the ZVZCS PWM DC/DC Converter 1273
Shen Hong, Wan Jianru, Yang Xiaobo, Wu Weiyang, Wang Xiaohuan

ZVS DC-DC Converter with Parallel-Connected Current Doubler Rectifier 1278
Bor-Ren Lin, Shuh-Chuan Tsay, Chun-Sheng Yang, Chien-Lan Huang

Study on the Dynamical Model and Analytical Method for DC-DC Switching Converter 1283
Li-Li Wang, Yu-Fei Zhou, Jun-Ning Chen

A Novel Topology Family of Single-stage Parallel Mode Uninterruptible AC/DC Converter with PFC 1288
Xuejun Ma, Hongxia Wu, Congsheng Huang, Xuwen Huang

Analysis and Design of an Automatic-Current-Sharing Control Based on Average-Current Mode for Parallel Boost Converters 1293
Wenxun Xiao, Bo Zhang, Dongyuan Qiu

A Novel Digital Charge Control for DC-DC Converters 1298
Shi Wenqing, Xu Haiping, Wen Xuhui, Wen Wei

An Asymmetrical Switched Capacitor and Lossless Inductor Quasi-Resonant Snubber-Assisted ZCS-PWM DC-DC Converter with High frequency Link 1302
Khairy Fathy, Keiki Morimoto, Toshimitsu Doi, Hyun Woo Lee, Mutsuo Nakaoka

A Divided Voltage Half-Bridge High Frequency Soft-Switching PWM DC-DC Converter with High and Low Side DC Rail Active Edge Resonant Snubbers 1307
Khairy Fathy, Keiki Morimoto, Toshimitsu Doi, Hiroyuki Ogiwara, Hyun Woo Lee, Mutsuo Nakaoka

Dynamic Analysis of a Current Source Inductively Coupled Power Transfer System 1312
Wenqi Zhou, Hao Ma

A New Topology of Capacitor-Clamp Cascade Multilevel Converters 1318
Anees Abu Sneineh, Ming-Yan Wang, Kai Tian

Evaluation of Semiconductor Losses in Cryogenic DC-DC Converters 1323
C. Jia, A. J. Forsyth

Design and Performance Evaluation of a 10-kW Interleaved Boost Converter for a Fuel Cell Electric Vehicle 1328
G. Calderon-Lopez, A. J. Forsyth, D. R. Nuttall

Analysis of Abnormal Phenomenon in Common-Source-type Forward Converter with Self-driven Synchronous Rectifier 1333
Kentaro Fukushima, Takayoshi Hashimoto, Tamotsu Ninomiya, Takeshi Segawa

Power Quality Conditioning in Distributed Generation Systems 1338
R.K. Járdán, I. Nagy

xv

Table of Contents

Active Clamp Forward Converter Combined with Dither Voltage Generator for Poultry Stunning Applications..1343
S. -Y. Tseng, H.-T. Wen, H.-H. Chang, J. -S. Kuo

A Novel Zero-Voltage Switching Resonant Pole Inverter 1348
Sanbo Pan, Junmin Pan

Analysis of Three-Level ZVS PWM Inverter for Induction Heating Applications 1353
A. Jangwanitlert, J. Songboonkaew, W. Thammasiriroj, J.C. Balda

Dual Duty Cycle Controlled Voltage Source Soft-Switching High Frequency Inverter with AC Load Side Reverse Blocking Switched Resonant Capacitor 1358
Khairy Fathy, Ju-Sung Kang, Hiroyuki Ogiwara, Bin Eiuo, Hideki Omori, Hyun Woo Lee, Mutsuo Nakaoka

A Switched-Capacitor Lossless Inductor ZCS Snubber-Assisted Series Load Resonant High Frequency Inverter with Dual Mode Pulse Modulation Scheme..1363
Khairy Fathy, Takaaki Okude, Hideki Omori, Hyun Woo Lee, Mutsuo Nakaoka

Topologies of Switch-Linear Hybrid Power Conversion & Special Operation States....................1368
Lu-sheng Ge, Qian-zhi Zhou, Wu bin

Single Reverse Blocking Switch Type Pulse Density Modulation Controlled ZVS Inverter with Boost Transformer for Dielectric Barrier Discharge Lamp Dimmer..1372
Hisayuki Sugimura, Bishwajit Saha, Hideki Omori, Hyun-Woo Lee, Mutsuo Nakaoka

PDM Controlled Series Load Resonant Soft Switching High Frequency Inverter for Induction Heated Toner Fixing Outer Roller with Inner Cylindrical Working Coil Stator 1377
Hisayuki Sugimura, Hideki Omori, Hyun Woo Lee, Mutsuo Nakaoka

Zero-Voltage and Zero-Current Switching Two-Transformer Full-Bridge Converter Using the Output-Voltage-Doubler..1382
H.K. Yoon, E.S. Choi, S.K. Han, G.W. Moon, M.J. Youn

A Single-stage Boost-Flyback PFC Converter 1387
Zhao Qinglin, Wen Yi, Wu Weiyang, Chen Zhe

Control Bifurcation in PFC Boost Converter under Peak Current-Mode Control..1392
Yi-Jing Ke, Yu-Fei Zhou, Jun-Ning Chen

Analysis and Design of One-Cycle-Controlled Dual-Boost Power Factor Corrector 1397
Yue-feng Yao, Yuan-rui Chen

A Novel Single-phase Buck PFC Converter Based on One-cycle Control..1401
Chen Bing, Xie Yun-Xiang, Huang Feng, Chen Jiang-Hui

Modeling and Simulation of Three Phase High Power Factor PWM Rectifier factor correction. 1406
Yu Fang, Yong Xie, Yan Xing

Effect of the Ripple Current on Power Factor of CRM Boost APFC 1412
A. Abramovitz

Simulated Study of Three-Phase Single-Switch PFC Converter with Harmonic Injected PWM by MATLAB..1416
Zhanlong Li, Yupeng Tang

A Simple Digital Controller for Constant Instantaneous Input Power type Three-Phase Boost Rectifier under Unbalanced System..1421
Jin Ai-Juan, Li Hang-Tian, Li Shao-Long

An Improved and Digital Current Control Strategy for One Cycle Control Based Three-Phase Boost Rectifier under Unbalanced System..1426
Li Shao-Long, Jin Ai-Juan, Li Hang-Tian

xvi

Table of Contents

Control Method for Power Quality Compensation Based on Levenberg-Marquardt Optimized BP Neural Networks 1431
Zhou Ming, Wan Jian-Ru, Wei Zhi-Qiang, Cui Jian

A Nonlinear Method for Hybrid Electromagnetic Suspension 1436
Junwei Cui, Jianhui Wang

New topology of multi - level - converter for harmonic reduction 1442
Frank Grundmann, Jian Xie

Model Reference Adaptive Control based on Neural Network for Electrode System in Electric Arc Furnace 1447
Zhang Shi-feng, Zhang Shao-De, Li Kun, Zheng Xiao

STATCOM ETO Failure Analysis 1450
Zhong Du, Bin Chen, Chong Han, Zhaoning Yang, Wenchao Song, Subhashish Bhattacharya, Alex Q. Huang

Modeling and Control of Three-phase Voltage Source PWM Rectifier 1454
Yao Chen, Xin Min Jin

Mitigation of Electric Arc Furnace Voltage Flicker Using Static Synchronous Compensator 1458
Y.F. Wang, J.G. Jiang, L.S. Ge, X.J. Yang

Design of Distributed FACTS Controller and Considerations for Transient Characteristics 1463
Gaidi Ning, Shijie He, Yue Wang, Lei Yao, Zhaoan Wang

A Wind-Power Generation System Having a Function of Suppressing Line Voltage Deviation 1468
Y. Nakayama, S. Fukuda, M. Futami, M. Ichinose, S. Ohara, H. Kita

A Novel Active Islanding Detection Method of Grid-connected Photovoltaic Inverters Based on Current-Disturbing 1473
Zhang Chunjiang, Liu Wei, San Guocheng, Wu Weiyang

Grid Connection of Doubly-Fed Induction Generators in Wind Energy Conversion System 1477
Ahmed G. Abo-Khalil, Dong-Choon Lee, Se-Hyun Lee

Active and Reactive Power Control of DFIG for Wind Energy Conversion under Unbalanced Grid Voltage 1482
Jeong-Ik Jang, Young-Sin Kim, Dong-Choon Lee

A BASIC STUDY OF FUZZY-LOGIC-BASED POWER SYSTEM STABILIZATION WITH DOUBLY-FED ASYNCHRONOUS MACHINE 1487
Li Wu, Zhixin Wang

Quantitative Analysis on Different Modes of Energy Optimal Control for Series Power Quality Controllers 1492
Huang Xinming, Liu Jinjun, Zhang Hui

Resonance inverter power system for improving plasma sterilization effect 1497
Y.M Kim, J.Y Kim, M. C Jo, S.H Lee, S.P Mun, H.W Lee, S.K Kwon, K.Y Suh

Generic optimization for SMPS design with Smart Scan and Genetic Algorithm 1502
Heidi H.T. Yeung, N. K. Poon, Stephen L. Lai

Novel Single-Stage Isolated Buck-Boost Inverter Based on Improved SPWM Control Method 1507
Guang-Hui Tan, Fanpeng Zeng, Yanchao Ji, Xi Chen, Hua Wang

On the Effects of Voltage Loop in Paralleled Converters Under Master-Slave Current Sharing 1512
Yuehui Huang, Chi K. Tse

Improved Control for Parallel Inverter with Current-Sharing Control Scheme 1517
Zhao Qinglin, Chen Zhongying, Wu Weiyang

A Novel Digital Controlled battery charger for High power UPS application 1522
Fang Luo, Yong Kang, Shan Xu Duan, Xueliang Wei

xvii

Table of Contents

A Novel High Input Power Factor Single-Stage Single-Phase AC/AC Converter 1527
Chien-Ming Wang, Chien-Yeh Ho, Maoh-Chin Jiag

Research on the Power Sharing of the Parallel Inverters without Control Interconnection Basing on Droop Characteristic 1532
Kan Jiarong, Xie Shaojun

Analysis and Design of Repetitive controlled Inverter System with High Dynamic Performance 1537
Mingzhu Li, Zhongyi He, Yan Xing

Study on a large-volume high-performance programmable voltage disturbance source 1542
Zhan Qizhi, Zhuo Fang, Dong Wenjuan, Wang Zhao'an

1 KW Dual Interleaved Boost Converter for Low Voltage Applications 1546
Heinz van der Broeck, Ibrahim Tezcan

Control of Multilevel Flying Capacitor Inverters for High Performance 1551
L. Zhang, S. J. Watkins, Duan Qi Chang

Analysis of Harmonics in Input Line Current for Matrix Converter based on Double Input Line-toline Voltages 1557
Guo Yougui, Deng Wenlang, Zhu Jianlin

Research on Neutral-point Balancing Control for Three-level NPC Inverter Based on Correlation between Carrier-based PWM and SVPWM 1560
Wenxiang Song, Guocheng Chen, Xiaoyu Ding, Mantang Shu

Instantaneous Voltage Regulated Seamless Transfer Control Strategy for Utility-interconnected Fuel cell Inverters with an LCL-filter 1566
Guoqiao Shen, Dehong Xu, Xiaoming Yuan

An Anti-windup Design Method for Internal Model Control Based on H8 Optimization 1571
Hou Yansong, Li Hua

Study on Pwm Control Strategy of Photovoltaic Grid-connected Generation System 1576
Shi-cheng Zheng, Pei-zhen Wang, Lu-sheng Ge

Robust Sliding Model Control for Regenerative Braking of Electric Vehicle 1581
Min Ye, Zhifeng Bai, Binggang. Cao

A Self-adaptive Fuzzy Control Scheme of High Frequency Link SPWM Inverters 1585
Herong Gu, Deyu Wan, Weiyang Wu

Using Automatic Frequency Shifting Techniques for LLC-SRC Output Voltage Regulation 1590
Kuo-Kai Shyu, Ching-Ming Lai, Ko-Wen Jwo, Ming-Ho Pan, Chung-Ping Ku

Design and Test of Novel Programmable Digital Three Phases SPWM Chip 1595
Yang Yuan, Gao Yong, Chen Lijie

An Improved Performance of Five-Leg Inverter in Two Induction Motor Drives 1598
Ryuji Omata, Kazuo Oka, Atsushi Furuya, Shuji Matsumoto, Yusuke Nozawa, Kouki Matsuse

Adaptive Three Dimensional Space Vector Modulation in abc Coordinates for Three Phase Four Wire Split Capacitor Converter 1603
Xiao-bo Yang, Wei-yang Wu, Hong Shen

Inverters Parallel Operation Based on CAN 1608
Yong Wu, Xianglong Jiang, Jinbang Xu, Qingyi Wang, Shuyun Wan

EMI Reduction Method for a Single-Phase PWM Inverter by Suppressing Common-Mode Currents with Complementary Switching 1613
Toshiji Kato, Kaoru Inoue, Koji Akimasa

xviii

Table of Contents

Analysis and Design of a Novel Dual Secondary Winding and Dual Power Bridge High Frequency Link Inverter 1618
Zhang Zhe, Zhang Chunjiang, Wu Weiyang, Gu Herong, Shen Hong

Research of Complex Fuzzy Control on-off Magnetism Team Motor Speed-Adjusting System 1623
Zhao Ming-fu, Chen Yan, Zhang Zhi-yuan, Dong Chun, DongYu

A New BLDC Motor Drives Method Based on BUCK Converter for Torque Ripple Reduction 1626
Zhang Xiaofeng, Lu Zhengyu

Design of Wind Turbine Generator Control System 1630
Chen Guiyou, Zhou Li, Sun Tongjing, Wang Zhongmin

Non-touching Intelligent Control System of Water Intenerating Equipment Based on Sodion Exchange 1634
Chen Guiyou, Zhang Qingfan, Zhou Li, Luo Donghua

Investigation of Hybrid Modeling and Control for DC-DC Converters 1637
Hao Ma, Feng Qi, Wenqi Zhou

Effect of Peak Current Mode Control on Transient Response for VRM Application 1641
Seiya Abe, Tamotsu Ninomiya

Modulations for Voltage Source Rectification and Voltage Source Inversion Using Direct Space Vector Approach 1646
Keping You, M. F. Rahman

Synchronization of Voltage Waveforms in Basic Topologies of Dual Inverter-Fed Motor Drives 1651
V. Oleschuk, F. Profumo, A. Tenconi, R. Bojoi, A.M. Stankovic

Research on Fast Magnetic Valve Controllable Reactor 1657
Zhang Jian-wen, Cai Xu

Study and comparison of fault tolerant shunt threephase active filter topologies 1663
H. El Brouji, P. Poure, S. Saadate

Application of GA-BP in Fault Diagnosis of Power Circuit of SVC 1669
Zeng Guang, Xi Yu-fan, Su Yan-min, Zhang Jing-Gang

The Optimization-Sliding Mode Control For Three-Phase Three-Wire DSP-based Active Power Filter 1674
Zhou Wei-ping, Liu Da-ming, Wu Zheng-guo, Xia Li, and Yang Xuan-fang

Three-Phase DVR using a Single-Phase Structure with Combined Hysteresis/ Dead-band Control 1679
Seyyed Hossein Hosseini, Mehran Sabahi

Harmonic Detection Based on the TLS Estimation Algorithm 1684
Liu Kaipei, Zhang Junmin

Control Strategy Study of Hybrid Active Power Filter 1689
Jia Zhang, Guohong Zeng

Novel Harmonic Free Single Phase Variable Inductor Based on Active Power Filter Strategy 1693
Mu Xianmin, Wang Jianze, Ji Yanchao, Wei Xiaoxia, Fu Xiangyun

A Multi-Output Series Resonant Inverter with Asymmetrical Voltage-Cancellation Control for Induction-Heating Cooking Appliances 1697
S.H. Hosseini, A. Yazdanpanah Goharrizi, E. Karimi

Capacitor Voltage Control in a Cascaded Multilevel Inverter as a Static Var Generator 1703
M. Li, J. N. Chiasson, L. M. Tolbert

DC-link Pumping-up Voltage Suppression of a Series Active Voltage Regulator With Phase Shift Control 1708
G. C. Xiao, Z. L. Hu, C. H. Nan, Z. A. Wang

The Fuzzy Soft-startup Controller of Active Power Filter 1713
He Na, Wu Jian, Xu Dianguo

Table of Contents

A Novel Control Method for DSTATCOM Using Artificial Neural Network..1718
Yang Xiao-ping, Zhong Yan-ru, Wang Yan

A Detailed Analysis of Unexpected DC-side Voltage Boost in Series Power Quality Controllers 1722
Yuan Chang, Liu Jinjun, Wang Xiaoyu, Wang Zhaoan

Comparative Analysis of Popular Control Schemes for Parallel Active Power Filter and Experimental
Verification..1726
Xiaoyu Wang, Jinjun Liu, Chang Yuan, Zhaoan Wang

Accurate Modeling of the Three Phase Induction Motor Including Saturation Effects...................................1731
E. V. N. Souza, S. R. Naidu

A study on the reliability evaluation of driving parts for note handling units 1736
Joo Han Kim, Jung Kee Chung, Ha Kyeong Sung, Se Hyun Rhyu

Analysis on Toothless Permanent Magnet Machine with Halbach Array...1741
Xu Yanliang, Feng Kaijie

Improvement in Reliability of Doubly Salient Permanent Magnet Motor Drive..1746
Wenxiang Zhao, Ming Cheng, Xiaoyong Zhu, Wei Hua, Jianzhong Zhang

A New Approach of Modeling the Saturated Induction and Synchronous Salient Pole Machines 1751
A. Câmpeanu, M. Badica

Inductance characteristics of 3-phase fluxswitching permanent magnet machine with doubly-salient
structure 1758
Wei Hua, Cheng Ming

Performance Index Evaluations of a Micro Axialflux Switched-reluctance Motor..1763
Cheng-Tsung Liu, Yen-Ming Chen, Da-Chen Pang

Study of Variable Frequency Operation of Induction Generator for Wind Power...1768
Noriyuki Kimura, Mitsuhiro Hirao, Toshimitsu Morizane, Katsunori Taniguchi

Optimal Power Control Strategy of Maximizing Wind Energy Tracking and Conversion for VSCF
Doubly Fed Induction Generator System 1773
H. Li, Z. Chen, John K. Pedersen

Design and Evaluation of a Dual Mechanical Port Machine and System 1779
Longya Xu, Yuan Zhang

Characteristic Analysis on Overhang Effect in Axial Flux PM Synchronous Motors with Slotted Winding 1784
WonYoung Jo, YunHyun Cho, YonDo Chun, DaeHyun Koo

Design and Analysis of a Double-Stator Cup-Rotor Directly Driven Permanent Magnet Wind Power
Generator 1788
Dong Zhang, Shuangxia Niu, K. T. Chau, J. Z. Jiang, Yu Gong

Feasibility Analysis of Accelerometer Configuration of Non-gyro Micro Inertial Measurement Unit...................1793
Ding Mingli, Zhou Qingdong, Wang Qi, Wang Changhong

Design of Fractional-Order a PI Controller with two modes...1797
Wen Li, Yoichi Hori

Sliding Mode Robust Tracking Control Based on Learning Feedforward Compensation for High
Precision Linear Servo System 1802
Zhu Guoxin, Guo Qingding, Zhao Ximei

Application of Fuzzy Self-learning Sliding Mode Variable Structure Control in Linear AC Servo System...............1806
Qing Hu, Shuo Jie, Dongmei Yu

Dynamics Research of Robot Manipulator 1811
Zhibing Shu, Caizhong Yan, Hairong Zhang

xx

Table of Contents

Advanced Angle Control Schemes for Stator Hybrid Excited Doubly Salient Motor Drive 1815
Xiaoyong Zhu, Ming Cheng, Wenxiang Zhao, Wenguang Li

A Design Method of Reconfigurable Controller for AC Position Servo Systems 1820
Wu Qinmu, Qin Yi, Li Yesong

Position Sensorless Control of PMSM Based on a Novel Sliding Mode Observer over Wide Speed Range 1825
Song Chi, Student Member, Longya Xu,

Design of Motion Control System Used for Filter Rod Production Machine 1832
Yang Qingyu, Ge Sibo, Ye Kesong, Shi Ren

Analysis and Implementation of Sensorless Position Detection in a Permanent Magnet Generator 1836
Sebastian Rosado, Xiangfei Ma, Fred Wang, Jerry Francis, Dushan Boroyevich

Torque-Speed Characteristics of Interior-Magnet Machines in Brushless AC and DC Modes, with Particular Reference to Their Flux-Weakening Performance 1841
Y. F. Shi, Z. Q. Zhu, D. Howe

H8 Robust Control for Dual Linear Motors Servo System 1846
Zhao Ximei, Guo Qingding

Research on Linear Motor Driving System Based on Wavelet Transform 1849
Cui Jiefan, Zhao Lijun, Wang Hemin, Wan Junzhu, Jiang Lili

Study on Rotor Position Detection Error in Sensorless BLDC Motor Drives 1853
Li Qiang, Wang Ruixia

A New Scheme to Direct Torque Control of Interior Permanent Magnet Synchronous Machine Drives for Constant Inverter Switching Frequency and Low Torque Ripple 1858
Jun Zhang, M. Faz Rahman, Colin Grantham

A Modified Direct Toque Control for Interior Permanent Magnet Synchronous Motor Drive Without a Speed Sensor 1863
Yanping Xu, Yanru Zhong, Hui Yang

Direct Torque Control for Interior Permanent Magnet Synchronous Motors Using Matrix Converters 1867
D. Xiao, M. F. Rahman

A Neural Network Based Initial Position Detection Method To Permanent Magnet Synchronous Machines 1872
Mengjia Jin, P.C.K Luk, Jianqi Qiu, Cenwei Shi, Ruiguang Lin

A New Recurrent Fuzzy Neural Network Sliding Mode Position Controller Based on Vector Control of PMLSM Using SVM 1877
Junyou Yang, Ruijuan Chen, Naiguang Fa

DSP Implementation of Rotor Position Detection Method for Hybrid Stepper Motors 1882
M. Bendjedia, Y. Ait-Amirat, B. Walther, A. Berthon

An In-Wheel Switched Reluctance Motor for Electric Vehicles 1887
P.C.K. Luk, P. Jinupun

Speed Sensorless Vector Control of Induction Motor Based on Full-Order Flux Observer 1892
Shanshan Wu, Yongdong Li, Zedong Zheng

A Parameter Identification Method for General Inverter-fed Induction Motor Drive 1896
Xiaochun Jiang, Geng Yang, Yunfei Wang

Indirect Rotor Field Orientation Vector Control for Induction Motor Drives in the Absence of Current Sensors 1901
Z. S. WANG, S. L. HO

A Robust Adaptive Sliding-Mode Controller for Slip Power Recovery Induction Machine Drives 1906
J.Soltani, A. Farrokh Payam

xxi

Table of Contents

Identification of the Rotor Time Constant in Induction Machines without Speed Sensor..1912
M. Li, J.N. Chiasson, M. Bodson, L.M. Tolbert

Adaptive Control of Doubly Fed Field-Oriented Induction Machine Based On Recursive Least Squares Method Taking the Iron Loss Into account..1917
N. R. Abjadi, J. Askari, J. Soltani

Analysis and Design of PDM Converter with High Frequency Link for HEV Drive System..1922
Ma Xianmin

A Multi-Directional Power Converter for a Hybrid Renewable Energy Distributed Generation System with Battery Storage..1926
Mei Qiang, Wu Wei-Yang, Xu Zhen-lin

Four-bridge Multilevel Converters Based on Hybrid-clamped Techniques..1931
Xiaofeng Wang, Yan Deng, Xiangning He

Standardization of Input/Output Impedance Specifications of Buck Converters Based on the System Integration Concept..1936
Tao Wu, Xinbo Ruan

Research on The Magnetic Integration in Three-Level ZCS Quasi-Resonant Buck Converter..1942
Jiang Ying, Xiang Hui-jie, Yang Yu-gang, Liu Nan

Decoupling Control of Magnetically Levitated Induction Motor with Inverse System Theory..1947
Yang Zhou, Huangqiu Zhu, Tianbo Li

Fault Detection and Accommodation for Nonlinear Systems Using Fuzzy Neural Networks..1952
H. Xue, J.G. Jiang

A Novel Constant Power Control of High Frequency Electronic Ballast Applying the PLL Technique for a Metal Halide Lamp..1957
Chang-Hua Lin, Chung-Lun Ou, Tien-Shuo Liu, Ken-Chuan Hsu

The Voltage Stability Research of Ship Electric Power System..1962
Fanyinhai Zhaomin

Parasitic Gate Resistance and Switching Performance..1967
Alan Elbanhawy

PWM Rectifier with DC Reverse-Blocking Diode for High-Reliability Generating Apparatus and Its Application to Gas Heat Pump System..1971
Akio Toba, Toshihiro Maeda, Kouetsu Fujita, Tomohiko Kato

A Novel Stator Section Crossing Method of Long Stator Linear Synchronous Motor for Maglev Vehicles..1976
Qian Zhang, Fei Lin, Xiaojie You, Trillion Q. Zheng

Common Mode Current Suppression in Full-Bridge Converter Based on Simulated Annealing Algorithm..1981
Yonggao Zhang, Kai Zhang, Yunbin Zhou, Yong Kang

Summary of Distance Measurement Based on Vision in Localization Technology..1986
Handong Zhang, Gang Wang, Yuwan Cen

The studies of Single-phase Inverter Fault Diagnosis Based on D-S Evidential Theory and Fuzzy Logical Theory..1991
Wang Baocheng, Li Danhe, Sun Xiaofeng, Wu Weiyang

A Novel Single-Stage High-Power-Factor Electronic Ballast with Symmetrical Half-Bridge Topology..1995
Chien-Ming Wang, Chien-Yeh Ho

Smoothed-Power Output Supply System for Battery of Stand-alone Renewable Power System Using EDLC..2000
Y. Jia, R. Shibata, N. Yamamura, M. Ashida

Table of Contents

Supercapacitors characterization for hybrid vehicle applications 2005
F. Rafik, H. Gualous, R. Gallay, A. Crausaz, A. Berthon

Power Transfer Maximization and Di/Dt Based Extremum Tracking for a Swing Engine Based Portable Power System 2010
Satish Rajagopalan, Deepak M. Divan, Ronald G. Harley, J. Rhett Mayor

3D FEA of the Stator of the Linear Magnetic Flux Compression Generator 2015
Yanjie Cao, Chengxue Wang

The Effect of Current Control Strategies on Power Consumption of a Magnetically Levitated Turbomolecular Pump 2018
A.E. Hartavi, R.N. Tuncay, M.N. Sahinkaya

Direct Torque Control of an Interior Permanent Magnet Synchronous Machine fed by a Direct AC-AC Converter 2023
D. Xiao, M. F. Rahman

Control of Distributed Power Systems 2029
Z. Chen, Y. Hu, F. Blaaberg

Design Challenges For Distributed Power Systems

Fred C. Lee, Ming Xu, Shuo Wang and Bing Lu

Center for Power Electronics Systems
The Bradley Department of Electrical and Computer Engineering
Virginia Polytechnic Institute and State University
Blacksburg, VA 24061 USA
Email: fclee@vt.edu

Abstract: **Remarkable progresses have been made over the past decade in power conversion technologies, including advanced power semiconductor devices, power management ICs, innovative circuit topologies, and packaging and integrated system solutions. These technological advancements have been manifested in a wide range of products and applications with ever increasing performances, efficiency, and power density. This paper highlights some of the challenges and opportunities of power conversion technologies in the Distributed Power System (DPS) for computer, telecommunication and network products. Topics discussed in this paper include improved EMI filter design techniques to mitigate the detrimental effects of filter/converter parasitics; impacts of the operating frequency of PFC to the size and weight of EMI filter; power conversion architecture and potential simplification; high-frequency high-density AC/DC and DC/DC topologies and designs; bus converters; as well as non-isolated point- of -load converters**

Keywords-DPS, Power Architecture, EMI filter, AC/DC, DC/DC

I. INTRODUCTION

Widespread use of the internet and telecommunication requires infrastructure support using a more sophisticated, high-quality and reliable "power network" that naturally takes the form of distributed power generation, distribution and regulation. Such a system is expected to achieve fully controllable, fully reconfigurable, autonomous platforms and customized for ever changing applications. It is envisioned that these advanced systems will be required to provide on-demand power from a required source, and required load at any rate and in any desired form. A typical distributed power system (DPS) as shown in Fig. 1, is

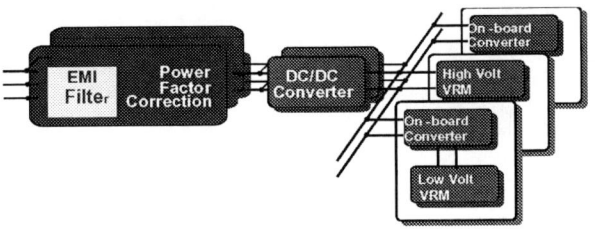

Fig. 1. Distributed power system

configured to be scalable and adaptable to ever changing power requirements of computers and telecommunication equipment and systems. The frond-end AC/DC converter is a standardized module with the paralleling capability to convert the AC voltage source into the 48V DC voltage bus. This DC bus voltage is often distributed via a back-plane into various circuit boards in a form of plug-in modules, or "circuit packs" in the telephone jargon, with additional point-of-load DC/DC modules locally supplying the need power levels at the appropriate voltages to the end users.

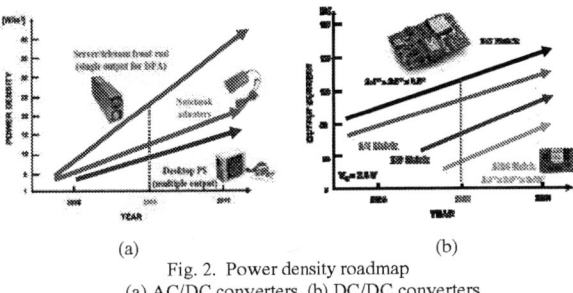

(a)   (b)
Fig. 2. Power density roadmap
(a) AC/DC converters, (b) DC/DC converters

With ever increasing in functionality and shrinking in size and weight of all forms of computer and telecommunication equipment, it is essential to pack the advanced power hungry processors onto each circuit boards together with the high quality customized and miniaturized power. The aggressive power density targets for AC/DC converters and DC/DC modules, as shown in Fig. 2 [1], respectively were met in the past decade and perhaps in the foreseeable future. The typical real estate utilization of the state-of-the-art 3kW 1U telecom AC/DC with 25W/inch^3 is shown in Fig.3. From this picture, it can be easily seen that EMI filter, PFC inductor and bulk capacitors and magnetics in the DC/DC converter represent the major portion of the system.

With rapidly growing computer, telecommunication, and internet technologies and applications, power supplies, other than demanding for higher power density, must achieve higher power conversion efficiency, especially at lighter load, in order to meet the stringent regulation for energy saving.

1-4244-0448-7/06/$25.00 ©2006 IEEE

In the following sections, design challenges and opportunities for EMI filter, front-end AC/DC, board mounted DC/DC will be presented. More distributed power architecture with potential cost saving and performance improvements is proposed. At the meantime, system packaging and integration technologies leading to further improvement of power density and performances are discussed.

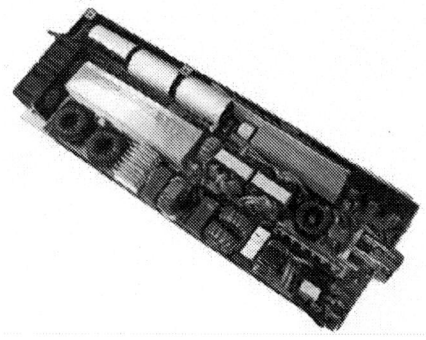

Fig. 3. State-of-the-art front-end converter layout

II. EMI FILTER DESIGN CONSIDERATIONS AND CHALLENGES

In power electronics applications, electromagnetic (EMI) filter is a necessary interface between the power line and power conversion equipment, such as AC/DC and DC/DC converters operating in high frequency switching mode. They generate conducted switching noise with a spectrum that ranges from the designed switching frequency up to 30MHz. EMI standards, such as EN55022 class A, specify the frequency range (150kHz-30MHz) and noise limits which all power supplies must meet. In order to satisfy these EMI standards, one or two stages of EMI filters are usually employed. A typical one-stage EMI filter used in power supplies is shown in Fig. 4.

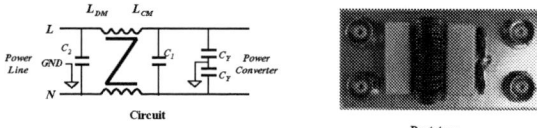

Fig. 4. One-stage EMI filter under investigation

Due to the self- and mutual parasitics, the EMI filters do not work as well as expected at high frequencies (HF). Fig. 5(a) compares three DM insertion voltage gain curves, the curve that represents filter with ideal components, the curve that includes components together with their associated self parasitics, and the curve including self parasitics and parasitics due to the coupling of electric field or electro-magnetic field referred to as mutual parasitics. It is shown that the self parasitics make DM filter's HF performance much worse than that of an ideal filter; however, the mutual parasitics finally determines filter's HF performance. For CM filter performance shown in Fig. 5 (b), the self parasitics, especially the winding capacitance (EPC) of

inductors degrades significantly the filter high frequency performances. Investigation [2, 3] shows the mutual coupling (mutual inductance) between inductor and trace loops, between inductor and capacitor, between two capacitors significantly affect differential mode (DM) EMI filter's high frequency performances. Equivalent series inductance (ESL) of capacitors also plays the role on the filter performances. The equivalent parallel capacitance (EPC), i.e. winding capacitance, of the inductor is a key factor detrimental to CM filter's high frequency performances [8]. These parasitics must be minimized to improve EMI filter's ability to attenuate high frequency noises.

The mutual inductance between inductor and trace loops can be easily reduced by reducing the loop areas. One simple method to minimize the coupling between the inductor and capacitors is to rotate the inductor winding by 90° as shown in Fig. 6 [3]. In so doing, the mutual inductance can be reduced by as high as 92% (89.3nH vs. 7.5nH). To reduce the mutual inductance between two capacitors, one simple method is simply to arrange physically the two capacitors to be perpendicular [4]. In this manner, the coupling between the two capacitors can be reduced by 66% as shown in the measurement result. An even better method is to place a ¾ turn in parallel with one of the capacitor as shown in Fig. 7 [7].

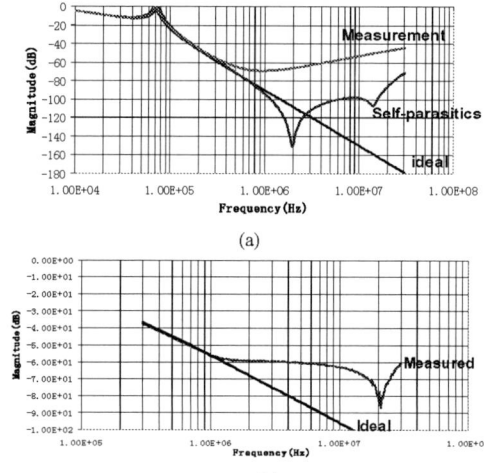

(a)

(b)

Fig. 5. Comparison of insertion voltage gains: (a) Differential Mode (DM) and (b) Common Mode (CM)

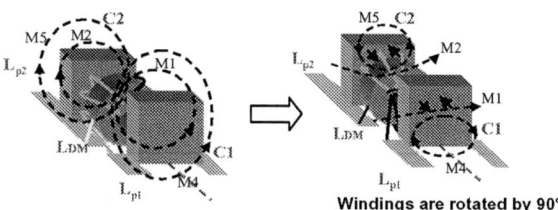

Fig. 6. Reducing the mutual inductance between the inductor and capacitor

2

In Fig. 7, the magnetic flux linking C2 also links cancellation turn, so the mutual inductance between two capacitors is cancelled. A 92.3% reduction (439pH vs. 19pH) is achieved in the measurement. The suggested methods illustrated in Fig. 6 and Fig. 7 can be employed at the same time to achieve even better results. Fig. 9 shows the measured insertion voltage gains with these proposed methods and compared them with the original EMI filter without incorporating any suggested methods for improvement.

Fig. 8 shows that the combination of methods suggested in Fig. 6 and Fig. 7 gives best result. A 40dB improvement achieved at 30MHz. It is well known that capacitor behaves like an inductor at high frequencies after the series resonant frequency between the capacitance and the ESL. For a 0.47μF/400V film capacitor, the measured series resonant frequency is around 2MHz, which means this capacitor behaves like an inductor beyond 2MHz. For an electrolytic capacitor (220μF/250V), the measured series resonant frequency is even below 100 kHz. If this electrolytic capacitor is used as a bulk capacitor for a converter with switching frequency above 100 kHz, the capacitor actually works like an inductor even for the fundamental component of the switching waveform.

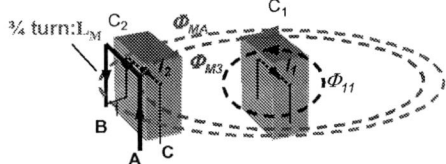

Fig. 7. Integrating cancellation turn with capacitor to cancel mutual inductance between two capacitors

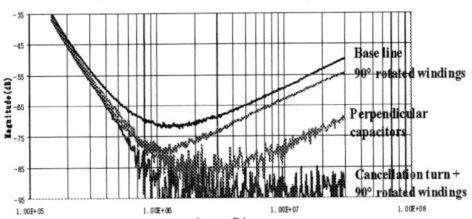

Fig. 8. Comparison of filter performance

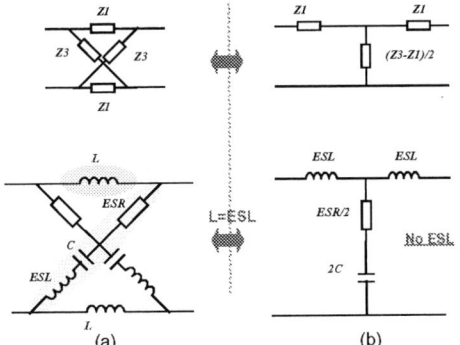

Fig. 9. Illustrating the idea of ESL cancellation: (a) network with cancellation inductor and (b) the equivalent network without ESL

Wang and Lee [9] further proposed a method to cancel the ESL of capacitors so as to significantly improve capacitor's filtering performance at high frequencies. The idea is shown in Fig. 9.

Fig. 9 shows that the two networks on the top are equivalent. If two capacitors are diagonally connected and the inductor L is chosen with a value equal to the ESL of the two capacitors, then the resultant network is a T filter with the capacitor free from ESL as shown in the bottom diagram. In the proposed concept, the L can be implemented with the PCB trace inductance. Fig. 10 shows the measured insertion voltage gains with and without ESL cancellation. It is shown that after ESL is canceled, the film capacitor's performance is improved significantly above 3MHz and a 27dB improvement is achieved at 30MHz when compared with two paralleled capacitors. For electrolytic capacitors, a 27dB improvement at 30MHz also demonstrated and performance improvement above 150 kHz is also noted. The performance improvement for both film and electrolytic capacitors is further verified in [9] by measuring the EMI noise of a converter using a noise separator [6] and a spectrum analyzer.

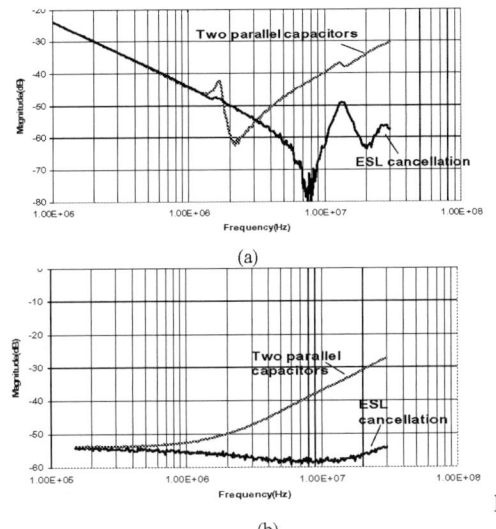

Fig. 10. Comparison of capacitor performance: (a) film (0.47μF/400V) and (b) electrolytic (220μF/250V)

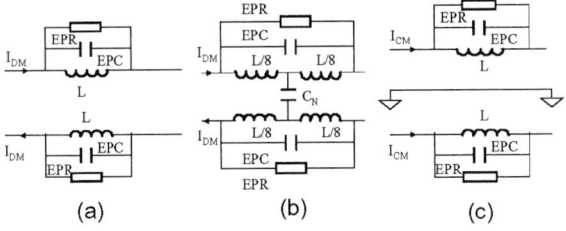

Fig. 11. Inductor model: (a) separate DM inductor model (b) model for two DM inductors in one core and (c) CM inductor model.

The method for cancellation of winding capacitance EPC of an inductor was proposed recently [10, 11]. Fig. 11

shows models for DM inductors and CM inductors taking into consideration of EPR and EPC. Fig. 12 shows the ideas of winding capacitance cancellation.

If the two DM inductors are separated, i.e. not on one core, the parasitic model can be shown in Fig. 11(a). If two DM inductors are on one core then there is equivalent parasitic capacitance C_N between two inductors as shown in Fig. 11(b), since the permittivity of the core would be relatively high and the distance between two inductor windings are relatively close. For CM inductor the model is shown in Fig. 11 (c).

In order to cancel the winding capacitance of DM inductors, two extra small capacitors are needed. For separate DM inductors, two small capacitors with values equal to EPC are diagonally connected to two inductors as shown in Fig. 12(a). The cancellation mechanism can be explained by network equivalence as well as mutual capacitance theory developed in [10]. For combined DM inductors, since two windings are on the same core, the effect of C_N can be equivalent to a negative capacitor with value of $C_N/2$ in parallel with each inductor [10, 11]. If the equivalent negative capacitance is smaller than original EPC, then the total winding capacitance is still positive. Two small capacitors with values of EPC-$C_N/2$ can be diagonally connected to two inductors to cancel it. This is shown in Fig. 12 (b). However, if the total winding capacitance is negative, then two small capacitors with values of $C_N/2$-EPC should be parallel with two DM inductors. Ironically, in order to cancel the winding capacitance, external capacitors are needed in parallel with windings. This is shown in Fig. 12 (c).

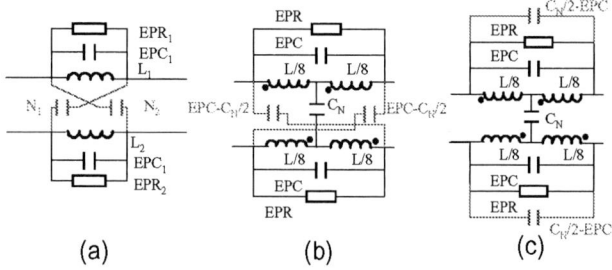

Fig. 12. Winding capacitance cancellation for DM inductors: (a) separate DM inductor, (b) two DM inductors on one core when equivalent winding capacitance is positive and (c) two DM inductors on one core when equivalent winding capacitance is negative

For CM inductors, the method proposed for DM inductors can also be applied if a winding is allowed in ground path. If the winding is not desirable on ground path, the method proposed in [4] can be considered. For the method in [4], the coupling coefficient between two half windings of one side must be as high as 0.9999, which is possible to be realized in an integrated inductor structure. However, for a conventional discrete CM inductor winding structure, the coupling coefficient is around 0.99. Therefore, the idea can not be applied, since a resonance between the leakage of

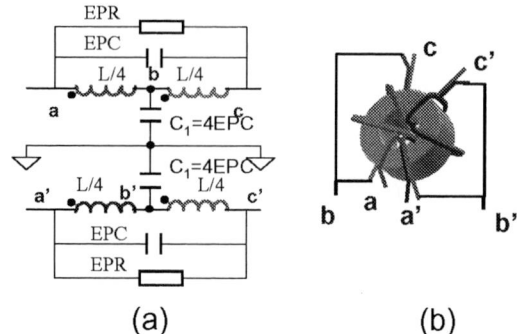

Fig. 13. The cancellation of winding capacitance for CM inductor: (a) Cancellation idea and (b) proposed bifilar winding structure

two winding halves and the cancellation capacitor can cause high frequency noises.

A bifilar winding structure, which is not used in conventional CM inductors due to large EPC, is proposed for high coupling coefficient. Because two winding halves are almost in the same position, the coupling coefficient is very high. The measured value is 0.99995. Although EPC is enlarged, the EPC cancellation techniques can be employed and the performance improved.

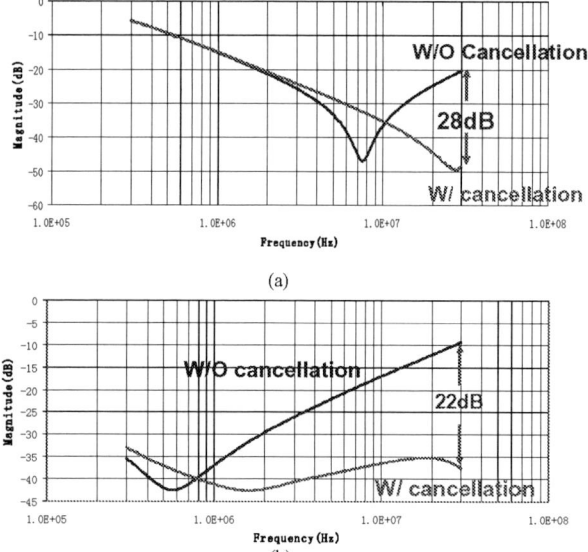

Fig. 14. Improvement on inductor impedance: (a) DM inductor and (b) CM inductor

Fig. 14 (a) and (b) show the measured impedance of the filter inductor. Impedance at high frequencies is significantly reduced for both DM and CM inductors due to winding capacitance cancellation. Both DM and CM inductor filtering performance is improved by more than 20dB. More elaborated EMI noise measurements on the proposed EPC cancellation techniques when applied to a power converter were reported in [10, 11] with significant improvements shown.

In this section, various techniques for cancellation of both mutual parasitics and self-parasitics are introduced. A much better understanding of the basic nature of EMI noises had led to development of in the practice. Continuous research in this area will eventually lead to an understanding and development of this subject matter as a branch of science in stead of art or "magic" as it is often referred to.

III. HIGH FREQUENCY HIGH DENSITY FRONT-END AC/DC

For the front-end AC/DC converters, normally they are implemented by two-stage approach. The first stage is the power factor correction (PFC) stage, which is able to provide power factor correction function and a constant voltage for the DC/DC stage input. The second stage is the DC/DC stage, which gives a regulated output voltage for the load. For the PFC stage, the single switch continuous current mode (CCM) PFC is the most widely used topology because of its simplicity and smaller EMI filter. For the whole front-end converter, the EMI noise major comes from the PFC stage. The differential mode and common mode EMI noise models for the single switch CCM PFC circuit have been derived and verified by CPES. Based on the derived the EMI noise model, the switching frequency impact on the EMI filter design of the PFC circuit is analyzed. The switching frequency selection guideline is given based on the analysis.

III-1. Benefits from High Frequency PFC:

Based on the filter attenuation requirement, we are able to develop the relationship between the corner-frequency of the filter and the switching frequencies of the converter, as shown in Fig. 15(a) [13~15]. In these curves, the zigzag nature of the curve is caused by the EMI standard which specifies the regulation requirement beginning at 150 kHz. Those vertical jumps at frequencies representing the various sub-harmonics of 150kHz, such as 50kHz, 75kHz, etc. When the switching frequency is slightly lower than those frequencies, the corner frequency of the DM filter is much higher. Beyond 150 kHz, the filter corner frequency will continue to increase as the switching frequency increases. From the curves, it can be seen that when the switching frequency is roughly higher than 400-500 kHz, the corner frequency of the DM filter will be higher than all the previous peaks. As it is know, higher the corner frequency means smaller the EMI filter size. Therefore, one can observe that the size reduction of the DM filter is achieved only if the switching frequency is higher than 400-500 kHz, in comparison with that at 150 kHz.

Similar observation can be made for the CM filter. From Fig 15(b), it can be seen that irrespective to what K value is (where K is related to the circuit parasitic, the switching speed of the MOSFET and the output voltage of the PFC circuit.), when the switching frequency is higher than 400kHz, the size of CM filter will be smaller than that previously achieved at 150kHz. Further increasing of the switching frequency will continue to reduce filter size.

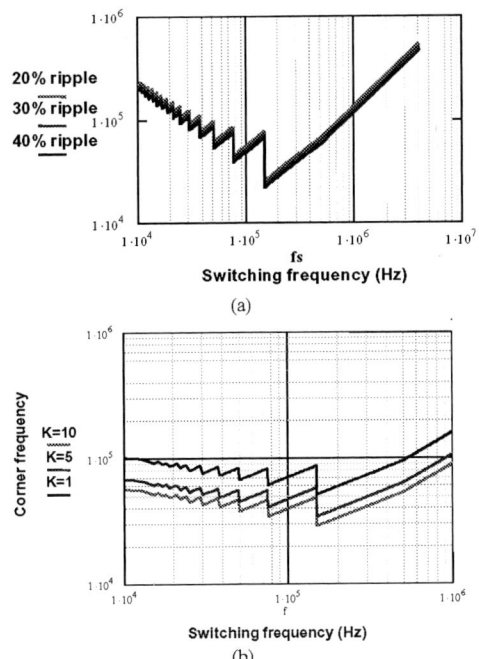

Fig. 15. Filter corner frequency vs. switching frequency:
(a) Differential Mode (b) Common Mode

From the previous analysis, we can conclude that the switching frequency has great impacts on the EMI filter design. Since the EMI spec begins at 150 kHz, certain frequency range should be avoided. Clearly, there is no filter size reduction achieved when the switching frequency is chosen between 150-400 kHz. Since the new generation of power devices, such as CoolMOS and SiC, enable the PFC circuit to operate at a much higher switching

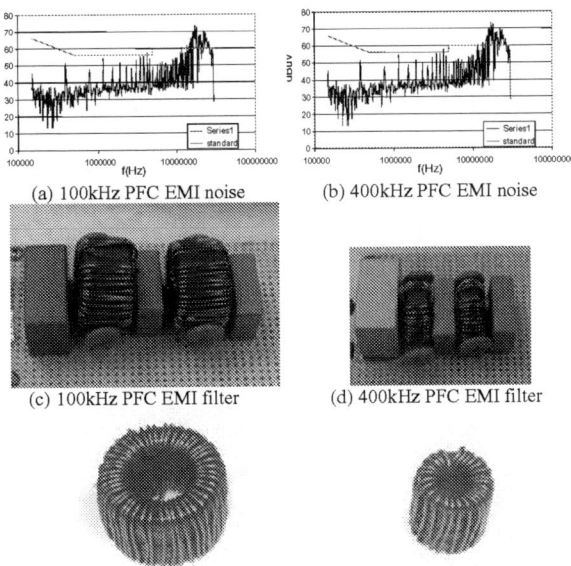

(a) 100kHz PFC EMI noise   (b) 400kHz PFC EMI noise

(c) 100kHz PFC EMI filter   (d) 400kHz PFC EMI filter

(e) 100kHz Boost inductor   (f) 400kHz Boost inductor

Fig. 16. EMI noise after filtering and EMI filter size comparison

frequency. It is possible to contemplating a switching frequency higher than 400 kHz. To illustrate this point, two CCM PFC circuits were built, one operated at 100 kHz and the other at 400 kHz. Two EMI filters were designed, respectively to meet the EMI standard EN55022 Class B. As shown in Fig. 16, both of the converters can meet the standard at the low frequency range. For the two different EMI filter and boost inductor, we can see that 35% and 55% volume reduction is achieved respectively by pushing the switching frequency from 100kHz to 400kHz. The excessive noises at high frequencies are due to parasitics of the EMI filter as discussed in the previous section. The problem can be solved by using the suggested techniques, grounding and shielding technologies.

In Fig. 17(a), the impedances of the boost inductors when designed for 100 kHz PFC, there is a resonant valley at the frequency about 17MHz. At this frequency, the impedance of the boost inductor is small. Therefore, the EMI noise at corresponding frequency is high, which can be seen in Fig. 17(b) [15, 16]. But for the 400 kHz boost inductor, the resonant valley is pushed beyond 30MHz, as discussed in the previous section; the filter's ability to attenuate high frequency noises is compromised with the presence of parasitics. Therefore, it is important to eliminate the high frequency noise peaks. From the experimental results, we can clear see that the benefit brought by pushing the switching frequency higher.

To get high switching frequency for PFC, CPES has evaluated different device combinations. As shown in Fig. 18, the revolutionary devices, CoolMOS[TM] and SiC diode enable much higher efficiency and higher frequency [17]. A

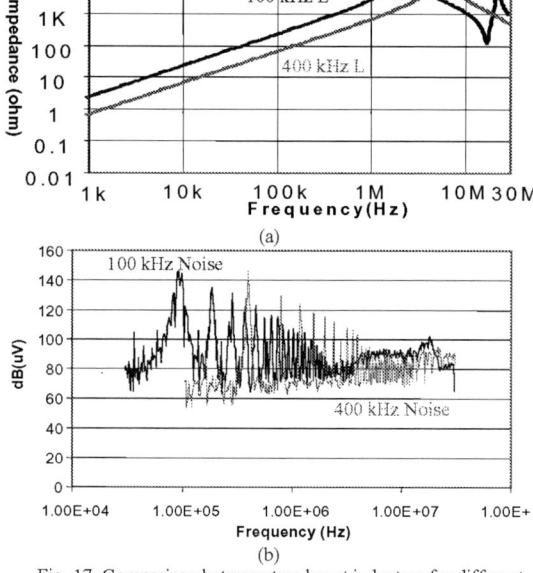

Fig. 17. Comparison between two boost inductors for different switching frequencies: (a) Impedance, (b) EMI noise

1kW CCM PFC using CoolMOS TM and SiC diode running at 400 KHz is demonstrated. It shows that by using the CoolMOS[TM] and SiC diode, the 400 KHz CCM PFC can achieve the similar efficiency by using conventional devices at 100 KHz switching frequency, as shown in Fig. 18.

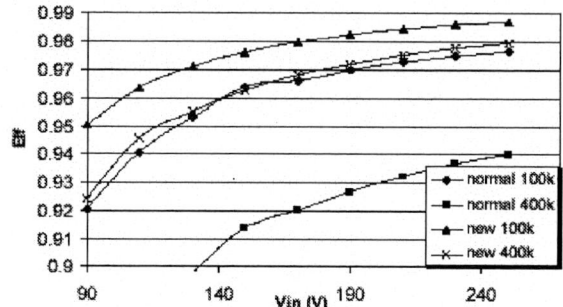

Fig. 18. Efficiency comparison between different PFC designs
Normal: IRFP460 + RHRP860, New: CoolMOS + SiC diode

III-2. High frequency DC/Converter:

To achieve high power density for the front-end AC/DC converters, increased switching frequency is desired. However, its effectiveness is limited by the large holdup capacitors. For the telecom and computer applications, system is required to maintain a regulated output voltage with full power for 20ms after AC input is lost. Therefore, bulky capacitors are employed to provide the needed energy during holdup time. The size of the selected bulk capacitors will directly impact the input voltage range of the down stream DC/DC converter, thus its conversion efficiency and power rating. To reduce the holdup time capacitors and improve converter power density, it is essential to select an appropriate converter topology that can achieve high efficiency with wide input voltage range, especially high efficiency at 400V input [18~20].

For conventional PWM DC/DC converters, a maximum duty cycle is designed for the minimum input (say, 300V) which occurs when the AC input power is lost. Consequently, the given circuit will operate at a smaller duty cycle when the input voltage is at around 400V. Thus, to realize a wide input range, it is inevitable that the converter efficiency is suffered at the nominal input of 400V. Fig. 19 demonstrates the efficiency of an asymmetrical half-bridge (AHB) optimally designed for a fixed input voltage at 400V. The circuit can achieve 94.5% efficiency. When the same circuit is designed to operate at 300V to 400V input range, the converter efficiency can only achieve 92% at 400V [20].

Instead of using PWM converters, certain class of resonant converters such as the LLC resonant converter as shown in Fig. 20(a) is able to operate with a wide input range without compromising circuit efficiency at the desired operating voltage, i.e. at 400V. Because the magnetizing inductor participates in resonant, converter voltage gain characteristic is change, as shown in Fig. 20(b).

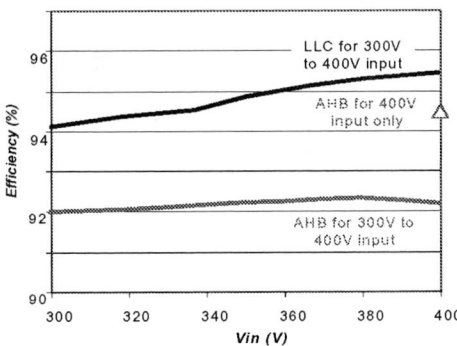

Fig. 19. Efficiency of AHB and LLC converters

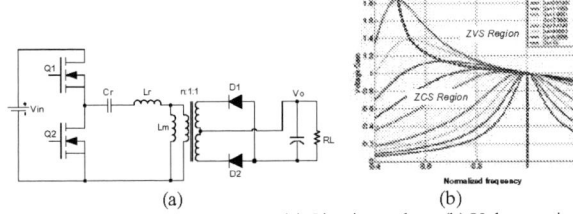

(a) (b)

Fig 20. LLC resonant converter (a) Circuit topology (b) Voltage gain

The LLC converter can achieve a voltage conversion ratio either higher or lower than unity. Moreover, the zero voltage switching (ZVS) can be achieved with switching frequency both lower and higher than the series resonant frequency determined by Lr and Cr. By choosing a suitable transformer turns-ratio, the converter can be targeted to operate right on top of the resonant frequency with optimal efficiency at normal operation condition, i.e.400V input. During holdup time, input voltage drops, and LLC resonant converter reduces its switching frequency and increase voltage gain to maintain regulated output voltage. Although LLC converter operates far away from the resonant point during the holdup time which means the circuit is less efficient, it only lasts for 20mS and will not cause extra thermal problem. Because LLC converter can operate at resonant frequency during normal operation condition, the circuit is operated at the most efficient point and its efficiency could be much higher in comparison with AHB or other PWM topologies. As shown in Fig. 19, with 200 kHz switching, LLC could achieve 2 to 3% efficiency improvement over that of AHB. Moreover, LLC resonant is operated with ZVS turn-on and relatively small turn-off current. These properties make   switching losses at the primary side switches very small. Besides, the secondary side diodes are also operated with ZCS thus reduces the diode reverse recovery loss. The much reduced switching losses enable LLC resonant converter to operate at much higher switching frequencies while maintaining high efficiency. A proto-type 1MHz LLC was developed with 94.5% efficiency at 1kW output. The efficiency of LLC resonant converter for different switching frequency is shown in Fig. 21. As shown in Fig. 22(b), 1MHz LLC achieves 76W/in³ power density [20, 22].

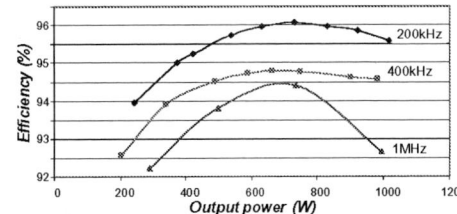

Fig. 21. Efficiencies for LLC converter with different switching frequency

(a) (b)

Fig. 22. Comparison among different DC/DC converter designs: (a) 200 kHz AHB, 12W/in³, (b) 1MHz LLC, 76W/in³

III-3. High density front-end AC/DC via IPEMs

Although higher density is achieved at higher frequency, some fundamental limitations prevent further improvement in power density. For example, the parasitic inductance in the switch commutation loop hampers the switching speed and causes more switching losses. Large voltage stresses appear on switching devices due to large parasitic inductance, which compromises reliability. The parasitic junction capacitances between high voltage transition points to the earth ground increases the common mode noise. Moreover, the parasitic components brought by the interconnection of the electrical layout played a negative role in the system electromagnetic interference noise (EMI) if it is not treated carefully.

The front-end ac/dc converters are essentially custom-designed and manufactured using discrete parts, which high labor content and high cost. To address the aforementioned performance issues and cost, the integrated power electronics module (IPEM) concept was proposed by Center of Power Electronics Systems (CPES). The IPEM concept is to explore the integration of discrete power devices to the extend that it is technologically practical and economically feasible. To this end, the active power semiconductor devices, with its associated drivers, protection circuits, sensors and controller are integrated together in the form of modular building block here referred to as active IPEM. Similarly, the passive power components such as inductors, capacitors and power transformers are integrated together into a Passive IPEM. One apparent benefit of integration is the size reduction. The integration of the switching devices together with their associated gate driver circuit, invariably will reduce parasitics associated with interconnects, thus resulting in smaller switching losses and voltage stresses. By integrating inductors, capacitors, together with power transformer, passive component size can be greatly reduced. Moreover, due to the integration, circuit component number can be greatly reduced. The assembly of such modules could be automated thus reduce labor content. Furthermore, by perfecting the process of integration, IPEMs can become standard building blocks to facilitate system integration. Therefore, the reliability, product cycle time and cost can be significantly reduced [23, 24].

To demonstrate the benefits of the IPEM concept under the system level, two 1kW front-end AC/DC converters were built using exactly the same topologies, one using discrete devices and the other one using IPEMs. As shown in Fig. 1 and 23, the system is constructed by the two stages: PFC stage and DC/DC stage. For the PFC stage, a 400 kHz single switch PFC using CoolMOS and SiC diode were chosen to reduce the boost inductor and EMI filter. The Asymmetrical Half Bridge operating at 200 kHz was used for the DC/DC stage. The converters are designed for the universal input line 90V~264V, and the power rating is de-rated to 600W when operated below 150V input. Two converters are show in Fig. 23. The discrete components have been replaced by the Active and Passive IPEMs. Through integration, density and form factor of the active and passive devices are much improved. Therefore, the power density at the system level is improved significantly. As for the discrete approach, the power density is 7.5 W/In^3, and for the IPEM-based converter, the power density is 11.4 W/In^3 and still has the room for improvement.

Fig. 23. Comparison between the discrete and integrated DPS System:
(a) Discrete devices DPS, (b) IPEM-based DPS

By replacing the discrete active and passive devices into the IPEMs, the number of components of the systems can be reduced from several hundreds part to about 20-30 parts. Not only the system power density has been improved, the system electrical performance has been improved as well. The system efficiency increases more than 2% at the high line voltage range, and more than 3% at 90V. Since the conduction losses are roughly the same for the same operation condition, the major improvement lies on the switching loss reduction by minimizing the circuit parasitics.

At the same time, due to the smaller parasitic inductance on critical path of the converter, less voltage stress is achieved. For the discrete PFC switch, when switch turn off occurs at 7Amp, the device voltage overshoot is 123V. But when the discrete components are replaced with an active IPEM, the voltage over-shoot is reduced to 72V even at 10Amp.

IV. HIGH FREQUENCY HIGH CURRENT DENSITY BOARD-MOUNTED ISOLATED DC/DC CONVERTERS

With rapidly growing computer and telecommunication applications, the point-of-load (POL) DC/DC module is becoming smaller and smaller, from non-isolated POL to isolated ¼ brick to 1/8 brick and even to 1/16 brick, and in the mean time, with continuous increasing in current demand and decreasing in output voltage. High power density and high efficiency are demanded by the customers Design engineers ,now a day are facing challenges in all fronts, including higher operating frequencies with reduced switching losses, conduction losses, body diode losses and even the gate driver losses; innovative packaging and thermal management; EMI and EMC containment and reduction.

This section will introduce two examples of circuit means of achieving higher operating frequencies and in the same time, higher efficiencies. These goals are realized by simultaneously reducing the primary side switching losses and conduction losses as well as the secondary side synchronous rectifier body diode conduction losses, reverse recovery losses, drive losses, and conduction losses.

IV-1. Single Stage isolated DC/DC:

A. State-of-the-art:
Fig. 24(a) shows a typical PWM hard-switching 48V input DC/DC topology. To get faster dynamic performance and higher power density, higher switching frequency is desired. However, as shown in Fig. 24(c), efficiency will suffer a lot at higher switching frequency mainly due to the switching loss, driving loss and SR body diode conduction loss. By far, the soft switching technique is a well-known approach to reduce the switching loss effectively.

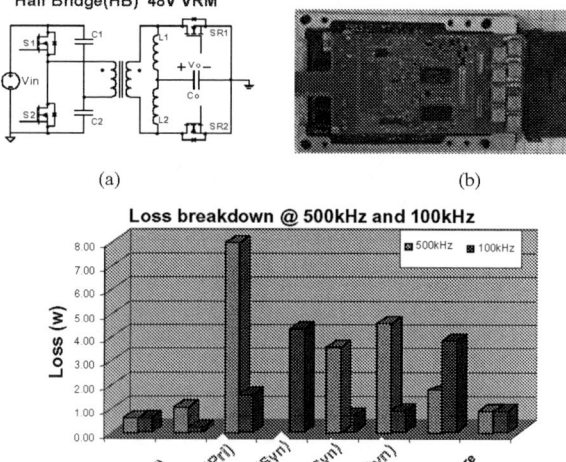

Fig. 24. State-of-the-art power pod in server

8

Therefore, Fig. 24(b) shows the state-of-the-art 48V DC/DC for server processor (power pod), running at 300KHz by employing phase-shift full-bridge ZVS topology. Its efficiency can be seen in Figure 30(b).

In the conventional synchronous rectifier, the dead time is necessary, accompanying with the body diode conduction loss and reverse recovery loss due to the current through the body diode during the dead time. There are some precise timing-control driving ICs to reduce the body diode conduction loss. However, they cannot effectively solve the body diode reverse recovery problem. And also, they are expensive and noise sensitive.

A. Self-Driven ZVS Full-Bridge DC/DC and Magnetic Integration:

The self-driven technique has been widely used in the industry practice due to their simplicity and low cost. However, for bridge-type symmetrical converters, implementation of self-driven capability is difficult because of the inherent dead time period. One possible solution is the level-shifted self-driven concept [25]. However, the proposed approach has several drawbacks: (a) Ringing occurs at the gate signal because the signal is coming from the main power transformer and severe ringing is coupled from the power stage, (b) there is extra body diode conduction loss, and (c) there is large amount of conduction loss due to the low driving voltage during dead time. These issues prevent the level-shifted self-driven concept from being used in high-frequency applications.

In order to overcome these issues, a self-driven ZVS full-bridge (FB) was proposed [26, 27]. The power stage is shown in Fig. 25. By simply rearranging the control strategy, it becomes very suitable for self-driven capability as well as achieving ZVS. It was further demonstrated that the self-driven scheme can save driving loss and body diode conduction loss and is very suitable for high-frequency applications where high power density is required.

A 1MHz prototype picture shown in Fig. 26(a) was built to verify this self-driven concept. Fig. 26(b) shows the efficiency comparison. Including the driver loss, the proposed self-driven ZVS full-bridge can achieve 81.7% efficiency. There is an efficiency improvement of 4.7% as compared with the conventional phase-shifted full-bridge with an external driver.

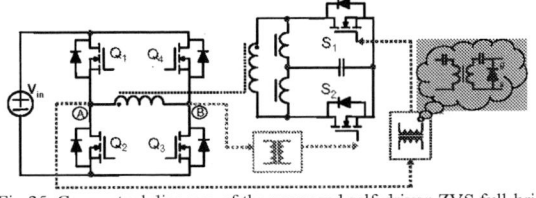

Fig.25. Conceptual diagram of the proposed self-driven ZVS full-bridge DC/DC converter

However, this topology requires a total of 3 discrete transformers, one main power transformer and two synchronous rectifier gate drive transformers, thus increasing the number of passive components. Further work

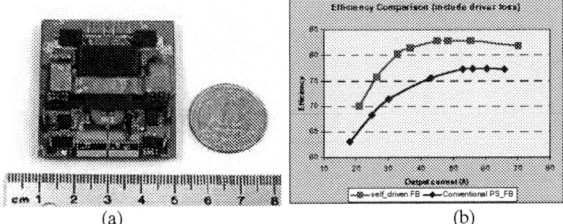

(a)   (b)

Fig. 26. (a) Hardware of the proposed DC/DC, (b) efficiency comparison

has been done to integrate all of the three magnetic components and two output inductors into a single core as shown in Fig. 27 [28]. This fully integrated magnetic has been adopted in a 1.2V/70A 1/8 brick prototype with 87% overall efficiency at 600 KHz switching frequency. Comparing to the state-of-the-art 1/8 brick product, it can deliver 40% more output current while having 2% higher efficiency.

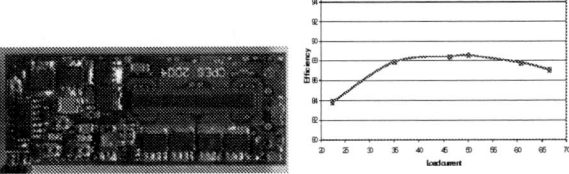

Fig. 27. 600 KHz 1.2V/70A1/8 brick hardware and measured efficiency

Generally, the gate driving loss can be reduced by the combination of the ZVS technique and Self-driven technique. As an example, the proposed self-driven ZVS full-bridge has the following advantages: (a) soft switching for primary switches; (b) clean gate signal and no level-shifting during dead time; (c) reduced gate driving loss; and (d) reduced body diode conduction loss. The experimental results verify that this topology is very promising in high-frequency applications. Furthermore, the proposed self-driven method can be applied to any bridge and non-bridge topologies employing complementary control.

B. Current Tripler Concept and Magnetic Integration:

For low-voltage and high-current applications, the secondary-side device switching and conduction losses have a major impact on system efficiency. To reduce the conduction loss in the secondary side, one solution is to reduce the on-resistances of the synchronous rectifiers and the transformer winding resistance. This can be realized by paralleling more synchronous rectifier switches (SRs) and enlarging the window area of the transformer. The drawbacks of this solution are the higher cost, larger gate driver loss and larger footprint.

Other than reducing the Rds (on) of the synchronous rectifier, proper secondary-side topologies should be selected to reduce the RMS current through the SRs. There are three major secondary-side topologies: forward rectifier, center-tapped rectifier and current-doubler rectifier. Among these three topologies, the current-doubler rectifier is the most suitable for high-current, low-voltage applications.

Because of its simpler transformer structure and halved inductor currents and transformer secondary currents, the current-doubler topology offers lower conduction losses than the conventional center-tapped topology [29].

The reason for the lower RMS current of the current-doubler rectifier is that during the freewheeling period (when there is no input-output energy transfer), both SR switches can conduct simultaneously to share the load current. As a result, the total rectifier conduction loss during the freewheeling period is reduced. To further reducing the conduction losses for higher current applications, often times, current-doubler with more semiconductor devices in parallel and distributed magnetics are used to reduce the transformer winding losses. However, those solutions have their limitations:

a) Increased cost,
b) Larger footprint and lower power density, and
c) More devices mean greater driver loss.

Recently, a novel current-Tripler topology was proposed, as shown in Fig. 28. The proposed topology can easily achieve ZVS for all MOSFETs, therefore switching loss is significantly reduced [30]. Through magnetic integration shown in Fig. 29, a three phase high frequency transformer can be used to greatly simplify the circuit.

Compared with the conventional current doubler, the proposed current Tripler can reduce the SR conduction loss and transformer winding loss by 20% and 12.5% respectively.

A 300 kHz prototype was developed to demonstrate the concept. Comparing to the typical industry design with the same spec and switching frequency, proposed current-tripler DC/DC converter can achieve 45% footprint reduction and 4% higher efficiency, as shown in Fig. 30.

IV-2. Two-Stage isolated DC/DC:

The 48V input DC/DC for high-end server and telecom applications requires higher voltage devices on the primary side and a transformer for isolation. To get fast dynamic response and regulation performance, the PWM type power conversion is preferred, whereas the efficiency and switching frequency is limited by the presence of the leakage inductances of the transformer. In order to achieve acceptable efficiency, lower switching frequencies, around 200~300 KHz, are normally adopted. Thus the size of the transformer and its passive components are bulky and the transient responses are slow. Excessive output capacitors are necessary to satisfy the dynamic transient requirement. In general, the isolated 48V DC/DC is normally customized design with higher cost, while its footprint and power density are significantly lower than that of the non-isolated POL converters.

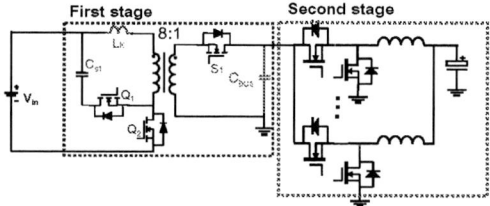

Fig. 31. Proposed two-stage solution

To leverage the 48V isolated DC/DC with standard high frequency non-isolated POL converter techniques, a superior two-stage approach was proposed [31], as shown in Fig. 31. The first stage utilizes a simple inductorless "DC/DC transformer operating at 1MHz switching frequency by adopting the resonant switching to minimize switching losses. The second stage employs the multi-phase buck capable of operating at multi-Mega Hertz, taking advantage of the already established infrastructure for low-voltage POL converters. Fig. 32 shows the two-stage prototype for the 48V power pod used in the server. Due to the MHz switching frequency in the second stage, the passive components can be greatly reduced. Table I lists the comparison between the CPES two-stage prototype and the industry single stage practice in Fig. 24(b), Based on the reduction of passive components, the power density of the power pod can be increased to around 150%. It is also apparent that since the output capacitors are significantly

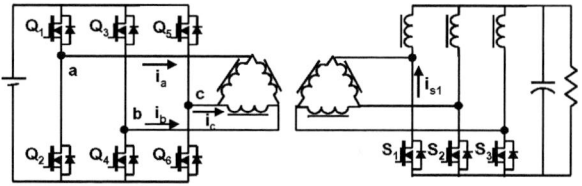

Fig. 28. Proposed Zero-Voltage-Switch current-Tripler DC/DC converter

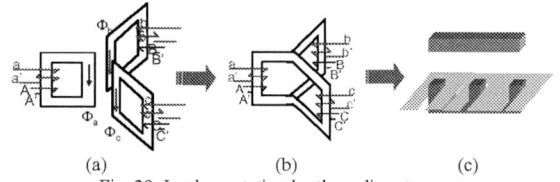

(a)   (b)   (c)

Fig. 29. Implementation by three discrete cores

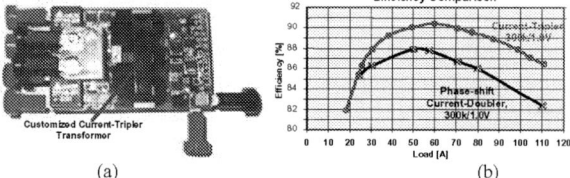

(a)   (b)

Fig. 30. (a) 300 KHz hardware of the proposed ZVS current tripler, (b) efficiency comparison with state-of-the-art

Fig. 32. 130W two-stage power pod prototype for server

10

TABLE I.   THE COMPARISON BETWEEN SINGLE-STAGE AND TWO-STAGE

	Switching frequency	Transformer	Inductor	Capacitor
Industry Practice	300kHz	Philips EI32	2 Philips EI22 (customized)	15*1mF Tantalum
CPES Prototype	1st stage: 300kHz, 2nd stage: 1MHz	Philips EI22	4 Vishay IHLP-5050FD 150nH	4*270uF ESRE
Volume Reduction		54%	90%	75%

reduced, the two-stage approach is less costly than the single-stage approach. This architecture has been quickly adopted by the industry and used in the current products.

IV-3. Intermediate Bus Architecture (IBA) and Bus Converter:

With the proliferation of low-voltage, high-current microprocessors/DSPs and high-voltage analog devices on a single circuit board, the number of different voltages encountered has mushroomed. All these voltages share a common ground so that it is unnecessary to use an isolated transformer for each of these loads, respectively. Therefore, the two-stage concept aforementioned was extended to the sub-system level, as shown in Fig. 33, in which an isolated bus converter steps down the 48V to an intermediate bus voltage to feed all the non-isolated point-of-load converters (POL) in the same board. This concept has been adopted by industry and is becoming a mainstream for high-end server and telecommunication applications, because it is more cost-effective, more flexible in terms of system structure.

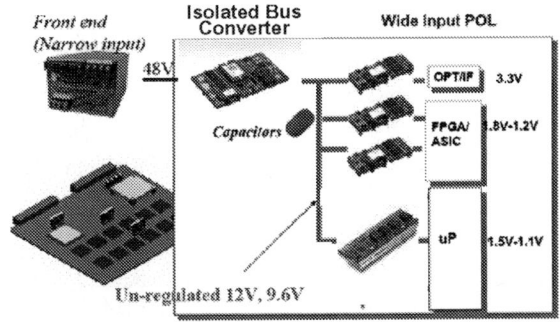

Fig. 33, Two-stage architecture—IBA

To minimize the size and cost of these bus converters, CPES proposed inductor-less bus converter family with ZVZCS switching behavior through the resonant between the transformer leakage inductor and output capacitor [32]. These proposed topologies can double the power capability compared with state-of-the-art products. As an example, Fig. 34(a) shows the full-bridge version of the proposed inductor-less bus converter family. Fig. 34(b) and (c) shows the 800 kHz 500W bus converter prototype with 96% efficiency at full load.

V. HIGH FREQUENCY HIGH CURRENT DENSITY NON-ISOLATED POINT-OF-LOAD (POL) CONVERTERS -- VRs

In 1997, CPES proposed a multi-phase buck converter, as

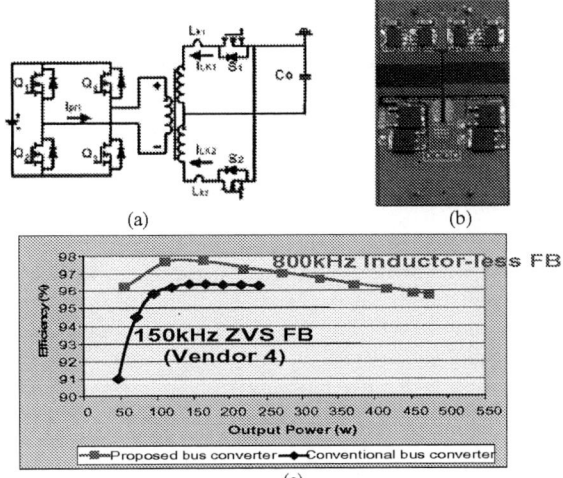

(a)   (b)

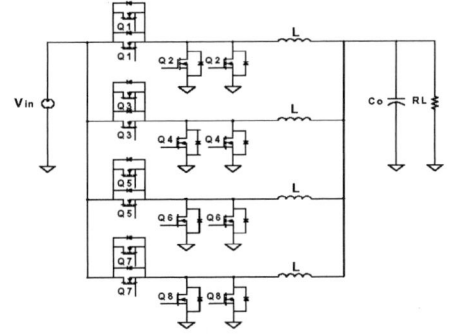

(c)

Fig. 34. Proposed inductor-less resonant FB DC/DC transformer

shown in Fig. 35, for the INTEL Pentium processor. This concept was quickly adopted by the industry. The VR is designed to operate at around 300 KHz and the control bandwidth is around 50 KHz. It is well known that the switching frequency can be increased to reduce the output capacitance. However, the efficiency of today's single stage suffers at higher switching frequencies. As shown in Fig. 36, the major loss factors are the switching losses and the body diode losses.

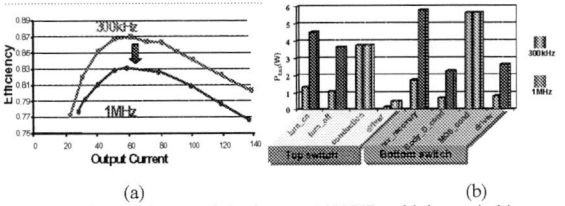

Fig. 35, State-of-the-art VR solution: Multi-phase Buck

(a)   (b)

Fig. 36. Issues of single stage 12V VR at higher switching frequency, (a) efficiencies, (b) Loss breakdown comparison

V-1. 1MHz ZVS Self-Driven Single-stage VRs:

From the loss breakdown in Fig. 36, the major factors affecting circuit efficiency at 1MHz are switching loss, body diode losses and gate driving loss.

11

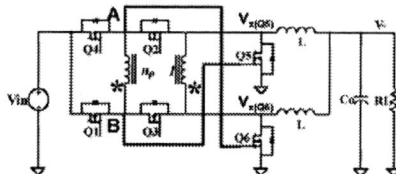

Fig. 37. Proposed ZVS self-driven full bridge Buck

To minimize the aforementioned high frequency losses, a novel self-driven dc/dc converter for non-isolated 12-V VR was proposed, as shown in Fig. 37 [33]. ZVS was realized for all the MOSFETs. By adding a transformer, the proposed topology extends its duty cycle so that both the switching loss and body diode reverse recovery loss are further reduced. This innovative self-driven concept eliminates the need for synchronous rectifier drivers which saves cost and driver loss. In addition, self-driven scheme reduces the body diode conduction losses. Furthermore, the magnetic integration has been realized to merge the output inductor into the transformer to further reduced the size. All of these benefits have been demonstrated in a 1MHz 100A 1U 12V VRM prototype shown in Fig. 38(a). Compared to the conventional Buck design, the proposed topology can elevate the efficiency by 6% at 1MHz switching frequency with the same active component setup, as shown in Fig.38(b).

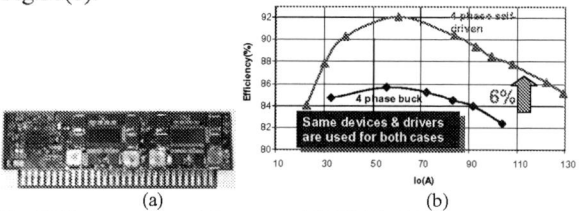

(a)   (b)

Fig. 38. Demonstration of the proposed DC/DC in 1U 100A 12V VRM
(a) Hardware, (b) efficiency comparison

Thought the proposed self-driven topology is compelling, the duty cycle loss induced by the transformer leakage inductor is one inherent limitation for higher than 1 MHz operation. Theoretical analysis reveals that the gain of this topology is diminishing with switching frequency increasing beyond 1MHz.

Thereby, a two-stage approach shown in Fig. 39 is proposed in [34]. The first stage can be designed at relatively low switching frequency to step down the input voltage from 12V to around 5V. With the lower input voltage, the switching losses and reverse recovery losses of the second stage, which are proportional to the input voltage, are dramatically reduced. Therefore, it was demonstrated that the second stage switching frequency can be pushed to 2MHz to achieve 350 kHz bandwidth at 83% efficiency [34, 35]. The prototype in Fig. 40 clearly shows that the two-stage approach can eliminate the output electrolytic capacitor entirely together with a 2% efficiency improvement over the single stage solution. This is achieved with a switching frequency is 4 times higher than the single

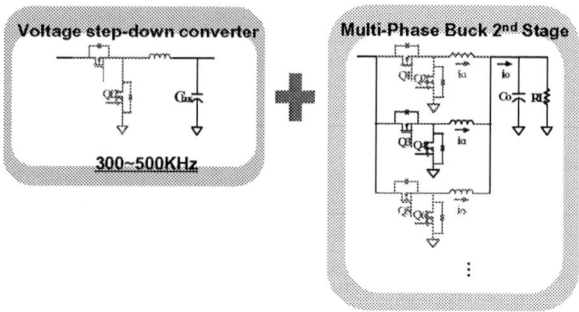

Fig. 39. The structure of the two-stage approach in [34]

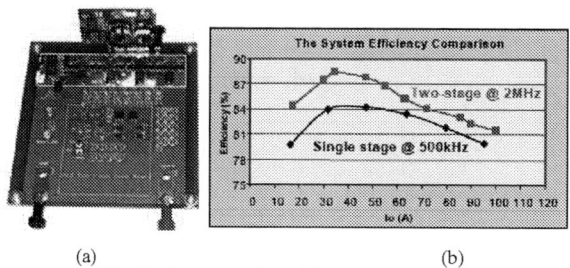

(a)   (b)

Fig. 40. Demonstration of the two-stage 12V VR
(a)   Hardware , (b) efficiency comparison

stage counterpart.

The conventional Buck converter was employed for the first stage initially, as shown in Fig. 39. It was designed to run at low switching frequency, e.g. 200-300 KHz, to attain high efficiency. However, this low switching frequency makes the first stage relatively large in size. Subsequently, a magnetic-less DC/DC converter was developed by adopting the switching capacitor technology [36]. Because no magnetic component is required, it can substantially boost

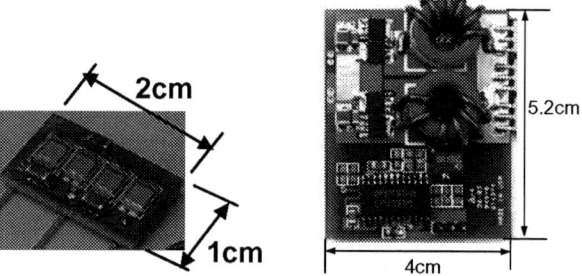

Fig. 41. Size comparison between voltage divider and Buck

the power density of the first stage up to 1KW/inch³. The proposed first stage with fixed 2:1 conversion ratio is essentially serves as a voltage divider. It can achieve 98% efficiency with 12V or even higher input voltage. Fig. 41 shows the comparison between the buck and voltage divider designs for the first stage.

Other than high power density, the proposed voltage divider can achieve ultra-high efficiency in the whole load range with capability to handle over load conditions. By adapting the switching frequency to the load, 98~99%

efficiency in whole load range has been demonstrated in a 70W prototype design as shown in Fig. 42. Even with 100% overload, the circuit can maintain 95.5% efficiency.

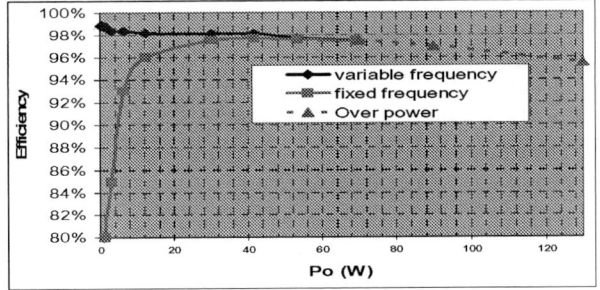

Fig. 42. Experimental efficiencies of 70W prototype

V-3. Non-isolated System Two-stage Architecture:

By extending this two-stage approach into system level, the non-isolated IBA structure is proposed, as shown in Fig. 43, in which the voltage divider is used as the non-isolated bus converter. This architecture has been investigated for the laptop with the configuration shown in Fig. 44 [37].

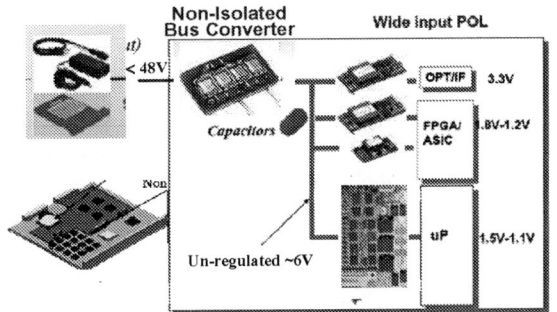

Fig. 43. Proposed non-isolated system two-stage architecture

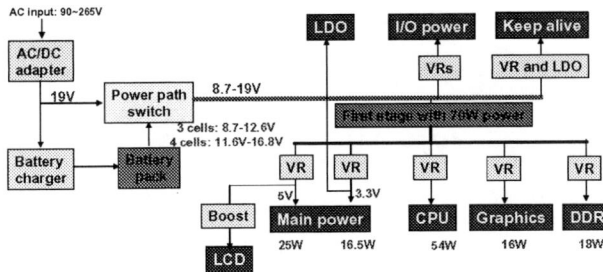

Fig. 44. Proposed system two-stage power architecture for laptop

Detailed analysis and experimental results have demonstrated that this architecture can achieve 45% total inductor size reduction, 35% total capacitor size reduction and 7% total cost reduction for major VRs inside the laptop. Furthermore, the proposed two-stage architecture can improve the light load efficiency for all down-stream VRs, which is critical to the battery life of the laptop. As an example, Fig. 45 shows 1~2% efficiency gain on the CPU VR at half load power and below.

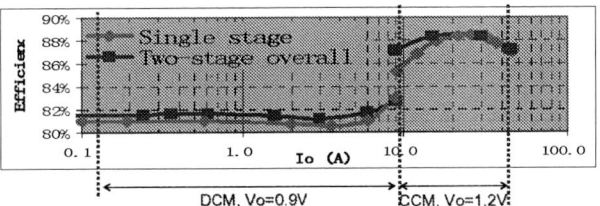

Fig. 45. Efficiency improvement from two-stage on CPU VR

VI. CONCLUSION

This paper provides an overview of some of the important design challenges and opportunities for power supplies, specifically using the example of a generic distributed power system for computer server, telecom and network applications.

First, the effects of mutual and self-paraitics of the filter capacitors and inductors on the performance of EMI filter are presented. Techniques for cancellation of both mutual parasitics and self-parasitics are introduced. Specifically, mutual coupling between inductors and capacitors as well as coupling between two capacitors can be minimized. Among the self-parasitics, the ESL of capacitors and the winding capacitance of the filter inductors are identified most detrimental to both DM and CM noises. The unwanted effects of self-parasitic can be neutralized by circuit means. The proposed technologies are proven by theory and verified by experiments. It is demonstrated that, EMI filter's HF performance can be greatly improved by applying proposed methods in practice.

To achieve higher power density of front-end converter, it is essential to using high switching frequency techniques to reduce passive component size. In the PFC stage, it is found that the impact of switching frequency to EMI filter size reduction is realized when the switching frequency is pushed beyond 400 kHz. The switching frequency between 150kHz to 400kHz should be avoid since the EMI filter size could be even larger than that at 70KHz due to the EMI regulation that specifies noise level from 150 KHz to 30MHz.

Besides the reduction on EMI filter and PFC inductor, it is essential to reduce the size of bulk capacitor for hold up time requirement. It is found that the LLC resonant converter offer special advantage for this application. It can work with wide input range without sacrificing the conversion efficiency at the normal operation condition. Furthermore, smaller switching loss allows LLC converter to operate at considerably higher switching frequency while maintaining high efficiency. A prototype LLC converter operated at 1 MHz was demonstrated with 76W/in3 power density.

System power density was further improved by means of integration using IPEM concept. Modular and integrated approach makes the system layout easy and manufacturing process more automated. Furthermore, it was shown that circuit interconnect parasitics were significantly reduced, thus further improving converter efficiency and reducing components stresses.

Products in the area of isolated and non-isolated point-of-load converters are fiercely competitive. The major driving forces are cost, efficiency and power density. Design challenges for high-frequency, high-density POLs involve innovative circuit means of reducing the primary side switching losses and conduction losses as well as the secondary side synchronous rectifier body diode conduction losses, reverse recovery losses, drive losses, and conduction losses. Packaging and thermal management are equally important. Some of the important recent developments are presented. A novel zero-voltage-switching (ZVS) current-Tripler converter was presented which offers great advantages compared with the popular current double configuration especially for low voltage and high current applications. A novel self-driven ZVS full bridge topology was introduced with significant saving of driver losses, switching losses, and body diode losses for isolated and non-isolated DC/DC converters. One of the potential applications is for powering the next generation of microprocessors. Prototype hardware is demonstrated operating at 1MHz 12V with an efficiency 6% above the comparable multi-phase Buck topology used in today's VRM design. Other than the topological innovations, CPES also proposed two-stage power architectures to further improve the system performance for isolated and non-isolated applications. Prototypes were developed to demonstrate significant improvements in efficiency, power density as well as potential cost saving compared to the state-of-the-art.

REFERENCE

[1] Milan.M Jovanovic "Technology Drivers and Trends for Power Supplies in Computer/Telecom Applications", IEEE APEC'2006, Plenary Session (presentation)

[2] Shuo Wang, Fred C. Lee, D. Y. Chen and W. G. Odendaal, "Effects of Parasitic Parameters on EMI Filter Performance," *Power Electronics, IEEE Transactions*, Volume 19, Issue 3, May 2004, pp. 869 – 877.

[3] Shuo Wang, Fred C. Lee and W. G. Odendaal, "Characterization and Parasitic Extraction of EMI Filters Using Scattering Parameters," *Power Electronics, IEEE Transactions*, Volume 20, Issue 2, March 2005, pp. 502 - 510.

[4] Rengang Chen, J. D. van Wyk, Shuo Wang and W. G. Odendaal, "Improving the Characteristics of Integrated EMI Filters by Embedded Conductive Layers," *Power Electronics, IEEE Transactions*, Volume 20, Issue 3, May 2005, pp. 611 - 619.

[5] J. D. van Wyk, Fred C. Lee, Zhenxian Liang, Rengang Chen, Shuo Wang and Bing Lu, "Development of Some Technologies for Integration in Power Electronics Systems," *Power Electronics, IEEE Transactions*, Volume 20, Issue 3, May 2005, pp. 523 - 536.

[6] Shuo Wang, Fred C. Lee and W. G. Odendaal, "Characterization, Evaluation and Design of Noise Separator for Conducted EMI Noise Diagnosis," *Power Electronics, IEEE Transactions*, Volume 20, Issue 4, Jul. 2005, pp. 974-982.

[7] Shuo Wang, Fred C. Lee and W. G. Odendaal and J. D. van Wyk, "Improvement of EMI Filter Performance with Parasitic Parameter

Cancellation," *Power Electronics, IEEE Transactions*, Volume 20, Issue 5, Sept. 2005, pp. 1221-1228.

[8] Shuo Wang, Rengang Chen, J. D. van Wyk, Fred C. Lee and W. G. Odendaal, "Developing Parasitic Cancellation Technologies to Improve EMI Filter Performance for Switching Mode Power Supplies," *Electromagnetic Compatibility, IEEE Transactions*, Volume 47, Issue 4, Nov. 2005, pp. 921- 929.

[9] Shuo Wang, Fred C. Lee and W. G. Odendaal, "Cancellation of Capacitor Parasitic Parameters for Noise Reduction Application," *Power Electronics, IEEE Transactions*, accepted to appear in 2006.

[10] Shuo Wang, Fred C. Lee and J.D. van Wyk, "Inductor Winding Capacitance Cancellation Using Mutual Capacitance Concept for Noise Reduction Application," *Electromagnetic Compatibility, IEEE Transactions*, accepted to appear in 2006.

[11] Shuo Wang, Fred C. Lee and J.D. van Wyk, "Design of Inductor Winding Capacitance Cancellation for EMI Suppress," *Power Electronics, IEEE Transactions*, under review.

[12] Rengang Chen, J.D. van Wyk, Shuo Wang and W.G. Odendaal, "Planar Electromagnetic Integration Technologies for Integrated EMI Filters," in *Proc. IEEE Industry Applications Conference*, 12-16, Oct. 2003, Volume 3, pp. 1582 -1588.

[13] Fu-Yuan Shih; Chen, D.Y.; Yan-Pei Wu; Yie-Tone Chen, "A procedure for designing EMI filters for AC line applications," Power Electronics, IEEE Transactions on, Volume: 11 Issue: 1, Jan. 1996 Page(s): 170 -181

[14] Ting Guo; Chen, D.Y.; Lee, F.C., "Separation of the common-mode- and differential-mode-conducted EMI noise" Power Electronics, IEEE Transactions on, Volume: 11 Issue: 3, May 1996 Page(s): 480 - 488

[15] Lu, B.; Dong, W.; Wang, S.; Lee, F.C.; "High frequency investigation of single-switch CCM power factor correction converter", in Proc. IEEE APEC 2004, pp.1481 - 1487

[16] Shuo Wang; Lee, F.C.; Odendaal, W.G.; "Single layer iron powder core inductor model and its effect on boost PFC EMI noise", in Proc. IEEE PESC 2003, Page(s):847 - 852

[17] Bing Lu, Wei Dong, Qun Zhao, F.C. Lee, "Performance Evaluation of CoolMOS™ and SiC Diode for Single-phase Power Factor Correction Applications," in Proc. IEEE-APEC 2003', pp.651-657

[18] Y. Jang; M.M. Jovanovic, D.L. Dillman, "Hold-up time extension circuit with integrated magnetics,", in Proc. IEEE APEC 2005. pp. 219 - 225

[19] Y. Xing; L. Huang; X. Cai; S. Sun; "A combined front end DC/DC converter," , in Proc. IEEE APEC 2003 , pp.1095 – 1099

[20] B. Yang, F.C. Lee, A.J. Zhang, G. Huang, "LLC resonant converter for front end DC/DC conversion," in IEEE-APEC 2002, pp. 1108-1112

[21] G. Ivensky, S. Bronstein, S. Ben-Yaakov, "Approximate analysis of the resonant LCL DC-DC converter," in IEEE Electrical and Electronics Engineers in Israel, 2004. Proceedings, pp. 44-47

[22] B. Lu, W. Liu, Y. Liang, F. C. Lee, J.D. van Wyk, "optimal design methodology for LLC resonant converter", in Proc. IEEE-APEC 2006, pp.533-538

[23] J.D. van Wyk, F.C. Lee, D. Boroyevich, Z. Liang; K. Yao, "A future approach to integration in power electronics systems," in Proc. IEEE-IECON '03. Volume: 1, pp. 1008-1019

[24] F.C. Lee, J.D. van Wyk, D.Boroyevich, G. Lu; Z. Liang; P.Barbosa, "Technology trends toward a system-in-a-module in power electronics," Circuits and Systems Magazine, IEEE , Volume: 2 Issue: 4 , 2002, Page(s): 4 –22

[25] Pedro Alou, Jose A. Cobos, et al., "A New Driving Scheme for Synchronous Rectifiers: Single Winding Self-Driven Synchronous Rectification," in IEEE Transaction on Power Electronics, Vol. 16, No. 6, Nov. 2001, pp. 803-811.

[26] Y. Ren, M. Xu, D. Sterk and F.C. Lee, "1MHz Self-Driven ZVS Full-Bridge Converter for 48V Power Pods," in Proc. IEEE- PESC 2003, Page(s): 1801 – 1806.

[27] Ming Xu, Yuancheng Ren, Jinghai Zhou, Lee, F.C.; "1MHz Self-Driven ZVS Full-Bridge Converter for 48V Power Pod and DC/DC Brick" IEEE Transaction on Power Electronics Volume: 20 , Issue: 5 , September 2005, Pages:997 – 1006

[28] D. Sterk, M. Xu, Y. Ren and F.C. Lee, "Novel Integrated Transformer Winding Scheme for Self-driven ZVS Interleaved Asymmetrical Half-Bridge for Telecommunications Quarter Brick" , in Proc. IEEE APEC 2004, pp.912-918

[29] Y. Panov and M. Jovanovic, "Design and Performance Evaluation of Low-Voltage/High-Current DC/DC On-Board Modules," IEEE Transactions on Power Electronics, Volume: 16 Issue:1,Jan.2001, pp. 26 –33.

[30] Ming Xu, Jinghai Zhou, Lee, F.C.; "A current-tripler dc/dc converter" IEEE Transaction on Power Electronics Volume: 19 , Issue: 3 , May 2004, Pages:693 – 700

[31] Y C.Ren, Ming Xu, Gary Yao, F C.Lee "Two-stage 48V power pod exploration for 64-bit microprocessor" in Proc. IEEE APEC'2003, Pages: 426 - 431

[32] Yuancheng Ren, Ming Xu, Lee, F.C.; "A Family of High Power Density Un-regulated Bus Converters", IEEE Transaction on Power Electronics , Volume: 20 , Issue: 5 , September 2005, Pages:1045 – 1054

[33] Jinghai Zhou, Ming Xu, Julu Sun, Lee, F.C.; "A Self-driven Soft-switching Voltage Regulator for Future Microprocessor", IEEE Transaction on Power Electronics Volume: 20, Issue: 4, July 2004, Pages:928 – 936

[34] Yuancheng Ren, Ming Xu, Kaiwei Yao,Yu Meng, Lee, F.C.; "Two-stage Approach for 12V VR", IEEE Transaction on Power Electronics , Volume: 19, Issue: 6, Nov. 2004, Pages:1498 – 1506

[35] Yuancheng Ren, Ming Xu, Kaiwei Yao, Lee, F.C.; "Analysis of the Power Delivery Path from 12V VR to Microprocessors", IEEE Transaction on Power Electronics, Volume:19, Issue: 6, Nov. 2004, Pages:1507 – 1514

[36] Ming X.;Julu Sun, and Lee, F.C.; "Voltage Divider and its Application in Two-stage Architecturess", in Proc. IEEE APEC'2006, Pages: 499 - 505

[37] Julu Sun, Ming Xu, and Lee, F.C.; "High Power Density, High Efficiency System Two-stage Power Architecture for Laptop Computers", in Proc. IEEE PESC'2006

A Smarter Grid for Improving System Reliability and Asset Utilization

D. Divan and H. Johal

Georgia Institute of Technology, School of Electrical and Computer Engineering, Atlanta, USA
deepak.divan@ece.gatech.edu

Abstract— The power grid is aging and under stress. Unlike other modern networked systems, the grid lacks intelligence and automation. This paper has presented a new look at the way a Smart Grid can be implemented. The conventional approach has been to first obtain real time information on critical parameters, and then by controlling VAR resources, tap changers, and FACTS devices to achieve the desired control. A simpler approach is presented here based on using highly interconnected meshed networks. Such networks have been used in high density urban areas for many years for the high reliability achievable, but suffer from poor line utilization and lack of flexibility under contingency or load growth conditions. The use of a large number of Current Limiting Conductor or CLiC modules provides a simple and cost-effective approach for realizing a controllable meshed network, maximizing network capacity under diverse contingencies and load growth scenarios. Using a low-tech approach, it is seen that basic network performance and reliability are dramatically increased. It is also seen that the distributed nature and inherent redundancy in the deployment of large numbers of CLiC modules, results in high system reliability.

Keywords- Transmission lines; Load flow control; Intelligent sensors

I. INTRODUCTION

The global electricity infrastructure represents perhaps the most complex edifice built by man. In the US, for instance, the grid is in excess of 50-60 years old, spans the entire continent with over 800,000 miles of high voltage transmission lines, and has a minimal level of monitoring and/or automation. Over the last two decades, electricity consumption and generation have continually grown at an annual rate of around 2.5% [1]. At the same time, investment in the T&D infrastructure has steadily declined. Further, it has become increasingly difficult and expensive to permit and build new power lines. As a result the aging power grid is congested and under stress, resulting in compromised reliability and higher energy costs. To compound the situation, the utilities often do not possess detailed information on the status and operating margins on their various geographically-distributed power-line assets, resulting in sub-optimal use. In today's competitive environment, the ability to use its assets efficiently becomes an important component of a utility's profitability.

While radical changes in infrastructure may seem attractive, and in line with transformations that have occurred in the telecom and Internet areas, the analogies are likely to be misleading at best. Unlike in telecom where society's bandwidth requirements have grown exponentially, no metrics can be identified in the power-delivery industry where performance has improved by several orders of magnitude over the last decade. With the exception of a few high growth opportunities, e.g. China and India, where building new infrastructure is a priority, enormous legacy investments exist in the existing power delivery infrastructure and one can only posit incremental improvements on the existing infrastructure.

A few critical requirements can clearly be defined. System reliability is sacrosanct and cannot be compromised. Utility system planners are moving away from radial systems towards networked systems to achieve higher reliability, especially under contingency conditions. While enhancing reliability, this degrades controllability of the network, as current flow along particular lines cannot easily be controlled. The situation is exacerbated when a contingency such as loss of a line or generator results in overload and tripping of lines, increasing the possibility of a cascading blackout. Finally, rapid load growth leads to congestion on key lines connecting low-cost generation to load centers, leading to an inefficient operation of energy markets and 'gaming' [2].

The answer seems to lie in the implementation of a 'Smart Grid', that is reliable, self-healing, fully controllable and asset efficient [3]. Continuous advances and cost-reductions in sensing, communications, power electronics and systems technology are at the heart of the Smart Grid of the future as envisioned. Components of tomorrow's grid include Flexible AC Transmission Systems (FACTS) devices rated at >100 MVA, HVDC Lite, Smart Wires, PMU's, and power line sensornetworks [4]-[5]. Of the various technologies under development, smart sensing is clearly an important component. However, much more critical is the ability to statically and dynamically control critical grid operating parameters, such as line current, voltage, phase angle, and power flow. This paper discusses some of the more promising technologies that allow real-time control of grid operations and that will help to create the Smart Grid.

This project has been done under the Intelligent Power Infrastructure Consortium at Georgia Tech under research funding by Tennessee Valley Authority

II. IMPROVING POWER GRID RELIABILITY AND UTILIZATION

System reliability is the paramount mission for a utility. Older systems are radial in structure, primarily because they provide a cost-effective and fully controllable system. However, radial systems suffer from poor reliability, because a fault results in an extended outage for all downstream customers, severely compromising system reliability. Utilities are moving from radial systems to meshed networks or 'networks' at the distribution, sub-transmission and transmission levels, in an effort to enhance system reliability. In a network, a fault results in the isolation of a single line segment, with alternate paths maintaining power to all other customers. This results in significantly higher reliability levels. Many urban centers, such as New York, have vast networks at the distribution level, and consistently deliver some of the most reliable power in the US [6]. Even at the transmission level, interconnections between major transmission lines are common, and provide alternate paths for power flow under contingency conditions.

The biggest drawback of networks is the inability to control how current flows on individual lines in the network. Balancing power flows dynamically under changing load and source conditions is not possible as the system is passive and has few 'handles' for control. Even if expensive phase angle controllers were used on each line, finding an optimal control strategy would be a daunting task. As a result, the reliability comes at a significant price. Inability to control power flow results in loop flows, congestion, and poor line utilization [7]. In a network, the first line that reaches a thermal limit constrains the power transfer capacity of the entire network, even though all the other lines may be operating substantially below their thermal capacity. Fig. 2 shows the utilization of a part of the IEEE 39 bus system (Fig. 1), when the first line in the system (Line 22_21) hits the thermal limit. It is observed that individual line utilization varies from 5% to 100%, with an average of ~59%. The real situation is even worse as it is driven by the need to have spare system capacity and to ensure system integrity and reliability under (N-1) or (N-2) contingency conditions. This reduces the allowed line current levels under normal operating conditions to well below nominal thermal limits, and further degrades system utilization.

Another compounding factor is the dynamic thermal limit of the line under prevailing weather conditions. System operators generally approach the line rating issue conservatively, with limits established under the highest ambient temperature and zero-wind conditions. Sometimes, prevailing winds and ambient temperature forecasts are factored into setting the limits. EPRI and others have proposed the use of line sensors located on a pre-identified 'critical span' to assess the dynamic thermal capacity of the line [8]. While this appears to be promising at first glance, lack of knowledge of micro-climate conditions, especially for long lines, raises a concern as to

whether the identified span is truly the critical span under current operating conditions [9]. This once again limits the system operator's ability to truly use any dynamic line capacity information that he may have.

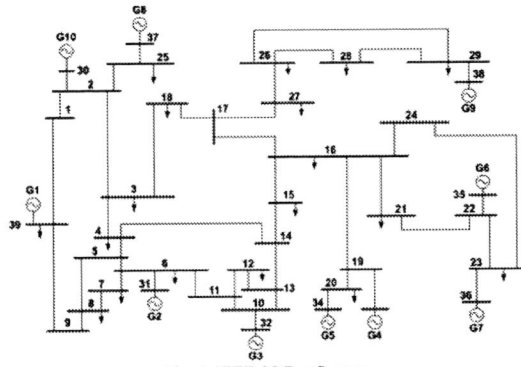

Fig. 1. IEEE 39 Bus System

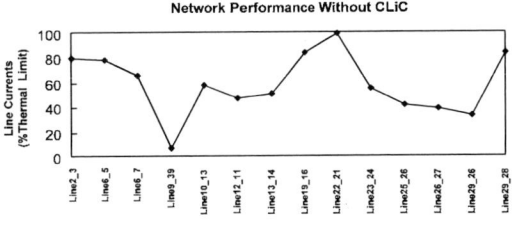

Fig. 2. Network utilization when the first line reaches thermal limit

Another important and related issue is transmission line congestion that limits the ability of end-users to freely access low-cost generation. Load growth and lack of incentives for transmission investment can result in increasing levels of congestion. When a transmission line reaches a constraint (thermal or stability), and limits the power that can be supplied, say to a load in region 2 from a low-cost generator in region 1, then an out-of-merit generator in region 2 is used to supply the power at a higher rate. This results in higher cost for all consumers in region 2. The insufficient transmission capacity also results in islanded networks that require higher level of generation reserves within each region to ensure system reliability. All these factors result in a cost increase for the consumers [10].

It is seen that system reliability and utilization seem to move power system designers in opposite directions. Cost and system-utilization can be more effectively controlled in radial systems, whereas reliability of meshed networks is substantially higher. Today's modern digital economy requires more reliable and higher quality power than ever before. Our total dependence on electronics appliances and industrial automation brings our everyday life to a halt, even for short interruptions in our electricity service. Except in remote rural areas, the trend is towards higher reliability achieved through an increasing use of networks. It is important to explore techniques for cost-effectively enhancing the reliability and utilization of meshed networks.

III. EXISTING SOLUTION

The traditional solution for overloaded lines and transmission congestion has been to build new lines. In a radial system, this is possibly the only solution. In a network, that may not be the case, as there are likely to be a number of under-utilized lines, as can be seen in Fig. 1(b). New lines are expensive to build, and are subject to delays in permitting, siting and obtaining rights of way (ROW). In the US, some transmission lines have been delayed for more than 20 years because of ROW issues. Further, it should be noted that a new line is likely to be lightly loaded and can make the overall grid utilization even worse [11].

Several important technologies are in use today for improving grid reliability and utilization. Some of these technologies are discussed below. A widely used approach is the use of shunt VAR compensation to provide voltage support. Shunt compensation techniques include electro-mechanically switched capacitor banks, static VAR compensators (SVC) and STATCOMs [12]. Capacitors are used to compensate for voltage drop along the line reactance, and provide only steady state VAR support. SVCs use thyristor controlled reactors (TCR) in parallel with shunt capacitors to realize reasonably fast control of shunt VARs, and to provide dynamic control of voltage. For sub-cycle response to prevent voltage collapse in the presence of faults on the system, it is necessary to have faster response capability using STATCOMs. STATCOMs are rated in the 20-100 MVA range, and utilize inverters to draw leading or lagging reactive current from the line. While shunt compensation is universally used, the value is primarily in voltage regulation on the system. The impact on system power flow control is extremely weak.

In order to achieve power flow control on the grid, series VAR compensation or phase angle control is required. Series capacitor compensation is often used to mitigate sub-synchronous resonance issues encountered with long lines, and to allow power flow over long haul transmission lines. Similarly, phase-angle controllers, essentially phase shifting transformers with tap-changers, are used to balance power flow on interconnected transmission systems. Both techniques are expensive and are typically applied only at 345 kV and higher. Also, neither technique offers dynamic control capability as far as grid power flow control is concerned. What is needed is the ability to dynamically control power flow, so that under contingency and/or overload conditions, the current can be redirected to under-utilized paths, thus averting a serious cascading blackout.

Several techniques are available that allow dynamic power flow control on the grid. These include Unified Power Flow Controllers (UPFC) and Synchronous Static Series Compensators (SSSC) [13]. Most of these fall into the category of Flexible AC Transmission or FACTS devices, as do SVCs and STATCOMs. Fig. 3. shows block schematics of various FACTS devices. The UPFC provides the highest level of flexibility, providing both series and shunt compensation, including implementation of STATCOM and SSSC functionality, but with the additional ability to exchange real power between the series and shunt inverters. The inverters use GTOs or IGCTs, are custom designed and built, are rated at up to 100 MVA, and are connected to the transmission line using transformers. The series connected transformer, in particular, has stringent design issues, including the ability to handle fault currents of up to 65,000 Amperes, and core saturation that can occur under certain types of transients and start-up conditions. The BIL issues related to a 345 kV line add further cost and complexity.

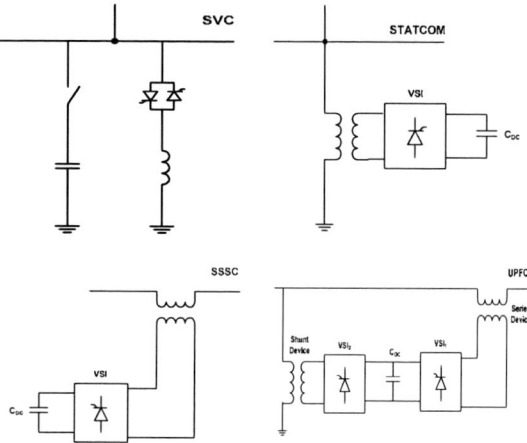

Fig. 3. Common FACTS devices: (a). SVC, (b). STATCOM, (c). SSSC, (d). UPFC

FACTS technology has been around for over 15 years, with several highly visible and successful demonstration projects that showcase the capability of the technology. The Marcy UPFC project in New York, representing 200 MVA of total control capability at the 345 kV level built at a cost of $54 million, exemplifies the capabilities of existing FACTS technology [14]. It is interesting to note that although FACTS devices have been commercially available for many years, there has been virtually no market penetration, especially in the area of grid power flow control.

Discussions with utility personnel suggest the following reasons. It is perceived that the first cost and life cycle costs for FACTS devices are high, while the uptime and reliability have not yet reached desired levels. Also, building a large centralized device results in susceptibility to a single point of failure, and the device complexity and unique components results in a mean time to repair that is much longer than desired. Finally, the utility personnel are not qualified to maintain and repair the FACTS devices.

Another interesting issue revolves around the cost differential between compensators that can influence steady state behavior, and those that can provide fast dynamic compensation. For example, STATCOMs are faster than capacitors or SVCs but cost a lot more. STATCOMs provide unique value only during infrequent transmission level system faults, while capacitors provide value on a continuous basis. This suggests that the return on investment from a market perspective for a steady state compensator may be easier than for a dynamic compensator that provides unique value only under occasional faults. This is particularly true for a steady state series VAR compensator that can enable additional MW flow along congested lines, that a customer is willing to pay for. However, there are few commercial technologies and solutions that are presently available that can provide cost-effective power flow control in networked systems, especially at the distribution level, so as to enhance system reliability and utilization.

IV. DISTRIBUTED SOLUTION FOR IMPROVING GRID RELIABILITY AND UTILIZATION

It is clear that a meshed network provides the highest level of reliability, albeit at apparently higher cost and with poor asset utilization. This results purely from the inability to control how current flows on individual lines in the network. Utility operations are based on an assumption that it is difficult to control current in individual lines in a network. This has resulted in a preference for radial networks wherever possible, and with vastly under-utilized systems where networks were unavoidable to meet urban distribution and reliability needs.

Fig. 4 shows an example of a meshed network, with controllable voltages at major buses. In an optimal situation, the amplitude and phase angle of the voltage at the various buses could be controlled so as to balance the individual line currents. This would require a real-time computation of the power flows in the network, with a calculation of voltage magnitude and angle at each of the controllable nodes to provide an optimal operating set point. This process would need to be continuously repeated as the load or any of the sources varied. For proper operation, this would also require a fast communication link between all nodes, with full visibility to the 'state' of the network/system. This is akin to the technique by which a larger power system is controlled, where system operators at the area control centers can dispatch generators, set VAR compensator tap changer operating points, etc. to optimize the load flow. Fast and reliable communications, information on the current in individual lines, and an accurate knowledge of the network topology at any given time is required for such a control strategy to work. Clearly, while this is theoretically possible, it would add substantially to the complexity and cost of the implementation. Further, under a contingency, such as an unanticipated line, generator or transformer outage, the set points have to be rapidly

recalculated if outages and the potential for cascading blackouts are to be avoided. While this approach may be acceptable for the transmission grid, it is too expensive and unwieldy to implement at the distribution level. As a result, interconnected and meshed networks are infrequently used in distribution, unless reliability is an overriding priority.

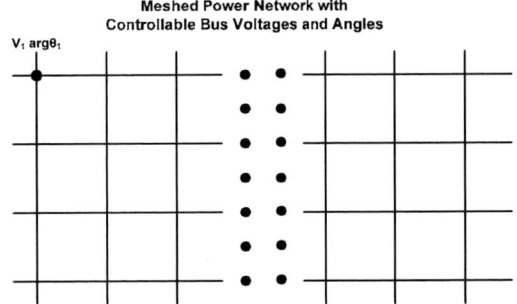

Meshed Power Network with Controllable Bus Voltages and Angles

Fig. 4. Meshed network with controllable voltages and phase angles

The fundamental issue with networks is that individual lines get overloaded as a result of load growth, line outages and source outages. The interconnectedness of the network, invaluable for reliability, now makes it difficult to predict how the current will distribute in the network, and which lines will be overloaded. Even if this information were known, there are no simple control handles that would allow us to limit the current without some form of load-shedding. What is clearly required is some form of 'throttle' control that would 'dial back' the current in an overloaded line. A recently proposed technique – 'Current Limiting Conductors'- provides a cost-effective method for implementing such control on lines that can potentially be subject to overload, offering a new approach to the implementation of meshed networks that promise high reliability, with high asset utilization and low cost [15].

V. CURRENT LIMITING CONDUCTORS

Recently, the concept of Current Limiting Conductor (CLiC) modules, a special implementation of the generic family of Distributed Series Impedance devices, has been proposed as a means of varying the impedance of existing transmission and distribution lines. CLiC modules clamp on to power lines, floating mechanically and electrically on the power line as shown in Fig. 5. Fig. 6 shows a schematic of a CLiC module, including a single turn transformer (STT) with a normally closed relay that bypasses the transformer impedance. The STT turns ratio is chosen to reduce the relay current, under nominal and fault conditions, to a reasonable value. A control circuit is powered parasitically off the line, and monitors the line current. When the current in the line reaches a predetermined threshold, the relay is opened inserting the magnetizing inductance of the transformer in series with the line impedance. Multiple CLiC modules are used together on a line, with individual module current trip thresholds tuned slightly apart from each other. At the line level, as the current rises, the impedance of the line gradually increases. If the current in other lines has not yet reached this threshold, then the increasing impedance will force the current to preferentially flow in other lines that have lower impedance.

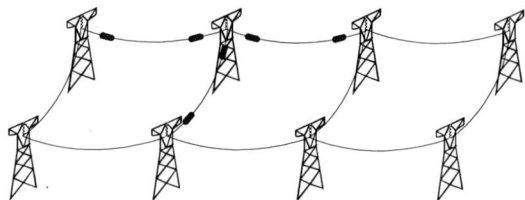

Fig. 5. CLiC modules attached to a meshed power grid

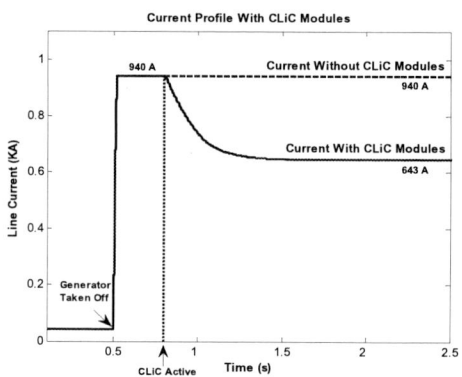

Fig. 6. Circuit schematic of CLiC

The CLiC module implementation is very simple, using readily available low-cost components. Control is purely based on locally measured parameters (line current), although communications can be used to augment the performance. A single CLiC module weighing approximately 55 kg can be suspended from most power lines, and can inject approximately 9-11 volts at ~600-1000 Amperes. The CLiC modules operate according to an algorithm that turns the device on or off with the appropriate current-delay characteristics.

One can explore what the impact of CLiC modules would be on system capacity and asset utilization. As current in a line increases to a value above the predefined threshold, as a result of load increase or a contingency, the impedance increases causing the current to redistribute to those lines where the impedance is unchanged – i.e. lines that are not seeing the same increase in current. This is a natural redistribution that does not require any coordinated control or action, or any communications. It is possible to analyze potential overloads under anticipated load increases and/or contingencies, and to only deploy the number of CLiC modules that would be needed to save the susceptible lines from overloads. Fig. 7 shows improvement in the network utilization of the IEEE 39 bus system, when operated with CLiC modules. It is seen that system utilization is increased from 59% to over 93.3%, showing an increase of more than 33% in system capacity without any addition of new lines and while ensuring that

all lines operate within their thermal limit. Further, it is seen that loss of lines and or sources does not result in system collapse. This is illustrated in Fig. 8, where a contingency condition is simulated by taking Gen. 7 (G7) off. The current through Line 22_23 is seen to jump to a very high value of 940 A from an initial operating point of 43 A. However, with the CLiC modules turned on, the current is brought down to a safer level of 643 A. Thus system reliability is ensured even under contingency conditions.

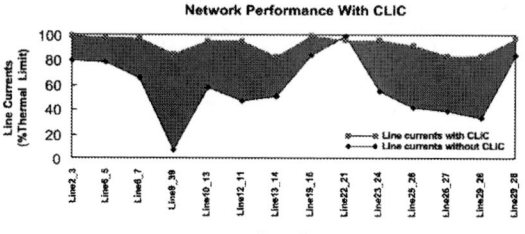

Fig. 7. Improvement in line utilization with CLiC modules

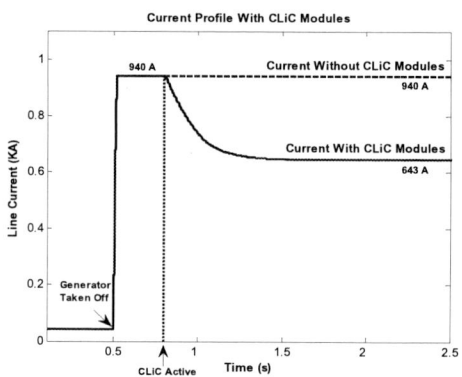

Fig. 8. Performance under contingency condition (generator outage)

The approach can be clearly applied to distribution as well as sub-transmission and transmission networks. Tightly meshed systems with short lines, typical at the distribution level, should show the most significant improvement in reliability and asset utilization. The CLiC modules can be incrementally deployed only on those lines where they are needed, and can handle a variety of contingencies without any need for real time computation. Further, the ability to keep operating even as a few of the modules fail indicates high reliability and availability at a system level. The ability to replace failed units in the field also promises a small mean time to repair. Finally, the use of standard CLiC modules promises low cost.

The introduction of new technology on the grid always raises a myriad of questions. Front and foremost are issues of fault current handling and impact on protective relaying. The CLiC module (Fig. 6) is designed with a thyristor pair in parallel to the NC relay. Under fault current conditions, the thyristor pair is rapidly triggered on (within ¼ cycle), effectively reverting the line to its normal impedance state. This serves two important functions. Firstly, it allows the CLiC module to ride-through the fault. Secondly, it allows existing protective relays to operate normally. This is clearly an issue that needs to be validated before significant level of deployment can occur. Other issues include the environmental susceptibility of the unit, and the ability to deploy or remove the CLiC module on a live line. These issues have been addressed in previous papers. An extension of the Distributed Series Impedance modules to realize a Distributed Series Static Compensator or DSSC, was reported in [16]. Fig. 9 shows that the device clamps on to power lines, and shows a schematic of the implementation. Fig. 10 shows operating waveforms, demonstrating the ability to inject leading or lagging impedance into the line.

Fig. 9. Laboratory demonstration of DSSC on a power line

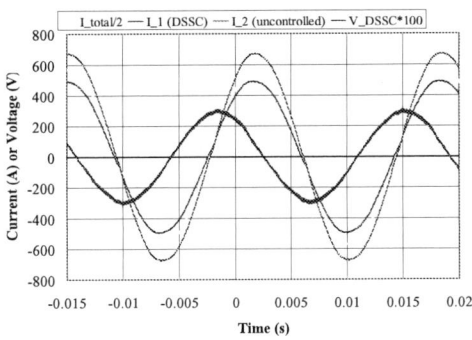

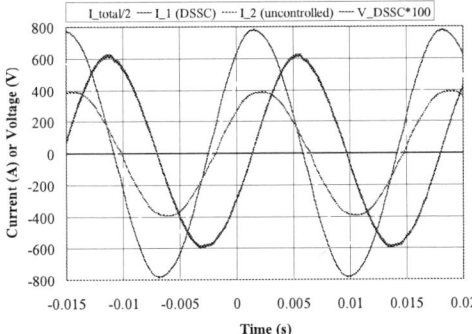

Fig. 10 Operating waveforms. (a). Injection of leading voltage (inductive impedance), (b). Injection of lagging voltage (capacitive impedance)

VI. POTENTIAL VALUE TO UTILITIES

A controllable meshed network can of course be implemented with existing networked systems. However, the more interesting possibility is that new power grid build-outs could be designed using a meshed network architecture to realize high reliability and asset utilization at low cost. In both cases, significant benefits are seen to accrue. The benefits are also seen to be realized for transmission as well as distribution networks. These include:

- Enhancement in reliability and operation of existing and new networks under (N-1) and (N-2) contingencies by automatically routing current from overloaded lines to lines with available capacity, without impacting system performance or losses under normal conditions.
- Improvement in line utilization significantly, especially for shorter length lines typical of distribution meshes, thus enhancing system capacity.
- Enhanced system performance based on local measurements, without the need for a communications link. Performance can be further enhanced with a low speed communications link.
- Ability to share generation reserves across wider part of the existing network.
- Distributed scalable solution allows strategic, targeted and incremental deployment.
- Reduction in the overall cost of energy by providing access to lower cost generation sources.
- Improvement in asset utilization of existing lines and deferment of investments in new lines.
- Maintaining and operating CLiC system using existing utility staff without the need for new skill sets.

VII. CONCLUSIONS

This paper has presented a new look at the way a Smart Grid can be implemented. The conventional approach has been to first obtain real time information on critical parameters, and then achieve the desired performance by using VAR resources, tap changers, and FACTS devices. A simpler approach is presented here based on using highly interconnected meshed networks. Such networks have been used in high density urban areas for many years for the need of high reliability, but suffer from poor line utilization and inflexibility under contingency or load growth conditions.

The use of Current Limiting Conductor or CLiC modules shows a simple and cost-effective approach for realizing a controllable meshed network, maximizing network capacity under diverse contingencies and load growth scenarios. Using a low-tech approach, it is seen that basic system performance and reliability are dramatically increased. While performance can be further enhanced using communications, the device operates based on locally measured parameters, and continues to deliver most of the improvement even without any communications. It is also seen that the distributed nature and inherent redundancy in the deployment of large numbers of CLiC modules, results in high system availability.

ACKNOWLEDGMENT

This project has been done under the Intelligent Power Infrastructure Consortium at Georgia Tech under research funding by Tennessee Valley Authority.

REFERENCES

[1] National Transmission Grid Study, US Department of Energy. [Online]. http://www.pi.energy.gov/pdf/library/TransmissionGrid.pdf.

[2] B. C. Lesieutre and J. H. Eto, "Electricity Transmission Congestion Costs: A Review of Recent Reports," [Online]. Available: http://certs.lbl.gov/pdf/54049.pdf.

[3] S. A. Massoud and B. F. Wollenberg, "Toward a smart grid: power delivery for the 21st century," *IEEE Power and Energy Magazine*, vol. 3, issue 5, Sept-Oct 2005, pp. 34-41.

[4] D. Divan, " Distributed Intelligent Power Networks – A New Concept for Improving T&D System Utilization and Performance," T&D Workshop, CMU, Dec 15-16, 2004. [Online]. Available.http://www.ece.cmu.edu/~electriconf/old2004/divan.pdf

[5] Y. Yang, D. Divan, R. G. Harley and T. Habetler, "Power Line Sensornet – A New Concept for Power Grid Monitoring," to be presented at the IEEE PES General Meeting 2006.

[6] M. J. Museler, President and CEO of the New York Independent System Operator (NYISO), presentation May 22, 2003. [Online]. Available: http://www.nyiso.com/topics/articles/news_releases/2003/pa3_presentation.pdf

[7] Kojo Ofori-Atta, Elliot Roseman, Bansari Saha, Scott Stuart, Marc Lipschultz and Jonathan Smidt, "Profiting From Transmission Investment," Public Utilities Fortnightly, Oct. 2004, pp 72-77.

[8] D.A. Douglass and A. A. Edris, "Real-time monitoring and dynamic thermal rating of power transmission circuits," Power Delivery, IEEE Transactions on, July 1996, vol. 11, issue: 3, pp: 1407-1418.

[9] S. D. Foss, S. H. Lin, R. A. Maraio and H. Schrayshuen, "Effect of variability in weather conditions on conductor temperature and the dynamic rating of transmission lines," Power Delivery, IEEE Transactions on, Oct. 1988, vol. 3, issue 4, pp:1832-1841.

[10] Harjeet Johal and Deepak Divan, "Calculating Economic Impact of Congestion Mitigation Using Distributed Series Impedances," under review for *IEEE PES, Power System Conference and Exposition 2006*, Atlanta

[11] W. W. Hogan, " Transmission Capacity Rights for the congested Highway: A Contract Network Proposal," Submitted to Federal Energy Regulatory Commission. [Online]. Available: http://ksghome.harvard.edu/~whogan/ferc691r.pdf

[12] N. G. Hingorani, "FACTS Technology and Opportunities," IEEE Colloquium on *Flexible AC Transmission (FACTS)- The Key to Increased Utilization of Power Systems*, Digest No.1994/005, 12 Jan. 1994, pp. 4/1-4/10.

[13] L. Gyugyi, C. D. Schauder, and K. K. Sen, "Static Series Compensator: a Solid-State Approach to the Series Compensation of Transmission Lines," *IEEE Transactions on Power Delivery*, Vol. 12, No. 1, Jan 1997, pp. 406-407

[14] NYPA MARCY FACTS Project - Phase II. [Online]. Available. http://www.nyiso.com/public/webdocs/market_data/reports_info/oper_studies_sys_perf_reports/ipfcmp6report_oc_final_approved.pdf

[15] Harjeet Johal and D. Divan, "Current Limiting Conductors: A Distributed Approach for Increasing T&D System Capacity and Reliability," *IEEE PES, Transmission and Distribution Conference* 2005-06, Dallas, Texas.R. Nicole, "Title of paper with only first word capitalized", *J. Name Stand. Abbrev.*, in press.

[16] D. Divan, W. Brumsickle, R. Schneider, B. Kranz, R. Gascoigne, D. Bradshaw, M. Ingram, and I. Grant, "A Distributed Static Series Compensator System for Realizing Active Power Flow Control on existing Power Lines," *IEEE PSCE Conference Records*, Oct 2004.

2006 5th International Power Electronics and Motion Control Conference

Medium-Voltage Power Conversion Systems in the Next Generation

Hirofumi Akagi, *Fellow, IEEE*, and Shigenori Inoue, *Student Member, IEEE*
Department of Electrical and Electronic Engineering
Tokyo Institute of Technology
S3-17, 2-12-1, O-okayama, Meguro, Tokyo, 152-8552, JAPAN
Phone/Fax: +81-3-5734-3549 E-mail: akagi@ee.titech.ac.jp

Abstract—This paper describes the next-generation medium-voltage power conversion systems characterized by using bi-directional isolated dc/dc converters. A 350-V, 10-kW and 20-kHz dc/dc converter is designed, constructed and tested as a core circuit of the medium-voltage power conversion systems. It consists of two single-phase full-bridge converters with the latest trench-gate Si-IGBTs and a 20-kHz transformer with a nano-crystalline soft-magnetic material core and litz wires. The transformer plays an essential role in achieving galvanic isolation between the two full-bridge converters. The overall efficiency from the dc-input to dc-output terminals is accurately measured to be as high as 97%, excluding gate drive circuit and control circuit losses from the whole loss. Moreover, loss analysis clarifies that the use of SiC-based power devices may bring a significant reduction in conducting and switching losses to the dc/dc converter. As a result, the overall efficiency may reach 99% or higher.

I. INTRODUCTION

Wide-band-gap semiconductors such as silicon carbide (SiC) and gallium nitride (GaN) have superior characteristics to silicon (Si). Many universities and/or manufacturers are now cooperating and competing to develop the next-generation, ultra low-loss, high-speed power devices using wide-band-gap semiconductors [1]-[10]. Infineon Technologies and Cree have already put 300-V, 600-V, and 1,200-V SiC-SBDs (schottky barrier diodes) on the market. These SiC-SBDs have been tested in power electroinic circuits to evaluate their effectiveness in reducing power loss [13]. In addition, SiC-JFETs (junction field-effect transistors) and SiC-MOSFETs (metal-oxide-semiconductor field-effect transistors) are now shifting from laboratory levels toward commercial levels. Mitsubishi Electric and Rohm in December 2004, and National Institute of Advanced Industrial Science and Technology of Japan (AIST) in March 2005, independently announced that they had developed SiC-MOSFETs with reduced channel resistances [9], [10].

Since SiC power devices with low conducting and switching losses can operate at higher temperature than Si counterparts can, smaller cooling devices such as heatsinks and fans are applicable. This makes a significant contribution to increasing the power density of power conversion systems.

State-of-the-art medium-voltage power conversion systems, however, usually require line-frequency (50 or 60 Hz) transformers to ensure galvanic isolation between the utility and the load. The size and weight of the transformer occupies a large part in the whole power conversion system. In other words, even if medium-voltage power conversion systems replaced Si power devices with SiC power devices, the transformer would impose limitations on the power density.

This paper describes a bi-directional isolated dc/dc converter [11], [12] considered as a core circuit of 3.3-kV/6.6-kV high-power-density power conversion systems in the next generation. Although the dc/dc converter has already been known as a technique to reduce the transformer size, few papers have dealt with this topology for many years. However, providing loss evaluation of the dc/dc converter in this paper may spur interest in this topology because new power devices and magnetic materials have been emerging. In this paper, the 350-V, 10-kW, 20-kHz dc/dc converter using latest trench-gate Si-IGBTs and soft magnetic material is designed, constructed, and tested as the core circuit of 3.3-kV, 270-kW adjustable-speed motor drives with regenerating braking. The tested circuit is unique in that the dc output terminals are connected back to the dc input terminals, so as to regenerate the dc output power to the dc voltage source. This special connection is useful to accurately measure the overall loss produced by the dc/dc converter. The overall efficiency from the dc-input to dc-output terminals in the experimental circuit is 96.8% at the rated power of 10 kW, and the maximum efficiency is 97.4% at 5.5 kW. Loss analysis carried out in this paper encourages to introduce SiC power devices to the dc/dc converter in terms of having the possibility of significantly reduced loss. As a result, the use of SiC-MOSFETs will improve the efficiency to 99% or higher.

II. TECHNICAL ISSUES IN MEDIUM-VOLTAGE POWER CONVERSIONS SYSTEMS

A. 6.6-kV BTB (Back-to-Back) Systems

The proliferation of distributed generation based on renewable energy and fuel cells in the near future encourages the Central Research Institute of Electric Power Industry of Japan in feasibility study of a 6.6-kV, 1-MW BTB system referred to as a loop power flow controller [14]. Fig. 1 shows the circuit configuration of the 6.6-kV BTB system that will be installed between two radial distribution feeders connected to the same primary distribution transformer. The BTB system forming

1-4244-0448-7/06/$25.00 ©2006 IEEE

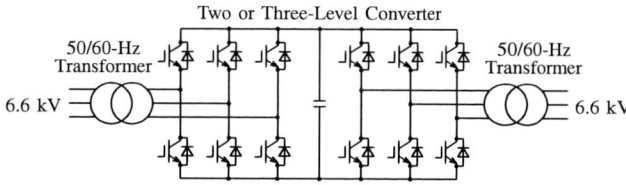

Fig. 1. The present BTB system for the 6.6-kV power distribution system.

a single looped distribution feeder actively controls active power as well as reactive power, so as to keep each voltage referred to 100 V in a range of 95 to 107 V throughout the looped feeder. The BTB system consists of two line-frequency transformers and two voltage-source PWM converters with a common dc-link capacitor. The transformers play an important role in preventing a zero-sequence current from circulating along the looped feeder. Unfortunately, each transformer rated at 6.6 kV and 1 MVA weighs around 4,000 kg, while each BTB converter ranges from 1,000 to 2,000 kg.

The use of leading-edge power devices such as IGCTs and IEGTs or IGBTs makes it possible to eliminate the two transformers from the 6.6-kV BTB system. However, the transformerless BTB system may cause a zero-sequence current circulating along the looped feeder. In the worst case, the zero-sequence current results in inappropriate operation of line-to-ground fault protection relays. To avoid the error accompanied by line-to-ground fault detection, the circulating zero-sequence current should be reduced less than 0.2 A in rms because the Japanese 6.6-kV utility power distribution system is a three-phase three-wire ungrounded circuit.

B. Medium-Voltage Motor Drive Systems

Fig. 2 shows a medium-voltage adjustable-speed motor drive system developed by Robicon Corp [15]. A three-phase diode rectifier and a single-phase H-bridge inverter form a power cell. The multi-winding line-frequency transformer supplies three-phase isolated voltages to the power cells. The ac output terminals of the power cells in each phase are connected in series to produce a sufficient voltage to drive the medium-voltage ac motor. The number of levels of the motor line-to-neutral voltage depends on how many cells are cascaded per phase. Moreover, a special multi-winding structure reduces harmonic currents flowing into the utility grid.

Although this unique system gains wide acceptance today in the field of industrial motor drives, the weight and volume of the transformer occupies a large part in the system, and it is impossible to regenerate power back to the utility grid when the motor is decelerated.

III. THE NEXT-GENERATION MEDIUM-VOLTAGE POWER CONVERSION SYSTEMS

This section deals with a medium-voltage power conversion system in the next generation, which is characterized by using bi-directional isolated dc/dc converters as the core circuit. This makes it possible to eliminate a line-frequency transformer

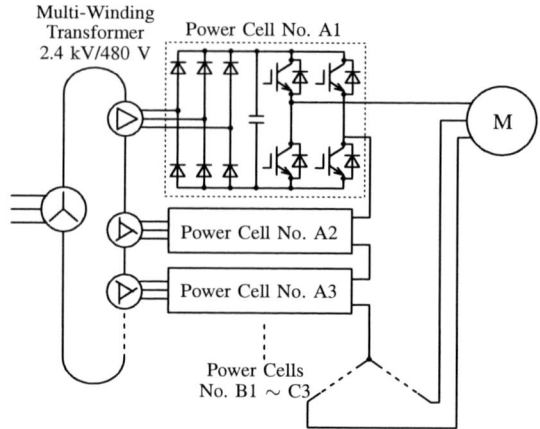

Fig. 2. The medium-voltage motor drive system proposed in [15].

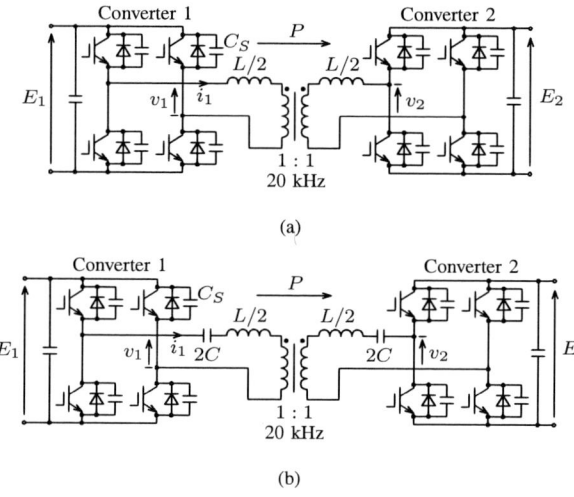

Fig. 3. The bi-directional isolated dc/dc converter: (a) based on a non-resonant circuit, (b) based on a series-resonant circuit.

A. The Bi-Directional Isolated DC/DC Converter

The bi-directional isolated dc/dc converters shown in Fig. 3 can galvanically isolate the output terminals from the input terminals, and can step up and down its output voltage by using a high-frequency transformer [11], [12].

Fig. 3 shows two possible circuits of a bi-directional isolated dc/dc converter. Fig. 3(a) is based on a non-resonant circuit while (b) is on a resonant circuit. In Fig. 3(b), the voltage across the transformer terminals becomes nearly sinusoidal by adjusting the operating frequency to the resonant frequency $(= 1/2\pi\sqrt{LC})$ so that the power devices can operate in ZCS (zero-current switching) manner.

This paper deals with 3.3-kV or 6.6-kV power conversion systems based on the bi-directional isolated dc/dc converter. In this case, the dc-link voltage in the dc/dc converter is less than 1 kV. The weight and volume of the resonant capacitors C in Fig. 3(b) is not negligible. Hence, it is more practical to employ the non-resonant dc/dc converter in Fig. 3(a), and so

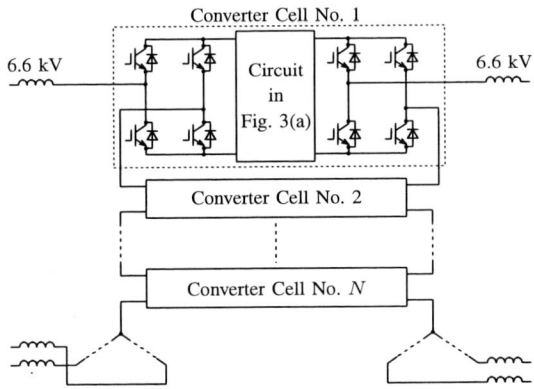

Fig. 4. A 6.6-kV BTB system in the next generation.

TABLE I

DESIGN EXAMPLES OF THE CONVERTER CELL USED FOR THE
NEXT-GENERATION BTB SYSTEM FOR THE 6.6-kV POWER DISTRIBUTION
SYSTEM.

N	Waveform	AC Input	DC Link	Device Rating
4	9 Level	952 V	1.52 kV	2.4 kV
5	11 Level	752 V	1.22 kV	2.0 kV
6	13 Level	635 V	1.02 kV	1.7 kV
7	15 Level	544 V	870 V	1.7 kV
8	17 Level	476 V	762 V	1.4 kV
9	19 Level	423 V	677 V	1.2 kV

When references [11] and [12] were published, the dc/dc converters did not draw much attention as a core circuit of medium-voltage power conversion systems because the first-generation IGBTs (insulated-gate bipolar transistors) at that time suffered from non-negligible conducting and switching losses. See Table IV. Fortunately, the advances of power device technology and magnetic material make the dc/dc converter feasible for elimination of bulky and heavy line-frequency transformers from power conversion systems.

Pavlovsky et al. have designed, constructed and experimentally verified a bi-directional isolated dc/dc converter rated at 750 V/600 V, 50 kW, and 25 kHz with a resonant ZVS and quasi-ZCS switching scheme that is different from the dc/dc converter in this paper [16]. The efficiency of their dc/dc converter was estimated to be as high as 97%, using trench-gate Si-IGBTs together with a transformer consisting of an amorphous core and foil windings. However, neither loss analysis was carried out, nor medium-voltage application was considered in [16].

B. The Next-Generation 6.6-kV BTB System

Fig. 4 depicts the 6.6-kV BTB (back-to-back) system in the next-generation. Each converter cell consists of a bi-directional isolated dc/dc converter and two single-phase PWM converters connected to the input and output terminals of the dc/dc converter. Cascade connection of N converter cells forms the BTB system.

Table I summarizes design examples of the converter cells.
The dc-link voltage of a converter cell depends on N, the

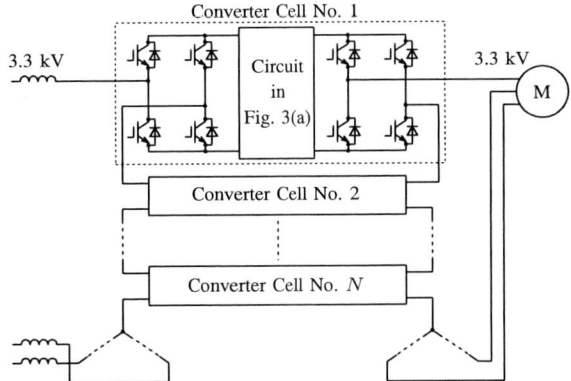

Fig. 5. A 3.3-kV motor drive in the next generation.

number of cascade connection. if N is designed to be nine, the ac-side rms voltage of a single converter cell is $6,600/9\sqrt{3} = 423$ V, and the dc-link voltage is 677 V, allowing to use 1.2-kV IGBTs. In this case, the nine cascaded converter cells in each phase produce nineteen-level voltage waveform at the ac side in each phase. Thus, switching ripples can be suppressed by only the ac-link inductors without any switching ripple filter. Since the switching frequency of the PWM converters is 1 kHz, their switching losses can be negligible.

C. A Medium-Voltage Motor Drive System in the Next Generation

Fig. 5 shows the 3.3-kV motor drive system in the next generation. The converter cells are cascaded and a 3.3-kV ac motor is connected to the ac output terminals. The system requires cascaded PWM converters at the line side because diode rectifiers instead of the PWM converters produce a large amount of harmonic current. Moreover, the PWM converters play an important role in balancing the dc-link voltages as well as providing regenerative braking capability.

If the rated power of the motor drive system is 270 kW with $N = 9$, the rated power of a single converter cell is 10 kW, and the dc-link voltage becomes 339 V, allowing to use 600-V IGBTs. Thus, this paper designs, constructs, and tests a bi-directional isolated dc/dc converter rated at 350 V, 10 kW, and 20 kHz, and analyzes the losses to estimate effectiveness in the use of SiC power devices.

In Germany, Austria, and Switzerland, the electrical railways are fed by single-phase 15-kV, 16 2/3-Hz power lines. This results in bulky and heavy low-frequency transformers on locomotives. Many reaserchers/engineeres in Europe have been tackling this issue and have proposed several different solutions to date [17]. Among them, the traction systems in [18] and [19] are similar to the system in Fig. 5, although they are different in application, voltage levels, and the use of the resonant dc/dc converter shown in Fig. 3(b). However, few experimental results have been reported yet about the traction system. Heinemann has designed, constructed and tested a prototype of a 3-kV, 350-kVA, 10-kHz transformer having an insulation capability of 35 kV [20], which has been intended to be used in the traction system in [19].

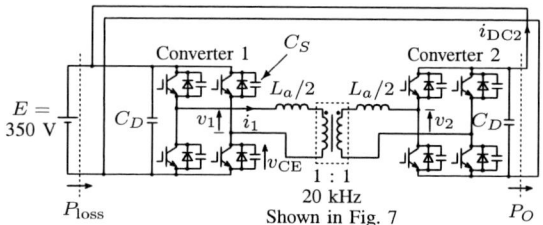

Fig. 6. The experimental circuit configuration for achieving precisely measuring the overall loss and loss analysis.

TABLE II

CIRCUIT PARAMETERS IN FIG. 6.

DC power supply voltage	E	350 V
Rated power	P	10 kW
DC capacitor	C_D	2,000 μF
Snubber capacitor	C_S	0.01 μF (1.5%)
Inductor	$L_a/2$	10.5 μH (11%)
Inductor resistance	$R_a/2$	4.5 mΩ (0.037%)
Transformer core material		FT-3M
Transformer turn number		17 : 17
Transformer leakage inductance	L_{trans}	1.6 μH (1.6%)
Transformer winding resistance	R_{trans}	17 mΩ (0.14%)

On a single-phase 350-V, 10-kW, 20-kHz base.

IV. DESIGN AND CHARACTERISTICS OF THE DC/DC CONVERTER AS THE CORE CIRCUIT

A. Circuit Configuration

Fig. 6 shows the circuit configuration of the 350-V, 10-kW, 20-kHz bi-directional isolated dc/dc converter for experiments. Table II summarizes the circuit parameters in Fig. 6. Fig. 7 shows the photograph of the transformer rated at 350 V, 10 kVA, and 20-kHz. Two full-bridge voltage-source converters (Converters 1 and 2) are connected symmetrically via the transformer. The output power P_O is regenerated back to the input terminals so that the overall loss of the circuit P_{loss} can be measured accurately and easily as the output power of the dc voltage source E. In the experiment, a digital power meter (Yokogawa WT130) is used to measure the overall loss P_{loss}. An analog ammeter (Yokogawa 2011 37) measures the average dc output current I_{DC2} so that the output power P_O can be obtained as $E \cdot I_{\text{DC2}}$.

The latest trench-gate IGBTs (Mitsubishi Electric CM200DY-12NF) are used in Converters 1 and 2. The core material of the transformer is "Finemet" FT-3M manufactured by Hitachi Metals. A set of litz wires is used as the windings to reduce the influence by the so-called skin effect. Two identical auxiliary inductors consisting of ferrite cores and litz wires are connected to adjust the inductance value that affect the power flow to be $L = L_{\text{trans}} + L_a = 22.6$ μH. The winding resistance of the transformer and the inductors are $R = R_{\text{trans}} + R_a = 26$ mΩ[1].

If all the power devices, the transformer, and the auxiliary inductors were ideal, the ac-side voltage of Converters 1 and 2, v_1 and v_2, would be rectangular voltage at $f = 20$ kHz.

[1]These inductance and resistance values (L_{trans}, L_a, R_{trans}, and R_a) were at 10 kHz and separately obtained by using an LCR meter (Agilent Technologies 4263B).

Fig. 7. The 350-V, 10-kVA and 20-kHz transformer.

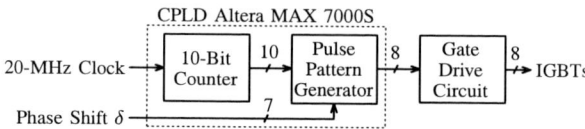

Fig. 8. The control circuit of the dc/dc converter.

The output power P_O is determined by the phase shift δ [rad] between v_1 and v_2 as follows: [11]

$$P = \frac{E^2}{\omega L}\left(\delta - \frac{\delta^2}{\pi}\right),\qquad (1)$$

where $\omega = 2\pi f$.

Connecting snubber capacitors C_S in parallel to the IGBTs in Fig. 6 realizes zero-voltage switching (ZVS) operation. The minimum required current I_{min} flowing in an IGBT to ensure ZVS operation is determined by: [12]

$$I_{\text{min}} = \frac{2E}{\sqrt{L/C_S}}.\qquad (2)$$

B. Control Circuit

Fig. 8 depicts the block diagram of the control and gate drive circuit. A clock signal at 20 MHz is fed to the CPLD (complex programmable logic device) Altera MAX 7000S (EPM7160SLC84-10). The 10-bit counter counts the clock pulse to create 10-bit digital sawtooth signal at 20 kHz. The pulse pattern generator receives the digital sawtooth signal and the 7-bit phase shift δ to generate switching patterns. A dead time of 500 ns is also created in the pulse pattern generator. The LSB (least significant bit) in the 7-bit phase shift δ corresponds to 0.36° $(2\pi/1,000$ [rad]) because the clock period of 50 ns (20 MHz) is 0.1% of the switching period of 50 μs (20 kHz) in the dc/dc converter. The gate drive circuit finally produces proper gate-emitter voltages (+15 V to be turned on and −15 V to be turned off) for the eight IGBTs.

Note that the control circuit in this paper is an open-loop system. Feedback control, for example, to control the power or the dc-link voltage constant, can be easily realized when installing analog-to-digital converters.

C. Conversion Efficiency and Overall Loss

Fig. 9 shows experimental waveforms at the rated power of

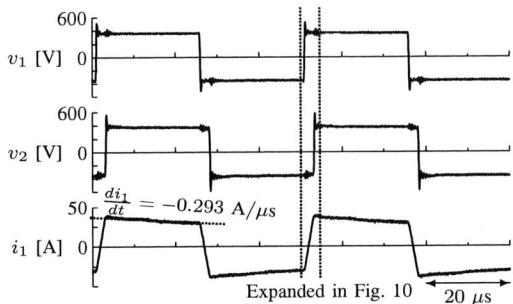

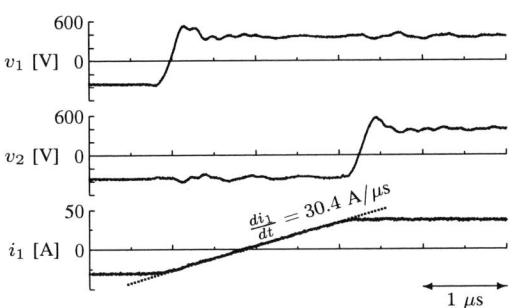

Fig. 9. Experimental waveforms when a power of 10 kW is delivered.

Fig. 10. Time-expanded waveforms of Fig. 9.

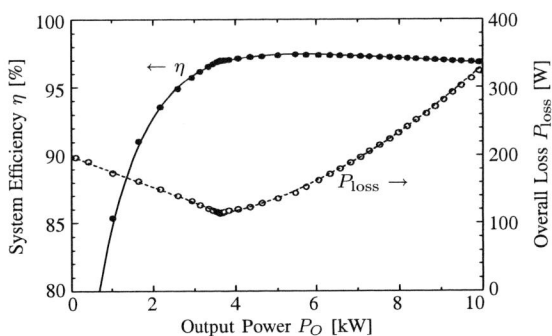

Fig. 11. Relationships between the output power P_O, the system efficiency η and the overall loss P_{loss}.

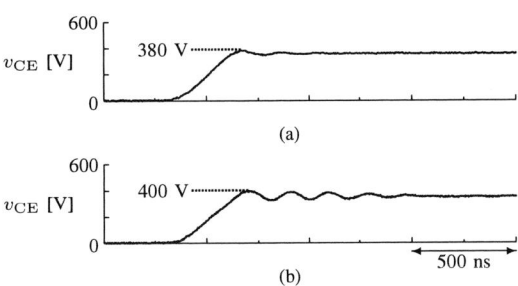

Fig. 12. Waveforms of the collector-emitter voltage of the IGBT: (a) when $P_O = 0$ kW, (b) when $P_O = 10$ kW.

Fig. 9. In this case, the phase shift δ was 17°, the rms value of the current i_1, I_1, was 32.6 A, and the average value of the absolute value of i_1, $\langle |i_1| \rangle$, was 32.0 A. The overall loss in the circuit P_{loss} was 335 W. Hence, the system efficiency η from dc input to dc output was $P_O/(P_O + P_{\text{loss}}) = 96.8\%$[2]. The di_1/dt values in Figs. 9 and 10 are used in the following loss analysis procedure.

Fig. 11 describes the relationships between the output power P_O, the system efficiency η, and the overall loss P_{loss}. The overall loss P_{loss} reaches its minimum at around $P_O = 3.8$ kW. When P_O is below 3.8 kW, incomplete ZVS operation results in forming a short circuit of the snubber capacitors that store some amount of energy so that the switching loss takes dominance. On the other hand, when P_O exceeds 3.8 kW, the conducting loss of the IGBTs becomes dominant. At $P_O = 5.5$ kW, the system efficiency η reaches its maximum, 97.4%.

D. Overvoltage across the IGBTs

Fig. 12 is the collector-emitter voltage across an IGBT in Converter 1, v_{CE}, when the IGBT is turned off. Note that v_{CE} is not the voltage on the IGBT chip but the voltage across the collector and emitter terminals of the IGBT module. Fig. 12(a) was taken when the output power P_O was zero, and (b) was

taken when P_O was 10 kW. Closely connecting the snubber capacitors C_S between the IGBT collectors and emitters and careful optimization of the gate resistance in the gate-drive circuit contributed to suppressing the maximum over voltage to be 380 V when $P_O = 0$ kW and 400 V $P_O = 10$ kW at $E = 350$ V.

E. Loss at $P_O = 0$ kW

When the output power P_O is zero, all the energy stored in a snubber capacitor C_S is lost when the IGBT is turned on. The loss in the experimental circuit from this mechanism P_{snub} can be obtained as:

$$
\begin{aligned}
P_{\text{snub}} &= 8 C_S E^2 f \\
&= 8 \times 0.01 \ \mu\text{F} \times (350 \ \text{V})^2 \times 20 \ \text{kHz} \\
&= 196 \ \text{W}.
\end{aligned} \tag{3}
$$

As will be stated in section V-B, the core loss in the transformer is estimated to be $P_{\text{core(tr)}} = 20$ W. Therefore, the theoretical loss at $P_O = 0$ kW can be calculated as $P_{\text{snub}} + P_{\text{core(tr)}} = 216$ W, agreeing well to the experimentally observed value $P_{\text{loss}} = 197$ W.

V. LOSS ANALYSIS

A. Conducing Loss in the IGBTs

From the di_1/dt values in Figs. 9 and 10, one can estimate the voltage drop $2(V_{\text{CE(sat)}} + V_F)$ across the power devices

---

[2]In the experiment, isolated dc power sources for the gate drive circuits were obtained from the single-phase 100-V utility outlet via small-rated 50-Hz transformers and diode rectifiers. When $P_O = 10$ kW, the loss of the gate drive circuit for the eight IGBTs was 8.9 W. The loss corresponds only to 2.7% of the overall loss of the main circuit $P_{\text{loss}} = 335$ W. Therefore, this

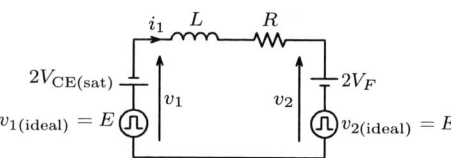

Fig. 13. Equivalent circuit of Fig. 6 where the winding resistance R of the transformer and the inductors, and the voltage drop across the power devices are taken into account.

$di_1/dt = 30.4$ A/μs when $v_1 = E$ and $v_2 = -E$ so that the following equation exists:

$$L\frac{di_1}{dt} = 2E - Ri_1. \qquad (4)$$

The second term in the above equation is negligible because it is obviously small compared to the first one (= 700 V). The inductance L can be calculated to be $L = 23.0$ μH from the above equation. The following loss analysis uses not the value $L = 22.6$ μH that appeared in section IV-A but the value $L = 23.0$ μH obtained here.

In Fig. 9, when $v_1 = v_2 = E$, the experimental circuit in Fig. 6 can be expressed by the equivalent circuit in Fig. 13. Note that Fig. 13 takes into account the winding resistance of the transformer and the inductors, and the voltage drop across the power devices. The voltages $v_{1(\text{ideal})}$ and $v_{2(\text{ideal})}$ would correspond to v_1 and v_2 if the power devices were ideal. In Fig. 9, $di_1/dt = -0.293$ A/μs. Thus, the voltage drop across the power devices can be calculated as:

$$2(V_{\text{CE(sat)}} + V_F) = -L\frac{di_1}{dt} - Ri_1 = 5.9 \text{ V}. \qquad (5)$$

If i_1 in the above equation is approximated to be its average value of the absolute value, $\langle |i_1| \rangle$, the conducting loss P_{cond} in the power devices can be estimated to be:

$$P_{\text{cond}} = 2(V_{\text{CE(sat)}} + V_F) \cdot \langle |i_1| \rangle = 189 \text{ W}. \qquad (6)$$

B. Losses in the Transformer and the Auxiliary Inductors

Table III summarizes the characteristics and losses of the transformer and each of the two identical auxiliary inductors at $P_O = 10$ kW . The maximum flux density in the transformer core $B_{\text{max(tr)}}$ can be calculated to be 0.86 T because the core effective cross-sectional area $A_e = 300$ mm^2, turn number (17 turns), the voltage $E = 350$ V, and the switching frequency $f = 20$ kHz are already known. The maximum flux density in the auxiliary inductor $B_{\text{max(ind)}}$ was obtained to be 0.174 T by using the rms value I_1 of the current i_1, 32.6 A. From the datasheets, the core losses can be estimated. As in Table III, the transformer core loss $P_{\text{core(tr)}}$ was 20 W, the inductor core loss $P_{\text{core(ind)}}$ was 9.0 W (4.5 W each) [3].

[3]The flux density in the transformer core becomes triangular because the rectangular voltages are applied to the transformer. On the other hand, trapezoidal current flows in the auxiliary inductors so that the flux densities in the auxiliary inductor cores also become trapezoidal. For the sake of simplicity, this paper deals only with the loss caused by the 20 kHz fundamental

TABLE III
CHARACTERISTICS AND THE LOSS OF THE TRANSFORMER AND THE INDUCTORS AT THE RATED POWER OF 10 KW.

Parameter	Transformer	Inductor
Core Material	Finemet FT-3M	Ferrite PC44
Effective Cross Section	300 mm^2	328 mm^2
Effective Volume	60.7 cm^3	37.2 cm^3
Turn Number	17 : 17	6
Maximum Flux Density	0.86 T	0.17 T
Winding Resistance	17 mΩ	4.5 mΩ
Core Loss P_{core}	20 W	4.5 W
Copper Loss P_{copp}	18 W	4.8 W

The core losses were calculated based on the datasheets. Two inductors were used in the experimental circuit.

Power Switching Devices — Magnetic Components

■ Conducting Loss: 189 W □ Switching Loss: 90 W
▨ Trans. Core Loss: 20 W ▥ Trans. Copper Loss: 18 W
▦ Inductor Core Loss: 9 W ▧ Inductor Copper Loss: 10 W

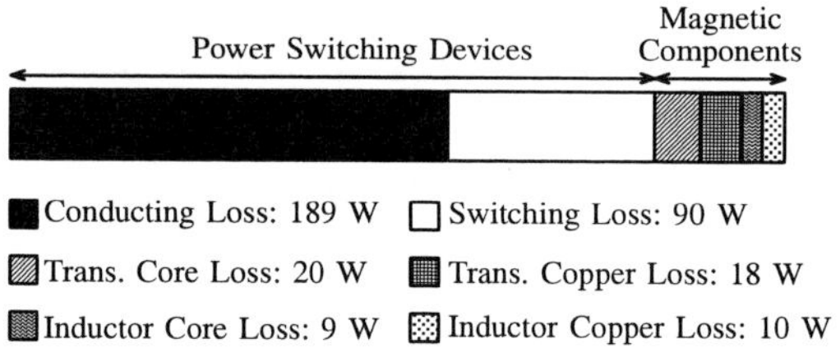

Fig. 14. Loss analysis in the experimental circuit.

The transformer winding resistance $R_{\text{trans}} = 17$ mΩ and the rms current $I_1 = 32.6$ A produced the copper loss of $P_{\text{cond(tr)}} = 17$ m$\Omega \times (32.6 \text{ A})^2 = 18$ W in the transformer. Similarly, the inductor copper loss was $P_{\text{cond(ind)}} = 4.5$ m$\Omega \times (32.6 \text{ A})^2 \times 2$ inductors $= 9.6$ W.

C. Loss Analysis Results

Fig. 14 summarizes the loss analysis results. The switching loss P_{sw} in the figure was obtained by subtracting the power device conducting loss and the transformer and inductor losses from the overall loss $P_{\text{loss}} = 335$ W. Although the "switching loss P_{sw}" in Fig. 14 may contain other losses than the real switching loss, the most dominant component in P_{sw} is the switching loss. Note that the snubber capacitor C_S theoretically does not contribute to the switching loss P_{sw} because each IGBT is operated in ZVS manner at 10 kW. The power loss produced by an equivalent series resistance existing in C_S is included in P_{sw}, but it may be negligible.

In Fig. 14, the loss in the power devices is 279 W and most dominant in the overall loss of 335 W. Therefore, significant loss reduction can be expected if the Si IGBTs are to be replaced by SiC power devices.

VI. THE POSSIBILITY OF LOSS REDUCTION BY SIC POWER DEVICES

A. Trends in SiC Power Devices

The breakdown field of SiC is ten times as high as that of Si so that high-voltage, low-loss, and high-speed unipolar devices can be realized. For example, while the specific on-state resistance in a 1.2-kV MOSFET drift region is theoretically

TABLE IV

TRENDS IN THE BI-DIRECTIONAL ISOLATED DC/DC CONVERTER.

Year	1991	2005 (This work)	2015?
Switching Devices	First-Generation Si-IGBT	Latest Trench-Gate Si-IGBT	SiC-MOSFET/JFET
Core Material	Ferrite	Finemet$^{\text{TM}\dagger}$	New Magnetic Material?
Efficiency (Loss)	around 91% (9%)*	97% (3%)	over 99% (1%)

* This value was estimated by the authors of this paper because the authors of [11] and [12] had made no description of efficiency.

† Nano-crystalline soft-magnetic material manufactured by Hitachi Metals

can reduce the theoretical limit to 0.248 mΩcm² [1]. In SiC-MOSFETs, the channel resistance is dominant because of low channel mobility. Thus, SiC-JFETs are also studied because their channel mobility is the same as that in the drift region [6], [7], [8].

Chang et al. have built a three-phase voltage-source inverter using SiC-JFETs (600 V, 25 A, 2.94 mΩcm²)⁴ and SiC-SBDs [6], and has reported that the on-state voltage across the JFET had been 38 mΩ × 20 A = 0.76 V. This value, 0.76 V, was less than the half of the voltage drop across the Si-IGBT used in this paper. Note that the SiC-JFETs in [6] were normally-on devices. Tone et al. have developed a normally-off SiC-JFET having a blocking voltage as high as 1,726 V and specific on-state resistance as low as 3.6 mΩcm² at 300 A/cm² [7]. Lai et al. have built single-phase and three-phase inverters using 500-V SiC-JFETs and have pointed out the synchronous rectification capability of the SiC-JFETs [8].

Arai and Yoshida have compared conducting and switching losses between 600-V Si-IGBTs and SiC-MOSFETs, and they have predicted both losses can be reduced to one-tenth [1]. SiC-MOSFETs, however, had suffered from relatively high channel resistance because high temperature (over 1,600°C) annealing process had deteriorated the carrier mobility on the SiC-SiO₂ surface after acceptor ion implantation to create the p-well. Thus, new technologies to increase the channel mobility, as well as to reduce the channel resistance, have been invented to fabricate the following novel SiC-MOSFETs. In December 2004, Mitsubishi Electric reported that they had developed a 1.2-kV, 12.9-mΩcm² SiC-MOSFET, using epitaxial growth to create the p-well. At the same time, Rohm announced that a 1-kV, 7.15-mΩcm² SiC-MOSFET based on a improved process to create the gate oxide. In March 2005, the National Institute of Advanced Industrial Science and Technology of Japan (AIST) reported a 1.1-kV, 4.3-mΩcm² SiC-MOSFET [10]. The AIST SiC-MOSFET used also epitaxial growth to create the p-well. The conducting loss can be reduced to about one-fifth, compared to the 1.2-kV Si-IGBTs when the 1.1-kV SiC-MOSFETs and the 1.2-kV Si-IGBTs have the same current density as 100 A/cm².

B. Prediction of Conversion Efficiency with Use of SiC-MOSFETs and SiC-SBDs

Replacing Si-IGBTs with SiC-MOSFETs in Fig. 6 can reduce the conducting loss to a half to one-fifth. Using SiC-SBD can significantly reduce the switching loss because both the switching devices and the diodes are unipolar devices.

In the dc/dc converter that operates in rectification mode (Converter 2 in Fig. 6), the ac current flows mainly in two diodes. So far, the 1.2-kV SiC-SBDs from Infineon Technologies produces a forward voltage drop ranging from 1.5 V to 2.1 V, which is much higher than the on-state voltage drop produced by the 1.2-kV SiC-MOSFET. However, synchronous rectification technique can be realized when using SiC-MOSFETs.

From the above-mentioned discussion, this paper assumes that the conducting loss P_{cond} can be reduced to one-fifth, and the switching loss P_{sw} can be reduced to one-tenth. As a result, the overall loss in the dc/dc converter at the rated power of 10 kW would be reduced from 335 W to 104 W, and so the efficiency of the dc/dc converter would reach $\eta = 99\%$.

Table IV summarizes trends in efficiency improvement and loss reduction of the dc/dc converter shown in Fig. 3(a). The dc/dc converter designed, constructed, and tested in this paper has reached the overall efficiency as high as 97%. As a result, the loss was reduced to one-third (from 9% to 3%) between 1991 and 2005. When SiC-MOSFETs/JFETs are available in the near future, the loss may be reduced to one-third again (3% to 1%).

In the 3.3-kV motor drive system in Fig. 5, the switching frequency of each cascaded PWM converter is 1 kHz so that the switching loss is small enough to be neglected, compared to that in the dc/dc converter with an overall efficiency of 99%. Therefore, the efficiency of each cascaded PWM converter may reach 99.5%. Thus, the overall efficiency of the 3.3-kV motor drive system is expected to reach 99.5% × 99% × 99.5% = 98% in 2015.

VII. CONCLUSIONS

This paper has described a technical issue in present medium-voltage power conversion systems, and has presented the next-generation medium-voltage power conversion systems intended to use SiC power devices. The next-generation systems are characterized by providing galvanic isolation with bi-directional high-frequency isolated dc/dc converters to significantly reduce their size and weight. In addition, this paper has designed and built a bi-directional isolated dc/dc converter rated at 350 V, 10 kW, and 20 kHz as the core circuit of a 3.3-kV, 270-kW motor drive system. The results of loss analysis has revealed that using SiC power devices would realize the dc/dc converter with an efficiency of 99%, and the 3.3-kV, 270-kW motor drive system with an efficiency of 98%. This loss evaluation would encourage power electronics researchers and engineers to proceed with further research on the key to put these power conversion systems into practical use. These issues remain as the next phases of research.

REFERENCES

[1] K. Arai and S. Yoshida, "Fundamentals and applications of SiC devices," Ohmsha, 2003 (in Japanese)

[2] B. J. Baliga, "Silicon carbide power devices," World Scientific Publishing, 2005

[3] M. Kawai and C. Horikiri, "The invisible strength of power control semiconductors," *Nikkei Electronics*, no. 893, pp. 79-97, 2005 (in Japanese)

[4] J. A. Cooper, JR. and A. Agarwal, "SiC power-switching devices – the second electronics revolution?" *Proc. of IEEE*, vol. 90, no. 6, pp.956-968, 2002

[5] P. Friedrichs and Roland Rupp, "Silicon carbide power devices – current developments and potential applications," *European Conference on Power Electronics and Applications (EPE)*, CD-ROM, 2005

[6] H.-R Chang, E. Hanna, and A. V. Radun, "Development and demonstration of silicon carbide (SiC) motor drive inverter modules," *IEEE Power Electronics Specialists Conference (PESC)*, vol. 1, pp. 211-216, 2003

[7] K. Tone, J. H. Zhao, L. Fursin, P. Alexandrov, and M. Weiner, "4H-SiC normally-off vertical junction field-effect transistor with high current density," *IEEE Electron Device Letters*, vol. 24, no. 7, pp. 463-465, 2003

[8] J.-S. Lai, H. Yu, J. Zhang, P. Alexandrov, Y. Li, J. H. Zhao, K. Sheng, and A. Hefner, "Characterization of normally-off SiC vertical JFET devices and inverter circuits," *IEEE IAS Annual Meeting 2005*, CD-ROM, 2005

[9] G. Majumdar, "Future of power semiconductors," *IEEE Power Electronics Specialists Conference (PESC)*, vol. 1, pp. 10-15, 2004

[10] S. Harada, M. Kato, M. Okamoto, T. Yatsuo, K. Fukuda, and K. Arai, "4.3 mΩcm^2, 1100 V normally-off IEMOSFET on SiC," *The Paper of Joint Technical Meeting on Electron Devices and Semiconductor Power Converter, IEE Japan*, EDD-05-49/SPC-05-74, pp. 27-31, 2005 (in Japanese)

[11] R. W. De Doncker, D. M. Divan, and M. H. Kheraluwala, "A three-phase soft-switched high-power density dc/dc converter for high-power applications," *IEEE Trans. Ind. Applicat.*, vol. 27, no. 1, pp. 63-73, 1991

[12] M. H. Kheraluwala, R. W. Gascoigne, and D. M. Divan, "Performance characterization of a high-power dual active bridge dc-to-dc converter," *IEEE Trans. Ind. Applicat.*, vol. 28, no. 6, pp. 1294-1301, 1992

[13] G. Spiazzi, S. Buso, M. Citron, M. Corradim, and R. Pierobon, "Performance evaluation of a schottky SiC power diode in a boost PFC application," *IEEE Trans. Power Electron.*, vol. 18, no. 6, pp. 1249-1253, 2003

[14] N. Okada, "Control of loop distribution network and result," *The Paper of Technical Meeting on Power Systems Engineering, IEEJ*, PSE-00-2, pp. 2-12, 2000 (in Japanese)

[15] P. W. Hammond, "A new approach to enhance power quality for medium voltage ac drives," *IEEE Trans. Ind. Applicat.*, vol. 33, no. 1, pp. 202-208, 1997

[16] M. Pavlovsky, S. W. H. de Haan, and J. A. Ferreira, "Concept of 50 kW DC/DC converter based on ZVS, quasi-ZCS topology and integrated thermal and electromagnetic design," *European Conference on Power Electronics and Applications (EPE)*, CD-ROM, 2005

[17] H. Stemmler, "State of the art and future trends in high power electronics," *International Power Electronics Conference (IPEC) Tokyo*, vol. 1, pp. 4-14, 2000

[18] N. Schibli and A. Rufer, "Single-phase and three-phase multilevel converters for traction systems 50 Hz/16 2/3 Hz," *European Conference on Power Electronics and Applications (EPE)*, vol. 4, pp. 210-215, 1997

[19] G. Kratz and H. Strasser, "Antriebskonzept für zukünftige elektrische Triebfahrzeuge," *Elektrische Bahnen*, vol. 96, pp. 333-337, 1998 (in German)

[20] L. Heinemann, "An actively cooled high power, high frequency transformer with high insulation capability," *IEEE Applied Power Electronics Conference and Exposition (APEC)*, vol. 1, pp. 352-357, 2002

Hirofumi Akagi (M'87-SM'94-F'96) was born in Okayama, Japan, in 1951. He received the B. S. degree from the Nagoya Institute of Technology, Nagoya, Japan, in 1974, and the M. S. and Ph. D. degrees from the Tokyo Institute of Technology, Tokyo, Japan, in 1976 and 1979, respectively, all in electrical engineering. In 1979, he joined the Nagaoka University of Technology, Nagaoka, Japan, as an Assistant and then Associate Professor in the department of electrical engineering. In 1987, he was a Visiting Scientist at the Massachusetts Institute of Technology for ten months. From 1991 to 1999, he was a Professor in the department of electrical engineering at Okayama University, Okayama, Japan. From March to August of 1996, he was a Visiting Professor at the University of Wisconsin-Madison and then the Massachusetts Institute of Technology. Since January 2000, he has been a Professor in the department of electrical and electronic engineering at the Tokyo Institute of Technology, Tokyo, Japan. His research interests include power conversion systems, ac motor drives, active and passive EMI filters, high-frequency resonant-inverters for induction heating and corona discharge treatment processes, and utility applications of power electronics such as active filters, self-commutated BTB systems, and FACTS devices. He has authored or coauthored some 70 IEEE Journal papers, including two invited *Proceedings of the IEEE* papers in 2001 and 2004. He has made presentations many times as a keynote or invited speaker internationally. He received two IEEE Industry Applications Society (IAS) Transactions Prize Paper Awards in 1991 and 2004, and two IEEE Power Electronics Society (PELS) Transactions Prize Paper Awards in 1999 and in 2003, along with nine IEEE IAS Committee Prize Paper Awards. He was elected as a Distinguished Lecturer of the IEEE IAS and PELS for 1998-1999. He was a recipient of the IEEE William E. Newell Power Electronics Award in 2001, and the IEEE IAS Outstanding Achievement Award in 2004.

Shigenori Inoue (S'02) was born in Fujimi, Saitama, Japan on January 29, 1979. He received B.S. and M.S. degrees from Tokyo Metropolitan University, Tokyo, Japan, in 2002 and 2004, respectively. Since April 2004, he has been working toward the Ph.D degree in Tokyo Institute of Technology, Tokyo, Japan. He currently is a JSPS (Japan Society for the Promotion of Science) research fellow. His research interests include medium-voltage power conversion systems, bi-directional isolated dc/dc converters, and active power filters.

2006 5th International Power Electronics and Motion Control Conference

Modern Electrical Drives:
Design and Future Trends

R. W. De Doncker, *Fellow, IEEE*

Institute for Power Electronics and Electrical Drives (ISEA)
RWTH-Aachen University, Aachen, Germany

Abstract—Electrical drives efficiently convert electrical power into mechanical power. As factory automation, comfortable lifestyle and energy conservation are growing businesses, the number of drives produced worldwide keeps growing. The increased use of information technology (computers, digital control) and communication systems not only has created new markets for drives, e.g. disc drives but also enforces more electrical drives to be used in systems as actuators and mechatronic systems. In this paper, the author reviews the present state of development of drive technology and probes into future application and technology trends.

Keywords- *electrical drives, electrical machines, power electronics, embedded control, rapid control prototyping*

I. INTRODUCTION

In general, as illustrated in Fig. 1, an electrical drive can be defined as a power conversion means characterized by its capability to efficiently convert electrical power from an electrical power source (voltage and current) into mechanical power (torque and speed) to control a mechanical load or process. In some cases, this power flow is reversed or can even be bi-directional. Today, modern drives make use of power electronic converters to (digitally) control this electro-mechanical energy conversion process. In addition, as drives are being integrated more and more in systems, communication links to higher level computer networks are essential to support commissioning, initialization, diagnostics and higher level process control. Consequently, the main drive components consist of an electro-mechanical energy converter (typically an electro-magnetic machine or actuator), a power electronic electrical-to-electrical power converter and an embedded digital control unit. The digital control unit directly controls the power electronic semiconductor switches of the power electronic converter. To this end not only suitable control hardware, sensors, high-speed digital logic devices and processors are needed but also suitable control algorithms. From this perspective, drive technology is a fairly modern development. Indeed, although electrical machines were first developed over 150 years ago, power electronic converter have been available for only 45 years, dynamic torque control algorithms for induction machines (field oriented control) have been around for

about 30 years and high-speed digital control using DSPs have been available for less than 25 years. Even today, with all components (machine, power electronics, control hardware and software) being developed, drive technology is still evolving at a rapid pace. Over the past 20 years, new machine types have been developed, optimized and investigated, such as linear machines, surface PM magnet and buried PM magnet machines, switched reluctance machines, transversal flux machines, axial flux machines, etc. Each machine type requires its specific control and sensors. During the past 10 years, (position) sensorless drives have been investigated to eliminate expensive sensors and make drives more robust (reliable).

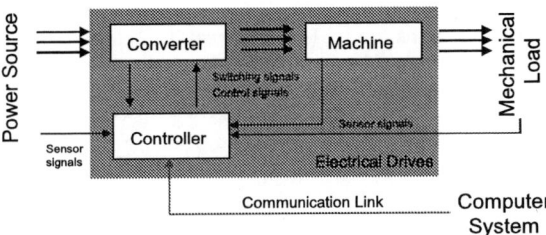

Figure 1. Electrical Drive System

The power range of modern drives spans many decades, from milliwatts up to hundreds of megawatts, which demonstrates the flexibility and the broad application of this technology.

Figure 2. Drive technology can scale from very small power (less than 1 W) to high power (more than 10 MW)

In the following several technology trends of state-of-the-art drives are being discussed. An attempt is made to derive future trends based on the development of drives over the past 25 years.

1-4244-0448-7/06/$25.00 ©2006 IEEE

II. MARKET TRENDS

A recently published report of ZVEI, illustrates the market of electrical drives in Germany [1]. Production technology, primarily driven by continued automation of industrial processes, energy efficiency and automotive applications has steadily increased over the past 10 years with a growth rate of 5-6 % annually. Sales are reaching 9 billion €, creating work for at least 60,000 people, not including maintenance and service personnel.

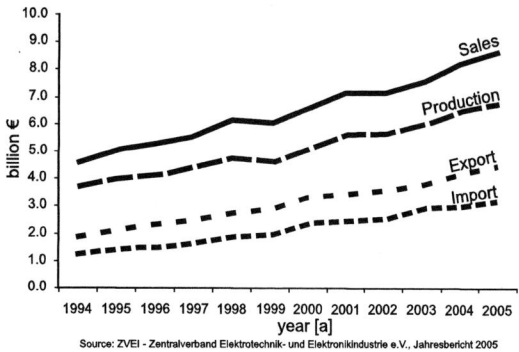

Figure 3.   Sales, production of electrical drives in Germany.

Similar growth rates can be noticed in the US [2]. Figure 4 shows a breakdown of the different drive technologies. Interestingly, the classical dc machine drives (with field excitation winding or with permanent magnets) maintain a constant (in absolute numbers) market share of around 1 billion $. The market increase is primarily due to the increased sales of ac induction machines but also PM synchronous (brushless DC) and switched reluctance machines.

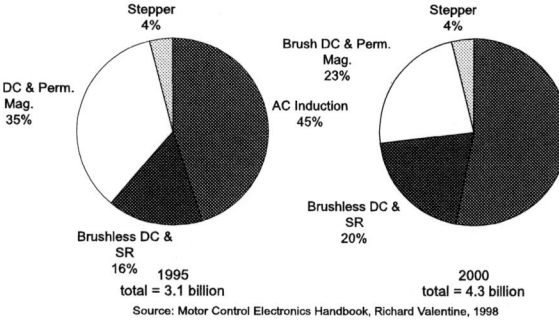

Figure 4.   Growth and market share of electrical drives in U.S.

Clearly, electrical drive technology represents growing markets, albeit less spectacular than recent IT and nano-technologies, but has proven to be a robust market segment which has been affected less by speculation and global market fluctuations or crisis. One can say that electrical drives literally are robust systems which keep the world's economy moving towards higher prosperity (more work done by machines) and more efficient use of primary energy (as variable speed drives are more efficient when production rates need to be adapted).

III. TECHNOLOGY TRENDS

A. Electrical machines

As was shown above, most drives sold today are based on induction machines, PM synchronous machines and increasingly switched reluctance machines (SRM) (see Fig. 5).

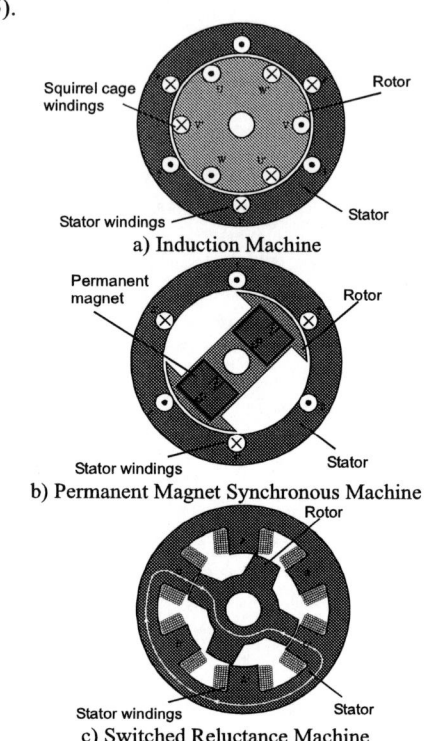

Figure 5.   Construction of most common electrical machine types

Back in 1980, it was thought that power density of electrical machines was saturating according to the classical S-curve, which is typical for a maturing technology. For example, Fig. 6, shows the evolution of power density of traction machines in 1980. This view was based on the fact that the materials used to build electrical machines were well developed and no innovations to improve power density (requires higher operating temperatures or less lossy materials) were to be expected. Furthermore, market demand had settled on standard machines with speeds up to 3,000 or 3,600 rpm as bearing life seemed to be limited at higher speeds.

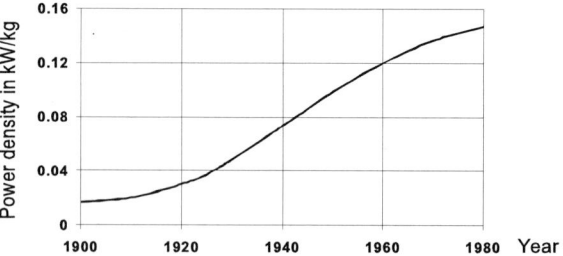

Figure 6.   Development of power density of electrical machines for traction applications.

32

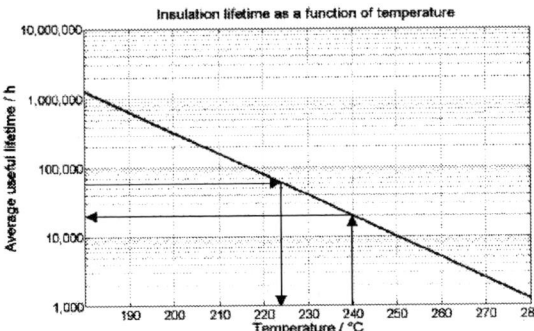

Figure 7. Insulation materials useful lifetime versus temperature

Note that power density of all electrical machines strongly depends on its maximum operating speed as power equals torque times speed and rated torque is the key variable which determines machine size in the first place. So, for constant power applications, higher speed machines have lower torque requirements and consequently can be built smaller and lighter.

Today, electrical machines still are built with the same silicon-steel alloys (max. induction still about 2 Tesla) for the laminations, copper or aluminum for the windings (same losses) and insulation materials, although the latter now reach somewhat higher maximum allowable hot spot temperatures of up to 225 °C for 50,000 hours operating life, as shown in Fig. 7. We can conclude that, in contrast to the first century of machine development, which relied greatly on material improvements, it were several engineering achievements which have allowed over the past 25 years to improve significantly power density of traction and industrial machines up to 1.2 kW/kg. These engineering developments can be situated in several areas:

- Improved quality control, automated production as well as new production techniques, for example copper injection (instead of aluminum) to form squirrel cage induction machines [3].
- Improved design tools (finite elements and physics based dynamic models linked to inverter and control simulations) allow the analysis of the drive under varying loading conditions and controls. Hence, they allow designs (synthesis) with less derating.
- Improved cooling avoids local hot spots, which allows for a better utilization of the available materials while keeping life constant (insulation materials with better thermal conductivity and improved thermal design of the cooling system)
- Reduced derating (used to be at least 15%) of inverter fed machines because less harmonics are being produced by power converters as higher switching frequencies are attainable today without compromising efficiency and cost of the inverter
- Applications have moved, with improved bearing and gear technology as well as improved converter technology, to higher speed applications. Fig. 8 shows an integrated inverter-machine drive with gear capable to operate above 3000 rpm [7].

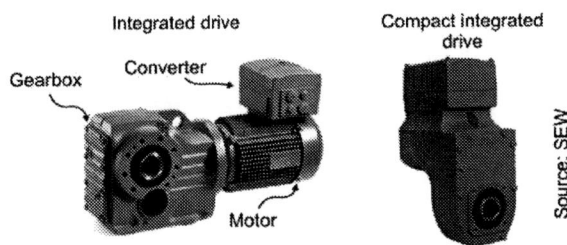

Figure 8. Integrated drives and gears using induction machines[7]

The latter can also be seen, for example, in modern traction applications which are now pushing machine speeds over 6,000 rpm. In electrical and hybrid vehicles electrical machines typically operate up to 16,000 rpm. Large high-power compressor drives operate up to 25,000 rpm. Recently, several manufacturers have implemented high-speed switched reluctance machines in vacuum cleaners at speeds reaching 100,000 rpm, well above the practical limits of universal dc machines. These machines reach, due to their high speed and inherent excellent cooling, power densities up to 3.5 kW/kg.

B. *Power converters*

It should be noted that power density of power converter systems (including auxiliaries, switchgear, cooling system) has improved even more dramatically over the past 25 years. This can be explained by the fact that power electronic converters are a more recent technology development having more opportunities to improve in several areas. Key to improve power density was the development of improved (less lossy) turn-off power semiconductors, improved heatsink technology, improved control minimizing losses, compact controller design and better design tools which allow the converter designer to "push for the limits" without compromising life of the inverter. However, in the author's opinion, deploying voltage source topologies (instead of current source topologies) and improving design and technology of passive components, in particular of capacitors, had an equally strong impact on power density of converter systems over the past decade.

Over the past 25 years volumetric power density of industrial air-cooled ac-to-ac converter systems improved from 30 kVA/m^3 up to 500 kVA/m^3. Several examples of commercially available converter units and converter systems are shown in Fig. 9 and Fig. 10.

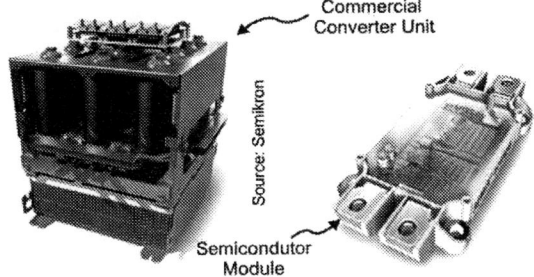

Figure 9. Universal commercial converter building block [8]

Figure 10. Modern vector controlled (FOC) inverters (source Lenze) with standard communication interfaces [9]

The most important limiting factor in designing high-power density converter modules is thermal, i.e. maximum operating temperature and thermal cycle life of power semiconductors and their packages. Significant difference can be noticed between disc type devices [27] and plastic modules with bond wires [28]. Hence, power converter density depends greatly on the specific losses of the devices (less losses requires less heatsinks), coolant temperature and operating conditions.

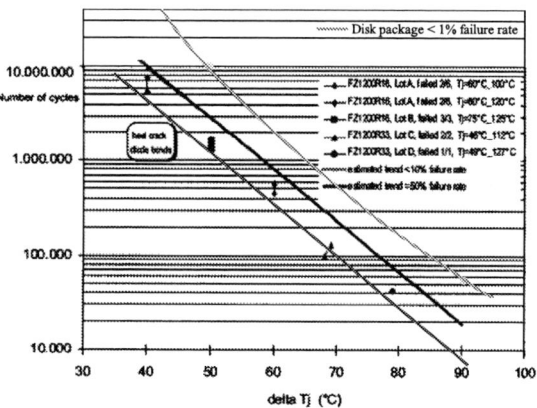

Figure 11. Thermal cycle life as function of temperature excursion of double sided cooled press pack devices (top curve) and power modules for 50 % failure rate (middle curve) and 10% failure rate (bottom).

In electric and hybrid vehicles, where power density plays an important role, it is not uncommon to see optimized liquid cooled dc-to-ac inverter designs with a power density of up to 6,000 kVA/m³ or 6 kVA/l, as illustrated by the 55 kW (peak 75 kW) propulsion unit for electric vehicles shown in Fig.12-14 [4,5]. In this example, a four-phase switched reluctance machine was developed for the propulsion system. The rated power density of the machine is around 1.2 kW/kg. Figure 15 illustrates the efficiency of the drive (tested only up to 10,000 rpm). This demonstrator proved that SRMs can be equally efficient as induction machines (IMs) or PM machines. Testing this propulsion system over different driving cycles showed that the energy consumption was better (up to 5%) than IMs or PM machines because partial load efficiency remains higher. In aerospace applications, power density and weight density are even more critical and converters with a power density of 20 kVA/l have been reported in literature [6].

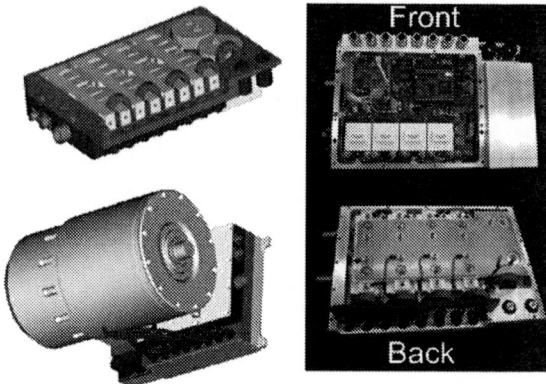

Figure 12. Tailor-made converter for a four-phase 75 kW SRM propulsion drive.

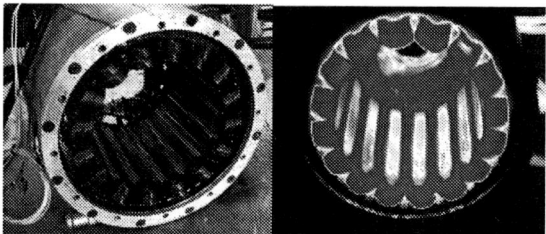

Figure 13. Stator construction of SRM, showing short end-turns and uniform heating of windings.

Figure 14. Test set-up showing mounting details of the propulsion unit.

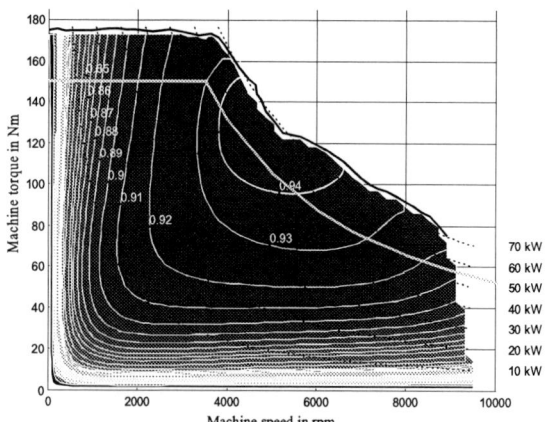

Figure 15. Speed-Torque diagram, showing efficiency contours of the 75 kW propulsion unit.

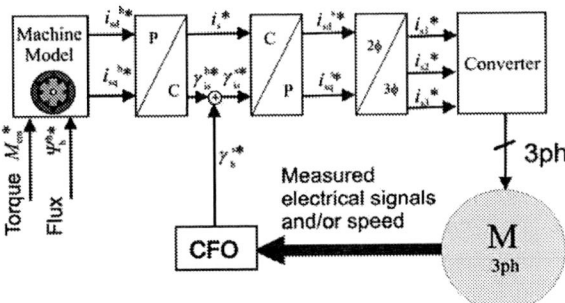

Figure 16. Field Oriented Control (FOC) for AC rotating filed machines

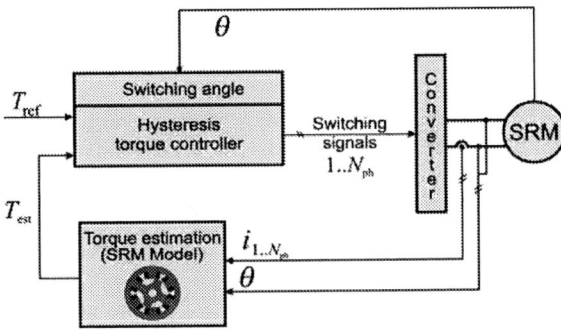

Figure 17. Direct Instantaneous Torque Control for switched reluctance machines (DITC)

C. Embedded control and communication links

As was shown above, most drives sold today are based on induction machines, PM synchronous machines and increasingly switched reluctance machines (SRM) (see Fig. 4). Each machine type requires dedicated control algorithms, some of which have been developed over the last decades. Note that SRMs were first introduced in the market in 1984. Since then they have been steadily improved upon so that SRM drives can be considered now for high quality servo drive systems, potentially offering lower cost. Control developers made several key contributions which made drives with servo performance possible and cost effective. Several control innovations (algorithms, software and hardware) developed over the past 30 years are worth mentioning:

- Principle of field oriented control (FOC) for ac rotating field machines maximizes torque per ampere of the drive [10-13] (Fig. 16)
- Space vector PWM modulation augmented with 15% the dc bus voltage utilization of three-phase inverters [14]
- Synchronized (space vector) PWM reduces low frequency harmonics and minimizes filter size
- Direct torque control [15-16], in particular of SRMs enables servo drive performance of this simple but highly non-linear drive [17-18], see Fig. 15.
- Position sensor elimination [19,22] or reducing number of current sensors (for example measuring output current via dc link current sensor) as well as sensor integration (integrating current sensor in

power devices) makes drives more robust and cheaper.

- Fast fixed point or floating point digital signal processors (DSPs) and programmable logic devices (PLDs) made implementation of complex control (FOC) and protection algorithms possible in a small space while offering high flexibility and avoiding expensive ASICs [23]
- Higher level programming languages C++ and tools shorten control development time, allowing flexible adaptation of the drive to new applications

As a result, drives have become programmable, tunable devices which can be adjusted for different applications. In the low to medium power level (0.5 to 50 kW), the cost reduction due to higher volume production can offset the cost overhead of the more flexible control hardware, especially when diagnostic functions need to be integrated in the drive. The classical distinction between adjustable speed drives and high performance drives is becoming less an issue as adjustable speed drives can be obtained by leaving out position sensors and fast communication ports while keeping control hardware identical. Plug-and-play concepts are becoming a reality for factory automation by integrating power, sensor wires and communication links in one cable and standardized connectors. Furthermore, more and more drive products can be easily integrated via fast communication links in a larger computer controlled environment.

IV. MODERN DRIVE DESIGN TOOLS

Design of new drive systems more and more uses powerful numerical design tools which can take specifications, fabrication rules and control algorithms into account. Design of drives requires an electromagnetic design of the electro-mechanical energy converter (machine, actuator), the power electronic converter and control methodology. In some cases, such as ac rotating filed machine, the design of the machine can be greatly decoupled form the design of the converter. However, design of switched reluctance drives requires a coupled design process which includes machine, converter and control aspects. The electromagnetic calculations of the machine are essential to calculate:

- winding and core losses
- machine efficiency
- hot spot temperatures, temperature cycles and life expectancy
- acoustic performance

Converter design involves:

- calculation of VA ratings of all components to determine weight and size and to select materials
- loss calculations to calculate efficiency, determine cooling concept
- temperature distribution calculation, temperature cycles, life of the converter
- EMI analysis and LF and EMI filter design

35

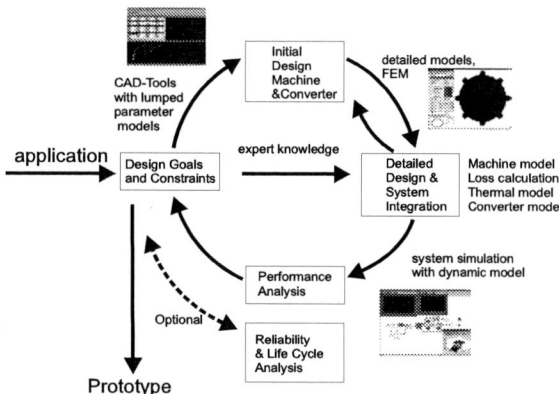

Figure 18. Typical design process of electrical drives

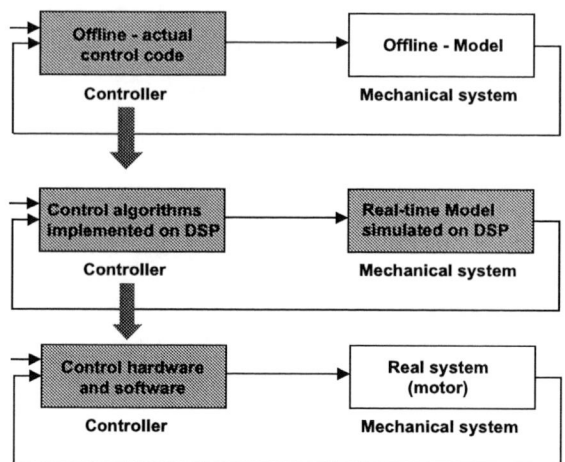

Figure 19. Rapid control prototyping: in all steps of the drive development the same control source can be used.

Control design and simulation is essential to assess stability, response performance of the drive or to optimize for specified design criteria. Clearly, control design can have a profound effect on inverter and machine efficiency, drive power density, EMI and acoustic noise. Usually, a closed loop design process which optimizes the entire drive system is needed. This design process requires iterative design steps, which is time consuming but allows different design tools to be used for each element of the drive and each step in the process as illustrated in Fig. 18.

To validate performance, thermal cycling, etc., the models need to be time domain, dynamic and physics based models which can be calculated quickly but still offer high accuracy and wide applicability. Detailed finite element simulations can be done to extract from complex geometries the parameters for these dynamic models. Most rotating machine can be modeled 2D, however more and more specialized products require 3D simulations. Embedding these FE simulations in dynamic models (including converter and control simulation) is possible but requires powerful computing environments and takes many resources (time). In some case, this coupling between dynamic models and FE is essential when parasitic effects are strongly coupled and difficult to model, for example eddy currents coupled with saturation and heating effects.

Control functions can be modeled in graphical programming environments (e.g. MatLab/Simulink). Pspice simulations are used to study the parasitic effects during switching of the inverter. Several companies offer idealized converter simulation tools which can be linked to the control simulation tools (MatLab/Simulink). These models offer much higher computation speed at the expense of HF details.

Device losses are calculated in postprocessing mode. Advanced control design concepts follow the rapid control prototyping principles, i.e. control software is coded in the off-line simulation the same way as in the actual control hardware. Time delays between inputs and outputs are simulated by the simulator. Real-time simulation of complex drives is becoming reality using high-speed DSP simulators (see Fig. 19) [29].

An example of a complex drive is shown in Fig. 20-21. A spherical machine was developed to integrate three-degrees of motion, providing very high stiffness for robot applications. The spherical machine has 96 independent phases to control the rotor with high torque and high precision (μm scale). Due to the tilt of the rotor, the pole pitch of each magnet appears to be variable from a stator perspective. Each phase was controlled using synchronized current regulators to avoid low-frequency beating and vibrations of the rotor. Six DSPs calculated the transformations and vector rotations such that each active phase was under field oriented control. Several search algorithms were developed to select quickly those phases which can produce the required force vectors most effectively (minimizing force per ampere) [25]. In this project, it was found that off-line simulations of the spherical machine (which was concurrently designed and constructed by other research groups) was too complex and too slow to test control algorithms for all practical purposes. A real-time simulation of the machine was built using the same hardware platform as the control hardware [23, 24]. Complete functionality of the control algorithms was proven within short time by working directly in this real-time environment. Due to small mechanical deviations from its initial design, only little tuning of parameters was required during initial testing when the spherical machine was first taken into service.

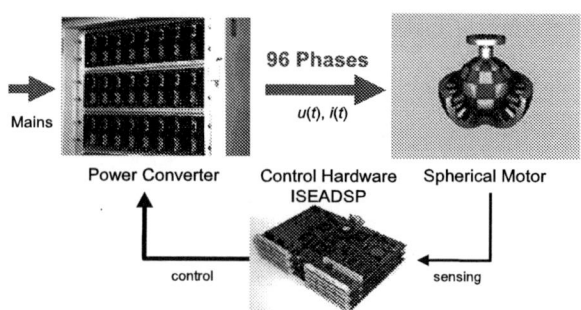

Figure 20. Spherical machine with 96-phase converter and control platform based on 6 DSPs in master-slave configuration.

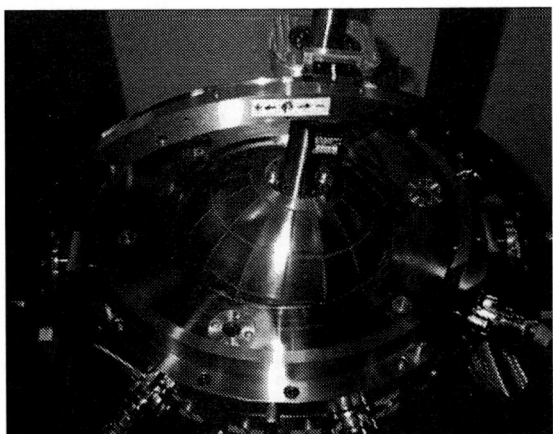

Figure 21. Spherical machine having 96 phases which need individual control due to the variable pole pitch of the magnets when rotor moves along trajectories and rotates

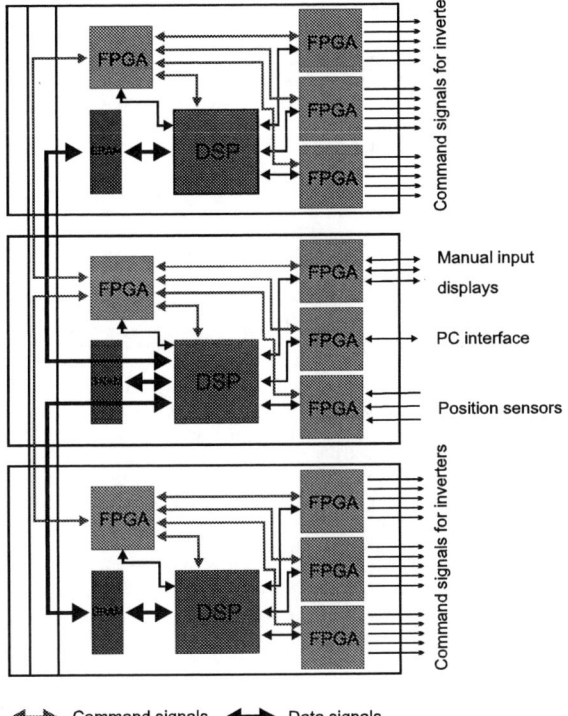

◀▬▶ Command signals   ◀▬▶ Data signals

Figure 22. DSP architecture showing control modularity. Flexibility is guaranteed by using programmable logic devices on all I/O.

Another example of productivity using modern rapid control prototyping is the design of a universal field oriented (UFO) controller for an induction machine, which is part of a students lab at ISEA. A target link in MatLab/Simulink emulates the control hardware and the machine. The DSP control software is entirely written in C-code. The developer (here a student training laboratory session) can focus on the control algorithms because the same C-code is used in the off-line simulation (target link) and the controller of the inverter. The execution time of the UFO controller is less than 50 μs. The total programming time takes less than a few hours including testing the performance of the drive. Just 20 years ago, this exercise would have taken an entire PhD program because none of these tools and algorithms existed!

V.   FUTURE DRIVE TRENDS

Although improvements in dynamic performance (torque response) and efficiency (already high) will be incremental (market does not demand major improvements in these areas), it is anticipated that in the next decade drive technology will offer:

- higher power densities. Indeed, converter designers are just starting to investigate more integrated packaging techniques as well as integrated passive components [26]. In addition, all auxiliaries, such as protective relays, switches, fuses, blowers, which all take considerable space, will be integrated or coordinated more effectively as converters designers are paying more attention to the overall system
- more diversity in machine types. More specialized drives require complete different designs. Amorphous powder metal instead of laminations offers more flexibility to adapt magnetic circuits to particular geometries and applications
- higher speed operation. More drives will operate at higher speeds as it reduces weight of the machine. One can speculate that, from this perspective, more SRM will come to market because SRM rotors can operate at higher speeds due to the fact that no magnets, windings or squirrel cages need to be supported in the rotor.
- high-temperature machines, power converters and control electronics. Machine winding isolation is limited to about 200 C
- propulsion systems with high-torque direct drives. High torque machines using powerful permanent magnets are very efficient at low speeds. Reactive power consumption at high-speed operation can be reduced by proper magnetic circuit design (saturation effects) or using so-called "memory effect" concepts.
- high-temperature superconductors will be used more and more in high power drives. To eliminate the maximum induction (2 Tesla) barrier of Si-steel lamination, large airgap machines or core-less machines are the only viable way to further reduce size of high power machines
- more intelligent drive controllers which takes away the burden of initialization and optimization of the drive performance. More diagnostic tools will be integrated in the drive as sensors at the system level can be avoided. High-speed field bus systems are already common place to transfer all data which can be extracted form a drive system. More drives will link to the Internet the same way as web cameras do today. Plug-and-play will become necessary to serve new applications which directly impact quality of life (domotica, mechatronics, automotive, alternative energy, power quality and medical).

VI. CONCLUSIONS

Judging form experience, publications and recent technology developments, drive technology is still an evolving technology. New applications are constantly emerging. These applications require continued engineering effort to adapt drives to the new requirements. Design tools are becoming mature and will improve over the next decades. New power semiconductor devices will enable higher power-frequency $P.f$ products (a measure of utilization of the devices in converters), while losses will continue to decrease.

In conclusion, drive technology is an exciting field of research and development for any young engineer: the more the world moves to automation, computer control, productivity improvements and improvement of the environment and life style, the more electrical energy needs to be converted to mechanical and vice versa. Another reason why drive technology is an attractive engineering field is the fact that it depends on many specialties, such as electrical machines, actuators, power electronics, passive and active components, control hardware and software, sensors technology and sophisticated algorithms and communication links.

REFERENCES

[1] ZVEI, Zentralverband Eletrotechnik- und Elektronikindustre e.V., Annual Report 2005

[2] R. Valentine, Motor Control Electronics Handbook, 1998

[3] Peters, D.T.; Cowie, J.G.; Brush, E.F., Jr.; Van Son, D.J., „Copper in the squirrel cage for improved motor performance," Electric Machines and Drives Conference, 2003. IEMDC'03. IEEE International, Volume 2, 1-4 June 2003 Page(s):1265 - 1271

[4] Inderka R.B., Altendorf J.-P., Sjöberg L., De Doncker R.W., "Design of a 75 kW Switched Reluctance Drive for Electric Vehicles," 18th International Electric Vehicle Symposium EVS18, 2001

[5] C. Carstensen, T. Schoenen, S. Bauer, R. De Doncker, „Highly Integrated 75kW Converter for Automotive Switched Reluctance Traction Drives with Robust 4-Quadrant Torque Control," 21th International Electric Vehicle Symposium EVS-21, Monaco, 2005

[6] Jahns, T.M.; De Doncker, R.W.; Radun, A.V.; Szczesny, P.M.; Turnbull, F.G., System design considerations for a high-power aerospace resonant link converter," Power Electronics, IEEE Transactions on, Volume 8, Issue 4, Oct. 1993 Page(s):663 - 672

[7] SEW-Eurodrive, product information, webpages 2006, http://corporate.sew-eurodrive.com/produkt/index.php

[8] Semikron, Semikube, product information, webpages 2006, http://www.semikron.com

[9] Lenze, product information, webpages 2006, http://www.lenze.de/en

[10] K. Hasse "Zur Dynamik drehzahlgeregelter Antriebe mit stromrichtergespeisten Asynchron-Kurzschlußläufermaschinen". Dissertation TH Darmstadt, Faculty of Electrical Energy, 1969

[11] Blaschke, F., "Das Verfahren der Feldorientierung zur Regelung der Asynchronmaschine," Siemens Forsch.- u. Entw.- Berichte 1 Nr. 1/72, 1972, S. 184-193

[12] Leonhard, W., "30 years space vectors, 20 years field orientation, 10 years digital signal processing with controlled AC drives," EPE-Journal 1991, pp. 13 and pp. 89

[13] De Doncker, R.W.; Novotny, D.W., „The universal field oriented controller," Industry Applications, IEEE Transactions on, Volume 30, Issue 1, Jan.-Feb. 1994 Page(s):92 - 100

[14] van der Broeck, H.W.; Skudelny, H.-C.; Stanke, G.V., "Analysis and realization of a pulsewidth modulator based on voltage space vectors" Industry Applications, IEEE Transactions on, Volume 24, Issue 1, Part 1, Jan.-Feb. 1988 Page(s):142 - 150

[15] M Depenbrock, "Direct self-control(DSC) of inverter-fed induction machine,"IEEE Transactions on Power Electronics, 1988 – Vol. 3, No. 4, Oct. 1988

[16] Takahashi, I. Ohmori, Y. "High-performance direct torque control of an induction motor," Industry Applications, IEEE Transactions, Mar/Apr 1989, Vol. 25, Issue: 2, pp. 257-264

[17] Inderka R.B., De Doncker R.W., "High Dynamic Direct Average Torque Control for Switched Reluctance Drives," 36st Annual Meeting of the Industrial Application Society IEEE-IAS, 2001, also IEEE Transaction of Industry Applications, Vol;39, No; 4 Juy/Aug. 2003, pp. 1040-1045

[18] Inderka R.B., R.W . De Doncker, "Direct Instantaneous Torque Control of Switched Reluctance Drives,", IEEE transactions of Industry Applications, Vol. 39, No. 4, July/Aug. 2003, pp. 1046-1051

[19] Ogasawara, S.; Akagi, H., „An approach to position sensorless drive for brushless DC motors," Industry Applications Society Annual Meeting, 1990., Conference Record of the 1990 IEEE, 7-12 Oct. 1990 Page(s):443 - 447 vol.1

[20] Bonnano, C.J.; Longya Xu; Xingyi Xu, „Robust, parameter insensitive position sensorless field orientation control of the induction machine," Power Electronics Specialists Conference, PESC '94 Record., 25th Annual IEEE, 20-25 June 1994 Page(s):752 - 757 vol.1

[21] Degner, M.W.; Lorenz, R.D., „Using multiple saliencies for the estimation of flux, position, and velocity in AC machines," Industry Applications, IEEE Transactions on, Volume 34, Issue 5, Sept.-Oct. 1998 Page(s):1097 - 1104

[22] Brosse, A.; Henneberger, G.; Schniedermeyer, M.; Lorenz, R.D.; Nagel, N., „Sensorless control of a SRM at low speeds and standstill based on signal power evaluation," Industrial Electronics Society, 1998. IECON '98. Proceedings of the 24th Annual Conference of the IEEE, Volume 3, 31 Aug.-4 Sept. 1998 Page(s):1538 - 1543

[23] De Doncker, R.W. „Twenty years of digital signal processing in power electronics and drives," Industrial Electronics Society, 2003. IECON '03. The 29th Annual Conference of the IEEE, Volume 1, 2-6 Nov. 2003 Page(s):957 - 960

[24] Fuengwarodsakul, N.H.; Radermacher, H.; De Doncker, R.W.; "Rapid prototyping tool for switched reluctance drive controls in traction applications," The Fifth International Conference on Power Electronics and Drive Systems, 2003. PEDS 2003,Volume 2, 17-20 Nov. 2003, p.927

[25] Kahlen, K.; Voss, I.; Priebe, C.; De Doncker, R.W., „Torque control of a spherical machine with variable pole pitch," Power Electronics, IEEE Transactions on, Volume 19, Issue 6, Nov. 2004 Page(s):1628 - 1634

[26] Smit, M.C.; Ferreira, J.A.; van Wyk, J.D.; Holm, M.F.K., „An integrated resonant DC-link converter, using planar ceramic capacitor technology," Power Electronics Specialists Conference, PESC '94 Record., 25th Annual IEEE, 20-25 June 1994 Page(s):664 - 670

[27] Somos, I.L.; Piccone, D.E.; Willinger, L.J.; Tobin, W.H.;"Power semiconductors empirical diagrams expressing life as a function of temperature excursion," Magnetics, IEEE Transactions on, Volume 29, Issue 1, Part 2, Jan 1993 Page(s):517 - 522

[28] Sommer K., Göttert J., Lefranc G., Spanke R.,"Multichip High Power IGBT-Modules for traction and Industrial Applications," *EPE Conference Record*, 1997, pp. 1.112-1.116.

[29] Aixcontrol, product information XCS-1000, webpages 2006, http://www.aixcontrol.de/

Power Semiconductors development trends

L. Lorenz [*]

[*] Infineon Technologies /Automotive Industrial and Multi-Market Marketing, Munich/Singapore, Germany/Singapore

Abstract—System integration and high power density of monolithic and multi-chip designs are the driving force for the progress in power electronic systems. The whole system has to be considered and optimized to meet this target and to keep the overall ruggedness, sensitivity towards EMI and long term reliability, Silicon utilization system reliability and power units miniaturization are the key factors.

Power semiconductors are primarily used to control the flow of energy between the energy source and the load, and to do so with great precision, with extremely fast control times and with low dissipated power.

The drive towards rational use of energy, miniaturization of electrical systems and power management has given impetus to the revolutionary development of power semiconductors and power electronic systems over the last 20 years.

In this paper new technologies, advanced devices concepts and future system aspect for industrial segments are discussed. In these fields of applications there are huge requirements towards system dynamic characteristic, overload capability, ruggedness behaviour and reliability.

High operating temperatures is required, in the industrial high blocking voltage capabilities are needed.

HIGH-VOLTAGE THYRISTORS DEVELOPMENT

High-power thyristors with direct light-triggering and integrated protection functions can be utilized advantageously for applications in which several thyristors are connected in series, because this enables a significant reduction in the number of components necessary for the construction of high-power thyristor converters. This in turn leads to greater reliability and lower fabrication costs. It applies in particular to High-Voltage DC (HVDC) transmission, Static Var Compensation (SVC), converters for medium voltage drives, and also to certain pulse power applications.

8-KV LIGHT-TRIGGERED THYRISTOR WITH INTEGRATED PROTECTION FUNCTIONS

Thyristors in HVDC converter stations have to be protected against several failures that may appear under Digitize or paste down figures.

standard operation. There are three classical failure events thyristors must be protected against :

• voltage pulses with a too high amplitude (overvoltage),
• voltage pulses with a too high voltage ramp dV/dt, and
• voltage pulses appearing during the forward recovery time.

A reliable protection of the thyristor can only be achieved, if the thyristor is safely turned on in failure case. A promising concept for an integration of protection functions is therefore to utilize the Amplifying Gate (AG) structure of the thyristor in such a way that in failure case a sufficiently large internal trigger current is generated turning on the device by means of the AG [1, 2].

Light-triggering and overvoltage protection

Figure 1 shows the central area of the thyristor including the AG structure, a part of the main cathode area and some peculiarities incorporated for the realization of the protection functions.

The innermost AG was adjusted such that the photogenerated current provided by the integrated diode triggers the thyristor when illuminated by a 40 mW light pulse with a duration of 10 μs. The breakdown voltage of the BOD can be controlled by the curvature of the junction and by the distance to the p⁻-ring below the optical gate and its doping concentration.

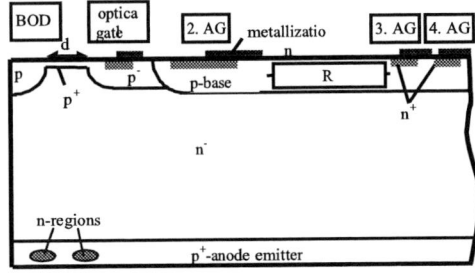

Figure 1. Central part of the thyristor (BOD & four-stage AG structure) and part of the main cathode area (from Ref. [1,2])

dV/dt protection

Integration of a dV/dt protection function can be achieved by designing the AG structure and the main cathode area such, that the innermost AG has a higher dV/dt sensitivity than the other AGs and the main cathode area.

Forward Recovery Protection

Commutation of a thyristor leads to an extraction of the free charge carriers in the n-base and the p-base and the formation of an anode-side depletion layer in the n-base. When large regions of the thyristor have completely recovered and an uncontrolled forward voltage pulse appears, the remaining small region with excess carriers in the main cathode area may turn on. To avoid this, the thyristor should also be turned on in a controlled way by the AG structure if such a failure occurs. However, an increase of the carrier concentration in the AG region can be achieved by modifying the carrier's lifetime distribution so that it is lower in the main cathode area than in the AG structure.

13-kV LIGHT-TRIGGERED TANDEM THYRISTOR WITH INTEGRATED OVERVOLTAGE PROTECTION

For blocking voltages of 13 kV, a new device concept consisting of a series connection of a 13-kV asymmetrical thyristor and a 13-kV diode has been developed.
Recently [3], it has been shown that the reverse recovery behaviour of 13-kV diodes can be improved significantly by applying a new field stop concept based on a deep buried field stop layer in front of the n^+ emitter. Fig. 3 shows typical time traces of the diode current and voltage during reverse recovery. The diodes were switched off under typical conditions appearing in High-Voltage DC (HVDC) converter stations with an RC-snubber circuit. The on-state current I_{on} and the current turn-off rate were 2 kA and -4 A/µs, respectively. The applied reverse voltage was -3.75 kV. Under these conditions, the diode shows a soft reverse recovery behaviour which is closely connected with the existence of the buried field stop layer:

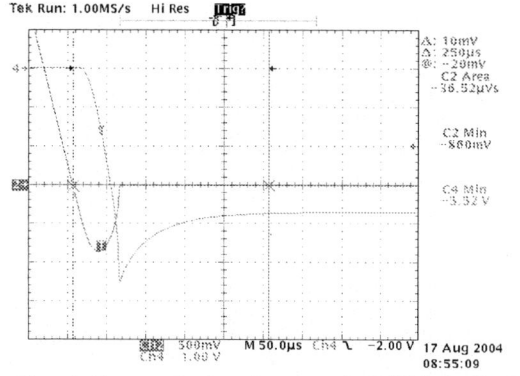

Figure 2. Current and voltage time trace of a 13-kV diode, $V_F = 3.7$ V, $V_r = -3.75$ kV, $R = 36$ Ω, $C = 2.4$ µF.

IGBT AND FAST SWITCHING DIODE FOR FUTURE POWER CONVERSION

For inverters as e. g. used in automation technology or traction there is an increasing demand for a matched system consisting of the control, intelligent drivers and the power switches. Key parameters for the power switches in inverters are ruggedness, on-state and switching losses and of course cost. Emerging new developments will focus more on the indirect contributors to losses of the power switches in the inverter systems. These topics become the more important the higher the load currents are. Key factors for current developments are inherent softness of the IGBTs and free wheeling diodes as well as enhanced controllability. Further improvement will be driven by optimized cell design and vertical structures, basically advanced field stop.

Talking about the IGBT as the key switch in these application fields in detail one must say that this technology has seen considerable innovation over the last decade regarding loss reduction as well as chip shrink resulting in more and more compact and cheaper packages as well as inverters.

Examples of this progress in Si development are shown in Fig. 3 in terms of on state voltage reduction as well as chip shrink path and in Fig. 4 regarding chip thickness reduction.

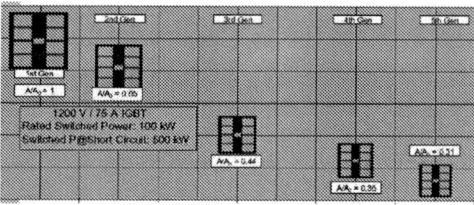

Figure 3. Shrink and on state voltage reduction of a 75A 1200V IGBT

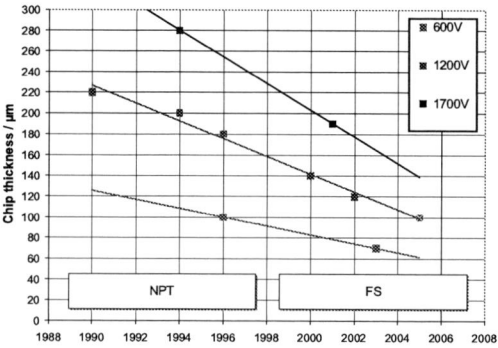

Figure 4. Decrease of chip thickness from first 1200V NPT IGBTs up to nowadays 600-1700V FS IGBTs.

Main enablers for these improvements were the implementation of modern IC technology into the IGBT process, e.g. small planar transistor cells as well as trench transistor cells [4, 5] – and also evolutional steps of the vertical structure from PT to NPT to FS concept (see Fig. 5), [6] by challenging ultrathin wafer technology.

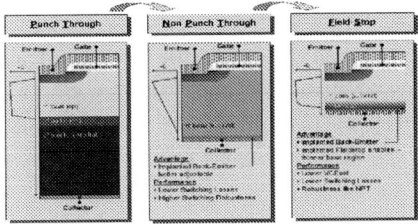

Figure 5. Evolution of vertical IGBT concept from PT (Punch Through) to NPT (Non Punch Through) to FS (Field Stop) IGBT

Some concerns might have come up what will come next: Is the silicon switch for low to medium to high power switches for drives and traction application now at its evolutionary end? Our definitive answer is no! On the one hand the roll out of modern chip technology to highest voltage class traction IGBTs (3.3kV up to 6.5kV) is still in progress, on the other hand for the lower voltage range (600, 1200, 1700V) the next IGBT generations are already in the development phase: for low power applications there might be possible some nice advantages by a reverse conducting IGBT concept with an integrated freewheeling diode [7, 8]. Especially in the medium to high power range, significant improvements can be foreseen in the switching performance by implementing the dynamic clamping feature [9, 10, 11] as well as more advanced sub-μ trench technology and next steps in the field stop technology for lower on state losses.

Also significantly higher maximum junction temperatures may become reachable (see short circuit test of a modern 1200V IGBT at $T_{junction}$ of 200°C in Fig. 6).

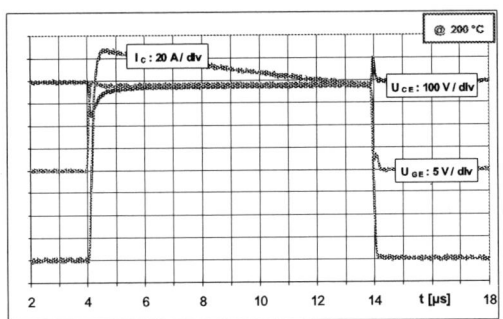

Figure 6. Short circuit pulse of a 25A 1200V IGBT at extreme temperature conditions: Tj=200°C, UGE=19V, UCE=800V

ULTRA FAST SWITCHING DEVICES WITH SUPER JUNCTION TECHNOLOGIES

System miniaturization, the reduction of system size and weight, is a strong driving factor in power electronics. The power supply is often a dominating part concerning dimensions and weight of the whole system. Together with robustness the most important key driver for improvement therefore is the efficiency.

COOLMOS C5 – THE LATEST GENERATION SUPER JUNCTION TECHNOLOGY IN VOLUME PRODUCTION

The efficiency of many power supply topologies is basically determined by the device capacities of the switching MOSFETs and, especially for low line conditions, the efficiency is significantly affected by the Rds,on of the switching transistor. For standard MOSFET technology these requirements are limited by the so called silicon limit. This basic conflict could be solved with the introduction of superjunction MOSFETs [12-16]. The on-resistance is now only a matter of technology performance and design but no longer subjected to the silicon limit. Fig. 7 shows the best commercial available standard transistor and CoolMOS C3 in comparison to its respective limits. (C5 is indicated by a blue star).

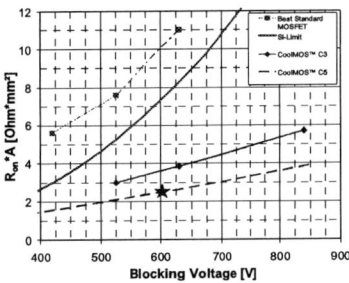

Figure 7. Best Std. HV MOSFET (Kobayashi), Si limit line, CoolMOS™ C3 & C5

Furthermore, this new superjunction technology combines the low area specific Rds,on with half the total gate charge Qg to achieve an outstanding figure of merit, Qg * Rds,on of 5 Ohm*nC, which is less than one tenth of the standard MOSFET value (Fig 8).

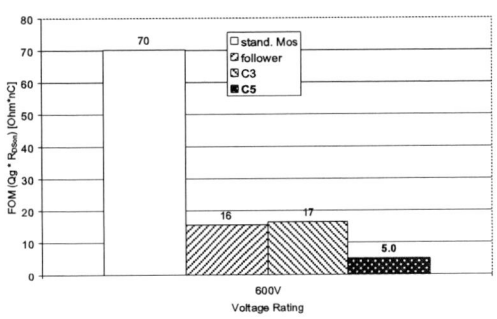

Figure 8. Figure of merit (Rds,on * Qg) for several

600V MOSFET technologies

Fig. 9 shows the development history of the superjunction technology. On the one side there was a continuous improvement in the loss reduction – dynamic and static losses – and on the other side an unusual improvement in the overall ruggedness performance. Overload capability, short circuit ruggedness and Avalanche ratings have been increased substantially.

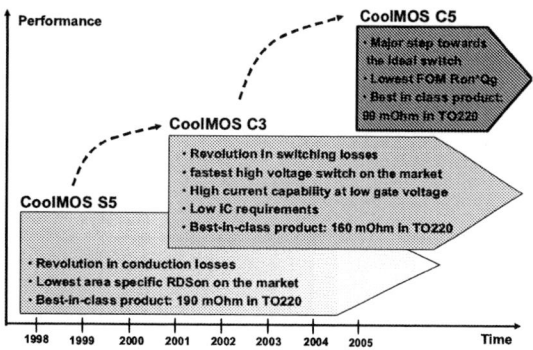

Figure 9. System leadership - CoolMOS history

Since 2001 Infineon is providing SiC Schottky diodes in the voltage classes of 300 V and 600 V. With the availability of these virtually switching loss free rectifiers the circuit developers have gained a new degree of freedom: frequency limitations due
to rapid increasing dynamic losses are out of the way when using this new device class. What remains are only capacitive losses which are more than one order of magnitude smaller compared to even the fastest Si bipolar diodes in this voltage range. Meanwhile power supply manufacturers have released new products including PFC stages with 200 kHz operation frequency based on this new technology.
Today SiC Schottky diodes are mainly used in high end power supply applications for servers and telecom base stations (> 500 W) and in less cost sensitive solar cell inverters. Today, frequently the most critical parameter for PFC design-ins is the surge current capability of this diode. This becomes very obvious when comparing the I^2t rating between SiC and standard diode (example: 6A SiC SDT06S60 → I^2t=2.3 A²s; 6A Si pn IDP06E60 → I^2t=4.3).
This issue is addressed in Infineon's upcoming new SiC diode generation by employing a merged pn-Schottky junction (see fig. 10).
In fact with this new approach the destructive surge current can be increased by a factor of 2 for a given current rating.

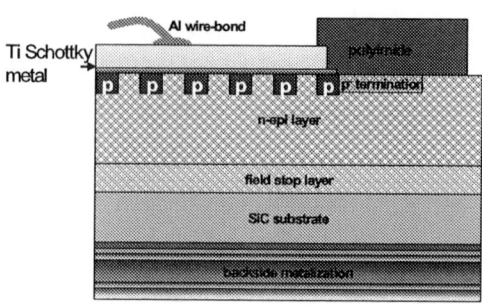

Figure 10. Schematic construction of a merged pn-Schottky diode. The p-wells have to be contacted by a separate ohmic contact (not shown) to get the desired bipolar injection.

Furthermore, this new diode will provide stable avalanche operation for a unipolar diode in SiC for the first time. Besides these performance improvements, an enlargement of our SiC Schottky diode product family will happen during 2005 towards larger breakdown voltage (1200 V) offering the benefits of lossless switching also for UPS, converter, drive applications, etc.

DRIVING TECHNOLOGIES FOR SYSTEM INTEGRATION IN THE INDUSTRY SEGMENT

MONOLITHIC HV-SYSTEM INTEGRATION

For high voltage applications, a thin-film SOI-technology with 600V-level-shift transistors and a wide range of low voltage devices integrated in a monolithic three phase driver IC is described. The benefits of junction insulation and silicon on insulator technology are combined using high dose SIMOX or wafer bonded material.

During the past few years, a trend from two level power conversion architectures to one level power conversion architecture for high-voltage applications has been observed. Galvanic insulation moves from the processor to the user interface. The microprocessor is being connected to the driver circuits directly, avoiding opto-coupler, transformer etc. As a response to this fundamental change, new semiconductor technologies have emerged integrating high voltage transistors mostly insulated with deep junctions. The main disadvantage of this junction-insulation concept (JI) is the presence of parasitic thyristor-like structures. In particular during negative transients and overheating these structures are prone to latch up. Such faults may lead to the destruction of complete driver units and to unpredictable error conditions in the controlled system.

Silicon on Insulator techniques (SOI) can exclude the latch up problem. However, many present SOI-technologies do not allow high blocking voltages or will cause unacceptable costs. Hence, a combined solution using the advantages of both insulation concepts (SOI and JI) may solve this problem.

The costs of SOI-devices are mainly driven by the used substrate material with its individual properties. Many common available SOI-wafers exhibit a buried

42

oxide layer with a thickness of only some hundred nanometers and an even thinner silicon film. These wafers are manufactured by SIMOX or Wafer-Bonding techniques. Instead of the typical p-wells or n-wells for CMOS transistors, the devices inside these thin silicon films are located in laterally and vertically insulated silicon islands. Therefore, higher packing densities and smaller chip sizes may be realized. But these materials are not appropriate for devices with vertical insulation voltages of 600V.

Beside the costs there are technical challenges for future devices. In some cases, controller and driver circuits have to be placed very close to the power devices, at a common direct copper bonding substrate (DCB) or a common lead frame in a molded package. They have to operate in harsh environments at elevated temperatures above 175°C. It is well known that SOI-devices in thin silicon films have a better temperature stability, smaller leakage currents and no latch-up effects. In total, our SOI-approach is one of the alternatives to bulk silicon to realize integrated smart power applications.

Device Structure of the HV-SOI-Transistor

Thin film SOI substrates were employed as the base material for the fabrication process. After processing they exhibit a buried oxide thickness of 400nm and a silicon film thickness of 150nm. The silicon body contains a phosphorus concentration of about 1e14/cm³.

The fairly thin buried oxide layer is not able to withstand voltage drops of 600V. Hence, the voltage drop across the buried oxide has to be reduced significantly.

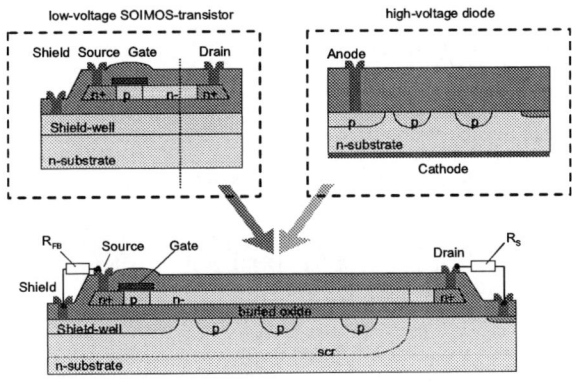

Figure 11. Cross section of a high voltage SOI-transistor combining a SOI-MOS-transistor and a high voltage diode

Therefore, a high voltage diode has been placed inside the silicon substrate below the SOI-transistor. Blocking junctions and drift regions of both devices lie on top of each other, only separated by the thin buried oxide. The outer contacts are electrically connected (figure 1). In case of a positive voltage drop between drain and source the space charge region in the silicon film will extend similar to the space charge region inside the silicon substrate. In the silicon film a primarily horizontal

electric field remains. The vertical voltage drop will occur inside the space

Coreless transformer a key technology for system integration

Although fields of application for insulated signal transfer can be found in such different areas as computer interfaces, sensors, control panels or gate drives for high power Transistors, its requirements can be reduced to a few major points. First of all insulation has to withstand not only the applied voltage stress over the whole life time but must also comply with the applicable standards.

To integrate an insulated signal transfer in an IC opto couplers or high voltage level shifter circuits are common. As both solutions have major draw backs, e.g. the transfer characteristic of opto couplers degrades over time and level shifter circuits tend to be sensitive to EMI, various new solutions are under development. Possible technical approaches include electromechanical transmission utilizing the piezo effects, dielectrical transmission via a capacitive coupler and magnetic transmission with receivers based on GMR-, Hall-effect or a secondary coil. For integration in a standard IC process a magnetic transmission with primary and secondary coil is favourable. With such an approach the advantages of a pulse transformer can be combined with the advantages of IC production technologies. Products which use this technology are being developed under the name coreless transformer. For realisation of a first coreless transformer the logic functions of the transmitter have been build into a primary die (figure 12).

The secondary die holds the receiver as well as primary and secondary windings of the transformer. Both dies are soldered to a split lead frame. Connection between transmitter and primary side of the transformer is made by wire-bonding. Afterwards the chip set is covered in a molding compound.

Figure 12. Sketch of coreless transformer technology

DRIVING TECHNOLOGIES FOR SYSTEM INTEGRATION IN THE AUTOMOTIVE SEGMENT

The major differences to other fields of applications technologies specified for automotive system integration are asked for an increased operation temperature range, particularly high ESD/ISO-pulse robustness and a high reliability. These requirements have a tremendous influence on the key features of the technology.

However, it is necessary to fit each process step to the automotive requirements. To cover all automotive segments from high- voltage to high current ratings along with integrated systems many technology approaches are used as shown in Fig. 23.

Figure13. Technology roadmap for automotive applications

A key issue for a cost effective system integration is the best chip partitioning. For this reason the CoC (chip on chip) CbC (chip by chip) and the monolithic integration (Fig. 24) has been developed and very successfully applied in the different automotive systems. For future system developments a new approach will be the embedded power system. As shown in Fig. 25 the power technology will be merged with the μC – technology on a common lead frame inside one package.

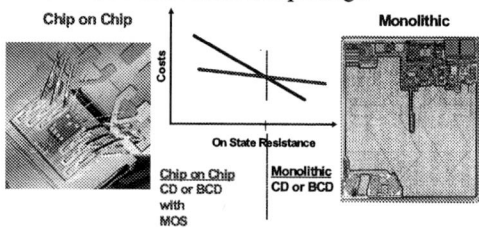

Figure 14. Multi Chip Packaging Chip on Chip

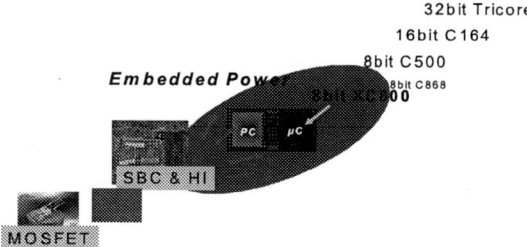

Figure15. Close a gap between μC and power with embedded power products

ACKNOWLEDGEMENT

The author would like to thank the whole team: G.Miller, F.J. Niedernostheide, H-J. Schulze, U. Kellner – Werdehauser T.Laska, A.Mauder, G. Deboy, H. Kapels, R. Rupp, J.Stengl, R. Rudolf, and Z.Gergintschew for their valuable discussions and technical support.

References:

[1]   F.-J. Niedernostheide, H.-J. Schulze, U. Kellner-Werdehausen, "Self-Protected High-Power Thyristors", Proc PCIM'2001, Power Conversion, Nuremberg, Germany, pp. 51-56, 2001.
[2]   H.-J. Schulze, F.-J. Niedernostheide, U. Kellner-Werdehausen, "Thyristor with Integrated Forward Recovery Protection Function", Proc. of 2001Int. Symp. on Power Semiconductor Devices and IC's, Osaka, Japan, pp. 199-202, 2001.
[3]   F.-J. Niedernostheide, H.-J. Schulze, U. Kellner-Werdehausen, R. Barthelmess, J. Przybilla, R. Keller, H. Schoof, D. Pikorz, "13-kV Rectifiers: Studies on Diodes and Asymmetric Thyristors", Proc. ISPSD'03, PR 122-125, 2003
[4]   J. Yamashita, C. Yoshida, C. Fujii, K. Takanashi, J. Mortani, "The 5th Generation Highly Rugged Planar IGBT Using Sub-micron Process Technology", *Proceedings of the 13th ISPSD*, pp.421, 424, 2001
[5]   T. Laska, F. Pfirsch, F. Hirler, J. Niedermeyr, C. Schaeffer, T. Schmidt"1200V-Trench-IGBT Study with Square Short Circuit SOA", *Proceedings of the 10th ISPSD*, pp.433-436, 1998
[6]   T. Laska, M. Münzer, F. Pfirsch, C.Schäffer, T. Schmidt, "The Field Stop IGBT (FS IGBT) – A New Power Device Concept with a Great Improvement Potential", *Proceedings of the 12th ISPSD*, pp.355-358, 2000
[7]   H. Takahashi, A. Yamamoto, S, Aono, T. Minato, "1200V Reverse Conducting IGBT", *Proceedings of the 16th ISPSD*, pp.133-136, 2004
[8]   E. Griebl, O. Hellmund, M. Herfurth, H. Hüsken, M. Pürschel, "LightMOS – IGBT with Integrated Diode for Lamp Ballast Applications", *PCIM Nürnberg Proceedings*, 2003
[9]   T.Laska, G.Miller, C.Schäffer, F.Umbach, "Field Stop IGBTs with Dynamic Clamping Capability – A New Degree of Freedom for Future Inverter Designs?", *to be published of EPE*, 2005
[10]   M. Otsuki, Y. Onozawa, S. Yoshiwatari, Y. Seki, "1200V FS-IGBT module with enhanced dynamic clamping capability", *Proceedings of the 16th ISPSD*, pp.339-342, 2004
[11]   M. Rahimo, A. Kopta, S. Eicher, U. Schlapbach, S. Linder, "Switching Self-Clamping-Mode "SSCM", a breakthrough in SOA performance for high voltage IGBTs and Diodes", *Proceedings of the 16th ISPSD*, pp.437-440, 2004
[12]   B. Murari, F. Bertotti, G. A. Vignola (eds); "Smart Power ICs: Technologies and Applications" ; Springer; 1995
[13]   L. Lorenz, M. März, "CoolMOS – A new approach towards high efficient power supplies", PCIM Hong Kong' 98 Proceedings
[14]   L. Lorenz, "CoolMOS – A New Approach Towards An Idealized Power Switch", EPE '99 Lausanne Proceedings
[15]   G. Deboy, M. März, J. –P. Stengl, H. Strack, J. Tihanyi and H. Weber, "A new generation of high voltage MOSFETs breaks the limit line of silicon", Tech. Digest IEDM 98, pp. 683-685, San Francisco 1998.
[16]   L. Lorenz, G. Deboy, A. Knapp and M. März, „ CoolMOS™ - a new milestone in high voltage power MOS", Proc. PCIM 98 Nürnberg
[17]   H. Mitlehner, W. Bartsch, K.-O. Dohnke, P. Friedrichs, R.Kaltschmidt, U. Weinert, B. Weis and D. Stephani,„Dynamic characteristics of high voltage 4H-SiC vertical JFETs", Proc. ISPSD 99, pp. 339-342, To-ronto 1999.
[18]   P. Friedrichs, H. Mitlehner, R. Kaltschmidt, U. Weinert, W.Bartsch, C. Hecht, K.-O. Dohnke, B. Weis and D.Stephanni, „ Static and dynamic characteristics of 4H-SiC JFETs designed for different

blocking categories", Proc. ICSCRM 99, 1999. T. Laska, M. Münzer, F. Pfirsch, C. Schaeffer and T.Schmidt, „The field Stop IGBT (FS-IGBT) – a new power device voncept with a great improvemnt potential", Proc. ISPSD 2000, pp. 355-358, Toulouse 2000.

[19] J.M. Hancock, "High Voltage SiC JFET switch enables New Points for Offline High Power SMPS
Power Systems World 2004 USA

2006 5th International Power Electronics and Motion Control Conference

Power Electronics in Wind Turbine Systems

F. Blaabjerg, Z. Chen, R. Teodorescu, F. Iov
Aalborg University, Institute of Energy Technology
Pontoppidanstraede 101, DK-9220 Aalborg East, Denmark
fbl@iet.aau.dk, zch@iet.aau.dk, ret@iet.aau.dk, fi@iet.aau.dk
www.iet.aau.dk

Abstract – **The global electrical energy consumption is still rising and there is a steady demand to increase the power capacity. The production, distribution and the use of the energy should be as technological efficient as possible and incentives to save energy at the end-user should be set up. The deregulation of energy has lowered the investment in larger power plants, which means the need for new electrical power sources may be very high in the near future. Two major technologies will play important roles to solve the future problems. One is to change the electrical power production sources from the conventional, fossil (and short term) based energy sources to renewable energy resources. The other is to use high efficient power electronics in power systems, power production and end-user application. This paper discuss the most emerging renewable energy source, wind energy, which by means of power electronics is changing from being a minor energy source to be acting as an important power source in the energy system. By that wind power is also getting an added value in the power system operation.**

I. INTRODUCTION

In classical power systems, large power generation plants located at adequate geographical places produce most of the power, which is then transferred towards large consumption centers over long distance transmission lines. The system control centers monitor and control the power system continuously to ensure the quality of the power, namely the frequency and the voltage. However, now the overall power system is changing, a large number of dispersed generation (DG) units, including both renewable and non-renewable sources such as wind turbines, wave generators, photovoltaic (PV) generators, small hydro, fuel cells and gas/steam powered Combined Heat and Power (CHP) stations, are being developed [1]-[2] and installed. A wide-spread use of renewable energy sources in distribution networks and a high penetration level will be seen in the near future many places. E.g. Denmark has a high penetration (> 20%) of wind energy in major areas of the country and today 18% of the whole electrical energy consumption is covered by wind energy. The main advantages of using renewable energy sources are the elimination of harmful emissions and the inexhaustible resources of the primary energy. However, the main disadvantage, apart from the higher costs, e.g. photovoltaic, is the uncontrollability. The availability of renewable energy sources has strong daily and seasonal patterns and the power demand by the consumers could have a very different characteristic. Therefore, it is difficult to operate a power system installed with only renewable generation units due to the characteristic differences and the high uncertainty in the availability of the renewable energy sources.

The wind turbine technology is one of the most emerging renewable technologies. It started in the 1980'es with a few tens of kW production power to today with Multi-MW range wind turbines that are being installed. This also means that wind power production in the beginning did not have any impact on the power system control but now due to their size they have to play an active part in the grid. The technology used in wind turbines was in the beginning based on a squirrel-cage induction generator connected directly to the grid. By that power pulsations in the wind are almost directly transferred to the electrical grid. Furthermore there is no control of the active and reactive power, which typically are important control parameters to regulate the frequency and the voltage. As the power range of the turbines increases those control parameters become more important and it is necessary to introduce power electronics [3] as an interface between the wind turbine and the grid. The power electronics is changing the basic characteristic of the wind turbine from being an energy source to be an active power source. The electrical technology used in wind turbine is not new. It has been discussed for several years [6]-[46] but now the price pr. produced kWh is so low, that solutions with power electronics are very attractive.

This paper will first discuss the basic development in power electronics and power electronic conversion. Then different wind turbine configurations will be explained both aerodynamically and electrically. Also different control methods will be explained for a turbine. Wind turbines are now more often installed in remote areas with good wind conditions (off-shore, on-shore) and different possible configurations are shown and compared. Finally, a general technology status of the wind power is presented demonstrating a still more efficient and attractive power source.

1-4244-0448-7/06/$25.00 ©2006 IEEE

II. MODERN POWER ELECTRONICS AND SYSTEMS

Power electronics has changed rapidly during the last thirty years and the number of applications has been increasing, mainly due to the developments of the semiconductor devices and the microprocessor technology. For both cases higher performance is steadily given for the same area of silicon, and at the same time they are continuously reducing the price. Fig. 1 shows a typical power electronic system consisting of a power converter, a load/source and a control unit.

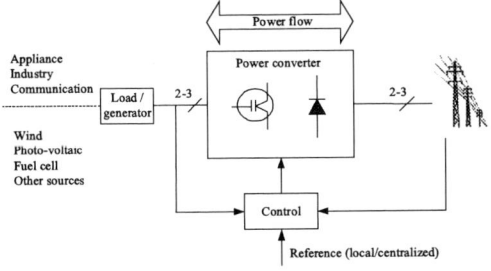

Fig. 1. Power electronic system with the grid, load/source, power

onverter and control.

The power converter is the interface between the load/generator and the grid. The power may flow in both directions, of course, dependent on topology and applications. Three important issues are of concern using such a system. The first one is reliability; the second is efficiency and the third one is cost. For the moment the cost of power semiconductor devices is decreasing 2-5 % every year for the same output performance and the price pr. kW for a power electronic system is also decreasing. A high competitive power electronic system is adjustable speed drives (ASD) and the trend of weight, size, number of components and functions in a standard Danfoss Drives A/S frequency converter can be seen in Fig. 2. It clearly shows that power electronic conversion is shrinking in volume and weight. It also shows that more integration is an important key to be competitive as well as more functions become available in such a product.

The key driver of this development is that the power electronic device technology is still undergoing important progress. Fig. 3 shows different key power devices and the areas where the development is still going on.

The only power device which is not under development any more is the silicon-based power bipolar transistor because MOS-gated devices are preferable in the sense of easy control. The breakdown voltage and/or current carrying capability of the components are also continuously increasing. Also important research is going on to change the material from silicon to silicon carbide. This may dramatically increase the power density of power converters but silicon carbide based

transistors on a commercial basis with a competitive price will still take some years to appear on the market.

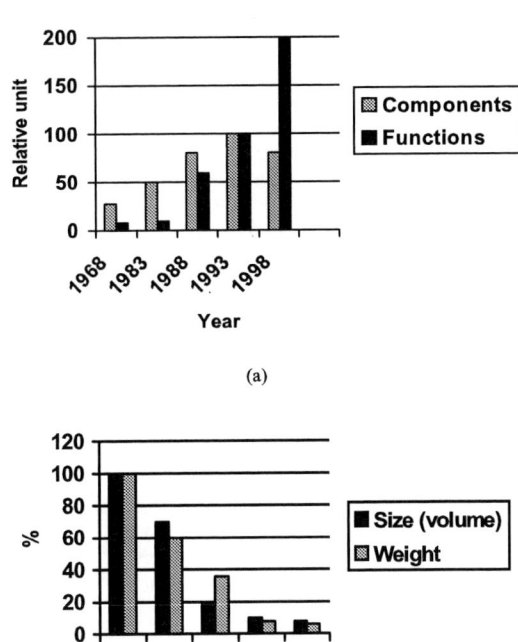

(a)

(b)

Fig. 2. Development of a 4 kW standard industrially adjustable speed

drive during the last 25 years [5].

a) Relative number of components and functions

b) Relative size and weight

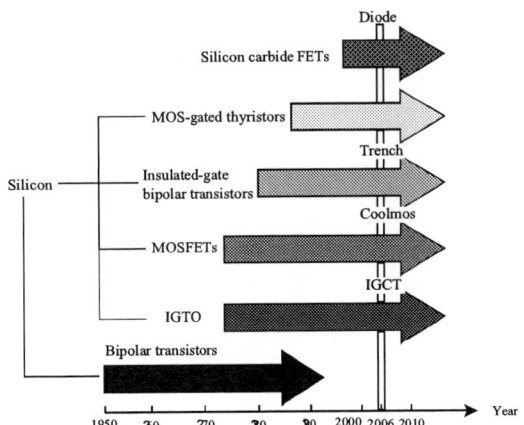

Fig. 3. Development of power semiconductor devices in the past and

in the future [34]

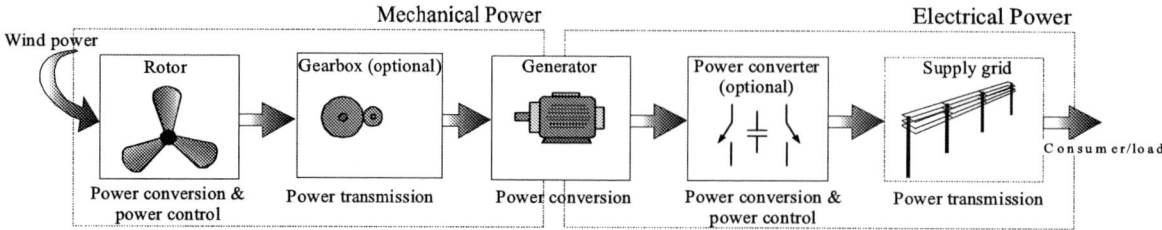

Fig. 4. Converting wind power to electrical power in a wind turbine [17].

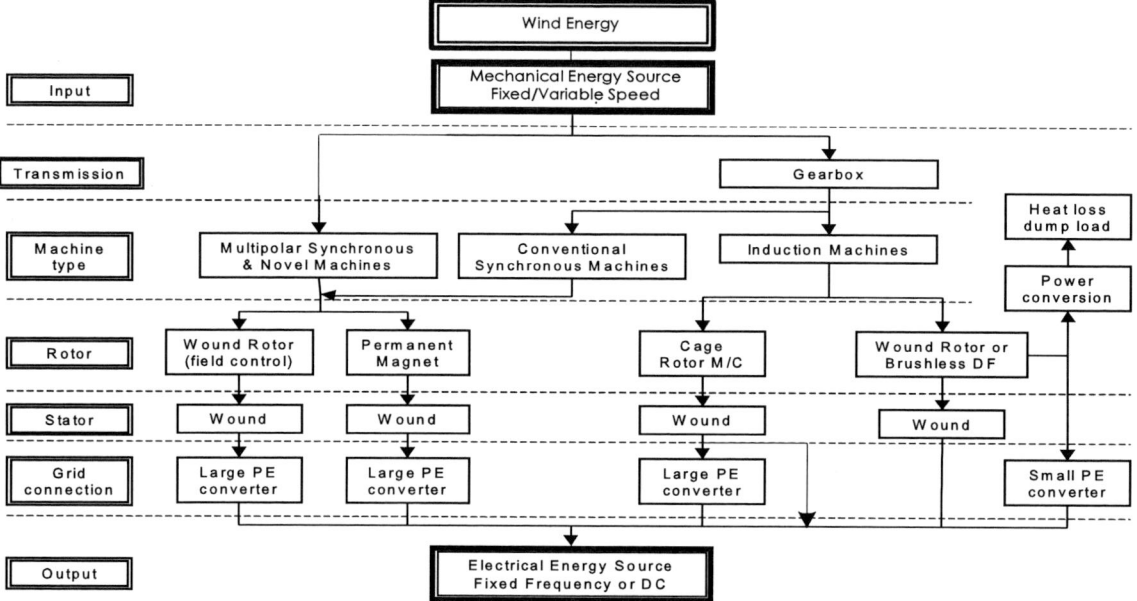

Fig. 5. Road-map for wind energy conversion. PE: Power Electronics. DF: Doubly-fed [15], [22].

III. WIND ENERGY CONVERSION

Wind turbines capture the power from the wind by means of aerodynamically designed blades and convert it to rotating mechanical power. The number of blades is normally three. As the blade tip-speed typically should be lower than half the speed of sound the rotational speed will decrease as the radius of the blade increases. For multi-MW wind turbines the rotational speed will be 10-15 rpm. The most weight efficient way to convert the low-speed, high-torque power to electrical power is using a gear-box and a standard fixed speed generator as illustrated in Fig. 4.

The gear-box is optional as multi-pole generator systems are possible solutions. Between the grid and the generator a power converter can be inserted.

The possible technical solutions are many and Fig. 5 shows a technological roadmap starting with wind energy/power and converting the mechanical power into electrical power. It involves solutions with and without gearbox as well as solutions with or without power electronic conversion. The electrical output can either be ac or dc. In the last case a power converter will be used as interface to the grid. In the following sections, some different wind turbine configurations will be presented and compared.

IV. FIXED SPEED WIND TURBINES

The development in wind turbine systems has been steady for the last 25 years and four to five generations of wind turbines exist. It is now proven technology. The conversion of wind power to mechanical power is as mentioned before done aerodynamically. It is important to be able to control and limit the converted mechanical power at higher wind speed, as the power in the wind is a cube of the wind speed. The power limitation may be done either by stall control (the blade position is fixed but stall of the wind appears along the blade at higher wind speed), active stall (the blade angle is adjusted in order to create stall along the blades) or pitch control (the blades are turned out of the wind at higher wind speed). The wind turbines technology can basically be divided into three categories: the first category is systems without power electronics, the second category is wind turbines with partially rated power electronics (small PE converter in Fig. 5) and the last is the full-scale power electronic interfaced wind turbine systems (large PE converter in Fig. 5). Fig. 6 shows different topologies for the first category of wind turbines where the wind turbine speed is fixed.

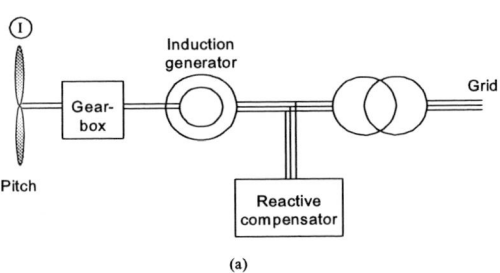

(a)

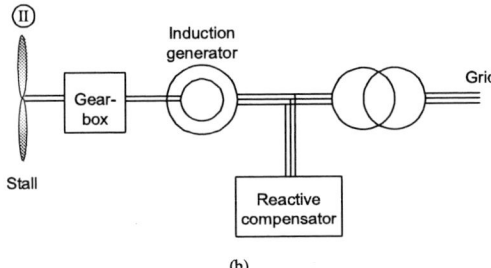

(b)

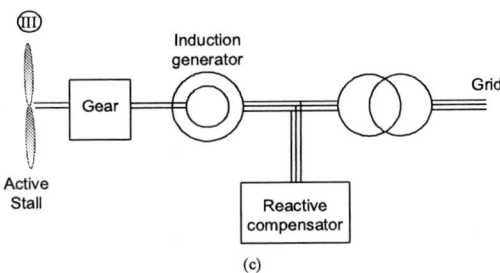

(c)

Fig. 6. Wind turbine systems without power converter but with

aerodynamic power control.

a)   Pitch controlled (System I)

b)   Stall controlled (System II)

c)   Active stall controlled (System III)

The wind turbine systems in Fig. 6 are using induction generators, which almost independent of torque variation operate at a fixed speed (variation of 1-2%). The power is limited aerodynamically either by stall, active stall or by pitch control. All three systems are using a soft-starter (not shown in Fig. 6) in order to reduce the inrush current and thereby limit flicker problems on the grid. They also need a reactive power compensator to reduce (almost eliminate) the reactive power demand from the turbine generators to the grid. It is usually done by continuously switching capacitor banks following the production variation (5-25 steps). Those solutions are attractive due to cost and reliability

but they are not able very fast (within a few ms) to control the active power. Furthermore wind-gusts may cause torque pulsations in the drive-drain and load the gear-box significantly. The basic power characteristics of the three different fixed speed concepts are shown in Fig. 7 where the power is limited aerodynamically.

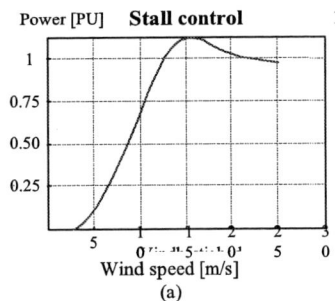

(a)

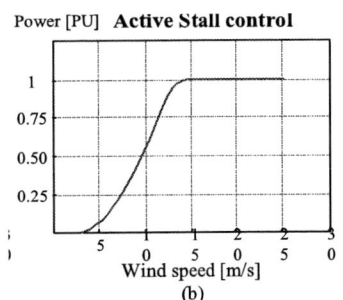

(b)

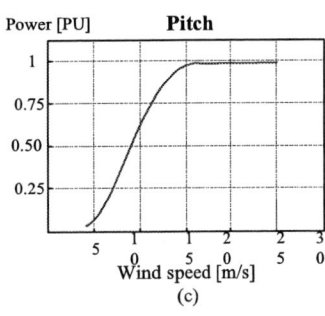

(c)

Fig. 7. Power characteristics of fixed speed wind turbines.

a)   Stall control b) Active stall control c) Pitch control

Fig. 7 shows that by rotating the blades either by pitch or active stall control it is possible precise to limit the power while the measured power for the stall controlled turbine shows a small overshoot. This depends a lot on the final aerodynamic design.

V. VARIABLE SPEED WIND TURBINES

The next category is wind turbines with partially rated power converters and by that improved control performance can be obtained. Fig. 8 shows two such solutions.

49

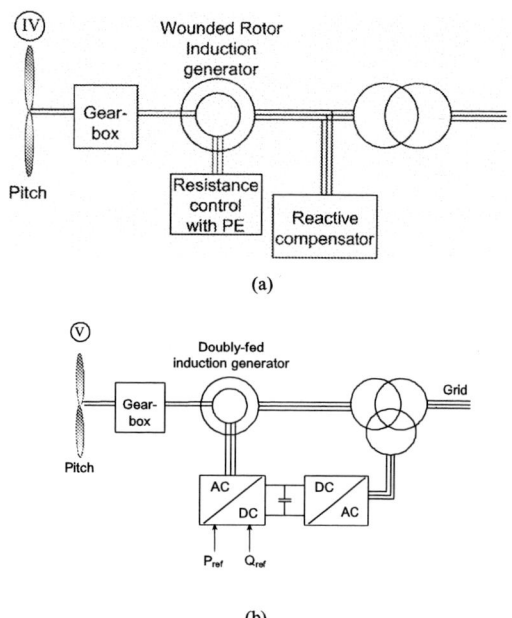

(a)

(b)

Fig. 8. Wind turbine topologies with partially rated power electronics

and limited speed range.

    a)   Rotor-resistance converter (System IV)

    b)   Doubly-fed induction generator (System V)

Fig. 8 shows a wind turbine system where the generator is an induction generator with a wounded rotor. An extra resistance is added in the rotor, which can be controlled by power electronics. This is a dynamic slip controller and it gives typically a speed range of 2 - 5 %. The power converter for the rotor resistance control is for low voltage but high currents. At the same time an

extra control freedom is obtained at higher wind speeds in order to keep the output power fixed. This solution still needs a softstarter and a reactive power compensator, which is in continuous operation.

A second solution of using a medium scale power converter with a wounded rotor induction generator is shown in Fig. 8b. Slip-rings are making the electrical connection to the rotor. A power converter controls the rotor currents.

If the generator is running super-synchronously electrical power is delivered through both the rotor and the stator. If the generator is running sub-synchronously electrical power is only delivered into the rotor from the grid. A speed variation of ±30 % around synchronous speed can be obtained by the use of a power converter of 30 % of nominal power. Furthermore, it is possible to control both active (Pref) and reactive power (Qref), which gives a better grid performance, and the power electronics is enabling the wind turbine to act as a more dynamic power source to the grid.

The last solution needs neither a soft-starter nor a reactive power compensator. The solution is naturally a little bit more expensive compared to the classical solutions shown before in Fig. 7 and Fig. 8a. However, it is possible to save money on the safety margin of gear, reactive power compensation units as well it is possible to capture more energy from the wind.

The third category is wind turbines with a full-scale power converter between the generator and grid, which are the ultimate solutions technically. It gives extra losses in the power conversion but it may be gained by the added technical performance. Fig. 9 shows four possible, but not exhaustive, solutions with full-scale power converters.

System comparison of wind turbines									
System	**I**	**II**	**III**	**IV**	**V**	**VI**	**VII**	**VIII**	**IX**
Variable speed	No	No	No	No	Yes	Yes	Yes	Yes	Yes
Control active power	Limited	No	Limited	Limited	Yes	Yes	Yes	Yes	Yes
Control reactive power	No	No	No	No	Yes	Yes	Yes	Yes	Yes
Short circuit (fault-active)	No	No	No	No	No/Yes	Yes	Yes	Yes	Yes
Short circuit power	contribute	contribute	contribute	contribute	contribute	limit	limit	limit	limit
Control bandwidth	1-10 s	1-10 s	1-10 s	100 ms	1 ms	0.5-1 ms	0.5-1 ms	0.5-1 ms	0.5-1 ms
Standby function	No	No	No	No	Yes +	Yes ++	Yes ++	Yes ++	Yes ++
Flicker (sensitive)	Yes	Yes	Yes	Yes	No	No	No	No	No
Softstarter needed	Yes	Yes	Yes	Yes	No	No	No	No	No
Rolling capacity on grid	Yes, partly	No	Yes, partly	Yes, partly	Yes	Yes	Yes	Yes	Yes
Reactive compensator (C)	Yes	Yes	Yes	Yes	No	No	No	No	No
Island operation	No	No	No	No	Yes/No	Yes/No	Yes/No	Yes/No	Yes

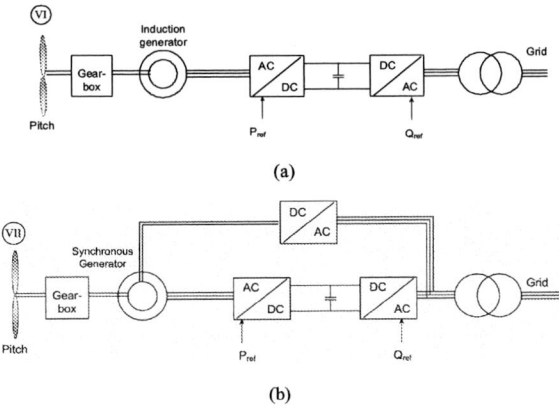

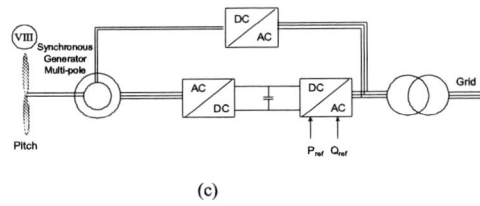

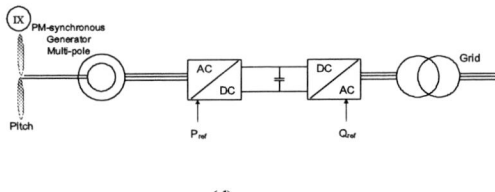

(d)

Fig. 9. Wind turbine systems with full-scale power converters.

a) Induction generator with gear (System VI)

b) Synchronous generator with gear (System VII)

c)   Multi-pole synchronous generator (System VIII)

d) Multi-pole permanent magnet synchronous generator (System IX)

The solutions shown in Fig. 9a and Fig. 9b are characterized by having a gear. A synchronous generator solution shown in Fig. 9b needs a small power converter for field excitation. Multi-pole systems with the synchronous generator without a gear are shown in Fig. 9c and Fig. 9d. The last solution is using permanent magnets, which are still becoming cheaper and thereby more attractive. All four solutions have the same controllable characteristics since the generator is decoupled from the grid by a dc-link.

The power converter to the grid enables the system to control active and reactive power very fast. However, the negative side is a more complex system with more sensitive electronic parts.

Comparing the different wind turbine systems in respect to performance shows a contradiction between

cost and the performance to the grid. Table I shows a technical comparison of the presented wind turbine systems, where issues on grid control, cost, maintenance, internal turbine performance are given. By introducing power electronics many of the wind turbine systems get a performance like a power plant. In respect to control performance they are faster but of course the produced real power depends on the available wind. The reactive power can in some solutions be delivered without having any wind producing active power.

Fig. 9 is also indicating other important issues for wind turbines in order to act as a real power source for the grid. They are able to be active when a fault appears at the grid and where it is necessary to build the grid voltage up again; having the possibility to lower the power production even though more power is available in the wind and thereby act as a rolling capacity for the power system. Finally, some systems are able to work in island operation in the case of a grid collapse. The market share in 2001 (Globally and in Germany) between the dominant system topologies is shown Table II.

TABLE II.
WIND TURBINE TOPOLOGIES MARKET IN 2002.

(Source: [4])

Turbine Concept	World-Market Share
Fixed speed (Stall or active stall, gearbox), System I, II, III	28%
Dynamic slip control (Limited variable speed, pitch, gearbox), System IV	5%
Doubly-fed generator (Variable speed operation, pitch control, gearbox), System V	47%
Direct-driven (variable speed operation, pitch control), System VIII	20%
TOTAL	**100%**

As it can be seen the most sold technology in 2001 is the doubly-fed induction generator system which occupies about 50% of the whole market. More than 75% of all sold wind turbines in 2001 are controlled by power electronics. That is even more in 2003.

VI. CONTROL OF WIND TURBINES

Controlling a wind turbine involves both fast and slow control. Overall the power has to be controlled by means of the aerodynamic system and has to react based on a set-point given by dispatched center or locally with the goal to maximize the production based on the available wind power.

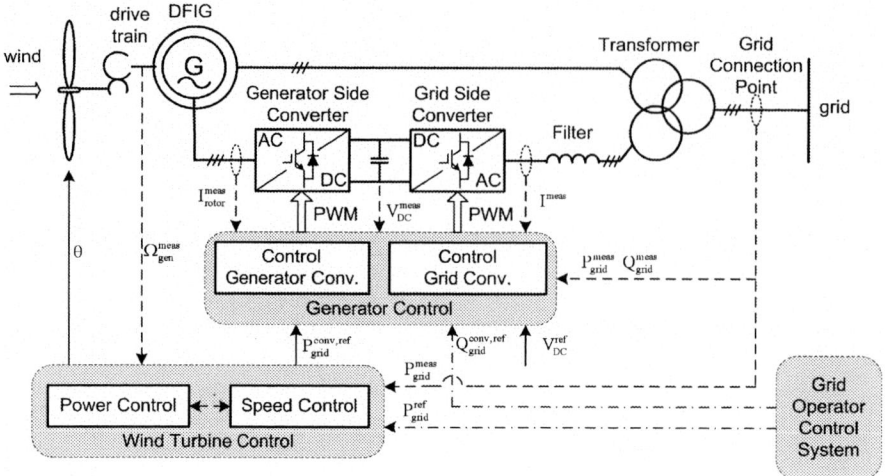

Fig. 10. Control of wind turbine with doubly-fed induction generator system [35].

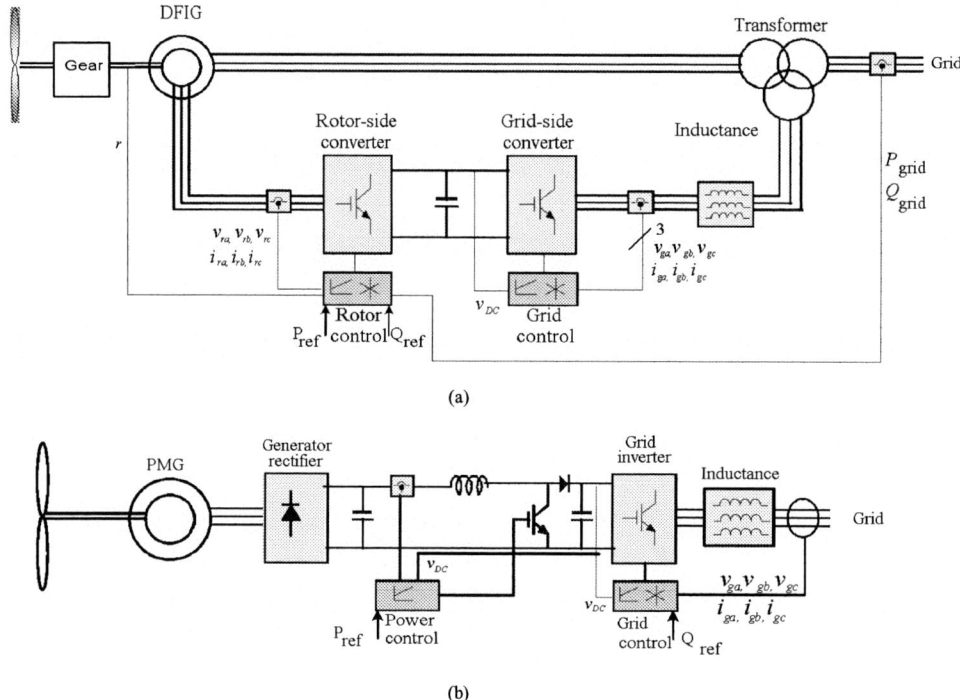

(a)

(b)

Fig. 11. Basic control of active and reactive power in a wind turbine [17].

a) Doubly-fed induction generator system (System V)

b) Multi-pole synchronous PM-generator system (System IX)

The power control system should also be able to limit the power. An example of an overall control scheme of a wind turbine with a doubly-fed generator system is shown in Fig. 10.

Below maximum power production the wind turbine will typically vary the speed proportional with the wind speed and keep the pitch angle □ fixed. At very low wind the speed of the turbine will be fixed at the maximum allowable slip in order not to have over voltage. A pitch angle controller will limit the power when the turbine reaches nominal power. The generated electrical power is done by controlling the doubly-fed generator through the rotor-side converter. The control of the grid-side converter is simply just keeping the dc-link voltage fixed. Internal current loops in both converters are used which typically are linear PI-controllers, as it is illustrated in Fig. 11a. The power converters to the grid-side and the rotor-side are voltage source inverters.

Another solution for the electrical power control is to use the multi-pole synchronous generator. A passive rectifier and a boost converter are used in order to boost the voltage at low speed. The system is industrially used today. It is possible to control the active power from the generator. The topology is shown in Fig. 11b. A grid inverter is interfacing the dc-link to the grid. Here it is also possible to control the reactive power to the grid. Common for both systems are they are able to control reactive to control the reactive power to the grid. Common for both systems are they are able to control reactive and active power very fast and thereby the turbine can take part in the power system control.

VII. OFFSHORE WIND FARM TOPOLOGIES

In many countries energy planning is going on with a high penetration of wind energy, which will be covered by large offshore wind farms. These wind farms may in the future present a significant power contribution to the national grid, and therefore, play an important role on the power quality and the control of power systems.

Consequently, very high technical demands are expected to be met by these generation units, such as to perform frequency and voltage control, regulation of active and reactive power, quick responses under power system transient and dynamic situations, for example, to reduce the power from the nominal power to 20 % power within 2 seconds. The power electronic technology is again an important part in both the system configurations and the control of the offshore wind farms in order to fulfill the future demands.

One off-shore wind farm equipped with power electronic converters can perform both real and reactive power control and also operate the wind turbines in variable speed to maximize the energy captured as well as reduce the mechanical stress and noise. This solution is shown in Fig. 12a and it is in operation in Denmark as a 160 MW off-shore wind power station.

For long distance transmission of power from off-shore wind farm, HVDC may be an interesting option. In an HVDC transmission, the low or medium AC voltage at the wind farm is converted into a high dc voltage on the transmission side and the dc power is transferred to the onshore system where the dc voltage is converted back into ac voltage as shown in Fig. 12c. For certain power level, an HVDC transmission system, based on voltage source converter technology, may be used in such a system instead of the conventional thyristor based HVDC technology.

The topology may even be able to vary the speed on the wind turbines in the complete wind farm.

Another possible dc transmission system configuration is shown in Fig. 12d, where each wind turbine has its own power electronic converter, so it is possible to operate each wind turbine at an individual optimal speed. A comparison of the topologies is given in Table III.

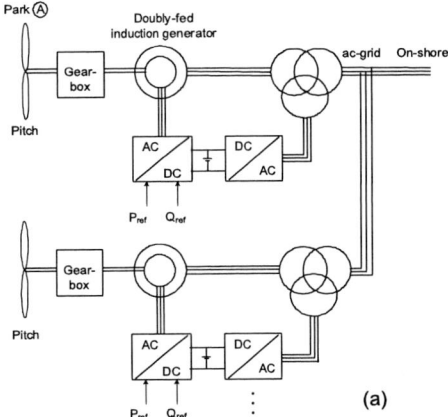

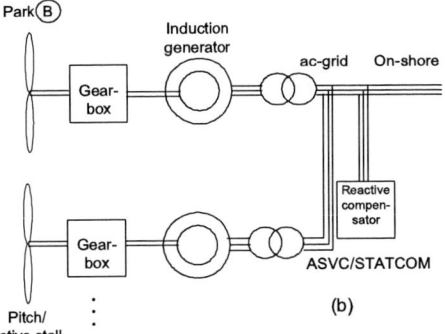

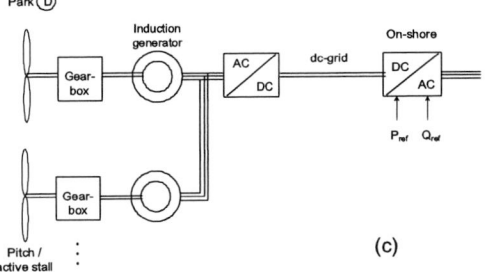

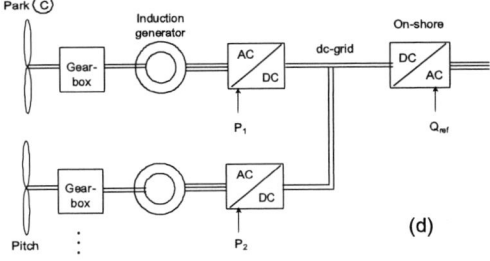

Fig. 12. Wind farm solutions.

a) Doubly-fed induction generator system with ac-grid (System A)

b) Induction generator with ac-grid (System B)

c) Speed controlled induction generator with common dc-bus and control of active and reactive power (System C)

d) Speed controlled induction generator with common ac-grid and dc transmission (System D)

As it can be seen the wind farms have interesting features in order to act as a power source to the grid. Some have better abilities than others. Bottom-line will always be a total cost scenario including production, investment, maintenance and reliability. This may be different depending on the planned site.

VIII. WIND POWER TRENDS

The installed power in wind energy has grown rapidly in many years. Today more than 45000 MW are installed globally with recently an annual market of 8000 MW. This is illustrated in Fig. 13.

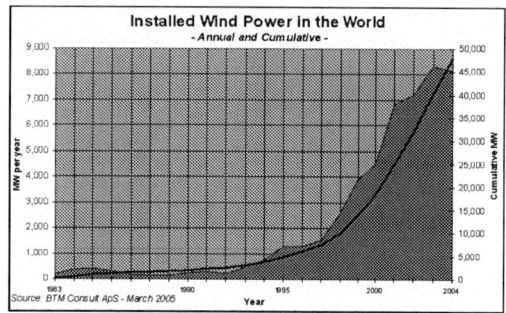

Fig. 13. Annually installed and accumulated wind power globally.

(Source: Risoe National Laboratory, Denmark)

The expectations for the future are also very positive as many countries have progressive plans. Table IV gives an estimate for the installed wind power in 2010 based on official statements from different European countries.

It can be seen that many countries will increase their wind power capacity in large scales. In Denmark the installed capacity is expected to approach saturation as the problems of a too high capacity compared to the load level are appearing. However, energy cost rise can

change this.

The power scaling has been an important tool to reduce the price pr. kWh. Fig. 14 shows the average size of the installed wind turbines in Denmark as well as their produced energy pr. m2 swept area pr. year. It can be seen that the technology is improving and it is possible to produce more than 900 kWh/m2/year. This depends of course on location and from experience off-shore wind-farms are able to produce much more energy

The influence on the power scaling can also be seen at the prices pr. kWh for different wind-turbine sizes in two different landscape classes and it is shown in Fig. 15. The key to reduce price is to increase the power and today prototype turbines of 4-5 MW are seen around the world being tested. Finally, the development of wind turbines is illustrated in Fig. 16. It is expected 10 MW wind turbines will be present in 2010.

IX. CONCLUSIONS

The paper discusses the applications of power electronic for the wind turbine technology. The development of modern power electronics has been briefly reviewed. The applications of power electronics in various kinds of wind turbine generation systems and offshore wind farms are also illustrated, showing that the wind turbine behavior/performance is very much improved by using power electronics. They are able to act as a contributor to the frequency and voltage control by means of active and reactive power control. Also it can be concluded the power scaling of wind turbines is important in order to be able to reduce the energy cost.

Tabel III. Comparison of Wind Farms				
Farm configuration (Fig. 12)	**A**	**B**	**C**	**D**
Individual speed control	Yes	No	Yes	No
Control active power electronically	Yes	No	Yes	Yes
Control reactive power	Yes	Centralized	Yes	Yes
Short circuit (active)	Partly	Partly	Yes	Yes
Short circuit power	Contribute	Contribute	No	No
Control bandwidth	10-100 ms	200ms - 2s	10 -100 ms	10 ms – 10 s
Standby-function	Yes	No	Yes	Yes
Softstarter needed	No	Yes	No	No
Rolling capacity on grid	Yes	Partly	Yes	Yes
Redundancy	Yes	Yes	No	No
Investment	+	++	+	+
Maintenance	+	++	+	+

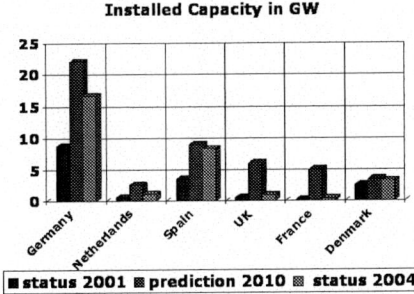

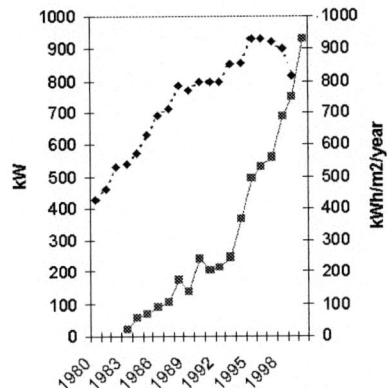

Fig. 14. Average size of wind turbines and produced energy pr.

m2 swept area pr. year in Denmark.

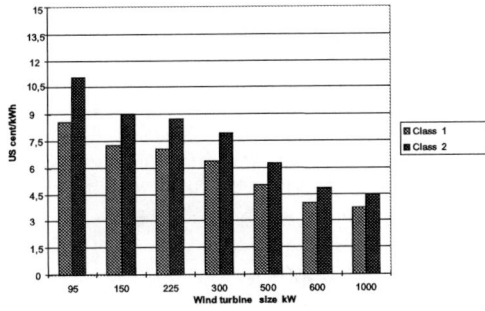

Fig. 15. Price pr. produced kWh at different landscape classes.

(Source: Risoe National Laboratory, Denmark)

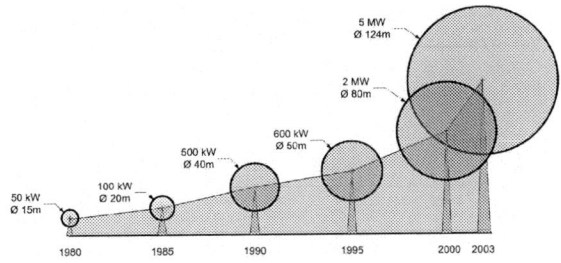

Fig. 16. Development of wind turbines during the last 25 years.

REFERENCES

[1] S. Heier, Grid integration of wind energy conversion systems/translated by Rachel Waddington. John Wiley, 1998. ISBN 0-47-197143x

[2] E. Bossanyi. Wind Energy Handbook. John Wiley, 2000.

[3] N. Mohan, T.M. Undeland, W.P. Robbins. Power Electronics-Converters, Applications and Design. 1'st Edition, John Wiley & Sons, 1989.

[4] A. D. Hansen, F. Iov, F. Blaabjerg, L. H. Hansen, "Review of Contemporary Wind Turbine Concepts and their Market Penetration". Journal of Wind Engineering, Vol. 28, No. 3, 2004, pp. 247-263.

[5] P. Thoegersen, F. Blaabjerg. "Adjustable Speed Drives in the Next Decade. Future Steps in Industry and Academia". Journal of Electric Power Components and Systems, Vol. 32, No. 1, 2004, pp. 13-32.

[6] Z. Chen, E. Spooner, "Voltage Source Inverters for High-Power, Variable-Voltage DC Power Sources", IEE Proc. –Generation, Transmission and Distributions, Vol. 148, No. 5, September 2001, pp. 439-447.

[7] F. Blaabjerg, Z. Chen, "Power Electronics as an enabling technology for Renewable Energy Integration", Journal of Power Electronics, Vol. 3, No. 2, April 2003, pp. 81-89.

[8] Z. Chen, E. Spooner, "Grid Power Quality with Variable-Speed Wind Turbines", IEEE Trans. on Energy Conversion, Vol. 16, No.2, June 2001, pp. 148-154.

[9] F. Iov, Z. Chen, F. Blaabjerg, A. Hansen, P. Sorensen, "A New Simulation Platform to Model, Optimize and Design Wind Turbine", Proc. of IECON '02, Vol. 1, pp. 561-566.

[10] S. Bolik, "Grid Requirements Challenges for Wind Turbines", Proc. of Fourth International Workshop on Large-Scale Integration of Wind Power and Transmission Networks for Offshore Windfarms, 2003.

[11] E. Bogalecka, "Power control of a doubly fed induction generator without speed or position sensor", Proc. of EPE '93, Vol.8, 1993, pp. 224-228.

[12] O. Carlson, J. Hylander, K. Thorborg, "Survey of variable speed operation of wind turbines", Proc. of European Union Wind Energy Conference, Sweden, 1996, pp. 406-409.

[13] M. Dahlgren, H. Frank, M. Leijon, F. Owman, L. Walfridsson, "Wind power goes large scale", ABB Review, 2000, Vol.3, pp.31-37.

[14] M.R. Dubois, H. Polinder, J.A. Ferreira, "Comparison of Generator Topologies for Direct-Drive Wind Turbines", IEEE Nordic Workshop on Power and Industrial Electronics (Norpie '2000), 2000, pp. 22-26.

[15] L.H. Hansen, P.H. Madsen, F. Blaabjerg, H.C. Christensen, U. Lindhard, K. Eskildsen, "Generators and power electronics technology for wind turbines", Proc. of IECON '01, Vol. 3, 2001, pp. 2000 – 2005.

[16] Z. Chen, E. Spooner, "Wind turbine power converters: a comparative study", Proc. of PEVD '98, 1998 pp. 471 – 476.

[17] M.P. Kazmierkowski, R. Krishnan, F. Blaabjerg. Control in Power Electronics-Selected problems. Academic Press, 2002. ISBN 0-12-402772-5

[18] Å. Larsson, The Power quality of Wind Turbines, Ph.D. report, Chalmers University of Technology, Göteborg, Sweden,2000.

[19] R. Pena, J.C. Clare, G.M. Asher, "Doubly fed induction generator using back-to-back PWM converters and its application to variable speed wind-energy generation". IEE proceedings on Electronic Power application, 1996, pp. 231-241.

[20] J. Rodriguez, L. Moran, A. Gonzalez, C. Silva, "High voltage multilevel converter with regeneration capability", Proc. of PESC '99, 1999, Vol.2, pp.1077-1082.

[21] P. Sørensen, B. Bak-Jensen, J. Kristian, A.D. Hansen, L. Janosi, J. Bech, " Power Plant Characteristics of Wind Farms", Proc. of the Int. Conf. in Wind Power for the 21st Century, 2000.

[22] A.K. Wallace, J.A. Oliver, "Variable-Speed Generation Controlled by Passive Elements", Proc. of ICEM '98, 1998.

[23] S. Bhowmik, R. Spee, J.H.R. Enslin, "Performance optimization for doubly fed wind power generation systems", IEEE Trans. on Industry Applications, Vol. 35, No. 4 , July-Aug. 1999, pp. 949 – 958.

[24] Z. Saad-Saoud, N. Jenkins, "The application of advanced static VAr compensators to wind farms", Power Electronics for Renewable Energy, 1997, pp. 6/1 - 6/5.

[25] J.B. Ekanayake, L. Holdsworth, W. XueGuang, N. Jenkins, "Dynamic modelling of doubly fed induction generator wind turbines", Trans. on Power Systems, Vol. 18 , No. 2 , May 2003 , pp.803- 809.

[26] D. Arsudis, "Doppeltgespeister Drehstromgenerator mit Spannungszwischenkreis Umrichter in Rotorkreis für Wind Kraftanlagen, Ph.D. Thesis, 1998, T.U. Braunschweig, Germany.

[27] D. Arsudis, "Sensorlose Regelung einer doppelt-gespeisten Asynchronmaschine mit geringen Netzrückwirkungen", Archiv für Elektrotechnik, Vol. 74, 1990, pp. 89-97.

[28] T. Matsuzaka, K. Trusliga, S. Yamada, H. Kitahara, "A variable speed wind generating system and its test results". Proc. of EWEC '89, Part Two, pp. 608-612, 1989.

[29] R.S. Barton, T.J. Horp, G.P. Schanzenback, "Control System Design for the MOD-5A 7.3 MW wind turbine generator". Proc. of DOE/NASA workshop on Horizontal-Axis Wind Turbine Technology Workshop, May 8-10, 1984, pp. 157-174.

[30] O. Warneke, "Einsatz einer doppeltgespeisten Asynchronmaschine in der Großen Windenergie-anlage Growian", Siemens-Energietechnik 5, Heft 6, 1983, pp. 364-367.

[31] L. Gertmar, "Power Electronics and Wind Power", Proc. of EPE 2003, paper 1205, CD-Rom.

[32] F. Blaabjerg, Z. Chen, "Power Electronics as Efficient Interface in Dispersed Power Generation Systems", IEEE Trans. on PE, Vol. 19, No. 4, 2004, pp. 1184-1194

[33] E.N. Hinrichsen, "Controls for variable pitch wind turbine generators", IEEE Trans. On Power Apparatus and Systems, Vol. 103, No. 4, 1984, pp. 886-892.

[34] B.J. Baliga, "Power IC's in the saddle", IEEE Spectrum, July 1995, pp. 34-49.

[35] A.D. Hansen, C. Jauch, P. Soerensen, F. Iov, F. Blaabjerg. "Dynamic Wind Turbine Models in Power System Simulation Tool DigSilent", Report Risoe-R-1400 (EN), Dec. 2003, ISBN 87-550-3198-6 (80 pages).

[36] T. A. Lipo, "Variable Speed Generator Technology Options for Wind Turbine Generators", NASA Workshop on HAWTT Technology, May 1984, pp. 214-220.

[37] K. Thorborg, "Asynchronous Machine with Variable Speed", Appendix G, Power Electronics, 1988, ISBN 0-13-686593-3, pp. G1.

[38] D. Arsudis, W. Vollstedt, "Sensorless Power control of a Double-Fed AC-Machine with nearly Sinusoidal Line Currents", Proc. of EPE '89, Aachen 1989, pp. 899-904.

[39] M. Yamamoto, O. Motoyoshi, "Active and Reactive Power control for Doubly-Fed Wound Rotor Induction Generator", Proc. of PESC '90, Vol. 1, pp. 455-460.

[40] O. Carlson, J. Hylander, S. Tsiolis, "Variable Speed AC-Generators Applied in WECS", European Wind Energy Association Conference and Exhibition, October 1986, pp. 685-690.

[41] J.D. van Wyk, J.H.R. Enslin, "A Study of Wind Power Converter with Microcomputer Based Maximal Power Control Utilising an Oversynchronous Electronic Schertives Cascade", Proc. of IPEC '83, Vol. I, 1983, pp. 766-777.

[42] T. Sun, Z. Chen, F. Blaabjerg, "Flicker Study on Variable Speed Wind Turbines With Doubly Fed Induction Generators". IEEE Trans. on Energy Conversion, Vol. 20, No. 4, 2005, pp. 896-905.

[43] T. Sun, Z. Chen, F. Blaabjerg, "Transient Stability of DFIG Wind Turbines at an External Short-circuit-Fault". Wind Energy, Vol. 8, 2005, pp. 345-360.

[44] L. Mihet-Popa, F. Blaabjerg, I. Boldea, "Wind Turbine Generator Modeling and Simulation Where Rotational Speed is the Controlled Variable". IEEE Transactions on Industry Applications, 2004, Vol. 40, No. 1. pp. 3-10.

[45] M. Liserre, R. Teodorescu, F. Blaabjerg, "Stability of Photovoltaic and Wind Turbine Grid-Connected Inverters for a Large Set of Grid Impedance Values", IEEE Trans. on PE, Vol. 21, No. 1, Jan. 2006, pp. 263-272.

[46] A. D. Hansen, P. Sørensen, F. Iov, F. Blaabjerg, "Centralised power control of wind farm with doubly fed induction generators", Journal of Renewable Energy, Vol. 31, 2006, pp. 935-951.

Sustainable Energy and Mobility, and Challenges to Power Electronics.

C.C.Chan

Harbin Institute of Technology, Harbin, China
University of Hong Kong, Hong Kong, China
Academician of Chinese Academy of Engineering
Fellow of Royal Academy of Engineering, U.K.
e-mail: ccchan@eee.hku.hk

Abstract—Review the issues of sustainable energy and mobility, energy development strategy and the challenges to power electronics.

Keywords - sustainable energy; sustainable mobility; challenges of power electronics.

I. INTRODUCTION

In 1987, the World Commission on Environment and Development developed a definition of sustainability as "Sustainable development meets the needs of the present without compromising the ability of future generations to meet their own needs." Sustainability implies long term social, economic and ecological health and vitality. In the development and application of energy, we need to take care, in particular, two important sustainability elements: the Energy Efficiency and the Environmental Protection.

The direct manifestations of a widespread and long-term trend toward warmer global temperatures include: the heat waves and periods of unusually warm weather, the sea-level rise and coastal flooding, the glaciers melting, and Arctic and Antarctic warming. On one hand, we need to have energy generation with reduced green house gas emission and on the other hand, we need to have more effective utilization of generated energy.

The growth of global population and the rapid growth of automobiles, particularly in developing countries, have had significant impacts to energy, environment and safety. We need to develop clean, efficient and intelligent vehicles and transportation system.

Sustainable energy and mobility development prompt opportunity and challenge for a wide range of power electronic devices and systems that posses high energy density, high power density, high efficiency, reliable, endurance, compatible EMI/EMC and competitive cost.

II. ENERGY STRATEGIES

Energy is a strategic resource for economy and society development. Therefore a clear energy development strategy is essential. Currently fossil fuels are still our main energy resource, however their reserve is limited and their impacts to environment and ecology are serious. It is also noted from the history of energy utilization that it will take long time to change the energy structure. Therefore we should promptly draw our energy development strategy, which may include energy resource survey, advanced energy utilization technology, impacts of energy to environment and ecology, and energy security assessment.

Energy sources can be described as renewable and non-renewable. Renewable energy sources are those which are continually being replaced such as energy from the sun. Renewable energy sources include solar, wind, tidal/wave, biomass, hydro and geothermal. If the energy resource is being used faster than it can be replaced, it is called non-renewable energy source, for example, fossil fuels. Non-renewable energy sources include coal, petroleum, gas and nuclear.

It is noted the global unbalance energy resources, which roughly can be classified into five categories countries:
1. Developed or big economic countries with rich energy resources; such as Russia, Canada.
2. Rich oil resources countries; such as OPEC countries.
3. Developed or big economic countries with reasonable energy resources; such as USA, China, France, Italy, Spain..
4. Developed or big economy countries with lag energy resources; such as Japan, Korea.
5. Underdeveloped countries with poor resources; such as some third world countries

In the World Energy Council Statement 2005, it highlighted a number of the strategies including:
- Keep all energy options open. No technology should be idolized or demonized and energy efficiency must be increased.
- Exploit the 'win-win' opportunities of emerging climate change responses. The mechanisms, whether voluntary or regulated, should embrace least cost

emissions reduction, encouraging transfer of clean technology from industrialized to developing countries.

- Ensure technical innovation. It is vital to reconciling development with environmental protection and calls for strong, sustained support for R&D.
- Foster and sustain public understanding and trust. This is in turn depends on energy section transparency and better public information, starting in particular with young people.

An energy structure should be comprehensive, balanced, secure, efficient and environmentally sound. The interaction among the electric generation technology, chemical engineering and environmental technology would hopefully lead to a new electric generation technology with possible new energy sources.

III.   TWO MAYOR RISKS AFFECTING SUSTAINABILITY

The two major environmental risks affecting sustainability in the production of electricity using fossil fuels are Green House Effect and Acid Rain.

Green House Effect

Fossil fuels are formed from over millions of years from the compressed decaying remains of plants and animals. Because of the heavy use over the past few decades, it has caused a significant rise in carbon dioxide in the atmosphere. The molecules of the greenhouse gases, after absorbing the infrared radiation from the sun, vibrate and hence the temperature of the atmosphere is warmed up.

The increase in temperature will mean everywhere will get warmer and the climate patterns will also be affected. The rise in temperature will melt the ice at the Arctic and Antarctic, thus resulting in flooding particularly in low-lying countries. The warmer atmosphere also causes convection current which may turn into hurricanes and violent storms.

Acid Rain

Acid rain is a broad term used to describe the ways that wet and dry acids fall out of the atmosphere. The wet deposition (refers to acidic rain, fog, and snow) flows over and through the ground hence affecting a variety of plants and animals. Dry deposition refers to acidic gases and particles that fall back to earth. The wind blows these acidic particles and gases onto buildings, cars, homes, and trees. Furthermore, the dry deposited gases and particles can be washed from trees and other surfaces by rainstorms, thus making the runoff water more acidic than the falling rain alone.

Sulfur dioxide (SO_2) and nitrogen oxides (NOx) are the primary causes of acid rain. Acid rain occurs when these gases react in the atmosphere with water, oxygen, and other chemicals to form various acidic compounds. Sunlight increases the rate of most of these reactions. The result is a mild solution of sulfuric

acid and nitric acid. This acid deposition has a variety of effects, including damage to forests and soils, fish and other living things, materials, and human health.

IV.   SUSTAINABILITY ENERGY SOLUTION

The followings are proposed solutions for sustainability in the production and utilization of electricity.

1. Renewable Energy

The most ideal method is to switch from fossil fuels to renewable energy which does not produce carbon dioxide. The use of renewable energy is also an overall social responsibility of the community and more proactive approach should be adopted for their long term application. It is believed that in the 22nd century, at least 30% of the energy should come from renewable energy.

Solar Energy
Solar energy is a clean and endless energy and they can be readily accessed. Photovoltaic and Solar Thermal are actively developed to convert solar radiation to electricity or thermal energy.

Wind Energy
Moving air turns the blades of the windmill to make electricity. Wind generators need a high wind speed for this electricity production.

Tidal/Wave Energy
If a dam is built across a river inlet, electricity can be generated by the flow of water through turbines in the dam as the tide rises and falls. The movement of wave can also drive air turbines to make electricity.

Biomass Energy
Biomass is plant and animal material used to generate electricity. This includes using wood from trees, waste from other plants and manure from livestock.

Hydro Energy
Fast flowing water released from dams in mountainous areas can produce electricity.

Geothermal Energy
Geothermal energy uses heat energy from beneath the surface of the earth.

The development of renewable energy has started but more innovative and technology breakthrough are required in order to achieve a commercial viable applications. For wider use of renewable energy, the government should play an leading role in promotion and setting up standard for renewable energy.

2. Energy Efficiency

As renewable energy and advanced emissions reduction technology are in fast development stage, it is essential to ensure energy efficient approach being adopted in both energy production and utilization.

Advances in gaining energy efficiency in the production of electricity are progressing satisfactorily. However, there are still a lot of opportunities to improve efficiency in the utilization of electricity. Energy labeling scheme is an energy scale given to the electrical appliances to show how efficiently the appliance uses electricity. Energy Efficiency Registration Scheme for Buildings is used to encourage the building owners adopted energy saving initiatives in their facilities. Governments are also progressively introducing energy efficiency codes, practices which are now being applied from an advisory to mandatory role. It is particularly important when replacing old and obsolete appliance and equipment.

The energy efficiency and conservation concept must also be promoted to gain acceptance and implementation. Remote control and monitoring system could help to affectively achieve the required saving targets. The purpose is to use the energy in an intelligent and effective way while avoiding or reducing the energy wasted to an absolute minimum.

Furthermore, the total life cycle energy concept is also to be promoted. For example, a list of Total Life Inventory is kept when selecting materials and process uses. In the material selection, we need to consider the energy used to produce such materials and selection is made to those where employing less energy. This embedded energy approach is also being developed. For example, we should use less aluminum than other metal as the production of conventional aluminum production process is costly and inefficient.

V. SUSTAINABLE MOBILITY

The key issues of sustainable mobility are: energy, environment and safety. Therefore we should develop clean, efficient and intelligent vehicles, including battery electric vehicles, hybrid electric vehicles and fuel cell electric vehicles. It is predicted that the global market of hybrid vehicles in year 2010 will be 2 million. This will give significant opportunity for power electronics devices and systems that can meet automobile environment. Figure 1 shows the requirement for next generation vehicles. Figure 2 shows the motor, power electronic controller and IGBT devices for electric and hybrid vehicles. The basic requirement of power electronic for electric, hybrid and fuel cell vehicles are: high energy density, high power density, high efficiency over wide speed and torque range, frequent start stop, continuous dynamic, capable working under severe ambient temperature and vibration environment, compatible EMI/EMC, high reliability, high endurance, and reasonable cost.

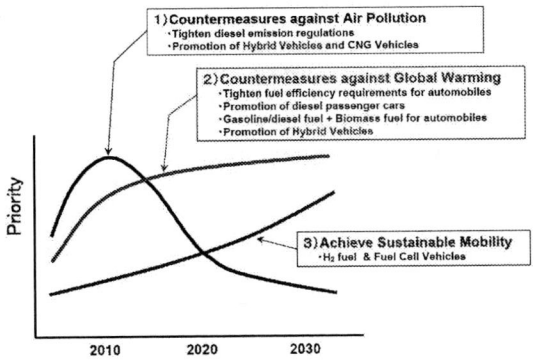

Figure 1. Requirement for next generation vehicles

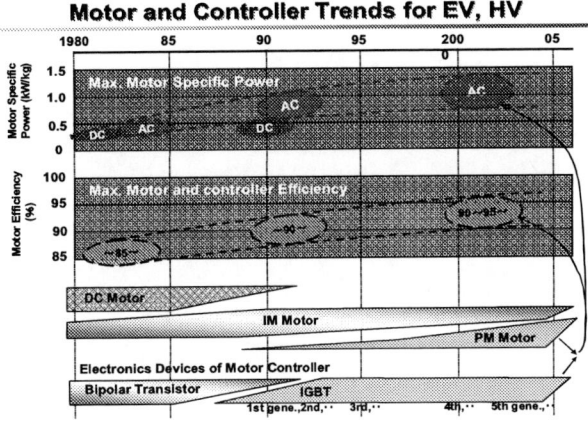

Figure 2. Motor and controller for electric and hybrid vehicles

VI. CHALLENGES TO POWER ELECTRONICS

1. Development of High Performance Devices

Figure 3 shows the historical development of power electronic devices. Currently silicon based devices are approaching the limit of the capability of their materials. Therefore we should research on new materials and new deice structure. Figure 4 shows the features of various devices. The product of multiplication of power and frequency is about $10^9 - 10^{10}$ WHz..

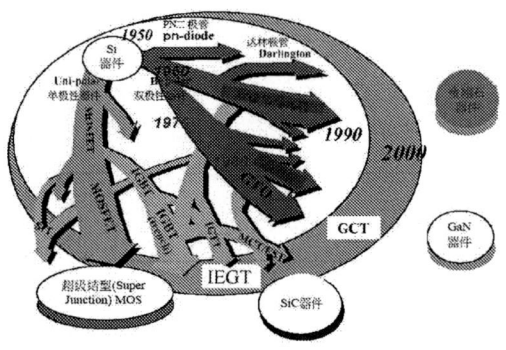

Figure 3. Historical development of power electronic devices

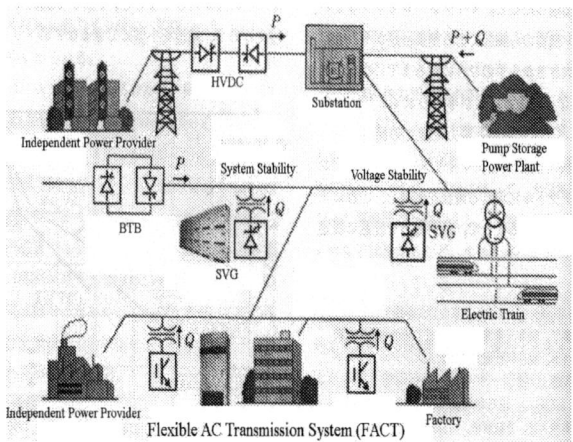

Flexible AC Transmission System (FACT)

Figure 5. Power quality control

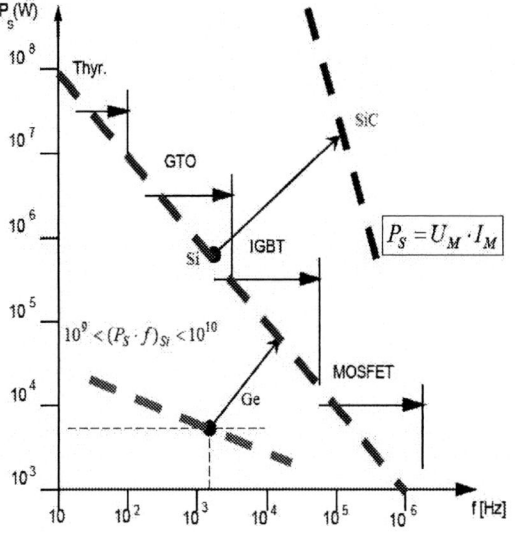

$$P_S = U_M \cdot I_M$$

$$10^9 < (P_S \cdot f)_{Si} < 10^{10}$$

Figure 4. Features of power electronic devices

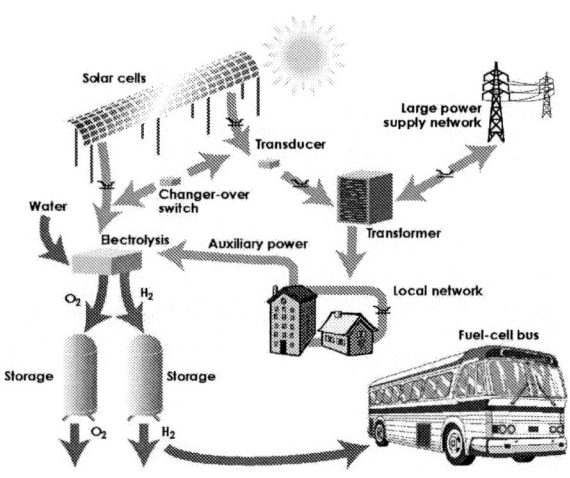

Figure 6. An integrated energy end –use system.

2. Integration

It is essential for the development of standardization, modularization and integration, such as power system on chip (PCOC), power system in package (PSIP), etc.

3. Converters Topologies and PWM Technology

It is noted the recent development of various converters topologies and PWM technology and the methods for their effective evaluation.

4. Examples in Sustainable Energy Systems

The followings are examples of sustainable energy systems. Figure 5 shows the application of power electronics for power quality control. . Figure 6 shows an integrated energy end-use system.

5. Examples in Sustainable Mobility

Hybrid vehicles play key role in sustainable mobility.

The followings are some examples of power electronics in hybrid vehicles

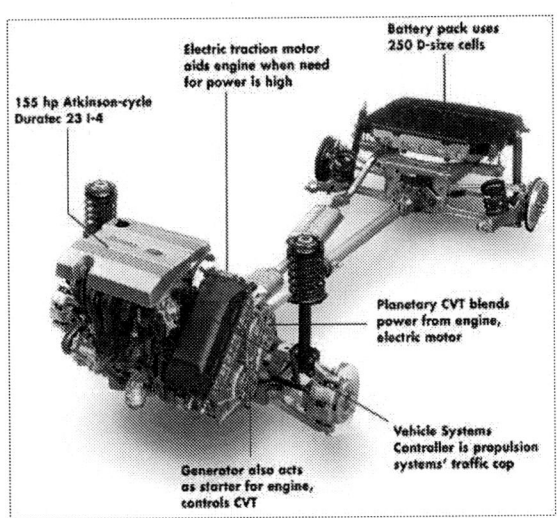

Figure 7. Hybrid electric vehicle powertrain

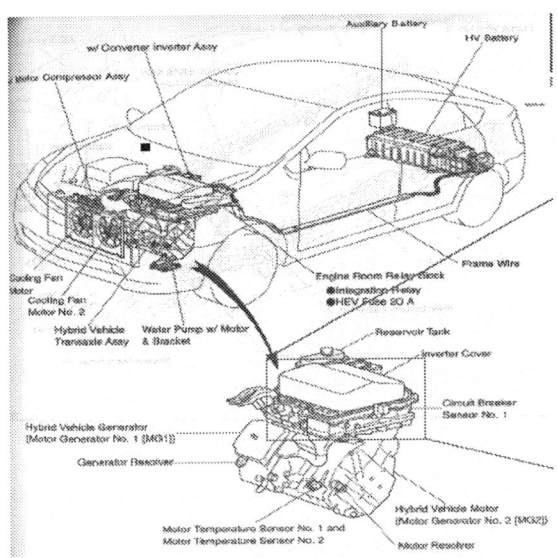

Electronics Control Architecture

Hybrid-ECU	Controls the drive force and other ECUs
Motor-ECU	Inverter control of 3-phase AC output
Engine-ECU	Electronic throttle, fuel, spark, VVTi, etc
Brake-ECU	Brake force strategy for regeneration
Battery-ECU	Monitors the battery SOC and DOD

Figure 8. Toyota Prius 2005 hybrid architecture

Figure 9. Toyota Prius Converter Inverter Module

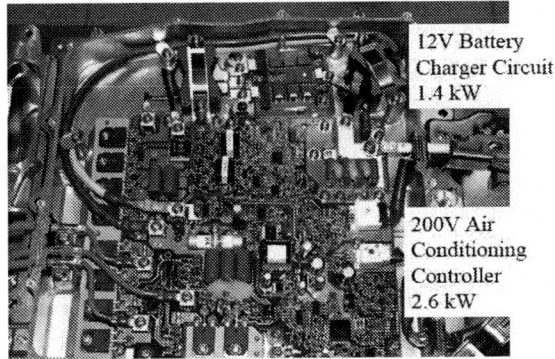

Figure 10. Toyota Prius Water Cooled Electronics for AC +12V Battery Charger

Figure 11. Toyota Prius Water Cooling Fins Between power modules

Figure 12. Toyota Prius Smart Power Driver and Current Sensor

Figure 13. Toyota Prius 22 Sets IGBTs & Power Diodes

VII. CONCLUSIONS

The sustainable energy and sustainable mobility prompt opportunity and challenges for power electronics developments. Power electronics engineers should not only further develop the fundamental theory, devices, circuit topologies, integration and optimization technology, but should also understand the emerging application requirements, to develop high energy density, high power density, high efficiency, reliable, endurance and EMI/EMC compatible at market acceptable cost for sustainable energy and sustainable mobility.

REFERENCES

[1] C.C. Chan, "The challenge of sustainable mobility", keynote at IEEE Vehicles Power and Propulsion Conference (VPPC 2005), Chicago, September 7-9, 2005.

[2] C.C. Chan, "Energy development strategy, industrial ecology and challenges to electrical machines systems", keynote paper, International Conference on Electric Machines Systems (ICEMS 2005), Nanjing, China, September 27-29, 2005.

[3] C.C. Chan and K.T. Chau , Modern Electric Vehicle Technology, Chapters 1,2,3,4,and 5, Oxford University Press, ISBN 0198504160, 2001.

Wind farms with increased transient stability margin provided by a STATCOM

Marta Molinas[*], Jon Are Suul[*] and Tore Undeland[*]
[*] Norwegian University of Science and Technology
Department of Electric Power Engineering, Trondheim, Norway
marta.molinas@elkraft.ntnu.no

Abstract— this paper analyzes the extent to which the transient stability margin of wind farms using squirrel cage generators can be increased by the use of a STATCOM. A simplified analytical approach is used first to quantify the effect of the STATCOM on the transient stability margin. An experimental validation of the calculated transient margin is then presented by measurements done in a laboratory prototype of 7.5 kW emulating a wind turbine or wind farms. Measurement results confirm that the STATCOM provides a clear transient stability margin increase and with adequate rating it becomes possible to ride through severe faults.

Keywords- Transient Stability Margin; Grid Code; Voltage Source Static Var Compensator; Voltage Sag; Squirrel Cage Wind Generator; Vector Control.

I. INTRODUCTION

The impact of the wind generation on the power systems is no longer negligible if high penetration levels are going to be reached. Significant barriers to interconnection are being perceived already with the severe requirements of the new emerged grid codes. Depending on the generator technologies, different solutions are found to support behavior in case of voltage sags. Voltage Source Static Var Compensator such as the STATCOM can be used to regulate voltage as shunt compensator with directly connected asynchronous wind generators.

In Norway, the potential for large scale generation is huge, but the extent to which it can be integrated into the power system without affecting the overall stable operation depends on the technology available to mitigate the negative impacts. There is a strong need for preparing the ground for large scale integration of wind power if the vast resources are to be used. There are several issues that need to be considered. Some are related to power quality; others like the fault ride through capability introduced with the new grid codes are new challenges [1,2]. Many countries in Europe and other parts of the world are developing or modifying interconnection rules and processes for wind power through a grid code [3,4]. Fig. 1 shows the low voltage ride-through profile in the Nordel Grid Code. The grid codes have identified many

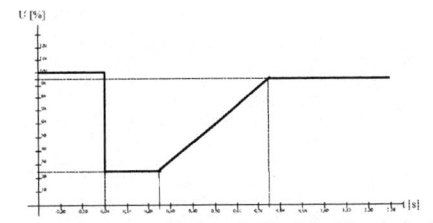

Figure 1 Ride through profile from the Nordel grid code for the Nordic countries Norway, Denmark, and Sweden.

potential adverse impacts of large scale integration. The impacts are wide ranging, but given the novelty of the industry and the scenarios, they are not fully understood. One of the issues addressed by these new grid codes is that wind turbines must be able to remain connected to the grid during and after network faults. This requires particular control strategies for wind turbines with squirrel cage induction generator as well as robust power electronics control for converter connected farms. Wind farms using squirrel cage induction generators directly connected to the network will more acutely suffer from the new demands, since they would usually disconnect from the power system when the voltage drops more than 30 % below rated value.

The STATCOM has been reported to have the capability to regulate voltage, control power factor, and stabilize power flow [5,6]. In this paper, the STATCOM is analyzed from the point of view of its potential for transient margin increase as reported in [7] to assess its capability for ride-through demanded by the new grid codes. This margin is the length of fault that the wind generation is capable of riding through without loosing its stable operating condition.

The paper concentrates on the directly connected asynchronous wind generator and analyses the improvement of transient stability margin in the event of a three phase-to ground fault when a STATCOM is used as the ride through solution. A torque-slip based analysis of transient stability limit with a STATCOM is first presented. Measurements on a lab set-up then verify the clear increase of the transient stability margin when a STATCOM is used.

II. System Model with the STATCOM

A. System Model

Figure 2 shows the schematic configuration of the system under consideration for compensation with a STATCOM. For this study it is assumed that the power system is subjected to a three phase fault along the transmission line. The STATCOM is a power electronics device based on the voltage source converter principle. The technology typically in use is a two level voltage source converter with a DC energy storage device, a coupling transformer connected in shunt with the power system, and DSP based control circuits. The main advantage of the STATCOM over thyristor type static var compensators is that the compensating current does not depend on the voltage level of the connecting point and thus the compensating current is not lowered as the voltage drops. This is an important feature now that the new grid codes will require wind turbines to supply reactive power variably depending on network demand and actual voltage level. However, regarding the fault ride through, the most relevant feature of the STATCOM will be its inherent capability to increase the transient stability margin. The model of the system is expressed by the equation of the system voltage at the point of compensation with the STATCOM. The three phase voltage is expressed as

$$v_{abc} = r \cdot i_{abc} + l \cdot \frac{di_{abc}}{dt} + v_{abc,STATCOM}, \qquad (1)$$

where v_{abc}, i_{abc}, $v_{abc,STATCOM}$ are grid voltages, grid currents, and voltage generated by the STATCOM, r and l are the per unit resistance and inductance between the converter and the grid.

B. Control of the STATCOM

The control strategy implemented is based on the vector control principle. Main advantage of this is the decoupled control of the DC link voltage and reactive current. Equation (1) is transformed into a dq reference frame rotating at the grid frequency ω.

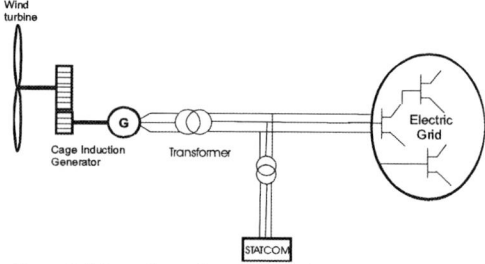

Figure 2 Schematic configuration of the system under study.

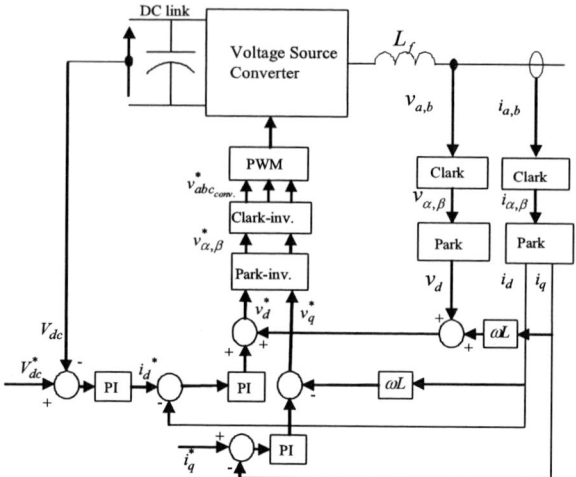

Figure 3. Block diagram of the control implemented in the STATCOM

$$v_{dq} = r \cdot i_{dq} + l \cdot \frac{di_{dq}}{dt} + l \begin{bmatrix} 0 & -\omega \\ \omega & 0 \end{bmatrix} i_{dq} + v_{dq_{STATCOM}} \qquad (2)$$

Aligning the d axis of the Park reference frame with the grid voltage vector we have that $v_q = 0$. The angular position of the supply voltage vector is computed as

$$\theta_v = \tan^{-1}\left(\frac{v_\beta}{v_\alpha}\right), \qquad (3)$$

where v_α and v_β are the components of the grid voltage in a stationary two axis reference frame (Clark transformation). The power balance between the DC link and the converter output gives

$$v_{dc} i_{dc} = \frac{3}{2} v_d i_d \qquad (4)$$

where v_{dc} is the DC link voltage and i_{dc} is the DC link current. In (4), v_d is determined by the depth of the voltage sag, and the DC link voltage can be controlled by modulating the converter direct axis current component i_d to compensate for the converter losses.

The converter quadrature current component i_q is used to modulate the flow of reactive power. Fig. 3 shows the block diagram of the control of the STATCOM. The reactive current injected is controlled so as to obtain 1 p.u. rated grid voltage before, during and after the fault.

III. Torque-slip analysis

The transient stability of a grid connected induction generator is analysed using a simplified approach based on torque-slip characteristics.

As argued in [9] the analysis is done neglecting stator and rotor transients of the induction machine. This leads to the use of a traditional per phase equivalent circuit representing the induction machine, and studying only the mechanical acceleration dynamics. For very fast transients this assumption will be too rough, but it is gives much simpler analysis than a 5th order model, and is generally used in power system studies involving voltage stability and load restoration [10].

A. System description and principal assumptions

With reference to the implemented control system of the STATCOM, the following premises are stated regarding analysis of the torque-slip curve before during and after a grid fault:

- Before the fault, the STATCOM is controlled to keep the terminal voltage constant. Initial operating conditions can be calculated from the induction generator equivalent circuit with terminal voltage $\mathbf{v}_1$ at its nominal value
- After the fault, the STATCOM will give maximum current, until nominal voltage is reached.

To simplify the analysis of rotor acceleration during the fault, it is also assumed that:

- During a three-phase fault the STATCOM current has negligible influence on the terminal voltage of the generator, and thus on the system behaviour.
- The accelerating torque during a three-phase fault is approximately constant and equal to the initial applied mechanical torque.

The equivalent circuit of the system after a fault is shown in Fig. 3, where $\mathbf{v}_g$, $\mathbf{v}_1$ are grid voltage and induction machine terminal voltage phasors, and $\mathbf{i}_l$, $\mathbf{i}_{STATCOM}$, $\mathbf{i}_1$, $\mathbf{i}_m$ and $\mathbf{i}_2$ are line, STATCOM, stator, magnetizing and rotor current phasors. Line impedance is $r_l + jx_l$ while r_1, r_2 are stator and rotor resistance and x_1, x_m, x_2 are stator, magnetizing and rotor reactance. The STATCOM is represented by a current source and the generator is connected to a stiff grid. Any series impedance between the STATCOM and the generator terminals, accounting for line- and/or transformer-impedance, can be added to the stator impedance in the figure.

If the system is stable after a fault, the torque as function of slip will follow a quasi-stationary equilibrium curve determined by the system in Fig. 3 until the speed is reduced enough for the STATCOM to bring the terminal voltage back to its nominal value. Thereafter the torque-slip curve will be given by this constant voltage, which is the same as used to find the initial conditions.

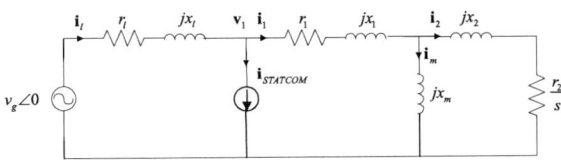

Figure 3 Quasi stationary equivalent circuit for the system under study, consisting of the traditional induction machine per phase equivalent, the STATCOM modelled as a current source and a grid equivalent.

B. System equations

With the current directions indicated in Fig 3, the relation between grid voltage and STATCOM voltage will be given as:

$$\mathbf{v}_g = \mathbf{v}_1 + \left(r_l + jx_l\right)\left(\mathbf{i}_1 + \mathbf{i}_{STATCOM}\right) \qquad (5)$$

The current $\mathbf{i}_1$ will depend on the voltage $\mathbf{v}_1$ and the slip. The Thevenin impedance for the rotor circuit and the magnetizing reactance for the induction machine, seen from the stator, will be

$$r_{Th,r} + jx_{Th,r} = \frac{jx_m\left(jx_2 + \dfrac{r_2}{s}\right)}{\dfrac{r_2}{s} + j\left(x_m + x_2\right)}, \qquad (6)$$

with the corresponding stator current:

$$\mathbf{i}_1 = \frac{\mathbf{v}_1}{r_1 + r_{Th,r} + j\left(x_1 + x_{Th,r}\right)} \qquad (7)$$

The STATOM current is purely reactive, and is always leading the voltage $\mathbf{v}_1$ by 90°. The STACOM current phasor can therefore be expressed as:

$$\mathbf{i}_{STATCOM} = j\frac{\mathbf{v}_1}{|\mathbf{v}_1|}\left|\mathbf{i}_{STATCOM}\right| \qquad (8)$$

Combining (7) and (8) with (5) gives:

$$\begin{aligned}
\frac{\mathbf{v}_g}{\mathbf{v}_1} &= 1 + \frac{r_l + jx_l}{r_1 + r_{Th,r} + j\left(x_1 + x_{Th,r}\right)} \\
&\quad + j\frac{r_l + jx_l}{|\mathbf{v}_1|}\left|\mathbf{i}_{STATCOM}\right|
\end{aligned} \qquad (9)$$

The grid voltage $\mathbf{v}_g$, is the constant reference voltage, and for a given STATCOM current and a given slip, this equation can, as shown in Appendix, be solved numerically to find the voltage $\mathbf{v}_1$. The corresponding stator current $\mathbf{i}_1$ is given by (7), and the per unit rotor current $\mathbf{i}_2$ and mechanical torque τ_{em} are given as:

$$\mathbf{i}_2 = \frac{jx_m}{\dfrac{r_2}{s} + j\left(x_2 + x_m\right)}\mathbf{i}_1 \qquad (10)$$

$$\tau_{em} = \frac{r_2}{s}\left|\mathbf{i}_2\right|^2 \qquad (11)$$

From these equations a torque-slip curve after the fault, for a given rating of maximum STATCOM current, can be established, and a critical clearing slip or speed for a given mechanical torque can be found,. This is similar to what is reported in [9] for induction generators with passive capacitor compensation. Such constant compensation can also be accounted for in this approach, by including the capacitance in the Thevenin impedance of (6).

As pointed out in [9], the critical clearing speed will in general not depend on the type of disturbance, since the stability of the induction machine depends only on the magnitudes of mechanical torque and reapplied electromechanical torque after the disturbance.

The mechanical equation is given by

$$T_a \frac{dn}{dt} = \tau_m - \tau_{em}(n),\qquad(12)$$

where T_a is the inertia constant, n is the speed, τ_m, τ_{em} are mechanical torque and electromagnetic torque. Assuming zero electromagnetic torque during a three-phase short circuit, and constant accelerating torque equal to mechanical torque, the critical clearing time (CCT) can be calculated from the critical speed and the initial speed, or the corresponding slip values as:

$$CCT_{3-phase} \approx T_a \frac{n_{cc} - n_{init}}{\tau_m}\qquad(13)$$

C. Torque-slip calculation example

The given equations are used for an example calculation with parameters given in Table 1. The resulting torque-slip characteristics for three different STATCOM current ratings are given as torque-speed curves in Fig. 4. The curves are plotted together with torque-slip curves for no compensation and for constant terminal voltage (ideal compensation).

TABLE 1
PARAMETERS USED FOR CALCULATION EXAMPLE

Asynchronous machine	
$r_1 = 0.050$	$r_2 = 0.040$
$x_1 = 0.095$	$x_2 = 0.063$
$x_m = 1.43$	
$T_a = 0.5$ s	$\tau_m = 1$
Grid	
$r_1 = 0.05$	$x_1 = 0.11$
$\mathbf{v}_g = 1e^{j0}$	

The initial speed with 1 pu torque is 1.043. Critical slip and corresponding speed for the uncompensated system, and for the different STATCOM current ratings, are given in Table 1. It is also given critical clearing time, assuming constant applied mechanical torque and zero electromagnetic torque during a three-phase short circuit.

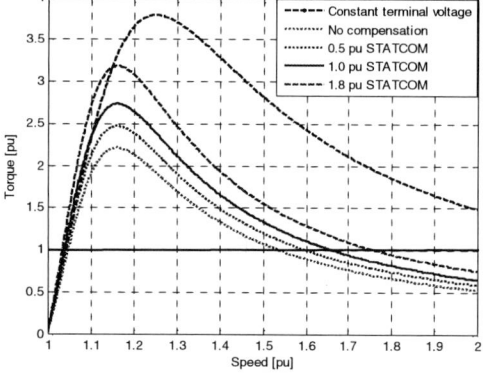

Figure 4 Torque-speed curves for uncompensated system, for different STATCOM current ratings, and for constant terminal voltage.

TABLE 2
CRITICAL CLEARING SLIP AND SPEED FOR DIFFERENT STATCOM CURRENT RATINGS WITH CORRESPONDING CRITICAL CLEARING TIME FOR A THREE-PHASE FAULT

$I_{STATCOM}$	Critical slip	Critical speed	Critical clearing time – $CCT_{3-phase}$
0 pu	-0.539	1.539 pu	0.248 s
0.5 pu	-0.601	1.601 pu	0.279 s
1.0 pu	-0.663	1.663 pu	0.310 s
1.8 pu	-0.764	1.764 pu	0.361 s

It is observed that the STATCOM gives a significant contribution to increase the critical speed, and thereby the stability limit of the induction generator, resulting in a corresponding increase of critical clearing time during a three phase fault.

D. Mechanical considerations

Following the suggested approach, neglecting electrical transients, the quasi-stationary torque is a function of slip and STATCOM current. When a fault is cleared, the decelerating torque for a stable case is the difference between applied mechanical torque and electromechanical torque for the particular slip. Subsequently, the electromechanical torque will follow the torque-slip curve for the given STATCOM rating, until the speed is reduced to the level where the stator voltage reaches its nominal value. As can be seen from Fig. 4, this will in general be for a lower speed than the maximum torque, and there will be a significant transient in mechanical torque before the speed has returned to its stationary valued. To limit the mechanical strain during recovery, the lowest possible STATCOM current that ensures stability would therefore be favourable. If the STATCOM current could be reduced before the nominal voltage is regained, but after stability is ensured, it would reduce the mechanical stress during speed recovery.

The calculation example is based on data of a machine similar to the available laboratory setup used for validation of the approach, based on a small induction machine with high loss and low inertia. For such a lossy machine, the pull-out torque with constant terminal voltage is very high when operating as a generator. That would not be the case for a real wind turbine. Low losses will give a much steeper torque-speed characteristic, and therefore lower critical speed. At the same time the mechanical time constant of a real wind turbine would be higher, up to several seconds, giving a much slower acceleration during a disturbance.

For a real wind turbine with induction generator, there would also be a gear with quite high speed ratio. This makes it necessary to account for the effects of shaft stiffness, and corresponding stored potential energy because of mechanical torsion, when estimating critical clearing time [11].

TABLE I
RATINGS OF DEVICES USED IN EXPERIMENTS

System Component	Description	Parameters
Wind Generator	Wound rotor induction machine with short circuited rotor windings	$P_n = 7.5$ kW
		$V_n = 230$ V
		$I_n = 27.5$ A
		$n_n = 1430$ r/min
		$r_1 = 0.05$ pu
		$r_2 = 0.04$ pu
		$x_\sigma = 0.15$ pu
		$x_m = 1.8$ pu
Wind Turbine	Separately excited DC machine	$P_n = 10.0$ kW
		$n_n = 2000$ r/min
Local Load	Induction Machine coupled to separately excited DC machine	Same as wind generator and wind turbine
STATCOM	IGBT-based, 3-ph inverter bridge PWM controlled	$V_{DC,MAX} = 600$V
		$I_{MAX} = 78$ A
Short Circuit Device	Short circuited 3-ph Thyristor bridge	$I_{Peak} = 2000$ A

IV. EXPERIMENTAL CONFIGURATION

The experimental setup is shown in Fig. 5. Ratings and characteristics of all system components are summarized in Table I. In all measurements, the pu system is based on the rated VA of the generator used to emulate the wind generator.

The wind generator is emulated with a 7.5 kW wound rotor induction machine with short-circuited rotor terminals. The wind turbine is emulated by a constant torque generated with a controllable DC motor. A constant torque is used to model the power from the wind, because transient phenomena are much faster than wind variations.

A local load, having induction motor characteristics is placed at the point of common coupling as shown in Fig. 5, in order to force the power system towards its stability limit. The STATCOM is an IGBT-based inverter bridge, whose maximum current is about twice the rated current of the wind generator. The vector control technique is implemented through a DSP and a host computer.

V. MEASUREMENT RESULTS

A sudden severe drop in voltage is caused at the point of common coupling (PCC) by simultaneously triggering all thyristors in the short circuit device (SCD) in Fig. 5. The depth of the voltage drop can be modified by properly adjusting the inductance in series with the SCD.

In the experiments shown in Fig. 6, a voltage drop of about 80 % is produced, and the fault is cleared after 300 ms. Before the fault, the wind generator is delivering about 1.3 pu of active power, of which 0.75 pu are absorbed by the local load, and the rest is sent to the weak grid. Figure 6. a, b, and c show the system responses for different control conditions.

When there is no STATCOM control, during the fault, the wind generator accelerates, since it is no longer able

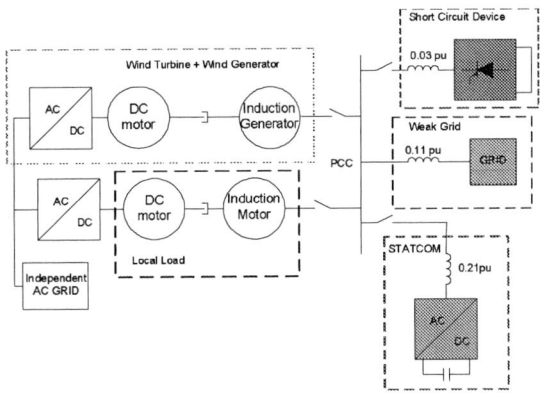

Figure 5 Experimental setup of the wind generation compensated with STATCOM

to generate enough electromagnetic torque to balance the torque coming from the wind, which is obviously unaffected by the grid fault. The voltage at the PCC does not recover its nominal value and remains very low (0.4 pu). When the fault is cleared, the generator speed is about 1.6 pu and, without fast reactive support; the generator is not able to produce enough braking torque to bring the speed back to its pre-fault value. The situation would have developed into voltage collapse, provided that we were not limited by the wind turbine emulator, which is unable to keep rated torque at such a high rotational speed. However, even if voltage collapse did not take place, the pseudo-stable post-fault operating condition is not a sustainable one, and will not be accepted by any grid code. The excessive currents in the system under this condition will also cause protections to trip very soon.

With a STATCOM connected at the PCC the system responses to the same fault condition, starting from the same initial operating point, are shown for different ratings of the STATCOM current (0.5, 1, and 1.8 pu.) When the maximum current the STATCOM can deliver is fixed to 0.5pu of wind generator rated current during the fault, the system behaves exactly like the one with no compensation, except for a small contribution of the STATCOM to the short circuit current. The voltage at the PCC and the generator speed are not affected by the STATCOM to any significant degree, during the fault time. However, it is clear that even with the relatively small rating of 0.5 pu, the STATCOM is able to stabilize the power system when the fault is cleared; bringing the PCC voltage, grid current and generator speed back to their pre-fault values. Complete voltage recovery process takes about 0.9 s from the instant of the fault clearance.

With an increased rating for the STATCOM current (1.0pu and 1.8pu, respectively), things do not change significantly during the fault time, but the voltage recovery process is considerably shortened as the STATCOM rating is increased. In particular, the wind generator is able to generate much more torque at the fault clearance, due to the increased reactive support at its

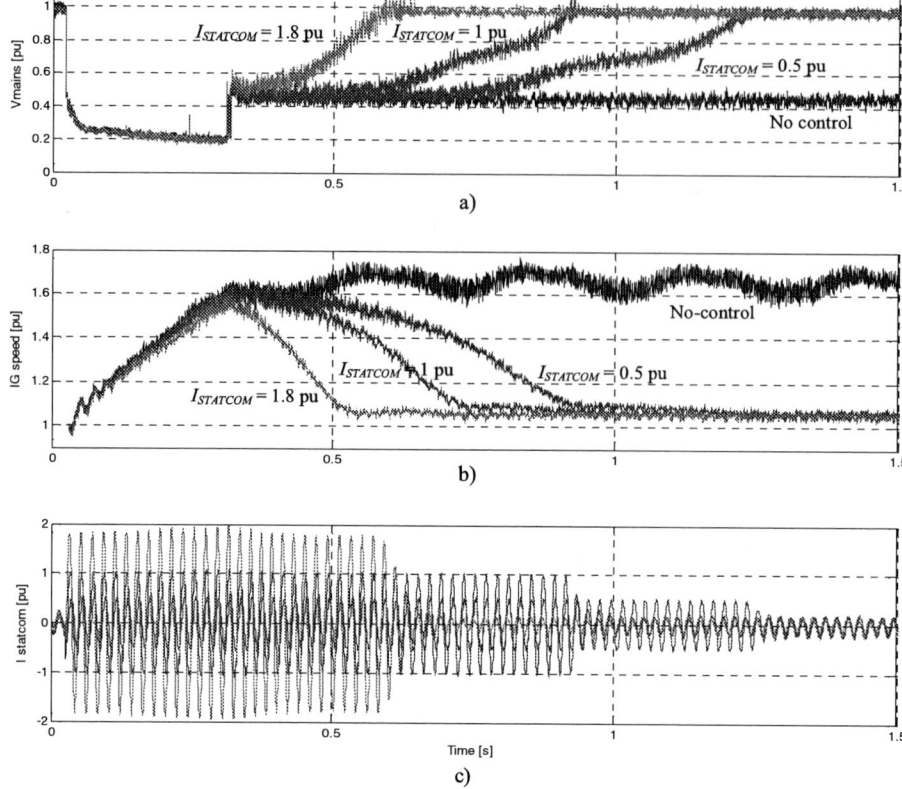

Figure 6 Measurement results of a 300ms, 80% voltage drop at PCC under several control conditions

terminals. This phenomenon is shown in Fig. 7, in which the accelerating torque for all the experimental conditions is calculated based on estimated speed.

It is clear that the higher the STATCOM current rating, the larger the accelerating torque, and thus the transient margin. This behavior shows an increased stability margin with the STATCOM ratings. That means that a system with higher rating STATCOM will be able to withstand longer short circuits and /or it will be able to withstand voltage drops while delivering higher amount of power. However, the higher rating of the STATCOM gives also higher transient torques, which represents a problem of mechanical stresses for the wind turbine. As noted in section III.D the lowest possible STATCOM rating that would keep the system stable is therefore favorable for the mechanical system. The torque-time curves in Fig. 7 also indicate that the torque during recovery is in accordance with the quasi-stationary torque-speed characteristics of Fig. 4.

VI. CONCLUSIONS

The improvement of ride through capability for an induction generator, achieved by the use of a STATCOM, is investigated with a simplified theoretic approach and with laboratory experiments. An estimate for the increase of CCT for different STATCOM current ratings after a three phase fault is calculated, and the laboratory measurements verify the increase of transient margin, with corresponding improvement of ride through capability.

The laboratory measurements indicate that the assumptions of the simplified theoretical calculations could be quite reasonable in this case. The maximum speed of the generator when the fault is cleared is not much affected by the STATCOM, and the torque curves of Fig. 7 indicate how the reapplied electromagnetic torque and the mechanical deceleration are influenced by the STATCOM.

The laboratory setup delivers 1.3 pu power before and after the fault, while the average accelerating torque during the fault is about 1 pu. This means that the machine is considerably more stable than the calculations indicates. Considering that electric transients are neglected, there is a quite good accordance between measurements and calculations, and the quantitative deviations could be mainly from parameter uncertainties and variation. A higher practical rotor resistance, for instance due to skin effect of the rotor windings at high speed, will make the induction generator more stable.

In terms of ride through capability, the measurement results show that with a reasonable rating of 0.5 pu and a fault of 80 % voltage drop for 300 ms, the STATCOM is an optimum candidate for providing ride through in wind farms equipped with asynchronous generators directly connected to the network. Increased STATCOM current rating will enlarge the transient margin of the system by providing higher decelerating torque, but will at the same time increase the maximum mechanical stress.

Considering that with the previous rules, wind turbine induction generators would trip when they detect a 30%

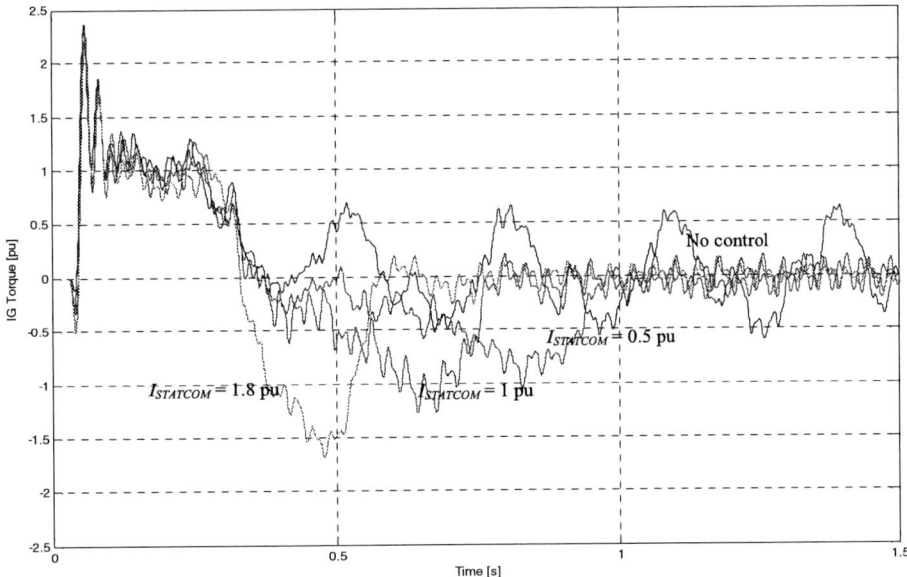

Figure 7 Induction Generator torque during and after the short circuit under different control conditions

voltage drop, it can be said that the STATCOM provides with a clear capability of handling fault ride through for an 80% voltage drop. The extent of this handling capability will depend on the power system configuration, the generator parameters and rating of the device itself.

In the electrical system analyzed in this paper, in spite of the additional cost of power electronics converters and control, with the new grid codes, the achievement of fault ride through capability is of relevance. Therefore, investment in a STATCOM will certainly be justified in the scenario presented in this paper.

REFERENCES

[1] E.ON Netz GmbH, "Grid code high and extra high voltage," E.ON Netz GmgH Bayreuth, August 2003, http://eon-netz.com

[2] I Erlich, U. Bachmann, "Grid code requirements concerning connection and operation of wind turbines in Germany," IEEE PES 2005 General Meeting, San Francisco.

[3] Nordisk Regelsamling (Nordic Grid Code), Nordel, 2004

[4] Eltra specifications, "Wind farms connected to the grid with voltages over 100 KV. Technical regulations for the properties and control of wind turbines," Eltra, Doc. NO. 17619 v6, in Danish, 2004.

[5] L. Gyugyi, , "Dynamic Compensation of AC Transmission Line by Solid-State Synchronous Voltage Sources," IEEE Power Engineering Society, Summer Meeting 1993, pp. 434 1-8.

[6] M.Molinas, J. Marvik, T. Undeland, "Impact of Large Scale Integration of Wind Power into the Electricity Grid," International Conference of Women Engineers and Scientists, Seoul, Korea., (2005).

[7] E. Larsen, N. Miller, S.Nilsson, S. Lindgren, "Benefits of GTO-based Compensation Systems For Electric Utility Application," IEEE Power Engineering Society, Summer Meeting 1991, pp. 397 1-8.

[8] M. Molinas, B. Nass, W. Gullvik, T.Undeland, "Control of Wind Turbines with Induction Generators Interfaced to the Grid with Power Electronics Converters," Proc. of the International Power Electronics Conference IPEC 05, Niigata, Japan

[9] K.C. Divya, P.S. Nagendra Rao, "Study of dynamic behaviour of grid connected induction generator," IEEE Power Engineering Society General Meeting, 6-10 June 2004, vol.2, pp. 2200-2205

[10] T. Van Cutsem, C. Vournas, "Voltage Stability of Electric Power Systems," Kluwer Academic Publishers, 1998

[11] S. K. Salman, A. L. J. Teo, "Investigation into the Estimation of the Critical Clearing Time of a Grid Connected Wind Power Based Embedded Generator", IEEE/PES Transmission and Distribution Conference and Exhibition, 6-10 Oct. 2002 vol. 2, pp. 975-980

APPENDIX

Equation (9) can be written as:

$$\mathbf{v}_g = \mathbf{v}_1 \left(1 + z_{l1} + \frac{z_{IS}}{|\mathbf{v}_1|} \right),$$

$$z_{l1} = \frac{r_l + j x_l}{r_1 + r_{Th,r} + j \left(x_1 + x_{Th,r} \right)}$$

$$z_{IS} = j \frac{r_l + j x_l}{|\mathbf{v}_1|} |\mathbf{i}_{STATCOM}|$$

Decomposed to real and imaginary part, this gives the following set of equations, with corresponding constants:

$$v_g = c_1 \cdot v_{Re} - c_2 \cdot v_{Im} + c_3 \frac{v_{Re}}{v_{Re}^2 + v_{Im}^2} - c_4 \frac{v_{Im}}{v_{Re}^2 + v_{Im}^2}$$

$$0 = c_2 \cdot v_{Re} + c_1 \cdot v_{Im} + c_4 \frac{v_{Re}}{v_{Re}^2 + v_{Im}^2} + c_3 \frac{v_{Im}}{v_{Re}^2 + v_{Im}^2}$$

$$c_1 = 1 + \text{Re}\left(z_{l1} \right)$$

$$c_2 = \text{Im}\left(z_{l1} \right)$$

$$c_3 = \text{Re}\left(z_{IS} \right)$$

$$c3 = \text{Im}\left(z_{IS} \right)$$

This set of equations can be solved numerically for varying slip, to find the real and imaginary part of the voltage $\mathbf{v}_1$. Using equations (7), (10) and (11), this gives as post-fault torque-slip characteristics of a induction generator with a given STATCOM current rating.

A New Super Junction LDMOS with N^+-Floating Layer

Baoxing Duan, Bo Zhang and Zhaoji Li

Research Institute of Micro-Electronics, University of Electronic Science and Technology of China, No. 4, Section 2, North Jianshe Road, Chengdu, 610054 Sichuan, People's Republic of China

FAX: 86-28-83202569; Tel: 86-28-83204101; Email: bxduan@163.com

Abstract— A new CMOS compatible Super Junction LDMSOT structure is designed with N^+-Floating Layer embedded in the high-resistance substrate, which suppresses charges imbalance effect resulting from substrate-assisted depletion N-type pillar, and the high electric field around the drain is reduced by N^+-Floating Layer which causes the redistribution of the bulk electric field in the drift region. The new structure features high breakdown voltage, low on resistance and charges balance in drift region due to N^+-Floating Layer.

Keywords: Super Junction; N^+-Floating Layer; substrate-assisted depletion

I. INTRODUCTION

Lateral double diffused MOSFETs based on the super junction concept [1-3] in which the N-type drift region of the conventional LDMOS is replaced by a set of alternating and highly doped N- and P-type semiconductor pillars were recently proposed to further improve the trade-off characteristics between the breakdown voltage (BV) and the on-resistance (R_{on}) which has always been a major issue in the design of power devices. In theory, the SJ structure results in high BV due to charge balance in the pillars and low R_{on} is achieved by highly doping in the N-pillars. However, the P-type pillar of the SJ-LDMOST implemented on a P-substrate can not be depleted completely before the electric field of silicon reaches the critical breakdown, which results from charges imbalanced effect due to substrate-assisted depletion N-type pillar, so this effect also be called the substrate-assisted depletion effects which degrade the BV of the device[4]. To eliminate this effect, several structures had been reported [5-8].

In this paper, a novel CMOS compatible SJ-LDMOST using N^+-Floating Layer (N^+-Floating SJ-LDMOS) embedded in the high-resistance substrate (Figure.1) is proposed to suppress the charges imbalance effect, and the high electric field around the drain is reduced by N^+-Floating Layer which causes the redistribution of the bulk electric field in the drift region due to REBULF (REduced BULk Field) effect [9].

II. Device Structure and Description

Figure.1 shows the proposed N^+-Floating SJ-LDMOST. The key feature in the structure is the use of the N^+-Floating layer of which concentration is more than $1 \times 10^{17} \text{cm}^{-3}$ in high-resistance substrate, and the distance from N^+-Floating layer to the bottom of drift region is W which must be less than the thickness of depletion layer in the substrate of conventional SJ-LDMOS. As the drain voltage increases in the off-state, the N-pillars of the N^+-Floating SJ-LDMOST are depleted by the neighboring P-pillars, as well as by the P-type substrate above N^+-Floating. The P-pillars start to be affected by the N^+-Floating layer after the P-type substrate above N^+-Floating is fully depleted at a high enough drain voltage. In the N^+-Floating SJ-LDMOST structure, both pillars are affected by vertical depletion effects, which causes the charges balance between the pillars, while in the conventional SJ structure only the N-pillars are affected by the P-type substrate.

The potential of N^+ layer is floated up when the depletion layer spreads into its region, thus the high electric field around the drain is reduced by the redistribution of the bulk electric field in the drift region, and the substrate supports more biases due to the parallel plane D_3 junction.

1-4244-0448-7/06/$25.00 ©2006 IEEE

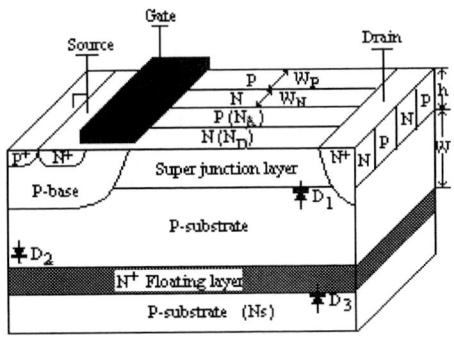

Fig.1.Three-dimentional view of the N^+-Floating SJ-LDMOST

III. Simulation Results

Figure.2 shows the simulated potential contours at breakdown using simulation software ISE [10] for the proposed N^+-Floating SJ-LDMOST and conventional SJ-LDMOST.

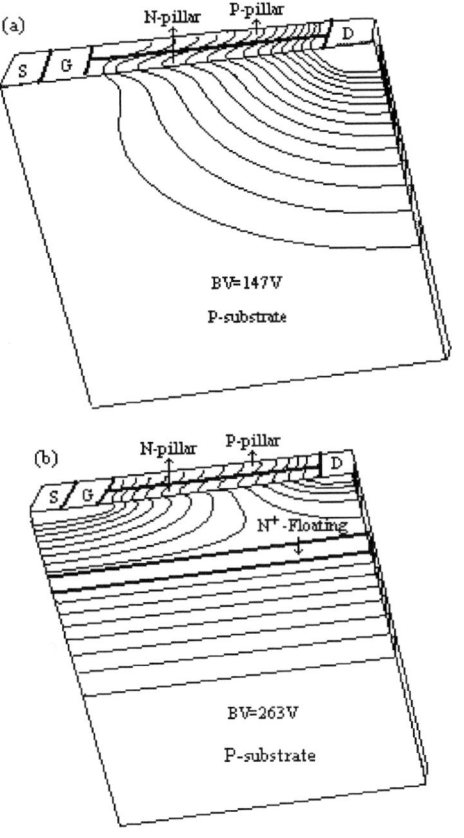

Fig.2 Potential distribution of conventional SJ-LDMOS (a) and N^+-Floating SJ-LDMOS (b): L_d=15µm, $N_D=N_A=5.0\times10^{16}$cm^{-3}, h=1µm,W=16µm

It is clear that the breakdown occurs at the surface of

the conventional SJ-LDMOST (Fig.2.a) due to field crowding and the P-type pillar cannot be depleted completely for charges imbalance effect, thus breakdown voltage is limited to 147V. However, comparing with conventional SJ-LDMOST, there are two significant differences for proposed N^+-Floating SJ-LDMOST (Fig.2.b). First, both P and N-type pillars are depleted completely at the breakdown due to the fact that charges balance is obtained in proposed structure. Second, a significant part of the potential contours gets pulled out toward the source region thanks to the potential acquired by the N^+ floating layer, by which the electric field around drain is reduced due to REBULF effect, thus the potential across the D_1 junction has been reduced significantly and the device breakdown occurs at the D_1 and D_2 junction simultaneously in the optimum case, its breakdown voltage reaches 263V.

The vertical electric field distributions around drain and source were shown in Fig.3 respectively. Under reverse bias condition, the electric field around drain of N^+-Floating SJ-LDMOST (Fig.3a) is divided into two parts, which were produced by D_1 and D_3 junctions. In this way, the maximal electric field around drain becomes lower than that of conventional SJ-LDMOST. It can be noticed from Fig.3b that the strength and area of electric field distribution near the source in N^+-Floating SJ-LDMOST are larger than that of conventional SJ-LDMOST. This fact is result from the depletion layer spreading into source region.

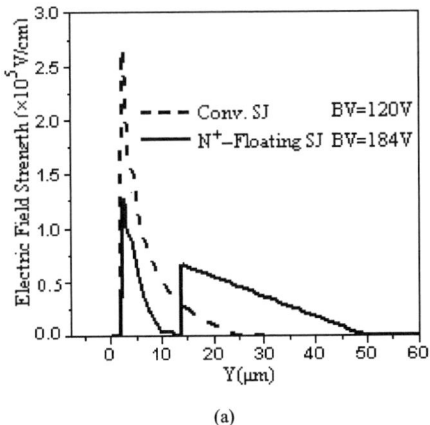

(a)

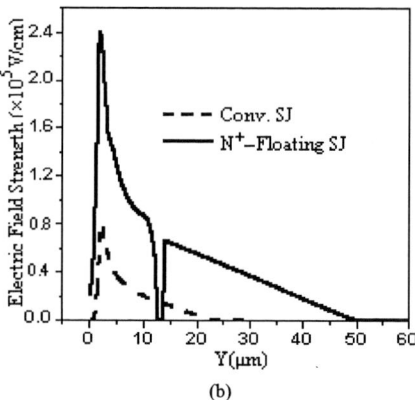

(b)

Fig. 3 Vertical electric field profiles (a) around the drain and (b) around the source for conventional and N^+-Floating SJ-LDMOS: h=1μm, L_d=10μm ,W=16μm

The effect of the doping imbalance on the BV for the N^+-Floating SJ-LDMOST and the conventional SJ-LDMOST are compared in Fig.4. The maximum breakdown voltage V_A in N^+-Floating SJ-LDMOST is increased drastically by $\triangle V_1$ compared with the V_B in the conventional SJ-LDMOS resulting from suppressing imbalance charge $\triangle Q$, and increased by $\triangle V_2$ compared with the V_C in the lateral unbalanced SJ-LDMOS [6] due to RESURF effect. The optimum W is applied in accordance with Figure.5 showing the relation the breakdown voltage and W in N^+-Floating SJ-LDMOS. It is clear that the breakdown voltage would be maximum when the REBULF effect is satisfied.

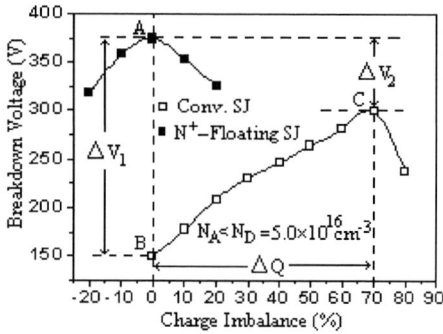

Fig.4 Breakdown Voltage versus doping imbalance for conventional and N^+-Floating SJ-LDMOS: h=1μm, L_d=30μm, W= 22μm

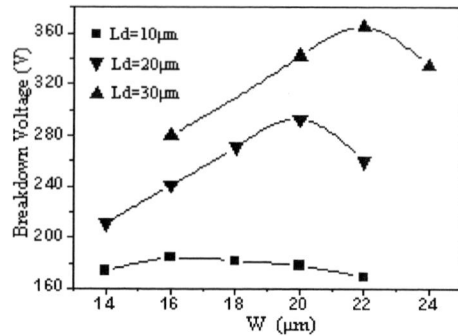

Fig.5 Breakdown Voltage versus W for N^+-Floating SJ-LDMOS: h=1μm

In general, the longer the drift region length is, the higher the breakdown voltage will be. But it should be noted that the device breakdown voltage would also saturate due to the saturation of the vertical breakdown voltage. This phenomenon is shown in Fig.6. In the N^+-Floating SJ-LDMOST structure, the saturated length of drift region is larger than that of the conventional SJ-LDMOST since the vertical breakdown voltage of N^+-Floating SJ-LDMOST is higher than that of conventional SJ-LDMOST thanks to REBULF effect. It's easily obtained from Fig.6 that the maximum breakdown voltages of N^+-Floating SJ-LDMOST reaches 417volts compared with that of 150volts in conventional SJ-LDMOST.

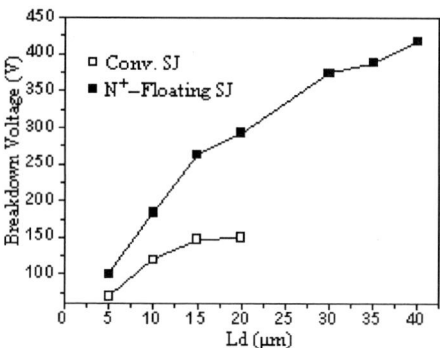

Fig.6 Breakdown Voltage Versus the length of drift region for conventional and N^+-Floating SJ-LDMOS: h=1μm

Figure.7 shows the on-state I_{DS}-V_{DS} characteristics of the conventional and N^+-Floating SJ-LDMOST with the various V_{GS} indicating that they have similar low on-resistance.

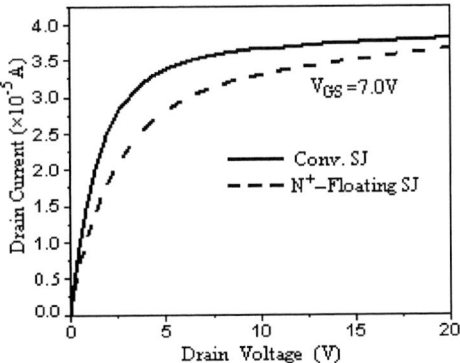

Fig.7 On-state characteristics of the conventional and N^+-Floating SJ-LDMOS: h=1μm, L_d=5μm, W=12μm

IV. Conclusions

A new SJ-LDMOS structure, which employs a N^+-Floating layer embedded in the high-resistance substrate, was analyzed. This structure eliminates charges imbalance effect to achieve high breakdown voltage while maintaining low on-resistance when the same length of drift region was applied. In addition, the maximum breakdown voltage in N^+-Floating SJ-LDMOS is higher than in conventional SJ-LDMOS due to suppressing charges imbalance and REBULF effect by N^+- Floating layer.

Acknowledgment

This work is partially supported by the projects of National Science Foundation of China (No.60436030 and No.60576052) and martial research foundation 514080609.

Reference

[1] F.Udrea, A. Popescu, and W. I. Milne, "The 3D RESURF double-gate MOSFET: A revolutionary power device concept," *Electron. Lett. Vol.* 34, pp. 808-809, 1998.

[2] L. Lorenz, G. Deboy, A. Knapp, and M. Marz, "COOL-MOSTM-a new milestone in high voltage power MOS," in *Proc. ISPSD*, pp. 3-10. 1999.

[3] Xing-Bi Chen and Johnny K. O. Sin, "Optimization of the Specific On-Resistance of the COOLMOS^{TM}," *IEEE Trans. Electron Devices, Vol.* 48, pp. 344-348, 2001.

[4] Sameh G. Nassif-Khalil and C. Andre T.Salama, "SJ/RESURF LDMOST," *IEEE Trans. Electron Devices, Vol.* 51, pp. 1185-1191, 2004.

[5] Sameh G. Nassif-Khalil and C. Andre T.Salama, "Super junction LDMOST in silicon-on-Sapphire technology (SJ-LDMOST)," Proc. *Int.Symp. Power Semiconductor Devices and ICs (ISPSD)*,

pp.81-84, 2002.

[6] R. Ng, F. Udrea, K. Sheng, K. Ueno, G. A. J. Amaratunga, and M. Nishiura, "Lateral unbalanced super junction (USJ)/3D-RESURF for high breakdown voltage on SOI," *Int. Symp Power Semiconductor Devices and ICs (ISPSD)*, pp.395-398, 2001.

[7] Il-Yong Park and C. Andre T. Salama, "CMOS Compatible Super Junction LDMOS with N-Buffer Layer," *Int. Symp. Power Semiconductor Devices and ICs (ISPSD)*, pp.163-166, 2005.

[8] Bo Zhang, Lin Chen, Jie Wu and Zhao Ji Li. SLOP-LDMOS–A Novel Super-Junction Concept LDMOS and Its Experimental Demonstration. *International conference on communications, circuits and system*, 2005. Volume 2, 1399-1402.

[9] Bo Zhang, Baoxing Duan and Zhaoji Li. Breakdown Voltage Analysis of REBULF LDMOS Structure with N^+-Floating Layer. Chinese Journal of Semiconductors, 2006. Volume 27(4), 730-734..

[10] ISE TCAD Manuals, release 8.5.

2006 5th International Power Electronics and Motion Control Conference

Unified Power Flow Controller: Comparison of Two Advanced Control Schemes and Performance Analysis for Power Flow Control

Liu Liming[1], Zhu Pengcheng[2], Kang Yong[3], Chen Jian[4], Senior Member, IEEE

Department of Applied Electronic Engineering, College of Electrical and Electronic Engineering, Huazhong University
of Science and Technology,1037 Luoyu Road, Wuhan, P. R. China, 430074
Email: [1]*hyt_llm@sohu.com*, [2]*zhu_pc1977@yahoo.com.cn*, [3]*Ykang@mail.hust.edu.cn*, [4]*Jchen@mail.hust.edu.cn*

Abstract—In this paper real, reactive power and voltage balance of the UPFC system is analyzed. Two advanced control schemes of UPFC have been proposed. One is the conventional coordination control scheme. In this scheme, the shunt converter of the UPFC controls the UPFC bus voltage and the DC-link voltage. The series converter controls the transmission line real and reactive power flow. By analysis, it is indicated that the shunt converter provides all the required reactive power during the power flow changes if the UPFC bus voltage is constant. The UPFC bus voltage can be control both from the sending side and from the receiving side. So a modified control scheme is presented. In this scheme, the shunt converter controls the transmission line reactive power flow and the DC-link voltage. The series converter controls the transmission line real power flow and the UPFC bus voltage. The conventional coordination controllers in the UPFC control system can get good performance, and the modified controller has an improved performance. Operation of UPFC using the proposed two control schemes is presented by experimental laboratory results.

Keywords- Unified Power Flow Controller, Coordination control scheme, Modified control scheme, Series converter, Shunt Converter

I. INTRODUCTION

The Unified Power Flow Controller (UPFC) is devised for real-time control and dynamic compensator of AC transmission systems. As the most comprehensive multivariable FACTS device, the UPFC is able to control, simultaneously or selectively, all the parameters affecting power flow in the transmission line (i.e., voltage, impedance, and phase angle). Alternatively, it can independently control both real and reactive power flow in the transmission line. With its multifunction, the control scheme of UPFC is complex and meets many problems. To solve these problems, some different types of control schemes are applied. In [1] the transient condition of a UPFC is analyzed and the DC-link capacitor design is discussed. In [2] a real and reactive power coordination controller has been designed. Although considerable UPFC research work has concentrated on developing

control system via simulation, there is a general lack of experimental verification of the proposed control schemes. In [3] a comprehend control scheme of both phase angle and cross-coupling control is proposed for power flow adjustment and gets good performance in a laboratory prototype. However, the analysis of the above control schemes is not comprehensive. On these work mentioned above, this paper has proposed and analyzed in detail a conventional coordination control scheme and a modified control scheme.

In contrast to the control schemes as listed in [1]-[8], the coordination control scheme is designed in detail in this paper. The shunt converter of the UPFC controls the UPFC bus voltage and the DC-link voltage. The series converter controls the transmission line real and reactive power flow. However, the shunt converter supplies increase/decrease reactive power in transmission line by analysis. So a modified control scheme of the UPFC can be developed. The shunt converter is controlled to make the DC-link voltage constant and output the reactive power for reactive power flow control, while the series converter is controlled to maintain the UPFC bus constant and adjust the transmission line real power flow. With this control scheme the UPFC bus voltage and transmission line reactive power can be controlled directly and a better performance is expected compared to conventional coordination control scheme.

In this paper, two advanced control schemes for UPFC that includes both the shunt converter and the series converter have been designed and their performance is evaluated. Section II gives the characteristic analysis of the UPFC system. Section III describes the conventional coordination control scheme for series and shunt converters of UPFC separately. Section IV describes the modified control scheme for series and shunt converters of UPFC separately. Section V provides experiment details of the UPFC and the results of the experimental works with the proposed two control systems. Brief conclusions are given in Section VI.

II. CHARACTERISTIC ANALYSIS OF UPFC

A. Voltage Balance of the UPFC system

Fig.1 shows the system configuration of a UPFC. The

1-4244-0448-7/06/$25.00 ©2006 IEEE

UPFC consists of two voltage source converters connected back to back with a common DC-link capacitor.

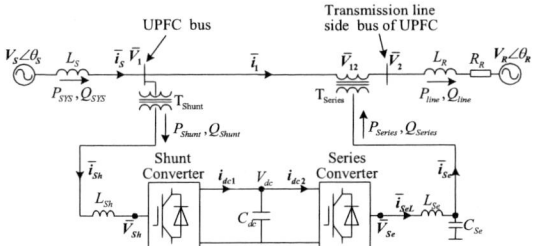

Fig.1 Configuration of a UPFC based on two back to back three-phase converters

According to Fig.1, the voltage balance can be established as below:

$$\overline{V_S} = \overline{V_1} + j\omega L_S \overline{i_S} \tag{1}$$

$$\overline{V_1} = \overline{V_{12}} + j\omega L_R \overline{i_1} + \overline{V_R} \tag{2}$$

If the d-axis is in phase with the UPFC bus voltage $\overline{V_1}$ and the q-axis is in quadrature with $\overline{V_1}$, a new voltage balance in synchronous rotating d-q frame can be expressed as follows:

$$\begin{cases} V_{Sd} = V_{1d} + L_S \, di_{Sd}/dt - \omega L_S i_{Sq} \\ V_{Sq} = L_S \, di_{Sq}/dt + \omega L_S i_{Sd} \end{cases} \tag{3}$$

$$\begin{cases} V_{1d} = V_{12d} + L_R \, di_{1d}/dt - \omega L_R i_{1q} + V_{Rd} \\ 0 = V_{12q} + L_R \, di_{1q}/dt + \omega L_R i_{1d} + V_{Rq} \end{cases} \tag{4}$$

where ω is utility angular frequency.

It can be inferred from (3) that i_{Sq} should not change to make V_{1d} constant. Because:

$$\overline{i_S} = \overline{i_1} + \overline{i_{Sh}} \tag{5}$$

And in the same d-q frame:

$$\begin{cases} i_{Sd} = i_{1d} + i_{Shd} \\ i_{Sq} = i_{1q} + i_{Shq} \end{cases} \tag{6}$$

So to make i_{Sq} constant when i_{1q} varies, it should meet the condition below:

$$\Delta i_{1q} + \Delta i_{Shq} = 0 \tag{7}$$

It means the shunt converter should provide the transmission line reactive power variation.

B. Real Power Balance Analysis

For real power balance:

$$P_{SYS} = P_{Shunt} - P_{Series} + P_{Line} \tag{8}$$

where P_{SYS} is the sending end output real power;

P_{Shunt} is the real power absorbed by the shunt converter;

P_{Series} is the real power output by the series converter;

P_{Line} is the real power flow from the UPFC to the receiving end.

According to the instant power theory, (8) can be transformed to:

$$\overline{V_S} \bullet \overline{i_S} = \overline{V_1} \bullet \overline{i_{Sh}} + \overline{V_{12}} \bullet \overline{i_1} + \overline{V_2} \bullet \overline{i_1} \tag{9}$$

where ' $\bullet$ ' means dot product of two vector.

To make the DC-link voltage stable, it should meet the condition in (10):

$$P_{Shunt} - P_{Series} = \overline{V_1} \bullet \overline{i_{Sh}} + \overline{V_{12}} \bullet \overline{i_1} = P_{dc} \tag{10}$$

where P_{dc} is the loss of the UPFC conversion system.

It can be seen from (8) to (10) that the sending end power source provides all the transmission line real power and system loss.

According to (8) and (9), it can be gotten:

$$P_{Line} = \overline{V_2} \bullet \overline{i_1} = \left(\overline{V_1} - \overline{V_{12}} \right) \bullet \left(\frac{\overline{V_1} - \overline{V_{12}} - \overline{V_R}}{j\omega L_R} \right) \tag{11}$$

And (11) can be simplified as:

$$P_{Line} = \left(\overline{V_1} - \overline{V_{12}} \right) \bullet \left(-\frac{\overline{V_R}}{j\omega L_R} \right) \tag{12}$$

If $\overline{V_1}$ and $\overline{V_R}$ do not vary, then the transmission line real power variation ΔP_{Line} in synchronous rotating d-q frame can be given as follows:

$$\Delta P_{Line} = \frac{\Delta V_{12d} V_{Rq}}{\omega L_R} - \frac{V_{Rd}}{\omega L_R} \left[\omega L_S \left(\Delta i_{1d} + \Delta i_{Shd} \right) + \Delta V_{12q} \right] \tag{13}$$

Equation (13) indicates that the series converter output voltage $\overline{V_{12}}$ can be utilized to control ΔP_{Line}. Because (4) indicates that ΔV_{12d} will cause variation of the amplitude of V_1, ΔV_{12q} is controlled to regulate ΔP_{Line}.

On the other hand, (10) can be transformed to (14) in the above mentioned d-q frame.

$$V_{1d} i_{Shd} = i_{dc2} V_{dc} + P_{dc} \tag{14}$$

where $\overline{V_{12}} \bullet \overline{i_1} = -i_{dc2} V_{dc}$ is the real power absorbed by the series converter. By determined by (14), i_{Shd} should vary with i_{dc2} to meet the active power balance of the UPFC. So a forward component including i_{dc2} can be added to the DC-link voltage controller of UPFC and make up a coordination controller.

C. Reactive Power Balance Analysis

According to Fig.1, the reactive power balance can be written as below.

$$Q_{SYS} - Q_{Xs} - Q_{Shunt} + Q_{Series} = Q_{Line} \tag{15}$$

where Q_{SYS} is the sending end output reactive power;

Q_{Xs} is the reactive power consumed by the sending transmission line;

Q_{Shunt} is the reactive power absorbed by the shunt converter;

Q_{Series} is the reactive power output by the series converter;

Q_{Line} is the reactive power flow from the UPFC to the receiving end.

According to instantaneous power theory, (15) can be transferred to (16)

$$\overline{i_S} \times \overline{V_S} = \overline{i_S} \times j\omega L_S \overline{i_S} + \overline{i_{Sh}} \times \overline{V_1} + \overline{i_1} \times \overline{V_{12}} + \overline{i_1} \times \overline{V_2} \qquad (16)$$

Through the same transformation as mentioned above, (16) can be transformed to (17):

$$i_{Shq}V_{1d} = i_{Sq}V_{1d} + i_{1d}V_{12q} - i_{1q}V_{12d} + Q_{Line} \qquad (17)$$

When there is power flow variation in the transmission line, (17) can be transformed to (18) by small signal analysis and (7):

$$\Delta i_{Shq}V_{1d} = \Delta i_{1d}V_{12q} + i_{1d}\Delta V_{12q} - \Delta i_{1q}V_{12d} - i_{1q}\Delta V_{12d} + \Delta Q_{Line} \qquad (18)$$

Equation (18) indicates that the reactive power flow in transmission line can be controlled by ΔV_{12d} or Δi_{Shq}.

D. Discussion

It can be concluded from the analysis above that there are two kinds of solution to reach the control goals of getting proper power flow and maintaining the UPFC bus voltage constant. One conventional method is to control the real power flow with V_{12q} and reactive power flow with V_{12d}, while maintain the UPFC bus voltage V_{1d} constant by adjusting the shunt converter output reactive current i_{Shq}. The real current i_{Shd} absorbed by shunt converter is controlled to make UPFC DC-link voltage constant and maintain the real power balance of the UPFC. The other modified method is that V_{12d} is controlled to maintain UPFC bus voltage V_{1d} constant, and i_{Shd} is controlled to regulate the reactive power in the transmission line. The rest of the modified method is the same as the traditional one. Details of the proposed two control schemes are given in the next two sections.

III. THE COORDINATION CONTROL SCHEME OF UPFC

A. Shunt Converter Control System

In the conventional control scheme, the shunt converter controls the DC-link voltage and UPFC bus voltage. The shunt converter is operated in the decoupled control and coordination control system. The control block for the shunt converter is presented in Fig.2. The control system is based on the synchronous rotating d-q frame as mentioned above. The control scheme is composed of two loops, the outer loop is voltage loop, and the inner loop is current loop. The d-axis component of shunt converter output voltage controls the DC-link voltage. The q-axis component controls the UPFC bus voltage. The shunt converter control system includes the coordination feedback between the series and the shunt converters. In the d-axis decoupled control, the equivalent d-axis additional current signal i_{dc2} is fed to inner control system with a gain $2V_{dc}/(3V_{1d})$. With the aid of two PI regulators, the references value (i_{shd}^{*}) for the internal current control loop can be obtained. Further, the shunt converter can provide enough real power demand by series converter and maintain DC-link voltage constant. In the q-axis decoupled control, the transmission line reactive power flow reference (Q^{*}) is added to outer loop output with a gain $2V_{1d}/3$. With the aid of two PI regulators, the references values (i_{shq}^{*}) for the internal current control loop can be obtained. So the shunt converter can generate appropriate amount of reactive power to maintain the UPFC bus voltage. The real and reactive power coordination between the series and shunt converter is applied in the control scheme to get better dynamic performance.

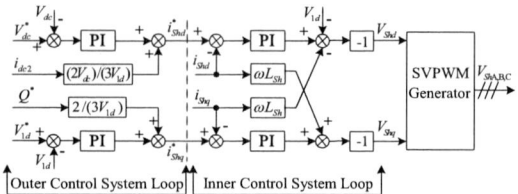

Fig.2 Shunt converter control system of UPFC in coordination control scheme

B. Series Converter Control System

The series converter provides simultaneous control of the transmission line real/reactive power flow. To do so, the series converter injected voltage is decomposed into two components in the above mentioned d-q frame. The d-axis component controls the transmission line reactive power flow. The q-axis component regulates the transmission line real power flow. Considering the variation of the parameters of the transmission line, a cross-coupling controller is designed. The control block of series converter is shown in Fig.3. The control system is composed of three loops, the power loop is outer loop, the voltage loop is middle loop and the current loop is inner loop. Every loop includes the corresponding state feedback decoupling terms. The control scheme can eliminate the inference which exists in voltage and current between the d-axis and the q-axis and has very good robust performance.

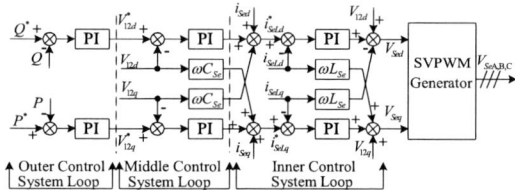

Fig.3 Series converter control system of UPFC in coordination control scheme

IV. THE MODIFIED CONTROL SCHEME OF UPFC

A. Shunt Converter Control System

In this modified control scheme, the shunt converter controls the DC-link voltage and the reactive power flow in the transmission line. The control diagram for the shunt converter is shown in Fig.4. The control system is composed of two loops, the outer loop and the inner loop. The outer loop generates the current reference, while the inner current loop make the shunt converter output the needed currents. The control system is based on the above mentioned d-q frame. The d-axis current is controlled to manage the DC-link voltage and balance the real power of the UPFC device. The q-axis current is controlled to manage the transmission line reactive power flow. The d-axis control scheme is the same as one in shunt converter control system of the above coordination control scheme. In the q-axis control scheme, the transmission line reactive power flow reference is added to the reactive power flow PI regulator output with a gain $2/(3V_{1d})$. So the shunt converter can fast generate appropriate amount of reactive power to compensate the reactive power flow change.

76

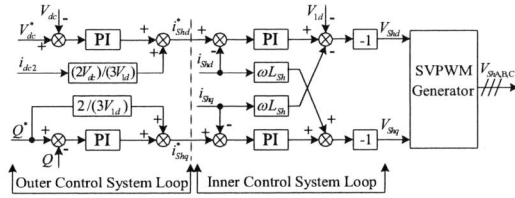

Fig.4 Shunt converter control system of UPFC in modified control scheme

B. Series Converter Control System

The series converter provides simultaneous control of UPFC bus voltage and the transmission line real power flow. To do so, the series converter injected voltage is decomposed into two components. The d-axis component controls the UPFC bus voltage. The q-axis component controls the transmission line real power flow. The detail of this control scheme is shown in Fig.5. The control system is composed of three loops. The outer loop generates the reference voltage signals for the middle loop. A current feed forward is added to the d-axis outer loop to compensate the voltage changes across the line impedance when there is a power flow change happening. The middle loop and inner loop are the same in above coordination control scheme and the modified control scheme. The UPFC bus voltage can be controlled directly with this method.

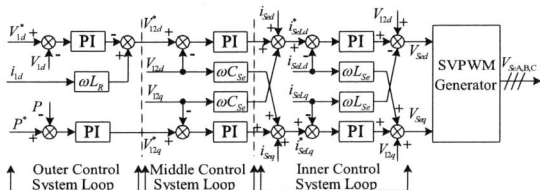

Fig.5 Series converter control system of UPFC in modified control scheme

V. EXPERIMENTAL RESULTS AND ANALYS

An experimental UPFC system rated at 15kVA has been constructed to verify the proposed control schemes as shown in Fig.1. The parameters of the whole system are shown in Table I. All the experimental result waveforms are recorded by a Yokogawa DL750 scope-recorder. The objectives for this experiment include the transmission line real power flow (P_{Line}), the transmission line reactive power flow (Q_{Line}), shunt converter real power (P_{Shunt}), shunt converter reactive power (Q_{Shunt}), series converter real power (P_{Series}), series converter reactive power (Q_{Series}), UPFC DC-link voltage (V_{dc}), UPFC bus voltage of phase A (V_{1A}), the transmission line voltage of phase A controlled by series converter (V_{2A}) respectively with UPFC in proposed two control schemes. The experimental results are carried out to confirm the effectiveness of UPFC and are analyzed in the following.

A. Power flow control performance of the UPFC in coordination control mode

The performance of the UPFC for real and reactive power control is demonstrated with the objectives of keeping the sending-end and receiving-end bus voltage and operating the UPFC in coordination control mode. As established previously, in this operating mode the UPFC regulates the real and reactive line power to given

TABLE I.
SYSTEM PARAMETERS OF THE UPFC

Sending End Voltage V_S	380V (line-line)
Receiving End Voltage V_R	380V (line-line)
The Transmission Angle θ_S-θ_R=δ	10 degree
Shunt Transformer Turning Ratio	2.5:1($Y-\Delta$)
Series Transformer Turning Ratio	6:8($Y-\Delta$)
Sending End Line Inductance L_S	18mH
Transmission Line Inductance L_R	60mH
Shunt Converter Output Filtering Inductance L_{Sh}	6mH
Series Converter Output Filtering Inductance L_{Se}	1mH
Series Converter Output Filtering Capacitor C_{Se}	10uF
DC-link Capacitor C_{dc}	9400uF
DC-link Voltage V_{dc}	400V

reference values. As illustrated in Fig.6, the initial real and reactive power flow (P_{Line}, Q_{Line}) in the transmission line are 1200W and 100Var respectively for the transmission angle δ=10° between the sending-end and receiving-end bus voltage source. Then the UPFC is instructed to perform a series step changes in rapid succession. First, P_{Line} is increased to +3kW in about 300ms, then Q_{Line} is increased to +3kVar, followed by a series of decreases, ending with zero value of P_{Line} and Q_{Line}. The waveforms illustrate clearly the operation of the UPFC. It can be seen from Fig.6 that the transmission line reactive power flow Q_{Line} is not affected significantly during the step changes in the transmission real power references. The series converter injected voltage makes the voltage V_{2A} leads V_{1A}. So the real power flows from the sending-end voltage source V_S to the receiving-end voltage source V_R. In the same way, the transmission line real power flow P_{Line} is not affected significantly during the step changes in the transmission reactive power references. It has also shown that the decrease/increase in the transmission line reactive power is balanced by an equal decrease/increase in the shunt converter reactive power (Q_{Shunt}) and series converter reactive power (Q_{Series}). So the V_{1A} which is controlled by the shunt converter does not deviate significantly from its reference value. In fact, the increase/decrease in the UPFC bus voltage causes the shunt converter to consume/absorb reactive power and bring the UPFC bus voltage back to its reference value. In addition, V_{dc} is controlled to 400V by the shunt converter.

B. Power flow control performance of the UPFC in modified control mode

Fig.7 shows UPFC in modified control scheme controls the power flow changes above. The initial real and reactive power flow (P_{Line}, Q_{Line}) in the transmission line are the same as those in coordination control mode. Then the UPFC is instructed to perform a series step changes in rapid succession. First, P_{Line} is increased to +3kW, then Q_{Line} is increased to +3kVar in about 100ms, followed by a series of decreases, ending with zero value of P_{Line} and Q_{Line}. The waveforms illustrate clearly the operation of the UPFC. It can be seen from Fig.7 that the transmission line reactive power flow Q_{Line} is not affected significantly during the step changes in the transmission real power references. Similarly the transmission line real power flow P_{Line} is not affected significantly during the step changes

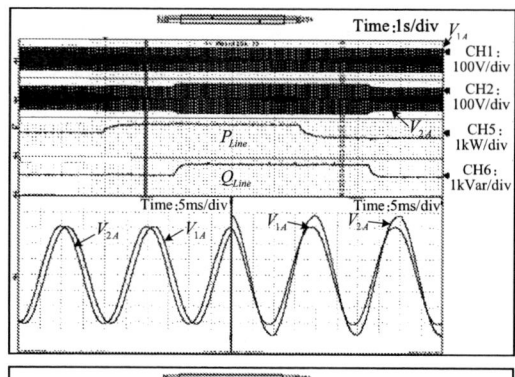

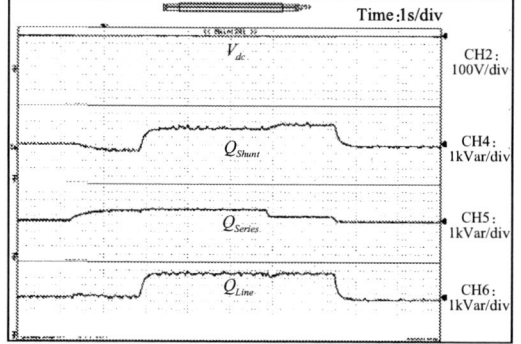

Fig.6 Performance of UPFC in coordination control mode for step changes in the line real power demands and reactive power demands

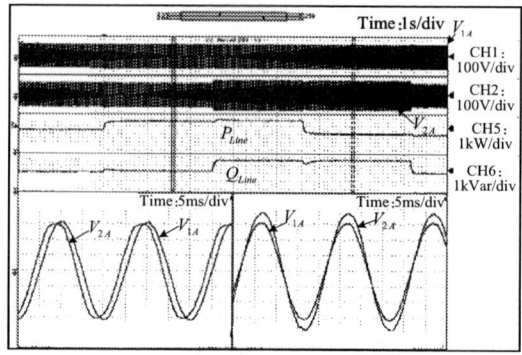

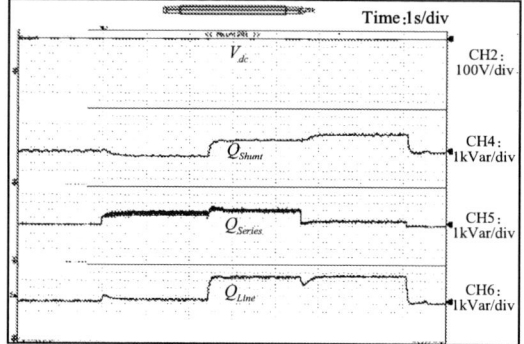

Fig.7 Performance of UPFC in modified control mode for step changes in the line real power demands and reactive power demands

in the transmission reactive power references. It has shown that the shunt converter almost supplies the whole decrease/increase in the transmission line reactive power.

Fig.7 has brought out the decoupled nature of the modified control scheme. Future, it has shown that it is possible to control the transmission line reactive power flow through the shunt converter. In addition, it has also shown that the UPFC bus voltage which is controlled by the series converter and UPFC DC-link voltage (V_{dc}) which is controlled by the shunt converter do not significantly deviate much from their reference value. It can be observed that with the modified control scheme the reactive power flow can get a better dynamic performance than real power flow, and that is the advantages compare to the coordination control scheme.

VI. CONCLUSION

The paper proposes two advanced control schemes of UPFC and analyzes dynamic performance of a UPFC in detail. In conventional coordination control scheme, the shunt converter controls the UPFC bus voltage and the DC-link voltage. The series converter controls the transmission line real and reactive power flow. In the modified control scheme, the shunt converter controls the DC-link voltage and regulates the transmission line reactive power flow, while the series converter controls the UPFC bus voltage and adjusts the transmission line real power flow. With this control scheme the UPFC bus voltage and transmission line reactive power can be controlled directly and a better performance is expected compared to conventional coordination control scheme.

The experimental results have verified that the proposed two control schemes are effective and feasible.

Further investigations of dynamic performance of a UPFC are necessary for a more complicated power system.

ACKNOWLEDGMENT

The author would like to thank Dr. Pengcheng Zhu at Huazhong University of Science and Technology, for his contribution to this work.

REFERENCES

[1] Hideaki Fujita, Yasuhiro Watanabe and Hirofumi Akagi "Transient Analysis of a Unified Power Flow Controller and its Application to Design of the DC-Link Capacitor," *IEEE* Power Electronics. *Trans.* Vol. 16, No.5, pp. 735–740, September 2001.

[2] S.Kannan, Shesha Jayaram, M.M.A.Salama "Real and Reactive power Coordination for a Unified Power Flow Controller," *IEEE* Power Systems. *Trans.* Vol.19,No.3, pp. 1454–1461, August 2004.

[3] Hideaki Fujita, Yasuhiro Watanabe and Hirofumi Akagi "Control and Analysis of a Unified Power Flow Controller a," *IEEE Power* Electronics. *Trans.* Vol. 14, No.6, pp. 1021–1027, November 1999.

[4] Yu Q., Round S., Norum L., Undeland T., "Dynamic Control of A Unified Power Flow Controller" *Proc. Of IEEE PESC'96*, pp.508-514, 1996

[5] I. Papič, P.Žunko, D.Povh, M.Weinhold, "Basic Control of Unified Power Flow Controller," *IEEE Trans. on Power Systems*, Vol. 12, No. 4, pp.1734-1739, Nov.1997

[6] K.R.Padiyar, K.Uma Rao, "Modeling and Control of Unified Power Flow Controller for Transient Stability," *Electrical power & Energy Systems*, Vol.3, pp.1-11, 1999.

[7] L. Dong, M.L.Crow, Z.Yang, C. Shen, L.Zhang and S.Aticitty "A Reconfigurable FACTS System for University Laboratories," *IEEE* Power Systems. *Trans.* Vol. 19, No.1, pp. 120–128, February 2004.

[8] Kataoka T., Hisa Y., Uchida Y., Ikeda Y., "A control strategy for a UPFC-based fast voltage compensator and its performance analysis," *2004 IEEE 35th Annual Power Electronics Specialists Conference*, Vol.2, pp.1513 – 1518, June 2004.

2006 5th International Power Electronics and Motion Control Conference

A New Analytical Model for the Surface Electrical Field Distribution of Double RESURF LDMOS

Qi Li*, Zhaoji Li

IC design Center, university of Electronic Science & Technology, Chengdu, 610054, China
*Email : lqphoenix@sina.com

Abstract—In this paper, a new analytical model for the surface electrical field distributions of double RESURF LDMOS is presented. Based on the 2-D Poisson solution, the model gives the influence on the surface electrical field in terms of the drain bias and structure parameters, such as the doping concentration, the depth and the position of the P-top region, the thickness and the doping concentration of the drift region and the substrate doping concentration; the dependence of breakdown voltage on the length of drift region is calculated. Further, an effectual way to gain the optimum high-voltage is also proposed. All analytical results are well verified by simulation results obtained by MEDICI and previous experimental data, showing the validity of the model presented here.

Keywords-Double RESURF; Surface electrical field; Model

I. INTRODUCTION

In recent years, RESURF (Reduced Surface Field) technology has been widely used in power integrated circuits devices[1~2]. Double RESURF is one of the most utilized methods to design high voltage device with low on-resistance. Many researches on numerical simulation and experimental results show that high breakdown voltage can be maintained while the drift region doping concentration is increased by twice as much as that in single RESURF devices[3~7]. Previously several analytical models have been introduced[8~11], but are all for the particular structures, providing little information about breakdown phenomena of bulk-silicon double RESURF device. To the best of our knowledge, there has not been any 2-D bulk-silicon analytical solution for the surface electrical field of double RESURF structure so for.

This work is to develop the 2-D analytical model for the surface electrical field and potential distributions of the bulk-silicon double RESURF devices based on the Poisson solution. The analytical results of the presented models show a good agreement with the numerical simulation results obtained by MEDICI. The dependence of the surface electrical field on the bias and structure parameters has been discussed in detail. The proposed models will be helpful for the designers to provide accurate first-order design schemes and afford an effective

way to improve the performance of the high voltage bulk-silicon double RESURF device.

II. ANALYTICAL MODLE

A schematic cross-section of the bulk- silicon double RESURF device is shown in Fig.1, where x measures the horizontal position relative to the left edge of the double diffused P$^+$N junction while y measures the vertical position relative to the surface. The drift region thickness is t_e with a uniform doping concentration of N_e, whereas P_{top} and t_{top} are the doping concentration and depth of the P-top region, negative concentration denotes P-type doping, and positive concentration is N-type doping. The whole drift region is divided into four regions along

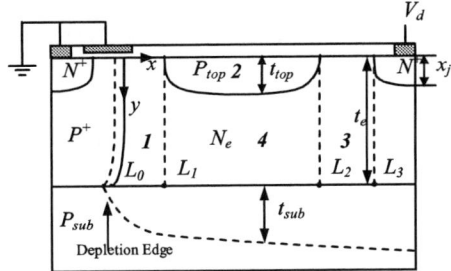

Fig.1. Cross-section of the double RESURF device

the edges of P-top region and the boundary positions of them are given by x=0, L_1, L_2, L_3 and y=0, t_{top}, t_e. The substrate depletion layer thickness is t_{sub} with doping concentration P_{sub}. The dielectric constant of bulk-silicon is ε_{si}. The device is biased in the off-state configuration; substrate, source and gate are grounded while the drain is biased to a positive voltage V_d, The potential function $\varphi_i(x,y)$ in the silicon film must be satisfied by 2-D Poisson's equation given by

$$\frac{\partial^2 \varphi_i(x,y)}{\partial x^2} + \frac{\partial^2 \varphi_i(x,y)}{\partial y^2} = -\frac{qN_i}{\varepsilon_{si}} \quad , i=1,2,3,4 \quad (1)$$

Assuming that the drift region is completely depleted and the potential functions $\varphi_i(x,y)$ can be approximated by two-order Taylor expanded formula of

1-4244-0448-7/06/$25.00 ©2006 IEEE

$$\varphi_i(x,y)=\varphi_i(x,0)+\frac{\partial \varphi_i(x,0)}{\partial x}y+\frac{\partial^2 \varphi_i(x,0)}{2\partial x^2}y^2 \quad ,i=1,2,3 \quad (2)$$

$$\varphi_4(x,y)=\varphi_4(x,t_{top})+\frac{\partial \varphi_4(x,t_{top})}{\partial x}(y-t_{top})+\frac{\partial^2 \varphi_4(x,t_{top})}{2\partial x^2}(y-t_{top})^2 \quad (3)$$

The boundary conditions for the potential functions are

$$\left.\frac{\partial \varphi_i(x,y)}{\partial y}\right|_{y=0}=0 \quad ,i=1,2,3 \qquad (4)$$

$$\left.\frac{\partial \varphi_i(x,y)}{\partial y}\right|_{y=t_e}=-\frac{2\varphi_i(x,t_e)}{t_{sub}} \quad ,i=1,3,4 \qquad (5)$$

$$\varphi_2(x,t_{top})=\varphi_4(x,t_{top}), \quad \left.\frac{\partial \varphi_2(x,y)}{\partial y}\right|_{y=t_{top}}=\left.\frac{\partial \varphi_4(x,y)}{\partial y}\right|_{y=t_{top}} \quad (6)$$

$$\varphi_1(L_1,0)=\varphi_2(L_1,0) \quad , \quad \left.\frac{\partial \varphi_1(L_1,y)}{\partial x}\right|_{y=0}=\left.\frac{\partial \varphi_2(L_1,y)}{\partial x}\right|_{y=0} \quad (7)$$

$$\varphi_2(L_2,0)=\varphi_3(L_2,0) \quad , \quad \left.\frac{\partial \varphi_2(L_2,y)}{\partial x}\right|_{y=0}=\left.\frac{\partial \varphi_3(L_2,y)}{\partial x}\right|_{y=0} \quad (8)$$

$$\varphi_1(0,0)=0, \qquad \varphi_3(L_3,0)=V_d \qquad (9)$$

where $N_1=N_3=N_4=N_e,N_2=P_{top}$, (4) is assumed that the electrical field at the surface may be minimized[12], (5) is obtained from a linear field variation along the vertical direction within the substrate depletion thickness t_{sub}, (6)-(8) are the continuity of the potential and electrical field along the boundary of 1-2, 2-3 and 2-4 regions, respectively, (9) is the voltage condition applied on the device. Substituting (2) and (3) into (1) under boundary conditions (4)-(6) leads to a general differential equation for the potential distribution function along the surface as

$$\frac{\partial^2 \varphi_i(x,0)}{\partial x^2}+\frac{\varphi_i(x,0)}{t^2}=-\frac{qN_{eff}^i}{\varepsilon_s}, \quad i=1,2,3 \qquad (10)$$

where $\quad N_{eff}^1=N_{eff}^3=N_e \quad , \quad t=\sqrt{\dfrac{t_e^2+t_e t_{sub}}{2}} \quad$ and

$$N_{eff}^2=(-P_{top}+\frac{P_{top}+N_e}{t_e^2+t_e t_{sub}}((t_e-t_{top})^2+t_{sub}(t_e-t_{top})))$$

We assume that t_{sub} is a constant at the first-order approximation, and on the general formulae for the double-sided junction may be taken as:

$$t_{sub}=\frac{1}{2}(\sqrt{(1+\frac{N_{eff}^3}{P_{sub}})t_e^2+\frac{2\varepsilon_s V_d}{qP_{sub}}}-t_e)$$

Solving (10) with the boundary condition $\varphi_1(L_1,0)=V_1$, $\varphi_2(L_2,0)=V_2$ and $\varphi_3(L_3,0)=V_3$ gives surface potential $\varphi_i(x,0)$ and electrical field $E_i(x,0)$ as:

$$\varphi_i(x,0)=\frac{qN_{eff}^i t^2}{\varepsilon_s}+(V_i-\frac{qN_{eff}^i t^2}{\varepsilon_s})\frac{\sinh((x-L_{i-1})/t)}{\sinh((L_i-L_{i-1})/t)}$$
$$+(V_{i-1}-\frac{qN_{eff}^i t^2}{\varepsilon_s})\frac{\sinh((L_i-x)/t)}{\sinh((L_i-L_{i-1})/t)}, \quad (L_{i-1}\le x<L_i) \quad (11)$$

$$E_i(x,0)=(V_i-\frac{qN_{eff}^i t^2}{\varepsilon_s})\frac{\cosh((x-L_{i-1})/t)}{t\sinh((L_i-L_{i-1})/t)}$$
$$-(V_{i-1}-\frac{qN_{eff}^i t^2}{\varepsilon_s})\frac{\cosh((L_i-x)/t)}{t\sinh((L_i-L_{i-1})/t)},(L_{i-1}\le x<L_i) \quad (12)$$

where i=1,2 and 3 are applied, V_1 and V_2 are the surface potential of two boundary between P-top and drift regions. The surface potential $\varphi_i(x,0)$ and surface electrical field $E_i(x,0)$ are now obtained from (11) and (12) respectively by finding V_1 and V_2 for given $V_0=0$ and $V_3=V_d$ using the continuity condition (7) and (8). Using $P_{top}=N_e$ in (12), the surface electrical field of single RESURF device can be obtained by

$$E(x,0)=(V_d-\frac{qN_e t^2}{\varepsilon_s})\frac{\cosh(x/t)}{t\sinh(L_3/t)}$$
$$+\frac{qN_e t}{\varepsilon_s}\frac{\cosh((L_3-x)/t)}{\sinh(L_3/t)} \quad ,(0\le x<L_3) \qquad (13)$$

III. RESULTS AND DISCUSSION

In order to verify the proposed model, the 2-D device simulation is performed using MEDICI for the same structure. In the following discussion, the curves denote the analytical results and the black points represent the numerical results, respectively.

Fig.2 shows the surface potential and electrical field distributions of single RESURF and double RESURF devices. A fair accordance between the analytical and numerical results may generally be found. The discrepancies between them are due to the penetration of the space charge region between two regions with different doping concentration in x=L_0, L_1, L_2, L_3. However, this kind of discrepancy has little effect on the

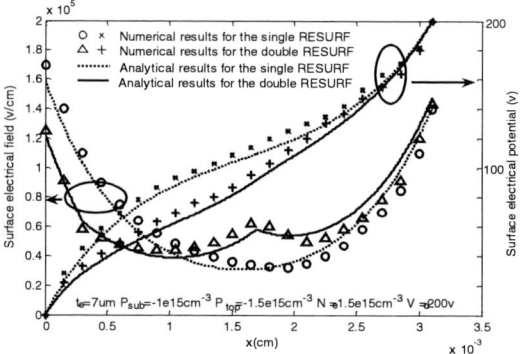

Fig.2 Surface potential and electrical field distributions of single RESURF and double RESURF devices

breakdown voltage analysis. One can see that a new electrical field peak appears at x=L_2 in the double RESURF device compared to the single RESURF structure. Because of the incorporation of the P-top region inside the drift region, the peaks of electrical field of double RESURF is decreased at x=L_0 while increased little at x=L_3. The potential of double RESURF is distributed linearly in most drift region, but the potential distribution of single RESURF shows a large curvature in the whole drift region to lead to a non-uniform surface field profile that may cause the degradation of breakdown voltage.

Fig.3 illustrates the surface electrical distributions for the different doping concentrations, thicknesses and positions of P-top region. It is evident that there are three surface electrical field peaks, which appear at x=L_0, L_2 and L_3, respectively and strongly depend on the parameters of the P-top region. In Fig.3 (a) and (b), the P_{top} and t_{top} have the same effect on the surface electrical field distribution. With an increase in the P_{top} or t_{top}, the electrical field peaks decrease at x=L_0 and increase at x=L_2, L_3, and there exists a electrical field value that is fixed between x =L_1 and L_2. It means that the highest electrical field may move from x=L_3 to x=L_0 with the decrease in P_{top} or t_{top}, which is responsible for the change of the potential distribution.

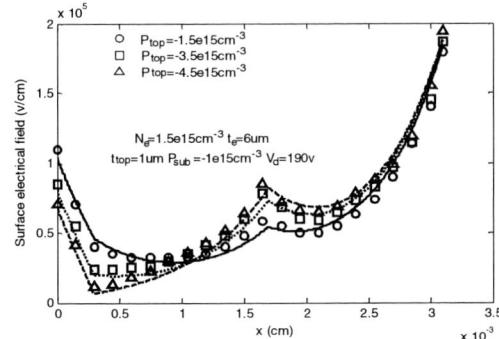

(a)

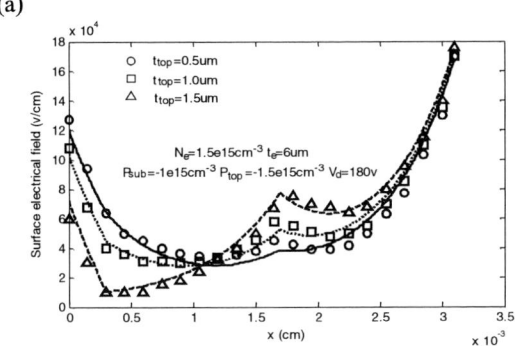

(b)

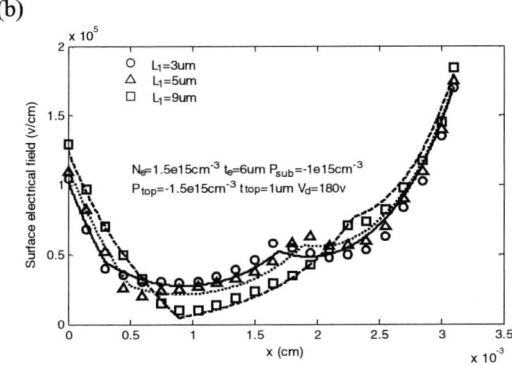

(c)

Fig.3 Surface electrical field distributions of double RESURF device along the drift region (a) for different P_{top} (b) for different t_{top} (c) for different L_1

Therefore the breakdown point moves from x=L_3 to x=L_0 as the P_{top} or t_{top} decreases below a critical value, at which the two electrical field peaks (x=L_0, L_3) equal each other. In Fig.3(c), the position of the electrical field peaks at x=L_2 moves with the change of the P-top region position and the positions of other electrical field peaks does not change. Because P-top region' charge can restrain the peaks at x=L_0, but enhance that of other place, one can find that with the increase of L_1, the value of electrical fields at x=L_0, L_2 and L_3 increase but the values of the electrical field decrease at x= L_1. So in order to obtain the ideal electrical field distribution at which the maximum breakdown is realized a smaller L_1 is required.

Fig.4 demonstrates the surface electrical field distributions for different drift region doping concentration, thickness and substrate doping

concentration. In Fig.4 (a) and (b), with the increase of the N_e or T_e, the magnitude of the electrical field rises at x=L_0, L_1, and reduces at x=L_3, but is fixed at x=L_2, so the maximum peak field point may translates from x=L_3 to x=L_0. The maximum breakdown point will change with the change of the position of the highest electrical field peak value. When the three electrical field peaks (x=L_0, L_2, L_3) have a uniform value that must be less than the critical value the maximum breakdown voltage will appear. In

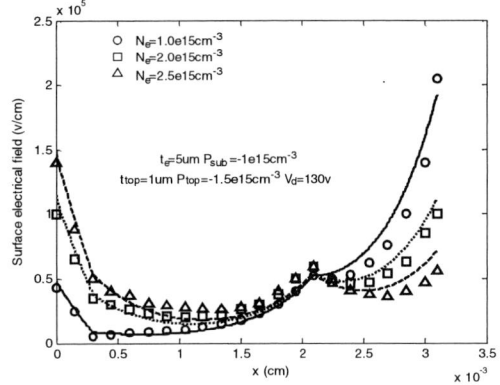

(a)

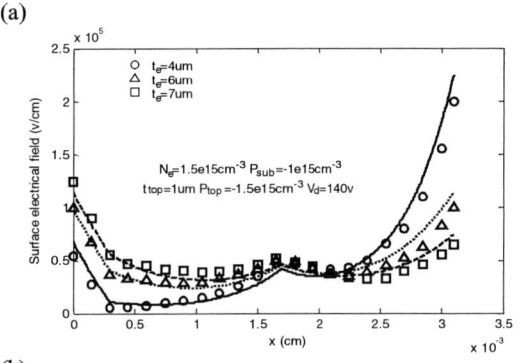

(b)

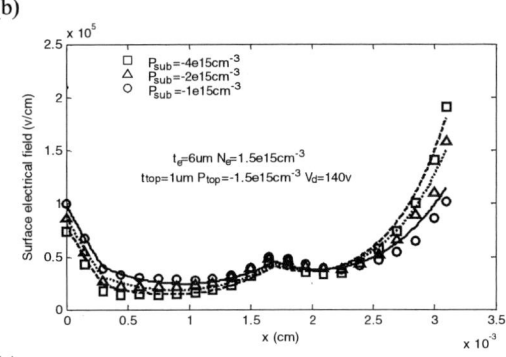

(c)

Fig.4 Surface electrical field distributions of double RESURF device along the drift region (a) for different drift region doping concentration (b) for different drift region thickness (c) for different substrate doping concentration

contrast, the substrate doing concentration of the substrate makes the magnitude of the peak field reduce at at x=L_0 and rise at x=L_3 with the increase in the substrate doping concentration. In the optimization of the devices, the substrate doping concentration is a very important parameter, which to a large extent determines the position and amplitude of the maximum peak electrical field.

The breakdown voltage of double RESURF device is determined by the minimum of the lateral breakdown voltage BV_{lat} and the vertical breakdown voltage BV_{ver} While BV_{lat}, due to the surface electrical field $E_i(x,0)$ can be calculated from the avalanche breakdown condition of the critical electrical field concept: the maximum surface electrical field reaches the critical value E_{clat}, that can be given as: $Max[E_i(x,0)] \leq E_{clat} = \dfrac{A}{1 - B \lg \dfrac{N_e}{10^{16}}}$, where

$A \cong 3.1 \times 10^5 V/cm, B \cong 0.5$ are related to the ionization rates[13]. BV_{ver} is determined by a solution of ionization integral with ionization rates for an abrupt two-sided p-n junction[14]:

$$BV_{ver} = 5.238 \times 10^{13} \left(\frac{P_{sub}}{1 + \frac{P_{sub}}{N_e}}\right)^{\frac{3}{4}} [1 - \frac{(1 - \frac{t_e N_e}{T_y P_{sub}})^2}{1 + \frac{N_e}{P_{sub}}}]$$

$$\text{with } T_y \approx [\frac{1.8 \times 10^{-35}}{8}(\frac{q P_{sub}}{\varepsilon_s})^7 (1 + \frac{P_{sub}}{N_e})]^{-\frac{1}{8}}$$

Analytical results for the breakdown voltage are showed in Fig.5 with simulation results as a function of

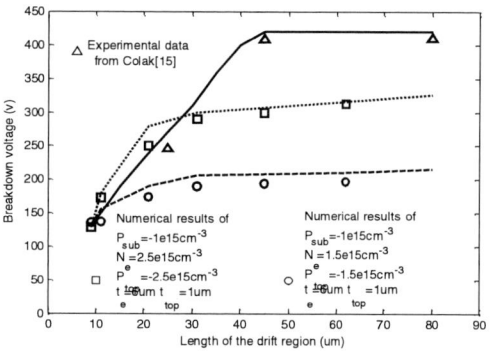

Fig.5 Breakdown voltage as a function of drift region length

drift region length for set of device parameters and compared with the experimental data obtained by Colak[15], one can observe both show good accordance. For a given set of parameters, with the increase of the drift region length, breakdown voltage reaches up to a constant value BV_{ver} limited by the vertical p-n junction breakdown.

IV. CONCLUSION

In this paper, the analytical model for the surface field distribution of double RESURF LDMOS has been studied. The dependence of the surface electrical field and potential distributions on the thickness and doping concentration of the drift region, depth, doping concentration and position of the P-top region and substrate doping concentration have been discussed. All

analytical results have been shown to be in agreement with the results obtained by the MEDICI simulation and previous experimental data. The analytical model proposed in this paper will be a good tool for a designer to optimize the double RESURF devices.

ACKNOWLEDGMENT

The author thanks Zhaoji Li for helpful discussions and critical reading of the manuscript.

REFERENCES

[1] J. Appels, H. Vaes, and J. verhoeven, "High voltage thin layer devices (RESURF devices)," IEDM Tech. Digest, pp. 238-241, 1979.

[2] B. J. Baliga, "An overview of smart power technology," IEEE Trans Electron Dev, vol. 38, pp. 1568-1573, 1991.

[3] M. M. D.Souza, and E. M. S. Narayanan, "Double RESURF technology for HVIC," Electron Lett, vol. 32, pp. 1092-1093, June 1996.

[4] S. Hardikar, M. M. D. Souza, Y. Z. Xu, T. J. Pease, and E. M. S. Narayanan, "A novel double RESURF LDMOS for HVIC's," Microelectronics, vol. 35, pp. 305-310, 2004.

[5] V. Parthasarathy, V. Khemka, R. Zhu, and A. Bose, "SOA improvement by a double RESURF LDMOS technique in a power IC technology," IEDM, pp. 75-78, 2000.

[6] Z. Hossain, M. Imam, J. Fulton,M. Tanaka, "Double-resuf 720V n-channel LDMOS with best-in-class on-resistance," ISPSD, pp. 137-140, 2002.

[7] M.Imam, Z.Hossain, M.Quddus, J. Adams, and C. Hoggatt el al, "Design and optimization of double-RESURF high-voltage lateral devices for a manufacturable process," IEEE trans Electron Devices, vol. 20, pp.1697-1701, 2003.

[8] Jian Fang, Kun Yi, and Zhaoji Li. On-State Breakdown Model for high voltage RESURF LDMOS. Chinese Journal of Semiconductors, vol. 26, pp. 436-442, Mar 2005.

[9] Qi Li, Zhaoji Li, and Bo Zhang, "A breakdown model of thin drift region LDMOS with a step doping profile," Chinese Journal of Semiconductors, vol. 26, pp. 2159-2163, Nov 2005.

[10] Yufeng Guo, Bo Zhang, Ping Mao, Zhaoji Li, and Quanwang Liu, "Unified breakdown model of SOI RESURF device with uniform/step/linear doping profile," Chinese Journal of Semiconductors, vol. 26, pp. 243-249, Feb 2005.

[11] M. Imam, M. Quddus, J. Adams, and Z. Hossain, "Efficacy of charge sharing in reshaping the surface electric field in high-voltage lateral RESURF Devices," IEEE Trans Electron Dev, vol. 51, pp.141-148, Jan 2004.

[12] S. Y. Han, H. W. Kim, S. K. Chung, " Surface field distribution and breakdown voltage of RESURF LDMOSFETs," Microelectronics Journal, vol.31, pp. 685-688, 2000.

[13] S. M. Sze, Physics of Semiconductor Devices, New York: Wiley, 1981.

[14] S.Y. Han, J. M. Na, Y. I. Choi, "An analytical model of the breakdown voltage and minimum EPI layer length for RESURF pn diodes," Solid-State Electronics, vol. 39, pp.1247-1248, 1996.

[15] S. Colak, "Effects of drift region parameters on the static properties of power LDMOST," IEEE Trans Electron Dev, vol. 12, pp. 1455-1466, Dec 1981.

2006 5th International Power Electronics and Motion Control Conference

A Novel Centralized HID Ballast System with Power-Bus

Xiaodong Lu, Bo Yang, Jiande Wu, Xiangning He
College of Electrical Engineering, Zhejiang University, Hangzhou 310027   China
E-mail: yangbo530@hotmail.com

Abstract- **In purpose of solving the problems on saving energy, resisting acoustic resonance and enhancing reliability in HID lamp ballasts, this paper addresses a novel centralized HID ballast system with power-bus. The system utilizes a central converter to convert the mains voltage to a particular waveform in power-bus, and then with the power-bus, simplified ballasts can be driven in parallel. According to the different voltages waveforms in power-bus, two centralized ballast systems are proposed: the centralized HID ballast system with DC bus and the centralized HID ballast system with AC bus. Compared with the traditional ballast, the centralized HID ballast system with power-bus is of great advantages such as remarkable energy and materials saving, easy dimming and control, high intelligence and reliability, etc. And besides, there is no acoustic resonance phenomenon in the centralized HID ballast system with AC bus.**

Keywords- HID ballast, AC bus, DC bus, Dimming, Energy-saving, Reliability;

I. INTRODUCTION

High Intensity Discharge (HID) lamp is an energy-saving light source of new generation. It is well known that the use of HID lamps has been widely extended, because of their much higher luminous efficacy and longer life than incandescent lamps. At present, HID lamp is ignited generally by the traditional 50Hz electromagnetic ballast or the single electronic ballast. However, different limitations of the two ballasts are of great existence. The traditional 50Hz electromagnetic ballast has the large size, a low power factor, a strong rush current and a flicker with 50Hz mains-frequency [3], which result in high metal consumption and clumsy in management, low energy efficiency and low lighting stability. As for the single electronic ballast, although it works with high efficiency, and a high power factor, still many technical problems remain unsolved. Firstly, harsh outdoor temperature condition causes little reliability and short life. Secondly, acoustic resonance phenomena appear in a frequency range from several to several hundred kHz [1].

To solve the problems mentioned above, this paper addresses a novel centralized HID ballast system with power-bus. The system utilizes a converter to convert the mains voltage to a particular waveform in power-bus, and then with the power-bus, simplified ballasts can be driven in parallel.

According to the different voltages waveforms in power-bus, two centralized ballast systems are proposed: the centralized HID ballast system with DC bus and the centralized HID ballast system with AC bus. Compared with the traditional ballast, the centralized HID ballast system with power-bus is of great advantages: The simplified ballast in the system is two to eight times lighter and smaller than the traditional one, thus the materials consumption is evidently reduced; The realization of dimming is more convenient, as a result, remarkable energy saving is easily achieved; The system is of higher level of integration, intelligence and reliability. No acoustic resonance phenomena appear in the centralized HID ballast system with AC bus.

II. THE CENTRALIZED HID BALLAST SYSTEM WITH DC BUS

This paper proposes a centralized HID ballast system with DC bus, comprised of a central rectifier supply and several units of DC/AC electronic ballasts. The central rectifier supply substitutes for the rectifiers in series of single ballasts, and converts 50Hz mains supply to 400 volts in DC bus, which is connected to several units of DC/AC electronic ballasts. The control on start, stop, and dimming, etc. over the whole system is available through control terminals. The system diagram is shown in Fig 1.

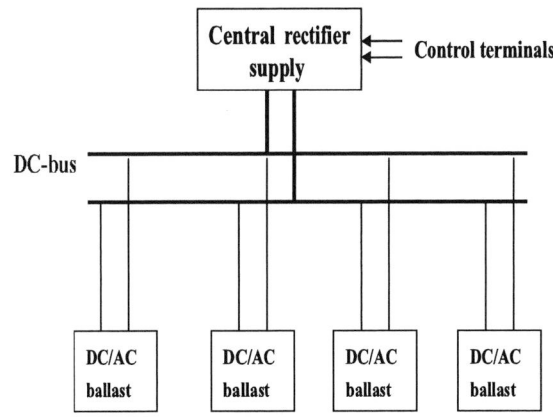

Fig 1. System diagram of the HID ballast with DC bus

The central rectifier supply in the system employs the commercialized products for DC drives, which is located in

Electrical Control Box. Its efficiency is more than 98%. All the DC/AC electronic ballasts are installed in the outdoor lamp holders, and are smaller than 1/2 the size of the single electronic ones. The DC/AC electronic ballast inverts 400 volts in DC bus to the appropriate ignition voltage and stable operation voltage for HID lamp, which consists of a half bridge inverter and high voltage starter circuit. The schematic diagram is shown in Fig 2.

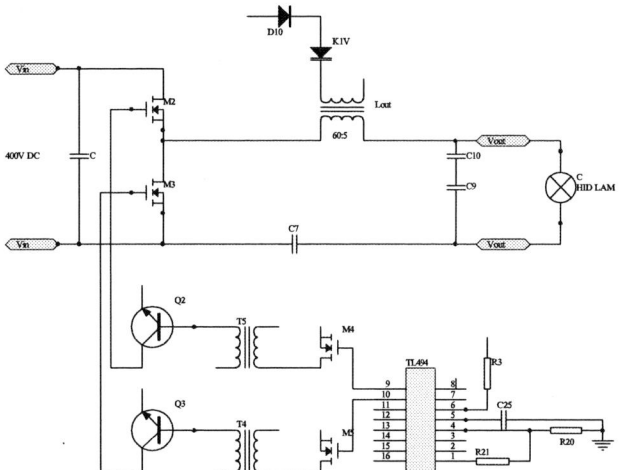

Fig 2. Schematic diagram of DC/AC ballast

The inverter uses the technology of fluctuated switching frequency to avoid generation of acoustic resonance, with working frequency of 50kHz±5kHz. The high voltage starter circuit generates 3000 volts ignition pulses by SIDAC and a transformer of 60:5. Fig 3 below describes the key waveforms of the DC/AC electronic ballast.

By voltage regulation in DC bus, smooth dimming and stepped dimming are achieved. The voltage of DC bus can be regulated between 300 volts and 400 volts through variable

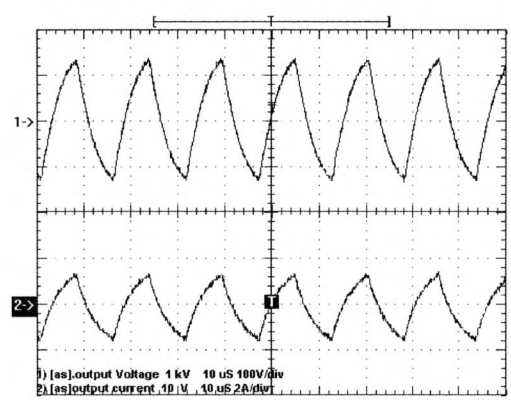

(a) Output waveforms of the DC/AC electronic ballast.
1: Voltage waveform; 2: current waveform

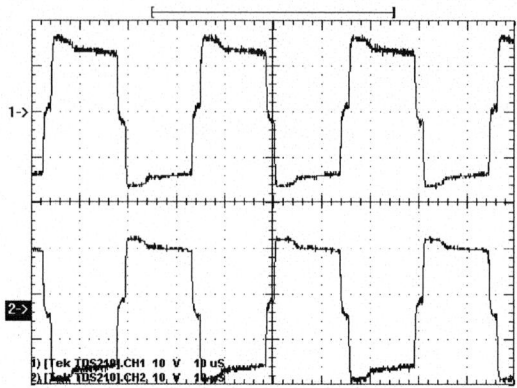

(b) Gate drive waveforms of the half bridge inverter in DC/AC electronic ballast
Fig 3. The key waveforms of the DC/AC electronic ballast

resistance and digital communication. The regulation method depends on AC/DC device adopted in the system. A caution should be given; dimming after HID lamp is in steady state is recommended, otherwise HID lamp gets nigrescent easily, and is caused to have a short serviceable life.

III. THE CENTRALIZED HID BALLAST SYSTEM WITH AC BUS

Furthermore, this paper proposes a centralized HID ballast system with AC bus, with a frequency higher than 50Hz(less than 400Hz is recommended), which consists of a VVVF, a 380/220V transformer with LC filter and several simplified ballasting components included reduced electromagnetic ballasts and high voltage starters. The system diagram is shown in Fig 4.

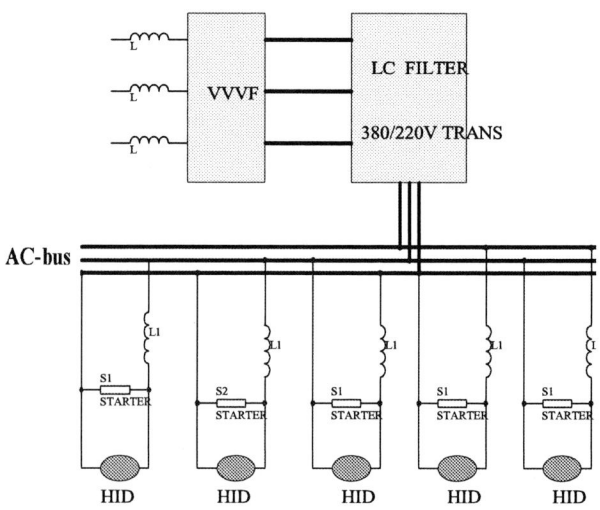

Fig 4. System diagram of the HID ballast with AC bus

Taking 400hz bus for example, the VVVF converts 380V/50Hz mains supply to 380V/400Hz PWM supply, and the PWM wave is further transmitted through the transformer with LC filter into 220V/400Hz sine waveform in AC bus, which is connected to several simplified ballasting components included electromagnetic ballasts and high voltage starters. Due to the increase of the frequency in AC bus, the traditional 50Hz electromagnetic ballast L0 is substituted with a comparatively small electromagnetic ballast L1. Where L1=L0/8. For example, L1=24mh when L0=190mh for 250W HID lamp. Accordingly, the size is reduced to 1/8 and a mass of materials is economized. The inductances in front of VVVF are for increasing the power factor of whole system and helping to the EMC solutions. The key waveforms of the HID ballast system can be seen from Fig 5.

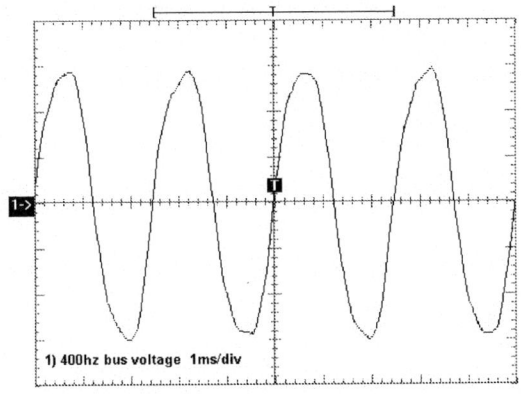

(a) Voltage of 400Hz bus

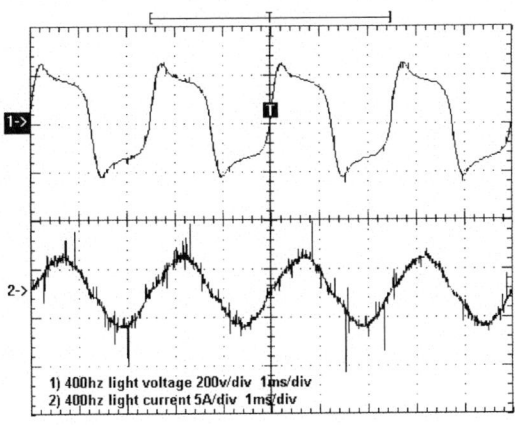

1) 400hz light voltage 200v/div 1ms/div
2) 400hz light current 5A/div 1ms/div

(b) Voltage and current of lamp at 400Hz
Fig 5. The key waveforms of the HID ballast system

The regulated voltage output of the VVVF excludes the impacts from the fluctuation of mains voltage. And owing to the usage of LC filter, the waveform in AC bus is a relatively exact sine wave, stably at 400Hz/220V as Fig.5 (a) describes. The waveforms of HID lamp at 400Hz are demonstrated in Fig.5 (b). The voltage waveform is of quasi square wave, and the current waveform is of similar triangular wave, both are

stable. The current crest factor of HID lamp is 1.5 approximately, less than the traditional 50Hz one, and of benefit to the lamp life [2]. There is no acoustic resonance phenomenon in the centralized HID ballast system with 400Hz bus.

Correlative power tests are executed in the ballast system with 400Hz bus. When the HID lamp output power is 183.3W, the total input power is 204.28W, so the ballasting efficiency at 400Hz is 183.3W/204.28W=89.7%. Compared with it, the ballasting efficiency of the traditional 50Hz electromagnetic ballast is 187.2/220.31=84.9%. Because the main power loss is caused by the reactor in ballast system, the power loss of 400Hz reactor is compared with 50Hz reactor's. In the experiment, seven 250W HID lamps of different kinds are tested at 400Hz and 50Hz, and Table 1 lists the test results. The conclusion can be drawn that the energy saving is advantaged using the ballast system with 400Hz bus, even if not dimming.

TABLE I
THE COMPARISON OF REACTOR POWER LOSS

HID lamp number	Freque-ncy	Lamp power (W)	Reactor power loss (W)	Reactor power loss rate	Decrease of reactor power loss rate
HID1	50Hz	187.30	33.20	17.7%	13.7%
	400Hz	183.30	7.34	4.0%	
HID2	50Hz	188.56	34.46	18.2%	14.1%
	400Hz	184.09	7.52	4.1%	
HID3	50Hz	190.59	33.87	17.8%	13.6%
	400Hz	185.88	7.81	4.2%	
HID4	50Hz	186.98	33.12	17.7%	13.7%
	400Hz	183.03	7.30	4.0%	
HID5	50Hz	198.32	35.96	18.1%	14.3%
	400Hz	194.25	7.39	3.8%	
HID6	50Hz	190.82	33.89	17.7%	13.6%
	400Hz	187.63	7.68	4.1%	
HID7	50Hz	188.47	33.77	17.9%	13.9%
	400Hz	182.34	7.35	4.0%	

For lamp dimming, there are many possible ways, e.g.: regulating the bus voltage, changing the frequency or phase control [2]. In the centralized HID ballast system, through VVVF, smooth lamp dimming and three-stepped lamp dimming are obtained by changing the frequency and voltage in AC bus. The VVVF utilizes the V/F control method, and the V/F curve is shown in Fig 6, where the VVVF switch frequency is set at 15kHz, and the rise time and fall time of the output frequency both are 0.1s. In addition, the digital communication function of the VVVF is available for the control on start, stop, and dimming, etc. over the whole system.

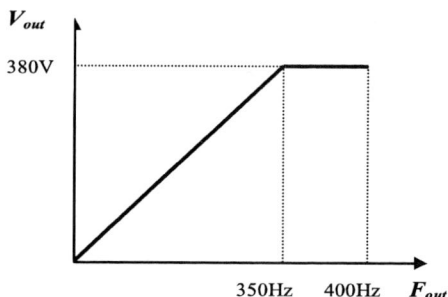

Fig 6. The V/F curve of VVVF output

According to the characteristic of the V/F curve, two types of lamp dimming tests were executed, one by changing the bus frequency from 350Hz to 400Hz without the variation of the bus voltage, the other by changing the bus frequency from 350Hz to 250Hz with the linear decrease of the bus voltage. The dimming power curve changing from 350Hz to 400Hz can be seen in Fig 7.

Frequency (Hz)	350	360	370	380	390	400
Lamp power (W)	215.0	208.4	203.8	197.2	189.2	183.3
Input power (W)	257.3	251.7	245.6	237.3	231.9	224.5

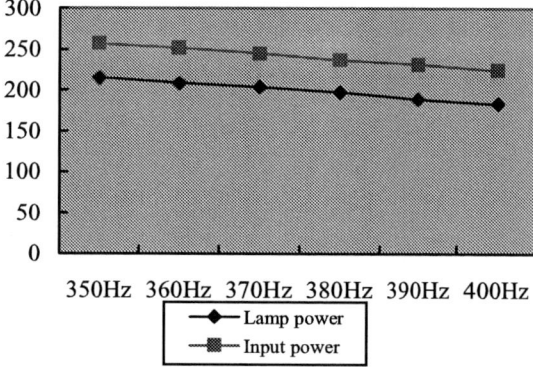

Fig 7. HID lamp dimming power curve from 350Hz to 400Hz

The dimming waveforms changing frequency from 350Hz to 250Hz can be seen from Fig 8.

Observing the dimming test results of HID lamps at different bus frequencies, this leads to the following conclusions:

- The HIP lamps can normally start, run up, and work in steady state with long time in the large frequency range from 250Hz to 400Hz.

- Smooth lamp dimming is obtained when changing the frequency form 350Hz to 400Hz. The dimming effect is obvious and the optical vision is steady during dimming. Fig.7 shows the HID lamp dimming power curve from 350Hz to 400Hz, and an average decrease of 15% lamp

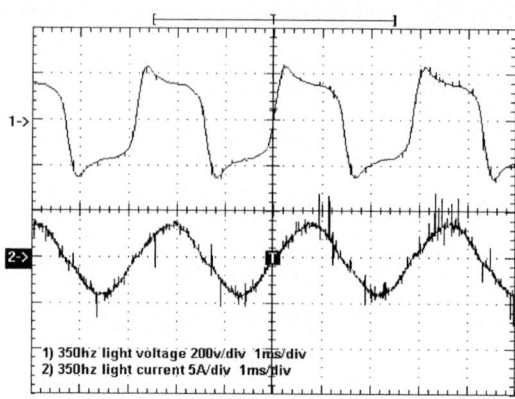

(a) Voltage and current of lamp at 350Hz

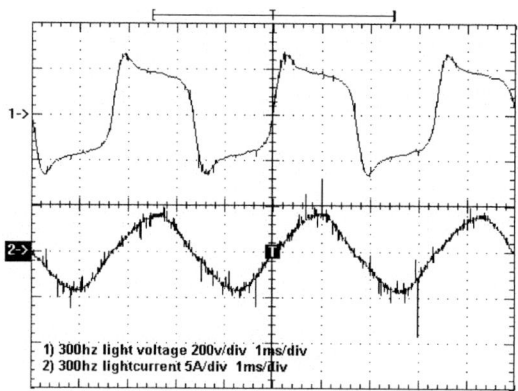

(b) Voltage and current of lamp at 300Hz

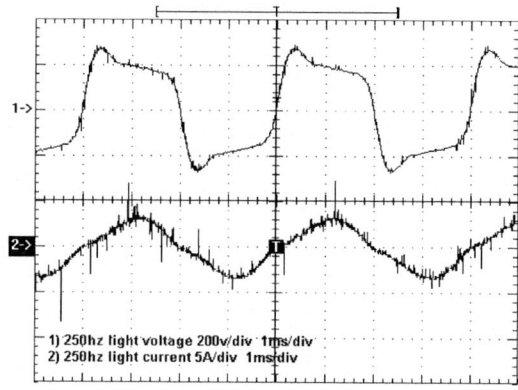

(c) Voltage and current of lamp at 250Hz
Fig 8. The dimming waveforms of HID lamp

power was measured between 350Hz and 400Hz.

- Stepped lamp dimming is obtained within the frequency range from 350Hz to 250Hz. Fig.8 shows the relative waveforms of lamp voltage and current by dimming at 350Hz, 300Hz and 250Hz. Because of the decrease of the bus voltage always with the linear decrease of the bus frequency, the dimming effect doesn't seem smooth, and then the stepped dimming generally is designed.

- During dimming, the current crest factor range of HID lamps at different bus frequencies is from 1.5 to 1.6, less than the traditional 50Hz electromagnetic one, and of much benefit to the lamp life. There is no acoustic resonance and lamp noise except for that from the inductances and the transformer.

In practice, many high pressure sodium (HPS) lamps made by some famous companies are most commonly tested. The test results validated the conclusions mentioned above.

IV. CONCLUSION

The paper addresses a centralized HID ballast system with power-bus. According to the different voltages waveforms in power-bus, the centralized HID ballast system with DC bus and the centralized HID ballast system with AC bus are distinguished. Both systems have been tested in various situations, the one with DC bus in port lighting and the one with AC bus in street lighting. And great advantages indicated in practice validated the theory in this paper.

Conclusions can be drawn that the centralized HID ballast system with power-bus is undoubtedly of energy and materials saving, easy dimming and control, high intelligence and reliability. It is of great practical value in wide-scale lighting such as centralized monitor for traffic lighting, multi-lamps in a higher point, etc, and there is much prospect of application and market. Two patents were obtained on this technology [4][5].

REFERENCES

[1] Laskai,L.; Enjeti,P.N.; Pitel,I.J.; "White-noise modulation of high-frequency high-intensity discharge lamp ballast," *IEEE Transitions on Industry application,* vol.34(3), pp. 597-604, May-June 1998.

[2] Van Tichelen, P.; Weyen, D.; Meynen, G.; "Test results from high intensity discharge lamps with current supplied at 50 Hz, 400 Hz and modulated between 15 and 35 kHz," *Industry Applications Conference, 1996. Thirty-First IAS Annual Meeting, IAS '96.,* Conference Record of the 1996 IEEE Volume 4, 6-10 Oct. 1996 Page(s):2225 - 2230 vol.4.

[3] Melis,J.; "An output unit for low frequency square wave electronic ballasts," *Southeastcon '94. 'Creative Technology Transfer - A Global Affair'., Proceedings of the 1994 IEEE 10-13,* April 1994, Page(s):106 – 108.

[4] Xiaodong Lu, Xiangning He. "A central HID ballast sysytem supplied by DC bus [P]". China patent: ZL 03 2 56145.8, 2004 - 10 - 13.

[5] Xiaodong Lu, Yang Deng, Xiangning He. "A central HID ballast sysytem supplied by AC bus [P]". China patent: ZL 03 2 10020.5, 2004 - 09 -08.

2006 5th International Power Electronics and Motion Control Conference

Research on a Novel Structure of SiGeC/Si Heterojunction Power Diodes[a]

Liu Jing[*], Gao Yong[*], Ma Li[**]

[*]Department of Electronic Engineering, Xi'an University of Technology, Xi'an China
[**]Department of Applied Physics, Xi'an University of Technology, Xi'an China
E-mail: liujing935@163.com

Abstract—A novel structure of p^+(SiGeC)-n^--n^+ power diodes is presented in this paper. On the basis of MEDICI, the physical parameter models applicable for SiGeC/Si power diodes are given and the effect on the device characteristics by incorporating smaller-sized carbon atoms substitutionally into SiGe system is simulated and analyzed. Compared to p^+(SiGe)-n^--n^+ diodes the new structure not only has the merits of faster and softer reverse recovery characteristics but also reduces the reverse leakage current largely. Besides, the dependence of devices characteristics on critical thickness is also reduced largely by adding smaller-sized carbon atoms to SiGe diodes. And therefore, the SiGeC alloys are more suitable for power devices than SiGe alloys.

Keywords–physical models, SiGeC alloys, power diodes, reverse leakage current.

□. INTROGUCTION

In recent years, group IV semiconductor alloys have attracted substantial attention for their potential applications in electronic and optoelectronic devices [1,2]. However, the 4.2% lattice mismatch between Si and Ge imposes several restrictions on the device structures [3], which limits the applications only to low Ge fractions, thin active layers and relatively lower process temperature windows. Incorporating smaller-sized carbon atoms substitutionally into SiGe system makes it possible to reduce the strain within the SiGeC system and to provide more flexible band-gap engineering. The application of SiGeC in metal-oxide-semiconductor (MOS) devices and optoelectronic devices has been studied by a number of groups [4], while that in power device has not been reported to date. A novel structure of SiGeC/Si heterojunction power diodes is presented in this paper. On the basis of analyzing the physical characteristics of SiGeC alloys, the physical parameter models applicable for SiGeC/Si power diodes are given and the effect on the device characteristics by adding carbon atoms to the SiGe system is simulated and analyzed.

II. DEVICE STRUCTURE

The structure of p^+(SiGeC)-n^--n^+ power diodes is shown in Fig. 1(a). Compared to p^+(SiGe)-n^--n^+ diodes, the new structure not only has more flexibility in the thickness of p^+(SiGeC) region which can be changed from tens of nanometers to hundreds of nanometers but also overcomes some limitations of SiGe on Si during further device processing steps. In addition, the thickness of n^- region is determined by the requirement of sustaining the reverse blocking voltage, which is analogous with that of p^+(SiGe)-n^--n^+ and Si p-i-n diodes, shown in Fig.1(b) and (c), respectively. Furthermore, the characteristics of p^+(SiGe)-n^--n^+ diodes have been previously reported by our research group [5].

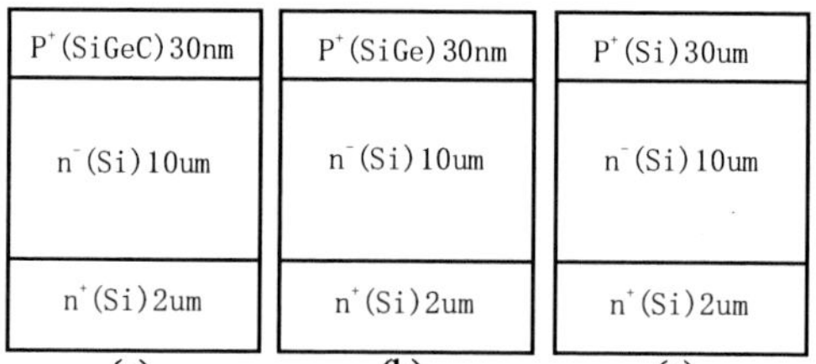

P⁺(SiGeC) 30nm	P⁺(SiGe) 30nm	P⁺(Si) 30um
n⁻(Si) 10um	n⁻(Si) 10um	n⁻(Si) 10um
n⁺(Si) 2um	n⁺(Si) 2um	n⁺(Si) 2um
(a)	(b)	(c)

Figure1. (a) Structure of SiGeC diodes (b) Structure of SiGe diodes (c) Structure of Si diodes

The strain within the SiGeC system can be compensated completely by altering the Ge:C ratio in theory. Actually, it is impossible to make all the carbon atoms into the substitutional sites completely for the SiGeC system with carbon content more than 1%, under all kinds of processing conditions compatible with Si [6]. That is, the SiGeC epitaxial layer can deteriorate quickly resulting from misfit dislocation, defects and Si-C precipitates when the carbon content is more than 1%. Thus, the carbon content is strictly limited to 1% in the novel structure of p^+(SiGeC)-n^--n^+ diodes presented in this paper.

III. MODEL

In order to achieve realistic results, several important physical models, used in MEDICI, applicable for SiGeC/Si power diodes are given in this paper.

A. Band Gap

---

[a] Project supported by the National Natural Science Foundation of China (Grant No. 50477012)
Project supported by the Specialized Research Fund for the Doctoral Program of Higher Education of China (Grant No. 20050700006)

The band offsets for SiGeC alloys strained on Si is calculated by considering hydrostatic and uniaxial strain and the intrinsic chemical effect of Ge and C. The effect is schematically illustrated [7] in Fig. 2

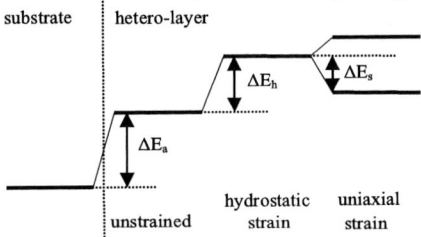

Figure2. Schematic band alignment between a coherently strained hetero-layer and the substrate.

Where ΔE_a stands for the alloy only effect for the unstrained material, ΔE_h represents the shift due to hydrostatic strain and ΔE_s is the possible splitting due to uniaxial strain. The total change in a band is expressed as

$$\Delta E = \Delta E_h + \Delta E_h + \Delta E_s \qquad (1)$$

Considering all the contributions, the band offsets for ternary SiGeC alloys with Ge and C contents limited to 50% and 3% respectively are finally given by

$$\Delta E_g = \begin{cases} \Delta E_{\Delta 4} - \Delta E_{hh} (y \leq x/8.2) \\ \Delta E_{\Delta 2} - \Delta E_{lh} (y \geq x/8.2) \end{cases} \qquad (2)$$

$$\Delta E_c = \min(\Delta E_{\Delta 2}, \Delta E_{\Delta 4}) \qquad (3)$$

$$\Delta E_v = \max(\Delta E_{lh}, \Delta E_{hh}) \qquad (4)$$

Where $\Delta E_{\Delta 2}(x,y) = 0.67x - (6.5 + 0.6x)y$ eV

$\Delta E_{\Delta 4}(x,y) = -(0.89 + 0.94x)y$ eV

$\Delta E_{hh}(x,y) = 0.74x - (3.37 + 0.56x)y$
$\qquad\qquad - (20.9 + 0.18x)y^2$ eV

$\Delta E_{lh}(x,y) = P_0(x) + P_1(x)y + P_2(x)y^2 + P_3(x)y^3$
$\qquad\qquad\qquad\qquad$ eV

$$\begin{cases} P_0(x) = 0.46x + 0.4x^2 - 0.4x^3 \\ P_1(x) = -(0.212 + 17.1x + 202.2x^2 + 245.6x^3) \\ \qquad\qquad \times(1 - 4.6x + 117.5x^2)^{-1} \\ P_2(x) = (26.8 + 2228x - 7349x^2 - 8594x^3) \\ \qquad\qquad \times(1 + 22.1x - 220.3x^2 + 1241x^3)^{-1} \\ P_3(x) = (-668.7 - 4.04\times10^4 x + 3.2\times10^5 x^2 \\ \qquad - 3.75\times10^5 x^3)(1 + 19.8x - 200x^2 + 929x^3)^{-1} \end{cases}$$

B. Mobility

The carrier mobility can be expressed as a function of the effective mass m^* and the scattering time τ :

$$\mu = q\tau / m^* \qquad (5)$$

The scattering time τ is a parameter that presents all scattering mechanisms at carrier experiences.

$$1/\tau = \sum 1/\tau_i \qquad (6)$$

Therefore, the mobility is dominated by the scattering mechanism with the smallest time constant.

On the basis of analysis mentioned above and the experimental data of Osten et al [8], we present the expression of carrier mobility as follows

$$\mu_p = \frac{\mu_{p.\max} - \mu_{p.\min}}{1 + (N_{total}/N_0')^\beta} + \mu_{p.\min} \qquad (7)$$

Where N_{total} is the doping content, $\beta = 0.7$,

$\mu_{p.\min} = 49.7\times(1 + 30x^2 - 17y - 900y^2) cm^2/V \cdot s$

$\mu_{p.\max} = 480 cm^2/V \cdot s$, $N_0' = 1.61\times10^{17} cm^{-3}$

$$\mu_n = \frac{\mu_{n.\max} - \mu_{n.\min}}{1 + (N_{total}/N_0)^\alpha}(1 + a_1 x + a_2 x^2 + a_3 x^3) + \mu_{n.\min}$$

$$(8)$$

Where N_{total} is also the doping content,

$\alpha = 0.625$, $N_0 = 1.1\times(1 + 14.15x)\times10^{17} cm^{-3}$

$\mu_{n.\min} = 175 cm^2/V \cdot s$, $a_1 = -3.02$, $a_2 = -7.08$,

$a_3 = 53.08$, $\mu_{n.\max} = 1350 cm^2/V \cdot s$

Furthermore, In order to achieve realistic results some other physical effects, such as bandgap narrowing and high-field saturation, are also considered in the simulation.

IV. SIMULATION RESULTS AND ANALYSIS

According to the models presented in this paper the characteristics of p^+(SiGeC)-n^--n^+ diodes are given by using the device simulator MEDICI.

A. Forward I-V Characteristics

Fig. 3 shows the comparison of the forward I-V characteristics of three kinds of diodes. It is observed that the forward current density of p^+(SiGeC)-n^--n^+ power diodes is the same with that of p^+(SiGe)-n^--n^+ diodes for $J_f < 1000 A/cm^2$, while has some enhancement for $J_F > 1000A/cm^2$. That is, the forward characteristics of $p^+(SiGeC)-n^--n^+$ diodes is better than $p^+(SiGe)-n^--n^+$ diodes for high current density, which indicates that the SiGeC alloys are more suitable for power devices than SiGe alloys. On the other hand, the natural shortcoming of SiGe alloys is the restriction against increasing the forward characteristics more. The SiGeC alloys reduce the lattice defect and increase the thermal stability by adding carbon atoms to the SiGe system, which makes the characteristics of the new structure superior to the $p^+(SiGe)-n^--n^+$ diodes for high current density.

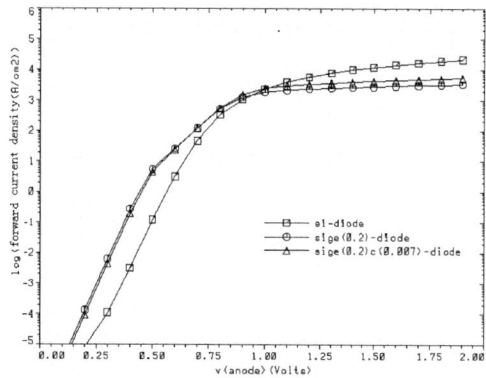

Figure3. Comparison curves of forward I-V characteristics

B. Reverse I-V Characteristics

The reverse I-V curve, shown in Fig. 4, exhibits that the $p^+(SiGeC)-n^--n^+$ diode is superior to the counterpart of SiGe alloys and the reverse leakage current of the new structure is largely reduced, with no deterioration for the reverse breakdown voltage.

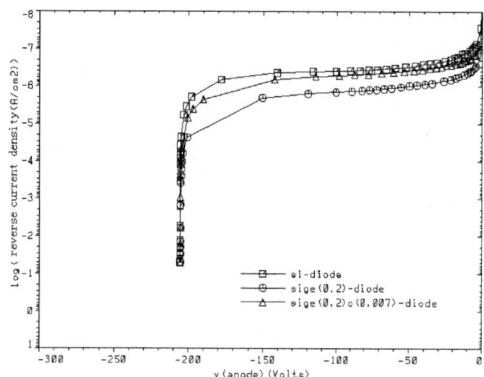

Figure4. Comparison curves of reverse I-V characteristics

Actually, we know that the 4.2% lattice mismatch between Si and Ge is the basic causes that the $p^+(SiGe)-n^--n^+$ diode has higher reverse leakage current. Thus, it can be said that the reduction of misfit dislocation in heterojunction interface is the basis of achieving stabilized I-V characteristics. The reverse leakage current of the new structure, presented in this paper, is largely reduced compared to $p^+(SiGe)-n^--n^+$

diodes by incorporating smaller-sized carbon atoms substitutionally into SiGe alloys, as demonstrated experimentally [9], which also proves the validity and correctness of the models presented in this paper.

C. Reverse Recovery Characteristics

Fig.5 shows the comparison of the reverse recovery characteristics.

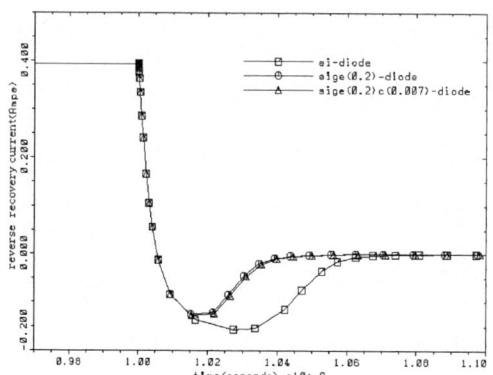

Figure5. Comparison curves of reverse recovery current

It can be seen that the reverse recovery characteristics of the new structure are almost unchanged compared to the $p^+(SiGe)-n^--n^+$ diode. In other words, the reduction of the reverse leakage current is not at the sacrifice of the reverse recovery characteristics in $p^+(SiGeC)-n^--n^+$ diodes, which also indicates that the new structure not only has smaller reverse leakage current but also has the merit of faster and softer reverse recovery characteristics.

D. Comparison of Reverse I-V Characteristics with Different p^+ Region Thickness

Fig. 6 shows the comparison of the reverse I-V characteristics between $p^+(SiGe)-n^--n^+$ diodes and $p^+(SiGeC)-n^--n^+$ diodes with different p^+ region thickness. It is observed that the reverse leakage current of $p^+(SiGe)-n^--n^+$ diodes deteriorates sharply while the leakage current of $p^+(SiGeC)-n^--n^+$ diodes has almost no change when the p^+ region thickness changes from 30nm to 150nm, which is consistent with the conclusion, presented above in this paper, that $p^+(SiGeC)-n^--n^+$ diodes has more flexibility in the thickness of p^+ region which can be changed from tens of nanometers to hundreds of nanometers. Actually, the 4.2% lattice mismatch between Si and Ge is the basic shortcoming which makes pseudomorphic SiGe epitaxial layer retain much compressive strain energy, and thus the film thickness must be below the critical thickness, or the quality of the layer will deteriorate by yielding misfit dislocation relaxation. Fig.6 also exhibits that it is possible to reduce the dependence of device characteristics on critical thickness by incorporating smaller-sized carbon atoms substitutionally into SiGe system. Besides, the consistency between simulation results and the theory mentioned above also indicates that the models presented in this paper is valid and correct.

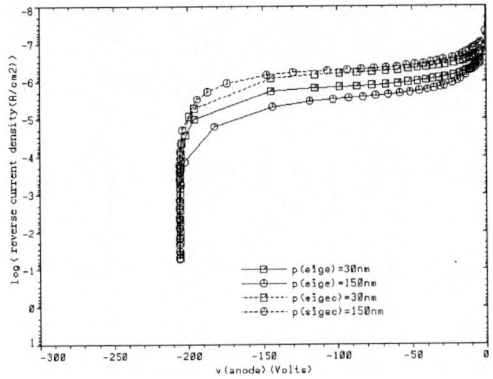

Figure6. Comparison curves of reverse I-V characteristics between p$^+$(SiGe)-n$^-$-n$^+$ and p$^+$(SiGeC)-n$^-$-n$^+$ diodes with different p$^+$ region thickness

V. CONCLUSION

In conclusion, a novel structure of SiGeC/Si heterojunction power diodes is presented in this paper. On the basis of MEDICI, the physical parameter models applicable for SiGeC/Si power diodes are given and the effect on the device characteristics by incorporation smaller-sized carbon atoms substitutionally into the SiGe system is simulated and analyzed. The simulation results show that the new structure not only has the merit of faster and softer reverse recovery characteristics but also reduces the reverse leakage current largely, compared to the p$^+$(SiGe)-n$^-$-n$^+$ diode. Furthermore, the forward characteristics of p$^+$(SiGeC)-n$^-$-n$^+$ diodes is better than p$^+$(SiGe)-n$^-$-n$^+$ diodes for high current density and the dependence of devices characteristics on critical thickness is also reduced largely by adding smaller-sized carbon atoms to SiGe diodes. And therefore, the SiGeC alloys are more suitable for power devices than SiGe alloys.

REFERENCE

[1] S C Jain, S Decoutere, M Willander, H E Maes, "SiGe HBT for application in BiCMOS technology: II. Design, technology and performance," Semicond Sci Technol. Vol. 16, pp. 67 - 85, 2001.

[2] Li Ma, Yong Gao, "Analysis and Optimal Design of a Novel SiGe/Si Power Diode for Fast and Soft Recovery," Chinese Physics Letters, vol. 21, pp. 414 - 417, 2004.

[3] Osten H J, "MBE growth and properties of super-saturated, carbon-containing silicon/germanium alloy on Si (001)" Thin Solid Films, vol. 367, pp.101 - 111, 2000.

[4] W. T. Hsieh, Y. K. Fang, S. F. Ting et al "Improving high temperature characteristic of SiGeC/Si heterojunction diode with propane carbon source", Solid-State Electronics, vol. 46, pp.1949 - 1952, 2002

[5] Gao Y, Cheng B T and Yang Y. "Analysis and Optimal Design of Novel SiGe/Si Heterojunction Switching Power Diodes," Chin. J. Semicond, vol. 23, pp. 735-740, 2002 (in chinese).

[6] V. Loup, J. M. Hartmann, G.. Rollend et al, "Growth temperature dependence of substitutional carbon incorporation in SiGeC/Si heterostructures", J. Vac, Sci. Technol. B, vol. 21 pp. 246-249, 2003.

[7] Sylvie Galdin, Philippe Dollfus, Valerie Aubry-Fortuna et al, "Band offset predictions for strained group IV alloys: $Si_{1-x-y}Ge_xC_y$ on Si(001) and $Si_{1-x}Ge_x$ on $Si_{1-z}Ge_z$(001)", Semicond. Sci. Technol, vol. 15, pp.565-572, 2000.

[8] H. J. Osten, P. Gaworzewski. "Charge transport in strained $Si_{1-y}C_y$ and $Si_{1-x-y}Ge_xC_y$ alloys on Si(001)", J. Appl. Phys, vol. 82(10), pp.4977-4981, 1997 .

[9] F. Chen, B.A.Orner, D.Guerin et al, "Current Transport Characteristics of SiGeC/Si Heterojunction Diode" IEEE. vol. 17 No.12, pp.589-591, 1996.

2006 5th International Power Electronics and Motion Control Conference

Gate driving of high power IGBT by wireless transmission

Stéphane Bréhaut, François Costa, member IEEE

SATIE UMR8029, ENS de Cachan. 61, Avenue du Président Wilson, 94235 Cachan France

stephane.brehaut@satie.ens-cachan.fr

Abstract— **In high power high voltage conversion, the technology of IGBT's drivers is a very sensitive point due to the need of a high reliability and a high degree of insulation to transmit driving information and energy to the gate driver. Many techniques exist that allow an insulated transmission of the driving signal. Some, like the level-shifter or the optocoupler are commonly used in medium voltage range but are totally unsuited to high voltage. In this last case, a classical solution leads to use, for each IGBT, a DC supply with a high insulation transformer for energy transmission and an optic fiber for the transmission of information. This solution is reliable, fast, and not sensitive to disturbances. However, in the case of numerous switches to drive, this solution becomes costly and bulky. Thus, in this paper, we propose a driving system well suited to high voltage, which can be insensitive to high-level disturbances. A wireless solution coupled to an energy supply loop is proposed, in order to reduce the cost and bulkiness of high power converters. The advantage of this solution lies in a very low propagation time, a strong immunity to noise if signal coding is well studied and a simplified supply system to drivers.**

I. INTRODUCTION

In electrical locomotive, the use of power converters directly fed by the catenary discloses the problem of the insulation of the drivers stage [1]. Indeed, with a 25 kV catenary voltage, standard technologies are not matched with usual insulation technologies, and new structures have to be developed for the realization of High-Voltage Insulated Drivers (HVID). The data an energy paths throughout the drivers need to be insulated from the IGBT and they must be reliable with respect to data transfer and to secured energy transfer. These systems must be compact because, in architectures of high-voltage converters, the number of switches can become significant as well as the quantity of connection's wires [2].

So, the signal transmission through a barrier of insulation is a technical challenge in a great number of applications [3]. The solutions generally adopted are the optic fiber, the opto-coupler, as well as the pulse transformer [4].

The requirements for a safe drive transmission are gathered in some points: firstly, the stage treating the driving signal must be immunized against disturbances (high dV/dt). Another significant point is that the signal transmission must be reliable and fast. Moreover, the transmitter must have low parasitic capacitances in order

to reduce the propagation impedance of the conducted Electromagnetic Interferences (EMI). At last, the energy supply system of the drivers must exhibit a high insulation while being compact.

All these specifications have led us to pay a great attention to wireless transmission. In fact, with this technology, there isn't any contact between the control and the converter, the propagation paths of EMI are greatly reduced with the distance between the power transistor and the low power drive. So, only the energy system supply has to be highly insulated. This can be obtained more easily than for the signal transmission, because these devices are more robust to EMI, as it will be shown hereunder.

In Radio Frequency Transmissions (RFT), the delay time depends on the band-pass of the transmitter/receiver. In order to avoid EMI disturbances [5], we must include a coding strategy in the driving signal [6].

This paper depicts this new high-voltage driver: firstly, we will present the diagram block of the system HVID (100 kV).

Secondly, we will propose a strategy of wireless transmission.

II. PRESENTATION OF THE ARCHITECTURE OF A HVID

A. General presentation

Many studies develop strategies of energy transmission between two distinct parts [7][8][9][10]. The suggested structure must take into account all the constraints previously detailed.

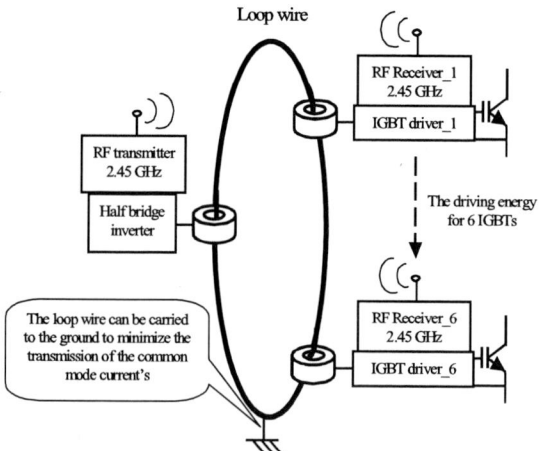

Figure 1. Concept of the HVID

1-4244-0448-7/06/$25.00 ©2006 IEEE

Fig. 1 shows the general working principle of our structure [11] [12]. At the low-voltage side of the driver, a half-bridge inverter induces a HF current in a Litz wire loop while a 2.45 GHz RF transducer transmits the encoded driving signal to all receivers. At high-voltage side, the driving energy is absorbed on the loop while a RF receiver and a decoder restore the driving signal.

A high insulation degree is achieved with this topology, while it enables to reduce the number of DC supplies and the wiring of the converter. Moreover, it offers an excellent galvanic insulation as well as an excellent immunity against EMI. In order to minimize the radiation of the HF loop, this one must be twisted.

B. Globlal model of the DGIT

The Double Galvanic Insulation Transformer (DGIT) is a key factor in our system due to all the constraints that it has to take into account. In order to have a better comprehension of the DGIT, we have decided to model and to compare the simulation with the measurement. The DGIT is constituted by two simple transformers connected together with a loop wire, fig. 2. Note that the turn ratio's product is equal to 1.

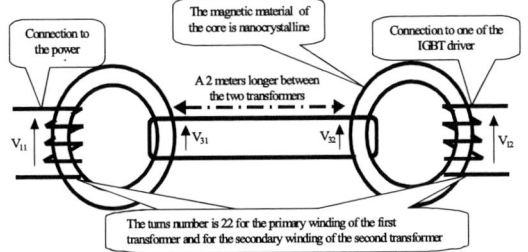

Figure 2. Concept of the DGIT with two windings

Each simple transformer can be represented by a typical model, as shown in fig. 3, [13]; however, it can be simplified regarding the frequency range and the design, as we will show hereunder.

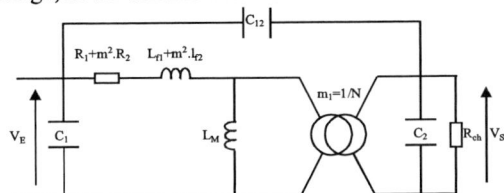

Figure 3. Simple transformer model with three spray capacitances

This one is used to describe the transformer behavior, where L_{f1} is the primary leakage inductance, L_{f2} is the secondary leakage inductance, L_M is the mutual inductance. C_1 and C_2 are the primary and secondary winding capacitances. C_{12} represents the capacitance between primary and secondary winding. m is the winding ratio.

We have measured the voltage gain in the first transformer, on the 10 Hz-10 MHz range, as shown in fig. 4. A network analyzer with high impedance ports was used. We got a typical 1st order voltage gain curve with a positive 20 dB slope from 10 Hz to 200 Hz. This is due to the mutual inductance and to the winding resistances. On the other hand, there is no influence of the stray capacitances. So, the transformer model can be represented as shown in fig. 4. If we compare the simulation with the measurement in the 10 Hz-10 MHz range, we can observe the same comportment on the full range.

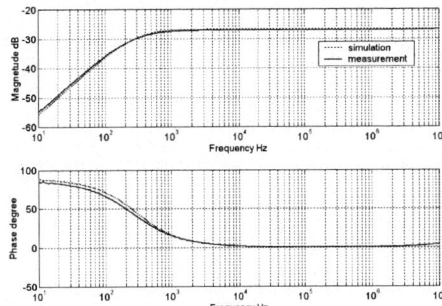

Figure 4. Measurement and simulation of voltage gain of the transformer connected to power supply, in the 10 Hz-10 MHz range

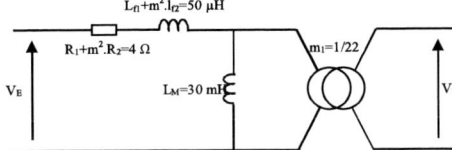

Figure 5. Final model of the transformer connected to power supply

Then, we have applied the same approach to the second transformer. In the 4 MHz-10 MHz frequency range, we found a –40 dB slope, as shown in fig. 6.

Thus, the stray capacitance effect is not longer negligible for this second transformer, so we have added it in the model as shown in fig. 7. The simulation curve matches with the measurement one in the full frequency range, fig. 6.

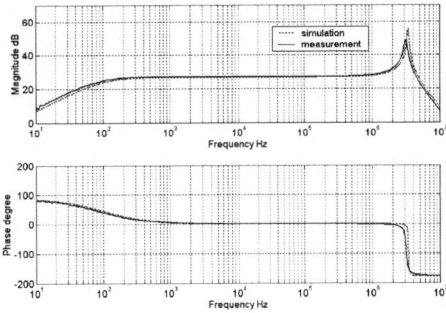

Figure 6. Measurement and simulation of voltage gain of the transformer connected to the IGBT driver, in the 10 Hz-10 MHz range

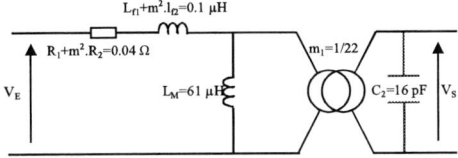

Figure 7. Final model of the transformer connected to the IGBT driver

The next step is to connect the two transformers to the loop wire. This one has stray elements composed by a resistance, R_c, in series with an inductance, L_c. It can be added to the leakage inductances, L_{fl} and L_{fl}', and to the winding resistances, R_1 and R_1', as shown in fig. 8.

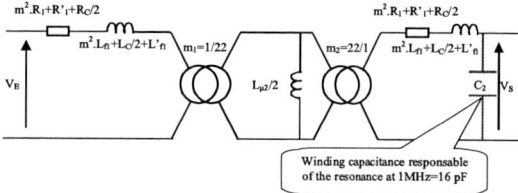

Figure 8.   Final model of the DGIT from the power point of view

The simulation and the measurement curves are superimposed, in the full 10 Hz-10 MHz range, as exhibited in fig. 9.

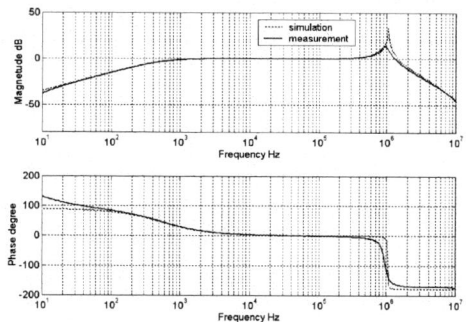

Figure 9.   Measurement and simulation of voltage gain of the DGIT from the energy transfer point of view, in the 10 Hz-10 MHz range

Now, we have a reliable model. Among all the stray capacitances, only the winding capacitance of the high voltage side transformer determines the −40dB slope. This is an important information in order to control the usable frequency range to transmit the energy.

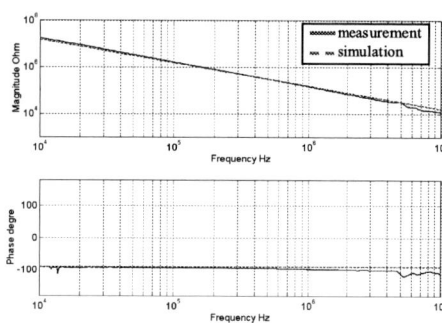

Figure 10.   Measurement and simulation of the parasitic capacitance in the 10kHz-10 MHz range

In order to verify the low capacitance between the primary and the secondary winding of the DGIT, we have measured its value as shown in fig. 10. The parasitic capacitance is very low, around 1 pF, which is efficient to limit the propagation of conducted EMI.

C.   The power supply efficiency

The converter, which generates the energy for the driver working is a half bridge inverter, as shown in fig. 12. It can deliver up to 30 Watts, the purpose being to supply up to 6 drivers with the same loop.

We have used Litz wire for the loop in order to reduce the losses due to skin effect. Fig. 11 shows the measurement setup of the power transferred across the system. Because neither of the load terminals is common to the converter's ground, differential voltage probe are used to measure voltage at the output of the half bridge and at the load terminals.

The power efficiency, μ, is defined as the ratio of the output power to the input power (1). Thus we get,

$$\mu = \frac{P_{out}}{P_{in}} \times 100\% = \frac{30\,Watts}{35\,Watts} = 84\,\% \qquad (1)$$

The primary voltage V_p, secondary voltage V_s, primary current I_p and secondary current I_s, of the transformer are shown in fig. 11. The rectified voltage at the secondary winding is conditioned by the standard specifications of drivers, typically 24 V_{DC}.

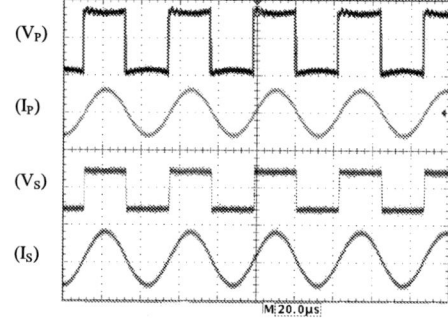

Figure 11. V_P (20 V/div), I_P (5 A/div), V_S (50 V/div), and I_S (1 A/Div), commutation frequency 22 kHz

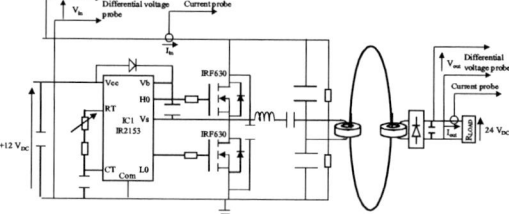

Figure 12.   Primary part of the energy supplier of the driver

III.   TWO STRATEGIES OF RF TRANSMISSION

Two strategies can be exploited to send out the RF driving signal. On one hand, we could directly transmit the numerical value of the duty cycle on the other hand, we could emit and encode the turning on/off order of the gate driver, this last solution has been chosen due to its simplest implementation. It can be considered as an asynchronous communication.

94

IV. THE ASYNCHRONOUS COMMUNICATION

A. Presentation

We consider this working as an asynchronous communication because the driver doesn't need a synchronization clock. In fact, only the electronic power switching orders are emitted, fig. 13.

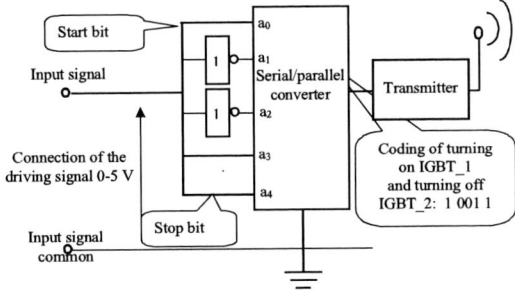

Figure 13. Reception of the electronic power switching orders, using 5 bits

The driving signal is connected to the input terminal of the emitter. This is coded on 5 bits serial/parallel converter and modulated by a transmitter. The choice of the coding is treated by several logic gates. As an example, table I shows how the driving of a 3 legs inverter can be operated by this method.

TABLE I.   Example of a driving code for a three phases inverter

Coded value of the IGBTs state	Address of the IGBTs
1 001 1	Set IGBT_1, Reset IGBT_2, arm_1
1 010 1	Set IGBT_3, Reset IGBT_4, arm_2
1 100 1	Set IGBT_5, Reset IGBT_6, arm_3
1 110 1	Reset IGBT_1, Set IGBT_2, arm_1
1 101 1	Reset IGBT_3, Set IGBT_4, arm_2
1 011 1	Reset IGBT_5, Set IGBT_6, arm_3

B. The RF reception

The receiver is made up very simply with just a decoder, a serial/parallel converter and a comparator, as depicted in fig. 14.

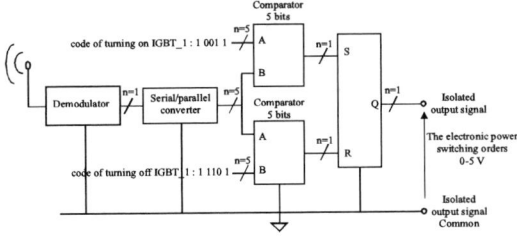

Figure 14. RF receptor for each switch

The asynchronous communication is very simple to achieve. Only 5 bits make it possible to control a three phases inverter.

V. DEVELOPMENT OF A PROTOTYPE

We must have a low delay time between the input signal and the output signal received by the gate driver. Thus, the data rate must be high in order to reduce the delay time.

A. Choice of the band pass

We are interested in the 2.45 GHz band-width. There are currently many techniques of wireless transmission in this range. "WIFI" as well as "Blue Tooth" is currently in full expansion. The main problem with "WIFI" like with "Blue tooth" is the use of complex implementation protocols. These require a consequent material and a considerable processing time.

We have chosen a RF video transmitter and receiver. These transmitters have a band-width in the order of 5 MHz. Moreover, several bandwidths can be selected. As we dispose of 5 bandwidths device, we can save one bit and reduce the code on 4 bits.

We have tested the RF asynchronous communication with the code presented in the previous section. The frequency of the transmitted signal is 2.5 kHz. We have selected a rate of 4 Mbytes/s. We have calculated the delay time of the transmission as expressed by (2).

$$Delay\_time = \frac{Number\_of\_command\_bits}{Flow\_of\_bits} * 2$$

$$Delay\_time = \frac{4\_bits}{4\,MHz} * 2 = 2\,\mu s \tag{2}$$

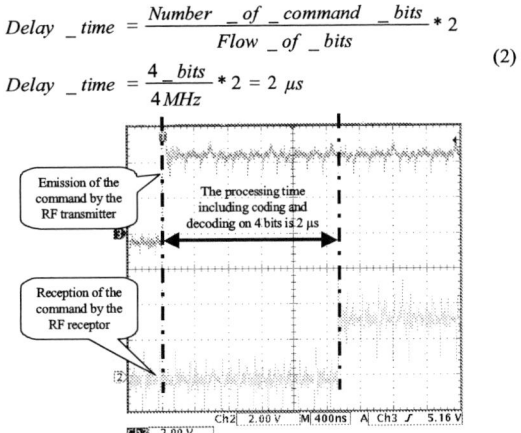

Figure 15. Delay time created by the 4 bits code

Fig. 15 shows that the signal received at the IGBT side is quite in phase with the input signal. Also, the effective delay time is very close to the calculated one. Now, we can consider that the RF transmission concept is possible.

VI. APPLICATION TO A LOW-VOLTAGE FIRST PROTOTYPE

In order to validate our concept, we have tested it in a MOSFET chopper with a RL load as shown in fig. 16. The gate driving of the MOSFET is transmitted by wireless communication.

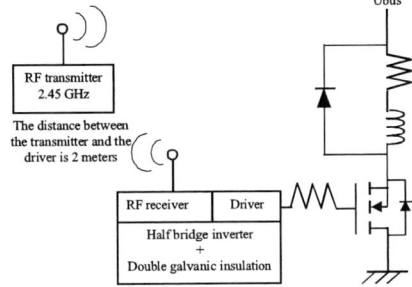

Figure 16. RF transmission on a chopper

The MOSFET has a max. reverse voltage of 200 V and a max. current of 50 A. The component's values of the load RL are selected in order to reduce the current ripple.

To validate the correct operation of the system, we have varied the commutation frequency between 100 Hz and 2.5 kHz. The information is well transmitted to the MOSFET, as fig. 17 and fig.18 show for two values of the duty cycle (50 and 20 %).

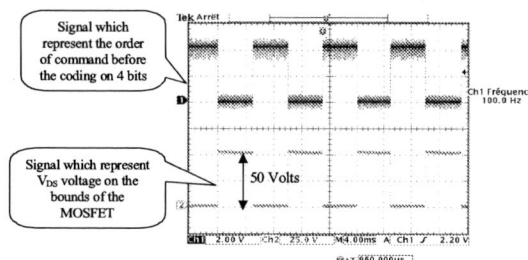

Figure 17. Control of chopper with a 2.5 kHz commutation frequency

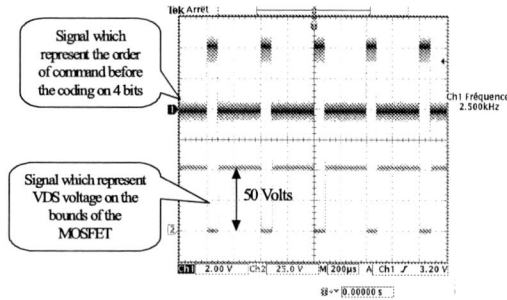

Figure 18. Duty cycle of 80 % with a 2.5 kHz commutation frequency

But now, a second prototype is to be realized in order to check the working of the full system under high-level constraints.

VII. CONCLUSION

We have treated in this article the gate driving of high power IGBT by wireless transmission. We have presented how to transmit the required energy thanks to a DGIT. A 30 Watts inverter supplies the power necessary for the drivers, with an efficiency higher than 80 %. For a first validation, we have driven a low-voltage chopper with our wireless communication system. The next step of our work will be to check it in a HV inverter leg.

REFERENCES

[1] A. Colasse, A. Dandoy, Ch. Delecluse, R. Maffei, Ph. Thomas, *Development of a multi-voltage locomotive with 6.5 kV IGBT*, Power electronics and applications, 2003, EPE 03

[2] J.Saiz, M. Mermet, D.Frey, P.O. Jeannin, JL. Schanen, P. Muzicki. *Optimisation and integration of an active clamping circuit for IGBT series association*, Industry Applications Conference, 2001. Thirty-Sixth IAS Annual Meeting. Conference Record of the 2001 IEEE Volume 2, 30 Sept.-4 Oct. 2001 Page(s):1046 - 1051 vol.2

[3] U. Schwarzer, R.W. De Doncker, *Design and implementation of a driver board for a high power and high frequency IGBT*

inverter, Power Electronics Specialists Conference, 2002. pesc 02. 2002 IEEE 33rd Annual Volume 4, 23-27 June 2002 Page(s):1907 – 1912

[4] M. Munzer, W. Ademmer, B. Strzalkowski, K.T. Kaschani, *Insulated signal transfer in a half bridge driver IC based on coreless transformer technology*, Power Electronics and Drive Systems, 2003. PEDS 2003. The Fifth International Conference on Volume 1, 17-20 Nov. 2003 Page(s):93 - 96 Vol.1

[5] M. Zimmermann, K. Dostert, *Analysis and modeling of impulsive noise in broad-band powerline communications*; Electromagnetic Compatibility, IEEE Transactions on Volume 44, Issue 1, Feb. 2002 Page(s):249 – 258

[6] G. Roberts, P. Hadfield, M.E. Humphries, F. Bauder, J.M.G. Izquierdo, *Design and evaluation of the power and data contactless transfer device*; Aerospace Conference, 1997. Proceedings., IEEE Volume 3, 1-8 Feb. 1997 Page(s):523 - 533 vol.3

[7] K.W. Klontz, D.M. Divan, D.W. Novotny, R.D. Lorenz, *Contactless power delivery system for mining applications*, Industry Applications, IEEE Transactions on Volume 31, Issue 1, Jan.-Feb. 1995 Page(s):27 – 35

[8] J.M. Barnard, J.A. Ferreira, J.D. Van Wyk, *Optimized linear contactless power transmission systems for different applications*, Applied Power Electronics Conference and Exposition, 1997. APEC '97 Conference Proceedings 1997., Twelfth Annual Volume 2, 23-27 Feb. 1997 Page(s):953 - 959 vol.2

[9] R. Mecke, C. Rathge, *High frequency resonant inverter for contactless energy transmission over large air gap* Power Electronics Specialists Conference, 2004. PESC 04. 2004 IEEE 35th Annual Volume 3, 20-25 June 2004 Page(s):1737 - 1743 Vol.3

[10] T. Bieler, M. Perrottet, V. Nguyen, Y. Perriard, *Contactless power and information transmission* Industry Applications, IEEE Transactions on Volume 38, Issue 5, Sept.-Oct. 2002 Page(s):1266 – 1272

[11] E. Bowles, T. Overett, T. Smith, R. Street, Advanced buck converter power supply "ABCPS" for APT.; Particle Accelerator Conference, 1999. Proceedings of the 1999 Volume 5, 27 March-2 April 1999 Page(s):3755 - 3757 vol.5

[12] Z. Hua, Y. Wang, K.D. Mueller-Glaser, O. Simon, Channel modeling for and performance of contactless power-line data transmission; Power Line Communications and Its Applications, 2005 International Symposium on 6-8 April, 2005 Page(s):305 – 309

[13] S. C. Tang, S. Y. Hui, H. Shu-Hung Chung, Coreless printed circuit board (PCB) transformers with multiple secondary windings for complementary gate drive circuits; Power Electronics, IEEE Transactions on Volume 14, Issue 3, May 1999 Page(s):431 – 437

The Characteristics of Thyristor Controlled Reactance Series Compensation by Adjustable Coupling[***]

Guo-rong Zhu [*], Min-zu Li [**] and Yong Kang [*]

* Huazhong University of Science and Technology, Wuhan, Hubei, China

** Guizhou University , Guiyang, Guizhou, China

zhuguorong@21cn.com, liminzu@163.com,Ykang@mail.hust.edu.cn

Abstract—This paper presents Thyristor Controlled Series Compensation by Adjustable Coupling, a novel method of thyristor controlled series compensation. The concrete device connection scheme and line power regulation characteristic of this method are obtained by analyzing the controlled reactance characteristic and adjustable principle of Thyristor Controlled Reactance by Adjustable Coupling. This paper analyzes not only the regularity of the concrete device's steady state current, voltage and the needed capacity with the variation of parameters but also the transient characteristic in the course of switching on/off the thyristors, on the basis of which the optimal switching controlling is obtained. A great deal of calculation data indicate that this method has good power regulation characteristic and transient characteristic. Compared with Thyristor Controlled Series Capacitor(TCSC), the total capacity of the concrete device can be considerably reduced if this method is used to regulate power flow. The result of the basic experiments proves the correctness of its theoretical analysis. In a word, this method can regulate the line reactance independently and quickly without high harmonic, and it is simple, economical and feasible in terms of thyristor controlled series compensation.

Key words —Thyristor Controlled Reactance by Adjustable Coupling; thyristor controlled series compensation; reactance characteristic; power regulation; transient characteristic

I. INTRODUCTION

Thyristor controlled series compensation is significantly practical among the various methods of regulating and controlling power flow in high voltage transmission line [1]. It can change its impedance quickly, regulate network power flow distribution effectively, increase the power transfer capabilities of transmission lines quickly[2], improve static stability and transient stability speedily[3], damp lower frequency oscillations and sub-synchronous resonance[4]and enhance steady state voltage stability margins. Thyristor Controlled Reactor(TCR) in parallel with capacitors constructs Thyristor Controlled Series Capacitor(TCSC) [2][3][4], which has been widely discussed and applied. TCSC can regulate circuit reactance, but there is a dead region of equivalent impedance (X_T) from inductive reactance to capacitive reactance and a resonance region ($X_T=\infty$) [5], which are both disadvantageous for series regulation. In order to avoid resonance and dead region, on the one hand, firing angle is required to leap, which will greatly change energy of capacitor and inductor, delay the transition period and reduce the speedy regulation characteristic；On the other hand, for the sake of narrowing or eliminating dead region, capacitor X_C and inductor X_L must be small, which will increase the capacity of capacitor and inductor. Therefore, narrowing or eliminating dead region and reducing regulation capacity are contradictory; that's to say, it is not ideal technically and economically to regulate circuit impedance with TCSC for power regulation.

This paper presents Thyristor Controlled Reactance by Adjustable Coupling(TCRAC) device, which in series with a capacitor structures Thyristor Controlled Series Compensation by Adjustable Coupling(TCSCAC) [6], a novel method of thyristor controlled series compensation. So long as the device is designed sound and has proper parameter, the line transmission power flow can achieve the evenly step control and realize long step control in the required scope, in which its despondence is quick, its transition period is short and it runs without high harmonic. The study shows that the total capacity of this method is much lower than that of TCSC employed for power regulation within the same regulation scope.

II. PRINCIPLE AND CHARACTERISTIC OF TCRAC DEVICE

The principle connection of TCRAC device is shown in Fig.1, where the first sides of the two three-phase or multi-phase series transformers BT_1 and BT_2 are in series, and the second sides are connected together by thyristor switches (VT). The VT can change the connection mode and winding circles of BT_1 and BT_2 so that it can adjust the magnetic coupling of the two transformers. When the current flows through the first sides of BT_1 and BT_2, self-induction magnetic flux will be produced in both transformers and mutual-induction magnetic flux will be produced by the induced current of the second sides . Provided the first sides current is fixed, that's to say, self-induction magnetic flux is fixed, the second sides current will be changed by the VT device, so mutual-induction and synthetic magnetic flux will be adjusted. In short, the equivalent impedance of the first side of each transformer is controllable [7].

Provided BT_1 and BT_2 are Y-connection-12, unloaded voltage ratios of the first side to the second side of BT_1 and BT_2 are K_1 and K_2 respectively, and wiring coefficient of VT is defined $\dot{K}_c = \Delta\dot{E}_2'/\Delta\dot{E}_1'$ [8], when this device runs symmetrically, the equivalent circuit of single-phase can be shown in Fig.2.

From Fig.2, when winding leak inductive reactance is neglected, we can get[6][7]:

$$\begin{cases} \Delta\dot{E}_1 - \Delta\dot{E}_2 = \dot{U}_1 - \dot{U}_2 \\ \Delta\dot{E}_1' - \Delta\dot{E}_{C1} = 0 \\ \Delta\dot{E}_2' - \Delta\dot{E}_{C2} = 0 \end{cases}, \qquad (1)$$

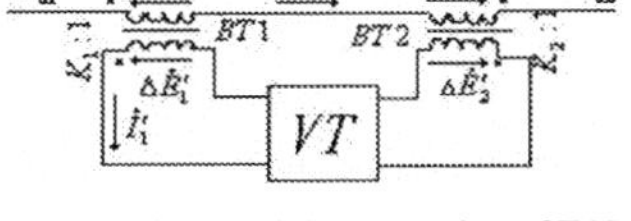

Figure 1. The principle connection of TCRAC

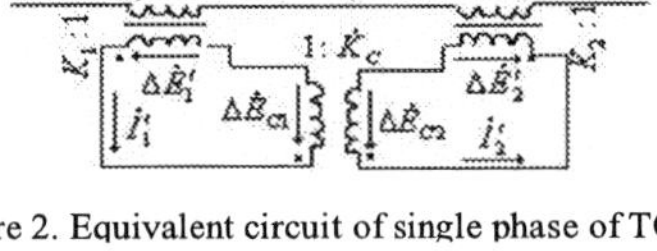

Figure 2. Equivalent circuit of single phase of TCRAC

*** This work is supported by NSFC Grant #50477056

$$\begin{cases} \Delta \dot{E}_1 = jX_{L1}\dot{I}_1 - jX_{m1}\dot{I}_1' \\ \Delta \dot{E}_2 = -jX_{L2}\dot{I}_1 + jX_{m2}\dot{I}_2' \end{cases}, \qquad (2)$$

$$\begin{cases} \Delta \dot{E}_1' = K_1 \Delta \dot{E}_1 = jX_{m1}\dot{I}_1 - jK_1 X_{m1}\dot{I}_1' \\ \Delta \dot{E}_2' = K_2 \Delta \dot{E}_2 = -jX_{m2}\dot{I}_1 + jK_2 X_{m2}\dot{I}_2' \end{cases} \qquad (3)$$

$$\begin{cases} \Delta \dot{E}_{C2} = \dot{K}_C \Delta \dot{E}_{C1} \\ \dot{I}_2' = \dfrac{\dot{I}_1'}{\dot{K}_C^*} \end{cases}, \qquad (4)$$

Where, $X_{L1} = \dfrac{X_{m1}}{K_1}$ and $X_{L2} = \dfrac{X_{m2}}{K_2}$ are respectively the self-induction reactance of BT_1 and BT_2, neglecting the leak induction reactance, which are fixed after the device is designed.

Upon the substitution of (2), (3), (4) in (1), then:

$$\begin{cases} A\dot{I}_1 - B\dot{I}_1' = \dot{U}_1 - \dot{U}_2 \\ C\dot{I}_1 - D\dot{I}_1' = 0 \end{cases}, \qquad (5)$$

Where， $A = j(X_{L1} + X_{L2})$， $B = j\left(X_{m1} + \dfrac{X_{m2}}{\dot{K}_C^*}\right)$，

$C = j(\dot{K}_C X_{m1} + X_{m2})$， $D = j\left(\dot{K}_C K_1 X_{m1} + \dfrac{K_2}{\dot{K}_C^*} X_{m2}\right)$.

From (5), the equivalent impedance of this device is: $Z_{equ} = \dfrac{\dot{U}_1 - \dot{U}_2}{\dot{I}_1} = A - \dfrac{BC}{D}$. $\qquad (6)$

Upon the substitution of A,B,C,D in (6), then：

$$Z_{equ} = j\frac{NK_C^2 + NK^2 - 2NKK_C \cos\alpha}{K_C^2 + NK^2} X_{L1} = jX_T . \qquad (7)$$

where ， $K = \dfrac{K_2}{K_1}, N = \dfrac{X_{L2}}{X_{L1}}, \dot{K}_C = K_C e^{j\alpha}$.

This device has pure reactance characteristic from (7). Adjusting K or $\dot{K}_C$ can change reactance characteristic of this device, so the equivalent reactance of this device can be expressed by controlled reactance X_T. This device can be regarded as controlled reactance which can be realized by thyristor switch adjusting magnetic coupling of two transformers BT_1 and BT_2, so this device is named Thyristor Controlled Reactance by Adjustable Coupling device.

From (7), given $N = 1, K < 1$, when the second sides of BT_1 and BT_2 are connected with Y-connection-12 ($K_C = 1, \alpha = 0^0$), and Y-connection-6 ($K_C = 1, \alpha = 180^0$), the equivalent reactance characteristic of the TCRAC device is shown in Fig.3.

From (7) and Fig.3, the reasons for choosing this characteristic are as follows: (1) adjust K or $\dot{K}_C$, then the equivalent reactance X_T of the TCRAC device can be obtained from 0 to $2X_{L1}$, that is to say, the adjustable reactance scope is wide; (2) it can make the best of the self-induction reactance of BT_1 and BT_2, so the need of transformers capacity is least, even when the second side windings of BT_1 and BT_2 are open, then follows the maximal device reactance which is still in normal work scope and will not affect transmission security; (3)

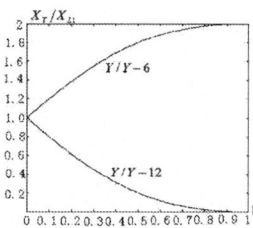

Figure 3. Equivalent reactance characteristic of TCRAC

although more thyristor switches capacity is needed, the cost of thyristor switches accounts for a small proportion of the TCRAC device.

The reactance curve in Fig.3 can be explained physically: (1) When $N = 1, K = 1$, neglecting the leak induction reactance and the second sides of BT_1 and BT_2 being connected with Y-connection-12 ($K_C = 1, \alpha = 0^0$), then the second side voltage of BT_1 and BT_2 are parallel and equal; therefore, the second circuit loop voltage is zero and the TCRAC device can be regard as the second sides of transformer in short circuit, then $X_T = 0$; (2) When $N = 1, K = 1$, neglecting the leak induction reactance and the second sides of BT_1 and BT_2 being connected with Y-connection-6 ($K_C = 1, \alpha = 180^0$), the second side voltage of BT_1 and BT_2 are in series and equal; therefore, the second circuit loop voltage is the sum of the second side voltage of BT_1 and BT_2 and the TCRAC device can be regard as the second sides of transformer in open circuit, then $X_T = 2X_{L1}$; (3) When $N = 1, K = 0$, namely, $K_2 = 0$, neglecting the leak induction reactance, the second winding of BT_2 can be regard as open circuit and the second winding of BT_1 can be regard as short circuit; therefore, $X_T = X_{L2}$, namely, $X_T / X_{L1} = 1$.

III. REASONABLE DETERMINATION OF THE CONNECTION SCHEME OF TCSCAC

A. Current, voltage, power characteristic

This TCRAC device has pure reactance characteristic and has no high harmonic while normally running. It must be in series with capacitor to structure Thyristor Controlled Series Compensation by Adjustable Coupling (TCSCAC) for regulation line reactance in a large scope. The connection scheme of applying this TCSCAC device on transmission line is shown in Fig.4a), and equivalent circuit is shown in Fig.4b).

Before installing this device, let's suppose that line transfer power is $\tilde{S}_0 = P_0 + jQ_0$, the voltage between two terminals is $d\dot{U}$, and line reactance is $Z = R + jX$. After installing this device, line reactance is $Z' = Z + jX_T - jX_C$; provided $d\dot{U}$ is fixed, then line transfer power:

$$\tilde{S} = \frac{U_N d\dot{U}^*}{Z'^*} = \frac{U_N d\dot{U}^*}{R - j(X + X_T - X_C)} . \qquad (8)$$

X_T is the function of K and $\dot{K}_C$, so adjusting K and $\dot{K}_C$ reasonably can regulate and control line transfer power. $\dot{I}_1$ can be obtained from the determined $\tilde{S}$, then from (4) and (5), the second side current is shown as following:

$$\dot{I}_1' = \frac{C}{D} \times \dot{I}_1 = \frac{K_C^2 + NK\dot{K}_C^*}{(K_C^2 + NK^2)K_1} \times \dot{I}_1 , \qquad (9)$$

$$\dot{I}_2' = \frac{\dot{I}_1'}{\dot{K}_C^*} . \qquad (10)$$

When the current of the first side and second side is determined, we can get the first side and second side voltage of BT_1 and BT_2 from (1)~(4).

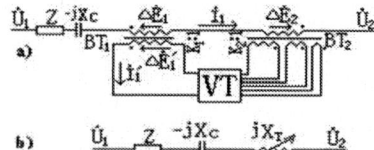

Figure 4. Transmission line and its equivalent circuit containing this device

The first side and second side voltage of BT_1:

$$\dot{U}_{T11} = \Delta\dot{E}_1 = jX_{L1}(\dot{I}_1 - K_1\dot{I}_1') , \qquad (11)$$

$$\dot{U}_{T12} = -\Delta\dot{E}_1' = jX_{L1}(K_1^2\dot{I}_1' - K_1\dot{I}_1) . \qquad (12)$$

The first side and second side voltage of BT_2:

$$\dot{U}_{T21} = -\Delta\dot{E}_2 = jX_{L2}(\dot{I}_1 - K_2\dot{I}_2') , \qquad (13)$$

$$\dot{U}_{T22} = \Delta\dot{E}_2' = jX_{L2}(K_2^2\dot{I}_2' - K_2\dot{I}_1) . \qquad (14)$$

Component power flow can be determined from current and voltage. Current, voltage and power are relevant to the X_T, thus they are the function of adjustable parameter K and $\dot{K}_C$.

B. Reasonable determination of optimal connection scheme

Connection scheme of applying for TCSCAC device must meet these demands: 1) providing the needed regulation scope and steps; 2) trying to evenly step control and meeting the needs of precision; 3) responding quickly, transition period must be short in the course of switching on/off the thyristors especially in realizing long step controlling; 4) component capacity should be small; 5) thyrisrtor working voltage must be low to decrease thyristor capacity and easy to realize; 6) connection must be simple and can be controlled easily.

Through analyzing various connection schemes, we obtain the practical connection scheme shown in Fig.5. Square 1~12 expresses thyristor switch, which is made up of a pair of bi-directional thyristors. K_1 is fixed and K_2 varies with the adjustment of w_{21}', w_{22}', w_{23}', w_{24}' positive–connection, opposite-connection or by-passing (switching on 2,3 and switching off 1,4 indicate w_{21}' positive–connection; switching on 1,4 and switching off 2,3 indicate w_{21}' opposite–connection; switching on 2,4 or switching on 1,3 indicate w_{21}' by-passing;). Given $w_{21}':w_{22}':w_{23}':w_{24}' = 1:3:9:15$, this device can construct 27 steps. Applying the TCSCAC device to a 380kM, 500kV double transmission line, we can make such supposition that equivalent parameter of the transmission line two terminals systems is $0.5KKZ_l$ (Z_l is the controlled single line reactance), assigning $KK=1.2$, fixed series compensation degree is 25%, and controllable series compensation degree is 25%; assigning $N=1, X_{L1}=13.25\Omega, X_C=53\Omega$, the transfer power on each line P_0=1GW is used as benchmark before installing this device. When one line is cut off, another line installs TCSCAC and the equivalent voltage of two terminal systems is fixed in the course of regulation; when Y-connection-12 and Y-connection-6, K=0~1, the curve of controllable transfer power is shown in Fig.6, which indicates that the scope of controllable transfer power is $(1.6\sim2.02)P_0$.

IV. TRANSIENT CHARACTERISTIC OF TCSCAC

Adjustment function of this device is realized by the thyristor switching on/off sequentially [9]. From the original state K' (adjustment parameter is K') to the later state K (adjustment parameter is K), there is a middle state in which either w_{21}', w_{22}' and w_{23}' are in short circuit or

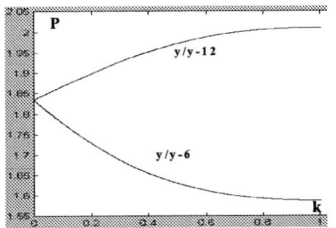

Figure 6. Adjusting characteristic of circuit power

all these are in open circuit. The elaborate analysis proves it is better that the middle state is in open circuit. In the middle state, $X_T = X_{T\max} = 2X_{L1}$ is comparatively big, but the open-circuit period is rather short, which will not affect the static stability and transient stability. Moreover, the instantaneous opening of the second side of device is beneficial for connecting all of the windings. The time of the second current i_2=0 is selected as the opening time, but the best time of entering K state can only be obtained from simulation, so we must build mathematical model.

The topology construction of K' state and K state is alike, which is shown in Fig.7, where $R1$ is the sum of the line resistance, and the first winding resistance of BT_1 and BT_2; $R2$ is the sum of the second wire resistance of BT_1 and BT_2; $L1$ is the sum of the line induction reactance and the leak induction reactance of the first windings of BT_1 and BT_2; $L2$ is the sum of the second windings leak induction reactance of BT_1 and BT_2; L_{11} and L_{21} are respectively the self-induction reactance of BT_1 and BT_2, neglecting leak induction.

The mathematical model can be obtained from the equivalent circuit as follows: (The inference is omitted.)

$$\begin{cases} (\dfrac{1}{pC} + R_1 + LL_1 p)i_1 - MMpi_2 = u , \\ -MMpi_1 + (R_2 + LL_2 p)i_2 = 0 \end{cases} \qquad (15)$$

Where, $p = \dfrac{d}{dt}$ -differential operator,

$LL_1 = L_1 + (1+N)L_{11}$, $MM = (1+NK)K_1 L_{11}$,

$LL_2 = L_2 + (1+K^2 N)K_1^2 L_{11}$, $u = u_1 - u_2$, $N = \dfrac{L_{21}}{L_{11}}$, $K = \dfrac{K_2}{K_1}$.

The middle state only has a circuit, and its mathematical model is: $(R_1 + \dfrac{1}{pC} + LL_1 p)i_1 = u$ $\qquad (16)$

Suppose t=0 is the time of entering K' state, and $t=t_1$=0.3s is the time of entering the middle state, in which i_2=0. Choosing $\Delta t = 0 \sim 0.02s$, and $t_2 = t_1 + \Delta t$, the values got in the middle state time t_2 are regarded as the starting values of K state, then we can build the relevant curve of free quantum's maximum of i_1 and i_2 (i_{1zymax} and i_{2zymax}), from which the time of the minimum of i_{2zymax} (Δt_m) can be got, so we can determine $t_{2m}=t_1+\Delta t_m$ is the best time of entering K state.

Based on Fig.6 transfer power characteristic curve, we can get the curve of $i_{2zymax}(\Delta t)$ shown in Fig.8. From Fig.8, we can conclude that: in order to reduce i_{2zymax} and optimize the switching transient characteristic, Δt =0s. Therefore when the current in K' state passing zero device goes to K state, we can obtain the optimal

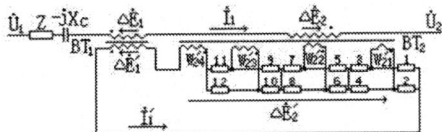

Figure 5. Single phase connect diagram of this device

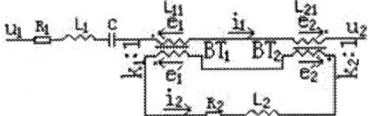

Figure7. Transient equivalent circuit

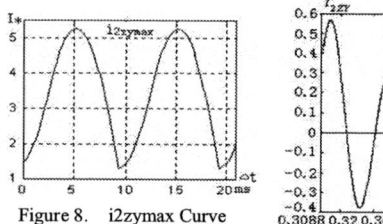

Figure 8. i2zymax Curve

Figure 9. Curve of i_{2zy} when the lowest step switching to the highest step

switching transient characteristic. In this state, when the two typical step changes, $i_{2zy}(t)$ is shown in Fig.9.

From Fig.9, it can be seen that: when K' state of the minimum power flow ($P' = 1.65P_0$) jumps to K state of the maximum power flow($P' = 2.0P_0$), i_{2zymax} =0.58, which is 9% of normal first harmonic current and drops to 1.8% in 5 cycles, thus the device transient characteristic is ideal.

V. BASIC CHARACTERISTIC EXPERIMENT OF TCSCAC

In order to prove the characteristic of TCSCAC, we did two basic characteristic experiments.

A. Reactance characteristic experiment of TCSCAC

We design and make two air gap iron cored transformer BT_1 and BT_2, in which $X_{L1} = 51.7\Omega$, $X_{L2} = 41.4\Omega$, then N=0.839. According to Fig.1, given Y-connection-12, I_1 is fixed, and we change K_2 manually. The basic reactance characteristic curve is shown in Fig.10. The first side winding voltage, the second side compositive voltage of BT_1, BT_2 are U_{A1X1}, U_{A2X2}, U_{A1X2}, respectively shown in Tab. I .

The result of reactance characteristic experiment indicates that: (1) the actual reactance characteristic is similar to the theoretical reactance characteristic of TCSCAC, that is to say, experiment can prove that the theoretical analysis of TCSCAC is right. The actual data are slightly high that of theoretical data because winding leak inductive reactance is neglected; (2) U_{A1X2} is not the sum of U_{A1X1} and U_{A2X2} but the value between U_{A1X1} and U_{A2X2}, which illuminates that U_{A2X2} mainly from self-induction is positive , U_{A1X1} mostly from mutual-induction is negative. The actual experimental data can prove that the theoretical analysis of TCSCAC in [9] is right. In actual data, U_{A1X2} is not the difference of U_{A1X1} and U_{A2X2} because winding leak inductive reactance is neglected.

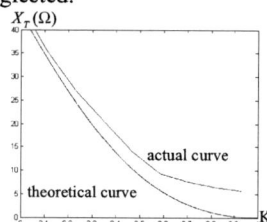

Figure10. Reactance characteristic of TCSCAC

TABLE I . RESULT OF REACTANCE CHARACTERISTIC EXPERIMENT

K	I_1/A	I_2/A	U_{A1X1}/V	U_{A2X2}/V	U_{A1X2}/V	X_T(actual)/Ω	X_T(theory)/Ω
0.0587							38.33
0.1174	4	8.5	22.9	151.8	142.4	35	33.404
0.2348	4	8.9	28.8	128.5	107.4	26.85	24.276
0.3523	4	9.1	34.1	105.2	81.9	20.42	16.487
0.4696	4	9.2	34.7	82.5	55.9	13.97	10.296
0.587	4	9.1	30.6	61.2	37.4	9.35	5.739
0.7044	4	8.85	25.1	42.8	30.2	7.55	2.676
0.8218	4	8.5	18.6	28.9	28.5	6.42	0.879
0.9329	4	8.1	11.7	15.7	22.9	5.72	0.092

B. Current regulation characteristic experiment of TCSCAC

We design and make the experiment circuit of TCSCAC, in which line impedance $Z_1 = (7 + j60)\Omega$, series capacitor $X_C = 32\Omega$, the controlled reactance $X_T = (3 \sim 25.7)\Omega$, constant $dU = 161V$. Using the optimal controlling thyristor switches, we measure the variable first current one step after another step and obtain the current curve in the course of thyristor switching. The result of current regulation characteristic experiment is shown in Tab. II , and the current characteristic curve when the lowest step (I1=3A) switches to the highest step (I1=5.1A) is shown in Fig.11.

From Tab. II and Fig.11, we can conclude that: (1) the difference between actual I_1 and theoretical I_1 is smaller, (2) the pulse current is little, which is only 1.22 times of steady state current and achieves steady state in 3 cycles.

From this simple basic characteristic experiment, we can see that the principle of TCSCAC is correct and the regulation characteristic and thyristor switching transient characteristic are good.

VI. TECHNICAL AND ECONOMIC PERFORMANCE OF TCSCAC

The basic technical performance of TCSCAC is similar to that of TCSC: (1) The two methods can change their reactance quickly, regulate network power flow distribution effectively, and increase the power transfer capabilities of transmission lines quickly. (2) The two methods can regulate continually. (3) TCSCAC has the same characteristic as TCSC when TCSC works at capacitance section [10]. When the thyristor device of TCSCAC is contingency open-loop, TCSCAC appears to be the lowest capacitance, which can partially increase the power transfer capabilities of transmission lines and ensure the transmission line to run safely. (4) Although TCSCAC device can only be regulated in stepwise fashion, we can obtain 27 steps which are enough to meet regulation demand. The step changes are quick, which cannot affect the static stability and transient stability. (5) TCSCAC mainly works at capacitance section, which is suited to domestic transmission system.

Economically comparing, TCSCAC and TCSC must meet the following demands: (1) the same system background, (2) the same power transmission before compensation, (3) during regulation, the equivalent voltage of two terminal systems is fixed and state parameters is in reliable scope. (4) controllable reactance is at (1.0~2.0~3.0)X_{C0} (X_{C0} is the fixed capacitor), in which (1.0~2.0)X_{C0} is applied to keep power regulation

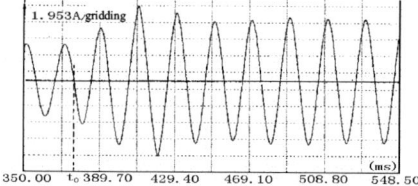

Figure 11. Transient current curve when the lowest step switching to the highest step

TABLE II . THE RESULT OF CURRENT REGULATION CHARACTERISTIC EXPERIMENT (K1=0.5,U=161V)

levels	1	2	3	4	5	6	7
K_2	0.51	0.34	0.17	0	0.17	0.34	0.51
K	-1.02	-0.68	-0.34	0	0.34	0.68	1.02
Actual I_1/A	3.0	3.1	3.3	3.7	4.4	4.9	5.1
Theoretical I_1/A	2.976	3.076	3.366	3.790	4.520	5.040	5.180
Error /%	-0.8	-0.7	2	2.4	2.7	2.8	1.5

for long running and determines normal running parameter of the device; $(2.0\sim3.0)X_{C0}$ makes use of overload capability of the capacitor and mainly is applied to improve transient stability and damp power oscillations.

TCSC device is made up of fixed capacitor C_1 (X_{C1}) in series with the paralleling TCR (X_{TR}) and capacitor C_2 (X_{C2})[11]. Its work mode is as follows: (1) the minimum capacitor and the maximum capacitor are respectively

$X_{T\min} = X_{C1} + X_{C2} = 2X_{C1} = 1.0X_{C0}$,

$X_{T\max1} = X_{C1} + X'_{T\max} = 2.0X_{C0}$, where $X'_{T\max}$ is the value of X_{C2} paralleling with some first harmonic reactance of TCR. $X_{T\max2}$ is applied to transient stability control and can achieve $3.0X_{C0}$.

TCSCAC device (X_T) is made up of fixed capacitor C (X_C) in series with TCRAC (X_{Tb}). Given $X_C=3.0X_{C0}$, $X_{Tb}=(2.0\sim1.0\sim0.0)X_{C0}$ is controllable, $X_T=(1.0\sim2.0)X_{C0}$ applied to power regulation determines the first harmonic maximum work parameter of the device.

In Yimin-Fengtun 500kV double transmission line power system, according to [4], the transfer power on each line is $P_0=1GW$ when one line is cut off, another line is installed with TCSC and TCSCAC respectively, and the equivalent voltage of two terminal systems is fixed. Through regulation calculation of the stable state, we can obtain the maximum work phase voltage U_{gmax}, maximum work current I_{gmax} and maximum work capacity S_{gmax} of two devices respectively. The result of the calculation between TCSCAC and TCSC is shown in Tab.3, and the power regulation scope of each line is $(1.488-1.997) P_0$.

From Tab.III, it can be seen that: the total capacity of capacitor in TCSCAC method is 42% lower than that of TCSC, and the capacity of transformer or inductor in TCRAC method is 44% lower than that of TCSC. These are the most important merits of the TCSCAC method. Even though the unit capacity price of transformer in TCSCAC is 20%-30% higher than the cost of reactor in TCSC, the aggregate price of TCSCAC is also much lower than that of TCSC.

In Yimin-Fengtun 500kV double transmission line power system, when TCSCAC is only applied to power regulation ($X_T=(1.0-2.0)X_{C0}$), through regulation calculation of the stable state, the first harmonic maximum work parameter of every component is shown in Tab.IV.

From Tab.III and Tab. IV, it can be seen that: When TCSCAC is only applied to power regulation, the total capacity of capacitor in TCSCAC method is 61% lower than that of TCSC, and the capacity of transformer or inductor in TCRAC method is 78.5% lower than that of TCSC, but the capacity of thyristor device in TCSCAC is 1.64 times that of TCSC.

VII. CONCLUSIONS

Based on the detailed study of the basic characteristic of TCSCAC, it can be concluded as following:

(1) TCRAC method can regulate reactance quickly within large scope by thyristor switches, its good regulation characteristic can be designed to meet demands, and the novel method can be interpreted with physical principle.

(2) TCRAC in series with capacitor constructs a novel transmission line thyristor controlled series compensation, which can also be applied to high voltage transmission line power regulation and increase the power transfer capabilities quickly.

(3) Transient characteristic of the long step control indicate that the transition period of this method is short and the transient characteristic is good.

(4) The principle of this method is simple, which can be realized easily and has no high harmonic in normal running.

(5) The basic technical performance of TCSCAC is the same as that of TCSC, but when TCSCAC is only applied to power regulation, it is much more economical than that of TCSC.

(6) If TCSCAC and TCSC are applied to series compensation together: TCSCAC is responsible for power regulation and TCSC is responsible for improving transient stability and damping power oscillations, we can obtain the optimal thyristor controlled series compensation.

REFERENCES

[1] N. G. Hingorani, and L. Gyugyi, *Understanding FACTS: concepts and technology of flexible AC transmission systems.* 1st ed., New York, U. S. A: Institute of Electrical and Electronics Engineers, Inc, 1999.

[2] O. Yan, and S. Chanan, "Improvement of total power transfer capability using TCSC and SVC," *IEEE Power Engineering Society Summer Meeting,* Vol.2, pp.944 – 948, 2001.

[3] C.Laurence Duarte, A.Silvio Cesar Braz, and F. Ednilton Bressan, "Stability analysis of power system including facts (TCSC) effects by direct method approach," *International Journal of Electrical Power and Energy System.* Vol. 27, Issue. 4, pp. 264-274, May. 2005.

[4] L. Fangcheng, L. Gengyin, Z. Jiancheng, C. Zhiye, and Y. Yihan, "The study on the VSC control of TCSC for damping power system low frequency oscillation," *POWERCON '98.* Vol.1, pp.349-352 , Aug. 1998

[5] X. Zheng, Z. Guibin, and L. Haifeng, "The controllable impedance range of TCSC and its TCR reactance constraints," *IEEE Power Engineering Society Summer Meeting,* vol.2, pp.939 – 943, 2001.

[6] L. Minzu, Z. Guorong, L. xiaodong, and W. xiaonan, "A method about power flow regulation with a novel controllable reactance," *Proceedings of CSEE,* Beijing, China,Vol.21,No.4, pp. 56-59, April 2001.

[7] L. Min-zu, Z. Guo-rong, L. Ying, and T. xiaoling, "Characteristic mechanism and application of thyristor-controlled reactance series compensation by adjustable coupling," *Power System Technology,* Beijing, China, Vol. 28, No. 10, pp.50 -54, 2004.

[8] Z. Guo-rong, L. Min-zu, L. xiaodong, and W. tingting, "Voltage regulation method in series Thyristor of distribution transformer," *Transformer ,*Shenyang, China, Vol. 39, No. 7, pp.22 -26, 2002.

[9] G. Jun, T. Luyuan, and G. Juncheng. "The mathematical model for describing transient characteristic of TCSC based on capacitor voltage synchronization mode," *Proceedings of CSEE,* Beijing, China,Vol.21,No.3, pp. 1-5, Mar 2001.

[10] Y. Besanger, S. Maginier, N. Hadjsaid, and R. Feuillet, "Thyristor controlled series compensation: some aspects of different circuit parameters and voltage stability margin, " *EMPD '95,* Vol. 2, pp.753 – 758, Nov. 1995.

[11] T. Jie, and Y. Jianhua. "Investigation of the thyristor controlled series compensation (I)," *Automation of Electric Power System,* Nanjing, China, Vol.21,No.10, pp.43-47, Oct 1997.

TABLE III. TECHNICAL AND ECONOMICAL COMPARESION FOR TCSCAC AND TCSC

method	Capacitor			Transformer or reactor			Thyristor device
TCSC	parameter	C1	C2	C1+C2	reactor		
	Ugmax/kV	32.52	97.9		97.9		26.43
	Igmax/kA	2.43	7.34		4.89		4.89
	Sgmax/MVA	230	2156	2386	1436		775
TCSCAC	parameter	C			BT1	BT2	BT1+BT2
	Ugmax/kV	191			46.6	66.38	33.2
	Igmax/kA	2.386			2.386	2.386	4.76
	Sgmax/MVA	1367			334	475	809	3792

TABLE IV. THE FIRST HARMONIC MAXIMUM WORK PARAMETER OF EVERY COMPONENT WHEN TCSCAC IS APPLIED TO POWER REGULATION

parameter	Capacitor	Transformer			Thyristor device
		BT_1	BT_2	BT_1+BT_2	
U_{gmax}/kV	127	23.3	30.9		15.5
I_{gmax}/kA	2.38	2.38	2.38		5.51
S_{gmax}/MVA	913	129	180	309	2051

2006 5th International Power Electronics and Motion Control Conference

Dual-Side Cooled Novel IPM and Improved Capability of Inverter for Elevated-Temperature Operations

Jie (Jay) Chang, Ph.D., Senior Member IEEE and Changming Liao, Ph.D.
Florida State University, Tallahassee, FL 32310
Email: jiechang@ieee.org

Abstract— Novel two-side cooling for integrated power modules (IPM) is developed for elevated-temperature operations. This approach applies and improves commercially off-the-shelf (COTS) products of IPM without significantly changing the original packaging design and manufacture's fabrication process. It can reduce the p-n junction temperature rise of the power devices inside by 20% at an equivalent load, thus increasing operating ambient temperatures, which is desirable for integrated motors and elevated-temperature applications. The size and volume associated with conventional cooling mechanism can be shrunk. Dc-link capacitor reduction with adaptive PWM control is also discussed.

Keywords- elevated-temperature operations; two-side cooling; integrated power modules; integrated AC motors and drives.

I. INTRODUCTION

In many industrial applications, such as integrated electric motors, automotive-grade power conversion systems and distributed alternative-energy generation systems, it is desirable to have new grade of power converters and AC motor drives that can operate safely in an elevated temperature environment, with reduced weight and size [1-2]. In the applications of integrated electric motors, electrical vehicles (EV) or hybrid electrical vehicles (HEV), the power converters or AC drives are often located in a hot housing or compartment. The local ambient temperatures of the converters can reach 90° C. It is significantly higher than the common industrial ambient temperature that is specified at 40° C for a power electronic product. This has significantly challenged today's power converter/inverter designs that use conventional power semiconductor devices or integrated power modules (IPM). This is because the maximum allowable p-n junction temperature of today's solid-state power devices is limited and gives insufficient margin over the local operating ambient temperatures.

This paper presents a novel approach of dual-side thermal interface and cooling [5]. This approach can effectively improve the thermal characteristics of the IPM and reduce their equivalent thermal impedance by about 20% with an equivalent load current. Furthermore, this approach can be adapted to improve today's commercially off-the-shelf (COTS) power devices and IPMs without changing their integration process.

II. ANALYSIS FOR ELEVATED TEMPERATURE OPERATIONS

Today's semiconductor power devices and IPM are commonly designed for industrial motor drives and power converters where the normal ambient temperature is specified at 40°C. At low to medium power level, Direct Bond Copper (DBC) has been the most popular substrate material for chip-on-board assembly for power device industry. Figure 1 presents a cross section structure, "device island" assembly, using the DBC substrate. The DBC consists of a ceramic isolator, Al2O3 (Alumina) or AIN (Aluminum-Nitride), onto which solid copper is bonded in a high temperature melting/diffusion process. In each island, IGBTs (and MOSFETs) are popular power devices for today's solid-state power inverters or converters [1-2, 8]. However, their maximum rated p-n junction temperature is specified for operation at or below 125°C. The power rating of the device unfortunately is derated to zero operating power at 150°C governed by a typical derating curve as shown in Figure 2.

The island structure can be extended to a pair of power devices of an IGBT and an anti-paralleled diode, which form the basic cell for integrated power modules including both totem-pole DC modules and bi-directional AC modules [7]. Figure 3 illustrates a cross-section structure of an IPM at low to medium power level. The silicon chips such as Diodes, IGBTs or MOSFETs are connected, according to a particular circuit configuration, to the copper conductors by aluminum thick wire bonds. A schematic of a three-phase bridge circuit of an IPM is shown in Figure 4.

At a full load operation in an elevated-temperature environment, the power devices in each island structure produced heat from both bottom and topside of the power chips. However, today's most commercially available IPM products provide only one thermal interface via the module's bottom copper plate. While the module transfers a large amount of heat from its base-plate thermal contact for

1-4244-0448-7/06/$25.00 ©2006 IEEE

external cooling and heat dissipation using an additional heat-sink, the top sides of the power chips are completely sealed by a thick layer of "gel" or soft encapsulation as shown in Figure 3. Between the gel and the plastic cap, perhaps more significantly, it forms a thick space layer of static air, 3-6 mm depending on the size of the module. Although the soft encapsulation provides the chip island assembly protection from exploring into the environment, it consists of high thermal impendence material. In fact, both the soft encapsulation and the air have high thermal impedance. In particular, the thermal conductivity of the air is as low as 0.028 W/mK. Because the static air is completely sealed in by the plastic case on the front side of the IPM, together with the gel layer, this forms a high-thermal impedance barrier blocking heat dissipation from the front side of the power module. On the other hand, the gate area of the IGBT devices and all wire-bounding solder joints are all on the topside of the devices. Our case study shows, as to be detailed in later sections, the heat emitted to the front side from the gate area can be a large percent of the total heat. The device switching characteristics and reliability are sensitive to temperature rise and periodical variations in the gate area and solder joints.

III. ARCHITECTURE OF DUAL-SIDE THERMAL MANAGEMENT

Now consider an example IPM circuit diagram in Figure 4 (a) that has a three-phase inverter bridge consisting of six IGBTs, each having an anti-paralleled diode in the totem-pole configuration. It also has six power diodes, in the front side, connected in a three-phase rectifier bridge configuration. Assuming the inverter is fed from an on-board alternator with balance three-phase voltages:

$$V_a(t) = \sqrt{2}\, V_1 \cos \omega t \qquad (1)$$

$$V_b(t) = \sqrt{2}\, V_1 \cos(\omega t - \frac{2\pi}{3}) \qquad (2)$$

$$V_c(t) = \sqrt{2}\, V_1 \cos(\omega t + \frac{2\pi}{3}) \qquad (3)$$

Where V1 is the rms value of input phase-to-neutral voltage, the average value of the Dc-link voltage is then given by:

$$V_{dc} = \frac{1}{\pi/3} \int_{-\pi/6}^{\pi/6} \sqrt{2}\, V_L \cos \omega t\, d\omega t = \frac{3}{\pi}\sqrt{2}\, V_L$$

$$\qquad (4)$$

Where VL = $\sqrt{3}\, V_1$ is the rms value of the input line-to-line voltage.

If VL = 230 V rms, the average Vdc = 311 V. The IPM power circuit under discussion can have eight different

voltage potentials that are changing rapidly with the PWM control waveform or patterns in a real-time operation. The voltages, for example, between the DC terminals, the output lines and the input lines can change periodically between 0 V and greater than over 400 V DC. If the input voltage doubles, the internal voltages between those circuit points, or between the internal DBC island assemblies can change periodically between 0 V and over 800 VDC. Simply extending the methods used for the low-voltage ICs might result in catastrophic failures of the power devices.

In this paper, we present a novel approach of dual-side thermal interfaces of IPM, containing IGBTs, MOSFETs or SiC devices [3-4]. The new approach employs an embedded thermal interface on the front-side of the power module package, in addition to a compact bottom-side thermal interface via a base-plate for external cooling. The dual-side interfacing approach can be applied directly to COTS products without inventing a fundamentally new design for the power module. A graphical illustration of the new concept is shown in Figure 5.

In the front-side thermal interface, our goal is to develop a practical cooling approach and structure that can significantly reduce the thermal impedance while meet the requirement of the high-voltage isolation and environmental protections as highlighted above. As shown in Figure 5, the new front thermal interface consists of a new encapsulation layer that uses modulated epoxy of Stycast material. The material has a high-thermal-conductivity coefficient of 1.4 W/mK that is 50 times higher than the thermal conductivity of static air in a conventional IPM design as shown in Figure 3, yet still possesses high-dielectric strength. While providing a low-thermal impedance, the new material is cast on the top of the power-chip-island assemblies via multiple layers of epoxy applications, so that it completely encapsulate the power chip and wire bonding joints. A set of specially designed metal cooling fins is then set closely contacting the top surface of the encapsulation layer, while a mini-scale cooling fan is embedded into the fin structure to blow the heat out. In fact, the fin's structure allows the micro-cooling fan to be embedded in the center of the fin's body, improving the efficiency of the heat exchange and reducing the volume. Key material properties of the dual-side cooling design given in Figure 5 and 6 are highlighted in Table 1. This new approach eliminates the airgap or static air compartment that has high thermal impedance between the top surfaces of the power chips and the heat dissipater. More detailed discussion will be included in a full paper. A bottom-side thermal interface with Micro-wind-channel heat-sink is also shown in Figure 5.

Table 1. Key material properties of the dual-side cooling design

Material Layer	Density (mg/cm³)	Thermal Conductivity (W/m*K)
SILICON	2.34	148
Solder	8.42	50
Cu (1)	8.9	400
Al₂O₃	3.8	21
Cu (2)	8.9	400
Al	2.707	200
Stycast	-	1.4
Air	1.17	0.028
Plastic	-	0.23

Table 2. Key parameters and material properties of an IPM

Material Layer	Thickness (mm)	Specific Heat (J/kg*K)	Thermal Conductivity (W/m*K)
SILICON	0.4	712	148
Solder	0.1	176	50
Cu (1)	0.3	385.1	400
Al₂O₃	0.635	795.35	21
Cu (2)	0.15	385.1	400
Al	-	896	200
Stycast	-	-	1.4
Air	-	1005.7	0.028
Plastic	-	-	0.23

IV. PRACTICAL IMPLEMENTATION OF DUAL-SIDE COOLING

To prove the concept and demonstrate the effectiveness, a laboratory prototype has been implemented. We selected a popular integrated power module for AC motor drives, CM25MD-24H, POWEREX [8], as a base-line IPM. The power module is rated at 25 A and 1200 V and a simplified schematic is shown in Figure 4. A 50-ampere IPM of this series has a similar circuit configuration and internal layout. This is a low-cost off-the-shelf product of the integrated power module.

Figure 6 illustrates the DBC layout of the three-phase IGBT inverter circuit. The power devices of the IGBTs and Diodes are labeled in cross-reference to the schematic in Figure 4 (a). The original gel layer that has high thermal impedance has been removed. As such, the prototype is now ready for applying the new encapsulation layer that uses modulated epoxy of Stycast material, as shown in Figure 5, for improved thermal conductivity as discussed in Section 3. This selected Stycast epoxy material provides high-service temperature characteristics and high-chemical resistance. The detailed fabrication of the micro-channel fin structure for the front-face heat dispassion can not be included in this paper due to the page limit.

The total power losses of this power module is estimated about 40 W, including the device conduction loss and switching losses, at an output of 3 horse power under a 4 kHz of PWM modulation. The key dimensional data of the micro-channel fin is computed and given in Figure 7 (c). The materials used in our prototype for dual-side thermal interfaces and their properties are summarized in Table 2.

V. EXPERIMENTAL IMPLEMENTATION AND RESULTS

Our first prototype of new two-side-cooled integrated power module (DSC-IPM) is shown in Figure 7. It is very compact measuring 4.53" x 2.36" x 1.25". In contrast to a flip-chip packaging technique [9], our new approach allows the manufacture to keep most of their original packaging design and fabrication process for the IPM production, but significantly reduces the equivalent thermal impedance and improves the compactness.

To assist our experimental test and evaluation for applications at elevated temperatures and requiring compactness and lightweight, the prototype of DSC-IPM has been integrated to a special power converter that is designed in a pancake shape. The physical shape of pancake converters makes it convenient for integration with AC motors, which is desirable for elevated-temperature power applications. Figure 8 presents the CFD evaluation result of the internal temperature distribution of new IPM module with dual-side cooling at a full load. The heat-sink or local ambient temperature is about 90 °C. The temperatures of the most area occupied by the DBC assembly and power chips are below 110°C. The simulation result indicates the highest temperatures of the IGBT chips, T3 and T4, reach about 122° ~ 125° C. The front-side thermal interface manages about 20.1% of heat dissipation out of the total heat to be dissipated, which helps to lower the IGBT's temperature.
The CFD analysis and evaluation of this complex IPM configuration has been validated by experimental tests and measurements.

Figure 9 illustrates the circuit assembly consisting of a DSC-IPM and it's mechanical interface in an integrated motor. DSC-IPM and power converter are tested at elevated heatsink/local ambient temperatures that surpass 90 degrees centigrade. This test has been conducted at full load in a sustained test routine and time. An experimental test record of elevated temperature operation of DSC-IPM at 90 degrees centigrade is shown in Figure 10. The recorded test duration shown in Figure 10 is 500 minutes. The DSC-IPM was functioning well as expected during the entire experimental test. The IGBT devices' switching waves under PWM control are again clean and healthy. The experimental results, together with the data given in Table 3, have proven our dual-side thermal interface approach and validated achieving our design objectives. A performance comparison of the new dual-side cooling and conventional one-side cooling is summarized and given in Table 3. The new approach reduces the equivalent thermal impedance of the power module over a conventional device by approximately 20%. This helps proportionally to reduce the rise of the p-n junction temperature of the power devices inside the IPM by 20% with an equivalent load current.

Table 3. Results comparison and performance improvement

Technolo-gies	Thermal dispassion front side (%)	Thermal dissipation back side (%)	Equivalent thermal impedance (pu)	T_j rise (pu)
Conven-tional 1- side interface	1.12	98.88	1	1
Proposed dual-side cooling	20.11	79.90	0.8	0.8

VI. DISCUSSION ON DC-LINK CAPACITOR AND REDUCTION BY ADAPTIVE PWM CONTROL

Traditional PWM controlled power switching assumes a DC input voltage is always constant on the DC power bus. This requires large (high voltage and high value) capacitors to filter out ripples components at six times the input frequency, commonly reducing the peak-to-peak DC voltage ripple to 2-3% at loaded conditions. The electrolytic capacitor provides energy storage capacity, but, unfortunately, it is also perhaps the most problematic power component in a high temperature environment. The electrolytic capacitors made with an insulator with a high dielectric constant exhibit unstable capacitance and high leakage current at elevated temperatures. Hot spots induced by surge currents coupled with dielectric breakdown at high temperatures can cause catastrophic failures.

After evaluating various techniques, an adaptive PWM controlled inverter scheme is developed for reducing the DC-link capacitance for integrated AC motor drives at elevated temperatures. The adaptive PWM scheme utilizes the circuit topology as shown in Figure 4, but the control scheme takes the DC-bus voltage fluctuations into account in each PWM cycle to produce improved performance at reduced DC capacitance. This improved control method can therefore accommodate a higher degree of DC-bus voltage fluctuation and allow the use of smaller DC-link capacitors. It also provides a reduced cost. When the adaptive PWM controlled inverter is combined with a high efficiency AC motor with permanent magnet rotor (which has low reactive current), the required DC-link capacitance can be reduced by 60-70%. Additional results will be given in a separate paper due to the page limit.

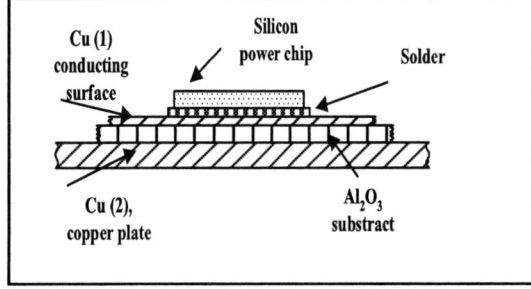

Figure 1. Cross-section island structure of power device packaging of IPM.

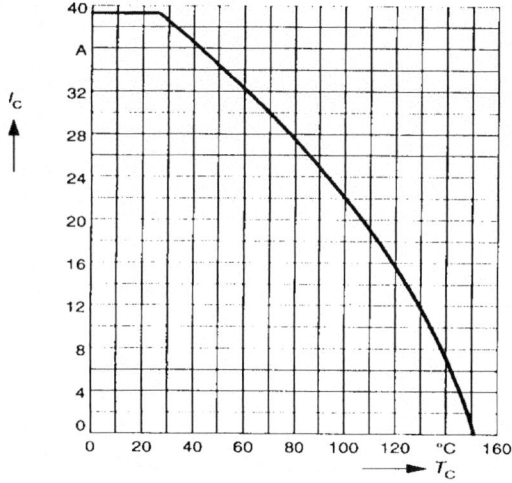

Figure. 2. A current derating characteristics of an IGBT (1200V, 20A)

105

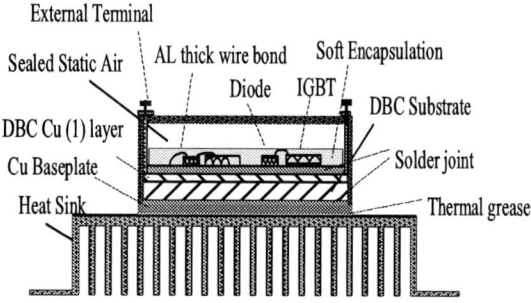

Figure 3. Cross-section structure of integrated power module.

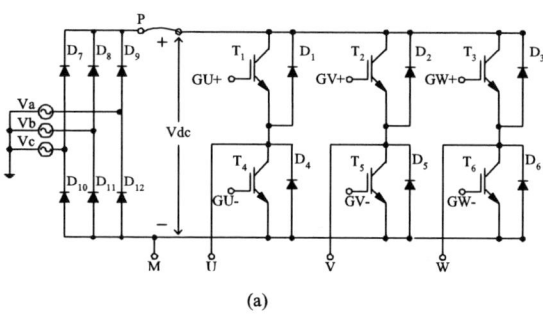

(a)

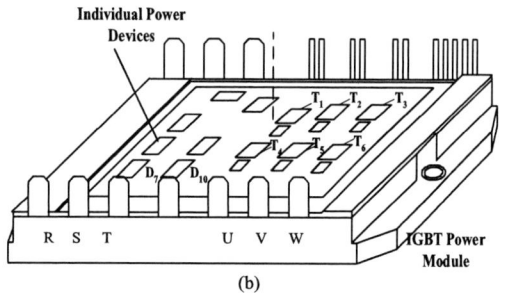

(b)

a) Circuit schematic
b) Top view of the circuit layout
Figure 4. Schematic diagram of three-phase bridge circuit of an IPM.

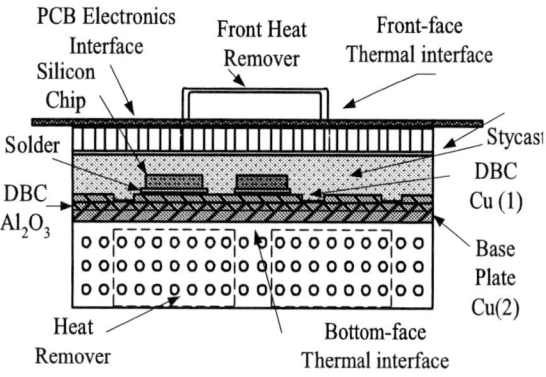

Figure 5. Architecture illustration of new dual-side ther

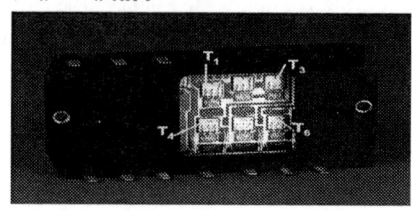

Figure 6. Converting off-the-shelf IPM of three-phase IGBT inverter to DSC-IPM

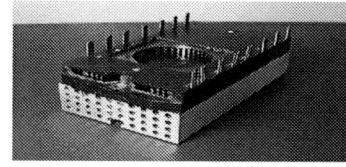

Figure. 7. Experimental prototype of dual-side thermal interfaces of IPM

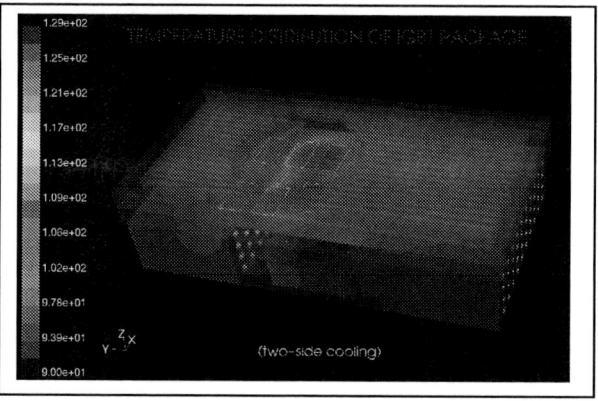

Figure 8. CFD thermal simulation results of Dual-side-cooled IPM prototype

(a)

(b)

(a) Circuit assembly consisting of a DSC-IPM
(b) Mechanical interface and integration

Figure 9. Illustration of the mechanical interface of DSC-IPM in an integrated motor.

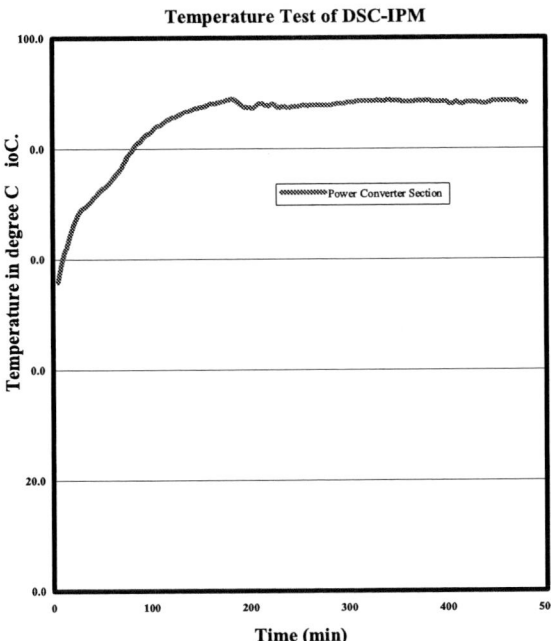

Temperature Test of DSC-IPM

Power Converter Section

Time (min)

Figure 10. An experimental test record of elevated local ambient temperature data of DSC-IPM

CONCLUSIONS

Novel two-side cooling for integrated power modules (IPM) has been developed for improved capability at elevated-temperature operations. It has been shown by our CFD analysis and hardware experimental results that the proposed approach can reduce the p-n junction temperature rise of the power devices inside the IPM by 20% at an equivalent load, thus increasing operating ambient temperatures. This is desirable for integrated motors, integrated automotive power circuits and many other applications at elevated-temperatures. This approach applies and improves commercially off-the-shelf (COTS) products of IPM without significantly changing the original packaging design and manufacture's fabrication process. The size and volume associated with conventional cooling mechanism can be shrunk proportionally. Dc-link capacitor reduction with adaptive PWM control is also discussed.

REFERENCES

[1] Jie Chang, "Design Tradeoffs and Technology Outlook of Highly Compact Integrated AC Motors", Power Conversion Conference, Nagaoka, Japan, Vol. 2, August 1997, pp.637-643.

[2] J. Miller, A. Emadi, A.V. Rajarathnam, M.Ehsani: "Current Status And Future Trends in More Electric Car power systems", 1999 IEEE 49th Vehicular Technology Conference, Volume: 2, 16-20 May 1999, Pages: 1380 – 1384.

[3] S. H. Ryu, A. K. Agarwal, R. Singh and J. Palmour, 1800 V, 3.8 A Bipolar Junction Transistors in 4H-SiC, presented at the Device Research Conference, University of Denver, Denver, CO, June, 2000.

[4] T. Yi; T. P. Chow, "High Gain Monolithic 4H-Sic Darlington Transistors", Power Semiconductor Devices and ICs, 2003. Proceedings. ISPSD '03. 2003 IEEE 15th International Symposium, 14-17 April 2003, Pages: 383 – 386

[5] Jie Chang, etc, "High Temperature Integrated Power Modules With Dual-Thermal Interfacing", U.S. Patent No. 6,989,592. January 24, 2006.

[6] Jie Chang and J. Hu, "Modular Design Of Soft-Switching Circuits For Two-Level And Three-Level Inverters", IEEE Transactions on Power Electronics, Vol. 21, No. 1, January, 2006, Page(s): 131 - 139.

[7] Jie Chang, X. Jin, A. Wang, J. Zhang: "High-Power Integrated Four-Quadrant AC Switch Modules with a Hybrid Topology of High-Current Switching Transistors and Arrays of Wide-Bandgap Diodes", Patent allowed. Patent issue fee received, Nov 2005. Publication, No. US 2004-0188706. Appl. #10/410,013.

[8] Data sheet, POWEREX CM25MD-24H.

[9] X. Liu, etc. "Low-cost 3D flip-chip packaging technology for integrated power electronics modules", U.S. Patent 6,442,033, August 27, 2002.

2006 5th International Power Electronics and Motion Control Conference

An Improved Current-Doubler with Coupled Inductors

T.-F. Wu, C.-T. Tsai, W.-C. Lin and Y.-M. Chen

Elegant Power Application Research Center
(EPARC)
Department of Electrical Engineering
National Chung Cheng University
Ming-Hsiung, Chia-Yi, Taiwan, China
E-mail: tfwu@ee.ccu.edu.tw
Tel: 886-5-2428159; Fax: 886-5-2720862

Abstract — In this paper, an improved current-doubler rectifier with coupled inductors is proposed, which can extend duty ratio to reduce the peak current through the isolation transformer and to lower output current ripple. Additionally, passive lossless snubbers are introduced to reduce reverse recovery loss and clamp the voltage spike across the rectifier diodes. In this study, a 500W prototype with a full-bridge phase-shift converter, the proposed rectifier, input voltage of 400V and output voltage of 12V was built, from which experimental measurements have shown that the efficiency can reach as high as 92% under full load condition.

I. INTRODUCTION

To reduce emission of exhausted gas, electric vehicles (EVs) and hybrid electric vehicles (HEVs) have been being developed rapidly. They need a charger to link an electric generator (EG) to batteries. Due to limited current density of its wire, an EG is always designed with high output voltage; thus, a converter with high step-down voltage ratio and high output current is usually required by this charger. Furthermore, an encapsulated environment of cars results in a serious thermal condition for the charger. The charger should be designed with high efficiency to reduce size and weight.

To achieve a high step-down voltage ratio and to increase power density and efficiency, a conventional full-bridge phase-shift current-doubler converter is usually a straightforward solution, because of its simple structure, constant frequency PWM control, high current capability and low switching loss [1]. However, it still has several limitations, such as for high step-down conversion, it requires a transformer with high turns ratio or it has to reduce the duty ratio of the switches. A high turns ratio will result in high winding loss and thermal problem, while a low duty ratio will increase input peak current and component stress. Additionally, a large circulation

current will flow through the primary winding of the transformer and the switches during a freewheeling interval [2]. As a result, conduction loss of the switches and copper loss of the transformer are significant. To release the above limitation of the conventional full-bridge current-doubler converter, a coupled-inductor current-doubler rectifier [3] and a current-tripler rectifier [4] were proposed. Although both of them can reduce the winding loss, they have a common disadvantage of complex transformer design. The phase-shift zero-voltage and zero-current-switching full-bridge converters can reduce circulation current flowing through the primary winding of the isolation transformer and switches during a freewheeling interval [5]-[6]. However, there still exist reverse recovery problems with the rectifier diodes, which result in voltage oscillation and spike.

In this paper, an improved current-doubler rectifier with coupled inductors is proposed. Unlike the conventional current doubler rectifier, there are two sets of coupled-inductors in the proposed rectifier, as shown in Fig. 1, which can extend the duty ratio to reduce the peak current through the isolation transformer and to lower output current ripple. Additionally, to reduce the ringing due to the resonance between the junction capacitor of the rectifier diodes and leakage inductors of the coupled inductors, the proposed rectifier is equipped with two sets of passive lossless snubbers, which can reduce circulation current flowing through the switches and the transformer windings and can clamp the voltage spike across the rectifier significantly [7]. In the paper, the operation principle of a full-bridge phase-shift dc-dc converter with the proposed current-doubler rectifier is analyzed in detail. The benefits of the proposed rectifier for high step-down voltage ratio and high output current applications are discussed. A 500W prototype has been designed and built, from which experimental results are used to verify the discussed merits.

1-4244-0448-7/06/$25.00 ©2006 IEEE

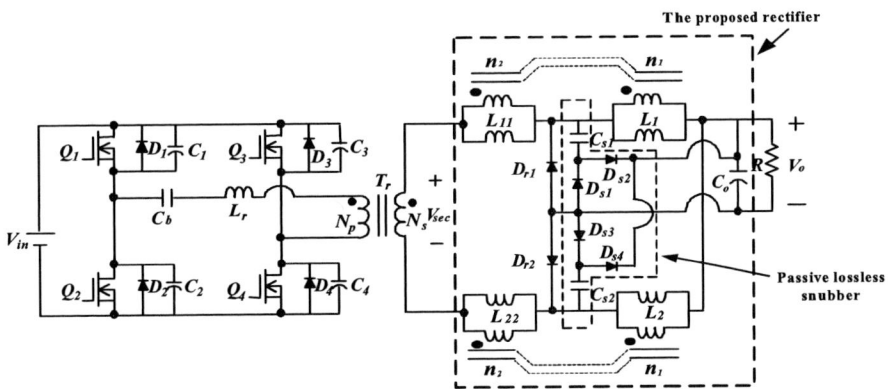

Fig. 1. The proposed current-doubler rectifier with a full-bridge phase-shift converter.

II. OPERATIONAL PRINCIPLE

The proposed rectifier with passive lossless snubbers and based on a full-bridge phase-shift converter is illustrated in Fig. 1. It mainly includes two sets of coupled inductors L_1, L_2, two sets of passive lossless snubbers, free-wheeling diodes D_{r1}, D_{r2}, and an output filter capacitor C_o. In Fig. 1, each set of the coupled inductors can be treated as an equivalent transformer with two magnetizing inductors. To simplify description of the operational modes, the following assumptions are made.

1) All of the switching devices, MOSFETs and diodes, are ideal.

2) Coupled inductors $L_1 = L_2$ and $L_{11} = L_{22}$, and leakage inductors are neglected.

3) Output filter capacitor C_o is large enough so as its voltage can be treated as a voltage source V_o.

Under continuous output current operation, five modes are identified during half a switching cycle. These modes are described below. Fig. 2 shows equivalent circuits of the operational modes and Fig. 3 shows conceptual voltage and current waveforms of the key components.

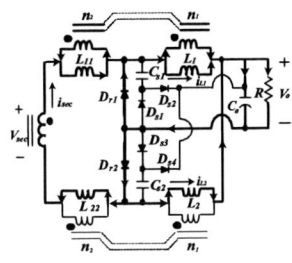

(e)

Fig. 2. Equivalent circuits of the five operational modes.

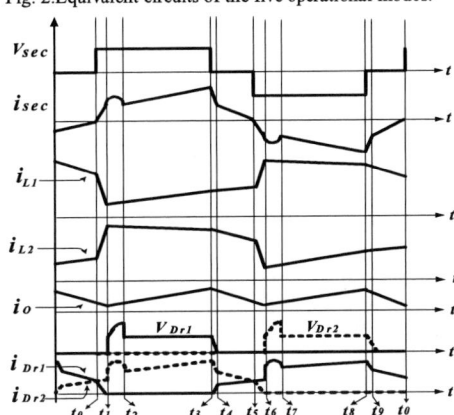

Fig. 3. Conceptual voltage and current waveforms of the key components in the proposed rectifier.

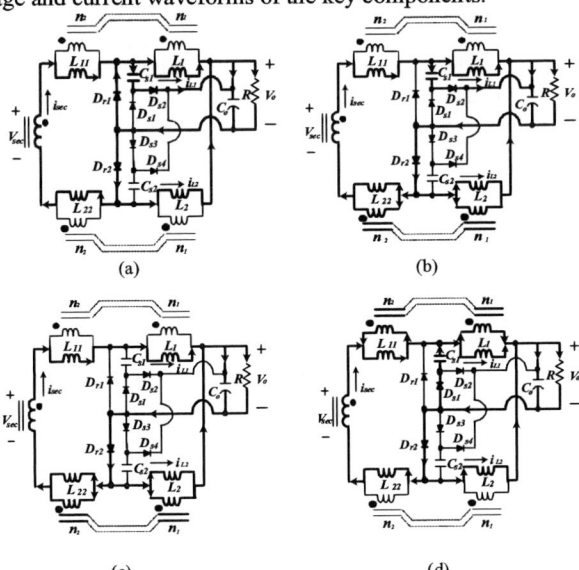

(a) (b)

(c) (d)

Mode 1 [**Fig. 2(a)**, $t_o \leq t < t_1$] :

With a positive voltage V_s across the secondary winding of transformer T_r at t_0, current i_{sec} flows through inductor L_{11} and charges clamping capacitor C_{s1}. During this interval, since current i_{sec} is not equal to inductor current i_{L1}, rectifier diodes D_{r1} and D_{r2} are conducting, while inductor L_1 and L_2 are discharged through diode D_{r1} and D_{r2}, respectively.

Mode 2 [**Fig. 2(b)**, $t_1 \leq t < t_2$] :

At time t_1, current i_{sec} is equal to inductor current i_{L1}, and diode D_{r1} is reversely biased. At this interval, clamping capacitor C_{s1} begins to resonate with inductor L_{11}, and inductor current i_{L1} flowing through the path V_o-D_{r2}-L_{22}-V_{sec}-L_{11}-L_1 is linearly increased. During this interval, the energy stored in inductor L_{22} will be coupled to inductor L_2 and inductor current i_{L2} flowing through the

109

path V_o-D_{r2} is decreased. Over this interval, inductors L_{11} and L_1 function as a tapped inductor, while inductors L_{22} and L_2 are coupled to function as a transformer. The time for charging capacitor C_{s1} from zero up to $2[nV_{sec}/(n+1) - V_o]$ can be determined as

$$t_{ch} = \frac{\pi}{2}\sqrt{L_{11}C_{s1}}. \tag{1}$$

In this mode, the inductor currents can be expressed as follows:

$$i_{L1}(t) = \frac{nV_{sec} - (n+1)V_o}{(n+1)L_1} \times (t - t_1) + i_{L1}(t_1) \tag{2}$$

and

$$i_{L2}(t) = i_{L2}(t_1) - \frac{nV_o}{L_2} \times (t - t_1), \tag{3}$$

where $n = n_1/n_2$ is the turns ratio of the coupled inductors.

Mode 3 [Fig. 2(c), $t_2 \leq t < t_3$] :

At time t_2, diode D_{s2} is in reverse bias and rectifier voltage V_{Dr1} drops to $nV_{sec}/(n+1)$. During this time interval, inductors L_1, L_{11} and L_{22} are still in the charging operation, while inductor L_2 is kept discharging. The inductor currents can be expressed as follows:

$$i_{L1}(t) = \frac{nV_{sec} - (n+1)V_o}{(n+1)L_1} \times (t - t_2) + i_{L1}(t_2) \tag{4}$$

and

$$i_{L2}(t) = i_{L2}(t_2) - \frac{nV_o}{L_2} \times (t - t_2). \tag{5}$$

Mode 4 [Fig. 2(d), $t_3 \leq t < t_4$] :

At time t_3, the secondary winding voltage V_{sec} of transformer T_r drops to zero, diode D_{s1} is conducting and the energy stored in clamping capacitor C_{s1} is transferred to the load. At time t_4, clamping capacitor C_{s1} is completely discharged, diode D_{s1} is turned off and rectifier diode D_{r1} is conducting. During this time interval, inductors L_1 and L_{11} are discharging and the energy stored in inductor L_{22} is no longer coupling to inductor L_2. Therefore, inductor current i_{L2} will decrease dramatically. The time for discharging C_{s1} from $nV_{sec}/(n+1)$ to zero can be determined as

$$t_{dis} = \frac{C_{s1}}{i_{L2}(t_3)} \frac{nV_{sec}}{(n+1)}, \tag{6}$$

and the inductor currents can be expressed as follows:

$$i_{L1}(t) = i_{L1}(t_3) - \frac{nV_o}{L_1}(t - t_3) \tag{7}$$

and

$$i_{L2}(t) = i_{L2}(t_3) - \frac{nV_o}{L_2}(t - t_3). \tag{8}$$

Mode 5 [Fig. 2(e), $t_4 \leq t < t_5$] :

Over this time interval, inductor currents i_{L1} and i_{L2} are still flowing through the load and through the rectifier diodes D_{r1} and D_{r2}, respectively. This ends a half-cycle operation.

III. DESIGN CONSIDERATIONS

This section discusses the voltage gain, diode voltage stress and output current ripple of the proposed rectifier and their effect on the operation of the circuit.

A. Voltage Gain

From Fig. 3 and the current equations of inductors L_1 and L_2 and by applying the volt-second balance principle, the voltage gain and diode voltage stress can be derived as follows:

$$\frac{V_o}{V_{sec}} = \frac{D}{\left(\frac{n+1}{n}\right)D + (n+1)(1-D)}, \tag{9}$$

and

$$V_{Dr1} = V_{Dr2} = \frac{nV_{sec}}{(n+1)}, \tag{10}$$

where D is the duty ratio of the active switches and $D < 0.5$.

B. Output Current Ripple

From Fig. 3 and by combining the currents of i_{L1} and i_{L2}, output current ripple can be determined as

$$i_{ripple} = n(1 - 2D)\frac{V_o}{L}T_s, \tag{11}$$

where T_s is the switching period of the active switches and L is the inductance of either inductor $L_1 (= L_2)$.

From (9) to (11), we can obtain a set of curves showing the relationship between duty ratio D and the voltage ratio of V_o/V_{sec} for different values of turns ratio n, as illustrated in Fig. 4. Fig 5 shows the plot of diode voltage stress versus n according to (10). Fig. 6 shows the plot of i_{ripple} versus n according to (11). From these plots, a proper duty ratio D and turns ratio n can be selected to yield high step-down output voltage, low voltage stress on the rectifier diode and low output current ripple. In considering the leakage inductance of the coupled inductors, the proposed current doubler rectifier is designed with turns ratio $n = 2$. To objectively judge the merits of the proposed current-doubler rectifier, performance comparison between the proposed rectifier and the conventional current-doubler rectifier (CCDR) is shown in Fig. 7. From these plots, it can be seen that the proposed current doubler yields higher voltage gain, lower current ripple and lower voltage stress on the rectifier diode over the conventional one.

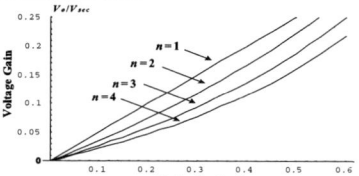

Fig. 4. Plots of the voltage gain (V_o/V_s) versus duty ratio D.

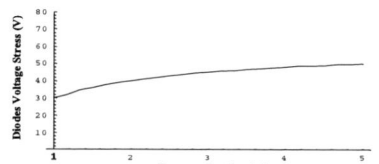

Fig. 5. Plots of diodes voltage stress versus turns ratio n of the coupled inductor.

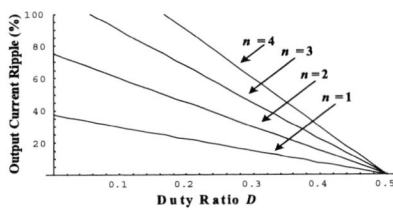

Fig. 6. Plots of output current ripple i_{ripple} (%) versus duty ratio D with n as a parameter.

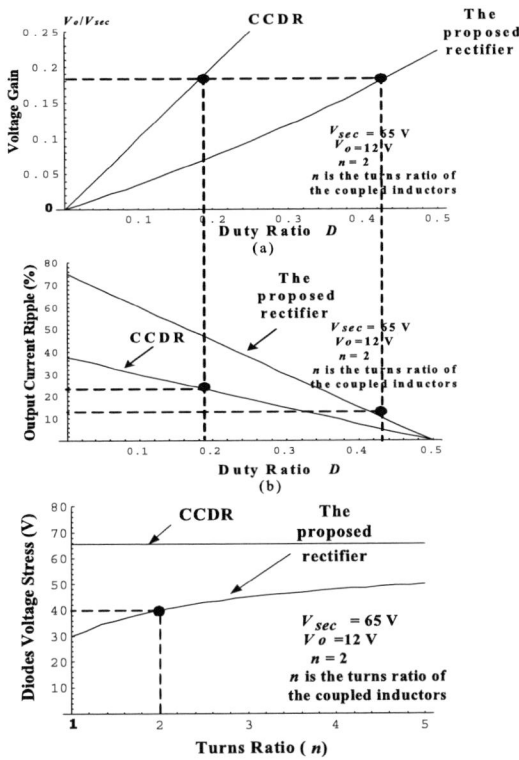

Fig. 7. Performance comparison between the proposed rectifier and the conventional current-doubler rectifier: (a) duty ratio, and (b) output current ripple, (c) diode voltage stress.

C. Determination of Clamping Capacitors

In the proposed rectifier, clamping capacitor C_{s1} or C_{s2} resonates with inductor L_{11} or L_{22} to clamp peak voltage at diode D_r ($= D_{r1} = D_{r2}$) turn-off transition and during the time intervals of t_0-t_2 and t_6-t_8. The energy W_{cs} stored in the capacitors must satisfy the following inequality:

$$W_{cs} = \frac{1}{2}C_s\left[\frac{nV_{sec}}{(n+1)}\right]^2 > \frac{1}{2}L_s i_{sec}^2, \quad (11)$$

where $C_s = C_{s1} = C_{s2}$ and $L_s = L_{11} = L_{22}$.

IV. EXPERIMENTAL RESULTS

To verify the performance of the proposed current-doubler rectifier, a 500W prototype with a full-bridge phase-shift converter and the proposed rectifier was built. Its specifications are listed as follows:

- input voltage: 400 VDC
- output voltage: 12 VDC
- output current: 40 A
- switching frequency: 50 kHz

The components of the power stage are designed as follows:

- $Q_1 \sim Q_4$: IRFP460
- C_b: 1 uF/600 V
- transformer core: TDK EE-55; $N_P = 13$ T; $N_S = 2$ T
- coupled inductors core: TDK ETD-44; $n_1 = 12$ T; $n_2 = 6$ T
- coupled inductors: $L_{11} = L_{22} = 5$ μH, $L_1 = L_2 = 20$ μH
- L_r: 12 μH
- D_{r1}, D_{r2}: C60P06Q
- $D_{s1} \sim D_{s4}$: 8TQ100
- C_{s1}, C_{s2}: 68 nF/250 V
- C_o: 1000 uF*6/25 V
- phase-shift PWM controller: UC3875

Fig. 8(a) shows measured current waveforms of inductors L_1 and L_2, and Fig. 8(b) shows full-load output current with low ripple. Fig. 9(a) shows measured voltage and current waveforms of the rectifier diodes without snubbers, where a severe voltage overshoot and ringing are observed. Fig. 9(b) shows those with RCD snubbers, where the spike voltage of the rectifier has been clamped, while the circulation current has not been minimized yet. Fig. 10(a) shows measured voltage and current waveforms of the rectifier diodes with passive lossless snubbers, where the circulation current has been reduced. Fig. 10(b) shows the spike voltage of the rectifier diodes has been properly clamped. Fig. 10(c) and (d) shows its extended waveforms and illustrates reduced reverse-recovery current at turn-off and ZCT feature at turn-on transitions. Fig. 11 shows efficiency measurements of the proposed rectifier with and without snubbers, from which it can be seen that the proposed rectifier with a passive lossless snubber can achieve as high as 92% efficiency under full load condition. The proposed rectifier is relatively feasible for the converter systems with high step-down voltage ratio, which has been verified by the experimental results.

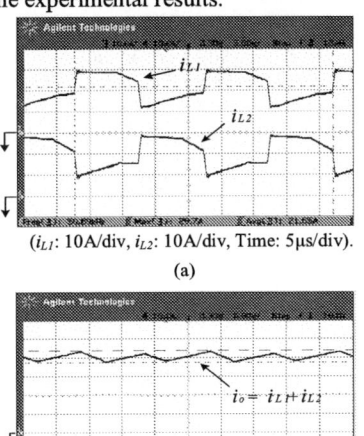

(i_{L1}: 10A/div, i_{L2}: 10A/div, Time: 5μs/div).

(a)

(i_o: 10A/div, Time: 5μs/div).

(b)

Fig. 8. Measured waveforms of (a) inductor currents i_{L1} and i_{L2}, and (b) output current $i_o = (i_{L1} + i_{L2})$.

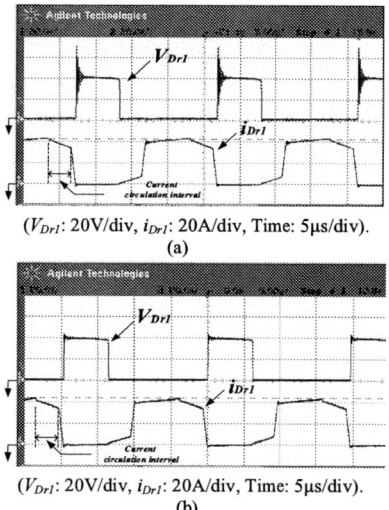

(V_{Dr1}: 20V/div, i_{Dr1}: 20A/div, Time: 5µs/div).

(a)

(V_{Dr1}: 20V/div, i_{Dr1}: 20A/div, Time: 5µs/div).

(b)

Fig. 9. Measured waveforms of the voltage and current of the rectifier diodes: (a) without snubber (b) with RCD snubber.

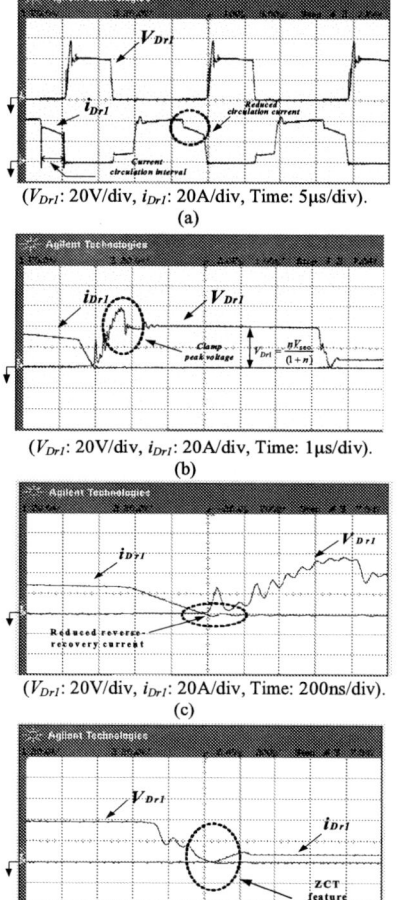

(V_{Dr1}: 20V/div, i_{Dr1}: 20A/div, Time: 5µs/div).

(a)

(V_{Dr1}: 20V/div, i_{Dr1}: 20A/div, Time: 1µs/div).

(b)

(V_{Dr1}: 20V/div, i_{Dr1}: 20A/div, Time: 200ns/div).

(c)

(V_{Dr1}: 20V/div, i_{Dr1}: 20A/div, Time: 200ns/div).

(d)

Fig. 10. Measured waveforms of the voltage and current of the rectifier diodes: (a) with lossless snubber (b) its extended waveforms (c) reduced reverse-recovery current at turn-off transition (d) ZCT feature at turn-on transition.

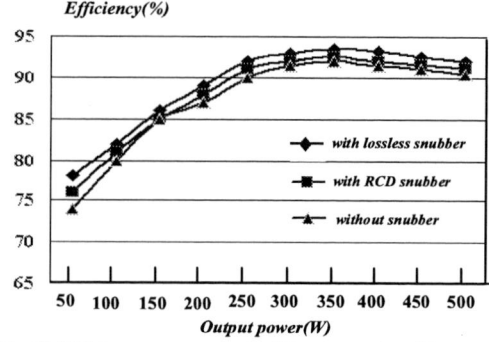

Fig. 11. Efficiency measurements of the proposed rectifier with and without passive snubbers.

V. CONCLUSIONS

In this paper, an improved current-doubler rectifier with coupled inductors has been proposed and analyzed. The proposed rectifier has the merits of low output current ripple and extended duty ratio, which can reduce the peak current through the isolation transformer, the conduction loss of the switches and the copper loss of the isolation transformer. When associated with a passive lossless snubber, it can clamp the peak voltage of the rectifier diodes and reduce the circulation current of the proposed converter significantly. Experimental results have verified that the proposed rectifier can achieve high efficiency over a wide load range. It is relatively suitable for high step-down voltage ratio and high output current applications.

REFERENCES

[1] O. D. Patterson and D. M. Divan, "Pseudo-Resonant Full Bridge DC/DC Converter," *Proceedings of Power Electronics Specialists Conf.*, June. 1987, pp. 424-430.

[2] J. G. Cho, *et al.*, "Zero-Voltage and Zero-Current-Switching Full-Bridge PWM Converter for High Power Application," *IEEE Trans. on Power Electronics*, July 1996, Vol. 11, No. 4, pp. 622-628.

[3] A. Pietkiewicz and D. Tollik, "Coupled-Inductor Currenr-Doubler Topology in Phase-Shift Full-Bridge DC-DC Converter," *Proceedings of Telecommunications Energy Conf.*, Oct. 1998, pp.41-48.

[4] M. Xu, J. Zhou and F. C. Lee, "A Current-Triple DC/DC Converter," *IEEE Trans. on Power Electronics*, May 2004, Vol. 19, No. 3, pp. 693-700.

[5] K. Chen and T. A. Stuart, "A 1.6 kW 110 kHz dc/dc Converters Optimized for IGBTs," *IEEE Trans. on Power Electronics*, 1993, Vol. 8, No. 1, pp.18-25.

[6] B. J. Masserant, J. L. Shriver and T. A. Stuart, "A 10 kW dc/dc Converter Using IGBTs with Active Snubbers," *IEEE Trans. on Aerospace and Electronic Systems*, 1993, Vol. 29, No. 3, pp.857-865.

[7] Jung-Goo Cho, *et al.*, "Novel Zero-Voltage and Zero-Current-Switching Full-Bridge PWM Converter Using a Simple Auxiliary Circuit," *IEEE Trans. on Industry Applications*, Jan.-Feb. 1999, Vol. 35, No.1, pp.15-20.

2006 5th International Power Electronics and Motion Control Conference

Monolithic Integration of Trench Power JFET with Schottky Diode

Yang Gao, Jie Chen*, Alex. Q. Huang
Semiconductor Power Electronics Center (SPEC)
North Carolina State University, Raleigh, NC 27695, USA
E-mail: ygao3@ncsu.edu
* College of Computer Science & Information Engineering, Zhejiang Gongshang University, P.R.China, Hangzhou 310018

Abstract—A monolithic integration of trench power JFET with schottky diode is proposed and analyzed. A unit JFET cell pitch of 1.1 um can be obtained. The specific on-resistance of the device is reduced to 14.4 mΩ·mm² which is close to state-of-art of power MOSFET. Two approaches for the integrated schottky diode – junction barrier schottky (JBS) and planar schottky diode (PSD) --are analyzed and compared. The integrated JBS diode shows 28% and 30% reduction while the integrated PSD shows 30% and 32% reduction on forward voltage drop and reverse recovery charge respectively compared with its p-n counterpart from the same integration technology.

Keywords-trench; JFET; power; schottky diode; monolithic

I. INTRUDUCTION

Higher power conversion efficiency at higher operating frequency is required for next generation microprocessors. Usually synchronous Buck converter is used to meet such requirements. It is important to reduce the power losses for the transistors used in the DC-DC converters. For the ControlFET, this implies the reduction of the on-resistance (Rds,on), and the gate-drain charge density Qgd. The Figure-Of-Merit: FOM= Rds,on*Qgd is generally used in order to describe the switching performance of the ControlFET. For the SyncFET it's more important to reduce the specific on-resistance Rds,on in order to reduce the conduction loss. Trench MOSFETs have been the dominant devices for the SyncFET due to its extremely low Rds,on that can be achieved today [1-3]. However, due to the inherent body diode of conventional trench DMOSFET, a significant portion of the power loss is generated during the dead time; especially when the switching frequency is pushed higher in order to reduce the output capacitance. One technique is to combine a schottky diode in parallel with MOSFET. Despite the dominant usage of MOSFET as low-side switch, trench JFET still could be an attractive alternative solution for the low-side switch due to its comparable low Rds,on and much simpler process technology. A trench JFET in parallel with a schottky diode has been developed [4]. A concern here is the parasitic inductance between the JFET and the schottky diode. To address this issue, we propose a novel structure to integrate the schottky diode onto a baseline trench power JFET process and layout. This monolithic solution also allows considerably more flexibility in implementing the schottky structure.

II. DEVICE STRUCTURE AND PROCESS DISCRIPTION

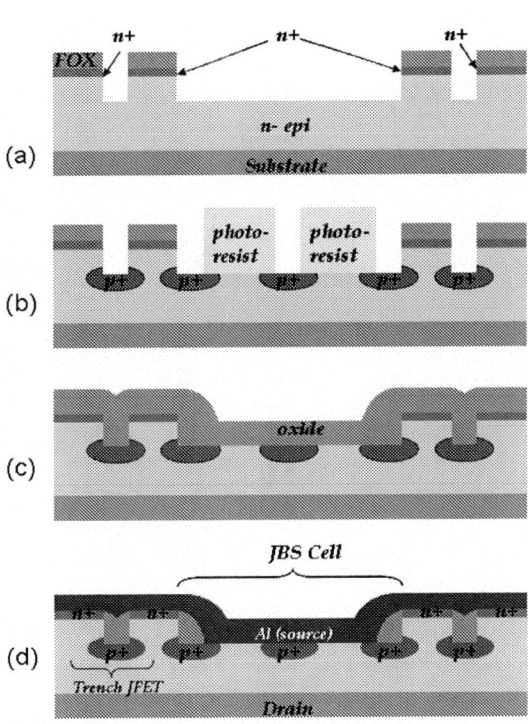

Figure 1. Key process steps for the proposed JBS structure

The reverse recovery charge Qrr in PSD is much smaller than its p-n counterpart because it is a unipolar device.

1-4244-0448-7/06/$25.00 ©2006 IEEE   113

However, the reverse bias leakage current in PSD interface is generally higher due to a lower junction barrier and surface interface state. JBS is a common method to alleviate the reverse leakage [5]. Fig. 1 illustrates the key process steps for the proposed trench JFET with integrated JBS structure. a) n+ source formation by blank Arsenic implantation with a dose of $10^{16}/cm^2$ and energy of 100 keV, trench formation by applying trench mask; b) photo resist (PR) spin and applying the 2nd mask for low energy p+ ($\sim 8 \times 10^{14}/cm^2$) implantation and nitrogen annealing at high temperature to form the gate as well as the p+ region in JBS cell; c) PR removal followed by trench refilling by LPCVD oxide deposition, d) source and gate contact opening followed by final front and backside metal deposition to form sources, drains and gates. Based on the same integration technology, the PSD and p-n diode structure can be fabricated by controlling the openings for p+ implantation in JBS cell area. The final cross-sections for these three structures are illustrated in Fig 2.

Fig. 3 shows the proposed layout of the trench JFET with integrated schottky diode. The active FET area is about 8.1 mm2. For the JFET cell, optimizations of dimension of cell pitch (CP), concentration of p+ gate and the n- epitaxy layer are the key issues to reduce Rds,on while keeping a high avalanche capability; while for schottky diode, design parameters need to be chosen carefully so as to balance the forward voltage drop, reverse leakage current as well as reverse recovery charge.

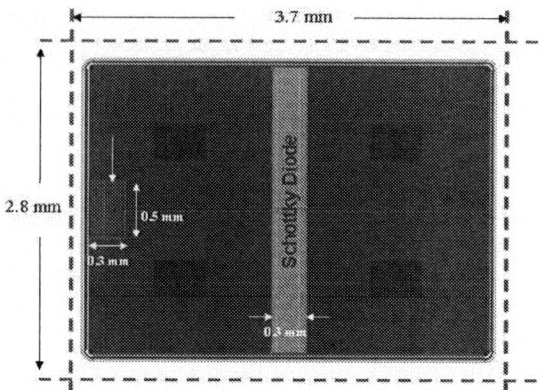

Figure 3. Layout of the proposed trench JFET with integrated schottky diode

III. SIMULATIONS AND RESULTS

2-D device and process simulations were carried out by using ISE-TCAD [6]. The cross section of JFET half cell with potential contours from process and device simulation is illustrated in figure 4 for the case of gate biases of 0 and -4V for a cell pitch of 1.1 µm. It can be seen that the n- region between two adjacent p+ gates is not fully depleted at zero gate bias. Under this condition, the drain current is determined by the resistance of un-depleted n-region between the source and drain. In our proposed trench JFET with integrated schottky diode, there are three mechanisms which will affect the breakdown of the whole device. The first mechanism is so called "gate barrier lowering effect". As shown in Fig. 4 (right), the n- region between the two adjacent gates is completely depleted at a gate bias of -4V. And a voltage barrier exists between the source and drain for the electrons due to the positive space charge in the depletion region. Fig. 5 shows the simulated energy band diagram as well as the gate barrier height Φ_B along the cut line for gate bias of -4V and drain to source bias of 0V. This barrier will become smaller if the drain bias increases due to the increased electric field in the depletion region. Fig. 6 illustrates this gate barrier lowering effect at CP = 1.3µm. The current density from source to drain under this gate barrier can be adequately described by the thermionic emission theory [7]:

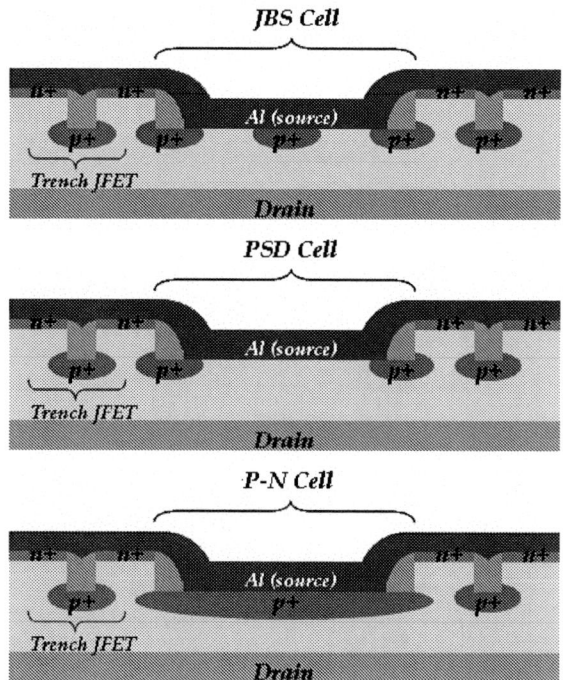

Figure 2. Cross sections of integrated trench JFET with JBS, PSD and p-n diode respectively under the same integration technology

$$J_{GB} = A^* T^2 \exp(-\frac{q\phi_B}{kT}) \qquad (1)$$

Where T is the temperature, and

$$A^* = \frac{4\pi \cdot qm^* k^2}{h^3} \qquad (2)$$

is the effective Richardson constant for thermionic emission, m* is the effective mass of the silicon. From equation (1), we can see that the J_{GB} changes exponentially in terms of Φ_B. A

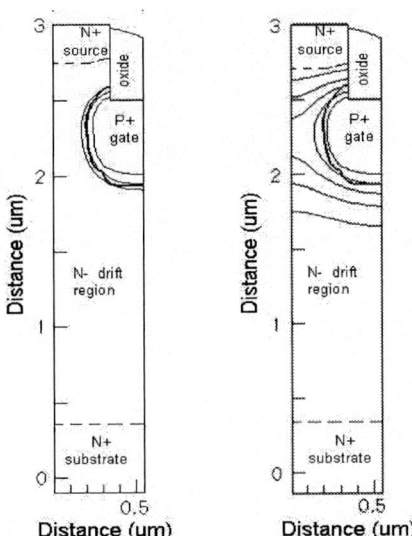

Figure 4. Potential contours obtained by simulation for gate bias of 0V(left), and gate bias of -4V(right)

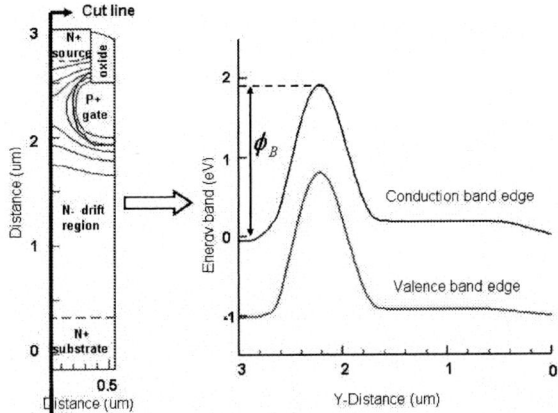

Figure 5. Simulated energy band diagram along the cut line of the JFET half cell with gate barrier defined

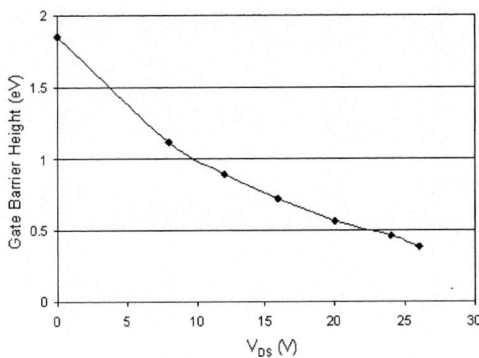

Figure 6. Influence of drain to source voltage on the gate barrier at CP = 1.3μm and V_G=-4V

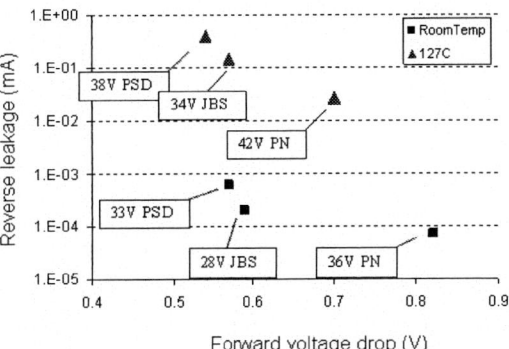

Figure 7. Trade-off between reverse leakage current at V_R=20V, V_{GS}=-4V and forward voltage drop at I_F=10A for PSD, JBS and P-N diode fabricated by the same integration technology under room and elevated temperatures. The breakdown for each structure is also marked in this graph

drastic increasing of J_{GB} happens when Φ_B is low enough. The breakdown current flows from drain to source at this breakdown mechanism. The second mechanism for the drain current surge is the avalanche breakdown between the gate and drain and the breakdown current will flow from drain to gate in this case. The third breakdown mechanism comes from the breakdown of the schottky diode. The smallest of these three voltages determines the overall breakdown voltage of the device.

It is also found that both Rdson and breakdown voltage will increase when the CP is reduced. Thus an optimal value for the CP exists by considering the trade-off between BV_{DS} and $R_{dson,sp}$. In our final design, CP = 1.1μm and BV_{DS} and $R_{dson,sp}$ are 27V and 14.4 mΩ-mm^2 respectively.

The on-state and off-state characteristics of the integrated schottky diode are illustrated in Fig.7. While to keep the reverse leakage smaller than 0.5mA (this corresponds to ~0.1% efficiency degradation in a typical DC/DC buck converter), the PSD and JBS show 30% and 28% reduction on forward drop respectively compared to their p-n counterpart.

Fig. 8 (a & b) shows the mix-mode circuit and the reverse recovery current waveforms for the PSD, JBS and their p-n counterpart at I_F = 20A and di/di = 100A/μs. Either schottky structure shows reduced trr and Irr over its p-n counterpart. The stored charge (Qrr) can be expressed as:

$$Q_{rr} = \int_{t_0}^{t_0+t_{rr}} i_{ds}(t)dt \qquad (4)$$

Based upon Fig.8, Qrr for the PSD and JBS structure are 2.9 nC and 3.0 nC respectively, which are 32% and 30% lower than the Qrr in p-n diode.

115

The electrical characteristics of the designed device are listed in Table 1, in comparison with a typical commercial available trench JFET with schottky diode in parallel [4]. It can be seen that the proposed trench JFET with implanted p+ trench gate shows a lower on-resistance and better blocking capability comparing to commercial power JFET. Both integrated schottky structures show more than 20% improvement on forward voltage drop and more than 60% reduction on Qrr compared to trench JFET with schottky diode in parallel, which translate to much lower power losses during the dead time in the DC/DC converter application.

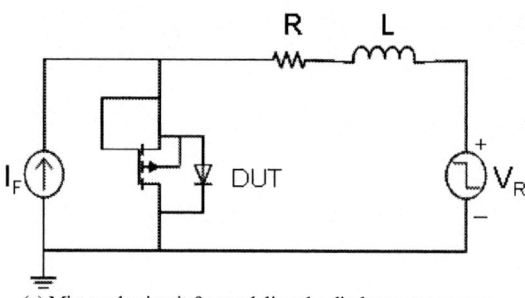

(a) Mix-mode circuit for modeling the diode reverse recovery

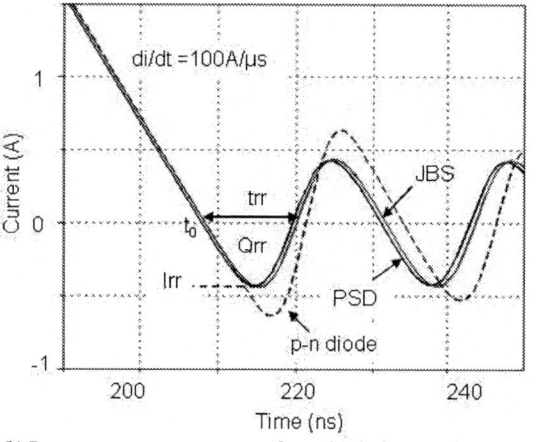

(b) Reverse recovery current waveforms for the integrated JBS and its p-n counterpart

Figure. 8 Test circuit and current waveforms for the diode reverse recovery

IV. CONCLUSIONS

A novel technique for monolithically integration of trench power JFET and schottky diode is proposed in order to achieve higher cell density, comparable on-resistance and blocking capability with the trench MOSFET. Two approaches for the integrated schottky diode structure are analyzed and compared to their p-n counterpart. Both schottky structures show a lower forward voltage drop and smaller stored charge than their p-n

counterpart. The overall performance of the proposed device shows that it is a good candidate for the low-side switch in power DC-DC converter.

TABLE 1

SUMMARY OF THE ELECTRIC CHARACTERISTICS OF THE PROPOSED STRUCTURES, IN COMPARASON WITH A TYPICAL COMMERCIAL AVAILABLE TRENCH JFET WITH SCHOTTKY DIODE IN PARALLEL

	Lovoltech Trench JFET[4]	This work	
Active FET Area	8.1 mm^2	8.1 mm^2	
BV$_{DSX}$ (I$_D$=0.5mA, V$_{GS}$=-4V)	24 V	27 V	
BV$_{GDO}$ (I$_G$=-50μA)	-28 V	-31 V	
BV$_{GSO}$ (I$_G$=-1mA)	-12 V	-12 V	
Gate Threshold Voltage	-0.8V	-0.84 V	
Rds,on (I$_G$=10mA, I$_D$=10A)	4.5 mΩ	1.8 mΩ	
Q$_{GD}$ (I$_D$=10A, V$_{DS}$=15V)	12 nC	17.8 nC	
Rds,on· Q$_{GD}$	54 mΩnC	37.1 mΩnC	
	Paralleled schottky diode	Integrated PSD	Integrated JBS
I$_R$ (V$_R$=20V, V$_{GS}$=-4V)	0.25 mA	0.4 mA	0.15 mA
V$_F$ (I$_F$=10A)	0.7 V	0.54 V	0.58 V
V$_F$ (I$_F$=20A)	0.9 V	0.66 V	0.72 V
Qrr(di/dt = 100A/μs)	8 nC	2.9 nC	3 nC

REFERENCE

[1] Jun Zeng et al "An Ultra Dense Trench-Gated Power MOSFET Technology Using A Self-Aligned Process" ISPSD2002, pp.147-150

[2] S.Ono et al "30V New Fine Trench MOSFET with Ultra Low On Resistance" ISPSD2003, pp.28-31

[3] M.Darwish et al "A New Power W-Gate Trench MOSFET with High Switching Performance" ISPSD2003, pp.28-31

[4] Lovoltech Data Sheet, "PWRLITE LD1003S"

[5] M.Mehrotra, B.J.Baliga,"Very low forward drop JBS rectifiers fabricated using submicron technology," IEEE Trans. on Elec. Dev, v.40, No 11, pp2131-2132, 1993

[6] The ISE-TCAD program from Synopsis Inc.

[7] S.M.Sze, Physics of Semiconductor Devices. 2nd Edition, New York, Wiley-Interscience,1985

2006 5th International Power Electronics and Motion Control Conference

Sequential Color LED Back-Light Driving System for LCD Panels

C.-C. Chen[*], C.-Y. Wu, P.-C. Lu, Y.-M. Chen and T.-F. Wu

Elegant Power Application Research Center
(EPARC)
Department of Electrical Engineering
National Chung Cheng University
Ming-Hsiung, Chia-Yi, Taiwan, China.
E-mail: tfwu@ee.ccu.edu.tw
Tel: 886-5-2428159; Fax: 886-5-2720862

[*]Department of Electrical Engineering
Nan-Jeon Institute of Technology
Yen-Shui, Tainan, Taiwan, China
E-mail: ccchen.johnson@msa.hinet.net
Tel: 886-6-6523111 ext. 241

Abstract—In this paper, a sequential color light emitting diode (LED) backlight driving system for liquid crystal display (LCD) panels is proposed. Due to improvement on luminous efficacy, long life and wide color gamut, LED has gradually substituted for cold cathode fluorescent lamp as backlight. The proposed driving system adopts sequential color scanning scheme to improve light utilization efficiency by removing color filter. To meet display performance requirement where color variation □uv needs to be limited below 0.002, this paper also proposes a family of output current ripple free topologies to reduce color variation caused by ripple current. In addition, a driving voltage reseter is introduced to resolve current-spike problem, improving backlight reliability and availability. A forward-type current ripple free converter has been built, from which experimental measurements have verified the discussed performance and feasibility of the proposed system.

Keywords-sequential color display; LED; LCD

I. INTRODUCTION

In an LCD TV, the backlight driving system consumes most of the power due to its extremely low light utilization efficiency (LUE). Fig. 1 shows the layout of an LCD display panel with backlight and the LUE of each component [1]. Without the color filter, LUE can be improved about 2.7 times. In literature [2]-[5], a sequential color scanning scheme is proposed to reduce power consumption by removing color filter. In a sequential color display, a frame is decomposed into RGB sub-frames, which are displayed sequentially at short time intervals so as the human eye cannot perceive the color latency. Recently, due to improvement on luminous efficacy, long life and wide color gamut, LED gradually substituted for cold cathode fluorescent lamp as backlight [6]-[9]. Furthermore, for their fast response, LEDs are very suitable for a sequential color display. Thus, this paper proposes a sequential color LED back-light driving system for LCD panels.

To achieve dimming feature for LCD TV, the current of backlight LEDs can be controlled with either variable current or current-type burst-mode control. The luminance of LED increases with its forward current,

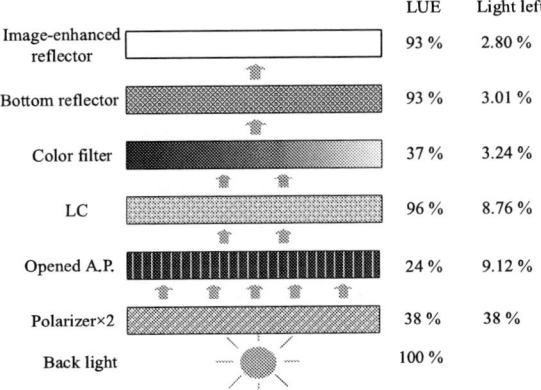

Fig. 1. Layout of an LCD display panel with back light and the LUE of each component.

but the chromaticity of the LED light output varies with the driving current levels, as shown in Fig. 2 [7]. Thus, it is difficult to maintain the desired white point at the chromaticity coordinate with variable current. Although, a burst-mode dimming approach can be readily implemented to control the desired chromaticity, the ripple current would influence the accuracy. In LCD display, a color variation of $\triangle uv > 0.002$ on a white screen may not meet performance requirement. Tracing the curves in Fig. 2, the ripple level of the driving currents for blue and green LEDs must be limited below 5 % to meet the requirement. In this paper, a family of current ripple free topologies is proposed to achieve low ripple current and accurate chromaticity.

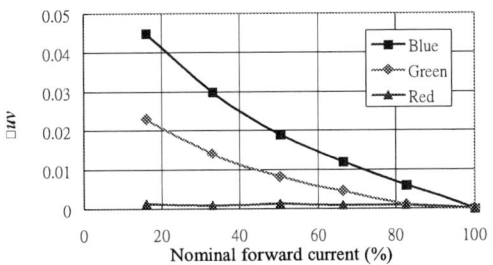

Fig. 2. Plots of the variation of RGB LED chromaticity versus the forward current under a constant temperature.

1-4244-0448-7/06/$25.00 ©2006 IEEE

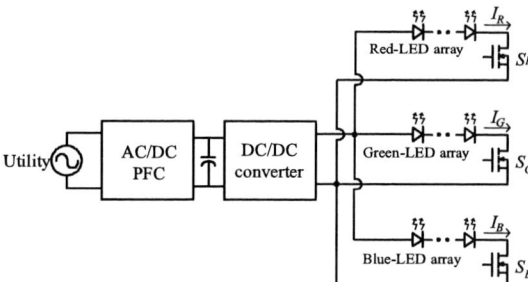

Fig. 3. The proposed driving system for sequential color display.

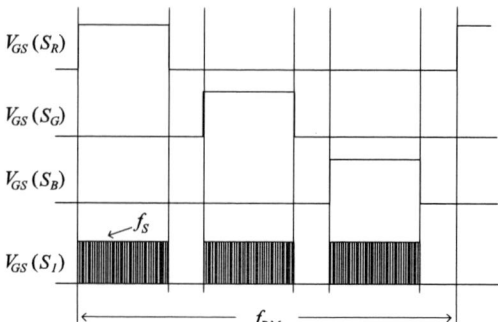

Fig. 4. Driving signals of the proposed system.

II. OPERATION OF THE PROPOSED SYSTEM

The proposed sequential color LED back-light driving system, as shown in Fig. 3, consists of a boost converter for power factor correction (PFC), a step-down dc/dc converter with isolation, and RGB LED arrays. In addition, a family of current ripple free topologies is introduced to the dc/dc converter to improve chromaticity accuracy. In this section, the emphasis is focused on the operation of the proposed driving system and ripple free topologies, which are explained briefly as follows:

A. Driving System

To achieve sequential color display, red, green and blue LEDs are driven with phase-shift manner, as illustrated by the gate signals shown in Fig. 4, in which $V_{GS}(S_R)$, $V_{GS}(S_G)$, $V_{GS}(S_B)$ are the gate signals of switches S_R, S_G and S_B, respectively, and $V_{GS}(S_I)$ is the signal for the switch in the dc/dc converter. The switching frequency f_{BM} of S_R, S_G and S_B is equal to the frame rate, such as 60, 120 or 180 Hz, and that of the switch in the dc/dc converter is f_s which is much higher than f_{BM}. With a sequential driving scheme, many components of the drivers can be shared each other so as its volume and cost can be reduced significantly. In the driving system, RGB LEDs are driven by the same circuits, except that switches S_R, S_G and S_B are controlled to be turned on sequentially to attain light-scanning feature.

B. Ripple Free Topologies

Fig. 5(a) shows a transformer model to illustrate a current ripple free concept [10]. With free current ripple, the voltage across L_2 must be zero; i.e., $V_{L2} = 0$. Thus, the model can be simplified to the one shown in Fig. 5(b), where

$$V_s = (L_1 + L_c)\frac{di_s}{dt} \qquad (1)$$

and

$$\left(\frac{N_p}{N_s}\right)aV_s = L_c\frac{di_s}{dt}. \qquad (2)$$

Combining these two equations yields

$$L_1 = L_c\left(\frac{N_s}{aN_p} - 1\right), \qquad (3)$$

which determines a condition for achieving free current ripple. This model can be implemented in the basic converter topologies, and a family of output current ripple free topologies can be therefore derived, as shown in Fig. 6, in which the output inductor of each basic converter is replaced with the secondary winding of the model shown in Fig. 5(a).

III. DESIGN AND PRACTICAL CONSIDERATION

In a practical circuit design, the following issues: choosing topologies, determining component values and designing driving voltage reseter should be considered and they are described in detail as follows:

A. Topologies

Fig. 6 shows a family of output current ripple free topologies, and each of which can be readily implemented in the proposed driving system to improve chromaticity accuracy for RGB LEDs. However, if non-isolation topologies are adopted, it needs an additional dc/dc converter with galvanic isolation. Thus, converter topologies with isolation are preferred. In this paper, a forward-type current ripple free topology is adopted, as shown in Fig. 6(d).

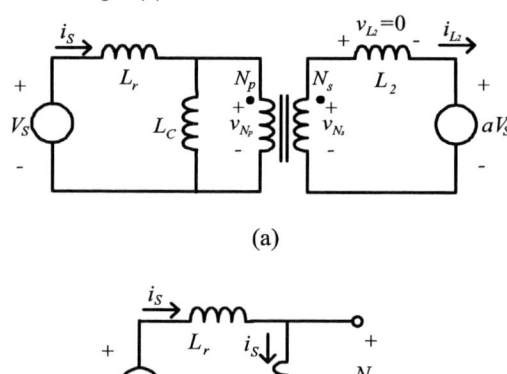

(a)

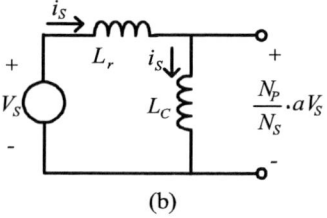

(b)

Fig. 5. Circuit model for a transformer with free current ripple.

118

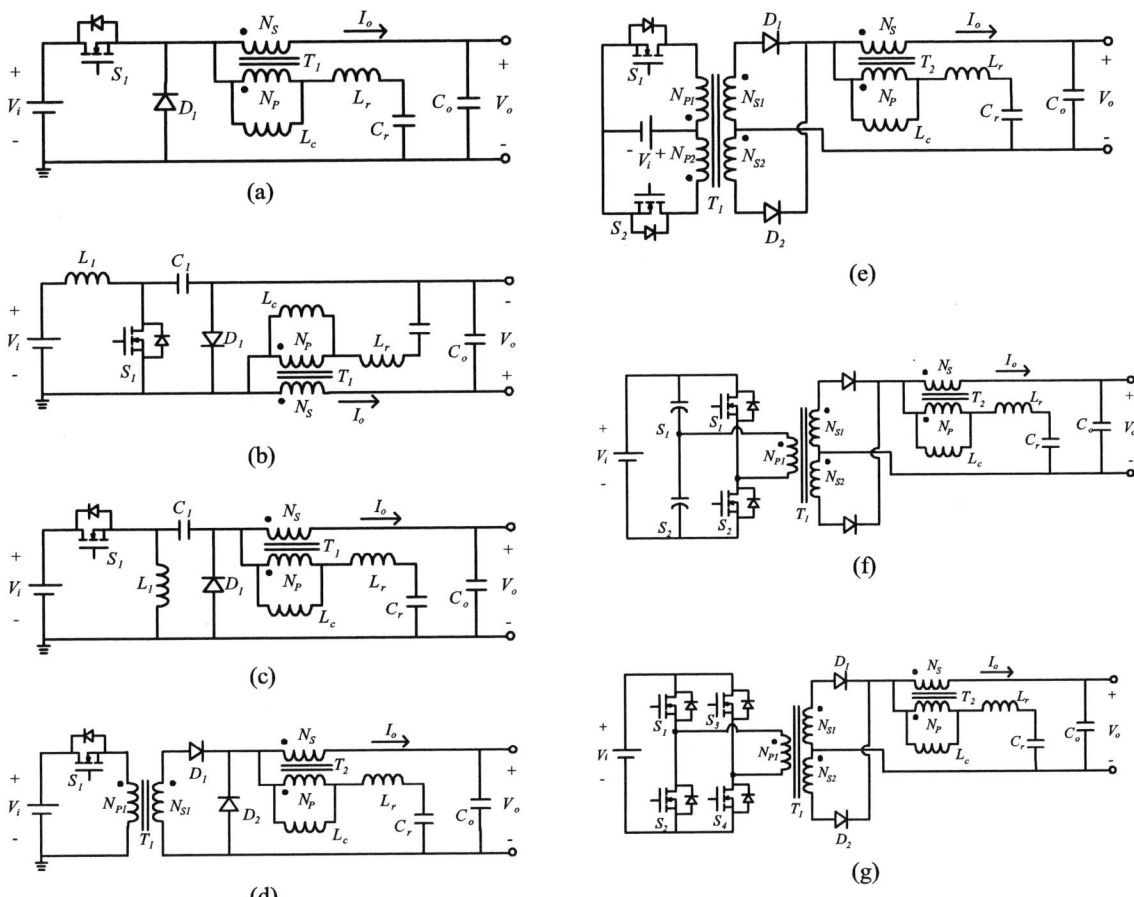

Fig. 6. A family of output current ripple free topologies: (a) Buck, (b) Cuk, (c) Zeta, (d) Forward, (e) Push-Pull, (f) Half-Bridge and (g) Full-Bridge.

B. Determination of Ripple Free Components

For a conventional forward converter, the output current ripple can be expressed as follows:

$$\Delta I_o = \frac{V_o T_s}{L}(1-D) \tag{4}$$

$$= (\frac{N_{s1}}{N_{p1}} \cdot DV_i) \times \frac{T_s}{L}(1-D) \cdot \tag{5}$$

With a specified current ripple, inductor L can be determined. For lowering current ripple, it needs to operate the converter at CCM. Thus, inductor L should satisfy the following inequality:

$$L > \frac{V_o^2 \cdot T_s}{2P_{O,min}}\left(1 - \frac{V_o}{V_{I,min}}\right) \cdot \tag{6}$$

To achieve extremely low current ripple, the inductance will be unreasonably large, which will increase the volume, weight and cost of the converter and worsen the response. This problem can be alleviated by adopting output current ripple free topologies. According (3), the turns ratio and inductance of the transformer with free output current ripple can be determined. Additionally, to make the voltage across capacitor C_c and C_o identical, parameter a in (3) must be designed to be unity. It is usually to select $N_s = 2N_p$, and L_c is therefore equal to L_1.

C. Driving Voltage Reseter

Since the LED arrays might have different forward voltages, a transition from a high voltage to a low one will cause current spike. For instance, it will happen at the transition between green and blue LED arrays. To solve this problem, the output voltage should be controlled to drop to a certain low voltage V_{set} in each R-G-B lighting transition. A driving voltage reseter, as shown in Fig. 7, is proposed to eliminate current spike induced at lighting transition. In the figure, SCR_1 is triggered when switches S_R, S_G and S_B are turned off. In a practical circuit design, the rising edges of the lighting synchronous signal can be used to trigger SCR_1 while its falling edges can be used to synchronize S_R, S_G and S_B gate signals sequentially. It should be noticed that the breakdown

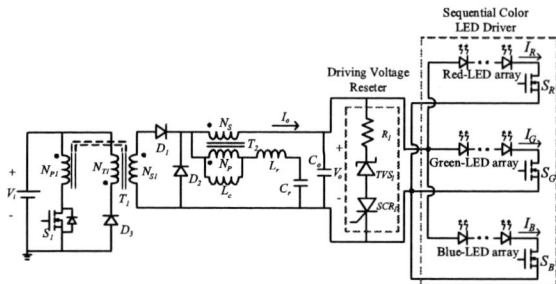

Fig. 7. The proposed driving system with the forward converter, current spike suppressor and the sequential color LED driver

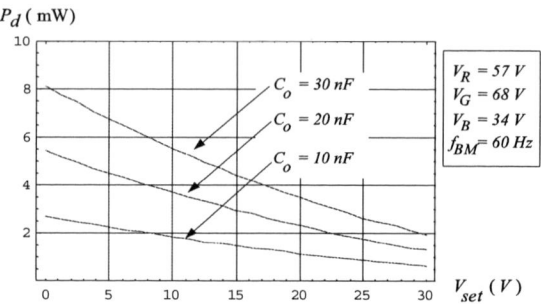

Fig. 8. Plots of the power dissipation P_d versus setting voltage V_{set} under different output capacitance C_o.

voltage of TVS_1 must be lower than the lowest forward voltage of LED arrays to avoid current spikes. The power dissipation P_d of the driving voltage reseter can be estimated as

$$P_d = \frac{1}{2} C_o \left[(V_R - V_{set})^2 + (V_G - V_{set})^2 + (V_B - V_{set})^2 \right] \times f_{BM}. \quad (18)$$

The curves showing the relationships between P_d and V_{set} for different values of output capacitance C_o are illustrated in Fig. 8, which reveal that P_d is negligible.

IV. EXPERIMENTAL RESULTS

A forward-type current ripple free converter shown in Fig. 7 was implemented to verify the theoretical analysis. The converter has been designed according to the following specifications:

- input voltage: $V_i = 400\ V_{dc}$,
- output current: $I_o = 350$ mA,
- switching frequency: $f_S = 100$ kHz,
- burst-mode dimming frequency: $f_{BM} = 60$ Hz.

The components of the power stage were designed as follows:

- S_1: FS7KM-16A,
- D_1 and D_2: ES3D,
- D_3: HFA08TP120,
- C_o: 10 nF,   • C_r: 100 nF,
- L_r: 500 μH,   • L_c: 500 μH,
- T_1 core: EI-33,   • T_2 core: ER-28,
- N_{P1}: N_{S1}: $N_{T1} = 2:1:3$,
- N_P: $N_S = 1:2$,
- LED array: Luxeon emitter LXHL-BM01 ×20,
                      LXHL-BD01 ×20, LXHL-BB01 ×10.

Fig. 9 shows waveforms of output current I_{o1}, illustrating the effects of the ripple free technique. With a conventional forward converter, the ripple current is around 41 mA, which is 11.7 % of the output current. With the forward-type current ripple free topology, the ripple current is only around 10 mA, which is 2.9 % of the output current. Since the ripple current is limited be-

low 5 % of the output current, the chromaticity variation $\triangle uv$ will not exceed 0.002, which can meet the display performance requirement. It reveals that the ripple free topology can reduce ripple current and chromaticity variation feasibly. Fig. 10 shows measured waveforms of output voltage V_o and current I_B flowing through blue LED arrays. Without the driving voltage reseter, the current spike is around 557mA, which would not only cause a chromaticity variation but also break LEDs. Fig. 10(c) illustrates that the driving voltage reseter can resolve the current-spike problem entirely.

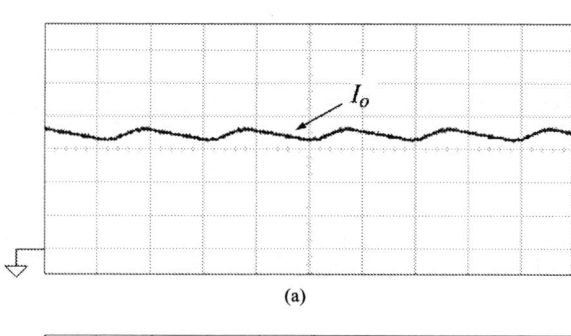

(a)

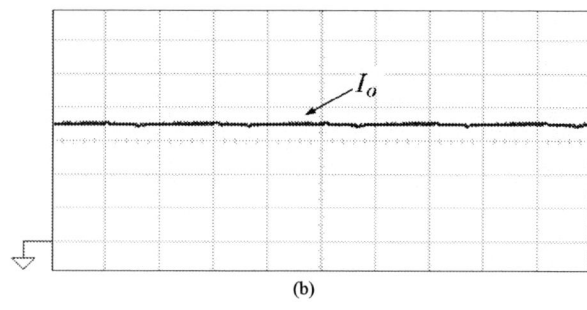

(b)

(I_o: 100 mA/div, Time: 5 μs/div)

Fig. 9. Measured waveforms of output current I_o illustrating the effects of the ripple free technique: (a) with a conventional forward converter and (b) with the forward-type current ripple free topology.

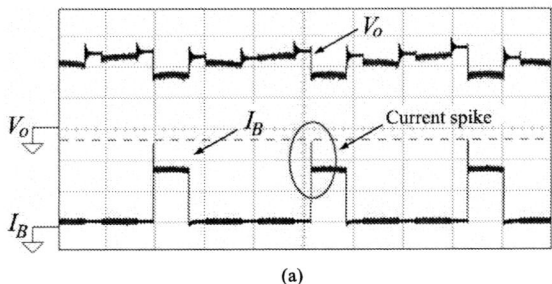

(a)

(V_o: 20 V/div, I_B: 200 mA/div, Time: 5 ms/div)

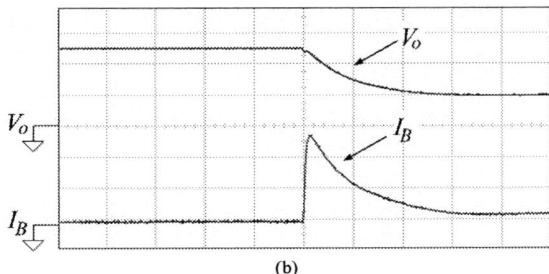

(b)

(V_o: 20 V/div, I_B: 200 mA/div, Time: 500 ns/div)

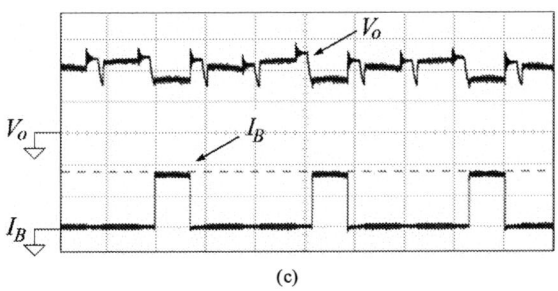

(c)

(V_o: 20 V/div, I_B: 200 mA/div, Time: 5 ms/div)

Fig. 10. Measured waveforms of output voltage V_o and current I_B illustrating the effects of the driving voltage reseter: (a) without the driving voltage reseter, (b) the expanded waveforms of (a), and (c) with the driving voltage reseter.

V. CONCLUSION

In this paper, a sequential color LED backlight driving system for LCD panels has been proposed. By adopting sequential color display, power consumption can be reduced significantly due to no need of color filter. A family of output current ripple free topologies has been proposed to reduce chromaticity variation caused by ripple current. In addition, a current-spike problem can be resolved by utilizing the proposed driving voltage reseter. In other words, the proposed driving system has the following features: low power consumption, accurate light output and high reliability. Operation and design consideration of the driving system and ripple free topologies have been also addressed in detail. Experimental results have verified the discussion and feasibility of the proposed driving system.

REFERENCES

[1] H.-P. D. Shieh, *et al.*, "Single Cell-Gap Transflective Color TFT-LCD by Using Image-enhanced Reflector," *Proceedings of the Sixth Chinese Symposium Optoelectronics*, 2003, pp. 270-272.

[2] W-C. Cheng, "Power Minimization of LED Backlight in a Color Sequential Display," *Proceedings of SID*, 2005, pp. 1384-1387.

[3] I. Choi, H. Shim and N. Chang, "Low-Power Color TFT LCD Display for Hand-Held Embedded Systems," *Proceedings of Symp. on Low Power Electronics and Design*, 2002, pp. 112-117.

[4] F. Gatti, *et al.*, "Low Power Control Technique for TFT LCD Display," *Proceedings of Intl. Conf. Compilers, Architecture, and Synthesis for Embedded System*, 2002, pp. 218-224.

[5] W-C. Cheng and M. Pedram, "Power Minimization in a Back-lit TFT-LCD Display by Concurrent Brightness and Contrast Scaling," *Proceedings of Design Automation and Test in Europe*, 2004, pp. 252-259.

[6] A. Konno, Y. Yamamoto and T. Inuzuka, "RGB Color Control System for LED Backlights in IPS-LCD TVs," *Proceedings of SID*, 2005, pp. 1380-1383.

[7] S. Muthu, F. J. Schuurmans and M. D. Pashley, "Red, Green, and Blue LED Based White Light Generation: Issues and Control," *Proceedings of Industry Applications Conf.*, 2002, pp. 327-333.

[8] S. Muthu, F. J. Schuurmans and M. D. Pashley, "Red, Green, and Blue LEDs for White Light Illumination," *IEEE Trans. on Quantum Electronics*, Vol. 8, pp. 333-338, 2002.

[9] S. Muthu and J. Gaines, "Red, Green, and Blue LED-Based White Light Source: Implementation Challenges and Control Design," *in Proc. Industry Applications Conf.*, 2003, pp. 515-522.

[10] P. Severns Rudolf and Bloom Gordon, Modern DC-TO-DC Switchmode Power Converter Circuit, Van Nostrand Reinhold Company, New York, 1985.

2006 5th International Power Electronics and Motion Control Conference

Development of Large Capacity Programmable Harmonic Current Generator Based on Three-phase-four-wire Configuration

LIU Tao, ZHUO Fang, CHEN Bo, ZHAI Xi, WANG Zhao-an
School of Electrical Engineering, Xi'an Jiaotong University, Xi'an, China
zffz@mail.xjtu.edu.cn lukeflow@gmail.com

Abstract—the large capacity harmonic current generator based on three-phase four-wire configuration uses four-leg converter as its main circuit. So it is different from the traditional small capacity linear current source. The whole equipment can output steady harmonic current include the selective order harmonic (below 25th order), reactive current and unbalanced current. Besides, dynamic change process of harmonic current can also be accurately simulated from the equipment. The output harmonic current can be set from communicating with computer or the EPROM. The harmonic current generator can be applied to testing electrical measure equipment or debugging the harmonic current elimination equipments. This paper first introduces the operation principle and main circuit; and then analyzes the control circuit structure and presents one control method with voltage feedforward control and the output current DFT feedback control; lastly the simulation results and the experiment results based on 50kVA harmonic current generator are given in this paper. The results of the simulation and experiment verify well performance of the device even though for the large capacity.

Keywords- harmonic current generator; three-phase four-wire; voltage feedforward control; DFT feedback control;

I. INTRODUCTION

With the development and the application of power electronics technology, the harmonic component in the power system increases progressively. Many harmonic elimination equipments such as passive filter and active power filter are applied to suppress the harmonic and improve power quality [1]. However, how to test the suppression performance of these equipments accurately before application is a problem. Usually the harmonic current can be produced by using thyristor or diode rectifier with the resistor load. In this way, typical order harmonic current can be generated with large capacity fundamental active power wasted in the power supply. Besides, the dynamic harmonic current change process is difficult to meet and large capacity cooling apparatus is necessary.

The harmonic current generator based on three-phase-four-wire power system can solve this problem. This generator is based on four-leg VSI configuration, which is applicable for the three-phase-three-wire system or three-phase-four-wire system by controlling the 4th leg. The output harmonic current can be set from communicating

with the computer or the EPROM. The whole equipment can output steady harmonic current include the selective order harmonic (below 25th order), reactive current and unbalanced current, and the amplitude and phase of each order harmonic current can be defined independently in order to meet the needs of different non-linear loads. The harmonic current generator can be applied to testing electrical measure equipment or debugging the harmonic current elimination equipments. The equipment has a low cost, easy operation or debugging contrast with the conventional debug method. This equipment output power is up to 50kVA and maximal phase harmonic current is 80A. Besides, the whole equipment has a steady precision and dynamic performance. The research in this paper is based on 50kVA harmonic current generator that has been developed out. The connecting configuration figure is shown in Fig.1:

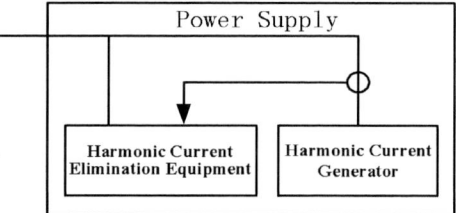

Fig.1 connecting configuration

II. OPERATION PRINCIPLE AND MAIN CIRCUIT CONFIGURATION

A. Operation Principle

The harmonic current generator is connected to the power system directly. The basic parameters include harmonic current orders, amplitude and phase angle of each phase can be programmed and the output harmonic current instruction can be generated from this information and the synchronous voltage signal. Later the output current instruction is disposed by the signal regulation circuit and transmitted to the current tracing circuit. The difference between the harmonic current instruction and the feedback of output current is compared with the triangle wave in order to generate the PWM drive pulse. The PWM converter receives the drive signal, and then the harmonic current is generated from the inductance between the power supply and the PWM converter. Because the switching frequency harmonic current also interfuses in the output harmonic current, the output filter is used to eliminate the switching frequency component.

1-4244-0448-7/06/$25.00 ©2006 IEEE   122

After filtering, the harmonic current injects directly into the power supply.

B. Main Circuit Configuration

The main circuit configuration based on three-phase four-wire system is divided into three-phase three-leg and three-phase four-leg structure according to the connection way[1]. The configurations are shown as the Fig.2:

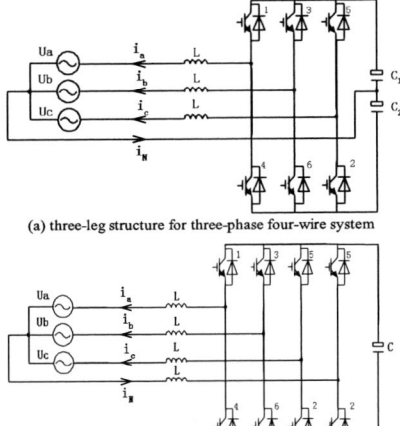

(a) three-leg structure for three-phase four-wire system

(b) four-leg structure for three-phase four-wire system

Fig2 two main circuit structures based on VSI

In Fig.2.(a), the voltage between two capacitors must be kept balance because the zero-sequence current is conducted by connecting N-wire with the midpoint of DC side. So the control system of this structure becomes more complex although it has fewer legs. Besides, the fluctuation of the DC side voltage resulting from the zero-sequence and negative-sequence current component makes capacity of the capacitors larger[2].

In Fig.2.(b), the additional leg is used to conduct the zero-sequence current, so the 4th leg can be controlled directly regardless of another three legs. And this configuration can be implemented easily and the capacity of the DC side capacitors is relatively small[3]. In this paper four-leg structure is adopted as the main circuit configuration of harmonic current generator. Fig.3 shows the main circuit configuration:

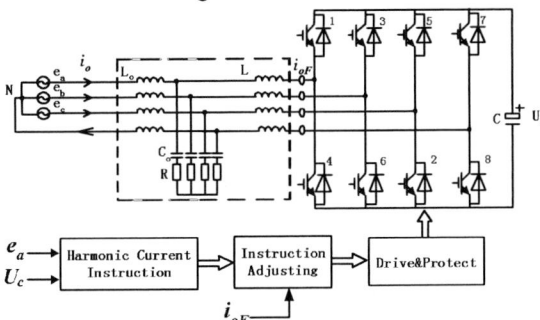

Fig.3 main circuit of the device

Moreover, maximum available peak voltage per leg (MVPL) ratio can be used as an index to express the effectiveness of the configuration of the VSI, where the utilization of power semiconductor capacity is evaluated to generate output voltages under the same dc side voltage condition. Therefore MVPL ratio can indicate the capability of voltage synthesis while considering the cost-effectiveness of the power circuit[4]. The two main circuit MVPL ratios are shown in the following table:

(a)	$\dfrac{V_{dc}}{2}\dfrac{3}{3}\dfrac{1}{V_{dc}} = 0.5$
(b)	$\left(\dfrac{V_{dc}}{\sqrt{3}} + \dfrac{V_{dc}}{2} - \dfrac{V_{dc}}{4\sqrt{3}}\right)\dfrac{3}{4}\dfrac{1}{V_{dc}} = 0.699$

In the table, (a) and (b) indicate three-leg and four-leg for three-phase four-wire system respectively. From the above table we can get that four-leg configuration has a higher MVPL ratio under the same DC side voltage, because the midpoint voltage of output ports can be controlled directly.

III. CONTROL CIRCUIT STRUCTURE

A. Control Strategy

The block diagram of control strategy of harmonic current generator is as shown in Fig.4.

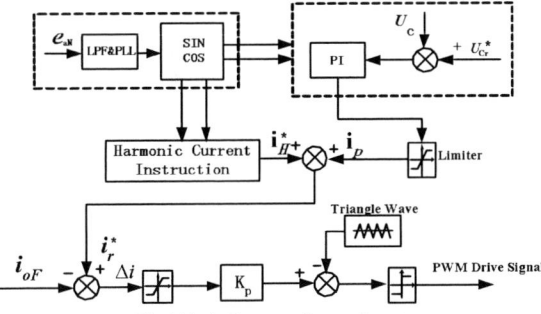

Fig.4 block diagram of control strategy

The programmable harmonic current generator obtains the sine signal that is synchronous to the power supply voltage in the synchronous circuit, and the corresponding sine table is produced consequently. After the amplitude and phase of output current are set in the program, the harmonic current instruction i_H* is produced. The variation DC side will have a bad influence on the stability and precision of the output harmonic current. In order to maintain the DC side at a constant voltage, the feedback control of DC side voltage is introduced in the control system. So i_p (positive-sequence of fundamental component) is included into the instruction current i_r*. Comparing with the output harmonic, the amplitude of i_p* is very small and no more than 5A in the steady state. After the difference between i_r* and i_{oF} (feedback signal of output current) is adjusted by K_p, PWM drive signal is produced by comparing the triangle wave with the difference signal in the current tracing circuit. Then the four-leg converter begins to generate harmonic current when the PWM drive signal is conducted to the gate of the IGBT.

The instruction circuit is based on the DSP digital circuit, and the address coding and protect function is realized from CPLD. All the current tracing function is completed in an analog circuit.

B. The Four-leg Instruction Generation and Realization Programmable Function

Three-phase harmonic current instruction can be obtained from defined output current information independently. Because the 4th leg is the channel of zero-sequence current component, so the 4th leg harmonic current instruction is related with the three-phase instruction. According to the symmetrical component law and the main circuit structure character, the 4th leg harmonic current instruction can be expressed in the following equation:

$$I_N = -(I_A + I_B + I_C)$$

Where I_N is the 4th leg instruction, $(I_A + I_B + I_C)$ is the zero-sequence component of the three-phase harmonic current instruction. Because of the reverse direction of the 4th leg current on the N wire, there is a minus in front of the $(I_A + I_B + I_C)$. So from computing the defined three-phase instruction, the 4th leg instruction can be produced; and the 4th leg instruction is not independent.

There are two ways to realize programmable function of the harmonic current generator. One is from the EPROM. All programs and the output current parameters are loaded into the fixed EPROM. This way is simple, but it's relatively complex when modifying output harmonic current parameters by erasing EPROM again and again. The other way of using series communication mode is adopted in this paper. The basic parameters of output harmonic current are loaded into DSP with the computer. This way is very convenient and effective; the communication interface between the harmonic current generator and computer is shown in Fig.5:

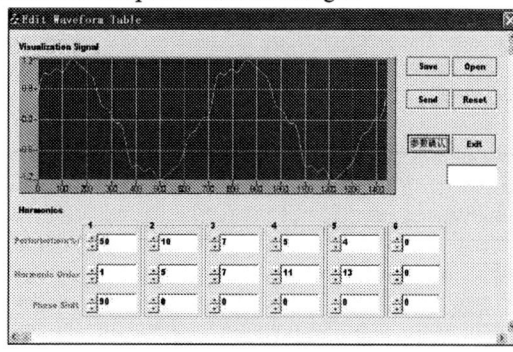

Fig.5 communication interface

C. Voltage Feedforward Control and Output Current DFT Feedback Control Method

When the switching frequency f_s of the harmonic current generator is far high than the modulation signal frequency even the signal isn't over-modified, there is a fixed proportional relationship between the modulation signal and the corresponding output voltage of converter. So the PWM converter can be regarded as a proportional adjuster [5], and its gain is related with DC side voltage

and the amplitude of the triangle wave. In this way, the equivalent model of the programmable harmonic current generator is shown in Fig.6:

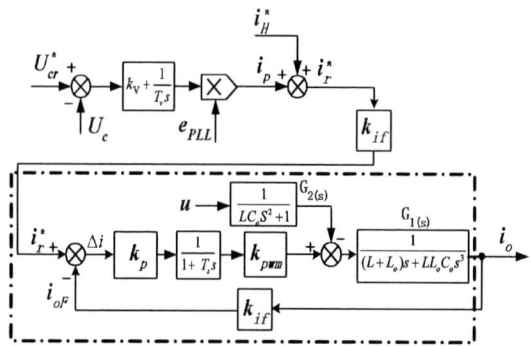

Fig.6 equivalent model of the harmonic generator

Where: u is phase voltage of power supply, U_c is DC side voltage, U_r^* is defined DC side voltage, e_{PLL} is standard sine fundamental positive-sequence component table, i_H^* defined harmonic instruction, i_r^* is final instruction, k_p and k_{if} is proportional coefficient, T_s is delay time of detecting and PWM switching [6]. The R and capacitor C_o of the switching ripple filter can be ignored, so the $G_1(s)$ and $G_2(s)$ are deduced as the following equation respectively:

$$G_1(s) = \frac{1}{(L + L_o)s}, G_2(s) = 1$$

From Fig.6 we can see that the current inner loop still includes power supply voltage u even if the DC side voltage is steady. The output current i_o is affected when the u fluctuates. When the influence of DC side and harmonic current instruction are not considered, the close-loop transfer function between the output current i_o and power supply voltage u can be presented in the following formula:

$$G_{iu}(s) = \frac{i_o}{u} = \frac{T_s + 1}{T_s(L + L_o)s^2 + (L + L_o)s + k_i k_{if} k_{pwm}}$$

The Bode diagram of $G_{iu}(s)$ is shown in Fig7, supposing that DC side voltage is constant.

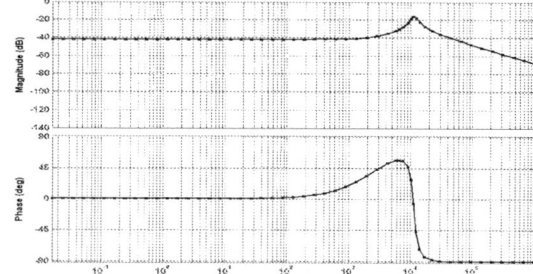

Fig.7 frequency response of output current

From Fig.7 we can see that the effect of suppressing for the fundamental frequency voltage is not good as for the high frequency harmonic voltage. Because the fluctuation of the power supply voltage affects the output current, power supply voltage feedforward control method is presented in this paper and the effect of power supply voltage u can be eliminated. This result can be validated by small signal analysis based on the transfer function $G_{iu}(s)$ of the system [2].

The results of simulation and experiment testify the realization of the mentioned strategy. The simulation waveforms for voltage feedforward control are shown in Fig.8 (based on PSIM simulation software):

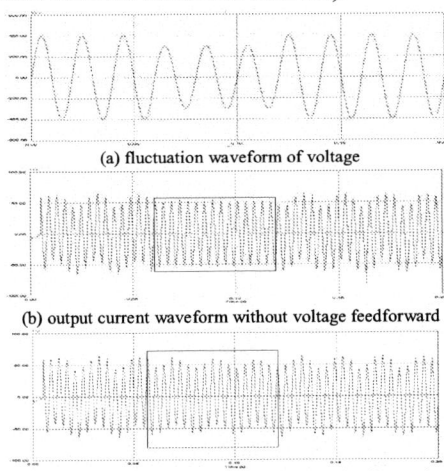

(a) fluctuation waveform of voltage

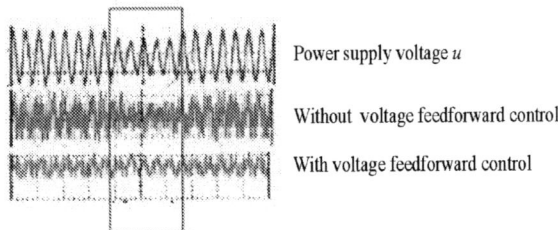

(b) output current waveform without voltage feedforward

(c) output current waveform with voltage feedforward

Fig.8 waveform of voltage feedforward control

The output current waveform with the voltage feedforward is different from without the voltage feedforward when power supply voltage u changes. It's obvious that the output current distortion resulting from the change of voltage u is attenuated when voltage feedforward control is introduced into the system. The experiment results are shown in Fig.9:

Power supply voltage u

Without voltage feedforward control

With voltage feedforward control

Fig.9 experiment results with voltage feedforward control

As a programmable harmonic current generator, the large capacity equipment presented in this paper must meet the need of output current precision. Although PI adjuster in the current loop can improve the steady output precision, the dynamic quality will become worse and the system stability will be affected badly. So the DFT feedback control of output current is introduced in this paper. From analyzing the output harmonic current with DFT, the available harmonic current component corresponding to the harmonic component defined in the instruction is introduced in the current loop to improve the output precision. It can be seen form the transfer function of this control method, appropriate DFT feedback coefficient (k_{f2} in the Fig.10) can improve output current precision and keep good performance on the system stability and dynamic character. The control block diagram of system is shown in Fig.10:

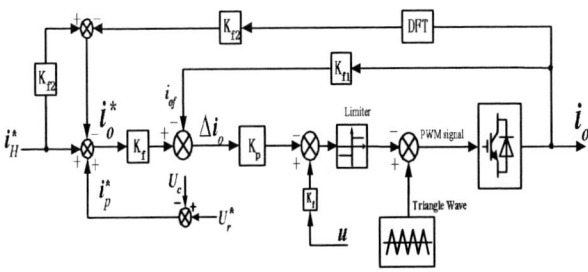

Fig.10 control block diagram of system

In order to verify the validity of this output current DFT feedback control, this control model in Fig.10 is simulated with Matlab6.5. The output harmonic current waveform and frequency spectrum analysis are shown in Fig.11:

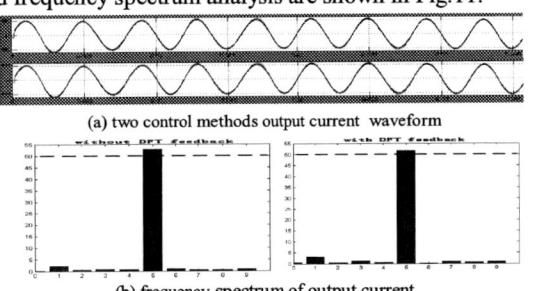

(a) two control methods output current waveform

(b) frequency spectrum of output current

Fig.11 waveform and spectrum of output current

In the Fig.11, the defined output current instruction is 5th harmonic current with amplitude 50A rms. Although the output wave with DFT feedback is almost the same as the one without DFT feedback, the frequency spectrum analysis of two control method are different. The 5th harmonic current component with DFT feedback control is closer to 50A than the one without DFT feedback.

IV. EXPERIENT RESULTS

All this experiments in this paper is based on the actual equipment 50kVA harmonic current generator that has been developed out. According to the design demand, the equipment can generate 2th~25th harmonic current. In the Fig.12, odd order harmonic current include 5th, 7th, 11th, 13th, 17th, 19th, 23th, 25th. The output current waveform and frequency spectrum analysis are shown as follows.

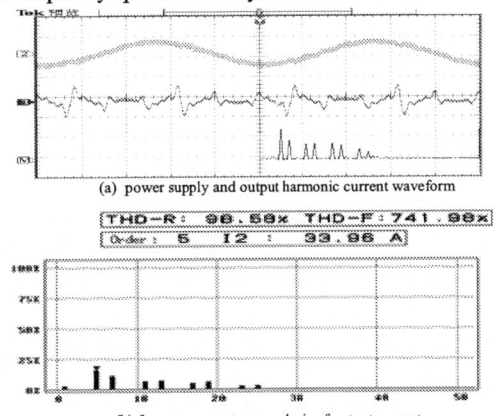

(a) power supply and output harmonic current waveform

(b) frequency spectrum analysis of output current

Fig.12 odd order harmonic current experimental results

125

Besides, even order harmonic current can also be generated by the equipment in this paper. The 4th, 5th, 6th harmonic current experiment results are shown in Fig.13.

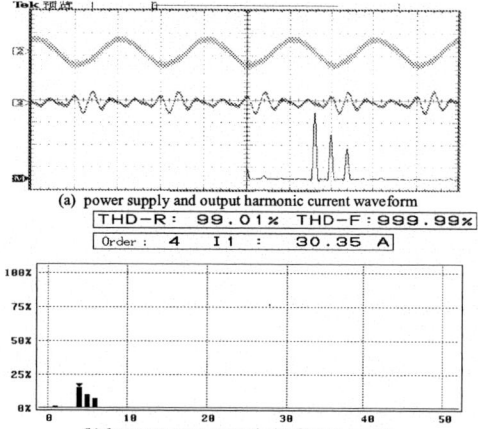

(a) power supply and output harmonic current waveform

THD-R: 99.01% THD-F: 999.99%

Order: 4 I1: 30.35 A

(b) frequency spectrum analysis of output current

Fig.13 even order harmonic current experimental results

Because of the four-leg main circuit, the harmonic current generator can produce the any unbalanced output current. The experiment results in the Fig.14 shows that the 4th leg current contains 3th, 5th, 7th, 11th harmonic current.

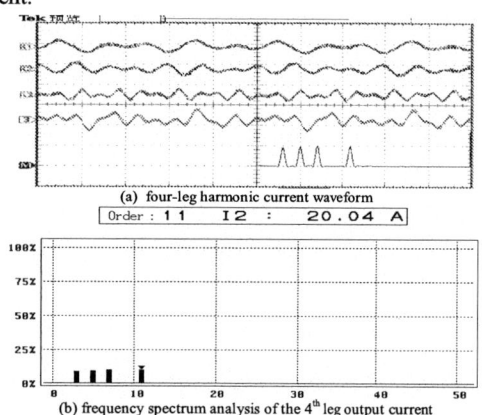

(a) four-leg harmonic current waveform

Order: 11 I2: 20.04 A

(b) frequency spectrum analysis of the 4th leg output current

Fig.14 unbalanced harmonic current experimental results

In order to realize the dynamic quality of programmable harmonic current generator, the variable process can be defined in the program in advance. The rise, sag and period change experiment results are shown in Fig.15.

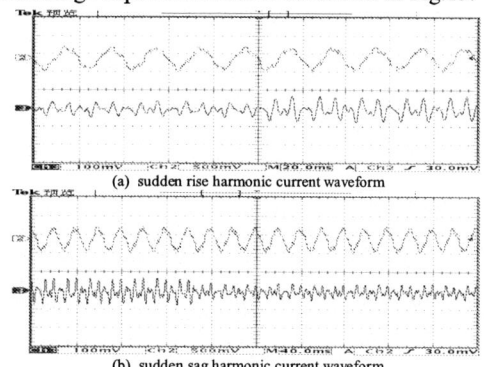

(a) sudden rise harmonic current waveform

(b) sudden sag harmonic current waveform

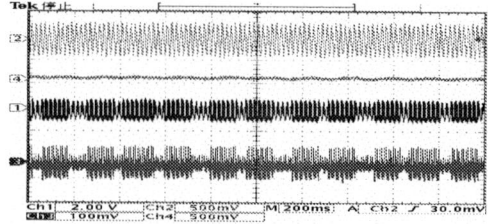

(c) period change harmonic current waveform

Fig.15 variable harmonic current experimental results t

V. CONCLUSION

The large capacity programmable harmonic generator based on three-phase-four-wire is presented in this paper. This equipment can produce selective order harmonic (below 25th order), reactive current and unbalanced current. Besides, the whole equipment has a good performance on dynamic quality. This paper presents the four-leg main circuit and control strategy in detail, and introduces voltage feedforward control and output current DFT feedback control to improve the system output quality. The detailed experience results verify the feasibility and validity of the system. Altogether, the large capacity programmable harmonic current generator can meet most debugging needs of electrical measurement equipment and harmonic current elimination equipment. It can be used as a standard harmonic current source for its good performance.

REFERENCES

[1] Zhaoan Wang; Jun Yang; and Jinjun Liu, Harmonic Elimination and Reactive Power Compensation. Beijing, China: China Machine Press, 1998.

[2] Chongwei Zhang; Xing Zhang. PWM Rectifier and Control Strategy. Beijing, China: China Machine Press, 2003.

[3] Wanjun Lei; Hongyu Li; Fang Zhuo; Longhui Wu; Zhaoan Wang. Development of large capacity active power filter based on DFT detection method. Advanced Technology of Electrical Engineering and Energy. vol. 23, no. 2, pp.69–72, Jan. 2004.

[4] Jang-Hwan Kim, Seung-Ki Sul.A Carrier-Based PWM Method for Three-Phase Four-Leg Voltage Source Converters, IEEE TRANSACTIONS ON POWER ELECTRONICS, VOL. 19, NO.1, JANUARY 2004

[5] Blasko V, Kaura V. A Novel Control to Actively Damp Resonance in Input LC Filter of a Three-phase Voltage Source Converter [J]. IEEE Trans on Ind Appl, 1997 , 3 3 (2):542-550.

[6] Maur'ıcio Aredes, *IEEE* J¨urgen H¨afner, and Klemens Heumann. Three-Phase Four-Wire Shunt Active Filter Control Strategies, IEEE TRANSACTIONS ON POWER ELECTRONICS, VOL. 12, NO. 2, MARCH 1997.

2006 5th International Power Electronics and Motion Control Conference

A Universal Digital Platform and Software Library for Power Electronic Systems Integration

Haibing HU, Tianjun Jin, Wenxi YAO, Zhengyu LU, Zhaoming Qian

National Key Lab. of Power Electronics, Zhejiang University, Hangzhou, China

huhaibing@163.com

Abstract—A new flexible digital platform based on the fixed-pointing DSP has been developed for power electronic systems integration, whose computational power can be easily expanded by floating-point units(FPU) which are configured by a FPGA. Some IP(Intellectual Property) cores for power electronic applications, coded in VHDL for modularity and portability, have been designed and validated in the digital platform. Software modules commonly used in power electronic applications are packaged in C language with fixed-point format for versatility and computation efficiency. All these IP cores and software modules are verified in real-time applications and can be easily integrated into power electronic systems.

Keywords-Digital platform; IP core; Software module; Power electronic systems integration

I. INTRODUCTION

With the advancement of power electronic systems integration, a universal digital platform for power electronics should be placed on the agenda, which will not only help to facilitate the research and development of power electronic applications, but also is an indispensable element for power electronic systems integration. Many forerunners have been continuing to design flexible platforms for their various applications. All these digital platforms can be categorized into three classes:(1)Very high performance digital platforms but complicated, such as dSPACE and ISEADSP[1][2]. (2) Platforms based on floating-point DSPs[3][4]. (3) Platforms based on fixed-point DSPs[5].

However all these platforms have their shortcomings as universal platforms for the purposes of power electronic systems integration. As for the first class platforms, the aim of these type platforms is mainly for research purposes due to the following special features:(1) Powerful computational power with multi floating-point DSPs to implement complex control algorithms; (2) Rich peripherals; (3) Friendly interface to high level simulation tools such as Matlab and Pspice, for example, their simulation code can be downloaded to the platforms without any modification in code. But the platforms with so many hardware resources are too expensive and complex to be candidates of universal platforms for systems integration. For the second class, these platforms

based on floating point DSPs have to expand peripheral circuits for power electronic applications such as A/D converters, D/A converters, PWM generator and so on. This kind platform not only increased the system cost greatly but also reduced the system reliability with so many "discrete" components added to the platforms. What is most important is that these platforms didn't take advantages of the state-of-art DSPs dedicated to power electronics and motor drive. As well known, the fixed-point DSPs have continued to be the mainstay of the industry, especially in the field of power electronics. The reason, of course, is the cost. But, unfortunately few attempts have been made to develop the platforms totally based on fixed-point DSPs. Even did, the flexibility and versatility of these platform are poor.

A new digital platform based on the fixed-pointing DSP is proposed for power electronic systems integration. Compared with above three kind platforms, the new digital platform have its own features, which are listed as follows: (1) Flexibility and versatility; (2) Low cost; (3) Powerful computational capability; (4) All level communication peripherals for systems integration. All these features ensure the platform is not only applicable to most power electronic systems such as motor drive, APF, PFC, multi-level inverter and so on, but also suitable to power electronic systems integration.

In order to provide a modular design on a software level, software standardization and modularization is necessary. Standardized software modules are basic elements for power electronic systems integration on software levels[6]. All these software modules, which should be transparent to users and irrelevant to the hardware platform, will accelerate the development of control algorithms in real-time control hardware. In this paper, some commonly used software modules for power electronic applications have been developed and validated in the digital platform.

II. HARDWAER SETUP

The platform was designed for power electronic systems integration, which is flexible and versatile enough for use in various topologies and with various switching devices. The computational power can be easily expanded by the floating-point units, which are configured by a FPGA. It is very convenient to accomplish the user-defined functions through configurable hardware and software modules. The block

Project supported by National Nature Science Foundation of China (50237030)

1-4244-0448-7/06/$25.00 ©2006 IEEE

diagram of the platform is illustrated in Fig.1.

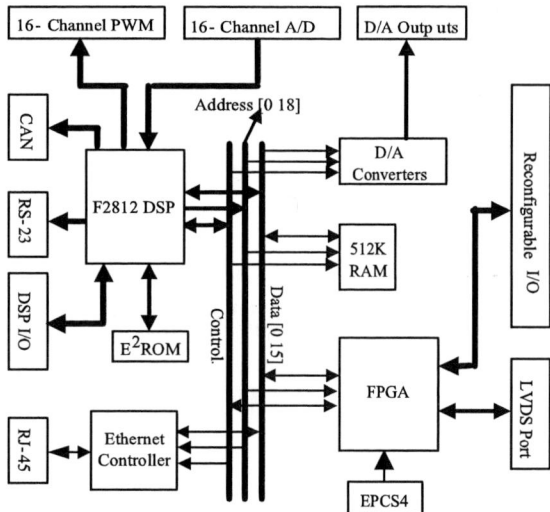

Figure 1. Diagram block of the platform

The platform consists of several main elements below:

- TMS320F2812 DSP: F2812 DSP, a fixed-point DSP from Texas Instrument, features 150-MPIS performance and complete system-on-a-chip integration dedicated to power electronics and motor drive[7]. The rich control peripherals include:(1)A 12-bit,16-channel ADC with conversion rate up to 80ns;(2)Optimized event managers, including flexible PWM generators, programmable general timers and glueless capture encoder interfaces;(3)Several standard communication controllers such as UART,CAN and SPI. The DSP is used as the main calculation unit in the platform.

- EP1C6: This ALTERA Field Programmable Gate Array(FPGA) is based on a 1.5V, 0.13um,all-layer copper SRAM process, featuring 5890 LEs(Logic Element)and a total embedded RAM of 92160 bits[8]. It is used for the following three major purposes:(1) To configure floating-point units to enhance the platform computational capacity;(2)To generate PWM signals, like the event manager of DSP. The platform can offer up to 48-channel PWM;(3)To expand the platform and make the platform more flexible via reconfigurable FPGA.

- Communication controllers: As a universal platform for power electronic systems integration, communication interfaces are indispensable, for this type platform can be used not only as a stand-alone controller but also as an open controller which can exchange information with other controllers. Therefore three communication ports are deliberately designed for their different purposes. The Ethernet controller serves as a high level communication port which can be accessed by Ethernet or even by Internet, that means these type platforms can easily construct a LAN network for high level system control via commonly-used twisted pair lines. The platform can

also be functioned as a host or slave node in a distributed power electronic systems via Controller Area Network(CAN) communication, which has a wide acceptance in the field of serial communication. While the RS-232 communication port is used for system monitoring and debugging. All these communication ports are isolated in electricity by a transformer or opticouplers for safety reason.

- Two channel D/A converter: The 12-bit D/A outputs are of great help for debugging during system development stage. They also can be fed out as a command reference for control systems.

- EPCS4: This chip configures the FPGA when power on.

- Memory elements: The platform is equipped with 1kx16 E^2ROM which is of vital importance to store the important data when power off, and 512Kx16 one wait state Static RAM(SRAM) for program and data memory.

The Fig. 2 depicts the actual platform board with the dimension 14x11cm^2.

Figure 2. The top view of proposed digital platform

III. STANDARDIZED SOFTWARE LIBRARY

The software library is one of the key components of the platform development system. It is divided into two categories: one is IP cores for FPGA, the other is software modules for DSP. The software library is an open source ware, to which the users can add their own software modules.

A. IP cores for FPGA

- Floating-point units

We have developed three type single floating-point units (adder, multiplier and divider) respectively for FPGA EC1C6 from ALTERA Cyclone family. All these three units can finish their arithmetic operations in 40 nanoseconds and the maximum absolute error of the results obtained by these units is less than 2ulp(Unit in Last Place). Fig.3 illustrated the IP modules for adder, multiplier and divider. OperandA and OperandB are 32-bit operands fed by DSP word by word and the result can be obtained from Result[31..0] port. While the Underflow port and Overflow port signify whether the operation is underflow or overflow. The operation result is Not A Number(NAN) if the output nan port is '1'. In Fig.3(c), a 10MHz clock input is required for the IP core.

128

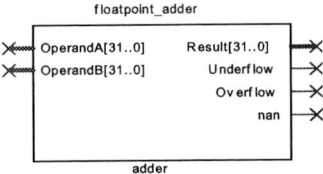

(a) IP core of floating point adder

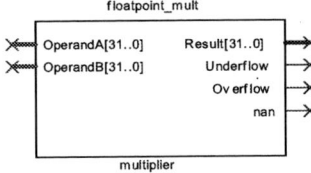

(b) IP core of floating point multiplier

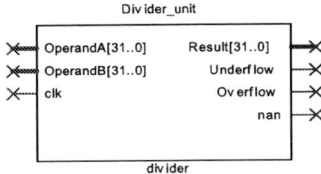

(c)IP core of floating point divider

Fig.3 IP cores of floating point units

♦ PWM generator

The function of PWM module, shown in Fig.4, is similar to that of PWM generator in DSP. It has a 16-bit Timer Period Register(TPR), a 16-bit Timer Count(TCNT),a 16-bit Timer Compare (TCMPR) and a Timer Control Register(TCON). All these registers can be read and written through 16-bit data bus. The PWM can be easily controlled using these registers, and its resolution can reach up to 10ns with the 100MHz input clock. Therefore it is also suitable to high frequency switch applications. Users can program their desired PWM signal using this module very conveniently.

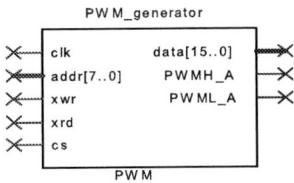

Figure 4. PWM IP core

The resources required by IP cores are listed in table 1.

Table 1. Resources required by IP cores

IP cores	LE	Memory bits	PLL
Adder	883	0	0
Multiplier	995	0	0
Divider	1505	51200	1
PWM generator	236	0	0

B. Software modules for DSP

These modules in this paper were written using fixed-point format in C language for its versatility. They were all verified in the platform.

♦ Two level SVPWM

SVPWM technique is widely used in three phase system such as motor drive, UPS for its merits in (1) less commutation losses; (2) 15 % higher output voltage utility than SPWM. In order to properly handle the over modulation area, the module adopted the Q14 format to express the range of inputs (Q14 is a number format with the point fixed in bit 14 position). Notice that the inputs(Ualpha,Ubeta) should be normalized . The module is shown in Fig.5. The outputs are calculated

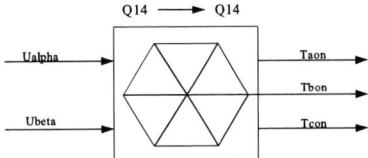

Figure 5. Two-level SVPWM module

on-time. The stator reference voltage is described by its (α,β) components, Ualpha and Ubeta. The meaning of terminals of the module is described in Tab.2.

Table 2. Terminal variables of two level SVPWM

	Name	Description	Format
Input	Ualpha	Component of reference stator voltage vector on alpha axis frame	Q14
	Ubeta	Component of reference stator voltage vector on beta axis frame	Q14
output	Taon	Duty ratio of PWM1	Q14
	Tbon	Duty ratio of PWM3	Q14
	Tcon	Duty ratio of PWM5	Q14

♦ Three-level SVPWM

Three-level SVPWM is a new modulation technology suitable to medium and high voltage applications. Some high power motor drives turned to the three level topologies due to their low dv/dt and high voltage rating. Three-level SVPWM is a prosperous technique for these applications. However three-level SVPWM is more complex than two-level SVPWM due to its too many freedoms and its neutral-point voltage unbalanced problem. The technique used in this module is only one of many control strategies[9]. But other strategies can be

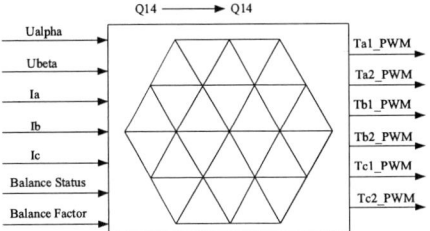

Figure 6. Module of three-level SVPWM

easily modified based on this software module. The module is illustrated in Fig.6. The meaning of terminals of the module is listed in Tab.3.

Table 3. Terminal variables of three level SVPWM

	Name	Description	Format
Input	Ualpha	Component of reference stator voltage vector on alpha axis frame	Q14
	Ubeta	Component of reference stator voltage vector on alpha axis frame	Q14
	Ia	Current of phase A	Q14
	Ib	Current of phase B	Q14
	Ic	Current of phase C	Q14
	Balance Status	the status of mid-point voltage	BOOL
	Balance Factor	a coefficient for redistributing vectors	Q15
Output	Ta1_PWM	Duty ratio of PWM1	Q14
	Ta2_PWM	Duty ratio of PWM3	Q14
	Tb1_PWM	Duty ratio of PWM5	Q14
	Tb2_PWM	Duty ratio of PWM7	Q14
	Tc1_PWM	Duty ratio of PWM9	Q14
	Tc2_PWM	Duty ratio of PWM11	Q14

◆ Transformation modules

Coordinate transformations such as (Park, Clarke and their inverse transforms) are frequently used in power electronics, especially in three phase systems.

◆ Lookup table module for sine and cosine

This module is widely used in coordinate transformation. In order to save storage size, only one-forth sine cycle with the size of 2K words is stored in the memory. Several code lines are required in mapping the first quadrant into other quadrants for looking up the right sine and cosine values in the table.

◆ PID module

Despite the advanced control techniques, such as adaptive control, fuzzy control and neural network control, the PID and its variations are still widely applied in power electronic systems. In this paper, a universal PID module was designed.

IV. MODULE VERIFICATIONS

A. Floating-point units

FFT algorithm was adopted to verify the correctness of the results obtained by the FPU and make a comparison between the FPU and DSP algorithms(supported by run time library) in computation efficiency. The results of FFT calculated by FPU were exported by using "data saving" feature available in CCS and then analyzed in Matlab. The results of FFT obtained by FPU are totally in consistence with those calculated by Matlab.

To study the computation efficiency between DSP algorithms and FPU, FFT algorithm with different points was adopted. The time consumed by DSP algorithms and FPU was measured by the interval between the very beginning of FFT calculation and the end using DSP T2 timer. Notice that all the operation environments were set the same for all these tests. Computation time of two methods was plotted in Fig.7. From this figure, it is very clear that the computation time by FPU is five times less than that of DSP algorithms. If the FPU were interfaced to the DSP with 32 bit data width or integrated into the DSP like other peripherals, the computation efficiency would further improved, since at least two-thirds time are

wasted in organizing, feeding and retrieving data to and from FPU.

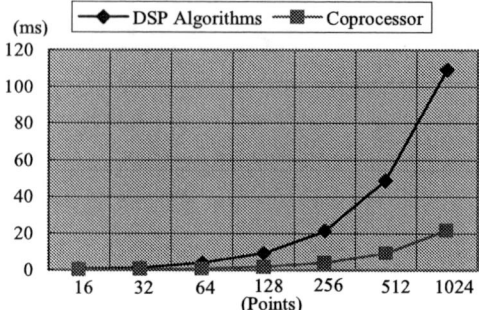

Figure 7. Computation time of FFT between FPU and DSP algorithms

B. Software modules

We used following two steps to verify the software modules. Step 1: Feeding results calculated by these modules to D/A converters to measure whether these modules properly functioned in all conditions. Step 2: Implement these software modules in the testing platform shown in Fig.(8) to make sure they really work. For the

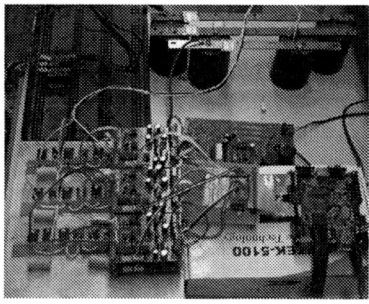

Figure 8. Setup for testing software modules

limited space, only parts of modules are presented here. Below are the verification waveforms of two-level and three-level modules. As can be seen from Fig9 and Fig10

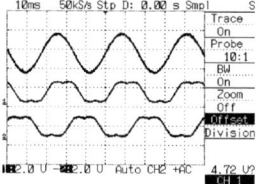

(a)Line and phase A,B voltages (m=0.855)

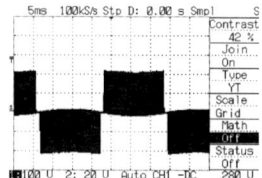

(c)Line PWM waveform(m=0.855)

(b) Line and phase A,B voltages (m=1.425)

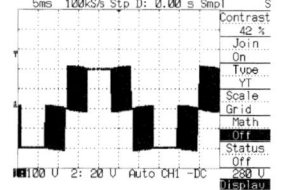

(d)Line PWM waveform(m=1.425)

Figure 9. Normal modulation and over-modulation testing for two-level SVPWM module

130

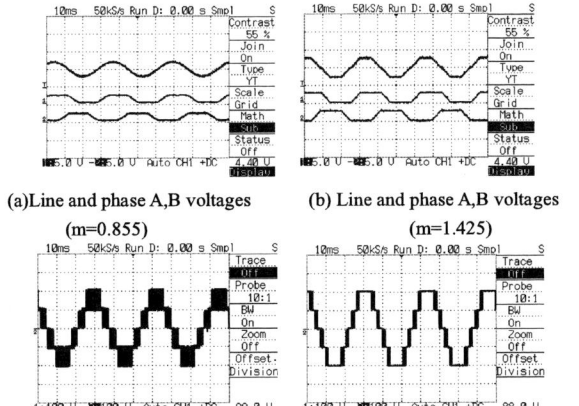

(a)Line and phase A,B voltages
(m=0.855)

(b) Line and phase A,B voltages
(m=1.425)

(c)Line PWM waveform(m=0.855) (d)Line PWM waveform(m=1.425)

Figure 10. Normal modulation and over-modulation
testing for three-level SVPWM module

both modules can work properly under normal condition and over-modulation condition.

Table 4 depicts the specifications of software modules mentioned in the Section III. Notice that all the figures in the table 4 are tested under the conditions that the DSP operates under 100MHz with software modules loaded in the internal RAM and lookup-table loaded in the external RAM with 5 wait states.

Table 4. Specifications of software modules

Software modules	Data ram (word)	Calculation Time(us)	Number of input	Number of output
Two level SVPWM	16	2.12	2	3
Three level SVPWM	34	5.35	7	6
Clarke transform	5	0.30	3	2
Inverse Clarke transform	5	0.48	2	3
Park transform	6	0.38	4	2
Inverse Park transform	6	0.38	4	2
PID module	21	0.94	1	1
8K-lookup table for sin(θ) and cos(θ)	7	1.15	1	2

V. CONCLUSION

A universal and flexible digital platform based on the TMS320F2812 fixed-point DSP and EP1C6 FPGA has been developed. It also can implement the floating-point operations using floating-point IP cores configured by the FPGA, whose speed is five times faster than that of DSP algorithms in computation efficiency. Software modules commonly used in power electronic applications are packaged in C language with fixed-point format for versatility and computation efficiency. All these IP cores and software modules are verified in the real-time applications using the digital platform. The platform packaged with software library is being used for systems integration in various power electronic applications.

REFERENCES

[1] Joep Jacob,Dirk Detjen, etc. "Rapid Prototyping Tools for Power Electronic Systems: Demonstration with Shunt Active Power Filters". IEEE Trans. on Power Electronics,Vol.19(2),2004,pp:500-507.

[2] DSPACE, Solutions for Control: Catalog 2005. http://www.dspace.de.

[3] D.D.Bester, J.A. du Toit J.H.R Enslin. "High Performance DSP/FPGA controller for Implementation of Computationally Intensive Algorithms".IEEE International Symposium on Industrial Electronics,Jul.1998,pp.240-244.

[4] Habib-ur Rehman, Richard J. Hampo. " A flexible high performance advanced controller for electric machines". Applied Power Electronics Conference and Exposition, Feb. 2000,pp.939-943.

[5] Mongkol Konghirun, Longya Xu Jennifer Skinner Gray. "Quantization Errors in Digital Motor Control Systems".Power Electronics and Motion Control Conference, Aug. 2004,pp.1421-1426.

[6] Ivan Celanovic."A Distributed Control Architecture for Power Electronics Systems", Master thesis, Virginia Polytechnic Institute and State University,2000.

[7] DSP TMS320F2812, Datasheet: http://www.ti.com.

[8] FPGA EP1C6,Datasheet: http://www.altera.com.

[9] Yao Wenxi, Lu Zhengyu,etc."Three-level SVPWM method based on two-level PWM cell in DSP". Power Electronics Conference and Exposition, 2004,pp.1720-1724.

2006 5th International Power Electronics and Motion Control Conference

Unipolar SiC Devices – Latest Achievements on the Way to a New Generation of High Voltage Power Semiconductors

Peter Friedrichs

SiCED Electronics Development GmbH & Co. KG, a Siemens Company
Paul-Gossen-Str. 100, D-91052 Erlangen, Germany
Tel. ++49 9131 734894   Fax ++49 9131 723046
Email: peter.friedrichs@siemens.com
http://www.siced.de

Abstract— Silicon carbide power devices are supposed to revolutionize certain parts of the power semiconductor business. After the successful product release of Schottky barrier diodes in 2001 and the development of a new generation of surge current stable diodes in 2005, the next logical step in the device chain should be a SiC switching device. Addressable are high voltage applications in energy systems as well as the huge market for low voltage power switches, where even in the blocking voltage range below 100V SiC is assumed to be one potential candidate for meeting the increasing demands on power density [1].

The paper will give an overview about latest developments of SiC power switches and diodes. In addition, some potential applications serving as drivers for the SiC power device development will be sketched. Finally, an outlook to near and long term perspectives for SiC power devices is given.

Keywords-silicon carbide, diodes, switches, power density

I. INTRODUCTION

A certain attention paid to wide band gap semiconductors was observed beginning in the early nineties. Regarding power devices, it was focused during the last years to the most interesting materials like GaN and SiC. Especially SiC has realized an impressive growth process due to its outstanding technological advantages. In contrast to the most other candidates among the wide band gap semiconductors SiC can be characterized by the following advantages

- o Indirect semiconductor, important for bipolar power electronics e.g.
- o Ability for selective doping of both, n- and p-type
- o Native thermal oxide SiO_2
- o Freestanding and high quality crystals available
- o Broad targeted range of applications (power electronics, high frequency electronics optoelectronics, high temperature electronics etc.)

At the moment, unipolar structures like Schottky barrier diodes and FETs are favored for main stream applications.

Bipolar structures are expected to be used for very high blocking voltages. Since additional technological challenges exist for these structures, the paper will focus predominantly on unipolar devices.

After only ten years from first attempts to grow device grade substrates, the introduction of first products (Schottky barrier diodes) occurred in 2001 by Infineon and Cree. Even considering the higher device price, the implementation in systems could be realized also from an economical point of view due to the achievable system advantages [2]. The use of SiC components is mainly triggered by the possibility to achieve higher power densities for power conversion systems via higher switching frequencies. The now released new generation of Schottky Barrier diodes represents a next step towards the really ideal power diode. Details will be outlined in the following paragraphs.

The development of SiC based switching devices is still at R&D level for MOSFET type structures, however, vertical JFETs are in a mature phase and close to be fabricated. The implementation into next generation power electronics will require a similar revolutionary way of mind changing in design of power systems as we have observed for the Schottky barrier diodes (SBDs). Nevertheless, the starting point for the switches might be somewhat easier then for the diodes since firstly, SiC is no longer suspicious in power electronics, and secondly, in contrast to the diodes, also the reduction of static losses is an issue, combined with a more relaxed cost ratio between the SiC devices and competing components.

A more sustaining development is mandatory in order to exploit the performance of SiC in the high voltage market. Even if technically attractive solutions are available, for instance in distributed energy components or for new generation traction systems, the technical and economical breakthrough cannot be fixed on a time scale based from a today's point of view. However, we will demonstrate some results which prove the huge technical advantages which can be gained from using SiC. New applications can come up which up to know were not imaginable with the abilities of existing power

1-4244-0448-7/06/$25.00 ©2006 IEEE

semiconductors. One of the scenarios could also be the use in pulse power applications.

In the following paragraphs, the mentioned topics will be discussed with respect to the today's status, near term development targets and application issues.

II. SCHOTTKYBARRIER DIODES

On a first glance, the easiest semiconductor device is the two terminal diode acting with its rectifying function often in combination with switches or as an individual component in power circuits. Regarding SiC, the ability to fabricate unipolar Schottky barrier diodes (SBDs) up to blocking voltages of 1700V or more offers new degrees of freedom in the design of power circuits. Due to actually zero reverse recovery, dynamic losses can be reduced drastically. First applications like power factor correction successfully exploited these features in server systems.

The optimum use of Schottky barrier diodes requires an understanding of the main differences compared to the established bipolar silicon counterparts (see Figure 1).

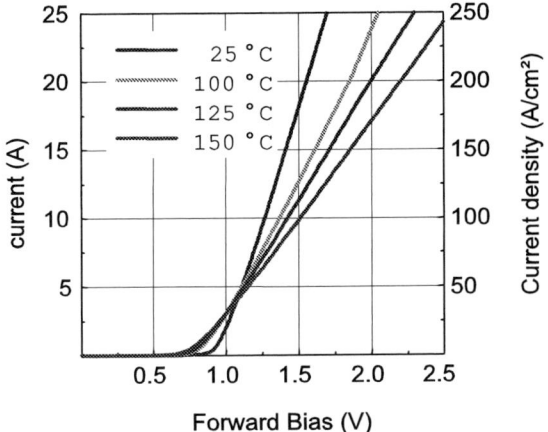

Figure 1. Characteristic forward I-V behavior for a standard SiC Schottky barrierdiode (rated current 25A)

Regarding the forward voltage drop, it increases comparable to MOSFETs with temperature for a given current rating. Furthermore, it has a resistive behavior after reaching the threshold of approximately 1V. Thus, peak currents through the devices can end up with high peak power densities and after reaching critical values, the device will be thermally destroyed.

In reverse direction, no defined avalanche is possible since electrons injected over and through the barrier dominate the reverse current at electric fields well below the critical field of SiC. Overvoltage spikes can cause local, destructive stress and cannot be managed by a stable avalanche like in well designed silicon bipolar diodes.

Meanwhile these devices are established as products with blocking voltage ranges from 300V up to 1200V. However, the range of applications is still restricted because of the cost performance ratio today. One of the limiting features is the described restricted surge-current handling capability which forces the application engineers to apply oversized semiconductors in order to meet the demands of functionality in the case of drop outs, start up etc. In order to address these limits, a new generation of surge current stable devices with avalanche capability was developed [3]. Infineon released these devices in 2006. The main feature is a bipolar part integrated grid like into the Schottky area. A careful design ensures purely unipolar operation at nominal currents and bipolar operation under surge conditions. The bipolar part restricts the forward current drop via conductivity modulation of the drift zone (see Figure 2).

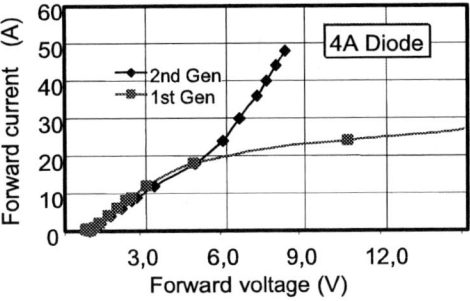

Figure 2. Behavior of first and second generation diodes under surge current conditions (nominal current 4A)

Additionally, the design can be tuned to establish avalanche conditions at every emitter region of the devices and thus, a stable and rugged avalanche with a positive temperature coefficient can be realized. Similar to the first generation standard Schottky diodes, these devices still have no reverse recovery, even if the unusual case of turning of the device in bipolar operation (over current) takes place (see Figure 3). The reason is that all injected carrier are quickly extracted via the neighboring Schottky islands and thus, these so called merged pin-Schottky barrier diodes behave from a dynamical point of view like pure Schottky diodes.

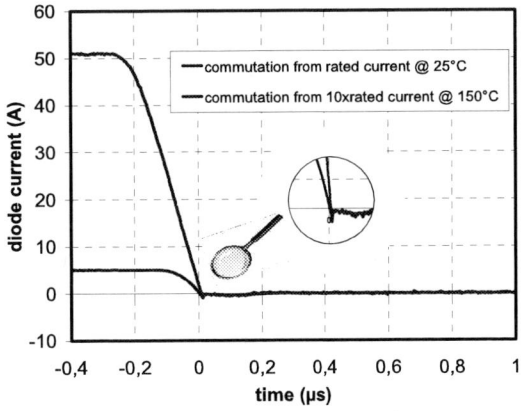

Figure 3. Turn off of a second generation diode from nominal and from 10x the nominal current

Future developments will address e.g. the suitability of SiC Schottky barrier diodes for high temperature operation, where today's technology shows certain limits with respect to leakage currents and static losses.

III. SWITCHING DEVICES

Currently the main focus of the development regarding switching devices in SiC is directed to unipolar switching devices for blocking voltages starting with 600V up to approximately 2kV. Further trends like bipolar junction transistor structures, thyristors or GTO's will not be discussed in this contribution since firstly, the non-military applications of these devices are still not clear and secondly, industrial needs clearly favor charge controlled devices today.

Investigated power switches are mostly MOSFET type devices and Junction Field Effect Transistors (JFET). In contrast to the diodes, these devices offer in addition to the superior dynamics also advantages regarding the static losses. Recently reported MOSFET structures are characterized by ultra short channels (0.5µm) and thus, high channel densities [4]. This approach allows achieving low specific on-resistances despite inferior channel characteristics.

JFETs with their bulk channel are able to offer very low on-resistances too. Japanese sources report about only 1mΩcm² for a 600V SIT structure [5], however, this was only a tiny chip and a positive gate bias is necessary to achieve this value. Regarding the area specific on-resistance at room temperature of a large area 600V unipolar JFET of the meanwhile third generation from SiCED e.g. [6], it is five times lower than the value achievable today in state of the art silicon based charge compensated MOSFET structures, at operating temperature of about 150°C at the junction of the device this ratio favors even more the use of SiC devices (see figure 4).

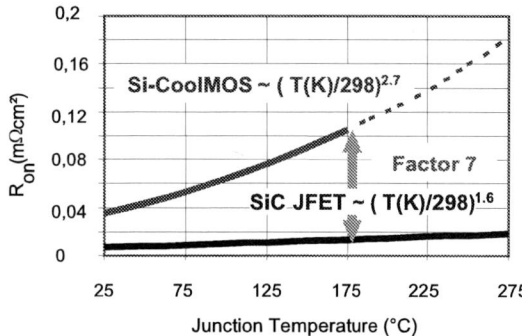

Figure 4. Increase of the on-resistance of unipolar 600V power switches, Si-Charge compensated MOSFET vs. SiC VJFET, please note that the silicon device is specified up to 150°C maximum

It clearly underlines the high potential hidden in the use of SiC devices. This advantage, however, is often not directly used in the applications, in reality the devices are benchmarked at a given current rating. Taking into account the thermal boundary conditions a comparison will show that for a given current rating or absolute on-resistance SiC devices are much smaller and most of the potential provided by lower area specific on-resistances will be shifted to additional dynamic improvements by achieving low input and output capacitances. Competition with today's powerful charge compensated devices at the 600V blocking voltage level is hard due to the mostly cost sensitive markets. However, especially in applications with extreme demands regarding ambient temperature and/or freewheeling functions, SiC JFETs can be an option even despite the higher cost at the device level.

The behavior of the internal body diode in a JFET as developed by SiCED is for most cases attractive enough to be used as an in-built freewheeling diode. An outstanding low reverse storage charge and extremely low reverse recovery time are obtained even for very high di/dt rates (at up to 1kA/µs only 40ns trr), changing only slightly with increasing temperature. It should be emphasized that classical VJFET devices do not offer a body diode between drain and source, thus, the developed power switch combines effectively the advantages of MOSFETs and JFETs.

The favorable use of SiC VJFETs at elevated temperatures is based on the following issues (besides the given wide band gap feature of a general high temperature suitability):

o The internal function is based only on pn-junctions which can operate in SiC theoretically up to several hundred degree Celsius. There is no issue regarding the stability of oxide interfaces under high field and high temperature stress like in MOS based silicon devices or even in SiC MOSFETs, where the electron injection under stress condition is enhanced due to the lower barrier between SiC and SiO_2.

o The threshold voltage is virtually independent of temperature, simply based on the fact that the pinch-off bias in a JFET is defined by the channel doping and its geometry. In MOSFETs (and also IGBT's), several temperature dependent factors result in a decrease of the threshold voltage with temperature with the danger of unwanted turn on at $T_J > 200°C$.

o The reverse recovery is also nearly independent on temperature, thus, de-rating with an increased operation temperature due to increasing dynamic losses is not required and provides more flexibility in designing circuits.

o Due to the independent design of the channel region, it is possible to tune and to reduce the increase of the on-resistance with temperature compared to standard unipolar devices made from silicon e.g. (see figure 4)

Parallel to the JFET, developments of MOSFETs gained new attention during the last two years due to improvements in channel mobility via special oxidation techniques as well as refined design methods using ultra short channels or even alternative crystal orientations which show reduces interface defects to thermally grown SiO_2 layers. These improvements clearly narrowed the gap between the achievable on-resistance in MOSFET and JFET structures. Figure 5 shows a comparison of the values achieved at SiCED, where also a SIT like structure is included (vertical channel instead of the lateral one in the JFET, no body diode).

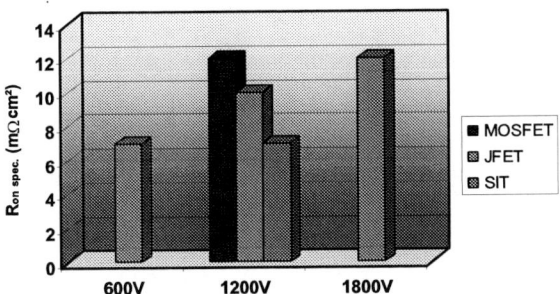

Figure 5. Area specific R_{on} for the latest generation of SiC VJFETs with lateral-vertical current flow (third generation devices), SiC SIT sans MOSFETs from SiCED, typical values for large area (>4mm² active area) at room temperature)

However, even taking into account these improvements and the advantage of being a normally-off device, still open issues regarding the long term reliability of MOS devices have to be investigated. Scattered publication show clear influences of the substrate preparation on the oxide lifetime [7], nevertheless, still the search for a suited insulator technology dominates today's development activities and we believe that reliable long term stability analysis can be carried out not before fixing the basic technological and structural solution for a future MOSFET device. Nevertheless, the potential for new generation power electronics provided by the normally off MOSFET device was proven by the latest results showing impressive on-resistance results of less than 10mΩcm² for a 1000V device [8].

Regarding switching devices for blocking voltages in the kV range, mostly bipolar and charge controlled devices like IGBT's are required by the system designers and thus, our focus is dedicated to these structures. This is partly in contrast to the activities currently being conducted in the United States where military applications are in the focus of high voltage SiC power switching devices and also alternative, current controlled devcies are investigated.

Basic challenges on the way to IGBT's exist due to the lack of highly conductive p-type substrates at the moment. Latest developments in crystal growth indicate that the conductivity can be improved, in addition, thin wafer technology could be applied to realize 10kV single chip devices. Since the success of IGBT's is strongly related to

the achievements in the MOSFET development, we demonstrated as an alternative approach a single die 4.5kV bipolar structure based on the JFET technology as well as a 9kV stacked solution based on a simple serial connection of unipolar JFETs (Fig. 6). The function of this arrangement is though roughly described in [9].

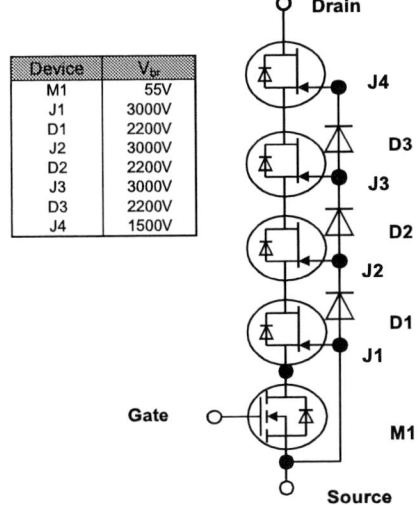

Device	V_{br}
M1	55V
J1	3000V
D1	2200V
J2	3000V
D2	2200V
J3	3000V
D3	2200V
J4	1500V

Figure 6. High voltage switch, based on a stack of SiC VJFETs and controlled by a simple low side silicon power MOSFET

Dynamical test showed that this arrangement can offer a very high speed even for a blocking voltage of several kV. Due to the modular structure (always via one diode and one VJFET the stack can be extended) theoretically there is no limitation for the maximum blocking voltage. Possible instabilities in the reverse current paths can be easily suppressed by simple passive elements [10]. Thus, this solution represents an interesting device for today still exotic semiconductor application like pulse power e.g.

IV. APPLICATION VISIONS

Early visions dealing with a widespread replacement of silicon power devices by SiC components clearly appeared to be and will be a dream. Especially the cost performance ratio and impressive technological improvements like thin wafer processing or charge compensated technologies strengthened the position of silicon as leading base material for power semiconductor components today and in the future. Nevertheless, despite this situation SiC found and will find its applications. Some guidelines for identifying or even creating these new systems could be the following:

o   Don't compare solutions at the device level; here silicon will be the better solution in most cases. At system level the judgment can result in a completely other situation, e.g. in case of higher switching frequencies passive components can be

135

shrunk and offer in sum an attractive cost – performance position.

○ try to forget about "Don't do" borders which may be originate from limits given by conventional semiconductors

○ try to use as high voltages as possible for achieving a certain power level since in SiC technology, higher voltages are in general achievable with less efforts than higher currents (it should be mentioned that creeping distances and isolation issues have to be taken into account by doing this)

Very promising are applications with high demands on power density or operating temperature. To fulfill "Moore's 2nd law" for energy transformation, the power density should be doubled every 4.5 years. The future goal of power density is 30W/cm^3 for such DC/DC converters in 10 years, which should be only possible by pure SiC design. On of the most frequently discussed issues in this frame are automotive applications, especially systems used in hybrid electric vehicles. The given preconditions are clearly in favor for SiC. Power density is crucial due to restricted space and weight considerations in modern cars, in addition an enlarged range of operating temperatures is recommended. The potential given by system advantages is very high since in today's solution a complete second cooling circuit only for keeping the power semiconductors at the desired temperature level is applied in existing hybrid cars. What have to be proven is whether the cost position achievable by using SiC is acceptable for the customers.

Furthermore, a smart combination with the right silicon device, e.g. a SiC diode and a CoolMOS switch for PFC or a well selected IGBT with a SiC freewheeling diode for bidirectional, hard switched converters or drives offer dramatic improvements compared to established components.

TABLE I.   COMPARISON OF DYNAMIC AND STATIC LOSSES FOR IBGT AND MOSFET TYPE SWITCHES, BOTH EQUIPPED WITH SiC FREEWHEELING SBD'S

		V_f	E_{on}	E_{off}
Si CoolMOS in series with a Schottky diode and bridged with a SiC Schottkydiode @125°C and 15A, 750A/μs		approx. 8V	255μJ	171μJ
Si IGBT with SiC Schottky freewheeling diode @175°C and 20A, 2000A/μs		approx. 4V	125μJ	160μJ

Table 1 compares the most prominent parameters of two different solutions of switches using a freewheeling diode. It can be clearly seen that as long as a IGBT can somehow manage the required switching frequency, it will be definitely the better solution for the overall system, especially with respect to cost and losses.

Classical switch mode power supplies already today take advantage of SiC e.g. in PFC circuits. Assuming switching devices based on SiC, it is possible to think about simple single switch solutions which can replace complex topologies in the field of high power converters (>1kW) [11]. Besides reduced costs due to less components the additional potential of increased reliability in the case of simpler circuits should be stressed. Finally, higher efficiencies can be achieved by the use of SiC switching devices [12]. This parameter is important for loss optimized energy conversion in order to meet the upcoming regulations by authorities in several parts of the world. If in addition energy saving is sponsored by the public like in the case of solar energy conversion, small advantages in efficiency can be quickly transferred into cost improvements.

References

[1]   Alex Lidow, in Power Electronics Europe, Issue 2/2003, pp44-45

[2]   L.Lorenz, M.März, A.Knapp and I.Zverev, IEEE IASConf. Rec. 2000, pp376-383

[3]   M. Treu, R. Rupp, C.S. Tai, P. Blaschitz, J. Hilsenbeck, H. Brunner, D. Peters, R. Elpelt, presented at ICSCRM in September 2005

[4]   S.H.Ryu, S.Krishnaswami, M.Das, J.Richmond, A.Agarwal, J.Palmour, J.Scofield, Material Science Forum, Volumes 483-485, pp. 797-800 (2005)

[5]   Refer to world wide web http://www.aist.go.jp/aist_e/latest_research/2005/20050407/20050407.html

[6]   P.Friedrichs, H.Mitlehner, R.Schörner, R.Elpelt, and D.Stephani, , Material Science Forum Vols. 457-460 (2004) pp.1201-1204

[7]   M. Treu, R. Schörner, P. Friedrichs, R. Rupp, A. Wiedenhofer, D. Stephani, H. Ryssel, , Material Science Forum, Volumes 338-342, pp. 1089-1092 (2000)

[8]   S.H.Ryu, S.Krishnaswami, M.Das, B. Hull, J.Richmond, A.Agarwal, J.Palmour, J.Scofield, to be presented at ISPSD 2005

[9]   H.Mitlehner, P.Friedrichs, R.Elpelt, K.Dohnke, R.Schörner, D.Stephani, Material Science Forum, Volumes 457-460, pp. 1245-1248 (2004)

[10]   R. Elpelt, P. Friedrichs, R. Schörner, K.-O. Dohnke, H. Mitlehner, and D. Stephani, "Serial connection of SiC VJFETs -features of a fast high voltage switch", presented at the EPE 2003 in Toulouse, ISBN 90-75815-07-7

[11]   Melkonyan A, Lorenz L., The 39th IEEE Industry Applications Society Annual Meeting. (IEEE / IAS'04), Seattle, USA, Oct. 3-7, 2004

[12]   Kuzumaki et al. The Papers of Technical Meeting on Semiconductor Power Converter, IEE Japan, SPC-04-96 (2004)(in Japanese)

2006 5th International Power Electronics and Motion Control Conference

Implementation of GA-trained GRNN for Intelligent Fast Charger for Ni-Cd Batteries

Panom Petchjatuporn[*] Noppadol Khaehintung[*] Khamron Sunat[**] Phaophak Sirisuk[**] Wiwat Kiranon[***]

[*] Dept. of Control and Instrumentation Engineering, Faculty of Engineering
[**]Dept. of Computer Engineering, Faculty of Engineering
[***]Dept. of Telecommunication Engineering, Faculty of Engineering
Mahanakorn University of Technology, Bangkok, THAILAND, 10530
e-mail *panom@mut.ac.th, noppadol@mut.ac.th, khamron@mut.ac.th, phaophak@mut.ac.th* and *wiwat@mut.ac.th*

Abstract— This paper presents the development of an intelligent genetic algorithm (GA) technique for training of a generalized regression neural network (GRNN) controller to achieve a compact network and to decrease battery charging time on a cost-effective RISC microcontroller. The suitable input-output data were selected from GA mechanism to establish GRNN. The computational complexity of GRNN can be reduced replaced by some simple polynomial forms. As a consequence, the fast charging device for Nickel-Cadmium (Ni-Cd) batteries can be efficiently implemented on a low-cost 16F876A RISC microcontroller. Experimental results are shown to demonstrate superiority of the proposed system.

Keywords-RISC microcontroller; genetic algorithm; fast charging; the generalized regression neural network; radial basis functions;

I. INTRODUCTION

The widely used secondary batteries are Nickel Cadmium (Ni-Cd), Nickel Metal Hydride (Ni-MH) and Lithium Ion (Li-Ion). These batteries can be recharged after being exhausted by passing current through them in the opposite direction to that of discharge current [1]. In principal, charging of a battery may be a straightforward task. In practical, however, it could involve complicated control algorithm, especially when high power demand needs to be supported by high performance while cost effective battery charger.

Recently, fast-charging algorithms have been proposed in many places. Their common purpose is to speed up the battery recharging time. In [2], for instance, the fast-charging algorithm implemented on microcrotroller based upon the relation between voltage, current and temperature of the battery has been presented. Several intelligent control techniques, such as Adaptive Neuro-Fuzzy Inference System (ANFIS) [3] and the Genetic Algorithm (GA) to optimize a fuzzy logic controller (FLC) [4] have also been applied for fast-battery chargers. However, these algorithms especially FLC, though very sophisticated and unchangeable to the environment, require relatively high performance processor and may not be appropriate for certain applications where cost is a prime concern [5].

Moreover, a fast charging technique using the Generalized Regression Neural Network (GRNN) has been introduced and successfully applied to Ni-Cd batteries [6]. With less computational burden, it has been found that its data selection performance is greater than some techniques, for example ANFIS. Moreover, the computational effort reduction of GRNN has been improved by further in [7]. Nonetheless, the data selection of construction GRNN still was not thoroughly considered. Although, the GA to select the optimal data for training of GRNN was present in [8] however this research work was implemented on expensive personal computer.

This paper proposes a cost-effective approach to implementation of GA-trained GRNN (GAGRNN) based on 16F876A RISC microcontroller [9]. Furthermore, the designed controller is integrated into an uncomplicated system. The remainder of the paper is organized as follows. Following this, a system configuration of the proposed fast-charging method is addressed in Section II. The GAGRNN controller for the fast-charging method is reviewed in Section III. In Section IV, the development of fast-charging is discussed and experimental results with emphasis on performance of the proposed GRNN controller are given. Finally, conclusion is drawn in Section V.

II. FAST CHARGING CIRCUIT CONFIGURATION

A. The Fast Charging Method

The maximum rate of charging current for Ni-Cd batteries is 8C [6], where C is capacity in Amp-Hour unit of the battery. In general, the battery will be charged at different rates with different durations. When charging at high rate, temperature inside the battery could change very rapidly, as shown by typical profiles of battery voltage and temperature during a charging period in Fig. 1. When the battery reaches its full charge level, the temperature gradient will increase significantly, and the charging rate should be slowed down in order to avoid the overcharge state. This can lead to significantly decreasing efficiency of the charging process, and, hence, the degradation or even serious damage of the battery. Therefore, this effect must be evaluated before application of a fast-charging process.

1-4244-0448-7/06/$25.00 ©2006 IEEE   137

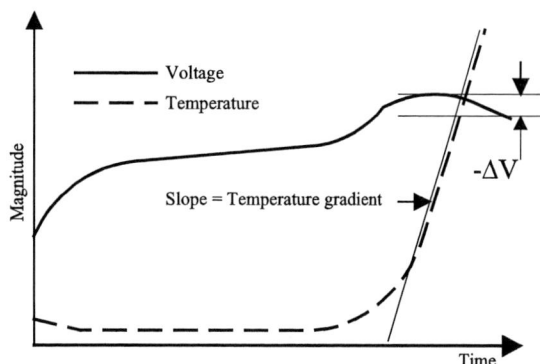

Figure 1. Typical profiles of battery temperature and voltage during fast-charging process

The battery temperature (T) and temperature gradient (dT/dt) can be used in the determination of the rate of charging current (C_t). Based on experimentation, input-output data consisting of 561 point as shown in Table I has been tabulated and taken from [10].

Fig. 2 depicts a surface plot that represents the relationship between the control inputs and corresponding charging current. Considering the figure implies that the charger should apply a relatively high charge current during the initial phase of battery charging. Then, the charge current should be properly reduced during the later phases of charging depending two control parameters T and dT/dt. Following this strategy, an optimal charging current within in the safety limit will always be supplied without any damage to the battery provided that the surface plot of Fig. 2 is well approximated.

B. Circuit Configuration

The proposed ultra fast charger using GAGRNN was realized on a low-cost RISC PIC16F876A microcontroller [9] with additional hardware circuitry. Its structure is illustrated in Fig. 3. In addition to 8kbyte programmable memory, the RISC-microcontroller features a 10-bit, 5-channel successive approximation Analog to Digital A/D. As for the hardware circuitry, it includes a voltage control current source, battery temperature and voltage detector, Digital to Analog (D/A) converter and two Voltage to Current (V/I) converters.

In the proposed charging system, the battery temperature, T, and the battery voltage, V, are measured by a temperature sensor, particularly a thermistor. The data is converted and fed to a microcontroller via a built-in A/D converter. Then, the temperature gradient, dT/dt, is computed using a unit delay. Both T and dT/dt are used as the control inputs of GAGRNN as discussed in the next sections. Computation of GAGRNN output, i.e. the charging current C_t, is performed in the microcontroller.

The D/A MAX503 is employed for conversion of the charging-current data to an analog voltage, which is fed through the V/I XTR110 in order to finally produce the proper charging current supplied to Ni-Cd battery. In

TABLE I. THE EXAMPLE OF REPRESENTATIVE INPUT-OUTPUT DATA FOR Ni-Cd BATTERY CHARGER

Sr.No.	T	dT/dt	C_t
1	0	0	8.00
⋮			
451	40	1	8.00
452	41	0	5.00
⋮			
484	43	1	5.00
485	44	0	4.00
⋮			
517	46	1	4.00
518	47	0	2.00
⋮			
528	47	1	2.00
529	48	0	0.1
⋮			
539	48	1	0.10
540	49	0	0.05
⋮			
561	50	1	0.05

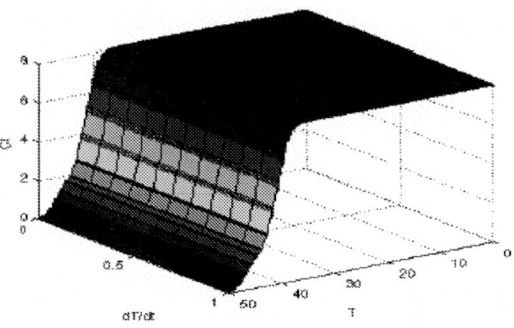

Figure 2. Surface plot of control inputs and a charging current

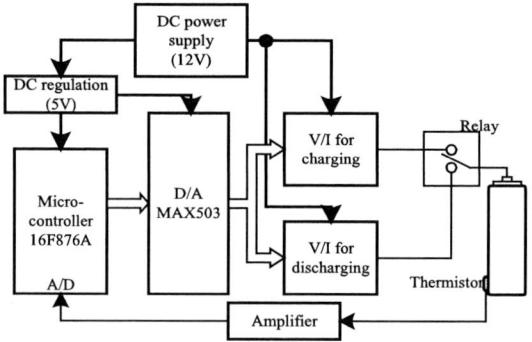

Figure 3. Hardware block diagram

addition to the temperature, the computed voltage gradient is monitored. As soon as the temperature approaches 50°C or the negative voltage gradient is detected, the charging process will be stopped, immediately. Fig. 4 depicts a hardware prototype of our proposed low-cost intelligent Fast Charger for Ni-Cd Battery. Moreover, the flowchart describing its operation is depicted in Fig. 5.

138

$$C_t(\mathbf{x}) = \frac{\sum_{i=1}^{n} \mathbf{Y}^i \mathrm{RBF}\left(\|\mathbf{x}, \mathbf{x_i}\|\right)}{\sum_{i=1}^{n} \mathrm{RBF}\left(\|\mathbf{x}, \mathbf{x_i}\|\right)} \qquad (1)$$

, where $\mathbf{x} \in R^n$ is input vector of dimension n, RBF(.) is radial basis function, $\|.\|$ denotes the Euclidean norm, $\mathbf{Y}^i$ are the weights or parameters of network, $\mathbf{x}_i \in R^n$ are known as the centers of RBF(.), and n is the number of centers.

Once the model training is completed, in order to evaluate the goodness the selected data set, its Mean-Square Error (MSE) of the validation set must be evaluated. MSE is defined by

$$MSE = \frac{1}{N} \sum_{n=1}^{N} \left(C_{td}(n) - C_t(n) \right)^2 \qquad (2)$$

, where $C_{td}(n)$ denotes the desired output of the validation set, $C_t(n)$ denotes the actual output of GRNN model, and N is denoted the number of data points taken for model validation.

B. GA-train GRNN

As similar to other neural network techniques, GRNN requires supervised training. There are various ways for training the network. Among those are genetic algorithms, which are probabilistic search techniques that emulate the mechanics of evolution [11]. They are capable of globally exploring a solution space, pursuing potentially fruitful paths while also examining additional random points to reduce the likelihood of settling for a local optimum.

In GA a set of variable for a given problem is encoded into a *string* (population) analogous to a *chromosome* in nature. In addition, each string contains a lot of *alleles* and each feature of the system located at a specific position in the string is called *gene*. Each string, therefore, contains a possible solution to the problem. The GA performs with a set of binary coding strings, representing a set of data having of 561 input-output data points given in [10] for training the GRNN. The searching of selected input-output data points i starts for from a created an initial population having randomly generating i values. For this case, the centers of RBF $\mathbf{x}_i$ can be decoded as:

$$\mathbf{x}_i = \text{input-output data point } (pop) \qquad (3)$$

, where *pop* is the population. The fitness (f) of each chromosome is evaluated as:

$$f = \text{MSE} \qquad (4)$$

The optimum solution can be obtained by minimizing a fitness function f. Thus, those with lower fitness values will be chosen to be the parents of the next generation while higher fitness values are rejected. Creating the new

Figure 4.   Hardware prototype

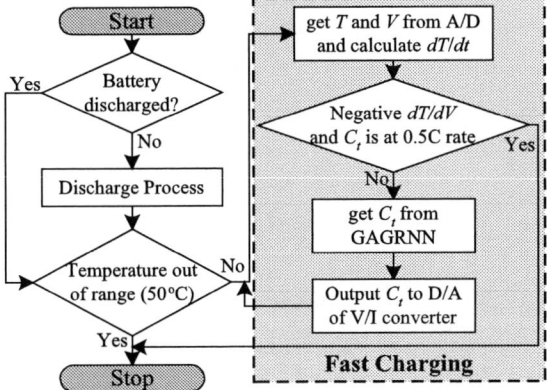

Figure 5.   The fast charging software flowchart

III.   THE PROPOSED FAST CHARGER DESIGN

A.   GRNN for fast-charging

The generalized regression neural network (GRNN) comprising of three layers is selected to perform this task. Once trained, GRNN will appropriately determine the charging current (C_t) given the temperature (T) and temperature gradient (dT/dt). In this research work, GRNN is employed for modeling the relationship among the observed data, namely T, dT/dt and C_t, obtained from experimentation on Ni-Cd battery [10].

Essentially, the output of GRNN given the inputs $\mathbf{x}$, is the weighted combination of the functional distance between the input data and the center of the basis function in the second (hidden) layer. The GRNN performance is sensitive to a choice of the basis function. The fast charger presented in [6] invoked Gaussian function in the GRNN to compute the charge current. However, it was subsequently reported in [7], [8] that with Radial Basis Function (RBF) being a basis function, GRNN can achieve better MSE performance. Thus, in the proposed fast charger RBF is invoked for computation of the GRNN output. That is, the charging current (C_t) is given by

offspring, the selected parents strings undergo a reproduction process such as crossover and mutation as described in [11]. By continuing such a procedure, the newer and fitter chromosome evolves until a predefined stopping condition is satisfied. In this work, the set of input-output pairs taken from [10] is used to train GRNN using GA as shown in Table I.

C. Computational Effort Reduction

The fast charger is intended to run on a very small microcontroller, the computational complexity of GRNN must be reduced. As a result, the RBF(), which is usually of an exponential function, **exp**(), is replaced by some simple polynomial forms presented in [7] which can be computed as the following pseudocode shown in Table II.

IV. EXPERIMENTAL RESULTS

A. Data Selection Results

The simulations were conducted to evaluate the performance of GA-trained GRNN controller for fast-charging using MATLAB/Neural Network Toolbox [12]. The population size used in each trial is 20. Each system is randomly assigned between 1 and 561 represented the order of the input-output data points taken from [10]. Table III shows the parameters used in the GA process. The GA selects 4 and 6 from 561 as the optimal data points to train GRNN as GAGRNN^1 and GAGRNN^2 for fast charging controller respectively. The result of MSE from both controllers is depicts in Table IV after 320 generations. Moreover, these controllers yield a very low MSE comparing with the previous proposed fast charging controller.

B. Charging Results

For performance evaluation, software was developed using C-programming language and was programmed into a low-cost PIC16F876A RISC microcontroller. In the experiment, we applied to charged a Panasonic Ni-Cd battery [13]. When both functions of **exp**(.) and the computational effort reduction algorithm were implemented, the results of memory consumption are compared as depicted in Table V. It can be observed that the computational effort reduction algorithm can be reduced the usage memory more than 69%. The battery voltage, current and temperature were observed and recorded via data acquisition system.

In order to determine the characteristic of normal charging this battery, the Ni-Cd battery was tested intensively at charging current rate at 0.5C and 3C at the environment temperature about 30ºC. Fig. 6 depicts the results of battery temperature and voltage during charging with constant current 0.5C. It indicates that charging process should be stopped within 8,500 seconds, because overcharge can be detected by positive dT/dt and negative dV/dt at this point. Fig. 7 shows the result of constant current charging at 3C. It indicates that charging process should be stopped within 1,300 seconds approximately.

TABLE II. THE PSEUDOCODE FOR COMPUTATIONAL EFFORT REDUCTION

BEGIN: Given distance value d, and a parameter $p \in \{0, 1, ...\}$
$\phi = 1 - d/2^{(p+1)}$
if $\phi \leq 0$ then *output* :=0
else *output* := $\phi \times \phi$
        for n :=1 to p do
                output := $\phi \times$ *output* $\times$ *output*
        endfor
endif
END {note: corresponding support interval is $[-2^{(p+1)}, 2^{(p+1)}]$}

TABLE III. GENETIC PARAMETER USED

Crossover	0.8
Mutation	0.1

TABLE IV. COMPARISON BETWEEN GRNN AND THE PROPOSED GAGRNN BATTERY CHARGER

	GRNN [6]	GAGRNN^1	GAGRNN^2
Number of Layers	3	3	3
Number of Nodes	13	9	13
MSE	0.0905	0.0249	0.0046

TABLE V. COMPARISON OF MEMORY CONSUMPTION

Function	Percentage of program code
exp(.)	16%
The computational effort reduction algorithm	11%

Although the results from Fig. 6 and Fig. 7 indicate that charging at 3C is faster than charging 0.5C, the former yields a high temperature gradient, the battery temperature is greater than 50°C which may cause batteries deterioration.

The results of implementation GAGRNN^1 and GAGRNN^2 with real time control are illustrated in Fig. 8 (a) and (b), respectively. Both proposed controllers decrease the charging current when temperature increases. From these results, by detecting the positive dT/dt and negative dV/dt, the Ni-Cd battery was reached the full charged within 1,500 and 1,070 seconds from GAGRNN^1 and GAGRNN^2, approximately. Furthermore, the maximum temperature during charging period was lower than the limited temperature comparing with charging result at 3C. To clearly represent the energy stored in a battery at a specific time. Fig. 9 shows the characteristic of battery discharge after each full charge. It can be observed that, the proposed fast charging GAGRNN provides the energy stored in a battery as normal battery charging rate.

To sum up, the GA-trained GRNN network for fast-charging battery can be compactly implemented on a low-cost PIC16F876A RISC microcontroller. Furthermore, the battery can be reached the full charge state in the short time without over limited temperature especially from GAGRNN^2.

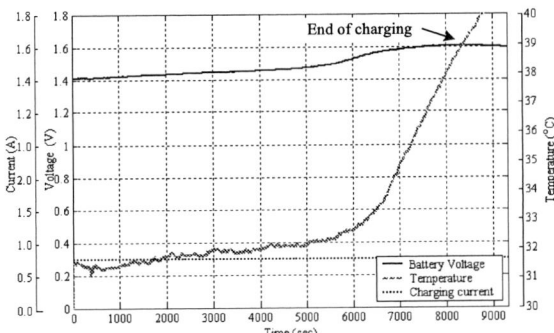

Figure 6. The battery voltage and temperature in a Ni-Cd battery (1.2V/600mAhr) at a constant 0.5C charging rate

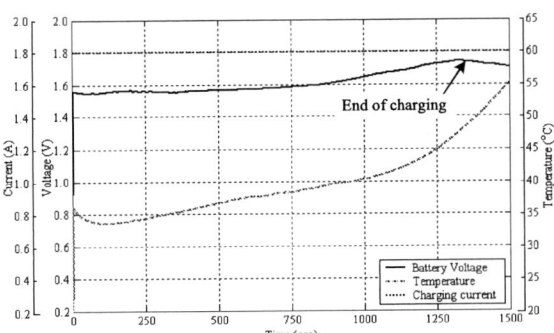

Figure 7. The battery voltage and temperature in a Ni-Cd battery at a constant 3C charging rate

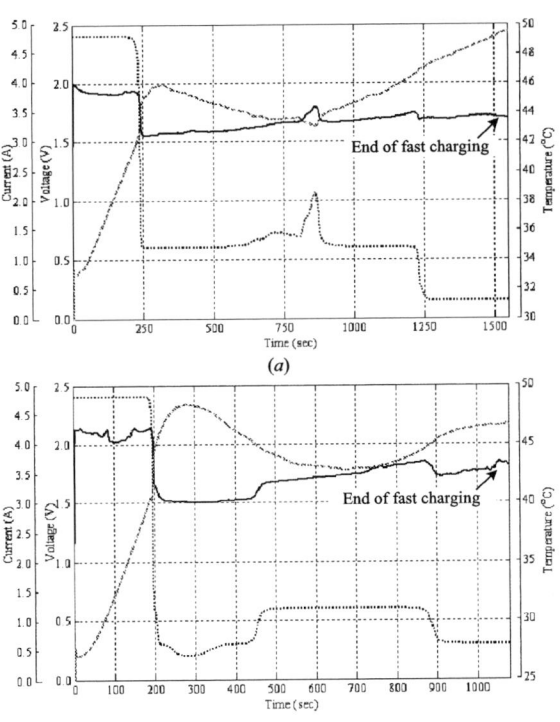

Figure 8. The battery voltage, current and temperature in a Ni-Cd battery of (a) GAGRNN[1] and (b) GAGRNN[2] for fast charging

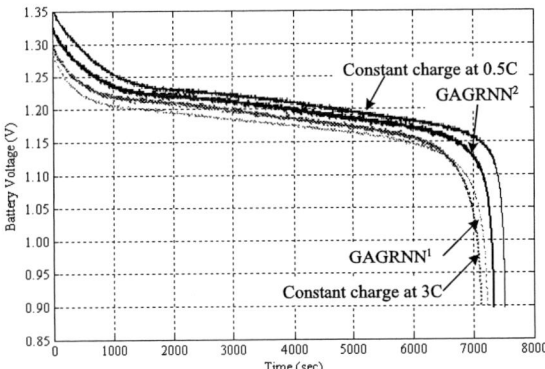

Figure 9. The battery discharge characteristic

V. CONCLUSIONS

In this paper, the fast charger for Ni-Cd battery has been developed. The proposed charger utilizes the GA trained GRNN (GAGRNN) controller to attain fast charging while avoiding battery damage. Efficient hardware implementation of GAGRNN based on a low-cost 16F876A RISC microcontroller was achieved. The experimental results showed and confirmed that the battery can be reached the full charge state in the short time without over limited temperature.

REFERENCES

[1] D. Linden, *Handbook of Batteries*, McGraw Hill Inc., 1995.

[2] J. Diaz, J. A. Martin-Ramos, A. M. Pernia, F. Nuno and F. F. Linera, "Intelligent and universal fast charger for Ni-Cd and Ni-MH batteries in portable applications," *IEEE Tran. on Industrial Electronics*, Vol. 51, No. 4, 2004, pp. 857-863.

[3] A. Khosla, S. Kumar, and K. K. Aggarwal, "Fuzzy controller for rapid nickel-cadmium batteries charger through adaptive neuro-fuzzy inference system (ANFIS) architecture," *Proc. of NAFIPS 2003 Conf.*, pp. 540-544, 2003.

[4] H. Surmann, "Genetic optimization of a fuzzy system for charging batteries," *IEEE Tran. on Industrial Electronics*, Vol. 43, No. 5, 1996, pp. 541-548.

[5] J. Binfet and B. M. Wilamowski, "Microprocessor implementation of fuzzy systems and neural networks," *Proc. of IJCNN '01 Conf.*, Vol. 1, pp.234-239, 2001.

[6] P. Petchjatuporn, P. Wicheanchote, N. Khaehintung, K. Sunat, P. Sookavatana and W. Kiranon, "Data Selection if a Compact GRNN for Ni-Cd Batteries fast charging," *Proc. of TENCON 2004 Conf.*, pp. 213-216, 2004.

[7] P. Petchjatuporn, P. Wicheanchote, N. Khaehintung, K. Sunat, W. Kiranon and S. Chiewchanwattana, "Intelligent ultra fast charger of Ni-Cd batteries," *Proc of ISCAS05 Conf.*, pp. 5162-5165, 2005.

[8] P. Petchjatuporn, N. Khaehintung, K. Sunat, W. Kiranon and P. Wicheanchote, "GA-trained GRNN for Intelligent Ultra Fast Charger for Ni-Cd Batteries," *Proc of PEDS' 2005 Conf.*, pp. 5162-5165, 2005.

[9] Microchip, PIC16F87XA Datasheet, 2003.

[10] http://www.research.4t.com.

[11] D. E. Goldberg, *Genetic Algorithm on Search Optimization and Machine Learning*, Reading, MA, Addision-Wesley, 1989.

[12] http://www.mathworks.com.

[13] Panasonic, Nickel Cadmium Handbook, August, 1998, p.51.

2006 5th International Power Electronics and Motion Control Conference

Modelling and Analysis of a Novel Transformer with Ability to Suppress Conducted interference

Zongxiang Chen*, Pengsheng Ye, Junmin Pan

Department of Electrical Engineering, Shanghai Jiao Tong University, Shanghai 200030, P.R.China

Abstract—A novel transformer with ability to suppress conduction-interference, named as Interference-free Transformer (IFT), is proposed in this paper. Comparing with the traditional transformers, it has a special magnetic structure, namely magnetic air-gap shunt, and adds a resonance coil on the secondary side. The modelling and analysis indicate that the magnetic air-gap shunt can be equivalent to a step-down inductor between the primary and secondary coils. It is in series with the parallel resonance-circuit to construct a filter-network, which makes the difference-mode (DM) interferences be attenuated obviously, while nearly doesn't affect the fundamental signal. In order to suppress the common-mode (CM) interferences, two coils with few turns and opposite direction, which are in series with the primary coil, are also added. A prototype of the IFT has been implemented in the lab. The simulation and experimental results prove the modelling, analysis and conclusion are right.

Keywords-IFT; magnetic air-gap shunt, DM interference suppression ; CM interference suppression component; formatting; style; styling; insert

I. INTRODUCTION

In recent years, with the development of modern industry, a lot of differential-mode (DM) and common-mode (CM) interferences are generated and transmitted into the power source system by the power line and ground. Such interferences are usually called conducted interferences. In order not to impose the interferences on the system, the EMI filters, including the passive and active EMI filters, are always used between the power and intermediate bus [1-6]. But those EMI filters are connected the two sides between the power sources and loads without any electrical isolation directly. Now the transformers have been used widely in order to supply the appropriate voltage and isolation.

Therefore it will be significant if the transformer has also the ability with the interferences suppression.

This paper proposes a new transformer, which has an interference-free ability to supply the power. The proposed Interference-free Transformer (IFT) is based on the resonance-circuit & magnetic air-gap shunt. With the help of special magnetic structure of the transformer, it is easy to form the inductance of DM interferences suppression. In the meanwhile, the CM inductor will exist by adding the CM windings on the iron core of the IFT. Under the prerequisite that the devices do not increase their bulks, electrical isolation and interferences suppression will be realized simultaneously. Simulation and experimental results prove that the IFT not only obviously restrains the interferences which includes the DM and CM, but also does not affect the fundamental wave signal.

The IFT can be used where the transformer-isolated is needed and interferences must be suppressed. Especially in some areas, such as national defense and telecommunications, the strong high-frequency signal exists.

II. IFT STRUCTURE

The IFT structure is shown in Figure1 (a). In the figure, R_{m1}, R_{m2}, and R_{mL} are magnetic reluctances of the primary, secondary and magnetic shunts of the IFT respectively. N_1, N_2 and N_c, are the numbers of turns of the primary, secondary and resonance windings of the IFT respectively. N_{c1} and N_{c2} are the numbers of turns of common-mode winding. C_{c1}, and C_{c2} are common mode capacitors.

The structure is similar to that of the constant voltage transformer (CVT). The major difference between these two structures is that the secondary iron

1-4244-0448-7/06/$25.00 ©2006 IEEE

core of CVT works in the saturation range of magnetization curve forming ferroresonance, while that of the IFT works in the range of linearity and resonates with parallel C_0. The efficiency of the CVT is low and its output voltage contains harmonics components. But the IFT can output the sinusoidal waveform, achieve higher efficiency, and suppress the interferences.

Because the secondary winding couples with the resonance winding closely, C_0 of the IFT can be converted into C of the secondary winding, as shown in Figure1 (b), where $C = (N_c/N_2)^2 \cdot C_0$.

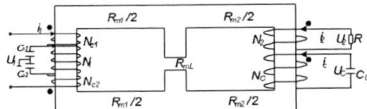

(a) Structure of IFT

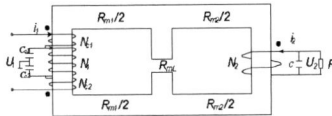

(b) Equivalent structure of IFT

Figure1. Structure & Equivalent structure of IFT

III. DM INTERFERENCE SUPPRESSION of IFT

A. Equivalent circuit

The IFT structure is similar to that of the CVT. Some studies have been done to analyze the magnetic circuit of the CVT, and obtained its equivalent circuit in the last few years [7-9]. This paper presents a simple method based on combination of the magnetic and electrical circuit, which adds an ideal transformer ingeniously.

The equivalent magnetic circuit of the DM part of IFT can be achieved (see Figure 2) from Figure1. By inspection of Figure 2, two equations can be written

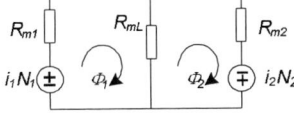

Figure2. Equivalent magnetic circuit

$$
\begin{cases}
i_1 N_1 = (R_{m1} + R_{mL}) \cdot \Phi_1 - R_{mL} \cdot \Phi_2 \\
i_2 N_2 = (R_{m2} + R_{mL}) \cdot \Phi_2 - R_{mL} \cdot \Phi_1
\end{cases}
\tag{1}
$$

From (1), we can obtain

$$
\begin{cases}
\Phi_1 = \dfrac{(R_{m1} + R_{mL}) \cdot i_1 N_1 + R_{mL} i_2 N_2}{R_M} \\
\Phi_2 = \dfrac{(R_{m2} + R_{mL}) \cdot i_2 N_2 + R_{mL} i_1 N_1}{R_M}
\end{cases}
\tag{2}
$$

where, $R_M = R_{m1}R_{m2} + R_{m1}R_{mL} + R_{m2}R_{mL}$.

According to the common transformer model, (3) can be written, regardless of the resistance of the primary and secondary windings.

$$
\begin{cases}
U_1 = j\omega N_1 \Phi_1 = j\omega L_1 i_1 + j\omega M i_2 \\
U_2 = j\omega N_2 \Phi_2 = j\omega L_2 i_2 + j\omega M i_1
\end{cases}
\tag{3}
$$

where, L_1 is the equivalent self-inductor of the primary winding; L_2 is the self-inductor of the secondary winding; M is the equivalent mutual induction between the primary and secondary windings.

Substitution of (2) into (3) gives:

$$
\begin{cases}
L_1 = (R_{m2} + R_{mL}) \cdot N_1^2 / R_M \\
L_2 = (R_{m1} + R_{mL}) \cdot N_2^2 / R_M \\
M = (R_{mL} N_1 N_2)/R_M
\end{cases}
\tag{4}
$$

Figure3 (a) is the IFT T-circuit. The distributed capacitance C_p is considered in the figure. Three inductors in the frame of the dotted line will be negative in some condition. So we add the ideal transformer to solve the problem of the negative inductors (see Figure3 (b)).

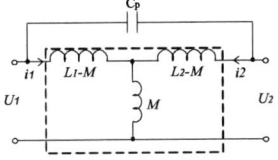

(a) T-circuit

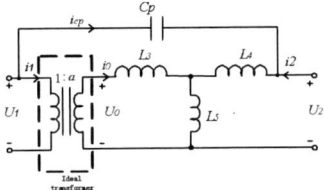

(b) Analytic circuit with ideal transformer

Figure3. Analytic circuit

In Figure3 (b), the turns ratio of ideal transformer is $a(a \in R^+)$, hence

$$
\begin{cases}
U_0 = aU_1 \\
i_0 = i_1/a
\end{cases}
\tag{5}
$$

$$U_1 = L_1 \frac{di_1}{dt} + M \frac{di_2}{dt} \qquad (6)$$

Substitution of (5) into (6) gives:

$$U_0 = a\left(L_1 \frac{a \cdot di_0}{dt} + M \frac{di_2}{dt} \right)$$

$$= \left(a^2 L_1 - Ma \right)\frac{di_0}{dt} + Ma\left(\frac{di_0}{dt} + \frac{di_2}{dt} \right) \quad (7)$$

According to Figure3 (b),

$$U_0 = L_3 \frac{di_0}{dt} + L_5\left(\frac{di_0}{dt} + \frac{di_2}{dt} \right) \qquad (8)$$

As the same derivation, the other equation can be written

$$U_2 = \left(L_2 - Ma \right)\frac{di_2}{dt} + Ma\left(\frac{di_0}{dt} + \frac{di_2}{dt} \right) \quad (9)$$

$$U_2 = L_4 \frac{di_2}{dt} + L_5\left(\frac{di_0}{dt} + \frac{di_2}{dt} \right) \qquad (10)$$

We can derive equations (7), (8), (9) and (10), and Equ.(11) can be obtained.

$$\begin{cases} L_3 = \left(a^2 L_1 - Ma \right) \\ L_4 = \left(L_2 - Ma \right) \\ L_5 = Ma \end{cases} \qquad (11)$$

For the special case of $a = \dfrac{L_2}{M} = \dfrac{R_{m1} + R_{mL}}{R_{mL}} \cdot \dfrac{N_2}{N_1}$, (12)

is an explicit solution of equation (11)

$$\begin{cases} L_0 = L_3 = \dfrac{\left(R_{m1} + R_{mL} \right) \cdot N_2^2}{R_{mL}^2} \\ L_4 = 0 \\ L_S = L_5 = \dfrac{\left(R_{m1} + R_{mL} \right) \cdot N_2^2}{R_{m1}R_{m2} + R_{m1}R_{mL} + R_{m2}R_{mL}} \end{cases} \quad (12)$$

Considering C, C_p and R, an equivalent circuit of the DM part of IFT is shown in Figure4.

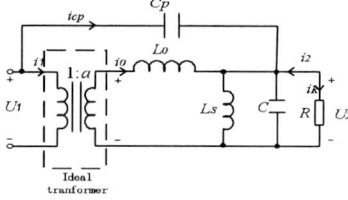

Figure4. Equivalent circuit of the DM part of IFT

The state equation of the part of DM interferences suppression, whose equivalent circuit is shown in Figure4, can be described as

$$\begin{cases} \dfrac{di_0}{dt} = \dfrac{U_0 - U_2}{L_0} = \dfrac{aU_1 - U_2}{L_0} \\ \qquad = \dfrac{R_{m1} + R_{mL}}{R_{mL}} \cdot \dfrac{N_2}{N_1} \cdot \dfrac{U_1}{L_0} - \dfrac{U_2}{L_0} \\ \dfrac{dU_2}{dt} = \left(i_0 + i_{cp} - i_R - i_s \right) \cdot \dfrac{1}{C} \\ i_{cp} = \left(\dfrac{dU_1}{dt} - \dfrac{dU_2}{dt} \right) \cdot C_P \end{cases} \quad (13)$$

From equation (13), the structural diagram is built, as shown in Figure5.

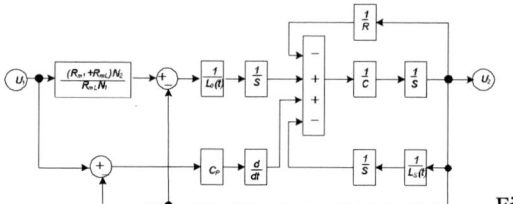

Figure

5. Systematic configuration of the DM part of IFT's equivalent circuit

B. Simulation Results

Using Matlab6.5/Simulink R13, the DM part of IFT will be simulated based on Figure5 with the following parameters:

Input Voltage U_1 (50Hz): 110VAC (RMS);

Output Voltage U_2 (50Hz): 20VAC (RMS);

Power: 100VA;

Resonance capacitance C_0: 3.6 μF

Rating load R: 4 Ω.

Figure6 (a) is simulated with the rating input & output voltage. Figure6(b) shows those of 250Hz (5th harmonic).

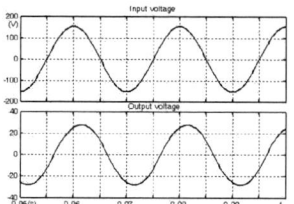

(a) 50Hz

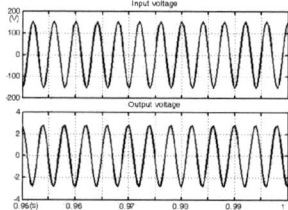

(b) 250Hz

Figure6.   Simulation waveform

C.   Experiment Results

144

An IFT prototype is made in the lab. The parameters of the simulation and experiment are the same. Figure7 (a)& (b) show the outputs in the different frequency, namely 50Hz, 250Hz. Their input voltages are all 110VAC (RMS) with rated loads. The top one (1st waveform) is the input voltage and the bottom one (2nd waveform) is the output voltage. (Figure8(a) has the same notation). In contrast to the simulation ones (see Figure6), the output voltage values and phases of experimental results are the same.

Figure8 (a) shows the waveform of triangle input voltage, which has the abundant odd harmonic, and waveform of output voltage. Figure8 (b) are the spectrum analysis of input and output voltage waveforms respectively. From Figure6,7 and 8, we can conclude that the suppression is more obvious with the frequency increasing.

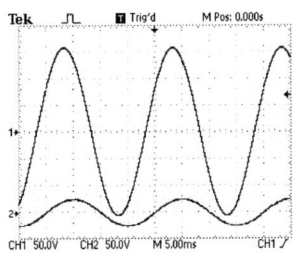

(a) 50Hz

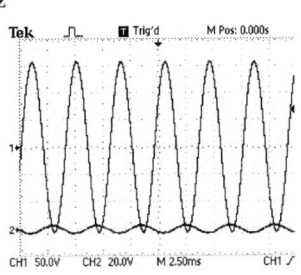

(b) 250Hz

Figure 7.   Experiment waveforms in different frequency with rated loads

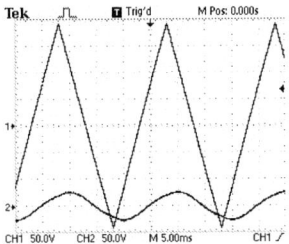

(a) input and output voltage waveform

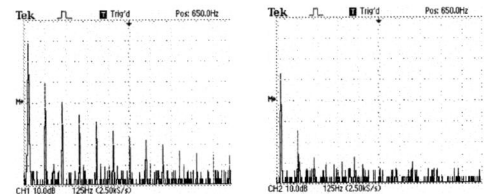

(b) input & output voltage spectrum

Figure8.   Experimental Results under triangle input

IV   CM INTERFERENCE SUPPRESSION of IFT

With the help of the magnetic structure of the IFT, it is easy to realize to suppress the CM interferences. As shown in Figure1, two CM windings, which have a few turns and opposite direction, are in series with the primary coil. Those windings can be equivalent to two common-mode inductors, named as L_{c1} and L_{c2}. Besides, two common-mode capacitors, namely C_{c1} and C_{c2}, are also added (see Figure9). By using the common-mode inductors and capacitors in the circuit, the CM interferences of power line can be filtered.

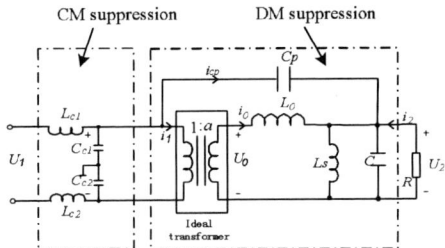

Figure9 Equivalent DM and CM circuits of IFT

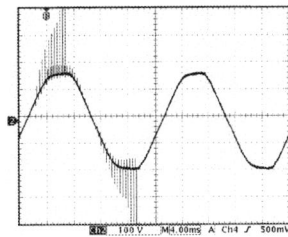

(a) Input voltage

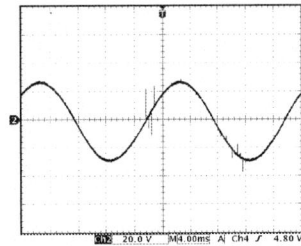

(b) Output voltage

Figure10.   Input and output voltages of IFT

In order to test the ability of CM interferences

suppression of IFT, some experiments have been done to test based on Electrical Fast Transient/burst (EFT/B) according to IEC61000-4-4 standard. The EFT/B is imposed on the loads via the lines and ground. It is always considered as the CM interferences test when the EFT/B is conducted on power lines [10]. In the experiment, NS64000-4A (EFT/B generator) is used, which is made in Shanghai Sanki Electronic Industries Co., LTD.

Figure10 (a) & (b) is the bursts input and output voltages of the IFT. The figure shows that both the value and number of the bursts have been attenuated and suppressed after the IFT transmission.

VI. CONCLUSIONS

In this paper, a new transformer (IFT) has been proposed and its equivalent circuit has been introduced. By using the proposed IFT, it is possible that electric isolation and CM&DM interferences suppression can be realized simultaneously. Therefore, it can be widely applied to power electronics systems. Its excellent effectiveness has been verified by the simulation and experiments. Meanwhile, this paper only studies the possibility of interferences suppression explanatorily. As to the optimized design of length of the air-gap and other parameters to improve the suppression abilities, this paper doesn't study. It is the authors' hope that the research of this special transformer will continue.

REFERENCES

[1] Jacobus. Daniel. Van. Wyk, Fred. C. Lee, "Integrating Active, Passive and EMI-Filter Functions in Power Electronics Systems: A Case Study of Some Technologies" IEEE Trans. Power Electronics, vol. 20,

no. 3, pp. 523-536, 2005.

[2] Sheng. Ye, Yan-Fei. Liu, "EMI filter design method for communication power sub-system". APEC'03. Eighteenth Annual IEEE, vol.1, pp.483 – 489, 2003.

[3] Shuo.Wang, Fred.C.Lee, "Improving the performance of boost PFC EMI filters". Applied Power Electronics Conference and Exposition, vol.1, pp. 368-374, 2003.

[4] Son. Y.-C, Seung-Ki, Sul. "A new active common-mode EMI filter for PWM inverter". IEEE Trans. Power Electronics, vol.18, no.3, pp. 1309 – 1314, 2003.

[5] Takahashi. I, Ogata. A, Kanazawa.H, "Active EMI filter for switching noise of high frequency inverters". Power Conversion Conference - Nagaoka 1997, vol.1, pp. 331-334, 1997.

[6] Dongbing. Zhang, Chen. D.Y., "A new method to characterize EMI filters". APEC '98. Conference Proceedings, vol.2, pp. 929-933, 1998.

[7] Lind. Magnus.G.J, Xiao. Weidong, "Modeling of a Constant Voltage Transformer". IEEE Trans. Circuits and Systems I: Regular Papers: Accepted for future publication, 2005.

[8] Fengren.Wang, "A Study of Nonlinear Property and Optimum". [Doctor Dissertation], Xi'an JiaoTong Univercity, 1995.

[9] Fuchs. E. F, "Large signal nonlinear model of anisotropic transformers for nonsinusoidal operation, Part □: $\lambda - I$ characteristic" IEEE Trans. Power Delivery, vol.6, no.1,pp.174-185,1991.

[10] Cormier. B., Boxleitner. W. "Electrical fast transient (EFT) testing-an overview". Electromagnetic Compatibility 1991.Symposium Record, pp. 291-296, 1991.

2006 5th International Power Electronics and Motion Control Conference

An Observer-Based Three-Phase Current Reconstruction using DC Link Measurement in PMAC Motors

Li Ying and Nesimi Ertugrul

Yunnan University, School of Information Science and Engineering
Kunming, Yunnan, E'mail: li_ying_km@163.com
Adelaide University, Department of Electrical and Electronics Engineering
Adelaide, Australia, E'mail: nesimi@elceng.adelaide.edu.au

Abstract—**This paper proposes a novel three-phase current reconstruction technique that utilises the measured DC link current and the switching signals of a three-phase inverter used in Permanent Magnet (PM) AC motors. The method is based on an adaptive current observer, which includes the model of a motor drive to estimate the three-phase currents. The relationship between the three-phase currents and the DC link current is used to estimate the DC link current first as an output of the observer. Then, this output is compared with the measured DC link current and the current error. The error is used to compensate the DC link voltage that can force the estimated phase currents to follow the actual winding currents. The paper provides a detailed analysis of the current conduction modes in a three-phase inverter and summarises the implementation details the method in real-time. A wide range of experimental results provided in this study demonstrates that the method can accurately reconstruct the phase currents both in the sinusoidal and in the rectangular current excitation modes of a PMAC motor. The real time experimental results are also given to demonstrate the closed-loop performance of the motor drive operating using a single current sensor.**

Keywords- Adaptive Observer, DC Link Current, Brushless Permanent Magnet AC Motors, Current Reconstruction

I. INTRODUCTION

Permanent magnet AC (PMAC) motors have a wide range of applications in industry and in electronic appliances, where the knowledge of winding currents is required to implement a closed-loop current control, and to limit the currents for the device protection. The conventional method is to measure the phase currents directly by using two or three low value resistors or Hall-Effect current sensors. The later is usually preferred since they provide electrical isolation at a relatively high frequency bandwidth. However, the current sensors and the associated accessories increase the complexity, the cost and the size of the motor drive, and reduce the reliability of the system. Therefore, the reduction of the number of sensors is desirable in motor drives.

A number of studies have been reported in the literature all utilize the dc link current to reconstruct the three-phase currents of an inverter. As reported in [1]-[4], the phase current reconstruction in a trapezoidal PMAC motor with rectangular current excitation is relatively easy. In this excitation mode, since only two of the three phase windings conduct currents at any instant of time, the value of the two-phase currents is equal to the DC link current, but their directions depend on the switching state of the inverter. However, during the current commutation periods, all three-phases conduct current simultaneously, which was not studied previously.

In the sinusoidal PMAC motors with sinusoidal current excitation, the three-phase current reconstruction from the DC link current measurement is also possible [5]-[12]. In this excitation mode, when all of the three phases conduct current simultaneously, there are two zero-states and six active-states in the inverter. During the active states, one of the phase currents flows through the DC link. However, during the zero-states, the phase currents circulate in the inverter bridge, which is not observable via the DC link. Under the normal PWM switching mode, there are two possible active-states in each PWM cycle. Hence, the two-phase currents can be derived from the DC link current at two consecutive points. If the PWM frequency is high enough, the phase currents vary slightly over one PWM cycle. Within this cycle, the derived values can represent the two-phase currents, and the third phase current can easily be calculated using the two-phase currents (assuming the motor is star-connected).

However, under certain operating conditions, one of the two active-states during a PWM cycle may last very short period of time. Due to the finite switching time of the power devices, the dead times, and the delays in the electronic circuits, the actual phase current may not be visible on the dc link measurement. Several different approaches have been proposed to solve this problem [11,12]. When the period of an active-state is too short, one typical solution is to replace the active-state with a zero-state and then adding the missing time to the subsequent PWM cycle. When the duty-cycles of two of the phases are almost equal, an alternative solution is to adjust the duty-cycle within one PWM cycle by shifting the switching signals one step forward and one backward. By this method, the dc link visible two active-states can be obtained, and the average voltages applied to the two

1-4244-0448-7/06/$25.00 ©2006 IEEE

phases are the same in this PWM cycle. A predictive state observer is found to be a possible solution for this problem.

Although the previous current reconstruction methods reported good results, they have a number of limitations when considered in the practical motor drives. Firstly, since the DC link current waveforms are mostly narrow current pulses, complex hardware circuits are required to detect the current values at specified points, which are usually distorted by the measurement noises. Secondly, the previous methods can only obtain the average values of the phase currents within one or several PWM cycles by an integration or an averaging method. Therefore, current variations within a PWM cycle cannot be obtained, which is required in the fast acting closed-loop current controllers. In addition, modified PWM signals, as in the previous methods, may affect the actual output currents of the inverter and may increase the total harmonic distortion and the switching losses.

This paper proposes a novel method of reconstruction of three-phase currents from the DC link current measurement, which can be used both in sinusoidal and rectangular current excited PMAC motor drives. The method is based on an adaptive current observer, which uses a model of the PMAC motor drive to estimate the three-phase currents. Using the relationship between the three-phase currents and the DC link current, the DC link current is reconstructed as the output of the observer. Then, this output is compared with the measured DC link current and the error is regulated by an adaptive scheme. Finally, using the symmetry of the three phases, the regulated errors are utilized to correct the estimated three-phase currents. To verify the method, a number of experimental results are provided in the paper. A set of experimental results is also given to demonstrate the performance of the motor drive under a closed-loop operation using a single current sensor on the DC link.

II. RECONSTRUCTION OF THE DC LINK CURRENT

Fig. 1 illustrates the power circuit of the motor drive. The conventional inverter in the figure consists of six power switches and six freewheeling diodes, in which the switches are controlled to provide desirable voltage waveforms to the motor windings. In the figure, the nominal current directions are indicated with the arrow signs, and I_{dc} is the instantaneous DC link current; I_a, I_b and I_c are the instantaneous three phase currents of the motor.

In a three-phase inverter, two switches on the same leg can be turned off but cannot be turned on at the same time. Therefore, a state variable, S_p with three different states can be defined to represent the states of the two switches on the same leg: if the upper switch of phase p (p = a, b, c) is closed $S_p=1$, if the lower switch of phase p is closed $S_p=-1$, and if both switches are open $S_p=0$. Using these explanations, it can be concluded that there are twenty-seven possible switch states in a three-phase inverter. However, since at least two of the three-phase windings should conduct currents at a given instant, only twenty switch states can be defined under the normal operating conditions. The operating modes and corresponding DC link current are listed in the table below.

Group No.	Switch States			DC Link current	Comment
	Sa	Sb	Sc		
1	1	-1	0	$i_a = -i_b$	two
	1	0	-1	$i_a = -i_c$	
	-1	1	0	$i_b = -i_a$	phases
	0	1	-1	$i_b = -i_c$	conduct
	-1	0	1	$i_c = -i_a$	
	0	-1	1	$i_c = -i_b$	current
2	1	-1	-1	$i_a = -i_b - i_c$	three
	1	1	-1	$i_a + i_b = -i_c$	
	1	-1	1	$i_a + i_c = -i_b$	phases
	-1	1	-1	$i_b = -i_a - i_c$	conduct
	-1	1	1	$i_b + i_c = -i_a$	
	-1	-1	1	$i_c = -i_a - i_b$	current
3	1	1	0	$i_a + i_b = 0$	
	-1	-1	0	$i_a + i_b = 0$	
	1	0	1	$i_a + i_c = 0$	zero
	-1	0	-1	$i_a + i_c = 0$	
	0	1	1	$i_b + i_c = 0$	
	0	-1	-1	$i_b + i_c = 0$	states
	1	1	1	$i_a + i_b + i_c = 0$	
	-1	-1	-1	$i_a + i_b + i_c = 0$	

From the table, the sum of the DC link current is equal to the sum of three-phase currents flowing via either positive or negative supply rail. Therefore, the instantaneous DC link current can be given as:

$$I_{dc} = \left(i_a \cdot S_a + i_b \cdot S_b + i_c \cdot S_c\right)/2 \qquad (1)$$

III. CURRENT OBSERVER

Figure 2 illustrates the block diagram of the current observer implemented in this study, which consists five main blocks. Phase Voltages block reconstructs the three-phase voltages of the motor using the switching signals

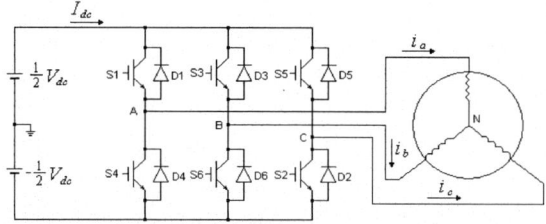

Figure 1. Three-phase inverter with a motor load

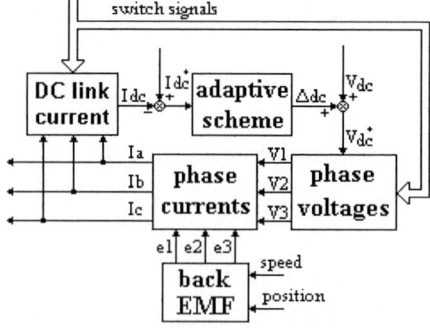

Figure 2. Block diagram of the current observer

148

and the measured DC link voltage. The phase currents of the motor are estimated using the voltage equations that are the outputs of the Phase Voltages. The block named DC Link Current reconstructs the DC link current of the motor drive using the estimated phase currents that are the output of the current observer. Then the reconstructed DC link current is compared with the measured DC link current and the result is regulated using an adaptive scheme.

A. Voltage Equation

$$\begin{bmatrix} v_a \\ v_b \\ v_c \end{bmatrix} = \begin{bmatrix} R & 0 & 0 \\ 0 & R & 0 \\ 0 & 0 & R \end{bmatrix} \begin{bmatrix} i_a \\ i_b \\ i_c \end{bmatrix} + \begin{bmatrix} L & 0 & 0 \\ 0 & L & 0 \\ 0 & 0 & L \end{bmatrix} \frac{d}{dt} \begin{bmatrix} i_a \\ i_b \\ i_c \end{bmatrix} + \begin{bmatrix} e_a \\ e_b \\ e_c \end{bmatrix}$$

(2)

Here v_a, v_b, and v_c are the phase voltages; R is the winding resistance, i_a, i_b, and i_c are the phase currents, L is the equivalent inductance of a winding, and e_a, e_b, and e_c are the back EMF voltages.

The three-phase back EMF voltages are the functions of the rotor position and angular speed, which can be expressed in matrix form as follow:

$$\begin{bmatrix} e_a \\ e_b \\ e_c \end{bmatrix} = k_e \cdot \omega_r \cdot \begin{bmatrix} e_1(\theta_e) \\ e_2(\theta_e) \\ e_3(\theta_e) \end{bmatrix}$$

(3)

Here k_e is the back EMF constant, ω_r is the angular speed of the motor, $e_1(\theta_e)$, $e_2(\theta_e)$ and $e_3(\theta_e)$ are the back EMF functions that vary with the rotor position as sinusoidal or trapezoidal waveforms, and θ_e is the electrical rotor position given by:

$$\theta_e = N_p \cdot \theta_r = N_p \cdot \int \omega_r dt$$

(4)

Here N_p is the number of pole pairs, and θ_r is the mechanical rotor position.

B. Phase Voltage Reconstruction

It is assumed that the rotor position and the speed are already known, and the three-phase voltages can be reconstructed using the DC link voltage measurement and the switching signals. A unified expression of the phase voltage is derived as:

$$v_p = \frac{1}{K} \begin{bmatrix} (K \cdot S_p - S_a - S_b - S_c) \cdot V_{dc} / 2 \\ + (|S_a| \cdot e_1 + |S_b| \cdot e_2 + |S_c| \cdot e_3) \end{bmatrix}$$

(5)

Where K = 3 when all three phases conduct currents and K= 2 for two-phase conduction.

C. Adaptive Scheme

In order to achieve a common modification method for all three phases of the motor drive, a novel correction algorithm is presented here. In the proposed algorithm, the DC link current error is not fed back directly to correct any phase current estimation as in the linear observer. Instead, it is used to compensate the DC link voltage. Then, the compensated value is modified by switch states of each phase, and is fed back to correct the current estimations. To eliminate the effects of the measurement noise in a practical system, a proportional and integral adaptive scheme is used to regulate the DC link current error, and the result is applied to compensate the DC link voltage in the current observer, which is given by

$$V_{dc}^* = V_{dc} + K_p \varepsilon_{dc} + K_i \int_0^t \varepsilon_{dc} dt$$

(6)

Here $\varepsilon_{dc} = I_{dc}^* - I_{dc}$, is the error between the measured and the reconstructed DC link currents, V_{dc} is the measured DC link voltage, V_{dc}^* is the corrected DC link voltage, K_p and K_i are the constants of the P and I current regulators respectively.

As seen from the above explanations, the observer based phase current reconstruction method depends on the DC link current measurement. As stated earlier, within the period of an active state, the DC link current measurement reflects one phase current and can be used to correct the phase current estimations. At zero state, however, the DC link current is equal to zero and cannot represent any phase current. Therefore, the current observer can be seen as an open-loop estimator. Fortunately, at least one valid DC link current value, which is equal to one phase current, is measurable in each PWM cycle under the normal operating condition. Therefore, the phase current estimation is corrected at least once per PWM cycle. Moreover, using Equation (1), a zero DC link current feedback at zero-states can correct the balance of three-phase estimated currents, which means that it forces the summation of the estimated three-phase currents to be zero.

IV. REAL-TIME IMPLEMENTATION AND EXPERIMENTAL RESULTS

To verify the validity of the phase current reconstruction and to demonstrate the single current sensor operation, a complete PMAC motor drive system has been implemented. The drive system includes an IRMDAC3 3-Phase 460VAC 3HP inverter module from International Rectifiers, which is controlled by ADMC

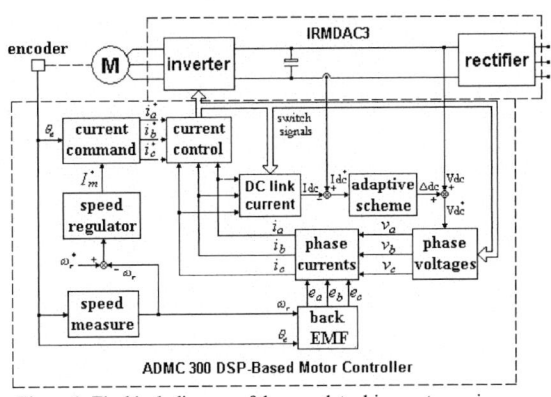

300 Evaluation Module, DSP-based motor controller from Analog Devices. The test motor used in the experiments has the motor parameters given in the appendix, and its back emf waveform is sinusoidal. However, to demonstrate the ability of the proposed reconstruction technique the motor is excited with both sinusoidal and rectangular currents. Figure 3 shows the block diagram of the implemented system, where two A/D converter channels of the DSP are used to measure the DC link current and the DC link voltage only.

Figure 4 shows the steady-state real-time experiment results of the single current sensor drive system under the sinusoidal current excitation. The measured DC link current and the reconstructed three-phase currents are given in the figure at two different speeds (43 Hz and 14 Hz). It can be seen in the figure that the reconstructed phase currents have small ripples caused by the low PWM frequency. In addition, it was observed by the additional tests results that at the low PWM frequency, the modulation index of PWM is high and the DC link current pulses are too narrow to be accurately detected by the available hardware. An inaccurate DC link current measurement causes unreliable phase current reconstruction and the motor drive may become unstable at the low PWM frequency, which may be improved using a higher speed DSP, such as ADMC 501.

When the motor is excited with the rectangular currents using a single DC link current sensor, the steady-state experimental results are given at two different speeds (43 Hz and 9Hz) in Figure 5. As can be observed in the figure,

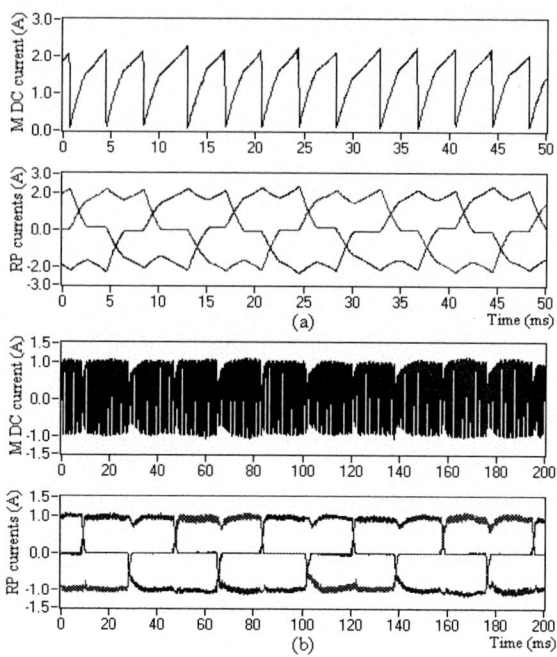

Figure 5. The measured (M) DC link current and the reconstructed phase (RP) currents in the real-time drive system under the rectangular current excitation: (a) High-speed operation without current control, (b) Low-speed operation with current control.

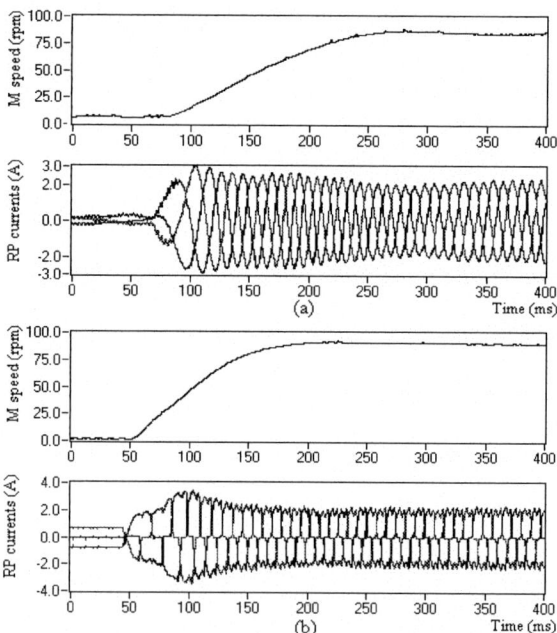

Figure 6. The measured (M) speed and the reconstructed phase (RP) currents under the dynamic states: (a) With a sinusoidal current excitation, a speed variation from 8 rpm to 80 rpm, (b) With a rectangular current excitation, a speed variation from 4 rpm to 85 rpm

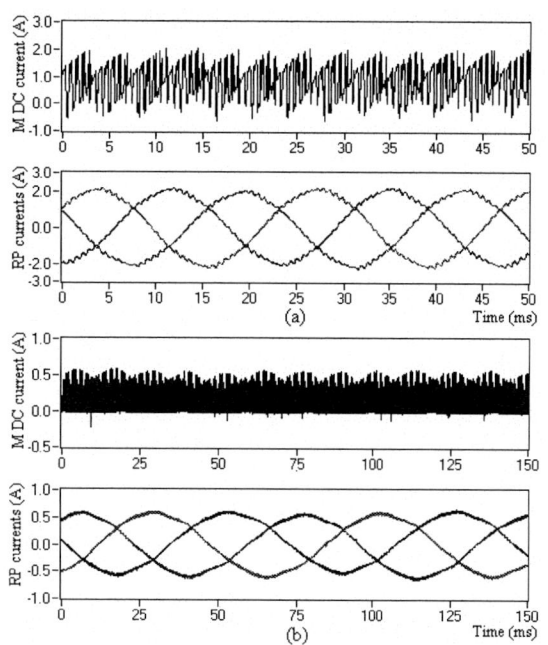

Figure 4. The measured (M) DC link current and the reconstructed phase (RP) currents in the real-time drive system under the sinusoidal current excitation:
(a) High-speed operation, (b) Low-speed operation.

since the PWM frequency in this type of excitation is high, no ripple appears on the reconstructed current waveforms. In the high-speed operation without current control (Figure 10a), the DC link current is easy to detect, and the reconstructed phase currents are accurate as the current feedback. As demonstrated, the motor drive can also operate steadily at very low speed. This is because the duty cycle of DC link current in this mode of excitation is around 50% at low speed range, so it can accurately be detected.

As shown in Figure 6a, when the motor is excited with a sinusoidal current, the phase currents are also reconstructed under the dynamic condition: a speed variation from 8 rpm to 80 rpm. The figure illustrates both the actual speed and the reconstructed current waveforms. Figure 6b shows the similar results when the motor is excited with the rectangular currents. A good dynamic performance is apparent in both tests.

V. CONCLUSIONS

The paper proposes a novel three-phase current reconstruction technique for both sinusoidal and rectangular current excited PMAC motors, which uses a single current sensor on the DC link. The technique is based on an adaptive current observer that effectively compares the reconstructed and the measured DC link currents and corrects the three-phase current estimates. The experimental results demonstrated that the reconstructed three-phase currents can follow the actual winding currents accurately, and can operate under steady-state and dynamic conditions. Although the motor parameters are required to estimate the phase currents in this method, it was studied by the off-line real data that the reconstructed phase currents are not sensitive to the parameter variations. In addition, the integration function in the adaptive scheme eliminates the fast transients that are observable via the DC link current sensor.

Compared with the previous phase current reconstruction methods, the method proposed requires only one valid DC link current measurement in each PWM cycle, which eliminates the use of complex hardware and does not require any modification on the PWM signals, as in the previous current reconstruction techniques.

As demonstrated in the sinusoidal excitation, the DC link current pulses are too narrow to be detected properly at low speeds. Therefore, in order to implement a practical system, a high speed DSP controller with a high performance A/D converter is required. If this is not possible, the PWM signals can be modified to improve the low speed performance of the reconstruction. However, no speed limitation is observed in the rectangular current excitation.

The principal purpose of the phase current reconstruction technique proposed here is also to reduce the number of current and voltage sensors in the PMAC motor drives. The future paper will integrate this study into the indirect position sensing technique that will use a single DC link sensor to implement a position sensorless PMAC motor drive.

APPENDIX

The parameters of the PMAC test motor:

Back EMF constant, k_e :	3.785 V/rad/s
Number of poles pair, P :	28
Winding resistance, R :	6.4 Ω
Equivalent winding inductance, L :	32.8 mH

REFERENCES

[1] Kavanagh R. C., Murphy J. M. D., and Egan M. G., "Innovative Current Sensing for Brushless DC Drives, " IEE PEVD Conference Publication, pp. 354-357, 1988.

[2] Acarnley P. P. and Eng C., "Current measurement in three-phase Brushless DC drives, " IEE Proceedings-B, Vol. 140, No.1, pp. 71-79, January 1993.

[3] Acarnley P. P., "Observability Criteria for Winding Currents in Three-Phase Brushless DC Drives, " IEEE Transactions on Power Electronics, Vol. 8, No.3, pp. 264-270, July 1993.

[4] Tan H. and Ho S. L., "A novel single current sensor technique suitable for BLDCM drives, " Record of PEDS`99, July 1999, Hong Kong.

[5] Green T.C. and Williams B.W., "Derivation of Motor Line-Current Waveforms from the DC-link Current of an Inverter, " IEE Proceedings-B, Vol. 136, No.4, pp. 196-204, July 1989.

[6] Xue Y., Xu X., et al., "A Stator Flux-oriented Voltage Source Variable-Speed Drive Based on DC Link Measurement," IEEE Transactions on Industry Application, Vol. 27, No.5, pp. 962-969, Sept/Oct, 1991.

[7] Mognihan J. F., Kavanagh R. C., et al., "Indirect Phase Current Detection for Field Oriented Control of a Permanent magnet Synchronous Motor Drive, " Proceedings of EPE`91, Vol. 3, pp. 641-646, 1991.

[8] Mognihan J. F., Bolognani S., et al., "Single Sensor Current Control of AC Servodrives Using Digital Signal Processors, " Proceedings of EPE`93, Vol. 4, pp. 415-421, 1993.

[9] Blaabjerg F., Pedersen J. K., "A New Low-cost Fully Fault Protected PWM-VSI Inverter with True Phase-current Information, " Proceedings of IPEC`95, Vol. 2, pp. 984-991, 1995.

[10] Riese M., "Phase Current Reconstruction of a Three-phase Voltage Source Inverter-fed Drive Using Sensor in the DC-link, " Proceedings of PCIM`96, pp. 95-101, 1996.

[11] Blaajerg F., Pedersen J. K., et al., "Single current sensor technique in the DC link of three-phase PWM-VS inverters: a review and a novel solution," IEEE Transactions on Industry Application, Vol. 33, No.5, pp. 1241-1253, Sept/Oct, 1997.

[12] Woo C. L., Taeck K. L and Dong S. H., "Comparison of Single-Sensor Current Control in the DC Link for Three-Phase Voltage-Source PWM Converters," IEEE Transactions on Industrial Electronics, Vol. 48, No.3, pp. 491-505, June, 2001.

[13] Chan C.C., Chau K.T., et al., "A Novel Dead-Time Vector Approach to Analysis of DC Link Current in PWM Inverter Drives," IEEE Applied Power Electronics Conference and Exposition, pp. 338-344, 1997.

[14] Ying L. and Ertugrul N., "A Novel, Robust DSP-Based Indirect Rotor Position Estimation for Permanent Magnet AC Motors without Rotor Saliency", IEEE Transactions on Power Electronics, Vol. 18, No.2, pp 539-546, March 2003.

2006 5th International Power Electronics and Motion Control Conference

Experiment Research of Chaotic PWM Suppressing EMI in Converter

R. YANG [*,**], B.ZHANG [*], F.LI [**] and J.J. JIANG [**]

* Electric Power College, South China University of Technology, Guangzhou, Guangdong Province, China
** School of Physics and Electronic Engineering, Guangzhou University, Guangzhou, Guangdong Province, China
E-mail: lisayang702@yahoo.com.cn

Abstract—Based on the analysis of constant frequency, periodical frequency-spreading scheme and chaotic frequency-spreading scheme, the article has discussed the mechanics and the characters of the chaotic PWM suppressing EMI in the converter. Experiments have been made on an AC-DC forward converter. The experiment results show that the EMI spectrum is decided completely by the PWM spectrum, and the chaotic frequency-spreading has an obvious impact on EMI suppressing, compared with the other schemes. The frequency character of the ripple has been changed but the peak value and the shape of the ripple have kept almost the same as before, and after the frequency is spread chaotically, the converter system works stable. The article has made up the lack of theory analysis and experiment analysis of chaotic PWM suppressing EMI.

Keywords- Chaotic frequency-spreading; Spectrum; EMI; Converter; PWM pulse process

I INTRODUCTION[1]

Principal method in power converter EMI suppressing has been so far passive noise filtering and shielding. Which have their limitations: size, weigh, design complexity, efficiency, costs, etc. Obviously the problem will be solved effectively by analysis the EMI mechanism of the converter. The soft-switching technique is one of the methods trying to solve EMI problem from the origin. Extensive research has been conducted on developing various soft-switching techniques to reduce both switching loss and EMI. But because it hasn't acted on the noise source, the effect is limited [1]-[2]. Suppressing EMI from the mechanism still is people's pursuit.

The main source of EMI emissions in power converters comes from the switching of DC voltages following a PWM scheme. So power converters generally operate in a periodic steady state, converter waveforms of interest are typically periodic functions of time. The experiments analysis shows that the peaks of converter EMI are mainly concentrated on the multiples of switch frequency.

Project supported by the National Natural Science Foundation of China (Grant No. 60474066) , and the Major Program of the Natural Science Foundation of Guangdong Province, China (Grant No. 05103540).

Power spectrum tends to be concentrated around the switching frequency and its harmonics.

Chaotic frequency-spreading technique has been recognized as an emerging technology for EMI suppressing [3]-[8] recently. It has been pointed out that chaos system poses a continuous spectrum [9]. Utilizing this characteristic, chaotic power converter's energy can be distributed around an acceptable rang. Then the EMI will be suppressed effectively.

Current research has been mostly settled on theory analysis and simulation. There is a lack of experiments' validating on commercial PWM control chips. Or after chaotic EMI suppressing , the dithering range of frequency is too large that the performance of converter has deteriorated, so there isn't commercial application value.

In this paper, based on the analysis of the relationship between PWM switch scheme and EMI spectrum, a series experiments have been made on a commercial PWM control chip, the experiments validate the theory analysis. The chaotic converter work stable and EMI is suppressed effectively, we can foresee the great potential prospect of the chaotic converter.

II DIFFERENT PWM FREQUENCY-SPREADING SCHEMES

The PWM frequency-spreading scheme of switch converter is defined by

$$f(t) = f_s + \Delta f = k_1 v_s + k_2 v_2 \qquad (1)$$

f_s denotes the PWM nominal frequency, and f_s is decided by the constant voltage v_s ; Δf denotes the additional PWM frequency, and Δf is decided by the frequency- spreading signal voltage v_2 .

Equation (1) shows that, when Δf is zero, the PWM scheme is a traditional constant frequency PWM; When Δf varies periodically, it is a periodical frequency-spreading PWM; When Δf varies chaotically, it is a chaotic frequency-spreading PWM. The circuit principle diagram is depicted in Fig.1.

From a statistical opinion, the PWM drive waveform can be taken as a PWM drive pulse process. In order to analysis the above three PWM frequency-spreading schemes, The PWM drive pulse process can be described with a unified equation

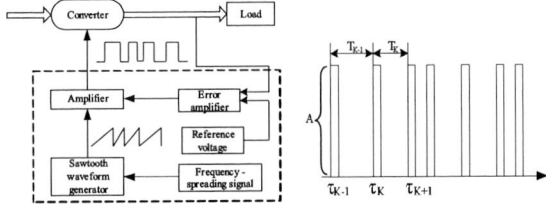

Fig.1 PWM frequency-spreading control   Fig.2 PWM driver pulse process

$$\xi(t) = \sum_{k=1}^{\infty} A\delta\left(t - \tau_{k-1}\right) \qquad (2)$$

Where τ_k is the start time of the kth PWM drive waveform. A is the amplitude of the PWM drive waveform, as marked in Fig.2.

Referring to Fig.2, under the above three PWM frequency-spreading schemes, the time intervals T_k keeps constant, varies periodically or chaotically respectively. τ_k can be represented as a cumulative sum of time intervals T_k. It is also keeps constant, varies periodically or chaotically respectively.

$$\tau_k = \tau_{k-1} + T_K \qquad (3)$$

III SPECTRUM UNDER PERIODICAL PWM FREQUENCY-SPREADING

In equation (1), supposing $\Delta f = f_A \sin(2\pi f_m t)$ is a sine periodical frequency-spreading signal, then the PWM switch frequency is

$$f = f_s + f_A \sin\left(2\pi f_m t\right) \qquad (4)$$

Where $f_m = 1/T_m$ is the frequency of the frequency -spreading signal. f_A represents the maximum PWM drive pulse frequency deviation in the switch frequency.

If $\Delta f = \begin{cases} f_A & nT_m \leq t < nT_m + T_m/2 \\ 0 & nT_m + T_m/2 \leq t < (n+1)T_m \end{cases}$ is a

square periodical frequency-spreading signal, the PWM switch frequency can be expressed as

$$f = \begin{cases} f_s + f_A & nT_m \leq t < nT_m + T_m/2 \\ f_s & nT_m + T_m/2 \leq t < (n+1)T_m \end{cases} \qquad (5)$$

If $\Delta f = \dfrac{f_A\left(t \bmod T_m\right)}{T_m}$ is a sawtooth periodical

frequency-spreading signal, the PWM switch frequency can be calculated as

$$f = f_s + \frac{f_A\left(t \bmod T_m\right)}{T_m} \qquad (6)$$

Using a sine waveform, a square waveform, and a sawtooth waveform as the frequency-spreading signal respectively, then the PWM drive waveforms are got as Fig.3 (a), (b), (c). The simulation and experiment circuit is depicted as Fig.7. All the frequency-spreading signals have a $f_m = 10kHz, T_m = 0.1ms$. The simulation software is SIMPLIS.

The PWM drive waveform under constant frequency is showed at Fig.3 (a). In Fig.3 (b), the frequency of the

PWM drive waveform varies every 0.1ms, and the variation rule is as same as the sine frequency-spreading voltage. Fig.3(c) shows that the frequency value of the PWM drive waveform decreases with the sawtooth frequency-spreading signal's decreases during every frequency-spreading signal period. In Fig3 (d), when the frequency-spreading signal is a square waveform with $f_m = 10KHz$, Duty=50%, the frequency variation of the MOSFET drive waveform is as same as the voltage variation of square frequency-spreading signal. Variation period is same, and half frequency-spreading signal period the frequency value of drive waveform is high, another half frequency-spreading signal period the PWM frequency value is low. The frequency of PWM drive changes with the shift of the square signal voltage.

In a word, Fig 3 illustrated the effective of the PWM frequency-spreading scheme formula (1).

The power spectrums of the periodical frequency-spreading are showed in Fig.4. Among the three periodical frequency-spreading schemes, sawthood frequency-spreading has the smoothest spectrum; square frequency-spreading signal has the most discrete spectrum. We can also observe that the spectrum of PWM drive consists of infinite discrete harmonics. The peak value of the spectrum can be decreased under periodical PWM frequency-spreading scheme compared with the constant frequency. But it is still a discrete spectrum, energy focus on the special frequency point $nf_s \pm kf_m$, and hasn't been spread completely.

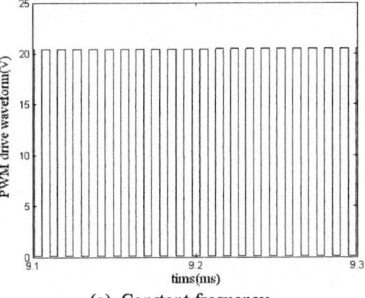

(a) Constant frequency

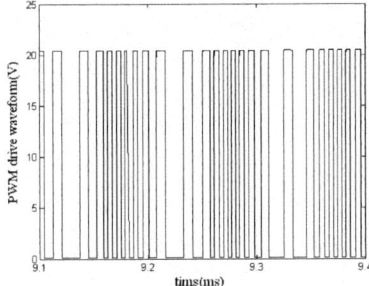

(b) Sine frequency-spreading

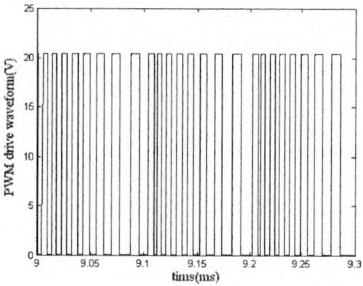

(c) Sawtooth frequency-spreading

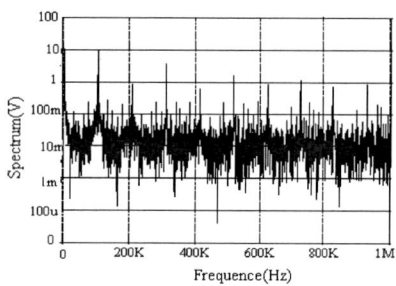

(d) Square frequency-spreading

Fig.3 PWM Drive waveform

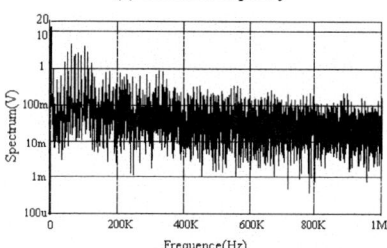

(a) Constant frequency

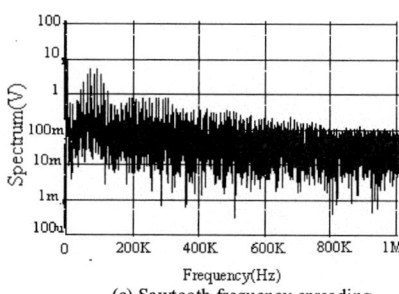

(b) Sine frequency-spreading

(c) Sawtooth frequency-spreading

(d) Square frequency-spreading

Fig.4 Spectrum of drive waveform

IV SPECTRUM UNDER CHAOTIC PWM SPREAD-FREQUENCY

Referring to Fig.2, under chaotic PWM frequency-spreading signal, the interval times T_k and the start time of the kth PWM drive waveform τ_k can be described with the chaotic map φ

$$T_{k+1} = \varphi(T_k) \qquad (7)$$

τ_k is also a chaotic sequence.

$$\tau_{k+1} = T_1 + \varphi(T_1) + \varphi^{(2)}(T_1) + \cdots + \varphi^{(K)}(T_1) \qquad (8)$$

According to the sequence τ_k （k=0,1,…）, a continuous-time process $N(t)$[10] is defined by,

$$N(t) = \max\{k : \tau_k \leq t\} \qquad (9)$$

Where $N(t)$ denotes the number of PWM drive pulse within the interval $[0,t)$, and is called the counting process of the PWM drive pulse process $\xi(t)$

For $\tau_{N(T)} \leq T \leq \tau_{N(T)+1}$, the FFT spectrum of the PWM chaotic pulse process is

$$F_T(j\omega) = \sum_{i=1}^{N(t)} Ae^{-j\omega\tau_i} \qquad (10)$$

Then the power spectrum density (PSD) of the chaotic PWM pulse process can be calculated by using [11]

$$S_\xi(\omega) = \lim_{T \to \infty} \frac{1}{T} E\left(\left|F_T(j\omega)\right|^2\right) \qquad (11)$$

According to the PWM frequency-spreading scheme formula (1). When Δf varies chaotically, the frequency value of the PWM drive pulse varies chaotically. The chaotic frequency-spreading signal source in Fig.7 is got from a Chua's circuit. Drive waveform and its spectrum under chaotic frequency-spreading scheme are shown on Fig.5and Fig.6 respectively. Comparing with Fig.4, the harmonic power of the drive spreads over the frequencies; the peak lever of the power spectra becomes less than that of the classical PWM scheme and the periodical frequency-spreading scheme.

Under the chaotic frequency-spreading, energy no longer concentrates on special frequency point; energy will spread over a wide frequency range of small magnitude. Effectively, the envelopes of the power

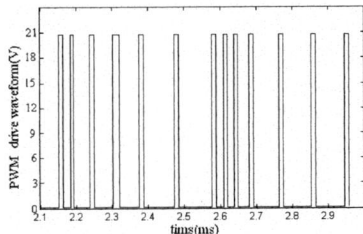

Fig.5 Drive waveform under chaotic frequency-spreading

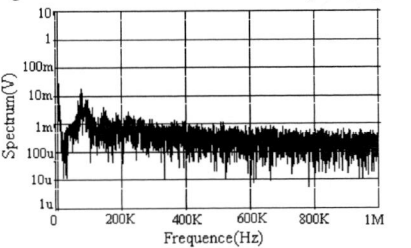

Fig.6 Spectrum of drive waveform's under chaotic frequency-spreading

spectra can in principle be reduced. Under chaotic frequency-spreading. So chaotic frequency-spreading becomes an effect scheme to solve switch converter EMI suppressing problem.

V EXPERIMENT VERIFICATIONS

A flyback converter has been built and tested. Fig.7 shows the schematic of the experiment circuit. The Power of the converter is 65W, input voltage is 220V50Hz, DC output is 12V5A, the switching frequency is 100 KHz and the variation frequency is 10 KHz.

The periodical frequency-spreading signal is achieved by the input alternating voltage. The chaotic frequency-spreading signal is got from a Chua's circuit. And all the PWM frequency schemes were examined at the same full load. The control circuit employs the highly integrated control chip SG6846, which is made by SYSTEM GENERAL LTD. Company, and has a wide commercial application.

Comparing with fig.8, Fig.9 (a) shows that spectrum has about 3dB peak reduction under periodical frequency-spreading. The shapes of the two spectrums are almost same. Almost power is densely concentrated around specific frequencies. From Fig.9 (b), we can conclude that spectrum is continuous under chaotic frequency-spreading, and at most frequency points there is almost 10dB peak reduction with respect to the constant PWM scheme.

Fig.10 shows the ripples under constant frequency and chaotic frequency-spreading respectively. It can be observed that the frequency of the ripple no longer keeps constant under chaotic frequency- spreading scheme, and varies with the frequency- spreading signal. But the peak-peak values of the ripple almost keep invariable; the shapes of ripple are also same under the two schemes. Due to the current and voltage feedback controls of the

PWM control chip, the converter system works stable. So it is realistic to suppressing EMI using chaotic frequency-spreading technology.

At last a conduct EMI test was done at the shielding room on the prototype. The conducted EMI measurements were carried out with a 150 kHz to 30MHz line impedance stabilization network (LISN) and an EMI analyzer Rohde&Schwarz. The EMI test criterion is CISPR. Practical measurements of the conducted EMI of the 65W AC-DC flyback converter are given in Fig.11, 12, 13. The top standard line is the quasi-peak (QP) value limit; the bottom standard line is the average value (AV) limit. According to the industry custom, the limits have 6dB margin comparing with the CISPR criteria.

Refer to Fig.11, under constant frequency, noise energy concentrated on nf_s, so in the middle frequency rang some discrete harmonics exceeds the QP and AV limits. The EMI test of the adapter is disqualified under constant PWM scheme.

Under periodical frequency-spreading (Fig.12), noise energy spread on some special frequency point, noise peak value decreases. The AV test values are within the limit, but the QP tested value exceeds the limit at the 3M frequency point. So the EMI test result is improved but still disqualified under periodical frequency-spreading.

Under the chaotic frequency-spreading (Fig.13), energy is spread over a wide frequency range. The QP and AV tested values are all within the limit, the adapter' EMI test is qualified now. Moreover there is an approximately 10dB reduction in the envelope of the spectrum. This can demonstrate the effectiveness of using chaotic frequency-spreading scheme in suppression EMI of the switch mode power supply.

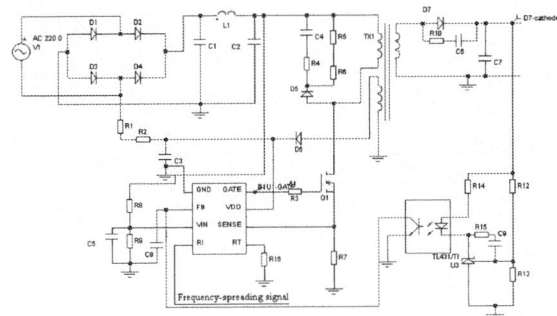

Fig.7 Experiment circuit

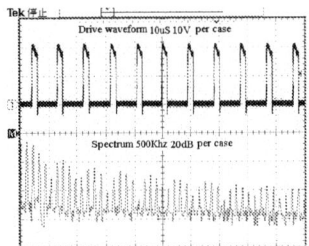

Fig.8 Driver waveform and spectrum under constant frequency

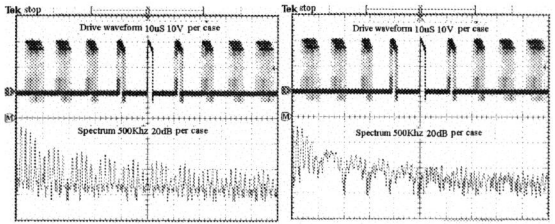

(a) Sinusoidal frequency- spreading   (b) chaotic frequency- spreading
Fig.9 Driver waveform and spectrum

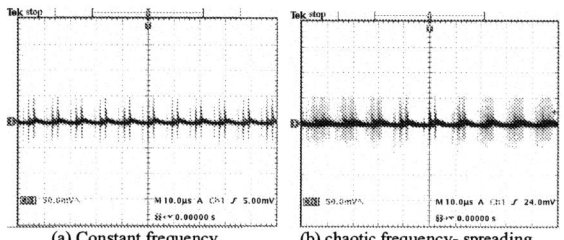

(a) Constant frequency   (b) chaotic frequency- spreading
Fig.10 Ripple waveform

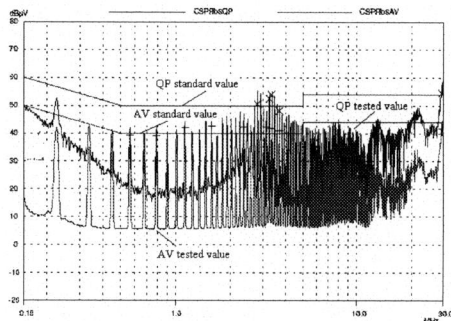

Fig.11 Conduct EMI test results under constant frequency

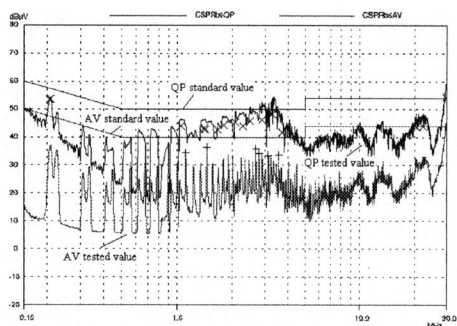

Fig.12 Conduct EMI test results under periodic frequency-spreading

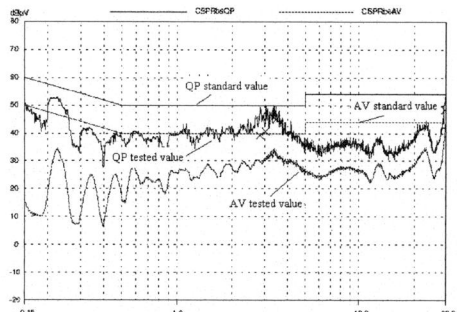

Fig.13 Conduct EMI test results under chaotic frequency-spreading

VI CONCLUSIONS

PWM drive pulse is the original source of switch converter EMI. Based on the analysis of the constant frequency scheme, the periodical frequency-spreading scheme and the chaotic frequency-spreading scheme, this paper has discussed the mechanics and characters of the chaotic PWM suppressing EMI in converter.

Experiments have been made on an AC-DC forward converter. The experiment results show that the EMI spectrum is completely decided by the PWM drive pulse. And after chaotic frequency-spreading, the converter system works stable under the original current and voltage feedback controls of the PWM chip. The frequency characteristic of the ripple has been changed, but the peak value and shape of the ripple have kept almost the same as before. Effectively, the envelopes of the power spectra of the conducted EMI can in principle be reduced. So the switch converter with chaotic frequency-spreading may become the next generation converter in the near future.

REFERENCES

[1] D.Zhang, D.Y.Chen and F.C.Lee., "Experimental Comparison of Conducted EMI emissions between a Zero-Voltage Transition Circuit and a Hard Switching circuits", 27th Annual IEEE Power Electronics Specialists Conference. Vol.2, pp. 1992-1997, PESC'96.

[2] Caldcira P, Liu R, Dalal D and Gu W. J.,"Comparison of EMI performance of PWM and resonant power converters", Power Electronics Specialists Conference, pp.134 – 140. June, PESC '93.

[3] Paramesh J and Jouanne A V, "Use of Sigma-Delta modulation to control EMI from switch-mode power supplies", IEEE Trans.on Industrial Electronics, vol.48 (1), pp.111-117, 2001.

[4] J.F. Gao, M.M.Huang and K.Zhao, "Suppressing Boost converter EMI with chaotic control", Power electronics, vol.38(3), pp.82-85, 2004.

[5] B.Zhang, "DC-DC converter Non-linear chaos phenomenon research", 5th annual meeting of Guangdong power electronics, Guangdong, Shunde, pp.17-20, 2003.

[6] M.Kuisma, "Variable frequency switching in power supplies EMI-control: an overview", IEEE AES systems magazine, pp.18-21, December 2003.

[7] A.Santolaria, J.Balcells, D.Gonzalez, "Evaluation of switching frequency modulation in EMI emission reduction applied to power converters", IEEE circuit and system, pp. 2306-2311, 2003.

[8] Tanaka T, Ninomiya T and Harada K, "Random-switching control in DC-to-DC converters", IEEE PESC Rec. pp.500-507, 1989.

[9] Q.CH.Zhang et al, "Theory and application of bifurcation and chaos", China, Tianjing university press 2004.

[10] Y.L.Deng, ZHI.SH.Liang, "Random point process and its application", China, Science press, 1992.

[11] Jorge crupper, Wolfgang Schwarz., "A performance estimation method for chaotic spread spectrum clock process", Circuits and Systems IEEE, Vol. 4, pp.3383 – 3386, 2005.

2006 5th International Power Electronics and Motion Control Conference

Emitter Size Effect in 4H-SiC BJT

Yan Gao, Alex Q. Huang, Sumi Krishnaswami*, Anant K. Agarwal*, Charles Scozzie[#]

Semiconductor Power Electronics Center (SPEC), North Carolina State University, Raleigh, U.S.

* Cree Inc, 4600 Silicon Drive, Durham, NC 27703

[#] US Army Research Laboratory, 2800 Powder Mill Road, Adelphi, MD 20783

Abstract—**SiC BJT with varying emitter width were investigated by numerical simulations as well as experiments. Emitter Size Effects (ESEs) are demonstrated in today's SiC BJT by comparing the characterization of BJT with different emitter width. Surface recombination current is found to be comparable with published result for Heterojunction Bipolar Transistors (HBTs) [9]. In SiC BJT design, this effect has to be considered. A good surface passivation is required to reduce the effect of surface recombination on the current gain.**

I. INTRODUCTION

SiC bipolar junction transistor (BJT) is a promising device candidate and considerable progress has been made in recent years [1][2][3][4]. Until now, achieving common emitter current gain is a major technical challenge in SiC BJT development. The investigation of the factors that affect the current gain is very important for SiC BJT application. Surface and space charge recombination are two main factors that affect the current gain. The base-emitter interface defect effects have been discussed in our previous work [5]. However, no detailed analysis of the Emitter Size Effect has been reported before[6]. In this paper, we analyze the effect of the emitter size by simulation as well as by experiments. Both show the emitter size to be an important factor affecting the current gain.

II. CURRENT GAIN Of SIC BJT

Figure 1 shows the structure and picture of the recently fabricated 1200V SiC NPN BJT by Cree with a cell pitch of 25 μm. The BJT has three epilayers that were grown on a 4H-SiC n+-type substrate with a two-layer metal system [4]. The total collector area is 0.0225 cm².

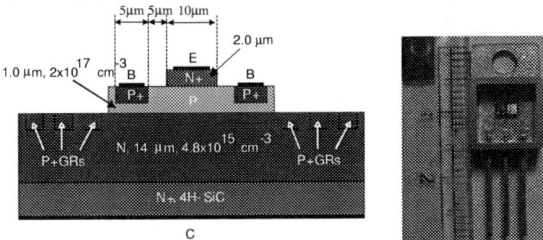

Figure 1. Cross-section of SiC BJT

Common emitter current gain as a function of collector current density was measured using the experimental setup shown in Figure 2: A pulse generator with rise and fall pulse time 5ns, and 5 μ s pulse duration was used to supply the base signal. The signal frequency was set as 20 Hz to minimize thermal effects. The BJT operates in its active region with V_{ce}=25V. The measured current gain as a function of collector current density is shown in Figure 3. Our simulation result which incorporates the interface defects and recombination is also shown in Figure 3. It can be seen from Figure 3 that our model can perfectly describe the experimental result. β decreases rapidly as Jc increases. This characteristic of SiC BJT will affect its application under high current density.

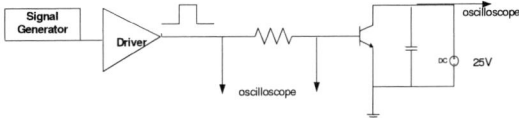

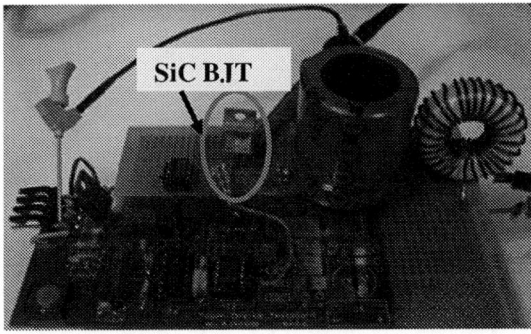

Figure 2. Experiment setup for measurement of ß as a function of collector current density

The device simulations were carried out using the commercial software ISE [7] to help us to investigate the reasons. Physical models used were doping, temperature and electric field dependent mobility, Shockley–Read–Hall recombination(SRH) and Auger recombination and bandgap narrowing.

1-4244-0448-7/06/$25.00 ©2006 IEEE

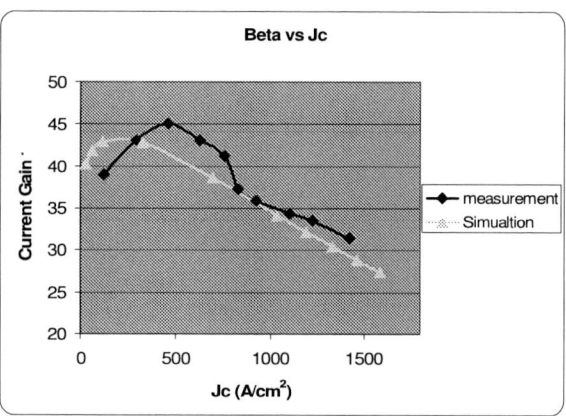

Figure 3. Measured and Simulated Current Gain vs the Collector Current Density at 25°C

In the emitter, complete ionization of $N_D^+ = N_D = 2 \times 10^{19} cm^{-3}$ was used while incomplete ionization of dopant was used in the base with the ionization energy $E_A = 0.191$ eV for Al. A constant density of states $D_{IT} = 7 \times 10^{10} cm^{-2} eV^{-1}$ at the interface of SiC/SiO$_2$ and a constant density of states $D_{IT} = 1.5 \times 10^{12} cm^{-2} eV^{-1}$ at the base-emitter interface were assumed at the mid-bandgap and a capture cross-section of $\sigma = 6 \times 10^{-15} cm^2$ was used for both of them.

III. EMITTER SIZE EFFECT

Under today's processing technology, the surface recombination current may be an important component that degrades the current gain. Some minority carriers injected from the emitter recombine with the base majority carriers at the surface. This surface recombination current, $I_{B,surf}$, has no contribution to the current gain. This base current component is proportional to the emitter periphery rather than the emitter area, unlike the collector current. This effect often appears in the Heterojunction Transistors and is not obvious in the Si BJT. In our study of the SiC BJT, we can clearly see this effect. This is because the SiO$_2$/SiC interface quality is not very good, and that the surface recombination current is high. With the Emitter Size Effect, the base current can be expressed as equation (1) [8]

$$\frac{Jc}{\beta} = (J_{BQN} + J_{BSCR} + J_{Bp}) + K_{Bsurf} \frac{P_E}{A_E} \quad (1)$$

Where Jc is the collector current density, J_{BQN}, and J_{BSCR} are the base bulk quasi-neutral, space charge recombination current densities respectively and J_{Bp} is the back-injection current density.

$K_{Bsurf} * P_E$ is the emitter periphery surface recombination current, A_E and P_E are the emitter area and periphery respectively [9]. K_{Bsurf} in A/cm is the normalized surface recombination current. Simulations were carried out to

study the emitter size effect on SiC BJT. Figure 4 shows the half-cell structure used in our simulations. Emitter length and width were defined in Figure 4.

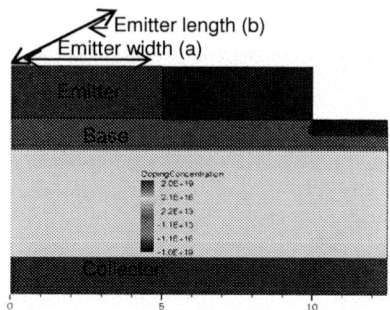

Figure 4. Half-Cell Structure for Emitter Size Effect Study

In the simulation, "b" is the emitter length and "a" is the emitter width. Decreasing "a" causes an increase in P_E/A_E. In the 2D simulation, only the effect of "a" can be simulated. So, P_E/A_E is expressed by 1/a in Figure 5, in which the Jc/β as a function of the P_E/A_E ratio is shown. From equation (1), the slope of the lines in Figure 5 is the

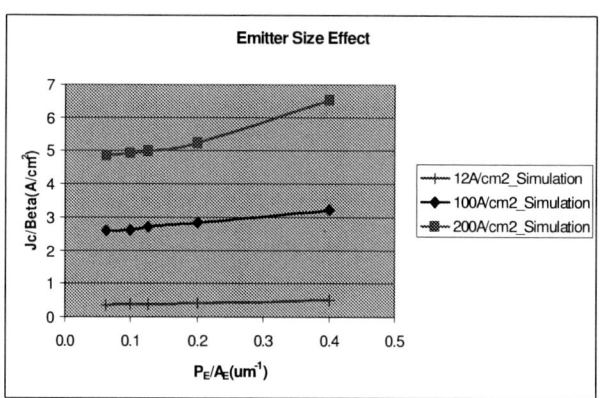

Figure 5. Jc/β as a function of P_E/A_E showing the Emitter Size

normalized periphery recombination base current $K_{B,surf}$ and is shown in Figure 10. It shows that as the collector current density increases from 12 to 200 A/cm^2, the slope, $K_{B,surf}$, also increases, which will cause a decrease of the current gain at high collector current density.

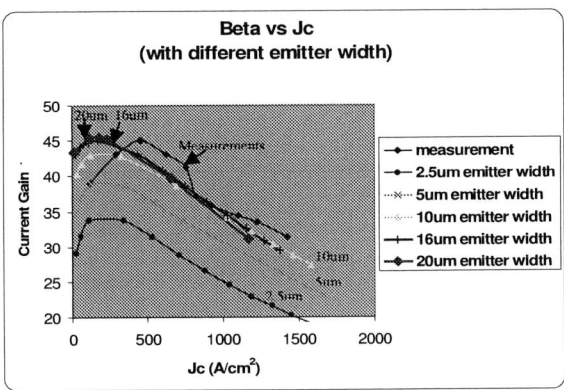

Figure 6.Current gain vs Jc at different Emitter Width

Figure 6 shows the beta as a function of current density at different emitter width. As the emitter size decreases, the gain shifts down. Improving the quality of SiC/SiO$_2$ interface is important to get a high gain especially for narrow emitter fingers. As emitter size increases above a certain value (16 μ m in our case), the gain stops increasing. This is due to the current crowding effect at the edges of the emitter, which starts to dominate.

IV. MEASUREMENT RESULTS

Devices with different emitter width on the same chip were fabricated at Cree. The three emitter widths were 6um, 8um and 10um respectively. I-V characteristics were measured by Tektronix 370A curve tracer. The common emitter current gain as a function of collector current density is shown in Figure 7.

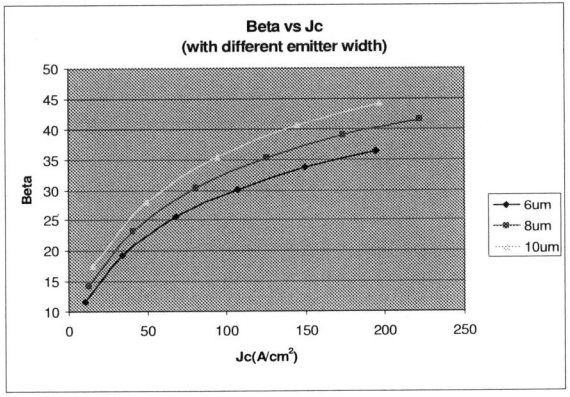

Figure 7. Current gain as a function of collector current density for different emitter width.

From Figure 7, the 10um emitter width BJT has the highest gain because it has the smallest emitter periphery over area (P$_E$/A$_E$) ratio. At I$_b$=100mA, V$_{ce}$=5V, the current gain increases from 34 to 44 as the emitter width goes from 6um to 10um as shown in Figure 8.

Figure 9 shows experimental plots of Jc/β at current densities of 12, 100, 200 A/cm^2. Similar to the simulation results shown in Figure 5, the result indicates an increasing effect of the surface current at higher current density.

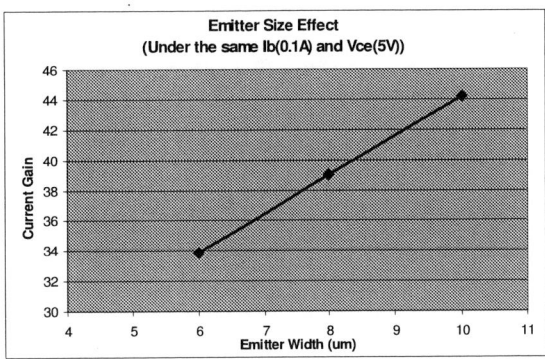

Figure 8. Current Gain as a function of Emitter Width

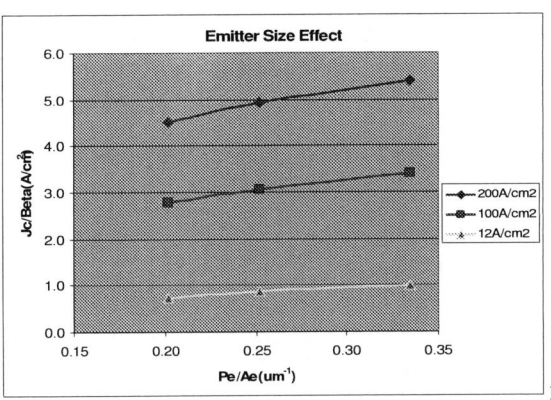

Figure 9. Emitter Size Effects for Jc=12, 100 and 200 A/cm^2(The slope in the graph is the normalized periphery recombination base current)

Surface recombination currents extracted by measurement and simulation data are shown together in Figure 10, which clearly shows the increase of the normalized periphery surface recombination current as a function of collector current density. Since this is a two-dimensional simulation, the value extracted by simulation data is smaller than that by the measured three-dimensional value.

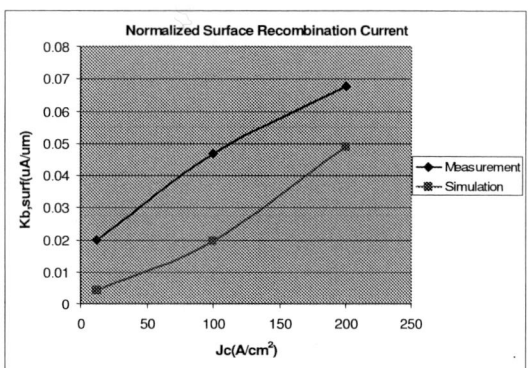

Figure 10. Normalized periphery surface recombination current as a function of collector current density Jc

But they have the same trend. Figure 10 also shows that the surface recombination current of today's SiC BJT is comparable to that of the Heterojunction Bipolar Transistors(HBTs)[9], and it is much larger than that of Si BJT , which has a very mature processing technology.

V. CONCLUSION

Both the simulation and experiment results show that in today's SiC BJT, Emitter Size Effect (ESE) plays a role in determining the common emitter current gain. As current density increases, the effect is more and more obvious. For the first time, surface recombination current is reported by the extraction from both measurement and simulation data for today's SiC BJT. Designers have to take this effect into account during device design. An optimum value in emitter size exists, that minimizes the surface recombination current effect and the emitter current crowding effect.

ACKNOWLEDGEMENT

This research was funded through the DAAD17-03-C-0117 program supported by the Army Research Laboratory in Adelphi, Md.

VI. REFERENCE

[1]. A. K. Agarwal, S-H. Ryu, J. Richmond, C. Capell, J.W. Palmour, S. Balachandran, T.P. Chow, B. Geil, S. Bayne, C. Scozzie and K.A. Jones, "Recent Progress in SiC Bipolar Junction Transistors," ISPSD Proceedings, pp. 361-364, 2004.
[2]. Sei-Hyung Ryu, A. K. Agarwal, J.W. Palmour, M.E.Levinshtein; "1.8 kV, 3.8 A bipolar junction transistors in 4HSiC", ISPSD, 2001 Page(s): 37 –40.
[3]. Sei-Hyung Ryu; A. K. Agarwal, R. Singh, J.W. Palmour, "1800 V NPN bipolar junction transistors in 4H-SiC", IEEE Electron Device Letters , Volume: 22 Issue: 3 , Mar 2001 Page(s): 124 –126.

[4]. Anant K. Agarwal, Sumi Krishnaswami, James Richmond, Craig Capell, Sei-Hyung Ryu, John W. Palmour, Santosh Balachandran1, T. Paul Chow1, Stephen Bayne2, Bruce Geil2, Charles Scozzie2 and Kenneth A. Jones2; " Evolution of the 1600 V, 20 A, SiC Bipolar Junction Transistors" ISPSD 2005 Page(s):271 – 274.
[5] Yan Gao; A. Q. Huang, A. K. Agarwal, Sumi Krishnaswami, Sei-Hyung Ryu," Characterization and modeling of 4H-SiC power BJTs" IECON 2005 Page(s):674 - 678
[6] M. Domeij, H.-S. Lee, E. Danielsson, C.-M. Zetterling, M. Östling, Fellow, IEEE, and A. Schöner," Geometrical Effects in High Current Gain 1100-V 4H-SiC BJTs" IEEE ELECTRON DEVICE LETTERS Volume 26, Issue 10, Oct. 2005 Page(s):743 - 745"
[7] ISE User's manual
[8] N. Hayama and K. Honjo, "Emitter size effect on current gain in fully self-aligned AlGaAs–GaAs HBTs AlGaAs surface passivation layer," IEEE Electron Device Lett., vol. 11, no. 4, pp. 388–390, Apr. 1990.
[9] Nick G. M. Tao, Honggang Liu, and C. R. Bolognesi, "Surface Recombination Currents in "Type-II" NpN InP–GaAsSb–InP Self-Aligned DHBTs" IEEE Trans.Electron Devices, vol.52 No. 6. June 2005.

2006 5th International Power Electronics and Motion Control Conference

PSIM and SIMULINK Co-simulation for Three-level Adjustable Speed Drive Systems

Zhang Yongchang, Zhao Zhengming, Baihua, Yuan Liqiang, Zhang Haitao
State Key Laboratory of Control and Simulation of Power System and Generation Equipment
Department of Electrical engineering, Tsinghua University, Beijing, P.R. China
zhangyongchang04@mails.tsinghua.edu.cn

Abstract—Applying co-simulation technology is a very important trend in power electronics simulation, which can make full use of the merits of different software to improve the efficiency of building up a system. In this paper, a simulation platform for three-level adjustable speed drive (ASD) based on co-simulation of PSIM and SIMULINK is presented, which includes the interface between PSIM and SIMULINK, main circuit in PSIM, space vector pulse width modulation (SVPWM), dead time and minimum pulse width, view of output voltage and its Fourier analysis in SIMULINK. Based on this platform, the cause of large current when motor starts is studied. Simulation and experimental results show that the minimum pulse width contributes a lot for the large starting current of the ASD. In conclusion, applying the simulation platform introduced in this paper, the dynamic behavior of the three-level ASD can be easily studied, which reduces the cost and improves efficiency at the same time compared to hardware investments, and may help to realize the optimal design of the ASD in practice.

Keywords: Simulation; SVPWM; Adjustable Speed Drive (ASD); Three-level inverter; Minimum pulse width

I. INTRODUCTION

Compared to conventional two-level inverter, the neutral point clamped (NPC) inverter provides significant advantages for high power applications, which can decrease the capacity of switch devices to get the same output voltage, and reduce the harmonic component of output waves, thus make the size of a filter smaller[1-3]. However, there are many factors to be considered when designing an ASD. In this condition, using computer simulation and CAD may help to simplify the analysis process and improve design quality while shortening the developing cycle. Computer simulation and CAD has been widely used in most research institutes and international companies such as ABB [4-5], and there are a variety of software packages available now for power electronics simulation [6], such as PSPICE, SABER, PSIM and MATLAB/SIMULINK, and so on. Each of them has its own merits and shortcomings. For example, PSPICE was originally developed for integrated electronics simulation, and the model of switch devices in

PSPICE are not perfect, besides there is a convergence problem for simulation. SABER is developed for mixed signal simulation and this makes it is very suitable to apply in power electronics simulation, but the high cost restricts its application. PSIM is suitable for power electronics circuit simulation, but it is not very convenient to build up a control circuit. MATLAB/SIMULINK is very popular in control simulation and CAD, but its power circuit simulation is far from perfect. In all, none of them can meet all needs in various fields. The recommended method is to try to develop interfacing technology between the existing software, so that we could make full use of the merits of different software [7]. In this paper, PSIM and SIMULINK are used to build the simulation platform and the starting characteristic of three-level ASD is studied. Simulation and experimental results are given to verify the validity of this simulation platform.

II. CO-SIMULATION OF PSIM AND SIMULINK

PSIM is developed especially for power electronics and motor control. With fast simulation, friendly user interface and powerful waveform processing, PSIM has been widely used in many areas [8-9]. MATLAB is one of the most popular software in control fields, and its graphical simulation environment SIMULINK is very suitable for dynamic system simulation because there are plenty of toolboxes and modules. However, it is awkward and cumbersome to simulate power electronic circuits in SIMULINK. Using SimCoupler, the main circuit can be implemented and simulated in PSIM, and the control part in MATLAB/SIMULINK, thus greatly shortening the time to set up a system which includes power electronic circuit and motor drives. In this part a condense introduction of SimCoupler will be given.

SimCoupler is an add-on module to the PSIM software, which provides interface between PSIM and SIMULINK for co-simulation. It has such key features: easy to use; minimum user input; fast simulation and waveform display in both PSIM and SIMULINK. There are three modules which can be seen in Fig.1. In PSIM, SLINK_IN receives signal from SIMULINK and SLINK_OUT output signal to SIMULINK; in SIMULINK, SimCoupler model block, which represents the main circuit defined in

---

1-4244-0448-7/06/$25.00 ©2006 IEEE

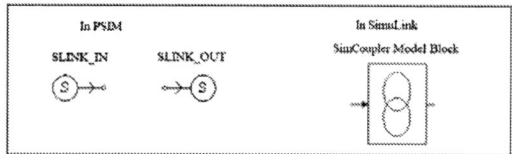

Figure 1. SimCoupler block

PSIM, interconnects with other part through input and output port.

In all, with the SimCoupler module, one can make full use of PSIM's capability in power simulation and SIMULINK's capability in control simulation in a complementary way.

III. MAIN CIRCUIT

Fig. 2 shows the main circuit for a three-level ASD, which employs AC-DC-AC topology. The main circuit consists of five parts: YYD step-down transformer, 12-pulse rectifier, IGCT clamping snubber circuit, NPC three-level inverter and integrated LC filter and step-up transformer. The 12-pulse rectifier provides more even DC link voltage and reduces the harmonics of the input side. The snubber circuit is introduced to suppress the di/dt of the system. The part of the inverter consists of twelve 4500V/630A IGCTs and six clamped diodes. In this system, the LC filter makes use of the leakage inductance of step-up transformer, which reduces the size of filter. In all, in PSIM one can build up a system just as the original form of circuit in practice, which is very easy and illustrative.

There are some points to be noted in practical simulation. Firstly, Because IGCT is developed recently, there is no corresponding model in PSIM, and in practice GTO is used in stead. One can build up a more precise model by using the ideal components in PSIM[10]; Secondly, the gating pulse is produced in SIMULINK, and it requires the interface SLINK_IN and SLINK_OUT be well defined; finally, make sure the simulation time step in PSIM should be made the same as that in SIMULINK to obtain high accuracy.

IV. CONTROL SYSTEM

A. The whole diagram

The main circuit built in PSIM works as a sub-circuit in SIMULINK, and it is SIMULINK that controls the start and stop of simulation. Fig. 3 shows the whole diagram of three-level ASD in SIMULINK. It includes mainly three parts: pulse block which produces the gating signals for IGCTs; SimCoupler block which receives the gating signals from pulse block and call a program called psim.dll to simulate the main circuit, and then sends the signals from PSIM to SIMULINK; measurement block which measure the output signals for view and analysis.

The pulse block is the core block in the system, and it is realized by a system function, which contains the SVPWM algorithm for the three-level inverter, curve of voltage vs. frequency, pre-excitation, minimum pulse and dead time, etc. Double click this block, and a dialog will appear from which all parameters of ASD can be set. Fig.4 shows the dialog and Fig. 5 shows the diagram of realization.

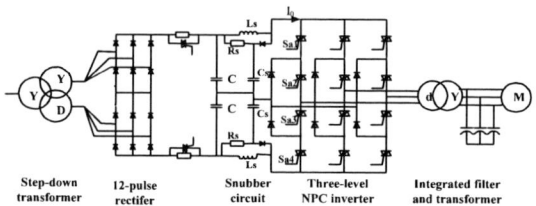

Figure 2. The main circuit of 3-level ASD

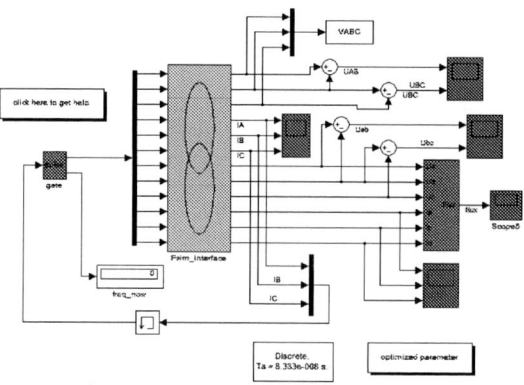

Figure 3. Three-level adjustable speed drive (ASD)

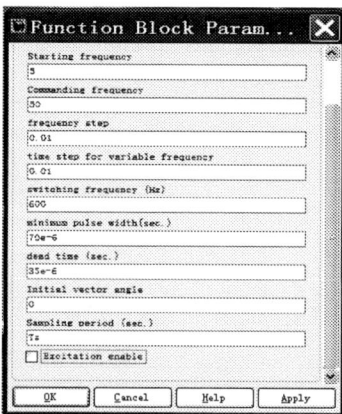

Figure 4. Parameter setting of pulse block

162

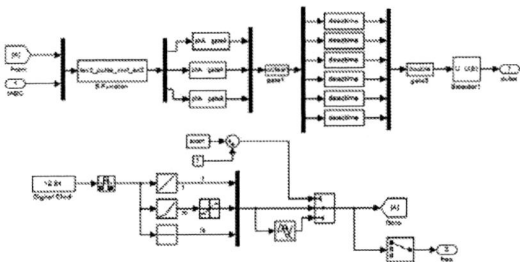

Figure 5. The diagram of pulse block

B. SVPWM algorithm

Fig. 6 shows the space vector diagram of the three-level inverter. It can be divided into 6 sectors and each sector is also divided into 6 regions. Fig. 7 shows the vector diagram of the first sector. When the reference vector falls to a certain region such as region 2, the duration of ta, tb and tc corresponding to space vectors Va, Vb and Vc can be calculated from the following equations [11]:

$$t_a + t_b + t_c = T_s \qquad (1)$$

$$V_1 t_a + V_2 t_b + V_3 t_c = V_r T_s \qquad (2)$$

Where

$$V_1 = \frac{1}{2}, V_2 = \frac{\sqrt{3}}{2} e^{j\frac{\pi}{6}}, V_3 = \frac{\sqrt{3}}{2} e^{j\frac{\pi}{6}}, V^* = V e^{j\theta}$$

and Ts is the sampling period. From the equations above the duration can be expressed as below:

$$t_a = \left(1 - 2k\sin\theta\right) \cdot T_s$$

$$t_b = \left[2k\sin\left(\theta + \frac{\pi}{3}\right) - 1\right] \cdot T_s \qquad (3)$$

$$t_c = \left[2k\sin\left(\theta - \frac{\pi}{3}\right) + 1\right] \cdot T_s$$

Here k represents the modulation index. To decrease the harmonic component and avoid the voltage jumping of output line voltage, the seven-segment output sequence of vectors is used. For example, the output vector sequence of region 2 in sector one is 211-210-110-100-110-210-211. The duration and output vector sequence of other sectors can be obtained in a similar way.

C. Minimum pulse and dead time

Usually the switching on and off of power electronics devices need time, especially for GTO and IGCT whose switching frequency is usually less than 1000Hz. For SVPWM, in some regions the reference vector is too near to a certain vector and the durations of other vectors are too short, which may cause the problem of minimum pulse. To ensure the safe operation of switching devices, it is necessary to limit the minimum pulse width, which

depends on the characteristic of switching devices and the snubber circuit. Here the minimum pulse width is limited to 50us. The restriction of minimum pulse width is realized in the pulse block by system function.

To avoid the upper and lower devices of the same leg to switch at the same time, a dead band time is needed to insert in between them. For three-level inverter, there are twelve gating signals. Here the gating signals for upper two devices of each phase are produced, and after the disposal of dead-time block, they are sent to IGCTs. Fig. 8 shows the diagram of dead time block and the dead band time is 30 us.

D. View of output wave and its Fourier analysis

The output wave can be viewed in both PSIM and SIMULINK. In the power system toolbox of SIMULINK, there is a very useful block called powergui, which provides excellent performance to simplify the analysis of output wave. In the simulation platform, it is used to discretize the system to improve the simulation speed, and to give Fourier analysis of the output wave. Fig. 9 shows the steady state line voltage at 50Hz and its corresponding frequency spectrum. The user only needs to select the wave to be analyzed and set the displaying format, and the results will be shown in the figure.

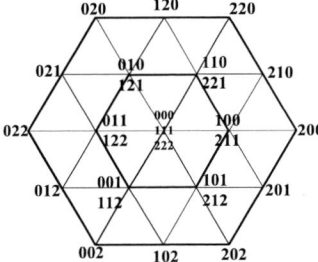

Figure 6. Space vectors for level-inverter

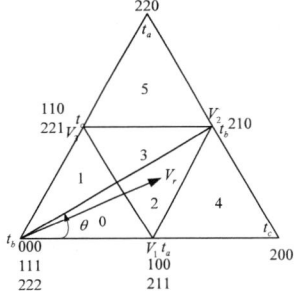

Figure 7. Vector synthesis for sector one

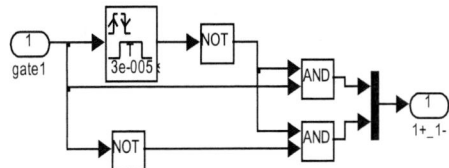

Figure 8. Dead-time block

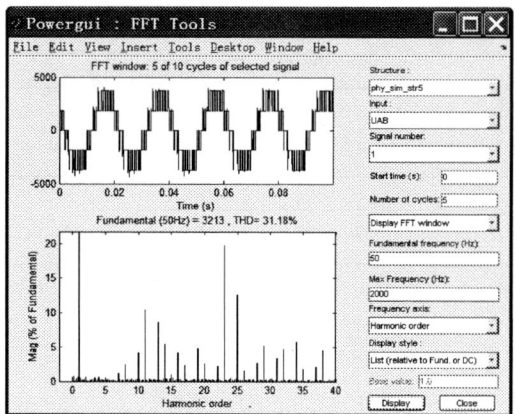

Figure 9. Fourier analysis of line voltage using powergui

V. STUDY OF STARTING FOR THREE-LEVEL ASD

Starting characteristic is one of the most important indices of ASD. In practice, it is found that for three-level ASD with 1250kW there is large starting current when the motor starts. Large starting current may do harm to the insulation of a motor and the switching devices. To ensure the safe operation of the system, usually devices with larger capacity are needed, which increases the cost.

Based on the simulation platform, helpful investigations

are done to find the cause of the overflowing. Analysis shows that the minimum pulse is the key point for overflowing. When the frequency is low, the influence of minimum pulse is even more serious. Fig. 10(a) shows the ideal reference vector at steady state of 5Hz. It is assumed that the reference space vector should distribute uniformly in the space, but in fact the amplitude and angle of true space vector are distorted because of the minimum pulse and dead time, which can be seen from Fig. 10(b).

Fig.11 shows the simulation results of starting process for the three-level ASD with 1250kW. Fig. 12 shows the experimental results of a 3kW motor with 380V. Both simulation and experimental results prove that after decreasing the width of minimum pulse, the peak value of the current is reduced obviously.

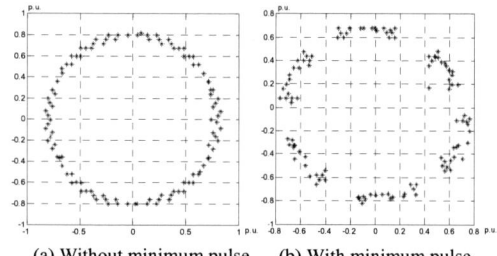

(a) Without minimum pulse   (b) With minimum pulse

Figure 10. Comparison of ideal and true space vector at 5Hz

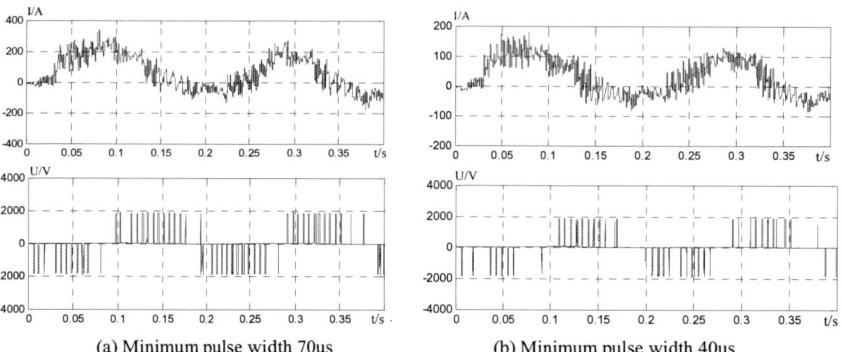

(a) Minimum pulse width 70us   (b) Minimum pulse width 40us

Figure 11. Line voltage and current (simulation results)

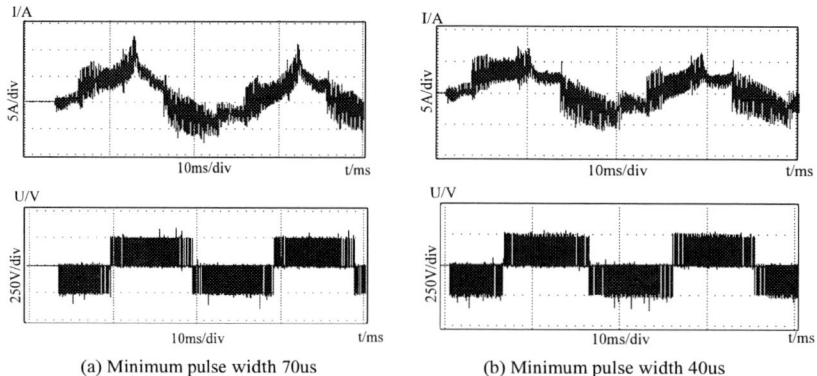

(a) Minimum pulse width 70us   (b) Minimum pulse width 40us

Figure 12. Line voltage and current (experimental results)

VI. CONCLUSIONS

A simulation platform for three-level ASD based on PSIM and SIMULINK co-simulation has been introduced in this paper. The co-simulation interface, main circuit and control circuit are all presented in detail. Applying the simulation platform, the overflowing of a three-level ASD with 1250kW/6000V is studied and the influence of minimum pulse is analyzed. The comparisons between simulation and experimental results verify the validity of the simulation platform.

REFERENCES

[1] Nabae A, Takahashi I, Akagi H. A new neutral-point-clamped PWM Inverter. IEEE Trans. on Ind. Appl., vol. 17, issue 5, pp. 518-523, 1981.

[2] TULBERT L M, PENG F Z. Multilevel converters for large electric drives. IEEE Trans. on Ind. Appl., vol. 35, issue 1, pp. 36-42, 1999.

[3] PENG F Z, LAI J S, MCKEEVER J W, el al. A multilevel voltage-source inverter with separate dc sources for static var generation. IEEE Trans Ind. Appl., vol. 32, issue 5, pp. 1130-1138, 1996.

[4] Yangjun, Wang Zhao'an. Modeling, simulation, CAD and synthesis of power electronics system. Transactions of China electrotechnical society, vol. 14(supplement), pp. 72-76, 1999.

[5] Gerhand B, Adrian W, Gerhand W. Simulation tools for the ACS 1000 standard AC drives. ABB Review, vol. 5, issue 5, pp. 43-51, 1998.

[6] Qian Zhaoming Dong Baofan He xiangning. The recent developments of power electronics and its applications. Proceedings of the CSEE, vol. 18, issue 3, pp. 158-159, 1998.

[7] Chen Jianye. The application of computer simulation in power electronics. Beijing: press of Tsinghua University, 2003, pp. 248-249

[8] Cui, P.; Zhu, J.G.; Ha, Q.P.; Hunter, G.P.; Ramsden, V.S. Simulation of non-linear switched reluctance motor drives with PSIM. ICEMS 2001, vol. 2, pp. 1061-1064

[9] Shigeru Onoda, Ali Emadi. PSIM-based modeling of automotive power systems: conventional, electric, and hybrid electric vehicles. IEEE Trans. on Vehicular Technology, vol. 53, issue 2, pp. 390-395, 2004.

[10] YUAN Liqiang; ZHAO Zhengming; BAI Hua; LI Chongjian; LI Yaohua. The functional model of IGCTs for the circuit simulation of high-voltage converters. Proceedings of CSEE, vol. 24, issue 6, pp. 65-69, 2004.

[11] Yo-Han Lee, Bum-seok Suh, Dong-seok Hyun. A novel PWM scheme for a three-level voltage source inverter with GTO thyristors. IEEE Trans. on Ind. Appl., vol. 32, issue 2, pp. 260-268, 1996.

2006 5th International Power Electronics and Motion Control Conference

Three-Phase Z-Source AC-AC Converter for Motor Drives

Xu-Peng Fang

Shandong University of Science and Technology, Qingdao, China, 266510

Xpfang69@yahoo.com.cn

Abstract— **This paper presents a Z-source ac-ac converter system and control for general-purpose motor drives. The Z-source ac-ac converter system employs a unique LC network to suppress the over-current and surge voltage due to the shoot-through or the open-circuit of the two switches. By controlling the duty cycle, the Z-source can provide buck-boost function, produce greater or less voltage than the line voltage. As a result, the new Z-source ac-ac converter system provides ride-through capability under voltage sags, reduces line harmonics, and improves power factor and reliability. Analysis, simulation and experimental results will be presented to demonstrate these new features.**

Keywords -Ac-ac converter; Z-source; PWM-converter.

I.   INTRODUCTION

The direct PWM ac-ac converter can perform voltage regulation in addition to conditioning, isolating, and filtering of the incoming power [1]. It can also be used for the general-purpose motor drive (or adjustable speed drive—ASD) system.

The use of self-commutated switches with PWM control can significantly improve the performances of ac-ac converters, which has been articulated in a number of technical publications [1] ~ [7]. Each paper proposed a different ac-ac converter and some simulation results were presented to illustrate their performance in the presence of voltage sags, surges, and load fluctuations.

Z-source ac-ac converter is a novel topology [11], it overcomes the conceptual and theoretical barriers and limitations of the traditional ac-ac PWM converters and provides a novel power conversion concept. It also could be directly extended to motor drive applications.

In this paper, the novel three-phase Z-source ac-ac converter-fed ASD system will be proposed, its operating principle will be described, and simulation and experimental results will be presented.

II.   Z-SOURCE AC-AC CONVERTER-FED ASD SYSTEM

Figs.1 (a) (b) show the proposed topology of three-phase Z-source ac-ac converter-fed ASD system. They utilize only two active devices (S_1 and S_2), and each is combined with a full diode bridge. They consist of a Z-network, which employs a unique LC network to provide the buck-boost function. Their operating principles are same, and we can analyze one of them to deduce the relations between input and output. In the paper, the circuit shown in Fig.1 (a) is analyzed.

The proposed ac-ac converter can be operated with PWM duty cycle control in exactly the same way as the conventional dc-dc converter. Fig.2 shows the control scheme, S_1 and S_2 are turned on and off in complement. By controlling the duty cycle, the output voltage can be regulated as desired. It could not destroy the two semiconductor devices when they are gated on or off synchronously due to the Z-network.

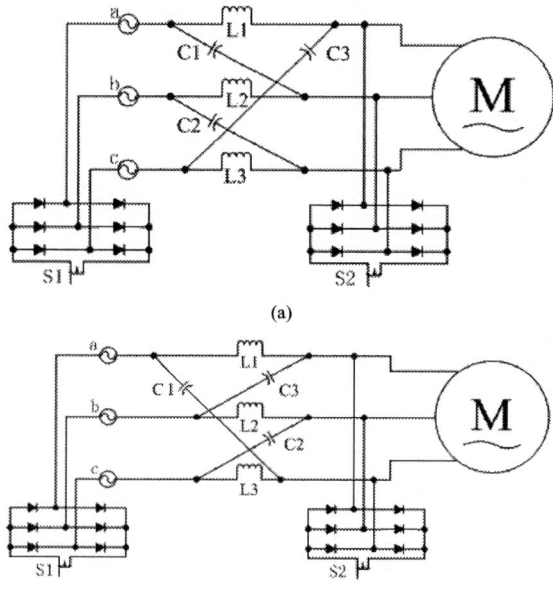

(a)

(b)

Fig.1. Three-phase Z-source ac-ac converter

1-4244-0448-7/06/$25.00 ©2006 IEEE

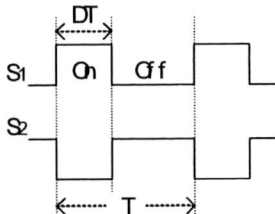

Fig.2. Duty cycle control of Z-source ac-ac converter

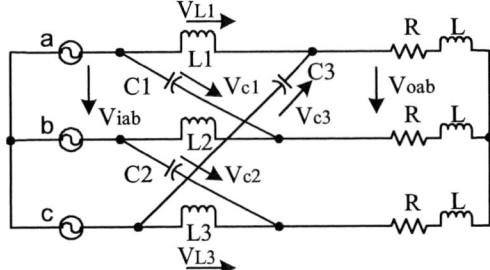

Fig.3. State 1: S_1 is on and S_2 is off.

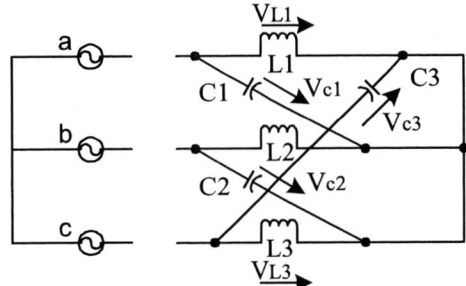

Fig.4. State 2: S_1 is off and S_2 is on.

III. EQUIVALENT CIRCUIT，OPERATING PRINCIPLE OF THE PROPOSED AC-AC CONVERTER

For the novel PWM ac-ac converter, the control scheme described in Fig.2 is simple and easy to implement. S_1 and S_2 are gated on and off in complement as shown in Fig.2. Since the switching frequency is much higher than the frequency of the input voltage, in a switching cycle, we can look the input voltage as a constant, and the motor can be modeled as a resistor in series with a inductor, then two equivalent circuits of the two states can be achieved, as shown in Fig.3 and Fig.4.

Since the inductors and capacitors of the Z-network have the same inductances (L) and capacitances (C) in Fig.3 and Fig.4 respectively, the Z-source networks among three phases become symmetrical. Suppose the load of motor is symmetrical, and

$$\begin{bmatrix} v_{ia} \\ v_{ib} \\ v_{ic} \end{bmatrix} = V_i \begin{bmatrix} \sin(\omega t) \\ \sin(\omega t - 120^\circ) \\ \sin(\omega t + 120^\circ) \end{bmatrix},$$

$$\begin{bmatrix} v_{L1} \\ v_{L2} \\ v_{L3} \end{bmatrix} = \omega L I_L \begin{bmatrix} \sin(\omega t + \phi_L) \\ \sin(\omega t + \phi_L - 120^\circ) \\ \sin(\omega t + \phi_L + 120^\circ) \end{bmatrix},$$

$$\begin{bmatrix} v_{C1} \\ v_{C2} \\ v_{C3} \end{bmatrix} = V_C \begin{bmatrix} \sin(\omega t + \phi_C) \\ \sin(\omega t + \phi_C - 120^\circ) \\ \sin(\omega t + \phi_C + 120^\circ) \end{bmatrix},$$

$$\begin{bmatrix} v_{oa} \\ v_{ob} \\ v_{oc} \end{bmatrix} = V_o \begin{bmatrix} \sin(\omega t + \phi_o) \\ \sin(\omega t + \phi_o - 120^\circ) \\ \sin(\omega t + \phi_o + 120^\circ) \end{bmatrix} \quad (1)$$

In state 1, the bi-directional switch S_1 is on and S_2 is off, and the interval of the converter operating in this state is DT, where D is the duty cycle of switch S_1 and T is the switching cycle, shown in Fig.3, then one has,

$$\begin{bmatrix} v_{iab} \\ v_{ibc} \\ v_{ica} \end{bmatrix} = \begin{bmatrix} v_{C1} \\ v_{C2} \\ v_{C3} \end{bmatrix} - \begin{bmatrix} v_{L2} \\ v_{L3} \\ v_{L1} \end{bmatrix}, \begin{bmatrix} v_{oab} \\ v_{obc} \\ v_{oca} \end{bmatrix} = \begin{bmatrix} v_{C1} \\ v_{C2} \\ v_{C3} \end{bmatrix} - \begin{bmatrix} v_{L1} \\ v_{L2} \\ v_{L3} \end{bmatrix} \quad (2)$$

In state 2, the bi-directional switch S_1 is off and S_2 is on, and the interval of the converter operating in this state is (1-D)T, shown in Fig.4, one has,

$$\begin{bmatrix} v_{C1} \\ v_{C2} \\ v_{C3} \end{bmatrix} = \begin{bmatrix} v_{L1} \\ v_{L2} \\ v_{L3} \end{bmatrix}, \quad (3)$$

The average voltage of the inductors over one ac line period in steady state should be zero, from (2) and (3), we have

$$V_C \begin{bmatrix} D\cos(\omega t + \phi_C) + (1-D)\cos(\omega t + \phi_C - 120°) \\ D\cos(\omega t + \phi_C - 120°) + (1-D)\cos(\omega t + \phi_C + 120°) \\ D\cos(\omega t + \phi_C + 120°) + (1-D)\cos(\omega t + \phi_C) \end{bmatrix}$$

$$= \sqrt{3} D V_i \begin{bmatrix} \cos(\omega t + 30°) \\ \cos(\omega t - 90°) \\ \cos(\omega t + 150°) \end{bmatrix}$$

$$\text{Or} \quad \frac{V_C}{\sqrt{3}V_i} = \frac{D}{\sqrt{3D^2 - 3D + 1}} \tag{4}$$

Assume that the inductor in the Z-network are very small and there is no line frequency voltage drop across the inductor, thus the output line-to-line voltage should equal to V_C, the voltage across the capacitor in the Z-network, that is:

$$\frac{V_o}{V_i} = \frac{D}{\sqrt{3D^2 - 3D + 1}} \tag{5}$$

Evidently, by controlling the duty cycle D, the output voltage of the proposed ac-ac converter can be bucked or boosted. Fig.5 shows the voltage gain versus the duty cycle. It clearly shows that there are two operating regions. When the duty cycle is less than 0.5, the converter operates in the buck mode; and when the duty cycle is greater than 0.5, it operates in the boost mode.

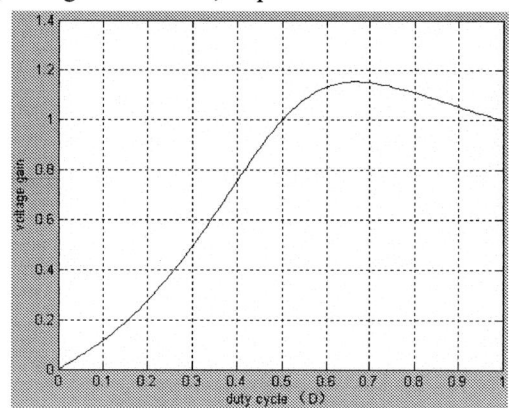

Fig.5 The relationship between voltage gain and duty cycle

IV. SIMULATION AND EXPERIMENTAL VERIFICATION OF THE Z-SOURCE ASD SYSTEM

To confirm the operating principle of the new ASD system, simulations have been carried out and a 3-kVA prototype has been built. In order to show clearly the output voltage obtained from the ac-ac converter, an LC filter with 1-kHz cutoff frequency is placed in between the converter and the motor. The simulation and experimental system is setup with the following Parameters.

1) Three-phase line voltage: 220 V, line impedance: 3%.
2) Load: three-phase 220V3-kW induction motor.
3) Z-source network: L=0.1mH and C=10uF.
4) Switching frequency: 10 kHz.

Fig.6 shows simulation waveforms. The ac phase input voltage has voltage sag during the time [0.12s, 0.18s] from the nominal voltage of 220V rms to 200V rms, and has a surge during the time [0.18s, 0.24s] from 200V rms to 240V rms. By PWM duty cycle control, we can keep the output voltage constant at 220V rms. The proposed converter operates in the boost mode during the voltage sag, and in the buck mode during the voltage surge. Fig.7 shows the frequency spectrum of the input phase current, output line-to-line voltage and Z-network capacitor voltage. The simulation results are well consistent with above theoretical analysis. The simulation proved the Z-source converter concept.

Figs.8 shows the experimental results. Fig.8 (a) shows the results when V_{irms}=220V and D=0.1, the proposed converter operates in the buck mode in the condition, and Fig.8 (b) shows the results when V_{irms}=165V and D=0.9, the proposed converter operates in the boost mode in the condition. The waveforms verify the theoretical analysis and simulation results.

V. CONCLUSIONS

This paper has presented an impedance-source power converter for implementing ac-ac power conversion. The Z-source converter employs a unique impedance network (or circuit) to couple the converter main circuit to the power source, thus providing unique features that cannot be observed in the traditional ac-ac converters. It can buck-boost the input voltage, minimized component count, increase efficiency and reduce cost.

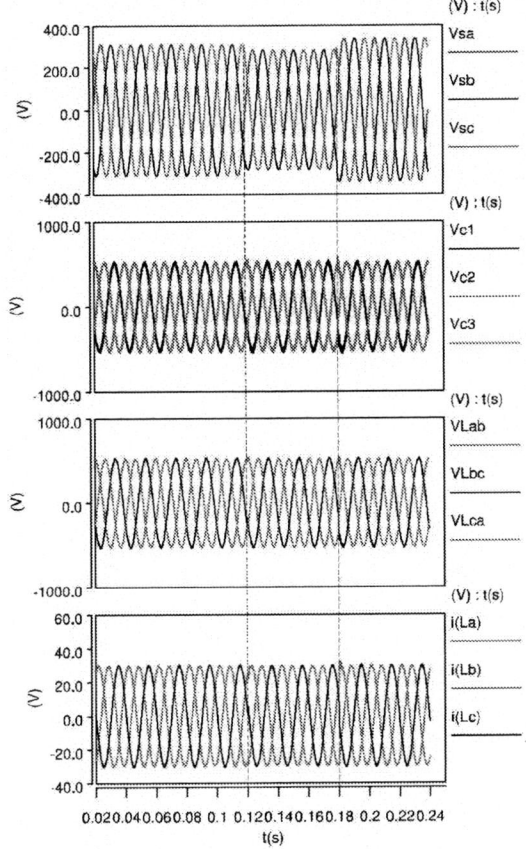

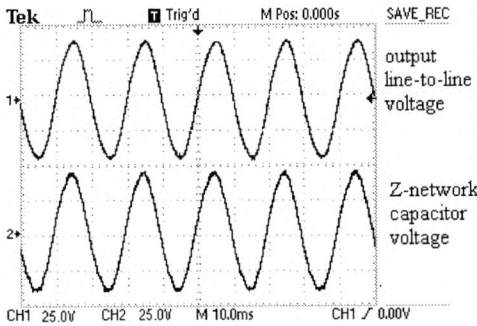

(a)V_{irms}=220V and D=0.1

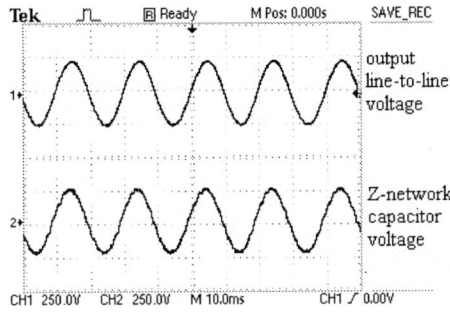

(b) V_{irms}=165V and D=0.9

Fig.8. Experimental results of the proposed converter.

Fig. 6 Simulation results (the input voltage, Z-source network capacitor voltage, output line-to-line voltage and Z-source network inductor current).

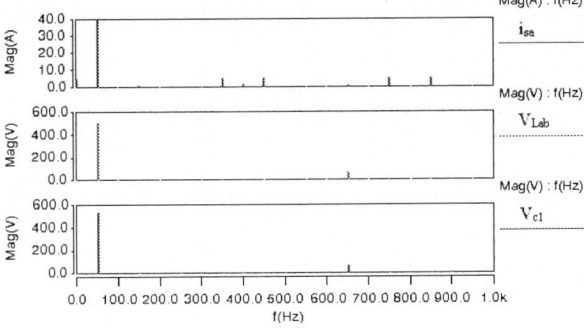

Fig.7. Frequency spectrum (the input current, output line-to-line voltage and Z-source network capacitor voltage).

REFERENCES

[1] Fang Zheng Peng, Lihua Chen, Fan Zhang, "Simple topologies of PWM AC-AC converters", / IEEE, Power Electronics Letters/, Vol.1, Issue: 1, pp. 10–13, March 2003.

[2] Lin, B.R., Yang, T.Y., Wei, T.C., "Single-phase ac/ac converter based on half-bridge NPC topology", in/Record, Circuits and Systems, International Symposium/, 2003, Vol.3, pp. III-340 - III-343.

[3] Jie Chang, Sun, T, Anhua Wang, "High energy-efficient and compact AC-AC converters", in/Record, IEEE Power Electronics and Drive Systems Conf/. 2001, Vol.2, pp.543 - 548.

[4] Jacobina, C.B., de R. Correa, M.B., Ribeiro, R.L.A., Oliveira, T.M., da Silva, E.R.C., Lima, A.M.N., "AC/AC converters with a reduced number of switches", in/Record, IEEE Industry Applications Conf/., 2001, Vol.3, pp.1755- 1762.

[5] Kazerani, M., "A direct AC/AC converter based on current-source converter modules", /IEEE Trans. Power Electronics/, Vol. 18, pp.1168 – 1175, Sept. 2003.

[6] Bor-Ren Lin, Ta-Chang Wei, "Implementation of a single-phase AC/AC converter based on neutral-point-clamped topology", /IEEE Trans. Aerospace and Electronic Systems/, Vol. 39, pp.625 – 634, April 2003.

[7] Shenkman, A., Axelrod, B., Berkovich, Y., "Improved modification of the single-switch AC-AC converter for induction heating applications", /IEE Proc. Electric Power Applications/, Vol.151, pp.1 – 4, 9 Jan. 2004.

[8] Bor-Ren Lin, Chen, D.-J., "Implementation of a single-phase three-leg AC/AC converter with neutral-point diode-clamped scheme", /IEE Proc. Electric Power Applications/, Vol. 149, pp.423 – 432, Nov. 2002.

[9] Xu-peng Fang, Zhao-ming Qian, Qi-Gao, Bin-Gu, Fang-zheng Peng, Xiao-ming Yuan, "Current mode Z-source inverter-fed ASD system", in/Record, IEEE Power Electronics Specialist Conf/., 2004, pp.2805-2809.

[10] Peng F.Z., Xiao-ming Yuan, Xu-peng Fang, Zhao-ming Qian, "Z-source inverter for adjustable speed drives", in/Record, IEEE Power Electronics Specialist Conf/., 2003, pp. 33 –35.

[11] Xu-peng Fang, Zhao-ming Qian, Fang-zheng Peng, "Single-phase Z-source ac-ac converters", Power Electronics Letters, Vol.3, Issue: 4, pp. 121–124, Dec 2005.

2006 5th International Power Electronics and Motion Control Conference

Construction and Application of Macro Model for ZVS Resonant Mode Controller MC34067

Wei Chen, Yilei Gu, Zhengyu Lu (Senior member, IEEE), Zhaoming Qian (Senior member, IEEE)
College of Electrical Engineering
Zhejiang University
Hangzhou, Zhejiang, China
lionhcw@hotmail.com

Abstract—To solve the low efficiency issue in cyber-simulation for high frequency resonant converters in power electronic system integration, improve the speed of the simulation for the resonant mode controller and make the design process for high frequency switching power supply to be more efficient, proposed the construction process of the macro model for ZVS resonant mode controller MC34067 followed by the analysis for the internal structure and main characteristics. This macro model maintained the main functions of MC34067 while some simplification was made to suit for simulation. Based on the macro model, A 250~400V input, 54V/300W output LLC resonant mode converter was designed and simulated. Whereafter a prototype with the same circuit electrical parameters was made to verify the authenticity and validity of this macro model. The experimental results show the proposed macro model can truly imitate the external characteristics of MC34067 while maintains highly precision as well. Therefore the proposed macro model is attractive for the power electronic system integration applications, where the high frequency is intensively required, to improve the efficiency of the design process for the high frequency resonant power supply.

Keywords-Resonant mode; Macro model; LLC topology simulation

I. INTRODUCTION

Along with the electronic technique developing apace, high power density power supply set is required to achieve higher efficiency, reduce volume and lower the cost [1]. To realize this object, increasing the switching frequency is the most feasible method. However, with the ascent of the switching frequency, the switching loss of the power switches increases as well in the conventional PWM power converters, where the further increase for the switching frequency is limited. Therefore, the resonant mode power converters, which employs soft switching technique to reduce switching loss, are very attractive for the high frequency applications [2] [3]. However, in the design process for a resonant converter, the simulation stage is at very low efficient for the complex structure and colossal electrical circuit of the controller microchip special for resonant mode converters, which induce the slow speed of the

simulation caused by the limitation of computer's memory and convergence of calculation. Therefore the simulation must be operated by employing the simplified macro model of the controller, not just directly by the original controller's internal circuit.

According to previous analysis, it is necessary to research the resonant converter controller's macro model for computer simulation to improve the efficiency of the design for high frequency converters. This paper presents the construction process of the macro model for MC34067 yielded by Motorola based on the microchip's main characteristics, simulates a ZVS LLC resonant power converter by employing the constructed macro model, finally a prototype with the same electrical parameters with the simulation verifies the validity of the proposed macro model.

II. MACRO MODEL CONSTRUCTION CONSIDERATIONS

The overall structure and each main function block are described in detail in Ref. [4], which indicate that it is not suitable for simulation by directly duplicating the circuit. What we need is the simplified model basically with the same external characteristics. According to this conception, the macro model of MC34067, which is suitable for transient analysis simulation for power system when some moderate simplification was made, is presented in this paper. The scheme is shown in Fig. [1].

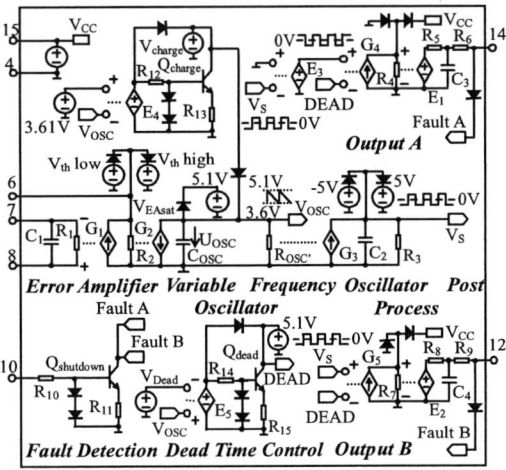

Figure. 1   Scheme of the macro model for MC34067

Project supported by National Natural Science Foundation of China (50237030ZD)

1-4244-0448-7/06/$25.00 ©2006 IEEE

A. Error Amplifier

The *Error Amplifier* function block is simulated by the VCCS G_1, R_1 and C_1, where the latter two represent input impedance respectively. Assuming the conductance G_1 is 2mS, paralleled resistance R_1 is 5MΩ, so the open loop gain can be obtained:

$$Ao = G_1 \times R_1 = 10000 \ (80dB)$$

The bandwidth BW of this error amplifier is determined by capacitor C_1, which can be calculated by the equation below in case C_1 is 50pF for example:

$$BW = 1/(2\pi R_1 C_1) = 600Hz$$

B. Variable Frequency Oscillator

According to the detailed circuit scheme and structure in Ref. [4], the frequency of MC34067 is modulated by the time of the whole time period for capacitor C_{OSC} to be charged and discharged. However, since C_{OSC} charges from 3.6V to 5.1V is less than 50ns, this time period can be omitted, which means the frequency is mainly controlled by the discharge process from 5.1V to 3.6V for C_{OSC}. $I_{C_{OSC}}$, the current of the capacitor C_{OSC}, can be expressed as follows by Fig. [2]:

$$I_{C_{OSC}} = I_{R_{OSC}} + I_{OSC} \tag{1}$$

Where $I_{R_{OSC}}$ and I_{OSC} are elicited by Eq. [2] and Eq. [3] separately from Ref. [4]:

$$I_{R_{OSC}} = \frac{1.5}{R_{OSC}} \varepsilon^{\left(-\frac{1}{f_{min} R_{OSC} C_{OSC}}\right)} \tag{2}$$

$$I_{OSC} = \frac{2.5 - V_{EAsat}}{R_{VFO}} \tag{3}$$

Where in the two equations presented above, R_{VFO}, which is controlled by the maximum switching frequency, and R_{OSC}, which is controlled by the minimum switching frequency, are both constant when the parameters of the power supply are determined. Use Eq. [4], R_{OSC} can be calculated if the minimum frequency was already known:

$$R_{OSC} = \frac{\frac{1}{f_{min}} - 70ns}{0.348 C_{OSC}} \tag{4}$$

Since the switching period is comparatively longer than the delay time 70ns, therefore the delay time can be omitted for simplification, and the Eq. [5] can be obtained from Eq. [4]:

$$R_{OSC} = \frac{1}{0.348 f_{min} C_{OSC}} \tag{5}$$

Put [5] into [2]:

$$I_{R_{OSC}} = \frac{1.5 \times \varepsilon^{-0.348}}{R_{OSC}} \tag{6}$$

Therefore $I_{R_{OSC}}$ is determined by R_{OSC} as well as the

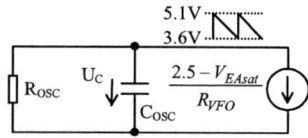

Figure. 2   Equivalent circuit for variable frequency oscillator

discharge current $I_{C_{OSC}}$ of the capacitor C_{OSC}, which determines the frequency of oscillator f_{OSC} by Eq. [7]:

$$I_{C_{OSC}} = 1.5 C_{OSC} f_{OSC} \tag{7}$$

Put [3], [6] and [7] into [1], the approximative voltage-frequency equation for the variable frequency oscillator can be obtained:

$$f_{OSC} = \frac{\varepsilon^{-0.348}}{R_{OSC} C_{OSC}} + \frac{2.5 - V_{EAsat}}{1.5 R_{VFO} C_{OSC}} \tag{8}$$

The output frequency of the variable frequency oscillator changes in reverse along with the output of the error amplifier V_{EAsat} changes. In fact, the relationship between the output frequency of MC34067 f_{OUT} and the frequency of the oscillator f_{OSC} can be expressed in Eq. [9]:

$$f_{OSC} = 2 f_{OUT} \tag{9}$$

For the purpose of simplifying the simulation circuit, the macro model adopts the design that just making the frequency of the oscillator f_{OSC}' be equal to the final output frequency, which can be expressed as follows:

$$f_{OSC} = 2 f_{OSC}' \tag{10}$$

Revise Eq. [8] to obtain Eq. [11]:

$$f_{OSC}' = \frac{\varepsilon^{-0.348}}{2 R_{OSC} C_{OSC}} + \frac{2.5 - V_{EAsat}}{2 \times 1.5 R_{VFO} C_{OSC}} \tag{11}$$

Which is the very mathematical model employed by the macro model inherently. Therefore the parameter for the VCCS G_2 can be obtained:

$$G_2 = \frac{2.5 - V_{EAsat}}{2 R_{VFO}} \tag{12}$$

By Eq. [12], the current I_{OSC}' is also obtained by $I_{OSC}' = V_{EAsat} \times G_2$. At the same time, the frequency output of the oscillator can meet Eq. [11] by setting the R_{OSC}' of Fig. [1] to $2R_{OSC}$. In addition, the charge circuit, which ensures the capacitor C_{OSC} can be charged to 5.1V after being discharged, is adopted based on the VCVS E_4 and the transistor Q_{charge}. Therefore all the functions of the *Variable Frequency Oscillator* are achieved.

C. Oscillator Post Process

The main function of this block is to transfer the output voltage of the variable frequency oscillator V_{OSC} to a square wave V_S. Then the VCCS G_3 is adopted to divide V_{OSC} by 4.35V, where the parameter of G_3 is :

$$G_3 = k \times (Vosc - 4.35V) \quad k \rightarrow +\infty \tag{13}$$

The voltage across R_3, which is also clamped, is a square wave V_S which has the duty ratio of 50%.

D. Output Driver

To realize the dead time control, the *Variable Frequency Oscillator* output voltage V_S is compared with the dead time pulse, which yields two ways of complementary pulses with the dead time added. Voltage source V_{Dead} is the reference voltage to adjust the dead time, which value is between 3.6V and 5.1V. After the processing of dead time control, be amplified and clamped, the final dual output driver A&B is created. Also another VCVS, which is inserted between the macro model and the exterior circuit to isolate the two parts, can prevent the exterior impendence and electrical characteristics affecting the internal macro model characteristics.

E. Shutdown Circuit (Fault Detection Circuit)

The main function of Transistor $Q_{shutdown}$ is to shutdown the output of the macro model. When voltage level of pin 10 is high, $Q_{shutdown}$ is conduct, which can bring down the voltage level of pin 14 and pin 12 and then shutdown the macro model output.

According to previous analysis, the proposed macro model, which maintains the main functions for MC34067, can be used in the system level simulation for a power supply system analysis and design process.

III. EXPERIMENTAL RESULTS

To make this macro model more comprehensive and can be properly utilized, a close loop simulation for a 250~400V input, 54V/300W output LLC resonant converter given by Fig. [3], which based on the proposed macro model, was performed. The operation waveforms are shown as Fig. [4]~[8]. The output drive waveform of the macro model for MC34067, whose duty ratio is about 50% with dead time, is shown in Fig. [4]. The converter's transient simulation output waveform with the time length of 8ms is shown as Fig. [5], wherein the stead-state error ratio is less than 2%. The output capacitor Co is initialized with 25V to save the total simulation time. Fig. [6] shows the ZVS for main switches is achieved. The simulated voltage and current waveforms for the primary side of the transformer are shown in Fig. [7], and the simulated current waveforms for both primary and secondary side (after rectified) of the transformer are given by Fig. [8]. All waveforms shown in Fig. [7] and Fig. [8] are accordant to the LLC resonant topology operation principles [2] [3] [11] [12].

To verify the coherence between the simulation results and the true circuit, a prototype of the same electrical parameters used in simulation is made and Fig. [9]~[11] are the experimental results. The measured driving signal and drain to source voltage of S_2 are shown in Fig. [9], which shows the ZVS for the main switches is well realized. Fig. [10] shows the measured waveforms of the voltage and current for the primary side of the

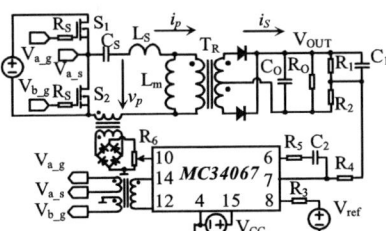

Figure. 3 LLC resonant converter scheme with MC34067 macro model block

transformer, and Fig. [11] shows the measured waveforms of the current for both the primary and

secondary side (after rectified) of the transformer. Actually, Fig. [9]~[11] are the same electrical parameter waveforms compared with Fig. [6]~[8], but just measured in the true circuit. The comparison between Fig. [6]~[8] and Fig. [9]~[11] can show under the same condition (311V input), the proposed macro model can truly imitate the external characteristics of MC34067 while maintains highly precision. Table. [1] shows the simulation results at varying input voltages.

TABLE I. SIMULATION RESULTS AT VARYING INPUT VOLTAGES

	Output Voltage (V)	Load Current (A)	Output Power (W)	Input Power (W)	Efficiency
V_{in}=250V	53.5	5.50	294.3	308.2	95.5%
V_{in}=311V	53.9	5.54	298.6	310.1	96.3%
V_{in}=400V	54.1	5.56	300.8	309.8	97.1%

IV. CONCLUSION

The macro model presented in this paper, which compared with the actual circuit scheme, maintains the main functions featured in the resonant mode controller MC34067 and a high imitating precision, while the internal structure of the macro model is greatly simplified. By employing the proposed macro model, the time for simulation in the design for a high frequency resonant power converter can be reduced, which results in the improvement of efficiency for the design simulation. Therefore, the proposed macro model is attractive for the simulation process in the power electronic system integration applications, where the high frequency is intensively required to realize the miniaturization and to improve efficiency for the resonant mode power supply set .

REFERENCE

[1] Fred C. Lee, Denming Peng, "Power Electronics Building Block and System Integration" PIEMC 2000, The Third International Volume 1, 15-18 Aug. 2000 Page(s):1-8 vol.1

[2] Bo Yang; Lee, F.C.; Zhang, A.J.; et al, "LLC resonant converter for front end DC/DC conversion", APEC '02. Seventeenth Annual IEEE Volume 2, 10-14 March 2002 Page(s):1108-1112 vol.2

[3] Yanjun Zhang; Dehong Xu; Min Chen; et al, "LLC resonant converter for 48 V to 0.9 V VRM", PESC '04. 2004 IEEE 35th Annual Volume 3, 20-25 June 2004 Page(s):1848-1854 Vol.3

[4] Motorola Analog IC Device Data Book 1999

[5] Rong Luo, Huazhong Yang, Yu Shen, et al, "A macromodel for operational amplifiers with adjustable offset voltage", Communications, Circuits and Systems and West Sino Expositions, IEEE 2002 International Conference on Volume 2, 29 June-1 July 2002 Page(s):1326 - 1329 vol.2

[6] Mutnury, B.; Swaminathan, M.; Cases, M.; et al "Macro-modeling of transistor level receiver circuits", Electrical Performance of Electronic Packaging, 2004. IEEE 13th Topical Meeting on 2004 Page(s): 243-246

[7] Mutnury, B.; Swaminathan, M.; Libous, J.; "Modeling of power supply noise using efficient macro-model of non-linear driver", Electromagnetic Compatibility, 2004. EMC 2004. 2004 International Symposium on Volume 3, 9-13 Aug. 2004 Page(s): 988-993 vol.3

[8] Zou Gang; Chen Xiangxun; Zheng Jianchao; et al, "A macro-model of SCR for transient analysis in power electronic system", Power System Technology, 1998. Proceedings.

POWERCON '98. 1998 International Conference on Volume 1, 18-21 Aug. 1998 Page(s): 659-663 vol.1

[9] Schmid, R.; "Creating high frequency current feedback macro models with enhanced performance characteristics", Circuits and Systems, 1993., Proceedings of the 36th Midwest Symposium on 16-18 Aug. 1993 Page(s): 1345-1348 vol.2

[10] J. Alvin; pyung choi; Macromodeling with spice; Eglewood cliffs, N, J.I, Prentice Hall c1992

[11] Yilei Gu, Zhengyu Lu, Zhaoming Qian, Three Level LLC Series Resonant DC/DC Converter, APEC '04. Nineteenth Annual IEEE Volume 3, 2004 Page(s):1647-1652 Vol.3

[12] Yilei Gu, Lijun Hang, Huiming Chen, "A simple structure of LLC resonant DC-DC converter for multi-output applications", APEC 2005. Twentieth Annual IEEE Volume 3, 6-10 March 2005 Page(s): 1485-1490 Vol. 3

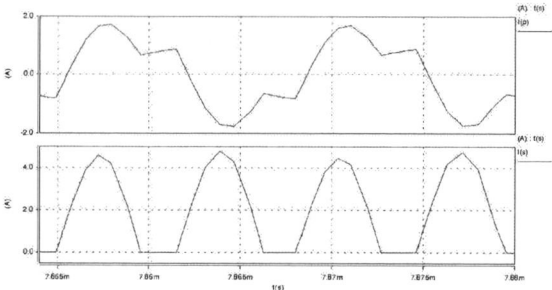

Figure. 8   Simulated current waveforms for both primary and secondary side of the transformer

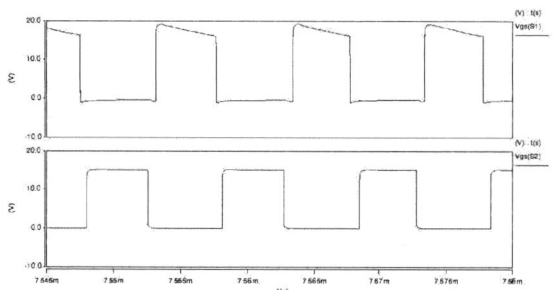

Figure. 4   Output waveform of MC34067 macro model

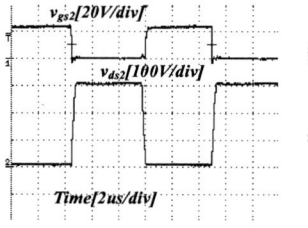

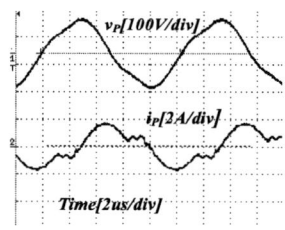

Figure. 9 Measured ZVS waveforms for main switches

Figure. 10 Measured voltage and current waveforms for primary side of the transformer

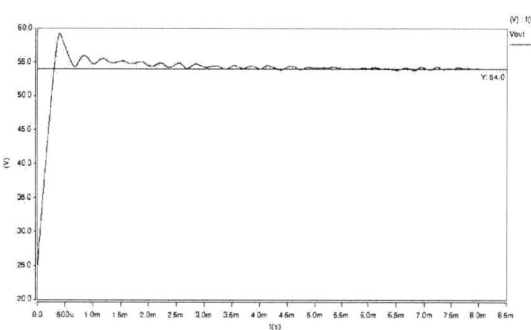

Figure. 5   Output waveform of LLC resonant converter

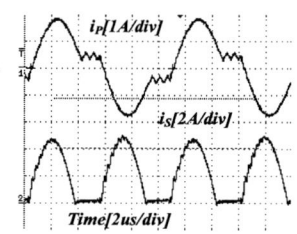

Figure. 11 Measured current waveforms for both primary and secondary side of the transformer

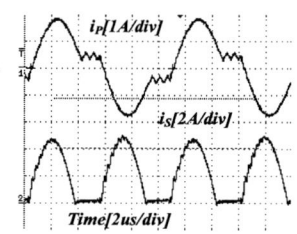

Figure. 6   ZVS for main switches in simulation

2006 5th International Power Electronics and Motion Control Conference

Optimum Design of Hollow Conductor in Stator Winding for Large Evaporative Hydro-generator

Z. Wen , L. Ruan, G. Gu

Institute of Electrical Engineering/Evaporative Cooling Technology Research Center,
Chinese Academy of Sciences, Beijing, China, 100080
Phone: +86-010-62635883   Fax: +86-010-62622056   E-mail: wenzhw@mail.iee.ac.cn

Abstract—A new type of evaporative inner-cooling system used in large hydro-generator has exceeded over other traditional cooling methods that is characterized by two-phase flow and boiling heat transfer. For this reason, in the case of the optimum design of generator, some design variables different from an water-cooling or air-cooling machine should be considered, which will lead to substantial improvement on the temperature rise of stator winding and core. The major design variables that are considered by using 3D electromagnetic Finite Element Method (FEM) coupled with evaporative-cooling heat transfer analysis in this paper are dimension of hollow conductor used in the stator bar, efficiency of machine and ultimate capacity of machine. The analysis results show that, based on the existing design of water-cooling system of Three Gorges' hydro-generator installed on the left bank, increase the dimension of hollow conductor will result in a whole better temperature distribution and thermal load capability in the stator of evaporative cooling hydro-generator than in water-cooling hydro-generator. This study presents an effective dimension of hollow conductor for the Three Gorges' hydro-generators that satisfy the request of operating efficiency. Conclusion of this paper can be provided as academic foundation for improving design and manufacture high-capacity hydro-generator with evaporative inner-cooling system.

Keywords- hydro-generator; evaporative cooling; stator winding; Optimum Design

I.   INTRODUCTION

At the present time, the inner cooling system with water is the main type of cooling system used in stator windings of large hydro-generator, but the possible leakage of water may cause a serious hazard to insulation and electrical accident may happen. The self-circulating evaporative cooling system developed by the Institute of Electrical Engineering of Chinese Academy of Science (IEECAS) used in the stator of hydro-generator has the same cooling performance as traditional inner cooling system such as water cooling system, also has advantage in operation, installation and maintenance. Hydro-generator rated 10MW, 50MW and 400MW with the evaporative cooling system has been put into operation successfully since

1983, which is the evidence that the evaporative cooling technology has been an important application technology in the power industry [1-3].

Compared with water-cooling system, economic and reliability considerations are of the most important concern in the design of large hydro-generator with evaporative system. Because characteristic of heat transfer of boiling two-phase fluid in evaporative cooling system is different from single-phase convection heat transfer in water-cooling system. Hollow conductor used in the water-cooling hydro-generator is not the best choice for evaporative cooling machine. Increase of area of hollow conductor could benefit for the heat transfer in the conductor and increase the ultimate output power of machine with evaporative cooling system has been testified [4]. Because increase of area of hollow conductor could also cause the increase of armature loss, it is worthy to have a deep research on optimization of cooling system and maximize the ultimate output power based on assurance of efficiency. With the advent of large-scale multiprocessing computers new techniques have been developed that provide economical, multidimensional design optimizations. These methods are considered very important for engineering design. Although optimization methods have been widely used in the design of a large number of engineering systems, it is only recently that they were applied to the design of large electrical machinery.

The purpose of this paper is to present the design philosophy and implementation procedure in the optimum design of stator winding of large evaporative cooling hydro-generator such as the Three Gorges' hydro-generator. Sizing equations of the machines are derived using generalized equations. Optimum machine design is illustrated by choosing the dimension of hollow conductor. Using the optimum design data, electromagnetic and two-phase flow temperature's field analysis of the hydro-generator are investigated. The outline of this analysis is indicated in Fig.1.

In this paper, the design variables of machine are analyzed by using 3D FEM including actual stator and rotor winding topology at a rated-load condition shown as Fig.2. Considering symmetry, the analysis models are set as 1/10 of its full model. Except windings, other structures of

1-4244-0448-7/06/$25.00 ©2006 IEEE

hydro-generator including camper winding, end links also be contained.

Three models of tube conductor width and four models of hollow conductor height are selected because its mechanical size is limited. Fig. 3 shows hollow conductor coil used in the stator of the machine.

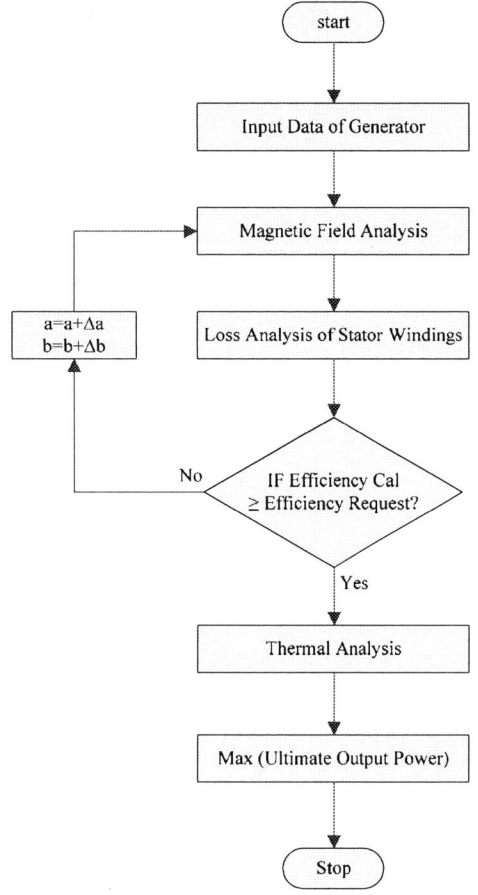

Figure 1. Simplified chart for optimization procedure

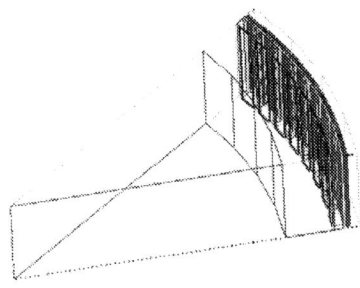

Figure 2. Magnetic analysis model of the hydro-generator

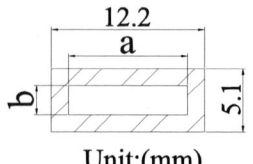

Unit:(mm)

Figure 3. Figure of hollow conductor

Table I. shows specifications of the hydro-generator. The request minimize efficiency of the machine is 98.75%.

TABLE I.   GENERATOR CHARACTERISTIC

Rated power	778	MVA
Rated voltage	20000	V
Rated current	22453	A
Rated power factor	0.9	
Nominal frequency	50	Hz
Nominal speed	75	rpm
Number of stator slots	510	
Stator bore diameter	18500	mm
Stator stacking height	3130	mm

II.   FEM FORMULA AND THERMAL ANALYSIS

In deriving the formulations for the solution of electromagnetic and temperature fields in hydro-generator, the following assumptions have been adopted to keep the computational algorithm and computer CPU times within reasonable limits:

(1) The influence of transposition on each conductor of coil bar is equal, and average value of resistance losses of the coil bar is taken into account.

(2) In the end region the coil is taken into consideration by a connection with external resistance and induction.

(3) The coolant in the evaporative cooling system takes all the heat generated by the coils away.

A.   3D FEM Formulation

In order to describe the static magnetic field using the scalar potential, its rotational part should be modified by the magnetic field intensity.

$$\vec{H} = \vec{T} - \nabla\Omega \tag{1}$$

where $\vec{T}$ is an arbitrary function describing input current density $\vec{J}$, and Ω is magnetic scalar potential. Equation (1) satisfies Ampere'law and the solenoidal condition of flux density.

$$\nabla \cdot \mu\left(\vec{T} - \nabla\Omega\right) = 0 \tag{2}$$

where μ is the permeability of components.

After equation (2) discrete by FEM, the loss of component is calculated by

$$P = \sqrt{\frac{\omega\mu}{8\sigma}}\int_{s} H_t \cdot H_t^* dS \tag{3}$$

where ω is angular velocity, σ is the conductivity.

176

B. Thermal analysis

The formulation of temperature rise analysis is based on the heat conduction equation shown as equation (4):

$$k\nabla^2 T + Q - \rho c \frac{\partial T}{\partial t} = 0 \tag{4}$$

where k is thermal conductivity, ρ is density, Q is internal heat generation rate, and c is heat capacity. Considering the convective heat exchange on the inner surface of conductor [5], the convection boundary condition is applied as equation (5):

$$-k\frac{\partial T}{\partial n}\bigg|_s = \alpha(T - T_f) \tag{5}$$

where T and T_f are the surface of conductor and coolant temperature, respectively, and α denotes heat transfer coefficient.

C. Design constraints

The values assigned to the design variables are usually restricted by a number of constraints imposed on the stator as dictated by the actual design problem. However, in the evaporative cooling system, the related problems of two-phase flow in the evaporative cooling loop in micro channel are very complicated and no common design and calculation methods are in use up to the present and technological criteria for such methods have yet been developed. In order to improve the safety of system, sufficient margins must be considered during the design process, and to hold the design between fixed limits. Based on the present research results evaporative cooling system, constraints are introduced using the inequality type

$$T_{out\,min} \le T_{out} \le T_{out\,max} \tag{6}$$

where T_{out} is temperature at the output of cooling channel in the stator winding.

Another important performance constraint which must be considered is the limitation of the maximum temperature of coolant as result of the temperature distribution along the stator winding which is given by eqn.4. Also, the constraints are introduced using the inequality type

$$T \le T_{max} \tag{7}$$

D. Design objective function

The objective function in a general optimization problem represents a basis for the choice between alternate acceptable designs. In this case the design objective function is to maximize the output power of the generator. However, it can not be calculated directly by some present formula based on eqn. 6 and eqn. 7. the final result of the design must be tested by experiments.

III. RESULTS AND DISCUSSION

A. Efficiency analysis

The analysis of the machine is carried out by using 3D FEM. Fig.4 and Fig.5 shows that the characteristics of efficiency according to the variation of dimension of hollow conductor.

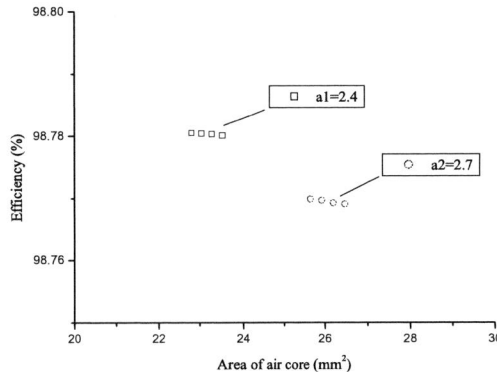

Figure 4. Calculating efficiency of hydro-generator

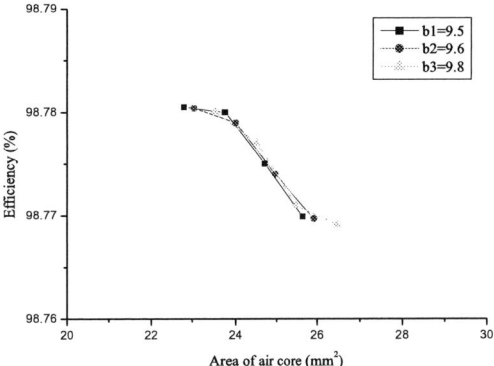

Figure 5. Calculating efficiency of hydro-generator

B. Temperture field

We used the CFC-113 as the coolant of the experimental model. The height of infusing coolant is set as the height of stator's bar. The hollow conductors are employed direct current to simulate thermal load. In the rated working condition, copper losses of a hollow conductor can be calculated on the different copper losses of real hydro-generator. The calculated temperature distribution in the stator with the initial inner size of hollow conductor 2.4mm*9.5mm and the optimization result with the size of hollow conductor 2.7mm*9.8mm is shown as Fig.6.

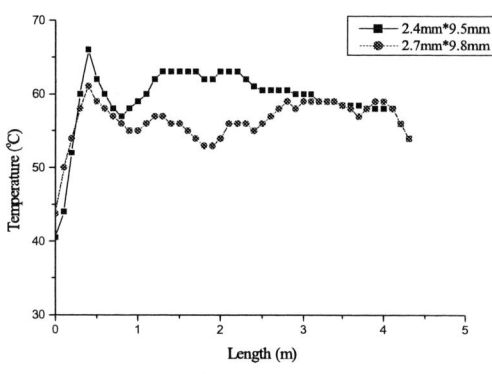

Figure 6. Temperature distribution in the stator of hydro-generator

C. Ultimate output power

Ultimate output power of the machine with design efficiency meet the request is estimated by Empirical Model Analysis (EMA) [6]. Fig.7 shows calculated utmost thermal load with different area of hollow conductor. This result shows that by using the evaporative cooling system large hydro-generator can have even bigger heat-dissipating capability and it will be beneficial to the overloading operation.

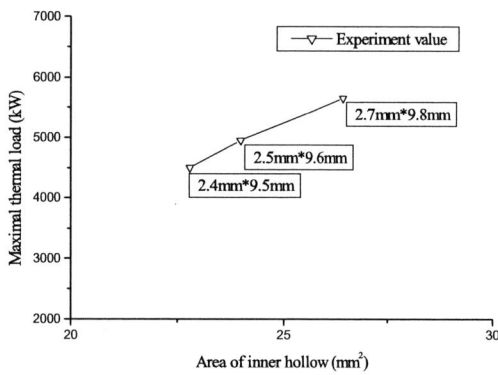

Figure 7. Maximal thermal loads with different hollow conductors

IV. CONCLUSIONS

The major design variables considered in this paper is the inner size of the hollow conductor- width and height. The variation of inner width is more important than height in terms of efficiency. The results show that the increase of the inner size of the hollow conductor can be testified to improve large hydro-generator performance on ultimate power output.

In this paper, we have presented an effective dimension of hollow conductor for the Three Gorges' hydro-generator that satisfy the request of operating efficiency and at the same time obtain the better over-load capacity. Conclusion of this paper can be provided as academic foundation for improving design and manufacture high-capacity hydro-generator with evaporative inner-cooling system.

REFERENCES

[1] G. Gu, Z. Cheng, X. Tian, W. Zeng, "Calculation and experiment of hydro-generator stator bar with CLSC evaporative cooling system", *Proceeding of the International Conference on Electrical Machines Manchester Machines Research Group*, UK, pp.990-994, 1992.

[2] L. Ruan, G. Gu, "Analysis of the key techniques in hydro-generator with inner evaporative cooling", *ICEMS'2003*, Beijing.

[3] L. Ruan, G. Gu, and X. Tian, "Numerical simulation for circulating systems and experimental comparison of the closed-loop, self-circulating evaporative cooling of hydro-generator", *Electrical Engineering*, vol. 86, pp. 127–134, 2004.

[4] L. Ruan, G. Gu, and X. Tian, and Z. chang, "The experimental research of new coolant in evaporative cooling technology of large hydro-generator", *Advanced Technology of Electrical Engineer and Energy*, vol.21, pp.21–24, 2002.

[5] J. G. Collier and J. R. Thome, *Convective boiling and condensation*, Third edition, Oxford, Oxford University Press, 1994.

[6] G. Gu, L. Ruan, "A few questions about using the evaporative cooling method as the substitute for the water-cooling method in the Three Gorges' hydro-generator", *ICEMS'2005*, Nanjing. China

2006 5th International Power Electronics and Motion Control Conference

Rotor Suspension Principle and Decoupling Control for Self-bearing Induction Motors

Tengchao Zhang, Huangqiu Zhu and Yuxin Sun

Jiangsu University/School of Electrical and Information Engineering, Zhenjiang 212013, China

tengchao311@126.com

Abstract—**In this paper, a self-bearing induction motor with additional windings in the stator slots is proposed. The self-bearing induction motor is a strong coupled, multivariable and nonlinear system. The torque and radial suspension forces are coupled each other through airgap flux linkages. The decoupling control of the torque and radial suspension forces is the base of the stable operation of the motor. In this paper, the principle of radial suspension forces is introduced. The mathematics models of radial suspension forces and the rotation part of the motor are deduced. A control system based on rotor magnetic field oriented control strategy is designed. The control system is simulated using Matlab/Simulink toolbox. Simulation results have shown that the rotor can suspend steadily, torque and radial suspension subsystems can be controlled independently, the control method is valid and the whole control system has satisfactory static and dynamic performance.**

Keywords-self-bearing induction motor; mathematics model; radial suspension force; decoupling; simulation

I. INTRODUCTION

The self-bearing motor which combines function of motor and magnetic bearing can provide both torque and radial suspension forces. Two sets of three-phase windings that a special one acts as bearings and a conventional one for rotation, are wound in the same stator slots [1]-[4]. The self-bearing motors have all advantages of magnetic bearing, such as no mechanical contact, no wear, no need of lubrication and high speed. In addition, it has compact structure, higher reliability and shorter shaft length. For this reason, they have good application prospects in high speed drives, such as special motors in the areas of biotechnology, chemical, medical, semiconductor and so on. There are many kinds of self-bearing motors and the self-bearing induction motor is one of the focuses in this research field because of its simple structure and high reliability.

When the motor is operating, the radial suspension forces must be controlled in real time. Therefore, accurate calculation of the radial suspension forces is important to the control of the motor. The coupling of the torque and radial suspension forces through the airgap flux linkages

exists. The decoupling control of the torque and radial suspension forces is important to stable operation of the motor. In this paper, the rotor magnetic field oriented vector controller is used to realize the decoupling control of this nonlinear system, and the control system has satisfactory performance.

II. WORKING PRINCIPLE OF SELF-BEARING INDUCTION MOTOR

Two sets of three-phase windings are wound in the same stator slots. One is the motor windings with the pole pairs p_1 for the production of motoring torque. The other is the suspension force windings with the pole pairs $p_2=p_1\pm1$ for controlling the rotor radial position in the airgap. When the suspension force windings are introduced, the airgap flux distribution is asymmetrical. The flux which is generated by suspension force windings strengthens the airgap flux in some area and weakens on the opposite area. Then, the Maxwell-Force caused by asymmetrical of the airgap flux points to the flux strengthened area. Fig. 1 ($p_1=1$, $p_2=2$) shows that the addition of two kinds of flux linkages ψ_2 and ψ_4 results in the asymmetrical of the airgap flux distribution, and Fig. 1(a) the Maxwell-Force points to the positive *x*-direction and Fig. 1(b) points to the positive *y*-direction, respectively. By controlling the current in the suspension force windings, the rotor can suspend steadily. The torque of the self-bearing induction motor is based on Lorentz-Force, the same as conventional induction motor.

III. RADIAL SUSPENSION FORCES IN SELF-BEARING INDUCTION MOTOR

A. Lorentz-Force

The Lorentz-Force acts on a conductor which is in

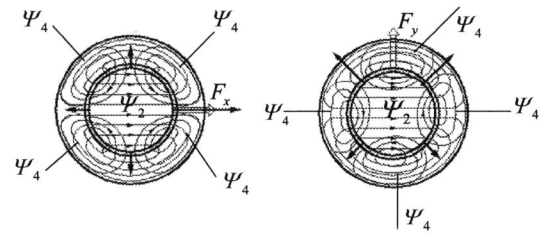

(a)　　　　(b)

Figure 1.　Principle of suspension force generation

Project supported by National Natural Science Foundation of China (50575099)

magnetic field with a current flow, and the Lorentz-Force is tangental to the rotor surface. Let l is the active length of the motor, r is rotor radius, $B(\theta)$ is airgap flux density and $J(\theta)$ is current density in suspension force windings, the Lorentz-force element which acts on the rotor surface is

$$dF_l = -J(\theta) \times B(\theta) \cdot l \cdot r \cdot d\theta \qquad (1)$$

where θ is space position angle.

By integrating (1) in the whole rotor, the Lorentz-Force ($p_2 = p_1 \pm 1$) can be expressed as

When $p_2 = p_1 - 1$:

$$F_{lx} = -F_l \cdot \cos(\theta_2 - \theta_1), F_{ly} = F_l \cdot \sin(\theta_2 - \theta_1) \qquad (2)$$

When $p_2 = p_1 + 1$:

$$F_{lx} = F_l \cdot \cos(\theta_2 - \theta_1), F_{ly} = F_l \cdot \sin(\theta_2 - \theta_1) \qquad (3)$$

where $F_l = \dfrac{\pi r l J_{max} B_{max}}{2}$;

J_{max} is amplitude of current density; B_{max} is amplitude of airgap flux density; θ_1 is initial phase angle of motor windings; θ_2 is initial phase angle of suspension force windings.

Fig. 2 shows the vector graph of Lorentz-Force based on (2), (3).

Based on the vector multiplication operation, in a synchronous rotating reference frame (3) can be expressed as follows

$$\begin{cases} F_{lx} = K_l(i_{s2d}\Psi_{1d} + i_{s2q}\Psi_{1q}) \\ F_{ly} = K_l(i_{s2q}\Psi_{1d} - i_{s2d}\Psi_{1q}) \end{cases} \qquad (4)$$

where $K_l = \dfrac{\pi p_1 N_2}{8 r N_1}$;

N_1 and N_2 are number of turns for motor windings and suspension force windings, respectively; Subscripts expression "1" and "2" correspond to motor windings and suspension force windings, respectively; Subscripts expression "s" and "r" correspond to stator and rotor, respectively; Ψ_{1d} and Ψ_{1q} are airgap flux linkages components of motor windings; i_{s2d} and i_{s2q} are current components of suspension force windings.

B. Maxwell-Force

The Maxwell-Force is produced in magnetic circuits at the boundary layers of materials with different permeability. Their directions are right angled to the rotor surface.

Let airgap flux density is B, the Maxwell-Force which acts on a rotor surface element $dA = lrd\theta$ is

$$dF_m = \frac{B^2 dA}{2\mu_0} \qquad (5)$$

where μ_0 is the vacuum permeability.

Taking no account of the magnetic saturation, the total airgap flux density produced by motor windings and suspension force windings is

$$B(\theta) = B_{1max}\cos(p_1\theta - \theta_1) + B_{2max}\cos(p_2\theta - \theta_2) \qquad (6)$$

Substituting (6) into (5) and integrating (5) in the whole rotor, the Maxwell-Force ($p_2 = p_1 \pm 1$) can be expressed as

When $p_2 = p_1 - 1$:

$$F_{mx} = F_m \cdot \cos(\theta_2 - \theta_1), F_{my} = -F_m \cdot \sin(\theta_2 - \theta_1) \qquad (7)$$

When $p_2 = p_1 + 1$:

$$F_{mx} = F_m \cdot \cos(\theta_2 - \theta_1), F_{my} = F_m \cdot \sin(\theta_2 - \theta_1) \qquad (8)$$

where $F_m = \dfrac{\pi r l B_{1max} B_{2max}}{2\mu_0} \qquad (9)$

Fig. 3 shows the vector graph of Maxwell-Force based on (7), (8).

Per pole airgap flux

$$\Phi_1 = \frac{2rlB_{1max}}{p_1}, \quad \Phi_2 = \frac{2rlB_{2max}}{p_2} \qquad (10)$$

Per phase airgap flux linkages

$$\Psi_{1m} = \Phi_1 N_1, \quad \Psi_{2m} = \Phi_2 N_2 \qquad (11)$$

The amplitude of airgap flux linkages

$$\Psi_1 = \frac{3}{2}\Psi_{1m}, \quad \Psi_2 = \frac{3}{2}\Psi_{2m} \qquad (12)$$

Neglecting the induced current in the rotor produced by suspension force windings

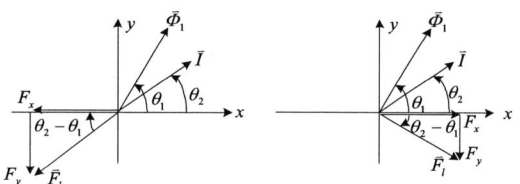

$p_2 = p_1 - 1$ $p_2 = p_1 + 1$

Figure 2. Vector graph of Lorentz-Force

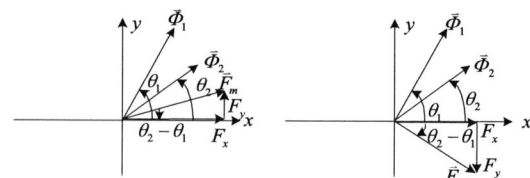

$p_2 = p_1 - 1$ $p_2 = p_1 + 1$

Figure 3. Vector graph of Maxwell-Force

$$\Psi_2 = L_{m2}i_{s2} \tag{13}$$

where L_{m2} is mutual inductance of suspension force windings.

Substituting (10) ~ (13) into (9), one can obtain

$$F_m = \frac{\pi p_1 p_2 L_{m2} \Psi_1 i_{s2}}{18\mu_0 lr N_1 N_2}$$

Based on the vector multiplication operation, in a synchronous rotating reference frame (8) can be expressed as

$$\begin{cases} F_{mx} = K_m(i_{s2d}\Psi_{1d} + i_{s2q}\Psi_{1q}) \\ F_{my} = K_m(i_{s2q}\Psi_{1d} - i_{s2d}\Psi_{1q}) \end{cases} \tag{14}$$

where $K_m = \dfrac{\pi p_1 p_2 L_{m2}}{18\mu_0 lr N_1 N_2}$.

IV. Mathematics Models of Self-Bearing Induction Motor

A. Radial Force Equation

Fig. 2 and Fig. 3 show that when $p_2=p_1+1$ the Maxwell-Force and Lorentz-Force are in the same direction and when $p_2=p_1-1$ the two forces are in the opposite direction. Therefore, $p_2=p_1+1$ will be chosen to obtain more effective radial suspension forces [5].

Then, the controllable radial forces is $F=F_m+F_l$

$$\begin{cases} F_x = K(i_{s2d}\Psi_{1d} + i_{s2q}\Psi_{1q}) \\ F_y = K(i_{s2q}\Psi_{1d} - i_{s2d}\Psi_{1q}) \end{cases} \tag{15}$$

where $K=K_m+K_l$.

When the rotor is out of the center, another radial force will exist. Only a displacement of the rotor out of the center gives an asymmetrical flux distribution and a radial force points to the direction of the displacement. This effect is known as the magnetic tensile force in the theory of the electrical motor. The radial force can be expressed as

$$\begin{cases} F_{sx} = k_s x \\ F_{sy} = k_s y \end{cases} \tag{16}$$

where $k_s = k\dfrac{\pi rlB^2}{\mu_0 \delta}$;

k_s is the force-displacement coefficient; δ is the length of the airgap; k is the attenuation gene, $k \approx 0.3$ [5].

B. Basic Equations of Rotation

When the motor is a self-bearing induction motor with a squirrel cage rotor, the magnetic field produced by suspension force windings also can induce current (with the pole pairs p_2) in the rotor and the interaction of the magnetic field and current can produce the torque too. Then, the self-bearing induction motor with a squirrel cage rotor can be described as two motors with the pole pairs p_1 and p_2, respectively. The torque of the self-bearing

motor is the sum of the two motors. Because the current in suspension force windings is small, the torque and the flux linkages produced by suspension force windings can be ignored. Therefore, the basic rotating equations of the self-bearing induction motor can be expressed as follows [6]

Flux linkages equation

$$\begin{cases} \Psi_{1d} = L_{m1}(i_{s1d} + i_{r1d}) \\ \Psi_{1q} = L_{m1}(i_{s1q} + i_{r1q}) \end{cases} \tag{17}$$

Rotor voltage equation

$$\begin{cases} U_{r1d} = R_{r1}i_{r1d} + p\Psi_{r1d} - \omega_s\Psi_{r1q} = 0 \\ U_{r1q} = R_{r1}i_{r1q} + p\Psi_{r1q} + \omega_s\Psi_{r1d} = 0 \end{cases} \tag{18}$$

where ω_s is the slip angular frequency; R_{r1} is rotor resistance of motor windings.

Torque equation

$$T_e = p_1(i_{s1q}\Psi_{1d} - i_{s1d}\Psi_{1q}) \tag{19}$$

C. Motion Equation

Let m is the mass of the rotor, J is the rotor moment of inertia, T_L is the load torque, F_{dx}, F_{dy} are the disturbance forces in x- and y-direction, respectively. The motion equation can be expressed as

$$\begin{cases} F_{dx} + F_{sx} - F_x = m\ddot{x} \\ F_{dy} + F_{sy} - F_y = m\ddot{y} \\ T_e - T_L = \dfrac{J}{p_1} \cdot \dfrac{d\omega_r}{dt} \end{cases} \tag{20}$$

where ω_r is the rotor angular frequency.

V. Design and Simulation of Control System

It can be obtained from (15) and (19) that the suspension forces and torque are coupled each other through the airgap flux linkages. The decoupling of suspension forces and torque can be realized by airgap magnetic field oriented vector controller [7]. Compared with the rotor magnetic field oriented vector controller, many limitations exist in the control strategy based on the airgap magnetic field oriented vector controller, such as great amount control computation, inherent pull-out torque and difficulty in realizing the adaptive control. In this case, a nonlinear control strategy based on rotor magnetic field oriented vector controller is proposed. The stable suspension can also be obtained because the information about the airgap flux linkages can be obtained by system identification.

When the rotor magnetic field oriented vector controller is adopted, one can obtain

181

$$\Psi_{r1d} = \Psi_{r1}, \; \Psi_{r1q} = 0 \tag{21}$$

Then the mathematics models of rotation can be expressed as

$$\begin{cases} \Psi_{r1} = \dfrac{L_{m1}}{T_{r1}p+1} i_{s1d} \\[2mm] \omega_s = \dfrac{L_{m1}}{T_{r1}\Psi_{r1}} i_{s1q} \\[2mm] T_e = p_1 \dfrac{L_{m1}}{L_{r1}} i_{s1q}\Psi_{r1} \end{cases} \tag{22}$$

where $T_{r1} = L_{r1}/R_{r1}$;

L_{r1} is self-inductance of motor windings.

Using the relationship between airgap flux linkages and rotor flux linkages, the airgap flux linkages components can be obtained

$$\begin{cases} \Psi_{1d} = (1 + \dfrac{L_{r1l}p}{R_{r1}})\Psi_{r1} \\[2mm] \Psi_{1q} = \dfrac{L_{m1}}{L_{r1}} L_{r1l} i_{s1q} \end{cases} \tag{23}$$

The mathematics models of radial suspension forces are established by (15) and (23).

Fig. 4 shows the control block diagram of self-bearing induction motor.

In order to verify the control strategy proposed in this paper, the control system is simulated using Matlab/Simulink [8]. The parameters of the prototype motor are given as follows:

Rated power P_N=1 000 W, rated speed n=6 000 r/min, mass of the rotor m=2.85 kg, rotor moment of inertia J=0.00769 kg·m^2, length of the airgap δ=0.375 mm, radial airgap of outboard bearing δ_1=0.25 mm; Motor windings: stator resistance R_{s1}=2.01 Ω, rotor resistance R_{r1}=11.48 Ω, mutual inductance L_{m1}=158.56 mH, stator leakage inductance L_{s1l}=4.54 mH, rotor leakage inductance L_{r1l}=9.22 mH, pole pairs p_1=2; Radial suspension force

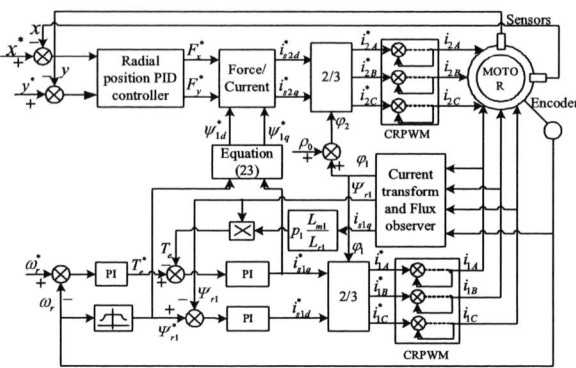

Figure 4. A control block diagram of self-bearing induction motor

windings: stator resistance R_{s2}=1.03 Ω, rotor resistance R_{r2}=0.075 Ω, mutual inductance L_{m2}=9.32 mH, pole pairs p_2=3.

Fig. 5 shows the simulation diagram of the control system for self-bearing induction motor. Fig. 6 shows the simulation model of self-bearing induction motor.

The speed command is 6 000 r/min. The load torque is 5N·m at the starting, and at the 0.3 s, the load torque turns to 15 N·m. The initial displacement in x-direction and y-direction is 150 μm and 250 μm, respectively. The radial disturbance force is 100 N in x-direction. Fig. 7 shows the simulation results of the control system.

Fig. 7(a) shows the results of speed regulation. The speed overshoot is under 1%, and the fluctuating error of speed in steady state is less than 0.5 r/min. Fig. 7(b) shows the performance of the torque. The pulsating movement of torque is less than 5%, and the startup torque is great. The results of A-phase motor windings current in Fig. 7(c) have shown that the current changes with the torque. It is obvious that the system has fine speed-regulation performance.

Fig. 7 (d) shows the A-phase current in suspension force windings. The current changes with the radial displacement. When the rotor is stable suspension in the center, the current becomes small. The track for the mass center of the suspended rotor is shown in Fig. 7(e). The rotor approaches to the center in a helix track, and suspends stably in the center. Fig. 7(f) shows the radial

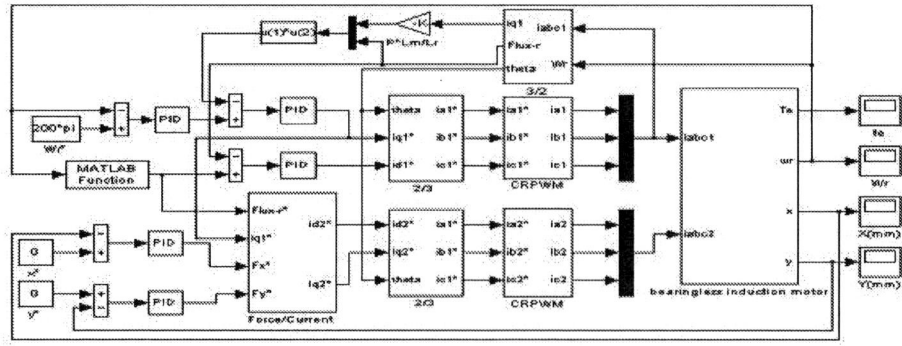

Figure 5. Simulation diagram of the control system for self-bearing induction motor

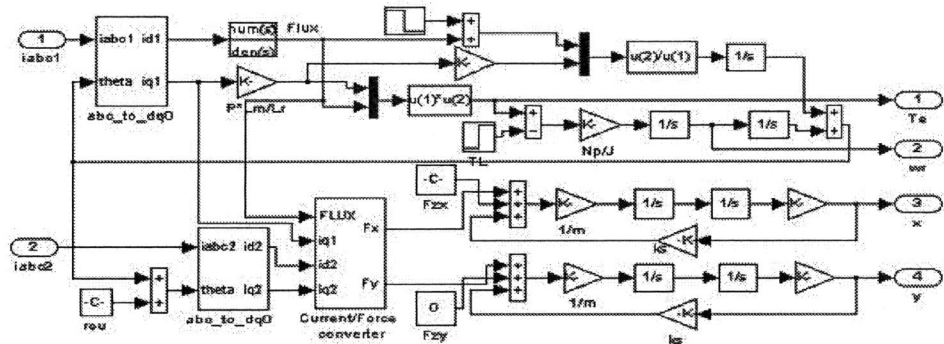

Figure 6. Simulation model of self-bearing induction motor

displacement in *x*-direction when the radial disturbance force is 100 N. The stable suspension is realized with radial displacement error less than 10 μm. It can be seen that the system has satisfactory suspension performance.

From Fig. 7(f), it can be concluded that the radial displacement is not interfered by the load torque. Fig. 7 (b) shows that the torque is not disturbed by radial interference force. Therefore, the successful decoupling of them is realized.

VI. CONCLUSIONS

In this paper the detailed mathematics models of suspension forces and torque on the self-bearing induction motor are deduced. The main problem in the closed loop control lies in the coupling of the torque and the radial suspension forces through the airgap flux linkages. The rotor magnetic field oriented vector controller is used to succeed in realizing decoupling control of this motor, and the control system is not complicated. Simulation results have shown that the control strategy is valid and the control system has satisfactory performance.

REFERENCES

[1] H. Zhu, Z. Deng, Y. Yan, S. Yuan, "Principles of bearingless motors and research status," *Micromotors* , vol. 33, no. 6, pp. 29-31, June 2000.

[2] D. Chiba, T. Power and M. A. Rahman, "Analysis of no-load characteristics of a bearingless induction motor," *IEEE Trans. Industry Application*, vol. 31, no. 1, pp. 77-83, Jan./Feb. 1995.

[3] Y. Okada, K. Dejima and T. Ohishi, "Analysis and comparison of PM synchronous motor and induction motor type magnetic bearings," *IEEE Trans. Industry Application,* vol. 31, no. 2, pp. 1047-1053, Mar./Apr. 1995.

[4] R. Schöb and J. Bichsel, "Vector control of the bearingless motor.," in *Proc. 4th Int. Symp. Magnetic Bearings*, ETH Zürich, Switzerland, pp. 327-332, Aug. 1994.

[5] R. Schöb, "Beiträge zur lagerlosen Asynchronmachine," Ph.D. dissertation, ETH Zürich, Switzerland,1993.

[6] B. Chen, *Electrical Towage Automatic Control Systems*. Beijing: China Machine Press , 2003.

[7] T. Suzuki, A. Chiba, M. A. Rahman, and T. Fukao, "An air-gap flux oriented vector controller for stable operation of bearingless induction motors," *IEEE Trans. Industry Application*, vol. 36, no. 4, pp. 1069-1076, Jul./Aug. 2000.

[8] L. Sun, *Matlab and Simulation for Control System*. Beijing: Beijing University of Technology Press, 2002.

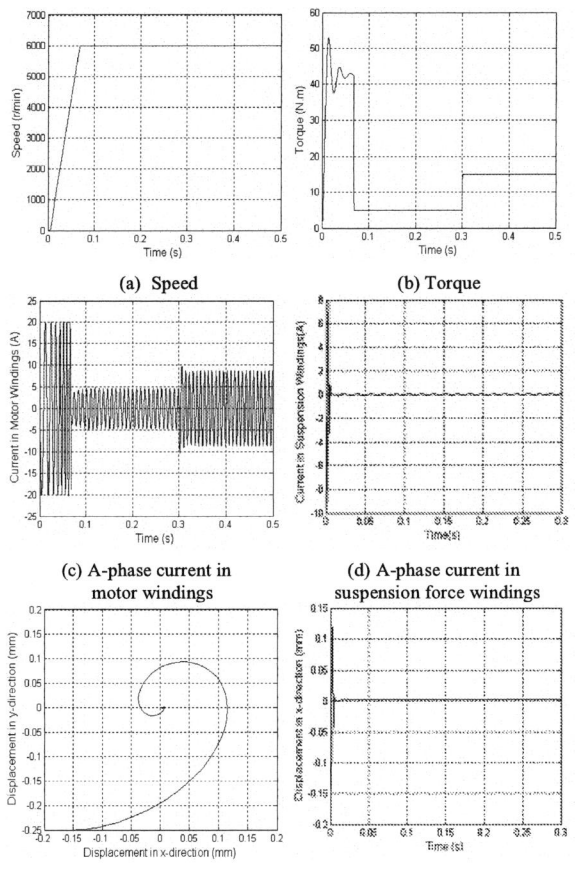

(a) Speed   (b) Torque

(c) A-phase current in motor windings   (d) A-phase current in suspension force windings

(e) Rotor suspension track   (f) Displacement in *x*-direction

Figure 7. Simulation results of the control system

2006 5th International Power Electronics and Motion Control Conference

Field Oriented Control of Linear Induction Motor Considering Attraction Force & End-Effects

Jianqiang Liu, Fei Lin, Zhongping Yang, Trillion Q. Zheng

School of Electrical Engineering, Beijing Jiaotong University, Beijing 100044, China

Abstract—This paper presents a field oriented control scheme of the linear induction motor (LIM) in order to achieve high performance of the machine. The dynamic model of LIM taking into account end-effects is derived. Then the modified mechanical load equation considering the effect of the attraction force on the performance of linear induction motor is analyzed. Finally the control strategy based on field oriented control is presented. The simulation results are presented to show the control system has the quick speed response and the robust thrust characteristics.

Keywords-linear induction motor (LIM); end effects; attraction force; field oriented control

I. INTRODUCTION

Linear induction motor (LIM) has been developed as electrical drives for a new urban transit system with cruising speed 40-100km/h, in some countries. Now in china this new urban transit system will be built in Guangzhou and Beijing. This system has the advantage that the construction costs of the subway is reduced as the cross-sectional area of tunnel become smaller because of lower level of the floor of the vehicle, compared with the conventional railcar which has rotating motors. Furthermore, the linear motor subway has the following special advantages because of the absence of adhesive drive: the length of track is shorted because the vehicles can climb steep gradients and turn sharp curves, and in addition the ride quality is improved and noise is lowered [1].

In order to obtain high performance of linear motor vehicle, it is important to develop vector control for LIM. The basic idea behind the field oriented control is uncoupling the flux and the torque of an induction motor in order to achieve the torque response similar to that of a separately excited direct current machine. Similar is in case of linear induction motor except the torque has been replaced by thrust force (propulsion force) and the rotational speed by linear speed. Thus, the field oriented control, can be adopted to decouple the dynamics of the thrust force and the linor (secondary) flux amplitude of the LIM.

The aim of the field oriented control is to maintain constant the d-axis linor flux and making null the q-axis linor flux. The dynamic model of the LIM is analyzed by using the $d-q$ model of the equivalent electrical circuit with end effects included. A speed inverse function factor is determined to express the effects that the LIM speed causes in the magnetization branch of the equivalent electrical circuit [2-4].

The linear induction motor comes into the category of special electromechanical device as it is associated with special phenomenon due to its extra-ordinary structure compared to conventional rotary induction motor. As a unique and distinct feature, linear induction motor produces attraction force between its primary and linor members due to magnetic discontinuities in the magnetic circuit of LIM.

In this paper we first give the dynamic model of the linear induction motor, taking into account the end effects. Next, we discuss the effect of attraction force between the primary and linor members of the LIM by considering it as a load, aiding to the virtual load of the linear induction motor. Lastly, simulation results are presented to verify the validity of the proposed field oriented control strategy.

II. DYNAMIC MODEL OF LINEAR INDUCTION MOTOR

Fig.1 shows a conceptual construction of a LIM used in this research. The short primary is moveable and the infinite linor is fixed. In a LIM, the primary voltage or current excitation produces a magnetic field, which moves from the front to the back of the primary. This magnetic field induces currents in the conducting layer on the surface of the linor, which produces a second traveling magnetic field. The interaction of these two magnetic fields produces a force, which tends to move the primary along the surface of the linor. When the primary moves, the linor is continuously replaced by a new linoric region. This new linoric region tends to oppose to the sudden increase in the penetration of the magnetization flux allowing a gradual accumulation of the magnetization field density in the air gap. The arising of a new linoric region and its influence in the magnetic field modifies the LIM performance when compared to the traditional induction motor.

The dynamic model of the linear induction motor is analyzed by using the $d-q$ model of the equivalent electrical circuit with end effects included [2]. The q-axis equivalent circuit of the LIM is identical to the q-axis equivalent circuit of the induction motor (RIM), i.e. the parameters do not vary with the end effects. However, the d-axis entry linoric currents affect the air gap flux by decreasing ψ_{dr}^e. Therefore, the d-axis equivalent circuit of

1-4244-0448-7/06/$25.00 ©2006 IEEE

the RIM cannot be used in the LIM analysis when the end effects are considered.

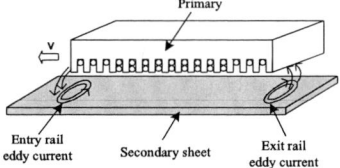

Fig.1. Linor currents at the entry and exit ends for a given velocity.

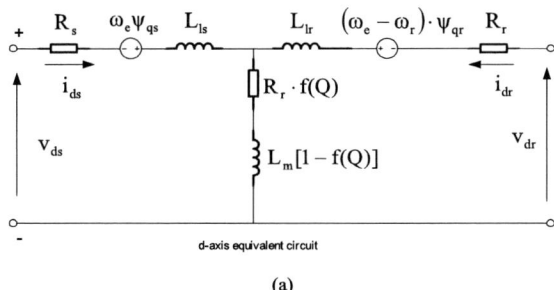

(a)

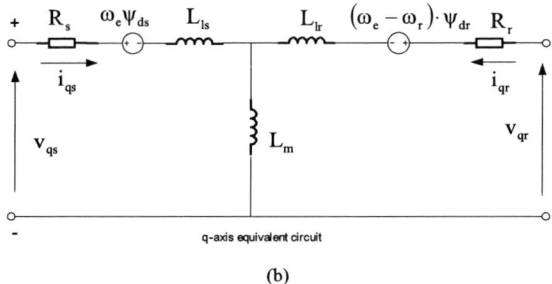

(b)

Fig.2. The LIM equivalent circuits taking into account the end effects.

Fig.2 (a) shows the d-axis equivalent which magnetization branch is different from the traditional induction motor. In Fig.2 (b) the equivalent circuit is the same that the traditional induction motor. From the dq equivalent circuit of the LIM (Fig.2), the primary and linor voltage equations in a synchronous reference system (superscript "e") aligned with the linor flux are given by:

$$v_{ds}^{e} = R_s i_{ds}^{e} + R_r f(Q)(i_{ds}^{e} + i_{dr}^{e}) + p\psi_{ds}^{e} - \omega_e \psi_{qs}^{e} \quad (1)$$

$$v_{qs}^{e} = R_s i_{qs}^{e} + p\psi_{qs}^{e} + \omega_e \psi_{ds}^{e} \quad (2)$$

$$v_{dr}^{e} = R_r i_{dr}^{e} + R_r f(Q)(i_{ds}^{e} + i_{dr}^{e}) + p\psi_{dr}^{e} - (\omega_e - \omega_r)\psi_{qr}^{e} = 0 \quad (3)$$

$$v_{qr}^{e} = R_r i_{qr}^{e} + (\omega_e - \omega_r)\psi_{dr}^{e} + p\psi_{qr}^{e} = 0 \quad (4)$$

The linkage fluxes are given by the following equations:

$$\psi_{ds}^{e} = L_{ls} i_{ds}^{e} + L_m(1 - f(Q))(i_{ds}^{e} + i_{dr}^{e}) \quad (5)$$

$$\psi_{qs}^{e} = L_{ls} i_{qs}^{e} + L_m(i_{qs}^{e} + i_{qr}^{e}) \quad (6)$$

$$\psi_{dr}^{e} = L_{lr} i_{dr}^{e} + L_m(1 - f(Q))(i_{ds}^{e} + i_{dr}^{e}) \quad (7)$$

$$\psi_{qr}^{e} = L_{lr} i_{qr}^{e} + L_m(i_{qs}^{e} + i_{qr}^{e}) \quad (8)$$

$$Q = \frac{DR_r}{L_r v} \quad (9)$$

$$f(Q) = \frac{1 - e^{-Q}}{Q} \quad (10)$$

$$L_m' = L_m(1 - f(Q)) \quad (11)$$

The thrust force is expressed as:

$$F_e = \frac{3\pi}{2\tau} \cdot \frac{n_p}{2} \left(\psi_{ds}^{e} i_{qs}^{e} - \psi_{qs}^{e} i_{ds}^{e} \right) \quad (12)$$

Where $v_s = v_{ds} + jv_{qs}$ is the primary voltage, $v_r = v_{dr} + jv_{qr}$ is the linor voltage, $L_s = L_{ls} + L_m$ is the primary inductance, $L_r = L_{lr} + L_m$ is the linor inductance, L_{ls} is the primary leakage inductance, L_{lr} is the linor leakage inductance, and L_m is the magnetizing inductance. Subscripts "s" and "r" denote the primary and linor values, respectively. v_{ds}, v_{qs} are the primary voltages; v_{dr}, v_{qr} are the linor voltages; i_{ds}, i_{qs} are the primary electrical currents; i_{dr}, i_{qr} are the linor electrical currents; $\psi_s = \psi_{ds} + j\psi_{qs}$, $\psi_r = \psi_{dr} + j\psi_{qr}$ are the primary and linor linkage flux; R_s, R_r are the primary and linor resistances; ω_r is the linor speed; n_p is the pole number; p is the mathematical derivative operator; v is the linear speed in m/s; τ is the pole pitch; D is the primary length in meters; Q is a factor related to the primary length, which quantifies the end effects as a function of the speed (v).

III. EFFECT OF ATTRACTION FORCE

The attraction force between the primary and secondary members of the linear induction motor provides ability to negotiate steep and sharp turns, reduction in braking time (transportation applications), stability on track thereby minimizing the problem of derailment etc. but at the same time imposes control problem as well as brake the motion [3]. Braking of motion increases the response time during acceleration, as it is acting vertically and virtually aiding to the load of the drive. The analysis of LIM has been done by considering the attraction force as an additional load acting vertically and adding the load thrust force to be transported. In this case the modified mechanical load equation becomes:

$$F_{em} = (M\frac{dv}{dt} + Bv)k + F_L \qquad (13)$$

Where F_L is the load thrust force, B is the viscous friction coefficient of the LIM, M is the mass of the LIM, k is the compensated coefficient considering the effect of the attraction force on the performance of linear induction motor.

Hence, the attraction force causes a virtual change in the load on the drive and the performance of the LIM drive is affected accordingly.

IV. FIELD ORIENTED CONTROL FOR LINEAR INDUCTION MOTOR

The aim of the field oriented control is maintaining constant the d-axis linor flux and making null the q-axis linor flux. The field oriented control for LIM can be analyzed in the same way that the traditional induction motor. The field oriented control utilizes rotor flux and torque feedback, obtained through direct sensing or an estimator. The rotor flux angle permits the decoupling of primary current into its force-producing component and its rotor-flux component [3-5].

In order to determine the slip angular speed, the linor q-axis flux must be considered zero (ψ_{dr}^e). Then, the slip frequency can be determined from (4) and (8) giving the following equation:

$$\omega_{sl} = -\frac{R_r i_{qr}^e}{\psi_{dr}^e} = \frac{1}{T_r}\frac{L_m}{\psi_{dr}^e}i_{qs}^e \qquad (14)$$

Where $T_r = L_r / R_r$ is the linor time constant.

The slip frequency equation ω_{sl} is the same that the conventional induction motor. The main difference between a LIM and a RIM is the transient characteristic of ψ_{dr}^e. We obtain from (3) and (7) that

$$\psi_{dr}^e = \frac{[L_m - L_r f(Q)]R_r}{[L_r - L_m f(Q)]p + R_r(1 + f(Q))}i_{ds}^e \qquad (15)$$

Where p is the mathematical derivative operator.

The primary angular frequency can be determined by aiding the linor angular speed with the slip frequency. This will give as:

$$\omega_e = \omega_r + \omega_{sl} \qquad (16)$$

The transformation of linear speed of the LIM to an angular speed is given by:

$$\omega_r = \frac{\pi}{\tau}v \qquad (17)$$

Where ω_r represents the linor angular speed and τ is the pole pitch.

The linor flux position angle related to the primary can be determined by:

$$\theta_e = \int\omega_r dt + \int\omega_{sl}dt \qquad (18)$$

The linor flux instantaneous position angle has to be used in the transformation of the stationary reference systems to synchronous and vice versa, as can be seen in the block diagram of the field oriented control in Fig.3.

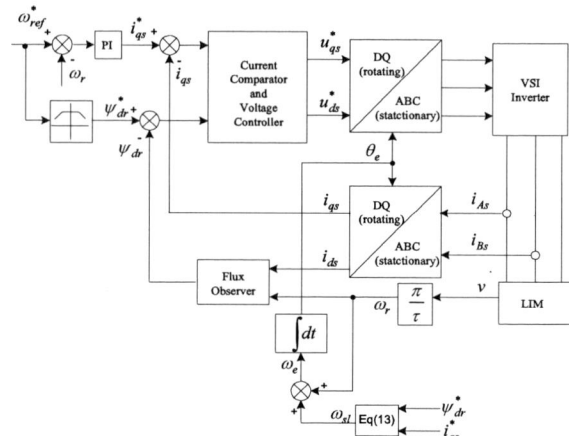

Fig.3. Block diagram of the proposed field oriented control for LIM

In the speed control loop of field oriented control for LIM, commonly, there is saturation-type nonlinearity, i.e. windup phenomenon, which causes large overshoots, slow setting time and even instability in the speed response. To overcome the windup, we adopt an antiwindup technique [6] in the speed PI controller as shown in Fig.4.

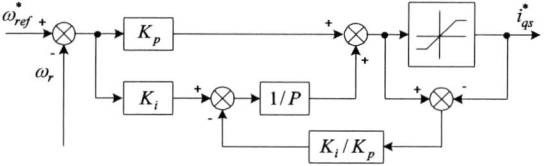

Fig.4. Block diagram of speed controller with antiwindup.

V. SIMULATION RESULTS

In this paper, the simulation of the presented field oriented control for LIM is performed in MATLAB/SIMULINK. The linear induction motor parameters used in the dynamic simulation are given in table Ⅰ. For all the simulations the linor flux is 0.4 Wb and k (the compensated coefficient considering the effect of the attraction force) is 1.2.

TABLE I.
LINEAR INDUCTION MOTOR PARAMETERS

Parameters	Values(Units)
Primary resisitance- R_s	1.25 Ω
Linor resistance- R_r	2.7 Ω
Primary inductance- L_s	40.1 mH
Linor inductance- L_r	32.6 mH
Magnetizing inductance- L_m	33.1 mH

Fig.5 and Fig.6 shows speed, thrust force and primary current respectively. The references used in simulation (case1): Load force =0N [t=0s to 5.0s]; Speed =1.05m/s [t=0s to 5.0s].

Fig.7 and Fig.8 shows speed, thrust force and primary current respectively. The references used in simulation (case2): Load force =100N [t=2.0s to 5.0s]; Speed =1.05m/s [t=0s to 5.0s].

Fig.9 and Fig.10 shows speed, thrust force and primary current respectively. The references used in simulation (case3): Load force =150N [t=2.5s to 4.0s]; Speed =1.05m/s [t=0s to 1.5s], Speed =2.1m/s [t=1.5s to 5.0s].

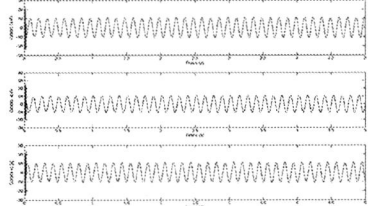

Fig.5. Simulations: Primary current for case1

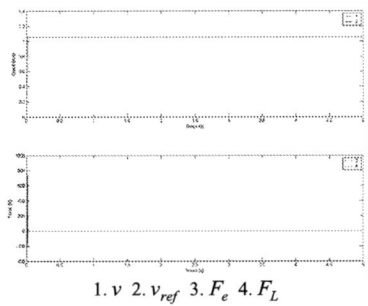

1. v 2. v_{ref} 3. F_e 4. F_L

Fig.6. Simulations: Speed and thrust force for case1

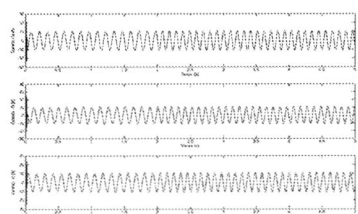

Fig.7. Simulations: Primary current for case2

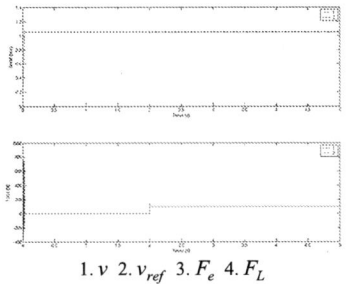

1. v 2. v_{ref} 3. F_e 4. F_L

Fig.8. Simulations: Speed and thrust force for case2

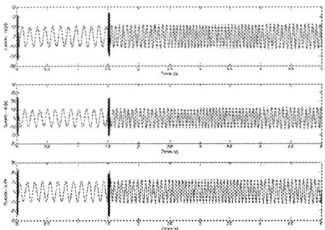

Fig.9. Simulations: Primary current for case3

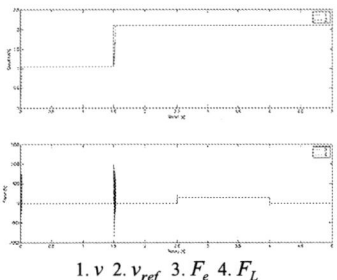

1. v 2. v_{ref} 3. F_e 4. F_L

Fig.10. Simulations: Speed and thrust force for case3

All the results from simulation present response very close to that expected from a field oriented control system. The reference forces were reached with high performance and fast response. The speed was reached satisfactorily without oscillating when a load force is applied. From simulation results we can see the validity of the proposed control strategy. The method used has been satisfactory and will be applied in real situation using a DSP system and a power converter.

VI. CONCLUSION

In this paper, LIM has been discussed with focus on its generalized mathematical modeling along with the end-effects and the effect of attraction force and besides all, the methodology for linor flux orientation for linear induction motor for high performance control. The effect of attraction force is analyzed by giving the compensated coefficient. Dynamic model and control strategies for LIM have been presented. Simulation results are given to show that the control system has the fast and precise speed response and the robust thrust characteristics.

REFERENCES

[1] Higuchi T, Naonak S, Ando M. "On the design of high-efficiency linear induction motors for linear metro," Electrical Engineering in Japan, vol. 137, no. 2, 2001, pp.36-43.

[2] K. Nam, J. H Sung, "A new approach to vector control for linear induction motor considering end effects," IEEE IAS annual meeting, 3-7 Oct, in Phoenix, Arizona, 1999, pp.2284-2289.

[3] Rathore, A.K., Mahendra, S.N., "Simulation of secondary flux oriented control of linear induction motor considering attraction force & transverse edge effect," Power Electronics Congress, 2004. CIEP 2004. 9th IEEE International, 17-22 OCT. 2004, pp.158-163.

[4] Ezio Fernandes da Silva, Charles C. dos Santos, "Field oriented control of linear induction motor taking into account end-effects,"

187

The 8th IEEE International Workshop on 25-28 March 2004, pp.689 – 694.

[5] Da Silva, E.F., dos Santos, E.B., Machado, P.C.M, de Oliveria, M.A.A., "Vector control for linear induction motor, " Industrial

Technology, 2003 IEEE International Conference on Volume1, 10-12 Dec.2003, pp.518-523.

[6] Y. Peng, D. Vrancic and R. Hanus, "Anti-windup, brumpless, and conditioned transfer techniques for pid controllers," IEEE Control Systems Magazine, vol.16, no.4, Aug.1996, pp.48-57.

2006 5th International Power Electronics and Motion Control Conference

Series Resonant High Frequency Link Sine-wave Inverter System Modeling using Sampled Data

Jin Xiaoyi, Dong Wei, Sun Xiaofeng and Wu Weiyang
College of Electrical Engineering, Yanshan University, Qinhuangdao China
jinyi_3@tom.com

Abstract—Based on the detailed analysis of the series resonant high frequency link sine-wave inverter operational principle, a system modeling methodology is presented using the sampled data. For this special modeling application, a black-box approach is described based on the data which are gotten from simulation or experimental results. And this approach aims at generating discrete-time small-signal linear equivalent model which describes the inverter as a linear time-invariant system, so the knowledge of the linear system can be used to the identified model for constructed feedback controller. The identified model is verified by the simulation and experimental results.

Keywords-system modeling; sampled data; series resonant; high frequency link; black-box

I.   INTRODUCTION

Resonant converter topologies are being widely used in power processing systems because of their soft-switching characteristics at high frequencies [1]. The advantages of high frequency operation include smaller size and lighter weight for the passive components. The main problem of operating at high frequency is ensuring the reliability of the feedback control system. For such a resonant link power converter system, from the view of analytical and designing, the design of LC tanks [2] and the control to maintain the stable oscillation of the tanks during the transient response are the most significant points.

For a nonlinear switching system, linear resistors, inductors, capacitors, nonlinear magnetic components and semiconductor switches are mixed together, and it is hard to design stable feedback controllers using exact mathematical model, as there are too many complex nonlinear differential equations to be solved.

Therefore, linearization and averaging become the main tools to realize and predigest analysis of nonlinear switching system. But this also causes some issues which concern absence of high frequency information and switching dynamic characteristic. Using small signal model, the switching frequency or sampling frequency information can be contained in analysis. Also from this small signal model, important performances such as loop gain, output impedance and stabilization margin are calcu-

This work is supported by National Natural Science Foundation of China (No. 50237020, 50407012).

lated calculated and can be easily measured and verified by experiments.

Based on the probability switching function, state space averaging is a commonly used modeling method for small-signal modeling of switching systems [3, 4]. And this averaging technique is based on an assumption that the natural frequency of the inverter power stage is well below the switching frequency. However, the ZCS series resonant inerter has the same switching frequency and inherent resonant frequency and their interaction determines the dynamic response mainly. By averaging, the interaction between switching frequency and resonant frequency is eliminated.

Discrete-time or sampled-data modeling method is anther systematic modeling way.

Utilizing the sampled-data either through a time-domain simulation or a hardware measurement, the system identification of a series resonant high frequency link sine-wave inverter is presented. Using a black-box approach, a model-free identification process is described. The paper is organized as follow: in section II, the detailed operational principle analysis of the series resonant inverter is presented to construct the modeling base; the modeling method and arithmetic is described in section III as well as the inverter system equivalent model identification utilizing this black-box technique; in section IV, a comparison between the resulting model and experimental results is performed, also the explanation of difference is given; a conclusion is drawn in section V.

II.   OPERATION PRINCIPLE

The interested topology of series resonant high frequency link sine-wave inverter is shown in Fig.1.

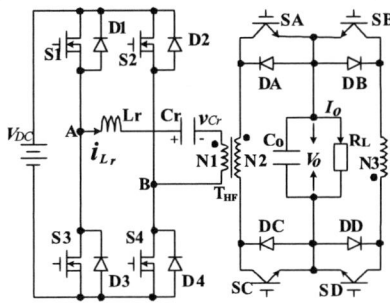

Figure 1.   Series resonant high frequency link sine-wave inverter

As detailed analyzed in referenced paper [5], six controllable switching modes are combined to realize the stable sine-wave output. Table I shows the switching sequence according to the corresponding resonant mode.

TABLE I.
SWITCHING SEQUENCE

Consequence		PM	FRM	PFM	PRM
S1	PHP	on	off	off	off
S2	NHP	on	on	on	off
S3	NHP	on	off	off	off
S4	PHP	on	on	on	off
SA	PHP	on	on	off	off
	NHP	on	on	off	off
SB	PHP	on	on	off	off
	NHP	on	on	off	off
SC	PHP	off	off	off	off
	NHP	off	off	on	on
SD	PHP	off	off	on	on
	NHP	off	off	off	off
Quadrant		1st quadrant		2nd quadrant	

There are four essential resonant modes in table I which are used to ensure the inverter system working properly. And anther two modes are used to improve the output waveform quantity as mentioned in [5]. Besides these six controllable resonant modes, there are three non-controllable resonant modes during the inverter operation. Using the equivalent circuit and corresponding waveform, the working principle and condition are presented.

Zero-Free Resonant Mode (ZFRM) is a special non-controllable FRM. Fig.2 (a) shows the equivalent circuit and corresponding waveform. Switches (S1, D2) and (S3, D4) pairs, or (S4, D3) and (S2, D1) pairs, or (S1, D2) and (S2, D1) pairs, or (S4, D3) and (S3, D4) pairs are turned on and off alternatively in synchronization with the resonant current zero crossing points. But for the reason of v_{Crmax} is less than u_o ($u_o=v_o/N$), the oscillation of series resonant tank does not work, so i_{Lr} maintains zero until the next controllable resonant mode.

Like ZFRM, Zero-Regenerating Mode (ZRM) is a special non-controllable RM. Four switches (S1-S4) are turned off, as $v_{Crmax}<v_{DC}+u_o$, i_{Lr} also maintains zero. Fig.2 (b) shows the same as Fig.2 (a).

As two special modes described above in 1st quadrant, there is a non-controllable resonant mode in 2nd quadrant too. Zero-Power Regenerating Mode (ZPRM) is a special PRM. Four switches (S1-S4) are turned off, SC and SD are turned on and off alternatively in synchronization with the inherent resonant frequency. For $u_o+v_{Crmax}<v_{DC}$, and the oscillation condition does not satisfied, i_{Lr} also maintains zero, Fig.2 (c) shows the equivalent circuit and corresponding waveform.

Note that these three special modes are not desirable during the inverter operation for their non-controllable.

And they appear nearby the output voltage crossing point mainly, but randomly. For $i_{Lr}=0$ during these resonant modes, the inverter is departed into two parts which are series resonant high frequency inverter and cycloconverter, and the HF transformer does not carry the energy flow. In 1st quadrant, the filter capacitor offers output power; and in 2nd quadrant, the capacitor absorbs the regenerating energy from load.

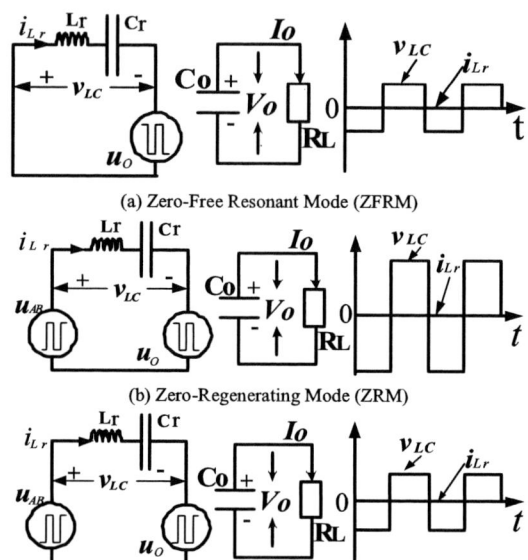

(a) Zero-Free Resonant Mode (ZFRM)

(b) Zero-Regenerating Mode (ZRM)

(c) Zero-Power Regenerating Mode (ZPRM)
Figure 2. Equivalent circuit and corresponding waveform

As mentioned above, i_{Lr} maintains zero until the next controllable resonant mode, but these non-controllable modes will influence the initial condition and movement trajectory of next controllable mode.

III. SYSTEM MODELING

A. Identification Basic Arithmetic

Considering the above three non-controllable modes and six controllable modes which are described in [5], there are so many complex differential equations based on the operational mechanism. In order to set a useful identification method to analyze and design the inverter operation principle and close-loop feed-back control, a black-box system analyze arithmetic is advanced [6]. Using the powerful simulation tools such as Pspice and Matlab, the sampled data can be gotten easily with a Zero-Order Holder (ZOH) from the time-domain waveforms.

In order to discuss canonical forms for highly non-linear MIMO systems, it is first necessary to define the concept of observability indices. Beginning with an assumed discrete-time state-space model:

$$\begin{cases} x(k+1) = A_o x(k) + B_o u(k) \\ y(k) = C_o x(k) + D_o u(k) \end{cases} \quad (1)$$

For a given m inputs, p outputs and n orders MIMO system, $(n \times n)$ state matrix A, full column rank $(n \times m)$ input matrix B and full row rank $(p \times n)$ output matrix C describing an observable system. The observability matrix Qo has $(np \times n)$. Since, by assumption the system is observable; there must be n linearly independent lows in Qo. In each case, a nonsingular $n \times n$ transformation matrix T may be formed and used to derive the corresponding observable canonical forms.

Define the observability index as follow:

$$Q_k = \left[C^T \vdots (CA)^T \vdots \cdots \cdots \vdots \left(CA^{k-1}\right)^T \right]^T \qquad (2)$$

Now increase k from 1, if $k=v$ and rank$Q_k=n$, then the smallest integer v, is the observability index of given model.

Define the observability indexes set $\{v_i\}$ $(1 \le i \le p)$ as follow:

$$Q_v = \left[c_1^T \vdots c_2^T \cdots c_p^T \vdots (c_1 A)^T \vdots (c_2 A)^T \cdots (c_p A)^T \right.$$

$$\left. \cdots \cdots (c_1 A^{v-1})^T \vdots (c_2 A^{v-1})^T \cdots (c_p A^{v-1})^T \right]^T \qquad (3)$$

Represent Q_v as above, and search the n linear independent rows from Q_v, then rearrange n rows as:

$$\left[c_1^T \vdots (c_1 A)^T \cdots (c_1 A^{v_1-1})^T \vdots c_2^T \vdots (c_2 A)^T \cdots (c_2 A^{v_2-1})^T \right.$$

$$\left. \cdots \cdots c_p^T \vdots (c_p A)^T \cdots (c_p A^{v_p-1})^T \right]^T \qquad (4)$$

Set $\{v_i\} = \{v_1, v_2 ... v_p\}$ is the observability indexes set, and has characteristics as:

$$v_1 + v_2 + \cdots + v_i + \cdots + v_p = n \qquad (5)$$

$$v = max\{v_1, v_2, \cdots v_i, \cdots v_p\} \qquad (6)$$

For a given system, set $\{v_i\}$ has the total number of combinations as

$$Num = \binom{n-1}{p-1} = \frac{(n-1)!}{(p-1)!(n-p)!} \qquad (7)$$

For example, a 3 inputs 3 outputs 6 orders system has 10 indexes set combinations totally as:

$$\begin{pmatrix} v_1 \\ v_2 \\ v_3 \end{pmatrix} = \begin{pmatrix} 4 & 3 & 3 & 2 & 2 & 2 & 1 & 1 & 1 & 1 \\ 1 & 2 & 1 & 3 & 2 & 1 & 4 & 3 & 2 & 1 \\ 1 & 1 & 2 & 1 & 2 & 3 & 1 & 2 & 3 & 4 \end{pmatrix} \qquad (8)$$

Utilizing the crate diagram which is simply a graphical method of visualizing the selection of linearly independent

rows from the given observability matrix, choose $\{v_i\}=\{2,3,1\}$ for following analysis.

{2,3,1}			
c1	c2	c3	
1	1	1	A^0
1	1	0	A^1
0	1		A^2
	0		A^3

Figure 3. Crate diagram for {2,3,1}

Among the 18 $(n*p)$ rows in Qo, the 6 (n) linearly independent elements are the rows (search the crate by row): c_1, c_2, c_3, c_1A, c_2A and c_2A^2. For further analysis and system identification, some selector vectors and matrices are defined from the crate diagram.

By omitting the first row and selecting the non-black elements from the crate, the vector v_a in row-wise:

$$v_a = \begin{bmatrix} 1 & 1 & 0 & 0 & 1 & 0 \end{bmatrix} \qquad (9)$$

Denoted v_b has the binary inverse components of v_a :

$$v_b = \begin{bmatrix} 0 & 0 & 1 & 1 & 0 & 1 \end{bmatrix} \qquad (10)$$

Considering the black elements to be zeros, and including the first row:

$$v_{1a} = \begin{bmatrix} 1 & 1 & 1 & 1 & 1 & 0 & 0 & 1 & 0 & 0 & 0 & 0 \end{bmatrix} \qquad (11)$$

Also including the first row, but now taking the black elements to be unit valued, and taking the binary complement, resulting in:

$$v_{1b} = \begin{bmatrix} 0 & 0 & 0 & 0 & 0 & 1 & 1 & 0 & 0 & 0 & 1 & 0 \end{bmatrix} \qquad (12)$$

Once the observability indexes set $\{v_i\}$ is decided, the above selector vectors are determined uniquely. And the "selector" matrices which are used in obtaining the observable form identification are derived from the associated selector vectors, shown as:

$$S_a = \begin{bmatrix} 1 & 0 & 0 & 0 & 0 & 0 \\ 0 & 1 & 0 & 0 & 0 & 0 \\ 0 & 0 & 0 & 0 & 1 & 0 \end{bmatrix}, \ S_b = \begin{bmatrix} 0 & 0 & 1 & 0 & 0 & 0 \\ 0 & 0 & 0 & 1 & 0 & 0 \\ 0 & 0 & 0 & 0 & 0 & 1 \end{bmatrix}$$

$$S_{1a}^T = \begin{bmatrix} 1 & 0 & 0 & 0 & 0 & 0 & 0 & 0 & 0 & 0 & 0 & 0 \\ 0 & 1 & 0 & 0 & 0 & 0 & 0 & 0 & 0 & 0 & 0 & 0 \\ 0 & 0 & 1 & 0 & 0 & 0 & 0 & 0 & 0 & 0 & 0 & 0 \\ 0 & 0 & 0 & 1 & 0 & 0 & 0 & 0 & 0 & 0 & 0 & 0 \\ 0 & 0 & 0 & 0 & 1 & 0 & 0 & 0 & 0 & 0 & 0 & 0 \\ 0 & 0 & 0 & 0 & 0 & 0 & 0 & 1 & 0 & 0 & 0 & 0 \end{bmatrix}$$

$$S_{lb}^T = \begin{bmatrix} 0 & 0 & 0 & 0 & 0 & 1 & 0 & 0 & 0 & 0 & 0 & 0 \\ 0 & 0 & 0 & 0 & 0 & 0 & 1 & 0 & 0 & 0 & 0 & 0 \\ 0 & 0 & 0 & 0 & 0 & 0 & 0 & 0 & 0 & 0 & 1 & 0 \end{bmatrix} \quad (13)$$

In the subsequent discussion, a state space observable form $R_o = \{A_o, B_o, C_o, D_o\}$ will be structured based on the above selector matrixes and sampled data.

For a system represented as (1), $v=max\ (v_i)$, system can rewrite as:

$$\begin{bmatrix} y(k) \\ y(k+1) \\ \vdots \\ y(k+v) \end{bmatrix} = \begin{bmatrix} C_o \\ C_o A_o \\ \vdots \\ C_o A_o^v \end{bmatrix} \cdot x(k)$$

$$+ \begin{bmatrix} D_o & \cdots & 0 & 0 \\ C_o B_o & \cdots & 0 & 0 \\ \vdots & \ddots & \vdots & \vdots \\ C_o A_o^{v-1} B_o & \cdots & C_o B_o & D_o \end{bmatrix} \begin{bmatrix} u(k) \\ u(k+1) \\ \vdots \\ u(k+v) \end{bmatrix} \quad (14)$$

Rewrite (14) as:

$$y_k = Q_{oo} \cdot x(k) + H \cdot u_k \quad (15)$$

Where y_k and u_k are $(v+1)p$ and $(v+1)m$ dimensional columns containing output and input vectors. In order to get the relationship between output and input, the state variables $x(k)$ must be eliminated from (15).

Premultiplying (15) by the selector matrices S_{la}^T and S_{lb}^T which are defined in (13):

$$y_{1k} = x(k) + H_1 \cdot u_k \quad y_{2k} = A_r \cdot x(k) + H_2 \cdot u_k \quad (16)$$

here

$$y_{1k} = S_{la}^T \cdot y_k \qquad y_{2k} = S_{lb}^T \cdot y_k \quad (17)$$

$$H_1 = S_{la}^T \cdot H \qquad H_2 = S_{lb}^T \cdot H \quad (18)$$

Eliminating $x(k)$ from (16):

$$y_{2k} = \begin{bmatrix} (H_2 - A_r \cdot H_1) & A_r \end{bmatrix} \cdot \begin{bmatrix} u_k \\ y_{1k} \end{bmatrix} \quad (19)$$

For a concise form, (19) can be written as:

$$y_{2k} = \begin{bmatrix} N_r & A_r \end{bmatrix} \cdot z_k \quad (20)$$

Since the relationship between output and input has been derived, the next step is solving N_r and A_r from (20). Link the vectors y_{2k} and z_k corresponding to samples $k=0$, $1...$, q into $(p \times q)$ and $(h \times q)$ matrices Y_2 and Z respectively, (where it is assumed that $h \leq q$, and $h=(v+1)m+n$) yielding:

$$Y_2 = \begin{bmatrix} N_r & A_r \end{bmatrix} \cdot Z \quad (21)$$

where

$$Z = \begin{bmatrix} U \\ Y_1 \end{bmatrix} \begin{matrix} \}(v+1)m \\ \}n \end{matrix} \quad (22)$$

If and only if *rank Z=h*, the observability indexes set $\{v_i\}$ is admissible. There are too many Z satisfying the condition. Chosen a different set $\{v_i\}$ could lead to a better conditioned Z. Solving (21), then get $[N_r\ A_r]$ which contain the parameter information for $R_o=\{A_o, B_o, C_o, D_o\}$.

$$\begin{bmatrix} N_r & A_r \end{bmatrix} = Y_2 Z^T (ZZ^T)^{-1} \quad (23)$$

which reduces to $Y_2 Z^{-1}$ if the matrix Z is square.

B. System identification

As mentioned in part II, the parameters of series resonant high frequency link sine-wave inverter are shown here (though they are not necessary during identification):

Input DC source: $V_{DC} = 48 \pm 5V$;

Switching frequency: $f_s = 50KHz$;

Output voltage: $v_o = 220V / 50Hz$;

Power capability: $P_{cap} = 1KW$;

Holistic efficiency: $\eta \geq 88\%$;

Resonant inductor: $L_r = 6.5\mu H$;

Resonant capacitor: $C_r = 1.5\mu F$;

Turn ratio: $N=N_2 /N_1 =N_3 /N_1 =8;$

As the identification procedure has been deduced, the steps should be followed:

Step 1: select reasonable output and input variables. Choose the variation of input voltage δv_{DC}, the variation of output current δi_o and the switching signal α ($\alpha=-1$, 0 or 1) as input variables and choose the input current i_{DC}, the maximum of resonant current i_{Lr} and the output voltage v_o as output variables.

Step 2: inject a small range of elaborate input perturbations. The perturbations should ensure that the sampled data satisfy *rank Z =h*.

Step 3: determine a nominal range of system order with the restriction that the inputs are "persistently exciting". For a black-box system identification method, the order of the system is unknown. Since the topology is given in Fig.1, the order of inverter system $n=3$.

Step 4: sample the corresponding data to construct (21).

Step 5: calculate an observable form state-space model $R_o=\{A_o, B_o, C_o, D_o\}$ from (23).

Using the resonant modes which are included in Table I, also covering the non-controllable resonant modes which are presented in part II, a small signal state-space linear model is constructed utilizing the sampled data and the above arithmetic.

$$A_o = \begin{bmatrix} 3.5226 & -1.3484 & -0.6214 \\ 3.3840 & -1.9715 & 0.3764 \\ 0.1543 & -0.1977 & 1.1037 \end{bmatrix}$$

$$B_o = 10^3 * \begin{bmatrix} -0.3345 & -3.2977 & -0.0802 \\ -0.2502 & -2.7746 & -0.0720 \\ -0.0012 & -0.0284 & -0.0059 \end{bmatrix}$$

$$C_o = \begin{bmatrix} 1 & 0 & 0 \\ 0 & 1 & 0 \\ 0 & 0 & 1 \end{bmatrix}$$

$$D_o = 10^3 * \begin{bmatrix} 0.6898 & -1.9620 & 0.7344 \\ 0.2309 & -0.8246 & 0.2942 \\ 0.0153 & 0.0258 & 0.0126 \end{bmatrix} \quad (24)$$

Having the identified model, a predicted controller is constructed to realize real-time sine-wave output and Fig.4 just shows it. For selecting optimal switching signal $a(k)$, a rule which is ensuring the error between $v_{ref}(k+1)$ and $v_{pre}(k+1)$ is minimum must be obeyed.

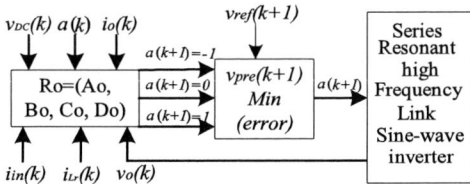

Figure 4. Block diagram of predicted controller

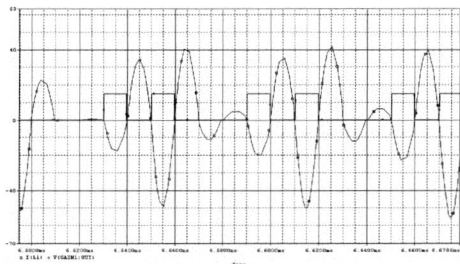

(a) simulation results

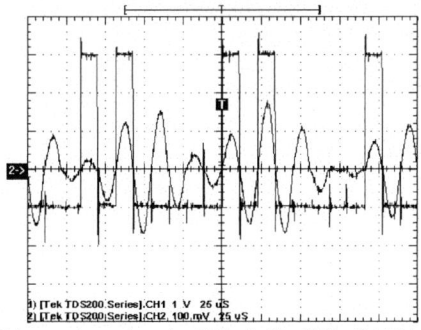

(b) experimental results (CH1: 5v/div; CH2: 40A/div)

Figure 5. switching signal and correponding variaty of i_Lr

Fig.5 shows the switching signal and corresponding variety of i_{Lr}. Both the simulation and experimental results conclude the following resonant modes: PM, FRM and ZFRM. In 2nd quadrant, there is a similar working

waveform as Fig.5 and the corresponding modes are PFM, PRM and ZPRM.

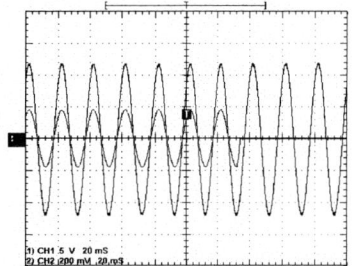

Figure 6. output voltage response (CH1: 8A/div CH2: 150v/div)

Fig.6 shows the instantaneous response of the inverter system with the load abrupt change from 100%~0% and the opposite change is omitted for page limit. The rapid response also confirms the validity of identified model and predicted controller.

IV. CONCLUSION

For a highly nonlinear power electronics converter, the paper presents a system identification method which is based on the sampled data. The detailed modeling procedure and arithmetic are derived and analyzed in an observable form. The operational principle of series resonant high frequency link sine-wave inverter is depicted and the system model is identified by utilizing this black-box tool. The resulting model is compared with experiments and the validity is confirmed. With this method, a small-signal state-space model can be constructed directly without the knowledge of the system order and the parameters of components.

REFERENCES

[1] Yong-Ho Chung, Bong-Soo Shin, Gyu-Hyeong Cho, "Bilateral Series Resonant Inverter for High Frequency Link UPS", Power Electronics Specialists Conference, 1989. PESC '89 Record, 20th Annual IEEE, 26-29 June 1989, vol.1Pages: 83-90

[2] Jin Xiaoyi, Wu Weiyang, Sun Xiaofeng, and Liu Jun "Resonant Tank and Transformer Design in series resonant converter" in IEEE IAS'05 Rec., Hong Kong, 2-6 Oct. 2005 Pages:1475-1482

[3] Mikihiko Matsui and Yong-Xia Yang "Macroscopic modeling of a parallel resonant link converter and its application to closed-loop link voltage control" in IEEE IAS'97 Rec., Volume 2, 5-9 Oct. 1997 Page(s):1421-1428

[4] Foster, M.P.; Sewell, H.I.; Bingham, C.M.; Stone, D.A.; Hente, D.; and Howe, D.;" Cyclic-averaging for high-speed analysis of resonant converters" in IEEE PE Transactions July 2003 Page(s):985-993

[5] Wu Weiyang, Jin Xiaoyi, and Sun Xiaofeng "A Novel Series Resonant High-Frequency Link Sine-wave Inverter Family" in IEEE PESC'05 Rec., Brazil, 12-16 June 2005, Page(s): 650-655

[6] S. Bingulac and H. F. VanLandingham, Algorithms for Computer-Aided Design of Multivariable Control Systems. New York, NY: Marcel Dekker, 1993

2006 5th International Power Electronics and Motion Control Conference

Maximal Power Point Tracking under Speed-Mode Control for Wind Energy Generation System with Doubly Fed Introduction Generator

Y. Zhao[*],[**], X. D. Zou[*], Y. N. Xu[*], Y. Kang[*] and J. Chen[*]

[*]Huazhong University of Science and Technology / Electrical and Electronic College, Wuhan, China
[**]Hubei University of Technology / School of Electrical and Electronic Engineering, Wuhan, China

Abstract—**The Doubly Fed Induction Generator (DFIG) with back-to-back four-quadrant converters between its rotor winding and the grid can realize the maximal wind energy capture in variable-speed wind energy generation system. Based on the stator flux-oriented excitation vector control strategy, this paper presents that the system dynamic characteristics under speed-mode control are superior to the dynamic characteristics under current-mode control in the process of Maximum Power Point Tracking (MPPT) and brings forward the explanation based on state space theory. And it designs the gains of two typical PI controllers in the current inner loop and the speed outer loop under speed-mode control based on Internal Model Control (IMC) theory and pole assignment method, respectively. At last, simulation results in MATLAB/ SIMULINK are put forward to verify the good static and dynamic performance of the designed closed-loop control system.**

Keywords-Doubly Fed Induction Generator (DFIG); current-mode control; speed-mode control; PI controller; Internal Model Control (IMC); pole assignment

I. INTRODUCTION

The Doubly Fed Induction Generator (DFIG) produces constant frequency power to the grid with variable rotor speed in variable-speed wind energy generation system. It can realize the maximal wind energy capture with the rotor speed varying from subsynchronous to super-synchronous speed when the wind speed is above the cut-in speed and under the rating speed; the generated active and reactive power can be controlled independently; the four-quadrant ac-to-ac converter connected between the rotor winding and the grid handles only a fraction of the total power to achieve the full control of the system. So with the rapid increase in capacity of single generator and wind farm, the research and application of the DFIG progress at a rapid rate in wind power generating application. The frequency conversion circuit between the rotor winding and the grid often adopts the configuration of two back-to-back four-quadrant voltage source converters, which is popular at present, because of its powerful function. The rotor-side converter adopts stator flux-oriented vector control strategy to realize the decouple control of the active and reactive power, and the grid-side converter adopts grid voltage-oriented vector control strategy to control the dc-link voltage and the grid power factor. This paper discusses only the active power control of the system.

The excitation control strategy adopted in the rotor-side converter based on the stator flux-oriented vector control enables Maximum Power Point Tracking (MPPT) to improve energy conversion efficiency by controlling the q-axis rotor current i_{qr}, namely the active component of rotor current. The basic theory of MPPT is that the wind turbine can obtain the maximum wind energy when it rotates in a certain speed at a certain wind speed. Therefore, the $P_{opt} - \omega_r$ characteristics have been given out [1], where P_{opt} is the maximum energy that the power system obtains from wind energy and ω_r is the rotor speed. The MPPT control strategy can be divided into two hierarchical controls [2], i.e., the setting of reference value and the tracing of reference value. Based on the MPPT principle, the reference value is generally the active power P_s [3]-[5], the electromagnetic torque T_e [1],[6]-[7] or the rotor speed ω_r [1],[8]-[9]. The former two may be termed 'current-mode control', and the last may be termed 'speed-mode control'[1].

Based on a third-order model related to the electromagnetic and mechanical aspects of DFIG, Section II presents a decoupled active and reactive power control strategy. Comparing the current-mode control and the speed-mode control, Section III presents that the system dynamic characteristics under speed-mode control are superior to the dynamic characteristics under current-mode control and brings forward the state space explanation. Section IV designs the gains of two typical PI controllers in the current inner loop and the speed outer loop under speed-mode control based on Internal Model Control (IMC) theory and pole assignment

1-4244-0448-7/06/$25.00 ©2006 IEEE

method, respectively. Simulation results in MATLAB/SIMULINK are brought forward to verify the good static and dynamic performance of the designed closed-loop control system in Section V.

II. MODEL AND CONTROL STRATEGY OF DFIG

The DFIG equations depicted in a two axis d-q reference frame rotating at synchronous speed are derived from Park's equations. The stator side of DFIG uses generator convention and the rotor side of DFIG uses motor convention.

The voltage equations can be written as

$$
\begin{bmatrix} u_{ds} \\ u_{qs} \\ u_{dr} \\ u_{qr} \end{bmatrix} = \begin{bmatrix} -R_s & 0 & 0 & 0 \\ 0 & -R_s & 0 & 0 \\ 0 & 0 & R_r & 0 \\ 0 & 0 & 0 & R_r \end{bmatrix} \begin{bmatrix} i_{ds} \\ i_{qs} \\ i_{dr} \\ i_{qr} \end{bmatrix} + p \begin{bmatrix} \psi_{ds} \\ \psi_{qs} \\ \psi_{dr} \\ \psi_{qr} \end{bmatrix} + \begin{bmatrix} -\omega_1 \psi_{qs} \\ \omega_1 \psi_{ds} \\ -\omega_2 \psi_{qr} \\ \omega_2 \psi_{dr} \end{bmatrix} \quad (1)
$$

The stator and rotor flux linkages in (1) are

$$
\begin{bmatrix} \psi_{ds} \\ \psi_{qs} \\ \psi_{dr} \\ \psi_{qr} \end{bmatrix} = \begin{bmatrix} -L_s & 0 & L_0 & 0 \\ 0 & -L_s & 0 & L_0 \\ -L_0 & 0 & L_r & 0 \\ 0 & -L_0 & 0 & L_r \end{bmatrix} \cdot \begin{bmatrix} i_{ds} \\ i_{qs} \\ i_{dr} \\ i_{qr} \end{bmatrix} \quad (2)
$$

The motion equation is given by

$$
J \frac{d\omega_r}{dt} + B\omega_r = n_p \cdot (T_m - T_e) \quad (3)
$$

The electromagnetic torque in (3) is

$$
T_e = n_p L_0 (i_{qs} i_{dr} - i_{ds} i_{qr}) \quad (4)
$$

Equations (1) to (4) are set of differential equations making up of a fifth-order model which describes the dynamic behavior of DFIG. The voltages, currents and flux linkages are expressed by the d-axis and q-axis components in synchronous rotating reference frame.

It is assumed that (a). Neglecting the influence of the stator flux linkage's transient state and orienting the d-axis of the synchronous frame to the direction of the stator flux vector. This implies that the differential of the stator flux linkage is zero and its vector with constant magnitude rotates in synchronous velocity. (b). Omitting R_s because the voltage drop on it is small.

From the assumption (a)

$$
p\psi_s = 0 \quad \text{and} \quad \begin{array}{l} \psi_{ds} = \psi_s \\ \psi_{qs} = 0 \end{array} \quad (5)
$$

The following voltage equations can be derived from (1) to (5) with the assumption (a) and (b).

$$
\begin{aligned}
u_{ds} &= 0 \\
u_{qs} &= U_s \\
u_{dr} &= (R_r + \sigma L_r p) i_{dr} - \omega_2 \sigma L_r i_{qr} \\
u_{qr} &= (R_r + \sigma L_r p) i_{qr} + \omega_2 (\sigma L_r i_{dr} + L_0^2 i_{ms} / L_s)
\end{aligned} \quad (6)
$$

Where σ represents the leakage coefficient, $\sigma = 1 - L_0^2 / (L_s \cdot L_r)$; i_{ms} is the stator flux magnetizing current, $i_{ms} = \Psi_s / L_0 = U_s / (\omega_1 \cdot L_0)$.

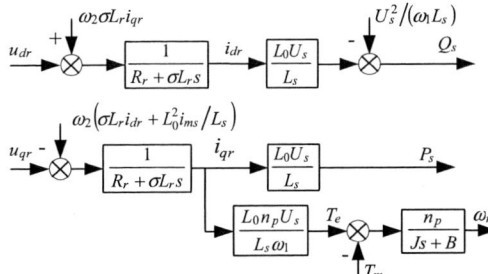

Figure 1. Block diagram of the DFIG model

The excitation control strategy based on the assumption (a) and (b) is often termed stator flux-oriented vector control strategy. Then the stator voltage vector is oriented to the q-axis of the reference frame. Thus realizes the stator voltage vector orientation, and the order of the DFIG model decreases from fifth to third which is beneficial to simplify the excitation control system of DFIG. The rotor voltage equations in (6) and (3) make up of a new third-order model.

As mentioned in section I, the reference value could be active power P_s, electromagnetic torque T_e or the rotor speed ω_r in order to realize the MPPT. The relationship between the reference values and the other variables is

$$
T_e = n_p L_0 (i_{qs} i_{dr} - i_{ds} i_{qr}) = \frac{L_0 n_p U_s}{L_s \omega_1} i_{qr} \quad (7)
$$

$$
P_s = u_{ds} i_{ds} + u_{qs} i_{qs} = U_s i_{qs} = U_s \frac{L_0}{L_s} i_{qr} \quad (8)
$$

Eq. (3), (7), (8) and the rotor voltage equations in (6) compose the full DFIG model as shown in Fig.1.

III. CURRENT-MODE CONTROL AND SPEED-MODE CONTROL

Under current-mode control, active power P_s and electromagnetic torque T_e both have the proportional relationship with the q-axis rotor current i_{qr} as (7) and (8) show. So, the reference value i_{qr}^* for the current inner loop can derived from P_s^* or T_e^* directly according to the certain proportion. Thus the active power and electromagnetic torque adopt the open-loop control as shown in Fig.2., where $K_1 = L_s / (L_0 \cdot U_s)$, $K_2 = L_s \omega_1 / (L_0 n_p U_s)$. The cascaded control scheme under speed-mode control is depicted in Fig.3., where $K_3 = L_0 \cdot n_p \cdot U_s / (L_s \cdot \omega_1)$. The controller in the speed outer loop is a PI controller. The reference value of the active power, electromagnetic torque and rotor speed in Fig.2. and Fig.3. are determined by the MPPT theory.

The controllers in the current inner loop under current-mode control or speed-mode control are both typical PI controllers. The cross-relation between the d-axis and q-axis rotor current components in (6) is feed-forward compensated in both Fig.2. and Fig.3.

The active power, electromagnetic torque and rotor speed have the definite relationship at a certain wind

195

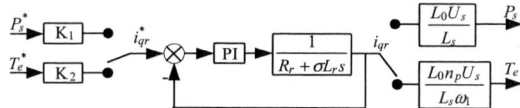

Figure 2. Block diagram of the controller in current-control mode

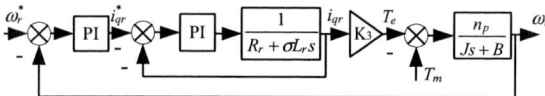

Figure 3. Block diagram of the controller in speed-control mode

speed based on the MPPT theory, which implies that with anyone of them reaching its reference value the system can obtain the maximal wind energy in static state.

However, the current-mode control and the speed-mode control have great differences in dynamic characteristics. The responsive rate of the system is low under current-mode control and at worst the system can not realize MPPT. On the contrary, the responsive rate of the system is fast under speed-mode control and can realize MPPT. This can be explained as below. The stator of DFIG can acquire the expected active power and electromagnetic torque rapidly because of the good response performance of the current loop under current-mode control. But this doesn't mean that the system reaches static state at the moment the rotor current reaches its desired value. The error between the mechanical power input and the active power the stator fed to the grid is divided into two parts: one part for rotor speedup and the other part for the rotor-side converter fed to the grid. The part for acceleration decreases and the other part increases with the rotor rotating velocity varying until the system comes to equilibrium. This transient process is determined by the DFIG motion characteristics and out of control. So, with little influence on the steady-state precision, the current-mode control doesn't improve the system dynamic performance whether the reference value is the active power or electromagnetic torque, whether the control scheme is the open-loop control in Fig.2. or the closed-loop control. The speed-mode control adds a speed outer loop to the current inner loop to form a cascaded closed loop control system, which including the motion part of the DFIG model as shown in Fig.3. That means the rotor speed varying process, the balance of active power and the balance of torque on the rotor shaft are in control.

When analyzed with state-space theory, the q-axis rotor voltage u_{qr} is regarded as input variables; the q-axis rotor current i_{qr} and the rotor speed ω_r are regarded as state variables; the active power, electro-magnetic torque or rotor speed is regarded as output variable; the mechanical torque T_m is regarded as a disturbance. The control scheme of the current-mode control comprises only one state variable, the q-axis current i_{qr}, which is fed back to form a current closed

loop, hence some electromagnetic quantities such as i_{qr}, P_s and T_e can be controlled. But the scheme doesn't contain the other state viable ω_r and the motion characteristics of DFIG, which reflect on the uncontrollable of the motion response. The control scheme of the speed-mode control comprises both state variables, the q-axis current i_{qr} and the rotor speed ω_r, which are fed back to form a current inner closed loop and a speed outer closed loop. Thus the closed-loop control system holds the full system state information and can control all the variables, not only the electromagnetic ones but also the mechanic ones. Therefore, the dynamic characteristics under the speed-mode control are superior to the ones under current-mode control.

IV. CONTROLLERS

In order to achieve the excellent static and dynamic performance, this paper adopts the excitation control strategy under speed-mode control with two typical PI controllers used in the current inner loop and the speed outer loop, respectively. The gains of the two PI controllers are designed based on IMC theory and pole assignment method, respectively.

A. PI Controller in Current Inner Loop

The controller in current closed loop can be easily designed to get good dynamic performance of tracking introduction and restraining disturbance by using the IMC method [9],[10]. For the first-order system, the controller becomes a simple PI controller.   .

The PI controller transfer function is

$$C(s) = K_{pc} + K_{ic}/s \qquad (9)$$

With the model of the object being $1/(R_r + \sigma L_r s)$, the gains of the controller can obtained from below

$$\begin{aligned} K_{pc} &= \sigma L_r/\tau \\ K_{ic} &= R_r/\tau \end{aligned} \qquad (10)$$

Thus the current inner loop equivalents an inertial link with filter parameter τ. So, it is easy to design the gains of the PI controller by adjusting τ to achieve nice dynamic performance and robust control of the inner-loop control system.

B. PI Controller in Speed Outer Loop

The controller in speed outer loop is a PI controller too. The reference value for the current inner loop which offered by the outer loop controller represents below [3]

$$i_{qr}^* = K_{ps}\left(b\omega_r^* - \omega_r\right) + K_{is}\int\left(\omega_r^* - \omega_r\right)dt \qquad (11)$$

Where b is a parameter introduced to the typical PI controller.

The simplified block diagram of the control system is shown in Fig.4, where $K_4 = L_0 \cdot n_p^2 \cdot U_s/(L_s \cdot \omega_1 \cdot B)$, $\tau_1 = J/B$.

196

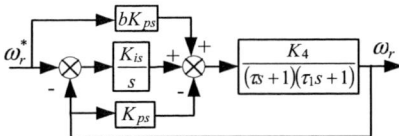

Figure 4. Block diagram of the system under speed-mode control

From Fig.4., the system closed-loop transfer function is derived as below

$$G(s) = \frac{K_4\left(bK_{ps}s + K_{is}\right)}{\tau\tau_1 s^3 + (\tau + \tau_1)s^2 + (K_{ps}K_4 + 1)s + K_{is}K_4} \quad (12)$$

The closed-loop characteristic equation is

$$D(s) = \tau\tau_1 s^3 + (\tau + \tau_1)s^2 + (K_{ps}K_4 + 1)s + K_{is}K_4 \quad (13)$$

Since the closed-loop control system is a third-order system, the gains of the PI controller are designed based on the pole assignment method to ensure good dynamic characteristics by assigning the closed-loop poles correctly [11]. The dynamic characteristic of a high order system are mainly determined by closed-loop dominant poles. The dominant poles of the system is

$$s_{1,2} = -\zeta_r\omega_r \pm j\omega_r\sqrt{1 - \xi_r^2} \quad (14)$$

And the non-dominant pole of the closed-loop control system is

$$s_3 = -n\zeta_r\omega_r \quad (15)$$

Where n is a parameter which decides the third pole's location in complex plane. The bigger value it has, the more approximate the response characteristics of the third-order system decided by $s_{1,2,3}$ are to the response characteristics of the two-order system decided by the dominant poles $s_{1,2}$. Usually, $n = 5 - 10$.

The characteristic equation of the third-order system decided by $s_{1,2,3}$ is

$$\begin{aligned} D_r(s) &= \left(s^2 + 2\zeta_r\omega_r s + \omega_r^2\right)\left(s + n\zeta_r\omega_r\right) \\ &= s^3 + (n+2)\zeta_r\omega_r s^2 + \omega_r^2\left(2n\zeta_r^2 + 1\right)s + n\zeta_r\omega_r^3 \end{aligned} \quad (16)$$

The gains of the controller are obtained by the comparison of (13) and (16)

$$\begin{aligned} K_{ps} &= \left[\omega_r^2\tau\tau_1\left(2n\zeta_r^2 + 1\right) - 1\right]/K_4 \\ K_{is} &= n\zeta_r\tau\tau_1\omega_r^3 / K_4 \end{aligned} \quad (17)$$

Where n, ζ_r, ω_r, τ and τ_1 have the relationship below

$$(n+2)\zeta_r\omega_r = (\tau + \tau_1)/(\tau\tau_1) \quad (18)$$

Simulation results indicate that there exists overshoot even if ζ_r is larger than 1 because of the influence of the zero point which the PI controller introduced into the closed-loop control system. The parameter b in (12) allows placing independently the unique zero.

V. SIMULATION RESULTS

The characteristics of the rounded induction machine that have been used in the simulations are shown in Table

I.

In the simulation, filter parameter τ in the current loop is 2ms, parameter n is 5 and b is 50 in speed controller. The damping ratio and the natural frequency that the closed-loop control system expected is 1.2 and 60Hz. Controllers parameters are designed as $K_{pc} = 6.9$, $K_{ic} = 408$, $K_{ps} = 0.512$, $K_{is} = 11.88$.

TABLE I.
CHARACTERISTICS OF DFIG

Machine Characteristic	Value
rating active power (kW)	4
mutual inductance (H)	0.28
stator inductance (H)	0.287
rotor inductance (H)	0.287
stator resistance (Ω)	0.5
rotor resistance (Ω)	0.816
number of pole pair	3
Inertia ($kg \cdot m^2$)	0.05

Fig.5. shows the simulation results under current-mode control. The mechanical torque fed into the rotor shaft is constant power torque, $P_m = 4$ kW. The active power as the reference value for the control system changes from 5kW to 3.33kW at 1.1s. The stator-side active power P_s in Fig.5 (a) and the q-axis rotor current i_{qr} in Fig.5(b) have fast response within 5ms because of the good dynamic performance of the current inner loop. While at the same time, the rotor-side active power P_r in Fig.5 (a) and the rotor mechanical rotating speed ω_r in Fig.5 (c) reach their static value at 1.5s. The transient process lasts at least 0.4s and it will last longer with bigger rotor inertia. Fig.5 (d) shows the rotor current in rotor-side reference frame.

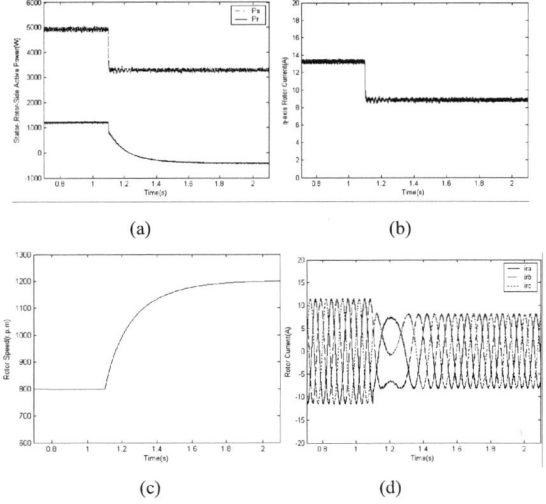

Fig. 5. Simulation results under current-mode control

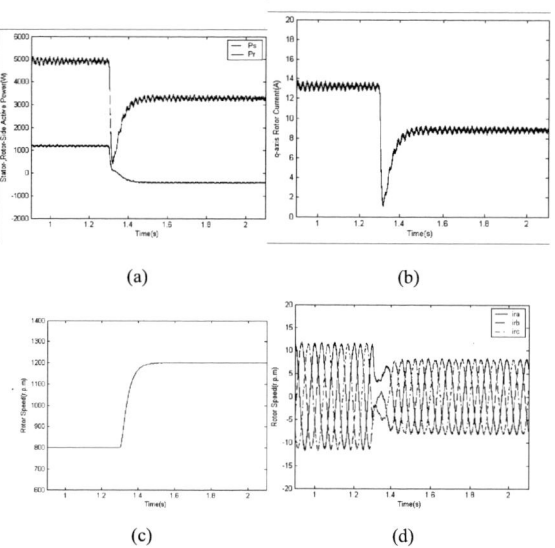

(a)   (b)

(c)   (d)

Fig.6 Simulation results under speed-mode control

Fig.6. shows the simulation results under speed-mode control. The PI controller is designed based on pole assignment method in Section IV. The mechanical torque fed into the rotor shaft is a constant power torque, $P_m = 4$ kW. The rotor speed as the reference value for the control system changes from 800rpm to 1000rpm at 1.3s. The stator-side active power P_s and the rotor-side active power P_r in Fig.6(a), the q-axis rotor current i_{qr} in Fig.6(b) and the rotor mechanical rotating speed in Fig.6(c) reach their static value at 1.4s. Fig.6 (d) shows the rotor current in rotor-side reference frame. The transient process lasts only 0.1s. There is a dip in the stator-side active power curve and the q-axis rotor current curve as shown in Fig.6(a) and (b). This implies that, at the beginning of the transient process, the error between the mechanical power input and the active power the stator fed to the grid is larger than the error under current-mode control, and the part of the error for rotor speedup is larger than that under current-mode control too. The rotor accelerates rapidly with the appropriate change of the i_{qr}. So, the transient process with all the electrical and mechanical variables varying does not last long.

VI. CONCLUSIONS

This contribution presents that the speed-mode control is superior to the current-mode control because of the controllable of all the state variables under the speed-mode control. To get good static and dynamic performance, two typical PI controllers are designed based on the IMC and pole assignment method under speed-mode control, respectively. The simulation results show the good performance of the control system.

REFERENCE

[1] R. Pena, J. C. Clare, and G. M. Asher, "Doubly fed induction generator using back-to-back PWM converters and its application to variable-speed wind-energy generation," *Proc. Int. Elec. Eng., Electr. Power Appl.*, vol. 143, no. 3, pp.231-241, May 1996.

[2] L. Jing, F. Yong, S. Jiahua, and W. Weisheng, "Research on subsection and layer control strategy of doubly-fed variable speed wind turbine," *Proc. Power System Technology*, Vol. 29, No.9, pp.15-21, May 2005.

[3] A. Tapia, G. Tapia, J. X. Ostolaza, and J. R. Sanenz, "Modeling and control of a wind turbine driven doubly fed induction generator," *IEEE Trans. Energy Convers.*, vol. 18, no. 2, pp.194-204, June 2003.

[4] C. Abbey, and G. Joos, "Optimal reactive power allocation in a wind powered doubly-fed induction generator," in *Proc. IEEE Power Engineering Society General Meeting*, Vol.2, pp. 1491 – 1495, June 2004.

[5] Park, J.W.; Lee, K.W.; Lee, H.J.; "Control of active power in a doubly-fed induction generator taking into account the rotor side apparent power," in *Proc. IEEE 35th Power Electronics Specialists Conference*, Vol. 3, pp. 2060 – 2064, June 2004.

[6] B. Rabelo, and W. Hofmann, "Optimal active and reactive power control with the doubly-fed induction generator in the MW-Class wind-turbines," in *Proc. IEEE Power Electronics and Drive Systems*, vol.1, pp. 53 – 58, Oct. 2001.

[7] A. Mullane, and M. O'Malley. "The inertial response of induction-machine-based wind turbines," *IEEE Trans. Power Syst.*, vol. 20, no. 3, pp.1496-1503, Aug. 2005.

[8] R. Datta, and V. T. Ranganathan, "A method of tracking the peak power points for a variable speed wind energy conversion system," *IEEE Trans. Energy Convers.*, vol. 18, no. 1, pp. 163-168, Mar. 2003.

[9] J. Morren, S. W. H. de Haan. "Ridethrough of wind turbines with doubly-fed induction generator during a voltage dip," *IEEE Trans. Energy Convers.*, vol. 20, no. 2, pp.435-441, June 2005.

[10] X. D. Zou , *Research on VSCF AC Excitation Doubly Fed Wind Energy Generation System and Its Control Technology*, Doctor Dissertation. Wuhan: Huazhong University of Science and Technology, 2005.

[11] L. Peng, *Research on Control Technique for PWM Inverters Based on State-Space*, Doctor Dissertation. Wuhan: Huazhong University of Science and Technology, 2004.

2006 5th International Power Electronics and Motion Control Conference

Effective Mobility in Nano-Scaled n-MOSFETs[*]

Yue-Hua Dai, Jun-Ning Chen, Dao-Ming Ke and Jia-E Sun

School of Electronic Science and Technology, Anhui University, Hefei, China, 230039

jnchen@ahu.edu.cn

Abstract—**In this work, we present a methodology for calculating mobility of nano-scaled MOSFET's from the Boltzmann transport equation(BTE). Approximate solution of the BTE for electrons in nano-scaled MOSFET's is given, and the improved distribution function of the carriers is used to model the mobility of carriers. A new model is presented for two-dimensional characteristic field-dependent mobility. Comparing the theoretical curves with an extensive set of simulation ones has validated this model.**

Keywords - Boltzmann transport equation; modeling of carrier mobility; effective electrical field;2D MOSFET.

I. INTRODUCTION

It is imperative to have accurate models for the mobility in nano-scaled ULSI MOSFET's so that useful simulations can be carried out.[1] Existing models are largely empirical [2-8], and their agreement with experiments is not quite satisfactory when dealing with nano-scaled ULSI devices. On the other hand, there are also some analytical mobility models based on physical mechanisms[9-13], and among all studies, the dependence of the channel mobility on the transverse field (the gate field) has been of much interest[10-12]. Most studies investigated the dependence at low longitudinal field (the drain field) to separate the drain field influence from the gate field. The later approach ignores the inversion layer as an entity and simply includes in Matthiessen's rule five or six scattering mechanisms[9]. Expressions for mobility obtained from such calculations differ by 50 percent or more from one another [1]. A careful modeling of electron mobility in nano-scaled ULSI MOSFET's must account for the strong two dimensionality of the mobile carriers, and it will be important for understanding and designing nano-scaled devices where starting with the correct physics may be crucial to success. The objective of this work is the development of a new mobility model for channel electrons in nano-scaled n-MOSFET's biased at strong inversion.

Deep sub-micron MOS processes require gate oxides less than 10nm thick, and channel doping of the order of $10^{18} cm^{-3}$. Due to these features, the electric field in the channel reaches values much higher than $10^5 V/cm$ even when the device is biased at threshold. So for nano-scaled MOS, the electrical field in the channel becomes even higher. Accurate calculations of mobility in high electric fields require knowledge of the distribution function of the carriers. The most straightforward way of determining the distribution function is to solve the Boltzmann equation[14]. This is frequently difficult, particularly for high fields. Thus the basic suggestion of near isotropy[15], although it can be regarded as a fruitful approach to the problem, would probably give numerical results of less value for electric fields of interests ($\sim 5 \times 10^5 V/cm$)[16]. However, simply and analytical expressions for mobility can be obtained if we use the anisotropy approximation and the relaxation time approximation for the collision term in the BTE.

In the next section, the improved distribution function of the carriers is given along with the simplifying assumptions used to derive them from the BTE. The solution strategy for the electron mobility in nano-scaled ULSI MOSFET's are explained in the third section. In the fourth section, calculated results about the new mobility model are compared with the simulation results for a MOS transistor with the channel length of 80 nm.

II. IMPROVED DISTRIBUTION FUNCTION

The Boltzmann Transport Equation (BTE) is a complex integral-differential equation that is based on both quantum-mechanical and classical laws of dynamics. As such, the BTE in its original form does not yield a closed form solution for mobility, and simplifying assumptions are necessary to make the solution tractable.

Assuming a nearly isotropic distribution function f_S in low or zero fields and change of the distribution due to the acceleration of the carriers by the high field as f_A (the anisotropic part of distribution function), the distribution function f in high fields may be written as

$$f(x, y, E) = f_S(E) + f_A(x, y, E) \quad , f_A << f_S \quad (1)$$

And usually solve BTE under the relaxation time approximation (RTA),

$$\frac{\partial f}{\partial t} + \frac{\hbar}{m^*} \overline{K} \cdot \nabla_r f + \frac{-qF}{\hbar} \cdot \nabla_K f = -\frac{f - f_0}{\tau_0} \quad (2)$$

For steady-state transport under a uniform electric field,

$$\frac{\partial f}{\partial t} = 0 \quad and \quad \nabla_r f = 0 \quad (3)$$

*This work was supported by the National Natural Science Foundation of China. (60276042)

1-4244-0448-7/06/$25.00 ©2006 IEEE

For low field transport,

$$f_S(E) \approx f_0(E) \qquad (4)$$

BTE reduces to,

$$\frac{-qF}{\hbar} \cdot \nabla_K (f_0 + f_A) = \frac{-f_A}{\tau_0} \qquad (5)$$

Accounting for $f_A \ll f_S (\approx f_0)$, we replace the term of $(f_0 + f_A)$ in the equation of (5) with f_0, then the anisotropic part of distribution function f_A can be obtained.

$$f_A(x, y, E) = q\tau_0 \frac{df_0}{dE}(v \cdot F) \qquad (6)$$

Where f_0 indicates the (local) equilibrium distribution function, and in equilibrium we may use the Maxwell-Boltzmann distribution function. E is the energy of a carrier, and τ_0 is a microscopic relaxation time. q is the electronic charge, m^* is the effective mass of carriers, and $\hbar$ is the Planker constant. F is the eletrical field in the channel of nano-scaled MOSFETs, and v is the carriers' velocity.

For advance MOSFETs, it is very necessary to account for the velocity and the eletrical field distribution in two dimensions, especially in the condition that the transverse field (the gate field) and the longitudinal field (the drain field) are simultaneously high. In fact, the both fields may not be separated rigorously from each other, or it is difficult to separate the drain field influence from the gate field. On the other hand, the drift of these "highly mobile" electrons increases with higher electric fields in nano-scaled MOSFETs, and the rate of energy loss of a electron to the crystal lattice must be smaller than the rate of energy gain of an electron due to the acceleration by the high field. So we deal with the term of $(v \cdot F)$ in the expression (6) as following:

$$v \cdot F = (v_{x0} + \frac{qF_x}{m^*}\tau_x)F_x + (v_{y0} + \frac{qF_y}{m^*}\tau_y)F_y \qquad (7)$$

Where the subscript x stand for the direction along the channel and y for the direction normal to the channel, respectively v_{x0} and v_{y0} are the beginning drift velocity along the channel and normal to the channel, respectively. In this work, both of the transverse field and the longitudinal field are very high. So we can suppose that $v_{x0} = v_{y0} = v_0$ and $\tau_x = \tau_y \approx \tau_0$, then

$$v \cdot F \approx \alpha v_0 \sqrt{F_x^2 + F_y^2} + \frac{q\tau_0}{m^*}(F_x^2 + F_y^2)$$

$$= \alpha v_0 F_{eff} + \frac{q\tau_0}{m^*}F_{eff}^2 \qquad (8)$$

Here α is a proportionality constant from the mathematics processing, F_{eff} is the effective electrical field of the channel from the point of view of regarding the inversion layer as an entity and described as

$$F_{eff} = \sqrt{F_x^2 + F_y^2}$$

As we known, $f_0 \propto \exp(-E/(K_0 T))$, so

$$\frac{df_0}{dE} = -\frac{1}{K_0 T}f_0 \qquad (9)$$

Substitute the equations (8) and (9) into (6), and we obtain the expression of f_A finally.

$$f_A(x, y, E) = -\frac{q\tau_0}{K_0 T}(\alpha v_0 F_{eff} + \frac{q\tau_0}{m^*}F_{eff}^2)f_0 \qquad (10)$$

Where K_0 is Boltzmann's constant, and T is the absolute temperature.

So, BTE solution is just a shifted version of the equilibrium occupancy.

$$f(x, y, E) = f_0 + \frac{-q\tau_0}{K_0 T}(\alpha v_0 F_{eff} + \frac{q\tau_0}{m^*}F_{eff}^2)f_0 \qquad (11)$$

III. NEW MOBILITY MODEL

BTE solution for mobility defines relationship between distribution function and mobility as

$$\mu = \frac{q}{m^*}\frac{<\tau v^2>}{<v^2>} = \frac{q}{m^*}\frac{\int \tau v^2 fdK}{\int v^2 fdK} \qquad (12)$$

Inserting the expression (11) into the numerator of the above formula, and replacing f_0 with the term of f in the denominator due to $f_A \ll f_S (\approx f_0)$, we can describe the mobility as

$$\mu \approx \mu_0 \left[1 - \frac{q\tau_0}{K_0 T}\left(\alpha v_0 F_{eff} + \frac{q\tau_0}{m^*}F_{eff}^2\right)\right] \qquad (13)$$

Where K is wave vector, and we assume that τ_0, α, v_0 and F_{eff} are all independent of K. μ_0 is the mobility in zero or very low electrical field, and is the function of doping concentrations.

$$\mu_0 = \frac{q}{m^*} \frac{\int \tau v^2 f_0 dK}{\int v^2 f_0 dK} \approx \frac{q\tau_0}{m^*} \qquad (14)$$

In the formula (13), v_0 is the accumulation of drift velocity in high fields comparing with the near equilibrium state. So we make the below approximation,

$$\alpha v_0 \approx \beta \mu \, F_{eff} - \gamma \mu_0 F_{eff} \qquad (15)$$

Where β and γ are the parameters to modify the velocity expression of $v = \mu F$, which is reasonable only under certain conditions.

Combining the expression (13), (14) and (15), we can easily deduce the relationship between mobility and effective electrical field in nano-scaled MOSFETs.

$$\mu(F_{eff}) \approx \mu_0 \left(1 + \eta \frac{K_0 T}{m^*} \mu_0^{-2} F_{eff}^{-2} \right) \qquad (16)$$

with η defined as,

$$\eta = \frac{\gamma - \beta - 1}{\beta^2} \qquad (17)$$

IV. RESULTS AND DISCUSSION

For the 2D MOSFET, we assumed a given temperature at 300K, and simulated numerically the $80nm$ NMOSFETs using the device simulation software of Medici. The MOSFET's were simulated with thermal oxides 1.6nm thick, the source/drain junctions 16nm depth, $V_{ds} = 2.5V$ and different channel doping concentrations of $3.0 \times 10^{17} cm^{-3}$, $5.0 \times 10^{17} cm^{-3}$, $1.0 \times 10^{18} cm^{-3}$, and $3.0 \times 10^{18} cm^{-3}$, respectively. The effective mass of carriers used in our work are $m^* = 1.08 m_0$.

Figure1 shows the basic structure of the $80nm$ NMOSFET simulated in this work. When we change the channel doping concentrations during simulation, the mobility of carriers is different due to the variational effective electrical fields. According the new model described above, there is a relationship between mobility and effective electrical field such as the formula (16). The calculated results are shown in Figure 2, 3, 4 and 5 for different channel doping, and the values of the parameter η in different conditions are given in Table 1.

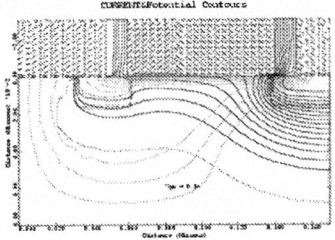

Figure 1. The basic structure of the **80nm** NMOSFET

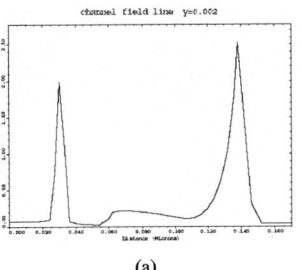

(a)

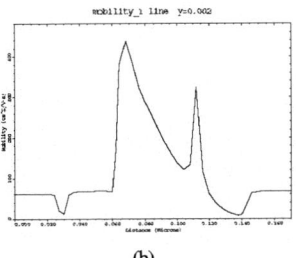

(b)

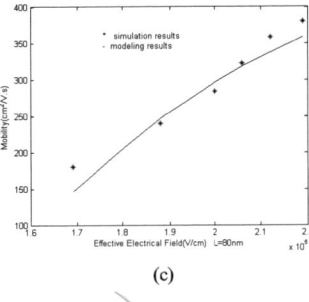

(c)

Figure 2. The simulation and calculation results of the 80nm NMOSFET with **3.0*10^17 cm^-3** channel doping concentration: (a) the electrical field distribution along the channel; (b) the mobility distribution along the channel; (c) the mobility dependence of the electrical field.

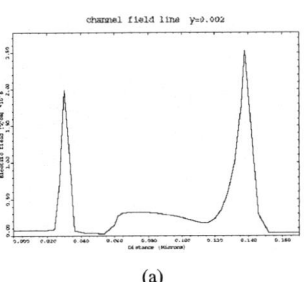

(a)

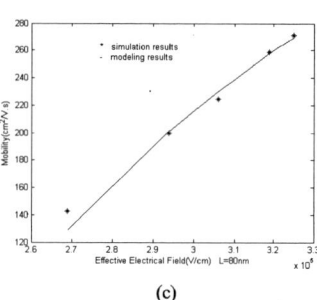

(b)

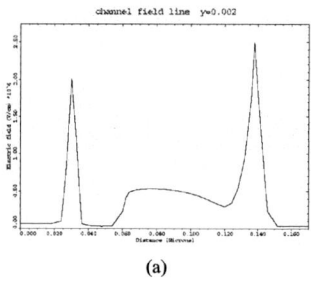

(c)

Figure 3. The simulation and calculation results of the 80nm NMOSFET with **5.0*10^{17}cm^{-3}** channel doping concentration: (a) the electrical field distribution along the channel; (b) the mobility distribution along the channel; (c) the mobility dependence of the electrical field.

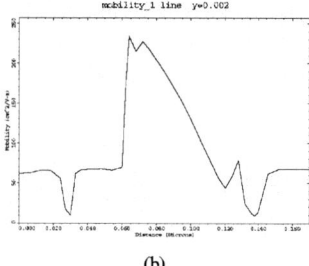

(a)

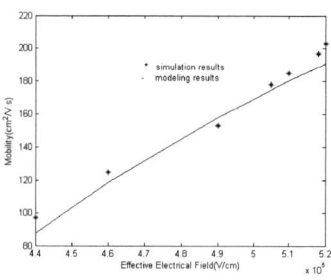

(b)

(c)

Figure 4. The simulation and calculation results of the 80nm NMOSFET with **1.0*10^{18}cm^{-3}** channel doping concentration: (a) the electrical field distribution along the channel; (b) the mobility distribution along the channel; (c) the mobility dependence of the electrical field.

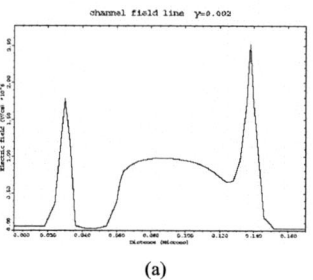

(a)

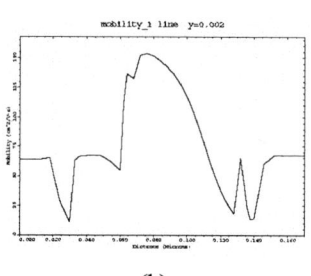

(b)

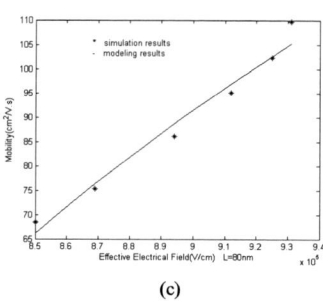

(c)

Figure 5. The simulation and calculation results of the 80nm NMOSFET with **3.0*10^{18}cm^{-3}** channel doping concentration: (a) the electrical field distribution along the channel; (b) the mobility distribution along the channel; (c) the mobility dependence of the electrical field.

TABLE I.   DIFFERENT VALUES OF η AND μ_0 IN DIFFERENT STATES (80NM)

	3.0×10^{17} (cm^{-3})	5.0×10^{17} (cm^{-3})	1.0×10^{18} (cm^{-3})	3.0×10^{18} (cm^{-3})
η	-238	-441	-750	-1204
μ_0	670 (cm^2/V.s)	575 (cm^2/V.s)	450 (cm^2/V.s)	300 (cm^2/V.s)

From the figures we can see the reasonable agreement between our model and the simulation results. Only by choosing different values of η in different states, our model can do well. But there are small errors due to not accounting for the gredience of electrical field in the channel. On the other hand, the effective field in this work is different from the electrical field of Medici, in which the interactive between transverse and longitudinal field is not considered.

V. CONCLUSION

By solving the Boltzmann Transport Equation, an improved distribution function of carriers is determined. From which, a new mobility model is presented in this work. Reasonable agreement with the simulation results makes our model quite satisfactory when dealing with submicron ULSI devices.

REFERENCES

[1] Yutao Ma, Litian Liu, and Zhijian Li, "A Discussion on the Universality of Inversion Layer Mobility in MOSFET's," IEEE Transactions On Electron Devices, VOL. 46, NO. 9, pp.1920-1922, SEPTEMBER 1999.

[2] Kostis Michelakis, Antonio Vilches, Christos Papavassiliou, Solon Despotopoulos, Kristel Fobelets, and Chris Toumazou, "Average Drift Mobility and Apparent Sheet-Electron Density Profiles in Strained-Si–SiGe Buried-Channel Depletion-Mode n-MOSFETs," IEEE Transactions On Electron Devices, VOL. 51, NO. 8, pp.1309-1314, AUGUST 2004.

[3] Masaki Kondo and Hiroyoshi Tanimoto, " An Accurate Coulomb Mobility Model for MOS Inversion Layer and Its Application to NO-Oxynitride Devices," IEEE Transactions On Electron Devices, VOL. 48, NO. 2, pp.265-270, FEBRUARY 2001.

[4] Haitao Gan and Ting-Wei Tang, "A New Method for Extracting Carrier Mobility from Monte Carlo Device Simulation," IEEE Transactions On Electron Devices, VOL. 48, NO. 2, pp.399-401, FEBRUARY 2001.

[5] Seonghearn Lee and Hyun Kyu Yu, " A New Technique to Extract Channel Mobility in Submicron MOSFETs Using Inversion Charge Slope Obtained from Measured S-Parameters," IEEE Transactions On Electron Devices, VOL. 48, NO. 4, pp.784-788, APRIL 2001.

[6] Agostino Pirovano Andrea L. Lacaita, Günther Zandler, and Ralph Oberhuber, "Explaining the Dependences of the Hole and Electron Mobilities in Si Inversion Layers," IEEE Transactions On Electron Devices, VOL. 47, NO. 4, pp.718-724, APRIL 2000.

[7] Shin-ichi Takagi, Akira Toriumi, Masao Iwase, and Hiroyuki Tango, "On the Universality of Inversion Layer Mobility in Si MOSFET's: Part I-Effects of Substrate Impurity Concentration,"

IEEE Transactions On Electron Devices, VOL. 41, NO. 12, pp.2357-2362, DECEMBER 1994.

[8] Shin-ichi Takagi, Akira Toriumi, Masao Iwase, and Hiroyuki Tango, "On the Universality of Inversion Layer Mobility in Si MOSFET's: Part 11-Effects of Surface Orientation," IEEE Transactions On Electron Devices, VOL. 41, NO. 12, pp.2363-2368, DECEMBER 1994.

[9] Syed Aon Mujtaba, "Advanced Mobility Models For Design And Simulation Of Deep Submicrometer Mosfets"(A DISSERTATION), The Department Of Electrical Engineering And The Committee On Graduate Studies Of Stanford University, December 1995.

[10] Hyungsoon Shin, Al F. Tasch, JR., Christine M. Maziar, and Sanjay K. Banerjee, "A New Approach to Verify and Derive a Transverse-Field-Dependent Mobility Model for Electrons in MOS Inversion Layers," IEEE Transactlons On Electron Devices, VOL. 36, NO. 6, pp.1117-1124, JUNE 1989.

[11] Ronald van Langevelde and Fran‚cois M. Klaassen, "Effect of Gate-Field Dependent Mobility Degradation on Distortion Analysis in MOSFET's," IEEE Transactions On Electron Devices, VOL. 44, NO. 11, pp.2044-2052, NOVEMBER 1997.

[12] J. A. M. Otten and F. M. Klaassen, "A novel technique to determine the gate and drain bias dependent series resistance in drain engineered MOSFET's using one single device," *IEEE Trans. Electron Devices*, vol. 43, pp. 1478–1488, Sept. 1996.

[13] Shunji Seki, Osamu Kogure, and Bunjiro Tsujiyama, "A Semi-Empirical Model for the Field-Effect Mobility of Hydrogenated Polycrystalline-Silicon MOSFET's," IEEE Transactions On Electron Devices, Vol. 35, No. 5, Pp.669-674, May 1988.

[14] Kausar Banoo, "Direct Solution Of The Boltzmann Transport Equation In Nanoscale Si Devices" (A Thesis), Purdue University, December 2000.

[15] Ting-wei Tang and Haitao Gan, "Two Formulations of Semiconductor Transport Equations Based on Spherical Harmonic Expansion of the Boltzmann Transport Equation," IEEE Transactions On Electron Devices, VOL. 47, NO. 9, pp.1726-1732, SEPTEMBER 2000.

[16] Bernd Meinerzhagen and Walter L. Engl, "The Influence of the Thermal Equilibrium Approximation on the Accuracy Classical Tw o-Dimensional Numerical Modeling of Silicon Submicrometer MOS Transistors," IEEE Transactions On Electron Devices, VOL. 35, NO. 5, pp.689-697, MAY 1988.

2006 5th International Power Electronics and Motion Control Conference

Investigation on the Factors Affecting Inrush Current of Transformers Based on Finite Element Modeling

Dr. M. Reza Feyzi and Dr. M. B. B. Sharifian
Faculty of Electrical and Computer Engineering, University of Tabriz, Tabriz, Iran
feyzi@tabrizu.ac.ir   sharifian@tabrizu.ac.ir

Abstract— A finite element model of a single phase transformer coupled with a voltage fed electric circuit is presented. The model can be used for various purposes. It is also capable to include the eddy-current effect in the core and/or in solid bars, if needed. A time-stepping finite element solution is used to study the influence of some major parameters on the inrush current level, including the primary winding distance from the core, the primary winding resistance, residual magnetic flux density in the core and its polarity. Simulation results showed a good agreement with corresponding test results..

Keywords-finite element method; inrush current; transformer

I.   INTRODUCTION

Inrush current is one of the power system transients that may occur when a transformer is switched on, even if the transformer is not loaded. It is one of the serious problems in both power systems and domestic networks. It not only may result in unnecessary tripping of the protective relays, but causes a voltage dip in the power system or the network. The peak value of inrush current may exceed ten times that of rated current. This value depends on various factors including the B-H characteristics of the iron core, the peak voltage and in particular its phase angle at the instant of switching, the resistance of the primary winding, the internal impedance of the power supply, and the magnitude, and more importantly, the polarity of the residual magnetic flux density in the core at the instant of switching. Many researchers have studied the inrush current from different points of views.

Paul C. Y. Ling and Amitava Basak [1] conducted a set of laboratory tests on a 2.5kVA transformer and investigated the effect of phase angle and the residual flux density on the inrush current level. A similar work was carried out by C. E. Lin *et. al* [2] who proposed an analytical method for the calculation of the inrush current. Yacamini and Bronzeado [3] developed a transformer model based on the physical dimensions of the transformer. They divided the ferromagnetic core and air space within the transformer into several portions in order to improve the accuracy of the model. However, this may not be accurate enough in particular when the saturation occurs.

N. Richard and N. Szylowicz [4] carried out a set of simulations and compared corresponding results from an analytical method by using a permeance model, a two-dimensional (2-D) and a three-dimensional (3-D) finite element model. Choosing the 3-D model as reference, surprisingly they concluded that the permeance model could be more accurate with %25 error compared to the %40 error of the 2-D model. André Gaudreau *et. al.* [5] conducted a set of comprehensive tests along with analyses and developed a flux density dependant model with an instantaneous magnetizing resistance for power transformers. The developed model can be used to predict the transformer losses during over excitation and inrush current. Sng Yeow *et. al.* [6] used a wavelet-based model to distinguish the inrush current from internal fault currents of transformers. Using the waveform character of spiry pulse in the interior half cycle of the transformer current, Hu Yufeng *et. al.* [7] proposed a novel theory to identify the inrush current from the high fault currents.

In a recent work, Cheng *et. al* [8] showed that the inrush current can be reduced by increasing the distance of the primary winding from the core. This, in fact increases the radius of the primary coil and results in the increase of the resistance and the leakage reactance of the winding. They ignored the primary winding resistance and also assuming an extremely saturation in the core area, neglected the influence of iron core and evaluated inductances from classic formula in an isotropic field. Although this may be acceptable in some parts of the core, but it can not be accurate enough for entire core region. More over, it can not be valid after a few cycles when the saturation level is damped down with time.

In this paper, the influence of the distance of primary winding from the core, the effect of residual magnetic flux density in the iron and also some other factors are studied by using a finite element (FE) model. The effect of residual magnetic flux has been ignored in most of the works in the literature. A 5kVA, 220/110, 50Hz transformer has been modeled and the simulation results have been compared with the corresponding test results.

II.   THEORETICAL BACKGROUND

The non-linear nature of the transformer, in particular during the switching on transients, dictates the application

1-4244-0448-7/06/$25.00 ©2006 IEEE   204

of a finite element model. The transformer is a three-dimensional system and therefore, a 3-D model is usually required in order to include all influencing parameters, such as the effect of laminations, the induced eddy current in the core, the effect of the winding taps and so forth. On the other hand, a 3-D model needs high computer resources. Some simplifications, such as part modeling and/or choosing coarse mesh sizes, may be used to overcome this disadvantage.

The variation of the field parameters is usually very low along the direction perpendicular to the laminations (z axis). Therefore, a 2-D model can be constructed for the transformer. Most of the essential solution results can be extracted from such a model, including the magnetic field distribution in the core and the surrounding air/oil, the non-uniform distribution of the current in case of solid conductors (eddy current effect in the conductors), leakage and mutual inductances of the windings. It also can be used for further investigations such as thermal analysis and study of internal forces in the transformer. The generated model is capable to handle all of aforementioned specifications. However, only the inrush current related factors are investigated in this work.

The governing equation in a 2-D magnetic field can be expressed as (1).

$$\frac{\partial}{\partial x}(\nu\frac{\partial \mathbf{A}}{\partial x}) + \frac{\partial}{\partial y}(\nu\frac{\partial \mathbf{A}}{\partial y}) = -\mathbf{J_s} + \sigma\frac{\partial \mathbf{A}}{\partial t} \qquad (1)$$

where:

$\mathbf{A}$	The magnetic vector potential
ν	Reluctivity of iron
σ	Conductivity of the conducting regions
t	Time
$\mathbf{J_s}$	Current density

In this work, the circuit equation of a voltage fed winding (KVL) is combined to the field equations and solved simultaneously. This allows more realistic modeling of the transformer compared to a current fed winding with pre-determined current. The developed program allows nonlinear isotropic or anisotropic magnetic characteristics for the iron core, such as that of the grain oriented materials. However, the anisotropy along z axis, i.e. the effect of laminations cannot be included in the 2-D model.

III. MODEL GENERATION AND SIMULATIONS

A single phase, 5kVA, 220V/110V, 50Hz dry-type transformer is modeled. Fig. 1 shows different regions of the transformer as well as some important dimensional data.

As shown, the secondary winding is split into two parts. A fraction of x of the secondary winding is located

between the core and the primary winding. Therefore, the distance of the primary winding from the core will increased as x increases. Consequently, the mean length of the primary coil turns and the primary winding resistance will increase with x.

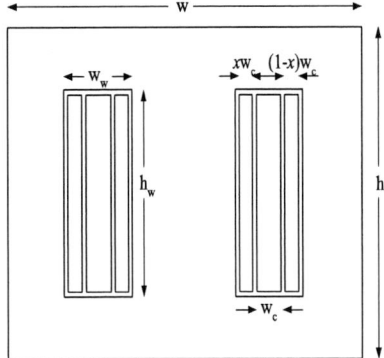

Fig. 1. Dimensions of the transformer

The meshed areas of the model as well as its details in the coil region are shown in Fig. 2.

a. The meshed regions

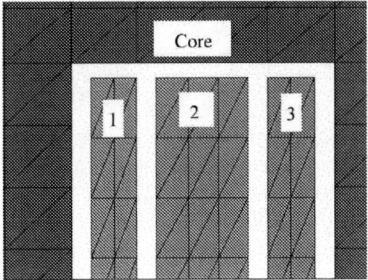

1. LV coils, part1; 2. HV coil; 3. LV coils, part 2

b. Zoomed view of a part of coil region

Fig. 2. Finite element model of the transformer

As shown, triangular mapped elements are used for the core and the winding areas. A large surrounding area is also included around the model in order to include the far-field effect; this part is not shown in the figures. A total of 3160 first order elements with 1618 nodes were used in the model. No boundary conditions are imposed around the core; otherwise, the leakage flux would not be simulated correctly.

Since the secondary is open-circuited within this study; this winding could be omitted from the FE model without loosing the accuracy of the results. However, it is still shown for the sake of clarity. Some of the required parameters of the model are given in table 1. The magnetization curve of the transformer core is shown in Fig. 3. The magnetization curve data was extracted from a set of test results on the modeled transformer. The extracted data was processed for the smoothing purpose; otherwise, the non-linear finite element model would not converge.

TABLE 1. SOME PARAMETERS OF THE MODELED TRANSFORMER

S	5kVA	h	200mm
V_1	220V	w	210mm
V_2	110V	h_w	130mm
N_1	109 Turns	w_w	35mm
N_2	55 Turns	w_c	12mm
Depth	128mm		

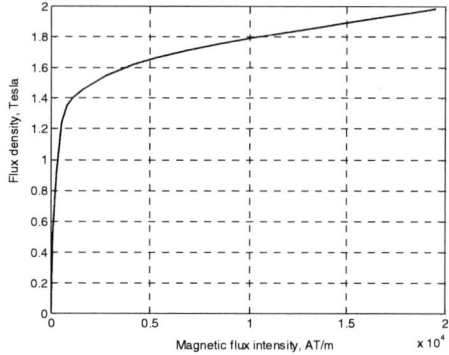

Fig. 3. Magnetization curve of the core material

IV. SIMULATION RESULTS

The influences of the distance of the primary winding from the core (d_1), the residual magnetic flux density in the core (B_r) and its polarity and the resistance of the primary winding (R_1) are individually investigated in this work. To insure the worst case is investigated, the rated sinusoidal voltage is assumed to be applied to the primary winding with zero phase angles in all cases.

Effect of d_1 and the primary resistance

As shown in Fig. 1, the secondary winding is split in two parts to allow larger distance between the primary winding and the core without loosing the winding area. x pu of the secondary winding is located next to the core to produce a total of d_1 [mm] distance between the inner surface of the primary winding from the core. This can be

simply performed in low-voltage transformers. Changing x results in different values of d_1 and consequently the leakage reactance of the primary winding. x was varied from zero to 1 in 5 steps and the peak values of the inrush current at each step was recorded. This was carried out in two different conditions:

a. the winding resistance was frozen at its value with $x = 0$

b. the winding resistance is changed as its mean length is varied with d_1

Some fictitious positions of the primary winding with $x = 0$ were also simulated in order to investigate the effect of the distance from core on the inrush current peak.

Table 2 shows the simulation results for the aforementioned cases. Fig. 4 shows the variation of I_{peak} with d_1 for the same cases, normalized to their values at $d_1 = 0.1\,\text{mm}$. As shown, the peak value of the inrush current reduces by about 7.8% when the variation of the primary resistance is included.

TABLE 2. SIMULATION RESULTS FOR DIFFERENT VALUES OF d_1

x [mm]	d_1 [mm]	$I_{inrush,peak}$ [A]	
		Case a	Case b
Additional Data with x = 0 (just simulation)	0.1	150.6	155.2
	1	148.8	151.5
	3	143.2	144. 6
0	5	138.1	138.1
1/6	7	135.4	134.8
2/6	9	132.0	129.2
3/6	11	130.3	125.1
4/6	13	126.7	121.6
5/6	15	124.4	116.8
6/6	17	121.9	113. 5

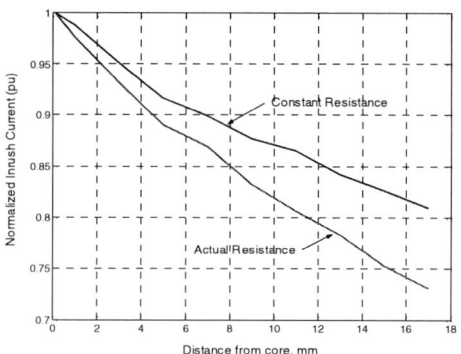

Fig. 4. Variation of I_{peak} with d_1 when:

a. the primary resistance is assumed to be constant

b. the primary resistance is varied with x

Variation of the inrush current with time and magnetic flux lines at $t = 0.01\,\text{Sec}$. are given in figures 5 and 6 respectively. A zoomed part of the magnetization current after 2.1 seconds is also shown in Fig. 5. The details of the magnetic flux lines inside the window area are also shown as a part of Fig 6. As expected, despite an extreme saturation of the core, the major part of the flux passes through the core.

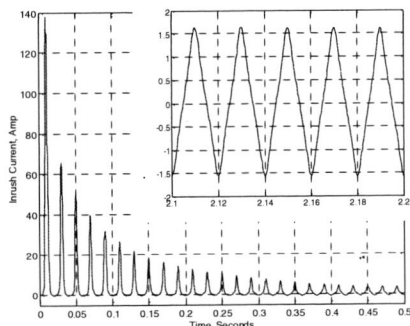

Fig. 5.   Variation of I_{inrush} with time

Fig. 6.   Magnetic flux lines in the core at $t = 0.01$ Sec.

Effect of $B_{residual}$ and its polarity

The transformer was exited with low DC currents in order to produce an initial flux density in the core prior to the application of the rated sinusoidal voltage. This was carried out at several DC current levels in two different directions. The peak values of the resulted inrush currents are tabulated in table 3 and their normalized values are plotted in Fig. 7.

As expected, the inrush current increases considerably as $B_{residual}$ increases with positive polarity. At the same time, it reduced significantly for the negative values of $B_{residual}$.

V.   VERIFICATION OF THE RESULTS

Some laboratory tests were carried out on the modeled transformer in order to check the validity of the simulation results. The tests were carried out at two different values of d_1, 5mm and 11mm within the simulated cases (see table 2). The instantaneous values of the inrush current and primary voltage were stored in a computer via a data-acquisition card in different test conditions while the switching time of the applied sinusoidal voltage was adjusted to occur at zero phase angles in all tests. The

corresponding no-load currents at each step were also recorded.

TABLE 3. PEAK VALUES OF THE INRUSH CURRENTS WITH RESIDUAL MAGNETIC FLUX DENSITY

$|B_r|$ [T]	$I_{p,inrush}$ [A] when B_r is	
	Negative	Positive
0	138.1	138.1
0.052	75.9	138.3
0.104	65.3	163.9
0.156	59.19	182.6
0.208	55.0	195.6
0.260	51.4	208.9
0.312	49.4	216.6
0.364	47.76	224.23
0.418	45.82	229.14

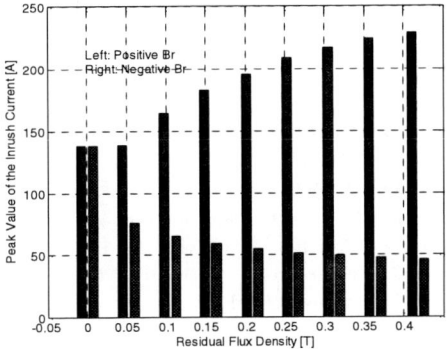

Fig. 7.   Variation of the inrush current with $B_{residual}$

The 110-turn (220V) winding was energized during the first case with $d_1 = 5$mm. However, for the sake of test setup and time saving, the existing 55-turn (110V) winding of the transformer was energized as the primary for $d_1 = 11$mm case and the results were converted to equivalent 110-turn winding by using the winding turn ratio. A time delay switching circuit was used to ensure the switching occurs at zero phase angle of the applied sinusoidal voltage. This was double checked by using the recorded data for the applied voltage across the winding.

The peak values of the inrush current and also the no-load currents at $d_1 = 5$mm, $d_1 = 11$mm were measured. The test results, the corresponding simulation results and the percentage of the error for each case are tabulated in table 4.

As can be seen, the measured values are slightly smaller than that of corresponding simulation results. This can be attributed to the internal impedance of the voltage source which is not included during the simulation; it is not unexpected to have a larger error in the case of the evaluated inrush current compared to that of the no-load currents. Never the less, the simulation results seem to be satisfactory.

TABLE 4 COMPARISON OF SIMULATION RESULTS AND THE TEST RESULTS WHEN D_1 IS 5 AND 11mm

	Peak Current	Simulation results	Test results	Percent Error
$d_1 = 5$mm	Inrush	138.1	128.4	7.02
	No-load	1.60	1.55	3.13
$d_1 = 11$mm	Inrush	125.1	117.3	6.23
	No-load	1.60	1.55	3.13

VI. CONCLUSION

The influence of some of the major parameters on the peak value of inrush current in a 5kVA single phase transformer was investigated. A coupled electric circuit with a sinusoidal voltage source and a finite element model was used for this purpose. It was shown that the peak value of the inrush current decreases as the distance of the primary winding increases from the core. This is attributed to the increase of the leakage inductance of the winding. It was also shown that the winding resistance considerably contributes to the total input impedance of the primary winding during the the first few cycles just after switching on the transformer. Note that the primary inductance reduces significantly during this period as a result of highly saturated core. The influence of the residual magnetic flux density in the core and its polarity was also studied. A significant increase or reduction occurs in the inrush current level when the transformer is switched on with negative or positive residual flux densities in the core respectively. Variation of this current is highly dependant on the magnetization characteristics of the core material. Some of the simulation results were compared with test results and showed a good agreement.

REFERENCES

[1] Paul C. Y. Ling and Amitava Basak, "Investigation of Magnetizing Inrush Current in a Single-Phase Transformer", IEEE Trans. on Magnetics, Vol. 24, No. 6, Nov. 1988, pp. 3217-3222.

[2] Lin C. E., Cheng C. L., Huang C. L. and Yeh J. C., January 1993, "Investigation of Magnetizing Inrush Current in Transformers", IEEE Trans. on Power Delivery, Vol. 8, No. 1, pp. 246-254.

[3] Yacamini R. and Bronzeado H. Nov. 1994, "Transformer Inrush Calculations Using a Coupled Electromagnetic Model", IEE Proc., Science, Measurements and Technology, Vol. 141, No. 6, pp. 491-498.

[4] Richard N. and Szylowicz N., September 1994, "Comparison Between a Permeance Network *Model* and a 2D Finite Element Model for the Inrush Current Computation in a Three Phase Transformer", IEEE Trans. on Magnetics, Vol. 30, No. 5, pp. 3232-3235.

[5] Gaudreau A., Picher P., Bolduc L. and Coutu A., Oct. 2002, "No-Load Losses in Transformer Under Overexcitation/Inrush-Current Conditions: Tests and a New Model", IEEE Trans. on Power Delivery, Vol. 17, No. 4, pp. 1009-1017.

[6] Sng Y H and Wang Qin, December 2000, "A Wavelet-based Method to Discriminate Between Inrush Current and Internal Fault", PowerCon 2000, Australia, pp. 927-931.

[7] Yufeng Hu, Deshu Chen, Xianggen Yin and Zhe Zhang, Sept. 2003, "A Novel Theory for Identifying Transformer Magnetizing Inrush Current", Transmission and Distribution Conference and Exposition, 2003 IEEE PES, Vol. 1, pp. 274-278.

[8] Cheng C. K., Liang T. J., Chen J. F., Chen S. D. and Yang W. H., May 2004, "*Novel* approach to reducing the inrush current of a power transformer", IEE Proc. Electr. Power Appl., Vol.151, No.3, pp. 289-295.

2006 5th International Power Electronics and Motion Control Conference

An Improved Support Vector Machine Method for Harmonic and Inter-harmonic Detecting

Ma Li, Liu Kaipei and Lei Xiao

School of Electrical Engineering, Wuhan University, Wuhan, China

mali99532@126.com

Abstract—Based on the linear regressive model of support vector machines, iterative re-weighted least squares of the support vector machines for signal spectral analysis is introduced by virtue of abstract expressions of cost functions, and the uniform mathematical analytical expression of support vector machine for solving the problem of signal spectral analysis is obtained. And a new method is presented by changing resolution, namely defining the range of spectrum by taking advantage of wide range with rough resolution at first, and then using narrow range with high resolution. The simulation has showed that this method solves the problem of large calculation and slow calculating speed.

Keywords-Discrete Fourier Transform; iterative re-weighted least squares; support vector machine

I. INTRODUCTION

Discrete Fourier Transform (DFT) is the radical method for harmonic and inter-harmonic detecting in the power system. The precondition of its application is that the detected signal is periodic or close to periodic, and the sampling data must cover integral period of the original signal, otherwise, spectrum leakage and spectrum aliasing in sampling data will occur. When harmonic and inter-harmonic appear at the same time, the hypothesis of signal periodicity may not come into existence, or the period is very long, even unknown. Then it leads that the hypothesis of signal integral periodicity is impossible (or hard) to come true, and it results in the phenomena of spectrum leakage, aliasing and the fence effect, or the number of sampled data is very large in an integral period (satisfying integral period as well as Shannon sampling theory), and the computation load will be increased. If the inter-harmonic frequency is changeable, it is far more difficult to analyze spectrum by DFT. Support Vector Machines are based on VC dimension of statistical learning theory and structural risk minimization principle. According to the information of limited samples, the optimal compromise between the complicacy of the model (learning precision of given training samples) and learning ability (the ability to recognize random samples faultlessly) has been sought and expected to get the best capacity of generalization. Support Vector Machines have been successfully applied to many areas in the last years,

especially for speech recognition, character recognition, and forecast of time series. This statistical learning approach can avoid the problem of constructing network structures that are demanded by some algorithms such as neural networks. Its results have been proved more precise than those of other methods of pattern recognition and regression forecast [1]. This paper will discuss the applied problem of SVM for harmonic and inter-harmonic detecting in the power system.

II. LINEAR REGRESSION OF SVM

Given a labeled training data set $\{(x_1, y_1), (x_2, y_2),..., (x_l, y_l)\}$, where $x_i \in R^n$, $y_i \in R$, SVM defines a kind of machines that can make sure the mapping relation between x and y, namely $x \to f(x, \alpha)$, with α as alterable parameter which can be sure by learning (training) known data set. And the given ε is required as the largest error for $f(x, \alpha)$ towards y, at the same time the regressive curve is smooth enough.

The linear regression defines mapping function $f(x) = \langle \omega, x \rangle + b, \omega \in R^n$, $f(x) \in R$, and the smallest α is required to make sure that the curve is smooth. The least Euclidean norm of vector α is a common method with its mapping error acceptable to fall into the scope of ε. It can be expressed as the following model [8]:

$$Min \quad \frac{1}{2}\|\omega\|^2 \qquad (1)$$

$$st. \quad y_i - \langle \omega, x_i \rangle - b \le \varepsilon \quad , \quad \langle \omega, x_i \rangle + b - y_i \le \varepsilon \qquad (2)$$

It is very difficult to satisfy (2) sometimes. Considering that regressive errors could be permitted, the slack variables $\xi_i^{(*)}$ are introduced and $\xi_i^{(*)} \ge 0$; cost functions $\sum L(\xi_i^{(*)})$ are constructed, where $\xi_i^{(*)}$ stand for ξ_i and ξ_i^*. Expressions (1) and (2) can be rewritten as:

$$Min \quad \frac{1}{2}\|\omega\|^2 + C\sum_{i=1}^{l} L(\xi_i^{(*)})$$
$$st. \quad y_i - \langle \omega, x_i \rangle - b \le \varepsilon + \xi_i \qquad (3)$$
$$\langle \omega, x_i \rangle + b - y_i \le \varepsilon + \xi_i^*$$
$$\xi_i^{(*)} \ge 0$$

1-4244-0448-7/06/$25.00 ©2006 IEEE

Where C is the penalization constant of such deviations when regressive precision exceeds permitted value, and $C>0$, $\xi_i^{(*)}$ are the regressive slack variables introduced in SVM to balance the empirical risk and the capability of generalization. Also $\xi_i^{(*)}$ serve as a solution for the regressive algorithm when errors are larger than ε in some points.

III. HARMONIC AND INTER-HARMONIC ANALYTICAL MODEL

A discrete time series $\{y_{t_k}\}$ can be expressed as a Fourier series, $k=1,2,\ldots N$. The sinusoidal approximation is given by

$$y_{t_k} = \sum_{i=1}^{N_w} A_i Cos(w_i t_k - \phi_i) + e_{t_k} \qquad (4)$$

Where A_i and ϕ_i is respectively the amplitude and the phase that correspond with frequency w_i, e_{t_k} is the model error for the kth sampling.

Suppose $c_i = A_i Cos(\phi_i), d_i = A_i Sin(\phi_i)$, and introduce them to (4), one can get:

$$y_{t_k} = \sum_{i=1}^{N_w} [c_i Cos(w_i t_k) + d_i Sin(w_i t_k)] + e_{t_k} \qquad (5)$$

In the linear regression, the regressive function is $f(x) = \langle \omega, x \rangle + b, \omega \in R^n, f(x) \in R$. Comparing with (5), we can get the following equations:

$$x_i = [Cos(w_1 t_i), \ldots, Cos(w_{N_w} t_i), Sin(w_1 t_i), \ldots, Sin(w_{N_w} t_i)]^T$$

$$c = [c_1, c_2, \ldots, c_{N_w}]^T, d = [d_1, d_2, \ldots, d_{N_w}]^T, \omega = [c^T, d^T]^T \qquad (6)$$

With the available (6), (3) can be arranged as the form of (7) that is the goal functions of spectrum evaluation in SVM.

$$Min \quad \frac{1}{2} \sum_{j=1}^{N_w} (c_j^2 + d_j^2) + C \sum_{k=1}^{N} L(\xi_k^{(*)})$$

$$st. \quad y_{t_k} - \sum_{i=1}^{N_w} c_i Cos(w_i t_k) - \sum_{i=1}^{N_w} d_i Sin(w_i t_k) \le \varepsilon + \xi_k \qquad (7)$$

$$-y_{t_k} + \sum_{i=1}^{N_w} c_i Cos(w_i t_k) + \sum_{i=1}^{N_w} d_i Sin(w_i t_k) \le \varepsilon + \xi_k^*$$

$$\xi_k, \xi_k^* \ge 0$$

Thus obtained ones are the goal functions of spectrum evaluation algorithm based on SVM. To work out the value of c_j, d_j from (7), we find that the amplitudes of harmonic and inter-harmonic w_j are $\sqrt{c_j^2 + d_j^2}$, and the phases are $arctan(d_j/c_j)$. If the harmonic or inter-harmonic signal with frequency w_k occurs in the analyzed signal, the corresponding amplitude of this frequency spectrum $A_k \ne 0$, otherwise it equals to zero. This is the ground on which the algorithm carries out spectrum evaluations.

IV. SVM METHOD FOR HARMONIC AND INTER-HARMONIC DETECTING

A. The Model of SVM

Expression (7) can be rewritten as:

$$Min \ L_{PD} = \frac{1}{2}\|\omega\|^2 + \sum_{k=1}^{N} \alpha_k (y_{t_k} - \omega^T x_k - \varepsilon)$$
$$+ \sum_{k=1}^{N} \alpha_k^* (-y_{t_k} + \omega^T x_k - \varepsilon) \qquad (8)$$
$$+ C \sum_{k=1}^{N} L(\xi_k^{(*)}) - \sum_{k=1}^{N} (\beta_k \xi_k + \beta_k^* \xi_k^*) + \alpha_k \xi_k + \alpha_k^* \xi_k^*]$$

where α_k、α_k^*、β_k、β_k^* are Lagrange multipliers .

For (8), a general method by using norm quadratic optimization technique to settle the dual problem will make the operation speed extraordinarily slow. The main reason is that the SVM needs to calculate and memorize kernel function matrix when sampling number is quite large, and it demands huge memory. For example, while the sampling number is beyond 4000, 128 million of memory will be required. What's more, the SVM will need a mass of matrix computation during the course of quadratic optimization. In many cases, optimization algorithm is the main part of costing the whole time. And the training speed of the SVM method is also the major problem to limit its application. In recent years, there are many algorithms brought out for this method to figure out dual optimal problem, and a basic idea in most of these algorithms is iterative, i.e. the primary problem is decomposed into some sub-problems according to certain iterative method.

To avoid the problems which occur in solving norm quadratic optimal issues to deal with dual problem, iterative re-weighted least squares(IRWLS) procedure can be used [2][3] . It can be described as follows:

Firstly suppose:

$$\begin{cases} e_k = y_{t_k} - \omega^T x_k & e_k^* = -y_{t_k} + \omega^T x_k \\ \lambda_k = \dfrac{2\alpha_k}{y_{t_k} - x_k^T \omega - \varepsilon} & \lambda_k^* = \dfrac{2\alpha_k^*}{-y_{t_k} + x_k^T \omega - \varepsilon} \end{cases} \qquad (9)$$

Then put (9) into (8), we can get:

$$Min \ L_{PD} = \frac{1}{2}\|\omega\|^2 + \frac{1}{2}\sum_{k=1}^{N} \lambda_k (e_k - \varepsilon)^2 + \frac{1}{2}\sum_{k=1}^{N} \lambda_k^* (e_k^* - \varepsilon)^2$$
$$+ C \sum_{k=1}^{N} L(\xi_k^{(*)}) - \sum_{k=1}^{N} (\beta_k \xi_k + \beta_k^* \xi_k^*) + \alpha_k \xi_k + \alpha_k^* \xi_k^*] \qquad (10)$$

In terms of KKT conditions [4], the product of Lagrange multipliers and its corresponding constraints is zero on the corresponding point of gained maximum. If KKT conditions are introduced, then the result derived from (10) can be expressed as:

$$\frac{\partial L_{PD}}{\partial \omega} = \omega - X^T D_\lambda [Y - X\omega - I\varepsilon] + X^T D_{\lambda^*} [-Y + X\omega - I\varepsilon] = 0 \quad (11)$$

$$\frac{\partial L_{PD}}{\partial \xi_k^{(*)}} = C\frac{dL(\xi_k^{(*)})}{d(\xi_k^{(*)})} - \alpha_k^{(*)} - \beta_k^{(*)} = 0 \qquad \forall k=1,\cdots,n \quad (12)$$

$$\beta_k^{(*)}\xi_k^{(*)} = 0 \qquad \forall k=1,\cdots,n \quad (13)$$

Where $X = [x_1,...,x_N]$, and (11) can be written as:

$$\omega = [I + X^T D_{\lambda+\lambda^*} X]^{-1}[X^T D_{\lambda+\lambda^*} Y - X^T D_{\lambda-\lambda^*} I\varepsilon] \quad (14)$$

Where (14) is the iterative equation of ω, and Y is the vector of sampling. $D_{\lambda+\lambda^*}$ denotes the diagonal matrix whose main diagonal kth element is $\lambda_k + \lambda_k^*$; $D_{\lambda-\lambda^*}$ denotes the diagonal matrix whose main diagonal kth element is $\lambda_k - \lambda_k^*$.

According to (9), (12) and (13), we can get:

$$\lambda_k^{(*)} = \begin{cases} 0 & e_k^{(*)} - \varepsilon < 0 \\ \dfrac{2C}{e_k^{(*)} - \varepsilon}\left.\dfrac{dL(\xi^{(*)})}{d(\xi^{(*)})}\right|_{\xi^{(*)}} & e_k^{(*)} - \varepsilon \geq 0 \end{cases} \quad (15)$$

In summary, the IRWLS procedure consists of the following steps.

(1) Start with an arbitrary ω_0, set n = 1;

(2) Calculate errors $e_k = y_{t_k} - \omega_{n-1}^T x_k$ $\quad e_k^* = -y_{t_k} + \omega_{n-1}^T x_k$;

(3) Calculate λ_k, λ_k^* as given in (15);

(4) Solve (14) to obtain ω_n;

(5) Come to an end if $\|\omega_n - \omega_{n-1}\|$ is smaller than a value or times reach the maximum; otherwise, set n=n+1, then return to step (2).

B. The Confirmation of Cost Functions

Figure 1 illustrates the general cost functions [7], among which the cost function in figure 1(a) corresponds to the norm of least square error; another cost function in figure 1(b) is a Laplacian one which is less sensitive than figure 1(a) to outliers; when the data distribution is unknown, the cost function in figure 1(c) presented by Huber is optimal in performance. However, these three cost functions above cannot make the support vector sparse. To solve this problem, Vapnik [5] presents the cost function in figure 1(d), which is approximate to Huber cost function but capable of making support vector sparse.

In the regressive method, it is necessary to choose an appropriate cost function and other additive constraint possibly needed, thus better understanding of the problem and a good knowledge of noise distribution is required. Without the knowledge, Huber function could be the best choose. ε -insensitive cost function, an eclectic cost function was thereby introduced by Vapnik to share parts of the characteristics of Huber cost function. Though capable of making the obtained support vector spare, it invites heavy computational load, and limitations also exist [6].

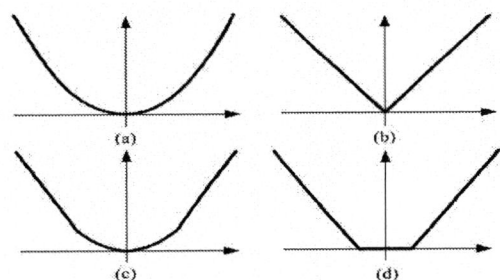

Figure 1. Cost functions

Huber cost function is used in this paper with its expression as follows:

$$L(\xi) = \begin{cases} \dfrac{1}{2\sigma}\xi_k^2 & 0 < \xi_k < \sigma \\ \xi_k - \dfrac{1}{2}\sigma & \xi_k \geq \sigma \end{cases} \quad (16)$$

Considering $\xi_k = e_k - \varepsilon$, we can get:

$$\lambda_k = \begin{cases} \dfrac{2C}{e_k - \varepsilon} & e_k > \sigma + \varepsilon \\ \dfrac{2C}{\sigma} & \varepsilon \leq e_k \leq \sigma + \varepsilon \\ 0 & elsewhere \end{cases} \quad \lambda_k^* = \begin{cases} \dfrac{2C}{e_k^* - \varepsilon} & e_k^* > \sigma + \varepsilon \\ \dfrac{2C}{\sigma} & \varepsilon \leq e_k^* \leq \sigma + \varepsilon \\ 0 & elsewhere \end{cases} \quad (17)$$

C. Simulation Example Research

Spectrum analysis for the following signal:

$$y(t) = \cos(2\pi \times 50 \times t) + v \quad (18)$$

Where v is a Gaussian, white noise. In the simulation, the sampling frequency is 200Hz, and sampling times are 200. Here, the sampling series are computed and analyzed respectively by DFT algorithm and the new one proposed in this paper. For the latter, the resolution in frequency domain is set as 1Hz, the analytical domain is $0 \sim 100$Hz, and the parameters of cost functions are $\varepsilon = 0$, $C = 0.02$, $\sigma = 0.2$ (other simulations in this paper are also conducted on this condition).

From figure 2, as for SVM, the amplitude-frequency figure of the signal can be accurately illustrated whether the samplings cover integral periods or not. While the amplitude-frequency figure of the signal given by DFT is accurate only under the condition that the samplings cover integral periods and the sampling rates satisfy the Shannon sampling theory, otherwise spectrum leak will occur. To prove it, we can set the signal as:

$$y = \cos(2\pi \times 50 \times t) \quad (19)$$

Sample at intervals of 0.005 sec. during 1.55 sec. (samplings can not cover integral periods), and 310 samplings can be collected. Then the sampling series are analyzed respectively by SVM and DFT. Figure 3 has showed the spectrums: the new algorithm brings out correct spectrum while spectrum leak does occur under the second circumstance (under DFT circumstance).

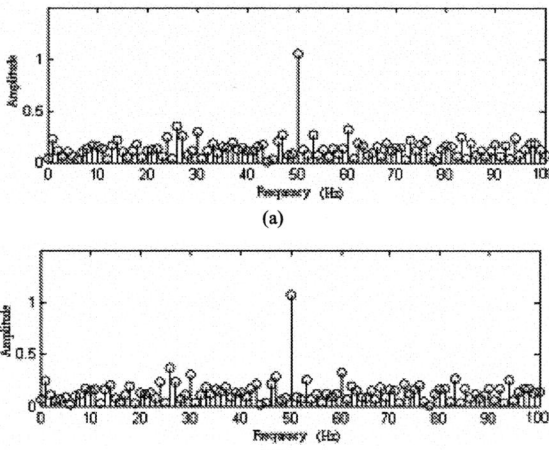

(a)

(b)

Figure 2. Spectrum analysis for the same signal by SVM and DFT.
(a) New algorithm. (b) DFT.

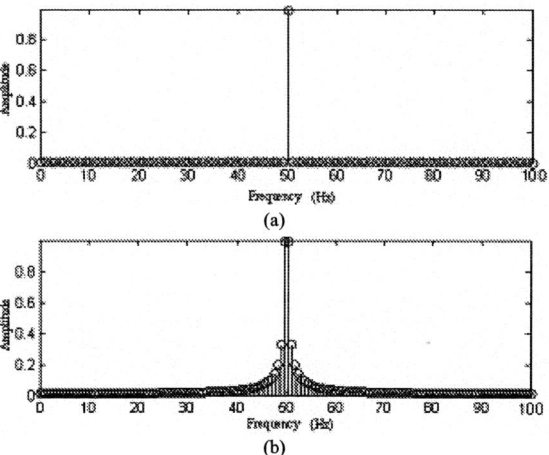

(a)

(b)

Figure 3. The influence of sampling covering the non-integral periods.
(a) New algorithm. (b) DFT.

Compared to SVM algorithm, which is more flexible in setting resolutions in frequency domain, i.e. the resolution can be arbitrarily chosen, resolutions of DFT in frequency domain are the ratio of sampling frequency and counting numbers. Both of them are limited by integral period sampling, which results in limitation of resolutions and disadvantage to analysis, while resolutions of SVM are comparatively more unrestricted in frequency domain, so the application holds much adaptability in application.

D. An Improved Support Vector Machine Method for Harmonic and Inter-harmonic Detecting

As indicated by the above analysis SVM regressive algorithm based on IRWLS enjoys more flexibility in setting resolutions, so it avoids constrains of resolutions in DFT. However, although using IRWLS has greatly reduced computations, it's still heavier than that of DFT.

The analysis of algorithm deducing shows that the leading factor determining computation load is X matrix.

The number of columns of X matrix corresponds to the length of entire sampling series, and correspondingly the number of rows of X matrix is twice as large as the number of points to be counted in frequency domain for the algorithm. When the rank of X matrix is very high, it takes much time to create this matrix, and expression (14) will be more complex as well. Hence, it is proposed that reducing samplings or shortening the domain of frequency can reduce computation load. In view of this, application of this algorithm has been improved so that computation load can be cut down.

Detailed algorithm follows:

Supposing the sampling frequency of signal is f_s and signal sampling aggregate is R.

Give data $\{y_1, y_2, \cdots, y_{n-1}, y_n\}$ from aggregate R, and set the analytical domain of frequency as $0 \sim 0.5 f_s$ for the algorithm, with a frequency resolution of f_R.

Enter algorithm based on IRWLS flow and get the signal spectrum. Frequencies whose amplitudes are larger than other ones are gathered as aggregate f_T and the analytical domain of frequencies are limited close to elements of the aggregate f_T, namely $[f_T - \Delta f, f_T + \Delta f]$.

Judge whether f_R reaches the demand or not, if so, go to step (4), otherwise confirm the new f_R, get the length of data extracted from R according to the demand of aggregate f_T and frequency resolutions, form new input series, return to step (2).

Stop the computation.

We now illustrate the signal:

$$y = \cos(2\pi \times 556 \times t) \qquad (20)$$

Supposing sampling frequency is 2000 Hz, and sampling time is $0 \sim 1$ sec. So, 2000 sampling points have been collected to create a sampling aggregate R. And the steps are as follows:

(1) Get the preceding 100 points with calculating domain within $0 \sim 1000$Hz and resolution to be 10Hz. Through IRWLS procedure, obtain signal spectrum in figure 4. Solve the mean of amplitudes which are larger than surrounding ones. Then note the points that are larger than the mean and regard them as the elements in the aggregate f_T.

(2) The analytical domain of frequency is obtain as [500, 600] from step (1). The resolution is 1Hz. And get 2000 sampling points from aggregate R, return to IRWLS procedure. After that we can have new signal spectrum in figure 5.

The new method has remarkable effect to reduce computations. Before the new method is adopted, if the sampling number is $0.5/0.0005 = 1000$, 2000×1000 X matrix is needed to make computations with the domain within $0 \sim 1000$Hz and the resolution to be 1Hz. After its adoption, during the first step when the samplings are

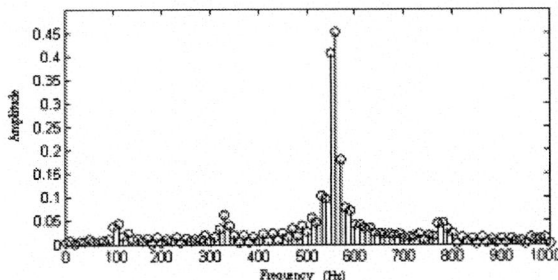

Figure 4. Large-scale sampling in rough resolution

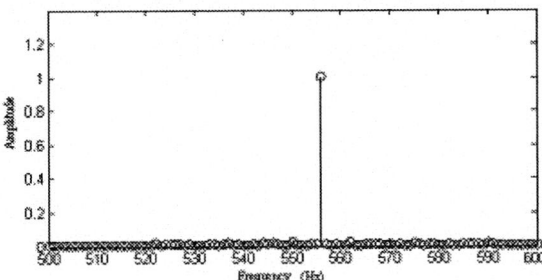

Figure 5. Small-scale sampling in high resolution

0.05/0.0005＝100 and the number in frequency domain is 100, the needed X matrix are 200×100; during the second step when the sampling number is 2000 and the computational number in frequency domain is 600-500＝100, 200×2000 X matrix is needed. Therefore, the method of confirming domains by applying rough resolution is proved to reduce computations greatly.

It is to point out that the amplitude of spectrum plot gained by the new algorithm has great practical significance. The previous deduction shows that the amplitude in the last spectrum figure is the actual signal one. But in simulation, signal amplitude may not be 1 due to lack of samplings. If the sampling number is large enough, the signal amplitude can reach to 1. The reason that causes this situation can be analyzed mathematically:

for the regressive function, the correct curve can be regressed only when the computational number is large enough.

V. CONCLUSION

The SVM linear regressive spectrum analytical algorithm based on IRWLS can be regarded as a supplement to nonparametric spectrum analysis and enrichment for its practical application. The algorithm advanced in this paper shows great significance in inaugurating a new way to analyze signal spectrum by introducing nerve algorithm into the analysis and simultaneously facilitating practical application by setting flexible resolution of frequency in the analysis. What's more, much improvement in the application of the above-mentioned new algorithm is proposed to cut down computations effectively and to improve computational speed, which accordingly improves the efficiency of this algorithm and its application.

REFERENCES

[1] Keerthi S S, Shevade S K, Bhattacharyya C, Murthy K R K. Improvements to platt's SMO algorithm for SVM classifier design [J]. Neural Computation, 2001, (13):637-649.

[2] F. Pérez-Cruz, A. Navia-Vázquez, P. L. Alarcón-Diana, and A. Artés-Rodríguez, "An IRWLS procedure for SVR," in Proc. EUSIPCO 2000,Tampere, Finland, Sept. 2000.

[3] P. W. Holland and R. E. Welch, .Robust regression using iterative re-weighted least squares. Comm. of Stat. Theory Methods, vol. A6, no. 9, pp. 813.27, 1977.

[4] R. Fletcher, Practical Methods of Optimization, 2nd ed. Chichester, U.K. Wiley, 1987.

[5] VAPNIK V. Statistical learning theory [M]. New York: New York Wiley, 1998.

[6] A. J. Smola, B. Schölkopf, and K.-R. Müller, "General cost functions for support vector regression," in Proc. 9th Australian Conf. Neural Networks, Brisbane, Australia, 1998, pp. 79–83.

[7] José Luis Rojo-Álvarez, Manel Martínez-Ramón, Aníbal R. Figueiras-Vidal, Ana García-Armada, Antonio Artés-Rodríguez, "A Robust Support Vector Algorithm for Nonparametric Spectral Analysis," IEEE, Nov. 2003，10（11）：320-323

[8] ZHANG Gongxue, "Introduction to statistical learning theory and Support Vector Machines". ACTA AUTOMATIC SINICA, 2000, 26(1):32-42.

A Common Mode and Differential Mode Integrated EMI Filter

Liu Nan   Yang Yugang

(Department of Electrical Engineering, Liaoning Technical University, Fu Xin, China)

E-mail address: nan9905@126.com

Abstract— **As the trend of Power electronics equipment is miniaturization and modularization, it is necessary to reduce the volume of each device in power electronics equipment. Traditional EMI filter, which has many devices, has become the major obstacle to reduce the volume power electronics equipment. Based on the worldwide research of integrated filter, a novel EMI integrated filter constructed with two different types of magnetic core is proposed. Equivalent wiring of Common Mode (CM) and Differential Mode (DM) current are analyzed. CM and DM equivalent circuit model are established. At last, the validity of the structure is certified by Saber software simulation and experiment.**

Keywords —*power electronics; EMI filter; integrated filter; Common Mode (CM); Differential Mode (DM)*

I. INTRODUCTION

Switch-mode power supplies generate high electromagnetic inference (EMI) because of their fast switching action. Because EMI usually exceeds acceptable levels, the emission must be reduced. At present, in most practical cases, there are two ways to solve EMI problem in power electronics: one way is to find the EMI source in power electronics circuit and reduce or eliminate EMI by developing new device and designing new circuit topology. The other way is accomplished by using EMI filter to attenuate high frequency noise. And the second way is also the most common and effective method.

Because the trend of power electronics equipment is miniaturization and modularization, it is necessary to reduce the volume of each device in power electronics equipment. Traditional EMI filter, which has many

---

Project Supported by National Natural Science Foundation of China (50207004)

devices, has become the major obstacle to reduce the volume power electronics equipment. At present, there are many kinds of integrated filter [2] [3] [4]. In literature [2], the integrated filter include one E type magnetic core and one I type magnetic core, two windings are placed in I type magnetic core. The advantages of this structure are that two windings can use one framework and CM and DM inductance value can be adjusted by regulating air-gap of E type magnetic core. But the magnetic core having air-gap affects the CM inductance value. Furthermore, the effective permeability would be reduced. In reference [3], the integrated filter is realized by two U type magnetic core of high permeability and one I type magnetic core of low permeability. The flaw of this structure is that the length of I type magnetic core is processed relatively accurate, besides, DM inductance is not easily adjusted. In reference [4], the I type magnetic core are transversely put on the window of the □ type and 日 type magnetic core. The flaw of this structure is that CM and DM magnetic flux share largely common path, therefore which affect the CM inductance value.

In order to overcome above the flaws, this paper proposes a new structure, which is easily implemented, no air-gap in CM magnetic path, small influence between CM and DM inductance, and has favorable suppression EMI performance. This structure is come from reference [5].

II. PROPOSED STRUCTURE

The new integrated filter is realized by one big loop type magnetic core of high permeability and two small loop type magnetic cores of low permeability. Two small loop type magnetic cores are symmetrical transversely placed on big loop type magnetic core. The two identical windings are placed in big and small loop type magnetic cores. The new integrated filter's structure is shown in Fig.1.

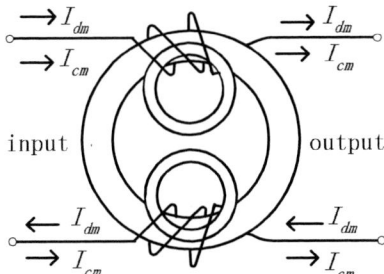

Figure1. A novel CM and DM integrated EMI filter

DM and CM current direction is illustrated in Fig.1. The big loop type magnetic core is to suppress CM noise. According to CM current direction and right-hand screw rule, the big loop magnetic core exhibits high impedance for CM signals but low impedance for DM signals. So the small loop type magnetic core is to suppress DM noise.

For integrated EMI filter, in CM and DM inductance's common path, superposition of CM and DM flux could reduce the effective permeability and CM inductance value. In reference [2] [3] [4], common magnetic path of CM and DM is very long which reduce the CM inductance value and seriously affects filtering performance of CM inductance. However, the new structure in this paper greatly reduces the common path of CM and DM magnetic flux, which improve the filtering performance especially for CM suppressing performance.

III. THEORY ANALYSES

Because of the different source and conducted path of CM noise and DM noise, CM and DM filter should be designed and analyzed respectively.

A. The theory analysis of CM filter

CM noise is come from between alternating current (AC) line (L.N) and ground (E). According to CM current direction, CM current equivalent wiring of this new CM and DM integrated EMI filter is shown in Fig.2. [6-7]

In Fig.2, because two windings are symmetrically placed in the big loop type magnetic core, $N_1=N_2=N$, $L_1=L_2=L$; N_1 and N_2 are the number of windings; L_1 and L_2 are each winding inductance; Φ_1 and Φ_2 are flux; L_{AB} is the inductance between terminal A and B; R is the magnetic resistance.

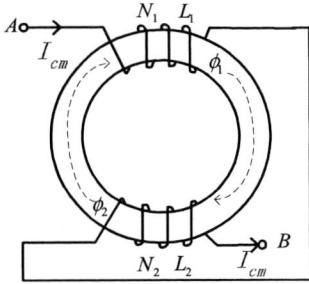

Figure2. Equivalent wiring of CM current

$$L = N^2\big/R \qquad (1)$$

$$L_{AB} = \frac{(N_1+N_2)^2}{R} = \frac{(2N)^2}{R} = \frac{4N^2}{R} = 4L \quad (2)$$

But because of the existence of leakage inductance, in practices,

$$L_{AB} < 4L \qquad (3)$$

From the reference [8], CM equivalent circuit is a LC low-pass filter, which is shown in Fig.3. L_{cm} is CM inductance, C_{cm} is CM capacitance.

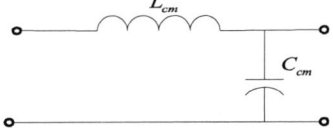

Figure3. CM equivalent circuit

$$L_{cm} = L_{AB} \qquad (4)$$

$$C_{cm} = 2C_y \qquad (5)$$

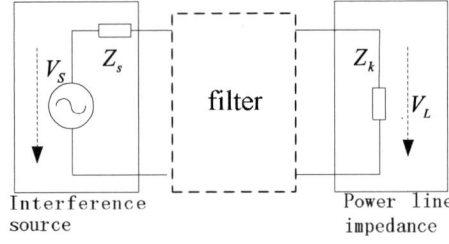

Figure4. Filter design equivalent circuit

$$\left|H(\omega)\right|_{LC} = \left|\frac{V_L}{V_S}\right| = \frac{1}{\omega^2 L_{AB} 2C_y} \qquad (6)$$

Above equation is the approximate transfer function of

CM equivalent circuit at high frequency. Input and output impedance don't affect the CM filtering performance and performance will be much better if the devices' parameter turn bigger. But in fact, C_y has the relation with the leakage current, so C_y is not too big.

B. The theory analysis of DM filter

DM noise exists in between AC lines L and N. According to DM current direction, DM current equivalent wiring of this new CM and DM integrated EMI filter is shown in Fig.5.

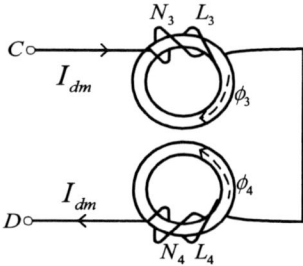

Figure5. Equivalent wiring of DM current

The two identical windings are feed through. Moreover, $N_3=N_4=N$, $L_3=L_4=L_d$; N_3 and N_4 are the number of windings; L_3 and L_4 are each winding inductance; Φ_3 and Φ_4 is flux; L_{CD} is the inductance between terminal C and D.

$$L_{CD} = L_3 + L_4 = 2L_d \qquad (7)$$

But because of the existence of leakage inductance, in practices,

$$L_{CD} < 2 L_d \qquad (8)$$

DM equivalent circuit is a Π type filter, which is shown in Fig.6, L_{dm} is DM inductance, C_{dm} is DM capacitance.

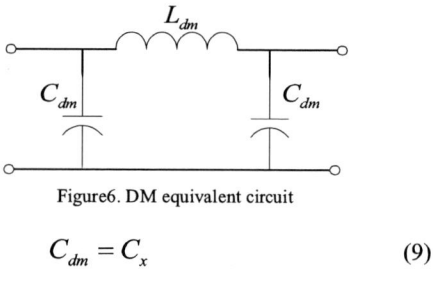

Figure6. DM equivalent circuit

$$C_{dm} = C_x \qquad (9)$$

$$L_{dm} = L_{CD} \qquad (10)$$

Equation (11) is the approximate transfer function of DM equivalent circuit at high frequency. Input

impedance will influence DM filtering performance.

$$\left| H\left(\omega\right)\right|_{LC} = \left|\frac{V_L}{V_S}\right| = \frac{1}{\omega^3 C_x{}^2 L_{CD} Z_S} \qquad (11)$$

IV. RESULTS

A. Simulated results

Based on above analysis, CM and DM equivalent circuit will be simulated by Saber software. Parameter is given by : L_{cm}=1.8mF, C_{cm}= 3nF, L_{dm} = 52.5 μ F, C_{dm}=71nF。The Source and load impedance are 50Ω .1V AC small signal is applied to the input of the filter. The amplitude change of output signal of filter with the input signal frequency is analyzed. Simulated waveforms are shown in Fig.7 and 8.

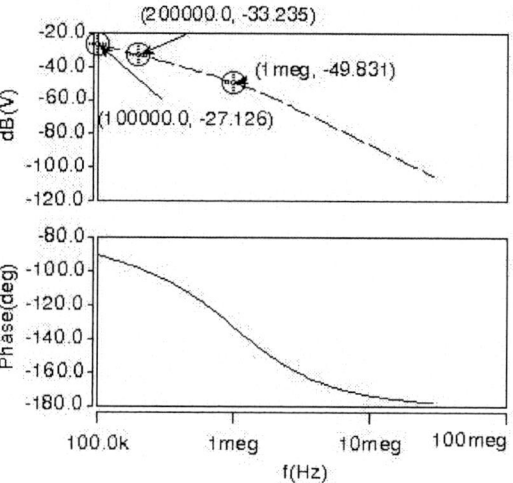

Figure7. Simulated output waveform of CM equivalent circuit

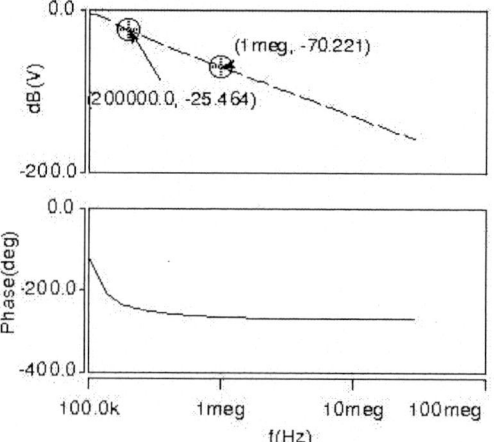

Figure8. Simulated output waveform of DM equivalent circuit

Seen from the simulated waveform, output amplitude is continually reduced when the frequency is becoming higher, and the curve is very smooth. But in fact, the measured curve would divert from the ideal curve because of the impacts of parasitic parameter on high frequency. The parasitic parameter is not in domain of this paper.

B. Experimental results [9-10]

In widely accepted practices, EMI filter can be evaluated by insertion loss (IL). A four-terminal network can represent EMI filter. Its insertion loss is expressed by the ratio of two powers or voltages.

$$IL=20\lg(U_1/U_2) \qquad (12)$$

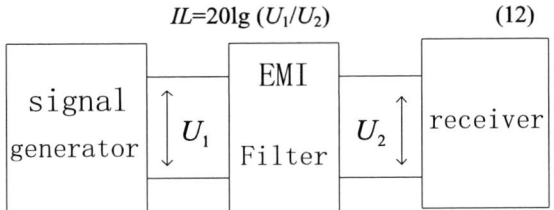

Figure9. Measurement of insertion loss of EMI filter

According to the definition of the insertion loss, designing of experiments is as followed: big loop type magnetic core is PHILIPS Company ferrite TX74/39/13, small loop type magnetic core is MICROMETALS Company ferrocart core T130-26. The 3.5V sine-wave signal is applied to input terminal of the new integrated filter. Output waveform is observed by oscilloscope. The waveforms are shown in Fig.10.

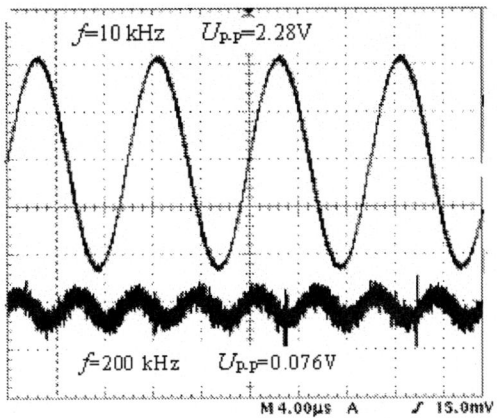

Figure10. (a) Measured output waveform of filter

f—frequency; U_{P-P}—peak-peak value of output

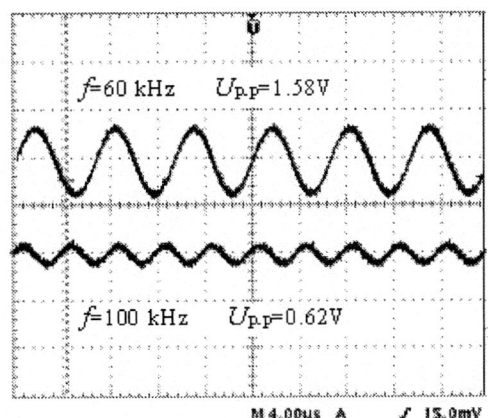

Figure10. (b) Measured output waveform of filter

f—frequency; U_{P-P}—peak-peak value of output

Seen from the measured waveform, the output amplitude is attenuated at 2.28V when frequency is 10 kHz; when frequency is 200 kHz, the output amplitude is attenuated at 76mV. So the output voltage attenuation is very obviously. From the equation (12), the calculated the insertion loss is 33.26 dB at 200 kHz. Compared with the simulation result, the measured result is little bad. There are three main reasons: ①there are distributed inductor and capacitor in breadboard. ②there are difference between theory simulation model and actual model of CM choke. ③ source and load impedance in actual circuit is different with that of simulation model. But, the attenuation of output voltage is still obviously.

V. CONCLUSIONS

A novel CM and DM integrated EMI filter is proposed in this paper, which is composed of one big loop type magnetic core of high permeability and two small loop type magnetic cores of low permeability. This filter has good filtering performance validated by simulation and experiment and partial experimental results is given. Because the research of integrated filter is needed to study, more work will be finished later.

REFERENCES

[1] DONG Ji-qing, CHEN Wei. A Novel EMI Integrated Filter [J]. Journal of Jimei University(Natural Science),2005(2);173-176.

[2] Motorola Company. Integrated Common Mode and Differential Mode Device [P]. US Patent 5313176.

[3] Magnetek Company. Integrated Common Mode and Differential Mode Filter [P]. US Patent 5731666.

[4] TAI Da Company. A Common Mode and Differential Mode
   Integrated Filter [P].Chinese Patent 01121117.2.

[5] YANG Yu-gang. A Suppression Common Mode and Differential
   Mode EMI Integrated Filter. Chinese Utility Model Patent.
   Number: ZL03214078.9，2003.7.

[6] Caponet Marco Chiado, Profumo Francesco, Ferrais L,el al.
   Common and Differential Mode Noise: Separation: Comparison
   of Two Different Approaches[A]. Conference Record IEEE
   PESC[C].2001.

[7] Caponet Marco Chiado, Profumo Francesco. Device for the
   Separation of the Common and Differential Mode Noise: Design
   and Realization [A]. Conference Record IEEE APEC[C].2002.

[8] WEI Ying-dong, WU Bian-hua. Theory and Design Research of
   EMI Filter in Switched Power Supply [J]. Power supply
   technologies and application, 2005(2)，36-40.

[9] YANG Yu-gang. Magnetic Technologies of Modern Power
   Electronics[M].Beijing: Science Press, 2003，173-180.

[10] ZHOU Li-fu, LIN Ming-yao. Design and Simulation of EMI
Source Filter [J]. Low-Voltage Apparatus, 2004, (4): 7-9

2006 5th International Power Electronics and Motion Control Conference

Electromagnetism Model and Characteristic Simulation of Novel Claw Pole Generator with Permanent Magnet Outer Rotor

Fengge Zhang*, Haijun Bai*, Shifu Zhang*, Hans Pert Gruenberger**, Eugen Nolle**

* Shenyang University of Technology, Shenyang, 110023,China
** Esslingen University of Applied Sciences, Esslingen, 73732,Germany
e-mail: zhangfg@sut.edu.cn

Abstract—This paper introduces a new type of permanent magnet claw pole generator (PMCPG) with outer rotor. The magnetic circuit and electrical circuit model of this kind of machine have been built, and the effect of armature current on magnetic field is also considered in the models. The operation characteristics are simulated by means of powerful simulation software on the basis of coupling between magnetic circuit and electric circuit. The computation results indicate that the PMCPG have good operation characteristics.

Keywords-claw pole machine; permanent magnet generator; electromagnetism model

I. INTRODUCTION

A claw pole generator (CPG) is the main source of electric energy in today's internal combustion engine automobile. It is widely used as car alternators because of its simpler structure than conventional machines [1]. At present, the stator of claw pole generator is made up of insulated laminations and the stator slots carry a three-phase winding. The rotor consists of two claw pole wheels, the core with circumferential exciting coil and the shaft. The performance of this electric machine is hard to improve and the manufacture cost is relatively high. Because price of permanent magnet (PM) is lower and performance of new soft magnetic composite (SMC) material is also very well, a new permanent magnet claw pole generator with outer rotor has been put forward. SMC material has isotropic magnetic property. Its outside rotor is covered by PMs and claw pole inner stator is made of new soft magnetic material and generating winding, so this is a kind of brushless machine. The cost of the PMCPG will be much lower than the CPG, and power electronics components which are used to rectify is much fewer. The use of the same low-cost coil and solid steel claws as the inner stator of a brushless generator is limited to very small size [2][3]. Because its structure is much different with conventional CPM, therefore, it is very

necessary to investigate the PMCPG.

II. STRUCTURE AND MAGNETIC CIRCUIT MODEL OF THE PMCPG

A. Structure of the PMCPG

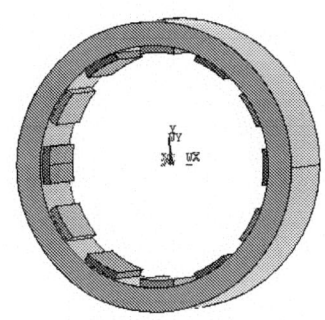

Figure 1. The outside rotor of the PMCPG

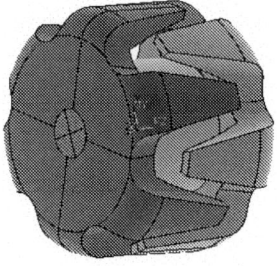

Figure 2. The claw pole inner stator of the PMCPG

The rotor of the PMCPG is similar to a conventional PM outer rotor and comprises 12 surface-mounted PMs and a mild steel cylinder. Fig.1 shows the outside rotor model of the PMCPG. The stator structure is similar to the claw pole inner rotor of a conventional claw pole generator and consists of two claw pole pieces mounted

1-4244-0448-7/06/$25.00 ©2006 IEEE   219

on a shaft with a single-phase concentrated winding between two pieces. Fig.2 shows the inner stator model of the PMCPG.

When the outer rotor with PM pieces is rotated by other mechanism, the rotating magnetic field will be produced and magnetic loop circuit is formed by means of stator claws. In this case, two sets of stator claws face alternately N or S Pole of the outer rotor, so that a single phase AC electromotive force (EMF) will be produced in the stator winding to output electrical energy.

B. Magnetic circuit model of PMCPG

The essential Electromagnetism relationships of any electrical machine rely on magnetic coupling, and the magnetic circuit model is very important for the qualitative analysis and quantitative calculation of new type of electric machine. The cross-section structure schematic diagram and equivalent magnetic circuit of the PMCPG are shown in Fig.3 and Fig.4 respectively.

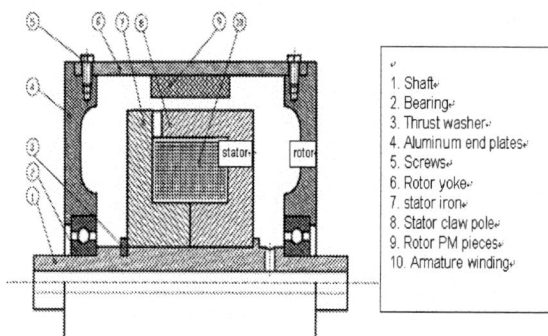

Figure 3. Schematic structure diagram of PMCPG

1. Shaft
2. Bearing
3. Thrust washer
4. Aluminum end plates
5. Screws
6. Rotor yoke
7. stator iron
8. Stator claw pole
9. Rotor PM pieces
10. Armature winding

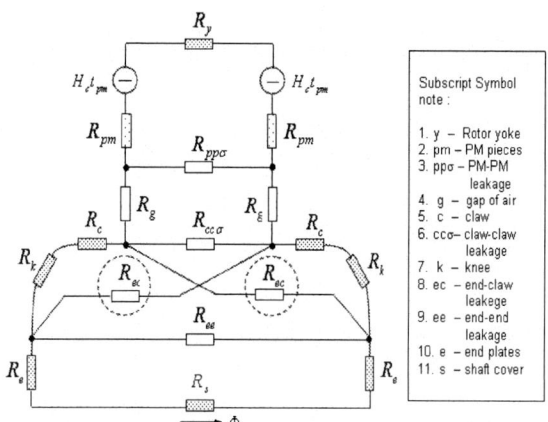

Figure 4. Equivalent magnetic circuit of PMCPG

Subscript Symbol note :

1. y – Rotor yoke
2. pm – PM pieces
3. ppσ – PM-PM leakage
4. g – gap of air
5. c – claw
6. ccσ– claw-claw leakage
7. k – knee
8. ec – end-claw leakege
9. ee – end-end leakage
10. e – end plates
11. s – shaft cover

In Fig.4, symbol R represents the reluctance and its subscript symbol is explained by the right remark of Fig.4. The magnetomotive force of PM can be calculated by

$$F_m = H_c t_{pm} \tag{1}$$

Where, H_c is the coercive force of the PM, and t_{pm} is the thickness of the PM pieces.

It can be seen from the Fig.4 hat the magnetic circuit of the PMCPG is truly three-dimensional structure. The PM pieces is the source of magnetomotive force, magnetic circuit starts from N pole of PMs and go through air gap, one claw pole, iron core, another claw pole, air gap, S pole of PMs, yoke of rotor, and return to start point. Due to the special structure of claw pole, transverse-flux will exist in the part of inner stator.

The key problem of magnetic circuit analysis is to calculate the reluctances of various parts. Here, calculation of three key reluctances is presented as follows.

The leakage reluctance between the end and the end of the stator is calculated by

$$R_{ee} = \frac{1}{\mu_0} \frac{L_r}{\pi \frac{D_e^2 - D_{so}^2}{4} \frac{1}{p}} = \frac{4 p L_r}{\mu_0 \pi (D_e^2 - D_{s0}^2)} \tag{2}$$

Where L_r is the axis length of the stator, and p is the pole pair number, and De, Ds0 is the outer and inner diameter of armature winding respectively.

The leakage reluctance between the claw and the claw can be calculated by

$$R_{cc\sigma} = 1/(2 \int_{\frac{\tau_r - b_{cm}}{2}}^{\frac{\tau_r}{2}} \frac{\mu_0 L_c dr}{\pi r \cos \alpha} + \frac{2\mu_0 L_r h_{cm}}{L_{cc} \cos \alpha}) \tag{3}$$

Where, all of variables are described by Fig.5, which is the schematic diagram of claw pole.

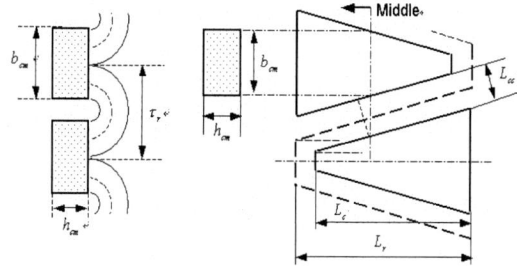

Figure 5. The schematic diagram of claw pole

The reluctance of the air gap can be calculated by

$$R_g = \frac{\delta}{\mu_0 \tau_1 L_{ef}} \tag{4}$$

Where, L_{ef} is the effective axial length of iron, and δ is the length of air gap, and τ_1 is the stator pole pitch as shown in Fig.5, and μ_0 is the permeability of free space.

Other reluctance may be calculated by reluctance calculation formula just as a conventional electric machine. Thus, the modeling of magnetic circuit calculation may be got.

220

III. ELECTRIC CIRCUIT MODEL AND PARAMETER

A. Equivalent electric circuit model

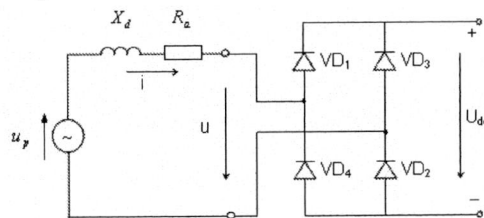

Figure 6. The equivalent electric circuit of PMCPG

When the rotor of generator is rotated by other mechanism, the stator winding will output single-phase AC voltage. Output voltage is commutated to DC current by the single-phase bridge rectifier and is transferred to charge storage battery and other load. The scheme diagram of PMCPG system is shown in Fig.6.

When outer rotor of the machine rotates, the EMF, which is represented by u_p, will be produced in single-phase stator winding. Its instantaneous value can be wrote as follow formula

$$u_p = \sqrt{2} U_p \cos(\omega_1 t) \qquad (5)$$

Where $U_p = 4.44 f \phi N_1 k_{N1}$ is effective value of the EMF, and $\omega_1 = 2\pi f$ is the angle frequency of this ac voltage, and f is the frequency of the current, and ϕ is the magnetic flux per pole, N_1 is the series turn number of the armature winding, and k_{N1} is the winding factor, for the PMCPG $k_{N1} = 1$.

The voltage equation of the stator winding may be expressed as

$$\dot{U}_p = \dot{U} + \dot{I} R_a + j \dot{I}_d X_d \qquad (6)$$

Where $\dot{U}$ and $\dot{U}_p$ are respectively the source voltage and the EMF produced by outside permanent magnetic piece, R_a is the stator winding resistance, and X_d is the synchronous reactance of the PMCPG.

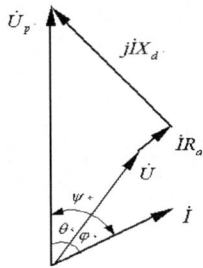

Figure 7. The electrical vector graph of PMCPG

According to the equation (5) and (6), the electrical vector graph of the PMCPG can be expressed in Fig.7.

The phase current of the PMCPG can be obtained by means of the geometrical relationship shown in Fig.7.

$$I_d = \frac{U_p X_d - U X_d \cos\theta - U R_a \sin\theta}{R_a^2 + X_d^2} \qquad (7)$$

$$I_q = \frac{U \sin\theta}{X_d} + \frac{R_a}{X_d} I_d \qquad (8)$$

Where I_d and I_q are the current components of d-axis and q-axis respectively.

B. Calculation of the circuit parameter

The resistance of armature winding can be calculated through the formula as follows.

$$R_a = \frac{1}{N_{ap}} \rho_{Cu} \frac{L_{aw}}{\pi (0.5 d_a)^2} \qquad (9)$$

Where N_{ap} is the parallel conductor number of armature winding, and ρ_{Cu} is the resistivity of the copper, and L_{aw} is the total length of the armature copper lead, and d_a is the nude diameter of copper line.

The leakage reactance of the armature winding is calculated by

$$X_\sigma = 2\pi f N_1^2 \left(\frac{1}{R_{ee}} + \frac{1}{R_{cc\sigma}} \right) \qquad (10)$$

Where, N_1 is turn number of the armature winding in series. R_{ee} and $R_{cc\sigma}$ are the leakage reluctance respectively acquired by magnetic circuit calculation.

The armature reaction reactance of the winding is calculated by

$$X_{hd} = 2\pi f_1 \frac{1}{K_s} N_1^2 \mu_0 \frac{\tau_1 L_{ef}}{\delta + \dfrac{t_{pm}}{\mu_{rpm}}} \qquad (11)$$

Where K_s is the saturation factor. T_{pm} and μ_{rpm} is the thickness and relative permeability of PM respectively.

Thus, the synchronous reactance of the armature winding can be calculated by

$$X_d = X_{hd} + X_\sigma \qquad (12)$$

IV. CONSIDERATION FOR THE ARMATURE REACTION

When the PMCPG operates with load, the armature current would be produced in the armature winding of the stator and magnetic field of armature reaction is formed at the same time.

After the effect of the armature reaction is considered, the compositive magnetic field of air gap will be different from no load magnetic field, and both magnetic flux and saturation degree of the magnetic circuit are changed. Meantime, iron core loss, inductance parameter and armature current will also vary. So the magnetic circuit of the generator with load should be calculated again. It is similar to a conventional PM synchronous machine with the conceal pole rotor and its current and MMF vector graph are shown in Fig.8

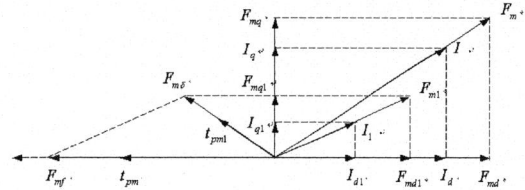

Figure 8.   The current and MMF vector graph of generator

By means of the vector graph, the relationship between the equivalent thickness of PM and the armature current can be shown as follows

$$(t_{pm1}H_c)^2 = \left(t_{pm}H_c - I_{d1}\frac{N_1}{2}\right)^2 + \left(I_{q1}\frac{N_1}{2}\right)^2 \quad (13)$$

$$t_{pm1} = \sqrt{\left(t_{pm} - I_{d1}\frac{N_1}{2H_c}\right)^2 + \left(I_{q1}\frac{N_1}{2H_c}\right)^2} \quad (14)$$

Because the PMCPG is similar to synchronous generator with conceal pole rotor, therefore, $I_{d1} = I_d$ and $I_{q1} = I_q$. Where, t_{pm1} is the equivalent thickness of PM piece corresponding to compositive magnetic field of air gap considering the armature reaction, and N_1 is the turn number of the armature winding.

V.   SIMULATION OF ORPERATION CHARACTERISTICS

In order to get the operation characteristics, the prototype machines of the PMCPG has been designed. The major parameters and dimensions of the machine are presented at Table I.

TABLE I.
DIMENSIONS AND MAJOR PARAMETERS OF PROTOTYPE MACHINES

Dimensions and parameters	Quantities
Length of rotor iron core along the axis	37mm
Length of the PM along axis direction	31mm
Width of the PM piece	15mm
Width of the outer rotor yoke	10mm
Length of air gap	0.7mm
Outer diameter of stator	92.6mm
Outer diameter of rotor	120mm
Turn number of the armature winding	16

There is one point to emphasize that the stator of the PMCPG is made of soft magnetic composite material, which is made by powder metallurgy techniques. This type of material is magnetically isotropic due to its powdered nature and is suitable for electrical machine of three-dimensional magnetic flux [4][5].

In order to calculate operation characteristics of prototype machine, MATLAB simulation program is compiled according to the above the equation and analysis theory. Because of the importance of PM, simulation result will be obtained with different PM thickness.

Fig.9 shows the electromotive force of the PMCPG. It can be seen that the EMF will be varied as the thickness of PM and rotate speed.

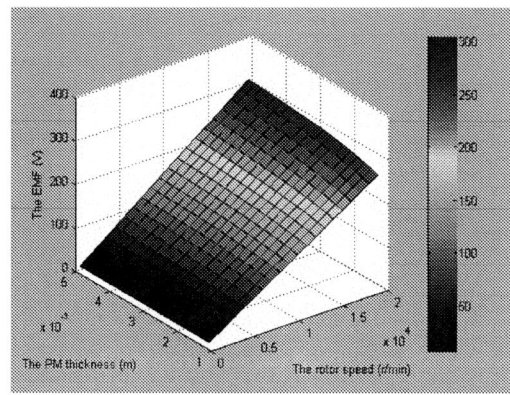

Figure 9. Relations of EMF, PM thickness and rotate speed

Fig.10 shows variation of the output power with PM thickness and rotor speed. Maximum of output power occurs at the speed of 5000 rev/min, and this output power can satisfy the needs of load.

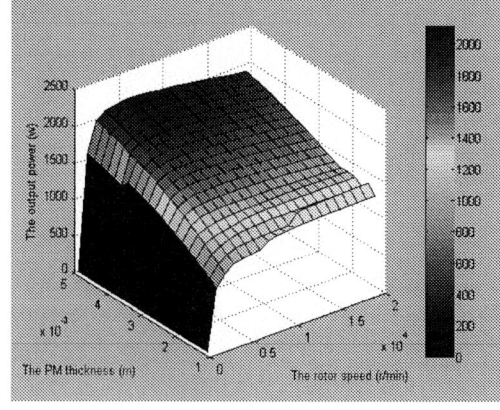

Figure 10.   The relations of output power, PM thickness and rotate speed

Because the permanent magnet material is used, this generator would have higher efficiency than conventional claw pole machine. Fig.11 shows the variation of the efficiency with rotor speed and PM thickness.

The loss of iron core with loaded is relative to the frequency of current. The simulation results of the relationship between the loss of iron core, PM thickness and the frequency is shown in Fig.12.

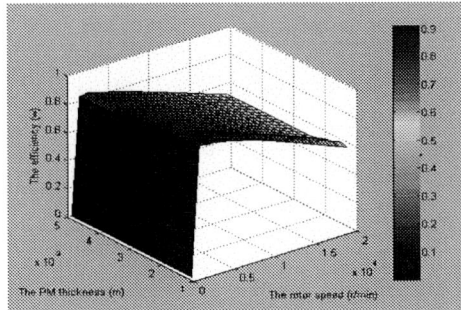

Figure 11. The relations of efficiency, PM thickness and rotate speed

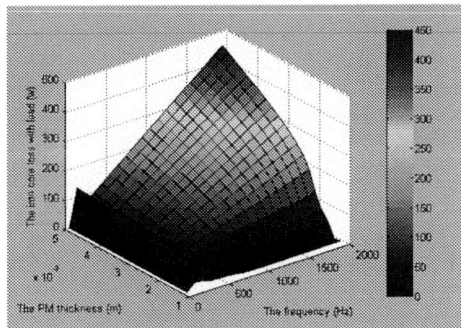

Figure 12. The relations of iron core loss, PM thickness and frequency

VI. CONCLUSION

In this paper, a novel permanent magnet claw pole generator with outer rotor is presented. Due to its simplicity of structure and low manufacturing costs, the novel generator may be used as an electrical source of electrical energy in vehicles. The efficiency of this generator will be increased because PMs is used as magnetic excitation. Compared to the other electric machine with the same size, the PMCPG will produce the bigger output power. The computation results indicate that the PMCPG have good operation characteristics.

Because its structure is much different with conventional CPM, therefore, the further research should be done, specially in the aspect of magnetic field analysis.

REFERENCES

[1] H.C.Lai, D.Rodger, "Three-dimensional Finite Element Modelling of a Claw Pole Type Car Alternator," IEEE Transactions on Industry Applications, 1995, 457~451

[2] Y.G.Guo, J.G. Zhu, P. A. Watterson, "Comparative Study of 3-D Flux Electrical Machines With Soft Magnetic Composite Cores," IEEE Transactions on Industry Applications, 2003, Vol. 39, pp.1696-1703

[3] Y.G.Guo, and J.G.Zhu, "Magnetic field calculation of claw pole permanent magnet machines using magnetic network method" proceedings of AUPEC2001, 2001, pp.199-204.

[4] Wen Ouyang, Surong Huang, Anne Good, T.A. Lipo, "Modular permanent magnet machine based on soft magnetic composite," unpublished

[5] A.G. Jack, "Experience with the use of soft magnetic composites in electrical machines," Proceeding of International Conference on Electrical Machines, Istanbul, Turkey, 1998, pp.1441–1448.

2006 5th International Power Electronics and Motion Control Conference

An Improved Adaptive Filter for Voltage and Current Reference Extraction

A. Abedini and A. Nasiri

Power Electronics and Motor Drives Laboratory
University of Wisconsin-Milwaukee
Milwaukee, WI 53211
Email: nasiri@uwm.du, URL: www.uwm.edu/~nasiri

Abstract—A novel improved adaptive filter is presented is this paper for extraction of fundamental components of voltage and current signals. The filter uses the concept of Least Square Method to cancel the distortions. Mathematical analysis of the proposed filter is described. The application of this filter includes active filters, UPS systems, and other power quality conditioners. The simulation and experimental results of the system are also presented to verify the viability of this filter.

Keywords-adaptive filter; distortions; least mean square; reference extraction;

I. INTRODUCTION

In recent years, with the increase of non-linear loads drawing non-sinusoidal currents, power quality has become a serious problem in power systems. Power distortions in electrical systems include voltage sag/swell, voltage and current harmonics, voltage spikes, and voltage outages. Many power electronics devices have been designed and implanted to suppress and mitigate power distortions. For instance, active filters have been known as the best tool for harmonic mitigation as well as reactive power compensation, load balancing, voltage regulation, and voltage flicker compensation. Active filters have been designed, improved, and commercialized in last 25 years. They are applicable to compensate current-based distortions such as current harmonics, reactive power, and neutral current. They are also used for voltage-based distortions such as voltage harmonics, voltage flickers, voltage sags and swells and voltage unbalances. Uninterruptible Power Supply (UPS) systems also have been designed to provide uninterrupted, reliable, and high quality power for vital loads. They protect sensitive loads against power outrages as wells as over voltage and under voltage conditions. UPS systems also suppress line transients and harmonic disturbance. They have widespread applications in medical facilities, life supporting systems, data storage and computer systems, emergency equipment, telecommunications, industrial processing, and on-line management systems. Other devices such as Static VAR Compensator (SVC), Unified Power Quality Conditioner (UPQC), and Dynamic Voltage Regulator (DVR) have also been used to compensate voltage sags and improve power quality [1]-[3].

To compensate voltage and current distortions, all of these devices need to calculate current and voltage reference with minimum time delay. In this paper, a new algorithm for detection of power distortion using Least Mean Square (LMS) method is introduced. Applications of this method in calculating the maximum of input signal and extraction of fundamental component is discussed. Derivation of reactive component from line current is also explained using this method. The main characteristic of this method is that the parameters of the system can be adjusted to achieve higher speed and better precision. Additionally, using this method the reference components are calculated with no time delay and phase shift.

II. LEAST MEAN SQUARE METHOD

The electric power signals consist of a fundamental sinusoidal component and disturbances. Power distortions include sag, swell, noise, harmonic and etc. Some of these distortions such as noise and harmonics are present in the system all the time but some of them are short term disturbances such as sag and swell. An electric power signal, current or voltage, can be written as:

$$x(t) = a\sin(\omega_0 t + \theta) + \sum_{i=2}^{j} a_i \sin(i\omega_0 t + \theta_i) \qquad (1)$$
$$+ g(t) + s(t)$$

Where:

$a\sin(\omega_0 t + \theta)$ is the fundamental component and

$a_i \sin(i\omega_0 t + \theta_i)$ represents harmonics.

$g(t)$ represents sudden power disturbances such as power interruptions and $s(t)$ represents the other distortions such as noise [5]-[6].

A discrete-time version of the above equation can be shown as below:

$$x(n) = a\sin(\Omega_0 n + \theta) + \sum_{i=2}^{j} a_i \sin(i\Omega_0 n + \theta_i) + \qquad (2)$$
$$g(n) + s(n)$$

1-4244-0448-7/06/$25.00 ©2006 IEEE

The proposed algorithm draws the fundamental component from the signal and then the distortion is achieved by subtracting the input signal from fundamental component. The general diagram of this algorithm is shown in Figure 1.

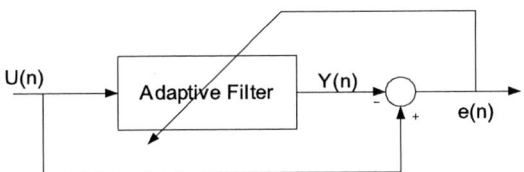

Figure 1. General block diagram of the adaptive filter.

In Figure 1, u(n) is the input signal, y(n) is the output of the adaptive filter, and e(n) is the disturbances. The output signal of the adaptive filter is:

$$y(n) = w_0(n)\sin(\Omega_0 n) + w_1(n)\cos(\Omega_0 n) \quad (3)$$

The adaptive algorithm changes the coefficients $w_0(n)$ and $w_1(n)$ according to LMS method to generate the fundamental component of the input signal. According to this method the coefficients change according to following equation:

$$W(n+1) = W(n) + \mu \begin{bmatrix} \sin(\Omega_0 n) \\ \cos(\Omega_0 n) \end{bmatrix} e(n) \quad (4)$$

Where: $W(n) = \begin{bmatrix} w_0(n) \\ w_1(n) \end{bmatrix}$

In the above equation, μ is a coefficient called step size, which represents the convergence rate of the filter. The higher value for μ results in faster convergence but if μ is selected very high, the system moves to unstable region. To ensure the stability of the filter, μ always must follow equation (5) [6]-[8].

$$0 < \mu < \frac{1}{3P} \quad (5)$$

Where: P is the power of the input signal.
The block diagram of the adaptive filter is shown in Figure 2.

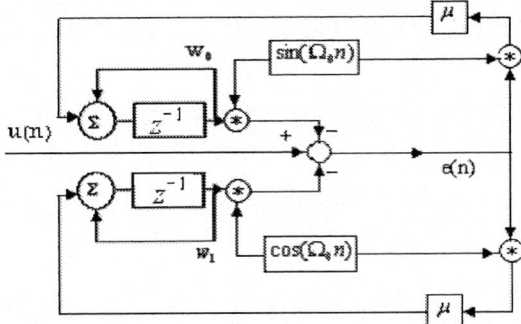

Figure 2. Block diagram of the LMS adaptive filter.

This filter generates the fundamental component of an input signal but the convergence rate is constant. In most applications, it is desired to achieve the fundamental component of the input signal in less than two cycles. Due to constant step size, this filter cannot provide fast response in the presence of distortions and non-linear loads. In the next section, the above algorithm is modified to change the convergence rate and improve filter speed.

III.   MODIFIED ADAPTIVE ALGORITHM

The discussed algorithm follows the fundamental component of the input signal but its precision depends on the convergence rate. For large distortions, the step size, μ, can be selected very high to speed up the response of the filter. If the ratio of the distortion to signal is small, then the convergence factor must be

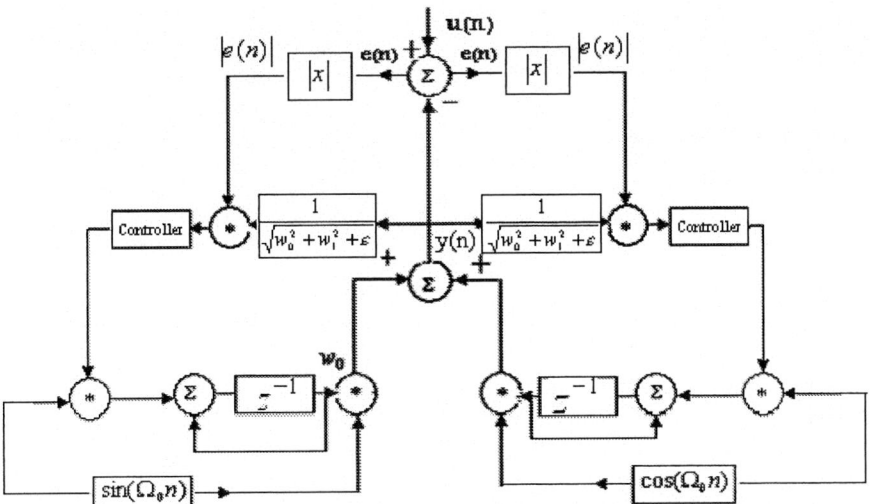

Figure 3. The block diagram of the improved adaptive filter.

225

lowered to prevent the instability of the filter and provide more precision. Considering above statements, the following equation is presented for μ.

$$\mu = k \frac{|e(n)|}{\sqrt{w_0^2 + w_1^2 + \varepsilon}} \qquad (6)$$

In (6), ε is a small positive constant to avoid dividing by zero during iterations. Its value should be selected small enough so that it does not affect on the performance of the algorithm. When the distortion occurs on the input signal, the error between output signal and fundamental component of input signal increases. At this stage, the step size, μ, is chosen high to increase the convergence speed of the filter. When the error decreases, this value should be decreased as well to increase the precision and minimize the error. Figure 3 shows the block diagram of modified filter. Two controllers in this figure include limiter and low pass filters.

IV. SIMULATION RESULTS

In this section, utilization of the adaptive control method for different applications is discussed.

A. Extracting Sine Wave

The algorithm is used to extract the fundamental component of an input signal. The input signal is a sine wave:

$$u(n) = w_0^o \sin(\Omega_0 n) + w_1^o \cos(\Omega_0 n) \qquad (7)$$

and $w_0^o = 1$ and $w_1^o = 0$

Figure 4 shows the simulation results including the input and output signals. The filter is converged in less than one period..

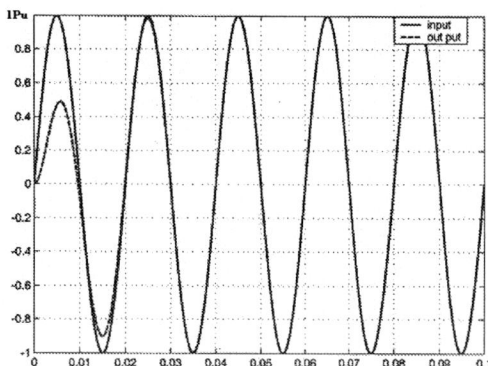

Figure 4. Output of the filter when input is a sine wave.

Figure 5 shows the variation of the coefficients for the output of the filter during this time. The initial values for both sine and cosine functions are set to zero.

Figure 6 shows the variation of μ during this time. In the first period, μ is set high to converge to fundamental component of the input signal. Then it is

lowered to increase the precision of the system and minimize the error.

In another example, the input signal is considered as:

$$u(n) = \sin(100\pi n) + 0.2\sin(200\pi n) + 0.1\sin(300\pi n) + 0.09\sin(500\pi n) \qquad (8)$$

Figure 7 shows the input signal and output signal of the filter. After nearly two cycles, the filter converges to fundamental component of the input signal.

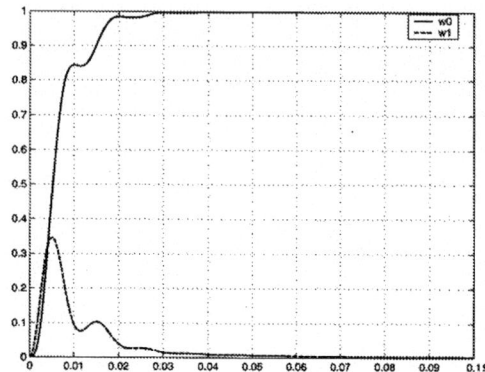

Figure 5. Variation of coefficients for signal of equation (7).

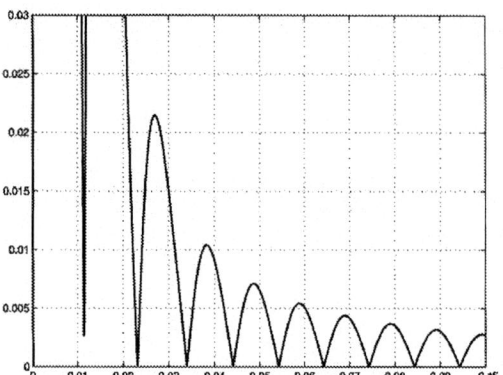

Figure 6. Variation of μ for signal of equation (7).

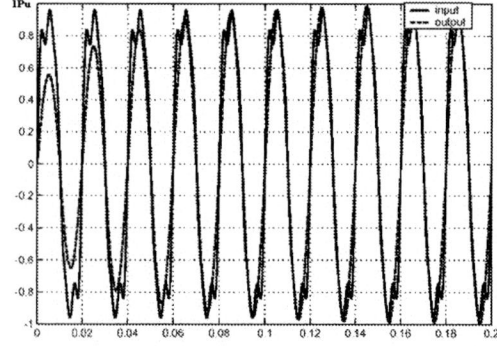

Figure 7. The output of the filter when The Input signal is a harmonic signal.

The frequency spectrums of the input and output signals are shown in figure 8. The output signal consists of only fundamental component of the input signal.

226

B. Maximum Voltage Detection.

In some power quality applications, the maximum voltage of a signal is used as an input to control the system[7]. In this part, the maximum voltages of input signals are measured using of this algorithm.

If u(n) is considered as:

$$u(n) = w_1^o \sin(\Omega_0 n) + w_2^o \cos(\Omega_0 n) \qquad (9)$$

The maximum voltage of this signal is:

$$Max\{u(n)\} = \sqrt{w_1^{o^2} + w_2^{o^2}} \qquad (10)$$

By using this algorithm, the peak detection is accelerated. Figure 9 shows the input and output signals. The input signal is changing as follows:

In the first 5 cycles, it is a sine wave with amplitude 1 pu. In the next 5 cycles, its amplitude is decreased to 0.5 pu and in the 10 next cycles, the amplitude oscillates with amplitude of 0.1 pu.

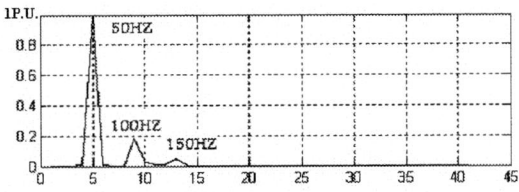

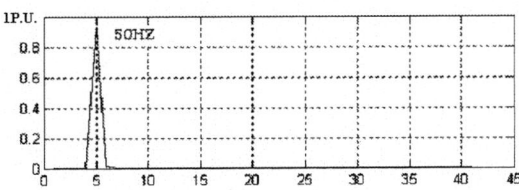

Figure 8.   The components of the Input and output signal.

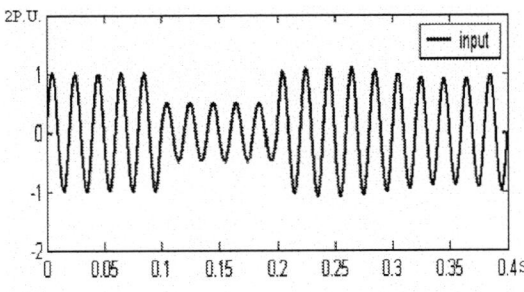

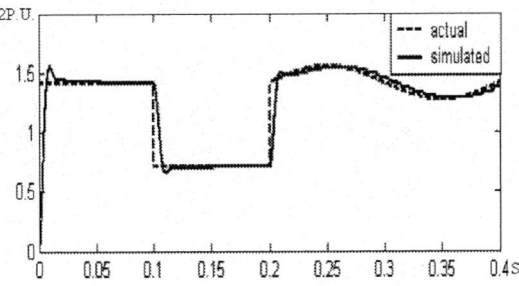

Figure 9.   Detection of maximum of the input signal.

Results of the measurement and simulation using the adaptive filter are presented and compared if figure 9 as well.

C. Reactive power Detection

Detection of reactive power component is needed for power factor correction[8]. In this part, using this algorithm, the reactive part of the input power is measured.

The input voltage is considered as:

$$v_1(n) = v_1 \sin(\Omega_0 n) + v_2 \cos(\Omega_0 n) \qquad (11)$$

and the line current is:

$$i(n) = i_l(n) + i_h(n) \qquad (12)$$

Where i_l is the fundamental component of the input signal and i_h is other distortions imposed on the line current including harmonics.

$$i_1(n) = i_1 \sin(\Omega_0 n) + i_2 \cos(\Omega_0 n) \qquad (13)$$

The line current signal can be written as:

$$i_1(n) = I_1 \sin(\Omega_0 n + \delta_i) = i_a(n) +$$
$$i_r(n) = I_1 \cos(\delta_i - \delta_v) \sin(\Omega_0 n + \delta_v) \qquad (14)$$
$$+ I_1 \sin(\delta_i - \delta_v) \cos(\Omega_0 n + \delta_v)$$

where:

$$\delta_i = \tan^{-1}(\frac{i_2}{i_1}) \qquad (15)$$

and

$$\delta_v = \tan^{-1}(\frac{v_2}{v_1}) \qquad (16)$$

i_a is the active current and i_r is the reactive current. The total current injected to the system to remove the harmonics and reactive part is as following.

$$i_f(n) = i_h(n) + i_r(n) \qquad (17)$$

The simulation results performed by MATLAB are shown in Figure 10. The network voltage is considered as:

$$v(t) = \sin(100\pi t) \qquad (18)$$

and the line current is:

$$i(t) = \sin(100\pi t + \frac{\pi}{3}) \qquad (19)$$

The comparison between actual reactive current and results of simulation are also shown in Figure 10.

V.   EXPERIMENTAL RESULTS:

To validate the results generated by MATLAB, an experimental setup was built to investigate the performance of the filter. The block diagram of the experimental setup is shown in Figure 11.

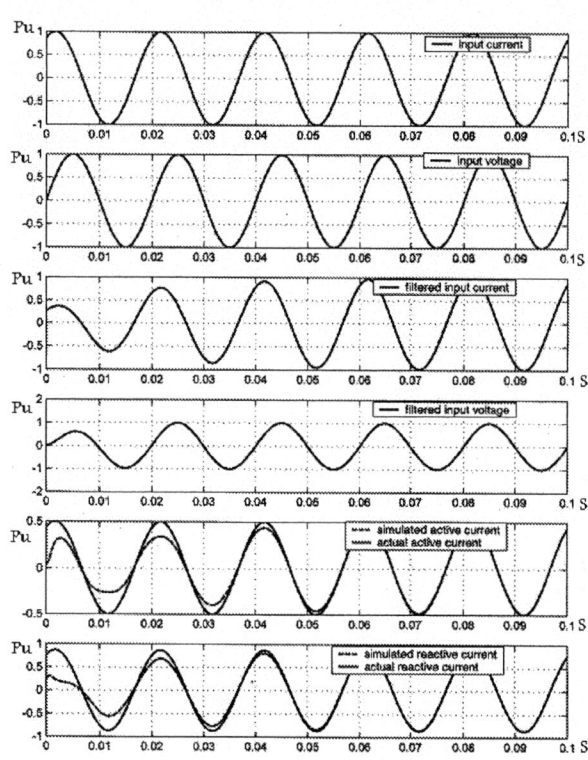

Figure 10. Simulation result for reactive current extraction; from top: waveforms of line current, line voltage, filtered line current and voltage, active current, and reactive current.

The signal generator provides the input signal and the interface block gathers the input data and sends to the computer. Primarily, the system was tested for a sine wave. Figure 12 shows the input signal and the output of the filter.

In the second test, a triangle waveform was used as input signal. Input signal and output signal are shown in figure 13 and 14 respectively. In figure 15, the output signal of the filter and actual fundamental component of the input signal are presented for comparison.

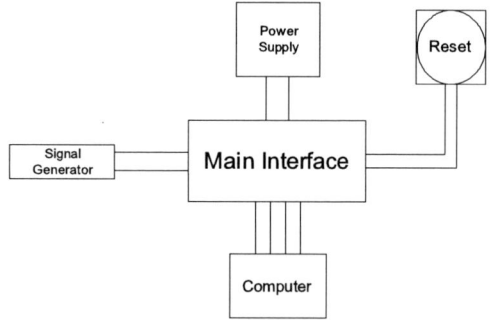

Figure 11. The hardware set up for experiments.

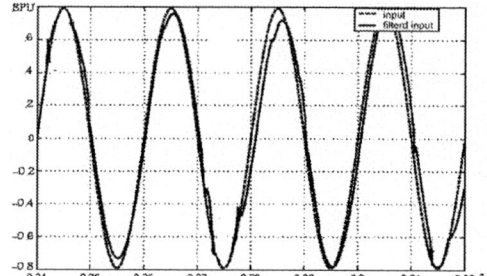

Figure 12. Input signal and output signal of the filter with sinusoidal input.

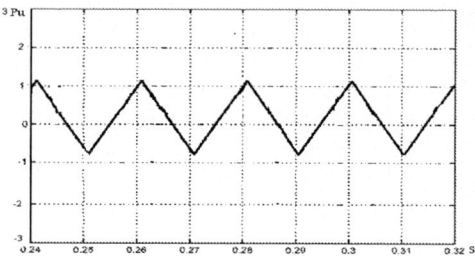

Figure 13. Input signal for the experimental setup.

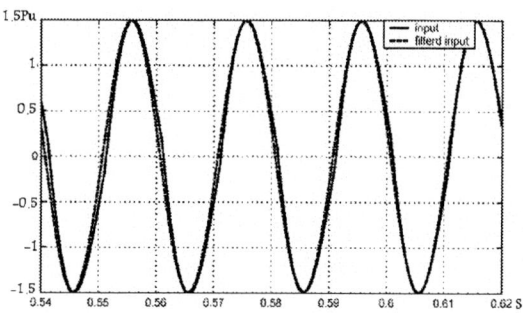

Figure 14. The fundamental component of triangle input and output of the adaptive filter.

REFERENCES

[1] M. Bollen, "Understanding Power Quality Problems," *IEEE Power engineering Review*, Vol. 21, no. 9, pp. 47-47, 2001

[2] R. Dugan, M. McGranagha, and H. Beaty, "Electrical Power System Quality," Mac Graw Hill, Newyork, 1996.

[3] G. T. Heydt, "Electric Power Quality," Weat Lafayette, Stars in Circle, 1991.

[4] B. Farhang-Boroujeny, "Adaptive Filter Theory and Application," John wiley & sons, 1999.

[5] Simon Haykin, "Adaptive Filter Theory," 3rd edition, Prentice Hall, International Editions, 1996.

[6] M. Karimi-Ghartemani, and M. Iravani, "A Nonlinear Adaptive Filter for Online Signal Analysis in Power System: Applications," *IEEE Transaction on Power Delivery*, vol. 17, no 2, pp. 617-622, 2002.

[7] F. Z. Peng, L. M. Tolbert, and Z. Qian, "Definitions and Compensation of Non-Active Current in Power Systems," in *Proc. 33rd IEEE Power Electronics Specialists Conference*, vol. 4 2002 pp. 1779 – 1784.

[8] M. Karimi, H. Mokhtari, M. Iravani, "Wavelet Based On-Line Disturbance Detection for Power Quality Applications." *IEEE Transaction on Power Delivery*, vol. 15, no. 4, pp. 1212–1220. 2000.

2006 5th International Power Electronics and Motion Control Conference

Simulation Analysis on Current SVM Algorithm of Matrix Rectifier

Xi-jun Yang, Peng-sheng Ye, Xiang Liu, Xing-hua Yang, Jian-quan Wang, Luan-guo Zhang

（Electrical Engineering Dept., Shanghai Jiaotong University, Shanghai 200030, China）

yangxijun@sjtu.edu.cn, psye@sjtu.edu.cn

Abstract—**Matrix rectifier (MR) is a buck rectifier in nature, which can be reduced from the traditional matrix converter. Like other converters interfaced with the power lines, it hardly avoids the problems caused by the abnormal cases of three-phase input voltages. The existing basic PWM algorithms of MR include current SVM algorithm and switching function algorithm, suitable for normal power supply. In view of this, after deriving the input current SVM algorithm under balanced condition, a modified input current SVM algorithm is derived under imbalanced conditions based on the concept of switching function algorithm. The practical simulation model of MR is then built up and its working principle is realized by employing simulation software SIMULINK6.0. The simulation analysis results of MR are also offered, including input voltage waveforms, input current waveforms and output DC voltage waveforms, which are consistent with the theoretical analysis results.**

Keywords-matrix rectifier; current space vector modulation; balanced input voltage; unbalanced input voltage

I. INTRODUCTION

Three-phase to three-phase matrix converter (MC) can be derived into three-phase to two-phase MC by removing any three bidirectional switches with responsibility for yielding one phase output. When the desired output frequency is set to zero, the three-phase to two-phase MC is a three-phase AC-DC MC, i.e. DC matrix converter, named matrix rectifier (MR) for convenience here[1-6]. It is a generalized buck three-phase AC-DC converter with four-quadrant operation capability. MR has a mature input current space vector algorithm, but it hasn't been studied enough especially under unbalanced power supply conditions. If three-phase input voltages appear abnormal, the constraints of switching function algorithm will be destroyed, thereby the former input current SVM algorithm should be modified. Otherwise the current commutation will fail, and the output DC voltage will contain even-order harmonic component. Thus this paper is to derive MR's

current SVM algorithm based on switching function algorithms to make it suitable for balanced and unbalanced power supply to some extent[7-10], and simulate it with SIMULINK6.0 after establishing much more practical system of MR.

II. SWITCHING FUNCTIONS ALGORITHM UNDER BALANCED CONDITION

The topology of traditional MC is shown in Fig.1, and that of MR in Fig.2, powered by three-phase AC voltages and outputting a DC voltage, where the bidirectional switches are common-emitter anti-paralleled type. In addition, MR can also be the front-stage circuit of AC-DC-AC equivalent structure of MC and that of quasi-sparse matrix converter. At this point, the further study on MR is significant. Fig.3 shows MR's input current space vectors and input current vector synthesis, where I, II, III, IV, V and VI stand for six vector sectors, and I1~I6 stand for active current vector, I7~I9 stand for zero current vectors.

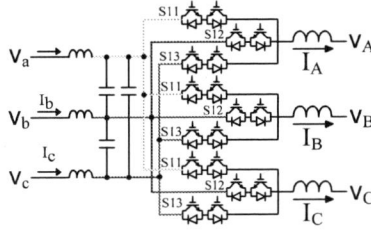

Figure 1. The topology of traditional MC

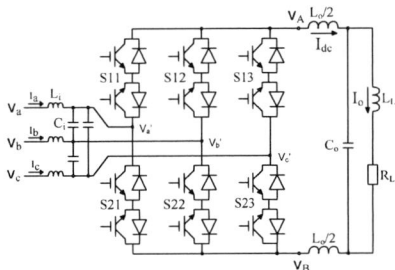

Figure 2. The topology of MR

The Project is supported by National Natural Science Foundation of China (No. 50377025)

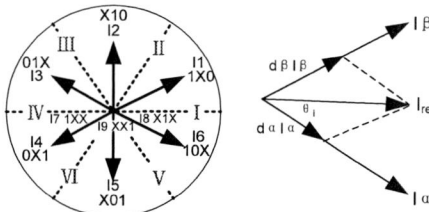

(a) Input current space vectors (b) Input current vector synthesis

Figure 3.   Input current space vectors and its vector synthesis

Fig.4 shows input voltage zones and corresponding active current vector selection. There are six such zones in a mains period, which are divided according to input voltage zero-crossing.

When MR is powered by balanced three-phase input voltages, supposing that the DC current is constant, the amplitude of input current is I_{im}, the DC current is I_{dc}, the modulus $|\mathbf{I}_j| = 2I_{dc}/\sqrt{3}$, the input voltage and input current are in positive phase sequence, Park transformation is in negative phase sequence, then the input current vector reference $\vec{I}_r = 2(i_a + i_b e^{j2\pi/3} + i_c e^{-j2\pi/3})/3 = I_{im}e^{j(\omega_i t + \varphi_i)}$, and the resultant input current vector $\mathbf{I}_R = (T_\alpha/T_s)\mathbf{I}_\alpha + (T_\beta/T_s)\mathbf{I}_\beta + (T_o/T_s)\mathbf{I}_o$, where $\mathbf{I}_\alpha$ and $\mathbf{I}_\beta$ denote active current vectors, and $\mathbf{I}_o$ denotes zero current vector. Their duty cycle expressions are

$$d_\alpha = \frac{T_\alpha}{T_s} = \frac{2}{\sqrt{3}}\frac{|\vec{I}_s|}{|\vec{I}_\alpha|}\sin(60° - \theta_k) = \frac{I_{im}}{I_{dc}}\sin(60° - \theta_k) = m_c\sin(60° - \theta_k) \quad (1)$$

$$d_\beta = \frac{T_\beta}{T_s} = \frac{2}{\sqrt{3}}\frac{|\vec{I}_R|}{|\vec{I}_Y|}\sin\theta_k = \frac{I_{im}}{I_{dc}}\sin\theta_k = m_c\sin\theta_k \quad (2)$$

$$d_o = 1 - d_\alpha - d_\beta \quad (3)$$

where θ_k varies from zero to $60°$, $\theta_k = \theta_0 + k\omega_i T_s$. Additionally $m_c = I_{im}/I_{dc}$, obviously $m_c \in [0,1]$, then in a switching period, the conduction times of $\mathbf{I}_\alpha$, $\mathbf{I}_\beta$ and $\mathbf{I}_o$ are

$$T_\alpha = d_\alpha T_s = m_c T_s \sin(60° - \theta_k) \quad (4)$$

$$T_\beta = d_\beta T_s = m_c T_s \sin(\theta_k) \quad (5)$$

$$T_0 = d_o T_s = T_s - (T_\alpha - T_\beta) \quad (6)$$

Supposing the input current vector is operating in zone I, then $-30° < k\omega_i t < +30°$, $\theta_i = \omega_i t + 30°$. The averaged input currents in a switching period is

$$\begin{bmatrix} i_a \\ i_b \\ i_c \end{bmatrix} = \frac{I_{dc}}{T_s}\left(\begin{bmatrix} T_\alpha \\ -T_\alpha \\ 0 \end{bmatrix} + \begin{bmatrix} T_\beta \\ -T_\alpha \\ -T_\beta \end{bmatrix} \right) \quad (7)$$

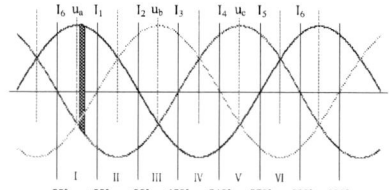

Figure 4.   Input voltage zones and current vector selection

$$\begin{bmatrix} i_a \\ i_b \\ i_c \end{bmatrix} = m_c \begin{bmatrix} \sin[60° - (\omega_i t + 30°)] + \sin(\omega_i t + 30°) \\ -\sin[60° - (\omega_i t + 30°)] \\ -\sin(\omega_i t + 30°) \end{bmatrix} = m_c \begin{bmatrix} \cos(\omega_i t) \\ \cos(\omega_i t - 120°) \\ \cos(\omega_i t + 120°) \end{bmatrix} \quad (8)$$

Evidently, the three-phase current instantaneous values vary with three-phase symmetry sinusoidal voltage waveforms, and the conclusion is the same with other zones.

Also supposing input current vector operates in zone I, then $V_{dc} = V_{ab}$ when I6 acts, $V_{dc} = V_{ac}$ when I1 acts, and $V_{dc} = 0$ when zero current vector I7, I8 or I9, The averaged output voltage in a switching period is

$$\bar{V}_{AB} = [T_\alpha V_{ab} + T_\beta V_{ac}]/T_s = m_c V_{ab}\sin(60° - \theta_k) + m_c V_{ac}\sin(\theta_k)$$
$$= m_c V_{ab}\sin[60° - (\omega_i t + 30°)] + m_c V_{ac}\sin(\omega_i t + 30°) = \frac{3}{2}m_c V_{im} \quad (9)$$

It shows clearly that the maximum voltage gains are identical for input current SVM algorithm and switching function algorithm.

III.   CURRENT SVM ALGORITHM AND SWITCHING FUNCTION ALGORITHM

A.   Switching Function Algorithm When Balanced

Considering maximum voltage transfer ratio is $\sqrt{3}/2$, initial phase is φ_{ip} , and the three-phase input voltages are in positive-sequence and symmetrical, and the desired DC output voltages is

$$\begin{bmatrix} V_{Ap} \\ V_{Bp} \\ V_{Cp} \end{bmatrix} = V_{omp}\begin{bmatrix} \sqrt{3}/2 \\ -\sqrt{3}/2 \\ 0 \end{bmatrix} + \frac{V_{omp}}{2\sqrt{3}}\begin{bmatrix} \cos(3\omega_i t + 3\varphi_{ip}) \\ \cos(3\omega_i t + 3\varphi_{ip}) \\ \cos(3\omega_i t + 3\varphi_{ip}) \end{bmatrix} \quad (10)$$

The general switching function expression of MR is

$$f_{lkp}(t) = \frac{1}{3}\{1 + q_{vp}\cos[\omega_i t - (k-1)\alpha + (l-1)\alpha + \varphi_{ip} + 30°]$$
$$+ q_{vp}\cos[\omega_i t - (k-1)\alpha - (l-1)\alpha + \varphi_{ip} - 30°]$$
$$+ \frac{7}{18}q_{vp}\cos[2\omega_i t + (k-1)\alpha + 2\varphi_{ip}]$$
$$- \frac{1}{18}q_{vp}\cos[4\omega_i t - (k-1)\alpha + 4\varphi_{ip}]\} \quad (11)$$

where voltage transfer ratio $q_{vp} = V_{omp}/V_{imp}$, $q_{vp} \in [0, \sqrt{3}/2]$, $q_{vp\max} = \sqrt{3}/2$, letter "p" denotes positive-sequence, and different phase sequence corresponds to different expression. Row $l = 1,2,3$ stand for output phases (A,B,C) , and column $k = 1,2,3$ stand for input phases (a,b,c) . The constraints are $0 \le f_{lk} \le 1$ and $\sum_{k=1}^{3} m_{lk}(t) = 1$, and the maximum DC output voltage is $V_{AB} = V_{Ap} - V_{Bp} = \sqrt{3} \cdot (V_{imp} \cdot q_{vp\max}) = 1.5V_{imp}$. Since output phase C voltage is zero, and there is no instantaneous current flowing through it, the switching function matrix can be composed by the first two rows in traditional MC's switching function matrix. The first row produces positive voltage V_{Ap} , and the second row produces negative voltage $V_{Ap} = -V_{Bp}$. Based on the matrix conversion theory, the input current of MR is

$$\begin{bmatrix} I_{ap} \\ I_{bp} \\ I_{cp} \end{bmatrix} = I_{dc} \begin{bmatrix} f_{11P}(t) - f_{21P}(t) \\ f_{12P}(t) - f_{22P}(t) \\ f_{13P}(t) - f_{23P}(t) \end{bmatrix} = \frac{2q_{vp}I_{dc}}{\sqrt{3}} \begin{bmatrix} \cos(\omega_i t + \varphi_{ip}) \\ \cos(\omega_i t + \varphi_{ip} - \alpha) \\ \cos(\omega_i t + \varphi_{ip} + \alpha) \end{bmatrix} \quad (12)$$

which indicates the input currents are sinusoidal and in phase with input phase voltages likewise.

B. Switching Function Algorithm When Unbalanced

When the three-phase input voltages are imbalanced or distorted due to some facts, using (11) to calculate MR's instantaneous duty cycles will bring about great deviation, owing to the dissatisfied constraints of switching function algorithm. One straightforward thought to solve such problem is to add the detected instantaneous values of input voltages to the original expressions[6-10], in order to adapt to the input voltages' unbalance and distortion.

Also considering maximum voltage transfer ratio is $\sqrt{3}/2$, the initial phase φ_{ip}, then the general switching function of MC can be rearranged as

$$f_{lk}(t) = \frac{1}{3}\{1 + 2q_v \cos(\omega_i t - (k-1)\alpha + \varphi_{ip})$$
$$\times [\cos(\omega_o t - (l-1)\alpha + \varphi_{ov}) - \frac{1}{6}\cos(3\omega_o t + 3\varphi_{ov}) + \frac{1}{2\sqrt{3}}\cos(3\omega_i t + 3\varphi_{ip})] \quad (13)$$
$$- \frac{2q_v}{3\sqrt{3}}[\cos(4\omega_i t - (k-1)\alpha + 4\varphi_{ip}) - \cos(2\omega_i t + (k-1)\alpha + 2\varphi_{ip})]\}$$

Substituting the instantaneous voltage values of the input voltages and the desired output voltages, then (13) turns into

$$f_{lk}(t) = \frac{1}{3} + \frac{2V_k V_l}{3V_{imeqv}^2} - \frac{2q_v}{9\sqrt{3}}\{\cos[4\omega_i t - (k-1)\alpha + 4\varphi_{ip}] \quad (14)$$
$$- \cos[2\omega_i t + (k-1)\alpha + 2\varphi_{ip}]\}$$

where $V_{imeqv}^2 = 2(V_a^2 + V_b^2 + V_c^2)/3$, it is no longer a constant. V_k stands for input phase voltage instantaneous value, $k \in \{a,b,c\}$. V_l stands for the desired output phase voltage instantaneous values, $l \in \{A,B,C\}$. The above equation states that MR's switching functions are the functions of input instantaneous voltages, which is helpful for obtaining desired output voltage.

C. SVM Algorithm When Balanced

Under the conditions that the output line is never opened and the input inter-phase is never shorted, MR has in total nine switching combinations, shown in Table 1, where the first six are active vectors, and the last three are null vector or zero vectors. DC voltage output is transmitted from the input side when one active vector acts. Whereas there isn't DC voltage output conducted when one inactive vector acts, but the output lines are shorted to provide load current commutation. Table 1 also gives the relations among output voltage, input currents and switching combinations. MR's active vector conduction durations are given in Table 2 after rearrangement[6].

D. SVM Algorithm When Unbalanced

Due to without consideration of the abnormal power conditions and the varying instantaneous values of input voltages and voltage zones, If the former current SVM

algorithm is not modified, the DC output voltage will be distorted, the input current will also deformed. According to (14), it is possible to establish the active vector conduction duration under unbalanced or distorted, conditions, shown in Table 3. The input voltage zones need redivision, shown in Fig.5. It is obvious that Table 3 has the generality, inclusive of Table 2.

In Fig. 5 and Table 3, $\varphi_{I1} \sim \varphi_{I6}$ represent the starting phases of $I1 \sim I6$. Apparently when V_{omp} is constant, m_c is the function of V_{imeqv}^2 and V_a, and V_{omp} will be greater than one sometimes. Since V_{imeqv}^2 is not constant, the input current will be distorted and different from that of input voltages. When using Table 3 to implement the current SVM algorithm, the active and zero vector conduction duration calculation are no longer same among the six sectors, needing the detection and calculation of input voltages in real time, and needing repartition of each sectors. As for amplitude unbalance and initial phase

TABLE I.
INPUT CURRENT VECTORS, SWITCHING COMBINATIONS, INPUT CURRENTS AND OUTPUT VOLTAGE

N	V_{pn-p}	i_{ap}	i_{bp}	i_{cp}	S11,S12,S13	S21,S22,S23	Vector
1	v_{acp}	I_{dcp}	0	$-I_{dcp}$	1 0 0	0 0 1	I1
2	v_{bcp}	0	I_{dcp}	$-I_{dcp}$	0 1 0	0 0 1	I2
3	$-v_{abp}$	$-I_{dcp}$	I_{dcp}	0	0 1 0	1 0 0	I3
4	$-v_{acp}$	$-I_{dcp}$	0	I_{dcp}	0 0 1	1 0 0	I4
5	$-v_{bcp}$	0	$-I_{dcp}$	I_{dcp}	0 0 1	0 1 0	I5
6	v_{abp}	I_{dcp}	$-I_{dcp}$	0	1 0 0	0 1 0	I6
7	0	0	0	0	1 0 0	1 0 0	I7
8	0	0	0	0	0 1 0	0 1 0	I8
9	0	0	0	0	0 0 1	0 0 1	I9

TABLE II.
ACTIVE VECTOR CONDUCTION DURATIONS WHEN BALANCED

zones	Active Vector	Conduction Duration (t_α and t_β)
I $-30° < \omega_i t + \varphi_{ip} < 30°$	I6	$f_{22p} - f_{12p} = m_c \sin(\omega_i t + 5\pi/6 + \varphi_{ip})$
	I1	$f_{23p} - f_{13p} = m_c \sin(\omega_i t + \pi/6 + \varphi_{ip})$
II $30° < \omega_i t + \varphi_{ip} < 90°$	I1	$f_{11p} - f_{21p} = m_c \sin(\omega_i t + \pi/2 + \varphi_{ip})$
	I2	$f_{12p} - f_{22p} = m_c \sin(\omega_i t - \pi/6 + \varphi_{ip})$
III $90° < \omega_i t + \varphi_{ip} < 150°$	I2	$f_{23p} - f_{13p} = m_c \sin(\omega_i t + \pi/6 + \varphi_{ip})$
	I3	$f_{21p} - f_{11p} = m_c \sin(\omega_i t - \pi/2 + \varphi_{ip})$
IV $150° < \omega_i t + \varphi_{ip} < 210°$	I3	$f_{12p} - f_{22p} = m_c \sin(\omega_i t + \pi/6 + \varphi_{ip})$
	I4	$f_{13p} - f_{23p} = m_c \sin(\omega_i t - 5\pi/6 + \varphi_{ip})$
V $210° < \omega_i t + \varphi_{ip} < 270°$	I4	$f_{21p} - f_{11p} = m_c \sin(\omega_i t - \pi/2 + \varphi_{ip})$
	I5	$f_{22p} - f_{12p} = m_c \sin(\omega_i t + 5\pi/6 + \varphi_{ip})$
VI $270° < \omega_i t + \varphi_{ip} < -30°$	I5	$f_{13p} - f_{23p} = m_c \sin(\omega_i t - 5\pi/6 + \varphi_{ip})$
	I6	$f_{11p} - f_{21p} = m_c \sin(\omega_i t + \pi/2 + \varphi_{ip})$

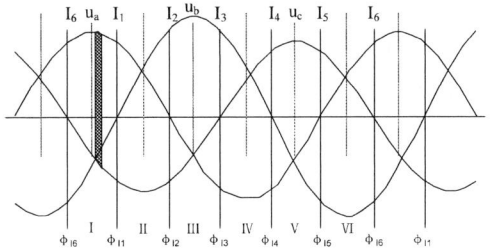

Figure 5. Input voltage zones redivision when unbalanced

TABLE III.
ACTIVE VECTOR CONDUCTION DURATIONS WHEN UNBALANCED

zones	Active Vector	Conduction Durations (t_α and t_β)
I $\varphi_{I6} < \omega_i t + \varphi_{ip} < \varphi_{I1}$	I6	$f_{22} - f_{12} = 2V_b(V_B^* - V_A^*)/3V_{imeqv}^2 = -\dfrac{2}{\sqrt{3}}\dfrac{V_{omp}}{V_{imeqv}^2}V_b$
	I1	$f_{23} - f_{13} = 2V_c(V_B^* - V_A^*)/3V_{imeqv}^2 = -\dfrac{2}{\sqrt{3}}\dfrac{V_{omp}}{V_{imeqv}^2}V_c$
II $\varphi_{I1} < \omega_i t + \varphi_{ip} < \varphi_{I2}$	I1	$f_{11} - f_{21} = 2V_a(V_A^* - V_B^*)/3V_{imeqv}^2 = \dfrac{2}{\sqrt{3}}\dfrac{V_{omp}}{V_{imeqv}^2}V_a$
	I2	$f_{12} - f_{22} = 2V_b(V_A^* - V_B^*)/3V_{imeqv}^2 = \dfrac{2}{\sqrt{3}}\dfrac{V_{omp}}{V_{imeqv}^2}V_b$
III $\varphi_{I2} < \omega_i t + \varphi_{ip} < \varphi_{I3}$	I2	$f_{23} - f_{13} = 2V_c(V_B^* - V_A^*)/3V_{imeqv}^2 = -\dfrac{2}{\sqrt{3}}\dfrac{V_{omp}}{V_{imeqv}^2}V_c$
	I3	$f_{21} - f_{11} = 2V_a(V_B^* - V_A^*)/3V_{imeqv}^2 = -\dfrac{2}{\sqrt{3}}\dfrac{V_{omp}}{V_{imeqv}^2}V_a$
IV $\varphi_{I3} < \omega_i t + \varphi_{ip} < \varphi_{I4}$	I3	$f_{12} - f_{22} = 2V_b(V_A^* - V_B^*)/3V_{imeqv}^2 = \dfrac{2}{\sqrt{3}}\dfrac{V_{omp}}{V_{imeqv}^2}V_b$
	I4	$f_{13} - f_{23} = 2V_c(V_A^* - V_B^*)/3V_{imeqv}^2 = \dfrac{2}{\sqrt{3}}\dfrac{V_{omp}}{V_{imeqv}^2}V_c$
V $\varphi_{I4} < \omega_i t + \varphi_{ip} < \varphi_{I5}$	I4	$f_{21} - f_{11} = 2V_a(V_B^* - V_A^*)/3V_{imeqv}^2 = -\dfrac{2}{\sqrt{3}}\dfrac{V_{omp}}{V_{imeqv}^2}V_a$
	I5	$f_{22} - f_{12} = 2V_b(V_B^* - V_A^*)/3V_{imeqv}^2 = -\dfrac{2}{\sqrt{3}}\dfrac{V_{omp}}{V_{imeqv}^2}V_b$
VI $\varphi_{I5} < \omega_i t + \varphi_{ip} < \varphi_{I6}$	I5	$f_{13} - f_{23} = 2V_c(V_A^* - V_B^*)/3V_{imeqv}^2 = \dfrac{2}{\sqrt{3}}\dfrac{V_{omp}}{V_{imeqv}^2}V_c$
	I6	$f_{11} - f_{21} = 2V_a(V_A^* - V_B^*)/3V_{imeqv}^2 = \dfrac{2}{\sqrt{3}}\dfrac{V_{omp}}{V_{imeqv}^2}V_a$

unbalance, when V_{omp} is low, dissatisfying $\sum_{k=1}^{3} f_{ik}(t)=1$, and when V_{omp} is high, dissatisfying $0 \le f_{ik} \le 1$ and $\sum_{k=1}^{3} f_{ik}(t)=1$. As for distortion unbalance, satisfying $\sum_{k=1}^{3} f_{ik}(t)=1$ constantly. But when V_{omp} is high enough, dissatisfying $0 \le f_{ik} \le 1$. Therefore the maximum voltage utilization will be lowered under such conditions, dependent on the level of unbalance. A simple method is put forwards to solve the problem when the desired V_{omp} is low, which means that $0 \le f_{ik} \le 1$ is satisfactory. That is, according the actual input phase voltage zones, to get two relative input phase voltages, then subtract from them one third the instantaneous sum of three-phase input voltages as new input phase voltages to calculate the active vectors' conduction durations.

IV. SIMULATION ANALYSIS

A. *Simulation When Balanced*

Simulation circuits of one balanced case and three unbalanced cases mentioned above are established, including power stage and respective control stage. Simulation parameters are listed below: the desired output voltage $V_{omp}=220V$, equivalent to $q_v=\sqrt{2}/2$ and $m_c=\sqrt{2}/\sqrt{3}$, i.e. the desired DC output voltage $V_{PN}=220\sqrt{3}$ at null load. Switching frequency is 8kHz, and the control is in open loop. Only the power feeding state of MR is simulated without considering regenerative state, and the related input voltage waveforms are scaled down ten times. The simulation algorithm is ode4 (Runge-Kutta) with fixed step of 1e-6S and free sampling time. For simulation under balanced conditions, let the desired leading and lagging displacement of input current is 60°, equivalently to displace the input voltages at the same lagging and leading 60°.

The power stage is shown in Fig.6, where input filtering inductance L_i=1mH and capacitance C_i=5uF, power switches consists of six ideal bilateral switches, PS1~PS6 correspond to S11, S12, S13, S21, S22 and S23 in Fig.2. Output filtering inductance L_{o1}= L_{o2}=2.5mH and capacitance C_o=220µF. The load is formed by a resistor and an inductor in series connection with R_o=20Ω and L_o=5mH. Also in Fig.6, VM1~6 stand for voltage meters, CM1~5 for current meters, Scope1~6 for scopes, Mux1~3 for multiplexes, Fcn"x" for maths functions, Demux for demultiplexes with its six output driving signals corresponding to the gates of PS1~6 in Fig.2.

The control circuit includes four parts, they are, current vectors duration calculation circuit shown in Fig.7, time into pulse conversion circuit in Fig.8, current vector pulse generation circuit in Fig.9 and current vector pulse selection circuit in Fig.10. In Fig.7, Fcn8 and Fcn9 generate saw-tooth signal with a period of 1/6 mains period, ZOH1's sampling time is 10/3mS, Const1 indicates voltage transfer ratio. Fcn5 and Fcn6 are used to figure out active vectors T_α and T_β, Fcn7 is responsible

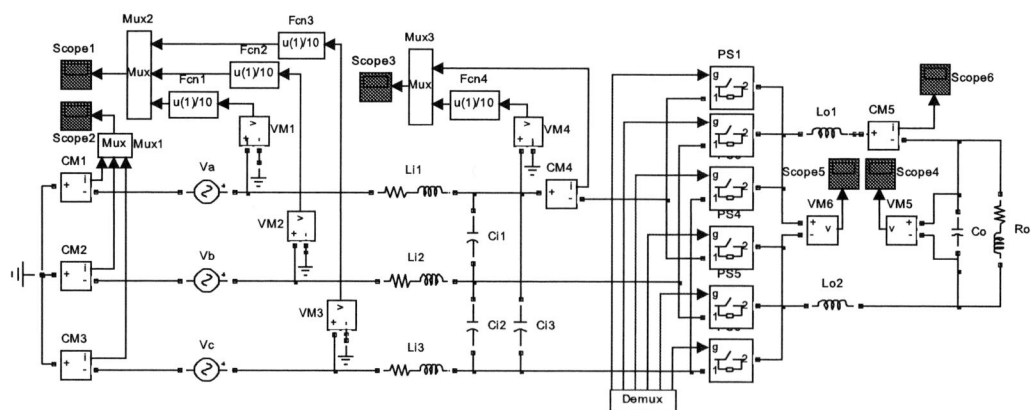

Figure 6. Power stage of simulated matrix rectifier

for calculating T_0 and divide it into three equal divisions. ZOH2's sampling time is 1.25e-4S, generating sawtooth–carrier signal with a frequency of 8kHz. Thus a switching period contains five time segments, they are zero vector→active vector 1→zero vector→active vector 2→zero vector in sequence.

In Fig.8, at first Fcn10~Fcn14 are to calculate the time length from the beginning a switching period to the end of the current vector, after a single step delay through Mem1~5 and trimming relay with upper limit and lower limit of ±eps, the above obtained time length are changed into unitary pulse series. After further via Fcn15~Fcn18, the positions and the durations of each vector pulse series are attained.

In Fig.9, Switch1~5's thresholds are $\pi/3$, $2\pi/3, \pi$, $4\pi/3$ and $5\pi/3$, and ZOH3's sampling time is 20mS, thus Fcn19 output sawtooth signal with a period of 20mS (mains period) and amplitude of 2π. According to Table 1 and Table 2, subsystems Subsys1~6 are designed to gain six pulse series from five vector pulse series by proper synthesis, and the six pulse series are used to drive six bidirectional switches.

Each subsystem includes six maths functions, shown in Fig.10. Eventually, after selected through conversion switches Switch1~5, the six pulse series satisfying the current fan sector are transferred to power stage.

Referring to (10), standard three-phase 380VAC/50Hz co-sinusoidal AC power supply are utilized and the initial phase is -30°, then the input current vectors, switching combinations and input currents and output voltages are employed in term of Table 1. Take input voltage phase "a" as example, to give the input voltage and input current simulated results. Those in Fig.11 (a) correspond to $\varphi_{ip}=0$, the filtered averaged DC output value is 379V, peak to peak voltage ripple is 4V, and the output value is 381V at near null load. Those in Fig.11 (b) correspond to $\varphi_{ip}=60°$, the filtered averaged DC output value is 189V, peak to peak voltage ripple is 4V, and output value is 190V at near null load. Those in Fig.11 (c) correspond to $\varphi_{ip}=-60°$, the filtered averaged DC output value is 189V, peak to peak voltage ripple is 4V, and output value is 190V at near null load. And Fig.11 (d) shows the filtered averaged DC output waveform corresponding to $\varphi_{ip}=0$.

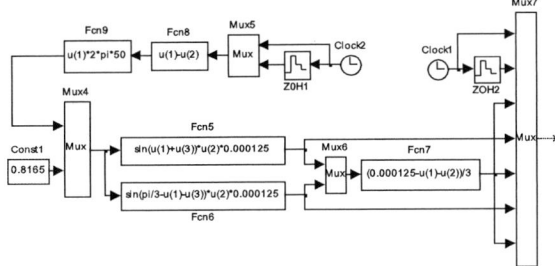

Figure 7.   Current vectors duration calculation when unbalanced

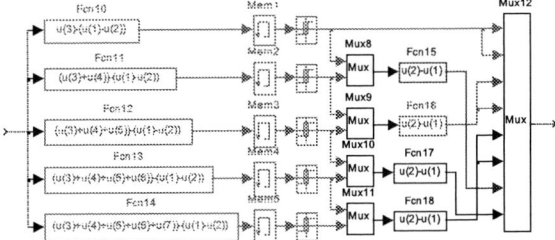

Figure 8.   Time to pulse conversion circuit

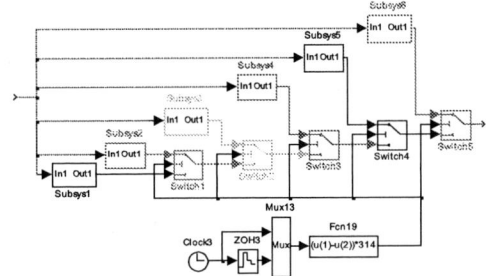

Figure 9.   Current vector pulse generation circuit

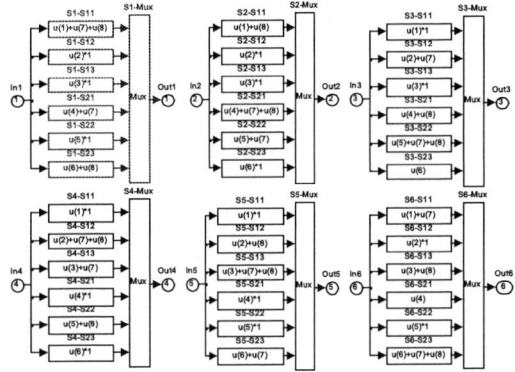

Figure 10. Current vector pulse selection circuit

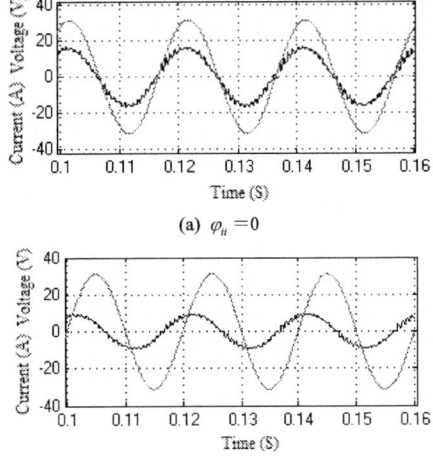

(a) $\varphi_{ii}=0$

(b) $\varphi_{ii}=+60°$

233

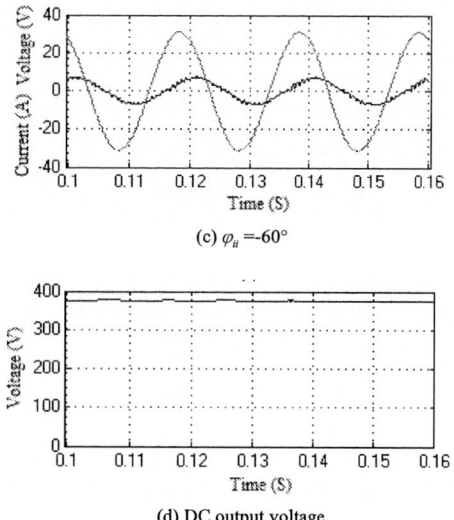

(c) $\varphi_{ii} = -60°$

(d) DC output voltage

Figure 11. Input voltage and input current waveforms when balanced

B. Simulation when Unbalanced

Here let Fig.7 substitute for Fig.12, Fig.7, Fig.8 and Fig.9 are kept unchanged. Function Fcn1 produces triangular signal with period of 20mS and amplitude of 2π, meaning the circulation of mains period. Multiplexes Mux1~6 and switches Switch1~5 yield two phase input voltages according to the current fan sector, used for calculating active vector durations. In addition, the thresholds of Switch1~5 correspond to the six rising edges and trailing edges outputted by function Fcn2. Signs Sign1~3 are used for detection of the zero-crossing points of input phase voltage. Fcn2 produces pulse signals, and the level transition reflects the entrance into different input voltage zone. Fcn4 calculates v_{imsqv}^2, and constant v_{omp} set the desired output voltage $220\sqrt{3}$. Fcn6 and Fcn7 calculate active vectors τ_α and τ_β. Sat1~4 are amplitude-limiters, their lower limits are zero. Fcn3 calculates the sum of three phase input voltages, and divide it by three and negate it, used for compensating the active vectors, provided in Fcn6 and Fcn7. the remainders are the same as those in Fig.7.

C. Simulation when Distorted

Here three-phase co-sinusoidal input voltages 380VAC/50Hz with initial phase -30° are superimposed by three-phase co-sinusoidal five order harmonic input voltages 30VAC/50Hz. The active vector conduction durations can refer to Table 3. Switch1~5's thresholds are 0.365π, 0.635π, π, 1.365π, 1.635π and 2π. Input phase voltage and current waveforms are shown in Fig.13, indicating the input currents are distorted also, notwithstanding in phase with input voltages, considering the phase-shifting effect of input filter in advance. The filtered averaged DC output value is 380V, peak to peak voltage ripple is 4V, and its waveform resembles that in Fig.11(d).

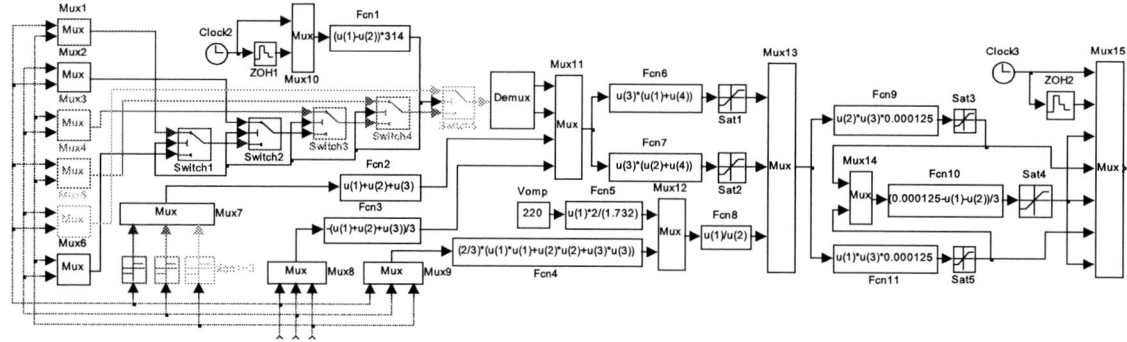

Figure 12. Current vectors duration calculation circuit when unbalanced

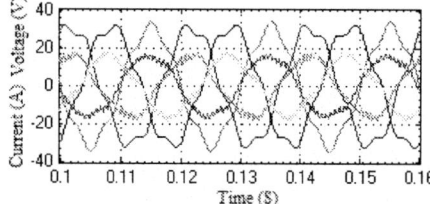

Figure 13. Input voltage and input current waveforms when balanced

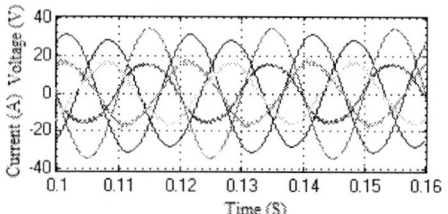

Figure 14. Input voltage and input current waveforms when balance

D. Simulation when Amplitude-Unbalanced

Here the amplitudes of three-phase input voltages are respective 311V, 331V and 291V with initial phase -30°. The active vector conduction durations can refer to Table 3. Switch1~5's thresholds are $\pi/3$, $2\pi/3$, π, $4\pi/3$, $5\pi/3$ and 2π. Input phase voltage and current waveforms are shown in Fig.13. The filtered averaged DC output value is 380V, peak to peak voltage ripple is 4V, and its waveform resembles that in Fig.11(d).

E. Simulation when Initial Phase-Distorted

Here the amplitudes of three-phase are 311V with respective initial phase+5°, 0° and -5°. The active vector conduction durations can refer to Table 3. Switch1~5's thresholds are 0.3611π, 0.6389π, π, 1.36π and 1.64π. Input phase voltage and current waveforms are shown in Fig.15. The filtered averaged DC output value is 378V, voltage ripple peak to peak value is 10V, and its waveform resembles that in Fig.11(d).

Note that unmodified current space vector algorithm when unbalanced will destroy the constraints of switching function. Modified one can acquire stable and continuously adjustable DC output voltage, but the maximum voltage utilization decrease duo to the unbalance.

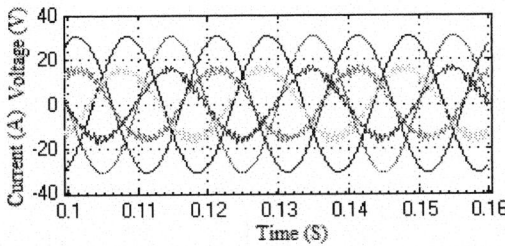

Figure 15. Input voltage and input current waveforms when balance

V. CONCLUSIONS

The current SVM algorithm of MR, under balanced and unbalanced power supply conditions, are in-depth discussed and simulated by SIMULINK6.0. The algorithm is more adaptable under balanced power supply conditions, and the input current displacement can be regulated continuously. But it isn't convenient to regulate input current displacement under unbalanced power supply conditions.

REFERENCES

[1] D. J. Holmes, T. A. Lipo, "Implementation of A Controlled Rectifier Using AC-AC Matrix Converter Theory," *IEEE Trans. on Power Electronics*, vol. 7, no.1, pp.240–249, 1992.

[2] Christopher S. Czerwinski, "Reduced common voltage in a DC matrix converter," *United States Patent, Patent Number: 6166930*, Date of Patent: Dec. 26, 2000.

[3] J. B. Ejea, E. Sanchis-Kilders, J. A. Carrasco, J. M. Espi and A. Ferreres, "Implementation of bi-directional AC-DC matrix converter," *IEE Electronics Letters*, vol. 38, no. 16, pp. 933–934, 2002.

[4] Xi-jun Yang, Chao-ping Deng, zhi-bin Ling and Peng-sheng Ye, "Three-phase controlled rectifier employing matrix converter theory," *Journal of Shanghai Jiaotong University*, vol. 38, no. 8, pp. 1291-1295, 2004 (in Chinese) .

[5] Hai-hui Lu. *Researches on AC-AC matrix power converter and its control system*. Shanghai: Shanghai University, 1998 (in Chinese).

[6] Christian Klumpner and Frede Blaabjerg, "Modulation method for a multiple drive system based on a two-stage direct power conversion topology with reduced input current ripple," *IEEE Trans. on Power Electronics*, vol.20, no.4, pp. 922–929, July 2005.

[7] L. Zhang, C. Watthanasarn and W. Shepherd, "Analysis and comparison of control techniques for AC-AC matrix converters," *IEE Proc. Electr. Power. Appl.*, vol. 145, no. 4, pp.284–294, 1998.

[8] C. Watthanasarn, *Optimal control and application of AC-AC matrix converters*, UK: University of Bradford, 1997.

[9] Akio Ishiguro, Takeshi Furuhashi and Shigeru Okuma, "A novel control method for forced commutated cyclo-converters using instantaneous values of input line-to-line voltages," *IEEE Trans. on Industrial Electronics*, vol. 38, no. 3, pp.166–172, July 1991.

[10] Aklo Ishiguro, Katsuhisa Inagaki, Muneaki Ishida, Shigeru Okuma, Voshlkl Uchikawa and Kojl Iwata, "A new method of PWM control for forced commutated cyclo-converters using microprocessors," *IEEE IAS*, vol. 1, pp. 712–721, 1988.

2006 5th International Power Electronics and Motion Control Conference

Study of Measurement Approach of Loop Gain of Converter

Weiping Zhang , Yunpeng Chen , Yuanchao Liu , Dongyan Zhang and Zheng Meng

North China University of Technology/Lab of Green Power & Energy System, Beijing, China

Email: zwp@ncut.edu.cn

Abstract—**In this paper, an approach by using the Agilent 4395A to measure the loop gain of the power supply has been deeply developed. The main issues have been investigated as the following: (1). Measurement techniques about the parasitic parameters of the filtering components will be studied. One is how to test of the electrolyte capacitor by Agilent 4395A, another is how to test dynamic inductance of filter inductor by Tektronix TDS5052 with TDSPWR3 software; (2). A measurement approach for loop gain has been put forward; (3). A prototype has been made up and its loop gain has been tested to verify the proposed approaches; (4). The experimental result has first revealed that the biggest analytic error of the loop gain occurs nearby about the resonance frequency of low pass filter. It can be explained that the error results from the equivalent resistor of the power diode and MOSFET; (5). Based on considering the equivalent resistor, a novel accurate small-signal model of Buck converter has been proposed to reduce the theoretic error. The proposed approaches, model and other results have provided a powerful tool to analyze and design a power supply.**

Key Words—*loop gain; model; power components stability*

I. Introduction

A general power switching regulation system includes power circuit (viz. switching converter) and control circuit. Both parts interact and work together. Switching network and LC low-pass filter constitute power circuit, and the control circuit is made up of sampling network, error amplifier, compensator, Pulse-Width Modulation (Abbrev. PWM) and driver etc[1]. In general, voltage-mode control and current-mode control are usually adopted. The voltage-mode control is to compare the sampling of output voltage with the reference, amplify the error voltage, and convert it into the impulse width D (the duty cycle) through PWM, and then drive the switch to regulate the output voltage. This mode is now extensively applied in switching power supply. The current-mode control has two modes: average-current-mode control and current-programmed-mode control, and can realize the regulation of the output current. In order to improve the performance of the system, the double-loop

Project supported by Beijing Natural Science foundation (No. 4052011)
Project supported by National Natural Science foundation of China (N0.50477054)

controlling approach has been employed. In general, current control is an internal loop and the voltage control is an external loop.

Fig.1 shows a system configuration schematic of Buck converter with voltage-mode control, the part in broken line frame is power circuit. The feedback control loop plays an important role in all linear power supply and switching power supply [2].

Currently many engineers think that modeling is the most powerful tool in the design of control loop, and it is enough to apply the ac small-signal model to predict the performance of system by professional simulation software. Some other engineers only copy mature compensation networks that have good performances in some typical circuits when they design the control circuit. Although the ac small-signal model is useful, it is vitally necessary to measure the loop gain by a frequency response analyzer for several reasons: (1). As well known, the elements are neither pure resistance nor pure reactance, and should be the combination of these components. So the elements have parasitic parameters in reality, for instance, there is inductance in resistor, resistance in capacitor, and capacitance in inductor. The values of parasitic parameters are different because of different materials and manufacture technologies as well as different condition. So if parasitic parameters of depositing energy elements cannot be confirmed, the analysis to model or the simulation results may be deviate from the fact. However, the parasitic parameters have been usually neglected when we model the system and do simulation. So the reality will deviate from theoretic results. The parasitic parameters of the elements should be measured and employed to reduce analytic errors. (2). The higher di/dt and dv/dt as well as reverse recovery current of power diodes in the system cause a serious

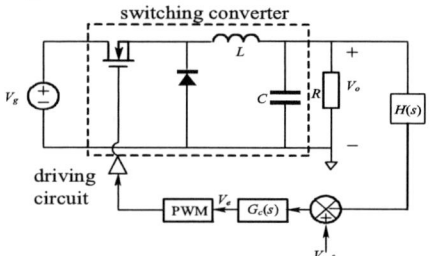

Figure 1. System configuration schematic of Buck converter with voltage feedback

1-4244-0448-7/06/$25.00 ©2006 IEEE

EMI and be influence of testing the frequency response. So, some special measurement equipments have to be employed to avoid this EMI while we are trying to measure the frequency response accurately. In a word, to minimize the risk, we have to measure the stability of control loop in the entire frequency range by employing special measurement equipments after the design of control circuit is completed.

Recently, many papers and books have been published and concentrate on theory and modeling, but the measurement approach for frequency response of the power switching regulation system has been rarely mentioned [3]. In the section II, a few of useful measurement approaches for the parasitic parameters have been introduced; In the section III, the measurement technique for loop gain has been developed by 4395A Network /Spectrum/Impedance Analyzer; In the section IV, a measurement example has been given out to verify the testing approaches proposed in this paper. At the same time, a new special phenomenal of loop gain has been discovered and the reason for that has been revealed.

II. MEASUREMENT APPROACH FOR THE PARASITIC PARAMETER

Our experiments show that the bigger analytic errors will occur if the parasitic parameters are neglected and the ac small-signal model is employed to analyze the frequency response. The maximum analytic error can reach to 30%, so that the theoretic analysis almost loss the sense. So, parasitical parameters of elements in a switching power supply can have dramatic effects on the overall system operation, especially on stability .It is, therefore, crucial to know how to get the values of the parasitic parameters . In this section, both measurement techniques will be developed. One is how to test the parasitic parameters of the electrolyte capacitor by Agilent 4395A, another is how to test dynamic inductance of filter inductor by Tektronix TDS5052 with TDSPWR3 software.

Agilent 4395A is a very powerful tool for characterizing the impedance of components, and finding the critical parasitical values. This equipment also provides a function that automatically calculates approximate values of specific parameters of an equivalent circuit that corresponds to an element. This function supports five circuit models, as shown in Table I . In addition, the resulting parameter values can be used to simulate the frequency-based characteristics of the equivalent circuit; this allows you to compare the simulated characteristics with the actually measured characteristics. We may also enter values of equivalent circuit parameter to simulate the frequency-based characteristics of equivalent circuit, consequently find exact parameter values.

Fig.2 shows the impedance characteristics of a 47uF electrolyte capacitor. The curve ① is simulation result,

TABLE I .
FIVE EQUIVALENT CIRCUIT MODELS

Equivalent Circuit	Type of Devices
	Inductors with high core loss
	Inductors and resistors
	High-value resistors
	Capacitors
	Resonators

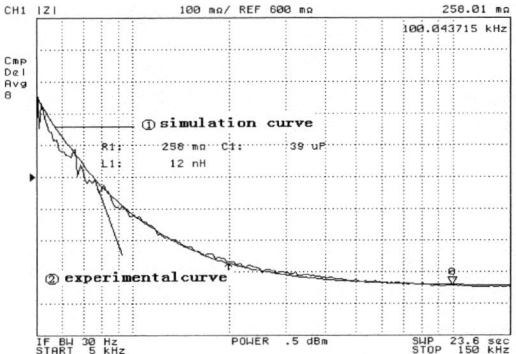

Figure 2. Impedance characteristics of a 47uF electrolysis capacitor

and the curve ② is experimental result. The measurement results of 4395A tell that the values of ESR, parasitic inductance and capacitance respectively are 0.258Ω, 12nH, and 39μF. In the following section, these parameters' values will be employed to analyze the frequency response properties.

For an inductor, we can make a measurement by Handheld LCR Meter, but the measured result should be regarded as static inductance, because the dc magnetizing force is so low that the inductance has little sense. The curve of increment permeability Vs. dc magnetizing force is shown in Fig.3. The curve displays that the inductance will reduce as the working current increases. So, in order to derive the exact inductance, the dynamic inductance should be measured.

Tektronix TDS5052 oscilloscope with TDSPWR3

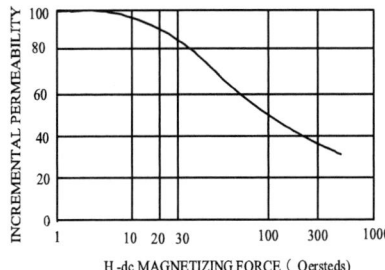

Figure 3. Incremental permeability Vs dc magnetizing force

software-power measurement and analysis application software-can do B-H analysis, dynamic inductance, permeability, maximum magnetic flux density and etc. The inductance of an inductor applied in our prototype was test by TDS5052, and Fig.4 shows that, $L = 68.62\mu H$.

If the increments (ΔI, Δt) and the voltage (V) of the inductor are measured by a oscilloscope, the dynamical inductance can be also calculated by the following formula,

$$L = V\Delta t / \Delta I \ . \qquad (1)$$

III. Measurement Approach of Loop Gain

We can employ Agilent 4395A to measure frequency response. If the loop gain of a closed-loop system has been measured, phase and magnitude margin can be found out through analyzing the magnitude and phase Bode plot, the crossover frequency. Based on these data, we can judge whether the design of feedback loop can meet the requirements or not. In this section, Buck converter with voltage feedback shown in Fig.1 will be taken as an example to introduce how to use the 4395A to measure the loop gain of the switching power supply. Then comparing the experimental results with the simulation waveforms, the correctness of the experimental method will be verified.

To make a measurement, we should inject a disturbance signal into the loop. The output impedance of the disturbance signal is asked to be too low to have influence upon loop gain. Furthermore, the signal should be injected into the place where the feedback signal has only one path in the loop. Usually, the feedback path that connects to the output of the power supply or the error amplifier is the preferable place to the signal. Here, we choose the former.

Fig.5 shows the equipment setup for loop gain measurement. The injected signal is generated by the radio frequency (RF) signals output from the RF OUT port of 4395A, and goes into the circuit through an isolation transformer T_1. It is important to check the following points:

（1）The signals detected at the B and R ports should

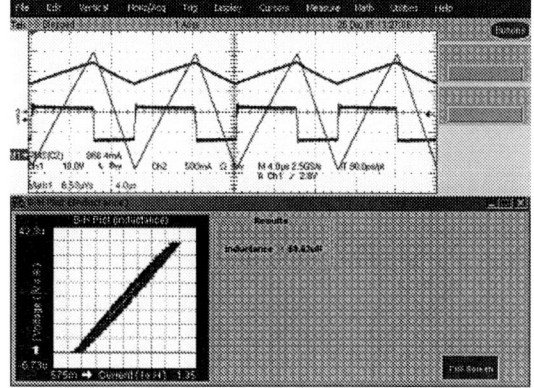

Figure 4. Inductance measured by Tektronix TDS5052

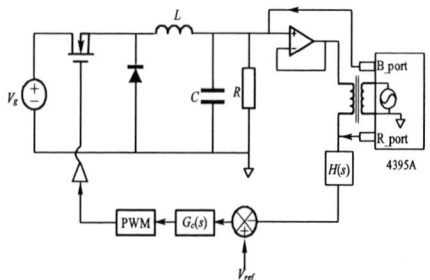

Figure 5. Equipment setup for loop gain measurement

be enough higher than the 4395A's noise background to avoid that it is submerged, and this signal should be enough to make sure of that it is a small-signal. So it is key issue to properly select amplitude of the injected disturbance signal. In Fig.5, the signal should be injected into the loop by the transformer T_1 because it is a difference signal. To obtain the maximum signal power, the magnetizing inductance should be far greater than the output impendence (50Ω) of 4395A RF-OUT and the current though it could be neglected and the load resistance R_T of the transformer should match its output impendence of 4395A RF-OUT when the input impendence of under test circuit in the secondary side of the transformer. TDK's 3701/3702 transformer can be used to fulfill impedance matching.

(2) Injecting the isolation transformer T_1 and both probes should not lead to any resonance in the loop in the measurement frequency range because T_1 must exhibit inductance behavior in this range in order to inject the RF signal.

To reduce the influence of the voltage sampling network to the load, we connect a voltage follower between the output of the power supply and the isolation transformer T_1, the follower not only has high precision, the error between input voltage and output voltage is about one millionth, but also the input resistance is high and the output resistance is low, the input resistance can reach $10^{12}\Omega$ in theory. In generally, the voltage follower is not necessary in the measurement of loop gain. It is much better if T_1 is injected into the input port of the voltage follower.

In Fig.5, a monitor's signal is injected into sampling network at the R-port and the signal output from output of the power supply is measured at the B-port. We make use of 41802A (1M Input Adapter) and 10441B that is one kind of Agilent 10400B series miniature passive oscilloscope probe to get the monitor signals. The 41802A is used to fulfill impedance matching. When 41802A is employed, make sure that the sum of ac and dc voltages at the measure point can't exceed 50V. In Fig.5, the ratio of B/R is directly equivalent to loop gain of the switching power supply.

In order to observe its frequency response, the scanning frequency range of the injected signal should be enough wide to cover the operational frequency of under

238

test power supply. For example, if the operational frequency of a switching converter is 100 kHz, the measure frequency should be between 10Hz and 100 kHz. The 4395A can measure the input and output monitor signals in each point of the entire frequency range, then make the phase and magnitude Bode plot on the screen. We may set frequency and magnitude to the form of logarithm .when the vertical axis is set to phase and magnitude, the Bode plot will be shown on the screen. We can also use the dual channel display function to merge the channels into one screen or split the screen into the two channels. By doing this, the manipulator may conveniently use the marker read the phase and magnitude numerical values.

By the way, we should note that the measurement curve is not stable in the low frequency range; this is because the influence of noise on the measurement is big in this frequency range.

IV. MEASUREMENT EXAMPLE OF LOOP GAIN AND A SMALL−SIGNAL ACCURATE MODEL

Based on the circuit shown in Fig.1, a prototype has been made up to exhibit the measurement of loop gain. The experimental conditions are as following: Input line voltage: V_g=15.4V, Output voltage: V= 7.3V, inductor :L =68.62μH, the ESR of L is 0.025Ω (the value is measured by Handheld LCR Meter), capacitor :C = 47μF, the peak to peak of saw-tooth waveform: V_M =2.4V. Single pole and single zero circuit shown in Fig.6 is used as the compensation network, R_1=29.8kΩ, R_2=29.8kΩ, C_1=0.104nF, C_2=31.1nF, the sampling network is shown in Fig.7, R_3=4.27kΩ, R_4=8.14kΩ, R_5=5.53kΩ.

Fig.8 shows the block diagram of voltage-control-mode switching regulation system. The definitions of all of the symbols are as follows:

$\hat{v}_{ref}(s)$ —— image function of the reference voltage

$\hat{v}_e(s)$ —— image function of the error

$G_c(s)$ —— transfer function of the controller

$\hat{v}_c(s)$ ——image function of the output of the controller

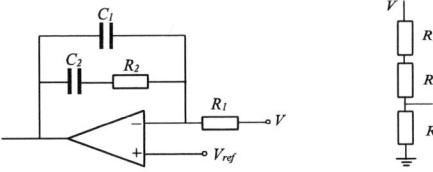

Figure 6. Compensation network   Figure 7. Sampling network

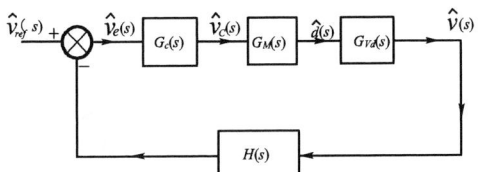

Figure 8. The block diagram of voltage-control-mode switching converter

$G_M(s)(1/V_M)$—— transfer function of the PWM

V_M—— peak to peak of the saw-tooth waveform

$\hat{d}(s)$ —— image function of the duty cycle

$G_{vd}(s)$—— transfer function of the control-output

$\hat{v}(s)$ —— image function of the output voltage

$H(s)$—— transfer function of the sampling network

The loop gain can be written in the following form:

$$A_{loop} = H(s)G_c(s)G_{vd}(s)G_M(s) \cdot \qquad (2)$$

If the parasitic resistor (R_L) of the filter inductor(L) and the filter capacitor(C)'s ESR (R_c) and parasitic inductor (L_c) is considered, the transfer function of the control-output is

$$G_{vd}(s) = \frac{V_g}{1+(sL+R_L)\cdot(\frac{1}{R}+\frac{1}{R_c+sL_c+\frac{1}{sC}})} \cdot \qquad (3)$$

The expression of loop gain can be written,

$$A_{loop}(s) = \frac{\frac{V_g}{V_M}H(s)G_c(s)}{1+(sL+R_L)\cdot(\frac{1}{R}+\frac{1}{R_c+sL_c+\frac{1}{sC}})} \cdot \qquad (4)$$

Based on the above formula, the simulation has been done by by Mathcad program. Fig.9 shows the simulation result and experimental result of magnitude Bode plot of the loop gain. Fig.10 shows the simulation result and experimental result of phase Bode plot of the loop gain. The curve ① is simulation result, and the curve ② is experimental waveform. The experimental results are well agreed with theoretic results except resonance frequency and its neighborhood. What is reason about the error?

The measurement of loop gain can be divided into

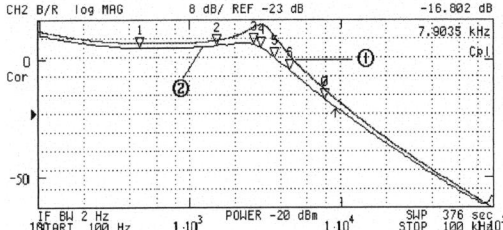

Figure 9. The comparison of the simulation curve of magnitude Bode plot of the loop gain and experimental one

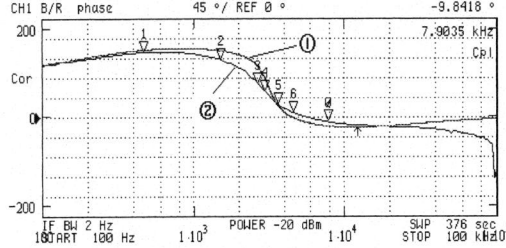

Figure 10. The comparison of the simulation curve of phase Bode plot of the loop gain and experimental one

three parts in terms of the expression: one is $H(s)G_c(s)$, another is $V_g G_M(s)$, the last is $G_{vd}(s)/V_g$. The total value should be the sum of three parts. We respectively measure the three parts, and make seven points that have the same frequency in the three curves. The sum of the three parts and the value of loop gain at the corresponding frequency points are shown in Table II.

The magnitude Bode plot of $V_g G_M(s)$ is shown in Fig.11. This curve is not smooth and has a sunken near the resonance frequency. However, the simulation results and theoretic results show that this magnitude $V_g G_M(s)$ always keeps a constant in whole frequency range. Similarly, Fig.12 shows the phase Bode plot of $V_g G_M(s)$. We may also notice that the curve is not zero near the resonance frequency. But when we make emulation, we regard the phase value as zero in the entire frequency range. So, this part maybe produces the bigger analytic error of the loop gain near the resonance frequency.

In order to verify our above opinion, we change the resonance frequency by changing the value of the filter capacitor C, and find the above phenomena still exits and the sunken point on the amplitude Bode plot varies as the value of the filter capacitor C varies. Hence, our conclusion is that $V_g G_M(s)$ is not a constant nearby the resonance frequency; this is the reason why the experimental curve deviates from the simulation one nearby the resonance frequency.

Why the experimental curve deviates from the simulation one nearby the resonance frequency? Ideally the value of $V_g G_M(s)$ is a constant, V_g /V_M, if an equivalent interior resistor R_s for the diode and MOSFET is neglected. Unfortunately, the reality is that the interior resistance R_s can not be neglected. So the accurate model of this part should be a voltage source V_1 and an interior resistance R_s, as shown in Fig.13. The value of V_1 is equal to V_g/V_M. In fact, the measurement value in the experiment should be the value of V_{ab}, not V_1.

The inductance, capacitance and load can be seen as an equivalent input impendence R_1. At the resonance frequency, Fig.13 can be simplified to Fig.14. So the following formula can be obtained:

$$R_1 = sL + R_L + \frac{R(R_c + sL_c + 1/sC)}{(R_c + sL_c + 1/sC + R)} \quad (5)$$

$$V_{ab} = V_g R_1/(R_1 + R_s) = V_g/(1 + R_s/R_1) . \quad (6)$$

The value of R_1 reaches minimum if the circuit works at the state of resonance. V_{ab} may reach minimum in the state of resonance because V_{ab} is in proportion to the value of R_1. This is the reason why the value of $V_g G_M(s)$ is not a constant, or decreases dramatically nearby the resonance frequency.

V. SUMMARY

The analysis to model is not enough in the design and production of switching power supply. In order to reduce the adventure and save time or money, the control loop stability must be measured within the operation frequency range. In this paper we introduce how to use the 4395A to measure the loop gain of the switching power supply. what's more, the results obtained from experiment are compared with the ones from simulation. The measurement approach is verified by comparison. At the same time we may use impedance performance provided by the 4395A to measure the impedance characteristics of power components, because many manufactures of power components do not give adequate data for their parts for proper converter design.

TABLE II.

THE SUM OF THREE PARTS AND VALUE OF LOOP GAIN

N	SWP PARAM	Vg/Vx	Gvd/Vg	H(s)G(s)	SUM OF THE THREE PARTS	LOOP GAIN	ERROR
1	465.39 Hz	15.18 dB	.533 dB	-11.374 dB	4.339 dB	4.8742 dB	0.5352 dB
2	1.5375 kHz	14.709 dB	2.8682 dB	-12.115 dB	5.462 dB	5.9954 dB	0.5334 dB
3	2.6535 kHz	12.952 dB	5.5913 dB	-12.77 dB	5.773 dB	6.4407 dB	0.3677 dB
4	2.9432 kHz	13.246 dB	4.3985 dB	-12.936 dB	4.709 dB	5.1635 dB	0.4545 dB
5	3.6376 kHz	14.367 dB	-.0971 dB	-13.361 dB	0.909 dB	.606 dB	0.303 dB
6	4.5795 kHz	14.999 dB	-5.2718 dB	-14.057 dB	-4.33 dB	-4.5287 dB	0.1987 dB
7	7.9035 kHz	15.465 dB	-15.448 dB	-16.558 dB	-16.541 dB	-16.802 dB	0.261 dB

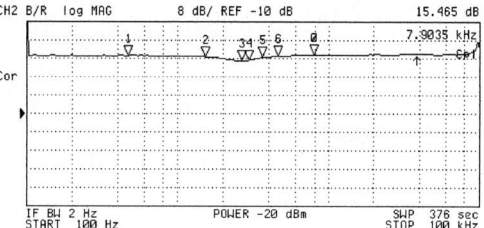

Figure 11. Magnitude Bode plot of $V_g G_M(s)$

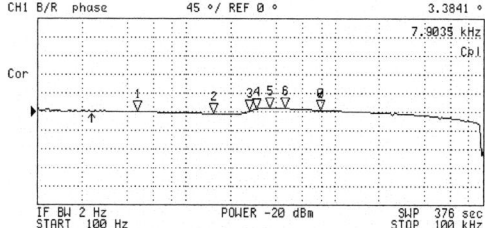

Figure 12. Phase Bode plot of $V_g G_M(s)$

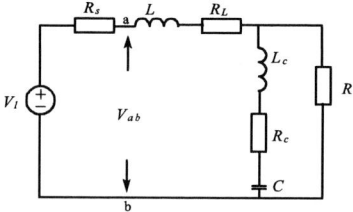

Figure 13. Accurate model of Buck converter

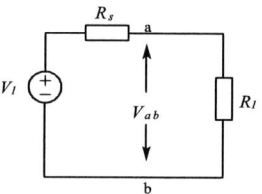

Figure 14. The equivalent circuit at resonance frequency

REFERENCES

[1] Weiping Zhang, *Modeling and Controlling of Switching Converter*, 1st ed., China: China Electric Power Press, January, 2006, pp.88–90.

[2] Robert W . Erickson and Dragon Maksimovic, *Fundamentals of Power Electronics,* 2nd ed., U.S.A: Kluwer Academic Publishers , 2001, pp.331–334.

[3] Dr. Ray Ridley, *"Frequency Response Measurements for Switching Supplies"*,Sem1300-Unitrode Design Seminar Archive, Appendix A-1.

[4] Agilent, *4395A Network/Spectrum/Impedance Analyzer Operation Manual*, Malaysia ,May ,2003.

[5] Tektronix, *power measurement and analysis application software*, 2004.

2006 5th International Power Electronics and Motion Control Conference

A Stand-Alone Hybrid Generation System Combining Solar Photovoltaic and Wind Turbine with Simple Maximum Power Point Tracking Control

Nabil A. Ahmed and Masafumi Miyatake

Sophia University, Tokyo, Japan
Email: nabil@power.ee.sophia.ac.jp

Abstract—This paper proposes a hybrid energy system combing solar photovoltaic and wind turbine as a small-scale alternative source of electrical energy where conventional generation is not practical. A simple and cost effective control technique has been proposed for maximum power point tracking from the photovoltaic array and wind turbine under varying climatic conditions without measuring the irradiance of the photovoltaic or the wind speed. The proposed system is attractive owing to its simplicity, ease of control and low cost. A complete description of the proposed hybrid system along with detailed simulation results which ascertain its feasibility are given to demonstrate the availability of the proposed system in this paper. Simulation of the hybrid system under investigation was carried out using PSIM software.

Keywords-Hybrid energy system, solar photovoltaic, wind turbine, stand alone applications, boost dc-dc converter, maximum power point trackin.

I. INTRODUCTION

Renewable energy from wind turbine and solar photovoltaic are the most environment-friendly type of energy to use. They have come of age and are global phenomenon, the world's fastest growing energy resources, a clean and effective modern technology that provides a beacon of hope for a future based on sustainable, pollution-free technology. Today's wind turbines are state-of-the-art of modern technology-modular and very quick to install. The importance of utilizing the renewable energy system, including solar photovoltaic (PV) and wind turbine (WT) generation systems have been attracted greatly in these days because the electricity demand is growing rapidly all over the world. Therefore, there is an urgent need for the renewable energy resources and it has formulated as a national strategy for the development of renewable energy applications and energy conservation measures. For this purpose, continuous effort to develop more attracting systems with lower-cost, higher-performance and multi-functions are required. Sensor-less approaches and combined generators explained in this paper are one of such key aspects.

Small-scale stand-alone power generation systems are an important alternative source of electrical energy, finding applications in locations where conventional generation is not practical. Consider, for example, remote villages in developing countries or ranches located far away from main power lines. It has been shown that a remote load has only to be a matter of a few miles away from a main power line for a stand-alone wind generator to be cost-effective [1]-[3].

The certainty of load demands at all times is greatly enhanced by hybrid generation systems, which use more than one power source. It is possible to achieve much higher generating capacity factors by combining wind turbine and photovoltaic generators with a storage technology to overcome the fluctuations in plant output. An efficient energy storage system is required, to get constant power and the electrical energy delivered by the wind turbine and photovoltaic has to be easy converted into storage energy. This conversion might be realized by a battery bank or energy capacitor system (ECS). The battery bank or ECS meets the daily load fluctuations [4]-[5].

In this paper a hybrid energy system combining variable speed WT and PV array generating system is presented to supply continuous power to the stand-alone load. The wind and PV are used as main energy sources, while the battery is used as back-up energy source. Two individual dc-dc boost converters are used to control the power flow to the load. A simple and cost effective control with dc-dc converter is used for maximum power point tracking (MPPT) and hence maximum power extracting from the WT and the PV array.

II. PROPOSED HYBRID ENERGY SYSTEM

Fig. 1 depicts the topology of hybrid energy system consisting of variable speed WT coupled to a permanent magnet generator (PMG) and PV array. The two energy sources are connected in parallel to a common dc bus line through their individual dc-dc converters. The load may be dc connected to the dc bus line or may include a PWM voltage source inverter to convert the dc power into ac at 50 or 60 Hz. The load configuration is beyond the scope of this paper.

Each source has its individual control. The Diodes *D1* and *D2* allow only unidirectional current flow from the source to the dc bus line, thus keeping each source from acting as a load on each other or on the battery. Therefore in the event of malfunctioning of any of the energy sources, the respective diode will automatically disconnect that source from the system.

The output of the hybrid generating system goes to the dc bus line to feed the isolating dc load or to the inverter, which converts the dc into ac. A battery charger is used to keep the battery fully charged at a constant dc bus line voltage. When the output of the system is not available, the battery powers the dc load or discharged to the inverter to

1-4244-0448-7/06/$25.00 ©2006 IEEE

power ac loads, through a discharge diode *Db*. A battery discharge diode *Db* is to prevent the battery from being charged when the charger is opened after a full charge. A dump load may be required, if excessive power is still available after fully charging the battery. As depicted in the system configuration represented in Fig. 1, the Vdc is st to a fixed dc bus line voltage and the output dc voltage from each source is controlled independently for both generation systems to get maximum power point tracking.

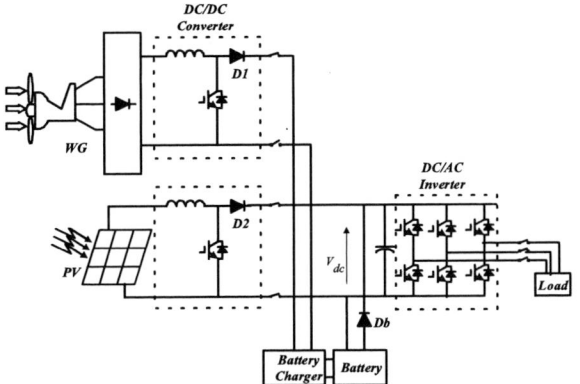

Figure 1.   Equivalent circuit of PV module.

III.   SOLAR PHOTOVOLTAIC SYSTEM

The European PV industry Association reported that the total global PV cell production world wide in 2002 was over 560 MW and has been growing about 30% annually in recent years.

The physical of PV cell is very similar to that of the classical diode with a pn junction formed by semiconductor material. When the junction absorbs light, the energy of absorbed photon is transferred to the electron-proton system of the material, creating charge carriers that are separated at the junction. The charge carriers in the junction region create a potential gradient, get accelerated under the electric field, and circulate as current through an external circuit. The solar cell is the basic building of the PV power system it produces about 1 W of power. To obtain high power, numerous such cell are connected in series and parallel circuits on a panel (module), The solar array or panel is a group of a several modules electrically connected in series-parallel combination to generate the required current and voltage. The electrical characteristics of the PV module are generally represented by the current vs. voltage (I-V) and the current vs. power (P-V) curves. Figs. and   show the (I-V) and (P-V) characteristics of the used photovoltaic module at different solar illumination intensities.

Using the equivalent circuit of solar cells shown in Fig. 2, the radiation dependent V-I characteristic of ns series cell and np parallel modules can be represented by:

$$V = n_s \left(\frac{AkT}{q} \right) \text{lin} \left[\frac{n_p I_{sc} - I + n_p I_D}{n_p I_D} \right] - \frac{n_s}{n_p} I R_s \qquad (1)$$

where I_{sc} : Short circuit current per cell (A)

$\quad I_D$: Diode saturation current (A)

$\quad q$: Electron charge ($1.6e^{-19}$ C)

k : Boltzmann constant ($1.38e^{-23}$ J/°K)

A : pn junction material factor

T : Temperature (°K)

R_s : Series resistance.

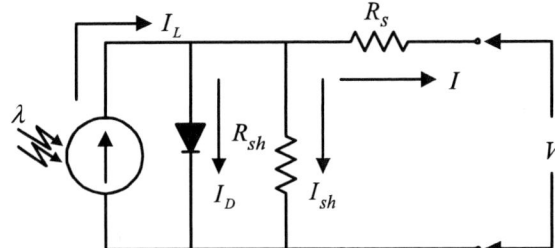

Figure 2.   Equivalent circuit of PV module.

For an ELR615 160Z, 750 W, Fuji Electric solar panel (n_s =3, n_p =5) used in this paper and neglecting the series and shun resistances, Eq. (1) can be written as:

$$V = \frac{3}{0.482} lin \left[\frac{5*2.281 - I + 2*8.66e^{-5}}{5*8.66e^{-5}} \right] \qquad (2)$$

Fig. 3 and 4 shows the strong non linearity of the I-V and P-V characteristics of the used solar wt different insolation levels. The I-V characteristic of the solar PV decreases gradually as the voltage goes up and when the voltage is low the current is almost constant. The power output of the panel is the product of the voltage and current outputs. The PV module must operate electrically at a certain voltage that corresponds to the peak power point under a given operation conditions.

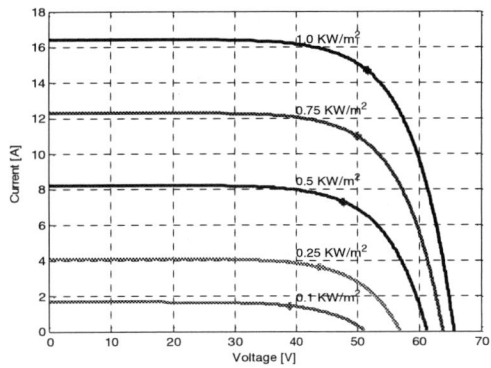

Figure 3.   I-V characteristics of PV module.

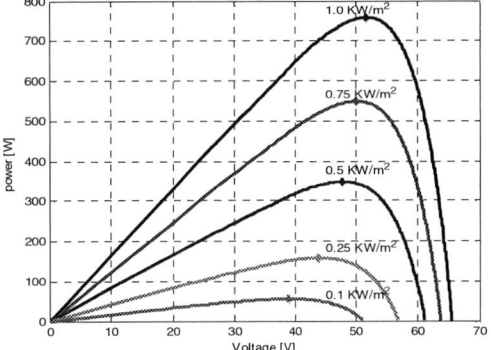

Figure 4.   P-V characteristics of PV module

Various techniques of maximum power tracking have been considered in PV power applications. Among these, the perturbation and observation (P&O) method, which moves the operation point toward the maximum power point by periodically increasing or decreasing the array voltage, is often used in many PV systems. The advantage of this method is that it works well when the irradiation does not vary quickly with time, however, the P&O method fails to quickly track the maximum power points [6]. The incremental conductance (IncCond) method is also often used in PV systems. The IncCond method tracks the maximum power points by comparing the incremental and instantaneous conductance of the PV array. This incremental conductance method offers good performance under rapidly changing atmospheric conditions [7]. However, it has two divisions and the structure is similar with P&O algorithm because the condition, $dP/dV=0$, is rarely happen.

For most PV modules, the ratio of the voltage at the maximum power point for different insolation levels to the open circuit voltage (V_{mp}/V_{oc}) is approximately constant. Also, the ratio of the current at the maximum power point for different insolation levels to the short circuit current (I_{mp}/I_{sc}) is constant [8], [9]. Figs. 5 and 6 indicate the linear relation $V_{mp}=0.77V_{oc}$ and $I_{mp} = 0.89I_{sc}$ with the computed (almost linear) dependency shown by "*" signs. Therefore, if unloaded cell is installed on the array and kept in the same environment as the power producing cells, and its open circuit voltage or short circuit current are periodically measured. The operating voltage or the current of the power producing array are then set to the required values, which corresponding to maximum power as shown in Figs. 5 and 6. The MPPT technique proposed in this work makes use of a predetermined relationship between the operating voltage or current and the open circuit voltage/short circuit current to obtain MPPT at any operating conditions.

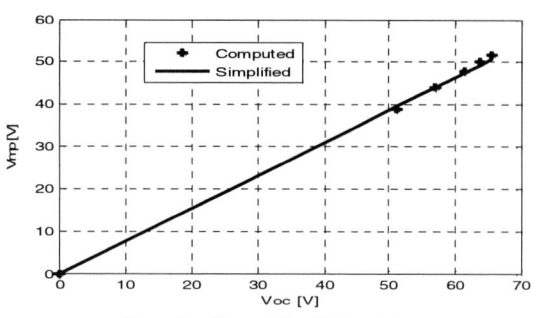

Figure 5. V_{mp} and V_{oc} of PV module.

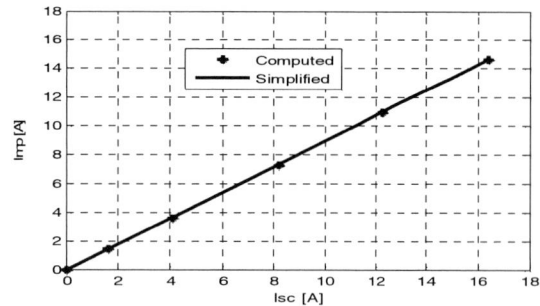

Figure 6. I_{mp} and I_{sc} of PV module.

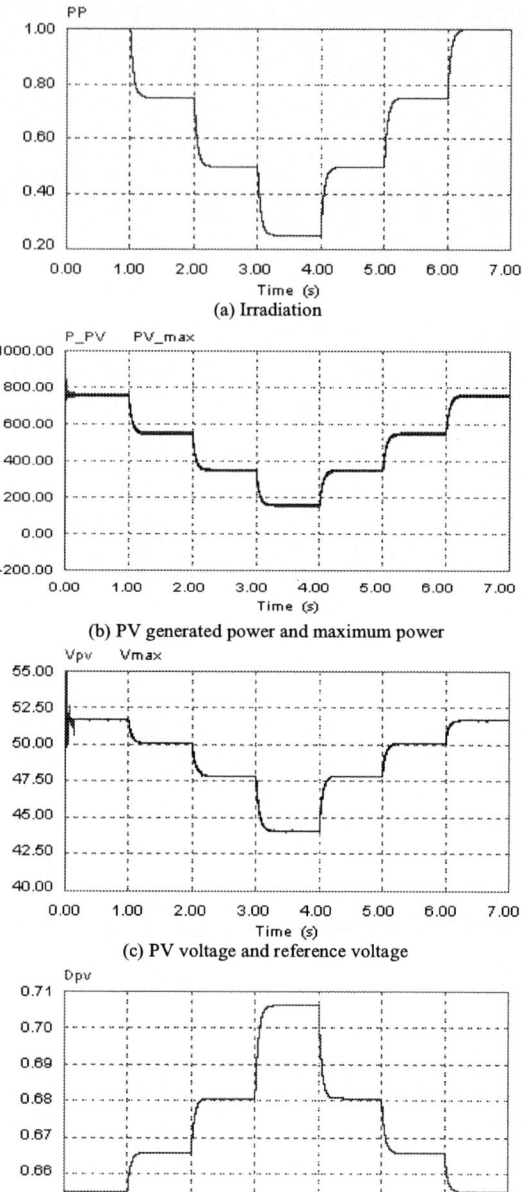

(a) Irradiation

(b) PV generated power and maximum power

(c) PV voltage and reference voltage

(d) Duty cycle

Figure 7. PV generation system characteristics under MPPT.

Simulation of the PV system under investigation was carried out using PSIM software. The simulation results of the dynamic performance, which validates the efficient MPPT of PV generation system when the irradiance changes dramatically are presented. Fig. 7 shows the irradiation, the power and maximum power, PV voltage and reference voltage and the PV DC-DC boost converter duty cycle, respectively of the voltage-based maximum power point tracking technique when the irradiation changes dramatically from 1 kW/m2 to 0.25 kW/m2 and again to 1 kW/m2 at a step of 0.25 kW/m2 and at a time step of 1s. The proposed simple MPPT is efficiently able to capture the maximum power corresponds to each irradiance. The PV generated power is not constant and it depends on the irradiance conditions.

IV. WIND ENERGY SYSTEM

Because wind energy has become the least expensive source of new renewable energy that is also compatible with environment preservation programs, many countries promote wind power technology by means of national programs and market incentives. The wind turbine captures the wind's kinetic energy in a rotor consisting of two or more blades mechanically coupled to an electrical generator.

The fundamental equation governing the mechanical power capture of the wind turbine rotor blades, which drives the electrical generator, is given by:

$$P = \frac{1}{2}\rho A C_p V^3 \tag{3}$$

where ρ : Air density (kg/m^3)

A : Area swept by the rotor blades

V : Velocity of air (m/sec),

C_p : Power coefficient of the wind turbine.

The theoretical maximum value of the power coefficient C_p is 0.59 and it is often expressed as function of the rotor tip-speed to wind-speed ratio (*TSR*). *TSR* is defined as the linear speed of the rotor to the wind speed.

$$TSR = \frac{\omega R}{V} \tag{4}$$

where R and ω are the turbine radius and the angular speed, respectively. In practical designs, the maximum achievable C_p ranges between 0.4 to 0.5 for modern high speed turbines and between 0.2 to 0.4 for slow speed turbines. Attaining C_p above 0.4 is considered good. Whatever maximum value is attainable with a given wind turbine, it must be maintained constant at that value for the efficient capture of maximum wind power. A relatively small deviation on either side of the *TSR* will result in a significant reduction of the power available for conversion to electrical energy. Fig. 8 exhibits the poor C_p performance at different TSR for various types of wind turbines [10]. Fig. 9 illustrates the typical power coefficient C_p curve for a 503 series WINDSEEKER by Southwest wind power, which is used for the analysis and simulations discussed in this paper. Fig. 9 shows that C_p has its maximum value ($C_{p\max}$) at a certain optimum value of tip-speed to wind-speed ratio called TSRopt. It is clear that (for this case) the maximum power captured by the wind turbine will occur when *TSR* is approximately 9. The typical turbine torque and power vs. rotor speed are plotted in Figs.10 and 11. The maximum power for different wind speeds is generated at a different rotor speeds. Therefore, the turbine speed should be controlled to follow the ideal TSR, with an optimal operating point which is different for every wind speed. This is achieved by incorporating a speed control in the system design to run the rotor at high speed in high wind and at low speed in low wind. Employing control of the rotational speed of the turbine allows the *TSR* to be controlled and the coefficient of performance to be maximized. Thus, in turn, the generated electrical energy may be maximized. Unfortunately, accurate wind speed measurement in the rotor of the turbine is difficult and requires the use of a relatively expensive anemometer if it is to be used for system control. Based on Eq. (4), the optimum speed of the rotor can be estimated as:

$$\omega_{opt} = \frac{TSR_{opt}V}{R} \tag{5}$$

Combining Eqs. (3) and (5), the output torque of the turbine can be written as:

$$T = \frac{1}{2}\frac{\rho A C_{p\max}}{\omega_{opt}}\{\frac{R\omega_{opt}}{TSR_{opt}}\}^3 \tag{6}$$

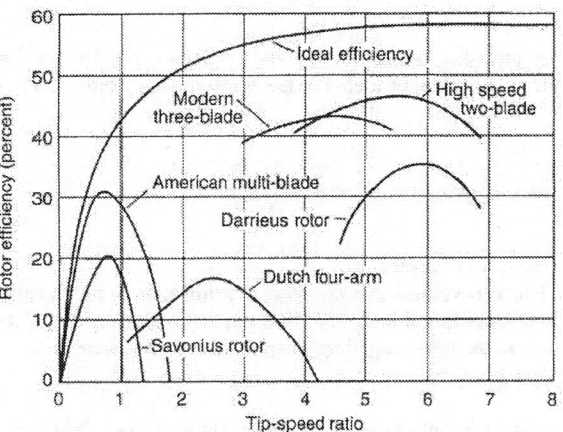

Figure 8. C_p vs. *TSR* for various types of wind turbines.

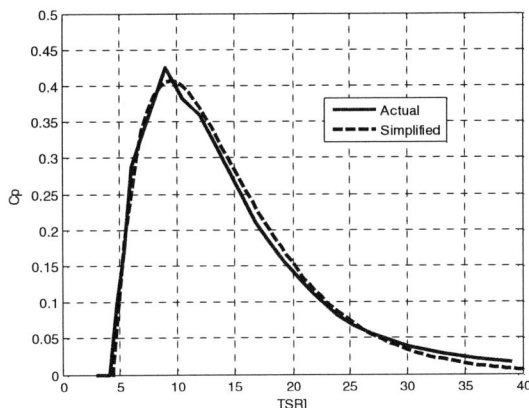

Figure 9.   Typical C_p curve used for the analysis and simulation.

A typical, small-scale, stand-alone, wind electric system is composed of a variable speed wind turbine, a permanent-magnet generator (PMG) and a diode bridge rectifier. In many small-scale systems, the dc system is set at a constant dc voltage and is usually comprised of a battery bank, allowing energy storage; a controller to keep the batteries from overcharging; and a load. The load may be dc or may include an inverter to an ac system. Connecting a wind generator to a constant dc voltage has significant problems due to the mismatching the poor impedance matching between the generator and the constant dc voltage (battery), which will limit power transfer to the dc system. In response to these problems, researchers have investigated incorporating a dc–dc converter in the dc link [11], [12].Adjusting the voltage on the dc rectifier will change the generator terminal voltage and thereby provide control over the current flowing out of the generator. Since the current is proportional to torque, the dc–dc converter will provide control over the speed of the turbine. Control of the dc–dc converter can be achieved by means of a predetermined relationship between rotor speed and rectifier dc voltage to achieve maximum power point tracking or by means of a predetermined relationship between generator electrical frequency and dc-link voltage [13].

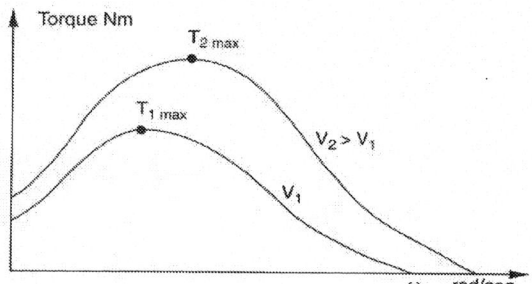

Figure 10. Wind turbine torque vs. rotor speed.

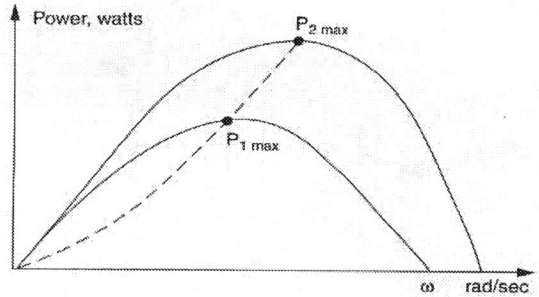

Figure 11. Wind turbine power vs. rotor speed.

A. Permanent Magnet Synchronous Generator

An analytical model of a small PMSM is used to investigate the effect of controlling the dc link voltage on the capture of maximum power. The model relates the dc link voltage of the machine to its rotor speed. It neglects magnetic saturation. The effective air gap in a PMSM with magnets mounted on the rotor surface can be considered constant and relatively large. This is due to the relative permeability of the PM material being close to unity. The d and q-axis synchronous reactances are consequently identical. The generator armature current can be related to the torque and induced voltage as follows:

$$T = K_t I_a \qquad (7)$$

$$E = K_e I_a \qquad (8)$$

Control over the rotor speed can be achieved simply by varying the generator terminal voltage. The steady state terminal voltage of the generator can be determined for a machine with negligible saliency can be expressed as:

$$V_a = \sqrt{E^2 - (I_a X_s \cos\phi + I_a R_a \sin\phi)^2} + I_a X_s \sin\phi - I_a R_a \cos\phi \qquad (9)$$

It is assumed that the generator is connected to a diode rectifier and assumed that the phase voltage and fundamental component of the armature current of the generator are in phase. Then Eq. 9 may be written as

$$V_a = \sqrt{E^2 - (I_a \omega L_s)^2} - I_a R_a \qquad (10)$$

The rectified dc-link voltage may be obtained using the standard equations for a three-phase full-bridge diode rectifier taking the effect of commutation overlap into account as [14]

$$V_{DC} = \frac{3\sqrt{6}}{\pi} V - 2V_{diode} - \frac{3}{\sqrt{6}} \omega L_s I_a \qquad (11)$$

Using Eqs. (6)-(11), it is possible to obtain a prediction for the dc-link voltage as a function of the terminal phase voltage or mechanical speed and TSR. Fig. 12 shows the optimum relation between the dc voltage and the rotor speed for the capture of maximum power when the generator operates at the peak power coefficient $C_{p\max}$ and TSR_{opt}.

Considering Eq. (4)-(6) and Fig. 12, a sudden increase in wind speed will decrease both TSR and C_p. According to Eq. (6), an increase in the wind speed will result an increase in the torque transmitted from the turbine to the generator. Then, the turbine will try to accelerate in response to an increase in wind speed. An acceleration of the turbine will result in an increase in the commanded dc-link voltage (i.e., dc-link voltage will increase in response to an increase in wind speed). Increasing the dc link voltage increases the difference between the generated voltage and he dc-link voltage. Thus, the armature current decreases which decreases the braking torque. This will continue until the speed is increased such that torque is balanced. When the wind speed falls rapidly, a sudden decrease in wind speed will result in a high TSR and C_p will decrease, decreasing the torque. With low applied torque to the generator, the inductance and inertia of the system will result in a braking torque being applied, slowing the generator and turbine. The reduction in speed will lower the dc–link voltage. As the dc voltage falls, the difference between generated voltage and dc-link voltage will remain high, maintaining current flow and applied braking torque. This process will continue until the speed is reduced such that the TSR is low enough that the turbine increases and torque is balanced.

In order to evaluate the dynamic performance of the wind generation system, an example wind speed variation was developed, defined as

$$v_w = 9 + |6\sin(4t) + 0.6\sin(36t)| \qquad (12)$$

The choice of (12) allows the investigation of the system response to a fast and continuous change in wind speed. The development of the control relationship is based on the ideal steady-state relationship of the wind speed and rotational (turbine) speed given by Eq. 4. In case where the

246

wind speed is continuously changing, the system inertia will introduce a time lag between a change in wind speed and a noticeable change in rotational speed. This time lag is neglected in this study.

Fig. 13 depicts the simulation results of the dynamic performance which validates the efficient MPPT of WT generation system when the wind speed changes rapidly and continuously. Fig. 12 plots the variation in wind speed, power coefficient C_p, tip-to-speed ratio TSR, rectified dc-link voltage, wind power, generated power, turbine speed and dc-dc converter duty cycle. By controlling the dc-link link voltage according to Fig. 11, the TSR can be kept closer to the ideal value of 9 and the power coefficient is almost constant at its maximum value of 0.42. Therefore, the wind turbine generated power increases with wind speed. The output power from the wind system is not constant and varies with wind speed.

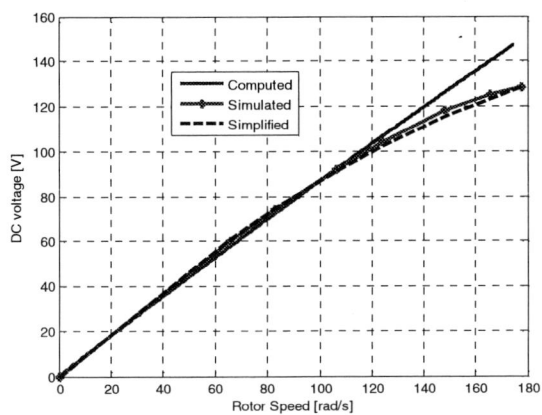

Figure 12. Optimum dc voltage vs. rotor speed characteristic.

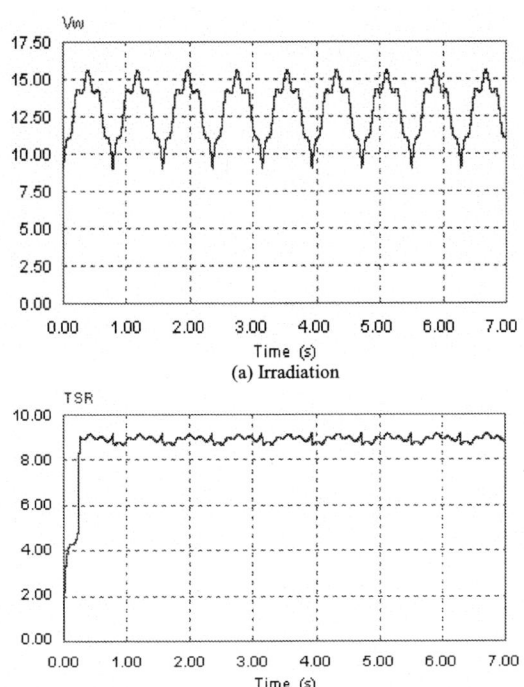

(a) Irradiation

(b) Tip-speed-ratio

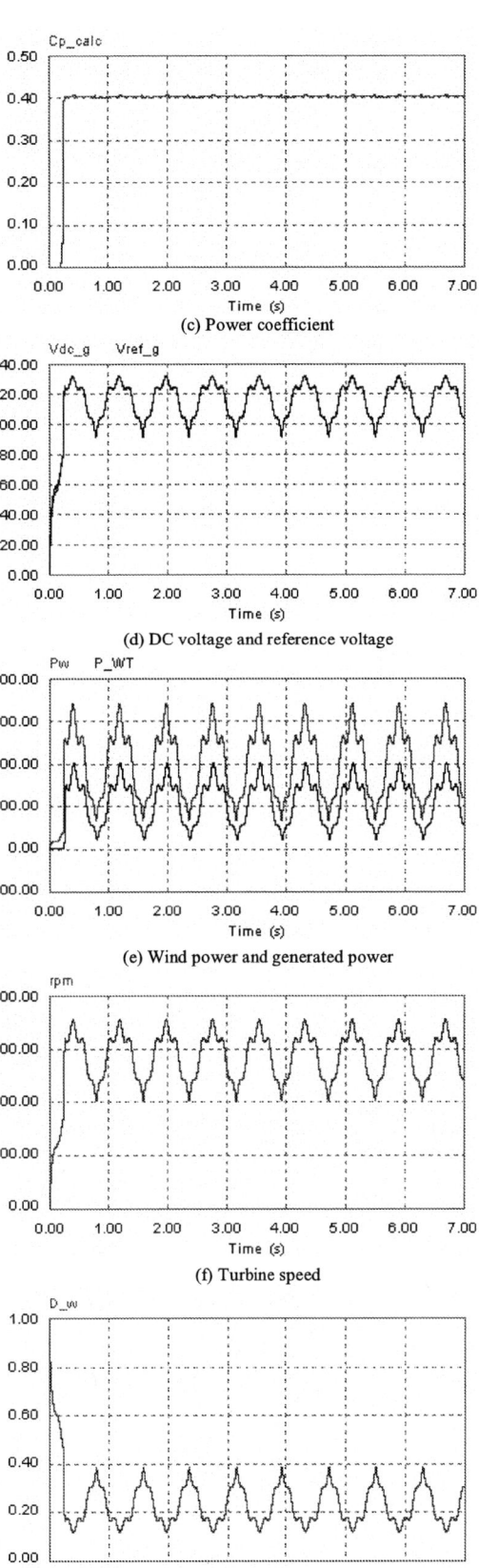

(c) Power coefficient

(d) DC voltage and reference voltage

(e) Wind power and generated power

(f) Turbine speed

(g) Duty cycle

Figure 13. Wind generation system characteristics under MPPT.

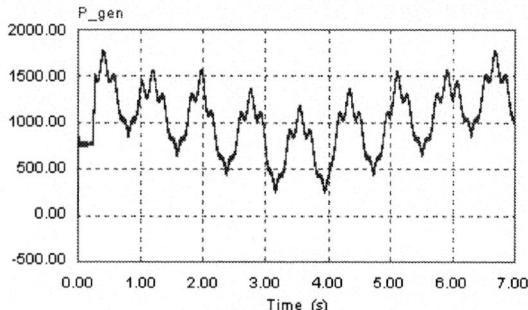

Figure 14. Generated power of hybrid system.

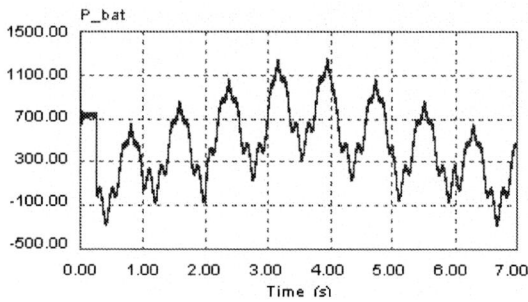

Figure 15. Power supplied by battery.

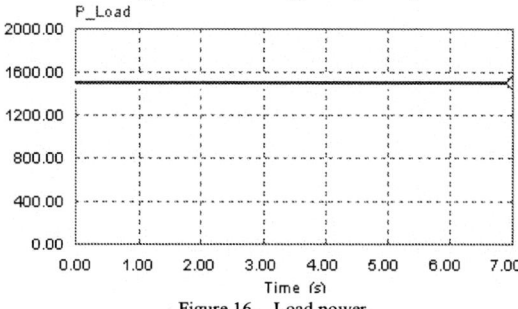

Figure 16. Load power.

Fig. 14 illustrates the total generated power of the hybrid system. The output power of hybrid system is mostly fluctuating and the fluctuation has an effect on system frequency. From Fig. 13, it is clear to note that the power fluctuation of the hybrid system is less dependent on the irradiance conditions and wind speed variations as compared to the power generated of individual PV and WG systems shown in Figs. 7(b) and 13(e).

However, this fluctuation must be suppressed. One existing method to solve these issues is to install batteries which absorb power from the system as shown in Fig. 1. The other method is to install a dump load, which dissipates fluctuating power. Using these methods the PV/WT hybrid generation system can supply almost good quality power as shown in Fig. 15. Fig. 16 shows the power supplied by the battery. However, these methods have disadvantages that they require batteries, which are costly and the installation of dump load is not an efficient method to dissipate fluctuating power. Moreover, they can not guarantee certainty of load demands at all times especially at bad environmental conditions, where there is no power from the PV and WG systems. In the future work, the authors suggest a new hybrid generation system which combining Solar PV, WG and fuel cell generation systems.

V. CONCLUSIONS

This paper describes a renewable energy hybrid generation system combining solar photovoltaic and variable speed wind turbine. A simple and cost effective maximum power point tracking technique is proposed for the photovoltaic and wind turbine without measuring the environmental conditions. This is based on controlling the photovoltaic terminal voltage or current according to the open circuit voltage or short circuit current and the control relationship between the turbine speed and the dc-link voltage is obtained using simple calculations. More expensive and complex control algorithms are not required. A complete description of the hybrid system has been presented along with its detailed simulation results which ascertain its feasibility. The power fluctuation of the hybrid system is less dependent on the environmental conditions as compared to the power generated of individual PV and WG systems. This power fluctuation has been suppressed using a battery in this paper and it will be the subject of future work.

REFERENCES

[1] W. D. Kellogg, M. H. Nehrir, G. Venkataramanan, and V. Gerez, "Generation unit sizing and cost analysis for stand-alone wind, photovoltaic, and hybrid wind/PV systems," IEEE Trans. Energy Conversion., vol. 13, no. 1, pp. 70–75, Mar. 1998.

[2] F. Valenciaga and P. F. Puleston, "Supervisor Control for a Stand-Alone Hybrid Generation System Using Wind and Photovoltaic Energy," IEEE Trans. Energy Conversion, vol. 20, no. 2, pp. 398-405, June 2005.

[3] T. Senjyu, T. Nakaji, K. Uezato and T. Funabashi, "A hybrid System Using Alternative Energy Facilities in Isolated Island," IEEE Trans. Energy Conversion, vol. 20, no. 2, pp. 406-414 June 2005.

[4] S. J. Chiang, K. T. Chang and C. Y. Yen, "Residental Photovoltaic Energy storage System," Trans. Ind. Elec., vol. 45, no. 3, pp. 385-394, June 1998.

[5] C. C. Hua and P. K. Ku, "Implementation of a Stand-Alone Photovoltaic Lighting System with MPPT, Battery Charger and High Brightness LEDS," Proceedings of the IEEE Sixth International Conference on Power Electronics and Drive Systems, 28 Nov - 1 Dec 2005, Kuala Lumpur, Malaysia.

[6] C. Hua, J. Lin, and C. Shen, "Implementation of a DSP Controlled Photovoltaic System with Peak Power Tracking," IEEE Trans. Ind. Elec., vol. 45, no. 1, pp. 99-107, Feb. 1998.

[7] K.H. Hussion, and G. Zhao, "Maximum Photovoltaic Power Tracking: An Algorithm for Rapidly Changing Atmospheric Conditions", Proceedings of the IEE, vol. 142, no. 1, pp. 59-64, 1995.

[8] M. A. S. Maoum, S. M. M. Badejani and E. F. Fuchs, "Microprocessor -Controlled New Class of Optimal Battery Chargers for Photovoltaic Applications", IEEE Trans. Energy Conversion, vol. 19, no.3, pp. 599-606, Sep., 2004.

[9] M. A. S. Maoum, H. Dehbone and E. F. Fuchs, "Theoretical and Experimental Analyses of Photovoltaic Systems with Voltage- and Current-Bases Maximum Power-Point Tracking, IEEE Trans. Energy Conversion, vol. 17, no.4, pp. 514-522, Dec. 2002.

[10] M. R. Patel, "Wind and Solar Power systems, Design, Analysis and Operation", 2nd ed. Taylor & Francis, New York, 2006.

[11] N. Yamamura, M. Ishida, and T. Hori, "A simple wind power generating system with permanent magnet type synchronous generator," in Proc. IEEE Int. Conf. Power Electronics Drive Systems, vol. 2, pp. 849–854, Hong Kong, July 27–29, 1999.

[12] A. M. De Broe, S. Drouilhet, and V. Gevorgian, "A Peak Power Tracker for Small Wind Turbines in Battery Charging Applications", IEEE Trans Energy Conversion, vol. 14, no. 4, pp. 1630–1635, Dec. 1999.

[13] A. M. Knight, and G. E. Peters, "Simple Wind Energy Controller for an Expanded Operating Range", IEEE Trans Energy Conversion, vol. 20, no. 2, pp.459-466, June 2005.

[14] N. Mohan, T. M. Undeland, and W. P. Robbins, Power Electronics, Converts, Applications and Design, 2nd edition, New York, Wiley, 1995.

2006 5th International Power Electronics and Motion Control Conference

Design Optimization of Industrial Motor Drive Power Stage Using Genetic Algorithms

F. Wang[1], W. Shen[1], D. Boroyevich[1], S. Ragon[2], V. Stefanovic[3], M. Arpilliere[4]

[1] Center for Power Electronics Systems
Virginia Polytechnic Institute & State University
Blacksburg, VA 24061-0111 USA
Email: f.wang@ieee.org

[2] Phoenix Integration, Inc.
1715 Pratt Drive, Suite
2000, Blacksburg, VA
24060, USA

[3] V-S Drives
Afton, VA
22920, USA

[4] Schneider Toshiba Inverter Europe
F-27120 Pacy sur Eure, France

Abstract— A design optimization tool of motor drive power stage has been developed. Through analyzing and modeling three major blocks, including front-end harmonic filters, IGBT inverters, and EMI filters, in a general-purpose drive, optimization programs have been implemented by using Genetic Algorithm (GA) engine. The optimizer can be used as design tools for engineers who might not have deep insights of all aspects like thermal, electromagnetic, and PWM controls. Even experts can use the optimizer to quickly and conveniently search all possible designs, which would be impossible for manual calculations and simulations. The design results obtained from the optimizer have been implemented and tested, and the experimental results have verified the models and programs.

Keywords – Motor Drive, Optimization, Genetic Algorithm

I. INTRODUCTION

The IGBT based PWM voltage source inverter with diode front-end rectifier has become the converter of choice for three-phase AC-fed general-purpose industrial motor drives. The motor drive power stage, which mainly consists of the front-end rectifier, the inverter, the DC link capacitor, the harmonic and EMI filter, and the thermal management system, as shown in Fig. 1, is the primary contributor to the overall converter cost and size. Since there are interdependencies and tradeoffs among components or subsystems in the power converter, it is very desirable to have a systematic methodology and tool for achieving a cost and/or size optimized converter design while meeting the system performance requirement. A design tool will also reduce the development cycle and effort.

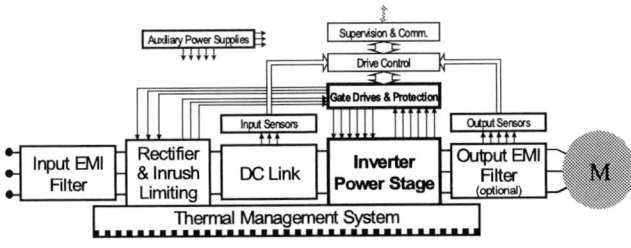

Figure 1. Functional configuration of an industrial motor drive

There have been numerous previous work on converter optimization, including work on AC motor drive optimization [1-5]. However, most of the work focused on optimization considering limited aspects of the converter power stage design, such as losses and thermal management in [3,4]. Some focused on the optimization of specific aspects, such as minimizing harmonics or losses via topology and control design [3]. There

are some work on computer-aided design and optimization tool, with some of them requiring detailed simulation [4, 5], and some for a specific type of converter or a subsystem (e.g., filter) of a converter.

This paper develops a new design optimization methodology and tool for the motor drive converter power stage, considering all major components and subsystems. Recognizing the relative functional independence of the front-end diode rectifier, the inverter and its thermal management system, and the EMI filter, the approach is to first establish three separate design optimizers for the three relatively independent subsystems, and then consider the interactions and trade-offs. Fig. 2 shows the three optimizations and their links. The front-end rectifier shares the thermal management system with the IGBT inverter and also acts as the propagation path of the EMI emission; IGBT inverter and heatsink impacts EMI through switching, propagation, and gate control.

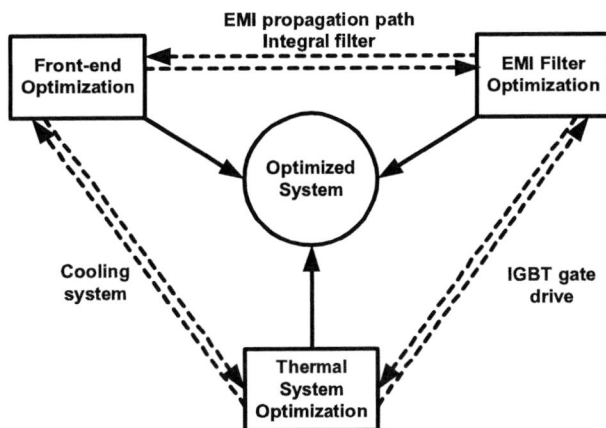

Figure 2. Approach for system optimization for motor drive

To solve the optimization problems, a genetic algorithm (GA) based discrete variable optimizer called DARWIN [6, 7] was chosen. In addition to discrete variables (such as capacitors, cores, and wires), the optimizer also has the capability of handling a small number of continuous variables. Continuous variables are modeled directly by using a single real value, and specifying a lower and upper bound for the variable. The algorithms have been successfully used in some previous work [1, 2].

Using the genetic algorithm to carry out the design optimization is another important reason for first developing

---

This work made use of ERC Shared Facilities supported by the National Science Foundation under Award Number EEC-9731677.

1-4244-0448-7/06/$25.00 ©2006 IEEE

three separate subsystem optimization programs. In order to achieve an efficient design, it is critical to use analytical relations between variables, constraints, and system responses, rather than detailed system simulations, in optimization. Decoupling of relatively independent subsystem and functions can greatly simplify establishing the analytical relationships.

II. ANALYTICAL RELATIONS

The first step for design optimization is to understand system specifications, variables, constraints, and design objectives. The key to system optimization is to establish analytical relationships among all these parameters and variables.

A. Front-end Rectifier

The basic function of the front-end rectifier is to provide required DC voltage for a given power. In addition, it will have to meet requirement for harmonics. The key to the front-end rectifier design is selection of passive components since they often account for a significant portion of cost and size of the total converter. DC link capacitor, AC line inductor (or DC link inductor) together provide voltage support, and limit the harmonics and inrush current during transients. Some of the more challenging analytical relationships include: 1) AC line harmonics and peak current; 2) DC voltage ripple (also required for capacitor selection); 3) inrush current considering the inductor magnetic saturation. The development details have been reported in a previous paper [8].

Fig. 3 shows design constraints, specs and conditions, and variables for the front-end passive components (*L* and *C*) optimization for the lowest cost or size. The design constraints are the performance specifications such as maximum RMS current as well as the component physical limits such as temperature rise. The design variables are the selection of the actual, commercially available components associated with capacitors and inductors. The optimizer provides for a systematic and efficient search of the databases of preferred components, which contain all (datasheet-based) component parameters necessary for evaluating the constraints and objective function.

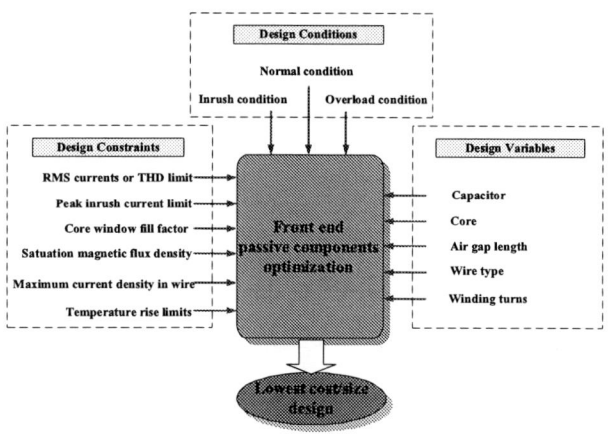

Figure 3. Front-end design optimization

The front-end optimization part can be used as a stand-alone optimizer as well as part of the system design tool. It is interesting that we can examine the optimal design result by the program to identify which operating conditions are more critical in determining the parameter value, cost, or size of the front-end passives. The studied case, shown in [8], actually indicates that the optimal design based on nominal conditions would have much lower inrush current compared with the specification.

Another important application of the optimizer is that design tradeoffs between 3-phase AC inductor and DC inductor schemes can be found.

B. Inverter and Thermal Management System

For three-phase voltage source inverter and thermal design optimization, the focus is on IGBT selection and heatsink design. The most important analytical relationships are device loss models and system thermal models considering design constraints, such as maximum junction temperature of the IGBT module.

Device Loss Model

Device selection often involves a tradeoff between fast switching characteristics (i.e., lower switching losses) and lower on-state conduction losses. In addition, many other parameters must be considered, including gate drive circuit parameters, temperature and switching pattern.

It is also very useful to express the device losses as analytical functions of design parameters so that they can be used in the overall system design optimization to facilitate device selection and thermal system design. Both linear and polynomial curve fittings have been used to represent losses in IGBTs. These previous work, however, focused mostly on a single variable relationship, for example, the effect of loss dependency on temperature or currents but not on the gate drive circuit. In this paper, multi-variable behavioral models for IGBT losses are used, which are functions of current, voltage, temperature, and gate drive circuit parameters.

Here, the conduction loss and switching loss characteristics are implemented by a simple direct curve-input method for flexibility. For example, Fig.4 shows the switching energy loss (including turn-on energy loss *Eon*, turn-off energy loss *Eoff*, and diode reverse-current-effect energy loss *Erec*) vs. gate resistance characteristics (under the rated current) of one IGBT in the IGBT module. Three data points on each curve are put into the optimization database. The actual conduction loss and switching loss are obtained through interpolation based on the design and operating point. Taking into account the gate resistance effect and DC voltage effect, the switching energy losses at the operating point V are:

$$Eon = Eon\_i \cdot \frac{Eon\_rg}{Eon\_rgrated} \cdot \frac{V}{V_{test}},$$

$$Eoff = Eoff\_i \cdot \frac{Eoff\_rg}{Eoff\_rgrated} \cdot \frac{V}{V_{test}},$$

$$Erec = Erec\_i \cdot \frac{Err\_rg}{Err\_rgrated} \cdot \frac{V}{V_{test}}.$$

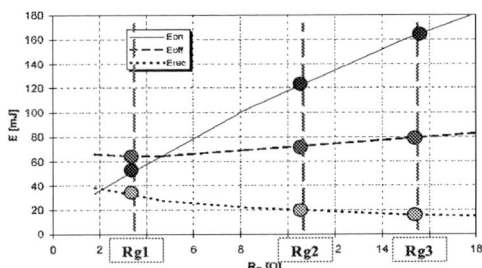

Figure 4. Gate resistance effects on switching energy loss of IGBT

With instantaneous loss model and considering the effect of modulator, the total loss can be calculated.

Thermal Model

Thermal impedance network model is needed to account for both the steady-sate and transient (e.g., overload) conditions. The thermal impedance corresponding to fundamental frequency f_{out} can be calculated by

$$Zjc\_igbt = \sum_{k=1}^{i} \frac{1}{\dfrac{1}{r_{k\_igbt}} + j \cdot 2 \cdot \pi \cdot f_{out} \cdot \dfrac{\tau_{k\_igbt}}{r_{k\_igbt}}} \quad \left(r_{i\_igbt} > 0 \right),$$

and

$$Zjc\_diode = \sum_{k=1}^{i} \frac{1}{\dfrac{1}{r_{k\_diode}} + j \cdot 2 \cdot \pi \cdot f_{out} \cdot \dfrac{\tau_{k\_diode}}{r_{k\_diode}}} \quad \left(r_{i\_diode} > 0 \right)$$

Thermal Analysis

The flow charts for the losses and thermal calculations are shown in Fig. 5. Due to a large difference between the thermal time constants of the IGBT module and the heat sink, there are two main parts in the thermal calculation: One part is the junction-to-case temperature rise calculation, as shown in Fig. 5(a); the other part is the case-to-ambient temperature rise calculation, as shown in Fig. 5(b).

The key idea of the loss and thermal analysis in this paper is to calculate the power losses of the IGBT module just for a single cycle in the time domain, and to obtain the thermal prediction results in the frequency domain. For each thermal calculation case, all the information can be obtained from just a single-cycle loss calculation according to the output frequency of the VSI. By means of the Fast Fourier Transformation (FFT), the average power loss and the power loss harmonics for each device are calculated. Inserting the loss harmonics into the IGBT module thermal impedance networks will result in the temperature harmonics. The sum of the average junction-to-case temperature rise and the junction-to-case temperature harmonics is the time-domain junction-to-case temperature rise.

Fig. 6 shows the structure of the inverter and thermal optimizer.

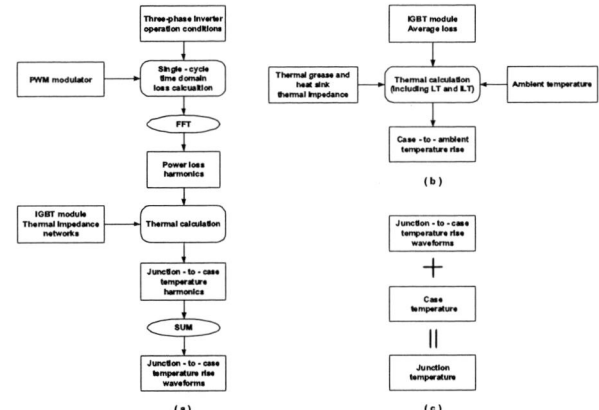

Figure 5. Flow charts for the losses and thermal calculations

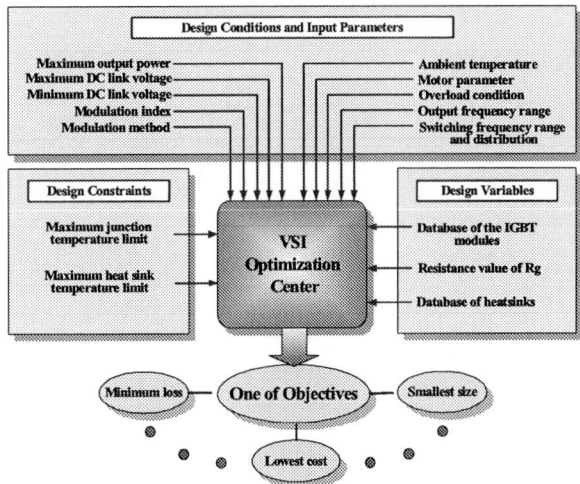

Figure 6. Inverter and thermal system optimization

C. EMI Filter

To minimize the filter cost and size in a motor drive, the EMI noise generation and propagation characteristics must be understood first. There are three possible EMI sources in a motor drive: diode commutation, high-frequency power supply switching, and IGBT switching, with IGBT being the dominating source. The EMI noise modeling and characterization of IGBT switching in a motor drive has been presented in previous work [9]. The EMI filter optimizer focuses on input EMI filter itself assuming the bare noise characteristics at the input terminal is an input variable.

The filter topologies considered in the current optimization are the passive filter shown in Fig. 7 and its two-stage counterpart. Other topologies can be included with the same approach.

The EMI design optimization approach is illustrated in Fig. 8. The key analytical relations are inductor and capacitor high-frequency impedance, the magnetic core thermal model, and the EMI attenuation model. Fig. 9 shows the measured and

251

calculated temperature rise comparison for the design CM choke.

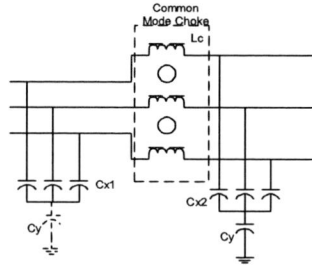

Figure 7. One-stage EMI filter topology

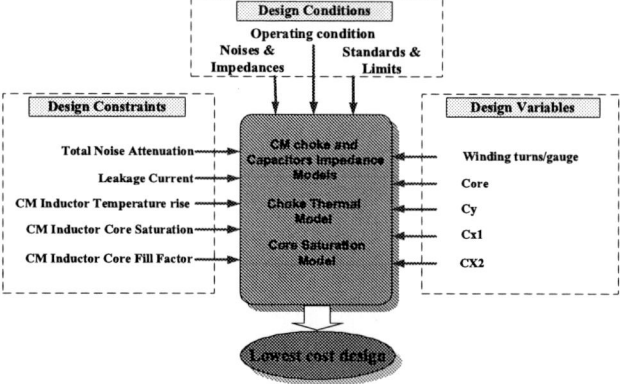

Figure 8. Structure of the EMI filter optimizer

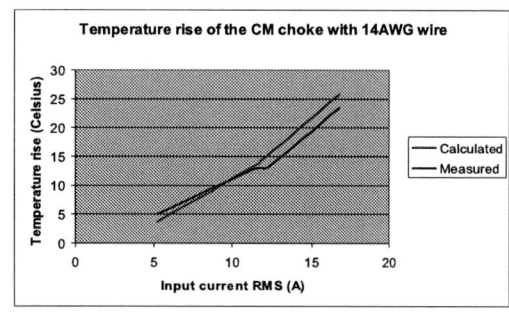

Figure 9. Measured and calculated temperature rises of the designed CM choke.

III. OPTIMIZER DEVELOPMENT AND APPLICATION EXAMPLES

The optimization algorithm used is GA tailored specifically for engineering system design. GA's are one of the few optimization algorithms that work directly with discrete design variables, are also excellent all-purpose discrete optimization algorithms. Compared to traditional gradient-based optimizers, genetic optimizers are more likely to find the overall best (globally optimal) design. In addition to finding the overall best design, GA's are also capable of finding many near-optimal

designs as well, providing the user with many options when selecting a final design configuration.

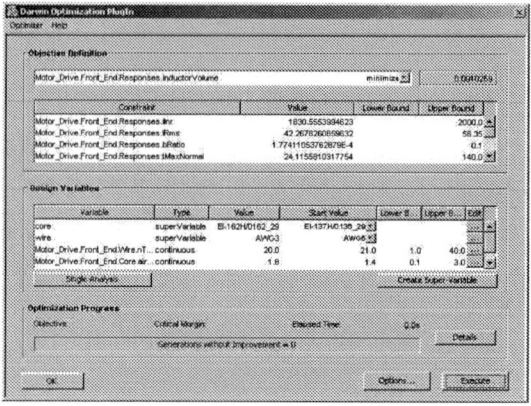

Figure 10. Optimization Software Interface

Table 1. Design Comparison of Harmonic Inductors

Parameters	Manual design (Three-phase)	Optimum design	
		Three-phase	DC
Capacitor (uF)	2400	2400	2400
Core (Silicon steel)	3_8_AC	2_8_AC	EI-225S/0221
Winding turns N	12	10	19
Air gap lg (mm)	2.3	1.6	2.7
Inductance (uH)	152.2	145.9	274.0
B_{rated} (T)	1.20	1.48	1.49
I_{rms} (A)	92.5	94	89.0
I_{pk} (A)	183	188	169
I_{inrush} (A)	611	637	672
Wire type	AWG1	AWG1	AWG1
J_{rated} (A/mm^2)	2.18	2.22	2.57
ΔT (°C)	26.6	24.1	77.6
$\Delta T_{transient}$ (°C)	29.3	26.7	80.8
Volume of the inductor (cm^3)	3952	2863	2199

Table 2. Three Phase Optimization Results (7.5kW)

Parameters	Single Analysis	Optimized
IGBT Module	FS35R12KE3G	FS25R12KE3G
Heat Sink	Heat Sink 2	Heat Sink 2
Gate resistance (Ohm)	27.0	36.0
TjcIGBTMax (°C)	17.45	24.51
TjcDiodeMax (°C)	14.11	23.74
TjIGBTMax (°C)	100.17	107.43
TjDiodeMax (°C)	96.83	106.66
TjcIGBTOverloadMax (°C)	23.82	33.91
TjcDiodeOverloadMax (°C)	19.20	32.84
TjIGBTOverloadMax (°C)	110.00	120.54
TjDiodeOverloadMax (°C)	105.39	119.47
THSSteadyState (°C)	78.58	78.75
THSOverloadMax (°C)	80.57	80.87
Pmodule (W)	206.49	208.33
PmoduleOverload (W)	280.96	287.90
Total cost of IGBT and heat sink	100%	82.6%

Fig. 10 shows the optimization software interface. Tables 1 and 2 show design examples, using front-end and inverter /thermal optimizers, in comparison with the manual designs. In both cases, the optimizers yield significant superior results. Fig. 11 shows the optimized filter layout with illustration of CM choke, Cx2, and Cy. In Fig. 12, the measured noise of the drive with the filter is below the standard limit, and the filter performance margin is small as expected, which means small and necessary size of the filter achieved.

252

Figure 11. EMI filter optimizer result implementation illustration

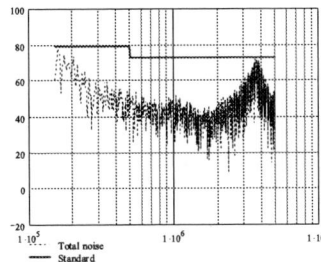

Figure 12. Measured EMI noise with the optimized filter meet standard

IV. DISCUSSIONS ON SYSTEM OPTIMIZATION

The optimization tool described in this paper primarily focus on optimization of three separate parts of the drive power stage. The goal is to link these three optimizers together to achieve the total optimization. As shown in Fig. 2, physical relationships between any two of the three blocks, the front-end passive, inverter loss and thermal, and conducted EMI noise level, have been identified. Using the relationships, any two sets of optimizations results can be combined to achieve a new optimal design considering two parts together. Similarly, it can be extended to all three.

As an example, the gate resistance effects on both EMI noise and inverter losses have been studied. For certain IGBT device SKM150GB123D, four different gate resistances of $R_g = 3.9$ ohm, 4.8 ohm, 13.8 ohm and 31.8 ohm, have been used, and the device rise and falling times are measured as shown in Fig. 13, under the test conditions of load current $I_{load} = 100A$ and bus voltage $V_{bus} = 100A$. Also the conducted DM and CM EMI noises are measured for these Rg, as in Fig. 14. It is clear that the higher gate resistance, the slower IGBT switching, the larger switching loss, but smaller CM and DM noises. Therefore, the value of Rg could be used as a common design variable to optimization programs of thermal and EMI filter parts.

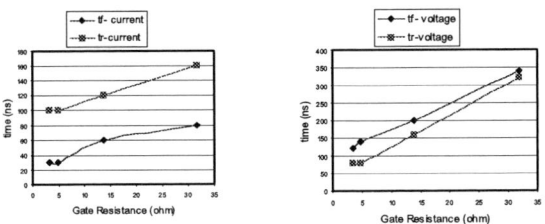

Figure 12. *Rg* effect on IGBT switching turn-on and turn-off times

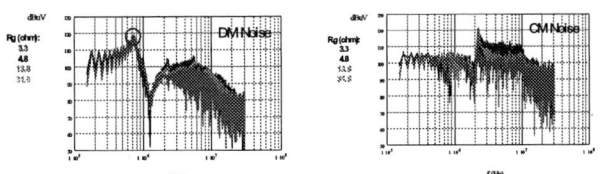

Figure 13. *Rg* effect on CM and DM noise levels

The developed optimizers are flexible to add more design variables, such as gate resistance, switching frequency. With the common variable as the linkage, the optimizer will give systematic optimal result. An overall objective function, such as total cost, can be applied to govern the use of results from all three optimizers.

V. CONCLUSION

This paper presents a genetic algorithm based design optimization tool for AC-fed industrial motor drive power stage considering all major subsystems - front-end rectifier, inverter and thermal management system, and EMI filter. Analytical relationships are developed and implemented into three optimizers. Global optimization is achieved by considering the relations between the subsystems. Design examples verified the usefulness and correctness of the design methodology and tool.

REFERENCES

[1] S. Busquets-Monge, J-C Crebier, S. Ragon, E. Hertz, D. Boroyevich, Z. Gurdal, M. Arpilliere, D. Lindner, "Design of a boost power factor correction converter using optimization techniques", IEEE Transactions on Power Electronics, Vol. 19, No. 6, Nov. 2004 pp: 1388 – 1396

[2] S. Busquets-Monge, G. Soremekun, E. Hertz, C Crebier, S. Ragon, , D. Boroyevich, Z. Gurdal, M. Arpilliere, D. Lindner, "Power converter design optimization", IEEE Industry Applications Magazine, Vol. 10, No. 1, Jan-Feb 2004 pp:32 – 38

[3] B. Ozpineci, L.M. Tolbert, J.N. Chiasson, "Harmonic optimization of multilevel converters using genetic algorithms", IEEE Power Electronics Letters, Vol. 3, Nov. 3, Sept. 2005 pp: 92 – 95

[4] F. Blaabjerg, J. K. Pedersen, "Optimized design of a complete three-phase PWM-VS inverter", IEEE Transactions on Power Electronics, Volume 12, No. 3, May 1997 pp: 567 – 577

[5] H. Kragh, F. Blaabjerg, J.K. Pedersen, "An advanced tool for optimised design of power electronic circuits", The Thirty-Third IEEE IAS Annual Meeting., Oct. 1998, Vol. 2 pp:991 - 998

[6] *DARWIN* User's *Manual*, Advanced Design and Optimization Technologies, Inc., 1999.

[7] G. Soremekun and Z. Gürdal, "Stacking Sequence Blending of Multiple Composite Laminates Using Genetic Algorithms," AIAA Paper No. 2001-1203. 42nd AIAA/ASME/ASCE/AHS/ASC Structures, Structural Dynamics, and Materials Conference, Seattle, WA, 2001.

[8] G. Chen, M. Rentzch, F. Wang, D. Boroyevich, S. Ragon, V. Stefanovic, M. Arpilliere, "Analysis and design optimization of front-end passive components for voltage source inverters", IEEE APEC '03, Vol. 2, Feb. 2003 pp: 1170 - 1176

[9] Q. Liu, F. Wang, D. Boroyevich, "Model conducted EMI emission of switching modules for converter system EMI characterization and prediction", 39th IEEE IAS Annual Meeting. Conference Record, Vol. 3, Oct. 2004 pp: 1817 – 1823.

2006 5th International Power Electronics and Motion Control Conference

FEM Based Simulation of a Permanent Magnet Synchronous Motor Performance Characteristics

L. Petkovska, Senior Member IEEE and G. Cvetkovski, Member IEEE
Ss. Cyril and Methodius University, Faculty of Electrical Engineering
Karpos II b.b., P.O.Box 574, Skopje, MACEDONIA
lidijap@etf.ukim.edu.mk

Abstract—The paper deals with simulation and analysis of both steady state and dynamic performance characteristics of a synchronous motor with surface mounted permanent magnets. For this purpose, the machine parameters are determined as exact as possible. Different methods, including classical two-axes theory, Finite Element Method (FEM) and Simulation method are applied. The complex modelling of the motor, in order to obtain the most convenient models for application of the proposed methods, is carried out. The results of analytical and numerical calculations, as well as the simulations, are compared with the experimentally obtained ones; they show a very good agreement. On the basis of the analysis of the results, each of the methods is evaluated.

Keywords - Permanent Magnet Synchronous Motor; Finite Element Method; Steady-state and dynamic characteristics.

I. INTRODUCTION

In the paper, performance characteristics of permanent magnet synchronous motor (PMSM) are determined and analysed on the basis of FEM computational results.

The recent development of high energy magnets has enhanced the application of PMSM in wide range of areas. The built-in of permanent magnets as an excitation, and in particular the use of neodymium-boron-iron or samarium-cobalt magnets has challenged innovations in the PMSM design and analysis.

The steady state and especially the dynamic analysis of the permanent magnet motors is rather complicated issue. The main task is always to determine both the steady state and the transient performance characteristics. It is obvious that the main stress should be put on the exact determination of the parameters, as they are "playing" an important role in the accuracy with which mathematical models of the motor under consideration will be derived.

II. OBJECT OF STUDY

The object of investigation is a KONCAR motor type EKM 90M-6, with rated data: 18 A, 10 Nm, 1000 rpm. The motor is supplied from an AC source, by current or voltage sine waves at 50 Hz. Six permanent magnet poles made of SmCo5 are surface mounted. The motor view and its geometrical cross section are presented in Fig. 1 and Fig. 2, respectively.

Figure 1. Permanent magnet synchronous motor type EKM 90M-6

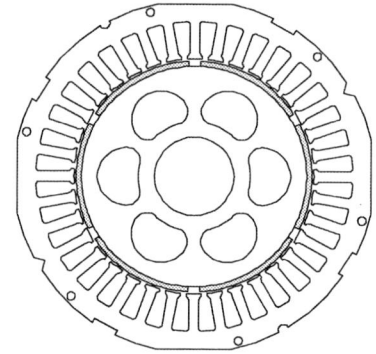

Figure 2. Cross-section of the motor EKM 90M-6

III. FEM ANALYSIS OF PMSM

The Finite Element Method, either as two-dimensional or three-dimensional field problems solver, has been used extensively in the numerical analysis of electric machines, in general. Many researchers all over the world, including the authors of the paper [1-4] have done a lot of work in this area. Nowadays, many different software packages exist. When applying FEM for analysis of the permanent magnet synchronous motor, the magnetic problem is considered to be two-dimensional time dependent harmonic problem. Hence, calculations of the magnetic field are performed at rated frequency f_n=50Hz.

The presented results in the paper are computed by using software package FEMM [5]. In the first step, the mesh of finite elements is generated over the whole cross section of the motor, as can be seen in Fig. 3.

1-4244-0448-7/06/$25.00 ©2006 IEEE   254

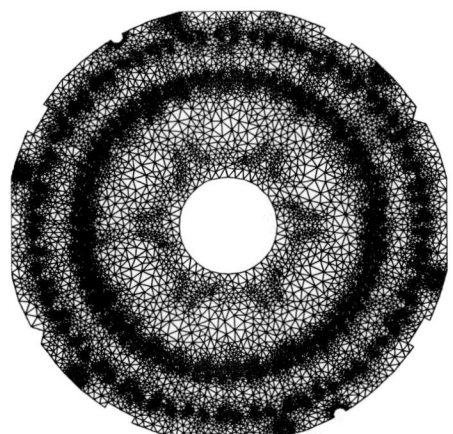

Figure 3. Finite elements mesh in PMSM

In the preprocessor are included all requested input data: • exact geometrical cross section of stator and rotor core; • current density in the stator windings; • boundary conditions of the region which is analysed; • all material characteristics of the motor (permanent magnets, copper wire, B-H curves). The numerical FEM model of the PMSM, being completed, is ready for practical use.

The field solutions are obtained by running the FEMM several times. As the whole cross section of the motor is used, only the first order Dirichlet's boundary conditions are applied: on the outer stator line and inner rotor line it is set A=0. Armature currents in stator windings are varied from I=0 to the rated value I_n=18A. Rotor is freely moving (rotating) in the air-gap, continuously changing position; d-axis of the rotor is taking different angles θ against the referential axis of the stator, firmly linked with one of the winding axes (usually phase A).

After the processing step is executed, the values of magnetic vector potential in every node of the motor domain are obtained. Later, one can use them for many purposes. The postprocessor in the FEMM package is offering user friendly calculations and graphical presentations of the most important electromagnetic and electromechanical quantities.

A. Magnetic Field Distribution

The best way to understand the phenomena in the investigated motor is "to get inside and to see" magnetic field distribution. Graphical presentation and visualization of the FEM results sketch out magnetic flux distribution in the cross-section of permanent magnet synchronous motor at various working conditions.

In Fig. 4, a part of the most interesting results of the field computations is given. The distribution of magnetic field is presented as follows: (a) no-load, i.e. zero armature current, when magnetic field is obtained by the permanent magnets only; (b) rated-load at rated stator winding current I_n=18A and rated loading angle δ_n=39⁰ [deg.el.]; (c) load at pull-out (maximum) torque, meaning load angle δ_{max}=90⁰ [deg.el.], i.e. θ=30⁰ [deg.mech.].

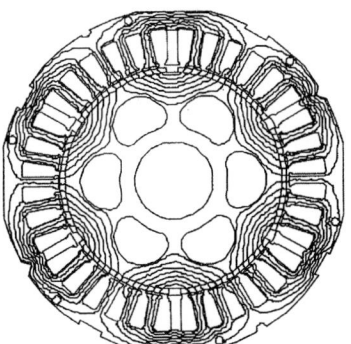

(a) no-load at I=0 and θ=0 deg.

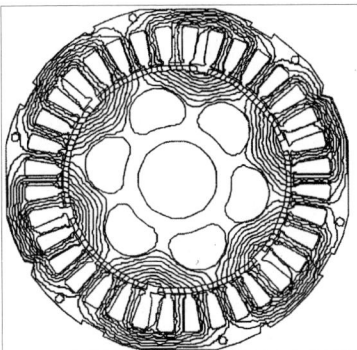

(b) rated load at I_n=18 A and θ=13 deg.

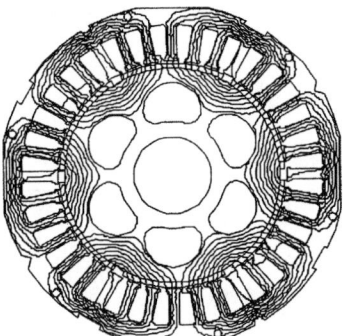

(c) pull-out load at I_n=18 A and θ=30 deg.

Figure 4. Magnetic flux plots in the middle cross-section of PMSM

B. Air-gap Flux

The numerical calculation of flux is performed per pair of excited poles, basing on field theory and by using the auxiliary quantity **A** (magnetic vector potential):

$$\Phi_g = \int_{\Sigma} \text{rot} A \cdot dS = \oint_C A \cdot d\,r = \int_{\Sigma} B \cdot dS \qquad (1)$$

The postprocessor in FEMM enables presentation of the magnetic flux density spatial distribution along an arbitrary selected line, as well. The results along the mid-gap line are presented in Fig. 5 (a), (b) and (c), in the same way as preceding explanation.

255

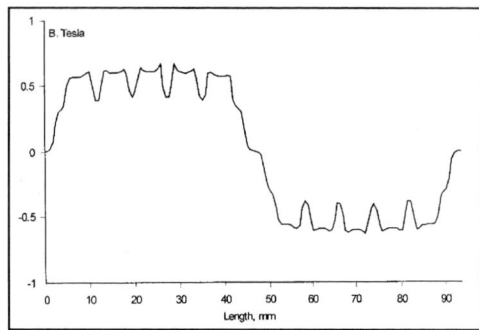

(a) no-load at I=0 and θ=0 deg.

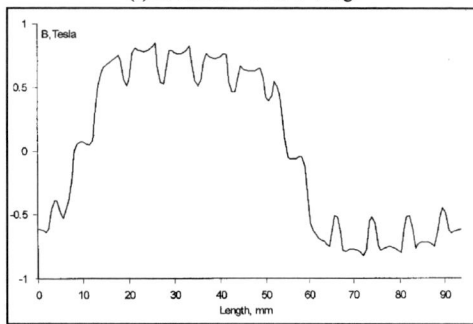

(b) rated load at I_n=18 A and θ=13 deg. (δ_n=39 deg.el.)

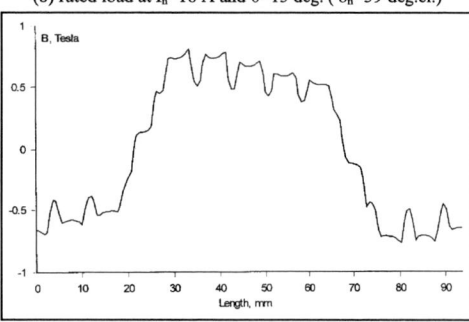

(c) pull-out load at I_n=18 A and θ=30 deg. (δ_{max}=90 deg.el.)

Figure 5.   Mid-gap magnetic flux density spatial distribution

IV.   MOTOR PARAMETERS

The two-axes model of the synchronous machines is well established classical approach for an analytical investigation of their behaviour. Many researchers widely use this method for fast prediction of the initial data for further profound analysis [6]. The basic idea is to develop a set of equations, describing the motor performance in d,q reference frame and in terms of the load angle δ. The only request is to have available the motor parameters.

Some of the parameters of PMSM could be easily measured; some are available from the producer's data. But very often their exact values are unknown, and it is presumed an experience and good skill to apply existing and well known numerical or analytical methods for calculations. Different approaches are possible.

The armature winding resistance R_a per phase, as well as leakage inductance L_{sa} are available from experimental

investigation of the PMSM type EKM 90M-6. Their measured values are given bellow:

$$R_a = 0.1242 \,[\Omega]; \quad L_{sa} = 2.2 \,[mH] \Rightarrow X_{sa} = 0.691 \,[\Omega]$$

Starting with the numerical procedure, d,q parameters of the PMSM under consideration are calcualted [7]. By using the computational FEM results, one can determine d,q reactance of PMSM, at 50 Hz as:

$$X_d = 1.827 \,[\Omega] \quad \text{and} \quad X_q = 1.823 \,[\Omega]$$

The already known fact that, in synchronous motors with surface mounted permanent magnets, there is almost no difference between reactance along d- and q-axis has been proved in the PMSM under consideration.

Having available all parameters of the PMSM a phasor diagram at rated operating conditions is constructed, and the rated load angle was found to be:

$$\delta_n = 39.2 \,[deg.el.]$$

V.   STEADY–STATE CHARACTERISTICS

The intention of researchers, producers and users is always focussed to analysis and estimation of the electric machine behaviour. For that purpose, it is requested to have computed both steady-state and dynamic performance characteristics, as accurate as possible.

The considered motor is analysed at different operating conditions. Numerical calculations of the most relevant electromagnetic and electromechanical quantities, based on the FEM results, are presented below. The armature currents I and rotor positions θ along one pole pitch, are arbitrary selected. The reference axis is selected to be the A-phase axis of stator windings; the initial rotor position and θ=0 deg.mech. is defined when the stator A-axis and the axis of rotor N-pole (d-axis) are in accordance [8]. The rotation is supposed to be counterclockwise.

A.   A. Magnetic Flux Density

The flux density B is calculated from the basic relation used in the definition of the magnetic vector potential **A**:

$$\nabla \times \mathbf{A} = \mathbf{B} \tag{2}$$

Applying the numerical procedure for its solution in the air-gap domain, magnetic flux density B_g per a pair of poles is computed. The set of characteristics, presenting the magnetic flux density at different armature currents I and different rotor positions θ along one pole pitch of the motor ($0 \leq \theta \leq 60$ deg.), are given in Fig. 6.

B.   Magnetic Field Coenergy

In linear magnetic field problems, the magnetic energy W and the coenergy W' are equal. But, in the most cases the problem is non-linear and they are different; magnetic coenergy is computed starting with the expression:

$$W' = \frac{1}{2} \int_V \mathbf{J} \cdot \mathbf{A} \, dV \tag{3}$$

256

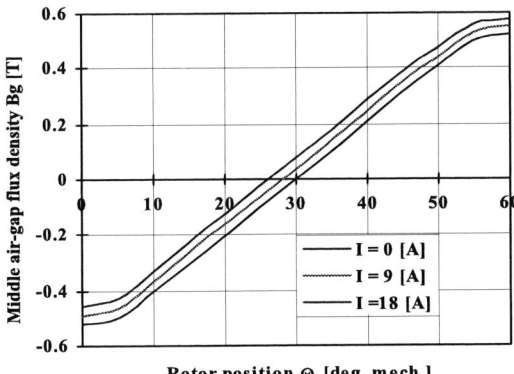

Figure 6. Magnetic flux density characteristics $B_g = f(\theta, I)$

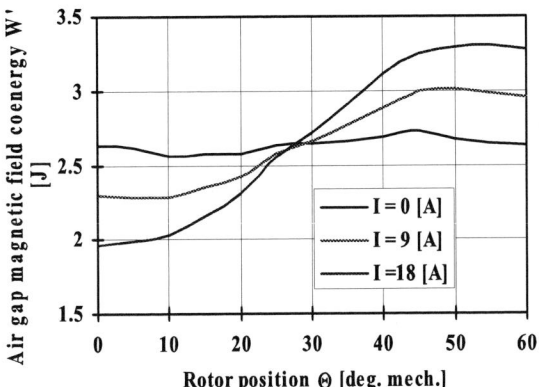

Figure 7. Magnetic coenergy characteristics $W' = f(\theta, I)$

In fact, this quantity has no physical explanation, but it is very useful for calculation of the electro-magneto-mechanical quantities when an energy concept is applied.

For the quasi static model of the analysed PMSM, the coenergy is calculated numerically from the expression:

$$W'(\theta, I) = \int_0^I \psi(I, \theta) dI \Big|_{\theta = const} \tag{4}$$

The magnetic coenergy is calculated in dependence of the position of moving parts in the domain (the rotor) at arbitrary selected armature current. The characteristics are presented in Fig. 7.

C. Electromagnetic Torque

The knowledge of the static torque characteristics is very important issue for the performance analysis and behaviour of electric motors. For its calculation, various approaches exist. In theory, the torque is computed from the field solution in a number of various ways. Three approaches for calculation are in practical use: Flux-Current Method (FCM), Maxwell Stress Method (MSM) and Virtual Work Method (VWM). Recently, torque and forces in electromagnetic topologies are computed using Weighted Stress Tensor (WST). In this paper, the energy concept for numerical calculation of torque is applied.

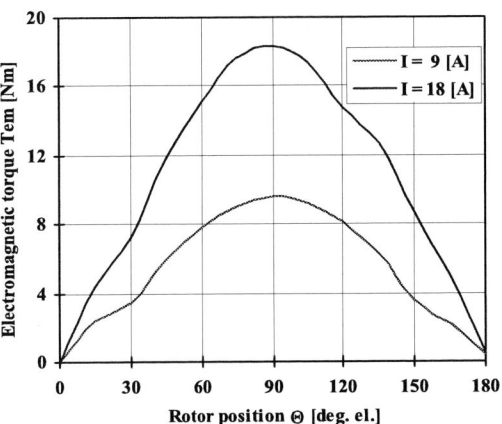

Figure 8. Electromagnetic torque characteristics $T_{em} = f(\theta, I)$

The electromagnetic torque T_{em} is effected by the variation of magnetic field coenergy in the motor air-gap, at virtual displacement of the rotor, while the armature current is forced to be constant. For calculation we use:

$$T_{em}(\theta, I) = \frac{\partial W'(\theta, I)}{\partial \theta} \Big|_{I = cons.} \tag{5}$$

The results of calculations, performed at rated current $I_n = 18$ A and $I_n/2 = 9$ A, are presented in Fig. 8.

VI. DYNAMIC CHARACTERISTICS

Matlab/Simulink® is widely accepted simulation tool which enables "to view" dynamic characteristics and to analyse transient performance of the motor. The model of the permanent magnet synchronous motor, suitable for implementation of the Matlab/Simulink Method bases on d,q transformations known from the reference frame theory of electrical machinery [9].

A. Motor Model Development

When using simulation methods the main attention has to be put on development the corresponding mathematical model of the analysed motor which will represent the physical phenomena as close as possible. The flow-chart of the PMSM simulation model is presented in Fig. 9.

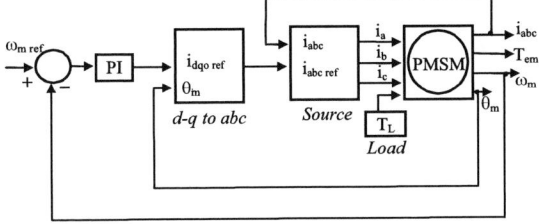

Figure 9. Flow-chart of PMSM simulation model

The d,q components of the armature current of the PMSM, in terms of the flux linkage λ_m and load angle δ, are introduced. After the relations between currents and flux linkages are implemented, the equations for electromagnetic torque T_{em} are developed in a form:

$$T_{em} = \frac{3}{2} p \left[\lambda_m i_q + \left(L_d - L_q \right) i_d i_q \right] \qquad (6)$$

$$T_{em} = T_L + B \cdot \omega_m + J \cdot p \omega_m \qquad (7)$$

where: p is number of pole pairs; T_L is loading torque; B is damping effect of friction loss; J is rotor inertia.

The angular velocity of the rotor ω_m is expressed by the time derivation of its displacement θ_m:

$$\omega_m = \frac{d\theta_m}{dt} \qquad (8)$$

B. Simulation results

On the basis of derived simulation model of PMSM, the simulation of transient performance characteristics is carried out. Flux linkage λ_m as well as synchronous d- and q- reactance of the motor are obtained from the FEM computations; stator windings parameters are measured; the load torque is given as input data. The simulation procedure is performed for rated supply conditions and at two the most common operating modes: no load start-up; motor starting at the rated load $T_L = T_n$.

The results of simulation and dynamic characteristics for both analysed cases are presented in Fig. 10 and Fig. 11, respectively. After the transients are suppressed, the obtained values of the simulated quantities, as speed, current and torque are corresponding to the quantities at the steady-state operating mode. Once determined from the diagrams, later they could be used for the comparative analysis of the investigated motor.

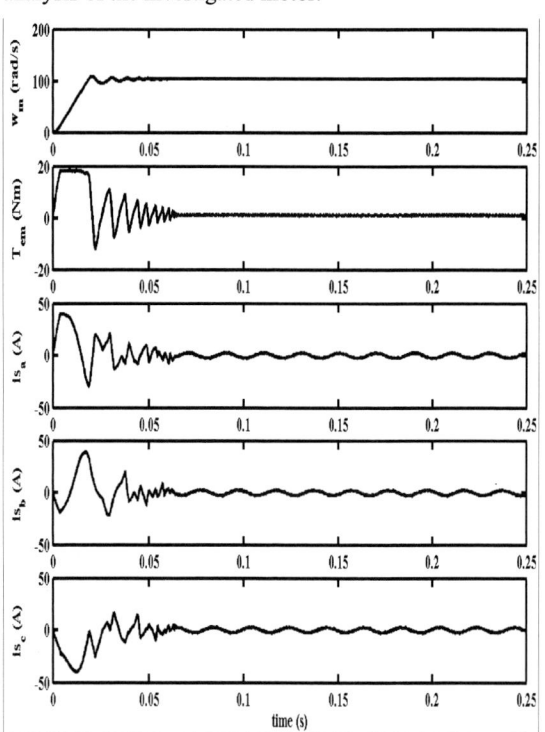

Figure 10. Dynamic characteristics of PMSM at no-load start-up

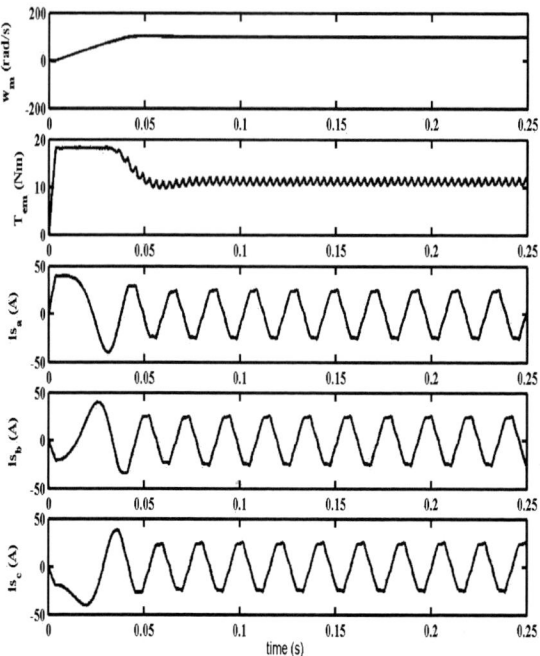

Figure 11. Dynamic characteristics of PMSM at rated load start-up

The performance characteristics of the considered PMSM are verified in two ways: computed results are compared with measured values and/or with the producer data [7]. Showing a very good agreement, they prove the applied methodologies as accurate and reliable.

REFERENCES

[1] S. Low and W. H. Lee, "Characteristics and performance analysis of a permanent magnet motor with multistacked imbricated rotor", *IEEE Trans. on Energy Conversion,* vol. EC-2, No.3, pp. 450-457, October 1987.

[2] L. Petkovska, G. Cvetkovski and V. Sarac, "Different aspects of magnetic field computation in electrical machines", Book of Abstracts 10th International Symposium on Numerical Field Calculation in Electrical Engineering - IGTE'2002, p.p. 73; on CD pp. 1-6, Graz, Austria, 2002.

[3] Z. Kolondzovski and L. Petkovska, "Identification of a Synchronous Generator Parameters Via Finite Element Analysis", Book of Abstracts 11th International Symposium on Numerical Field Calculation in Electrical Engineering - IGTE'2004, p.p. 96; on CD pp. 1-6, Seggauberg (Graz), Austria, 2004.

[4] S. J. Salon, *"Finite Element Analysis of Electrical Machines"*, Kluwer Academic Publishers, Norwell, MA, USA, 1995.

[5] D. Meeker, *"Finite Element Method Magnetics - FEMM"*, User's Manual, Ver. 4.0, Foster-Miller, MA, USA, 2005.

[6] G. Cvetkovski, L. Petkovska and S. Gair, "Performance Analysis of a Permanent Magnet Disc Motor for Direct Electric Vehicle Drive", *Journal Electromotion*, vol. 10, No. 3, p.p. 353-358, July-September 2003.

[7] L. Petkovska and G. Cvetkovski, "Identification of a PM Synchronous Motor Parameters Based on Finite Element Analysis", *Journal Przeglad Elektrotechniczny*, Vol. 81, No.12, p.p. 10-15, Warsaw, Poland, December 2005.

[8] L. Petkovska, M. Cundev and G. Cvetkovski, "Transient Performance Simulation of Permanent Magnet Motor", Proceedings 9th IEEE Mediterranean Electrotechnical Conference - MELECON '98, pp. 1135-1139, Tel Aviv, Israel, 1998.

[9] Chee-Mun Ong, *"Dynamic Simulation of Electric Machinery"* – *Using MATLAB/SIMIULINK*, Edition Prentice Hall, PTR, New Jersey, USA, 1998.

2006 5th International Power Electronics and Motion Control Conference

Analytical Modeling of Semiconductor Losses in Matrix Converters

Bingsen Wang   Giri Venkataramanan

Department of Electrical and Computer Engineering
University of Wisconsin-Madison
1415 Engineering Drive
Madison, WI 53706 USA
bingsen@cae.wisc.edu; giri@engr.wisc.edu

Abstract— **Analytical models for estimating semiconductor losses are commonly used for heatsink selection in the design process of power converters. While such models are established and widely known for different dc-dc converters, rectifiers and inverters, they have not been developed for matrix converters. Therefore, one has to resort to the use of numerical simulation for this purpose. Although numerical simulation is a straightforward approach as long as the power switching devices are properly modeled, it is typically time consuming and requires accurate physical models for the device. In this paper, an analytical approach to characterizing the semiconductor losses of the conventional matrix converter (CMC) and the indirect matrix converter (IMC) is presented. The analytical results are verified against the simulation results from a detailed numerical model under a wide variety of operating conditions.**

Keywords- conduction loss; conventional matrix converter (CMC); indirect matrix converter (IMC); simulation; switching loss;

I. INTRODUCTION

Since the introduction of high frequency synthesis approach for ac-ac power conversion proposed by Venturini in 1980 [1-3], significant research effort has been dedicated to the matrix converter due to its attractive features, such as high quality input and output waveforms and high power density [4-12]. The focus of much of the work has been mainly on topology, modulation, and commutation aspects, while definitive models for loss characterization have not been fully developed. Since the converter topology eliminates internal energy storage elements, the converter footprint is predominantly determined by the heat sink for the semiconductors. And in order to properly size the heat sink, estimation of the power loss/dissipation with acceptable accuracy is a key step.

Common practice of power loss estimation has been through the use of numerical time series simulation. Although the numerical simulation is a very straightforward procedure as long as the power switching

devices are properly modeled, it is typically time consuming and this is particularly true if various different operating conditions need to consider. Furthermore, the loss predictions are a strong function of the accuracy of the physical models available for simulation, while most widely available semiconductor loss models are behavioral. It is known that the fidelity of loss simulations may vary by quite a wide factor in repeating actual thermal performance [13]. In contrast, the analytical approach to predicting power losses is a much more favored alternative due to its ability of fast profiling, based on readily available behavioral models for semiconductors.

Estimation of conduction loss and switching loss for voltage source and current source converters have been established [14-18]. In this paper, an analytical approach to power loss calculation for conventional matrix converter (CMC) and the indirect matrix converter (IMC) are explored. This paper is organized as follows. In Section II, the simplified loss model of the power devices, namely insulated gate bipolar transistors (IGBTs) and diodes, are reviewed. In Section III, the conduction loss and switching loss are derived for the CMC topology. The conduction loss and switching loss are derived for the IMC topology in Section IV. In Section V, analytical results are verified against the simulation results for CMC and IMC topology followed by a comparison. The contributions of the paper are summarized in the concluding Section VI.

II. SEMICONDUCTOR LOSS MODEL

Typically, three operating modes can be categorized for the semiconductor power devices in a power converter: 'on', 'off' and 'transition' modes. The power loss when the *device* is off is usually negligible compared to the 'on' mode and 'transition' modes, in which the associated losses are commonly denoted as conduction and switching losses, respectively. The conduction loss can be modeled by power dissipation caused by the voltage drop across the device and the current through the device. The voltage drop is approximated by the linear dependence on the current.

1-4244-0448-7/06/$25.00 ©2006 IEEE

$$v_{CE}(i_C) = V_{CE0} + r_{CE} i_C$$
$$v_F(i_F) = V_{F0} + r_F i_F \tag{1}$$

where v_{CE} and v_F are the voltage drop across the IGBT and the diode for current i_C and i_F, respectively. r_{CE} and r_F are the incremental resistance of the IGBT and the diode. V_{CE0} and V_{F0} are the forward voltages when the current is zero (or quite small) [18].

The switching loss is characterized by the switching loss energy associated with each switching event. The switching energy is generally assumed proportional to the blocking voltage and the conducting current at the instant of switching event [18].

$$E_{sw} = E_{swR} \frac{v}{V_R} \frac{i}{I_R} \tag{2}$$

where E_{swR} is the switching energy at the reference voltage V_R and reference current I_R, which are typically the test conditions of in device datasheets; and v and i are the actual operating voltage and current in the particular application.

For IGBT, there are switching losses, E_{on} and E_{off}, associated with both turn-on and turn-off processes. While for diodes, the switching loss is typically cause by the reverse recovery (E_{rr}) mechanism, which only occurs during turn-off of a diode.

III. CMC Loss Calculation

As shown in Fig. 1, the topology of the CMC is represented by three single-pole-triple-throw (SPTT) switches. Each bidirectional throw S_{mn} (m, n = 1,2,3) is realized by two IGBTs and two diodes.

The input voltages and output currents are given by

$$v_{ik}(t) = V_i \cos\left(\alpha_i(t) - (k-1)\frac{2\pi}{3} \right)$$
$$i_{ok}(t) = I_o \cos\left(\beta_o(t) - (k-1)\frac{2\pi}{3} \right) \tag{3}$$

where k = 1, 2, 3; The phase angle of the input voltage and output currents can be expressed as $\alpha_i(t) = \omega_i t + \alpha_{i0}$ and $\beta_o(t) = \omega_o t + \beta_{00}$, with ω_i and ω_o being the fundamental frequency of the input voltage and output current, respectively.

A. Conduction Losses

At any instant, the output currents flow through one IGBT and one diode. Based on the balanced three phase output currents given in (3), only one SPTT is considered and the total conduction loss is calculated by multiplying three times. Only a quarter of the period is needed to integrate due to the output current waveform symmetry.

$$P_{c\_CMC} = 3\frac{2}{\pi} \int_0^{\pi/2} \left[v_{CE}(i_{o1}) + v_F(i_{o1}) \right] i_{o1} d\beta_o \tag{4}$$

Substituting the current i_{o1} in (3) and the diode/IGBT characteristics in (1) into (4) results in the closed-form expression for the conduction loss of the CMC

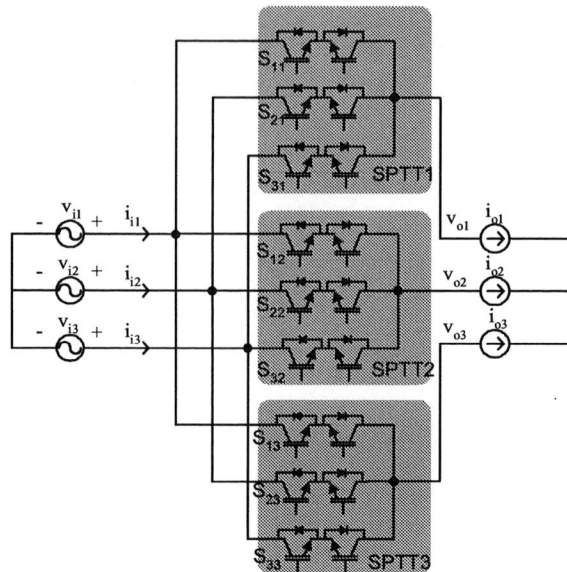

Fig. 1 Simplified schematic of the conventional matrix converter topology represented by three SPTT switches

$$P_{c\_CMC} = \frac{6}{\pi}(V_{CE0} + V_{F0})I_o + \frac{3}{2}I_o^2(r_{CE} + r_F) \tag{5}$$

It can be observed that the conduction loss is only determined by the peak value of the output current and is not affected by the operating conditions such as modulation index or power factor.

B. Switching Losses

Since the switching loss depends on the alternating blocking voltage of each throw, the commutation sequence will affect the switching loss. In this analytical development, a typical double-sided switching pattern is assumed [19]. Due to the symmetry of the balanced three-phase systems at input and output terminals, only one SPTT switch is considered. In Fig. 2, the switching functions h_{11}, h_{21} and h_{31} for throws S_{11}, S_{21} and S_{31} of switch SPTT1 are illustrated. There are four commutation events in one switching cycle T_s, which is assumed constant. The commutation voltage associated with each commutation event is labeled in Fig. 2.

For the topology shown in Fig. 1, during the commutation between throws, neither overlap nor dead-time is allowed since overlap means short circuit to the input voltages and dead-time means open circuit to the output currents. One way to work around this is to make the commutation happen in several steps [20]. The current-based four-step commutation scheme is assume in deriving the analytical expression for the switching losses although the methodology adopted here can be extended to other schemes, such as voltage-based commutation. The current-based four-step commutation scheme is explained here for completeness.

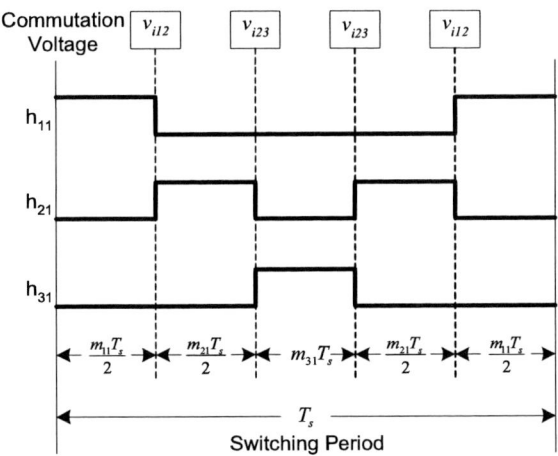

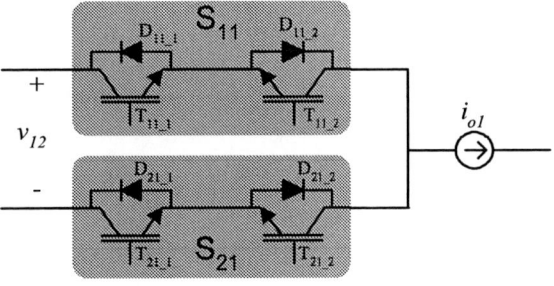

Fig. 2 Illustration of switching functions for throws S_{11}, S_{21}, and S_{31} with double-sided switching pattern.

The explanation proceeds with the example of commutating current i_{o1} from throw S_{11} to S_{21}. As illustrated in Fig. 3, if the current i_{o1} is positive, the current initially flows through the IGBT T_{11_1} and Diode D_{11_2}. The IGBT T_{11_1} is first turned off and no energy loss is involved since T_{11_1} is previously not carrying any current. Then T_{21_1} is turned on. If the commutation voltage v_{12} is negative, there will be turn-on loss with T_{21_1} and reverse recovery loss with D_{11_2}. Otherwise, nothing will happen until the next instant when T_{11_1} is turned off, there will be turn-off energy loss with T_{11_1}. The final step is to turn on T_{21_2}. Thus, the exact instant when actual commutation happens depends on the polarity of the commutation voltage. As indicated in Fig. 3, for positive v_{12} the commutation occurs when T_{21_1} is switched on. For native v_{12}, the commutation occurs when T_{11_1} is switched off.

Similar analysis can be carried out for the case of negative i_{o1}. Furthermore, the commutation process from S_{21} to S_{11} has been carried out to complete one commutation cycle between S_{11} and S_{21}. The switching energy losses for the commutation between S_{11} and S_{21} for different pole current and throw voltage polarities are summarized in TABLE I.

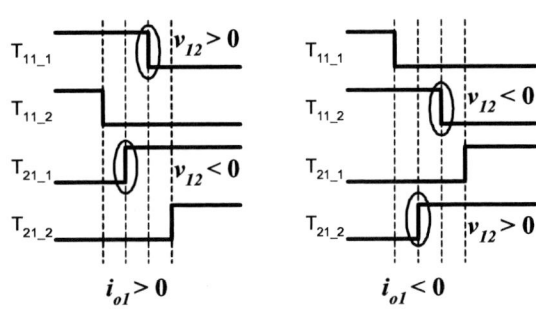

Fig. 3 Circuit schematic and actual switching function waveforms illustrating four step commutation from S_{11} to S_{21}

The observations from the table can be generalized that, for the two commutation events comprising a complete switching cycle, there are (i) one turn-on loss transient, and (ii) one turn-off loss transient for the IGBT; and (iii) one recovery energy loss for the diode. A similar table is possible for the commutation of between S_{21} and S_{31}, difference being the commutation voltage v_{23} instead of v_{12}. Thus, the total switching energy loss of one SPTT in one switching cycle can be calculated by

$$E_{sw/Ts} = \left(E_{on\_T} + E_{off\_T} + E_{rr\_D}\right)\frac{(|v_{12}| + |v_{23}|)|i_{o1}|}{V_R I_R} \quad (6)$$

where E_{on_T}, E_{off_T}, E_{rr_D} are the turn-on and turn off energy of the IGBT and the reverse recovery energy of the diode per switching event, respectively, with the loss reference conditions being V_R and I_R.

If the switching frequency is much higher than the fundamental frequency of input voltage and output current, the average switching power loss over time interval T can be estimated using the integral

$$P_{sw/T} = f_s \left(\frac{E_{on\_T} + E_{off\_T} + E_{rr\_D}}{V_R I_R}\right)\frac{1}{T}\int_0^T \left[(|v_{i12}| + |v_{i23}|)|i_{o1}|\right]dt \quad (7)$$

The fundamental frequency of the input is typically different from the output frequency. It follows that the integration interval must be sufficiently long to overcome possible low "beating" frequency of the integrand in (7). So, any attempt to characterize the loss of the matrix converter via simulation will be challenging unless particular input-output frequency ratio has been chosen, which does not result in extreme low "beating" frequency. Notice that $|v_{12}|$, $|v_{23}|$ and $|i_{o1}|$ can be expanded to Fourier

TABLE I. SWITCHING ENERGY LOSSES FOR THE COMMUTATION BETWEEN S_{11} AND S_{21}

	$I_{o1}>0$		$I_{o1}<0$	
	$S_{11}{\to}S_{21}$	$S_{21}{\to}S_{11}$	$S_{11}{\to}S_{21}$	$S_{21}{\to}S_{11}$
$v_{12}>0$	$E_{off\_T11\_1}$	$E_{on\_T11\_1}$ $E_{rr\_D21\_2}$	$E_{on\_T21\_2}$ $E_{rr\_D11\_1}$	$E_{off\_T21\_2}$
$v_{12}<0$	$E_{on\_T21\_1}$ $E_{rr\_D11\_2}$	$E_{off\_T21\_1}$	$E_{off\_T11\_2}$	$E_{on\_T11\_2}$ $E_{rr\_D21\_1}$

series and the average loss coming from harmonics can be neglected if the input and output frequency ratio is not integer, which is generally true for variable frequency drive applications. Using this process the total switching loss of the converter can be determined to be

$$P_{sw\_CMC} = \frac{24\sqrt{3}}{\pi^2} f_s \left(E_{on\_T} + E_{off\_T} + E_{rr\_D} \right) \frac{V_I I_o}{V_R I_R} \quad (8)$$

IV. IMC LOSS CALCULATION

As shown in Fig. 4, the IMC (also called the dual bridge matrix converter) is composed of two bridges, which are treated as the current source bridge (CSB) at the input terminals and the voltage source bridge (VSB) at the output terminals. One important advantage with this topology is the simple commutation of the bidirectional switches as identified in [21]. By utilizing the zero link current created by the VSB, the CSB can commutate without overlap-time or dead-time. In this discussion, the CSB is modulated in such way that one of the three input phases a, b or c with maximum reference current amplitude is connected to the dc link for the entire switching period and the return dc link current is split between the other two phases [22]. The VSB modulation is very similar to the ordinary VSI modulation except that the amplitude of the modulation functions of its throws appropriately modified to result in sinusoidal input currents and output voltages [22]. The input voltages and output currents still follow the definitions described in (3).

A. Conduction Losses

It is relatively easy to estimate the conduction loss of the CSB since the link current always flows through two IGBTs and two diodes in the CSB. Noticing that the link current i_p is discontinuous due to the VSB switching as illustrated in Fig. 5, we may first calculate the average conduction loss of the CSB over one switching period as

$$P_{c\_avg/Ts} = 2\frac{1}{T_s} \left[\int_{Ts} (V_{CE0} + V_{F0}) i_p dt + \int_{Ts} (r_{CE} + r_F) i_p^2 dt \right] \quad (9)$$

$$= 2(V_{CE0} + V_{F0}) i_{p\_avg/Ts} + 2(r_{CE} + r_F) i_{p\_rms/Ts}^2$$

in terms of average and rms dc link currents over one switching period, which can be derived for $0 < \alpha_o(t) < \pi/3$

$$i_{p\_avg/Ts} = \frac{3}{4} M(t) I_o \cos(\phi_o); \quad (10)$$

$$i_{p\_rms/Ts}^2 = \frac{3M(t) I_o^2}{4} \left[\cos\left(\alpha_o(t) + \frac{\pi}{3} \right) + \frac{1}{2} \sin(3\alpha_o(t) + 2\phi_o) \right]$$

where $M(t)$ is the the amplitude of the modulating functions of the VSB expressed as m_1, m_2, and m_3 which defined by

$$m_k = \frac{1 + M(t) \cos(\alpha_o(t) - (k-1)2\pi/3)}{2} \quad (11)$$

where $\alpha_o(t) = \omega_o t + \alpha_{o0}$, $k \in \{1,2,3\}$. $M(t)$ is the modulation function amplitude, which is time varying at frequency of $6\omega_i$ to synthesize the sinusoidal input current. The detailed derivation of (10) is found in Appendix A.

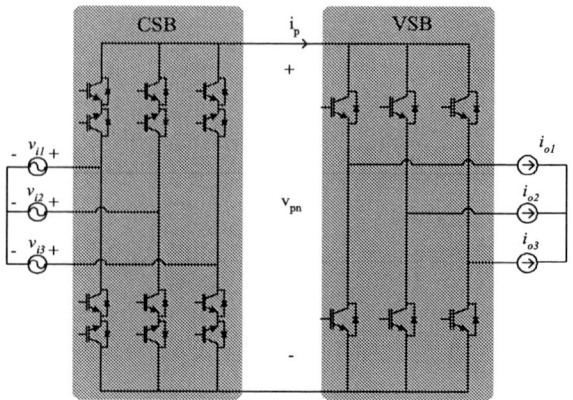

Fig. 4 Simplified schematic of the IMC topology using IGBTs and diodes.

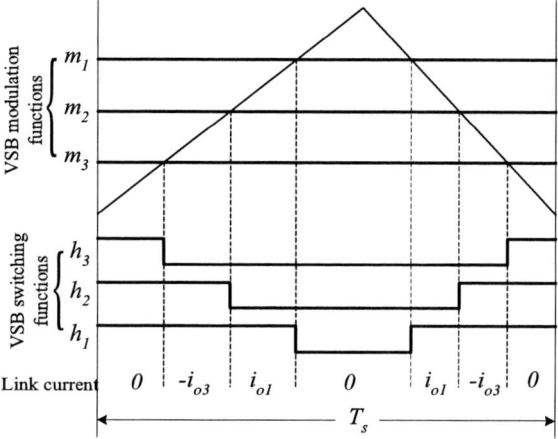

Fig. 5 Carrier based modulation for the VSB. $m_{1,2,3}$ are modulation functions of the upper throws in the VSB.

Combining (9), (10) and (11), the average conduction loss of the CSB may be determined as

$$P_{c\_CSB} = \frac{9}{2\pi} (V_{CE0} + V_{F0}) M_o I_o \cos\phi_o + \frac{3\sqrt{3}}{2\pi^2} (r_{CE} + r_F) M_o I_o^2 \left(1 + 4\cos^2\phi_o\right) \quad (12)$$

To calculate the conduction loss of the VSB, the top IGBT and diode in VSB phase-leg "1" is considered. The conduction loss of the top IGBT can be written as

$$P_{c\_Tp} = \frac{1}{T} \int_0^T i_{o1} v_{CE}(i_{o1}) \frac{[sign(i_{o1})+1]}{2} \frac{[m_{o1}(t)+1]}{2} dt \quad (13)$$

The conduction loss of the top diode in the same phase-leg is

$$P_{c\_Dp} = \frac{1}{T} \int_0^T -i_{o1} v_F(-i_{o1}) \frac{[sign(-i_{o1})+1]}{2} \frac{[m_{o1}(t)+1]}{2} dt \quad (14)$$

The conduction loss of the VSB can be calculated by

$$P_{c\_VSB} = 6 \left[\begin{array}{l} \dfrac{(V_{CE0} + V_{F0}) I_o}{2\pi} + \dfrac{(r_{CE} + r_F) I_o^2}{8} + \\[2ex] \dfrac{3}{8\pi} (V_{CE0} - V_{F0}) I_o M_o \cos\phi_o + \dfrac{(r_{CE} - r_F) I_o^2}{\pi^2} M_o \cos\phi_o \end{array} \right] \quad (15)$$

The expression (15) says the conduction loss of the VSB is affected by the modulation index and the output power factor angle due to the different characteristics of the IGBT and the diode, which is reasonable because the modulation index and the power factor angle will affect the relative duty ratios of the IGBTs and the diodes.

B. Switching Losses

Since the CSB only commutates only when the link current i_p is zero, there is no switching loss in the CSB. As shown in Fig. 5, there are six commutation events associated with the VSB in each switching cycle T_s. Due to the three phase symmetry of the output current, the loss in one phase leg alone needs to be determined and the total switching loss is estimated to be three times as much. During each switching period, the VSB commutation voltage (link voltage) will take one of the values of the two different input line-line voltages shown in bold in Fig. 6.

Using this formulation, the switching loss of the VSB can be expressed as

$$P_{sw\_VSB} = 3f_s \frac{1}{\theta} \int_0^\theta \left(E_{on} + E_{off} + E_{rr} \right) \frac{\left(v_{pn1} + v_{pn2} \right) i_{o1}}{V_R I_R} \frac{1 + sign\left(i_{o1} \right)}{2} d\omega_o t \quad (16)$$

Evaluating the integral, (16) can be simplified to

$$P_{sw\_VSB} = \frac{27}{\pi^2} f_s \left(E_{on} + E_{off} + E_{rr} \right) \frac{V_i}{V_R} \frac{I_o}{I_R} \cos(\phi_i) \quad (17)$$

It can be seen the IMC switching loss varies with the input power factor. The switching losses peak at unity input power factor.

V. SIMULATION VERIFICATION AND DISCUSSION

To verify the analytical solutions derived in proceeding sections, the detail numerical model has been built using Matlab SimuLink™. The key parameters used in simulation are listed in TABLE II. The device parameters are based on the datasheet of SEMIKRON IGBT module SKM 75GB123D.

Selected quantities of the current and voltage waveforms for CMC and IMC obtained from the simulations are illustrated in Fig. 7 and Fig. 8 respectively.

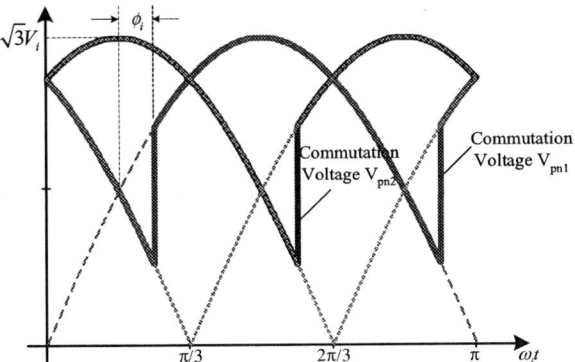

Fig. 6 Commutation voltage of the VSB with input power factor of ϕ_i

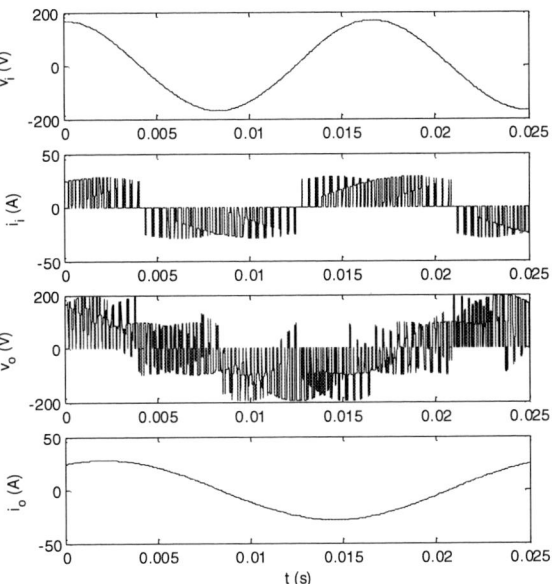

Fig. 7 Simulation waveforms of CMC: from top to bottom are input voltage v_i, input current i_i, output voltage (phase-to-load-neutral) v_o and output current i_o. Switching frequency of 3 kHz is chosen to show the waveforms with legible resolution.

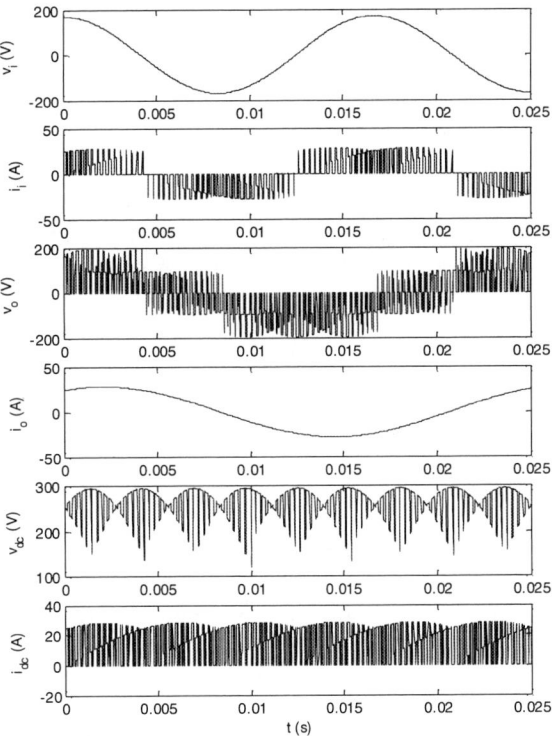

Fig. 8 Simulation waveforms of IMC: from top to bottom are input voltage v_i, input current i_i, output voltage (phase-to-load-neutral) v_o, output current i_o, link voltage v_{dc}, and link current i_{dc}. Switching frequency of 3 kHz is chosen to show the waveforms with legible resolution.

TABLE II.   Switching Energy Losses for the Commutation between S_{11} and S_{21}

Reference conditions	V_R (V)	600	I_R (A)	50
IGBT	V_{CE0} (V)	1.6	E_{on_T} (mJ)	8
	r_{CE} (mΩ)	30	E_{off_T} (mJ)	5
Diode	V_{F0} (V)	1.2	E_{rr_D} (mJ)	2.5
	r_F (mΩ)	18		
Input (l-n)	V_i (V_{rms})	120	ω_i (rad/s)	$2\pi\times60$
Output	I_o (A_{rms})	20	ω_o (rad/s)	$2\pi\times40$
Switching frequency	f_s (kHz)	10		

The semiconductor losses of IMC and CMC are plotted in Fig. 9 as functions of the modulation index. The per-unitized quantities are used to provide the common basis for comparison between the two different topologies. In the figure, the thin solid line, dashed line, dotted line and dash-dotted line are plots of analytical solution of the IMC losses with different input/output power factor angles. The markers on those lines are plots of the simulation results under the corresponding operation condition. The thick solid line and the markers are analytical and simulated results of the CMC losses, which are not affected by any of the operating conditions. The tight match between the analytical solution and simulation results validates the effectiveness of analytical approach developed in Section III and IV.

It can be observed from Fig. 9 at low modulation indices, the IMC has lower losses than CMC. Of course the break point relates to the different operating conditions of the IMC as indicated in the figure. For IMC, among the four different operating conditions, the losses with the $\phi_i = 0$ and $\phi_o = 0$ are maximum.

To better illustrate the relation, a separate 3-D plot of the IMC losses as function of the input/output power factor angle is presented in Fig. 10 using the analytical solution. It can be clearly seen the IMC losses peak at the operating point of $\phi_i, \phi_o = 0$.

In Fig. 9, all the plots are computed at the same switching frequency of 10 kHz. For different switching frequencies, the power losses of IMC and CMC are plotted in Fig. 11 to show the differences in dependency to switching frequency variations. In this comparison, the IMC power losses are plotted for different modulation indices with $\phi_i, \phi_o = 0$. It is observed that for high modulation indices, the IMC has higher losses than CMC. For case of modulation index = 0.75 as plotted in the figure, there is break point where the IMC losses become lower than CMC as the frequency goes higher since the CMC losses are more 'sensitive' to the switching frequency.

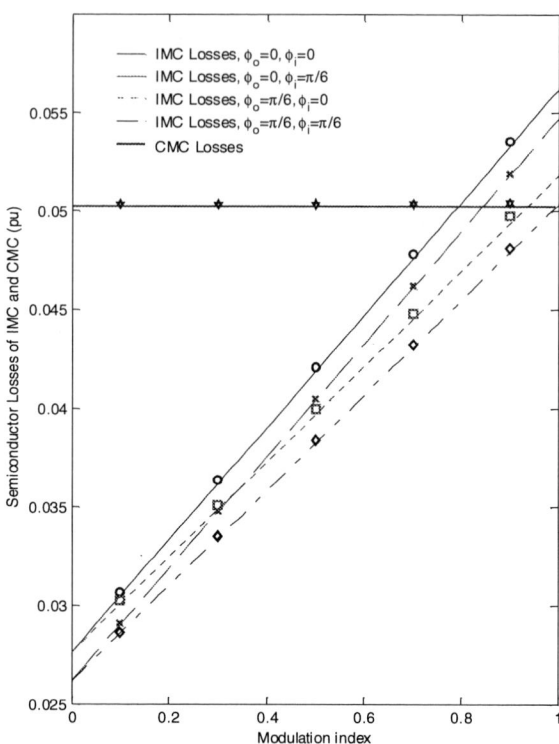

Fig. 9 Simulated and analytical power losses for CMC and IMC for different operating conditions. The lines are plots of analytical solutions and the markers on lines are the simulation results for the corresponding operating conditions.

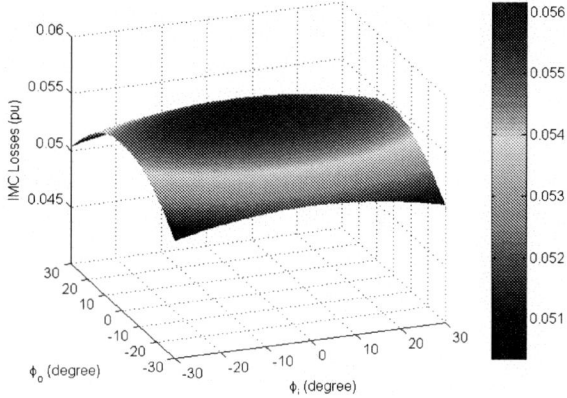

Fig. 10 Variation of the IMC losses vs input and output power factor angle ϕ_i and ϕ_o at modulation index = 1

VI. Conclusions

In this paper, an analytical approach to characterizing the semiconductor losses for two classes of matrix converters, namely CMC and IMC, has been developed. The analytical solutions are verified against numerical results from detailed simulation built on the SimuLink model. A close match between the analytical and simulated results under various operating conditions validates the effectiveness of the proposed approach.

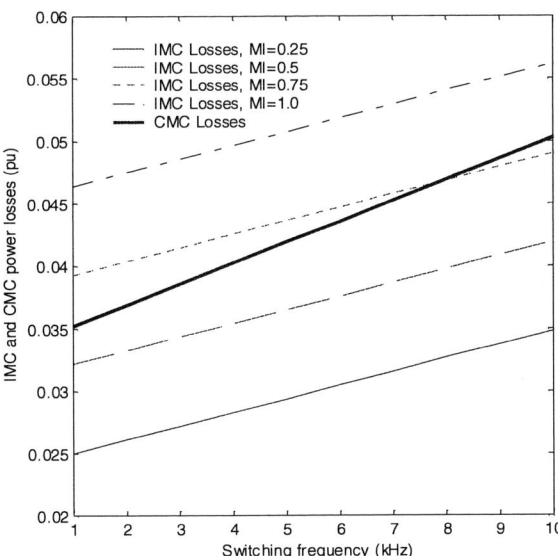

Fig. 11 Variation of power losses IMC and CMC at different switching frequencies.

It has been found power losses of the CMC and IMC respond quite differently to different operating conditions. The power losses of CMC increase as switching frequency increases while they remain invariant regardless the input and output power factors and voltage transfer ratio (or modulation index) variation. The power losses of IMC increase as the switching frequency increases, but at a smaller slope compared to the CMC. Furthermore, the

IMC power losses depend on the modulation index, and input/output power factor angle and peak when modulation index is 1.0 and the input and output power factor angles are zeros. In summary the IMC tends to perform better at higher switching frequency and lower modulation index than CMC in terms of the power losses.

The analytical model presented in the paper can be readily used in computer mathematical calculation programs like a spreadsheet to estimate loss calculations in matrix converter as a rapid computer aided design tool. They may also be used in system level optimization for evaluating the effect of variations in operating conditions during thermal cycling, startup conditions, load transients, etc. in sinusoidal input/output conditions. The systematic methodology of loss estimation can be further extended to study system performance under alternate modulation algorithms, input and output waveforms and non-sinusoidal input/output conditions.

ACKNOWLEDGMENT

The authors would like to acknowledge support from the Wisconsin Electric Machine and Power Electronics Consortium (WEMPEC) at the University of Wisconsin-Madison. The work made use of ERC shared facilities supported by the National Science Foundation (NSF) under AWARD EEC-9731677.

APPENDIX A

With reference to Fig. 5, the averaged link current over one switching cycle is

$$
\begin{aligned}
i_{p\_avg} &= -(m_2 - m_3)i_{o3} + (m_1 - m_2)i_{o1} \\
&= \frac{\sqrt{3}MI_o}{2}\left[\begin{array}{l}\cos(\alpha_o(t) + \pi/2)\cos(\beta_o(t) + 2\pi/3) + \\ \cos(\alpha_o(t) + \pi/6)\cos(\beta_o(t))\end{array}\right] \quad (18) \\
&= \frac{\sqrt{3}MI_o}{2}\cos(\phi)\cos(\pi/6) \\
&= \frac{3}{4}MI_o\cos(\phi)
\end{aligned}
$$

The rms value current of the link current over one switching cycle is

$$
\begin{aligned}
i_{dc\_rms}^2 &= (m_2 - m_3)i_{o3}^2 + (m_1 - m_2)i_{o1}^2 \\
&= \frac{\sqrt{3}M(t)I_o^2}{2}\left[\begin{array}{l}\cos(\alpha_o(t) + \pi/2)\cos^2(\beta_o(t) + 2\pi/3) + \\ \cos(\alpha_o(t) + \pi/6)\cos^2(\beta_o(t))\end{array}\right] \\
&= \frac{\sqrt{3}M(t)I_o^2}{4}\left[\begin{array}{l}\sqrt{3}\cos\left(\alpha_o(t) + \dfrac{\pi}{3}\right) + \\ \dfrac{1}{2}\left[\cos\left(3\alpha_o(t) + 2\phi_o - \dfrac{\pi}{6}\right) + \cos\left(\alpha_o(t) + 2\phi_o + \dfrac{5\pi}{6}\right)\right] + \\ \dfrac{1}{2}\left[\cos\left(3\alpha_o(t) + 2\phi_o + \dfrac{\pi}{6}\right) + \cos\left(\alpha_o(t) + 2\phi_o - \dfrac{\pi}{6}\right)\right]\end{array}\right] \\
&= \frac{\sqrt{3}M(t)I_o^2}{4}\left[\begin{array}{l}\sqrt{3}\cos\left(\alpha_o(t) + \dfrac{\pi}{3}\right) \\ \dfrac{1}{2}\left[\cos\left(3\alpha_o(t) + 2\phi_o - \dfrac{\pi}{6}\right) + \dfrac{1}{2}\cos\left(3\alpha_o(t) + 2\phi_o + \dfrac{\pi}{6}\right)\right]\end{array}\right] \\
&= \frac{3M(t)I_o^2}{4}\left[\cos\left(\alpha_o(t) + \dfrac{\pi}{3}\right) + \dfrac{1}{2}\sin(3\alpha_o(t) + 2\phi_o)\right]
\end{aligned}
$$

$$(19)$$

REFERENCES

[1] M. Venturini, "A new sinewave in, sinewave out conversion technique eliminates reactive elements " in *Proc. POWERCON'80*, 1980, pp. E3-1.

[2] A. Alesina and M. G. B. Venturini, "Solid-state power conversion; a Fourier analysis approach to generalized transformer synthesis," *IEEE Transactions on Circuits and Systems*, vol. CAS-28, no. 4, pp. 319, 1981.

[3] M. Venturini and A. Alestina, "The generalised transformer: a new bidirectional sinusoidal waveform frequency converter with continuously adjustable input power factor," in *PESC '80 Record. IEEE Power Electronics Specialists Conference*, Atlanta, GA, USA, 1980, pp. 242.

[4] P. W. Wheeler, J. C. Clare, L. Empringham, M. Bland, and K. G. Kerris, "Matrix converters," *IEEE Industry Applications Magazine*, vol. 10, no. 1, pp. 59, 2004.

[5] P. W. Wheeler, J. Rodriguez, J. C. Clare, L. Empringham, and A. Weinstein, "Matrix converters: a technology review," *IEEE Transactions on Industrial Electronics*, vol. 49, no. 2, pp. 276, 2002.

[6] P. W. Wheeler, J. C. Clare, D. Katsis, L. Empringham, M. Bland, and T. Podlesak, "Design and construction of a 150 KVA matrix converter induction motor drive," in *Second IEE International Conference on Power Electronics, Machines and Drives (Conference Publication No.498)*, Edinburgh, UK, 2004, pp. 719.

[7] S. Angkititrakul and R. W. Erickson, "Control and implementation of a new modular matrix converter," in *APEC 2004. Nineteenth*

Annual IEEE Applied Power Electronics Conference and Exposition (IEEE Cat. No.04CH37520), Anaheim, CA, USA, 2004, pp. 813.

[8] M. Aten, C. Whitley, G. Towers, P. Wheeler, J. Clare, and K. Bradley, "Dynamic performance of a matrix converter driven electro-mechanical actuator for an aircraft rudder," in *Second IEE International Conference on Power Electronics, Machines and Drives (Conference Publication No.498)*, Edinburgh, UK, 2004, pp. 326.

[9] E. H. Miliani, D. Depernet, J. M. Kauffmann, and A. Lacaze, "Experimental control of matrix converter for active generator," in *2004 IEEE 35th Annual Power Electronics Specialists Conference (IEEE Cat. No.04CH37551)*, Aachen, Germany, 2004, pp. 2899.

[10] R. Kazemzadeh and J. M. Kauffmann, "Power flow study in connection a generator to a power grid via a natural commutated matrix converter (NCMC)," in *2004 IEEE 35th Annual Power Electronics Specialists Conference (IEEE Cat. No.04CH37551)*, Aachen, Germany, 2004, pp. 2894.

[11] D. Katsis, P. Wheeler, J. Clare, and P. Zanchetta, "A three-phase utility power supply based on the matrix converter," in *Conference Record of the 2004 IEEE Industry Applications Conference. 39th IAS Annual Meeting (IEEE Cat. No.04CH37569)*, Seattle, WA, USA, 2004, pp. 1447.

[12] L. S. Large, A. Green, S. C. Mason, S. Bhatia, J. C. Clare, P. Zanchetta, L. Empringham, and P. W. Wheeler, "Matrix converter solution for aircraft starting," in *IEE Seminar Matrix Converters (Digest No.03/10100)*, Birmingham, UK, 2003, pp. 5.

[13] M. T. Aydemir, A. Bendre, and G. Venkataramanan, "A critical evaluation of high power hard and soft switched isolated DC-DC converters," in *Conference Record of the 2002 IEEE Industry Applications Conference. 37th IAS Annual Meeting (Cat. No.02CH37344)*, Pittsburgh, PA, USA, 2002, pp. 1338-45.

[14] L. K. Mestha and P. D. Evans, "Analysis of on-state losses in PWM inverters," *IEE Proceedings B (Electric Power Applications)*, vol. 136, no. 4, pp. 189-95, 1989.

[15] J. Bergauer, L. Schindele, and M. Braun, "Optimised space vector control reducing switching losses in current source inverters," in *9th International Conference and Exhibition on Power Electronics and Motion Control. EPE - PEMC 2000 Kosice. Proceedings*, Kosice, Slovakia, 2000, pp. 3-40.

[16] J. W. Kolar, F. C. Zach, and F. Casanellas, "Losses in PWM inverters using IGBTs," *IEE Proceedings-Electric Power Applications*, vol. 142, no. 4, pp. 285-8, 1995.

[17] A. M. Trzynadlowski and S. Legowski, "Minimum-loss vector PWM strategy for three-phase inverters," *IEEE Transactions on Power Electronics*, vol. 9, no. 1, pp. 26-34, 1994.

[18] M. H. Bierhoff and F. W. Fuchs, "Semiconductor losses in voltage source and current source IGBT converters based on analytical derivation," in *PESC Record - IEEE Annual Power Electronics Specialists Conference*, Aachen, Germany, 2004, pp. 2836-2842.

[19] D. Casadei, G. Serra, A. Tani, and L. Zarri, "Matrix converter modulation strategies: a new general approach based on space-vector representation of the switch state," *IEEE Transactions on Industrial Electronics*, vol. 49, no. 2, pp. 370-81, 2002.

[20] N. Burany, "Safe control of four-quadrant switches," in *Conference Record of the IEEE Industry Applications Society Annual Meeting (Cat. No.89CH2792-0)*, San Diego, CA, USA, 1989, pp. 1190.

[21] L. Wei and T. A. Lipo, "A novel matrix converter topology with simple commutation," in *Conference Record of the 2001 IEEE Industry Applications Conference. 36th IAS Annual Meeting (Cat. No.01CH37248)*, Chicago, IL, USA, 2001, pp. 1749.

[22] B. Wang and G. Venkataramanan, "A carrier based PWM algorithm for indirect matrix converters," in *Proceedings of 37th IEEE Power Electronics Specialists Conference*, Jeju, Korea, June 18-22, 2006.

2006 5th International Power Electronics and Motion Control Conference

Nonlinear Robust Sliding Mode Control for PM Linear Synchronous Motors

Xi Zhang, Junmin Pan

Department of Electrical Engineering, Shanghai Jiao Tong University, Shanghai 200030, P.R.China

braver1980@sjtu.edu.cn

Abstract—A new nonlinear robust control scheme of permanent magnet linear synchronous motors (PMLSM) is proposed in this paper. A quasi-linearized and decoupled model with uncertainties is derived by the mathematical model of PMLSM according to its characteristic. A fixed-boundary sliding mode controller using the *m*sat function is designed to guarantee the robustness and remove the chattering which usually exists in normal sliding mode control. Design of a force observer is given to estimate the load force unknown in the new model. The validity of the proposed algorithm compared with the conventional PID control scheme is proved by MATLAB simulation results.

Keywords-PMLSM; feedback linearization; sliding mode controller; robustness; *m*sat function; force observer

I. INTRODUCTION

Unlike rotary motors, high-performance linear motors can give machine tools linear motion directly without indirect coupling mechanisms such as gear boxes, chains and screws. The advantages of linear motors include high speed, high acceleration and the most importantly, high motion precision. With the increasing demand in industrial applications, some manufacturing factories and corporations have produced some types of linear motors [1]. Conventional control methods for linear motors are faced with challenge because of the existence of system parameter perturbation and edge effect etc.[2]. New advanced control schemes were proposed for disturbance reduction. Ref.[3] proposed a feedforward-assisted feedback controller based on neural network. A precision motion control with disturbance observer for pulse-width-modulated-driven permanent magnet linear motors was proposed in [4]. An artificial relay tuning and zero-phase filtering-based iterative learning control (ILC) for high-precision motion control firstly appeared in [5]. But no specific models with uncertainties which can be directly controlled were proposed in these papers.

In this paper, the conception of feedback linearization is used according to the special PMLSM mathematical model. Firstly, the three-closed-loop PID control scheme for PMSLM is introduced for comparison in simulation results. Then, the feedback linearization technique is applied to obtain a linearized and decoupled model, and the linear design technique is used to complete the control design based on Ref.[6]. But the linear design technique can not guarantee the system robustness due to the disturbance factors mentioned in the previous paragraph. Therefore, a quasi-linearized and decoupled model with all kinds of disturbances is presented to overcome this trouble. A fixed-boundary sliding mode controller is designed to improve the system performance. The chattering which exists in normal sliding mode control can be removed by the *m*sat function. Design of a force observer is given to estimate the load force unknown in the quasi-linearized and decoupled model. For this proposed control scheme, it is not necessary to measure system parameters precisely. Only the estimated variation bounds of system parameters of PMLSM are required. The good robustness can be obtained in spite of variations of system parameters. Finally, compared with the PID control scheme, the speed behaviors under the step response with uncertainties is analyzed, and the displacement error under the displacement sine response is also analyzed in the simulation. The validity of the proposed control algorithm is proved by simulation results in this paper.

II. MATHEMATICAL MODEL OF PMLSM

The dynamics of PMLSM can be described as follows [7]:

$$\frac{di_d}{dt} = -\frac{R}{L_d}i_d + \frac{L_q\pi}{L_d\tau}vi_q + \frac{1}{L_d}U_d \quad (1)$$

$$\frac{di_q}{dt} = -\frac{R}{L_q}i_q - \frac{L_d\pi}{L_q\tau}vi_d - \frac{\psi\pi}{\tau L_q}v + \frac{1}{L_q}U_q \quad (2)$$

$$\frac{dv}{dt} = \frac{3\pi\psi}{2\tau M}i_q + \frac{3\pi}{2\tau M}(L_d - L_q)i_d i_q - \frac{B}{M}v - \frac{1}{M}(F_l + F_d) \quad (3)$$

where i_d, i_q and v are the state variables which represent direct-axis current, quadrature-axis current and linear speed, respectively, and U_d, U_q the direct-axis and quadrature-axis primary voltage components, respectively, M the total mass of load, B the viscous damping coefficient, R the primary winding resistance, L_d, L_q the direct-axis and quadrature-axis primary inductors, respectively, ψ the permanent magnet flux, τ the polar pitch, F_l the load force, and F_d the edge effect force.

The quadrature-axis primary inductor is usually

1-4244-0448-7/06/$25.00 ©2006 IEEE

equivalent to the direct-axis inductor, namely, $L_q = L_d = L$. Thus, (1)-(3) become :

$$\frac{di_d}{dt} = -\frac{R}{L}i_d + \frac{\pi}{\tau}vi_q + \frac{1}{L}U_d \qquad (4a)$$

$$\frac{di_q}{dt} = -\frac{R}{L}i_q - \frac{\pi}{\tau}vi_d - \frac{\psi\pi}{\tau L}v + \frac{1}{L}U_q \qquad (4b)$$

$$\frac{dv}{dt} = \frac{3\pi\psi}{2\tau M}i_q - \frac{B}{M}v - \frac{1}{M}(F_l + F_d) \qquad (4c)$$

III. MODEL TRANSFORMATION AND PROPOSITION OF NEW STRATEGY

A. Nonlinear Control Scheme for PMLSM Using Feedback Linearization

For the sake of terseness, we note that

$$X = \begin{pmatrix} x_1 \\ x_2 \\ x_3 \end{pmatrix} = \begin{pmatrix} i_d \\ i_q \\ v \end{pmatrix}, \quad U = \begin{pmatrix} U_d \\ U_q \end{pmatrix}, \quad K = \begin{pmatrix} \frac{1}{L} & 0 \\ 0 & \frac{1}{L} \\ 0 & 0 \end{pmatrix},$$

$$F_L = F_l + F_d$$

$$f(X) = \begin{pmatrix} f_1(X) \\ f_2(X) \\ f_3(X) \end{pmatrix} = \begin{pmatrix} -\dfrac{R}{L}x_1 + \dfrac{\pi}{\tau}x_2 x_3 \\ -\dfrac{R}{L}x_2 - \dfrac{\pi}{\tau}x_1 x_3 - \dfrac{\psi\pi}{L\tau}x_3 \\ \dfrac{3\pi\psi}{2\tau M}x_2 - \dfrac{B}{M}x_3 - \dfrac{1}{M}F_L \end{pmatrix}$$

Thus (4) becomes

$$\frac{dX}{dt} = f(X) + KU \qquad (5)$$

in which i_d and v are chosen as outputs of the state equation above so as to obtain a total feedback linearization, U_d, U_q are inputs. Because the electromagnetic time constant is much smaller than the mechanical time constant, the variation of the generalized load force F_L is very slow compared with electrical variations. Thus F_L is considered as a constant during a small time range [7]. From (4), a new state equation can be obtained as follows:

$$\begin{pmatrix} \dfrac{di_d}{dt} \\ \dfrac{d^2v}{dt^2} \end{pmatrix} = C + DU = \begin{pmatrix} f_1(X) \\ \dfrac{3\pi\psi}{2\tau M}f_2(X) - \dfrac{B}{M}f_3(X) \end{pmatrix} + \frac{1}{L}\begin{pmatrix} 1 & 0 \\ 0 & \dfrac{3\pi\psi}{2\tau M} \end{pmatrix}\begin{pmatrix} U_d \\ U_q \end{pmatrix}$$

$$(6)$$

where

$$C = \begin{pmatrix} f_1(X) \\ \dfrac{3\pi\psi}{2\tau M}f_2(X) - \dfrac{B}{M}f_3(X) \end{pmatrix}, \quad D = \frac{1}{L}\begin{pmatrix} 1 & 0 \\ 0 & \dfrac{3\pi\psi}{2\tau M} \end{pmatrix} \qquad (7)$$

Assume

$$\begin{pmatrix} U_d \\ U_q \end{pmatrix} = D^{-1}(\begin{pmatrix} u_1 \\ u_2 \end{pmatrix} - C) \qquad (8)$$

where u_1 and u_2 are new inputs. From (6)-(8), we can derive the following relations:

$$u_1 = \frac{di_d}{dt}, \quad u_2 = \frac{d^2v}{dt^2}$$

Take $e_1 = i_d^* - i_d$, $e_2 = v^* - v$, where i_d^* and v^* are the reference values of the direct-axis current and linear speed, respectively. A new controller can be designed as follows:

$$u_1 = \frac{di_d^*}{dt} + K_1 e_1 \qquad (9)$$

$$u_2 = \frac{d^2v^*}{dt^2} + K_2\frac{de_2}{dt} + K_3 e_2 = \frac{d^3S^*}{dt^3} + K_2\frac{d^2e_3}{dt^2} + K_3\frac{de_3}{dt} \qquad (10)$$

where S^* is the reference value of the displacement, $e_3 = S^* - S$ is the displacement error and $K_i, i = 1, 2, 3$ is a constant. From (9) and (10), it can be obtained that

$$\frac{de_1}{dt} + K_1 e_1 = 0 \qquad (11)$$

$$\frac{d^2e_2}{dt^2} + K_2\frac{de_2}{dt} + K_3 e_2 = 0 \qquad (12)$$

By selecting proper K_i, $i = 1, 2, 3$, the dynamic behaviors can satisfy requirements [8]. U_d and U_q achieved by this method above are inputs of PMLSM.

B. Sliding Mode Control Scheme for PMLSM

Due to the system parameter perturbation during running, the voltage inputs are given by

$$\begin{pmatrix} U_d \\ U_q \end{pmatrix} = D_n^{-1}(\begin{pmatrix} u_1' \\ u_2' \end{pmatrix} - C_n) \qquad (13)$$

where C_n and D_n are obtained by substituting the nominal parameter values into (7). The generalized load force F_L in (8) and (13) is unknown and replaced by its estimated value. The specific design methodology of an observer will be introduced in Sec.3.3.

By substituting (13) into (16) and letting $y_1 = i_d, y_2 = v$ and $Y = (y_1, y_2)^T$, a quasi-linearized and decoupled model can be obtained as shown in the following equations:

$$\frac{dy_1}{dt} = \frac{R_n - R}{L}y_1 + u_1' = f_{e_1}(Y) + u_1' \qquad (14)$$

$$\frac{d^2y_2}{dt^2} = \frac{R_n - R}{L}\frac{3\pi\psi}{2\tau M}i_q + \frac{3\pi^2\psi}{2\tau^2 M}y_2(\psi_n - \psi) + \frac{3\pi\psi}{2\tau M}i_q(\frac{B_n}{M_n} - \frac{B}{M})$$

$$-(\frac{M_n\psi}{M\psi_n}\frac{B_n^2}{M_n^2} - \frac{B^2}{M^2})y_2 - (\frac{\psi M_n}{\psi_n M}\frac{B_n}{M_n^2}\hat{F}_L - \frac{B}{M^2}F_L) + \frac{\psi M_n}{\psi_n M}u_2'$$

$$= f_{e_2}(Y) + \sigma u_2' \qquad (15)$$

where the subscript 'n' represents the nominal parameter values. $\hat{F}_L$ is the estimated value of F_L. $f_{e_1}(Y)$, $f_{e_2}(Y)$ and σ can be estimated as $\hat{f}_{e_1}(Y)$, $\hat{f}_{e_2}(Y)$ and $\hat{\sigma}$, respectively. Let

$$\left|\hat{f}_{e_1}(Y) - f_{e_1}(Y)\right| \le F_1(Y),$$

$$\left|\hat{f}_{e_2}(Y) - f_{e_2}(Y)\right| \le F_2(Y) \quad (16)$$

and assume $\psi = \delta\psi_n$, $M = \gamma M_n$, $B = hB_n$, $R = qR_n$. The bounds of these coefficients are shown as follows:

$$0.7 \le \delta \le 1.3, \quad 0.95 \le h \le 1.05, \quad 1 \le q \le 1.4, \quad 1 \le \gamma \le 3$$

Assume $\sigma_{\min} \le \sigma \le \sigma_{\max}$. Calculate the maximum and the minimum value of $f_{e_1}(Y)$, and $\hat{f}_{e_1}(Y)$ is the average value of them. It is same that $\hat{f}_{e_2}(Y)$ can be obtained. $F_1(Y)$ is the difference between the maximum value and $\hat{f}_{e_1}(Y)$. It is same that $F_2(Y)$ can be achieved. The estimated value of σ can be calculated as $\hat{\sigma} = \sqrt{\sigma_{\min}\sigma_{\max}}$. Then the controller is designed, and the values of u_1' and u_2' are calculated by the following proposed scheme.

The sliding surface s_1 can be chosen as follows:

$$s_1 = e_1 + \lambda_1 \int_0^t e_1 dt \quad (17)$$

To satisfy $s_1\dot{s}_1 < -\eta_1|s_1|$, the control law is chosen as

$$u_1' = \hat{u}_1 - k_1(Y)\operatorname{sgn}(s_1) \quad (18)$$

where $\hat{u}_1 = -\hat{f}_{e_1}(Y) + \dot{y}_{1d} + \lambda_1 e_1$, $k_1(Y) = F_1(Y) + \eta_1$,

$$\operatorname{sgn}(s_1) = \begin{cases} 1 & \text{if } s_1 \ge 0; \\ -1 & \text{if } s_1 < 0, \end{cases} \quad (19)$$

But the chattering under such a control law is serious and does harm to the PMLSM. Ref.[9] proposed a boundary sliding mode control scheme using the $m$sat function to remove chattering. The control law is expressed as:

$$u_1' = \hat{u}_1 - k_1(Y)m\operatorname{sat}(\alpha_1(Y), s_1, \phi_1) \quad (20)$$

where

$$m\operatorname{sat}(\alpha_1(Y), s_1, \phi_1) = \begin{cases} \alpha_1(Y)s_1/\phi_1 & \text{if } |s_1| \le \phi_1; \\ \operatorname{sgn}(s_1) & \text{otherwise,} \end{cases} \quad (21)$$

$$\alpha_1(Y) = \frac{\lambda_1\phi_1}{k_1(Y_d)} \quad (22)$$

$Y_d = (y_{1d}, y_{2d})^T$ is the expected value of Y, namely, $(y_{1d}, y_{2d})^T = (i_d^*, v^*)^T$.

The sliding surface s_2 is chosen as follows:

$$s_2 = \frac{de_2}{dt} + 2\lambda_2 e_2 + \lambda^2 \int_0^t e_2 dt \quad (23)$$

To ensure $s_2\dot{s}_2 < -\eta_2|s_2|$ and remove chattering, the control law is designed as follows:

$$u_2' = \hat{\sigma}^{-1}(\hat{u}_2 - k_2(Y)m\operatorname{sat}(\alpha_2(Y), s_2, \phi_2)) \quad (24)$$

where

$$\hat{u}_2 = -\hat{f}_{e_2}(Y) + \ddot{y}_{2d} + 2\lambda_2\frac{de_2}{dt} + \lambda_2^2 e_2$$

$$= -\hat{f}_{e_2}(Y) + \ddot{S}^* + 2\lambda_2\frac{d^2 e_3}{dt^2} + \lambda_2^2\frac{de_3}{dt} \quad (25)$$

$$k_2(Y) = \hat{\beta}[F_2(Y) + \eta_2] + (\hat{\beta} - 1)|\hat{u}_2| \quad (26)$$

$$\hat{\beta} = \sqrt{\sigma_{\max}/\sigma_{\min}} \quad (27)$$

S^* is the reference value of the displacement, and $e_3 = S^* - S$ is the displacement error. The proposed control system is shown in Figure 1.

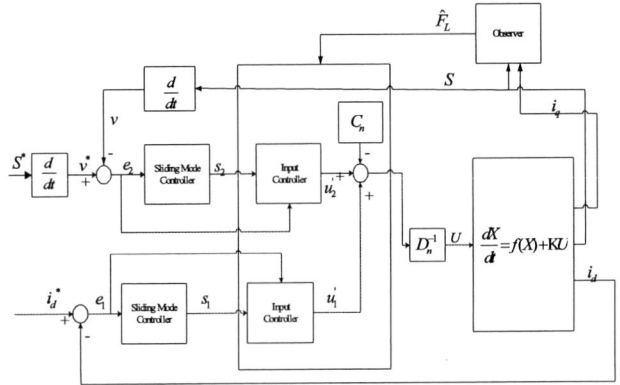

Figure 1. Diagram of the proposed control system

C. Design of Force Observer

Because the variation of the generalized load force F_L is very slow compared with electrical variations, F_L is considered as a constant in a small time range. F_L and v are chosen as new state variables. From (5), it is obtained that

$$\frac{d\xi}{dt} = H\xi + Aw \quad (28)$$

where

$$\xi = (F_L \quad v)^T, \quad H = \begin{pmatrix} 0 & 0 \\ \dfrac{1}{M} & \dfrac{B}{M} \end{pmatrix}, \quad A = \begin{pmatrix} 0 \\ \dfrac{3\pi\psi}{2\tau M} \end{pmatrix}, \quad w = i_q$$

Obviously, this system can be observed. An observer can be designed as follows:

$$\frac{d\hat{\xi}}{dt} = H\hat{\xi} + Aw + P(v - E\hat{\xi}) \quad (29)$$

where $P = (p_1 \quad p_2)^T$ is a constant vector, $E = (0 \quad 1)$.

From (29), the estimated value $\hat{F}_L$ of the generalized load force F_L can be obtained.

IV. SIMULATION RESULTS

The simulation results are obtained by MATLAB 6.5. The edge effect force is expressed as follows [10]:

$$F_d = F_{dm}\sin(S \cdot 2\pi/\tau + \theta_0) \tag{30}$$

where S represents the linear displacement, and θ_0 represents the initial phase.

The values of the PMLSM system parameters are shown in TABLE I.

The maximum permissible value of the quadrature current in the systems is taken as 100A. For the PID control scheme using the partial model matching method introduced in Ref. [7] , take the proportional gain of the current loop $K_{pC} = 60$, the gains of the speed loop $K_I = 0.5$, $K_P = 100$, $K_D = 0.1$ and the proportional gain of the position loop $K_{pP} = 500$. For the proposed fixed-boundary sliding mode control scheme, parameters are selected as $\lambda_1 = 3000, \lambda_2 = 1000$, $\phi_1 = 0.001$, $\phi_2 = 10$ .We choose $l_1 = -10000, l_2 = 500$ for the force observer and take $i_d^* = 0$.

TABLE I. PMLSM SYSTEM PARAMETERS

Primary Winding Resistance	$1.32\,\Omega$
Direct-Axis Primary Inductance	11mH
Quadrature-Axis Primary Inductance	11mH
Permanent Magnet Flux	0.65Wb
Mass of the Primary Part	20kg
Polar Pitch	30mm
Viscous Damping Coefficient	2Ns/m

The dynamic speed step responses in various conditions are shown in Figs. 2–4. As shown in Figure 2(a), because of the edge effect, the fluctuation exists in the speed waveform when the PID control scheme is used for the case that there are no parameter variations. However the proposed control scheme is not influenced by the edge effect as shown in Figure 2(b). When the total mass is increased ($M = 2M_n$), the settling time in the PID control scheme is increased up to 55ms as shown in Figure 3(a). On the other hand, the speed waveform using the robust control scheme proposed above is nearly not influenced as shown in Figure 3(b). For the case of +30% flux variation, the PID control scheme shows increased

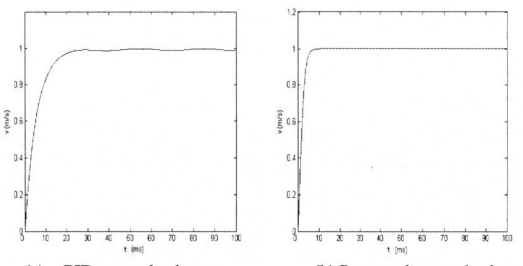

(a)   PID control scheme   (b) Proposed control scheme
Figure 2. Speed step response in the normal condition

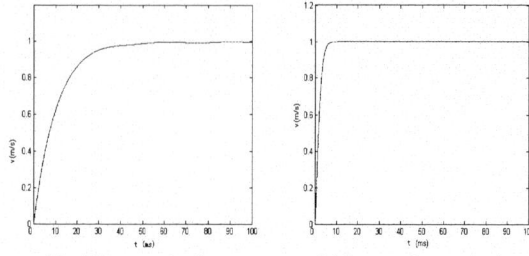

(a)   PID control scheme   (b) Proposed control scheme
Figure 3. Speed step response under +100% mass variation

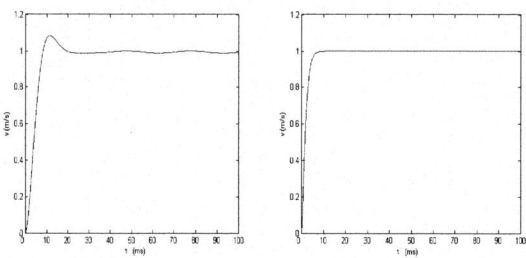

(a) PID control scheme   (b) Proposed control scheme
Figure 4. Speed step response under +30% PM flux variation

overshoot of +10% and settling time of 25ms as shown in Figure 4(a). However, the proposed robust control scheme is nearly not influenced as shown in Figure 4(b).

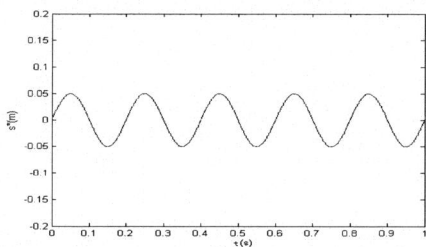

Figure 5. Expected displacement sine waveform

The expected displacement sine waveform is shown in Figure 5. The frequency is 5Hz, and the amplitude is 0.05m.

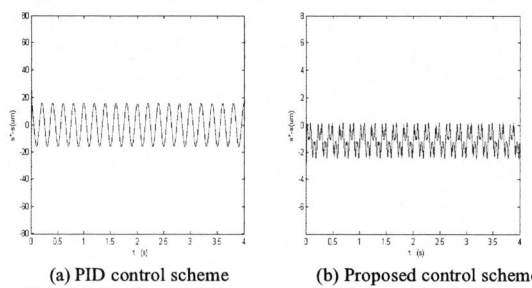

(a) PID control scheme   (b) Proposed control scheme
Figure 6. Displacement error under displacement sine response

The displacement errors of two control schemes under displacement sine response with +100% mass and +30% PM flux variations are shown in Figure 6. As shown in Figure 6(a), when using the PID control scheme the displacement error is within the range of $\pm 16\,\mu m$, and the control resolution is $16\,\mu m$. The displacement error when using the proposed control scheme is within the range of $\pm 3\,\mu m$, and the control resolution is $3\,\mu m$. The motion precision of the proposed control system is higher than that of the PID control system. When using the conventional sliding-mode control scheme with the sgn function, sliding errors under the displacement sine response are shown in Figure 7. The left figure is s_1 defined in (17), and the right one is s_2 defined in (23). When using the sliding-mode control scheme with the $m$sat function, sliding errors under the displacement sine response are shown in Figure 8. The left figure is s_1 defined in (17), and the right one is s_2 defined in (23). By comparison, it can be seen that the chattering is very serious when using the conventional sliding-mode control scheme with the sgn function. However, the proposed control scheme using the $m$sat function removes the chattering.

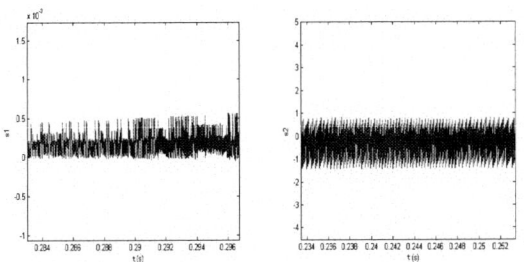

Figure 7. Sliding error using the sgn function under displacement sine response

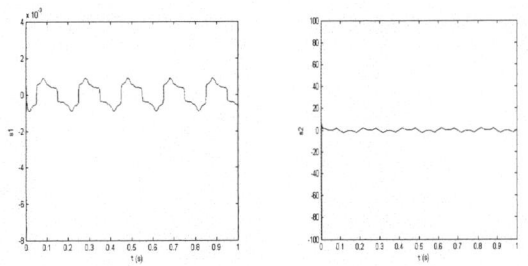

Figure 8. Sliding error using the $m$sat function under displacement sine response

V. CONCLUSIONS

Conventional control schemes may not satisfy the precision positioning requirements due to the uncertainties including the edge effect, parameter perturbation and external load. This paper creates a quasi-linearized and decoupled model derived from the mathematical model of PMLSM using the conception of feedback linearization. A fixed-boundary sliding control algorithm using the $m$sat function is employed to guarantee robustness.

In the simulation, the speed step response and displacement sine response are analyzed. The proposed control results show the strong robustness and good motion precision compared with the PID control results under disturbances. The results also show the proposed control scheme can remove the chattering compared with the conventional sliding mode control strategy. The validity of the proposed control scheme is confirmed.

REFERENCES

[1] C.Stephen and E.Boaz, "Linear motors: The future of high performance machine tools," *American Machinist*, vol.140, no.9, pp.44-48, 1996

[2] Alter, D.M. and Tsao, Tsu-Chin, "Control of linear motors for machine tool feed drives," *ASME Journal of Dynamic Systems, Measurement and Control*, vol.118, pp.649-656, 1996

[3] G. Otten, T. J. A. de Vries, J. van Amerongen, A. M. Rankers, and E.W. Gaal, "Linear motor motion control using a learning feedforward controller," *IEEE/ASME Trans. Mechatron.*, vol. 2, pp. 161–170, 1997.

[4] K.K.Tan, T.H.Lee, H.F.Dou, S.J.Chin, and S.Zhao, "Precision motion control with disturbance observer for pulsewidth-modulated-driven permanent-magnet linear motors," *IEEE Trans. Magnetics*, vol.39, no.3, pp.1813-1818, 2003

[5] K.K.Tan, T.H.Lee, S.Y.Lim, and H.F.Dou, "Learning enhanced motion control of permanent magnet linear motor," *in Proc. 3rd IFAC Int. Workshop Motion Control*, Grenoble, France, pp. 397–402,1998

[6] Isidori and Alberto, *Nonlinear Control Systems*, Berlin, New York : Springer-Verlag, 1995

[7] Q.D.Guo, C.Y.Chen, M.W.Zhou and T.Y.Sun, "Precision Motion Control Technology of Linear AC Servo Systems," *China Machine Press*, Beijing, China, 2000

[8] K.Ogata, Modern Control Engineering, *Prentice Hall PTR*, 1998

[9] P.Kachroo and M.Tomizuka, "Chattering reduction and error convergence in the sliding-mode control of a class of nonlinear systems," *IEEE Trans. Automatic Control*, vol.21, no.7, pp.1063-1068, 1996

[10] Q.L.Lee and X.K.Wang, "Design of permanent-magnet AC linear synchronic motor position servo control system," *China Mechanical Engineering*, vol.12, no.5, pp.577-581, 2001

2006 5th International Power Electronics and Motion Control Conference

Dynamic Analysis of PWM Switching DC-DC Converters

Liu Jian*, Wang Yuanbin *

* Xi'an University of Science & Technology/ School. of Electrical & Control Engineering, Xi'an, China
wangyb998@163.com

Abstract—The dynamic behavior of DC-DC converters is analyzed on the basis of a state-space averaged model including the feed back loop. The low frequency ripple behavior of DC-DC converters is analyzed in s filed. A buck DC-DC converter is used as an example. The condition causing oscillation and impulse during starting period and in case of input perturbation is obtained. The output to input voltage transfer function in the feedback loop is deduced. The stability requirement is obtained based on Hurwitz Stability Criterion. The frequency and amplitude of the low frequency ripple voltage are determined by using zero-pole analysis method. The proposed approach is verified by simulation and experiment results.

Keywords—switching converters; Pulse Width Modulation (PWM); dynamic analysis; stability; low frequency ripple voltage

I. Introduction

Switching DC-DC converters can be analyzed in steady state with averaged method [1]. In the case that the interruption is of low frequency and small magnitude, dynamic behavior of switching DC-DC converters can also be analyzed by averaged method [2][3]. Design consideration of a switching DC-DC converter is also based on above analysis [4].

Although great achievements have been made in switching DC-DC converter analysis and design, there are some problems that can't be solved by the published approaches. For instance, in some isolated switching DC-DC converters, a Voltage-Frequency-Converter (VFC) is used to convert the feed back voltage into a sequence of pulses whose density is modulated by the feed back voltage. A high frequency transformer is implanted to transfer the modulated pulses to the isolated side where a low-pass filter restores the feed back voltage. It is obvious that a delay is introduced in such feed back loop and sometimes results in a large ripple of much lower frequency in the output voltage of the converter making the converter noisy. It is not easy to predict whether above phenomena will occur by traditional method, let alone the frequency and amplitude of the ripple voltage and the way to avoid such phenomenon.

Further more, oscillation and impulse sometimes occurs on the output voltage during starting period and in case of input perturbation. It is necessary to determine whether oscillation and impulse will occur for designers.

Simulation is a feasible approach [5], but it can only obtain the dynamic behavior under certain parameters. The inherent regularities cannot be found.

In this paper, a Pulse-Width-Modulated (PWM) buck DC-DC converter is used as an example to discuss such problems.

II. Dynamic behavior in starting period and input perturbation

A model of the PWM buck DC-DC converter is shown in Fig.1, where the parasitic parameters of the inductor and capacitor are not considered. The PWM regulator is made up of a comparator (OA2) and a saw tooth voltage generator (TR) with the amplitude of the saw tooth voltage of V_p. OA1 is an amplifier with the gain of e. V_{ref} is the reference voltage.

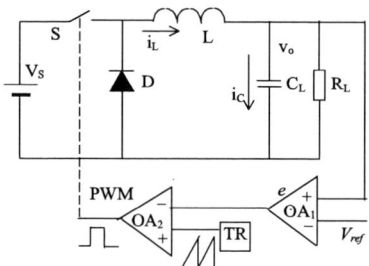

Fig1. Modeling a PWM buck DC-DC Converters

The duty ratio d is
$$d = 1 - e(v_0 - V_{ref})/V_p \qquad (1)$$

In Continuous Conduction Mode (CCM), when S is on and D is off, the increment of inductor current Δi_{L1} is
$$\Delta i_{L1} = (V_S - v_o)dT_S / L \qquad (2)$$
where T_s is switching cycle.

when S is off and D is on, the decrease of inductor current Δi_{L2} is
$$\Delta i_{L2} = v_o(1-d)T_S / L \qquad (3)$$

1-4244-0448-7/06/$25.00 ©2006 IEEE   272

Since i_L, i.e., the envelope of inductor current, varies much more slowly than a switching cycle, it can be written as

$$\frac{\partial i_L}{\partial t} = (\Delta i_{L1} - \Delta i_{L2})/T_S = (V_S d - v_o)/L \quad (4)$$

According to **KCL**, it can be have

$$i_L = i_C + v_0/R_L = C_L \frac{\partial v_o}{\partial t} + v_o/R_L \quad (5)$$

On the base of (5) and by derivation, it can be have

$$\frac{\partial i_L}{\partial t} = C_L \frac{\partial^2 v_o}{\partial t^2} + \frac{1}{R_L}\frac{\partial v_o}{\partial t} \quad (6)$$

From (1), (4) and (6), it can be have

$$LC_L \frac{\partial^2 v_o}{\partial t^2} + \frac{L}{R_L}\frac{\partial v_o}{\partial t} + \left(1 + \frac{eV_S}{V_P}\right)v_o = \left(1 + \frac{eV_{ref}}{V_P}\right)V_S \quad (7)$$

When the converter is in its stable state, $\frac{\partial^2 v_o}{\partial t^2} = \frac{\partial v_o}{\partial t} = 0$. Thus the output voltage in stable state $V_o(\infty)$ can be written as

$$V_o(\infty) = V_s(V_p + eV_{ref})/(V_p + eV_s) \quad (8)$$

The solution of (8) is

$$v_o(t) = k_1 e^{\lambda_1 t} + k_2 e^{\lambda_2 t} + V_o(\infty) \quad (9)$$

where λ_1 and λ_2 are characteristic value, k_1 and k_2 are constants.

In case of $4R_L^2 C_L(1 + eV_s/V_p) \le L$, λ_1 and λ_2 are of real values. Therefore, there is no oscillation in $v_o(t)$. In other words, neither up-impulse nor down-impulse occurs on $v_o(t)$ during starting period and in case of input perturbation.

In case of $4R_L^2 C_L(1 + eV_s/V_p) > L$, λ_1 and λ_2 are of complex values. Therefore, there must be oscillation and impulse on $v_o(t)$ during starting period and in case of input perturbation. In this case, it can be have

$$\lambda_1 = \lambda_2 = -\alpha \pm j\alpha \quad (10\text{-a})$$

$$\alpha = 1/2R_L C_L \quad (10\text{-b})$$

$$\omega = \frac{1}{2R_L C_L}\sqrt{\frac{4R_L^2 C_L(1 + eV_s/V_p)}{L} - 1} \quad (10\text{-c})$$

According to (10-c), the frequency of the oscillation on $v_o(t)$ during starting period and in case of input perturbation can be determined.

III. LOW FREQUENCY RIPPLE BEHAVIOR

Once a delay is inserted into the feed back loop of the converter, the output voltage sometimes is not stable, but with a remarkable low frequency ripple even in its stable state.

On the basis of a large number of experiment results, it can be found the followings: 1) There is no ripple voltage when the converter is in Discontinuous Conduction Mode (DCM). 2) The ripple voltage is due to the delay in feed back loop. 3) The envelopes of inductor current and ripple voltage are approximately sine. Thus the two parameters can be analyzed in s filed.

Let τ be the time constant of the delay in the feed back loop, the duty ratio d is

$$d = 1 - ev_o/V_p(\tau s + 1) \quad (11)$$

According to (7), we have

$$LC_L s^2 v_o(s) + \frac{L}{R_L}s v_o(s) = V_s\{1 - \frac{e[v_o(s) - V_{ref}]/(\tau s + 1)}{V_p}\} - v_o(s) \quad (12)$$

Based on (12), the s filed voltage transfer function can be made out, the characteristic equation of which is

$$LC_L V_p \tau s^3 + (LC_L V_p + \frac{L}{R_L}\tau V_p)s^2 + \frac{L}{R_L}V_p s + V_s e = 0 \quad (13)$$

According to Hurwitz Stability Criterion, the converter is stable if and only if

$$(LC_L V_p + \frac{L}{R_L}\tau V_p)(\frac{L}{R_L} + \tau)V_p - LC_L V_p \tau(V_s e + V_p) > 0 \quad (14\text{-a})$$

Conversely, the condition when the low-frequency output voltage ripple exists is

$$(LC_L V_p + \frac{L}{R_L}\tau V_p)(\frac{L}{R_L} + \tau)V_p - LC_L V_p \tau(V_s e + V_p) \le 0 \quad (14\text{-b})$$

In addition, whether the converter is stable may also be determined by the poles in s field. The converter is stable if and only if all of its poles are in left plane. In other words, if there is at least one pole in right plane, the low-frequency output voltage ripple exists.

When there is a low-frequency output voltage ripple, the converter has three poles as the following

$$s_1 = \alpha + j\omega \quad s_2 = \alpha - j\omega \quad s_3 = S \quad (15)$$

where, s_1, s_2 and s_3 can be made out by solving (13) with Cardano's Formula. Therefore the frequency of ripple voltage is ω(rad/s).

The output voltage with ripple can be approximately written as

$$v_o(t) = V_1 + V_2 \sin \omega t \quad (16)$$

where, V_1 is the DC component of the output voltage, V_2 is the amplitude of ripple voltage.

The feed back voltage v_f can be written as

$$v_f = k[V_1 + V_2 \sin(\omega t + \varphi)] \quad (17)$$

where φ and k are the phase delay and the gain of amplitude due to the feed back loop, respectively. So, it can be have

$$tg\varphi = -\omega\tau \quad (18)$$

$$k = e / \sqrt{1 + \omega^2 \tau^2} \qquad (19)$$

$$d = 1 - v_f / v_p \qquad (20)$$

According to (4) and (16)~(20), it can be have

$$\frac{\partial i_L}{\partial t} = \frac{V_s}{L}\{1 - \frac{k[V_1 + V_2 \sin(\omega t + \varphi)]}{V_p}\} - \frac{1}{L}(V_1 + V_2 \sin \omega t)$$

(21)

It can be seen from (21) that the envelope of inductor current i_L must meet

$$i_L = \beta_1 \cos \omega t + \beta_2 \cos(\omega t + \varphi) + \beta_3 \qquad (22)$$

and

$$\frac{\partial i_L}{\partial t} = -\omega \beta_1 \sin \omega t - \omega \beta_2 \sin(\omega t + \varphi) \qquad (23)$$

According to (21) and (23), it can be have

$$\beta_1 = V_2 / L\omega \qquad (24\text{-a})$$

$$\beta_2 = V_s k V_2 / LV_p \omega \qquad (24\text{-b})$$

$$\beta_3 = \frac{1}{\tau}\int_0^\tau i_L dt = V_1 / R_L \qquad (24\text{-c})$$

Since the inductor current envelope is near a sine in the stable state, the DC component in (21) must be zero. Thus, it can be have

$$V_1 = \frac{V_s}{1 + \frac{eV_s}{V_p \sqrt{1 + \omega^2 \tau^2}}} \qquad (25)$$

In theory, if there are poles in right plane, the amplitude of output ripple voltage will infinitely increase. In the light of (5), with the increment of the amplitude of output ripple voltage, a negative value of iL should appear. But according to Fig.1, the current through the inductor can not be negative since there is a diode D. In other words, iL will remain zero in some period and therefore restrain the increment of the amplitude of output ripple voltage. Finally, the amplitude of ripple voltage keeps at a certain value while the wave trough of the envelope of the current through the inductor just reaches zero, i.e., $\omega t = \pi$. In this case, substituting β1, β2 and β3 into (22), it can be have

$$-\frac{V_2}{L\omega} - \frac{V_s k V_2}{LV_p \omega}\cos \varphi \approx -\frac{V_1}{R_L} \qquad (26)$$

From (26), it can be have

$$V_2 = \frac{V_1 L\omega}{R_L(1 + \frac{eV_s}{V_p \sqrt{1 + \omega^2 \tau^2}}\cos \varphi)}$$

(27)

As for the case when the converter is stable, the output voltage v_o can be considered as a constant value, which is described in (8).

IV. EXPERIMENT AND SIMULATION RESULTS

The proposed analysis is verified by SPICE simulation and experiments, respectively. In the experiments and simulation, e=1.0, VP=3V, fs=100kHz. In the experiment, TL494 is used as the PWM regulator, IRF9640 is used as the power switch S.

Experiment results of the dynamic behavior during the starting period are shown in Fig.2(a) and (b), respectively.

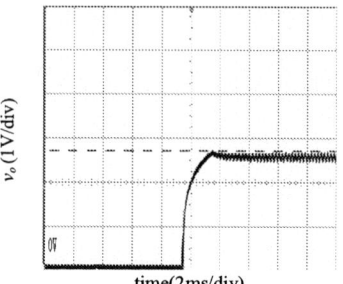

time(2ms/div)

(a) output voltage without oscillation

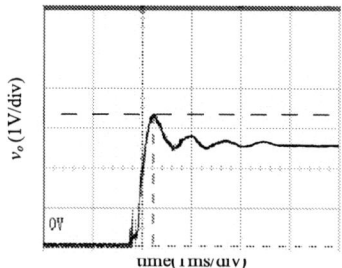

time(1ms/div)

(b) output voltage with oscillation

Fig.2 Experiment results of the dynamic behavior during the starting period

In Fig.2(a), V_S=10V, L=900μH, C_L=50μF, R_L=1Ω, V_{ref} =0V. It can be seen by calculation that $4R_L^2 C_L(1 + eV_s/V_p) < L$. There should be no oscillation and impulse on the output voltage. The experiment and simulation results are consistent with the analysis.

In Fig.2(b), V_S=8V, L=400μH, C_L=100μF, R_L=5Ω, V_{ref} =0V. It can be seen by calculation that $4R_L^2 C_L(1 + eV_s/V_p) > L$. There should be oscillation and impulse on the output voltage, which is proved by the experiment and simulation results. The frequency of oscillation can be calculated by (10-c). The theoretical value is 9521rad/s. The experiment and simulation results are 9091 rad/s and 9751 rad/s, respectively, which are consistent with the analysis.

Experiment results of the low frequency ripple voltage are shown in Fig.3(a) and (b), respectively.

In Fig.3(a), V_S=10V, L=330μH, C_L=330μF, R_L=5Ω, V_{ref}=5V, τ=12μs. It can be seen by calculation that (14-a) is satisfied. Thus the converter is stable without ripple voltage. Results of simulation and experiment are consistent with the analysis.

In Fig.3(b), V_S=10V, L=330μH, C_L=330μF, R_L=5Ω, V_{ref}=5V, τ=30μs. It can be seen by calculation that (14-a) is not satisfied. Thus the converter is unstable with remarkable ripple voltage on the output voltage. Results of simulation and experiment are consistent with the analysis. The theoretical value of DC component in output voltage is 6.15V, while the simulated result is 6.07V and the experiment result is 5.80V. The theoretical ripple frequency is 6224rad/s. The experiment and simulation results are 6798 rad/s and 6389 rad/s, respectively,.

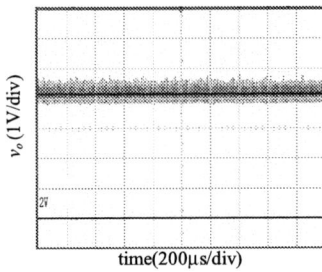

(a) output voltage without low-frequency ripple

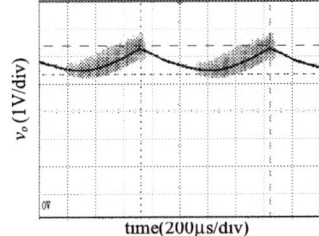

(b)   output voltage with low-frequency ripple

Fig.3 Experiment results of low frequency ripple voltage

V.   CONCLUSIONS

The dynamic behavior of DC-DC converters can be analyzed on the basis of a state-space averaged model including the feed back loop.

The low frequency ripple behavior of DC-DC converters can be analyzed in s filed.

As for a buck DC-DC converter in CCM, neither oscillation nor impulse occurs on the output voltage during starting period and in case of input perturbation if $4R_L^2 C_L(1+eV_s/V_p) \leq L$. Otherwise, there must be

oscillation and impulse with the frequency described in (10-c).

The delay in feed back loop sometimes results in a large low frequency voltage ripple in the output of the converter. As for a buck DC-DC converter, the low-frequency output voltage ripple occurs in case of

$$(LC_L V_p + \frac{L}{R_L}\tau V_p)(\frac{L}{R_L}+\tau)V_p - LC_L V_p \tau(V_s e + V_p) \leq 0$$

The proposed analysis is verified by the results of simulation and experiments.

REFERENCE

[1]   R.D.Middle brook and S. Cuk, A general unified approach to modeling switching—converters power stage, IEEE, PESC Rec. 1976, pp.18 -34

[2]   B. Lehman and R. M. Bass, Extensions of averaging theory for power electronic systems. IEEE trans. On Power Electronics, Vol.11, No.4, 1996, pp.542-553

[3]   F.Guinjoan, J.Calvente, A.Poveda, et al, Large signal modeling and simulation of switching DC-DC converters, IEEE trans. On Power

[4]   Cai Xuansan and Gong Shaowen，High Frequency Power Electronics，Bei Jing，Science Press，1993, pp361—363, in Chinese

[5]   Siu Chung Wong and Yim Shu Lee, SPICE modeling and simulation of hysteretic current controlled Cuk converter, IEEE trans. on Power Electronics, Vol.8, No.4, 1993, pp.580-587.

[6]   Hu Shousong, Principle of Automatic Control，National Defense Industry Press, 1994, Beijing, pp.105-107, in Chinese

A BRIEF INTRODUCTION OF THE AUTHORS

LIU JIAN (M'96, SM'01) received his Ph.D degree in Electrical Engineering in 1997. He is a professor and the chairman of academic committee of Xi'an University of Science and Technology . He is also a visiting professor of other three universities including North-East University , North-west University and Xi'an University of Technology. He has published six books and over 130 papers. His research interest focus on electrical engineering.

WANG YUANBIN received her master degree on Electrical Engineering in 2003. Her research interest is power electronics. Now she is a Ph.D candidate.

2006 5th International Power Electronics and Motion Control Conference

A Novel LLC Resonant Converter Topology: Voltage Stresses of All Components in Secondary Side Being Half of Output Voltage

Yilei Gu, Zhengyu Lu(Senior member, IEEE), Zhaoming Qian(Senior member, IEEE)
College of Electrical Engineering, Zhejiang University, Hangzhou 310027, China
Email: guyilei9602@hotmail.com

Abstract—Paper presents a novel LLC resonant converter topology, where voltage stresses of all components in secondary side are half of output voltage. The advantages, which are featured in conventional LLC converter, i.e. being easy to obtain full range ZVS for primary side switches, being easy to obtain ZCS for secondary side diodes, resonant inductors and transformer being easy to be integrated and wide range application, are taken on in the presented topology as well, furthermore the prominent advantages that voltage stresses of secondary side diodes and secondary side capacitors are half of output voltages are taken on. Therefore this converter is fit for application with the characteristic of high output voltage. Evolvement of this topology, detailed operation principle and design considerations of key parameters are detailed analyzed. Finally, an experimental prototype of 1100W is built to verify the reliability and practicability of the proposed topology. Efficiency of the prototype with full load is above 96%.

I. INTRODUCTION

Due to disaffinity of numbers of components for various types of converter topologies and moreover, difference of voltage and current stress, magnet induction intensity and temperature stress for each component, each types of topologies are appropriate to various application occasion [1~3]. Forward, flyback and etc, voltage platform of diodes included on secondary side for such types of topologies is about twice of output voltage, in addition, considering the voltage peak resulted by reverse recovering process, voltage stress is considered to be 4 times of output value. Besides for control type soft-switching topologies, for example, phase-shifted full-bridge and asynchronous half-bridge converter, voltage stress of diodes for secondary side is about 4 times of output value as well (considering reverse recovering process). If bridge type of rectifier is adopted on secondary side of phase-shifted full-bridge converter and asynchronous half-bridge converter, voltage stress of diode for secondary side can be reduced to twice of output value.

Switches of LLC resonant converter operates on control type soft-switching state, furthermore, diodes of secondary side turn off under state of zero current, therefore no voltage peak appears. Once, bridge type of rectifier is adopted on secondary side, voltage stress of diode equals to output voltage. Compare with the above mentioned topologies, the very type of topologies is appreciate to such application with

the characteristic of high output voltage.

a novel LLC resonant converter topology, where voltage stresses of all components in secondary side are half of output voltage is a preferred candidate for such application with characteristic of high output voltage, is presented. Appropriate diode and output capacitor can be picked out in such application with high output voltage. Reliability and practicability of the very topologies is verified by a experimental prototype of 1100w.

II. TOPOLOGY EVOLVEMENT

Diodes of secondary side turn off under zero current state, therefore problem of reverse recovery does not occur, and correspondingly, no voltage peak appears. LLC resonant converter is a preferred candidate for application with characteristic of high output. However, for being adopted in application of high output, optimization of LLC resonant topology is required for reducing the voltage stress of secondary component.

Current type zero-mode rectifier, which is shown in Fig.1, is commonly adopted in secondary configuration of LLC resonant converter. The gain of current type bridge-mode rectifier, which is shown in Fig.2, compared with the above rectifier, increases by one time and is much more applicable to application of high output. Trapezoid type voltage-double rectifier, which is shown in Fig.1 (c), can be adopted in secondary circuit of LLC resonant converter as well, compared with current type bridge-mode rectifier, gain of whole topology increases one time. In addition, voltage stress of C2 and C3 is respectively Vo/2 and Vo, voltage stress of each diode is Vo, compared with the above two rectifier configuration, the very type is furthermore appropriate to application of high output. However, once voltage stress of C3 and every diode can be reduced to half of output voltage, the converter can be adopted in application of further high output. The evolution process of topology is shown in Fig.1 (d) and (e). First of all, two components are in series to substitute the original one, then, D4 is removed to the bottom line, finally, the respective midpoints of C3 and C4, D1 and D3 are connected with each other. The novel topology can be named rectifier configuration with the characteristic of minimized voltage stress. Two parameters of voltage stress and gain of the four topologies are compared in Table.1. The whole novel LLC resonant topology is shown in Fig.2.

Project supported by National Natural Science Foundation of China (50237030ZD)

1-4244-0448-7/06/$25.00 ©2006 IEEE

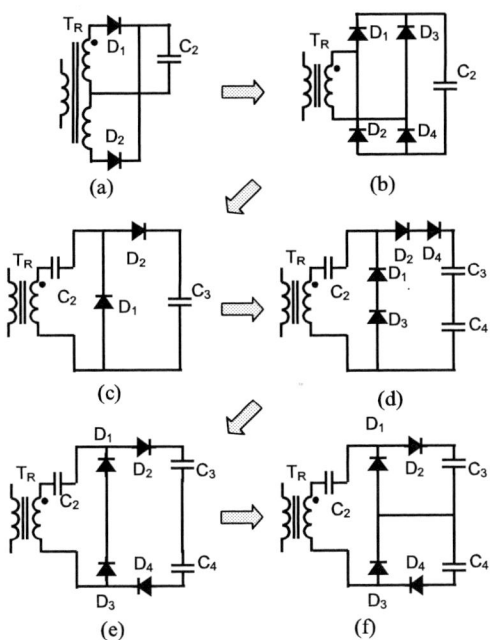

Fig.1 Evolvement of the topologies

Tab.1 Voltage stress and DC gain of various LLC resonant converters

	Voltage stress of Capacitor	Voltage stress of Diode	Gain
a	Vo	2Vo	M
b	Vo	Vo	2M
c	Vo、Vo/2	Vo	4M
f	Vo/2	Vo/2	4M

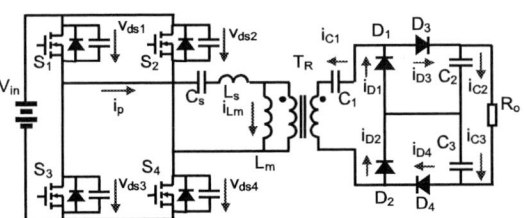

Fig.2 Topology of LLC resonant converter with minimized voltage stress for secondary side

III. OPERATION PRINCIPLE

Operation process of the proposed converter can be divided into eight stages during a switching cycle. The timing diagram is shown in Fig.3 and the equivalent circuit for each stage of operation is shown in Fig.43. These stages are described below.

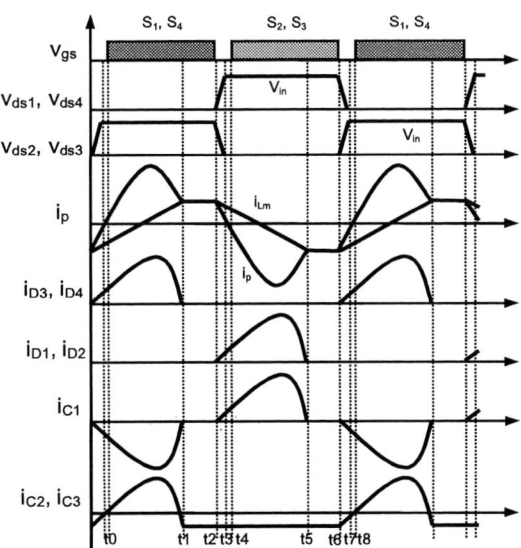

Fig.3 Principle waveform of the proposed converter

Stage 1 [t0, t1]: at the time of t0, S1 and S4 are turned on simultaneously at t0. The primary resonant current ip flows through S1 and S4 and increases by sine-wave type due to the resonance between Ls and Cs. D3 and D4 are turned on and secondary side winding of transformer is clamped by the sum value of voltage across C1,C2, and C3. So, the magnetizing current iLm increases linearly from negative to positive peak. The output current is a ratio of the difference between resonant current ip and magnetizing current iLM. At the time of t1, ip and iLm equals to each other, the period finishes.

Stage 2 [t1, t2]: due to equalization of ip and iLm, current circulating through D3 and D4 naturally decreases to zero, the diodes turn off under condition of zero current, reverse recovery does not occur. Correspondingly, voltage sum across C1, C2 and C3 does not clamp the transformer, therefore, Lm becomes free resonant inductor, resonance occurs between the resonant tank of Lm, Ls and Cs. This resonant cycle is much longer than the previous one because Lm is much larger than Ls. It can be recognized that Lm is a constant current source. The stage is over once S1 and S4 are turned off.

Stage 3 [t2, t3]: At the moment of t2, S1 and S4 are turned off, primary current ip charges parasitic capacitors of S1 and S4, simultaneously, discharges the parasitic capacitors of S2 and S3. Therefore, D1 and D2 turn on at the output end, correspondingly, voltage across C1 clamps the transformer. Once, voltage across the parasitic capacitors of S2 and S3 decreases to zero, the very stage is over.

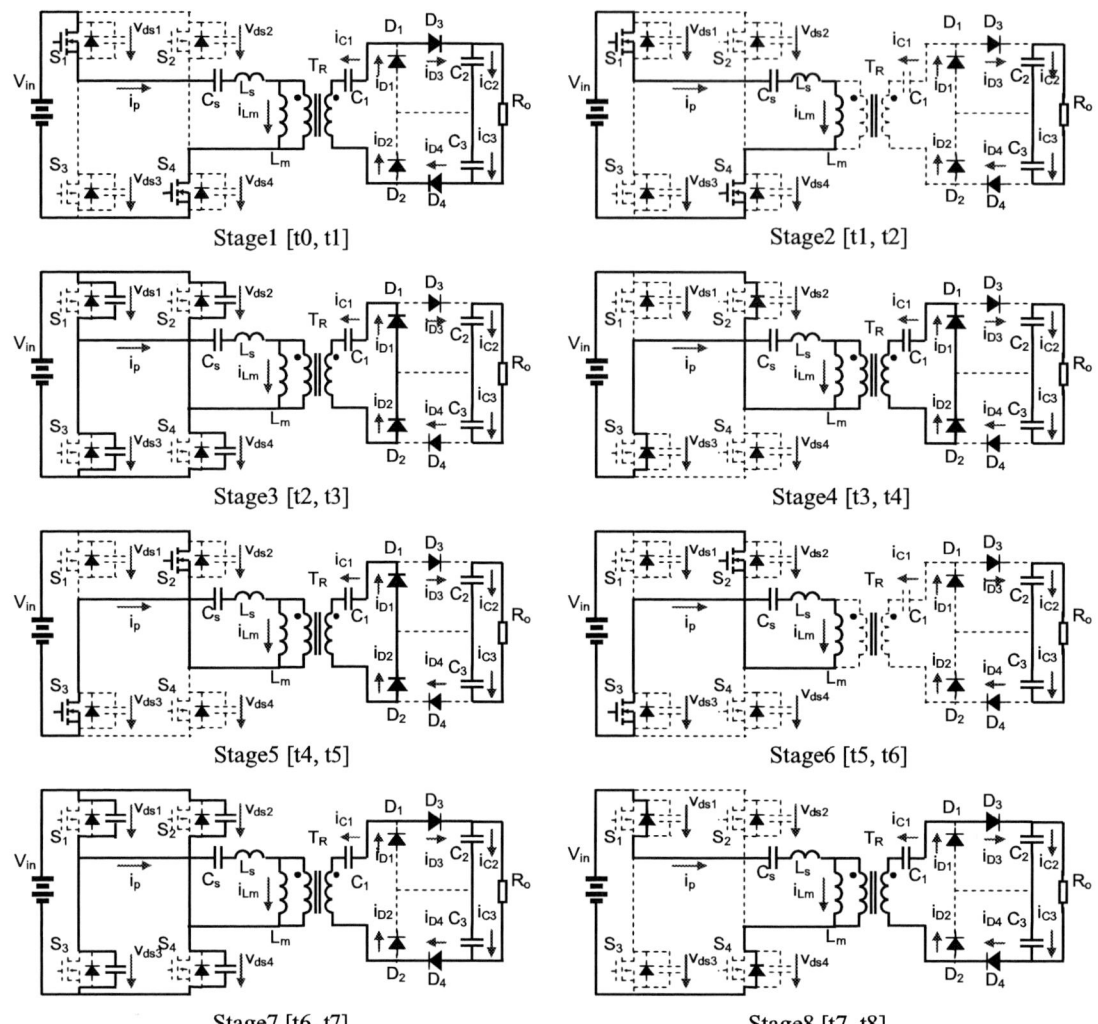

Fig.4 Equivalent circuit for each stage

Stage 4 [t3, t4]: Voltages across S2 and S3 are already decreased to zero in the previous stage, therefore ZVS is achieved, primary current is carried by body diodes of S2 and S3, and reversely increases as sinusoid. Due to the conduction of body diodes, voltage across drain to source keeps zero, ZVS is achieved for S2 and S3. Once, S2 and S3 turn on, this stage is over.

Stage 5 [t4, t5]: At the moment of t4, S2 and S3 turn on, primary resonant current ip, which circulates through S2 and S3, increases reversely as sinusoid mode. At the output end, D1 and D2 turn on, transformer is clamped by the voltage across C1. Therefore, resonance occurs between Ls and Cs, magnetizing current iLm, which circulates through Lm, decreases linearly. Once ip equals to iLm, the stage is over.

Stage 6 [t5, t6]: at the moment of t5, due to equation between ip and iLm, current circulating through D1 and D2 decrease to zero naturally, therefore, almost no reverse

recovery appears. At the moment, voltage across C1 does not clamp the transformer, and Lm becomes free inductor, correspondingly, resonance occurs between Lm, Ls and Cs. Once S2 and S3 are turned off, the stage is over.

Stage 7 [t6, t7]: at the moment of t6, S2 and S3 is turned off, primary current ip charges parasitic capacitors of S2 and S3, simultaneously, discharges the parasitic capacitors of S1 and S4. D3 and D4 turn on at the output end. Once, voltage across the parasitic capacitors of S1 and S4 decreases to zero, the very stage is over..

Stage 8 [t7, t8]: Voltages across S1 and S4 are already decreased to zero in the previous stage, therefore ZVS is achieved, primary current is carried by body diodes of S1 and S4, and reversely increases as sinusoid. Due to the conduction of body diodes, voltage across drain to source keeps zero, ZVS is achieved for S1 and S4. Once, S1 and S4 turn on, this stage is over.

According to the previous analysis, both ZVS for switches of primary side and ZCS for diodes of secondary is achieved. Voltage sum across C2 and C3 equals to output, according to waveform symmetry, voltage across C1 equals to half of output.

Provided that reverse recovery occurs for D1~D4, moreover, reverse recovery current and time for every one is discrete, then voltage balance between C2 and C3 is broken. However, secondary diode operates under the condition of ZCS and reverser recovery does not occur, therefore, voltage balance for C2 and C3 can be easily achieved, a relatively large resistor is required to respectively parallel with the capacitor, and that conversion efficiency will not be influenced by power loss of resistors(power loss of resistors in the following experiment is less than 0.2% of output power).

IV. DESIGN CONSIDERATION

For primary side, operating process of the proposed converter is similar to the conventional LLC resonant converter, therefore, design consideration, which is specified in [4~9], is similar. Design consideration of secondary side is analyzed in the following text.

a) SELECTION OF SECONDARY CAPACITOR

Resonant period of Ls and Cs, Ts can be expressed as follows:

$$T_s = 2\pi\sqrt{L_s C_s} \qquad (1)$$

Three capacitors are involved in LLC resonant converter with characteristic of minimized voltage stress on secondary side, and voltage stress of them is respective half of output, which makes it more applicable to application of high output. Current waveform of the capacitors is shown in Fig.3.

According to Fig.3, relation of switching ripple peak-peak value of C1, i.e. ΔU and its capacitance C can be expressed as follows:

$$\Delta U_1 = \frac{T I_o}{C} \qquad (2)$$

Relation of switching ripple peak-peak value of C2 and C3, i.e. ΔU and their capacitance C can be expressed as follow:

$$\Delta U_{2,3} = \frac{(T - T_s/2)I_o}{C} \qquad (3)$$

b) SELECTION OF SECONDARY DIODES

According to the previous analysis, no voltage peak occurs due to the diodes on secondary side operates on ZCS state, therefore, voltage stress is just the voltage platform, which equals to half of output and that contributes to high output. Analysis of current stress for diode is relatively complex. Current circulating through diodes equals to difference between a half of sinusoid current and the magnetizing current. For convenience of calculation, the current is assumed to be a half sinusoid.

Average current of diode equals to output current Io. Peak current value can be expressed as

$$I_{Dm} = \frac{\pi T I_o}{T_s} \qquad (4)$$

V. EXPERIMENTAL RESULT

Tab.2 Conversion efficiency at different load current under various input voltage

	0	2A	4A	6A	8A	10A
400V	8W	95.7%	97.1%	96.8%	96.6%	96.4%
380V	8.3W	95.8%	97%	97%	96.8%	96.4%
360V	8.6W	95.6%	97%	97%	96.4%	96.1%
340V	8.6W	95.6%	97%	96.8%	96.2%	96.1%
320V	8.8W	95.5%	96.8%	96.9%	96.2%	95.6%
300V	9W	95.5%	97.2%	96.8%	96.2%	95.7%
280V	9W	95.4%	96.9%	96.5%	95.5%	94.9%
260V	9.3W	95.2%	97.1%	96.4%	95.6%	94.4%
240V	9.6W	95.4%	97.1%	96.4%	95.2%	93.2%

Tab.3 Switching frequency at different load current under various input voltage

	0	2A	4A	6A	8A	10A
400V	247	201.8	199.6	197.5	194.6	193.1
380V	218	173.6	171.5	170.6	168.2	167.3
360V	170.7	154.6	153.2	152.2	150.1	148.8
340V	149.8	140.4	139.2	137.6	136.2	134.7
320V	136.3	129.2	127.7	126.6	125	123.2
300V	126	121.8	118.5	116.5	114.5	113.3
280V	118	114.7	110.6	108.6	106.4	104.6
260V	110.8	108.1	104.1	101.3	99.8	97.6
240V	104.8	102.6	98.3	95.6	93.6	92.3

An experimental prototype, which is designed according to specification of power operation supply by adopting the LLC resonant converter with characteristic of minimized voltage stress on the secondary side, verifies the above analysis. Specification and main parameters are as follows:
Input voltage Vin: 240V~400V
Output voltage Vo: 220V;
Output current Io: 0~5A
Main switches S1, S2, S3, S4 : IRF460
Rectifier diodes D1~D4: B20150
Transformer TR: n=46:12,
Magnetizing inductance: 260uH,
Leakage inductance: 55uH,
Resonant inductance Lm and Ls are integrated in TR.
Resonant capacitor C1: 19.8nF

For keeping voltage balance between the output capacitors C2, C3 and limiting the operating frequency without load, 10K resistor is respectively parallel with C2 and C3. Total power dissipation of the two resistors is about 0.2% of full load and that will not result in reduction of total conversion efficiency.

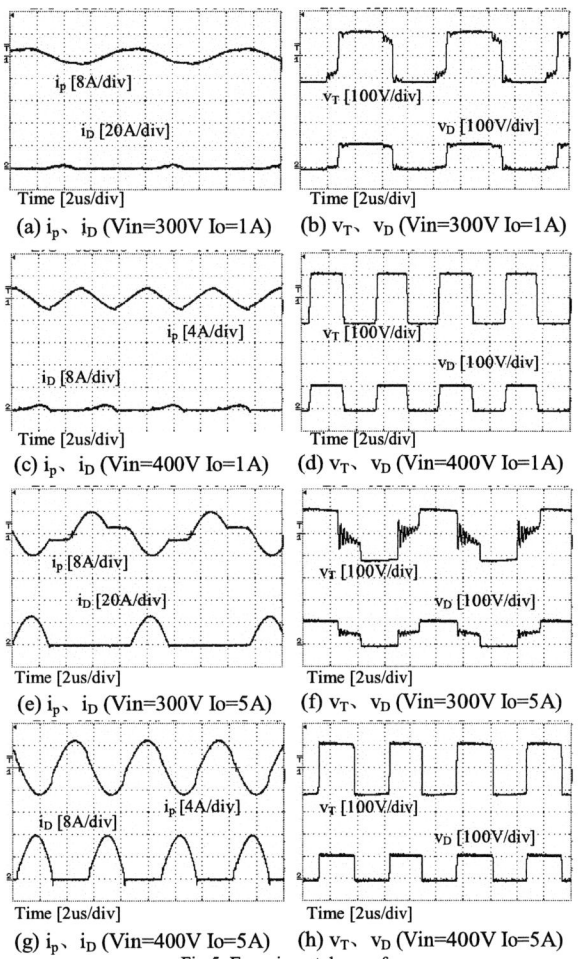

(a) i_p、i_D (Vin=300V Io=1A)　　(b) v_T、v_D (Vin=300V Io=1A)

(c) i_p、i_D (Vin=400V Io=1A)　　(d) v_T、v_D (Vin=400V Io=1A)

(e) i_p、i_D (Vin=300V Io=5A)　　(f) v_T、v_D (Vin=300V Io=5A)

(g) i_p、i_D (Vin=400V Io=5A)　　(h) v_T、v_D (Vin=400V Io=5A)

Fig.5 Experimental waveform

Experimental value of conversion efficiency of different input and different load is detail shown in Tabel.2 (wherein, when the load is empty, the data shown is the empty loss). It

is indicated that the full load efficiency reaches 96% at high line input. Experimental value of switching frequency of different load and different input is shown in Table.3.

Experimental waveforms of different input and different load are shown in Fig.5. It is obvious that experimental waveform matches the analysis perfectly. ZVS for primary side and ZCS for secondary side are both achieved, moreover, voltage stress of all components on the secondary side is half of output.

VI.　CONCLUSION

The proposed novel LLC resonant converter topology, possesses the merits of conventional LLC resonant converter, furthermore, voltage stress of diodes and capacitors on the secondary side is half of output, which makes it to be further applicable to high output. In addition, soft switching operation for all of the semiconductor devices is achieved, which contributes to make the very topology to be the preferred candidate in application with the characteristic of high frequency and high power density.

REFERENCES

[1]　Y. L. Gu, Z. Y. Lu, Z. M. Qian, "DC/DC Topology Selection Criterion," *in record, IEEE IPEMC,* 2004, pp.508-512
[2]　Tan F. D, "The forward converter: from the classic to the contemporary," IEEE APEC 2002, pp. 857 -863 vol.2.
[3]　S. Hamada, M. Nakaoka, "Analysis and Design of a Saturable Reactor Assisted Soft-Switching Full-Bridge DC-DC Converter," IEEE PESC Proceedings, 1991: 155-162.
[4]　Guisong Huang, Alpha J Zhang, Yilei Gu. "LLC series resonant DC-to-DC converter," US Patent, No.: 6344979, Feb. 5, 2002.
[5]　Bo Yang, Yuancheng Ren, Fred C. Lee, "Integrated magnetic for LLC resonant converter," IEEE APEC Proceedings, 2002, pp. 346-351.
[6]　Bo Yang, Fred C. Lee, Alpha J. Zhang, Guisong Huang,"LLC resonant converter for front end DC/DC conversion," IEEE APEC Proceedings, 2002, pp. 1108-1112.
[7]　Bo Yang, Fred C. Lee, Matthew Cancannon, "Over current protection methods for LLC resonant converter," IEEE APEC Proceedings, 2003, pp. 605-609.
[8]　E.G. Schmidtner, "A high frequency resonant converter topology", In Conference record of High Frequency Power Conversion 1988, pp.390-403.
[9]　Ashoka K. S. Bhat, "Analysis and design of LCL-type series resonant converter", IEEE Transactions on Industrial Electronics, 1994, Vol.41 No.1, pp.118-124.

2006 5th International Power Electronics and Motion Control Conference

On the hybrid automaton models and control synthesis of a single inductor, double output boost converter

Sreekumar C* and Vivek Agarwal**

* IDP in Systems and Control Engg, IIT Bombay, Mumbai, India- 400 076.
** Electrical Engg. Department, IIT Bombay, Mumbai, India- 400 076.

Abstract— In a Single Inductor Double Output (SIDO) boost converter, power is fed to both the outputs through the same inductor and hence the control requirements on output voltages are inter-dependent, causing complexity in modeling and control design of such converters. But, there is some flexibility in disguise due to the increased number of controlled devices and storage elements. This is utilized in this paper, to arrive at different hybrid automaton models for representing the SIDO boost converter. The features of each representation are highlighted. For these models, a safe spherical region around the required set point is identified using the stability requirements for a given set of disturbances. Inside the safe region, hybrid control laws are proposed to regulate the two different outputs at their required levels. Simulation studies are carried out in MATLAB/SIMULINK and the results are presented. The response of the SIDO converter, under various disturbances, is found to be satisfactory under the proposed control scheme.

Keywords-SIDO boost converter; voltage regulation; stability; hybrid automaton; Inner product; continuous current mode; discontinuous current mode.

I. INTRODUCTION

Inductors are the most critical elements in determining the bulkiness of any Switched Mode Power Supply (SMPS). Therefore, in the design of any SMPS, considerable efforts are made to reduce the inductor size to get considerable savings in cost, weight and size. The same idea of miniaturization is used in a Single Inductor Double Output (SIDO) boost converter, which is an ideal choice for applications where two outputs at different, but, higher voltage levels than the input voltage are required. A family of Single Inductor Multiple Output (SIMO) switching converters has been proposed as a means of miniaturization in power processing applications and their configurations are presented in the literature [1-3]. Unfortunately, the merits of reduced bulkiness get undermined due to their modeling and control design complexities. Though the literature covers several topologies of SIMO converters, they all suffer from a lack of good model for control design or a good control algorithm. A hybrid automaton model has been

proposed by Matthew Senesky *et al.* [4], as an extension of their work on hybrid control of boost power converter. But the paper does not explore the various topologies used in SIDO converters. Also the analysis is restricted to Continuous Current Mode (CCM) operation [5] and the robustness and stability of the converter under various disturbances are not studied. A more general model including the Continuous Current Mode (CCM) and Discontinuous Current Mode (DCM) operation of the converter and the control design for a boost converter is given in [6].

In this paper, various switching possibilities, among the different configurations of the SIDO boost converter, are investigated. This leads to different hybrid automaton representations for the converter. Control design using such a model of a CCM operated dual boost converter is already reported [4]. Our approach is to extend a similar concept to the other possible hybrid automaton models. Both, CCM and DCM operation of converter are considered for modeling and control design. At first, a safe set, which defines an area around the steady state point in the feasible state space, is determined. Then the controller design is carried out for each representation based on the stability and voltage regulation requirement. The proposed scheme is implemented using SIMULINK and the results are presented.

II. HYBRID MODELING AND AUTOMATON REPRESENTATION

The hybrid modeling of any DC-DC power converter preserves the identity of the individual configurations. The hybrid model captures the exact behavior of the circuit, without any approximation . Consider a DC-DC SIDO boost converter shown in Fig.1. As there are three switching elements in the circuit, eight discrete states are possible. They are: q1 (SW1 on, SW2 off, SW3 off), q2 (SW1 off, SW2 on, SW3 off), q3 (SW1 off, SW2 off, SW3 on) , q4 (SW1 off, SW2 off, SW3 off), q5 (SW1 on, SW2 on,SW3 on), q6 (SW1 on, SW2 on, SW3 off), q7 (SW1 on, SW2 off, SW3 on) and q8 (SW1 off, SW2 on, SW3 on). Out of these, q5, q6, q7 and q8 are not feasible. So the set of possible discrete states is, Q = (q1, q2, q3, q4). The various possibilities for the set of events are: E_1 = [(q1, q2), (q2, q1), (q1, q3), (q3, q1)], E_2 = [(q1, q2),

1-4244-0448-7/06/$25.00 ©2006 IEEE   281

(q2, q1), (q1, q3), (q3, q1), (q2, q3), (q3, q2)], E_3 = [(q1, q2), (q2, q4), (q4, q1), (q1, q3), (q3, q4)] and E_4 = [(q1, q2), (q2, q4), (q4, q1), (q1, q3), (q3, q4), (q2, q3), (q3, q2)]. Out of these, E_1 and E_2 represent the CCM operation. E_1 defines the multiplexed form of independent operation of two boost converters. In E_2, two additional events are permitted for achieving the desired control objective of regulating the output voltage. Similarly, E_3 and E_4 represent the counterparts of E_1 and E_2 in DCM case. $E_1 \cup E_2$ can be considered as the set of events for describing both the CCM and DCM operation of a SIDO boost converter, employing multiplexed independent converter operation. Similarly, $E_3 \cup E_4$ renders a more general and flexible set of events, capturing the CCM and DCM dynamics in converter operation. In this paper, two hybrid models are developed, one with $E_1 \cup E_2$ and the other with $E_3 \cup E_4$ as the feasible events of the system. The control algorithm is set to decide on the various switching possibilities in the converter.

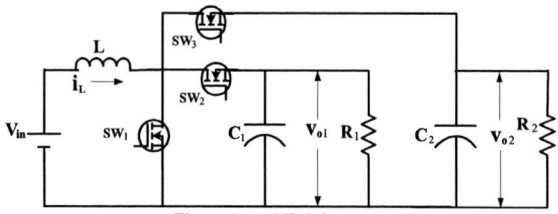

Figure 1. A SIDO boost converter.

For the analysis and design, let the states of the system be defined as $x = [i_L, v_{o1}, v_{o2}]$ where i_L is the inductor current, v_{o1} and v_{o2} are the two different output voltages of the converter appearing across R_1 and R_2. Then, the system can be represented in terms of three state equations corresponding to q_i, (i = 1, 2, 3, 4) as:

$$\dot{x}(t) = A_i x(t) + B_i = f_i(x(t)) \qquad (1)$$

where the matrices A_i and B_i for i=1, 2, 3, 4 are as shown in Table-1.

Table 1: System matrices under different modes.

Operating mode	A_i	B_i
1 (for q_1)	$\begin{pmatrix} 0 & 0 & 0 \\ 0 & -1/R_1C_1 & 0 \\ 0 & 0 & -1/R_2C_2 \end{pmatrix}$	$\begin{bmatrix} V_{in}/L & 0 & 0 \end{bmatrix}^T$
2 (for q_2)	$\begin{pmatrix} 0 & -1/L & 0 \\ 1/C_1 & -1/R_1C_1 & 0 \\ 0 & 0 & -1/R_2C_2 \end{pmatrix}$	$\begin{bmatrix} V_{in}/L & 0 & 0 \end{bmatrix}^T$
3 (for q_3)	$\begin{pmatrix} 0 & 0 & -1/L \\ 0 & -1/R_1C_1 & 0 \\ 1/C_2 & 0 & -1/R_2C_2 \end{pmatrix}$	$\begin{bmatrix} V_{in}/L & 0 & 0 \end{bmatrix}^T$
4 (for q_4)	$\begin{pmatrix} 0 & 0 & 0 \\ 0 & -1/R_1C_1 & 0 \\ 0 & 0 & -1/R_2C_2 \end{pmatrix}$	$\begin{bmatrix} 0 & 0 & 0 \end{bmatrix}^T$

In the present work, the two hybrid automaton models developed are used independently for the analysis and control design of a SIDO boost converter. The hybrid automaton [4,6-8] is a 6 tuple collection, $H = (Q, X, f, I, E, G)$ where $Q = q_1, \ldots q_N$ is a set of discrete states; $X \subset \mathbb{R}^n$ is the continuous state space; $f: Q \rightarrow (X \rightarrow \mathbb{R}^n)$ assigns to every discrete state a Lipschitz continuous vector field on X ; $I: Q \rightarrow 2^X$ assigns each $q \in Q$ an invariant set; $E \subseteq Q \times Q'$ is a collection of discrete transitions; $G: E \rightarrow 2^X$, $e = (q,q') \in E$, is a guard.

During the converter operation, let the inductor current 'ramping-up' phase be defined as mode-1 (σ_1), the current 'ramping-down' phase through R_1 as mode-2 (σ_2), the current 'ramping-down' phase through R_2 as mode-3 (σ_3), and the zero current phase as mode-4 (σ_4). Hence, $\Sigma = (\sigma_1, \ldots \sigma_4)$ represents the discrete symbol showing the discrete states of operation of the SIDO boost converter. The transition from mode-1 to mode-2, (called event-1), occurs on reaching the guard condition, G_{12}. Similarly, transition from mode-2 to mode-1 (event-2) in CCM, is achieved on satisfying the guard condition, G_{21}. Similarly, the other guard conditions, G_{13}, G_{31}, G_{24}, G_{41}, G_{34}, G_{32}, and G_{23} can also be defined. Using the hybrid automaton definitions [4, 6-8], a more generalized model representing a SIDO boost converter, incorporating the DCM operation is shown in Figs. 2 and 3. The hybrid control problem is to find the guards which satisfy the regulation requirements of the converter simultaneously assuring the stability of operation. The control design problem is discussed in the next section.

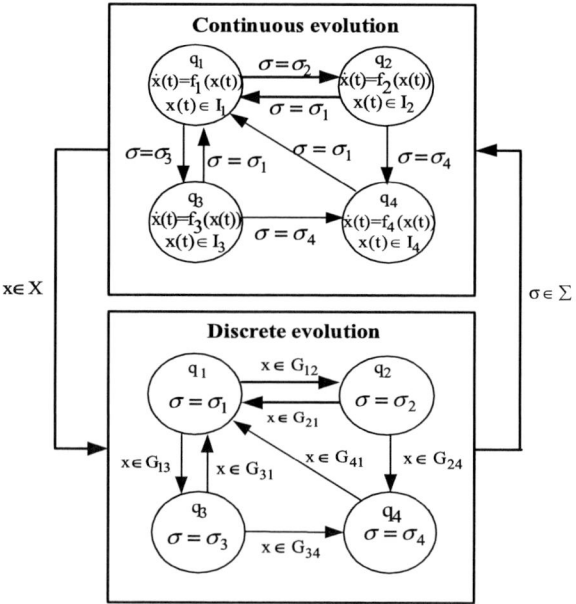

Figure 2. Hybrid automaton representation of the SIDO boost converter with event set $E_1 \cup E_2$.

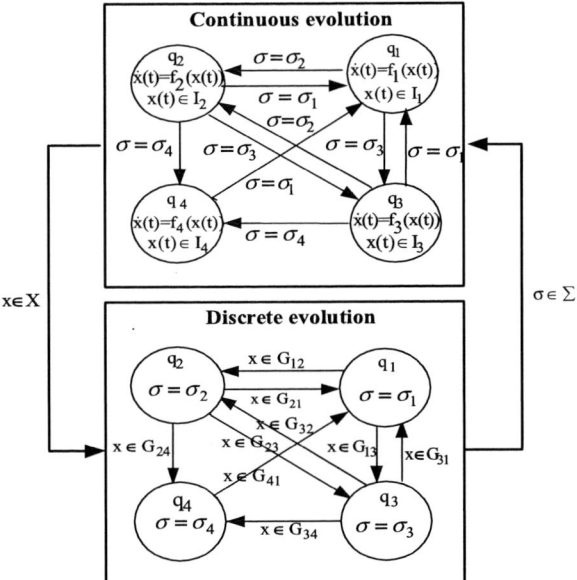

Figure.3. Hybrid automaton representation of the SIDO boost converter with event set $E_3 \cup E_4$.

III. THE CONTROL PROBLEM

For a SIDO boost converter, the control problem is to regulate the output voltages under various disturbance conditions. Also, it is required to ensure that the switching transitions are stable. First, a safe area in state-space which ensures switching stability is determined for a given load and line variation. Then, a controller is designed inside this safe area to regulate the output voltage.

In order to mathematically track the problem, the safe area is assumed to have a spherical boundary around the steady state operating point, x_d. To accommodate DCM operation, a zero current surface is super-imposed over this spherical boundary to decide on the admissible region of operation of the converter.. The value of maximum radius around the steady-state operating point is computed by proceeding outwards from the equilibrium point and checking for at least one element of the vector field pointing inwards, by computing the inner product $< x - x_d, f_i(x) >$ for each i. Whenever the value of the instantaneous current becomes zero, the current is constrained to be zero and the inner product is evaluated at the zero current boundary. This is done by incrementing the parameters representing the sphere, one by one in steps proceeding towards the boundary of the admissible set. Or, the radius vector is allowed to grow in linear as well as in angular directions of the three dimensional space to describe a ball or sliced ball to cover the entire admissible region of operation. The sphere of largest radius contained in the admissible region where at least one vector field pointing towards the steady state point describes the safe set.

As the stability of the system is guaranteed inside the safe boundary, controllers for the interior of the safe set can be designed based on some performance requirement. In our design, the system states are allowed to move inside the safe set, if the guard conditions hold true. When the guard condition is false, the control is chosen so as to minimize the cosine of the angle between $x - x_d$ and $f_i(x)$ as:

$$\sigma_i = \arg(\min_{i \in \Lambda} \frac{<x - x_d, f_i(x)>}{\|f_i(x)\|}) \qquad (2)$$

where, the term $\|(x - x_d)\|$ in the denominator is omitted, because it is independent of i. Or, in short, the system will move to the mode dictated by σ_i.

The system will work in just DCM when the steady-state current is exactly equal to the radius of the spherical shaped safe boundary. Beyond this operating point, the system will work in DCM and the corresponding guard conditions are applicable. The constraint of zero current acts along with a reference voltage profile and inner product condition to form the guard condition for transition from q_4 to q_1. A controller based on regulation requirement is defined and simulated in MATLAB/SIMULINK to test the suitability of the proposed algorithm. Some of the details of the design and computer simulations are included in the next section.

IV. SAFE BOUNDARY SYNTHESIS AND SIMULATIONS

The specifications and parameters of the example SIDO converter used for the present study are $v_{in}=10V$, $L=100\mu H$, $C_1=740\mu F$, $C_2=140\mu F$, $R_1=20\Omega$ and $R_2=90\Omega$. The control objective is to regulate the output voltages at $V_{o1}=18V$ and $V_{o2}=22V$ with a tolerance of $\pm0.5V$. For a load disturbance in R_1 from 20Ω to 650Ω, and that in R_2 from 90Ω to 500Ω, input disturbance from 5V to 15V requires inductor current variation in the range [0, 5A], which implies an admissible set $F = x \in R^3$: $0 \leq x_1 \leq 5$; $17.5 \leq x_2 \leq 18.5$; $21.5 \leq x_3 \leq 22.5$. Steady-state operation requires that $I_L = V_{o1}^2 / (V_{in}R_1) + V_{o2}^2 / (V_{in}R_2)$, where I_L, V_{o1} and V_{o2} are the steady state inductor current and the two output voltages respectively. With the help of a MATLAB program, as described in section III, by proceeding outwards from the desired state point corresponding to each operating conditions for the specified range, it is found that the radius of the safe set is 0.5.

The proposed control algorithm is simulated in MATLAB-SIMULINK using the variable step ode45 algorithm with a time step of 1 μs and a tolerance of 1e-6. The controller action is studied for a wide range of load and input voltage disturbance. The disturbances and the resulting variation in output voltages and inductor current are plotted in Figs. 4 and 5. It is observed that the output voltage remains constant and stable without any overshoot. The controller responds quickly to settle the output voltages at their set values, V_{o1} or V_{o2}. The change in the safe set under a large change in load resistances is shown in Fig. 6. With small load resistances, the converter operates in CCM and the safe set is spherical, where as the safe set takes a new shape while working in DCM with large load resistances.

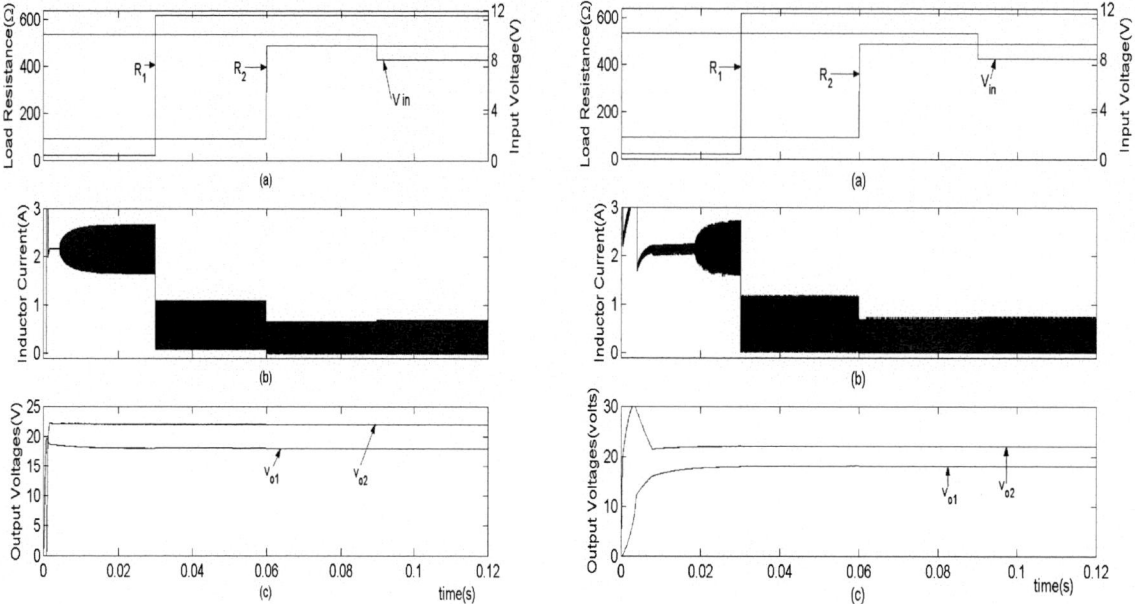

Figure.4. Variation of R, v_o and i_L with hybrid automaton model 1.

Figure.5. Variation of R, v_o and i_L with hybrid automaton model 2.

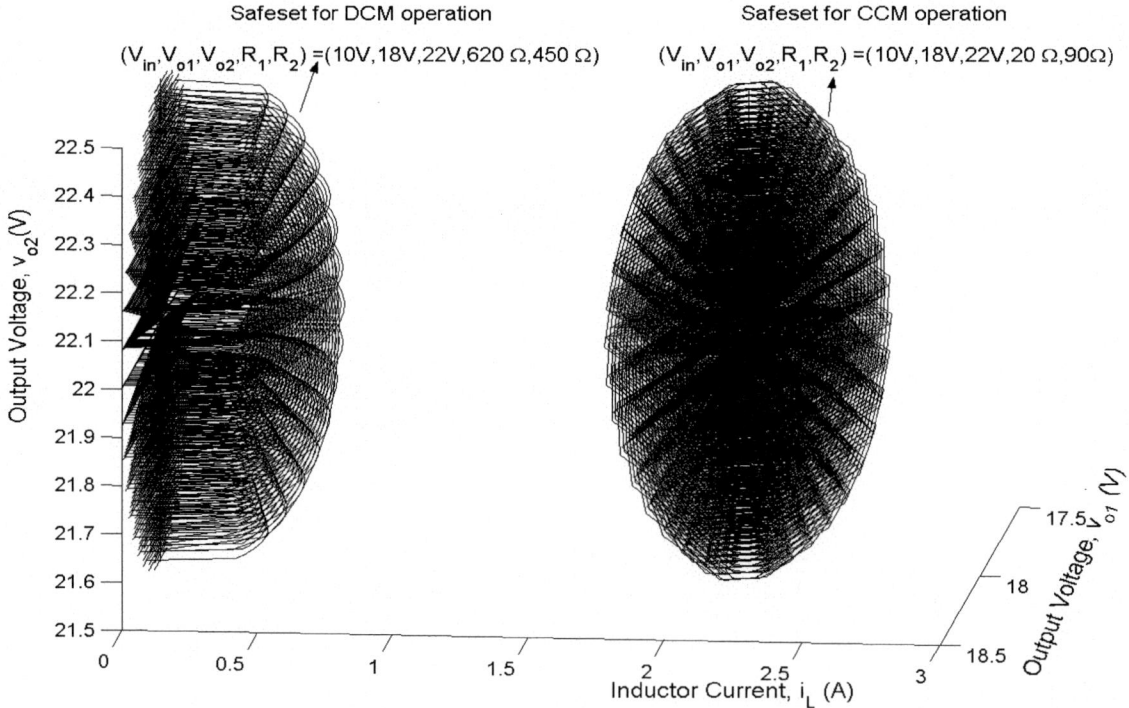

Figure.6. Change in switching surface for a load change (CCM to DCM).

284

From Figures 4 and 5 it is clear that both the model gives tight regulation of the outputs of the converter. But with model 2, it is found that there is an overshoot in one of the output whereas the other has more delay to reach the steady state from rest. Hence model 1 gives a better performance than model 2.

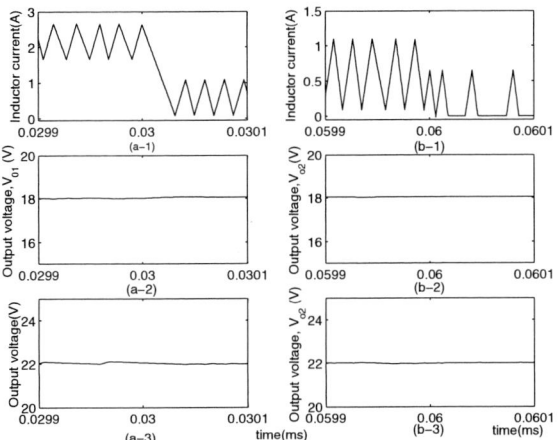

Figure7. Transient conditions for a load change (a) corresponding to CCM operation.(b) corresponding to CCM-DCM operation for model 1.

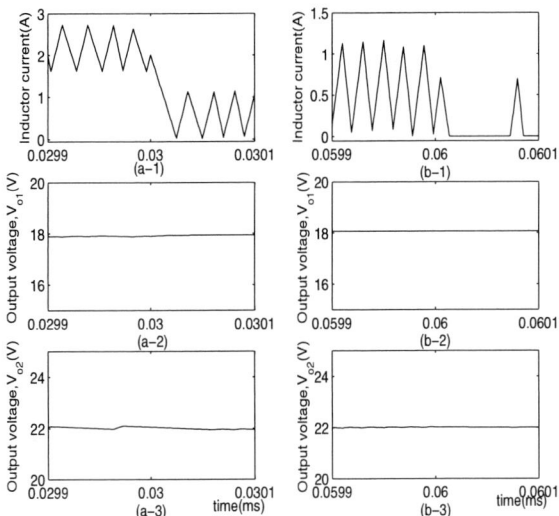

Figure.8. Transient conditions for a load change (a) corresponding to CCM operation.(b) corresponding to CCM-DCM operation for model 2.

An expanded view of the transients in inductor current and output voltages with the above two model under the defined control law for two different load and line conditions are shown in Fig.7. The transient behaviors for both the models are almost identical for similar disturbances occurring at 3ms and 6ms.

V. CONCLUSION

This paper describes the different hybrid automaton representations of a single inductor multiple output boost converter. After formulating the voltage regulation problem in a single inductor multiple output boost converter as a hybrid control problem, the guard conditions for the control of the different output voltages under various load conditions corresponding to CCM and DCM operation are derived. Though there is additional flexibility in the modes of operation in the second model, it is found that there is no significant advantage in using the operations in this model as far as the voltage regulation is considered. Also the transient behavior with the second model is found to be comparatively poor than the corresponding behavior with model 1.

REFERENCES

[1] Wing Hung Ki and Dongsheng Ma. "Single Inductor Multiple Output Switching Converters", *IEEE Power Electronics specialist's Conference,* Vancouver, pp. 226-231, June 2001.

[2] Dongsheng Ma, Wing Hung Ki and Chi-Ying Tsui, " A Pseudo-CCM/DCM SIMO Switching Converter with Freewheel Switching", *IEEE Journal of solid state circuits,* Vol.38, No.6, pp. 1007-1014, June 2003.

[3] M.H.Rashid, *Power electronics, circuits, devices & applications,* 3rd.ed., Pearson Education, Singapore, 2004.

[4] Matthew S., Gabriel E. and T. J. Koo, "Hybrid modeling and Control of Power Electronics", *Hybrid Systems: Computation and Control, Lecture Notes in Computer Science,* Springer, pp. 450-465, 2003.

[5] R W. Erickson, D. Maksimovic , *Fundamentals of Power Electronics,* 2nded., Kluwer Academic publishers Group, Massachusetts, USA, 2001.

[6] Sreekumar C, Vivek agarwal, "Hybrid control of Boost Converter Operating in Discontinuous Current Mode" to appear in the Proceedings of the Power electronics Specialist's Conference (PESC'06), to be held in Jeju, Korea, June 2006.

[7] A. Van der schaft and M. Schumacher, *Introduction to hybrid dynamical systems,* Springer Verlag, 2000.

[8] Geyer T., Papafotiou G., Morari M., "On the Optimal Control of Switch Mode Dc-Dc Converters", *Hybrid systems: Computation and Control, Lecture Notes in Computer Science.* Springer, 2004, pp 342-356.

2006 5th International Power Electronics and Motion Control Conference

Complex Intermittency in Voltage-Mode Controlled Buck Converter

Zheng-Ping Li, Yu-Fei Zhou* and Jun-Ning Chen
Department of Microelectronics, Anhui University, Hefei, P. R. China
* zhouyf@ahu.edu.cn

Abstract—Intermittency are commonly observed by practicing power supply engineers in their design workbenches. It shows to be a symmetrical period-doubling bifurcation in time domain with fixed long intermittent period. Sometimes it is called "breathing". Such kind of operation mode is usually undesired and can be avoided by means of screen technique or modifying the circuit parameters. This paper explore the mechanism and condition for the emergence of intermittency in a voltage-mode controlled Buck converter, and thus conclude that intermittency is a common and complex phenomenon by the fact that any interference, with frequency approaching switching frequency, or switching frequency's multiples or fractions, will induce the intermittency. Furthermore, the type and period of intermittency, and also the initial operation state have relations with the interference strength and frequency separately.

Keywords- switching power converter; voltage-mode control; intermittency; chaos; subharmonics

I. INTRODUCTION

Intermittency is a particular phenomenon frequently found in periodically driven nonlinear systems. It is sometimes called "breathing" in the physics literature [1]. Such intermittent operation has been shown to exhibit period-doubling bifurcation in two symmetrical directions over the time domain, i.e., system will bifurcate from the initial regular (or subharmonic) operation to the higher subharmonic operation, or eventually to chaos, and then come back to the initial operation through the same bifurcation route in reverse. In order to separate from "parameter-bifurcation", here we name the intermittency with "time-bifurcation" as it can provide information about the change of the qualitative behavior of the system as time elapses [2].

Intermittency may arise in periodically driven nonlinear systems, when the frequency of the accessional period signal is not consistent to system's driving frequency. PWM switching power converter is a kind of periodically driven nonlinear system. There have abundant periodic interference, which is coupled to the converter via

This work was supported by the National Natural Science Foundation of China(60402001) and the Natural Science Research Foundation of the Education Committee of Anhui Province (Grant No. 2005KJ054)

unintended paths (e.g., conducted or radiated EMI) [3]. When the interference frequency differs from the switching frequency, and the interference is strong enough, intermittent operation occurs. It heard as "frizzle" with a rather long period, so we can distinguish it clearly from switching sound (it has a high frequency) or irregular noise. Generally intermittency is an abnormal operation state should be avoided, and the emergence of intermittency is common and complicated. It is demanded to explore the condition and the way of its occurrence, so as to design switching power converters working stably and reliably.

This paper will construct a model coupling with intrusion interference signal base on a voltage-mode controlled Buck converter. By investigate this model, the condition for the emergence of intermittency is further expanded. Simulation and experiment all prove that the intermittency is complex, because the appearance of intermittency is varied, and any interference with frequency approaching switching frequency, or switching frequency's multiples or fractions, may cause the emergence of intermittency. The relative research results can also be extended to other switching power converters.

II. VOLTAGE-MODE CONTROLLED BUCK CONVERTER COUPLING WITH INTRUSION INTERFERENCE

A. Overview of Circuit Operation

The Buck converter consists of an inductor, a switch, a diode and a resistor load, which are connected as shown in Fig. 1(a). When operated under a common voltage-mode control, the buck converter has been shown to exhibit varied behavior [2, 4]. Fig. 1(b) gives the key waveforms about the control. In the practical occasion, the intrusion interference can take the form of coupling via conducted or radiated paths. Sometimes, the intruders can live on the same circuit board or be present at a very close proximity. Suppose the interference is injected to the reference voltage of the converter, we can model this coupling as an additive process which adds the spurious signal v_s directly to the V_{ref}, as shown in Fig. 1 (a), and now the perturbed reference voltage is:

$$V_{ref}^* = V_{ref} + v_s \qquad (1)$$

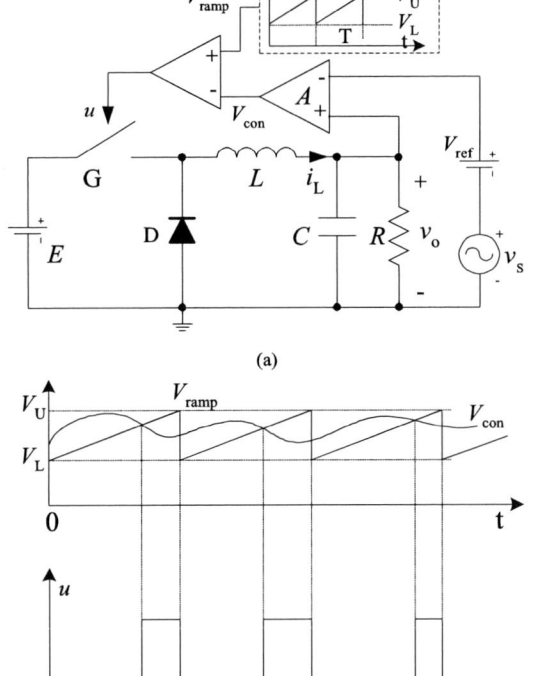

(a)

(b)

Figure 1. Voltage-mode controlled Buck converter coupled with spurious interference. (a) Schematic diagram; (b) key operation waveforms.

When the circuit parameters are chosen as follows: $L =$ 20mH, $C = 47$ μF, $R = 22$ Ω, $1/T = f_o = 2500$ Hz, V_{ref} =11.3 V, A = 8.4, $V_L = 3.8$ V, $V_U = 8.2$ V, and without the interference ($v_s = 0$), the Buck converter will experience a typical period-doubling bifurcation with input voltage E varying from 12 V to 34 V [2]. The first bifurcation of operating state occurs when $E \approx 24.6$ V, and the Buck converter eventually enter chaos at about $E \approx 32.3$ V.

B. Response Induced by Weak Resonant Interference

Still considering the circuit parameters afore, the original unperturbed Buck converter will operate in regular period-1 when E is selected to be 22 V (Fig. 2). However if the interference v_s is a periodic signal, e.g., sinusoidal signal with amplitude $\hat{v}_s$, then the perturbed

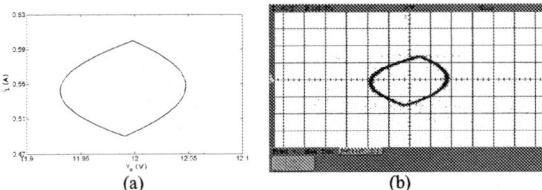

(a)   (b)

Figure 2. Regular period-1. (a) Simulation phase portrait; (b) measured phase portrait.

reference voltage is:

$$V_{ref}^* = V_{ref} + \hat{v}_s \sin(2\pi f_s t) = V_{ref}\left[1 + \alpha_v \sin(2\pi f_s t)\right] \quad (2)$$

where f_s is the frequency of the interference, and α_v is the strength of the interference which is defined as the ratio of the $\hat{v}_s$ to V_{ref}, i.e., $\alpha_v = \hat{v}_s/V_{ref}$. Coupling with this intrusion interference signal, the converter will exhibit varied resonant response, such as quasi-periodicity, period-1, period-2, ... according to the frequency ratio α_f ($\alpha_f = f_s / f_o$). We can conclude them in two occasions as follow.

(1) Frequency ratio α_f is a irrational number

On this occasion, the operating state of the Buck converter is quasi-periodicity, because there are two incompatible frequency in it, i.e., switching frequency f_o and interference frequency f_s. And the quasi-periodicity is characterized by a torus on the Poincaré section. Fig. 3 (a) shows an example with $\alpha_f = \sqrt{3}/2$ and $\alpha_v = 0.001$.

(2) Frequency ratio α_f is a rational number

If α_f is a rational number, we thus have:

$$\alpha_f = f_s/f_o = m/n \quad (3)$$

where m, n are positive integers. On this occasion, the Buck converter operates periodically with period number equaling n, i.e., have an operation of period-n subharmonics. The follows are two particular examples.

● $\alpha_f = 1 / n$

According to the former deduction, the Buck converter will operate in period-n subharmonics. Fig. 3(b), (c) give three examples with $n = 3, 2, 1$, which correspond to period-3, period-2 and period-1 separately, and characterized by 3, 2, 1 intersections on Poincaré section.

● $\alpha_f = m$

In this case, the Buck converter will operate in period-1 definitely. Fig. 3(c), (d) give three examples with $m = 1, 2,$

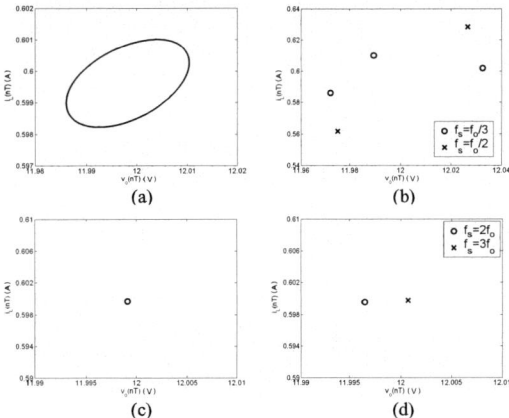

(a)   (b)

(c)   (d)

Figure 3. Response induced by resonant interferences. (a) $\alpha_f = \sqrt{3}/2$, $\alpha_v = 0.001$; (b) $\alpha_f = 1 / n$, $\alpha_v = 0.0014$; (c) $\alpha_f = 1$, $\alpha_v = 0.0003$; (d) $\alpha_f = m$, $\alpha_v = 0.0003$.

3, all corresponding to period-1, and characterized by one intersection on Poincaré section.

III. INTERMITTENCY: TIME BIFURCATION PANORAMA

The spurious coupling in last section can occur in switching power converters that operate in an RF environment [5]. Because the coupling intrusion interferences are imported unintended, it is impossible for its frequency f_s equaling switching frequency f_o, or switching frequency's multiples or fractions exactly. So there is little probability of the emergence of the resonant response given in Fig. 3. But it is possible that the interference frequency f_s approaches switching frequency f_o, or switching frequency's multiples or fractions. We can express this occasion as $f_s = \alpha_f \cdot f_o + \hat{f}$, here $\hat{f}$ is a small number compared to f_o. This kind of interference will induce the occurrence of intermittency, with varied appearance (such as intermittent subharmonics, intermittent chaos) determined by interference frequency f_s and strength α_v.

On this occasion of interference frequency approaching switching frequency, or switching frequency's multiples or fractions, interference signal (2) can be rewritten as:

$$V_{ref}^* = V_{ref}\left[1 + \alpha_v \cdot \sin(2\pi f_s t)\right]$$
$$= V_{ref}\left\{1 + \alpha_v \cdot \sin\left[2\pi(\alpha_f \cdot f_o + \hat{f})t\right]\right\} \quad (4)$$

Assuming $\hat{f} = 1$, and the frequency ratio α_f being 1/2, 1, 2 respectively, we can gain the time bifurcation diagrams of intermittency by simulation as shown in Fig. 4-6, from which we observe the following.

- When the strength of the interference is very weak, the converter can still maintain its original resonant response of period-n operation, though the average operating point fluctuates. The effect is not significant, and intermittency does not arise yet. Fig. 4(a)-6(a) describe this occasion.

- As the interference signal strength increases, the converter experiences higher subharmonic operation intermittently with the original resonant response of period-n. For a rather low interference signal strength, period-$2n$ subharmonics are observed intermittently with period-n. Fig. 4(b)-6(b) are the occasion.

- Further increase in interference signal strength causes period-2^2n subharmonics to occur intermittently with period-$2n$ subharmonics and the original period-n, as shown in Fig. 4(c)-6(c).

- Still further increasing the strength of interference to a certain value, the converter start to experience chaotic operation intermittently with period-$2^i n$ subharmonics and initial period-n, as shown in Fig. 4(d)-6(d).

- The intermittent period T_{int} is equal to the $1/\left|f_s - \alpha_f \cdot f_o\right| = 1/\hat{f}$. Thus, if the interference signal frequency is very close to the switching frequency of the converter, or switching frequency's multiples or

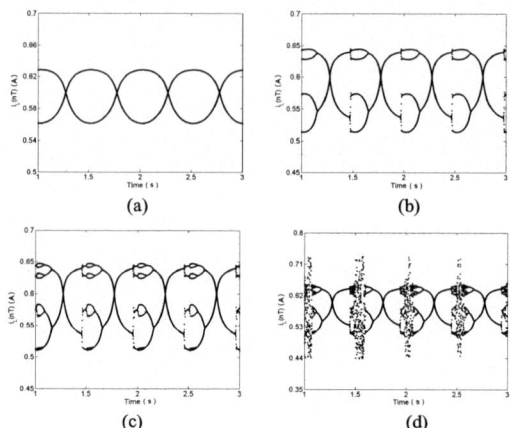

Figure 4. Sampled inductor current waveforms for the Buck converter coupling with interference of different α_v and $f_s = 1251$ Hz. (a) $\alpha_v = 0.0014$; (b) $\alpha_v = 0.0043$; (c) $\alpha_v = 0.0045$; (d) $\alpha_v = 0.0049$.

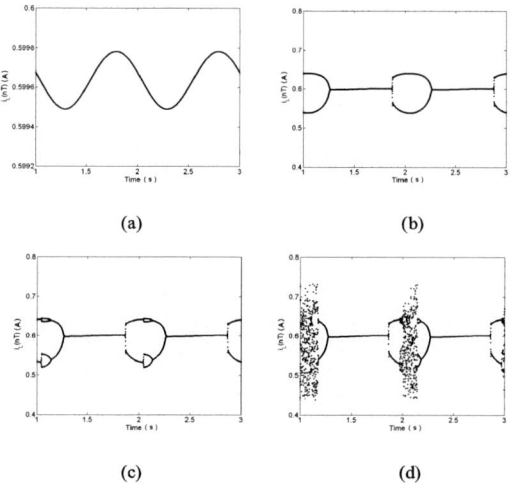

Figure 5. Sampled inductor current waveforms for the Buck converter coupling with interference of different α_v and $f_s = 2501$ Hz. (a) $\alpha_v = 0.0003$; (b) $\alpha_v = 0.003$; (c) $\alpha_v = 0.0035$; (d) $\alpha_v = 0.0045$.

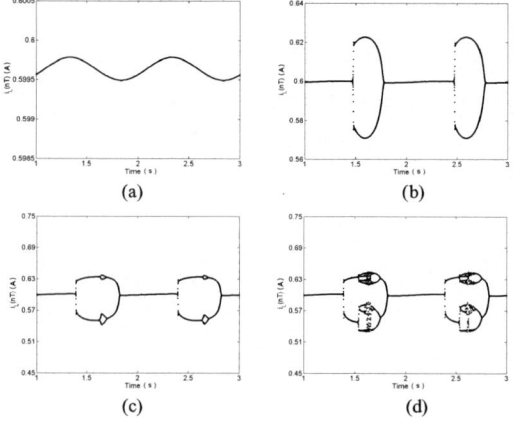

Figure 6. Sampled inductor current waveforms for the Buck converter coupling with interference of different α_v and $f_s = 5001$ Hz. (a) $\alpha_v = 0.0003$; (b) $\alpha_v = 0.0008$; (c) $\alpha_v = 0.0027$; (d) $\alpha_v = 0.00343$.

fractions, the intermittency is long. In the examples here, $T_{int} = 1$ s with $\hat{f} = 1$Hz .

- For the case of $\alpha_f \leq 1$, intermittent period T_{int} will be reduced to $\alpha_f / \hat{f}$ because the interference signal period is $1/\alpha_f$ times of switching period T. Hence, in Fig.4, intermittent period is reduced to 1/2 s from the original 1 s.

IV. EXPERIMENTAL OBSERVATIONS

We have built a circuit prototype to emulate the proposed coupling circuit. The circuit parameters are the same as those used in the simulations. Sinusoidal intrusion interference signals have been used in the experiment. Limited by space, here we only provide three typical cases. By varying the interference signal strength, we observe intermittent subharmonics and chaos in turn, which is consistent with the simulations. Fig. 7-9 show a few time-bifurcation diagrams obtained experimentally (Upper trace: output voltage v_o; lower trace: inductor current i_L.).

- Fig. 7 is the case of interference frequency approaching $f_o/2$. We observe fluctuating period-2 (Fig. 7(a), v_o: 200 mV/div, i_L: 100 mA/div), intermittent subharmonics (Fig. 7(b), v_o: 200 mV/div, i_L: 100 mA/div) and chaos (Fig. 7(c), v_o: 500 mV/div, i_L: 250 mA/div) in turn with interference strength α_v increasing.

- Fig. 8 is the case of interference frequency approaching f_o. We observe intermittent

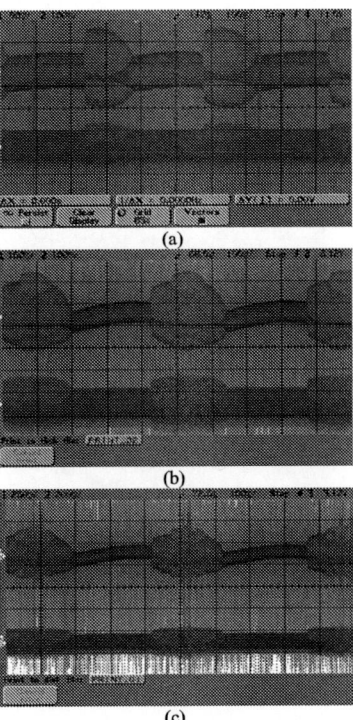

Figure 8. Measured time-bifurcation diagrams for buck converter coupling with sinusoidal interference of different strength α_v and $\alpha_f = 1$. (a) $\alpha_v = 0.001$; (b) $\alpha_v = 0.004$; (c) $\alpha_v = 0.0042$.

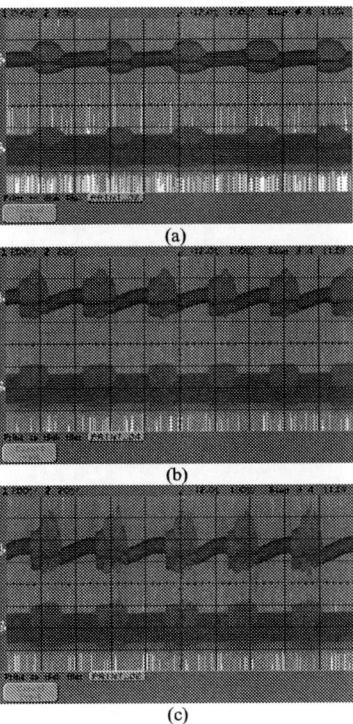

Figure 9. Measured time-bifurcation diagrams for buck converter coupling with sinusoidal interference of different strength α_v and $\alpha_f = 2$. (a) $\alpha_v = 0.0008$; (b) $\alpha_v = 0.0034$; (c) $\alpha_v = 0.004$.

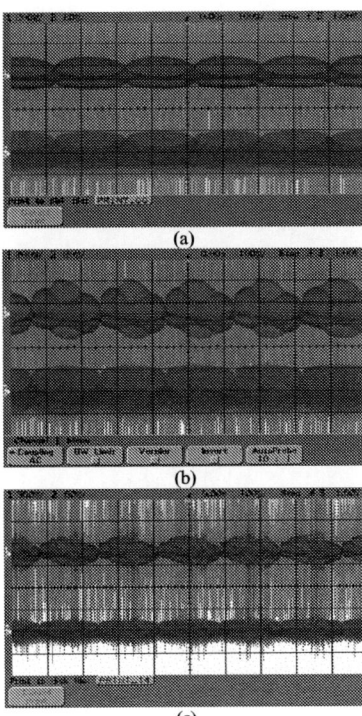

Figure 7. Measured time-bifurcation diagrams for buck converter coupling with sinusoidal interference of different strength α_v and $\alpha_f = 1/2$. (a) $\alpha_v = 0.0014$; (b) $\alpha_v = 0.0044$; (c) $\alpha_v = 0.0061$.

subharmonics (Fig. 8(a), v_o: 90 mV/div, i_L: 100 mA/div and Fig. 8(b), v_o: 150 mV/div, i_L: 100 mA/div) and chaos (Fig. 8(c), v_o: 250 mV/div, i_L: 200 mA/div) in turn with interference strength increasing. But fluctuating period-1 cannot be observed, because the minimum output of the signal generator will still induce the intermittent subharmonics.

● Fig. 9 is the case of interference frequency approaching $2f_o$. Similar to the occasion in Fig. 8, we observe intermittent subharmonics (Fig. 9(a, b), v_o: 200 mV/div, i_L: 100 mA/div) and chaos (Fig. 9(c), v_o: 200 mV/div, i_L: 100 mA/div) in turn with interference strength increasing, but still cannot see fluctuating period-1, despite of its coming forth in simulations.

V.   CONCLUSIONS

Switching power converters are nonlinear systems which have been shown to exhibit a variety of complex behavior. In this paper we attempt to explore a commonly observed but rarely explained phenomenon in power supply design. It is concluded that, when the interference frequency approaching the switching frequency, or rational multiples of the switching frequency, intermittency will occur with initial state being period-n subharmonics. This paper shows that the signal strength and frequency of the interference signal are vital parameters that affect the type and the period of intermittency. The same analysis can be used to study the intermittency in other types of converters.

REFERENCES

[1]   Z. Qu, G. Hu, G. Yang and G. Qin, "Phase effect taming nonautonomous chaos by weak harmonic perturbations," *Phys. Rev. Lett.*, vol. 74, no. 10, pp. 1736–1739, 1995.

[2]   Y. Zhou, C. K. Tse, S. S. Qiu and F. C. M. Lau, "Applying resonant parametric perturbation to control chaos in the buck dc/dc converter with phase shift and frequency mismatch considerations," *Int. J. of Bifurcation and Chaos*, vol. 13, no. 11, pp. 3459–3472, 2003.

[3]   T. Williams, *EMC for product designers* in *Magnetism.* Oxford, England, Butterworth-Heinemann Ltd., 1994.

[4]   S. Banerjeeand G. Verghese, *Nonlinear phenomena in power electronics: attractors, bifurcations, chaos, and nonlinear control.* New York, IEEE Press, 2001.

[5]   S. C. Wong, C. K. Tse and K. C. Tam, "Intermittent chaotic operation in switching power converters," *Int. J. of Bifurcation and Chaos*, vol. 14, no. 8, pp. 2971–2978, 2004.

2006 5th International Power Electronics and Motion Control Conference

Dual Mode Control Multiphase DC/DC Converter for CPU Power

Li-Wei Lin*, Chung-Hsing Chang*, Huang-Jen Chiu*, Member, *IEEE*, Shann-Chyi Mou**

*Dept. of Electrical Engineering, Chung-Yuan Christian Univ., Chung-Li, Taiwan, China
**Dept. of Mechanical Engineering, Ching-Yun Univ., Taiwan, China

Abstract—High performance voltage-regulator- modules (VRMs) for the new generation of microprocessors have many strict and challenging specifications that include high power-density, high output-current capability, low output-voltage deviation and fast transient-response. In this paper, a dual mode control multiphase DC/DC converter is presented. The proposed scheme combines voltage and hysteretic control modes and has the merits of high efficiency, simple control and fast transient response. It can be also used for single-phase and built in integrated circuit without additional cost and space. The operating principles and design considerations are analyzed and discussed in detail. A 1.3V/ 80A four-phase laboratory prototype was simulated and implemented to verify the feasibility of the proposed scheme. The results are satisfactory.

Keywords— Voltage Regulator Module, Multiphase DC/DC Converter, Dual Mode Control

I. Introduction

According to Intel's roadmap, over one million transistors will be integrated into one processor by 2010. The consumption current will be increased to 200A while the voltage is down to 0.8V [1, 2]. In addition to efficiency considerations, the transient response feature is a major challenge in power supply design for such applications because of the stricter transient current slew-rate and output voltage deviation requirements. For example, Intel's VRD 10.1 specification requires a ±19mV static regulation and less than 50mV overshoot with 120A transient load in one microsecond [3, 4]. Intel also defined the voltage droop specifications according to different CPU current. The CPU voltage, V_{core} will decrease while the I_{core} current increases and must be within the window between the maximum and minimum voltage lines. This voltage droop function is helpful in reducing CPU power consumption under heavy load conditions. Many researches have proposed high performance voltage-regulator-modules (VRMs) for the new generation of microprocessors. The most popular VRM topology is a multiphase buck DC/DC converter shown in Fig. 1(a). From the size and cost saving viewpoints, the transient response features of the DC/DC converters must be satisfied without using too many output capacitors especially in mobile applications. With traditional voltage and current-mode control schemes, the transient response is not fast enough except when a high switching frequency is adopted [5]. However, high-frequency operation could cause undesired switching loss. As shown in Fig. 1(b), a hysteretic control scheme has many advantages

including near instantaneous transient response [6, 7]. However, its switching frequency is influenced by the load condition and component tolerance. Operating in a wide frequency range causes difficulties in efficiency optimization and EMI treatments. In this paper, a dual mode control multiphase DC/DC converter is presented. The proposed scheme has high efficiency, simple control and fast transient response. The operating principles and design considerations will be discussed in the following sections. A laboratory prototype was also simulated and implemented to verify the feasibility.

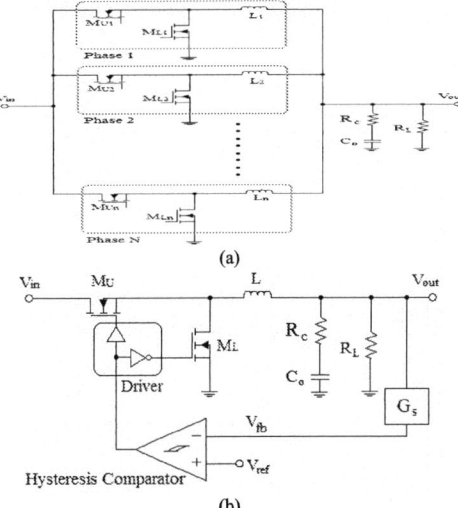

Fig. 1(a) Multiphase buck DC/DC converter and (b) a hysteretic control scheme

II. Multiphase Dual Mode Control

A faster response speed can reduce the number of output capacitors to save cost and space [8, 9]. It is important to choose a suitable control method for VRMs. To solve the disadvantage of the hysteretic and voltage modes, a dual mode control scheme shown in Fig. 2 is studied. In steady-state operation, the switching frequency is fixed as a normal voltage mode. In transient-state operation, it can keep the fastest response as a hysteretic mode. This control method combines the advantages of the hysteretic and voltage mode control methods. In Figure 2, the output voltage is fed back by Z_1 and compared with V_{ref} that comes from CPU VID information and is adjusted based on load current for load line requirements. Z_1 and Z_2 are used for loop compensation for voltage mode control. The feedback

1-4244-0448-7/06/$25.00 ©2006 IEEE

voltage is compared with a sawtooth signal from the oscillator for duty control. In steady state, the dual mode logic circuit just passes PWM signal to driver circuit so that the operation is same as that of the voltage mode. When V_{fb} is higher than $(V_{ref}+V_h)$, M_U will be turned-off immediately by the dual mode control circuit. Conversely, V_{fb} is lower than $(V_{ref}-V_h)$, M_U will be turned-on. The operation is the same as a hysteretic mode in transient state. Table 1 shows all conditions including normal PWM signal, over-voltage (OV) and under-voltage (UV) conditions. We don't need to consider the conditions that meet UV and OV at same time. The logic function can be expressed as Equation (1) and be implemented as shown in Fig. 2.

$$Q = UV + PWM \cdot \overline{OV}, \tag{1}$$

During a transient state, the MOSFETs can be turned on or off simultaneously to supply total inductor current more quickly without interleaving limitation. Figure 3 shows an unexpected operation waveform in hysteretic mode. The output voltage V_o still keeps increasing even though the high-side MOSFET M_U has been turned-off. The phenomenon is occurred because the inductor current keeps charging to the output capacitor C_o. The charging current $i_c(t)$ and the output voltage $v_o(t)$ can be expressed as follows.

$$i_c(t) = i_L(t) - I_o$$

$$= \begin{cases} \dfrac{I_p}{t_{on}}(t-t_o) - \dfrac{I_p}{2}, & t_0 \leq t \leq t_1 \\ -\dfrac{I_p}{t_{off}}(t-t_1) + \dfrac{I_p}{2}, & t_1 \leq t \leq t_2 \end{cases} \tag{2}$$

$$v_o(t) = i_c(t) \cdot R_c + C_o \int_0^t i_c(t)dt + V_o$$

$$= \begin{cases} \dfrac{I_p}{2C_o t_{on}}(t-t_o)^2 + (\dfrac{I_p \cdot R_c}{t_{on}} - \dfrac{I_p}{2C_o})(t-t_o) - \dfrac{I_p \cdot R_c}{2} + V_o, & t_0 \leq t \leq t_1 \\ -\dfrac{I_p}{2C_o t_{off}}(t-t_1)^2 + (\dfrac{I_p}{2C_o} - \dfrac{I_p \cdot R_c}{t_{off}})(t-t_1) + \dfrac{I_p \cdot R_c}{2} + V_o, & t_1 \leq t \leq t_2 \end{cases} \tag{3}$$

Where Ip is the peak-to-peak value of the inductor current ripple and Rc is the ESR of output capacitor. We can then get the time points, t_{min} and t_{max} at the lowest voltage and highest voltage.

$$t_{min} = t_0 + \frac{t_{on}}{2} - C_o \cdot R_c, \tag{4}$$

$$t_{max} = t_1 + \frac{t_{off}}{2} - C_o \cdot R_c, \tag{5}$$

For the requirement of hysteretic mode control, the high-side MOSFET M_U should be turned-off at the highest output voltage and it should be turned-on at the lowest output voltage. It means that the Equation (6) and (7) must be satisfied. In general, the CPU voltage is much lower than the input voltage so that t_{off} is longer. The ESR limitation of output capacitors should be determined by Equation (7). If this condition is met, the ripple voltage equals (I_pR_c).

$$R_c \geq \frac{t_{on}}{2C_o} = \frac{\delta}{2f_s C_o}, \tag{6}$$

$$R_c \geq \frac{t_{off}}{2C_o} = \frac{1-\delta}{2f_s C_o}, \tag{7}$$

Where δ is the duty cycle of the high-side MOSFET M_U and fs is the switching frequency.

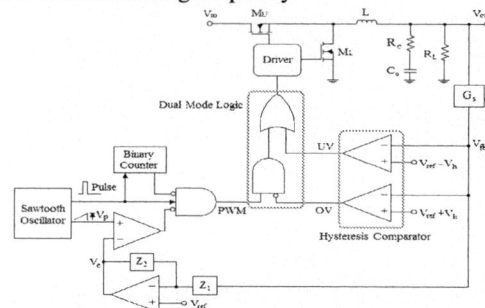

Fig. 2 The proposed dual mode control scheme

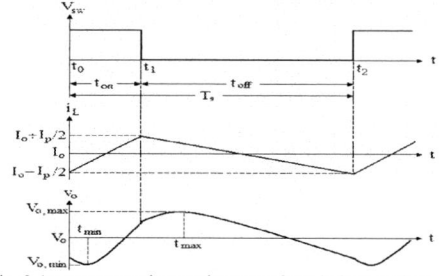

Fig. 3 An unexpected operation waveform in hysteretic mode

III. Design Considerations

A. Current Balance Circuit

Current balance can make thermal balance each phase and we do not need to reserve too much design margin for power components. The unbalanced current will increase the conduction loss and output ripple voltage. Figure 4 shows a current balance circuit adopted in this paper. The V_{IL} pins are connected to the output side of the current sense resistors and the V_{IH} pins are connected to the inductor side of the sensing resistors. All of the V_{IH} and V_{IL} signals are summed so that the signal $V_{I(AVG)}$ represents the average current of all phases. Amplifier A2 converts inductor current information of phase A. Amplifier A3 acts as an error amplifier and drives an adjustable current source, I_{s2}. Current source I_{s1} provides a constant 20μA for charging the internal ramp capacitor, C_T. If current sharing is perfect, the charge current keeps 20μA to C_T. If 1mV difference is detected on sensing resistor between phase A and average value, I_{s2} will inject 3.4μA into the capacitor. It will increase capacitor voltage faster and turn off MOSFET earlier to reduce duty cycle when the ramp signal is compared with V_{comp} of the voltage feedback loop.

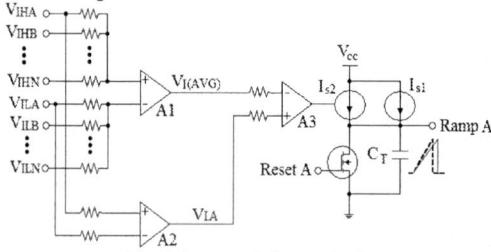

Fig. 4 A current balance circuit

B. Multiphase Interleaving Circuit

The multiphase interleaving function can be made with the D-type flip-flop circuit shown in Fig. 5(a). This circuit is popular and is widely used for generating four-phase pulses. Figure 5(b) shows the signal waveforms of the interleaving circuit. The clock frequency of the signal Pulse_IN is set as 4 times of the switching frequency in each phase. The four-phase pulses can be generated sequentially according to the signal Pulse_IN.

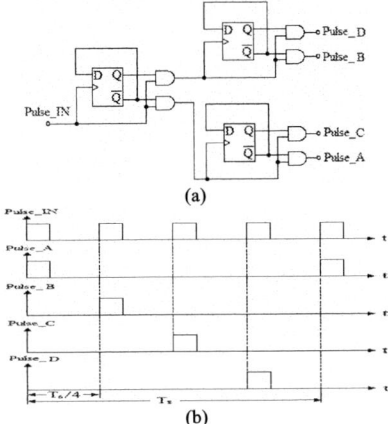

(a)

(b)

Fig. 5 (a) The multiphase interleaving circuit and (b) operating waveforms

IV. Simulation and Experimental Verifications

To verify the feasibility of the proposed dual mode control multiphase DC/DC converter, a four-phase prototype with following specifications was simulated and implemented.

- Source Voltage: Vin = 9V (Battery)/ 20V (Adaptor)
- Output Voltage: V_o = 1.3V
- Output Current: I_o =80A
- Switching Frequency: f_s=300kHz

To prevent out of voltage specification during a transient stage, the output capacitor ESR shall be lower than the load line specification of 1.5mV/A. We chose 6 Sanyo POSCAP 330μF/2.5V/9mΩ capacitors as the output capacitors. Let the current ripples be 30 percent of the inductor currents under full load. The inductance of the output inductor L_o can be determined using Equations (8) as follows.

$$L_o = \frac{V_o \times (1-\delta)}{\Delta I_L \times f_s} = \frac{1.3 \times (1-1.3/20)}{0.3 \times 20 \times 300k} \approx 0.7\mu H, \tag{8}$$

We can calculate output ripple voltage based on output ripple current and ESR directly as follows [10].

$$\Delta I_c = \frac{V_o \times (1-4\delta)}{L_o \times f_s} = \frac{1.3 \times (1-4 \times 1.3/20)}{0.7 \times 300} = 4.6A, \tag{9}$$

$$\Delta V_o = \Delta I_c \times R_c = 4.6 \times 1.5m = 6.9mV \tag{10}$$

The hysteretic voltage setting must be higher than half the ripple voltage for dual mode control. Considering capacitor ESL, trace impedance, output specification and noise immunity, we assume 15mV for hysteretic voltage setting. It will not influence normal operation at steady state but will respond immediately

when load change is over 10A. Figure 6 shows the simulated waveforms for the multiphase converter with conventional voltage mode control. The inductor currents and the charging current of output capacitor are shown in Figure 6(a). The inductor current are interleaved and balanced due to the addition of a current balance circuit. The current ripple cancellation on the output capacitor can be also observed. Figure 6(b) shows the simulated waveforms of the gating signals, output voltage during load change transient. The on-times are adjusted sequentially after load change. Output voltage does not recover within one phase period due to compensation and current limitation. Figure 7 is the simulation circuit with the proposed dual mode control. We can see faster response as shown in Fig. 8 because the high-side MOSFETs of 4 phases are turned on at same time after adding a dual mode logic circuit. Four times inductor current is provided to output capacitors to speed up voltage recovery. In Figure 9(a), we can see the measured waveforms of four-phase converter under steady state condition. The duty cycle is same and it is operated at voltage mode. Figure 9(b) shows the inductor current waveforms with interleaving function. The current per phase is similar due to current balance circuit. Figure 9(c) indicates the response with the dual mode control. The phases are turned-on simultaneously during transient. V_o is measured for showing the relationship with the duty adjustment during transient interval. The fast transient response can be achieved by using the proposed dual mode control method. Both the static regulation and transient response can satisfy the specifications. The voltage keeps at lower level after transient because load line function is supported. Figure 9(d) is the measured data about load line. For full load variations, the output voltage is always within the window between the maximum and minimum voltage lines. The efficiencies based on 20V and 9V input voltages are shown in Fig. 9(e).

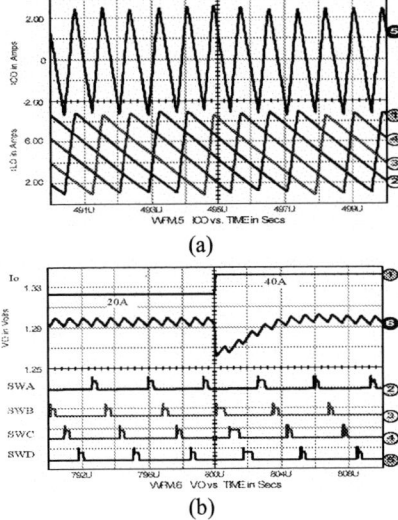

(a)

(b)

Fig. 6 Simulated waveforms for the simulation circuit with voltage mode control

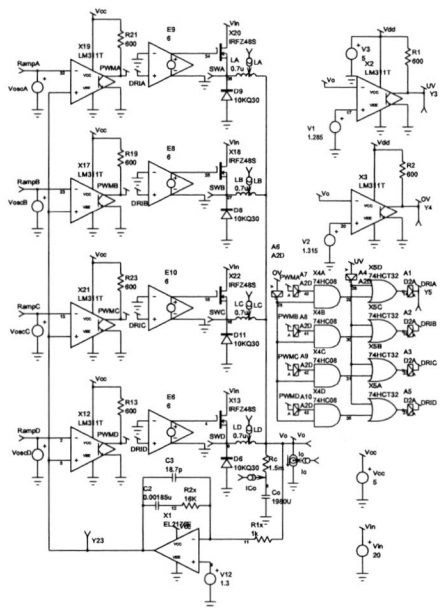

Fig. 7 Simulation circuit with the proposed dual mode control

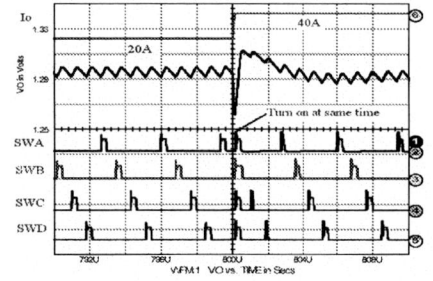

Fig. 8 Transient response of the circuit with the proposed dual mode control

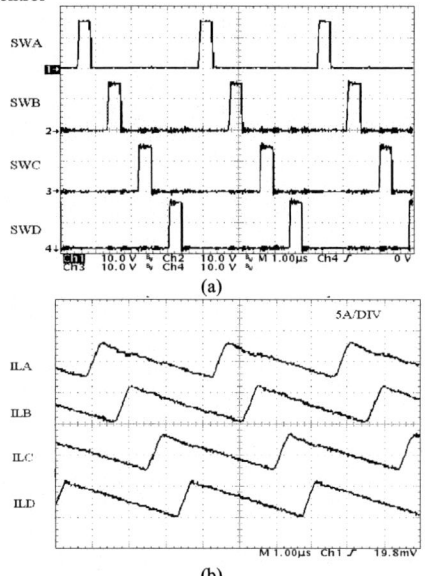

(a)

(b)

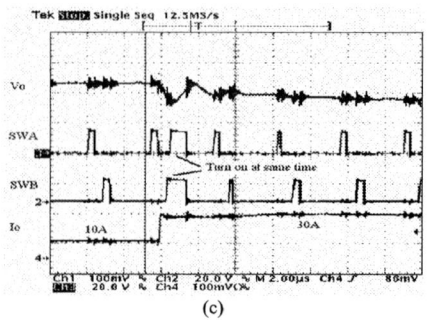

(c)

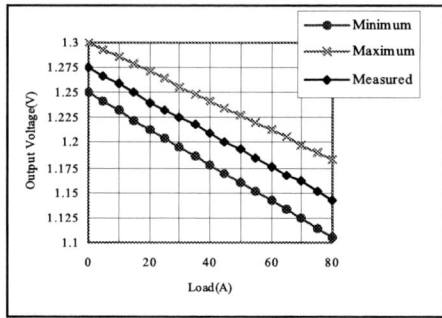

(d)

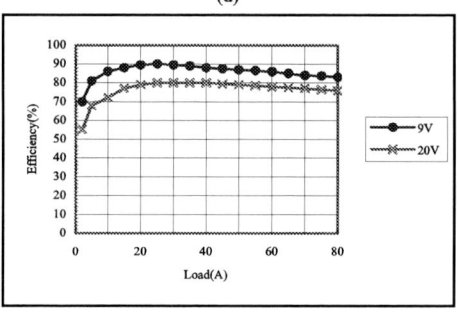

(e)

Fig. 9 Measured circuit results with the proposed dual mode control

Table 1 Truth table for dual mode control

UV	PWM	OV	Q
0	0	0	0
0	0	1	0
0	1	0	1
0	1	1	0
1	0	0	1
1	0	1	X
1	1	0	1
1	1	1	X

V. Conclusion

This paper presented a dual mode control multiphase DC/DC converter. The dual mode control method combines the voltage and hysteretic mode to have a fixed frequency operation in steady state and the fast response in transient state. It can be used for some chips which require low voltage and high current. The operating principles and design considerations are analyzed and discussed in detail. A four-phase laboratory prototype was simulated and implemented to verify the feasibility of the proposed scheme. The results were satisfactory. The dual mode control circuit is simple and easy to be built in integrated circuit without additional cost and space. Thus, it has the potential to be used widely with vast benefits in the future.

Acknowledgment

The authors would like to acknowledge the financial support from the National Science Council of Taiwan, China. through grant number NSC 94-2622-E-033-009–CC3.

Reference

[1] E. Stanford, "Power Technology Roadmap for Microprocessor Voltage Regulators," in Proc. Power Sources Manufactures Associations Conf., Feb. 2004.

[2] E. Stanford, "New Processors Will Require New Powering Technologies," Mag. Power Electronics Technology, pp. 32-42, Feb. 2002.

[3] Intel Corp., "Voltage Regulator-Down (VRD) 10.1 Design Guide," July 2004.

[4] Intel Corp., "Intel Pentium 4 Processor 660,650,640 and 630 Datasheet," Feb. 2005.

[5] K. Yao, Y. Ren, F. C. Lee, "Critical Bandwidth for The Load Transient Response of Voltage Regulator Modules," IEEE Trans. on Power Electronics, Vol. 19, No. 6, pp. 1454-1461, Nov. 2004.

[6] J. Abu-Qahouq, H. Mao and I. Batarseh, "Multiphase Voltage-Mode Hysteretic Controlled DC-DC Converter with Novel Current Sharing," IEEE Trans. on Power Electronics, Vol. 19, No. 6, pp. 1397-1407, Nov. 2004.

[7] T. Nabeshima, T. Sato, S. Yoshida, S. Chiba and K. Onda, "Analysis and Design Considerations of a Buck Converter with a Hysteretic PWM Controller," in Proc. IEEE PESC'04, Vol. 2, June 2004, pp. 1711-1716.

[8] N. Chabini, I. Chabini, E. M. Aboulhamid and Y. Savaria, Noureddine, "Methods for Minimizing Dynamic Power Consumption in Synchronous Designs with Multiple Supply Voltages," IEEE Trans. on Computer-Aided Design of Integrated Circuits and Systems, Vol. 22, No. 3, pp. 346-351, March 2003.

[9] C. Qiao, J. Zhang, P. Parto and D. Jauregui, "Output Capacitor Comparison for Low Voltage High Current Applications," in Proc. IEEE PESC'04, Vol. 1, June 2004, pp. 622-628.

[10] P. L. Wong, "Performance Improvements of Multi-channel Interleaving Voltage Regulator Modules with Integrated Coupling Inductors," Ph.D. Dissertation, Virginia Polytech. Insit. State Univ., Blacksburg, Mar. 2001.

2006 5th International Power Electronics and Motion Control Conference

An Analog Implementation of Pulse-Width-Modulation Based Sliding Mode Controller for DC–DC Boost Converters

Siew-Chong Tan*, Y. M. Lai, and Chi K. Tse
Department of Electronic and Information Engineering
Hong Kong Polytechnic University, Hong Kong, China
*E-mail: ensctan@polyu.edu.hk

Abstract— **This paper addresses the issues concerning the implementation of a pulse-width-modulation based sliding mode controller for boost converters. The methods of modelling the system and translation of the SM control equations for the PWM implementation are illustrated. It is shown that the control technique is easily realized with simple analog circuitries. The proposed system is verified by experimental measurements.**

I. INTRODUCTION

The advantages of employing sliding mode (SM) controllers for applications in nonlinear control systems are well discussed and understood [1]. In power converters requiring wide operating range, SM controllers are understandably better candidates than conventional linear PWM controllers due to their excellent robust and stability properties in handling large-signal perturbations [2]. This spurs numerous researches in the area. However, among the various proposed systems, the fixed-frequency SM controllers are particularly suited for practical implementation in power converters [3]–[10].

Study of fixed-frequency SM controllers has been initiated by the practical constraint of power converters in requiring to prevent excessive power losses and EMI noise generation, and also to simplify the design of input and output filters [11]. However, the nature of the SM controller is to ideally operate at infinite, varying, and self-oscillating switching frequency such that the controlled variables can track a certain reference path to achieve steady-state operation [1].

There are numerous methods proposed to constrict the switching frequency of SM controllers [3]–[10]. Those that employ the hysteresis-modulation (HM) (or delta-modulation) as the medium for implementing the control law, will require either constant timer circuits to be incorporated into the hysteretic SM controller to ensure constant switching frequency [3], [4], or the use of an adaptive hysteresis band that varies with parameter changes to control and fix the switching frequency [5], [10]. However, these solutions require additional components and are unattractive for low cost voltage conversion applications. Moreover, some of these converter systems suffer from deteriorated transient response.

Alternatively, the switching frequency of SM controllers can be constricted (fixated) by changing the modulation method

from HM to pulse-width-modulation (PWM) [2], [6], [7], [9]. This idea is originated from one of the earliest papers on SM controlled power converters [2], which suggests that under SM control operation, the control signal of *equivalent control approach* u_{eq} in SM control is equivalent to the *duty cycle control signal* d of a PWM controller. The proof was later provided in the papers [12], [13]. It has been shown that at a high switching frequency, the *control action* of a sliding mode controller is equivalent to the *duty cycle control action* of a PWM controller. Hence, the use of PWM techniques in lieu of HM methods in SM control is possible under these principles.

Lately, this idea is revisited and experimentally demonstrated on a sliding mode voltage controlled (SMVC) buck converter [9]. Building on the work of [2], [6], [12], it is demonstrated in [9] how PWM based SM controller can be easily realized with simple analog ICs. However, the discussion does not cover the design methodology of other converters. It was also unknown if the proposed PWM based SM controller can be as easily realized in converters that are more complex.

Therefore, in this paper, we extend the work in [9], by exploring into the possible application of the PWM based SM voltage controller on the boost converter. Specifically, we illustrate the method of modelling the system and the translation of the SM control equations for the PWM implementation. Finally, with an experimental prototype, we validate that the PWM based SM controller under a different circuit architecture, is also applicable for the control of boost converters.

II. THE DESIGN APPROACH

A. System Modelling

A second order PID type of SM voltage controller is adopted. Fig. 1 shows the schematic description of the proposed sliding mode voltage controlled (SMVC) boost converters, where C, L, and r_{L} are the capacitance, inductance, and load resistance of the converters respectively; i_C, i_L, and i_r are the capacitor, inductor, and load currents respectively; V_{ref}, v_{i}, and βv_{o} are the reference, input, and sensed output voltage respectively; and $u = 0$ or 1 is the switching state of power switch S_{W}.

1-4244-0448-7/06/$25.00 ©2006 IEEE

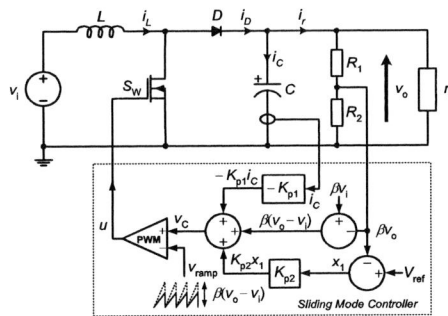

Fig. 1. Schematic diagrams of the PID SMVC boost converter.

For any PID SMVC converters type, the control variable x may be expressed in the general form:

$$x = \begin{bmatrix} x_1 \\ x_2 \\ x_3 \end{bmatrix} = \begin{bmatrix} V_{\text{ref}} - \beta v_o \\ \frac{d(V_{\text{ref}} - \beta v_o)}{dt} \\ \int (V_{\text{ref}} - \beta v_o)dt \end{bmatrix} \quad (1)$$

where x_1, x_2, and x_3 represents the *voltage error*, the *voltage error dynamics* (or the rate of change of voltage error), and the *integral of voltage error*, respectively. Substitution of the boost converter's behavioral models under continuous conduction mode (CCM) of operation into (1) produces the following control variable description:

$$x_{\text{boost}} = \begin{bmatrix} V_{\text{ref}} - \beta v_o \\ \frac{\beta v_o}{r_L C} + \int \frac{\beta(v_o - v_i)\bar{u}}{LC}dt \\ \int x_1 dt \end{bmatrix}. \quad (2)$$

where $\bar{u} = 1 - u$ is the inverse logic of u. Next, the time differentiation of equation (2) produces the state space description

$$\dot{x}_{\text{boost}} = \mathbf{A}x_{\text{boost}} + \mathbf{B}\bar{u} \quad (3)$$

where

$$\mathbf{A} = \begin{bmatrix} 0 & 1 & 0 \\ 0 & -\frac{1}{r_L C} & 0 \\ 1 & 0 & 0 \end{bmatrix} \quad \text{and} \quad \mathbf{B} = \begin{bmatrix} 0 \\ \frac{\beta v_o}{LC} - \frac{\beta v_i}{LC} \\ 0 \end{bmatrix}. \quad (4)$$

B. Controller Design

For this system, it is appropriate to have a general SM control law that adopts a switching function such as

$$u = \begin{cases} 1 & \text{when } S > 0 \\ 0 & \text{when } S < 0 \end{cases} \quad (5)$$

where S is the instantaneous state variable's trajectory, and is described as

$$S = \alpha_1 x_1 + \alpha_2 x_2 + \alpha_3 x_3 = \mathbf{J}^T x, \quad (6)$$

with $\mathbf{J}^T = [\alpha_1 \ \alpha_2 \ \alpha_3]$ and $\alpha_1, \alpha_2,$ and α_3 representing the control parameters termed as sliding coefficients.

1) Derivation of Existence Conditions: To ensure the existence[1] of SM operation, the local reachability condition

$$\lim_{S \to 0} S \cdot \dot{S} < 0, \quad (7)$$

must be satisfied. For the proposed converter, this can be expressed as

$$\begin{cases} \dot{S}_{S \to 0^+} = \mathbf{J}^T \mathbf{A} x_{\text{boost}} + \mathbf{J}^T \mathbf{B} u_{S \to 0^+} < 0 \\ \dot{S}_{S \to 0^-} = \mathbf{J}^T \mathbf{A} x_{\text{boost}} + \mathbf{J}^T \mathbf{B} u_{S \to 0^-} > 0 \end{cases}. \quad (8)$$

The specific conditions for the existence of SM control operation for the boost converters are

- Case 1: $S \to 0^+$, $\dot{S} < 0$ – substitution of $u_{S \to 0^+} = \bar{u} = 0$, and the matrices in (2) and (4) into (8) gives $-\alpha_1 \frac{\beta i_C}{C} + \alpha_2 \frac{\beta i_C}{r_L C^2} + \alpha_3(V_{\text{ref}} - \beta v_o) < 0$.

- Case 2: $S \to 0^-$, $\dot{S} > 0$ – substitution of $u_{S \to 0^-} = \bar{u} = 1$, and the matrices in (2) and (4) into (8) gives $-\alpha_1 \frac{\beta i_C}{C} + \alpha_2 \frac{\beta i_C}{r_L C^2} + \alpha_3(V_{\text{ref}} - \beta v_o) - \alpha_2 \frac{\beta v_i}{LC} + \alpha_2 \frac{\beta v_o}{LC} > 0$.

2) Derivation of Control Equations for PWM Based Controller: The conventional SM controller implementation based on HM [8] requires only control equations (5) and (6). However, if the PWM based SM voltage controller is to be adopted, an indirect translation of the SM control law is required so that pulse-width modulation can be used in lieu of hysteresis modulation [9]. The procedure for the PWM design can be summarized in two steps. Firstly, the equivalent control signal u_{eq}, which is a smooth function of the discrete input function u, is formulated using the *invariance conditions* by setting the time differentiation of (6) as $\dot{S} = 0$ [1]. Secondly, the equivalent control function is mapped onto the duty cycle function of the pulse-width modulator [9]. For the PWM based SMVC boost converter, the derivations are as illustrated.

- Equating $\dot{S} = \mathbf{J}^T \mathbf{A} x + \mathbf{J}^T \mathbf{B} \bar{u}_{\text{eq}} = 0$ yields the equivalent control function

$$\bar{u}_{\text{eq}} = \frac{\beta L}{\beta(v_o - v_i)}\left(\frac{\alpha_1}{\alpha_2} - \frac{1}{r_L C}\right) i_C$$
$$- \frac{\alpha_3 LC}{\alpha_2 \beta(v_o - v_i)}(V_{\text{ref}} - \beta v_o)$$

where $\bar{u}_{\text{eq}}$ is continuous and $0 < \bar{u}_{\text{eq}} < 1$. Since $u = 1 - \bar{u}$, which also implies $u_{\text{eq}} = 1 - \bar{u}_{\text{eq}}$, the substitution of (9) into the inequality and a multiplication by $\beta(v_o - v_i)$ gives

$$0 < u_{\text{eq}}^* = -\beta L\left(\frac{\alpha_1}{\alpha_2} - \frac{1}{r_L C}\right) i_C \quad (9)$$
$$+ LC\frac{\alpha_3}{\alpha_2}(V_{\text{ref}} - \beta v_o) + \beta(v_o - v_i) < \beta(v_o - v_i).$$

- Finally, the mapping of the equivalent control function (9) onto the duty ratio control d, where $0 < d = \frac{v_c}{\hat{v}_{\text{ramp}}} < 1$,

---

[1]Satisfaction of the existence condition is one of the three necessary conditions for SM control operation to occur. It ensures that the state trajectory at locations near the sliding surface will always be directed towards the sliding surface. The other two necessary conditions are the hitting condition, which is satisfied by the control law in eqn. (5), and the stability condition, which is satisfied through the assignment of sliding coefficients [14].

gives the following relationships for the control signal v_c and ramp signal $\hat{v}_{\text{ramp}}$ where

$$v_c = u_{\text{eq}}{}^* = -\beta L \left(\frac{\alpha_1}{\alpha_2} - \frac{1}{r_L C} \right) i_C \qquad (10)$$
$$+ LC \frac{\alpha_3}{\alpha_2} \left(V_{\text{ref}} - \beta v_o \right) + \beta \left(v_o - v_i \right)$$

and

$$\hat{v}_{\text{ramp}} = \beta \left(v_o - v_i \right) \qquad (11)$$

for the practical implementation of the PWM based SM controller.

III. IMPLEMENTATION OF THE PWM BASED SM CONTROLLER

A. Conversion of Control Equations to Circuit Form

1) Control Signal Computation: The computation of the control signal v_c in (10) can be performed using simple gain amplification and summing functions. In our prototype, we realize the equation using only three analog gain amplifiers and a summer circuit (LM318). The parameters of these circuitries can be easily calculated using known values of L, C, r_L, and β, and proper choices of α_1, α_2, and α_3.

2) Ramp Signal Generation: The peak magnitude of the variable ramp signal $\hat{v}_{\text{ramp}}$ is to follow description (11). In our prototype, a transistor configuration of multiple current mirror circuitries (9015 and 9016) and a charging capacitor are employed to realize the ramp generation. Since the desired output voltage is normally constant with little deviation, only the input voltage change is considered in the design. As for the frequency of the ramp signal, it is controlled by an impulse generator (LMC555 and CD4049).

3) Duty Cycle Protection: The incorporation of the control and ramp signal circuitries into the pulse-width-modulator (LM311) forms the basic architecture of the PWM based SM controller. However, recalling that the boost type converter cannot operate with a switching signal u that has a duty cycle $d = 1$, a small protective circuitry is required to ensure that the duty cycle of the controller's output is always $d < 1$. In our prototype, this is satisfied by multiplying the logic state u_{PWM} of the pulse-width-modulator with the logic state u_{CLK} of the impulse generator using a logic AND IC chip (CD4081). By doing so, the maximum duty cycle of the controller is clamped by the duty cycle of the impulse generator.

B. Experimental Prototype

The derived controller is verified through an experimental prototype developed with the specification shown in Table I. Figure 2 shows the schematic diagram of the proposed PWM based SMVC boost converter. The enumerated points on the diagram represent different test locations in the controller where waveforms are captured and analyzed. The theoretical description of the signals are derived and shown in Table I.

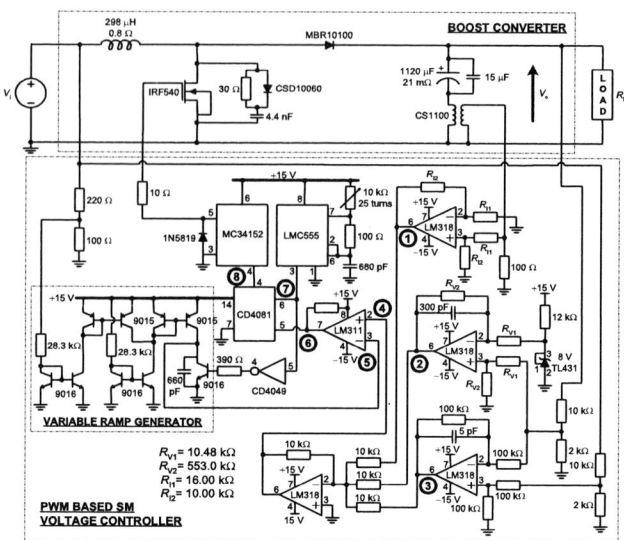

Fig. 2. Schematic diagram of the proposed PWM based SMVC boost converter.

TABLE I

THEORETICAL DESCRIPTION OF SIGNALS

Test Location	Description
1	$\beta L \left(\frac{\alpha_1}{\alpha_2} - \frac{1}{r_L C} \right) i_C$
2	$-LC \frac{\alpha_3}{\alpha_2} \left(V_{\text{ref}} - \beta v_o \right)$
3	$-\beta \left(v_o - v_i \right)$
4	v_c
5	$\hat{v}_{\text{ramp}}$
6	u_{PWM}
7	u_{CLK}
8	$u = u_{\text{PWM}} \bullet u_{\text{CLK}}$

IV. EXPERIMENTAL RESULTS AND DISCUSSIONS

A. Measured Signal of Test Locations

Fig. 3 shows the steady-state waveforms of the test locations when the converter is operating at full-load condition. The captured results are consistent with the theoretical expectation.

Figs. 4(a) and 4(b) show respectively the DC output voltage versus the input voltage and operating load resistance. For line variation test, it can be observed that v_o decreases with increasing v_i. Specifically, the output voltage deviation is -0.15 V (i.e. -0.31 % of V_{od}) for the entire input range $18 \text{ V} \leq v_i \leq 30 \text{ V}$, i.e., line regulation $\frac{dv_o}{dv_i}$ averages at -12.5 mV/V. For load variation test, it can be concluded that voltage regulation of the converter is robust to load changes, with only a 0.08 V deviation (i.e. 0.17 % of V_{od}) in v_o for the entire load range $48 \ \Omega \leq r_L \leq 900 \ \Omega$, i.e., load regulation $\frac{dv_o}{dr_L}$ averages at 0.09 mV/Ω.

B. Transient Performance

The dynamic behavior of the controller is studied using a load resistance that alternates between resistances of 240 Ω and 480 Ω, and an input voltage that alternates between

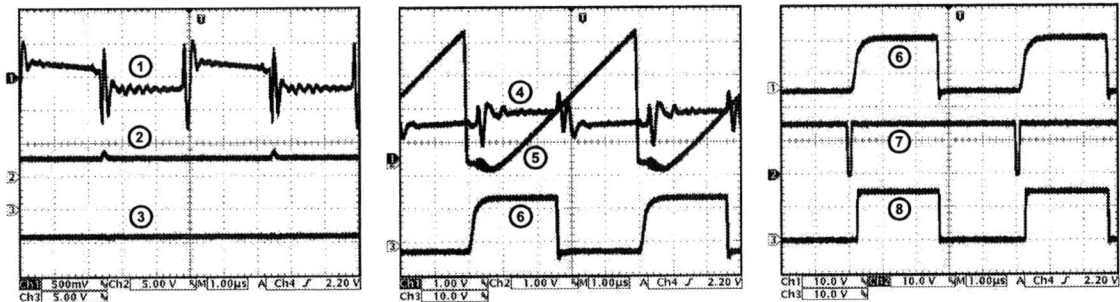

Fig. 3. Experimental waveforms of the test locations under full-load operation.

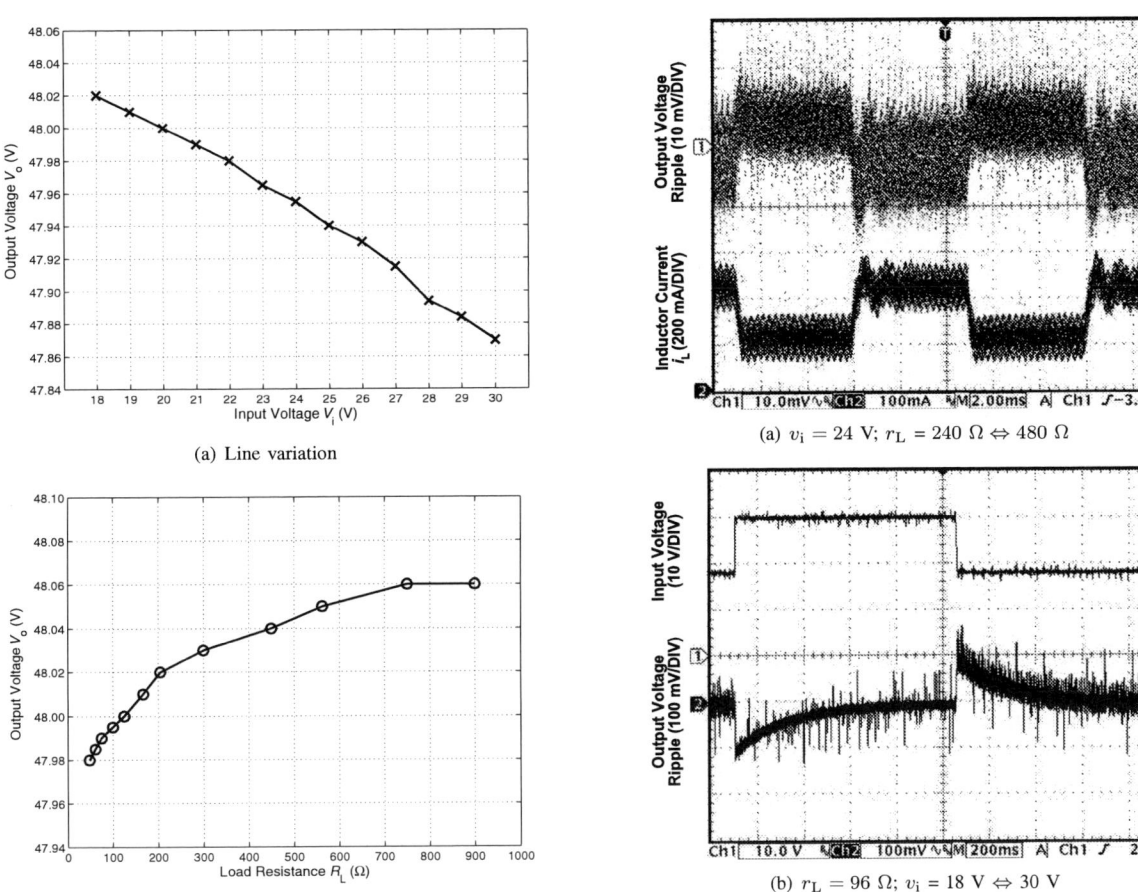

(a) Line variation

(b) Load variation

Fig. 4. Graphs of DC output voltage v_o against (a) input voltage v_i and (b) load resistance r_L for the PWM based SMVC boost converter.

(a) $v_i = 24$ V; $r_L = 240$ Ω $\Leftrightarrow$ 480 Ω

(b) $r_L = 96$ Ω; $v_i = 18$ V $\Leftrightarrow$ 30 V

Fig. 5. Waveforms of (a) output voltage v_o and inductor current i_L; and (b) input voltage v_i and output voltage v_o of the SMVC boost converter operating in dynamic line change and dynamic load change conditions.

voltages of 18 V and 30 V. Figs. 5(a) and 5(b) show respectively the output voltage ripple waveforms for both the dynamic load change and dynamic line change operations. The dynamic performances of the converter are adequate for common voltage conversion applications.

C. Operating in Discontinuous Conduction Mode

The PWM based SMVC boost converter, which is designed for operation in CCM, is also tested in discontinuous conduction mode (DCM). Fig. 6 shows the variation of the DC output voltage against different operating load resistances in the DCM. The voltage regulation of the converter has a 0.12 V

299

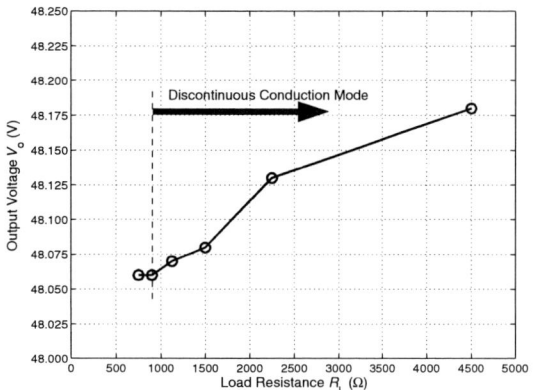

Fig. 6. Graphs of DC output voltage v_o against load resistance r_L for the PWM based SMVC boost converter in light load conditions.

deviation (i.e. 0.25 % of V_{od}) in v_o for the DCM load range $900 \ \Omega \leq r_L \leq 4500 \ \Omega$, i.e., load regulation $\frac{dv_o}{dr_L}$ averages at 0.033 mV/Ω.

V. CONCLUSION

A fixed-frequency PWM based SM voltage controller for boost converter is presented. The methods of modelling the system and translation of the SM control equations for the PWM implementation are illustrated. It is shown that the control technique is easily realized with simple analog circuitries. Different static and dynamic tests with line and load changes are also performed. It can be concluded from the results that the derived controller/converter system is feasible for common step-up conversion purposes.

REFERENCES

[1] V. Utkin, J. Guldner, and J.X. Shi , *Sliding Mode Control in Electromechanical Systems*. London, U.K.: Taylor and Francis, 1999.

[2] R. Venkataramanan, A. Sabanoivc, and S. Ćuk, "Sliding mode control of DC-to-DC converters," in *Proceedings, IEEE Conference on Industrial Electronics, Control and Instrumentations (IECON)*, pp. 251–258, 1985.

[3] B.J. Cardoso, A.F. Moreira, B.R. Menezes, and P.C. Cortizo, "Analysis of switching frequency reduction methods applied to sliding mode controlled dc-dc converters," in *Proceedings, IEEE Applied Power Electronics Conference and Exposition (APEC)*, pp. 403–410, Feb 1992.

[4] P. Mattavelli, L. Rossetto, G. Spiazzi, and P. Tenti, "General-purpose sliding-mode controller for dc/dc converter applications," in *IEEE Power Electronics Specialists Conference Record (PESC)*, pp. 609–615, June 1993.

[5] V.M. Nguyen and C.Q. Lee, "Tracking control of buck converter using sliding-mode with adaptive hysteresis," in *IEEE Power Electronics Specialists Conference Record (PESC)*, vol. 2, pp. 1086–1093, June 1995.

[6] V.M. Nguyen and C.Q. Lee, "Indirect implementations of sliding-mode control law in buck-type converters," in *Proceedings, IEEE Applied Power Electronics Conference and Exposition (APEC)*, vol. 1, pp. 111–115, March 1996.

[7] S.K. Mazumder and S.L. Kamisetty, "Design and experimental validation of a multiphase VRM controller", in press for final publication," in *Proceedings, IEE Electric Power Applications*, September 2005.

[8] S.C. Tan, Y.M. Lai, M.K.H. Cheung, and C.K. Tse, "On the practical design of a sliding mode voltage controlled buck converter", *IEEE Transactions on Power Electronics*, vol. 20, no. 2, Mar. 2005

[9] S.C. Tan, Y.M. Lai, C.K. Tse, and M.K.H. Cheung, "A fixed-frequency pulse-width-modulation based quasi-sliding mode controller for buck converters", *IEEE Transactions on Power Electronics* , vol. 20, no. 6, pp. 1379–1392, Nov. 2005.

[10] S.C. Tan, Y.M. Lai, and C.K. Tse, "Adaptive feedforward and feedback control schemes for sliding mode controlled power converters," *IEEE Transactions on Power Electronics*, vol. 21, no. 1, pp. 182–192, Jan. 2006.

[11] H.W. Whittington, B.W. Flynn, and D.E. Macpherson, *Switched Mode Power Supplies : Design and Construction*. New York: Wiley, 2nd ed., 1997.

[12] H. Sira-Ramirez, "A geometric approach to pulse-width modulated control in nonlinear dynamical systems," *IEEE Transactions on Automatic Control*, vol. 34 no. 3, pp. 184–187, Feb. 1989.

[13] L. Martinez, A. Poveda, J. Majo, L. Garcia-de-Vicuna, F. Guinjoan, J.C. Marpinard, and M. Valentin, "Lie algebras modelling of bidirectional switching converters," in *Proceedings, European Conference on Circuit Theory and Design (ECCTD)*, vol. 2, pp. 1425–1429, Sep. 1993.

[14] J. Ackermann and V. Utkin, "Sliding mode control design based on Ackermann's formula," *IEEE Transactions on Automatic Control*, vol. 43 no. 2, pp. 234–237, Feb. 1998.

2006 5th International Power Electronics and Motion Control Conference

Low Cost Electronic Ballast with Buck Converter as PFC Stage

Li Xiangrong, Xu Dianguo and Zhang Xiangjun
Department of Electrical Engineering, Harbin Inst of Tech, Harbin, P. R. China
leekoei@hotmail.com

Abstract—In this paper, a low cost buck converter was selected as the PFC stage of the electronic ballast for a 70W metal halide lamp. In this way, several advantages could be obtained, such as simple circuit topology, easy control strategy, especially decreased cost and low requirements for circuit components. All circuit parameters should be designed to ensure that the buck converter was operated in DCM. This paper presented the design of both the main circuit and control loop, analyzed the stability of buck converter, and discussed the function of PFC stage. Finally, the results of both simulation and experiment were given to testify the effectiveness of PFC.

Keywords-electronic ballast; PFC stage; buck converter; DCM

I. INTRODUCTION

Along with the increasing demands for higher lighting quality, traditional incandescent and fluorescent lamps can not meet these needs. Since the new generation of light sources, such as metal halide (MH) lamps and high pressure sodium (HPS) lamps, have high efficiency and long lifespan, they can be perfect alternatives for traditional lamps. Due to the negative resistance characteristic of MH lamps, they must be operated with proper electronic ballast to work effectively as demonstrated in Ref. [1]. Nowadays, the worldwide researches on high quality ballast are intensive, but because of some factors such as costs and reliability, most achievements are still in experimental stage. In order to popularizing MH lamps, electronic ballast must have low costs and high reliability. This paper sets a 70W MH lamp as the object, designs an electronic ballast and concentrates on the analysis of using buck converter as the PFC stage. Although the traditional passive PFC topology is more economical, they have intrinsic problems of crest factor and harmonic components, which can have negative effects on the power line. Therefore, a buck converter operated in DCM could be an option to settle the contradiction between costs and performance.

In section 2, the built prototype will be presented. The stability of the buck converter is analyzed in section 3. The function of this converter as the PFC stage is discussed in section 4 and the related testing results are given in section 5.

II. BUCK STAGE

The popular electronic ballast topology is divided into two stages: the first stage converts the power line voltage into DC bus voltage and also performs the function of PFC; the second stage applies a HF inverter to generate high frequency AC square wave to deliver power to MH lamp. Currently, there are many mature topologies of the first stage, for instance, the boost and buck-boost converters. The boost PFC circuit has the merits of low THD and high efficiency, but it generates high DC bus voltage, thus puts too much pressure on the power components and other passive components of the whole circuit. For the buck-boost circuit, though it has perfect performance as PFC stage, ordinary one-switch buck-boost circuit also puts high voltage stress on circuit components, thus increases the overall costs of the ballast. Consequently, a buck converter is chosen as the PFC stage of this design. It can achieve:

(a) No instable pole and zero points in open loop transfer function

(b) Easy design and thus lowered cost of control loop;

(c) Low DC bus voltage and low voltage stress on circuit components as stated in Ref. [1] [2].

A brief layout of the PFC stage is shown in Fig. 1.

In order to obtain constant DC bus voltage while retain low cost, the control circuit is a voltage feedback loop containing 555 and amplifiers. The scheme of control loop is shown in Fig. 2.

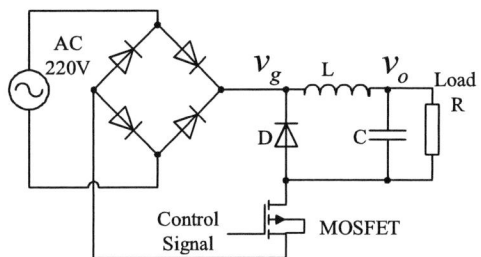

Figure 1. A brief layout of Buck PFC stage

1-4244-0448-7/06/$25.00 ©2006 IEEE

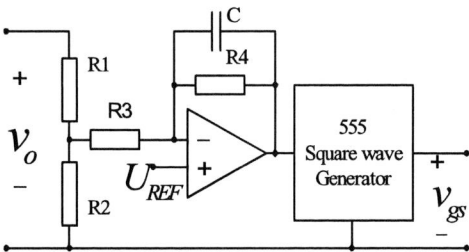

Figure 2. Scheme of control loop

In steady stage, R_1 and R_2 sample the DC bus voltage and serve as a voltage divider. Then, after compared with DC reference voltage and amplified by a PI regulator, the resulted differential voltage is delivered into the input end of the 555 square-wave generator. This generator operates in PFM to adjust the frequency as well as the duty ratio of the output control signal V_{gs}. Finally, this signal is sent to the gate of the MOSFET to form a complete voltage feedback loop.

Actually, there is an EMI filter between the power line and the PFC stage to eliminate the high frequency harmonic components superimposed on the input current, which is generated by the switching operation of power components. Furthermore, this filter can also keep the disturbance signal from passing in and out the ballast, which acts as an electromagnetic isolator.

III. STABILITY OF BUCK CONVERTER

In order to completely discuss the stability of Buck converter, a mathematic model of it must be established first. Since the transient time of the proposed ballast is relatively short, most emphasis is put on the steady state of the circuit. Therefore, the variation of the DC components in the circuit can be considered constant. In this paper, the small signal model of general buck converter is deduced. Then the transfer function and thus the relative analysis can be obtained easily as in Ref. [3].

Before any analysis, some important assumptions must be made:

(a) To ignore the influence from input side, assume that the input voltage of the converter is constant

(b) Suppose that in steady state, the duty ratio of the control signal is fixed;

(c) The amplitude of the small signal is far less than the fundamental components of all parameters;

(d) Assume the load is a constant resistor;

(e) Set the average value of all circuit parameters over one switching period as the instantaneous value of that period.

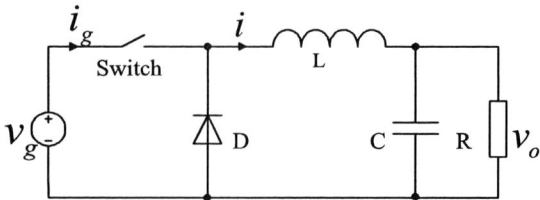

Figure 3. Scheme of buck converter and relative parameters

Based on the assumptions mentioned above, we can obtain the expressions of the input voltage, input current, inductor current, capacitor voltage as well as the duty ratio by superimposing small signal turbulences on the steady values respectively: $v_g = V_g + \hat{v}_g$, $i_g = I_g + \hat{i}_g$, $i = I + \hat{i}$, $v_o = V_o + \hat{v}_o$, $d = D + \hat{d}$. The parameters with superscripts are superimposed small signal components. The general scheme of buck converter is shown in Fig. 3 for the sake of easy analysis.

In continuous conducting mode (CCM), the Kirchhoff equations of buck converter can be obtained in reference to Fig 3. Then, substitute the small signal expressions of these parameters into the equations. In steady state, the DC components in both sides of each equation can be counteracted, and after ignoring the high-order items, the simplified version of circuit equations is obtained as (1).

To change the equations into small signal model, we can simply consider the small signal components in the equations as dependent and independent voltage sources as well as current sources, the scheme of the small signal model can be plotted as shown in Fig. 4. Then, after transforming all parameters into s domain, the transfer function of buck converter can be calculated.

$$\begin{cases} \hat{i}_g(t) = D\hat{i}(t) + I\hat{d}(t) \\ L\dfrac{d\hat{i}(t)}{dt} = D\hat{v}_g(t) - \hat{v}_o(t) + V_g\hat{d}(t) \\ C\dfrac{d\hat{v}_o(t)}{dt} = \hat{i}(t) - \dfrac{\hat{v}_o(t)}{R} \end{cases} \quad (1)$$

First, suppose the duty ratio to be constant and thus the small signal component $\hat{d}(s)$ is equal to zero. We can get the transfer function $G_{vg}(s)$ defined as the input voltage $\hat{v}_g(s)$ versus the output voltage $\hat{v}_o(s)$:

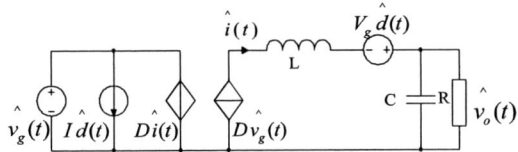

Figure 4. The small signal equivalent circuit of a buck converter operated in CCM

$$G_{vg}(s) = \frac{\hat{v}_o(s)}{\hat{v}_g(s)}\bigg|_{\hat{d}(s)=0} = G_{go}\frac{1}{1+\dfrac{s}{Q\omega_o}+\dfrac{s^2}{\omega_o^2}} \quad (2)$$

Where $G_{go} = D$, $Q = R\sqrt{\dfrac{C}{L}}$ and $\omega_o = \dfrac{1}{\sqrt{LC}}$

Similarly, we again assume the input voltage to be constant and get the transfer function $G_{vd}(s)$ as the duty ratio $\hat{d}(s)$ versus the output voltage $\hat{v}_o(s)$:

$$G_{vd}(s) = \frac{\hat{v}_o(s)}{\hat{d}(s)}\bigg|_{\hat{v}_g(s)=0} = G_{do}\frac{1}{1+\dfrac{s}{Q\omega_o}+\dfrac{s^2}{\omega_o^2}} \quad (3)$$

Where $G_{do} = V_g$ and the other two parameters are the same as (2).

In the similar method, the equivalent forms of these transfer functions in DCM can also be obtained. When operated in this mode, the two transfer functions above change into first-order and the numerators of which turn into constants. The specific approach of deduction is omitted here.

In consideration of the linear voltage feedback loop, the open loop transfer function of a buck converter is a second-order system in CCM but a first-order system in DCM. By properly setting circuit parameters, any small signal disturbance can be quickly eliminated. Therefore, with proper configuration of the circuit components, the stability of buck converter can be guaranteed.

IV. PFC FUNCTION OF BUCK CONVERTER

Currently, there are many choices of PFC topologies for electronic ballast, and the most mentioned one may be the active boost converter PFC circuit. By applying buck-boost converter, a power factor closed to 1 can be easily obtained. However, in order to operate buck-boost converter properly, additional control chips and relative ancillary circuit are needed, which certainly increases the overall costs of the ballast. Actually, in low wattage applications, the requirement of power factor is usually not so strict. Accordingly, a simple circuit topology can

be chosen to achieve acceptable power factor and other requirements. In this design, a buck converter operated in DCM is chosen to perform the PFC function of the proposed ballast in reference with Ref. [4]. Since the major part of the control circuit is a voltage feedback loop, this circuit is easy to design, install and debug. The following discussion will focus on the PFC function of buck converter.

Also, some presumptions must be made first:

(a) The main circuit is in steady state.

(b) All the components are ideal without any distributed parameters.

(c) The filter capacitor is large enough that the output voltage can be considered constant.

(d) The rectified voltage after the bridge can be considered constant in a given switching period. That is to say, the switching frequency is far higher than the line frequency.

Under these assumptions, we can get the waveforms of inductor current I_L and the current after the bridge I_g over a switching period as shown in Fig. 5.

In Fig. 5, it can be easily seen that when the instantaneous value of the rectified voltage U_g is greater than the DC bus voltage U_o, the operation of the buck converter during a switching period can be divided into three stages:

I: In this stage, the switch MOSFET is on. The rectified voltage charges the inductor and capacitor while supply the load at the same time. The current through the inductor is the same as the rectified current, and they rise linearly as time flows. The following equation can be obtained:

$$i_L(t) = \frac{U_g - U_o}{L}t \quad (4)$$

And $i_L(t)$ reaches its maximum value $I_{L\max}$ at DT_s.

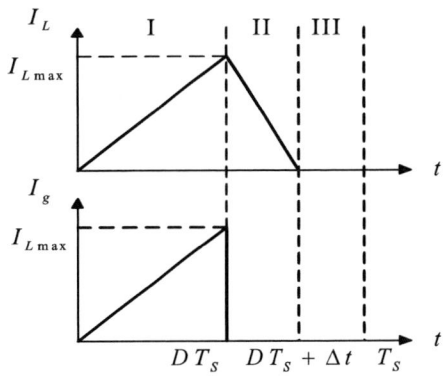

Figure 5. The waveform of inductor current I_L and the current after the bridge I_g over a switching period

II: The switch MOSFET is shut off in this stage; the magnetic energy stored in the inductor begins to charge the capacitor and supplies the load. Meanwhile, there is no input current to the converter and the diode D starts to offer channel for the inductor current. The inductor current decreases linearly:

$$i_L(t) = I_{L\max} - \frac{U_o}{L}(t - DT_s) \quad (5)$$

III: The energy stored in the inductor is expired, and the capacitor begins to supply the load. In this stage, both the inductor current and the rectified current are equal to zero.

Averaging the rectified current in a switching period, we can get:

$$I_g = \frac{1}{T_s}\int_0^{DT_s} i_g(t)dt = \frac{1}{T_s}\int_0^{DT_s} \frac{U_g - U_o}{L} t\,dt = \frac{D^2 T_s}{2L}(U_g - U_o) \quad (6)$$

According to (6), we can get the theoretical waveform of rectified voltage and current as shown in Fig. 6.

From the waveform below, when the rectified voltage U_g is greater than the output DC bus voltage U_o, the buck converter functions normally, and the rectified current I_g is in proportion with $(U_g - U_o)$. During this time period, the input voltage and current of the ballast are in-phase. On the other hand, when U_g is lower than U_o, which means that the input voltage of the buck converter is lower than its output voltage, though the control signal is still available, the converter stops working. Since in this stage, the rectifiers in the bridge are reversely biased, the input current is in dead-zone.

It must be emphasized that only when operating in DCM could the buck converter achieve PFC function. Conversely, if operated in CCM, at the beginning of each switching period, the inline current is not equal to zero. In this case, averaging the current over a switching period can not get a linear expression of current versus voltage, thus the PFC function could not be obtained. Moreover, if buck converter always works in CCM, the inductor current may be reversed in light load, and the on-state

loss will increase, causing the decrease in overall efficiency. In conclusion, in the whole course of designing, the converter must be guaranteed to operate in DCM.

V. EXPERIMENTAL AND SIMULATION RESULTS

According to the requirements of a Philips MHN-TD 70W lamp, the input side of the PFC stage is the rectified 220V AC voltage, while the output of this stage is a 120V DC bus voltage with low ripples.

After carefully adjusting the simulation parameter, some simulated results, which are very close to the practically tested ones, are obtained. Fig. 7 shows the waveforms of input voltage and current, from which we can see that the voltage and the current are in phase so that the power factor can be relatively high. Also in Fig. 8, the Fourier analysis of the input current is made to show that the high order harmonics are small in amplitude comparing with the fundamental wave.

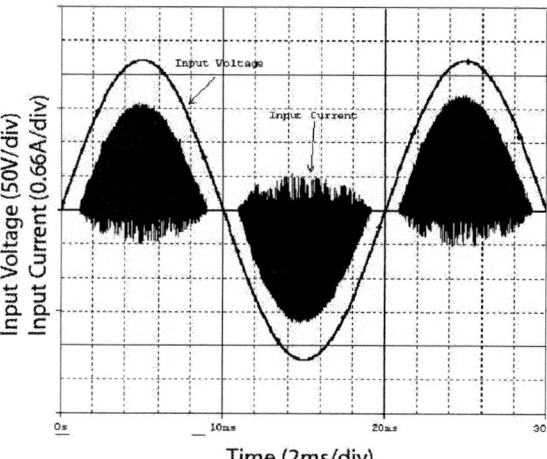

Figure 7. The simulated waveform of input voltage versus current in steady state

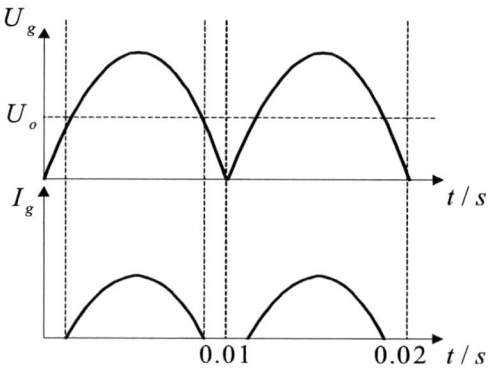

Figure 6. The waveform of rectified voltage and current

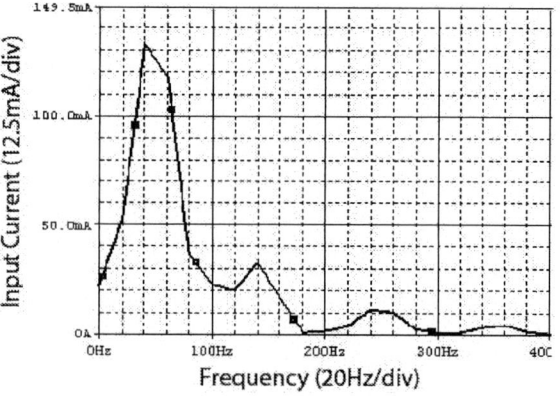

Figure 8. The Fourier analysis of the input current

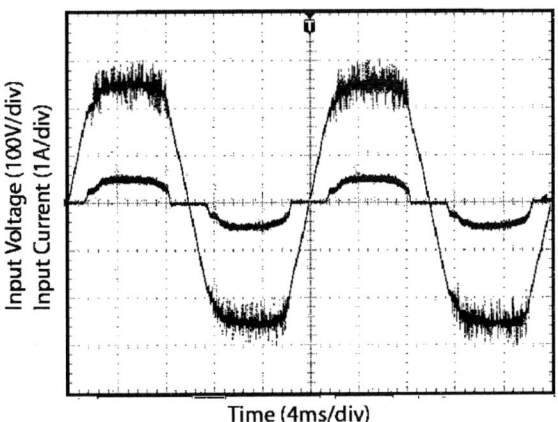

Figure 9. The waveform of input voltage versus current in steady state

TABLE I.

STATISTICS OF CIRCUIT PERFORMANCE WITH VARYING INPUT VOLTAGE

Input voltage (V)	Output power (W)	Power factor	Output voltage (v)
180.1	75.9	0.967	119.2
190.8	75.8	0.969	119.1
201.1	75.4	0.970	118.7
210.3	75.2	0.973	118.4
220.5	75.6	0.972	118.5
230.8	73.9	0.974	117.0
240.6	73.9	0.976	117.0
250.1	74.0	0.978	117.1

A Tektronix TDS3032B oscilloscope was used here to get both the input voltage and current when the whole circuit was in steady state, as shown in Fig. 9.

From the figure above, the input voltage is in phase with the input current, which is in accordance with the simulation result, and the measured power factor is fluctuating between 0.96~0.98. Moreover, due to the function of the DC bus voltage feedback loop, when the input voltage changes from 180V to 250V, the power factor, output power as well as the bus voltage change slightly, as listed in Table 1.

Since there is no filter capacitor between the rectifier and the buck converter, when the rectified voltage is lower than the DC bus voltage, the converter actually fails to operate, which can cause the dead-zone in the input current. In this case, although the input current and voltage is still in phase over the whole line period, due to the input current dead-zone, the THD of input current is relatively high, which directly have passive influence on the power factor. Moreover, in the input current dead-zone, the input energy is cut from the converter, and the capacitor discharges to supply the load. In this way, the voltage across the capacitor will inevitably decrease with time, which gives reason to the ripples in bus voltage. Although by increasing the filter capacitor and raising the switching frequency can lower the ripples to some extent, the ripples in the DC bus voltage can not be eliminated thoroughly since the input dead-zone is inevitable.

VI. CONCLUSION

The PFC stage of an electronic ballast for MH lamp is presented in this paper. With simple circuit topology and easy control strategy, the buck converter operated in DCM can achieve acceptable power factor, and the overall cost of the ballast is low. Specifically, the major advantages of this design are listed as below:

(a) Easy construction of both main circuit and control circuit. No specific PFC chip was used thus the overall costs can be kept low.

(b) Low DC bus voltage comparing with boost and buck-boost topologies. Lowered requirements on circuit components.

(c) Although the input current dead-zone is inevitable, through careful configuration of circuit parameter, a relatively high power factor can be obtained.

REFERENCES

[1] Garcia. J, Cardesin. J, and Alonso. M, "New HF Square-waveform Ballast for Low Wattage Metal Halide Lamps Free of Acoustic Resonance," in *Conference Record of the 2004 IEEE 39th IAS Annual Meeting*, 2004.

[2] Rico-Secades. M, Corominas. E. L, and Alonso. J. M, "Complete Low Cost Two-stage Electronic Ballast for 70W High Pressure Sodium Vapor Lamp Based on Current-mode-controlled Buck-boost Inverter," in *Conference Record of IEEE 37th IAS Annual Meeting*, 13-18 Oct. 2002, pp. 1841-1846.

[3] Erickson, Robert. W, *Fundamentals of Power Electronics.* 2nd ed. Kluwer Academic Publishers. 2000, Chap. 8, pp. 266-350

[4] Ned Mohan, Tore M. Undeland, William P. Robbins, *Power Electronics: Converters, Applications, and Design*, 3rd ed. John Wiley & Sons, Inc. 2003, Chap. 7, pp. 162-171.

2006 5th International Power Electronics and Motion Control Conference

A New Converter Architecture for Future Generations of Microprocessors

Dodi Garinto

Indonesia Power Electronics Center, Surakarta, Indonesia

Email: dodi.g@ieee.org

Abstract — **In this paper, a new converter architecture that is based upon the buck topology and that allows a multi-interleaving technique is proposed. The converter provides automatic current sharing and improving current ripple cancellation effect. Multiphase buck converters with multi-interleaving technique perform better than with interleaving technique because the multi-interleaving technique can extend the duty cycle, can improve the transient response without increasing current ripple in each cell, and can raise the switching frequency with low switching, gate drive and body diode losses. Moreover, a bypass LC filter is presented to achieve nonpulsating input and output currents. As a result, based on losses analysis and simulation results, high efficiency, high power density, fast transient response and low-cost 12 V input VRM with output voltage 0.5 to 1 V and output current 100 to 300 A to power future generations of microprocessors can be realized.**

Keywords–multi-interleaving; voltage regulator module; future microprocessors; switching frequency multiplication effect; new converter architecture; bypass LC filter

I. INTRODUCTION

Currently, Pentium-IV microprocessors run at 3 GHz. Future microprocessors will run at 20 GHz. In Fig. 1, according to the Intel Corporation, transistor account in microprocessor will increase from 200 millions in 2005 to 1 billion in 2010. To decrease the power consumption, supply voltage for future microprocessors must as low as possible. As shown in Fig. 2, to continue to extend Moore's Law, there is a necessity to reduce the power consumption of microprocessor. However, technical conflicts arise when the power supply, well known as Voltage Regulator Module (VRM), is operated at output voltage below 1 V. In 2010, future microprocessors are expected will draw current 150 A with 0.8 V supply voltage and ± 2 % output voltage tolerance. VRM design with output voltage below 1 V, fast transient response, high efficiency, and high power density pose serious technical challenges [1]. Today's 12 V input VRM widely used in industry based upon multiphase interleaving buck converter, as can be seen in Fig. 3. The topology is not only experimentally proven can reduce current ripple to the output filter capacitors and improve the transient response, but also increase the power density [2]. However, as shown in Fig. 4, the multiphase interleaving buck converter only produces current ripple cancellation

effect. The limitation is the inductor current ripple in each channel still large [3] and the inductor current ripple frequency unchanged.

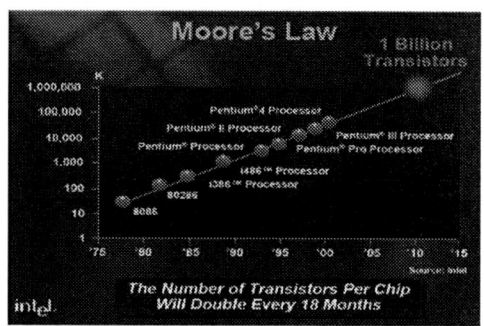

Fig. 1. Moore's Law

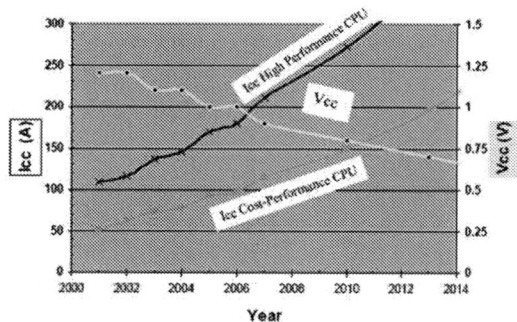

Fig. 2. Challenge to continue to extend Moore's Law

Therefore, the multiphase interleaving buck converter topology has three technical conflicts that can be formulated as follows. First, when the output voltage is decreased, then the duty cycle also decreased. In case of 12 V input VRM, to offer output voltage 0.5 to 1 V, this topology suffers very small duty cycle (D = Vo/Vin), that is below 10 %. It is not only yield poorer ripple cancellation effect, but also increase turn-off loss on the top switch and conduction loss on the bottom switch in each channel [4]. Second, when the switching frequency is increased, then the switching, gate drive and body diode losses also increased. With duty cycle below 10 %, raising the switching frequency to multi-MHz will reduce the efficiency less than 80 % [5]. Because the switching frequency is equal to the inductor current ripple

1-4244-0448-7/06/$25.00 ©2006 IEEE

frequency, then to keep high efficiency with very small duty cycle, the switching frequency is operated between 300 kHz to 500 kHz so that need the excessive amount of bulk output capacitors and decoupling capacitors [6]. Third, when the inductor current slew rate is increased with smaller inductance value to improve the transient response, then the inductor current ripple also increased. It is not only harmful action for the top switch due to larger turn-off loss, but also for the bottom switch due to larger conduction loss. It is also increase the inductor winding loss. This conflict limits the average inductor current in each channel. Moreover, there is a trade off between efficiency and transient response [7]. As a result, these technical conflicts are not only increase costs and sacrifice the power density, but also very difficult to meet the power requirements of future microprocessors before the technical conflicts are resolved [8].

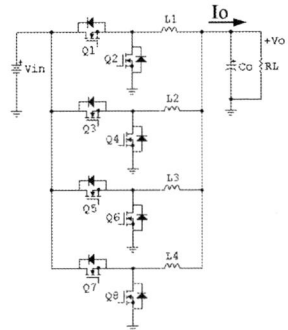

Fig. 3. Multiphase interleaving buck converter

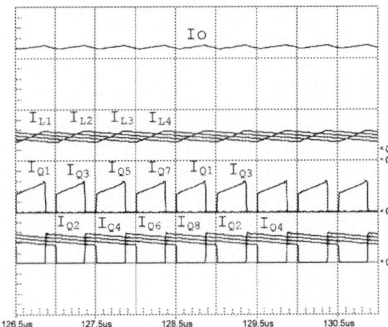

Fig. 4. Interleaving technique

In recent years, new converter topologies have been proposed to solve the above technical conflicts. Such as multiphase clamp coupled-buck converter [9], phase-shift buck converter [10], winding-coupled buck converter [11], etc. In fact, extending the duty cycle is not enough due to the technical conflicts are complex problems that can not be solved in one or two steps. In recent times, the two-stage solution was proposed [12], as can be seen in Fig. 5. However, the two-stage solution is an impractical approach because it is not removing the technical conflicts. How to remove the technical conflicts described above still a great challenge.

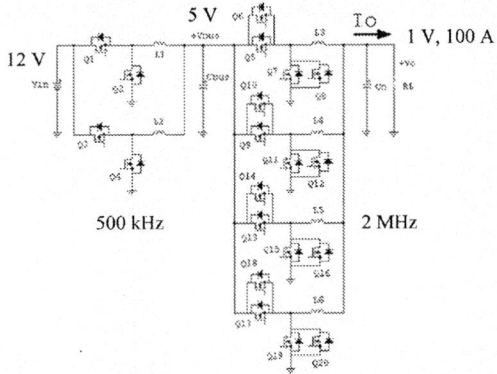

Fig. 5. Optimized duty cycle by using two-stage solution

In this paper, a new converter architecture that is based upon the buck topology and that allows a multi-interleaving technique to solve the technical conflicts discussed above is proposed. Multiphase buck converter with multi-interleaving technique perform better than with interleaving technique because the multi-interleaving technique is not only improving current ripple cancellation effect, but also can extend the duty cycle, can improve the transient response without increasing current ripple in each cell, and can raise the switching frequency with low switching, gate drive and body diode losses

II. MULTIPHASE CURRENT DOUBLER-BUCK CONVERTER – A NOVEL TOPOLOGY

A. Multiphase current doubler-buck converter

At this time, 12 V input VRM design uses multiphase buck converters with interleaving technique. The main benefit is current ripple cancellation effect. Based on the technical conflicts analysis, multiphase buck converters with interleaving technique and current ripple cancellation effect are not enough to meet the power requirements of future microprocessors. Therefore, as shown in Fig. 6, the multiphase current doubler-buck converter is proposed. Because the segmentation, phase shifting and merging principles are implemented, the converter architecture allows a multi-interleaving technique. Fig. 7 shows the operation waveform.

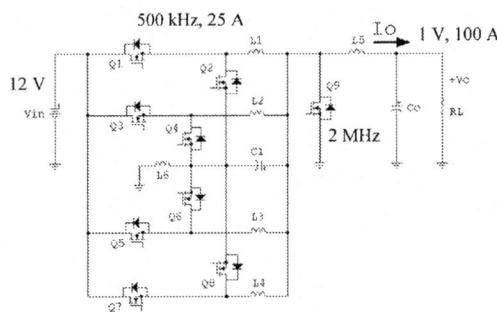

Fig. 6. Proposed multiphase current doubler-buck converter

307

In steady-state operation, when one of the top switches (Q1, Q3, Q5 or Q7) is turned-on, the current flows from voltage source to the load through the top switch, inductor cell (L1, L2, L3 or L4) and L5. In this condition, the energy stored in L6 is delivered to the load through C1 and L5. When the top switch (Q1, Q3, Q5 or Q7) is turned-off, the energy stored in the inductor cell is delivered to C1 through one of the bottom switches in one cell (Q2, Q4, Q6 or Q8). At the same time, the energy stored in L5 is transferred to the load through Q9. In this condition, the current also flows and stored in L6 through C1. The resonance inductor (L6), capacitor clamp (C1), the auxiliary output inductor (L5) and the auxiliary bottom switch (Q9) are added to produce current ripple cancellation effect, switching frequency multiplication effect and current multiplication effect.

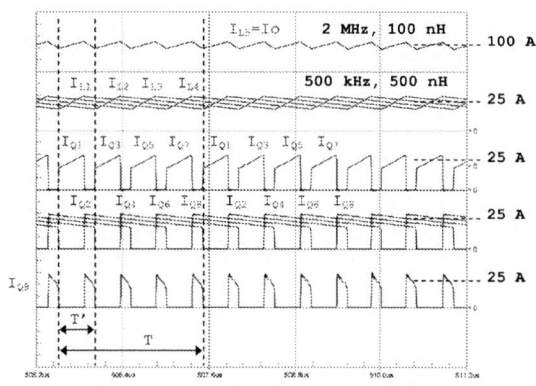

Fig. 7. The switching frequency can be raised with low switching losses

Simulation result shows that the operating frequency of Q9, as can be seen in Fig. 7, is naturally n times the switching frequency of buck converter cells, where n is amount of buck converter cells. In this case, n = 4. The gate drive signals for the top switch in each cell have a phase shift 360°/n. As can be seen in Fig. 7, from the operating frequency point of view of Q9, the switch Q3 replaces Q1, Q5 replaces Q3, Q7 replaces Q5, Q1 replaces Q7, and Q3 replaces Q1 again, so that the switching frequency can be raised with low switching, gate drive and body diode losses. Based on Fig. 7, T is the switching period of the top switch with regard to the bottom switch and T' is the switching period of the top switch relating to the auxiliary bottom switch. Since $T' = T/n$ and $T = 1/F_s$, where n is amount of buck converter cells and F_s is the switching frequency with regard to the bottom switch, then $1/F_s' = 1/n.F_s$ or $F_s' = n.F_s$ where F_s' is the switching frequency relating to the auxiliary bottom switch. The relationship between the pulse width of switch Q1, Q3, Q5 or Q7 and the pulse width of switch Q9 creates a new duty cycle ($D' = DT/T' = n.D$). This phenomenon can be viewed as a switching frequency multiplication effect. As shown in Fig. 7, when one of the top switches is turned-off, then Q9 is turned-on, and when one of the top switches is turned-on, then Q9 is

turned-off. Moreover, the auxiliary output filter inductor current $I_{L5} = I_{L1}+I_{L2}+I_{L3}+I_{L4}$, and peak current of the auxiliary bottom switch (Q9) is equal to peak current of the top switch (Q1, Q3, Q5 or Q7). It is due to when all of the top switches are turned-off, the inductor current of L5 also flows to the load through L6 and C1. As a result, multi-MHz switching frequency issue is solved when only one switch, that is the auxiliary bottom switch, in zero voltage switching condition, naturally operates at higher switching frequency. With this technique, energy storage costs can be reduced dramatically and transient response can be improved without sacrificing the efficiency. Furthermore, lower output voltage and higher output current can be achieved without duty cycle complication. However, high current flows into L5 and L6 due to current multiplication effect. Consequently, higher inductor current rating is required. It is interesting when all of the inductor cells and the auxiliary output inductor are integrated using coupled inductors.

B. Multiphase current doubler-buck converter with a bypass LC input filter

For that reason, to reduce the winding losses, as shown in Fig. 8, the multiphase current doubler-buck converter with a bypass LC input filter is proposed. As can be seen in Fig. 8, inductor L1 and capacitor C1 can be viewed as a bypass LC input filter. This topology uses inductor and capacitor to store and transfer energy from input to the output. Thus, the energy transfer mechanism occurs during one of the top switches is ON and OFF. Also, this configuration provides nonpulsating input and output currents. Fig. 9 shows the operation waveform.

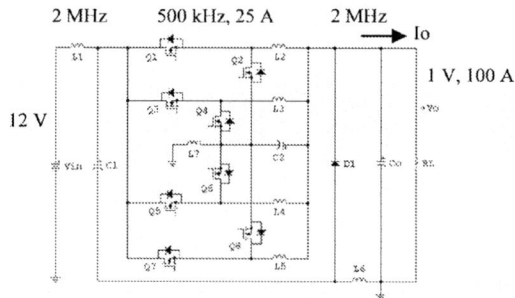

Fig. 8. Multiphase current doubler-buck converter with a bypass LC input filter

In steady-state operation, when one of the top switches (Q1, Q3, Q5 or Q7) is turned-on, then D1 is reverse biased and the current flows from voltage source to the load through L1, the top switch and inductor cell (L2, L3, L4 or L5). At the same time, the energy stored in C1 is discharged to the load through the top switch (Q1, Q3, Q5 or Q7), the inductor cell and L6. Consequently, the inductor currents L1 and L6 increase linearly and $I_{C1} = I_{L6}$. In this condition, the energy also flows from L7 to the load through C2. Also, peak current of the top switch is the sum of the peak currents L1 and L6. When the top switch (Q1, Q3, Q5 or Q7) is turned-off, the energy stored in the inductor cell is transferred to C2 through one of the bottom switches in one cell (Q2, Q4, Q6 or Q8). In this condition, D1 is forward biased and the current also flows from voltage source to the load through L1, C1 and D1. At the same time, the inductor current L6 is delivered to the load through D1. As a result, C1 is charged and peak current of D1 is $I_{L1}+I_{L6}$ or equal to peak current of the top switch. The output current Io = $I_{L2}+I_{L3}+I_{L4}+I_{L5}+I_{L6}$. It is clear that by using four buck-cells and 500 kHz switching frequency in each cell, this topology generates switching frequency multiplication effect as high as 2 MHz. Accordingly, the bulk output capacitor can be eliminated without increasing switching and gate drive losses.

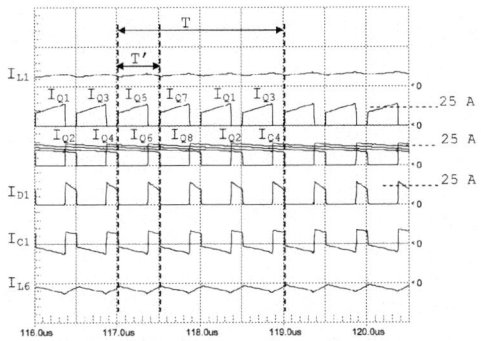

Fig. 9. Switching frequency multiplication effect

III. MULTIPHASE MULTI-INTERLEAVING CURRENT DOUBLER-BUCK CONVERTER

In low voltage and high current application, particularly for future microprocessors, it is clear that there is a motivation to achieve high bandwidth or multi-MHz switching frequency. In this paper, advanced multiphase multi-interleaving buck converter topology is proposed. As shown in Fig. 10, unlike interleaving technique, by using at least two modules and each module consist of n buck converter cells, where n is an integer greater than 2, the multi-interleaving technique can produce switching frequency multiplication effect as high as n times the switching frequency of buck converter cells in each module. Moreover, the capacitor clamp (C1, C2) and the resonance inductor (L6, L12) provide automatic current sharing, so that reduces the complexity of the controller.

Fig. 11 shows that interleaving occurs not only at buck converter cells in each module, but also inter-module. It means larger inductor current slew rate with smaller inductance value of the auxiliary output filter inductor in each module can improve the transient response. The auxiliary output inductor currents $I_{L5}=I_{L1}+I_{L2}+I_{L3}+I_{L4}$ and $I_{L11}=I_{L7}+I_{L8}+I_{L9}+I_{L10}$. The output current Io = $I_{L5}+I_{L11}$. To achieve multi-interleaving operation, at least required two modules and two buck converter cells in each module, and the gate drive signals for the top switch and the bottom switch in buck converter cells in each module have a phase shift 360° divided by amount of buck converter

cells in each module, and phase shift inter-module is 360° divided by amount of buck converter cells at all modules. Based on Fig. 11, the switching frequency can be raised from 1 MHz to 4 MHz without increasing switching and gate drive losses on the top switches. As a result, the output capacitor can be reduced dramatically, as well as high efficiency and fast transient response can be achieved.

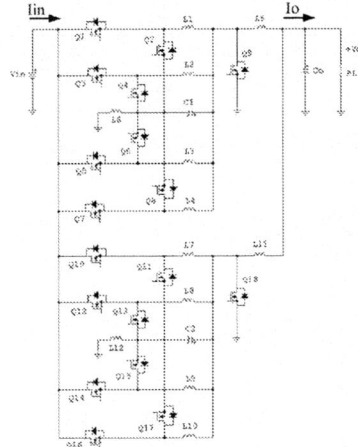

Fig. 10. Proposed multiphase multi-interleaving current doubler-buck converter

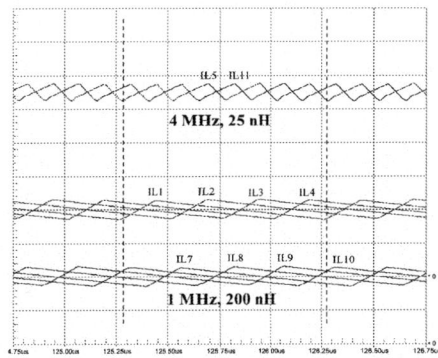

Fig. 11. Multi-interleaving technique

Fig. 12 shows advanced converter architecture with two modules and two buck converter cells in each module. However, these modules can be viewed as sub modules. The converter architecture of Fig. 12 can be viewed as one module, where two or more modules can be connected in parallel. Fig. 13 illustrates the operation waveform. As can be seen in Fig. 13, T_1 is the switching period of the top switches in relation to the bottom switches, T_2 is the switching period of the top switches in relation to the auxiliary bottom switches Q5 and Q10, and T' is the switching period of the top switches in relation to the extra bottom switch Q11. As $T_2 = T_1/n$, where n is the amount of buck converter cells in each sub-module, and T' = T_2/m, where m is the amount of sub-modules, then T' = $T_1/n.m$. Since T' = $1/F_s'$ and $T_1 = 1/F_{s1}$, then $1/F_s' = 1/F_{s1}.n.m$ or $F_s' = n.m.F_{s1}$ where F_s' is the switching frequency multiplication effect and F_{s1} is switching frequency of the top switches in relation to the bottom switches. In steady state operation, when one of the top switches is turned-on, the current flows from voltage source Vin to the load through top switch (Q1, Q3, Q6 or Q8), inductor cell (L1, L2, L5 or L6), auxiliary output inductor (L4 or L8) and extra output inductor L10. At the same time, the energy flows from the resonance inductor (L3 or L7) to the load through capacitor clamp (C1 or C2), auxiliary output inductor (L4 or L8) and L10. In this condition, the energy also flows from the resonance inductor L9 to the load through C3 and L10.

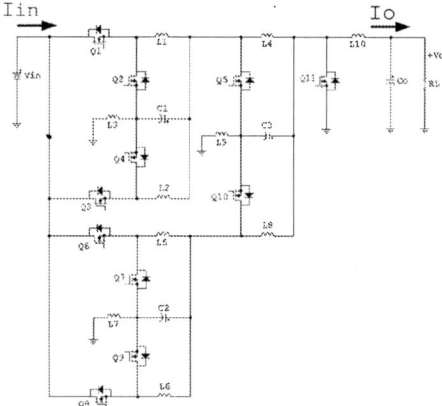

Fig. 12. Advanced converter architecture

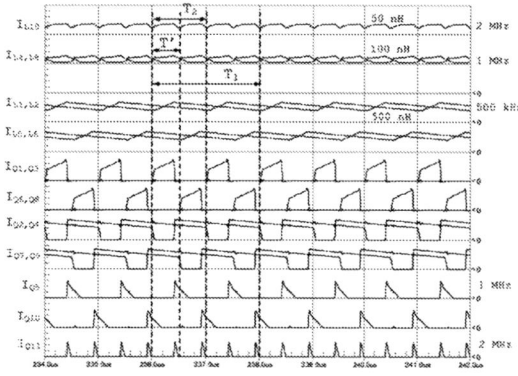

Fig. 13. Multi-interleaving technique and switching frequency multiplication effect

Due to the switching frequency multiplication effect, the transient response can be improved with low switching, gate drive and body diode losses by using smaller inductance value at the inductor cells, the auxiliary output inductor and the resonance inductor in each module or sub-module. In addition, to reduce the complexity of the controller, the switches Q5, Q10 and Q11 from Fig. 12 can be replaced by diodes. The switching frequency multiplication effect is a key to achieve high efficiency, fast transient response, high power density and low-cost VRM design for future generations of microprocessors.

IV. CONCLUSION

The multiphase buck converters with interleaving technique are not enough to meet the power requirements of future microprocessors. The technical conflicts are discussed. A new converter architecture that is based upon the buck topology and that allows a multi-interleaving operation is proposed as a candidate for future microprocessors. Simulation results show that the multi-interleaving technique is not only improving current ripple cancellation effect but also generating switching frequency multiplication effect. Therefore, the switching frequency can be raised with low switching losses. The converter architecture with multi-interleaving technique is possible with better performance compared to existing topologies. The experimental verification of the new converter architecture is under investigation. It is interesting when a comparison can be established with other solutions. However, a joint effort and advanced research are required.

REFERENCES

[1] Ed Stanford, "Power Technology Roadmap for Microprocessor Voltage Regulators," *IEEE* APEC 2004.

[2] X. Zhou, P. Xu, and F. C. Lee, "A high power density, high frequency and fast transient voltage regulator module with a novel current sensing and current sharing technique," *IEEE* APEC 1999.

[3] P. Wong, "Performance Improvements of Multi-Channel Interleaving Voltage Regulator Modules with Integrated Coupling Inductors," *Dissertation*, VPI&SU, Blacksburg, VA, March 2001.

[4] X.Zhou, P. Wong, P. Xu, F.C. Lee, and A. Q. Huang, "Investigation of candidate VRM topologies for future microprocessors," *IEEE Trans. Power Electron.*, Nov. 2000.

[5] Barry B, et al, "Comparison of two 12 V Voltage Regulator Module Topologies," *IEEE* APEC 2004.

[6] W. Huang, G.Schuellein, and D. Clavette, "A Scalable Multiphase Buck Converter with Average Current Share Bus," APEC 2003.

[7] Y. Panov and M. M. Jovanovic, "Design consideration for 12-V/1.5-V, 50-A voltage regulator modules," *IEEE Trans. Power Electron.*, pp. 776–783, Nov. 2001.

[8] Fred C. Lee and Xunwei Zhou, "Power Management Issue for Future Generation Microprocessors," *IEEE*, 1999.

[9] Peng Xu, "Multiphase Voltage Regulator Modules with Magnetic Integration to Power Microprocessors," *Ph.D. dissertation*, Virginia Tech, 2001.

[10] Jia Wei, "High Frequency High Efficiency Voltage Regulators for Future Microprocessors," *Ph.D. Dissertation*, Virginia Tech, 2004.

[11] Kaiwei Yao, "High Frequency and High Performance VRM Design for the Next Generations of Processors," *Ph. D. Dissertation*, Virginia Tech, 2004.

[12] Yuancheng Ren, "High Frequency, High Efficiency Two-Stage Approach for Future Microprocessors," *Ph.D. Dissertation*, Virginia Tech, 2005.

[13] G. Marcyk, "Breaking barriers to Moore's Law," Presentation at *Intel Developer Forum*, Spring 2002.

2006 5th International Power Electronics and Motion Control Conference

A Combined ZVS Converter with Naturally Sharing Input-Current and High Voltage Gain

Linbing Wang, Bo Yang

College of Electrical Engineering, Zhejiang University, Hangzhou, 310027， China

E-mail: WLB109@163.com

Abstract— **A new combined ZVS converter topology is studied in this paper to meet the demands of high power density of power supplies. By interleaving two-phase-shifted full-bridge converter modules, the presented converter has many advantages such as natural sharing input-current、 high voltage gain、 low voltage stress of the transformer second-side diodes、 small volume of input and output filters and so on. The simulated and experimental results show that the converter is suitable for high power and/or high voltage applications.**

Keywords -phase-shifted full bridge converter; interleaving; sharing current.

I. INDUCTION

Switching mode power supplies are widely used in industrial, residential and aerospace environments. They play an important role in modern telecommunication systems as well. The advancement in computer and communication technology requires that the power supplies have high efficiency and high power density. Over past few decades, great efforts have been made to develop isolated dc-dc converters which are capable of process the energy[1-6] effectively, and these converters exhibit the following characteristics: (i) high power density; (ii) low electromagnetic interference (EMI). In order to meet the item (i), it is necessary to decrease the size and increase the efficiency of switching mode power supplies. The sizes of filter components and isolated transformers are determined by the switching frequency. High switching frequency is necessary for achieving high power density. However, the increase of switching frequency results in an increase of switching losses, decreasing the efficiency of the conventional pulse width modulated converters , and increase of the electromagnetic interference (due to higher di/dt and dv/dt).

The soft switching is required to achieve high efficiency in PWM converters. Zero voltage switching technique and zero current switching technique are two commonly used soft switching methods. Phase-shifted Full-bridge (PSFB) converter is a favorite selection in the high voltage and large power applications. Although switches can be turned on at zero voltage or turned off at zero current, the rectifier diodes in the secondary side will suffer high voltage-stresses for their reverse voltages are

in direct ratio to the output voltage, which result in a serious reverse recovery problem and will damage the diodes. These restrain the PSFB's applications in the field of the high output voltage.

To meet the requirement of the high voltage and large current applications, a new combined converter topology with several PSFB cells parallel and/or series connecting is presented in this paper. This topology can resolve the problems of high voltage-stress and large reverse recovered current of the secondary-side diodes in PSFB. Furthermore, each PSFB cell can share the input current naturally, which simplifies the control circuit and extraordinarily improves the power density of the system. By interleaving the PSFB cells in the input-end, the presented converter can reduce the ripples of input current and the ripples of output voltage, which minimizes the size and weight of the input and output filters. Through series connecting in the output-end, the voltage-stresses of the second-side diodes are decreased to half, which decrease the reverse breakdown voltage of the diodes. On the other hand, in addition to physically distributing the magnetic components in the whole system, paralleling of the two converters also distributes power losses and thermal stresses of the semiconductors due to a smaller power processed through the individual paralleled power stages. As a result, paralleling is a popular approach to eliminating "hot spots" in power supplies. Besides, the switching frequencies of paralleled lower power stages may be higher than those of the corresponding single stage that processes high-power ,because faster semiconductor switches transferring low power can be used in each cell of paralleled power stages. Consequently, paralleling also offers an opportunity to reduce the size of the magnetic components.

Especially, compared with interleaving two-transistor forward converter [7-9], the new combined converter has such characteristics: the transformers operating in the first and third quadrant, small number of power components, ZVS, and so on.

II. ANALYSIS OF OPRATION PRINCIPLE

A. Description of the Proposed Power Circuit

The new combined converter topology presented in this paper is shown in Fig .1. Two PSFB cells are connected in parallel in the input-end and in series in the output-end.

1-4244-0448-7/06/$25.00 ©2006 IEEE 311

Each cell operates at the same switching frequency and same duty cycle, and the four groups of switches ((Q 1, Q 2), (Q 5, Q 6), (Q 3, Q 4) and (Q 7, Q 8)) are interleaved by introducing an equal phase shift of 1/4 switching period, shown in Fig 4. The parameters of the two transformers $T1$ and $T2$ are the same, so are those of the inductor $L1$ and $L2$. All semiconductors devices are the same too. In Fig.1，Q1、Q5、Q3、Q7、Q2、Q6、Q8、Q4 are MOSFET deviecs,D1~D8 are the diodes connected to the transformer secondary winding.L1, L2 are the output filter inductors, C1, C2 are the output filter capacitors. Lr1, Lr2 are the leakage inductors of the transformer T1 and T2.

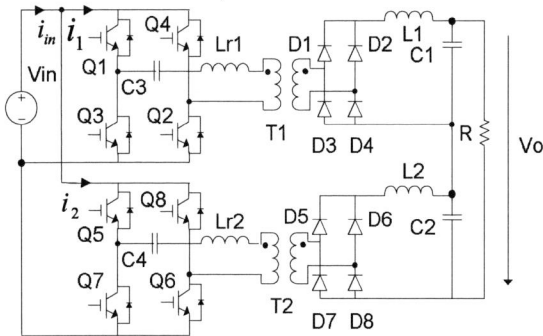

Fig 1. The combined converter schematic

B. Circuit Analysis

Both the two cells of the combined converter schematic in Fig.1 are ZVS phase-shifted full-bridge converters. To compare the characteristics of the converter cell with that of the combined converter, a full-bridge converter cell is shown in Fig.2.The converter in Fig.2 has such advantages: ZVS、 high transformer magnetic core utilization. The voltage gain of the single converter can be expressed as [1-6]:

$$Vo = 2 \cdot n_1 \cdot V_{in} \cdot D \qquad (1)$$

Where n_1 is the turns ratio of the transformer primary to secondary, D is the effective duty cycle when both Q1 and Q2 (or Q3 and Q4) are turned on. The new combined converter in Fig.1 is composed by interleaving two cells of the full-bridge converter in Fig.2. So, the voltage gain of the new combined converter is:

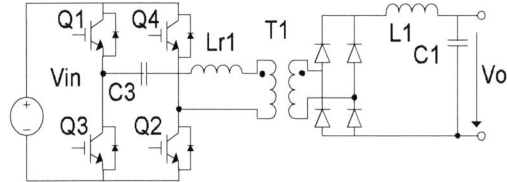

Fig 2. One cell of the combined converter schematic

$$Vo = 4 \cdot n_2 \cdot V_{in} \cdot D \qquad (2)$$

It can be seen from (2) that the effective duty cycle of the combined converter is 4D, which is two times that of the single full-bridge converter cell. So for the same input and output voltage, the turns ratio of the combined converter is half of the single converter cell's; the voltage stress of the diodes connected to the transformer secondary side in the combined converter is half of that in the single converter cell.

The input and output voltages of the two kinds of converters in Fig.1 and Fig.2 are considerably the same, the following equations can be derived from (1) and (2):

$$n_2 = 0.5 \cdot n_1 \qquad (3)$$

The diode voltage stresses of the transformer secondary sides in the two kinds of the converters respectively are:

$$V_{F2} = n_2 \cdot V_{in} \qquad (4)$$

$$V_{F1} = n_1 \cdot V_{in} \qquad (5)$$

Where V_{F2} and V_{F1} are the voltage stresses of the transformer secondary side diode in Fig.1 and Fig.2, and (6) can be derived from (3) ~ (5):

$$V_{F2} = 0.5 \cdot V_{F1} \qquad (6)$$

Obviously, the voltage stress of the transformer secondary side diode in the combined converter is half of that in the single converter cell. So, reliability can be acquired by using diodes with lower rated voltage and shorter reverse recovery time. The reverse recovery problem in the transformer secondary side diodes is alleviated.

In fact, the existing of drive pulse delay time as well as the parameters difference of the control chips and semiconductors devices, the conduction duty cycle of the

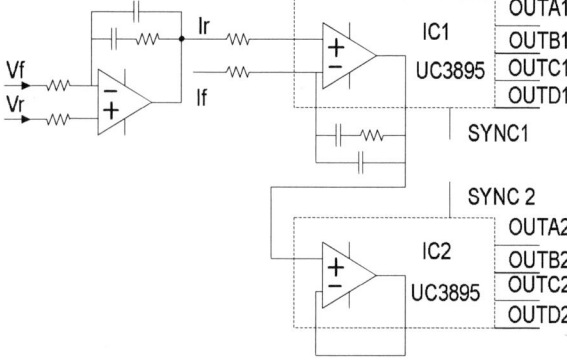

Fig 3. The control schematic of the combined converter

corresponding groups of semiconductor devices are not completely the same; but the switching power supplies can be considered as an ideal voltage sources with a very small output resistor[10],therefore the balanced output voltage can be acquired if the conduction duty cycle of the four groups of semiconductor devices are basically same (needn't completely same).In other words, the output voltage of the converter output filter capacitors C1 and C2 is basically same. The control schematic achieving basically the same conduction duty cycle is shown in Fig.3. Vf、 Vr、 If、 Ir denote the voltage feedback signal 、 the voltage reference signal、 the current feedback signal and the current reference signal respectively. SYNC1 and SYNC2 are synchronization signals.OUTA1~OUTB2 are driver signals. With the help of the synchronization pin of UC3895, the same switching frequency and the same conduction pulse width with a quart of one switching period phase difference among the four groups of the semiconductors devices can be achieved. So the interleaving driver pulses in the combined converter input port are obtained.

For the interleaving combined structure of the four groups of semiconductor devices in the converter input-end, the input current is all the time continuous under the condition that the conduction duty cycle D of each group of semiconductor devices is greater than 0.25(one group of semiconductor devices turns on at least at any instant time).The input current waveforms and the semiconductor devices conduction current waveform of the combined converter are shown in Fig.4 (a) when 0.25<D<0.5. Fig.4 (b) is the current waveform when D=0.25. The input current fluctuation amplitude of the combined converter is:

$$\Delta i_{in} = \frac{1-2D}{8L} \cdot Ts \cdot n_2 \cdot Vo \qquad (D=0.25) \quad (7)$$

Where L is the output filter inductor, D is effective duty cycle of each group semiconductor devices.

The input current fluctuation frequency of the combined converter is four times as the switching frequency .when the input current of the combined

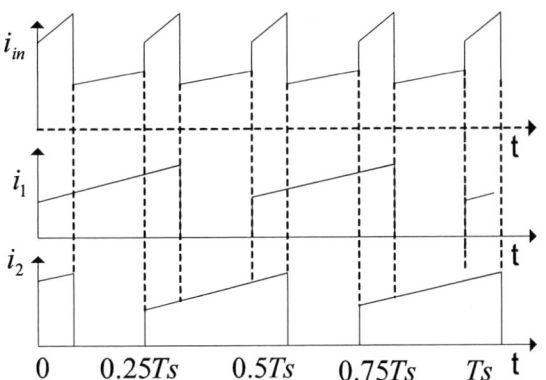

Fig 4 (a). The input current waveforms of the combined converter (0.25<D<0.5)

converter is continuous (0.5>D≥0.25), the input current fluctuation amplitude decreases, so do the volume and weight of the input filter. Because of the interleaving

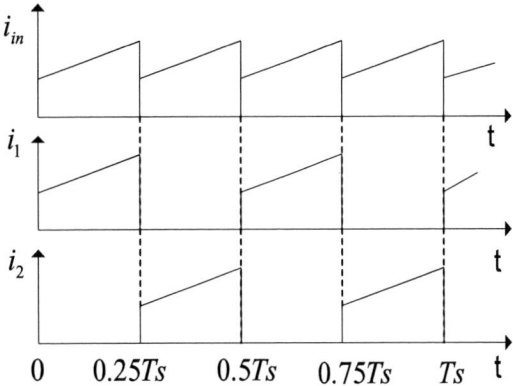

Fig 4(b). The input current waveforms of the combined converter (D=0.25)

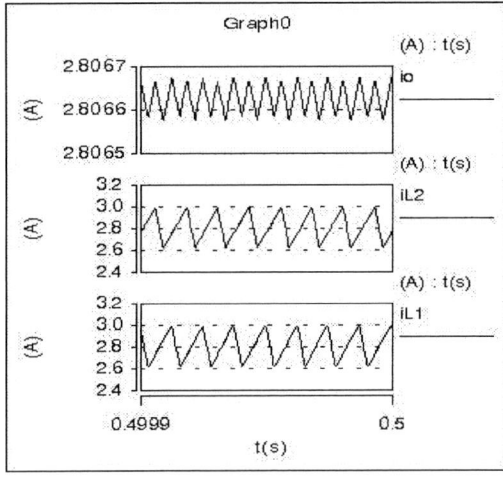

Fig 5. The simulation waveform of the output filter inductor current and the load current

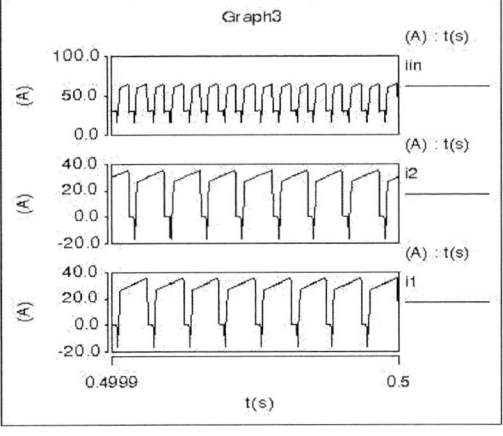

Fig 6. The input current simulation waveform of the combined converter

paralleled structure of the combined converter, the 180 degree phase difference between the two output filter inductors current ripples leads to the 180 degree phase difference between the two output filter capacitors voltage ripples, as shown in Fig.5 and Fig.7. The phase difference has an effect of the ripple cancellation of the output voltage and the output current, which is helpful for minimizing the volume and weight of the output filters. Owing to series connection in the output-end, the output filter inductor (L1 and L2) current values (average value) are the same. So the input current of the four groups of the semiconductor devices of the combined converter is the same, i.e. the input current is balanced naturally.

III. SIMULATION AND EXPERIMENTAL RESULTS

The simulation and experimental parameters of the combined converter are listed as follows:

fs=40kHz,Np:Ns=3:33,Vin=48V,Vo=730V,Ilaod=2.8A, C1=C2=940uF,L1=L2=3mH, Pout=2000W ,where fs 、 Np 、 Vin 、 Vo denote the switching frequency、 the primary side turns、 the secondary side turns、 the input dc Voltage、 the output dc voltage respectively. The input

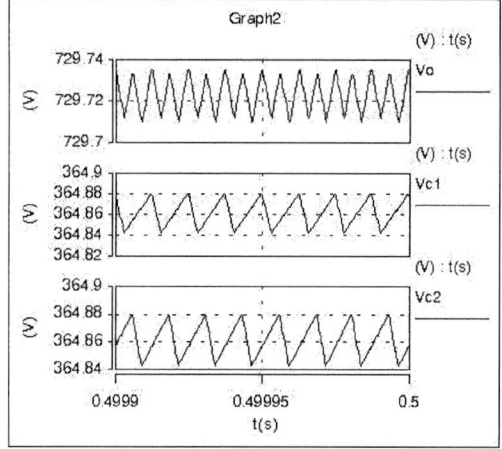

Fig 7. The simulation voltage waveforms of the output capacitor C1、C2 and the output Voltage Vo

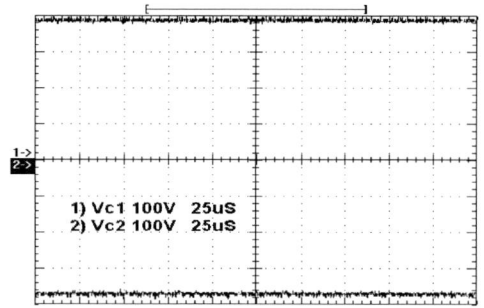

Fig 8. The experimental voltage waveforms of the output capacitor C1, C2.The voltage reference zero point lies in the connection point of the capacitor C1 and C2.

current simulation waveforms of the combined converter

are shown in Fig.6, where i1、 i2、 iin are the bus current of the two cells' circuit and the total bus input current. As Fig.6 shows, the input current is shared; the fluctuation frequency of the total input current is four times as the

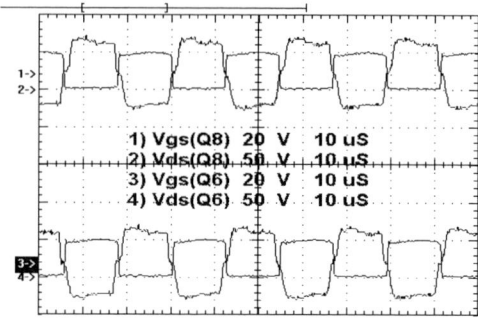

Fig 9. Switching waveforms in semiconductor devices (Upper: Q8 voltage, lower: Q6 voltage).

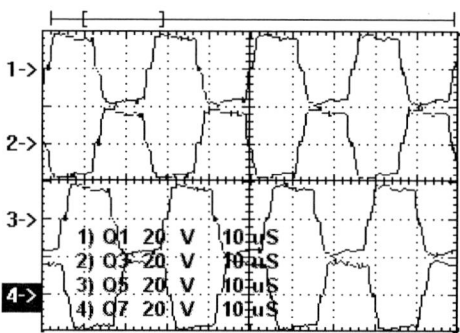

Fig 10. ch1:Q1 driver signal; ch2:Q3 driver signal; ch3:Q5 driver signal; ch4:Q7 driver signal

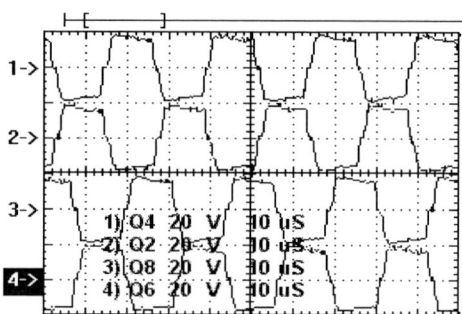

Fig 11. ch1:Q4 drive signal; ch2:Q2 driver signal; ch3:Q8 drive signal; ch4:Q6 driver signal。

switching frequency. The output filter inductor current and the load current waveforms are shown in Fig.5.The increase of the load current fluctuation frequency is resulted from the interleaving characteristics of the two groups of the output filter inductor current. The simulation waveforms of the output filter capacitor voltage and the total output voltage is shown in Fig.7, Where can be seen:

the balanced output voltage; the fluctuation cancellation of the output voltage; the high voltage gain.

The experimental waveforms are shown in Fig.8~Fig.10.The experimental voltage waveforms of the output filter capacitor C1 and C2 are shown in Fig.8, which are balanced in 365V.The experimental voltage waveforms of the lag leg semiconductor devices Q8, Q6 are shown in Fig.9. Fig.10 and Fig.11 are driver signals of the four groups of the semiconductor devicesQ1~Q8.The driver signals are interleaved to achieve the two cells of the combined converter interleaving paralleled connection. The driver signals reach the semiconductor devices through the high frequency magnetic loop.

Table 1 gives the output voltage values of Vc1 and Vc2 for the resistor load. Experimental results show that Vc1 and Vc2 are stable and balanced. The maximal error is 2.1V, which is caused by the control precision. Fig.12 is the picture of the combined converter.

TABLE I.
OUTPUT VOLTAGE EXPERIMENTAL DATASHEET OF THE COMBINED CONVERTER.

V_{in}/V	P_{out}/W	V_{C1}/V	V_{C2}/V
40	1000	365.8	366.3
40	1500	365.2	365.7
40	2000	364.5	364.3
48	1000	367.6	366.8
48	1500	366.9	366.1
48	2000	365.3	364.8
56	1000	367.1	366.9
56	1500	366.3	366.1
56	2000	365.1	364.9

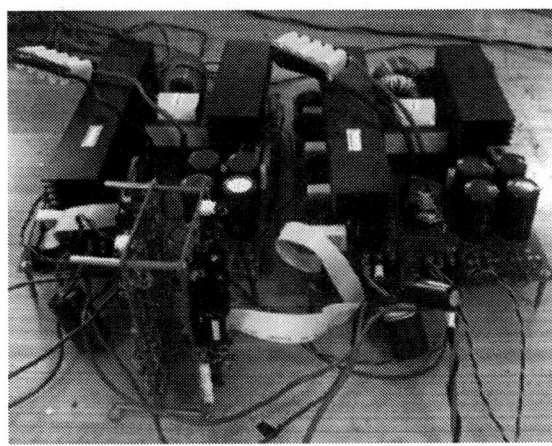

Fig.12 .the picture of the interleaving combined converter

IV.   CONCLUSIONS

Analysis, simulation and experiment results show that the interleaving combined converter presented in this paper have the following characteristics:

(1)   Naturally sharing input current and ZVS in the semiconductor devices;

(2)   High voltage gain ;

(3)   Naturally sharing output voltage;

(4)   Low voltage stress in transformer secondary side rectifier diodes;

(5)   Small volume of input and output filters.

REFERENCES

[1]   Kim, J.W, "Analysis and design considerations of zero-voltage and zero-current-switching (ZVZCS) full-bridge PWM converters,"[C]. 2002. PESC 02. 2002 Page(s):1835 – 1840.

[2]   Moschopoulos, G.; Jain, P, "Single-stage ZVS PWM full-bridge converter," [J]. Aerospace and Electronic Systems, IEEE Transactions on Volume 39, Issue 4, Oct. 2003 Page(s):1122 – 1133.

[3]   SongTingTing; HuangNianci, "A novel zero-voltage and zero-current-switching full-bridge PWM converter," [C]. Applied Power Electronics Conference and Exposition, 2003. APEC '03. Eighteenth Annual IEEE Volume 2, 9-13 Feb. 2003 Page(s):1088 - 1092.

[4]   Ayyanar, R.; Mohan, N, "A novel full-bridge DC-DC converter for battery charging using secondary-side control combines soft switching over the full load range and low magnetic requirement ,"[J]. Industry Applications, IEEE Transactions on Volume 37, Issue 2, March-April 2001 Page(s):559 – 565.

[5]   Chien-Ming Wang; Yen-Nien Wang; Chia-Hao Yang, "A new ZVS-PWM full-bridge step-up/down converter," [C]. Industrial Electronics Society, 2004. IECON 2004. 30th Annual Conference of IEEE Volume 1, 2-6 Nov. 2004 Page(s):908 - 913.

[6]   Xinbo Ruan; Yangguang Yan, "Soft-switching techniques for PWM full bridge converters," Power Electronics Specialists Conference, 2000. PESC 00. 2000 IEEE 31st Annual Volume 2, 18-23.

[7]   Jianping Xu, Xiaohong Cao, Qianchao Luo, "An improved two-transistor forward converter," [C]. PEDS '99. Proceedings of the IEEE 1999 International Conference on Power Electronics and Drive Systems. 1999, (1): 225 -228.

[8]   Michael T.Zhang,Milan M.Jovanovic,Fred C.Y.Lee, "Analysis and Evaluation of Interleaving Techniques in Forward Converters ," [J].IEEE Transaction on PE. 1998, 13(4): 690-698.

[9]   Shi jianjiang,He xiangning,Yong Yangguang, "Study on a parallel-series connection of dual two-transistor forward converter modules with output filter coupled-inductor ," [J]. Proceeding of the CSEE, 2004, 24(10). 157-161.

[10] John S. Glaser,and Arthur F. Witulski, " Analysis of Load-Sharing in Multiple-Module Converter Systems," [J]. IEEE Transaction on Power Electronics, Vol.9. No.1, January 1994:43-50.

2006 5th International Power Electronics and Motion Control Conference

Matrix Coefficient Polynomial Description Model of DC-DC Converters Based on Switched Linear Systems

Yongping ZHANG [*], Bo ZHANG [*], Zongbo HU [**], Dongyuan QIU [*], Guiping DU [*]

[*]Electrical Power College, South China University of Technology, Guangzhou , China
[**]ASTEC Power Supply (ShenZhen) Co.,Ltd.

e-mail: yptzhang@gmail.com

Abstract—**In this paper, DC-DC converters are viewed as one class of hybrid systems i.e. switched linear systems. A new mathematical model of switched linear systems-matrix coefficient polynomial description model is presented. An obvious description form of the relationship between the continuous variable and discrete variable is built, a new model of switched linear system of DC-DC converters, in which continuous variable and discrete variable interact on each other directly, is given. To some extent, the new model helps to reveal the essence of operation activity and analyze the performance of DC-DC converters. Both the results of steady-state analysis and closed-loop simulation of transient response have verified the validity and practicability of the new model.**

Keywords-switched linear systems; DC-DC converter; modeling

I. INTRODUCTION

Power electronic converters can be viewed as one class of switched linear systems, its continuous variable systems include each operation mode, while discrete event systems include switching action of power switch. These days, some scholars have tried to introduce switched linear systems to study the basic theory in power electronic converters, built up the switched linear system model and analyzed some basic control characteristic such as controllability, reachability and output controllability etc. of power electronic converters [1-4].

In this paper, based on the switched linear system model of DC-DC converters, we tried to apply a new mathematical modeling method (i.e. matrix coefficient polynomial description model) to the modeling analysis of DC-DC converters. Using the modeling approach, we obtained an obvious description form between the continuous variable and discrete variable. To some extent,

The Project Sponsored by Natural Science Foundation of China (60474066) and Natural Science Foundation of Guangdong (05103540).

the matrix coefficient polynomial description model in which the discrete variables and continuous variables interact on each other directly are expected to reveal the essence of DC-DC converters as switched linear systems much more deeply, furthermore, it helps to analyze the performance of this systems. The results of steady state analysis and closed-loop simulation of transient state response verified the validity and practicability of the new model. Meanwhile, the analysis of this paper showed that matrix coefficient polynomial description model is more general than the traditional state space averaging model.

II. MATRIX COEFFICIENT POLYNOMIAL MODEL

A. Switched linear systems

In general, switched linear system has the form as follows:

$$\begin{cases} \dot{x}(t) = A(r(t))x(t) + B(r(t))u(t) \\ y(t) = C(r(t))x(t) \end{cases} \quad (1)$$

where $x(t) \in \Re^n$ is continuous state vector, $u(t) \in \Re^n$ is control vector, $y(t) \in \Re^q$ is output vector, $A(r(t)), B(r(t)), C(r(t))$ and $r(t)$ are constant coefficient and discrete variable respectively, $Q = \{r_1, r_2 \cdots, r_N\}$ is a finite set.

As to the switched linear systems described by (1), although it is written as the form of $A(r(t))$ in system equation, $A(r(t)): Q \to \{A(r_1), A(r_2) \cdots, A(r_N)\}$ is just a formalization description, obvious relationship between the continuous variable and discrete variable can't be seen obviously from it. In other words, the description of direct relationship between them isn't obvious at all.

B. Matrix coefficient polynomial model

Generally speaking, owing to $Q = \{r_1, r_2 \cdots, r_N\}$ only represents N different discrete symbols, thus we may have many choices. We select $Q \in \Re$ i.e. select the value from real number set. Now apply polynomial interpolating technique to (1), we assume that all system matrix $A(r_1), A(r_2), \cdots, A(r_N)$ are known. Here, matrixes

1-4244-0448-7/06/$25.00 ©2006 IEEE

$A(r_1), A(r_2), \cdots, A(r_N)$ are corresponding to $r_1, r_2, \cdots, r_N$, according to the principle of polynomial interpolating technique, we can get a polynomial which satisfy the following condition

$$A(r_j) = \sum_{i=0}^{N-1} A^{[i]} r_j^i \qquad (2)$$

where $A^{[0]}, A^{[1]}, \cdots, A^{[N-1]}$ are ith matrix coefficient in interpolating polynomial respectively, we refer to it as matrix structure coefficient[5]. Simultaneously we get

$$B(r_j) = \sum_{i=0}^{N-1} B^{[i]} r_j^i \qquad (3)$$

$$C(r_j) = \sum_{i=0}^{N-1} C^{[i]} r_j^i \qquad (4)$$

where $B^{[0]}, B^{[1]}, \cdots, B^{[N-1]}$ and $C^{[0]}, C^{[1]}, \cdots, C^{[N-1]}$ are corresponding matrix structure coefficient. When the matrix dimension is greater than one, we deal with each element in system matrix applying the method shown by (2) (3) and (4). Now the switched system model indicated by (1) can be written as

$$\begin{cases} \dot{x}(t) = \sum_{i=0}^{N-1} A^{[i]} r^i(t) x(t) + \sum_{i=0}^{N-1} B^{[i]} r^i(t) u(t) \\ y(t) = \sum_{i=0}^{N-1} C^{[i]} r^i(t) x(t) \end{cases} \qquad (5)$$

Consequently, we get the interpolating polynomial model of switched linear systems by sampling interpolation technique. From (5) we can see that an obvious description form between the continuous part and discrete part is given, a mathematical model in which the discrete variable $r(t)$ and continuous variable $x(t)$ interacted on each other directly is obtained simultaneously. Compared to the system described by (1), a direct relationship between the discrete variable and continuous variable is given obviously in (5), so it helps to reveal the essence of switched systems much more deeply.

III. MATRIX COEFFICIENT POLYNOMIAL DESCRIPTION MODEL OF DC-DC CONVERTERS

According to references [1-4], considering Boost converter shown in Fig 1, we try to obtain the matrix coefficient polynomial model of Boost converter. It would hold under the following assumptions: (1) all components of converter are ideal; (2) converter is operated on continuous conduction mode (CCM). Here $x(t) = [i_L \ u_C]^T$ is state vector, $u(t) = V_{in}$ is control vector and $y(t) = u_C$ is output vector, $r(t)$ is discrete variable which describes the switching process and different operation mode. Since we consider the CCM of converter

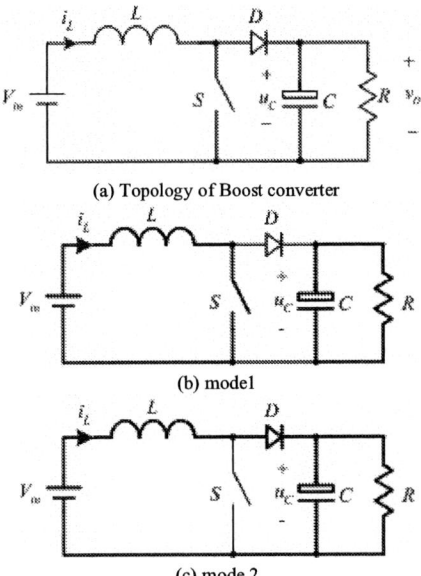

(a) Topology of Boost converter

(b) mode1

(c) mode 2

Fig. 1 Topology and operation modes of Boost converter

i.e. the set of $r(t)$ is $r(t) \in \{r_1, r_2\}$, and $N = 2$. Then the switched linear system equation of Boost converter is

$$\begin{cases} \dot{x}(t) = A(r_i) x(t) + B(r_i) u(t) \\ y(t) = C(r_i) x(t) \end{cases} \qquad (6)$$

Coefficient matrix of switched linear model is as follows:
Coefficient matrix of state vector

$$A(r_1) = \begin{bmatrix} 0 & 0 \\ 0 & -\dfrac{1}{RC} \end{bmatrix}; \quad A(r_2) = \begin{bmatrix} 0 & -\dfrac{1}{L} \\ \dfrac{1}{C} & -\dfrac{1}{RC} \end{bmatrix}.$$

Coefficient matrix of input vector

$$B(r_1) = B(r_2) = \begin{bmatrix} \dfrac{1}{L} \\ 0 \end{bmatrix}.$$

Coefficient matrix of output vector
$$C(r_1) = C(r_2) = \begin{bmatrix} 0 & 1 \end{bmatrix}.$$

Referring to (2), applying polynomial interpolating technique to coefficient matrix of state vector, we can get the matrix structure coefficient $A^{[0]}$ and $A^{[1]}$:

$$A^{[0]} = \frac{1}{r_1 - r_2} \begin{bmatrix} 0 & -\dfrac{r_1}{L} \\ \dfrac{r_1}{C} & \dfrac{r_2 - r_1}{RC} \end{bmatrix} \qquad (7\text{-}1)$$

$$A^{[1]} = \frac{1}{r_1 - r_2} \begin{bmatrix} 0 & \dfrac{1}{L} \\ -\dfrac{1}{C} & 0 \end{bmatrix} \qquad (7\text{-}2)$$

Then we get matrix structure coefficient:

$$A(r) = A^{[0]} + A^{[1]}r$$

$$= \frac{1}{r_1 - r_2} \begin{bmatrix} 0 & \dfrac{r - r_1}{L} \\ \dfrac{r_1 - r}{C} & \dfrac{r_2 - r_1}{RC} \end{bmatrix} \qquad (8)$$

Since coefficient matrix of input vector $B(r_1) = B(r_2)$, and coefficient matrix of output vector $C(r_1) = C(r_2)$, we simply get:

$$B(r) = B(r_1) = B(r_2) = \begin{bmatrix} \dfrac{1}{L} \\ 0 \end{bmatrix} \qquad (9)$$

$$C(r) = C(r_1) = C(r_2) = \begin{bmatrix} 0 & 1 \end{bmatrix} \qquad (10)$$

Consequently we get the matrix coefficient polynomial description model:

$$\begin{cases} \dot{x}(t) = A(r)x(t) + B(r)u(t) \\ \qquad y(t) = C(r)x(t) \end{cases} \qquad (11)$$

where coefficient matrix $A(r), B(r), C(r)$ is given by (8),(9) and (10) respectively. Compared (1) with (11), the new model obtained by using matrix coefficient polynomial interpolating technique contains not only the continuous variable but also the discrete variable, furthermore the resulting model is similar to linear system model.

As we can see from the principle of matrix coefficient polynomial interpolating technique, system discrete variable r_i is used to identify different switching mode, so it is random, during one switching period, $r_1 \in (0, d), r_2 \in (d, 1)$, where d is duty ratio and $d \in (0, 1)$. In order to illustrate the principle of matrix coefficient polynomial interpolating technique, we show $A(r)$ on a plane, seen Fig.2. $A(r)$ is a straight line which passed through point $(r_1, A(r_1))$ and $(r_2, A(r_2))$, where r_1, r_2 are optional. Basic state space averaging equation $\dot{x} = Ax + bv_s$, where $A = A(r_1)d + A(r_2)(1 - d)$, A is equal to the shadowed area. If we can find such a $A(r)$

that the area which is composed of $A(r)$ and two axis is equal to the shadowed area, that is $S_1 + S_3 = S_2 + S_4$, i.e.

$$\begin{aligned} &\big[A(r_1) - A(0)\big]r_1 + \big[A(r_2) - A(d)\big](r_2 - d) \\ &= \big[A(d) - A(r_1)\big](d - r_1) + \big[A(1) - A(r_2)\big](1 - r_2) \end{aligned} \qquad (12)$$

where $A(0), A(r_1), A(d), A(r_2), A(1)$ are corresponding value of $A(r)$ when $r = 0, r_1, d, r_2, 1$. If r_1, r_2 and d meet the following relationship

$$dr_1 + (1 - d)r_2 = \frac{1}{2} \qquad (13)$$

then matrix coefficient polynomial interpolating model is the same as the state space averaging model. So we can get state space averaging model from matrix coefficient polynomial interpolating technique, the latter is more general, it includes state space averaging method. In other words, state space averaging model is just a special case of the new model. Matrix coefficient polynomial interpolating model is expected to describe the characteristic of DC-DC converter better.

Referring to the modeling process of Boost converter, we can build the new model of Buck converter and Buck-Boost converter (seen Fig.3), the results are shown in table 1.

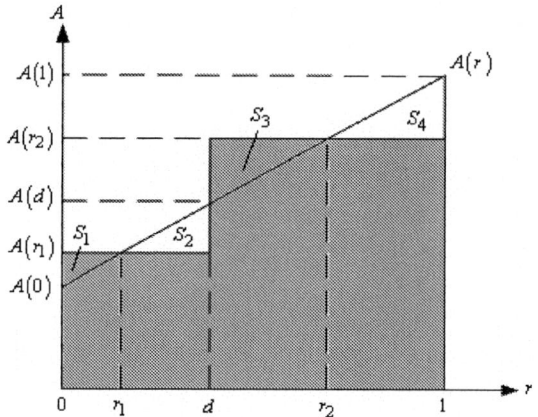

Fig. 2 Principle of matrix coefficient polynomial interpolating modeling

Tab. 1 Matrix coefficient polynomial description model of DC-DC converters

converter	A	B	C	Matrix coefficient polynomial model
Buck	$\begin{bmatrix} 0 & -\dfrac{1}{L} \\ \dfrac{1}{C} & -\dfrac{1}{RC} \end{bmatrix}$	$\dfrac{1}{r_1 - r_2}\begin{bmatrix} \dfrac{r - r_2}{L} \\ 0 \end{bmatrix}$	$\begin{bmatrix} 0 & 1 \end{bmatrix}$	(11)
Buck-Boost	$\dfrac{1}{r_1 - r_2}\begin{bmatrix} 0 & \dfrac{r - r_1}{L} \\ \dfrac{r_1 - r}{C} & \dfrac{r_2 - r_1}{RC} \end{bmatrix}$	$\dfrac{1}{r_1 - r_2}\begin{bmatrix} \dfrac{r - r_2}{L} \\ 0 \end{bmatrix}$	$\begin{bmatrix} 0 & 1 \end{bmatrix}$	(11)

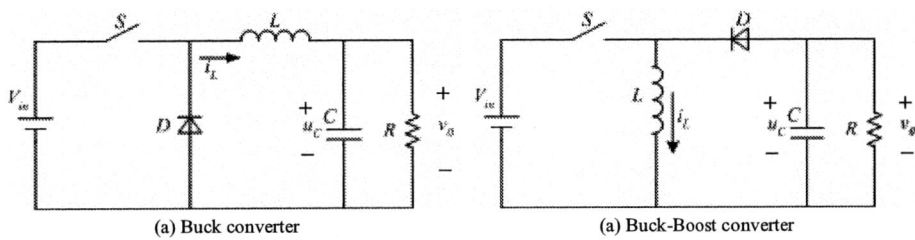

(a) Buck converter   (a) Buck-Boost converter

Fig. 3 Topologies of Buck converter and Buck-boost converter

IV. THE APPLICATION OF MATRIX COEFFICIENT POLYNOMIAL INTERPOLATING MODEL

Based on switched linear system model shown in (11), we can analyze the characteristic of basic DC-DC converters. Here we consider steady state analysis and closed loop transient state simulation of DC-DC converters, illustrate the application of the new model.

A. Analysis of Boost converter

Using the coefficient matrix parameter shown in tab.1, let $\dot{x}=0$ in (11), by simple calculation, we get

$$y(t) = -CA^{-1}Bu(t)$$

$$= -\frac{(r_1-r_2)LC}{(r-r_1)^2}[0 \quad 1]\begin{bmatrix} \dfrac{r_2-r_1}{RC} & \dfrac{r-r_1}{L} \\ \dfrac{r_1-r}{C} & 0 \end{bmatrix}\begin{bmatrix} \dfrac{1}{L} \\ 0 \end{bmatrix}u(t) \quad (14)$$

$$= \frac{r_1-r_2}{r-r_1}u(t)$$

i.e.

$$y = v_0 = \frac{r_1-r_2}{r_1-r}V_{in} \quad (15)$$

Likewise we can get the steady state analysis of Buck and Buck-Boost converter

$$v_0 = \frac{r-r_2}{r_1-r_2}V_{in} \quad (16)$$

$$v_0 = \frac{r_2-r}{r-r_1}V_{in} \quad (17)$$

If we select $r_1=0$, $r_2=1$, $r=1-D$, (15)、(16) and (17) have the following form respectively

$$v_0 = DV_{in} \quad (18)$$

$$v_0 = \frac{V_{in}}{1-D} \quad (19)$$

$$v_0 = \frac{D}{1-D}V_{in} \quad (20)$$

Note that we get the same result as conventional state space averaging technique.

B. Closed loop transient state simulation

Based on (11), we construct a new closed loop simulation model of DC-DC converters (Fig.4).

Select $r_1=0$, $r_2=1$, $r(t)=1-u$, where u is duty cycle, v_{ref} is reference voltage, use control law[6]

$$\dot{u} = -m\left(i_L v_C - \dot{v}_C i_L\right) - kv_{ref} \\ \cdot\left[Li_L + (1-u)\left(y-v_{ref}\right)\right] \quad (21)$$

Given parameters as follows: input voltage $V_{in}=12V$, reference voltage $V_{ref}=24V$, inductance $L=180uH$, capacitance $C=150uF$, load resistance $R=12\Omega$; $m=0.0005$, $k=1.8$. The results are shown in Fig.5.

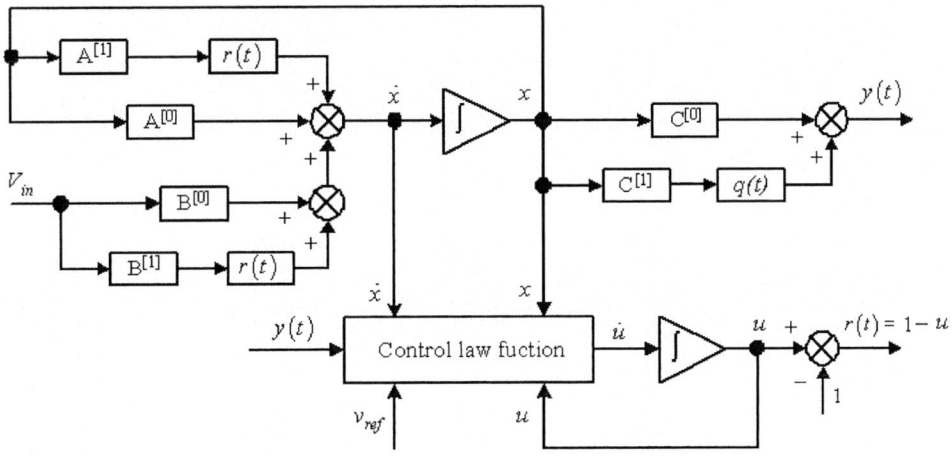

Fig. 4 Close loop simulation model of Boost converter

319

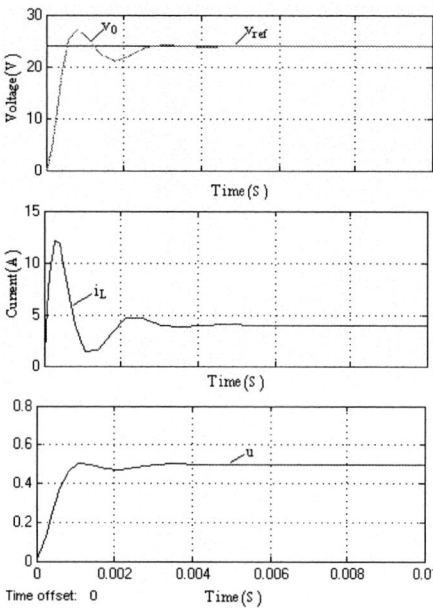

(a) Zero state response simulation

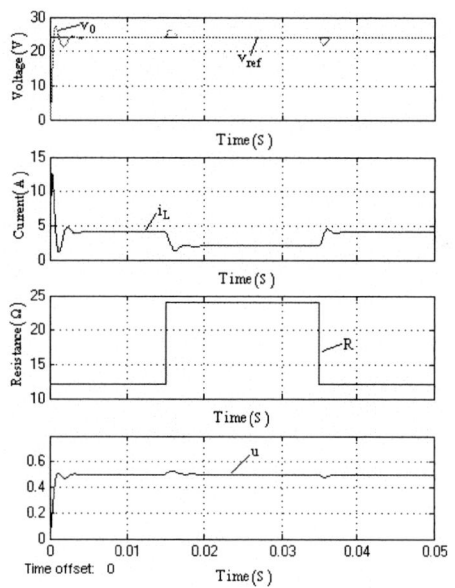

(b) Closed-loop simulation of transient state response

Fig. 5 Simulation results

V. CONCLUSION

In this paper, DC-DC converters are viewed as switched linear systems, matrix coefficient polynomial description model of DC-DC converters is obtained. Both the results of steady state analysis and transient state closed loop simulation have shown the validity and practicability of the new modeling approach. However, further research is needed in order to use the new model to study the characteristic of DC-DC converters much more deeply by selecting appropriate interpolating point.

REFERENCES

[1] Z Hu, B Zhang and W Deng, "Feasibility study on one cycle control for PWM switched converters," *C. 2004 35th Annual IEEE Power Electronics Specialists Conference, PESC'04*, Aachen, Germany, June, 2004: 3359~3365.

[2] Hu Zong-bo, Zang Bo, Deng Wei-hua, "Controllability and reachability of the hybrid dynamic system in Buck converter," *J. Journal of South China University of Technology—Natural Science Edition*, 2004, 32(7): 23~27.
胡宗波,张波,邓卫华, "Buck 变换器混杂动态系统的能控性和能达性," [J].华南理工大学学报（自然科学版）,2004, 32(7): 23~27.

[3] Zongbo HU, Bo ZHANG, Weihua DENG., "Output Controllability of Power Converters as Switched Linear Systems," *C. The 4th International Power Electronics and Motion Control Conference, IPEMC'04*, Xi'an, China, Aug. 2004: 1665-1668.

[4] Hu Zong-bo, Zang Bo, Deng Wei-hua, Zhang Yong-ping. "Controllability and reachability of DC-DC converters as switched linear systems," *J. Proceedings of the CSEE*, 2004, 24(12): 165-170.
胡宗波,张波,邓卫华,张涌萍, "基于切换线性系统理论的 DC-DC 变换器控制系统的能控性和能达性," [J]. 中国电机工程学报, 2004, 24(12): 165-170.

[5] Xie Guangming, Zheng Dazhong, "Matrix-coefficient Polynomial Description Model of Linear Switching Systems," *Control and Decision*, 15(3), 2000: 355-357.
谢广明,郑大钟, "线性切换系统的矩阵系数多项式描述模型," 控制与决策, 15(3), 2000: 355-357.

[6] Kawasaki N，Nomura H，Masuhiro M. "A new control law of bilinear DC-DC converters developed by direct application of Lyapunov" [J]. *IEEE Transactions on Power Electronics*，1995，10(3)：318~320.

2006 5th International Power Electronics and Motion Control Conference

Development of DC-DC Multiple Converter based on Push-pull Forward Topology

Weihao Hu, Yunqing Pei, Zhaoan Wang

Xi'an Jiaotong University, Xi'an 710049, China

Email: huweihao@mailst.xjtu.edu.cn

Abstract—The paper introduces a DC-DC multiple converter based on Push-pull Forward Topology and analyzes its operation principle in detail. It has some superiority in low voltage and high current situation. Then the design procedure of the power stage and control circuit is introduced. Finally 5kW experimental equipment is accomplished.

Keywords— push-pull forward; multiple; converter

I INTRODUCTION

Switching Mode Power Supply is a kind of power electronic equipment which is popular used in the field of telecommunication, electric, aviation and military. It has some advantages such as high efficiency, small size, high control precision and high response speed. Push-pull converter and forward converter are two kinds of popular DC-DC converter. But they have their inherent problems. Transformer in push-pull converter may meet unbalance magnetization and this will result in transformer saturation. Transformer in forward converter work in only one direction, so the core of transformer is not used sufficiently and the additional magnetism restoration circuit is needed.

Push-pull forward converter has the strongpoint of push-pull converter and forward converter at the same time. Its transformer works in two directions so it has a high usage factor of the transformer core. It has the low voltage spike of the switches and the additional magnetism restoration circuit is not needed. Because of these advantages, it has some superiority in low voltage and high current situation.

II OPERATION PRINCIPLE

Figure 1 shows the circuit diagram of the push-pull forward converter. The primary side of the converter consists of two MOSFET V1 and V2, two windings of the transformer N1 and N2, one capacitor C_1. Two diodes VD1 and VD2 are the body diode of the MOSFET V1 and V2. The secondary side of the converter is the full bridge rectifier. After the rectifier there is a LC filter connecting to the load R.

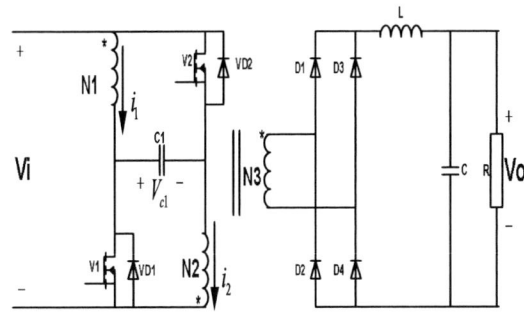

Figure 1. Push-pull forward converter topology.

In the steady state of the circuit, no matter which MOSFET turns on, the capacitor C_1 will parallel connect with one primary winding of the transformer. The voltage V_{c1} of the capacitor C_1 is positive. And its value equals to the input voltage approximately. So when one MOSFET turns on, the voltage of another MOSFET is: $V_{DS}=V_i+V_{c1}\approx 2V_i$.

In one cycle, the push-pull forward circuit has 8 operating modes. The key operating waveforms are given in Figure 2 and the equivalent circuit is shown in Figure 3.

(1) $[t_0 \sim t_1]$. Before t_0, two MOSFET are all in the off state. So the primary current freewheels through V_i, N1, C_1 and N2. All four diodes in the full bridge rectifier are in the on state. And the secondary winding of the transformer is in the short circuit state. At time of t_0, V1 turns on and V_i, N1, V1 consist of one loop. So the current i_1 in the winding N1 increases rapidly. C_1, N2, V1 consist of another loop. So the current i_2 in the winding N2 decreases rapidly and then increases in the reverse direction.

(2) $[t_1 \sim t_2]$. At the time of t_1, D2 and D3 turn off and the secondary winding of the transformer is not in the short circuit state. So the increase of the current i_1 becomes slowly. During this period, the converter works like two forward converters in parallel.

1-4244-0448-7/06/$25.00 ©2006 IEEE   321

(3) [$t_2 \sim t_3$]. At the time of t_2, V1 turns off. Because of the leakage inductance of the transformer, the current i_1 in the winding N1 can not change suddenly. So body diode of V2 is forced to turn on to continue the leakage inductance current. N1, C_1, VD2 consist of one loop and the current i_1 decreases rapidly. At the same time, V_i, VD2, N2 consist of another loop and the current i_2 increases rapidly. When $i_1 = i_2$, this mode ends.

(4) [$t_3 \sim t_4$]. During this period, two MOSFET V1 and V2 are all in the off state. V_i, N1, C_1, N2 consist of a loop. The voltage V_{c1} of the capacitor C_1 equals to the input voltage V_i approximately. So the current i_1 and i_2 of the two windings N1 and N2 keep fixed.

(5) [$t_4 \sim t_7$]. During this period, the circuit resembles the operation experienced in the previous half cycle.

Considering the voltage-seconds balance on the filter inductor L, the DC voltage relationship of the push-pull forward converter is:

$$V_o = nDV_i.$$

D is the switching duty cycle and n is the turn ratio of the transformer.

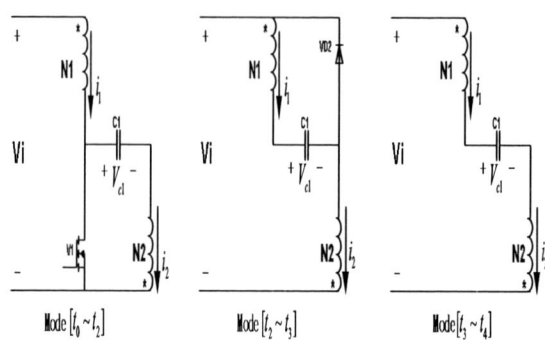

Figure 3. Equivalent circuit of each operation mode.

III DESIGN OF MAIN STAGE OF THE DC-DC MULTIPLE CONVERTER

A prototype was built to evaluate the performance of the DC-DC multiple converter based on push-pull forward topology. The input voltage is $48 \pm 10\%$ V, output voltage is 750 V, maximal power is 5 kW. Because of the high input current and high output power, we use multiple techniques. We use two 2.5 kW converters parallel connecting with each other in the primary side and series connecting with each other in the secondary side. And we make the two converters working in the different phase. The main stage of the prototype is shown in Figure 4. Multiple techniques can reduce the input current and output voltage of each converter. So it can increase the reliability of the converter. At the same time, it can reduce the ripple of the output voltage. So the output filter inductor L and capacitor C can be reduced observably.

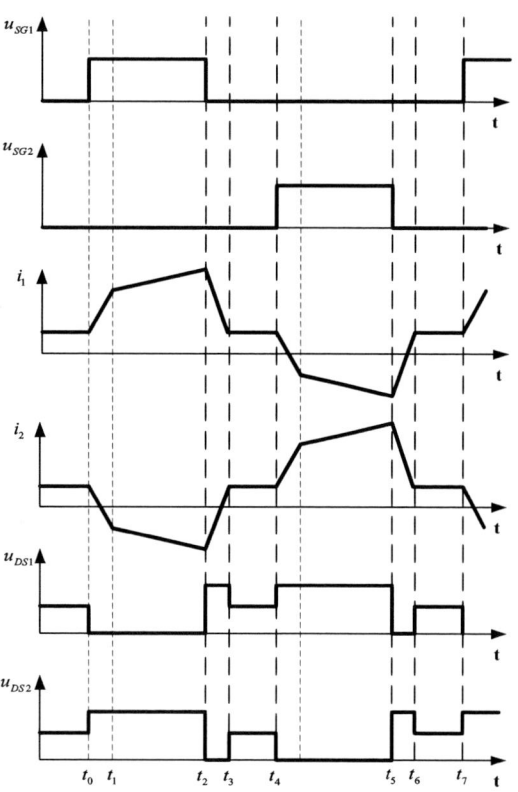

Figure 2. Operating waveform of the push-pull forward converter.

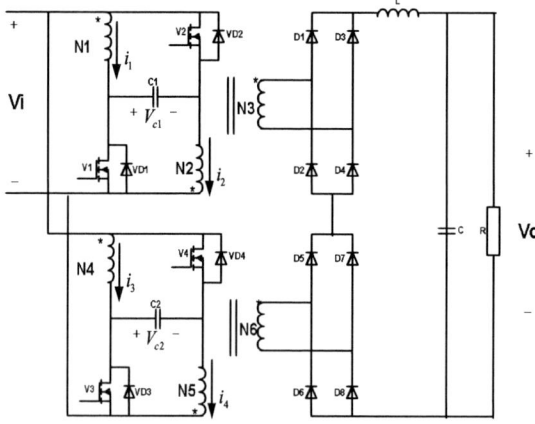

Figure 4. The main stage of the prototype.

Base on the analysis above, the highest voltage of MOSFET equals two times of the input voltage and the highest current of MOSFET equals the highest current in the filter inductor L divided the turn ratio of the transformer. So the highest voltage of MOSFET is about 100 V and the highest current is about 80 A. We choose IXFN180N20 (200 V, 180 A) form IXYS corporation as our experimental switch.

IV DESIGN OF CONTROL CIRCUIT FOR THE DC-DC MULTIPLE CONVERTER

We use SG3525 as the PWM control chip. This chip has two PWM signals to drive the two MOSFET of the push-pull forward converter. For realizing multiple techniques, the same MOSFET of the two converters should work in the different phase. So we must make the PWM control chip SG3525 of each converter work in synchronization state. The principle of control circuit is shown in Figure 5.

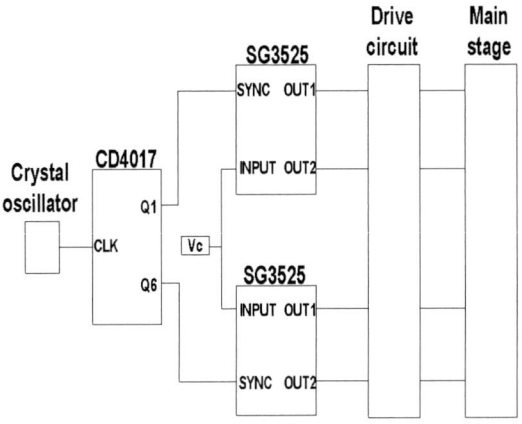

Figure 5. The principle of control circuit.

The crystal oscillator connects with the decade counter CD4017. CD4017 is used to make the input signal generated by the crystal oscillator change to a different phase. The output signal of CD4017 is used to synchronize the two PWM control chips SG3525. So the two chips can work in discrepant 90° phase. Because the control signal of the two chips is the same one, the output PWM signal has the same duty cycle. Finally we send the PWM signal to the driving circuit. Then the output of the driving circuit is sent to MOSFET of the main stage. So the two converters can work in the multiple states.

V EXPERIMENTAL RESULTS

Based on the principle above, we make 5kW experimental equipment. The input voltage is 48 V and the output voltage is 750 V. We choose IXFN180N20 as our experimental MOSFET and DSEP 12−12A as our diode of the full bridge rectifier. The turn ratio of the transformer is 0.1, the filter inductor L is 0.7mH, the filter capacitor C is $220\mu F$, the load is 110Ω and the switching frequency is 25 kHz. The waveform below is measured under the condition of 46 V input voltage, 750 V output voltage and 85% duty cycle.

From Figure 6, we can see the PWM control signals of MOSFET V1 and V3 have the same frequency and the same duty cycle. But these signals are generated in discrepant 90° phase. And the PWM control signals of MOSFET V2 and V4 are also generated in discrepant 90° phase. So the two converters can work in the multiple states. Figure 7 shows the current of the filter inductor L. The frequency of the inductor current is four times of the switching frequency. This also proves the two converters work in the multiple states. Figure 8 and Figure 9 show the operation waveforms of the multiple push-pull forward converter. These waveforms accord with the academic operation waveforms basically.

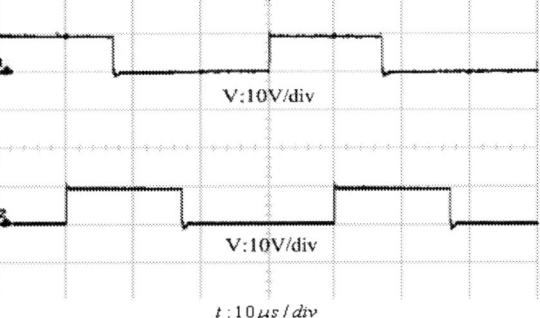

Figure 6. PWM control signals of MOSFET V1 and V3.

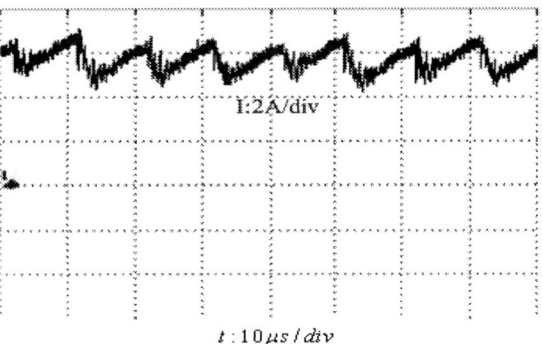

Figure 7. The current waveform of the filter inductor.

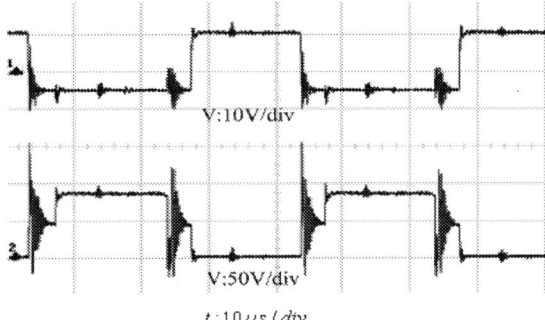

Figure 8. The GS voltage and the DS voltage.

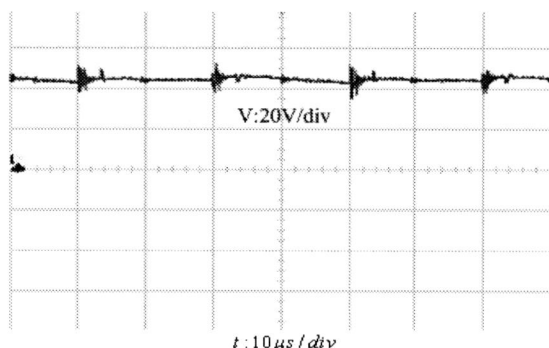

Figure 9.　The voltage of capacitor C_1.

VI　CONCLUSIONS

Push-pull forward converter has the strongpoint of push-pull converter and forward converter at the same time. Its transformer works in two directions so it has a high usage factor of the transformer core. It has the low voltage spike of the switches and the additional magnetism restoration circuit is not needed. Because of these advantages, it has some superiority in low voltage and high current situation. Multiple techniques can reduce the input current and output voltage of each converter. So it can increase the reliability of the converter. At the same time, it can reduce the ripple of the output voltage. So the output filter inductor L and capacitor C can be reduced observably. Finally 5kW experimental equipment is accomplished. Experimental results verify the analysis and the conclusion.

REFERENCES

[1] Zhou Xunwei, Yang Bo and Amoroso L, et al. "A Novel High-input-voltage, High Efficiency and Fast Transient Voltage Regulator Module — The Push-pull Forward Converter". IEEE APEC' 99[C], 1999:487~492.

[2] Mao Ye, Peng Xu. "Investigation of Topology Candidates for 48V VRM". IEEE 2002:699~705.

[3] J. Cobos, O. Garcia, J. Sebastian, J. Uceda, "Active clamp PWM forward converter with self driven synchronous rectification," IEEE International Telecommunication Energy Conference Proceedings, pp.200~206, 1990.

[4] Cai Xuansan, et al. The principle and design of switch-mode power supply [M]. Beijing: Electronic Industries Publishing Company, 2000.

[5] Wang Zhaoan, et al. Power electronic technology [M]. Beijing: mechanical Industries Publishing Company, 2000.

[6] Yang Xu, et al. The technology of switch-mode power supply [M]. Beijing: mechanical Industries Publishing Company, 2004.

2006 5th International Power Electronics and Motion Control Conference

Voltage Fed and Current Fed Full Bridge Converter for the Use in Three Phase Grid Connected Fuel Cell Systems

M. Mohr[1] and F.-W. Fuchs[2]

Christian-Albrechts-University of Kiel / Institute of Power Electronics and Electrical Drives; Kiel, Germany
[1]mam@tf.uni-kiel.de [2]fwf@tf.uni-kiel.de

Abstract—Fuel cells for low or medium power deliver comparatively low voltages compared to the mains voltage at high currents. A high dc-link voltage is needed to feed in electrical energy from fuel cells to the mains via a voltage source inverter. Several well known dc/dc converters can be used to transform the varying fuel cell voltage to the requested amplitude of the dc-link-voltage. Besides transformerless converters like the boost converter, converters with high frequency transformers can be used. In the considered higher power range, the full-bridge converter circuit is the appropriate solution. Depending on their input circuit, the converters are classified between voltage fed and current fed full bridge converter. In this paper, full bridge converters of the voltage type and of the current type as active front end for fuel cell inverters in the power range of 20 kW and higher are analysed and compared to each other. The focus is set on the operating behavior, the system complexity and the efficiency of the different converters.

Keywords—fuel cells, DC-DC power conversion, distributed energy

I. INTRODUCTION

Fuel cells can be an important component of future energy systems, enabling electricity and heat generation from hydrogen or similar substances with high efficiency and low or nearly zero emission. Their introduction is in the stage of beginning as their development has not yet been finished.

Fuel cells provide a variable dc current at variable fuel cell voltage. For feeding into the mains they have to be connected to the ac mains by means of inverters. Because of the comparatively low voltage of a fuel cell stack for low and medium power compared to the mains voltage, the inverter has to increase the voltage when feeding into the mains.

Several converters for fuel cell systems have been discussed in publications, see for example [1]-[3]. A very common solution to feed in electrical energy into the three phase mains is a voltage source inverter. To attain the high dc link voltage necessary for the voltage source inverter, a dc/dc converter is used between the fuel cell stack and the inverter.

For low ratios (i.e. 1:1 to 1:3) between fuel cell stack voltage and dc link voltage, a boost converter shows some advantages like high efficiency and low component

quantity [3], [4]. For higher ratios between fuel cell stack voltage and dc link voltage, transformerless converters become less applicable [4]. Dc/dc converters with high frequency transformers overcome these problems. They achieve the capability of high voltage ratios due to the magnetic coupling of the transformer and the different number of turns on the primary and the secondary side of the transformer. In addition they provide galvanic isolation between the fuel cell and the mains.

The proposed application is the feed in of electrical energy into the mains; therefore unidirectional converters are appropriate and an additional energy storage to supply peak power is not needed.

Many fuel cell inverters which have been published are designed for a power range of about 1 kW [1], [5]. In this power range, converters like push-pull converters or half bridge converters are suitable converter topologies utilizing high frequency transformers. For a power range of 20 kW and higher, the full bridge converter is an appropriate solution for dc/dc converters with controllable voltage ratio [1], [2]. Depending on their input circuit full bridge converters can be classified into voltage fed and current fed full bridge converters. Both types are used as converters for fuel cell applications.

In this paper, full bridge converters of the voltage type and of the current type are analysed and compared to each other with respect to their suitability as converters for fuel cell inverter systems in the mid power range. In addition to the general behaviour of the converters, the semiconductor losses and the complexity of the converter is investigated. Effects of the transformer's stray inductances and the dimensioning of the transformer's turns ratio are part of the analysis.

In chapter II, the basic performance of the fuel cell and the complete inverter system are introduced. Chapter III presents the operating principle of the voltage fed full bridge converter and chapter IV shows the operating principle of the current fed full bridge inverter. In chapter V, the results of the power loss calculation are shown. In chapter VI the two converters are compared to each other. In chapter VII experimental results of a laboratory setup of a current fed full bridge converter are presented. Finally chapter VIII presents the conclusion.

1-4244-0448-7/06/$25.00 ©2006 IEEE

II. BASIC SYSTEM PERFORMANCE

A. Fuel Cell characteristics

Figure 1 shows an example of the characteristic curve, of a single fuel cell [6]. Recommended operation of the cell is in the ohmic region (about a current density of 0.2 to 1 A/cm² in the curve below). Operation in the concentration region (declining section at about a current density of 1.1 to 1.2 A/cm² in the shown curve) of the fuel cell yields on the one hand to a bad efficiency of the fuel cell, on the other hand it may damage the fuel cell and has to be avoided. The fuel cell voltage typically specified in fig. 1 refers to one single fuel cell with a typical current density J. For a nominal operating point of 200 A at 100 V i.e. one has to take a series connection of 168 cells with a cell area of 200 cm² for the examplarily characteristic shown above. In general, there are curves that show a less stiff characteristic than the example shown in fig. 1. Due to the distinctive fuel cell characteristic curve, the voltage at nearly no load can reach values of much more than the nominal voltage, i.e. 168 V in the example above. Due to degradation effects of the fuel cell during its lifetime the voltage decreases. A degradation of 10 % corresponding to a lowest voltage of 90 V is assumed for the dimensioning of the converters. Thus, the converters have to be able to operate at input voltages from $V_1 = 90$ V… 200 V.

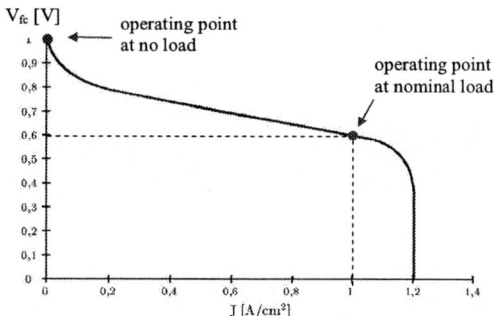

Fig. 1. Characteristic curve of a single fuel cell: Fuel cell voltage V_{fc} vs. current density J.

B. Fuel cell inverter system

Fig. 2 shows the whole schematic of the fuel cell inverter. The dc/dc converter elevates the fuel cell voltage V_1 to the the voltage source inverter's dc link voltage V_2 which has to be greater than the rectified line to line voltage [3]. The dc link capacitor decouples the voltage source inverter and the dc/dc converter and keeps the dc link voltage ripple at an adequate level. A feasible power

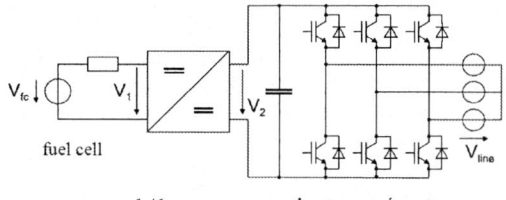

Fig. 2. Schematic fuel cell inverter system.

TABLE I
VOLTAGES OF THE FUEL CELL INVERTER SYSTEM FOR THE ANALYSIS

Fuel cell voltage	Dc-link voltage	Line to line voltage
V_1 = 90-200 V	V_2 = 700 V	V_{line} = 400 V +10% / - 15%

flow control method could be keeping the dc link voltage constant via the dc/dc converter. An additional energy storage is not required due to the fact that the inverter is connected in parallel to the mains and peak power is delivered from the mains. Tab. 1 shows the voltage levels of the inverter system taken for this analysis. Nominal operating point of the fuel cell is $V_1 = 100$ V at $I_1 = 200$ A.

III. VOLTAGE FED FULL BRIDGE CONVERTER

A. Operating principle

Voltage fed full bridge converters are well-known circuits for high power switch mode power supplies to convert high voltages at the input to low voltages at the output. They are used in fuel cell or photovoltaic inverters as well [1][2]. Fig. 3 shows the circuit diagram of the voltage fed full bridge converter. The transistors T_1–T_4 at the input can be MOSFETs due to the low input voltage V_1 of the fuel cell. The high frequency transformer is modelled by the following: the transformer has the turns ratio n, its total stray inductance L_σ is the sum of the primary stray inductance and the secondary stray inductance reflected across the transformer to the primary. The magnetizing inductance L_m is much bigger than the leakage inductance [7]. The rectification on the secondary side is realized with a full bridge rectifier consisting of the diodes D_1–D_4.

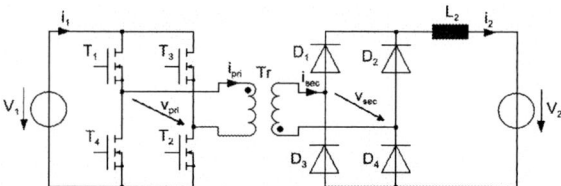

Fig. 3. Circuit diagram of voltage fed full bridge dc/dc converter

The waveforms and characteristic time instants are shown in fig. 4. The periodes of time are described in the following.

$t_1 - t_0$: At t_0 transitors T_1 and T_2 are switched on. The diodes D_1–D_4 are conducting, the secondary voltage of the transformer v_{sec} equals zero. The slope of the rising transformer current is determined by the transformer's leakage inductance L_σ and the Voltage V_1.

$t_2 - t_1$: At t_1 the primary current i_{pri} reflected over the transformer to the secondary has reached the output current i_2. The commutation of the diodes has finished, now only D_1 and D_4 are conducting.

Energy is transferred via the transformer to the secondary. The voltage at the secondary is $V_1 \cdot n$ minus the voltage drop over the leakage inductance L_σ. The current i_{pri} is furthermore rising due to the fact that the voltage at the transformer's secondary is higher than V_2.

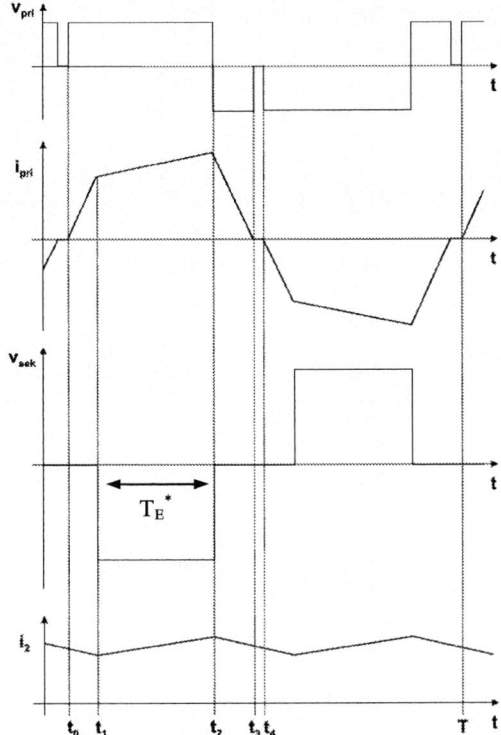

Fig. 4. Waveforms of voltage fed full bridge converter.

t₃ - t₂ : At t_2 the transistors T_1 and T_2 are switched off. There is a freewheeling path of the primary transformer current through the body diodes of T_3 and T_4. The transformer's primary voltage is negative ($-V_1$). The rectifier diodes D_2 and D_3 start to conduct in addition to D_1 and D_4 the secondary voltage of the transformer v_{sec} equals zero.

The slope of the falling current i_{pri} is determined by the voltage $-U_1$ and the leakage inductance L_σ.

t₄ - t₃ : At t_3 the current i_{pri} has become zero. Primary voltage and secondary voltage of the transformer are zero, too. The diodes D_1–D_4 are conducting, the transistors T_1–T_4 are off.

At t_4 the transistors T3 and T4 are switched on. The behavior and the waveforms in the second half period are like the ones shown above due to the symmetrical circuit of the converter.

B. Analysis

The duty cycle D^* of the voltage fed full bridge converter is defined here as follows:

$$D^* = \frac{2T_E^*}{T} \quad \text{with } T_E^* \text{ as shown in fig. 4.} \quad (1)$$

The ratio between the input voltage V_1 and the output voltage V_2 can be derived to:

$$V_2 = L_2 \, D^* n \frac{V_1}{(L_2 + L_\sigma \, n^2 - D^* L_\sigma \, n^2)} \quad (2)$$

It depends on L_2, L_σ, the transformer turns ratio n and the duty cycle as defined above.

The maximum possible duty cycle depends on the leakage inductance L_σ, the period T and the transformer turns ratio n. For a well smoothed output current i_2 it can be derived to:

$$D_{max}^* = 1 - \frac{4L_\sigma \, I_2 \, n}{V_1 \, T} \quad (3)$$

Eq. (2) inserted in eq. (1) yields to the maximum achieveable output voltage $V_{2,\,max}$ for given system parameters. From this it can be seen that the leakage inductance of the transformer L_σ has a great influence on the dimensioning of the converter. The leakage inductance limits the maximum output to input voltage ratio V_2/V_1 or the operation frequency $f = 1/T$ of the converter.

The blocking voltage of the rectifier diodes is $V_1 \cdot n$. It is very high at low load due to the high voltage at the fuel cell and the high turns ratio of the transformer.

IV. CURRENT FED FULL BRIDGE CONVERTER

A. Operating principle

The circuit of a current fed full bridge converter is shown in fig. 5. At the input there is the filter inductance L_1. As will be shown in the following, a clamping circuit is needed. This circuit consists of the diode D_{Cl} and the capacitance with assumed constant voltage V_{Cl}. The maximum blocking voltage of transistors T_1–T_4 is the voltage of the clamping circuit V_{Cl}. The transformer Tr has the turn ratio n. The rectification on the secondary side is realized with a full bridge rectifier consisting of the diodes D_1–D_4 directly connected to the output voltage V_2.

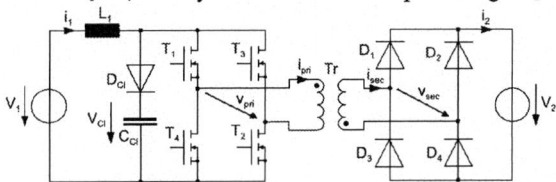

Fig. 5. Circuit diagram of current fed full bridge converter

Fig. 6 shows the waveforms and characteristic time instants of the current fed full bridge converter. In the following the converter's operation between the time instants is shown.

t₁ - t₀ : In the period before t_0, Transistors T_3 and T_4 have been conducting. At t_0 T_1 and T_2 are switched on in addition to T_3 and T_4 so all Transistors are conducting now. The primary voltage equals zero and the voltage at the transformer's secondary is V_2. The current through the transformer is decreasing due to the voltage V_2/n across the leakage inductance L_σ. The energy stored in the leakage inductance is transferred to the secondary through the diodes D_1 and D_4.

t₂ - t₁ : At t_1 the transformer current has become zero. It will be zero until the time instant t_2.

During the time interval t_2 - t_0 all transistors remain conducting, the input current i_1 is rising due to the voltage V_1 across the inductance L_1. Energy is stored in the inductance L_1.

t₃ - t₂ : At t_2 the transistors T_3 and T_4 are

327

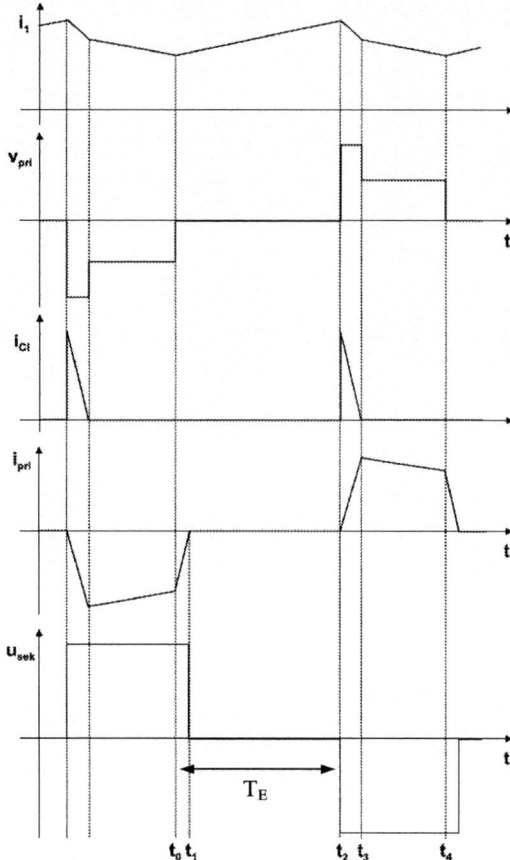

Fig. 6. Wave forms of the current fed full bridge converter.

switched off. The primary current i_1 is impressed into the transformer with its leakage inductance L_σ. Due to the fast current slope at L_σ, the voltage across the transformer's primary is rising so the clamping diode D_{Cl} starts to conduct. The primary voltage of the transformer is clamped to V_{Cl} which will be kept to a constant value. The falling slope of the input current i_1 is determined by the voltage V_1-V_{Cl} across the inductance L_1. The slope of the primary current i_{pri} is determined by the voltage $(V_{Cl}-V_2/n)$ across the leakage inductance L_σ.

$t_4 - t_3$:   At t_3 the transformer's primary current i_{pri} has reached the input current i_1. Energy is transferred via the transformer to the secondary. The current i_1 and i_{pri} are equal. The current slope is determined by the voltage (V_1-V_2/n) across (L_1+L_σ).

At t_4 transistors T_3 and T_4 are switched on, all transistors are conducting. Due to the symmetrical circuit of the converter the waveforms in the second half period are equal to the waveforms shown above.

B.   Analysis

The duty cyle D of the current fed full bridge converter is here defined as follows:

$$D = \frac{2T_E}{T} \quad \text{with } T_E \text{ as shown in fig. 6} \tag{4}$$

The ratio between output and input voltage V_1/V_2 depends on the mean input current I_1, L_1, n and the duty cycle D. It does not depend on stray inductance L_σ. The expression of the ratio between output and input voltage $V_1/V_2 = f(D, I_1, L_1, n)$ is rather huge. Showing it will go beyond the scope of this paper.

Due to the boost principle of the current fed full bridge converter, the input voltage is limited to a maximum amplitude. This yields to the maximum transformer turns ratio n_{max} depending on maximum input voltage $V_{1,max}$ and output voltage V_2:

$$n_{max} = \frac{V_2}{V_{1,max}} \tag{5}$$

As shown in chapter IV, section A, the current fed full bridge converter needs a clamping circuit for the energy stored in the transformer's leakage inductance. The power, which has to be dissipated through the clamping circuit, is given in eq. (6):

$$P_{Cl} = V_{Cl}\, I_1^2\, \frac{L_\sigma}{T}\, \frac{n}{(n\, V_{Cl} - V_2)} \tag{6}$$

It depends inter alia on the clamping voltage V_{Cl} and the leakage inductance L_σ.

In contrast to the voltage fed converter the duty cycle is not influenced by the leakage inductance, but the power in the clamping circuit is higher for a larger leakage inductance. In addition the blocking voltage of the rectifier diodes is V_2 at every operating condition.

V. SEMICONDUCTOR LOSSES

A.   Semiconductor losses of the voltage fed full bridge converter

In this section the calculation of the semiconductor losses is shown. The following simplifications have been assumed: the output current i_2 is assumed to be well smoothed. The output characteristic of the MOSFETs ($T_1 - T_4$) is modelled with the drain source resistance R_{DSon}, the output characteristic of the MOSFET's body diodes and the rectifier diodes are modelled with a forward threshold voltage V_{F0} and a differential resistance r_F ($V(I) = V_{F0} + r_F \cdot I$). The switching losses of the body diodes and the rectifier diodes are calculated [7] using their reverse recovery charge Q_{rr}, which depends on the slope of the falling diode current. Due to the slow current slopes the switching losses of the diodes are comparatively low. MOSFET's switching losses are estimated by multiplying idealized current and voltage waveforms (with rise time t_r and fall time t_f as defined in the respective datasheet) and integrating the product [7]. Due to the fact that the transistor current rises very slow, caused by the leakage inductance of the transformer, turn on losses are very small. The semiconductor losses in detail are as follows:

MOSFET conduction losses:

$$P_{C,MOS} = R_{DSon}\, \frac{n^3\, I_2^2}{3}\, \frac{L_\sigma\, I_2}{V_1\, T} + R_{DSon}\, n^2\, I_2^2\, \frac{D^*}{2} \tag{7}$$

MOSFET switching losses:

$$P_{S,MOS} = \left[\frac{V_1^2}{12 L_\sigma} t_r^2 + V_1 I_2 n^2 t_f + \left(V_1 + L_{streu} \frac{I_2 n}{2 t_f} \right) \right] \frac{1}{T} \qquad (8)$$

Conduction losses of MOSFET's body diode:

$$P_{C,Body} = V_{F0} \frac{n^2 I_2^2 L_\sigma}{2 V_1 T} + r_F \frac{n^3 I_2^3 L_\sigma}{3 V_1 T} \qquad (9)$$

Switching losses of MOSFET's body diode:

$$P_{S,Body} = \frac{Q_{rr} V_1}{4 T} \qquad (10)$$

Conduction losses of one rectifier diode:

$$P_{C,RF} = \frac{r_F n I_2^3 L_\sigma}{12 V_1 T} + \frac{V_{F0} I_2}{4} + \frac{r_F I_2^2}{8} + \frac{r_F}{T} \left(\frac{D^* T}{2} + \frac{n I_2 L_\sigma}{V_1} \right) \qquad (11)$$

Switching losses of one rectifier diode:

$$P_{S,RF} = \frac{Q_{rr} V_1 n}{2 T} \qquad (12)$$

B. Semiconductor losses of the current fed full bridge converter

For the loss calculation of the current fed full bridge converter the simplified conditions shown in the section above have been assumed, too. In contrast to the voltage fed converter, the input current i_1 is here well smoothed. In the current fed full bridge converter with the proposed modulation the body diodes are not used, but there are conduction and switching losses in the clamping diode. The losses in detail are as follows:

MOSFET conduction losses:

$$P_{C,MOS} = R_{DSon} \frac{I_1^2}{T} \left[\frac{8}{12} \left(\frac{n I_1 L_\sigma}{V_2} \right) + \frac{1}{2} \left(\frac{DT}{2} - \frac{n I_1 L_\sigma}{V_2} \right) \right] \dots$$
$$- R_{DSon} \frac{I_1^2}{T} \left[\frac{2 I_1 L_\sigma}{3 (V_{Cl} - V_2 / n)} + \frac{T}{2} - \frac{DT}{2} \right] \qquad (13)$$

MOSFET switching losses:

$$P_{S,MOS} = \left[\frac{V_2^2 t_r^2}{6 n^2 L_\sigma} + \frac{V_{Cl} I_1 t_f}{2} + \frac{I_1 t_f}{4} \left(V_{Cl} + \frac{L_{stray} I_1}{2 t_f} \right) \right] \frac{1}{T} \qquad (14)$$

Conduction losses of one rectifier diode:

$$P_{C,RF} = \left(V_{F0} \frac{I_1}{2n} + r_F \frac{I_1^2}{3 n^2} \right) \left(\frac{I_1 L_\sigma}{(V_{Cl} - V_2 / n)} + \frac{n I_1 L_\sigma}{V_2} \right) \frac{1}{T} \dots$$
$$+ \left(V_{F0} \frac{I_1}{n} + r_F \frac{I_1^2}{n^2} \right) \left[\frac{T}{2} - \frac{I_1 L_\sigma}{(V_{Cl} - V_2 / n)} - \frac{DT}{2} \right] \frac{1}{T} \qquad (15)$$

Switching losses of one rectifier diode:

$$P_{S,RF} = \frac{Q_{rr} V_2}{4 T} \qquad (16)$$

Conduction losses of the clamping diode:

$$P_{C,Cl} = V_{F0} \frac{I_1^2 L_\sigma}{(V_{Cl} - V_2 / n)} \frac{1}{T} + r_F \frac{2 I_1^3 L_\sigma}{3 (V_{Cl} - V_2 / n)} \frac{1}{T} \qquad (17)$$

Switching losses of the clamping diode:

$$P_{S,Cl} = Q_{rr} (V_{Cl} - V_2 / n) \frac{1}{T} \qquad (18)$$

VI. COMPARISON

A. Converter's dimensioning and operating conditions

In the following semiconductor losses for the voltage fed and the current fed full bridge converter are calculated for a certain dimensioning. Starting from the operating conditions shown in chapter II section B the converter's components are rated as follows. The projected MOSFETs are standard Infineon® Cool-MOS® IPW60R045CS ($I_D = 60$ A, $V_{DS} = 600$ V), the diodes are also standard ones: International Rectifier® fast, soft recovery diodes HFA30PB120 ($I_{F(AV)} = 30$ A, $V_R = 1200$ V). The numbers of the semiconductors in the respective design are chosen as follows: the junction temperature has been calculated for the worst case (maximum losses) by using the loss equations shown above for different numbers of semiconductors. The numbers of semiconductors is optimal dimensioned if the junction temperature of the used semiconductors is near to and does not exceed 125°C in the worst case. The assumed transformer leakage inductance is $L_\sigma = 0.75$ µH.

From this the dimensioning for the current fed full bridge converter is derived to: a = 4 MOSFETs per switch in parallel, a transformer turns ratio of n = 3.25, 4 rectifier diodes in total (one per place), two clamping diodes in parallel.

For the voltage fed full bridge converter: a = 7 MOSFETs per switch in parallel, a transformer turns ratio of n = 12 and 8 rectifier diodes in total (two of each in series due to the high blocking voltage).

B. Comparison of semiconductor losses

Using eq. (7)-(12) with given data from the semiconductor datasheet incorporating the fuel cell characteristic curve from fig. 1, chapter II, section A, we get the efficiency of the voltage fed full bridge converter. Equations (13)-(18) yield in the same way to the efficiency of the current fed full bridge converter. In the efficiency calculation for the current fed converter the energy of the clamping circuit has been taken into account: it is assumed that the energy that flows into the clamping capacitor will be transferred to the secondary via an additional auxiliary converter with an efficiency of $\eta = 90\%$.

Fig. 7. shows the efficiency based on semiconductor losses for the current fed and the voltage fed full bridge converter depending on the input current I_1. The current fed converter has significantly lower semiconductor

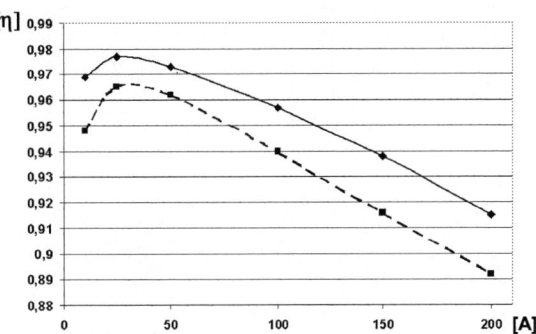

Fig. 7. Efficiency of voltage fed (dotted line) and current fed full bridge converter (solid line) connected to the fuel cell depending on fuel cell current. Optimal rating of semiconductors.

conducting losses because the left and the right leg are connected in parallel during T_E. The efficiency of the current fed converter is about 1-2% better than the efficiency of the voltage fed converter.

More than 80% of the semiconductor losses are conducting losses. From this, losses can be easily reduced by connecting more MOSFETs in parallel. Fig. 8. shows the efficiency of the two converters with an overdimensioning of the MOSFETs. It can be seen that even with more than twice as many MOSFETs per switch in parallel than the current fed converter, the voltage fed converter reaches just the efficiency of the optimal utilized current fed converter.

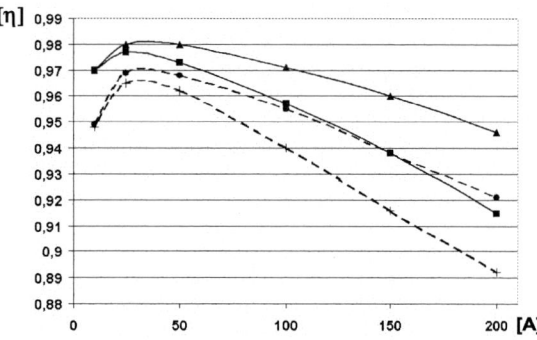

Fig. 8. Efficiency of voltage fed and current fed full bridge converter connected to the fuel cell. With overdimensioning of MOSFETs: voltage fed a = 7 (dotted line +), voltage fed a = 10 (dotted line •), current fed a = 4 (solid line ■), current fed a = 8 (solid line ▲).

C. Evaluation

This section gives a short outline of the characteristics of the voltage fed and current fed full bridge converter.

The working principle of the voltage fed full bridge converter could be characterized as "buck" principle. The series connected transformer generates the high voltages which will be averaged using the inductor L. Therefore the voltage fed converter needs a transformer with high turns ratio. The leakage inductance L_σ has a strong influence on the converter's dimensioning. The blocking voltage at the rectifier diodes is very high, particularly at low duty cylces (high input voltages at low fuel cell load). The start up of the converter from $V_2 = 0$ V can be easily obtained by the voltage fed converter. Due to the voltage fed input, the input current becomes temporarily negative. In order to provide negative currents from the fuel cell, additional filter elements are required. The transformer of the voltage fed converter is well utilized at nominal load due to high duty cycles. The voltage fed full bridge converter shows much higher losses and a higher amount of semiconductors compared to the current type converter. Switching losses of the MOSFETs contribute less than 4 % to the total losses and, in addition, turn on losses are low due to the slow current slope caused by the transformer's leakage inductance. Therefore phase shift modulation of the voltage fed full bridge converter [8] does not significantly reduce semiconductor losses.

The working principle of the current fed full bridge converter could be described as "boost" pinciple. The inductance L at the input of the converter boosts the input voltage to a higher level. In addition this voltage is increased by the transformer. Hence, the current fed converter needs a transformer with a comparatively low turns ratio. The value of the transformer's stray inductance influences the energy dissipated from the clamping circuit. The rectifier diode's blocking voltage is always the output voltage U_2. The current fed converter is not able to work with output voltages near zero. Thus starting-up the converter has to be done by loading the dc link from the mains voltage source inverter or using the additional converter at the clamping circuit. Due to the input inductance at the converter, the input current is well smoothed; no additional filtering towards the fuel cell is needed [9]. Semiconductor losses of the current fed converter are lower than for the voltage fed converter, hence the current fed converter needs less power semiconductors than the voltage fed converter. Another possibility to recycle the clamping energy could be to use an active clamp circuit [10] instead of the passive clamp circuit as described above .

VII. EXPERIMENTAL RESULTS

Fig. 9 shows the laboratory setup of a 20 kW current fed full bridge converter. CoolMOS® semiconductor bridge with gate drive circuits and fast clamping diodes with heat sink on the left, high frequency transformer with n = 3.25 consisting of two transformers in "power link" connection to reduce the stray inductances of the transformer (parallel connection of the primary windings, series connection of the secondary windings) in the middle. On the right hand side there is the full bridge rectifier. The MOSFETs T_1-T_4 and the clamping unit consisting of D_{Cl} and C_{Cl} are linked by cupper busbars to reduce the stray inductance of the clamping path. Below the heatsink of the MOSFETs there is the input inductance $L_1 = 10\,\mu\text{H}$.

Fig. 10 and 11 show experimental waveforms of the current fed full bridge converter. The energy stored in the clamping unit is here dissipated by a resistor. Fig. 10 shows the gate signals of the transistors, the input current of the converter i_1 and the transformer's primary current

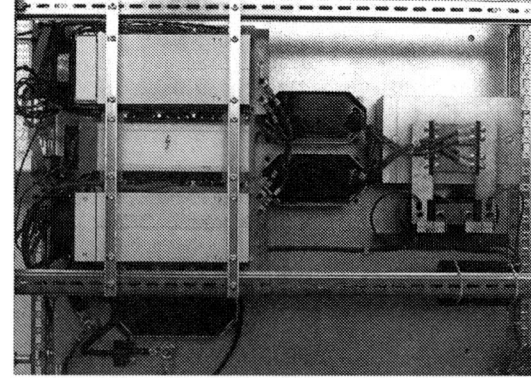

Fig. 9 Laboratory setup of the current fed full bridge converter with rated power of P = 20 kW.

330

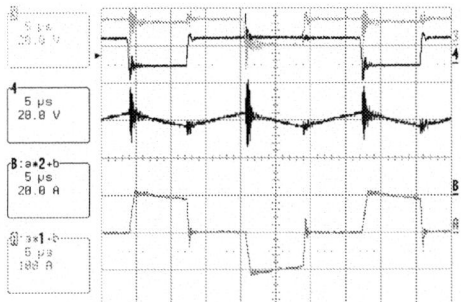

Fig. 10. Waveforms of current fed full bridge converter at an input power of P = 5 kW: gate signals of T_1/T_2 (Channel 3) and T_3/T_4 (Ch. 4), Input current I_1 (Ch. B), transformer current i_{pri} (Ch. A).

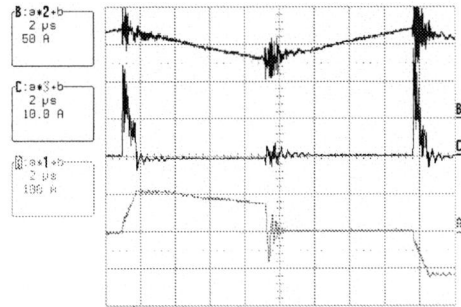

Fig. 11. Waveforms of current fed full bridge converter at an input power of P = 9 kW: input current I_1 (Ch. B), current of one of four paralleled clamping diodes (Ch. C), transformer current i_{pri} (Ch. A).

i_{pri} at an input power of 5 kW ($V_1 = 50$ V, $I_1 = 100$ A, $V_2 = 305$ V, $f_S = 30$ kHz).

Fig. 11 shows the input current i_1, the current through one of four paralleled clamping diodes and the primary current i_{pri} through the transformer at an input power of 9 kW ($V_1 = 90$ V, $I_1 = 100$ A, $V_2 = 665$ V). The clamping voltage here is $V_{Cl} = 225$ V, the power dissipated by the clamping unit at this operating point is $P_{Cl} = 505$ W.

Efficiency values for the converter will be measured by kalorimetric loss measurements to achieve adequate accuracy.

VIII. CONCLUSION

To feed in electrical energy from fuel cells into the three phase mains via a voltage source inverter, there is the need of a dc/dc converter which increases the fuel cell voltage to the dc link voltage of the inverter. Dc/dc converters with high frequency transformers meet the demands for larger input to output voltage ratios. In the considered power range of 20 kW and above, full bridge converters are the appropriate solution.

Two topologies, the current fed and the voltage fed full bridge converter, have been analysed in this paper.

The operating performance of the converters is shown, and their semiconductor losses have been calculated and compared. The effects of the tranformer's leakage inductance on the dimensioning of the converters and the dimensioning of the transformer turns ratio have been shown. The analysis has been done by analytical

calculation in steady state by average models.

The comparison between the two converter topologies is based on a converter design for a rated nominal input load of 20 kW at 100 V, 200 A, including standard fuel cell performance, and a dc link voltage of 700V for an inverter at the three phase mains with a line to line voltage of 400V. Standard power semiconductors have been selected.

The analysis shows that for this application the current fed converter has lower losses compared to the voltage fed converter. This is due to the fact that input MOSFETs are connected in parallel for the proposed modulation strategy. Due to the lower losses of the current fed full bridge converter, it can operate with less installed semiconductors compared to the voltage fed full bridge converter. However, the current fed converter needs an additonal clamping circuit and for further improving the performance an additional auxiliary converter to transfer the energy dissipated in the clamping circuit to the dc link of the inverter.

Regarding this analysis it can be concluded that current fed full bridge converters combine most advantages. Thus they are the favored solution for dc/dc converters with high frequency transformers for fuel cell applications in the medium power range.

REFERENCES

[1] Xue-Y.; Chang-L.; Kjær- S.-B.; Bordonau,-J.; Shimizu,-T. Topologies of single-phase inverters for small distributed power generators: an overview. *IEEE Transactions on Power Electronics*. Sept. 2004; 19(5): 1305-14

[2] Profumo,-F.; Tenconi,-A.; Cerchio,-M.; Bojoi,-R.; Gianolio,-G.: Fuel cells for electric power generation: peculiarities and dedicated solutions for power electronic conditioning systems *Proceedings of 11th International Power Electronics and Motion Control EPE-PEMC* 2004

[3] Mohr, M.; Fuchs, F.W.: Comparison of Three Phase Current Source Inverters and Voltage Source Inverters Linked with DC to DC Boost Converters for Fuel Cell Generation Systems, *Proceedings of the 11th European Conference on Power Electronics and Applications, EPE 2005*, Dresden

[4] Solmecke, H.: *Optimierte Stromrichter für Brennstoffzellenanlagen (Optimized power converters for fuel cell plants)*, phd-thesis, 1998, Fernuniversität-Gesamthochschule Hagen; in German

[5] Andersen, G. K.; Klumpner, Chr.; Kjær, S. B.; Blaabjerg F.: A New Green Power Inverter for Fuel Cells; *Proceedings of the Power Electronics Specialists Conference PESC 2002*, Cairns

[6] Ledjeff-Hey, K. et. al.: *Brennstoffzellen: Entwicklung, Technologie, Anwendung (Fuel Cells: Development, Technology, Application)*, edition 2, 2001, Müller Verlag, Heidelberg; in German

[7] Mohan, N.; Undeland, T.-M.; Robbins, W.-P.; *Power Electronics, Converters, Applications and Design;* Wiley&Sons, N. Y., 1995

[8] Sabate, J.-A.; Vlatkovic, V.; Ridley, R.-B.; Lee, F.-C.; Cho, B.-H.: Design considerations for high-voltage high-power full-bridge zero-voltage-switched PWM converter; *Proceedings of the 14th Applied Power Electronics Conference and Exposition APEC '90*.

[9] Schindele, L.; Braun, M.; Späth, H.: The Influence of Power Electronic Dynamics on PEM Fuel Cell-System; *Proceedings of the 11th European Conference on Power Electronics and Applications, EPE 2005*, Dresden

[10] Yakushev, V.; Meleshin, V.; Fraidlin, S.: Full-bridge isolated current fed converter with active clamp; *Proceedings of the 14th Applied Power Electronics Conference and Exposition APEC '99*.

2006 5th International Power Electronics and Motion Control Conference

Small-Signal Modeling of Asymmetrical Half Bridge Flyback Converter

Tso-Min Chen[*] and Chern-Lin Chen[**]

[*] System General Corp/System Engineering Department, Taiwan, China
[**] National Taiwan University/ Department of Electrical Engineering & Graduate Institute of Electronics Engineering, Taiwan, China
[*]Webber.Chen@sg.com.tw and [**]clchen@cc.ee.ntu.edu.tw

Abstract—A small-signal model for the asymmetrical half bridge flyback converter is presented. The effects of storage elements and peak current mode control on the transfer characteristics are investigated. The results show that the power stage of the studied converter is basically a fourth order system with two low-frequency poles and two high frequency poles. While the peak current mode control is applied, the transfer functions are reduced to three poles: one dominant pole dependent on the output capacitance and output load resistance, and two high-frequency poles determined by the resonant inductance and the DC blocking capacitance. The validity of the proposed models is verified by both PSPICE simulations and hardware experiments.

Keywords- asymmetrical half bridge; flyback converter; small-signal modeling

I. INTRODUCTION

The asymmetrical half bridge flyback converter, which can achieve ZVS operation of the power switches, is gaining popularity. There have been literatures [1-6] providing detailed steady-state analysis and design procedures. In addition to the steady-state characteristics, the dynamic behavior is equally important and critical when it comes to the design of a stable operation of the converter. However, there are no references exploring the dynamic characteristics of the studied converter. In fact, the dynamic characteristics of the studied converter are significantly different with the conventional flyback converter for its special resonant network consisting of the storage-elements. It will be shown in this paper that extra poles appear in the small-signal transfer functions, and these poles are dependent on the value of storage-elements in the studied converter. While the peak current mode control is applied to this converter, the control-to-output and input-to-output transfer functions are changed. Therefore, in order to optimize the dynamic performance of this converter, it is necessary to develop valid small signal models not only for its power stage, but also the applied control method.

There is a number of converter modeling techniques in the literatures. The state-space averaging method [7, 8] is employed in this work to derive the averaged state space equations. Based on these averaged state space equations, the small signal transfer functions of the power stage and the equivalent circuit model are obtained. This equivalent circuit model is then synthesized to facilitate the analysis using popular circuit analysis programs such as PSPICE. The small signal model of the studied converter with the peak current mode control is also presented. Effects of the values of the storage-elements on the dynamic characteristics are shown on the simulation results.

II. BASIC OPERATION REVIEW

The detailed operational principles of the studied converter have been discussed in the literatures [1-6]. Here, we give a simple review to facilitate the following analysis. Fig. 1 shows the simplified circuit diagram of the asymmetrical half bridge flyback converter. In the description, the following assumptions are made.

1. Dead times between conduction intervals are negligible.
2. C_{ds} accounts for any parasitic capacitance of the two switches and the transformer windings.
3. All elements of the circuit are lossless, i.e. there is no voltage drop across switches and other parasitic resistors.

As shown in Fig. 2, there are two stages of operation.

Stage 1 [$0 \leq t < DT_s$]: This stage begins when $S1$ is turned on at t=0. During this period, the DC input power source charges C_b, L_M, and L_r. $D1$ is reversed biased. During this period, the state equations can be written as follows:

$$(L_M + L_r)\frac{di_{Lr}}{dt} = V_{in} - v \cdot \tag{1}$$

$$C_b \frac{dv_b}{dt} = i_{Lr} \cdot \tag{2}$$

$$C_o \frac{dv_{CO}}{dt} = n(i_M - i_{Lr}) - \frac{V_o}{R_L} \cdot \tag{3}$$

Stage 2 [$DT_s \leq t < T_s$]: $S2$ is on and $S1$ is off. $D1$ is conducted, and i_{D1} is the difference between i_M and i_{Lr} multiplying by the turn ratios of T_X. C_b and L_r form a resonant network. In this stage, the state equations can be written as follows:

$$L_M \frac{di_M}{dt} = -nV_o \cdot \tag{4}$$

$$L_r \frac{di_{Lr}}{dt} = -v_b + nV_o \cdot \tag{5}$$

$$C_b \frac{dv_b}{dt} = i_{Lr} \cdot \tag{6}$$

1-4244-0448-7/06/$25.00 ©2006 IEEE

$$C_o \frac{dv_{CO}}{dt} = n(i_M - i_{Lr}) - \frac{V_o}{R_L} \tag{7}$$

The operations periodically repeat in each switching cycle. It is noted that the dead times between conduction intervals are rather short. They are ignored in our analysis.

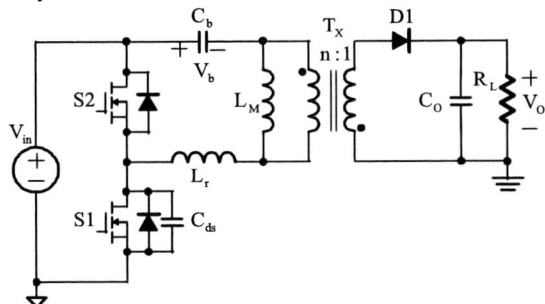

Fig. 1. Asymmetrical half bridge flyback converter.

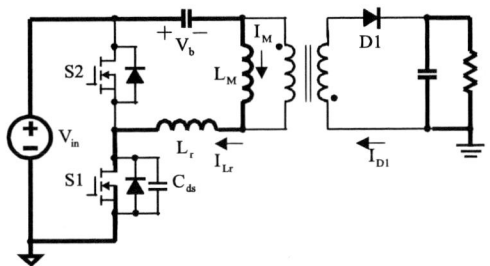

(a) For $0 \le t < DT_S$.

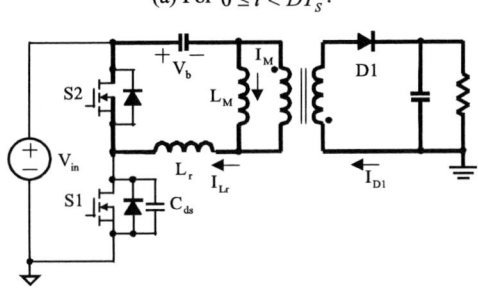

(b) For $DT_S \le t < T_S$

Fig. 2. Switching stages of the studied converter.

III. AVERAGED STATE SPACE EQUATION

The state space equations for the state vector $x = [i_M, i_{Lr}, v_b, v_{CO}]^T$ can be derived for each of the two operation intervals over one switching cycle and are give by A_i, B_i, and C_i for $i = 1,2$.

The averaged state space equations, $\dot{x} = Ax + BV_{in}$, $y = Cx$, can be obtained.

$$A = DA_1 + (1-D)A_2 \tag{8}$$
$$B = DB_1 + (1-D)B_2 \tag{9}$$
$$C = DC_1 + (1-D)C_2 \tag{10}$$

The results are given in (11)-(13).

$$A = \begin{bmatrix} 0 & 0 & -\dfrac{D}{L_M+L_r} & -\dfrac{n(1-D)}{L_M} \\ 0 & 0 & -\dfrac{D}{L_M+L_r} & -\dfrac{1-D}{L_r} & \dfrac{n(1-D)}{L_r} \\ 0 & \dfrac{1}{C_b} & 0 & 0 \\ \dfrac{n}{C_o} & -\dfrac{n}{C_o} & 0 & -\dfrac{1}{C_oR_L} \end{bmatrix} \tag{11}$$

$$B = \begin{bmatrix} \dfrac{D}{L_M+L_r} \\ \dfrac{D}{L_M+L_r} \\ 0 \\ 0 \end{bmatrix} \tag{12}$$

$$C = [0 \quad 0 \quad 0 \quad 1]. \tag{13}$$

DC characteristics can be obtained by letting $Ax + BV_{in} = 0$. This gives:

$$\frac{V_o}{V_{in}} = \frac{1}{n} \frac{L_M}{L_M+L_r} D \tag{14}$$

$$V_b = V_{in}D \tag{15}$$

$$I_M = \frac{I_o}{n}. \tag{16}$$

Equations (14)-(16) agree with the results shown in [6]. ((35), (36), and (38), respectively)

IV. SMALL SIGNAL MODEL FOR POWER STAGE

In a DC/DC converter, the small signal behavior is found by evaluating the large signal quantities at a given steady state point. The system is then perturbed around this operating point. In this Section, our objective is to formulate a small-signal model for the studied converter power stage.

A. Mathematical Model

The small-signal characteristics of the studied converter can be obtained by perturbing the steady-state operation point:

$$V_{in} \to V_{in} + \hat{V}_{in} \ , \ D \to D + \hat{D} \ , \ I_M \to I_M + \hat{I}_M \ , \ I_{Lr} \to I_{Lr} + \hat{I}_{Lr} \ ,$$
$$V_b \to V_b + \hat{V}_b \ , V_o \to V_o + \hat{V}_o \tag{17}$$

Inserting (17) into $\dot{x} = Ax + BV_{in}$ and $y = Cx$, and linearizing the result by neglecting second order terms, we find:

$$SL_M \hat{I}_M(s) = \left(\frac{L_M}{L_M+L_r}V_{in} - \frac{L_M}{L_M+L_r}V_C + nV_o \right)\hat{D}(s) -$$
$$\frac{L_M}{L_M+L_r}D\hat{V}_b(s) - n(1-D)\hat{V}_o(s) + \frac{L_M}{L_M+L_r}D\hat{V}_{in}(s) \tag{18}$$

$$SL_r \hat{I}_{Lr}(s) = \left(\frac{L_r}{L_M+L_r}V_{in} - \frac{L_M}{L_M+L_r}V_C + nV_o \right)\hat{D}(s) + n(1-D)\hat{V}_o(s) -$$
$$\left(\frac{L_M}{L_M+L_r}D - 1 \right)\hat{V}_b(s) + \frac{L_r}{L_M+L_r}D\hat{V}_{in}(s) \tag{19}$$

$$SC_b \hat{V}_b(s) = \hat{i}_{Lr}(s) \tag{20}$$

$$SC_o \hat{V}_o(s) = n\left(\hat{i}_M(s) - \hat{i}_{Lr}(s) \right) - \frac{1}{R_L}\hat{V}_o \tag{21}$$

where S is the Laplace transform of $\dfrac{d}{dt}$. Solving (18)-(21) gives

$$A(s)\hat{V}_o(s) = B(s)\hat{D}(s) + C(s)\hat{V}_{in}(s) \tag{22}$$

$$A(s) = a_4 s^4 + a_3 s^3 + a_2 s^2 + a_1 s + a_0, \quad B(s) = b_0, \quad C(s) = c_0 \tag{23}$$

$$a_4 = L_M L_r C_b C_O R_L (L_M + L_r) \tag{24}$$

$$a_3 = L_M L_r C_b (L_M + L_r) \tag{25}$$

$$a_2 = n^2 (1-D) C_b (L_M + L_r)^2 R_L + C_O L_M R_L [L_r + L_M (1-D)] \tag{26}$$

$$a_1 = L_M [L_r + L_M (1-D)] \tag{27}$$

$$a_0 = n^2 R_L (1-D)(L_M + L_r) \tag{28}$$

$$b_0 = n V_{in} L_M R_L (1-D) \tag{29}$$

$$c_0 = n D L_M R_L (1-D). \tag{30}$$

Therefore, we can get the small-signal transfer functions between the output voltage and the duty cycle and between the output voltage and the input voltage:

$$\frac{\hat{V}_o(s)}{\hat{D}(s)} = \frac{B(s)}{A(s)} \tag{31}$$

$$\frac{\hat{V}_o(s)}{\hat{V}_{in}(s)} = \frac{C(s)}{A(s)} \tag{32}$$

For the converter with 60W output power and the following set of circuit parameters: V_{in}=200V, V_O=12V, n=5, D=30%, L_M=373uH, L_r=16uH, C_b=1uF, C_O=1000uF, and f_S=100kHz,

$$\frac{\hat{V}_o(s)}{\hat{D}(s)} = \frac{1.12 \times 10^{20}}{s^4 + 4.17 \times 10^2 \cdot s^3 + 4.57 \times 10^{10} \cdot s^2 + 1.86 \times 10^{13} \cdot s + 2.93 \times 10^{18}} \tag{33}$$

$$\frac{\hat{V}_o(s)}{\hat{V}_{in}(s)} = \frac{1.69 \times 10^{17}}{s^4 + 4.17 \times 10^2 \cdot s^3 + 4.57 \times 10^{10} \cdot s^2 + 1.86 \times 10^{13} \cdot s + 2.93 \times 10^{18}} \tag{34}$$

Frequency responses of these two transfer functions are shown in Fig. 3(a) and (b), respectively. It is a four order system, having low-frequency and high-frequency double poles. These low-frequency and high-frequency poles are are located around 1.2kHz and 34kHz, respectively. The locations of these poles are dependent on the four storage elements in the studied converter. Fig. 4 shows the effects of the value of these storage elements on the transfer function characteristic. While the low-frequency double poles are dependent on L_M and C_O, the high-frequency double poles are dependent on L_r, and C_b. To simplify the feedback loop design, it is recommended to select a smaller C_b or a larger C_O to move the high-frequency poles at least one decade higher than the low-frequency poles. The converter stability can then be ensured by a compensation network with corner frequency sufficiently lower than the low-frequency double poles.

B. Equivalent Circuit Model

According to the averaged state space equations describing the studied converter shown in (11)~ (13), the equivalent circuit model can be obtained as shown in Fig. 5. This model can be implemented by Spice sub-circuits to analyze the DC and AC small signal characteristics.

V. PEAK CURRENT MODE CONTROL

The studied converter is susceptible to being controlled by the voltage or current mode control. While the voltage mode control is applied, the small signal characteristics of the studied converter are almost the same with that of the power stage. In this section, a peak current mode controller model is used to derive the small signal model for the studied converter under peak current mode control.

A. Mathematical Model

Peak current mode control may be applied to the studied converter as illustrated by Fig. 6. The converter switch is turned on every T_S seconds and turned off whenever $i_M(t)$ equals to the sum of a control signal $\dfrac{v_C}{R_S}$ and a compensating ramp of fixed slope $-m_c$ that restarts at 0 every cycle. The operation of the peak current mode control loop may therefore be represented by adding a constraint equation to the model.

$$I_M + m_{L1} \cdot \frac{DT_S}{2} = \frac{V_C}{R_S} - m_c \cdot DT_S \tag{35}$$

$$D = \frac{2}{(m_{L1} + 2m_c)T_S} \cdot \left(\frac{V_C}{R_S} - I_M \right) \tag{36}$$

By perturbation and linearization of (36), we get

$$\hat{d} = \frac{2}{(m_{L1} + 2m_c)T_S} \cdot \left(\frac{\hat{V}_C}{R_S} - \hat{I}_M \right) - \frac{D}{m_{L1} + 2m_c} \hat{m}_{L1} \tag{37}$$

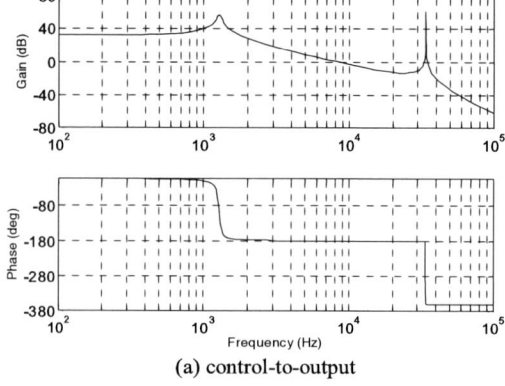

(a) control-to-output

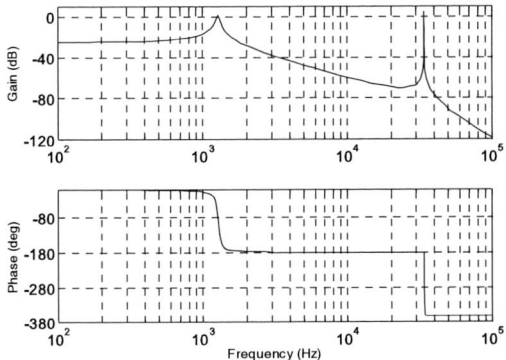

(b) input-to-output

Fig. 3. Frequency responses of the derived transfer functions.

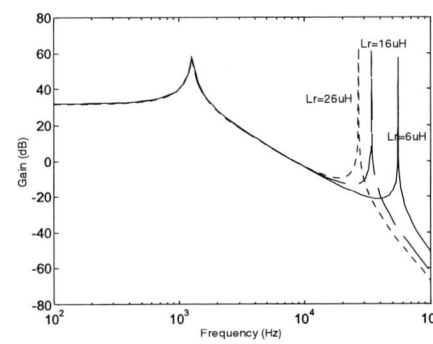

(d) resonant inductor

Fig. 4. Effects of storage-elements variations on the power stage small signal characteristics.

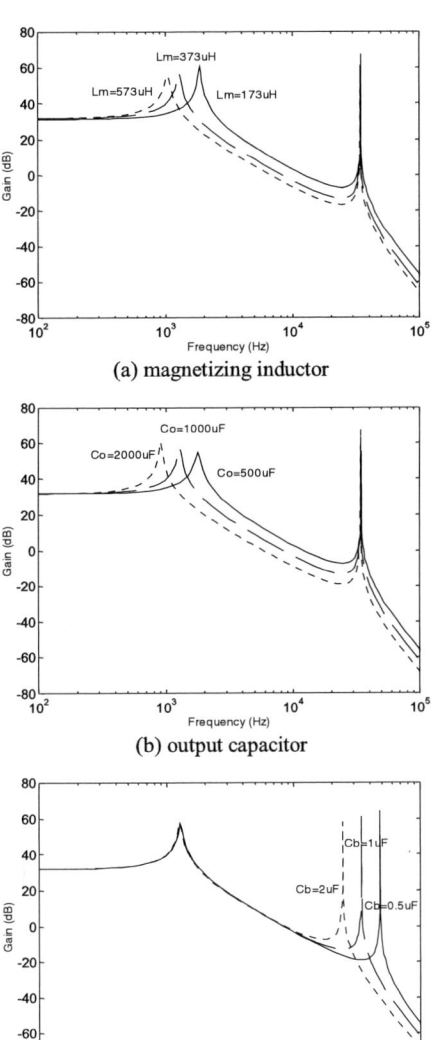

(a) magnetizing inductor

(b) output capacitor

(c) blocking capacitor

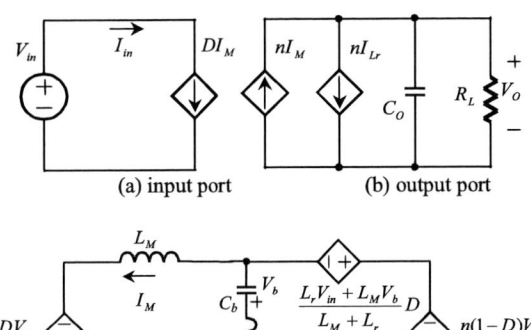

(a) input port   (b) output port

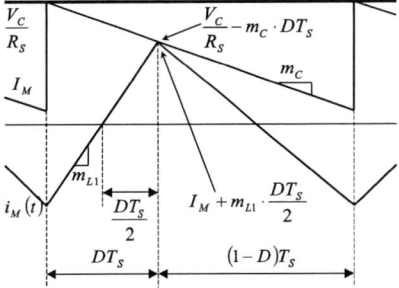

(c) inductor-capacitor loop

Fig. 5. Small-signal equivalent model.

Fig. 6. Controlled current $i_M(t)$ in peak current mode control.

The transfer function can now be obtained by inserting (37) into (22):

$$A_p(s)\hat{V}_O(s) = B_p(s)\hat{V}_C(s) + C_p(s)\hat{V}_{in}(s) \tag{38}$$

$$A_p(s) = a_{p4}s^4 + a_{p3}s^3 + a_{p2}s^2 + a_{p1}s + a_{p0} \tag{39}$$

$$B_p(s) = b_{p0} \tag{40}$$

$$C_p(s) = c_{p0} \tag{41}$$

$$a_{p4} = L_M L_r C_b C_O R_L (L_M + L_r)(L_M + L_r - \frac{V_{in}D}{m_{L1} + 2m_C}) \quad (42)$$

$$a_{p3} = L_M L_r C_b (L_M + L_r)(L_M + L_r - \frac{V_{in}D}{m_{L1} + 2m_C}) +$$
$$\frac{2L_M L_r C_b C_O R_O (L_M + L_r)V_{in}(1-D)}{(m_{L1} + 2m_C)T_S}$$

$$a_{p2} = \frac{2L_M L_r C_b (L_M + L_r)V_{in}(1-D)}{(m_{L1} + 2m_C)T_S} +$$
$$R_L (L_M + L_r - \frac{V_{in}D}{m_{L1} + 2m_C})C_O L_M [L_r + L_M(1-D)] +$$
$$R_L (L_M + L_r - \frac{V_{in}D}{m_{L1} + 2m_C})n^2(1-D)C_b (L_M + L_r)^2 \quad (44)$$

$$a_{p1} = L_M [L_r + L_M(1-D)](L_M + L_r - \frac{V_{in}D}{m_{L1} + 2m_C}) +$$
$$\frac{2n^2 C_b R_L V_{in}(1-D)^2(L_M + L_r)^2}{(m_{L1} + 2m_C)T_S} + \frac{2C_O R_L L_M V_{in}(1-D)[L_M(1-D) + L_r]}{(m_{L1} + 2m_C)T_S}$$

$$a_{p0} = n^2(1-D)(L_M + L_r)R_L(L_M + L_r - \frac{V_{in}D}{m_{L1} + 2m_C}) +$$
$$\frac{2L_M V_{in}(1-D)[L_M(1-D) + L_r]}{(m_{L1} + 2m_C)T_S}$$

$$b_{p0} = \frac{2nV_{in}(1-D)L_M(L_M + L_r)R_L}{R_S T_S(m_{L1} + 2m_C)} \quad (47)$$

$$c_{p0} = nV_{in}(1-D)L_M(L_M + L_r)R_L[\frac{D}{V_{in}} - \frac{D}{(L_M + L_r)(m_{L1} + 2m_C)}] \quad (48)$$

B. Equivalent Circuit Model

By merging the small signal peak current mode controller model presented in (37) into the power stage equivalent model derived in the previous section, the equivalent circuit model is obtained. This equivalent circuit model can be implemented by Spice sub-circuits to analyze the DC and AC small signal characteristics.

VI. MODEL VERIFICATION

A prototype converter with peak current mode control was constructed. The parameters are the same as that given in section IV. The results of the measurement and corresponding simulation are shown in Fig. 7. The star points and the solid line in Fig.7 show the magnitude and phase of the measured and simulated data, respectively. The measurement verifies the accuracy of the proposed model. There exists a dominant pole.

VII. CONCLUSIONS

In this paper, the continuous-time small-signal model of asymmetrical half bridge flyback converter is derived in an analytical form and the equivalent circuit model is developed. The small signal characteristic of the studied converter power stage is a four order system with two

low-frequency and two high-frequency poles. Among them, the low-frequency poles are associated with L_M and C_O, and the high-frequency poles are with L_r and C_b. Under voltage mode control, C_O and C_b may be designed to separate the low- and high-frequency poles at least one decade. The system stability can be guaranteed by compensation bandwidth sufficiently lower than the low-frequency double poles. The transfer function becomes first-order dominated when peak current mode control is applied. The frequency of this dominant pole is determined by the output capacitance and load resistance.

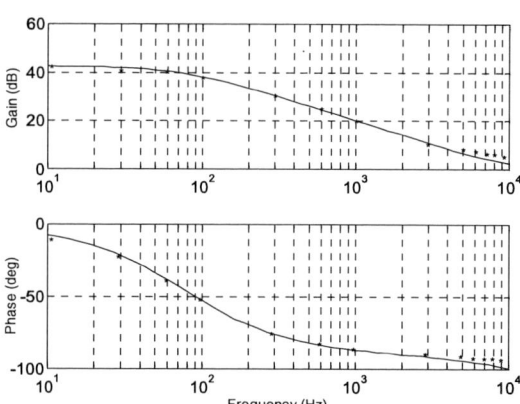

Fig. 7. Control-to-output transfer function $\frac{\hat{V}_O(s)}{\hat{V}_C(s)}$ under peak current mode control.

REFERENCES

[1] C. P. Henze, D. S. Lo, H. C. Martin, " Zero-voltage resonant transition switching power converter," U.S. Patent 5,057,986.

[2] P. C. Heng, R. Oruganti, "Family of two-switch soft-switched asymmetrical PWM DC/DC converters," *1994 Power Electronics Specialists Conference*, pp. 85-94.

[3] S. H. Lim, " Asymmetrical duty cycle flyback converter," U.S. Patent 5,959,850.

[4] D. H. Seo, O. J. Lee, S. H. Lim, J. S. Park, "Asymmetrical PWM flyback converter," *2000 Power Electronics Specialists Conference*, pp. 848-852.

[5] T. Tolle, T. Duerbaum, "Modeling of ZVS transitions in asymmetrical half-bridge PWM converters," *2001 Power Electronics Specialists Conference*, pp. 308-313.

[6] T. M. Chen, C. L. Chen, "Analysis and design of asymmetrical half-bridge flyback converter," *IEE Proc. Electric Power Application*, vol. 149, no. 6, Nov. 2002, pp. 433-440.

[7] R. D. Middlebrook, S. Cuk, "A general unified approach to modeling switching converter power stages," *1976 Power Electronics Specialists Conference*, pp. 18-34.

[8] R. D. Middlebrook, "Topics in multiple-loop regulators and current-mode programming," *IEEE Trans. Power Electronic*, vol. 2, no. 2, Apr. 1987, pp. 109-124.

2006 5th International Power Electronics and Motion Control Conference

A DSP Based Controller for High Power Dual-Phase DC-DC Converters

Xin Guo, Xuhui Wen and Ermin Qiao

Institute of Electrical Engineering, Chinese Academy of Sciences, Beijing, China
e-mail: guoxin@mail.iee.ac.cn

Abstract—**High power DC/DC converter has become the essential part of the distributed power system in fuel cell powered electric vehicles and stationary power systems. This paper proposes a topology of high power Dual-Phase Boost DC/DC Converter for Fuel Cell Power Supply. The principle of the converter operating in BOOST mode are analyzed thoroughly. A 150KW converter was designed. The converter was adopted DSP-based fully digital dual-loop control. The efficiency of the converter is over 97%, the experimental results realized perfect effect.**

Keywords-Dual-Phase DC/DC converter; Electrical Vehicle; Interleaved Boost converter; PWM

I. INTRODUCTION

Fuel Cell Electric Vehicle (FCEV) is believed to be the ultimate target in the development of EV in 21st century. In the FCEV, a distributed energy system consisting of fuel cell is used , high-power DC-DC converter is adopted to adjust the output voltage, current and power of Fuel Cell Engine (FCE) to meet the vehicle's requirements. As fig.1 shows, the power supply system of FCEV is composed of Fuel Cell (FC), Fuel Cell Compressor Motor, Traction Motor Driver, high-power DC-DC converter and Bi-directional power flow DC-DC converter.

In the application of FCEV and moveable power supply, the power converter devices especially demand small volume, light weight, stable and reliable properties, that is to say, DC/DC converter need high power density and excellent dynamic control characteristic. Boost converter has such good properties, simple system structure, high conversion efficiency, good stability and reliability, it is widely used in EV. In order to reduce the ripple current and to reduce the size of passive component, multiphase structure with interleaved control is adopted in the high-power DC-DC converter. In the high-power boost converter, the magnetic inductor is the important unit, which mainly composes the whole volume and weight of the device. It is a very hard and challenging work to design a high power inductor to store energy.

For such an interleaved controlled Boost converter, full digital control is a natural choice. In this paper, the TMS320LF2407A digital signal processor (DSP) is adopted to realize the full digital control for the high-power dual-phase Boost DC-DC converter[1][2]. The

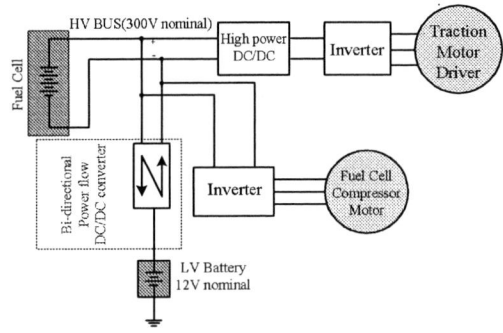

Figure 1. The power supply system of FCEV

experimental results show that the converter has excellent electrical characteristics.

II. THE PRINCIPLE OF DUAL-PHASE DC-DC CONVERTER

The main circuit of the Dual-Phase DC-DC converter is shown in fig 2. From fig 2 we can see that the circuit is composed of a bridge of power switches and storage energy inductor, it can work in Boost mode or in Buck mode. When S1u=S2u==OFF, S1d and S2d switch on and off, the system work in the Boost mode, as shown in table I.

From table I, we can see that in Boost mode, only the power device (S1d,S2d,D1u,D2u) have switching commutation, the power device (S1u,S2u,D1d,D2d) are in constant off state. The power switches S1d and S2d have 180-degree phase difference of driving pulses in a cycle from each other. The current ripple of input power supply reduced greatly because the two 180-degree phase difference inductor currents minify the fluctuation of each other[3][4]. In one switching cycle Ts, considering the commutation of power switches and diodes (S1d,S2d,D1u,D2u), there have eight kinds of running states, as shown in table II.

According to table II, the converter have eight equivalent sub-circuits of state1~state 8, as shown in fig. 3.[5][6]

When S1d=S2d==OFF, S1u and S2u switch on and off, the system can work in the Buck mode, as shown in Table III.

TABLE I.
THE STATE OF THE POWER DEVICE IN BOOST MODE

S1u=off	S2u=off	D1u=on/off	D2u=on/off
S1d=on/off	S2d=on/off	D1d=off	D2d=off

TABLE II.
THE EIGHT KINDS OF RUNNING STATES IN INTERLEAVED BOOST MODE

	S1d=on	S2d=on	D1u=on	D2u=on
S1d=on	State 2	State 7	*	State 1
S2d=on	State 7	State 5	State 4	*
D1u=on	*	State 4	State 3	State 8
D2u=on	State 1	*	State 8	State 6

From table III, we also can see that in Buck mode, only the power device (S1u,S2u,D1d,D2d) have switching commutation, the power device (S1d,S2d,D1u,D2u) are in constant off state. The power switches S1u and S2u have 180-degree phase difference of driving pulses in a cycle from each other. The current ripple of output power supply reduced greatly because the two 180-degree phase difference inductor currents minify the fluctuation of each other. In one switching cycle Ts, considering the commutation of power switches and diodes (S1u,S2u,D1d,D2d), there also have eight kinds of running states. The analysis method of Buck mode is similar to the Boost mode, we will not analyze the work station in detail in this paper.

From the analysis above, we can see that adopt appropriate control method , we can realize the conversion from Boost mode to Buck mode easily. Full digital control makes this conversion become easy.

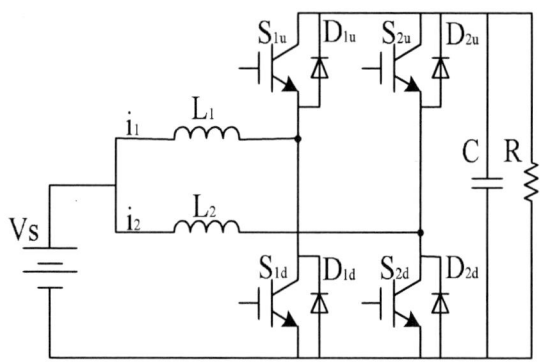

Figure 2.   The topology of the dual-phase DC-DC converter

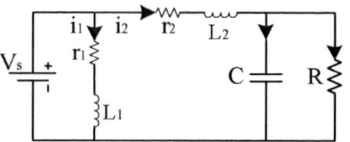

Figure 3a   The equivalent sub-circuits of state1

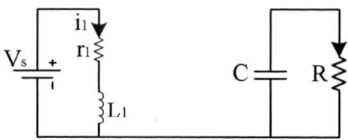

Figure 3b   The equivalent sub-circuits of state2

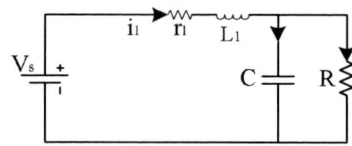

Figure 3c   The equivalent sub-circuits of state3

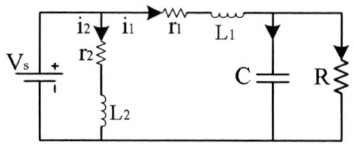

Figure 3d   The equivalent sub-circuits of state4

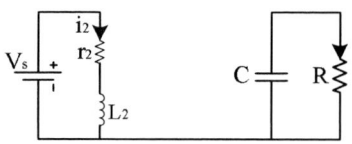

Figure 3e   The equivalent sub-circuits of state5

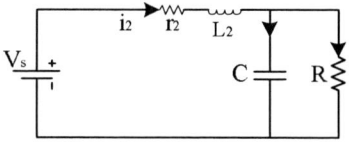

Figure 3f   The equivalent sub-circuits of state6

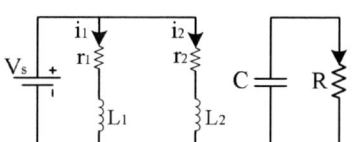

Figure 3g   The equivalent sub-circuits of state7

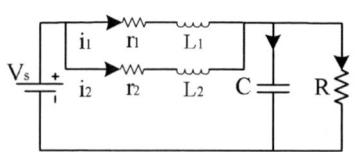

Figure 3h   The equivalent sub-circuits of state8

TABLE III.
THE STATE OF THE POWER DEVICE IN BUCK MODE

S1d=OFF	S2d=OFF	D1d=ON/OFF	D2d=ON/OFF
S1u=ON/OFF	S2u=ON/OFF	D1u=OFF	D2u=OFF

III. THE DESIGN OF THE CONTROLLER

The control scheme of the dual-phase DC/DC converter(Boost mode) is shown in fig. 4. From fig. 4 we can see that the converter adopt dual-loop control. In the dual-loop control structure, the outer loop is voltage loop, which provides the current reference for the inner current loop. The inner current loop can suffice the faster transient response of the DC/DC converter's requirement. This control requires the sampling of two variables: output voltage V_o, output current I_o, which are obtain through conventional Hall transducers.

A limiter on the output current reference is crucial in this controller. This is because during large signal transient like startup period, there is a big difference between the voltage reference and the DC output voltage. As a result, the output of the voltage compensator could give a command to the current loop higher than the maximum allowable current, resulting in the damaging the device. A limit on the output of the current loop is also crucial in this controller. This is because the TMS320LF2407A DSP is a fixed-point DSP, if we do not limit the maximum value of the output of the current loop, the value will be unpredictable when the converter work at some nonideal state, the PWM waveform will be abnormal, and these maybe result in damaging the device.

From fig. 4 we can see the control system of this converter is a double closed-loop system, outer voltage loop and inner current loop, full digital control with DSP. Now we draw the control graph of the dual-loop controller in fig. 5.

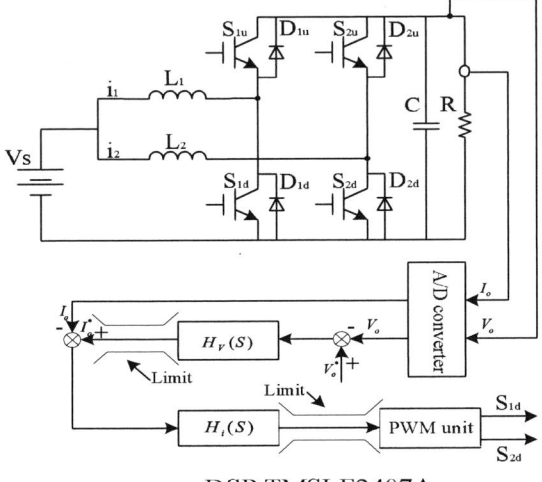

DSP TMSLF2407A

Figure 4　Dual-phase DC/DC converter (Boost mode) with digital control

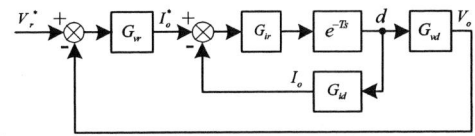

Figure 5　The dual-loop controller of the converter

The loop gains for inner current loop and outer voltage loop can be expressed as:

$$H_i(S) = G_{ir} * G_{id} * e^{-Ts} \qquad (1)$$

$$Hv(S) = \frac{G_{vr} * G_{ir} * e^{-Ts} * G_{vd}}{1 + H_i} \qquad (2)$$

H_i: Current loop gain, H_v: Voltage loop gain
G_{ir}: Current loop compensator, G_{vr}: Voltage loop compensator
e^{-Ts}: Delay caused by sampling period, computation time and duty cycle update

In the dual-loop system, the current compensator is designed for good loop dynamics, and the voltage compensator is designed for the desired crossover frequency and phase margin.

IV. EXPERIMENT RESULT

Based on the analysis above, using TMS320LF2407A DSP as main controller, a prototype of a fully digital controlled Dual-phase Boost converter was constructed with its basic technical specification as follows: P=150KW, 250V≤Vs≤450V, Vo=584V. The proposed converter weigh 50kg and hold 50L, which is lessen one-third of the volume and weight than single phase Boost converter, moreover, the fluctuation of current is less than 10%. Fig. 6 is the photo of the converter.

Figures 7 are the waveforms of the dual-phase DC/DC converter (Boost mode) which is working at 75KW. Fig. 7a is the driving signals of the two phase IGBT switches, which have equal duty and have 180-degree phase difference. Fig. 7b is the waveform of its output voltage and output current, which shows that the voltage fluctuation is less than 1%. Fig. 7c is the waveform of the inductor current of one phase and output voltage, it can be seen that the converter run at critical state between CCM mode and DCM mode. Fig. 7d is the waveform of voltage and current of inductor at half load. Fig. 7e is the waveform of two phase inductor current at half load.

Figure 6　The photo of the converter

Fig. 7f is the waveform of two inductor current and the sum of the two inductor current. Fig. 7g is the waveform of inductor current and input current. From the graph we can see that the ripple of inductor current can reduce the infection for input current obviously when we adopt interleaved control. The ripple coefficient of the input current is less than 10%. Fig.8 is the waveform of output voltage and output current when the system is soft start and soft stopped. The time of soft start and soft stopped can be redesigned easily. Fig. 9 is the system efficiency curve, which indicates that the whole efficiency is about 95~98%, and the efficiency is over 97% when the output power is large than the half of rated power.

The related parameters of the oscillograph are: voltage-200V/div, current-100A/div

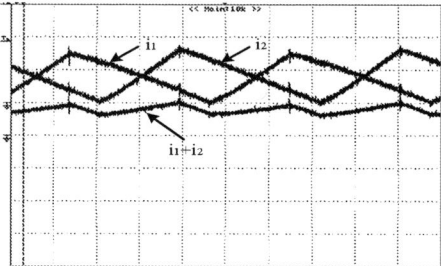

Figure 7e   two phase Inductor current at 75kW

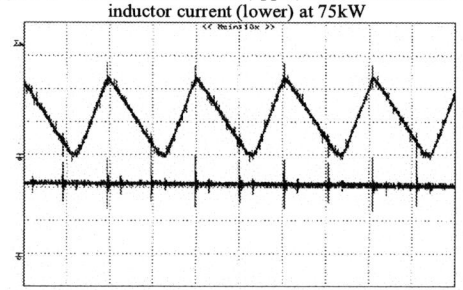

Figure 7f   Two Inductor current (upper) and the sum of the two inductor current (lower) at 75kW

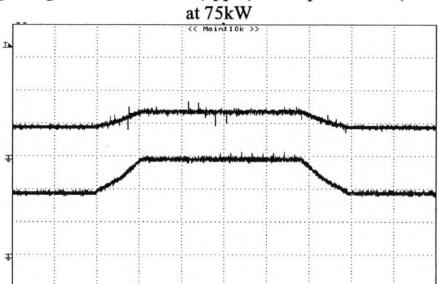

Figure 7g   Inductor current (upper) and input current (lower) at 75kW

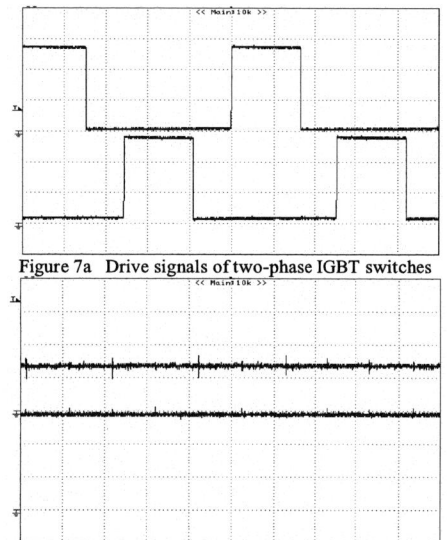

Figure 7a   Drive signals of two-phase IGBT switches

Figure 7b   Output current (upper) and output voltage (lower) at 75kW

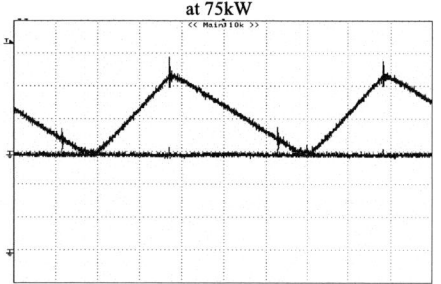

Figure 7c   Inductor current (upper) and output voltage (lower) at 75kW

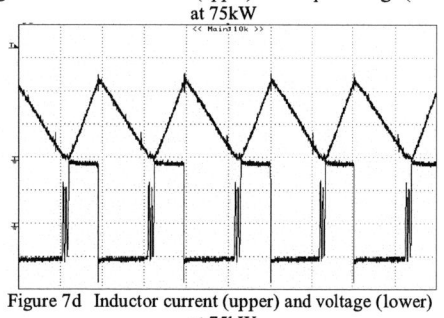

Figure 7d   Inductor current (upper) and voltage (lower) at 75kW

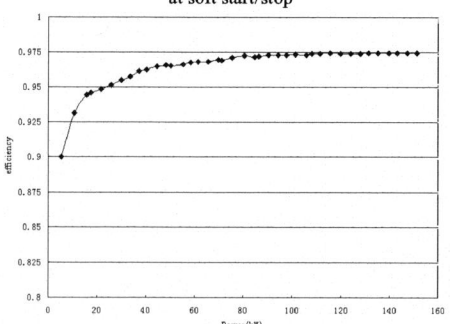

Figure 8   Output current (upper) and voltage (lower) at soft start/stop

Figure 9   the system efficiency curve

V. CONCLUSION

A DSP-based digital controller for a 150KW dual-phase DC/DC converter is presented in this paper. The converter can work at Boost mode and Buck mode. In this paper, analyze the work state of Boost mode in detail. the topology is interleaved Boost. The system adopt double closed-loop control, the outer loop is voltage loop, the inner loop is current loop. The experimental result shows that the converter has excellent dynamic characteristic, its efficiency is about 95~98%, its volume and weight is less than single Boost converter. The converter is suitable for Fuel Cell Electric Vehicle as well as the Fuel Cell Distributed Generation.

REFERENCES

[1] Xudong Huang, Troy Nergaard, Jih-Sheng Lai, Xingyi Xu and LiZhi Zhu, "A DSP Based Controller for High-Power Interleaved Boost Converters," APEC'03, vol. 1, pp.327-333, 9-13 February 2003.

[2] Xudong Huang, Xiaoyan Wang, Troy Nergaard, Jih-Sheng Lai, Xingyi Xu and Lizhi Zhu, "Parasitic Ringing and Design Issues of Digitally Controlled High Power Interleaved Boost Converters, " *IEEE Trans. On Power Electronics,* Vol.19, NO.5, pp. 1341-1352, September 2004.

[3] David J. Perreault and John G. Kassakian, "Distributed Interleaving of Paralleled Power Converters, " *IEEE Trans. On Circuits and Systems-I: Fundamental Theory and Applications,* Vol. 44, NO. 8, pp. 728-734, August 1997.

[4] Michael T. Zhang, Milan M. Jovanovic and Fred C. Y. Lee, "Analysis and Evaluation of Interleaving Techniques in Forward Converters, " *IEEE Trans. On Power Electronics,* Vol. 13, NO. 4, pp. 690-698, July 1998.

[5] Haiping Xu, Li Kong and Xuhui Wen, "Fuel Cell Power System and High Power DC-DC Converter, " *IEEE Trans. On Power Electronics,* Vol. 19(5), pp.1250-1255, 2004.

[6] Haiping Xu, Ermin Qiao, Xin Guo, Xuhui Wen and Li Kong, "Analysis and Design of High Power Interleaved Boost Converters for Fuel Cell distributed Generation System, " PESC 2005, pp. 140-145.

2006 5th International Power Electronics and Motion Control Conference

Effective Load Resistance; A New Method to Evaluate DC/DC converters Efficiency

Alan Elbanhawy
Fairchild Semiconductor, San Jose CA, USA
E-mail Address: aelbanhawy@fairchildsemi.com

Abstract—**This paper offers an alternative to efficiency measurements as an evaluation tool of power supplies. The proposed method works well with little dependency on the output voltage and temperature.**

Keywords-component; Power Efficiency; Effective Loss Resistance; DC-DC Converter power loss

I. INTRODUCTION

Power supplies have always been evaluated and compared based on the how efficient the conversion process is. Equations 1 and 2 are two of the most used formulae to calculate the power conversion efficiency η. The use of efficiency as a yardstick to evaluate DC-DC converters represents a very simple and in most cases effective way of comparison. Efficiency figure combined with the details of the converter parameters also allows the thermal engineer to calculate the thermal load of the converter and whether heatsink and/or cooling airflow is required. The efficiency tool works very well when we have established standard input and output voltages that are more or less fixed in value.

We will show that the efficiency alone is not the right tool to compare converters operating at different output conditions even with a fixed input voltage and switching frequency. We will explain why that is and propose an alternative methods that measures the performance "almost" independent of the output voltage that gives a much more accurate and unambiguous idea about the converter's performance so we introduce the concept of the Effective Loss Resistance, Rol

II. EFFICIENCY CALCULATIONS

The power conversion efficiency (ζ) may be calculated using the equation:

$$h = \frac{\sum_{i=1}^{n} Vouti \ Iouti}{\sum_{j=1}^{m} Vinj \ Iinj} \$100 \ \% \cdots (1)$$

This equation can also be written as follows:

$$h = \frac{\sum_{i=1}^{n} Vouti \ Iouti}{\left[\sum_{i=1}^{n} Vouti \ Iouti \ C \sum_{j=1}^{m} PLossj\right]} \$100 \ \%$$

..(2)

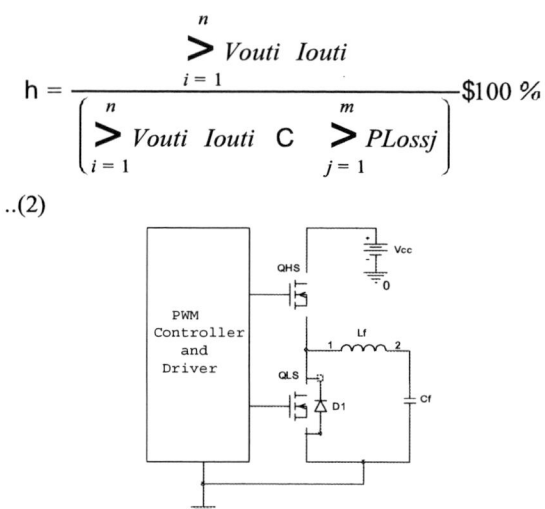

Figure 1. Synchronous Buck Converter

"Fig. 1," depicts a simplified block diagram of a synchronous buck converter used in the evaluation of this paper. It is worthwhile mentioning that this work may be adapted to almost all DC-DC converters

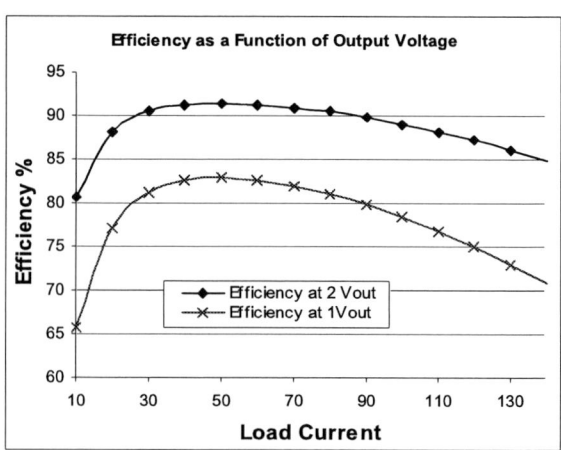

Figure 2. Synchronous buck converter's efficiency as a function of the output voltage

1-4244-0448-7/06/$25.00 ©2006 IEEE

"Fig. 2," shows the efficiency of a given converter at different output voltages. Although the power loss is mainly dependant on the load current at a fixed switching frequency and input voltage and is very lightly dependant on the output voltage within the range from 2V to 1V i.e. we have "almost" constant power loss in all of the out voltage conditions while the efficiency η varies dramatically. So even though the power loss is "almost" a constant, the very fact that we have a smaller output voltage will result in non realistic smaller efficiency figure.

III. SYNCHRONOUS BUCK CONVERTER LOSSES

Loss mechanisms in DC-DC converters can be divided into two major groups as follows:

A. Conduction or Ohmic losses, which is the loss due to I_{load}^2 x $R_{DS(ON)}$ x Δ where $R_{DS(ON)}$ is the on-resistance of the MOSFET, Iload is the load current and Δ is the duty cycle. Please note that this loss mechanism is mainly dependant on I_{load}^2 because of the quadratic relationship and to a lesser degree on the output voltage since Δ is a function of the output voltage that is topology dependant.

B. Dynamic or switching losses = I_{load} x Vin x ½ x fs x (tr+tf) where Vin is the input voltage and tr & tf are the rise and fall times and fs is the converter's switching frequency. Again you can see that the dynamic losses are not dependant on the output voltage.

This means that the losses are dependant on the output voltage in a secondary way. This leads immediately to the conclusion that we have more or less fixed losses regardless of the output voltage.

Why is this important? Because as can be seen in the efficiency equation (2) above, the smaller the output voltage the smaller the output power for the same output current. This results a lower efficiency for smaller output voltage as can clearly be seen in "Fig. 2,". One can see that there is a limiting condition as the output voltage goes to zero while maintaining the same output current at which point the efficiency η is theoretically zero:

$$h = \lim_{OutputPower \to 0} \left(\frac{OutputPower}{OutputPower + losses} \right) = 0$$

That is to say that the efficiency of a given converter is proportional to the output voltage for the same load current. This fact makes the comparison very difficult under different conditions of output voltage. "Fig 2," depicts this very case where you can see that the

efficiency at an output voltage of 2 volts is about 8% larger than that at 1V output at the same load current though the power dissipation is "almost" the same and the thermal load is also "almost" the same.

It can be shown that the efficiency is different for different heat sinking techniques for the same circuit and the same input and output voltages. In this case, there is less power dissipation where we have a heatsink and air flow compared to the same design in still air and no heatsink.

This leads to the following dilemma, as different DC-DC converter manufacturers show their efficiency results that optimally reflects their products; Starting with two converters A & B from two different vendors having the same efficiency figure, the engineer is left to compare them by guessing whether converter A tested at 30 Amps and a heatsink (of unknown performance generally) is better than converter B tested at 25Amp with no heatsink and 400LFM airflow?

The question that can be asked now is, "Is there a different way to evaluate converters independent of the output voltage and heat sinking techniques?"

IV. THE PROPOSED NEW APPROACH

In the above discussion, we have demonstrated the need for a different approach that can has a universal appeal both to vendors and buyers of converters and more importantly, to the circuit and thermal design engineer so that quick decisions regarding design issues can be crystal clear across different disciplines.

One can clearly see that the power loss is a better way to demonstrate the performance of power supplies within a limited span of output voltages, say 1V – 2V where the power loss may be considered constant. Needless to say if we widen the span of voltage say from 1V to 5V all the secondary effects will start becoming prominent and will not yield the same consistent results. "Fig. 3," below shows the losses as a function of the load current for different schemes of heat sinking.

To understand "Fig. 3," let us see what is happening here. It can be seen that the difference in losses is rather slight up to 60 Amps with less than two watts differential at this point. We can deliver currents up to 120 Amps when we have both heatsink and airflow from the same board because we can remove heat very efficiently from the board due to the use of the air and heatsink. The reader is unlikely to easily figure out that all four curves are taken for the very same VRM with heat sinking as the only difference though this is the case. In "Fig. 3," the testing was stopped when the board temperature reached $105°C – 110 °C$.

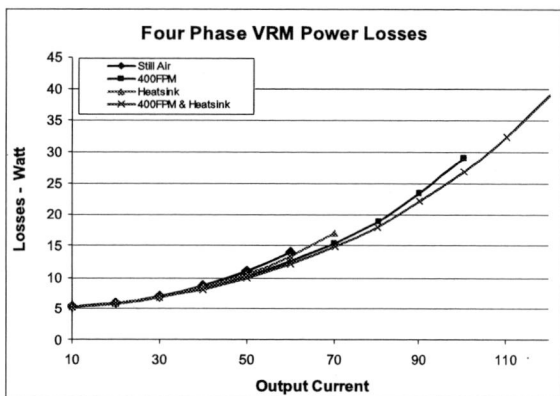

Figure 3. Power Loss vs. Load current

It is worthwhile mentioning that the difference in power dissipation between the fully heat sinked case and the still air case is mostly due to temperature differential since the fully heat sinked case is running at a much lower temperature and since the MOSFET on-resistance may be expressed as follows: $RDS(ON)_T = RDS(ON)_a \times (1 + \alpha \times \Delta T)$ where $RDS(ON)_T$ and $RDS(ON)_a$ are the MOSFET on-resistance at a temperature T and ambient temperature and ΔT is the temperature rise above ambient. The above equation indicates that at higher temperatures, the on-resistance is higher resulting in higher losses that can be observed in "Fig. 3,"

V. EFFECTIVE LOSS VOLTAGE

One way to explore the converter performance is to introduce the term "Effective Loss Voltage" which is equal to $\dfrac{Power\_dissipation}{I_{Load}}$. This represents a DC voltage in series with the converter output which dissipates power when Iload passes through it as can be seen in "Fig. 4,". On the y axes we have the "Effective Loss Voltage" and the x axes we have the load current.

Here are the benefits of this representation:
- We have direct evaluation of the losses as a function of heat sinking
- This graph translates the abstract efficiency curve into a actual performance as a function of the load current and power losses

VI. EFFECTIVE LOSS RESISTANCE

The second way is to propose the measuring of the "Effective Loss Resistance" Rol

$Rol = \dfrac{TotalConverterPowerLoss}{LoadCurrent^2}$. "Fig. 6," shows Rol for similar set of tests measured in "Fig. 2," with different output voltages and heat sinking scheme. It clearly shows that for a given current, the effective loss resistance, Rol is "almost" the same and "almost" independent of the heat sinking technique. Clearly Rol will differ slightly from one heat sinking technique to the other and the concept of "Rol Band" could be utilized to describe the difference in maximum and minimum Rol as will be shown later

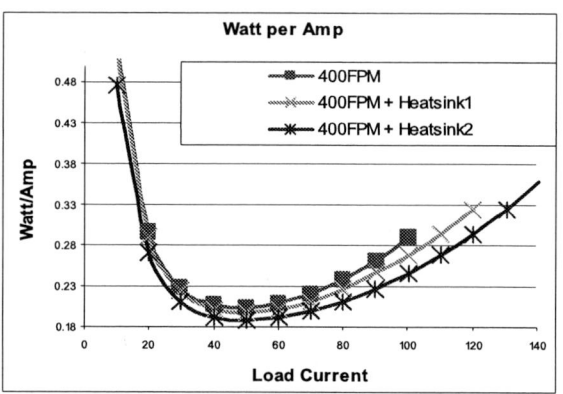

Figure 4. Loss voltage vs. load current

"Fig. 5," shows the effective loss resistance Rol in all four cases of heat sinking and clearly showing that the converter performance is "almost" independent of the cooling and output voltage leading to the conclusion that the effective loss resistance Rol, is a reliable means to evaluate the converter performance.

As mentioned above, there are some differences in the losses between a 1V and 2V output. A "Rol Band" could apply here too to fully describe the circuit performance. This may be presented as $Rol = Ro \pm \Delta R$. Where this equation applies to say ½ load to full load losses. Now we have a very simple parameter with spread that describes the performance of a given converter say a VRM operating between 1V and 2V. One may derive an equation for Rol that may help in converter comparison using an analytical approach such as a spreadsheet or a mathematical software analysis tool.

An unbiased comparison of different power supplies may now be done either using a set of curves as in "Fig. 5," or a set of equations in Rol and Iload mentioned above.

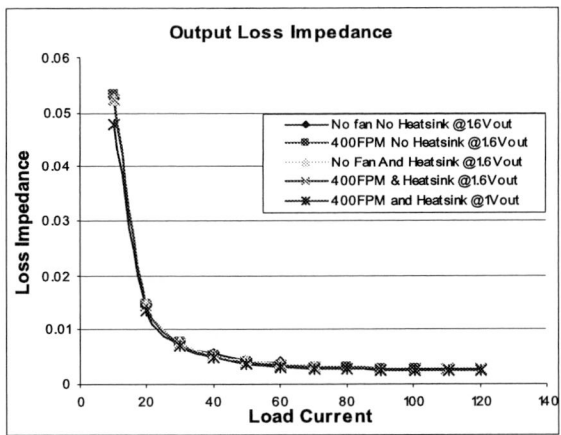

Figure 5. Effective Loss resistance of a converter under different conditions of output voltage and cooling

For completion, "Fig. 7," below shows Rol for different heat sinking conditions. As can be seen, Rol has a mean value and band at each current that represents the spread of the data range. In this particular case the data spread is due mainly to the temperature effects on the converter as explained above.

By knowing Rol and the ± spread one can immediately evaluate the effect of heat sinking on the total performance and would allow for correct decision making regarding whether the converter requires heat sinking.

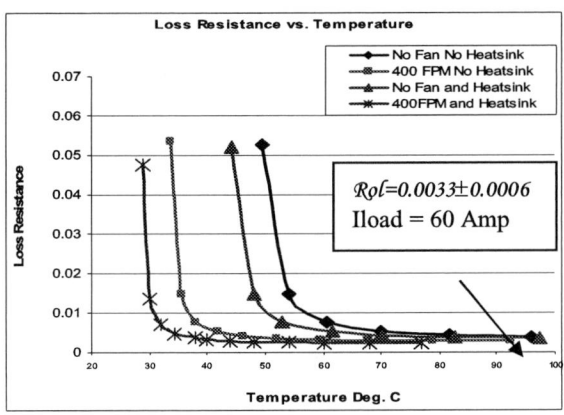

Figure 7. Effective Loss Resistance Vs. Temperature

VII. CONCLUSION

A. Some conditions must apply for power efficiency measurements to be used as a comparison tool between different converters. These conditions are the same input and output voltages, same switching frequency, same heat sinking and the same range of load current

B. A tool is needed that is independent of the above mentioned conditions. The Effective Loss Resistance, Rol, of a synchronous buck converter is "almost" independent of the output voltage and the heat sinking approach.

C. Effective Loss Resistance may be published in the data sheet of DC-DC converters allowing the design engineer to hold a very simple and accurate comparison between any number of converters

D. The mean value of Rol measured at the current range of interest may be used to determine the best converter for the application

REFERENCES

[1] A. Elbanhawy, "Effect of Parasitic Inductance on switching performance" in *Proc. PCIM Europe 2003, pp.251-255*
[2] A. Elbanhawy, "Effect of Parasitic inductance on switching performance of Synchronous Buck Converter" in *Proc. Intel Technology Symposium 2003*
[3] A. Elbanhawy, "Mathematical Treatment for HS MOSFET Turn off" in *Proc. PEDS 2003*
[4] A. Elbanhawy, "A quantum Leap in Semiconductor packaging" in Proc. *PCIM China, pp. 60-64*

2006 5th International Power Electronics and Motion Control Conference

Calculation of Power Loss in Output Diode of a Flyback Switching DC-DC Converter

Jiaxin Chen[*,**], Jianguo Zhu[**], and Youguang Guo[**]

[*]College of Electromechanical Engineering, Donghua University, Shanghai 200051, China
[**]Faculty of Engineering, University of Technology, Sydney, P.O. Box 123, Broadway, NSW 2007, Australia

Abstract—Based on the mathematical model of flyback switching DC-DC converter in the continuous current mode and the simulation model under heavy loads, this paper presents a study of the steady-state power loss in the output diode, including the loss calculation models under three operational modes (voltage control, under-voltage, and over-current modes), the relations of the diode loss versus the input and output voltages, and a comparison of the losses under different modes. By the heavy load simulation model, the effects of time delay on the diode loss are also investigated. Finally, the proposed models are applied to calculate the power losses of output diode of an existing flyback switching DC-DC converter.

Keywords-flyback switching DC-DC converter; output diode; power loss model; heavy load simulation model

I. INTRODUCTION

Flyback converters are commonly used as small power converters because of their simple structures. A great amount of work has been done on the operational principle, design methodology, modeling, and control of the flyback converter [1-5]. This paper investigates the power loss calculation in the output diode of a flyback switching DC-DC converter. The converter reliability mainly depends on two aspects: the over-voltage or over-current protection, which has been studied by using a heavy load simulation model based on state-machine [6], and the heat generation and dissipation, which has not been well studied because of the difficulty in appropriate determination of power loss in the output diode.

According to the state-machine-based simulation models under both the rated load and heavy load modes [6], this paper presents the calculation of output diode losses, including the models of average steady-state loss under different operational modes (voltage control, under-voltage, and over-current modes), the models of the maximum steady-state loss, and the relation of the diode power loss versus the input and output voltages. The effects of time delay on the diode loss are also investigated by the Matlab/Simulink-based simulation model. These models are applied to calculate the output diode loss of an existing flyback converter.

II. FLYBACK SWITCHING DC-DC CONVERTER

Fig. 1 shows the typical topology of a flyback converter with a single output port, where the dashed line connects control block to the main circuit, D1 is the output diode, Ro the output resistance, Co output capacitance, and Vout the output voltage.

To effectively analyze the flyback converter, the following assumptions and simplifications are made: (1) The power MOSFET in the ON state is modeled by a zero resistance and in the OFF state by an infinite resistance, R_{INF}. The output capacitance and inductance of the leading wires are ignored. (2) The diode in the ON state is modeled by a constant voltage source, V_F, with a constant forward resistance, R_F, and in the OFF state by an infinite resistance. The diode junction capacitance is negligible. (3) All the leakage inductances and stray capacitances of the transformer are neglected. (4) Passive components are linear, time invariant, and frequency independent. (5) UC3842 is selected as the PWM controller.

The system consists of the main input-output circuit block, the output voltage control block (VCB), the over-current protection (OCP) block, and the under-voltage protection (UVP) block. As the load increases, the last three blocks will be activated in order. When the converter is in the heavy load mode, the output voltage becomes lower than the rated value, so that the VCB block and the assistant winding N3 stop working and can be ignored in the analysis of heavy load operational modes. Figs. 2 (a) and (b) illustrate the equivalent electric circuits in the heavy load modes (OCP and/or UVP) and the normal operational mode (VCB), respectively.

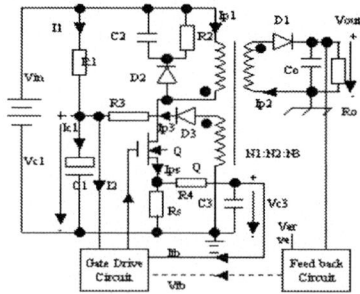

Figure 1.   Typical topology of a flyback converter

1-4244-0448-7/06/$25.00 ©2006 IEEE   346

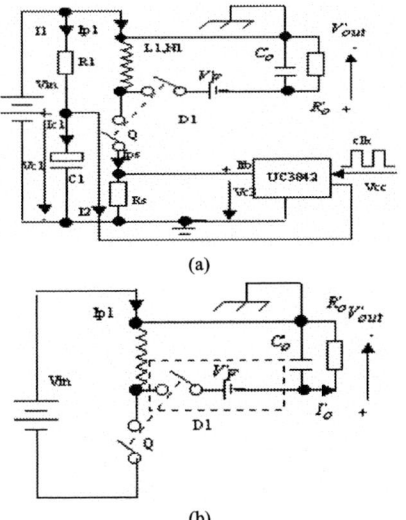

(a)

(b)

Figure 2. Equivalent electrical circuits of a flyback converter: (a) in heavy load modes; and (b) in normal operational mode

The parameters of the secondary winding are referred to the primary side by

$$V_F' = \frac{N1}{N2} V_F$$
$$C_o' = \left(\frac{N2}{N1}\right)^2 C_o \qquad (1)$$
$$V_{out}' = \frac{N1}{N2} V_{out}$$

where N1 and N2 are the numbers of turns of the primary and the secondary windings of the transformer, respectively.

III. STEADY-STATE POWER LOSS OF OUTPUT DIODE WITHOUT CONSIDERING TIME DELAY

In this section, the output diode loss is analyzed by ignoring the time delay caused by various components and the RC filter circuit (R4 and C3).

A. Output Diode Loss in VCB mode

When the converter operates in the VCB mode, the output voltage is a constant. According to [1] and Fig. 2(b), the following equation can be obtained:

$$dI_{p1} = \frac{Vin}{L1} DT = \frac{V_{out}' + V_F'}{L1}(1 - D)T \qquad (2)$$

where Ip1 is the current in the primary winding of the transformer, Vin the input voltage, L1 the primary winding inductance, D the duty ratio, and T the time period of a duty cycle. From (2), the duty ratio can be calculated by

$$D = \frac{V_{out}' + V_F'}{V_{out}' + V_F' + Vin} \approx \frac{V_{out}'}{V_{out}' + Vin} \qquad (3)$$

The duty ratio of PWM depends on the equivalent input and output voltages only and it is not affected by the load.

According to the output equation during one period, i.e. $dV_{out}' = 0$, one can get:

$$\frac{i_{p1} - i_o'}{C_o'}(1 - D)T = \frac{i_o'}{C_o'}DT \qquad (4)$$

$$i_{p1} = i_o'/(1 - D) \qquad (5)$$

$$P_{DT} = V_F' i_{p1}(1 - D) = V_F' i_o' \qquad (6)$$

The steady-state power loss in the output diode in the normal operational mode, P_{DT}, increases linearly with respect to the load.

As the converter may work at different voltages, the rated output power is designed as the maximum average power that the converter can deliver at any voltage within the whole range (Vmin to Vmax). The rated output power, Pe, corresponding to the highest output power when Vin=Vmin, and the maximum output power of the converter, Pem, which happens when Vin=Vmax, can be calculated by

$$\begin{cases} P_e = V_{min} D_{max} (I_{p1max} - \Delta I_{p1e}) \\ P_{em} = V_{max} D_{min} (I_{p1max} - \Delta I_{p1max}) \end{cases} \qquad (7)$$

where $Ip1max=1/Rs$ is the allowed maximum primary current, and

$$\begin{cases} \Delta I_{p1max} = \frac{V_{max}}{2L1} D_{min} T \\ \Delta I_{p1e} = \frac{V_{min}}{2L1} D_{max} T \\ D_{min} = \frac{V_{out}' + V_F'}{V_{out}' + V_F' + V max} \\ D_{max} = \frac{V_{out}' + V_F'}{V_{out}' + V_F' + V min} \end{cases}$$

It is shown that the converter output power increases when the input voltage increase, and Pem > Pe. To maximize the safety of the system in the VCB mode, the calculation of the maximum power loss is conducted at the highest input voltage. According to (6), the maximum average steady-state power loss of the output diode in the VCB mode can be computed by

$$P_{DTM} = \frac{V_F'}{V_{out}' + V_F'} P_{em} \qquad (8)$$

B. Output Diode Loss in OCP mode

The OCP mode can be considered as a normal operational mode (VCB) with Ip1=Ip1max and Vout ≠ constant. From Section III(A), it can be concluded that for a certain output voltage, the maximum average steady-state power loss of the output diode in the VCB mode happens with the highest input voltage. Therefore, the relation of the diode loss against the output voltage should be understood first. According to Fig. 2(a), the following equations can be obtained:

$$\begin{cases} P_o = V_{in} D (I_{p1max} - \Delta I_{p1}) \\ \Delta I_{p1} = \frac{V_{in}}{2L1} DT \end{cases} \qquad (9)$$

347

Then, the output diode loss in the OCP mode can be computed by

$$P_{DC} = \frac{V'_F}{V'_{out} + V'_F} P_o$$

$$= \frac{V'_F V_{in}}{V'_{out} + V'_F} D(I_{p1\max} - \frac{V_{in}}{2L1} DT) \qquad (10)$$

From (3), (9) and (10), it can be seen that for a certain input voltage, when the output voltage decreases, both the duty ratio of PWM and output power decrease as well, while the output diode loss increases. When the output is short circuited, the equivalent output current is close to Ip1max. In the OCP mode, when the system is with the highest input voltage and short circuited output, the power loss of the output diode reaches the maximum value as

$$P_{DCM} = V'_F I_{p1\max} \qquad (11)$$

Comparing (8) and (11) reveals:

$$P_{DCM} \gg P_{DTM}$$

C. Output Diode Loss in UVP mode

The UVP block is activated when the output voltage of the converter is lower than the rated value, so that the output voltage of the assistant winding N3 is too small to support the power consumption for the gate drive IC, such as UC3842. The UVP operation is an oscillation process with a large time period, as reported in [7]. Fig. 3 shows the simulated waveform of Vc1, where Ts is the oscillation period, T_C the rise time, and T_D the fall time. Assuming the highest operational voltage of the PWM control hysteresis loop is V_H, and the lowest voltage is V_L, referring to Fig. 2(a) and considering Vin $\gg$ Vc1, one can calculate T_C by

$$T_C = \frac{(V_H - V_L)R_1 C_1}{Vin} \qquad (12)$$

Since R1 is much larger than the equivalent resistance of UC3842, R_{IC}, the fall time T_D and the oscillating duty ratio D_s can be obtained by

$$T_D = R_{IC} C_1 \ln \frac{V_H}{V_L} \qquad (13)$$

$$D_s = \frac{T_D}{T_s} = \frac{T_D}{T_D + T_C} \qquad (14)$$

According to (9), (10) and (14), the average steady-state power loss of the output diode in the UVP mode is

$$P_{DV} = P_{DC} D_s$$

$$= \frac{V'_F V_{in}}{V'_{out} + V'_F + V_{in}} (I_{p1\max} - \frac{V_{in}}{2L1} DT) D_s \qquad (15)$$

From (12) – (15), it can be found that when the input voltage goes up, the system oscillating frequency and the oscillating duty ratio increase, as well as the power loss P_{DV}. When the input voltage reaches the allowed maximum value, the diode loss will also reach the maximum for a given output voltage, and if at the moment the output is short-circuited (zero output voltage), the diode loss is the highest by

$$P_{DVM} \approx V'_F I_{p1\max} D_{sm}$$

$$D_{sm} = \frac{T_D}{(V_H - V_L)R1C1/V\max + T_D} \qquad (16)$$

Comparing P_{DV} and P_{DC} reveals that it is very important to choose appropriate operational point of UVP and value of R1 for the high system efficiency. This can be done by the above models.

D. Calculation of Output Diode Loss in a Practical Flyback Converter

The proposed models are applied to calculate the power loss of the output diode in an existing flyback switching DC-DC converter. The major data of the converter include: the input voltage, Vin=102-370 VDC, the nominal output voltage, Vout=5 VDC, the rated output current, Io=3.6 A, and the switching frequency, f=60 kHz. When Vin=102 VDC, the system efficiency at the rated load is η≥72%, the maximum duty ratio Dmax=0.414, the peak value of the primary current, Ip1max=0.89 A (the transformer saturates when the current reaches 1.55 A), the corresponding maximum magnetic flux density, Bmax=290 mT, and the minimum primary current at conduction: Ip1min=0.30 A.

The transformer has a primary winding inductance of 1.186 mH, and the numbers of turns of three windings are 96:8:17. Other parameters include: R1=160 KΩ/1W, C1=47 μF/35V, R_S=1.3 Ω, R4=1.2 KΩ, C3=1.0 nF, MOSFET: SSS6N60A, R3=0 Ω, D1: MUR1620, D2: UF4006, D3: 1N4148, R2=100 KΩ/1W, C2=3.3 nF/1000V, and Co=1000 μF/25V.

According to (7) and (8), in the VCB mode, the relations of the duty ratio of PWM and the maximum steady-state power loss of output diode against the input voltage can be obtained, as shown in Fig. 4. When Vin=370 V, the power loss reaches the highest value as P_{DTM}=3.5 W.

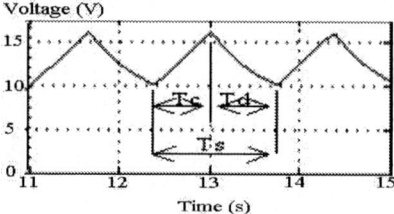

Figure 3.   Simulated waveform of Vc1 under the UVP mode

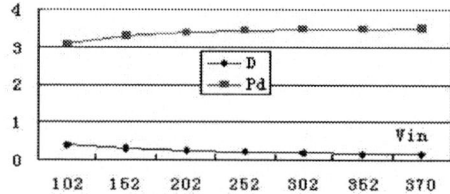

Figure 4.   Duty ratio and diode loss versus input voltage in VCB mode

According to (9) – (11), in the OCP mode, the relations of the maximum steady-state power loss of output diode against the input and output voltages are acquired and plotted in Fig. 5. The diode loss will increase if the input voltage increases or the output voltage decreases. When Vin=370 V, and Vout=0.05 V (almost short-circuited), the diode loss reaches the highest value P_{DCM}=6.87 W.

From (12) – (16), it can be found that when the converter switches from the OCP mode to the UVP mode, the output voltage at the corresponding operational point is Vout=4.0 V. At this operational point, the diode loss is studied with various input voltages, as illustrated in Fig. 6. Fig. 7 shows the diode loss against input voltage with the short-circuited output (Vout=0.05 V).

At the switching point (Vout=4 V) from the OCP mode to UVP mode, the maximum diode loss is found from Fig. 5 to be P_{DCM}=4.0 W. Because P_{DTM}=3.5 W (Fig. 4) and P_{DVM}=0.66 W (Fig. 7), the maximum diode losses in different operational modes satisfy: $P_{DCM} \geqslant P_{DTM} \geqslant P_{DVM}$.

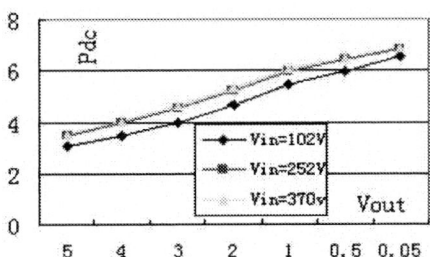

Figure 5. Diode loss versus input and output voltages in OCP mode

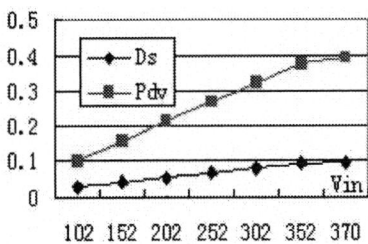

Figure 6. Diode loss versus input voltage at the switching point from OCP mode to UVP mode (Vout=4.0 V)

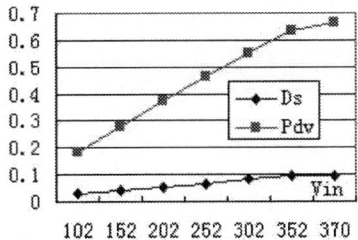

Figure 7. Diode loss versus input voltage in UVP mode with short-circuited output (Vout=0.05 V)

IV. STEADY-STATE POWER LOSS OF OUTPUT DIODE CONSIDERING TIME DELAY

Because of the high operational frequency, the time delay caused by converter components, such as the PWM control circuit and RC filter (R4 and C3) will affect the diode loss. This phenomenon will be studied in this paper by a Matlab/Simulink-based heavy load simulation model [6]. Fig. 8 shows the complete simulation model of the flyback converter and the major block simulation models.

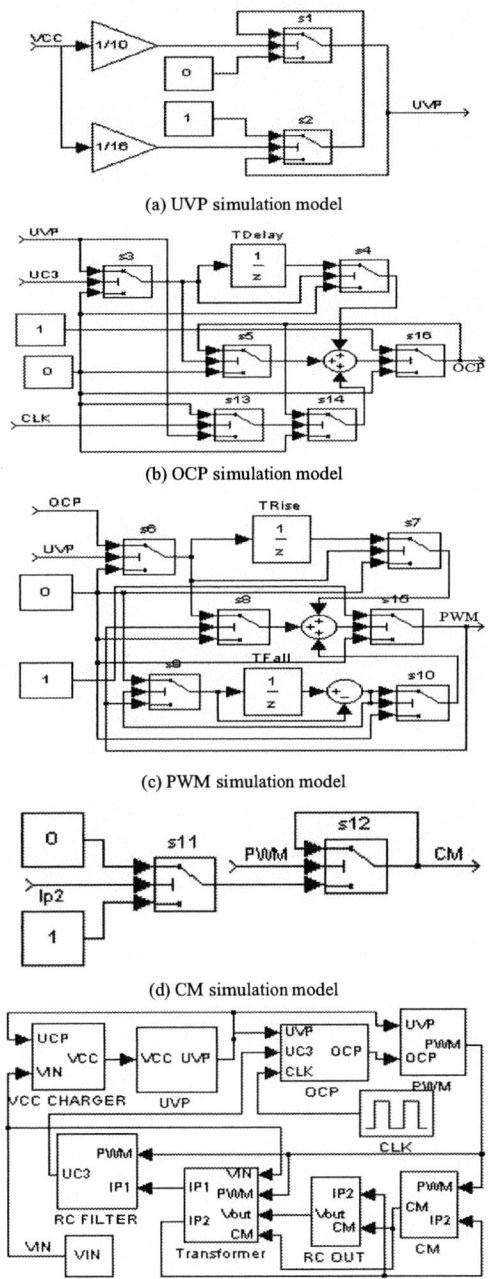

(a) UVP simulation model

(b) OCP simulation model

(c) PWM simulation model

(d) CM simulation model

(e) Complete simulation model of the flyback converter

Figure 8. Major block simulation models and complete simulation model of the flyback converter

The above model is applied to the existing flyback converter and the simulated results are shown in Figs. 9 – 11. The time delays of UC3842 are 300 ns for protection, 150 ns for switching off, and 50 ns for switching on, respectively.

In Fig. 9, Vout is the output voltage, Ip1 the primary current of the transformer, and PWM the pulse-width-modulated drive waveform of UC3842. The simulated results agree well with the previous design. Figs. 10 and 11 illustrate the relations between the steady-state diode loss and the input voltage in the VCB and OCP modes. Although the curves in Figs. 4 and 10 are similar, the maximum diode loss when considering time delay (4.0 W) is larger than that without time delay (3.5 W). The curves in Figs. 5 and 11 are also similar, but it can be shown that higher diode loss (4.44 W vs. 4.0 W) is caused by the time delay, when Vout=4.0 V. Therefore, in the thermal design, the calculation of the maximum diode loss in the OCP mode should take into account the effects of time delay. It is also noted that by choosing the appropriate switching point (Fig. 11), the diode loss can be reduced.

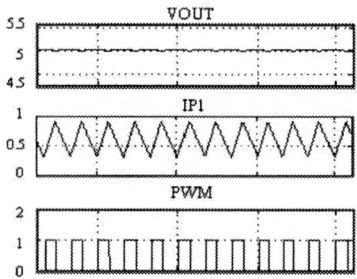

Figure 9. Simulated waveforms of Vout, Ip1 and PWM (Vin=102 V, switching frequency: 60 kHz)

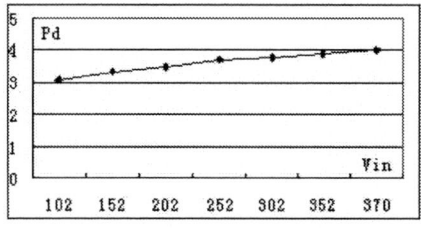

Figure 10. Diode loss versus input voltage in VCB mode (Vout=5 V)

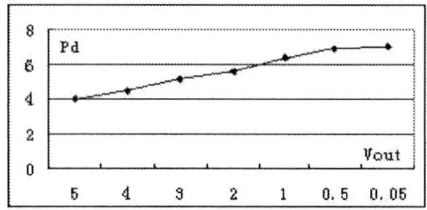

Figure 11. Diode loss versus output voltage in OCP mode (Vin=370 V)

V. CONCUSION

This paper investigates the steady-state power loss of output diode of flyback converters, such as the loss P_{DTM} with the allowed highest input voltage and rated output voltage in the normal voltage control mode, and the loss P_{DCM} when the converter switches from the over-current protection mode to under-voltage protection mode, with the highest input voltage. For thermal design, the larger one of P_{DTM} and P_{DCM} should be taken as the output diode power loss. In the under-voltage mode, the system has large time period and low duty ratio, so that the diode loss P_{DVM} is much smaller than P_{DCM} and can be neglected.

When considering the effects of time delay, the diode loss is larger than or equal to that without time delay. The thermal design should be conducted based on the power losses considering time delay.

REFERENCES

[1] Z. S. Zhang, *Theory and Design of Switching Converters* (in Chinese). Beijing: Electronics Industry Publishing Company, 1999.

[2] J. Liu and Y. B. Wang, "Low frequency oscillation of PWM switching DC-DC converters," *Proc. Chinese Society for Electrical Engineering* (in Chinese), Vol. 24, No. 4, April 2004.

[3] R. Tymerski and D. Li, "State-space models for current programmed PWM converters," *IEEE Trans. Power Electronics*, Vol. 8, No. 3, pp. 271-278, July 1993.

[4] Y. F. Liu and P. C. Sen, "A general unified large signal model for current programmed converters," *IEEE Trans. Power Electronics*, Vol. 9, No. 4, pp. 414-424, July 1994.

[5] T. H. Chen, W. L. Lin, and C. M. Liaw, "Dynamic modeling and controller design of fly back converter," *IEEE Trans. Aerospace and Electronics Systems*, Vol. 35, No. 4, pp. 1230-1234, Oct. 1999.

[6] J.X. Chen, J.G. Zhu, Y.G. Guo, and J.X. Jin, "Modeling and simulation of flyback DC-DC converter under heavy load," accepted to *the International Conference on Communication, Circuits and Systems*, 25-28 June 2006, Guilin, China.

[7] J. X. Chen, "Energy self-holding in flyback switching DC-DC converters," in *Proc. Int. Conf. Electrical Machines and Systems*, Nanjing, China, Sept. 2005, pp. 1194-1197.

2006 5th International Power Electronics and Motion Control Conference

A Multiple Output Forward Converter Adopting Weighted Time-Sharing Control and Switch-Linear Hybrid Scheme

Xiaodong Liu, Songqin Hu and Sizhou Sun

School of Electrical Engineering & Information Anhui University of Technology, Maanshan, P.R.China

e-mail:lxdong168@sina.com

Abstract—**A double forward multiple output converter with weighted time-sharing control and switch-linear hybrid (WTSC-SLH) technique is presented in this paper. Total regulation of   both output voltages is obtained by WTSC. Precise regulation and low ripple multiple output voltage can be achieved by a post regulator with SLH approach. The effects of WTSC-SLH is verified on an experimental two-output double forward converter.**

Keywords-switch-linear hybrid; weighted time-sharing control; multiple outputs; double forward

I. INTRODUCTION

Multiple outputs switching power supplies are widely used in electronic equipment for industrial, commercial and military applications. In high output power capability applications, Forward converter topology is usually used, because of its simple circuitry and high output current capability. However, one of its major drawbacks is the poor regulation in multiple output applications. When a good regulation in all their outputs is needed, the standard forward topology is no longer suitable for such applications.

In this circuit, the output voltage regulation is normally achieved with two different approaches. In one approach, the main output is regulated by a feedback control and the auxiliary outputs are controlled by a post regulator, such as a linear regulator[1], a magnetic amplifier(magamp) [3] or a buck converter[4] , this solution is usually called pre-regulator-post-regulator approach. In the other approach, all of the output voltages are summed and perhaps weighted differently from the feedback signal to control the transistor duty cycle to achieve regulation[2]. The former approach can obtain good regulation of all output voltages. But, the linear regulation option is cheap but lossy and the output current is limited below 1.5A.the saturated core of the magamp brings nonlinear properties into design. The buck converter post regulator spoils the natural simplicity of forward topology , and increased the costs. The later approach is economical and very often provides adequate

The project is sponsored by national nature science foundation of china, No: 50407017

regulation, it is consequently a popular approach for a low-costs power supply, but not suitable for the application requiring precise output voltage regulation.

The converter presented in this paper, named as WTSC-SLH multiple output forward converter, it is based on the weighted time-sharing control and switch-linear hybrid. WTSC can obtain adequate regulation, and SLH can obtain precise regulation and low output ripple voltage. WTSC-SLH converter present good regulation in all output voltages. It works at switching frequency and it can have three or more outputs.

In this paper, WTSC and design procedure for determining the weighing factors for a multiple-output converter are presented in section II .Section III present the SLH post regulator and the comparison with mostly used three terminal linear regulator. A design example and experimental results verify the proposed design procedure are given in section IV . Section V concludes the paper.

II.   WEIGHTED TIME-SHARING CONTROL

To explain the Basic operation of the WTSC control, a double forward converter with multiple output will be used like an example.

In this approach, all or a number of outputs are sensed, and the result of outputs WTSC is fed back to modulate the duty cycle d. Compared with the conventional single-

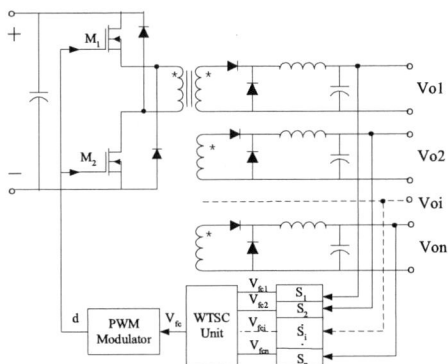

Figure 1. A multiple-output double forward converter with weighted time-sharing control

1-4244-0448-7/06/$25.00 ©2006 IEEE

output-sensing feedback control, the WTSC redistributes the DC regulation error among the sensed outputs. The basic structure of the WTSC forward converter with multiple insulated outputs without SLH post regulator is show in Fig. 1. In this figure both the power and the control stage have been separated.

The control stage presents these internal control signals: $V_{fc1}, V_{fc2} \cdots V_{fcn}$. They have been obtained by means of the feedback of output voltages ($V_{o1}, V_{o2} \cdots$ and V_{on}), Through WTSC, obtaining the end control voltage V_{fc}, and the control signal duty cycle d can be shaped by means o f the PWM modulator.

The function of the control block is given by

$$V_{fc}[T_c] = V_{fc1}[T_{c1}] + \cdots + V_{fcn}[T_{cn}]$$
$$= V_{fc1}[k_1 T_c] + \cdots + V_{fcn}[k_n T_c] \qquad (1)$$
$$= V_{fc1}[k_1 m T_{sw}] + \cdots + V_{fcn}[k_n m T_{sw}]$$

Where

$$T_c = m T_{sw} \quad (m \in N) \qquad (2)$$

$$k_1 + k_2 + \cdots + k_n = 1 \qquad (3)$$

$$[k_{i(i=1,n)} m T_{sw}] = \begin{cases} 1 & t \in (k_{i-1}T_c, k_i T_c) \\ k_0 = 0 & \\ 0 & t \notin (k_{i-1}T_c, k_i T_c) \end{cases} \qquad (4)$$

and T_c is the total control cycle, T_{sw} is the switching cycle of M_1 and M_2, V_{fc} is the end control voltage, k_i is the weighting factor of the i_{st} output.

The control stage firstly specified the control cycle T_c, then T_c is divided to different time interval $T_{c1}, T_{c2} \ldots T_{cn}$ by given weighting factors $k_1, k_2 \ldots k_n$. In different periods, the control power of PWM modulator is given to corresponding control branch. As can be seen from (1), in the period T_{c1} of T_c, V_{fc} is equal to the first output feedback control voltage V_{fc1}; in period T_{cn}, V_{fc} is equal to the n_{st} output feedback voltage V_{fcn}. This is base idea of WTSC scheme, it can be seen from Fig. 2.

For multiple output forward converter adopting

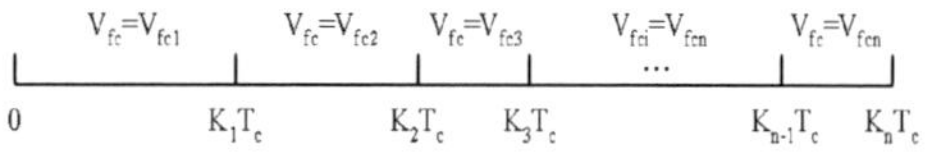

Figure 2. The Relation between internal control signals $V_{fc1}, V_{fc2} \cdots V_{fcn}$ and V_{fc}

WTSC, the selection of weighting factor is very important. The value of weighting factor will directly influence the stabilization of corresponding output voltage. The weighting factor k_i is big, the control signal

V_{fci} of this output has strong control ability on PWM modulator and the output voltage is more stabilization. Otherwise, weighting factor of one output is small, corresponding output voltage stabilization is relatively poor. The selection of time-sharing weighting factors should consider following principle:

1） According to the regulation requiement of each output voltage to determine the value of weighting factor. Different components of equipment have different seabilization requirements for power supply. Some devices(e.g. A/D, CPU) requirements are high, but others(e.g. relay, LCD) requirement are low. High seabilization requirement take big weighting factor, otherwise take the small weighting factor

2） In the total control cycle T_c division , minimum period Min($T_{ci,(i=1,n)}$)must be more than 3~5 times of T_{sw}.

3） third ,according to the power size of each output to determine the weighting factor, big size power takes big k_i.

4） Another, according to the output current capacity of each branch circuit to determine the weighting factor. High output current takes big k_i.

III. SLH POST REGULATOR

Converters with post regulator are typically used when a good regulation in all their outputs is needed. There are some kinds of post regulator. Their features have been introduced in section I. In this section, a SLH post regulator is presented. SLH post regulator has following features:

1） Independence and precise regulation of each output voltage by SLH.

2） Elimination of output ripple voltage.

3） High output current capability is achieved. The current is more than linear regulator option (1.5A), and the loss is far lower than linear regulator.

SLH power conversion technique, whose past name is CTA [5]. SLH's output waveforms just track the reference signal firmly in shapes , frequencies and phases, which do not depend on the switches power supply voltages' variations.

It has been proved that it got excellent characters in AC conversion system [6], [7]. Its application in DC power conversion has not been carefully studied. The present paper take the SLH scheme into forward converter system to stabilize output voltage and cancel the ripple voltage.

Fig. 3 shows the basic SLH post regulator circuit, in which the switch M is a power transistor with high rating current. In Fig. 3, the forward converter with weighted time-sharing control produces a ripple voltage V_c feeding the SLH unit, the reference voltage V_{ref} adding to the base of power transistor is a little lower than V_c, V_{ref} is followed voltage signal. The output voltage V_o of the SLH unit is also a little lower than that of the base voltage V_{ref} , V_o follows V_{ref} . V_{sat} is saturation voltage of power transistor. $V_c - V_o \approx V_{sat}$ drop on the power device can be limited small enough, the loss of

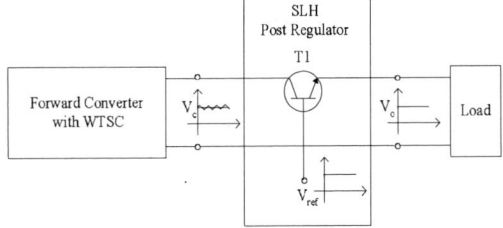

Figure 3. Basic SLH post regulator circuit

switch M also can be limited small. As long as the reference is standard enough, the output voltage waveform must be excellent by means of the voltage following function.

It is obvious that the power switch in the SLH post-regulator unit operate in a very special condition, obeying the regulation of linear amplifier but with loss consumptions like switch devices in on-state, this is the kernel idea of SLH technique. In this special state, the output voltage can be given by:

$$V_o = V_{ref} - V_{be} \qquad (5)$$

Considering the character of power transistor, V_{be} is a constant voltage when load current is changing, then V_o will keeps on a constant value. Output power is supplied by ripple voltage V_c , but the output voltage waveform that following V_{ref} voltage signal is excellent, ripple of V_c is cancelled.

Fig. 4 shows the simulation circuits of conventional three terminal linear regulator LM7815C[8] and presented one with ripple input voltage. The comparison of simulation results is shown in Fig. 5.The load operates

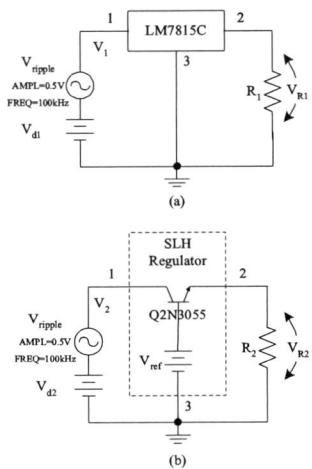

Figure 4. Simulation circuits of LM7815C and SLH regulator

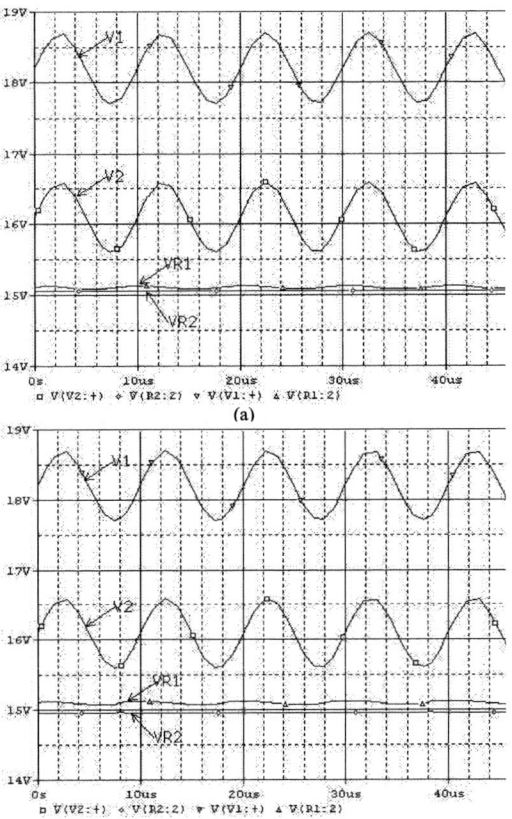

Figure 5. (a)The waveforms of input and output voltage, R1=150 Ω ,R2=150 Ω

.(b) The waveforms of input and output voltage, R1=15 Ω ,R2=3 Ω .

at 15V,0.1A ~5A..

Table I is the comparison of load regulation between three terminal linear regulator LM7815C and SLH regulator.

TABLE I.
OUTPUT VOLTAGE OF LM7815 AND SLH REGULATOR UNDER DIFFERENT LOAD CURRENT

I_{output}(A)	V_{output}(V)	
	LM7815C Regulator	SLH Regulator
0.1	15.11	15.06
1	15.09	15.02
3	×	14.99
5	×	14.96

The simulation results show that SLH regulator has sufficient ability to eliminate input ripple voltage and very good load regulation, load regulation is lower than 0.5%. SLH post regulator can output high current. The minimum drop voltage(about 1V) on SLH post regulator is lower than that of linear regulator（about 2.7V）,so the loss of SLH regulator is far smaller than linear regulator under the same output power size.

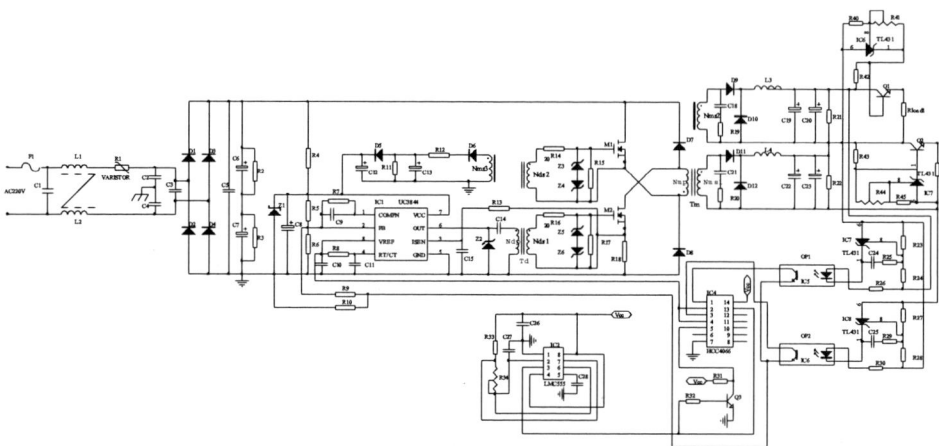

Figure 6. Schematic diagram of prototype

IV. EXPERIMENTAL RESULTS

To prove the effect of the proposed topology, a prototype converter is built according to above design criteria. The prototype has two outputs, 25V/4.5A and –25V/2.5A, it is a auxiliary power supply of frequency-converter employing double forward topology with TWSC as a pre-regulator and SLH as a post regulator. It is operated at 150kHz under a input AC voltage ranging between 175V~240V. Fig. 6 is the schematic diagram of prototype. Table II lists all the principle parameters and components of this prototype circuit.

To meet DC regulation specification , the ration of the weighted factors is chosen as k1:k2=7:3 , which yields k1= 0.7,k2=0.3.

The load regulation has been measured in the proposed prototype for different output power.

In table III is presented the output voltages for nominal

input voltage (AC V_i=220V) and for different output current, it presents the output results of two units, one is WTSC pre-regulator, the other is SLH post regulator. Drop voltage on SLH post regulator is considered in turns ratio selection of Tm.

The value of the line regulation for all outputs are shown in Fig. 7. The results of the line and load regulations are presented in Table IV. In all cases, the obtained regulation with SLH post regulator is lower than 0.3%.

TABLE II.
LIST OF THE CIRCUIT PARAMETERS

Components	Selection	Components	Selection
Nmp1	30	M1 /M2	IRF460
Nms1/Nms2	7/7	Q1/Q2	Q2N3055
Ndp1	18	D7/D8	MUR8100
Nds1/Nds2	28/28	OP1/OP2	TLP521

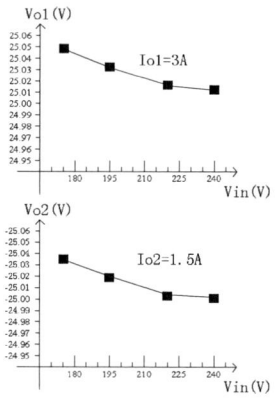

Figure 7. Line regulator with SLH post regulator , I_{01}=3A, I_{02}=1.5A.

TABLE III.
OUTPUT VOLTAGE WITH NORMAL INPUT VOLTAGE

I_{output}(A)		V_{output}(V)			
		WTSC		SLH	
I_{01}	I_{02}	V_{01}	V_{02}	V_{01}	V_{02}
1	0.5	26.7	-27.5	25.06	-25.05
2	1	26.4	-27.2	25.03	-25.02
3	1.5	26.2	-26.7	25.01	-25
4	2	26.2	-26.5	24.98	-24.99
4.5	2.5	26.1	-26.0	24.96	-24.98

TABLE IV.
LINE AND LOAD REGULATION WITH SLH POST REGULATOR

	Line Regulation	Load Regulation
Output 1	0.21%	0.24%
Output 2	0.18%	0.2%

V. CONCLUSIONS

In this paper, a new WTSC-SLH multiple output forward converter has been presented. The converter work at fixed switching frequency and they are regulated by means of WTSC and Switch-linear hybrid scheme. The operation principle of WTSC has been introduced, the selection criteria of weighted was given. If the weighting factors are properly designed, the regulation and dynamic characteristics of the sensed outputs can be significantly improved. Moreover, SLH post regulator has been presented, comparison simulation tests with three terminal linear regulator was done. Results show SLH having better performance than linear regulator. It can obtain independent and precise output voltage regulation and higher output current in each output. Also, a prototype with two outputs has been made to test the capabilities of WTSC-SLH. Verified with experimental results the proposed topology provides a promising solution in applications in which a very fixed voltage in the outputs is needed.

Acknowledgment

The authors thank to Pro. Qianzhi Zhou for her conduction about SLH technique with great enthusiasm.

References

[1] A. I. Pressman, *Switching Power Supply Design*. New York: McGraw-Hill, 1991, pp. 381–412.

[2] Q. Chen, F. C. Lee, and M. Jovanovic, "Analysis and design of weighted voltage-mode control for a multiple-output forward converter," in *Proc .IEEE APEC'93 Conf.*, 1993, pp. 449–455

[3] C. Jamerson and D. Chen, "Magamp post regulators for symmetrical topologies with emphasis on half bridge configuration," in *Proc. IEEE APEC'91 Conf.*, 1991.

[4] H. Matsuo and F. Kurokawa, "Precise regulation of multiple output voltages in a dc-dc converter," in *Proc. IEEE. PESC'80 Conf.*, 1980, pp.275–283.

[5] Qianzhi Zhou, et al., "Switch- linearity (CTA) conversion technique based on optimal waveform criterion," *in Proceedings of 1st International Power Electronics and Motion Control Conference*, Beijing, China, vol. 2 pp. 980--985, 1994.

[6] Zhou Qianzhi, "Switch-Linearity Hybrid Power Conversion and Its Application , " *Transactions of China Electrotechnical Society*, vol. 19, no. 8, pp. 28--33, 2004.

[7] Qianzhi Zhou and Luseng Ge, "Switch-linearity hybrid power conversion (SLH) with low output resistance," *in Proceedings of 4th International Power Electronics and Motion Control Conference*, Xian, China, vol.1, pp. 96--98, 2004.

[8] LM78XX voltage regulator datasheet, National Semiconductor Corp., USA.

2006 5th International Power Electronics and Motion Control Conference

A Novel Soft-Switching PWM Full-Bridge DC/DC Converter with DC Busline Series Switch-Parallel Capacitor Edge Resonant Snubber Assisted by High-Frequency Transformer Leakage Inductor

Khairy Fathy[1] , Toshimitsu Doi[2], Keiki Morimoto[2], Hyun Woo Lee[1] and Mutsuo Nakaoka[1],[3]

[1]Kyungnam University, Masan, Korea
[2]Daihen Corporation, Osaka, Japan
[3]Industrial College of Technology University, Hyogo, Japan
khairy@ieee.org

Abstract –This paper presents two new circuit topologies of DC busline side active edge resonant snubber assisted soft-switching PWM high frequency AC link full-bridge DC-DC converter acceptable for 200V-rms and 400V-rms utility AC mains. All the power switches in the full-bridge arms and DC busline can turns on with ZCS turn-on and turn-off with ZVS and the switching power losses of all the switches can be reduced for high-frequency switching PWM. Each operating principle of the proposed soft-switching PWM DC-DC converter with a high frequency planner transformer link is described, together with the experimental performances.

Keywords- DC-DC converter, , High-frequency transformer link, DC rail side series switch-parallel capacitor circuit, Active edge resonant snubber, Soft switching PWM, Low voltage large current output, TIG/MIG arc welding

I. INTRODUCTION

Recently, some soft-switching isolated topologies of high frequency transformer secondary side saturable inductor switches assisted phase shift ZVS-PWM full-bridge DC-DC converters [1] as well as lossless snubbing capacitors and transformer parasitic inductive components assisted DC-DC converter with phase-shift ZVS-PWM scheme in the secondary-side of high frequency transformer [2]-[4] have been developed by the authors and evaluated so far. These soft-switching PWM DC-DC converter circuit topologies are suitable for handling high output power more than several kW for high/medium voltage and low/medium current applications as new energy-related utility interactive boosted power conditioning supplies. However, secondary side power magnetic switches or power semiconductor switches in these phase shifted PWM DC-DC converters may cause large conduction losses in the secondary side of transformer when these circuit topologies are adopted for low voltage and large current applications as electro-plating, automotive DC feeding and arc welding power supplies. Therefore, for the low voltage and large current applications, a soft-switching

DC-DC converters with the main and auxiliary switches in the primary side of the high frequency transformer is considered to be more acceptable and cost effective.

Under the practical situation, this paper deals with two novel circuit topologies of voltage source full-bridge soft-switching PWM inverters suitable for 200V-rms or 400V-rms utility AC line, which are composed of typical voltage fed full-bridge inverter, high frequency planner transformer with secondary center-tapped windings and additional DC busline PWM series switches with the aid of a DC busline parallel lossless capacitive snubber. The operating principles of the soft-switching PWM full-bridge DC-DC converters topologies treated here are described using switch mode equivalent circuits, along with its remarkable features. The experimental results of these converters are illustrated including power loss analysis as compared with those of hard-switching PWM DC-DC converter. The practical effectiveness of the proposed converters acceptable for low voltage large current output is actually proved on the basis of experimental data.

II. Novel DC-DC Converter for Utility AC 200V-rms

A. Circuit Description

Fig. 1 shows a novel circuit topology of high frequency transformer linked zero voltage soft-switching PWM DC-DC converter with its primary side active edge resonant snubber which is designed for utility AC 200V-rms. This DC-DC power converter is composed of voltage source full-bridge inverter incorporating the time sharing operated active switches in series with DC busline and a single lossless snubbing capacitor in parallel with DC busline, a high frequency planner transformer with secondary side center-taped windings, DC filter and DC load as arc welder. The active edge resonant PWM switches; $Q_5(S_5/D_5)$ and $Q_6(S_6/D_6)$ in series with DC busline and a lossless capacitor in paralleled with DC busline are added in series with the

1-4244-0448-7/06/$25.00 ©2006 IEEE

DC busline connected to the full-bridge high frequency inverter composed of the bridge arm switches $Q_1(S_1/D_1)$, $Q_2(S_2/D_2)$, $Q_3(S_3/D_3)$ and $Q_4(S_4/D_4)$. In particular, it is noted that a single lossless snubbing capacitor C in DC input busline is inserted between active switches Q_5, Q_6 and the full-bridge inverter with high frequency transformer in order to achieve ZVS for all switches.

B. Gate Pulse Timing Sequences

Fig. 2 depicts the pulse pattern-timing sequences of the switching gate driving voltage signal to be provided to the IGBTs; Q_1-Q_4, and Q_5, Q_6. The gate voltage pulse signals with a certain dead time, which are delivered to Q_1 and Q_4 or Q_2 and Q_3 in the diagonal bridge arms of the full-bridge inverter arms, are the same as signal sequences of conventional one. Regarding the turn-on gate pulse voltage signals to the DC busline side series switches Q_5 and Q_6, the gate signals are applied to Q_5 or Q_6 at the same timing pulse period as the turn-on gate pulse signals to Q_1 and Q_4 or Q_2 and Q_3, respectively. As for the turn-off gate pulse voltage signals to Q_5 or Q_6, with a time sharing sequence the gate pulses are delivered to Q_5 or Q_6 before the predetermined specific length of time t_d on the basis of the time when the turn-off signals are applied to the switches Q_1 and Q_4 or Q_2 and Q_3 respectively.

C. Operation Principle of Converter Topology I

Fig. 3 illustrates the relevant voltage operating waveforms the DC-DC converter circuit for utility AC 200V-rms input mains in a complete switching period specified by the pulse pattern of gate drive timing sequences shown in Fig. 2. The operation modes of the converter circuit for the utility AC 200V-rms input are divided into seven switching modes from mode 0 to mode 6 in accordance with operation timing transitions from t_0 to t_6. The equivalent circuits corresponding to each switching mode are represented in Fig.4. The circuit operation in each mode in steady state is described below.

1) Mode 0: -t_0 before time t_0, the switches Q_1, Q_4 and Q_5 are turned on simultaneously. During this period, the primary side energy is supplied to the load R through the high frequency transformer secondary winding.

2) Mode 1:t_0-t_1 At time $t = t_0$, Q_5 is turned off with ZVS because i_{S5} through Q_5 is immediately cut off with the aid of the resonant components due to the lossless snubbing capacitor C assisted by high frequency transformer parasitic component. The supply DC voltage source is cut off from the DC-DC converter. After time t_0, the voltage v_C across C begins decreases toward zero from E. During this period, v_C across C can be approximately estimated as,

$$v_c(t) = E - (i_{t1}/C)\,t \qquad (1)$$

where, i_{t1} is a primary current of high frequency transformer. From eq. (1), the discharging time t_{dc} of C until the voltage v_C becomes zero is given by,

$$t_{dc} = CE/i_{t1} \qquad (2)$$

An appropriate delay time t_d indicated in Fig. 2 is designed so as to be a little longer than the time calculated from eq. (2) under the condition of the maximum output current. In this case, Q_1 and Q_4 or Q_2

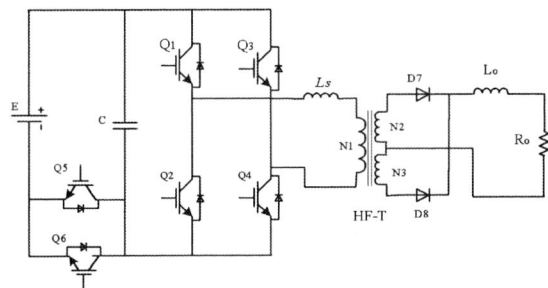

Fig. 1   A novel ZVS-PWM DC-DC converter for utility AC 200V-rms

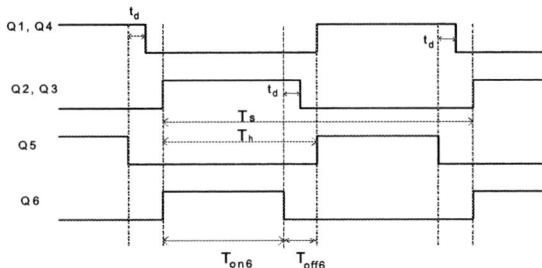

Fig. 2 Pattern sequences of switching gate driving pulses

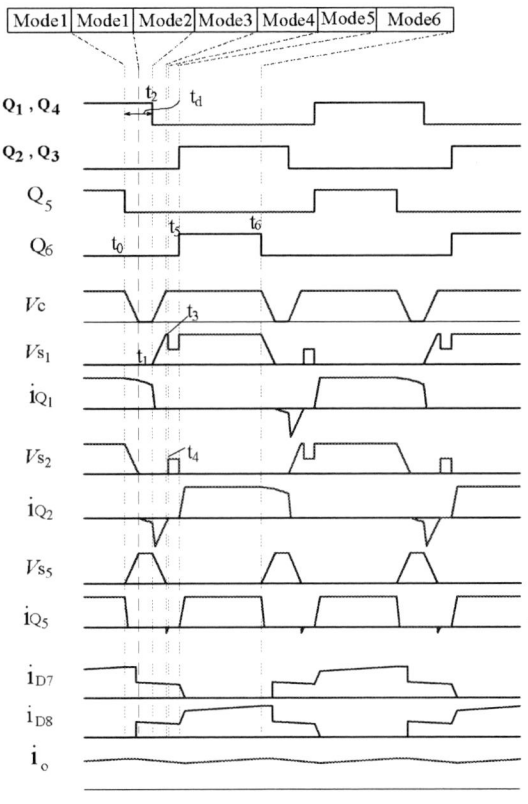

Fig. 3 Operating waveforms of the circuit for AC 200V-rms

and Q_3 can achieve ZVS transition completely. To implement a wider ZVS operation range at the turn-off commutation for t Q_1 and Q_4 or Q_2 and Q_3, the optimum

357

delay time t_d should be varied according to the value of the maximum current i_{t1} of high frequency transformer.

3) Mode 2:t_1-t_2: At time $t = t_1$, the voltage v_C is completely decreases to zero. In the interval from t_1 to t_2, D_2 of Q_2 and D_3 of Q_3 are turned on naturally. The current i_{t1} through the high frequency transformer primary winding flows through two closed loops; $L_S{\to}D_3{\to}S_1{\to}L_S$ and $L_S{\to}S_4{\to}D_2{\to}L_S$.

4) Mode 3:t_2-t_3 At time $t = t_2$, switches Q_1 and Q_4 are turned-off. At this time, because v_C has been already equal to zero. D_2 of Q_2 and D_3 of Q_3 immediately turn on, Q_1 and Q_4 can be turned off with ZVS. At this mode, the condition that capacitor C in parallel with DC busline has been just charged up to the same voltage as E can be estimated by eq. (3).

$$(1/2)CE^2 = (1/2)\,L_S\,(i_{t1})^2 \qquad (3)$$

However, as described after, in mode 6, the circuit parameters should be designed to meet the condition of $(1/2)CE^2 < (1/2)L_S(i_{t1})^2$ in order to achieve ZVS commutation at turn-on transition of Q_6.

5) Mode 4:t_3-t_4 Under a condition of $(1/2)CE^2 < (1/2)L_S(i_{t1})^2$, v_C across C is clamped to E after the voltage v_C reaches E, because of D_5 of Q_5 and D_6 of Q_6 are turned on and the energy stored into leakage inductance L_S of high frequency transformer is returned back to E.

6) Mode 5:t_4-t_5 In this mode, the circuit stops to operate in the primary winding of high frequency transformer, except the voltages across Q_1 and Q_4 decrease down to $(1/2)E$ and the voltages across Q_2 and Q_3 increase up to $(1/2)E$.

7) Mode 6:t_5-t_6 At time $t = t_5$, Q_2, Q_3 and Q_6 are turned on respectively. At this time, switches Q_2, Q_3 can be turned on with ZCS because of leakage inductance L_S of the transformer. And more, Q_6 can achieves ZVS/ZCS at a turn-on transition because v_C is the same as E. Thereafter, the aforementioned operating processes are repeated in sequence during each switching cycle.

I. NOVEL CONVERTER FOR UTILITY AC 400V-RMS GRID

Fig. 5 shows a high frequency transformer linked zero voltage soft-switching PWM DC-DC converter topology acceptable for 400V-rms utility mains. The DC busline voltage source can be selectively operated by the divided voltage sources E_1 and E_2. The voltages E_1 and E_2 ($E_1 = E_2$) are designed so as to be equal to E. It is noted that Q_5 in Fig. 1 is moved to the high side of DC busline in Fig. 5. The diodes D_9 and D_{10} in series are also inserted in parallel with the DC busline between Q_5 or Q_6 and full-bridge inverter arms. And the mid point between E_1 and E_2 is directly connected between D_9 and D_{10}.

Under the DC-DC converter topology II, when Q_5 or Q_6 are turned on and off alternately, half voltage E of DC busline voltage $2E$ is applied to C in parallel with DC busline and full-bridge inverter. Therefore, the same voltage and current rating as those in the converter circuit topology I for utility AC 200V-rms input can be used even in the circuit for the utility AC 400V-rms input grid.

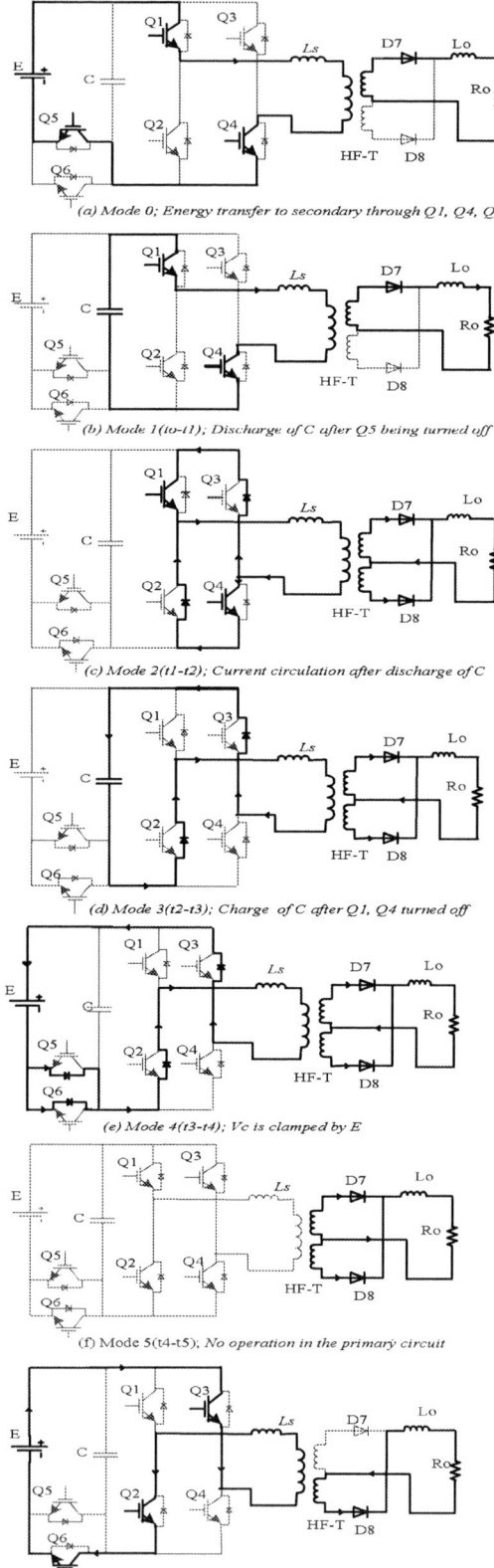

(a) Mode 0; Energy transfer to secondary through Q1, Q4, Q5

(b) Mode 1(t0-t1); Discharge of C after Q5 being turned off

(c) Mode 2(t1-t2); Current circulation after discharge of C

(d) Mode 3(t2-t3); Charge of C after Q1, Q4 turned off

(e) Mode 4(t3-t4); Vc is clamped by E

(f) Mode 5(t4-t5); No operation in the primary circuit

(g) Mode 6(t5-t6); Energy transfer to secondary through Q2, Q3, Q6

Fig. 4 Equivalent circuits for seven switching operation modes

In addition to inherent feature, when Q_1, Q_4 and Q_5 or Q_2, Q_3 and Q_6 are turned on and turned off alternately at the same timing pulse sequence as those for the switches in the converter topology I (Fig. 1.) employed for the utility grid AC 200V-rms shown in Fig. 1, all the switches can perform ZVS turn-off and perform ZCS or ZVS/ZCS turn-on transitions as all the switches in the DC-DC converter circuit topology I used for the utility grid AC 200V-rms.

The relevant operating waveforms of the converter circuit topology II are almost the same as that of the converter circuit topology I. The main difference between circuit I and circuit II is that v_C across C is not clamped to DC busline voltage source in case of circuit II.

IV. EXPERIMENTAL RESULTS AND DISCUSSIONS

A. System Implementations.

The experimental setups for the converters I and II are implemented, these circuit topologies in Fig. 7 are the same operating principle.

Under the converter circuits shown in Fig. 6 and Fig. 7, the 2in1 IGBT power modules 2MBI150TA-060 (I_C=150A, V_{CES}=600V) are used for all the switches. In Fig. 6, each IGBT with reverse conducting diode in the 2in1 IGBT power modules is used for switches $Q_5(S_5/D_5)$ and $Q_6(S_6/D_6)$ and another IGBT with reverse conducting diode in the 2in1 IGBT power modules is not in use. In Fig. 7, each IGBT with reverse conducting diode and one reverse conducting diode in the 2in1 IGBT power modules are used for the PWM switches $Q_5(S_5/D_5)$, D_9 and $Q_6(S_6/D_6)$, D_{10}. Fig. 8 demonstrates the whole appearance of the experimental setup using MAG arc welding power supply for both utility AC 200V-rms and AC 400V-rms grid. The maximum output rating of this experimental setup is 36V, 350A (12.6kW).

Fig. 9 represents the assembled PCB appearance in the transformer primary side of the main converter used in the experimental setup in Fig. 6.

B. Measured Switching Voltage and Current Waveforms

The observed operating voltage and current waveforms under maximum output specifications (36V, 350A) for utility AC 400V-rms grid when the switch Q_1 is turned on and turned off are depicted in Fig. 10 (a) and (b), respectively. Observing these switching waveforms Q_1 is turned on with ZCS and turned off with ZVS. are respectively as shown in Fig. 10 (c) and (d) when the switch Q_5 is turned on and turned off. Observing the operating switching waveforms, Q_5 is completely switches on with ZVS/ZCS and also turned off with ZVS. At the turn-off mode transition it is clear that Q_1 and Q_5, some turn off related power losses still exists due to inherent falling and tail current characteristics of IGBTs.

C. Power Loss Analysis

Considering power loss analysis in Fig. 11, the total power losses of all the switches in the bridge arms including Q_5 and Q_6 in DC busline for the newly-developed converter circuits I, II in Fig. 6 and Fig. 7 are compared with those of all the switches in

conventional hard-switching converters with a high frequency AC link. When the switching frequency is about 20 kHz, the total power losses for soft-switching converter and hard-switching converter are almost equal, So it has to be designed to be more than 20 kHz. The more the switching frequency of inverter increases, the more two newly-developed converter circuits I, II can have remarkable advantages from the view point of the efficiency and power density as compared with those of the conventional hard-switching converter.

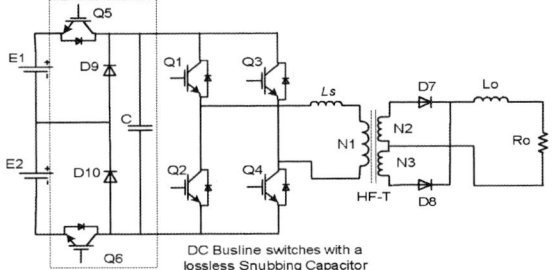

Fig. 5   A Novel ZVS PWM DC-DC power converter II

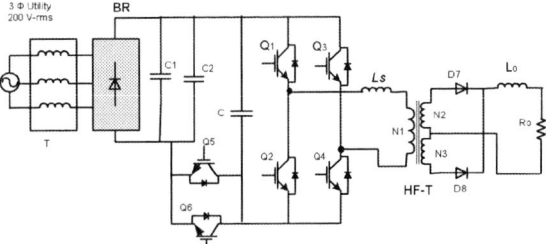

Fig. 6   Experimental setup for utility AC 200V-rms

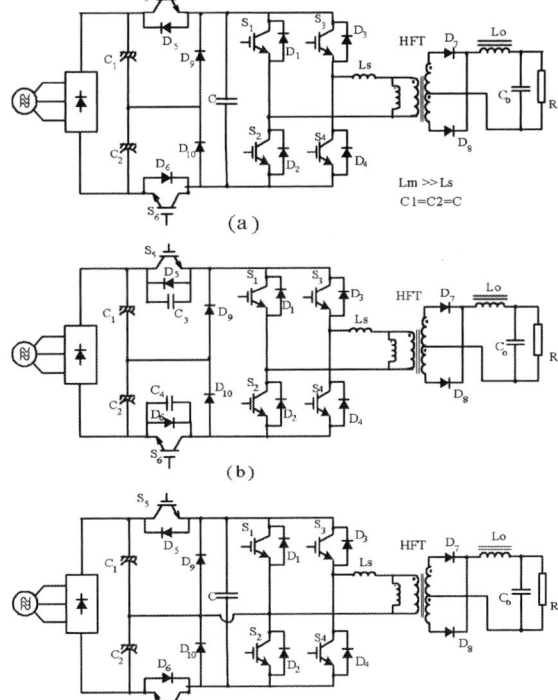

Fig. 7   Experimental setup for utility AC 400V-rms

Fig. 8   Whole appearance of experimental setup of I, II

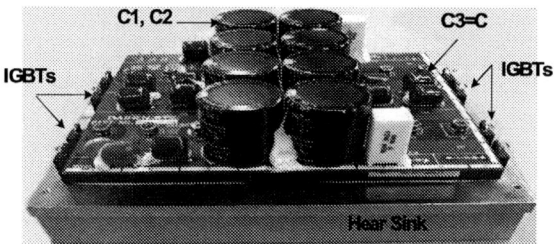

Fig. 9 Assembled appearance in transformer primary side

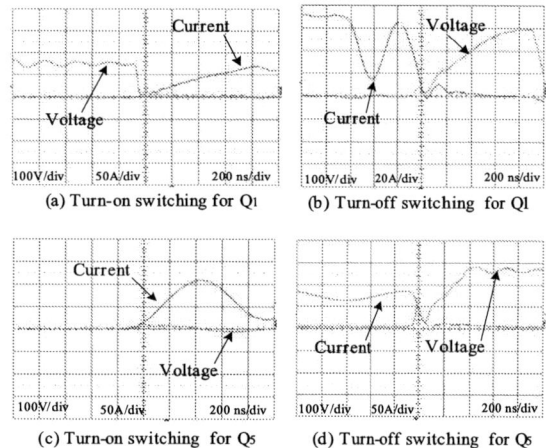

(a) Turn-on switching for Q_1

(b) Turn-off switching for Q_1

(c) Turn-on switching for Q_5

(d) Turn-off switching for Q_5

Fig. 10 Measured waveforms for Q_1, Q_5 for 400V-rms application

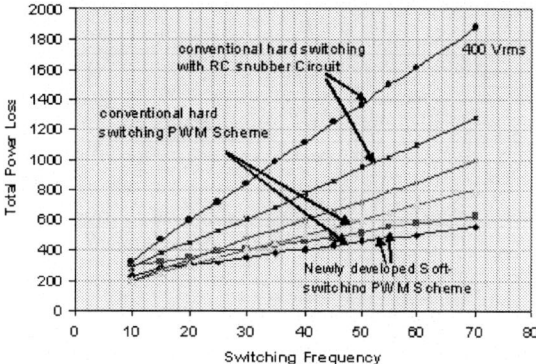

Fig. 11 Comparative power loss analysis between soft-switching and conventional hard-switching PWM converters

In case the switching frequency is designed for 40 kHz, the total power losses for conventional high frequency transformer link hard-switching PWM DC-DC converter including the power loss of RC snubber circuits are estimated as about 785 W in case of utility AC 200V-rms

grid and 1100 W in case of utility AC 400V-rms grid, respectively. These power losses are about two times more than the total power loss of newly-developed converter circuit I and two times more than that of newly-developed converter circuit II .

D. Arc Welding Products

Under the experimental setup of CO_2/MAG arc welding power equipment using the proposed power converter (see Fig. 8,9), the volumetric size is 59% less and its weight is 47% less than these of conventional arc welder using hard-switching PWM inverter, because the newly-developed converter circuits I, II enables 40 kHz switching frequency for the new generation MAG arc welding equipment without increasing the power loss of all power switches, while the inverter switching frequency of conventional one is designed so as to be 20 kHz for hard-switching PWM operation. In addition to this, the arc welding dynamic performance and power density can be improved by fast control responses in accordance with the high switching frequency.

V. CONCLUSIONS

In this paper, two new circuit topologies of active edge resonant snubber assisted zero voltage soft-switching PWM DC-DC power converters I, II with high frequency planner transformer acceptable for the dual voltage utility AC 200V-rms or 400V-rms. The design specifications of the voltage source-fed full-bridge inverter have been presented, The operating principle of these DC/DC converters have been described for MAG arc welding equipment in industry. The power loss analysis of newly proposed DC-DC converters with a high frequency transformer link have been discussed and evaluated as compared with that of hard-switching PWM DC-DC converter with a high frequency transformer link. The practical effectiveness of two DC-DC power converter topologies under a principle of soft-switching PWM scheme have been proved from a practical point of view.

ACKNOWLEDGMENT

Ministry of Commerce, Industry and Energy (MOCIE) financially supported this work through Industry and Energy Research Center (IERC) program.

REFERENCES

[1] S. Hamada, M. Nakaoka, "Saturable Inductor-Assisted ZVS-PWM Full-Bridge High-Frequency Link DC-DC Power Converter Operating and Conduction Losses", Proceedings of IEE-UK International Conference on Power Electronics and Variable-Speed Drives, pp.483-488, October, 1994.

[2] O. D .Patterson and D. M. Divan, "Pseudo-Resonant Full Bridge DC/DC Converter", Records of IEEE-PESC, pp.424-430, June,1987.

[3] M. Michihira, M. Nakaoka, "A Novel Quasi-Resonant DC-DC Converter using Phase-Shifted Modulation in Secondary-Side of High-Frequency Transformer", Records of IEEE-PELS Power Electronics Specialists Conference, Vol.1, pp100-105, June, 1996.

[4] S. Moisseev, S. Hamada, M. Nakaoka, "Novel Soft-Switching Phase-Shift PWM DC-DC Converter", Proceedings of Japan Society Power Electronics , Vol.28, pp107-116, 2003.

2006 5th International Power Electronics and Motion Control Conference

High-Efficiency Cascode Forward Converter of Low Power PEMFC System

Jiann-Fuh Chen *, Wei-Shih Liu*, Ray-Lee Lin*, Tsorng-Juu Liang* and Ching-Hsiung Liu**

* Department of Electrical Engineering, National Cheng Kung University, Taiwan, China
** Boyam Power System Co. Ltd., Taiwan, China
e-mail: N2893119@ccmail.ncku.edu.tw

Abstract - **This paper presents a topology of cascode scheme that is different from traditional cascade circuit. A high-efficiency cascode forward converter is proposed, with high voltage gain, low leakage impedance and same control method as conventional converter. In fuel cell system can use the characteristic that compensate the polarization loss of cascode circuit to achieve a steady output voltage. The operational principle and steady-state analysis are described. Finally, a simulation result with a 480W system output, 20V-38V fuel cell DC input and 48V DC output is presented to demonstrate the performance. It shows that the efficiency of the proposed converter is nearly 96%.**

Keywords-component; fuel cell; DC/DC converter; cascode; forward

I. INTRODUCTION

Recently, the development of 'green power' generation is becoming increasingly important to the global environment. Energy Conversion is an important to promote the application of recycles energy. Fuel cell attracts lots of attention, because of their high efficiency, low pollution, high reliability and high durability. A number of modified converters topologies for fuel cell system have been proposed [1-3].

The conventional cascade topology is adopted generally, the cascade DC/DC converter between the input source and output load for providing a smooth voltage. All of energy must be transformed through the DC/DC converter. However, the efficiency of the cascade circuit is limited by this converter and a part of energy is wasted. In view of efficiency point, modify the structure of cascade converter to propose the cascode scheme. The traditional cascade topology is shown in Fig. 1.

The proposition of cascode converter is to pile up a control voltage source on input source, the block diagram shown in Fig. 2. The main concept is to compensate the insufficient voltage of cascode system; however, the main power voltage supplied without any loss directly. The cascode converter topology with an isolation coupled transformer achieves high voltage gain and high conversion efficiency. Any isolation converter can use this topology including full bridge converter, half bridge

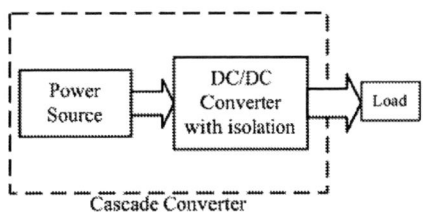

Fig. 1 Block diagram of traditional converter.

converter, push pull converter, fly-back converter and forward converter etc.

In this paper, use a simple forward converter to cascode on fuel cell generator source. The forward output terminal and fuel cell generator output terminal are serial to increase the output voltage gain [4].

The advantages of the proposed converter are:

- The control technique is simply of current mode PWM for duty cycle under 50%
- The leakage energy is recycled to the input source
- High voltage gain, high efficiency and low leakage impedance.
- Same control method as conventional converter.
- The design concept is easily extended to other converters similar to the basic topology with an isolation transformer.

II. CASCODE FRAME USE IN FUEL CELL

A. Efficiency analysis of DC/DC cascode converter

The structure in figure 2 is used, to proceed further into

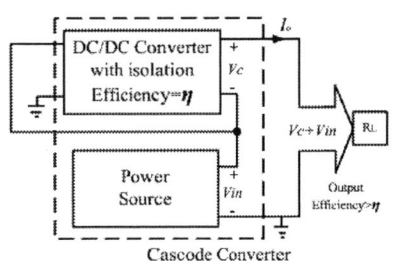

Fig. 2 Block diagram of DC/DC cascode converter

1-4244-0448-7/06/$25.00 ©2006 IEEE

an analysis of efficiency with cascode topology. Assume that efficiency of isolation converter is η, and the converter output and input source are serial to achieve cascode frame. Referring to Figure 2:

$$V_{in} + Vc = Vout \quad (1)$$

(1)

The output efficiency of cascode frame can be obtained:

$$Eff_T = \frac{Po}{Pin} = \frac{I_o^2 \times R_L}{I_o \times V_{in} + \dfrac{V_C \times I_o}{\eta}} \quad (2)$$

Rearranging the (2), equation 3 can be developed

$$Eff_T = \frac{Vout}{\dfrac{Vc}{\eta} + V_{in}} \geq \eta \quad (3)$$

From (3), part of the power from the main power source is directed to load without any loss. Therefore, the efficiency of the proposed is higher than that of the conventional circuit.

B. Fuel cell characteristic analysis

1). Fuel cell output voltage

Fuel cell is an electrochemical device that produces direct current electricity through the reaction of hydrogen and oxygen in the presence of an electrolyte. Recently, many research on simulating FCs and FC stack can be found in literatures [5-7].

The output voltage of a single cell can be gotten via the following expression

$$V_{fuel\_cell} = E_o - V_{act} - V_{conc} - V_{ohm} \quad (4)$$

In (4), E_o is the thermodynamic potential of the cell and it represents the reversible voltage; V_{act} is the voltage drop due to the activation of the anode and of the cathode; and V_{conc} represents the voltage drop resulting from concentration of mass transportation of the reacting gases; V_{ohm} is the ohmic voltage drop, measure of the ohmic voltage drop is associated with the conduction of the protons through the solid electrolyte and the electrons through the internal electronic resistances.

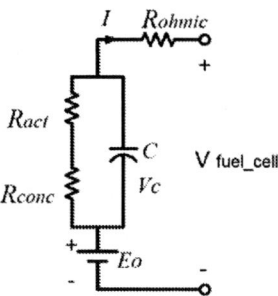

Fig. 3 Equation circuit of fuel cell stack

2). Double-Layer charging effect

In a PEM fuel cell, the two electrodes are separated by a solid membrane which only allows the H^+ ions to pass, but blocks the electron flow. The electrons will flow from the anode through the external load and gather at the surface of the cathode, to which the protons of hydrogen will be attracted at the same time. The layer of charge on or near the electrode-electrolyte interface is a store of electrical charge and energy, and as such behaves much like an electrical capacitor [8]. Considering this effect, the equivalent circuit of fuel cell is given in Fig. 3.

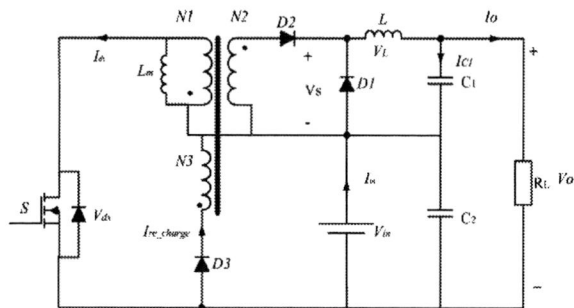

Fig. 4 Cascode topology of forward converter

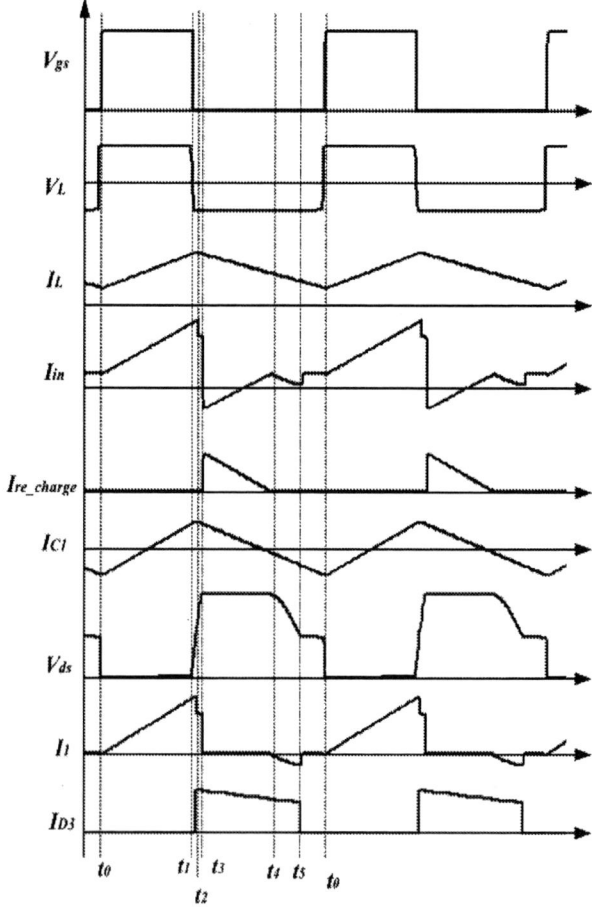

Fig. 5 Output waveforms of the cascode forward converter

362

III. CASCODE TOPOLOGY OF FORWARD CONVERTER

The proposed cascoded forward converter is shown in Fig. 4. Comparing to the conventional forward converter, the forward output terminal and input source are serial to increase the efficiency and output voltage gain. The magnetizing energy is recycled to the input sources or the capacitor C2, to utilize the input sources and increase the output voltage.

A. Operational principles analysis

The proposed cascode forward converter operating in continuous conduction mode (CCM) is now considered. To analyze the circuit, the coupled inductor is modeled as a magnetizing inductor L_m, the turn ratio is N1/N2/N3. The switching sequences and output waveforms of the converter in continuous conduction mode are illustrated in Fig. 5.

Figure 6 illustrates the states equivalent circuits in one switching cycle for the converter operated in CCM. The output capacitors C_1 and C_2 are assumed to be large enough so that the output voltages V_{C1} and V_{C2} across

them are considered as constant during the entire switching cycle, and become $V_o=V_{C1}+V_{C2}$. The six operational modes are briefly described as follows:

Mode 1 [t_0-t_1]: The switch S1 is turned on at t_0, the voltage across the magnetizing inductance is equal to *Vin* and the inductor current I_L will increase with a slope of V_{in}/L. Diode D1 and D3 are reversed-biased. Also, C1 and C2 supply the energy to load. The inductor current is continuous to increase at the time instant t_1 when the switch S1 is turn off.

Mode 2 [t_1-t_2]: During this period, the magnetizing current charges the parasitic capacitor *Cs* of MOSFET, and peak magnetizing current is obtained at the time instant t_2. D2 and D3 start to conduct.

Mode 3 [t_2-t_3]: When the diode *D3* starts to conduct, the V_{ds} voltage is pushed up to 2 time's *Vin*, and the capacitor *Cs* continues to increase to 2*Vin*. The reset winding starts to charge *Vin* at t_3.

Mode 4 [t_3-t_4]: During this time interval, the *D1* starts to conduct and large portion of the inductor current will flow into the capacitor *C1*, and supply the load current.

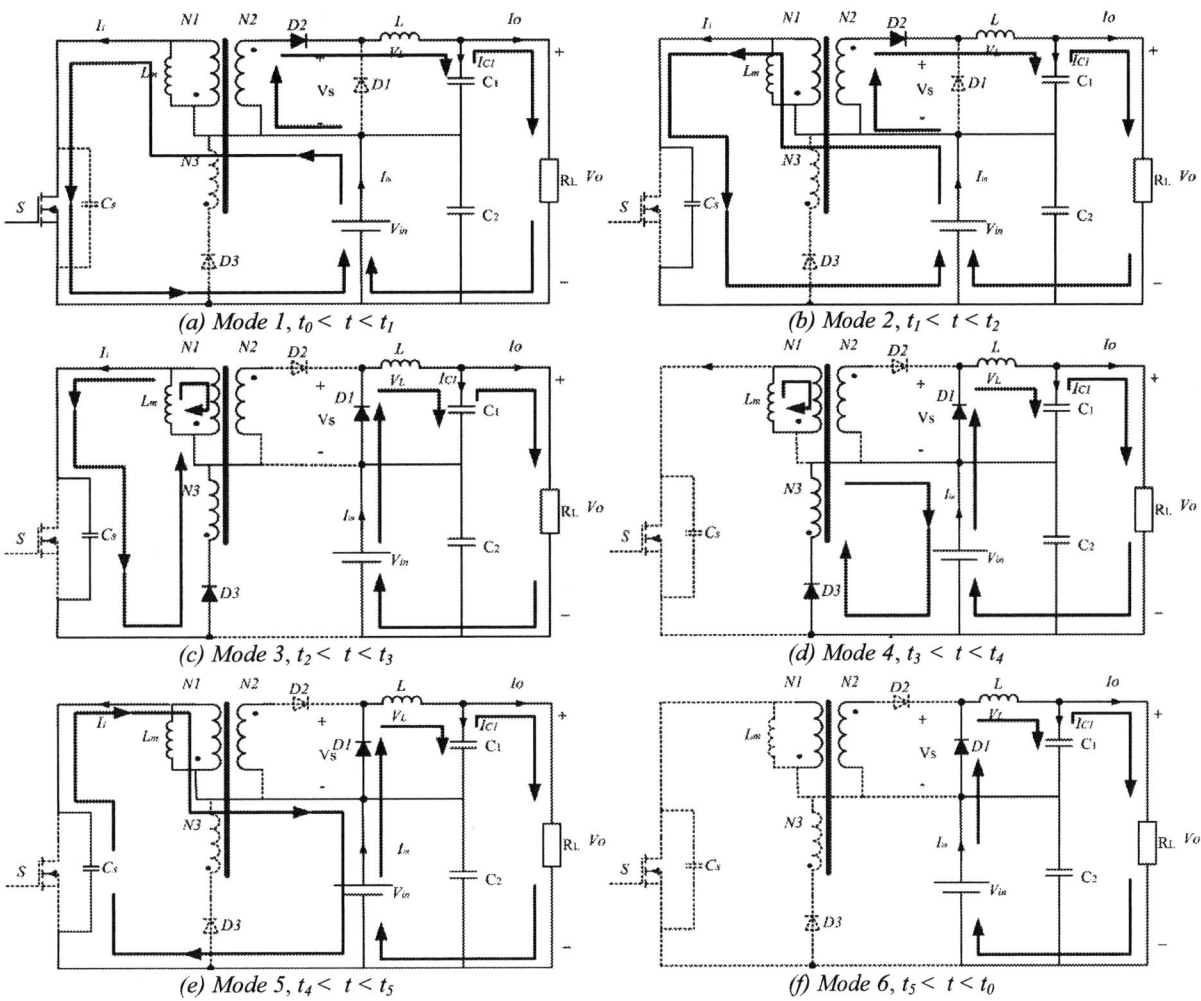

(a) Mode 1, $t_0 < t < t_1$

(b) Mode 2, $t_1 < t < t_2$

(c) Mode 3, $t_2 < t < t_3$

(d) Mode 4, $t_3 < t < t_4$

(e) Mode 5, $t_4 < t < t_5$

(f) Mode 6, $t_5 < t < t_0$

Fig. 6 Operational modes within one switching cycle

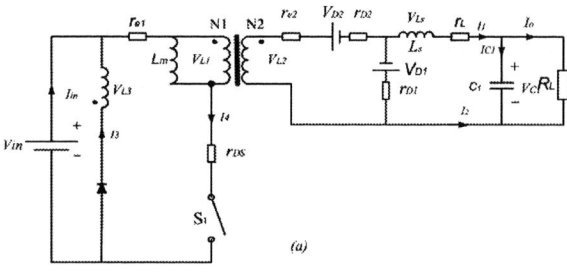

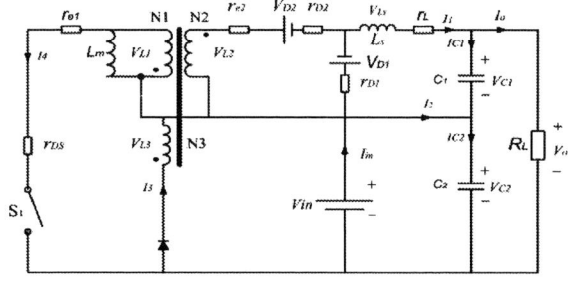

Fig. 7 Equivalent circuit with parasitic components
a. Conventional forward converter
b. Cascode forward converter

At the same time the reset winding current is decrease to zero at t_4, and the diode $D3$ is reversed-biased.

Mode 5 [t_4-t_5]: When the diode $D3$ is turn off, the V_{ds} voltage is clamp at $2Vin$, and V_{ds} starts to charge the Vin. The voltage of the V_{ds} will decreases to Vin at t_5.

Mode 6 [t_5-t_0]: At t_5, energy is discharged from C_1 to supply the load current. The input source is charging C_2 and supplying the load current at the same time.

B. Steady-state analysis

1) Conventional forward converter

Generally, the forward converter is similar to buck converter, it can operate at high voltage using turn ratio. Figure 7 (a) shows the equivalent circuit of the forward converter as follows: r_{DS} is the transistor on resistance, r_{e1} is the ESR of the primary winding of the transformer, r_L is the magnetic core series resistance, L is the magnetizing inductance of the transformer, r_2 is the resistance of the secondary winding of the transformer, V_D is the threshold voltage of the diode, r_D is the forward resistance of diode [9].

The DC voltage gain and efficiency can be expressed:

$$\frac{V_o}{Vin} = D\frac{N2}{N1} \times \frac{R_L - \frac{K}{\left(D\frac{N2}{N1}\right)}R_L}{D\left(\frac{N2}{N1}\right)^2(r_{e1}+r_{DS}) + D(r_{e2}+r_{D2}-r_{D1}) + (r_{D1}+r_L) + R_L} \quad (5)$$

$$\eta = \frac{Vin\left(D\frac{N2}{N1}\right)R_L - [(1-D)V_{D1}+DV_{D2}]R_L}{Vin\left(D\frac{N2}{N1}\right)\left[D\left(\frac{N2}{N1}\right)^2(r_{e1}+r_{DS}) + D(r_{e2}+r_{D2}-r_{D1}) + (r_{D1}+r_L) + R_L\right]} \quad (6)$$

Equation (5) shows very closely the effect of the ESR on the forward converter voltage gain. The inductor resistance also has an effect on the efficiency of the forward converter, as shown in (6). As the duty ratio and the resistance of the winding of the transformer affects the efficiency of the converter.

2) Cascode forward converter

Figure 7 (b) shows the corresponding equivalent circuit that includes copper inductances r_{e1} and r_{e2}, on-state resistances r_{D1} of diode D1, on-state resistance $r_{DS(on)}$ of the power switch.

During the turn-on period of the main switch S1, the following equation can be derived:

$$Vin + V_{L2} = I_1 \cdot r_{e2} + V_{D2} + I_1 \cdot r_{D2} + V_{Ls} + I_1 \cdot r_L + V_o \quad (7)$$

$$I_{C1(on)} = I_{1(on)} - \frac{V_o}{R_L} \quad (8)$$

$$I_{in(on)} = I_{1(on)}\frac{N2}{N1} + \frac{V_o}{R_L} \quad (9)$$

$$V_{L2} = \left(Vin\frac{N2}{N1} - I_{1(on)}\left(\frac{N2}{N1}\right)^2(r_{e1}+r_{DS})\right) \quad (10)$$

From (7) and (10) the V_{Ls} is determined as follows:

$$V_{Ls(on)} = Vin\left(1+\frac{N2}{N1}\right) - I_{1(on)}\left[\left(\frac{N2}{N1}\right)^2(r_{e1}+r_{DS}) + (r_{e2}+r_{D2}+r_L)\right] - V_o - V_{D2} \quad (11)$$

During the turn-off period of the main switch S1, the following equations can be derived:

$$I_{in(off)} = \frac{V_o}{R_L} \quad (12)$$

$$I_{C1(off)} = I_{1(off)} - \frac{V_o}{R_L} \quad (13)$$

$$Vin + V_{Ls} = I_1 \cdot r_{D1} + V_{D1} + I_1 \cdot r_L + V_o \quad (14)$$

From (14), obtain the following equation

$$V_{Ls(off)} = -Vin + I_{1(off)}(r_{D1}+r_L) + V_{D1} + V_o \quad (15)$$

The average inductor voltage flowing through Ls is zero in the steady state. From (7) and (9) the Volt-second balance equation can be derived for the steady-state condition of the inductor V_{Ls}, such that

$$D\left\{Vin\left(1+\frac{N2}{N1}\right) - I_{1(on)}\left[\left(\frac{N2}{N1}\right)^2(r_{e1}+r_{DS}) + (r_{e2}+r_{D2}+r_L)\right] - V_o - V_{D2}\right\}$$
$$= (1-D)\left[-Vin + I_{1(off)}(r_{D1}+r_L) + V_{D1} + V_o\right] \quad (16)$$

From (8) and (13) the Amp-second balance equation can be derived for the steady-state condition of the capacitor C1, such that

$$(1-D)\left(I_{1(off)} - \frac{V_o}{R_L}\right) + D\left(I_{1(on)} - \frac{V_o}{R_L}\right) = 0 \quad (17)$$

From (15), the DC component I_1 is determined as follows:

$$I_1 = \frac{V_o}{R_L} \qquad (18)$$

From (16) and (18), the DC voltage conversion ratio V_o/Vin is derived as follows:

$$\frac{V_o}{Vin} = \left(1 + D\frac{N2}{N1}\right)$$

$$\times \frac{R_L - \dfrac{R_L K}{\left(1 + D\dfrac{N2}{N1}\right)}}{R_L + (r_{D1} + r_L) + D\left(\dfrac{N2}{N1}\right)^2 (r_{e1} + r_{DS}) + D(r_{e2} + r_{D2} - r_{D1})} \qquad (19)$$

where $K = \dfrac{(1-D)V_{D1} + DV_{D2}}{Vin}$

If the ESRs are neglected, the ideal voltage gain of this converter is simplified as

$$\frac{V_o}{Vin} = \left(1 + D\frac{N2}{N1}\right) \qquad (20)$$

The input power and output power of the converter can be calculated as follows:

$$Pin = Vin \cdot I_{in(on)}D + Vin \cdot I_{in(off)}(1-D)$$
$$= Vin \cdot I_1 \left(1 + D\frac{N2}{N1}\right) \qquad (21)$$

$$Pout = Vo \cdot I_1 \qquad (22)$$

From (21) and (22), the converter efficiency is

$$\eta = \frac{Pout}{Pin} \qquad (23)$$
$$= \frac{Vo}{Vin\left(1 + D\dfrac{N2}{N1}\right)}$$

From (19) and (23), the efficiency can be derived as

$$\eta =$$

$$\frac{V_{in} \cdot R_L\left(1 + D\dfrac{N2}{N1}\right) - R_L\left[V_{D1} \cdot (1-D) + V_{D2}D\right]}{V_{in} \cdot \left(1 + D\dfrac{N2}{N1}\right) \cdot \left[R_L + (r_{D1} + r_L) + D\left(\dfrac{N2}{N1}\right)^2 (r_{e1} + r_{DS}) + D(r_{e2} + r_{D2} + r_L - r_{D1} - r_L)\right]} \qquad (24)$$

From (19), the cascode converter has a higher voltage gain than the conventional forward converter. Equation (24) reveals that the efficiency is affected by V_{D1}, r_{DS}, r_{e1}, r_{e2} and r_D.

The output voltage and efficiency for different load can

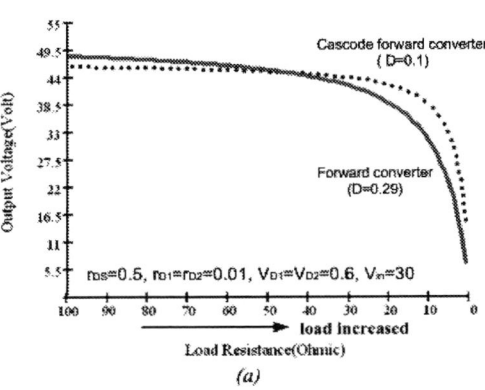

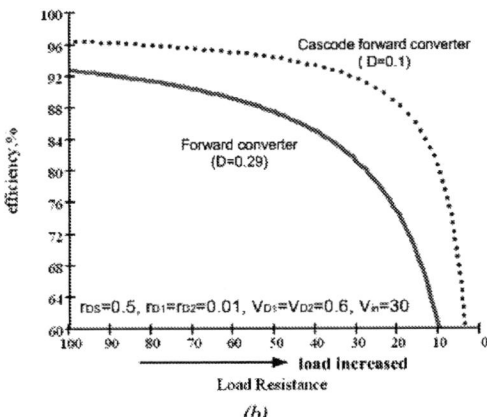

Fig. 8 Constant the turns ratio N2/N1=6, r_{e2}=0.6
(a) The output voltage performance for different load resistance
(b) Efficiency with different duty cycle and load resistance

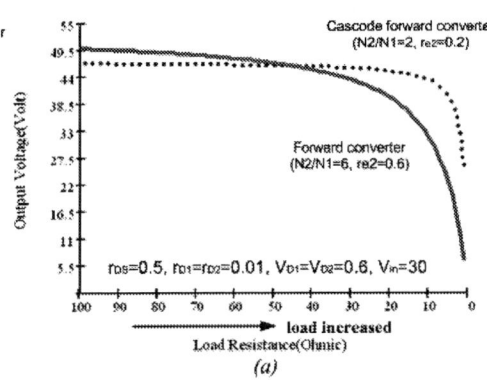

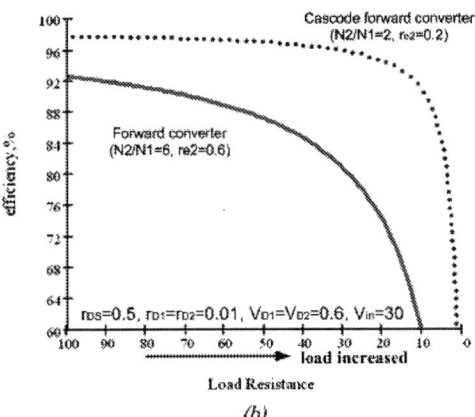

Fig. 9 Constant the duty ratio D=0.3
(a) The output voltage performance for different load resistance
(b) Efficiency with different turns ratio and load resistance

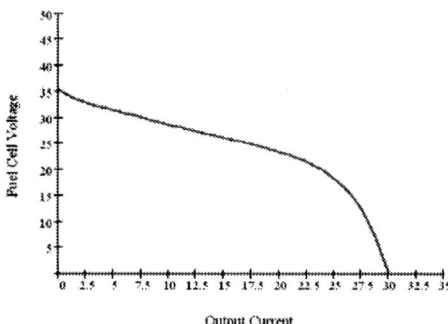

Fig. 10 Polarization curve of BPS stack

be described by equation (5), (6), (19) and (24), Figure 8 shows the efficiency of cascode circuit for the constant duty (D=0.3) and a similar output voltage. Figure 9 shows the efficiency of cascode circuit for the constant turns ratio (N2/N1=6, re2=0.6) and a similar output voltage. The efficiency of cascode forward converter is higher than the conventional converter.

IV. CASCODE FORWARD CONVERTER OF FUEL CELL SYSTEM SIMULATION RESULTS

The performance of the proposed converter is verified by SIMPLIS simulation. Based on the above analysis and design guidelines, the secondary-to-primary turns ratio is 2, the thirdly-to-primary turn ratio is 1. The inductance L is 1m; output filter capacitances C1and C2 =1mF; power MOSFET S1 is BUZ11; rectifier diodes D1, D2 and D3 are C92-02; the control current mode PWM IC is UC3842. The specifications of the circuit are:

Input fuel cell source = 20V~40V DC

Output voltage =48V DC

Load current =1A~10A

Switching frequency =20k Hz

The duty-ratio D of the PWM controller is nearly 0.5. According to the reference paper [5] and [6], the polarization curve presented in Fig. 10 was established for the 500W BPS stack, which also allows a comparison between the manufacturer and the simulated data [10].

The duty-ratio D of the PWM controller is smaller than 0.5. Figure 11 shows the simulation results for currents I_D, I_L, I_{in} and driving signal Vgs by light load operation with 40W. Figure 12 shows the simulation results for currents I_D, I_L, I_{in} and driving signal Vgs by full load operation with 400W. Figure 13 shows the driving signal and output voltage, the output voltage is kept steady at 48V DC.

V. CONCLUSIONS

This paper has presented the operation principles, theoretical analysis and design methodology of a cascode forward converter of fuel cell power system. The cascode converter achieves high efficiency without an extreme

duty ratio and high voltage gain using same control technique. The cascode scheme just a bit loss on cascode converter. By using the small-ripple approximation principle, the theoretical voltage gain and efficiency are derived. The features of this converter include high efficiency, high voltage gain, low inductance on the transformer and simplicity of design.

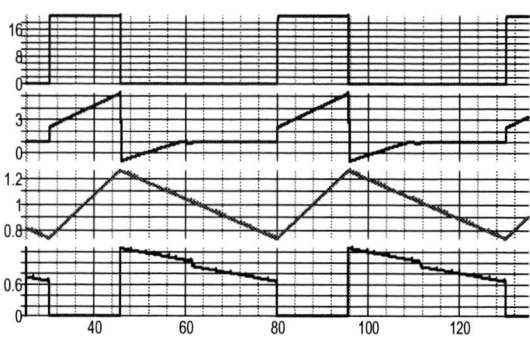

time/uSecs   20uSecs/div

Fig. 11 simulation results for current Iin, IL, ID and Vgs, at light output power

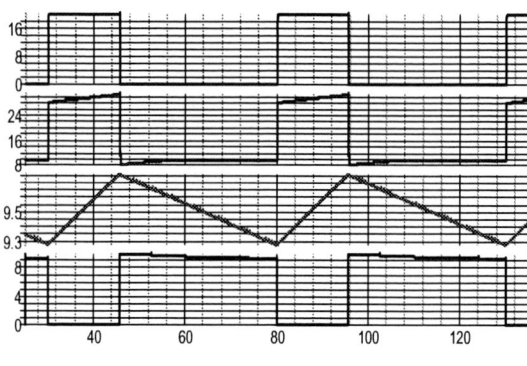

time/uSecs   20uSecs/div

Fig. 12 Simulation results for current Iin, IL, ID and Vgs, at light output power

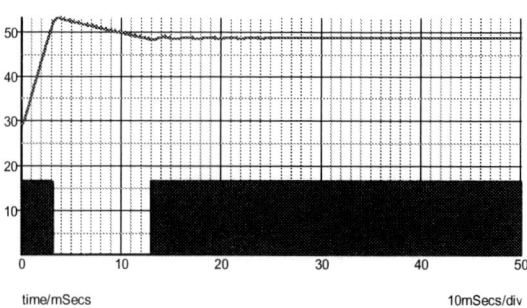

time/mSecs   10mSecs/div

Fig. 13 Cascode converter output voltage and driver signal.

ACKNOWLEDGMENT

The authors wish to thank the financial support of the BOYAM POWER SYSTEM CO. Ltd. And the National Science Council of Taiwan,China through grant number NSC 94-43010-7-0000-G-3058. The authors would also like to thank Kevin Liu for proof-read the paper.

REFERENCES

[1] R. W. Erickson and D. Maksimovic, "Fundamentals of power electronics", 2ned Edn. (John Wiley, New York, USA, 1950), pp. 39-60

[2] R. J. Wai, C. Y. Lin and C. C Chu, "High step-up DC-DC Converterfor Fuel Cell Generation System" in Annual . 30th Proc. IEEE IECON, PP. 57-62, NOV 2004.

[3] K. C. Tseng and T. J. Liang, "Novel High-efficiency step-up converter", IEE Proc, Vol. 151, pp. 182-190, No.2 March 2003

[4] T. J. Liang and K. C. Tseng, "Analysis of Integrated Boost-Flyback Step-up Converter", IEE Proc, Vol. 152, pp. 217-225, No.2 March 2003

[5] H. Xu, L. Kong and X. Wen, "Fuel Cell Power System and High Power DC-DC converter", IEE CAS, Vol 19, No.5, pp. 1250-1255, SEPT 2004

[6] J. M. Correa, F. A. Farret and J. R. Gomes, "Simulation of Fuel-Cell Stacks Using a Computer-Controlled Power Rectifier With the Purposes of Actual High-Power Injection Applications" IEEE Trans., Industry Application, vol. 39, pp. 1136-1142, JULY/AUGUST 2003.

[7] C. Wang, M. H. Nehrir and S. R. Shaw, "Dynamic Models and Model Validation for PEM Fuel Cell Using Electrical Circuits", IEEE Trans., Energy conversion, vol. 20, pp. 442-451, JUNE 2005.

[8] J. E. Larminie and A. Dicks, Fuel Cell Systems Explained. Chichester, U.K.: Wiley, 2000, p.308

[9] L. H. Dixon "Control Loop cookbook", Unitrode Power Supply Design Seminar, SHE-1200, 1997, pp. C4-1—C4-26

[10] "Data sheet of a 500W fuel cell stack", BPS Technology, Boyam, 2004

2006 5th International Power Electronics and Motion Control Conference

Control of Bifurcation by Fuzzy Logic Controller for Current-mode Boost Converters

Noppadol Khaehintung[*]   Phaophak Sirisuk[**]   Anantawat Kunakorn[*]

[*]Dept. of Electrical Engineering, Faculty of Engineering, King Mongkut's Institute of Technology Ladkrabang,
Bangkok, Thailand 10520
[**] Dept. of Computer Engineering, Faculty of Engineering, Mahanakorn University of Technology,
Bangkok, Thailand 10530
e-mail *noppadol@mut.ac.th, phaophak@mut.ac.th* and *kkananta@kmitl.ac.th*

Abstract— **This paper presents the design of a fuzzy logic controller for switching current-mode DC/DC boost converters. The proposed simple fuzzy logic controller, with nine rules, provides an optimal slope compensation to keep the system adequately remote from the first bifurcation point by means of reducing the current spectrum peak. In spite of nonlinear characteristics and instabilities of the converter, the performance of the closed-loop control system can be considerably improved to avoid bifurcation phenomena. It is found that the technique introduced in this paper gives satisfactory results with the regulating and tracking modes under changes in operating conditions.**

Keywords-a Current-mode Boost Converter; Fuzzy Logic Controller; Bifurcation Phenomena, Slope Compensation;

I.   INTRODUCTION

Nowadays, DC/DC converter is widely used, especially as a key part of a switching power supply. Generally, this is achieved by chopping and filtering the input voltage through an appropriate switching action, mostly implemented via Pulse Width Modulation (PWM) circuits. However, due to nonlinear characteristics of switching devices employed in the converter, the design of a control scheme for such a converter may be difficult with a rich variety of bifurcation [1].

Recently, the main antecedent in the study of bifurcations and chaos has been observed and analyzed for various kinds of power electronic circuits [2]. For a system that exhibits bifurcation when a certain parameter is changed, the key design problem is the *control of bifurcation* or *chaos control* [3]. These circuits are designed and guaranteed for stable operations. In most practical situations, the required stable operation is a period-1. Several researches have focused on the control of bifurcation for instance, the work presented in [4], with uses of the discrete time map to analyze and control the bifurcation for switching control converters. The proposed border bifurcation curves have been presented in [5] to provide useful information for circuit design and control. Although, a variable ramp compensation in a DC/DC

boost converter has been presented and implemented in [6], however, the design for automatic closed-loop operation has not been mentioned. In addition, applications of chaos anti-control to switch-mode power supplies to reduce spectral peaks have been shown in [7].

In the past decades, Fuzzy Logic Control (FLC) has become a popular candidate for applications in power electronic circuits, nonlinear control or optimal search problems [8]. The outmost performance of FLC comparing with a conventional controller has been illustrated in [9] and [10], when being applied to search the maximum power point tracking in solar arrays.

In this paper, the circuit configuration of the proposed a current-mode DC/DC boost converter (CMBC) with bifurcation control is introduced. A simple FLC is employed to search an optimal ramp compensation for CMBC. Apart from simplicity, the fuzzy rules, used in this research, occupy a very small memory space in a computation process, and are easy to implement. The performances of the control system are simulated using MATLAB/SIMULINK® [11] with various operating conditions of the system.

The paper is organized as follows. In Section II, the circuit configuration of CMBC is introduced. The proposed FLC for control bifurcations is then expressed in Section III. Simulation results from the control system are given in Section IV. Finally, conclusions are drawn in Section V.

II.   CIRCUIT CONFIGURATION

The proposed FLC for a CMBC with bifurcation control is depicted in Fig. 1. The system consists of a constant DC voltage source, a CMBC and a load. The current proportional plus integral (PI) controller is used to regulate the output voltage (v_{out}), accordingly to the desired output voltage (v_{ref}), in spite of the fluctuations in the input voltage (v_{in}) and the changes in load levels (R_{Load}).

In this system, the inductor current, i_L, is chosen as the programming variable which, by comparing with reference current, I_{ref}, from current controller, generates the turning-output signal for switch S. The switch S is

1-4244-0448-7/06/$25.00 ©2006 IEEE   368

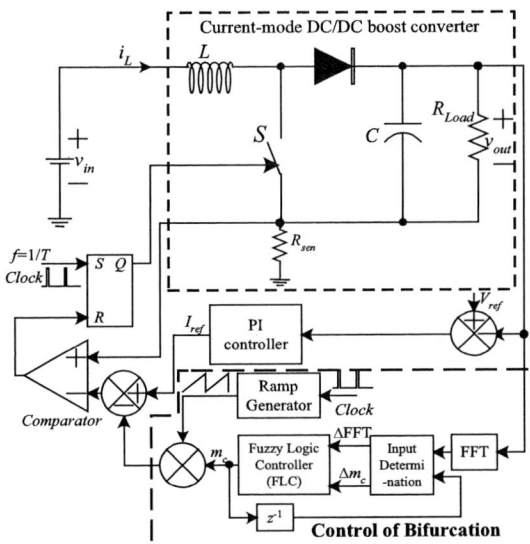

Figure 1. The proposed current-controlled mode DC/DC boost converter (CMBC)

turned on by a clock pulse signal at time equal to nT and turned off when i_L climbs up to the denoted value I_{ref} and remains in off position until the begin of the next cycle as shown in Fig. 2.

Typically, the system state equation can be represented by [5]

$$\begin{bmatrix} \dot{v}_{out} \\ \dot{i}_L \end{bmatrix} = \begin{bmatrix} -1/R_{Load}C & q/C \\ -q/L & 0 \end{bmatrix} \begin{bmatrix} v_{out} \\ i_L \end{bmatrix} + \begin{bmatrix} 0 \\ 1/L \end{bmatrix} v_{in} \quad (1)$$

, where q is a switching function which is one during the switch off interval while zero during on interval.

III. THE PROPOSED CONTROLLER

A. Control of nonlinear dynamics

Let us define $i_L(n)$ and $i_L(n+1)$ as the inductor current at a $t = nT$ and $t=(n+1)T$, respectively. In steady state condition, the inductor current can be derived from [6] as:

$$i_L(n+1) = I_{ref} + \frac{(v_{in} - v_{out})T}{L} t'_n \quad (2)$$

and

$$t'_n = T - T\left[\left(\frac{t_n}{T}\right) \bmod 1\right] \quad (3)$$

, where T is a clock pulse period and $\bmod(.)$ is the modulus function. The slope of inductor current can be expressed by inspecting as:

$$\frac{I_{ref} - i_L(n+1)}{(1-D)T} = \frac{v_{out} - v_{in}}{L} \quad (4)$$

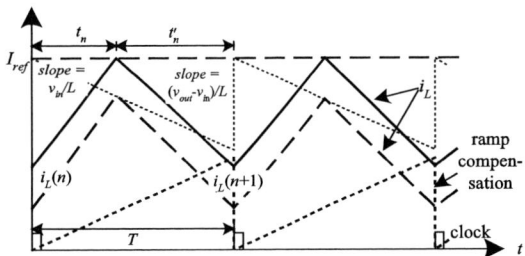

Figure 2. The inductor current i_L of CMBC

and

$$\frac{I_{ref} - i_L(n)}{DT} = \frac{v_{in}}{L} \quad (5)$$

In order to maintain a stable period-1 operation [6], a CMBC must operate with duty ratio D set below 0.5, so that the criterion of no bifurcation can be achieved. For the CMBC, the equivalent criterion of no bifurcation in terms of I_{ref} using the steady-state equation relating R_{Load} and D is given by:

$$i_{ref} < \frac{v_{in}}{R_{Load}}\left[\frac{DR_{Load}T}{2L} + \frac{1}{(1-D)^2}\right]\Bigg|_{D=0.5} \quad (6)$$

To prevent the operation in period-doubling when I_{ref} exceeds the above-stated limit, a variable compensated ramp is used to raise the upper bound of I_{ref} thereby widening the operation range as shown in Fig. 2, with dot and dash lines described in [6].

A variable ramp compensation for a CMBC needs only be controlled according to:

$$\frac{m_c L}{v_{in}} = M_c(v_{in}) \geq \frac{v_{out}}{2v_{in}} - 1 \quad (7)$$

, where m_c is a slope of compensating ramp which is appropriate for system environments.

B. An FLC for Control of Bifurcations

In order to decrease current spectral peaks, the FLC can be established to search an optimal value of m_c from (7). In particular, this optimal value of m_c can be achieved by the fuzzy algorithm based upon the meta-rule : "*If the last change in the Δm_c has caused the magnitude FFT to decline, keep moving the Δm_c in the same direction; otherwise, if it has caused the magnitude FFT to incline move it in the opposite direction.*" This can be translated into the following fuzzy control rule:

Rule (r) : **if** $\Delta FFT(k)$ is A_i **and** $\Delta m_c(k-1)$ is B_j
 then $\Delta m_c(k)$ is C^l (8)

369

, where A_i and B_j, are the fuzzy subsets in their universe of discourse including positive (PO), zero (ZE) and negative (NG), and C^l is the output fuzzy subsets, or fuzzy singleton for Sugeno fuzzy model [9]. It is noted that $\Delta FFT(k)$ and $\Delta m_c(k)$ are the change of magnitude FFT and the change of m_c at time k, respectively. The time instant k is corresponding to the sampling time of controller unit in system and is intentionally chosen to be different from n, which is corresponding to the switching time of the switch S.

It should be noted that the number of fuzzy function is optimally determined with considering the interrelation of control accuracy and calculation capacity. The membership function of each fuzzy set is selected based on trial-and-error such that the region of interest is covered appropriately. To avoid complicated calculation process, the triangular shape membership will be used. The membership functions in each universe of discourse are shown in Fig. 3, and the fuzzy rule base is shown in Table I.

For any input pair of $(\Delta FFT(k),\Delta m_c(k-1))$, the output can be calculated by fuzzy inference to determine the crisp value of $\Delta \bar{m}_c(k)$. When using the implication by production, aggregation with sum and defuzzification with centroid, the relation of $\Delta FFT(k)$, $\Delta m_c(k-1)$ and $\Delta m_c(k)$ is calculated by [9]

$$\Delta \bar{m}_c\left(k\right) = \frac{\sum_{l=1}^{9} \Delta m_c^l w_l}{\sum_{l=1}^{9} w_l} \qquad (9)$$

, where $w_l = \mu_{\Delta FFT\_k}(\Delta FFT(k)) \times \mu_{\Delta m_{c\_k\_1}}(\Delta m_c(k-1))$ is the compatibility (weighting factor) and $\Delta \bar{m}_c^l$ is a value corresponding to the membership function of $\Delta m_c(k)$.

The coefficient slope m_c of (7) for this nonlinear controller can be found as:

$$m_c\left(k\right) = m_c\left(k-1\right) + \Delta \bar{m}_c\left(k\right) \qquad (10)$$

IV. SIMULATION RESULTS

To evaluate performances of the proposed system, computer simulations were conducted using MATLAB/SIMULINK®[11] as shown in Fig. 4. The CMBC model described in the previous section with the parameters given in Table II was invoked.

The bifurcation diagram for the CMBC is used to study system nonlinear behaviours with the bifurcation parameter I_{ref} (horizontal axis) and I_L (vertical axis). Corresponding to the variation of I_{ref} from 0 to 1A, the sampled i_L is observed and recorded. While R_{Load} is equal to 80Ω, the bifurcation diagram of this system is shown in Fig. 5(a). In this figure, it can be seen that the period-1 is stable until I_{ref} is equal to 0.18A and 0.36A while v_{in} is fixed at constant values of 3V and 6V, respectively. When changing R_{Load} to 40Ω, this stable point is shifted to i_{ref} of 0.30A and 0.60A as shown in Fig. 5(b).

The variations of supply voltages and load resistances have been fictitiously set as follows. The supply voltage are denoted constant at 6V and declined to 3V at 0.15s, and the values of load resistance are 80Ω and 40Ω from 0 to 0.4s and 0.4s onwards, respectively. The output voltage converges from the starting point to 8.2V as desired for all cases

The output current and inductor current for the non-compensated CMBC are depicted in Fig. 6(a). Fig. 6(b) shows the frequency spectrum of inductor current. In this case, the system response is in chaos situations. After

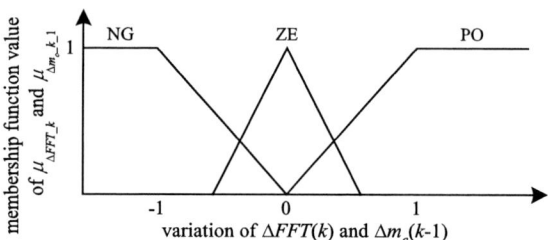

Figure 3. The membership functions of the proposed FLC

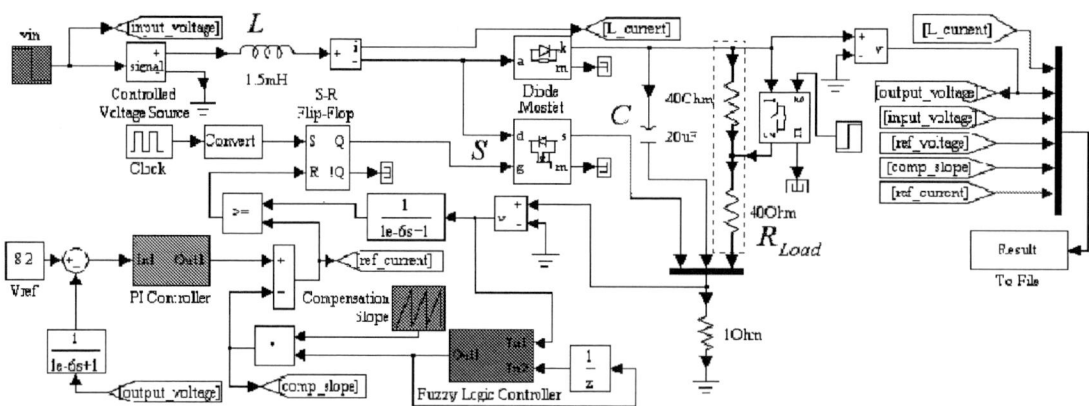

Figure 4. MATLAB/SIMULINK® model for CMBC with control of bifurcation

applying the proposed FLC, the output current and inductor current and frequency spectrum are depicted in Fig 7(a) and Fig 7(b) respectively. These results are confirmed that the proposed FLC can compensate effects of the bifurcation phenomena. Fig 8(a) and Fig 8(b) show the comparison of inductor current when the load resistance is changed for the system with and without FLC. It is found that the FLC can maintain the output at the period-1 in both decreases and increases of load resistance. The current ripples with R_{Load} of 80Ω and 40Ω, while v_{in} is kept constant at 3V, are 0.12 and 0.10A respectively. In addition, it is noticeable that, the output voltage and current response between with and without FLC are clearly different.

TABLE I. RULES OF THE PROPOSED FLC

$\Delta m(k-1)$ $\Delta FFT(k)$	NG	ZE	PO
NG	C^1 (-1)	C^2 (0)	C^3 (1)
ZE	C^4 (0)	C^5 (0)	C^6 (0)
PO	C^7 (1)	C^8 (0.5)	C^9 (-1)

TABLE II. THE CMBC PARAMETERS

Variable	Definitions
Inductor (L)	1.5mH
Capacitor (C)	20μF
Switching Freq ($1/T$).	10kHz
Output voltage (v_{out})	8.2V
Input voltage (v_{in})	3-6V

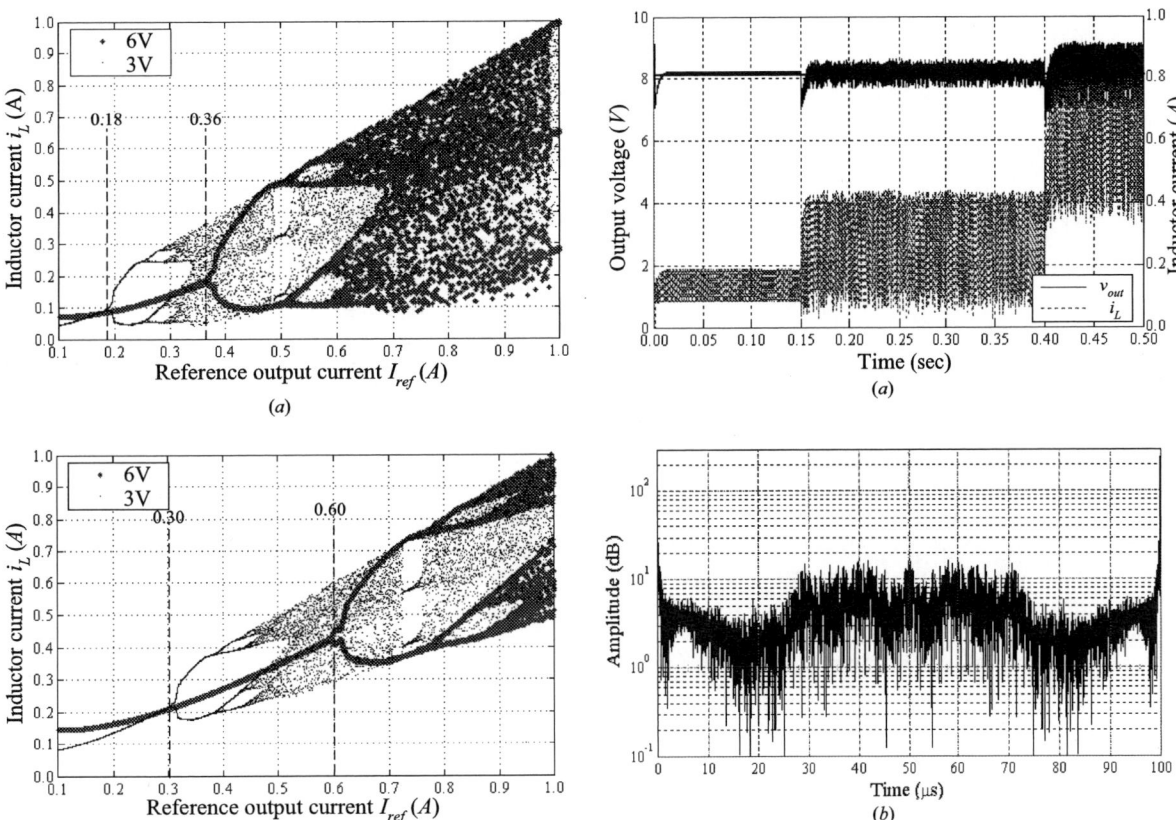

(a)

(b)

Figure 5. Bifurcation diagram in (I_{ref},i_L) plane with comparison of input voltage and load change R_{Load} (a) 80Ω and (b) 40Ω

(a)

(b)

Figure 6. System responses without bifurcation control (a) output voltage v_{out} and inductor current i_L and (b) current spectrum at 3V of v_{in}, R_{Load} of 80Ω

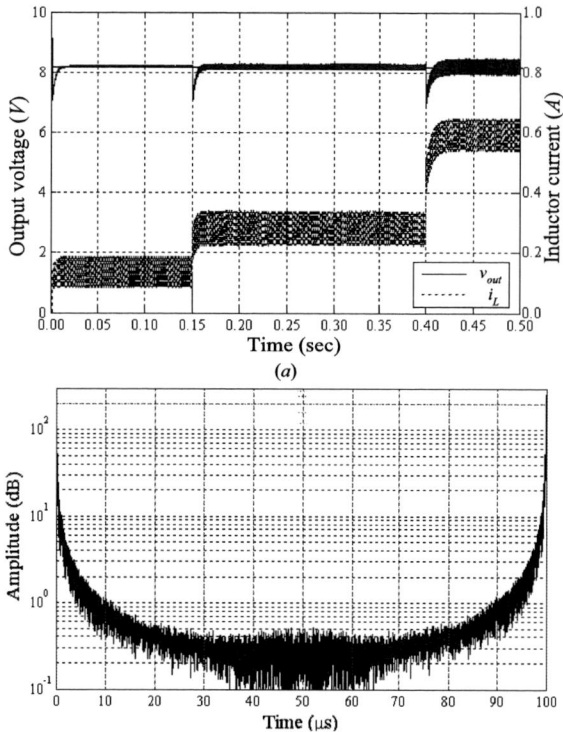

Figure 7. System response with bifurcation control (a) output voltage v_{out} and inductor current i_L and (b) current spectrum at v_{in} of 3V, R_{Load} of 80Ω

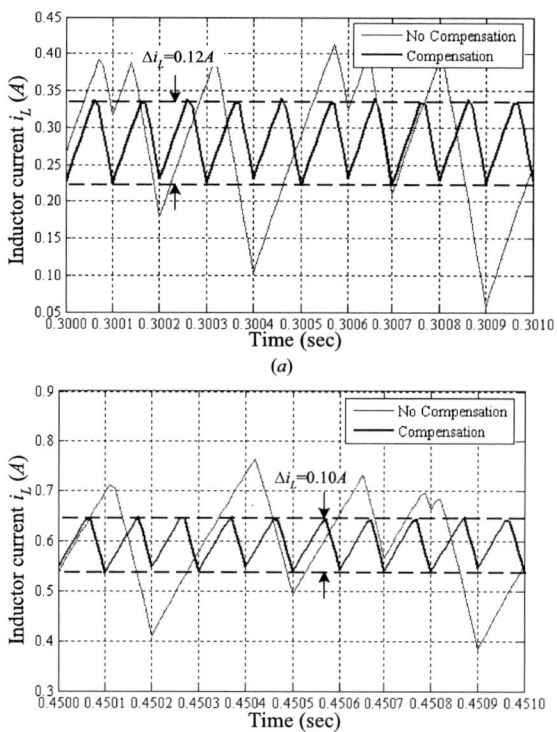

Figure 8. Comparison of inductor current when v_{in} is $3V$ (a) R_{Load} is 80Ω and (b) R_{Load} is 40Ω

V. CONCLUSIONS

In this paper, the advantages of a fuzzy logic controller for a current-mode boost converter with control of bifurcation have been demonstrated for stable system operations. The simulation results reveal that the proposed FLC controller can give the desired stable period-1 under changes in operating conditions. The technique introduced in this paper can be efficiently realized by means of a simple FLC, which can be easily implemented in a real-time control scheme.

REFERENCES

[1] S. Banerijee and G. C. Verghese, *Nonlinear Phenomena in Power Electronics*, IEEE press, New York, 2001.

[2] M. D. Bernardo, F. Garofalo, L. Glielmo and F. Vasca, "Switchings, Bifurcations, and Chaos in DC-DC Converters," *IEEE Tran. on CAS-I*, Vol. 45, No. 2, 1998, pp. 133–141.

[3] C. Morel, M. Bourcerie and F. Chapeau blondeau, "Extension of Chaos Anticontrol Applied to the Improvement of Switch-Mode Power Supply Electromagnetic Compatibility," *Proc. of the IEEE ISIE'04 Conf*, pp. 447–452, 2004.

[4] M. D. Bernardo and F. Vasca, "Discrete-Time Maps for the Analysis of Bifurcations and Chaos in DC-DC Converters," *IEEE Tran. on CAS-I*, Vol. 47, No. 2, 2000, pp. 130–143.

[5] M. Yue and H. Kawakami, "Control of bifurcation in DC/DC PWM switching converters," *Proc. of Int. ICARCV 2004 8th Conf.*, Vol. 2, pp. 1421 – 1426, 2004.

[6] C. K. Tse and Y. M. Lai, "Control Bifurcation in Current-Programmed DC/DC Converters: A reexamination of Slope Compensation," *Proc. of IEEE ISCAS's 2000 Conf.*, pp. I-671-674, 2000.

[7] W. C. Y. Chan and C. K. Tse, "Study of Bifurcations in Current-Programmed DC/DC Boost Converters: From Quasiperiodicty to Period-Doubling," *IEEE Tran. on CAS-I*, Vol. 43, No. 12, 1997, pp. 1129-1142.

[8] N. Khaehintung, P. Sirisuk and W. Kurutach, "A novel ANFIS controller for maximum power point tracking in photovoltaic systems," *Proc. of the IEEE PEDS'2003 Conf.*, Vol. 2, pp. 833-836, 2003.

[9] N. Khaehintung, and P. Sirisuk, "Implementation of Maximum Power Point Tracking Using Fuzzy Logic Controller for Solar-Powered Light-Flasher Applications," *Proc. of MWSCAS 2004 Conf.*, pp. III-171-174, 2004.

[10] N. Khaehintung, A. Kunakorn, M.Aorpimai1 and P. Sirisuk, "An Adaptive Fuzzy Logic Controller by Sliding Mode Control Method for DC/DC Converter," *Proc. of the IEEE PEDS'2005 Conf.*, Vol. 1, pp. 833-836, 2005.

[11] http://www.mathworks.com.

An Improved Three-Level Soft-Switching DC/DC Converter

Z. L. Lou , Z. S. Wang

Zhejiang University

Hangzhou ,310027, P. R. China

zjulzl@yahoo.com.cn

Abstract—This paper proposes an improved three–level zero-voltage zero-current soft-switching (ZVZCS) half-bridge type DC/DC converter. In order to turn off the inner switch in zero-current manner, a tapped inductor is designed to reset the primary circulating current during the freewheeling interval. Consequently, the resetting voltage and the time can be regulated conveniently by setting the turn ratios of the tapped inductor. The principle and operation modes are analyzed. Simulation and experimental results based on a 500W/125KHz prototype are given to verify the validity of the proposed converter.

Keywords-soft switching; three-level; tapped inductor; DC/DC converter

I. INTRODUCTION

The three-level converters have been attracting more and more interesting in high input voltage application cases due to their abilities to overcome the problems of high voltage stress on the switch devices: These converters can reduce the switch voltage stress down to half of the input DC link voltage [1]-[3]. In new generation power supply, many DC/DC converters have to operate in high voltage ranges: It is well known fact that power factor correction (PFC) technique has been widely used in order to meet the requirement of IEC61000-3-2 Class standard. The three-phase single-switch boost rectifier is a quite competitive option for this PFC stage, since it can easily comply with the aforementioned standard with simplicity, efficiency, reliability and low cost. However, in order to reduce the harmonic distortion in the three-phase boost rectifier, its output voltage must be increased with respect to the input voltage. In fact, its output voltage can be higher more than 800~1000V. Automatically, this increase in the output voltage also increases the voltage stress across the devices in the DC/DC step-down second-stage converter. On the hand, the zero-voltage zero-current switching (ZVZCS) technique can be implemented conveniently in three-level converter topology. Because this soft-switching technique can reset the primary current during the freewheeling stage to achieve zero-current switching (ZCS) for the inner switches, so that, the efficiency of the converter can be improved .Francisco C. etc. proposed a modified converter with an active switch and a clamped capacitor

in the secondary side [3]. When the auxiliary active switching device was turned on, the voltage of the clamed capacitor reflecting to the primary side of the transformer and applying to the leakage inductance. Thus the circulating current was forced to reduce until reach to zero, and the inner switches realize the zero current switching (ZCS). This topology has to increase an auxiliary active switching device and should design its corresponding control logical circuits. Reference [4] adopted auxiliary recoverable snubber capacitors to deliver and store energy. The lossless snubber capacitor can not only reset the primary current, but also absorb the high frequency ringing voltage due to the reverse recovery of the secondary rectification diodes. The main disadvantage was that there still was a high voltage on the rectifier diodes [5]. A zero current switching converter with the help of a tapped inductance and a snubber capacitor was presented in [6]. By setting an appropriate turn ratios of the tapped inductor, the ZVZCS condition can be extended to a very wide load ranges. However, this circuit has disadvantages such as the over-voltage on the secondary of the transformer by the resonant inductance of primary and the capacitor in secondary.

This paper proposes a new ZVZCS DC/DC converter through introducing a tapped inductance. The zero-current switching mechanism of the proposed converter is different from the traditional ZVZCS DC/DC converter. Because no recoverable snubber capacitors are used in the circuit, the voltage spike in the second side of the transformer can be avoided. The voltage and the time to reset the circulation current can be regulated conveniently by setting the turn ratios of the tapped inductor. No auxiliary active switching device is needed. The principle of this converter is analyzed and simulation, a 125KHz 500W experimental prototype has been built to verify the validity.

II. THE PROPSED ZVZCS THREE LEVEL DC/DC CONVERTER

The proposed ZVZCS phase-shifted DC/DC converter is shown in Fig. 1. In this circuit, a tapped inductor L_{d1}/L_{d2} is adopted to substitute normal output filter inductor. This tapped inductor functions both a filter and a transformer (to deliver the reset energy). The lossless snubber capacitors C_1 and C_4 are used to perform ZVS operation of the leading switches Q_1 , Q_4. The lagging switches Q_2 , Q_3 operate with ZCS at turn on due to effect of inductor L_K which can be substituted by leakage inductance of the high frequency transformer T_r. Tapped inductor filter L_{d1}/L_{d2} (electro-magnetically coupled inductors or autotransformer) is implemented to complete ZCS commutation for the switches Q_2 , Q_3 as well as to minimize the circulating current to zero during the freewheeling interval.

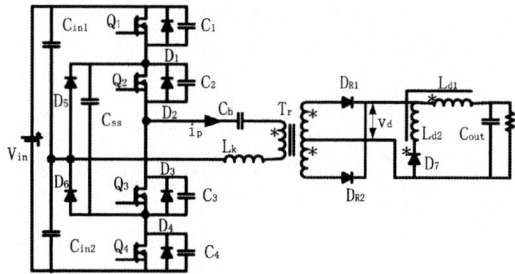

Fig. 1. ZVZCS three-level DC/DC converter

In order to simplify the analysis for the converter, it is assumed that the circuit operation is in steady state .The inductance of the tapped inductor L_{d1}/L_{d2} is large enough, so that its primary current can be treated as a constant level during commutation interval, The flying capacitor C_{ss} is so large that its voltage is constant. All the devices are ideal, The transformer magnetizing current is neglected. Fig.2 shows the operation waveforms. The proposed converter has six stages in each half of the operation period. These stages are described as follows.

Mode 0 ($t = t_0$): The outer switch Q_1 has been conducted before t_0 and the primary current i_p of the transformer flows from DC power supply V_{in} and capacitor C_{in}, via Q_1, Q_2 and L_k to the primary of the transformer. In this stage, the rectifier diodes D_{R1}, D_{R2} are conducting simultaneously, both primary and the secondary side voltage of the transformer is zero, and the input voltage ($V_{in}/2$) is applied to the inductor L_k. The primary current i_p increases linearly from zero to $(n_s/n_p)i_{Ld1}$, where n_p, n_s are the turns of the primary and secondary of the transformer respectively, and i_{Ld1} is the current of the tapped inductor L_{d1}.

$$i_p = \frac{1}{L_k} \frac{V_{in}}{2}(t_1 - t_0)$$ (1)

Mode 1 ($t_0 - t_1$): At time t_0, Q_1 is turned off under ZVS with the aid of lossless snubber capacitors C_1 . The rectified voltage v_d decreases as

$$v_d(t) = \frac{\frac{1}{2}v_{in} - v_{c_1}(t)}{\alpha_T}$$ (2)

where $\boldsymbol{a_T}$ is the turns ratio of the transformer. Freewheeling diode D_7 starts to conduct when v_d reaches value $a_L E_0$, where $\boldsymbol{a_L}$ is the turns ratio of the tapped inductor and is defined as $\boldsymbol{a_L = n_2/(n_1 + n_2)}$: n_1, n_2 are the number of turns of L_{d1} and L_{d2} respectively .

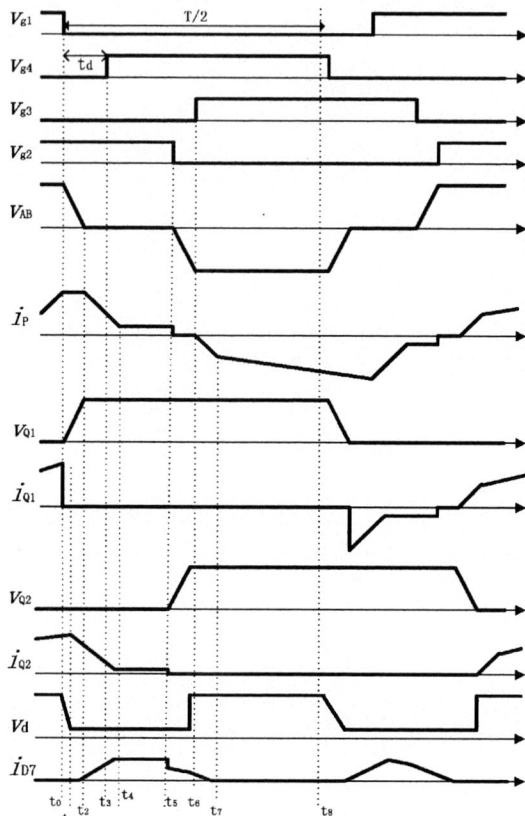

Fig. 2　Principal waveforms of the proposed converter

Mode 2 ($t_1 - t_2$): The output current starts flow through D_7, L_{d2} and L_{d1}. This interval ends when the voltage across C_1 rises to $(1/2)V_{in}$. At time $t = t_2$, $V_{C1}=(1/2)V_{in}$, the voltage of C_4 decreases to zero, and the clamped diode D_5 conducts spontaneously. This time $v_{AB} = 0$.

Mode 3 ($t_2 - t_4$): The diode D_5 starts to conduct and the voltage of C_4 is clamped to zero. The switch Q_4 is turned on under ZVS. The interval ($t_{Vg4} - t_{Vg1}$)>$t_2 - t_0$. In this Mode ,diode D_5 and the switch Q_2 are conducted, $v_{ab}=0$.The voltage of c_b is applied to the leakage inductance L_K, and the the primary current i_p of the transformer decreases. At the same time, the current i_{DR1} of the rectifier diode minimizes continuously. When the currents i_p and i_{DR1} reduce to zero ,This mode is ended during the time of $t = t_3$, the switch Q_4 is turned on under ZVS.

Because the voltage of C_b is large enough, so that the voltage v_{cb} can be treated as a constant level during commutation interval, and the current i_p decreases linearly, the current i_{DR1} decreases.

Mode 4 ($t_4 - t_5$): At time t_4, Whole output current flows through D_7, L_{d2} and L_{d1}. The output current is reflecting to the transformer primary side and the primary current i_p becomes zero. Ignoring of magnetizing current of the transformer T_r, the current of the transformer primary side i_p is keeping at zero level. At $t = t_5$, the switch Q_2 is turned off in ZCS manner. At this interval, the voltage V_B maintains $(1/2)V_{in}$, $V_A = (1/2)V_{in} + V_{cb}$.

Mode 5 ($t_5 - t_6$): At $t = t_5$, the switch Q_2 is turned off in ZCS manner. After a short time, the switch Q_2 is turned on. With the influence of the leakage, the increasing slope of the current i_p is restrained, as a consequence, the switch Q_3 is turned on under ZCS. The primary current of the transformer i_p is so small that it can't afford the current of the load. The leakage L_K on primary of the transformer supports the reversal voltage of $(1/2)V_{in} + V_{cb}$, as a results of i_p increases in reverse. On the other hand, when the current on the tapped inductor L_{d2} decreases to zero, the switch D_7 is turned off. Whole output current flows through D_7, L_{d2} and L_{d1}, This interval is defined by turn-off operation time of the switching power device Q_3.

Mode 6 ($t_6 - t_7$): The secondary current is delivering through D_6、L_{d1}, when i_p increases to the max value, this interval ends. The half cycle of operation ends at time t_7. The operation during the next half-cycle is symmetrical with half-cycle operation mentioned above. As described before Q_1 and Q_4 are turned on and turned off with ZVS, while Q_2 and Q_3 operate with ZCS at turn-on and turn-off. The circulating current is substantially lowered with no additional auxiliary circuits.

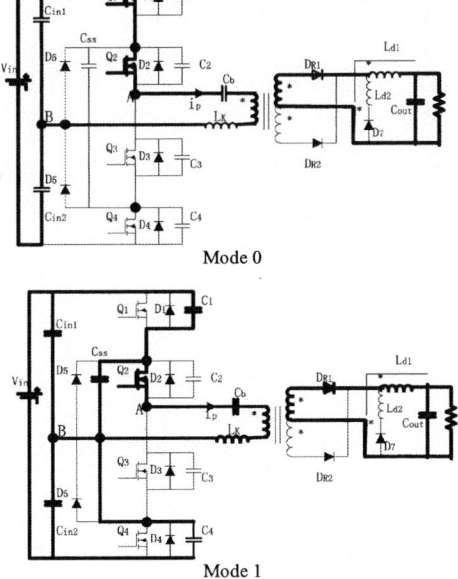

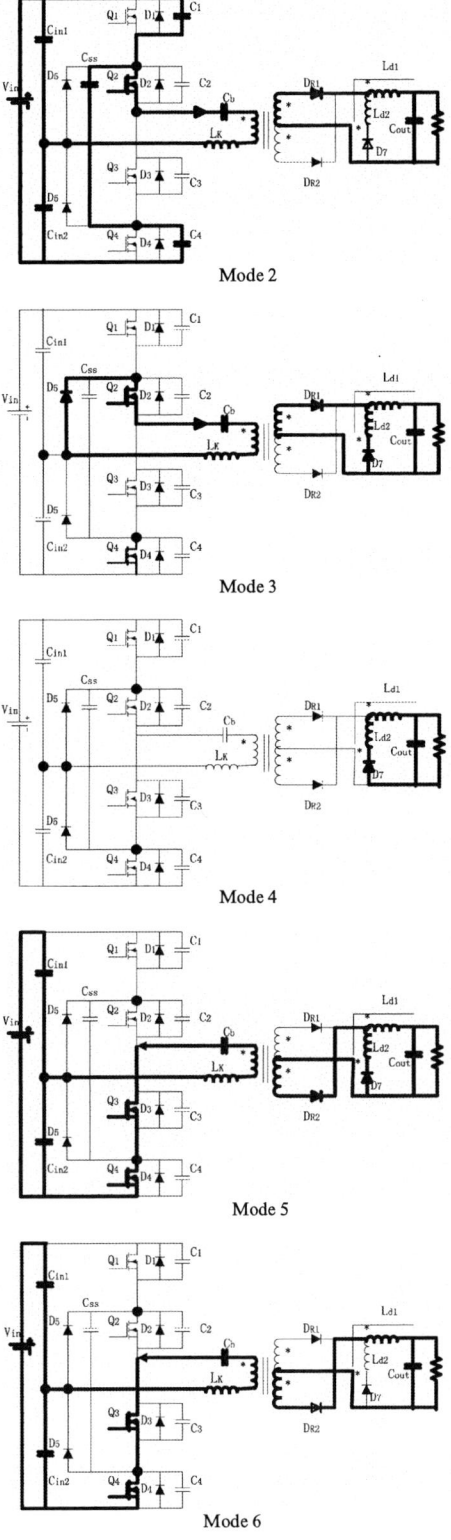

Fig. 3. Circuit configuration for operation mode in converter

III SIMULATION RESULTS

Circuit simulation parameters are as follows:

V_{in}=600V (input DC) , V_O=40V (output DC) , I_O=12.5A (output DC).C_{SS}=1 μ F (flying capacitor),

C_b=2 μ F (blocking capacitor), L_K=2 μ H(leakage inductance),n_{Ld1}/n_{Ld2}=100/2(tapped inductance).

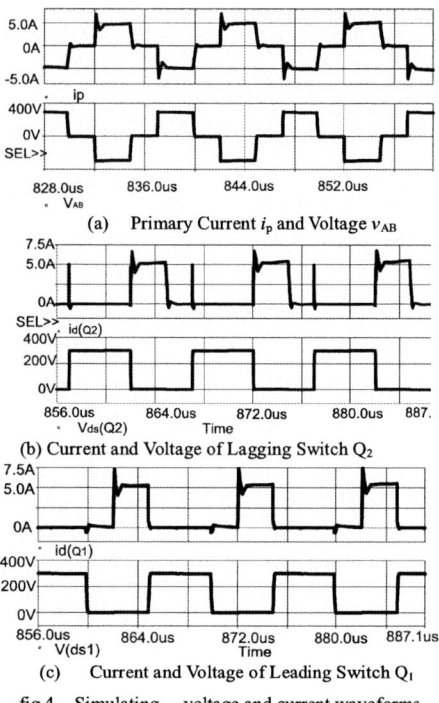

(a) Primary Current i_p and Voltage v_{AB}

(b) Current and Voltage of Lagging Switch Q_2

(c) Current and Voltage of Leading Switch Q_1

fig.4 Simulating voltage and current waveforms

Fig.4 shows the simulation wave forms of current and voltage . Fig. 4 (a) is the primary current i_p and voltage V_{AB}. It can be seen that ip decrease to zero swiftly when V_{AB} is zero. Fig. 4 (b) shows the current i_d and voltage V_{ds} of the inner switch Q_2. Obviously, the current i_d falls to zero prior to the voltage rising. Therefore a true ZCS action is achieved. Fig. 4 (c) shows the current i_d and voltage V_{ds} of outer switch Q_1, evidently the voltage V_{ds} drops to zero before current rises. Consequently, a true ZVS action is realized.

IV EXPERIMENT RESULTS

An experimental prototype was built to verify the validity of the proposed ZVZCS three-level DC/DC converter with the following specifications and parameters:

Input voltage V_{in} =530V, Output voltage/current: 40V/10A.

Switching frequency: 125kHz

$V_{Q1} - V_{Q4}$: (MOSFET) IRF840

$D_{R1} - D_{R2}$: MUR3060PT, $D_5 - D_{6}$: U860, D_7: IRO36X

$C_{in1} - C_{in2}$: 0.22 μ F/450V,Css : 0.33 μ F/250V,

Cb: 0.47 μ F/280V

a_T (n_P/n_s) : 18/5 = 3.6 (transformer turns ratio)

Lk: 11 μ H (leakage inductance) .

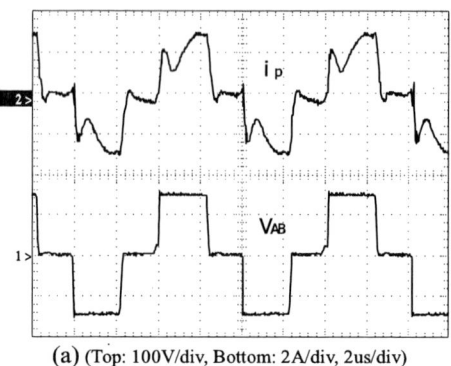

(a) (Top: 100V/div, Bottom: 2A/div, 2us/div)

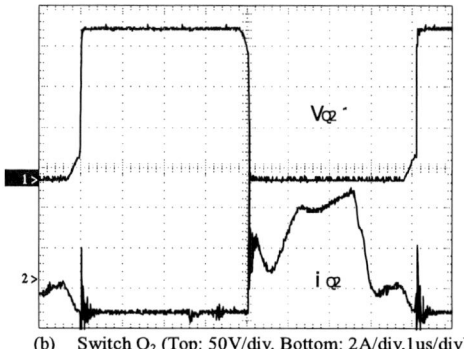

(b) Switch Q_2 (Top: 50V/div, Bottom: 2A/div,1us/div)

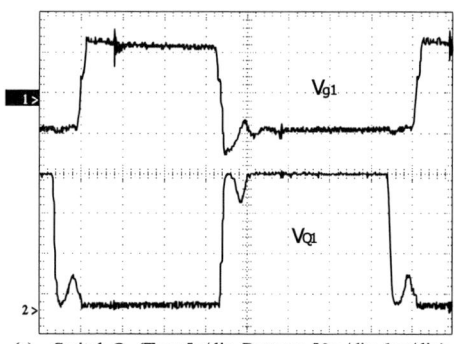

(c) Switch Q_1 (Top: 5v/div, Bottom: 50v /div, 1us/div)

Fig. 5 Experimental voltage and current waveforms

Experimental waveforms of the proposed converter are shown in Fig 5. Fig. 5 (a) is the primary voltage V_{AB} and current i_p. It is very similar to the one simulation that ip decrease to zero swiftly when V_{AB} is zero. Fig 5 (b) shows the voltage V_{Q2} and current i_{Q2} of the inner switch Q_2. It can be seen that the switch Q_2 is turned off with ZCS. Fig. 5 (c) shows the pulse signal V_{g1} and voltage V_{Q1} of outer switch Q_1, Obviously the voltage V_{Q1} drops to zero before pulse signal comes. Consequently, it is a true ZVS as described before.

IV. CONCLUSIONS

Zero-current switching can be achieved in three-level DC/DC converter through introducing a tapped inductor in the secondary side of transformer, and the circulating current can be reset to zero during the freewheeling stage. By regulating the turn ratio of the tapped inductor, the voltage, which is used to reset the primary current, can be adjusted easily. This converter realizes zero-voltage-switching for the leading switches and zero-current-switching for the lagging switches. The analysis of operation principle of the converter is presented in the paper, and experimental results based on a 125KHz 500W prototype are given to verify the validity.

REFERENCES

[1] Pinheiro J. R., Barbi I., "The three-level ZVS PWM converter, a new concept in high voltage-voltage DC-to-DC converter," IEEE Conference IECON'92, Nov. 1992, pp. 173-178.

[2] Pinheiro J. R. , Barbi I., "Wide load range rthree-level ZVS-PWN DC-to-DC converter," IEEE Annual Conference PESC'93 Record, 20-24 Jun 1993, pp.171-177.

[3] Francisco C., Peter M. Barbosa, Burdo J. S. etc., "A Zero Voltage Switching Three-level DC/DC Converter," *CPES* 2000, pp.366-371.

[4] E. S. Kim, K. Y. Joe, M. H. Kye, Y. H. Kim, etc., "An improved soft-switching PWM FB DC/DC converter for reducing conduction losses," *IEEE Transactions on Power Electronics*, Vol. 14, Issue, 2, March 1999 pp. 258 – 264.

[5] E. S. Kim; Y. B. Byun; T. G. Koo; etc., "An improved three level ZVZCS DC/DC converter using a tapped inductor and a snubber capacitor," *IEEE Power Conversion Conference,* Osaka 2002 Proceedings, Vol. 1, pp. 115 – 121.

[6] F. Canales, P. Barbosa, and F. C. Lee, "A zero-voltage and zero-current switching three-level DC/DC converter," *IEEE Transactions on Power Electronics*, Vol. 17, Nov. 2002, pp. 898 – 904.

A Novel Soft Switching Bidirectional DC/DC Converter and Design Consideration

Ma Gang   Qu Wenlong   Liu Yuanyuan

State Key Lab of Power System, Division of Power Electronics,
Department of Electrical Engineering, Tsinghua University, Beijing, 100084, China

magang03@mails.tsinghua.edu.cn   qwl@mail.tsinghua.edu.cn   liu-yy@mails.tsinghua.edu.cn

Abstract—This paper proposes a new soft switching bi-directional DC-DC converter. The described converter employs the symmetric topology with high frequency transformer. It gets soft switching with the designed pulse-width modulation control, which guarantees soft switching in wide load range without any additional circuit components. Operation principle and the ZVS condition of the proposed converter are also analyzed .Inductors L_1 and L_2 are proven to have effect on the ZVS condition. Therefore, the parameter design is taken into account to ensure the soft switching. Furthermore, simplified method to calculate the inductor's value is also introduced. Besides, the theoretical calculation, analysis and experimental results obtained from a 500W prototype are also presented, which substantiates the design consideration.

Key words-Bidirectional DC/DC converter, zero voltage switching (ZVS), design consideration

I. INTRODUCTION

The bi-directional DC-DC converter is a promising candidate for high frequency power conversion especially in the fuel cell application system and electric vehicle. High frequency endows the converter attractive features including high power density, high efficiency and small filter components. One of the most desirable characteristics of the converter applied above is the bi-directional power flow with soft switching which can lessen the device stress and switching losses [1]-[9].

For its special purpose of bi-directional power flow, bi-directional DC/DC converter requires more switches, more sophisticated control strategy and results in more complicated switching period [3]-[6]. It is more difficult for bi-directional DC/DC converter to achieve the soft switching [4].

The traditional soft switching bi-directional DC/DC converter has two defects: the one is the soft switching range being affected by load range. Accessory circuit components including inductance and capacitance should be added in the circuit to enforce the resonance effect in the light load. This measure will extend the soft switching range at the expense of considerable power losses, device stresses as well as the increasing of volume and cost.

The other is that the effect of rectifying diode reverse recovery can't be neglected. It leads to the power loss of

the diode especially in the high power application. As a result, it restricts the switching speed of the IGBT during the turn-on transient period. What's more, it could endanger the safe operation for the switches.

This paper presents a novel bi-directional DC/DC converter with soft switching characteristic as well as its operation principle. Then, theoretical analysis and the soft-switching character of the proposed converter are described in brief. And the effect of different parameters of the circuit components on the ZVS condition is discussed in detail. Finally, the experimental results are presented to verify the theoretical analysis.

II. OPERATION PRINCIPLE OF THE CONVERTER

The schematic of the proposed converter is shown as Fig.1.

Some assumptions are made as follows to simplify the analysis of the proposed converter:

1) The circuit operates in steady state.
2) All the switches are considered as ideal devices.
3) All energy storage components are free of loss.

The proposed converter can achieve bi-directional power flow because of the symmetric topology. Dependent on the energy flow direction, the proposed converter has two modes.

1. Forward mode: In this mode, power flow is transferred from input V_1 to output V_2 by conducting switches S_1 and S_2 complementarily. The gate drive signal of S_1 is leading to that of S_3. The duty cycle of S_2 modulates the output voltage V_2 of the converter.

2. Reverse mode: Energy transferred from input V_2 to output V_1, the gate drive signal of S_3 is leading to that of

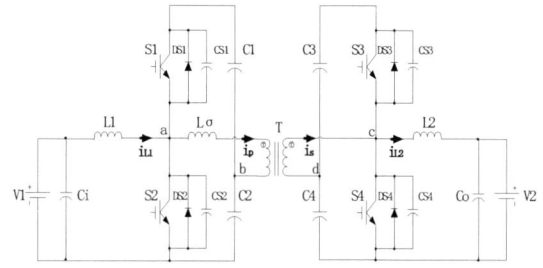

Figure 1. Bi-directional DC/DC converter

S_1 and the output voltage V_1 is mainly regulated by the duty cycle of switch S_4.

In the forward mode, the left part of the converter could be regarded as the combination of a Boost DC/DC converter and a Half-bridge DC/DC converter. By conducting S_1 and S_2 complementally, the left part of the converter not only boosts the input voltage of power supply but also inverts dc voltage to ac voltage with the help of the high frequency transformer. Compared to the left part of the converter, the principle of the right part is similar to the combination of a Half-bridge DC/DC converter and a Buck DC/DC converter. The ac voltage of the secondary windings is rectified to dc voltage by the freewheel diodes paralleled in S_3 and S_4. Owing to the big capacity of C_3 and C_4, the voltages of the capacitors can be supposed to be a constant, so the output voltage is mainly regulated by the duty cycle of S_2.

The soft switching principle of the proposed converter is described briefly as follows:

With the help of the paralleled and junction capacitors, the voltage over the conducting switches can not be changed suddenly and the switches turn off in ZVS.

The resonance, which occurs by the leakage inductance, the paralleled and junction capacitors in the short interval when the upper and lower switches are all off, provides the ZVS condition for turning on. When the voltage across the switch decreases to less than zero, the anti-parallel diode begins to conduct. Then the gate drive

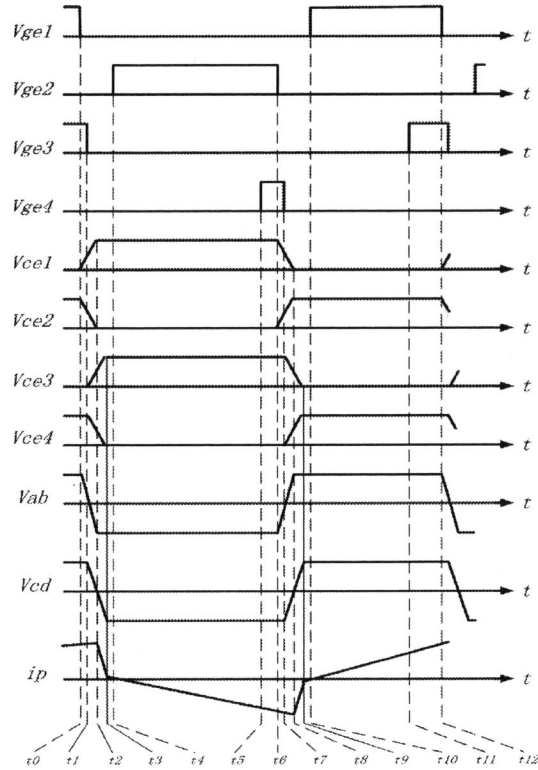

Figure 2. Theoretical voltage and current waveforms

signal is added, and the switch turns on in ZVS.

Similar to the above analysis, the switches of S_3 and S_4 also get their ZVS conditions by conducting of anti-parallel diode. Besides, conducting of S_3 and S_4 can also diminish the diode reverse recovery effect of D_{s3} and D_{s4}. For example: after S_3 turns on in the ZVS condition of D_{s3}'s conducting, the current flowing through D_{s3} decreases gradually and transfers into S_3. When it decreases to zero, D_{s3} will turn off in ZCS. Therefore, the diode reverse recovery effect will be weakened.

Due to the symmetric topology, the converter could also get soft switching in the reverse mode.

Fig. 2 shows the key waveforms of the proposed converter. Vge is the gate drive signal and Vce is the voltage across collector and emitter of the switches. Vab、Vcd are the voltages on primary and secondary windings of the transformer, and i_p is the primary current.

III. DESIGN CONSIDERATION

A. The ZVS condition of different switches

The ZVS condition is an important part of the steady state analysis, and it is different for S_1 and S_2 to get the ZVS operation.

(1) The ZVS condition of S_1

S_1 can achieve ZVS on the following conditions:

1. When S_2 is turned off, the transformer primary current i_p must be lower than the inductance current i_{L1}.

2. Before i_p ascend to i_{L1} (at this time D_{s1} is turned off with ZCS), S_1 must be triggered.

The first condition can be met easily. When S_2 is turned off, i_p will decrease to zero .Then i_p changes its direction and increases. In the interval when i_p is lower than i_{L1} (depicted in the direction as shown in Fig.1), D_{s1} is in conduct and the gate signal of S_1 should be added before i_p reaches to i_{L1}. Therefore, the D_{s1}'s freewheeling time needs to be longer than the dead time between the gate signals of S_2 and S_1.

(2) The ZVS condition of S_2

Similarly, S_2 can achieve ZVS on the following conditions:

1. When S_1 is turned off, the transformer primary current i_p should be lager than the inductance current i_{L1}.

2. S_2 must be triggered before i_p decreases to i_{L1}. If the triggering signal of S_2 is added after the current in D_{s2} decreases to zero, the voltage over S_2 will increase and S_2 will lose the ZVS condition.

B. The value of inductor L_1

The inductors play an important part in implementation of the soft switching: if inductance is too small, the current ripple of the inductor would be so large that the inductor current would not be continuous. But too large inductance of inductor L_1 will affect on the ZVS condition of S_3 in the reserve mode. The analysis could be referred to the analysis of L_2, and L_1 can be calculated by the similar way.

379

C. The value of inductor L_2

L_2 is also an important part of the energy transfer. In the previous analysis, L_2 is thought to be large enough to ensure the power supply as a current source. In fact, due to the limitation of volume, weight and power loss, the inductance cannot be too large, and the ripple of the output current cannot be neglected. What's more, too big inductance L_2 can do affect on the period of D_{S1}'s conducting especially in the heavy load. Heavy load also causes S_4 not to conduct as previously mentioned. Thus, the too big inductance L_2 will result in losing the ZVS condition of S_1 as the following reasons:

Due to the existence of leakage inductance, when S_2 is turned off, the primary winding current will keep its former direction. The sum of i_{L1} and i_p flows through D_{S1} and charges C_1. During this period, the secondary winding current flows through D_{S4} in the former direction and i_{L2} freewheels through D_{S4} too.

The voltage across the leakage inductance is the sum of V_{C1} and V_{C4}/n. V_{C4}/n is the referred voltage of V_{C4} that is converted to transformer primary winding and n is the turn ratio of transformer. Because the voltage across L_σ is large, the primary winding current changes its direction rapidly and increases in the opposite direction. As soon as i_p is equal to i_{L1}, the current flowing through D_{S1} will decrease to zero and D_{S1} will cease conducting. If the drive gate signal is still not sent at this moment, the voltage across S_1 will increase and S_1 will lose the ZVS condition. To ensure the ZVS condition of S_1, the interval between the gate drive signals of S_2 and S_1 should be lessened. Hence, the gate drive signal of S_1 is required to emerge before the primary winding current increases to i_{L1}.

The ZVS Condition of S_1 is easier to fulfill by lessening the interval of drive signals of S_1 and S_2. But the too small interval will endanger the safe operation of IGBTs. Furthermore, as D_{S4} is turned off under a certain reverse voltage, the reverse recovery effect of D_{S4} cannot be ignored.

On the basis of above analysis, the suitable value of L_2 is reasonable to guarantee i_{L2} decreasing to zero after S_2 is turned off. When i_{L2} drops to zero, D_{S4} will cease conducting. Both the primary and secondary windings voltage of transformer will change the polarity gradually and the voltage over the leakage inductance will decrease, changing from $V_{C1}+V_{C4}/n$ to $V_{C1}-V_{C3}/n$. Therefore, the changing rate of primary winding current will be reduced and the interval of D_{S1}'s conducting will be prolonged. Thus, the wider ZVS condition for S_1 is created.

The value of L_2 can be calculated as follows:

$$I_{L2\min}=I_{av}-\frac{1}{2}\Delta i_{L2}=\frac{P_2}{V_2}-\frac{P_2}{V_2}\cdot\frac{(1-D)T_sR}{2L_2}=\frac{P_2}{V_2}(1-\frac{1-D}{2\tau_{L2}}) \quad (1)$$

$$\tau_{L2}=\frac{L_2}{RT_s} \quad (2)$$

$$I_{L2\min}\leq 0 \quad (3)$$

By substituting (1) and (2) into (3), the required value of $I_{L2\min}$ is as:

$$L_2\leq\frac{R\cdot T_S\cdot(1-D)}{2} \quad (4)$$

In the above formulas, P_2 is the output power, and I_{av} is the average output current. D is the duty cycle of the switch while T_s is the switching period. $I_{l2\min}$ is the minimum of inductor current i_{l2}. The analysis is based on the condition of resistance load and R is the load resistance. As for other types of load, the requirement of the inductor can be expressed as:

$$L_2\leq\frac{V_2^2\cdot T_s\cdot(1-D)}{2\cdot P_2} \quad (5)$$

Taking the specification of the prototype for example:

Input voltage: 24 V;	Output voltage: 50 V;
Output power: 500 W;	Duty cycle: 50%;

Switching frequency: 20 kHz

The inductor L_2 can be calculated as:

$$L_2\leq\frac{V_2^2\cdot T_s\cdot(1-D)}{2\cdot P_2}=\frac{50^2\cdot 50\cdot 0.5}{2\cdot 300}=104\mu H \quad (6)$$

Adopted the above design procedure, the proposed converter can achieve the ZVS condition in the whole load range. Besides, D_{S4} is softly switched off due to the current decreasing to zero and avoids reverse recovery.

D. Diode reverse recovery

When the conductive diode is turned off by a high reverse voltage, the diode reverse recovery current will bring some problems especially in the high voltage, high power application. To some extents, it will endanger the safe operation of diode and IGBT. The proposed converter solves this problem by soft switching, for the integrated diodes turn off under the zero current. Therefore the diode reverse recovery effect is reduced greatly.

IV. EXPERIMENTAL RESULTS

A prototype of soft switching bi-directional DC/DC converter has been built to verify the analysis. Specifications of the experimental prototype are as follows:

Input voltage: 48Vdc;	Output voltage: 120Vdc;
Rated power: 500W;	Leakage inductance: 5uH;
Inductor L_1:138uH;	InductorL_2:98uH.

Switching frequency: 20 kHz;

Transformer turn ratio: 2:5;

Fig. 3 shows experimental waveforms of the proposed converter in 300W output condition. In Fig. 3, (a) shows gate drive signal and V_{ce} of S_1, and (b) shows the gate drive signal and V_{ce} of S_2.

Fig. 4 illustrates the waveforms in 15W output power. Being the magnification of Fig.4 (a), Fig.4 (b) shows S_2 is turned off in ZVS clearly. Fig. 3 and Fig. 4 manifest that

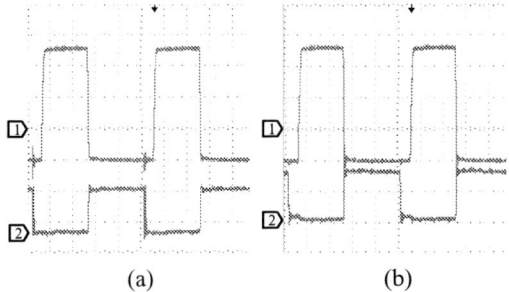

(a)   (b)

Figure. 3 Experimental waveforms when output power is 300W
and L_2 is 98uH.

(a)V_{ge1}: [Ch1: 5V/div], V_{ce1}: [Ch2:20V/div].

(b)V_{ge2}: [Ch1: 5V/div], V_{ce2}: [Ch2:20V/div]

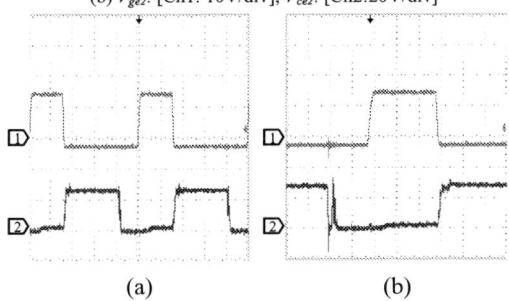

(a)   (b)

Figure. 4 Experimental waveforms when output power is 15W
and L_2 is 98uH

(a) V_{ge2}: [Ch1: 10V/div], V_{ce2}: [Ch2:20V/div],

(b) V_{ge2}: [Ch1: 10V/div], V_{ce2}: [Ch2:20V/div]

(a)   (b)

Figure. 5 Experimental waveforms when output power is 15W

(a)L_2=98uH, V_{ge1}: [Ch1: 10V/div], V_{ce1}: [Ch2:20V/div],

(b)L_2=238uH, V_{ge1}: [Ch1: 10V/div], V_{ce1}: [Ch2: 20V/div]

S_1 and S_2 can achieve ZVS in different output power with the proper value of inductor. Fig.5 (a) and (b) show the waveforms when L_2 is 98 *uH* and 238 *uH* respectively and demonstrate that the different value of inductance L_2 will affect on the interval of D_{s1}'s conducting. Fig 5 (b) shows that when L_2 is bigger than the calculated value, D_{s1} would turn off prior to the gate drive signal of S_1

becoming positive. Thus, the voltage across S_1 will increase and it will endanger the ZVS condition of S_1.

V. CONLUSION

A novel bi-directional DC/DC converter is presented. Adopting the proposed design consideration, the converter can operate with soft switching for all the power switches including power diode and lessens various losses as well as stresses. The experiment results also demonstrate that:

1. In terms of the proposed design consideration, all the switches of the proposed converter can achieve soft switching in a wide load range without any additional circuit components.

2. The new method of inductor value's calculation is convenient and effective.

VI. REFERENCES

[1] Jose,P; Mohan,N, "A novel ZVS bidirectional Cuk converter for dual voltage systems in automobiles",*Industrial Electronics Society,2003. IECON'03. The 29th Annual Conference of the IEEE*, Volume: 1, 2-6 Nov. 2003, pp: 117 - 122, vol.1.

[2]Yujin Song, Prasad N. Enjeti "A New Soft Switching Technique for Bi-directional Power Flow, Full-Bridge DC/DC Converter", *IEEE Trans. On Power Electronics, 2002*, pp: 2314-2319.

[3] G. Hua, F.C. Lee and M.M. Jovanovic,"An improved full-bridge zerovoltage-switched PWM converter using a saturable inductor", *IEEE Trans. On Power Electronics*, Vol. 8, Oct. 1993, pp530-534.

[4] F. Carichhi, F. Crescimbini, F. G. Capponi, and L. Solero, "Study of bidirectional buck-boost converter topologies for application in electrical vehicle motor drives," in *Proc. IEEE Applied Power Electronic specialist Conf. Expo*, 1998, pp. 287–293.

[5] E.S.Kim, K.Y.Joe, M.H.Kye and B.D. Yoon,"An improved soft-switching PWM FB DC/DC converter for reducing conduction losses", *IEEE Trans. On Power Electronics*, Vol.14, No.2, March 1999.

[6] J. Zeng, J. Ying, and Q. Zhang, "A novel DC/DC ZVS converter for battery input applications," in Proc. *IEEE APEC*, vol. 2, 2002, pp. 892–896.

[7] Z. R. Martinez and B. Ray, "Bidirectional dc/dc power conversion using constant frequency multi-resonant topology," in *Proc. APEC'94*, 1994, pp. 991–997.

[8] H.L. Chan, K.W.E. Cheng, and D.Sutanto,"A novel square-wave converter with bi-directional power flow", Power Electronics and Drive Systems, 1999PEDS'99. Proceedings of IEEE 1999 International Conference, Vol.2, pp966-971.

[9] H.L. Chan, K.W.E. Cheng, and D.Sutanto,"An extended load range ZCS_ZVS bi-directional phase-shifted DC-DC converter", Power Electronics and Variable Speed Drives, 2000. 8th International Conference, Vol.2, pp74-79.

2006 5th International Power Electronics and Motion Control Conference

State-Variable Description and Analysis of a DC-Rail ZVT Inverter Feeding a Permanent Magnet Synchronous Motor

MING Zhengfeng*, ZHONG yanru**
*Xidian University, Xi'an, China, 710071
**Xi'an University of Technology, Xi'an, China, 710048
E-mail: mzfxut@163.net

Abstract—**This paper proposes a state-variable description and analysis of a DC-Rail parallel resonant zero voltage transition (ZVT) voltage-source three-phase inverter feeding a permanent synchronous motor drive. In this DC-rail ZVT inverter, consists of two major parts: a ZVT resonant converter and an three-phase inverter. The DC-rail inverter is operated at a higher frequency and has higher efficiency than traditional inverters. A three-phase 1.1kW 20kHz DC-rail ZVT inverter has been constructed and successfully tested to feed a permanent magnet synchronous motor drive based on a field-oriented control. State Variable Description and Analysis, simulation and implementation of the whole system are described. The experimental results not only validate the theoretical development but also prove that this new system has better performance than traditional systems.**

Keywords—DC-rail ZVT, Inverter, Motor drive

I. INTRODUCTION

One of the most severe constraints on the performance of medium to high power transistorized pulse width modulated (PWM) inverters in uninterruptible power supply and motor drive application is the device switching loss which limits switching frequencies to, at most, a few kilohertz. Higher switching frequencies are the key to lowering harmonic distortion in the output waveforms, reducing losses in the load, and reducing acoustic noise in motor drive system.

This paper describes the state-variable description and analysis of a DC-rail ZVT three-phase inverter. The topic herein which differentiates it from previous paper is that it derives a state-variable description of the DC-rail ZVT converter. Based on the state-variable description, a systematic design of the resonant ZVT inverter is obtained. As a result, the steady-state characteristics of this inverter are derived, the parameters of the resonant elements and the control method of the resonant converter are determined, and a computer simulation program can be easily written. The simulated results are very close to the experimental waveforms. This shows that the state-variable description can precisely describe the behavior of the resonant ZVT inverter. To the best of author's knowledge, this

state-variable description of the DC-rail ZVT inverter has not yet been presented in other paper.

II. CIRCUIT TOPOLOGY AND STATE-VARIBLE DESCRIPTION

The DC-Rail ZVT three-phase voltage-source inverter is depicted in Fig. 1. The circuit consists of two major parts: a DC-rail ZVT converter and an three-phase inverter. The DC-rail ZVT converter is composed of a resonant circuit, and a voltage source clamp circuit. A capacitor Cr and an inductor Lr, the two major components of the resonant circuit, determine the resonant frequency of the circuit.

A switch S_L is located at upper DC-rail to allow the DC bus voltage to resonant; S_L is turned on and turned off under ZVS. A resonant inductor L_r is connected with two auxiliary switches S_a and S_b in series and connected between the top DC-rail and the center point of capacitors dividing DC-rail. S_a and S_b is turned on and off under ZCS .

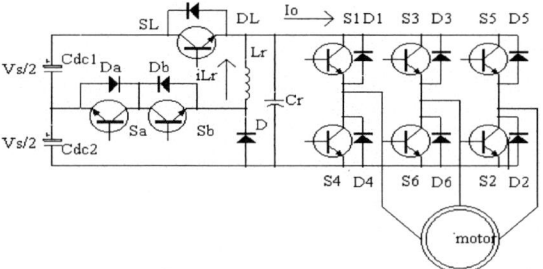

Figure1. Circuit topology of DC-rail ZVT three-phase inverter

A. *State-Variable Description of a DC-Rail ZVT Converter*

To simplify the illustrations of operation, the following assumptions are made:

1) All of the devices and components are ideal.

2) The load inductor Lo is much larger than the resonant inductor Lr so that the load current can be treated as a constant current source during a switching period.

Fig.2 shows the equivalent circuit of the DC-rail ZVT voltage-source three-phase inverter.

1-4244-0448-7/06/$25.00 ©2006 IEEE   382

Fig.3 shows the operation waveforms of the DC-rail ZVT inverter.

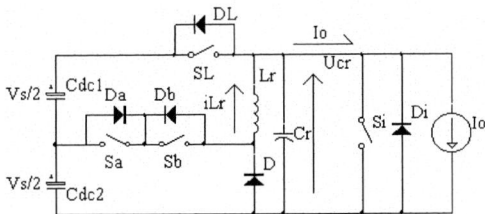

Figure2.   The equivalent Circuit

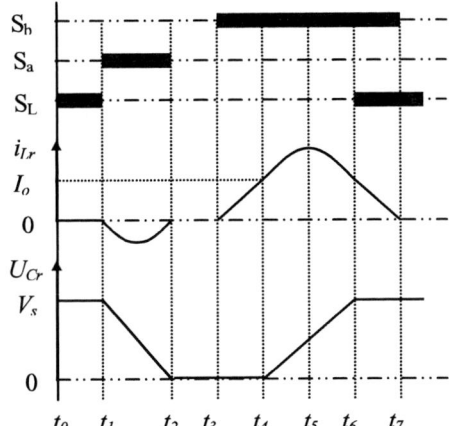

Figure4. The operation waveforms of switches' ZVT

From Fig.2, the dynamic equation of the circuit can be described by differential equations as follows:

$$
\begin{bmatrix} di_{Lr}/dt \\ dU_{Cr}/dt \end{bmatrix} = \begin{bmatrix} 0 & -1/L_r \\ 1/C_r & 0 \end{bmatrix} \begin{bmatrix} i_{Lr} \\ U_{Cr} \end{bmatrix} + \begin{bmatrix} 1/2L_r & 0 \\ 0 & -1/C_r \end{bmatrix} \begin{bmatrix} V_s \\ I_o \end{bmatrix} \quad (1)
$$

The output voltage is

$$
U_o = \begin{bmatrix} 0 & 1 \end{bmatrix} \begin{bmatrix} i_{Lr} \\ U_{Cr} \end{bmatrix} \quad (2)
$$

Where i_{Lr} is the current of the inductor L_r, U_{Cr} is the voltage of the capacitor C_r, U_o is the output voltage of the DC-rail ZVT converter, V_s is the input dc voltage, I_o is the equivalent load current, L_r is the inductor. There are six different modes to be operated in order to generate the DC-rail resonant ZVT source to the inverter.

During mode0 (t_0~t_1), the output load current I_o flows through either switch S_L or ant-parallel diode D_L of switch S_L and the auxiliary switches S_a and S_b are in off state.

During mode1 (t_1~t_2), if the switching status of the inverter needs to be changed, the auxiliary switch S_a is turned on with zero current condition at t_1, and dc-rail switch

S_L is turned off under zero voltage simultaneously. Hence, resonance occurs with dc-rail current Io. At t_2, i_{Lr} and U_{Cr} become zero simultaneously. During mode2 (t_2~t_3), S_a is turned off with zero current condition, and the inverter switch S_i can be turned on or off under zero voltage condition. During mode3 (t_3~t_4), at t_3, S_b is turned on under zero current condition, the inductor current is linearly increased with the slope of $V_s/2L_r$. During mode4 (t_4~t_5), at t_4, $i_L = I_o$, the freewheeling diode D_i is turned off softly. And a resonance occurs naturally with initial condition $i_{Lr}(t_4) = I_o$, $U_{Cr}(t_4) = 0$. During mode5 (t_5~t_6), at t_5, $U_{Cr} = V_s$, S_L is turned on under zero voltage condition, and the remnant current in L_r decays quickly to zero, switch S_b can be turned off under zero current condition. The next switching cycle is initiated.

B.   Steady-state Analysis

The steady-state analysis can be done by solving differential equations with bounding conditions or a balanced energy method [3]. We have used the balanced energy method in this paper to avoid a complicated computation. Fig.3 shows the steady-state voltage and current waveforms and identifies the various modes of operation. Mode0 occurs when the dc-rail short-circuits, and it changes to mode1 when the auxiliary switch S_a is turned on and switch S_L is turned off simultaneously. At mode0, the inductor current is zero and the voltage across C_r is $U_{Cr} = V_s$, the energy stored in the resonant circuit at the end of mode0 or the beginning of mode1 is

$$
E_{Cr} = \frac{1}{2} C_r U_{Cr}^2 = \frac{1}{2} C_r V_s^2
$$

$$
E_{Lr} = \frac{1}{2} L_r i_{Lr}^2 = 0 \quad (3)
$$

Mode1 begins as the auxiliary switch S_a is turned on and S_L is turned off simultaneously, and then the circuit reaches oscillation with the initial conditions, $i_{Lr}(t_1) = 0$, $U_{Cr}(t_1) = V_s$. According to equation (1), the results of analysis can be derived as

$$
U_o = \begin{bmatrix} 0 & 1 \end{bmatrix} \begin{bmatrix} i_{Lr} \\ U_{Cr} \end{bmatrix}
$$

$$
= \begin{bmatrix} 0 & 1 \end{bmatrix} \begin{bmatrix} I_o - A\sqrt{\dfrac{C_r}{L_r}} \sin[\omega_r(t-t_1) + \alpha] \\ \dfrac{1}{2}V_s + A\cos[\omega_r(t-t_1) + \alpha] \end{bmatrix} \quad (4)
$$

Where

$$
A = \sqrt{\frac{V_s^2}{4} + \frac{I_o^2 L_r}{C_r}} \qquad \alpha = \tan^{-1}\left(\frac{2I_o}{V_s}\sqrt{\frac{L_r}{C_r}}\right) \qquad \omega_r = \frac{1}{\sqrt{L_r C_r}}
$$

From (4), when

$$
t_2 = t_1 + \frac{\pi - 2\alpha}{\omega_r}
$$

$$
U_o = \begin{bmatrix} 0 & 1 \end{bmatrix} \begin{bmatrix} i_{Lr} \\ U_{Cr} \end{bmatrix} = \begin{bmatrix} 0 & 1 \end{bmatrix} \begin{bmatrix} 0 \\ 0 \end{bmatrix} = 0
$$

The energy stored in the resonant circuit at the end of mode1 and in mode2 is

$$E_{Cr} = \frac{1}{2}C_r U_{Cr}^2 = 0$$

$$E_{Lr} = \frac{1}{2}L_r i_{Lr}^2 = 0 \qquad (5)$$

At mode3 and according to equation (1), the inductor current can be expressed as

$$i_{Lr}(t) = \frac{V_s}{2L_r}(t_4 - t_3)$$

The energy stored in the resonant circuit at the end of mode3 is

$$E_{Cr} = \frac{1}{2}C_r U_{Cr}^2 = 0$$

$$E_{Lr} = \frac{1}{2}L_r i_{Lr}^2 = \frac{1}{2}L_r I_o^2 \qquad (6)$$

After i_{Lr} reaches I_o at t_4 of beginning of mode4, a resonance occurs naturally. $i_{Lr}(t_4)=I_o$, $U_{cr}(t_4)=0$. According to equation (1), the results of analysis can be derived as

$$U_o = \begin{bmatrix} 0 & 1 \end{bmatrix} \begin{bmatrix} i_{Lr} \\ U_{Cr} \end{bmatrix}$$

$$= \begin{bmatrix} 0 & 1 \end{bmatrix} \begin{bmatrix} I_o + \frac{1}{2}V_s \sqrt{\frac{C_r}{L_r}} \sin[\omega_r(t-t_4)] \\ \frac{1}{2}V_s[1 - \cos \omega_r(t-t_4)] \end{bmatrix} \qquad (7)$$

When

$$t_5 = t_4 + \pi\sqrt{L_r C_r}$$

$$U_o = \begin{bmatrix} 0 & 1 \end{bmatrix} \begin{bmatrix} i_{Lr} \\ U_{Cr} \end{bmatrix}$$

$$= \begin{bmatrix} 0 & 1 \end{bmatrix} \begin{bmatrix} I_o \\ V_s \end{bmatrix} = V_s$$

The energy stored in the resonant circuit at the end of mode4 is

$$E_{Cr} = \frac{1}{2}C_r U_{Cr}^2 = \frac{1}{2}C_r V_s^2$$

$$E_{Lr} = \frac{1}{2}L_r i_{Lr}^2 = \frac{1}{2}L_r I_o^2 \qquad (8)$$

In mode5, the remnant current in L_r decays quickly to zero due to the voltage $-V_s$ applied at t_6. The interval t_7-t_6 should be large enough for $i_{Lr}(t_6)$ to vanish completely.

$$\Delta T = t_7 - t_6 = \frac{L_r I_{o,\max}}{V_s}$$

At the end of mode5, the inductor current is zero and the voltage across C_r is V_s. The next switching cycle is initiated.

III. PM SYNCHRONOUS DRIVE SYSTEM

To validate the theoretical development, a DC-rail ZVT converter was used to drive a PM synchronous drives. The

block diagram for the speed-control system is shown in Fig.4. The principle of controlling the PM synchronous drive is a field-orientated control. For a PM synchronous motor, the magnetic flux generated from the rotor is fixed in relation to it, and the angular position of the flux is determined by the rotor position, which is in alignment with the d-axis. Thus, by properly controlling i_q, the generated torque T_e can be adjusted because the torque is linearly proportional to i_q, and them direct torque control of the motor can be achieved. The block diagram of a speed-controlled PM synchronous drive is shown in Fig.4. First, the i_q* is computed by a 16-bit 20MHz TI DSP2407 microcomputer. The microcomputer computes speed error, and executes a digital control law. Then, the 3Φ sin wave generator provides three-phase motor currents to the current controlled inverter by using an EPROM and a D/A converter. The three-phase reference currents are measured by Hall-effect current sensors and are fed back to the current controllers. The other feedback quantities are the absolute rotor position and the motor speed, which are measured by the resolve mounted on the motor shaft. The switching devices of a dc-rail ZVT scheme are subjected to very small losses duo to the soft commutation (zero voltage switching) conditions. The dc-rail ZVT inverter does not require snubber circuit because the inverter always changes state when the dc-rail voltage is zero. Thus, this inverter does not need to consider the losses caused by snubber circuits. The other losses can be calculated, and the efficiency of the dc-rail ZVT inverter can be easily obtained [4].

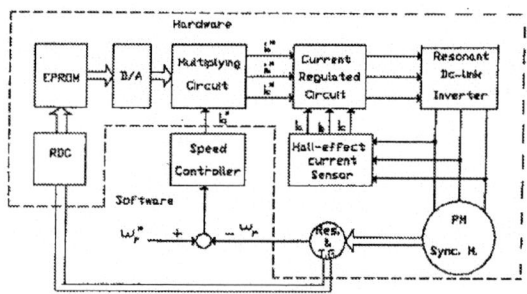

Figure4. Block diagram of a speed-controlled PM synchronous drive

IV. EXPERIMENTAL RESULTS

A high frequency dc-rail ZVT converter has been designed and tested using the method presented above. The unit is rated at 1.1kW. The component values of the resonant circuit are summarized in Table I. The input dc voltage of the dc-rail ZVT converter is 510V. In order to test the performance of the dc-rail ZVT converter, a PM synchronous drive system has been constructed. The resonant inductor current i_{Lr} and the resonant capacitor voltage U_{Cr}（DC-rail voltage）in experiment are shown in Fig.5. Fig.6 shows the line to line PWM voltage and the phase current when a synchronous motor runs at the speed of 500rpm. Remarkably clean current waveform is obtained with devices switching at high frequency. A microprocessor-based system is designed to measure the

384

efficiency of the converter, inverter, and motor. The currents are measured by Hall-effect sensors, and the voltages are measured by isolation-amplifiers. Through A/D converters, the measured data can be converted into digital values and then the efficiency of the inverter is computed. Fig.7 shows the efficiency of the dc-rail ZVT inverter. Its efficiency is between 84% and 93% from no load to full load. The efficiency of the traditional inverter, although operated at approximately 8KHz is less efficient. Fig.8 is the experimental speed response of the PM synchronous motor, which is driven by a dc-rail ZVT inverter with field-oriented control.

TABLE I
PARAMETERS OF THE DC-RAIL ZVT CONVERTER

L_r	22μh
C_r	0.05μf
V_s	510V

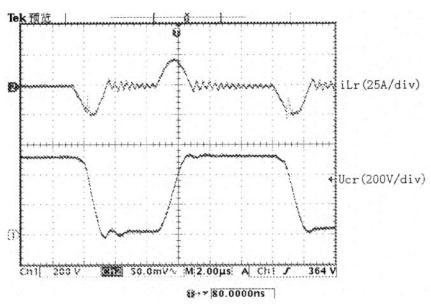

Figure5. Experimental link waveforms of i_{Lr} and U_{Cr}

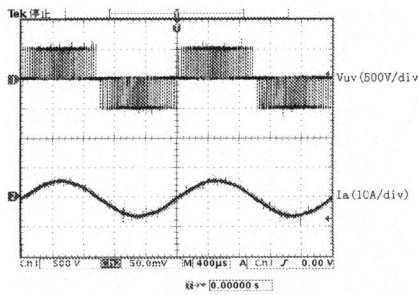

Figure6. Experimental phase voltage and phase current

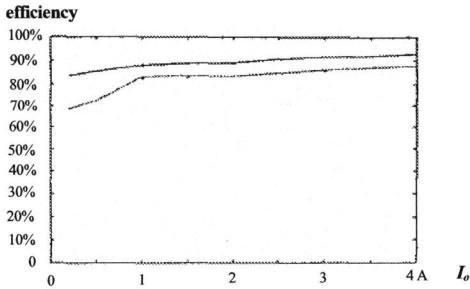

Figure7. Efficiency analysis of a dc-rail ZVT inverter

V. CONCLUDING REMARKS

The DC-Rail ZVT inverter is used to drive a PM synchronous motor, which is based on field oriented control.

Experimental results show that the new type of inverter has a higher switching frequency, lower switching losses and higher efficiency than a traditional hard switching inverter. The experimental results, which are very close to the computer simulated results, validate the theoretical development.

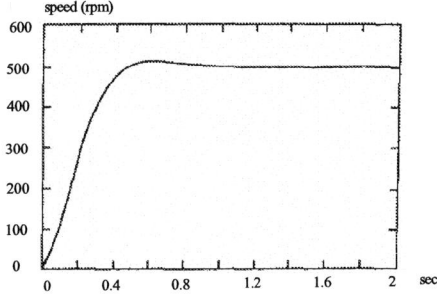

Figure8. Experimental speed response of the motor

REFERENCES

【1】 K.Wang, et., and F.C.Lee, "Novel DC-Rail Soft-Switched Three-Phase Voltage-Source Inverter," *IEEE Tran. on Indus. Appl.*,Vol.33（2）,1997

【2】 Ming Zhengfeng , Zhong Yanru ,"A Novel DC-Rail Parallel Resonant Zero Voltage Transition Three-phase PWM Voltage-Source Inverter " , *TRANSACTIONS OF CHINA ELECTROTECHNICAL SOCIETY* , No.6 , Vol.16 , 2001

【3】 Mertens and D. M. Divan, "A high frequency resonant DC Link inverter using IGBTs," Proceedings of 1990 International Power Electronics Conference, Tokyo, Japan

2006 5th International Power Electronics and Motion Control Conference

Analysis, Simulations and Experiments Of A Novel ZVS -ZCS Inverter With Pulse Current Feedback Transformer Auxiliary Commutation

Yaogang, Mahamnad Mansoor Khan and Chenchen
Department of Electrical Engineering
Shanghai Jiao Tong University
Huashan Road 1954#, Shanghai, 200030, China
yaogangth@sjtu.edu.cn

Abstract:- In order to enhance the performance of ARCPI (auxiliary resonant-commutated pole inverter), this paper proposes a novel ZVS-ZCS inverter with pulse current feedback transformer auxiliary commutation to solve the problem such as extra losses from auxiliary circuit, variable commutation duration, and poor reliability, which are common in existing ARCPIs. In this novel ARCPI, the application of commutation forcing voltage greater than half value of DC link voltage, fixed commutation duration, and protection for auxiliary switches furthest enhances the performance, and also extends its application. By the analysis, simulation, and experimentation, this new ARCPI is proved reasonable for use in high reliability application.

Key-Words:-Auxiliary resonant commutated pole inverter (ARCPI), Zero-voltage switching (ZVS), Zero-current switching (ZCS), Distribution flexible ac transmission system (DFACTS)

1 Introduction
Today, a variety of soft switching techniques have been applied in power inverters to solve the problems resulting from hard-switching in high-frequency application [1]. Among those soft switching techniques, the performance of auxiliary resonant commutated pole inverter (ARCPI) has been found relatively outstanding [2]-[6]. ARCPI takes advantage of auxiliary switch to achieve ZVS or ZCS for main switch, and it maximally enhances switching frequency, efficiency, and reliability. Although ARCPI techniques have widely been used in SPWM inverters to achieve obvious advantages, but still few of major issue are unsolved, for example extra switching losses from auxiliary circuit [6], variable commutation duration or duty

cycle loss and poor robustness. These problems seriously limit the performance of ARCPIs. A detail analysis about these limitations is described in following part of this section.

Fig.1 shows three kinds of ARCPIs, which all have achieved good performance in general applications, but in some application with severe current transients, the performance targets cannot be met. For example, the extra high resonant current of the topology shown in Fig.1a [4], resulting from high forcing voltage, can lead to severe current stresses in auxiliary switches, considerable power loss and severe thermal stress. The complex structure of the topology shown in Fig.1b [5], resulting from the advanced trigger control strategy, can lead to an unforeseen ZVS commutation failure and system oscillation [6]. As for as the topology shown in Fig.1c[6], one of major standing issues is the design of transformer, which is an over-designed transformer with larger leakage inductances. However in some high current and high voltage application such a transformer can be too inefficient and limit the application of this topology in high frequency range.

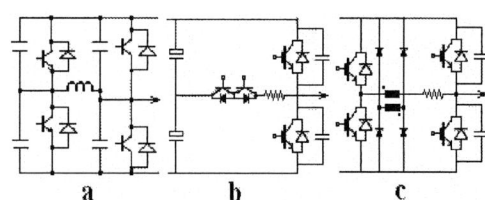

Fig.1 Configurations of some typical ARCPIs

1-4244-0448-7/06/$25.00 ©2006 IEEE   386

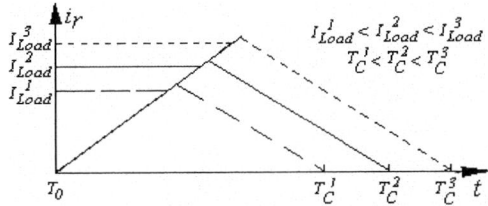

Fig.2 Variation of commutation duration VS. load current

Besides above-mentioned drawbacks, there is another drawback, which is common to all the ARCPIs, and that is the variation of commutation duration with load current as shown in Fig.2. It can be observed that, the commutation duration (Tc) is directly proportional to load current (ILoad). If an ARCPI is designed with fixed commutation duration of auxiliary switches, then at heavy load current, it can cause serious current stresses and extra switching losses to auxiliary switch. In order to ensure ZVS or ZCS operation of auxiliary switch and prevent auxiliary switch from getting damaged by the residual energy stored in Lx, the extra control strategy adjusting commutation duration following load current is always adopted [5],[6]. The solution of variable commutation duration complicates trigger strategy, and also limits switching frequency, duty ratio and capacity of ARCPI. And in some extreme situation, this solution also is helpless.

Since the requirements for switching frequency, output capacity, efficiency and reliability of an ARCPI are becoming more stringent, hence the robustness of the snubber circuit is also becoming more important. Although the performance of existing ARCPIs is excellent in standard inverter applications like motor driver systems, but still it is not suitable for the applications with high transient currents and asymmetrical operations like customer power quality application [4].

Therefore based on the comprehension of advantages and disadvantages among existing topologies, this paper

proposes an enhanced ARCPI as shown in Fig.3. It is named as ZVS-ZCS inverter with transformer-assisted commutation. This enhanced ARCPI caters the requirement of modern inverter technology and maximally enhances the performance of ARCPI. Furthermore, this ARCPI needn't keep track of load current to adjust commutation duration; therefore fixed commutation duration is adopted in this new ARCPI. The protection for auxiliary resonant circuit is considered by the structure of four auxiliary switches, and even in extreme over-current case, this special structure can well ensure the ZVS/ZCS commutation operation of the auxiliary switches. As a result, the stresses on auxiliary switch are much more reduced and there is very little risk of hard on/off commutation in auxiliary switches.

This paper is arranged in following order. Section 2 analyses commutation procedure. In section 3, detail analysis and simulation results under kinds of load condition are discussed. Section 4 addresses experimental results. Finally section 5 concludes in this paper.

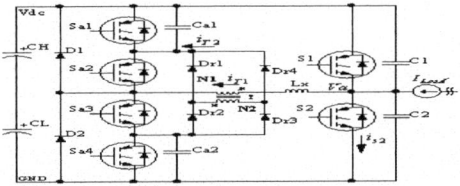

Fig.3 Configuration of the ZVS-ZCS inverter with transformer assisted commutation

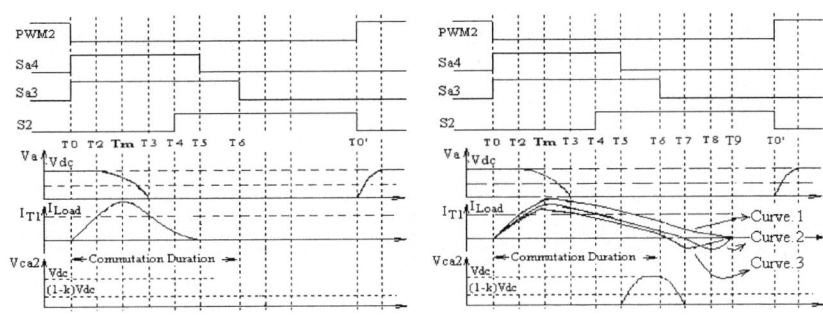

a. Under normal load current   b. Under abnormal load current

Fig.4 Predicted commutation wave during freewheeling diode of S1 to S2 under negative load current

387

2 Commutation Principle of The Novel ARCPI

The commutation sequence of the proposed ARCPI including two kinds of commutation processes, the commutation under normal load current and the commutation presented in Fig.4a under abnormal load current presented in Fig.4b, are supposed to occur. So the detail analysis of these two processes has been discussed separately. Since the case of fixed duration gating signals for auxiliary switches has been treated in this paper, the general switching sequence for all the auxiliary switches gate signals is same irrespective of commutation processes. Referring to Fig.3 and Fig.4, the general switching sequence is as follows. Considering, initially main switch S1 is in on state and all other switches including auxiliary switches are in off state. Now the target is to turn-off S1 under ZVS condition and turn-on S2 under ZVS/ZCS condition. To achieve this target, first S1 is turned off, Sa3 and Sa4 are turned on simultaneously; and then S2 is turned on if the voltage across S2 becomes zero in later part of the commutation. At the end of the of fixed commutation time for auxiliary circuit, first Sa4 is turned off at zero voltage and after a little delay, Sa3 is turned off

at zero current. Likewise the general switching sequence for Sa1, Sa2 and S1 can be inferred analogously. The separate analysis is presented in [7], and this paper doesn't further the process.

3 Mathematical Analysis of The Novel ARCPI

3.1 The Analysis of Commutation Transient

The expressions for the voltage across Ca, auxiliary inductor Lx current and auxiliary capacitor Ca2 voltage during various stages of commutation are presented. It can be observed from [7] that the commutation under extremely high load current contain all the possible modes of operation of this topology, however the successful commutation of the main switch cannot be guaranteed after a certain point in this range of operation, so the boundary of extremely high current mode and medium high current mode will be considered in rest of the section. Furthermore the expressions for this mode are enough to infer the various design parameters of this ARCPI. Due to this reason, the process for medium high load current is presented in this section. The equations for above–mentioned parameters are given in table I.

Table 1 The express of i_{T1}, Va, Vca2 for diode-to-switch commutation during the full commutation

$T_0 \le t < T_2$	$i_{T1}(t) = \dfrac{(1-k)V_{dc}}{L_x}(t-T_0) + I_{T1}(T_0)$ $v_{ca2}(t) = 0$ $v_{C2}(t) = V_{dc}$	$T_0 = 0.0s$ $I_{T1}(T_0) = 0A$ $t \in [T_0, T_2]$
$T_2 \le t < T_5$	$i_{T1}(t) = I_{Load}\cos[w(t-T_2)] + (1-k)V_{dc}\sqrt{\dfrac{2C}{L_x}}\sin[w(t-T_2)]$ $v_{ca2}(t) = 0$ $v_{C2}(t) = kV_{dc} + (1-k)V_{dc}\cos[w(t-T_2)] - I_{Load}\sqrt{\dfrac{2C}{L_x}}\sin[w(t-T_2)]$	$w = \sqrt{1/2CL_x}$ $t \in [T_2, T_5)$
$T_5 \le t < T_6$	$i_{T1}(t) = I_{T1}(T_5)\cos[w(t-T_5)] - \dfrac{kV_{dc}}{(1-k)}\sqrt{\dfrac{C_a}{L_x}}\sin[w(t-T_5)]$ $v_{ca2}(t) = \dfrac{kV_{dc}}{(1-k)}\{\cos[w(t-T_5)] - 1\} + I_{T1}(T_5)\sqrt{\dfrac{L_x}{C_a}}\sin[w(t-T_5)]$ $v_{C2}(t) = 0$	$w = \dfrac{(1-k)}{\sqrt{C_a L_x}}$ $t \in [T_5, T_6)$
$T_6 \le t < T_7$	$i_{T1}(t) = I_{T1}(T_6)\cos[w(t-T_6)] - \dfrac{V_{dc}}{k}\sqrt{\dfrac{C_a}{L_x}}\sin[w(t-T_6)]$ $v_{ca2}(t) = -\dfrac{V_{dc}}{k}\cos[w(t-T_6)] - I_{T1}(T_6)\sqrt{\dfrac{L_x}{C_a}}\sin[w(t-T_6)] + \dfrac{1+k}{k}V_{dc}$ $v_{C2}(t) = 0$	$w = \dfrac{k}{\sqrt{C_a L_x}}$ $t \in [T_6, T_7)$

$T_7 \le t < T_8$	$i_{T1}(t) = [\dfrac{k}{1+k}V_{dc} - V_{ca2}(T_7)]\sqrt{\dfrac{C_a}{L_x}}\sin[w(t - T_7)]$ $v_{ca2}(t) = [V_{ca2}(T_7) - \dfrac{k}{1+k}V_{dc}]\cos[w(t - T_7)] + \dfrac{k}{1+k}V_{dc}$ $v_{C2}(t) = 0$	$w = \dfrac{(1+k)}{\sqrt{C_a L_x}}$ $t \in [T_7, T_8)$
$T_8 \le t < T_9$	$i_{T1}(t) = I_{T1}(T_8) - \dfrac{kV_{dc}}{L}(t - T_8)$ $v_{ca2}(t) = 0$ $v_{C2}(t) = 0$	$i_{T1}(T_9) = 0$ $t \in [T_8, T_9)$

3.2 Simulation Results

The simulation waveforms for gating signal S2 (PWM2), voltages Va, Vca2, and current i_{T1} using PSCAD are shown in Fig.5. Referring to waveform Fig.5a for load current 50A, the commutation duration is less than 7us, and in waveform b with load current 200A, it also is about 7us. It means that fixed commutation duration strategy is feasible for the proposed circuit. By comparison of two waveforms, it can be observed that the time for main switch turn on at ZVS is delayed with the increase of load current. In practice, zero voltage detection in main IGBT driver ensures main switch to be turned on at zero voltage. It can also be observed that the voltage across Sa4 is always less than the Vdc in any case.

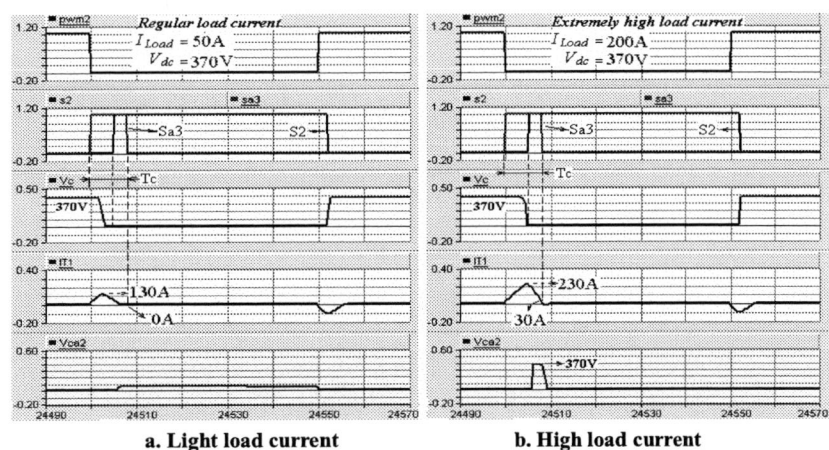

a. Light load current b. High load current

Fig.5 Simulation waveforms of PWM2, Va, i_{T1} and Vca2 (10us/Div)

By the analysis and simulation of the novel ARCPI, the introduction of fixed commutation duration has been proved feasible, even at high transient load current. To date, this enhanced ARCPI is most outstanding among all ARCPIs. And it is much more favorable in severe operating conditions.

4 Experimental Results

In order to check the feasibility of this ARCPI, a prototype half-bridge inverter of 10kw IGBT as show in Fig.6 has been built. In the prototype, the rated value of DC voltage is 370V, and DC link employs two 450V/10000uF capacitors to form the dc center tap; The main and auxiliary switches are SKM200GB128DN and SKM75GB128DN modules respectively and all the fast diodes are DSEI30-10A and the switching frequency of main switches reaches to 20kHz; The value of auxiliary resonant inductor is 5uH and the value of snubber capacitor C/Ca is 0.1uF/0.47uF; The ratio of auxilary transformer T is k (N1/N2), which is inside of 0.35~0.45; Six resistors (15Ω/20A) were connected in parallel as load; The driver of the main switch was interfaced with an extra zero-voltage detecting circuit, to ensures main IGBT ZVS turn-on.

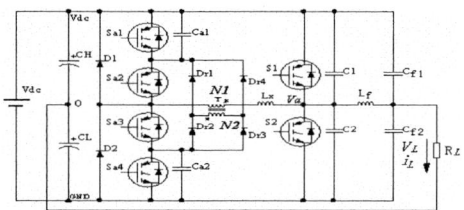

Fig.6 Configuration of the 10kW half-bridge IGBT ARCPI

Fig.7 gives the waveforms for main switch S2 turn-on and turn-off commutation under ZVS/ZCS operation for 50A load current. The values of dv/dt and di/dt during S2 turn-on are 70V/uS and 50A/uS respectively, and the values for dv/dt and di/dt during S2 turn off commutation are 175V/uS and 75A/uS respectively. All the values during switching transients are reasonable and inside the SOA of the IGBT.

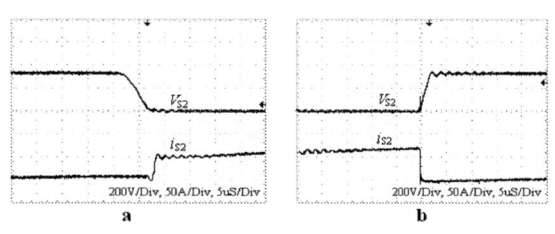

a. Turn-on of S2   b. Turn-off of S2

Fig.7 Experimental voltage and current waveforms

It can be observed in Fig.8 that all waveforms are obtained for fix commutation duration of the auxiliary switches Sa3/Sa4. So in any case the turn-on durations for the switches Sa4 and Sa3 are 5.5μS and 7μS respectively. Hence, the commutation duration from the start of the commutation cycle (turn-on of auxiliary switches Sa3/sa4) to the end (turn-off of auxiliary switches Sa3/sa4) is equal to7μs. These results indicate that the fixed commutation duration is feasible and the voltage Vca2 is never exceeds Vdc in any case. Besides, auxiliary Sa4 is always turned on and off at zero voltage; auxiliary switch Sa3 is turn on at zero voltage, and turned off at zero current under normal load current and at zero voltage under abnormal load current.

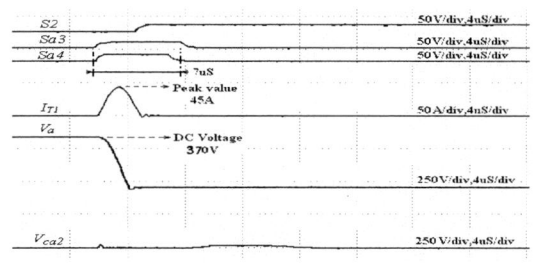

a. Under light load current about 27A

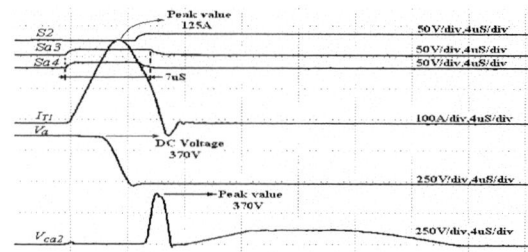

b. Under a normal high load current about 75A

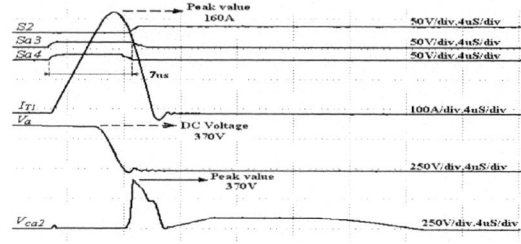

c. Under extremely high load current about 100A

Fig.8 Experimental waveforms with variable load current

Fig.8a shows the commutation waveforms for normal current range. Because of small resonant current, there is no peak in Vca2. All the waveforms are in accordance with the simulated results. It means that the procedure of commutation under light load current is same as predicted. Referring to the waveform shown in Fig.8b, there is an abnormal commutation under a little high load current 75A. Because of high resonant current in auxiliary circuit reaching to 125A, there is a peak in voltage across Ca2 almost touching the DC-link voltage value (370V). By comparing the commutation waveforms under abnormal load current and experimental waveforms given in Fig.8b, it can be inferred that the sequence of commutation is the same as sequence Step6 → Step8 → Step9 → Step10.

The waveform shown in Fig.8c represents the abnormal commutation under extremely high load current 100A. Referring to the waveform c, it can be seen that the peak value of Vca2 reaches to Vdc due to extremely high value of i_{T1}, simultaneously, the peak value of current i_{T1} reaches to 160A. Under extremely high load current 100A, the procedure of commutation is the same as sequence Step6 → Step 7→ Step10.

By comparison among predicated, simulated, and experimental waveforms, all key parameters and phenomenon are found consistent. With reasonable control logic, this topology can fulfill its task perfectly and achieves main switch and auxiliary switching at zero-voltage or zero-current, fixed commutation duration, and strong fault tolerance smoothly.

5 Conclusions

Based on above discussion, analysis and experimentation, the conclusions are inferred as following.

The voltage source $(1-k)$Vdc ($>\frac{1}{2}$Vdc) taking part in the commutation procedure that ensures zero-voltage switching of the main switch, and zero-voltage and zero-current switching of auxiliary switch without any extra detection and control circuit. So this novel ARCPI enhances reliability and efficiency.

Because of the special structure of four auxiliary switches, there is no need of extra protection for auxiliary switch and the fixed commutation duration can be adopted in it. And even if there is some fault at the resonant components, the stresses of auxiliary switches are in reasonable range. Therefore, the switching frequency is greatly enhanced and it also has strong fault tolerance.

By the simulation and experimentation, this new ARCPI is reasonable. The fixed commutation duration, zero-voltage switching for main switch and auxiliary circuit stress are validated.

By the experimentation, this new ARCPI is a prospective inverter to be used for high power, high efficiency, high switching frequency, and high reliability application, especially for DFACTS device.

Reference:

[1]. Maria D. Bellar, Tzong-Shiann Wu, and Aristide Tchamdjou, "A Review of Soft-Switched DC–AC Converters",in *IEEE Transactions on Industry Applications,* Vol.34, AUG. 1998, pp. 847-860.

[2]. Takano, H., Domoto, T., Takahashi, J., and Nakaoka, M., "Auxiliary resonant commutated soft-switching inverter with bidirectional active switches and voltage clamping diodes", *Industry Applications Conference,* 2001. Thirty-Sixth IAS Annual Meeting. Conference Record of the 2001 IEEE, Vol.3, Oct. 2001, pp.1441–1446.

[3]. Ahmed Elasser and David A.Torrey, "Soft Switching Active Snubbers for dc/ac Converter", in *IEEE Transactions on Power Electronics,* Vol.11, Sept. 1996. pp. 950-957.

[4]. Beukes, H.J., Enslin, J.H.R. and Spee, R., "Integrated Active Snubber for high Power IGBT Modules", in *Proceedings of IEEE APEC,* Vol.1, Feb. 1997, pp.161-167.

[5]. Turpin, C., Forest, F., Richardeau, F., Meynard, T.A., and Lacarnoy, A., "Switching Faults and Safe Control of an ARCP Multicell Flying Capacitor Inverter",in *IEEE Transactions on Power Electronics,* Vol.18, Sept. 2003, pp.1158-1167.

[6]. Xiaoming Yuan and Ivo Barbi, "Analysis, Designing, and Experimentation of a Transformer-Assisted PWM Zero-Voltage Switching Pole Inverter", in *IEEE Transactions on Power Electronics,* Vol.15, Jan. 2000, pp.72–82.

[7]. Yaogang, Mahamnad Mansoor Khan and Chenchen, "A Novel ZVS -ZCS Inverter With Auxiliary Resonant Snubber Using Pulse Current Feedback Transformer", WSWAS Trans. on Circuit and System, 4(11), pp.1770-1775, 2005

2006 5th International Power Electronics and Motion Control Conference

A Novel Eddy-Current Based Far-Infrared Rays Radiant Planner Heater using High-Frequency ZVT-PWM Inverter

Hisayuki Sugimura[1], Bishwajit Saha[1], Hideki Omori, Hyun Woo Lee[1] and Mutsuo Nakaoka[1][3]

[1]Electric Energy Saving Research Center, Graduate School of Electrical and Electronics Engineering,
Kyungnam University, Masan, Republic of Korea
[2]Home Appliance Company, Matsushita Electric Industrial Co., Ltd., Osaka, Japan
[3]Department of Electrical and Electronics Engineering, Industrial College of Technology-University, Hyogo, Japan

Abstract— **This paper presents an innovative prototype of a new conceptual electromagnetic induction heated type far infrared rays radiant heating appliance using the voltage-fed edge-resonant ZVS-PWM high frequency inverter using IGBTs for food cooking and processing which operates under a constant frequency variable power regulation scheme. This power electronic appliance with soft switching high frequency inverter using IGBTs has attracted special interest from some advantageous viewpoints of safety, cleanliness, compactness and rapid temperature response, which is more suitable for consumer power electronics applications.**

Keywords- Voltage source high frequency inverter; Active switched capacitor; Zero voltage soft switching (ZVS-PWM); Edge resonant soft commutation; Asymetrical PWM; Induction heating; Far-infrared rays radiant heater; Specially-designed planner stainless steel; Consumer power electronics

I. INTRODUCTION

Gas combustion heating system and sheath wired heating system for consumer home and business applications are usually used as the heating means for the food cooking and processing. Besides, low frequency and high frequency electromagnetic induction heating (IH) methods have been considered for food cooking and processing in business-use production plants and household kitchen applications. This actual heating efficiency is relatively high for consumer IH appliances as compared with traditional food cooking appliances. The induction heating appliances using power electronics circuits are more excellent in respect of energy saving, temperature control, clean environment, quick heating processing, safety and reliability. They are utilized in various power application fields for food cooking processing such as boiling, baking and steaming. The high frequency edge-resonant inverters and high frequency resonant cycloconverters are essentially indispensable for implementing the heating controllability, high efficient heating in addition to improvements of cleanliness, safety

and reliability of the heating processing work in the high efficient power conversion stages.

For a moment, the voltage source type single-ended edge-resonant zero voltage soft switching PFM high frequency inverter using a single IGBT for the commercial utility AC 100V grid has been developed so far for cost effective practical power applications. This type of high frequency zero voltage soft switching inverter has attracted special interest because of low cost, compact, high-efficiency, low noise. But, audible acoustic noise below 20kHz due to frequency difference cause in multi burner system composed of PFM control-based zero voltage soft switching inverter scheme, voltage peak stress across the power switching semiconductor device: IGBT

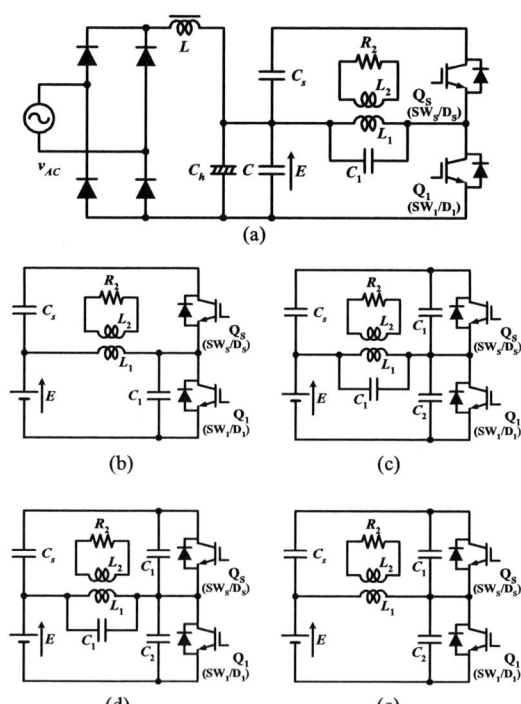

Figure 1. Total high frequency inverter system configuration

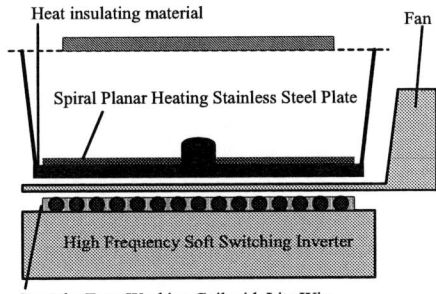

Figure 2. Griddle power appliance using electromagnetic induction heating ranges

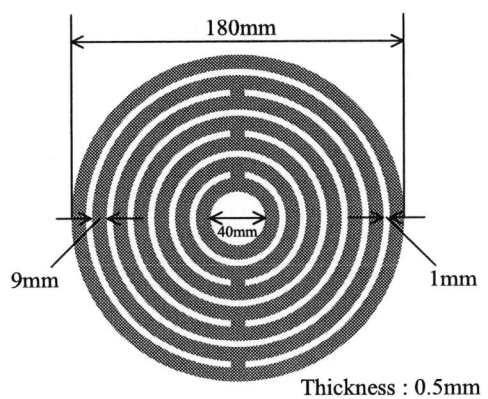

Figure 3.Eddy current-based heating plate shape

becomes extremely high. The operating range of this inverter is relatively small. In recent years, the modified topologies of zero voltage soft switching PWM high frequency inverter added an auxiliary active power switch connected in series with the capacitor for voltage clamping on the basis of conventional single ended high frequency inverter are practically developed for utility AC 100V or 200V grid consumer power applications. This paper presents an innovative prototype of the voltage source type high frequency edge-resonant soft switching inverter using IGBTs, which is based upon asymmetrical PWM control scheme due to Duty Cycle Time Ratio Control, which is newly developed for electromagnetic induction eddy current based far infrared rays radiant heating as a new generation griddle for food cooking processing. Its operation principle is described and unique salient features of this consumer power electronic appliance using a high frequency inverter are discussed and evaluated on the basis of computer-aided simulation and feasible experimental results, along with the development on the latest induction heated device as a griddle.

II. EDGE-RESONANT ZVS-PWM SOFT SWITCHING HIGH FREQUENCY INVERTER FOR INDUCTION HEATER

Figure 1(a) shows a schematic total system configuration including a high frequency edge-resonant

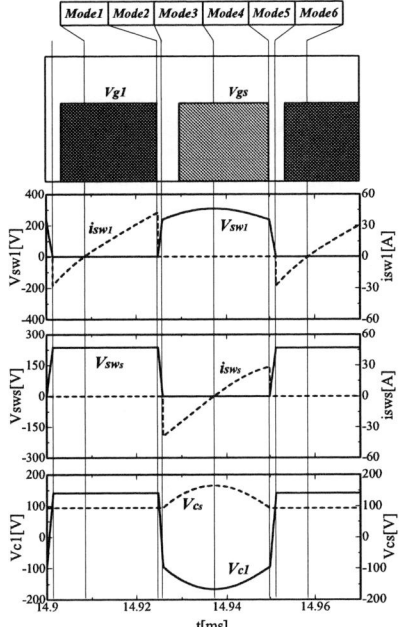

(a) In case of D=0.5

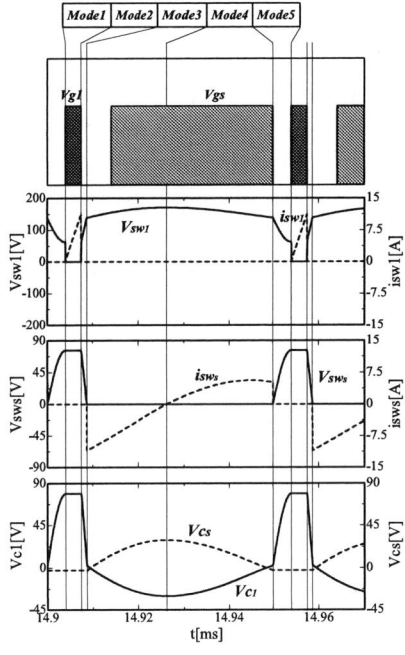

(b) In case of D=0.15

Figure 4. Steady state voltage and current switching waveforms for two duty cycle control conditions

(quasi-resonant) inverter circuit topology using two IGBTs that can operate under a principle of ZVS and constant frequency asymmetrical PWM control strategy. This high frequency inverter topology is newly applied for an electromagnetic induction eddy current based far

393

infrared rays radiant heating appliance as a new generation griddle for food cooking and processing.

The voltage source type high frequency zero voltage soft switching inverter with the duty cycle control-based variable power constant frequency (VPCF) function is connected to a single-phase utility AC 100V grid or 200V grid, the full bridge diode rectifier with the smoothing filter as shown in Fig.1(a).

Figure 1(b)-(e) illustrate modify of Total high frequency inverter system configuration

III. HIGH FREQUENCY EDDY CURRENT-BASED FAR-INFRARED RAYS RADIANT HEATING APPLIANCE UNITS

The produced induction heating griddle appliance using the high frequency inverter for consumer power electronic applications is shown in Fig.2. This consumer power appliance is composed of the eddy current-based planar spiral stainless steel heating plate: SUS304 ($70-80\mu\Omega cm$), wool board called ceramic fiber for heat insulating material, pancake type working coil, voltage source high frequency soft switching inverter without a matching transformer, forced air cooling fan driven by DC motor. The output current obtained from the voltage source high frequency soft switching PWM inverter flows through the working coil made of litz wire. The eddy current is directly induced into the spiral planar stainless steel plate by electromagnetic induction principle and its plate is directly heated on the basis of joule low. It is noted that this appliance is to make use of the radiant heat from

spiral planar stainless steel plate.

Figure 3 demonstrates a geometric structure of a new electromagnetic induction heating plate designed newly for consumer food cooking applications in household and business use. The condition required for the heating plate is specified that it heats rapidly when the output power is delivered to the load, on the other hand it rapidly stops to heat when the output power is shut off. In other words, the heat capacity of spiral planar heating plate is as small as possible. The heating material uses SUS304 of the non-magnetic material. The geometric structure of the diameter of the heating plate has the outer diameter: 180mm, the inside diameter: 40mm, the remainder width: 9mm, the chute width: 1mm. Its shape of groove connection is the special structure of the zigzag combination in order to prevent the thermal deformation when it is heated.

IV. INVERTER PERFORMANCE EVALUATIONS

Figure 4 illustrates the steady-state switching voltage and current waveforms of the main and auxiliary active power switching blocks; $Q_1(SW_1/D_1)$, $Q_S(SW_S/D_S)$, capacitor voltage V_{C1} and V_{CS} under two conditions of (a)D=0.5 and (b)D=0.15. Observing V_{SW1} in Fig.4(a), it is noted that zero voltage soft switching mode transition can be achieved for this inverter. On the other hand, from V_{SW1} in Fig.4(b), the hard switching mode transition can be observed in the case of D=0.15. This maximum duty cycle is determined by the maximum voltage value of the active power switch SW_1 and SW_S.

The simulation and experimental results of this IH griddle appliance for consumer cooking applications are illustrated and its steady-state operating performances are evaluated form a practical point of view on the basis of Duty Cycle Time Ratio Control defined in Fig.5 under 20kHz. Main active power switch SW_1 operates with ZVS when Duty Cycle is set to a value more than 0.2, and it becomes a hard switching operation mode, when Duty Cycle is set to a value smaller than 0.2. It is noted that the auxiliary active power switch SW_S can operate in ZVS in spite of the Duty Cycle: D control scheme and load circuit parameters. This high frequency edge-resonant soft switching PWM inverter can continuously control electric

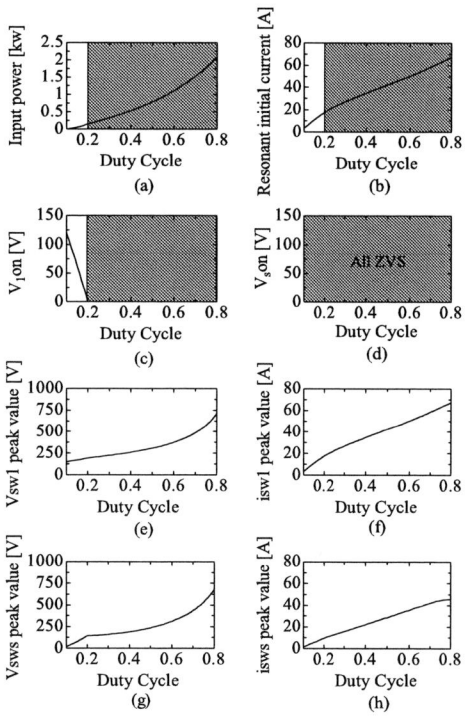

Figure 5. Various inverter characteristics for duty cycle control

TABLE I. DESIGN SPECIFICATIONS AND CIRCUIT PARAMETERS

Item	Symbol	Parameter Value
DC Source Voltage	V	141.4[V]
Switching Frequency	f	20[kHz]
Edge Resonant Lossless Snubber Capacitor	C_1	0.18[μF]
Active Voltage Clamped Capacitor	C_S	3.0[μF]
Working Coil	L_1	65.1[μH]
Electromagnetic Coupling Coefficient	k	0.652
Load Time Constant	τ	12.5[μsec]
Dead Time	t_d	3.0[μsec]

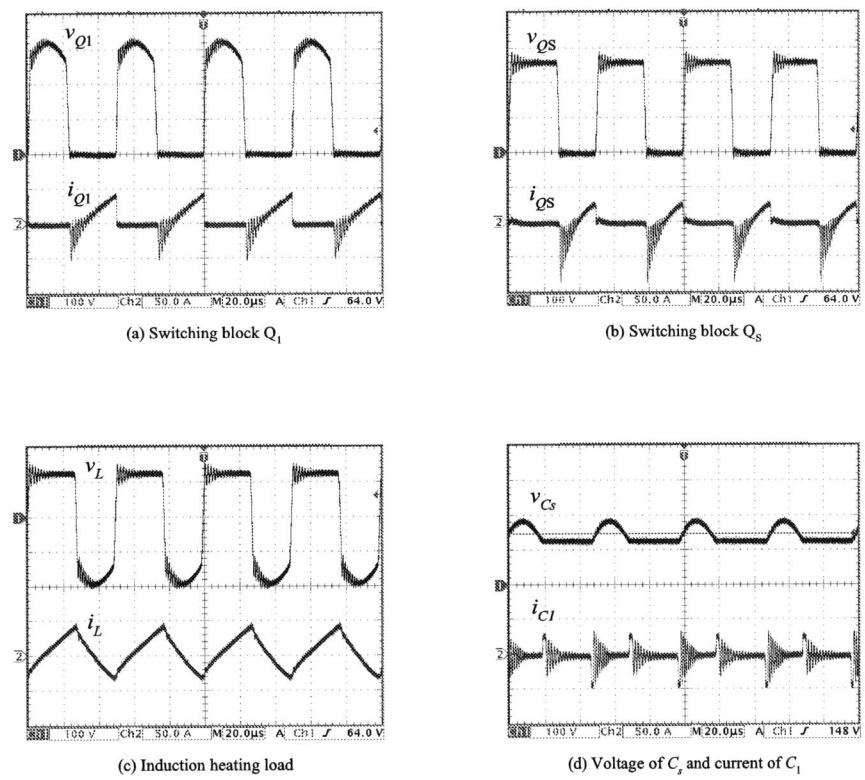

(a) Switching block Q_1

(b) Switching block Q_S

(c) Induction heating load

(d) Voltage of C_s and current of C_1

Figure 6. Observed voltage and current waveforms (100V/div, 50A/div, 20µsec/div, D=0.5)

power at the constant frequency. In addition, the soft switching operation of this inverter in this case; 20kHz (see Fig.1) is able to be done in 90% of the all Duty Cycle: D. The zero voltage soft switching of this inverter is not possible in the standby low electric power state when D is less than 0.2 under circuit parameters designed for induction heater as a griddle appliance.

V. EXPERIMENTAL RESULTS AND DISCUSSIONS

A. Comparative Operating Waveforms

Figure 6(a), Fig.6(b), Fig.6(c) and Fig.6(d) depict the measured voltage and current switching waveforms of the main active power switching block; $Q_1(SW_1/D_1)$, in addition to voltage and current waveforms of the auxiliary active power switching block; $Qs(SWs/Ds)$, the electromagnetic induction heater (see Fig.3) and the capacitor voltage and current waveforms under a condition of D=0.5. These measured voltage and current waveforms have a good agreement with simulated results, which are obtained from the simulation analysis developed by the authors. It is proved that this quasi-resonant ZVS-PWM high frequency inverter with VPCF (Variable Power Constant Frequency) scheme can completely work under a soft switching operation for a wide Duty Cycle control implementation.

Table 1 indicates the practical design specifications and circuit parameters of the feasible electromagnetic

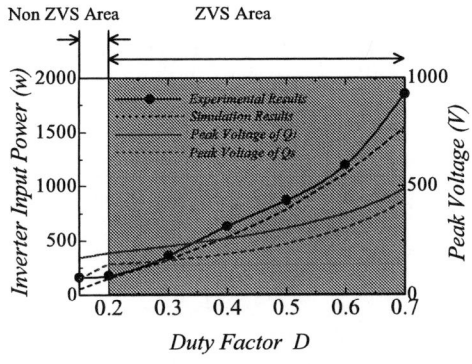

Figure 7. Duty cycle vs. input power characteristics

induction eddy current-based far infrared rays radiant heating appliance, which is built by edge-resonant ZVS-PWM high frequency inverter using the latest IGBT modules.

In addition, it is understood that this edge-resonant high frequency inverter using IGBT power module can clamp an excessive peak voltage applied to the main active power switch. Accordingly, the conduction power losses of the IGBTs and the current stresses of switching power semiconductor devices (IGBTs) can be effectively reduced by adopting this voltage-fed soft switching high frequency inverter.

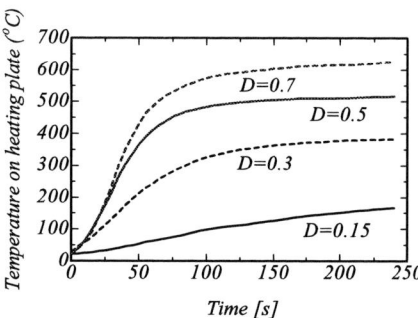

Figure 8. Temperature characteristics of induction heated stainless steel plate

B. Power Regulation Characteristics

Figure 7 represents Duty Cycle D vs. input power regulation characteristics under a fixed operation frequency (20kHz) asymmetrical PWM (Duty Cycle Time Ratio Control) strategy. Observing Fig.7, it is clearly proved that Duty Cycle as an independent control variable can be continuously adjusted in the accordance with the inverter output power.

C. Temperature Characteristics of Induction Heater

Figure 8 illustrates temperature characteristics of the spiral planar stainless heating plate using high frequency inverter shown in Fig.1 for the feasible setup in experiment. It is noted that this electromagnetic induction eddy current heated far infrared rays radiant heating processing scheme can more rapidly heat than the general purpose of gas combustion type or sheathed wired heating type heater for consumer food cooking and processing applications.

VI. CONCLUSIONS

In this paper, an innovative implementation of the electromagnetic induction eddy current-based far infrared rays radiant heating appliance using the specially designed spiral planar stainless steel heater has been successfully proposed as consumer household and business use products by using a voltage fed type active voltage clamped edge-resonant ZVS-PWM high frequency inverter using the latest IGBTs power modules, which can efficiently operate under the soft switching commutation on the basis of the asymmetrical PWM-Duty Cycle Time Ratio Control strategy and load parameter variations. Its steady state operation and power regulation characteristics have been evaluated in spite of simulation and experimental results.

Furthermore, this paper has been proved as a variety of industrial, automobile heat energy processing plants as well as consumer heat energy processing plants. The new and efficient induction heated far infrared rays radiant heating appliance using high frequency soft switching inverter could be cost effective than the gas combustion heating.

In the future, the power loss analysis of this soft switching high frequency soft switching inverter using the trench gate IGBTs and high conductivity IGBT (HiGT) should be done and the new generation consumer power electronics appliances for electromagnetic induction eddy current-heated far infrared rays radiant heating should be evaluated and discussed from a practical point of view. The computer aided design procedure of this power electronics appliances using a new inverter topology in the pipeline system should be studied from a theoretical point of view. The comparative studies between the zero voltage soft switching inverter treated here and zero current soft switching inverter has to be discussed in practical for induction heated griddle.

ACKNOWLEDGMENT

This work was financially supported by MOCIE through IERC program.

REFERENCES

[1] T.Miyauchi, I.Hirota, H.Omori and M.Nakaoka "Active Voltage-Clamped VPCF Soft Switching Inverter using 4th-Generation IGBTs for Induction-Heating Cooking Appliance and Practical Evaluations", Proceedings of IEE-J International Power Electronics Conference-Tokyo, Vol 3, pp1741-1746, April, 2000

[2] H.Tanaka, M.Kaneda, M.Ishitobi, E.Hiraki and M.Nakaoka "Electromagnetic Induction based Continuous Fluid Heating Appliance using Soft Switching PWM High Frequency Inverter", Proceedings of IEEE-IAS International Appliance Technical Conference-USA, pp11-20, May, 2000

[3] M.Kaneda, H.Tanaka, S.Muraoka, S.Hishikawa and M.Nakaoka "High Frequency Eddy Current-based Fluid Heater using Soft-Switching Inverter", Proceedings of IEEE Third International Power Electronics and Motion Control Conference-China, Vol 2, pp962-967, August, 2000

[4] H.Tanaka, M.Kaneda, S.Chandhakt, A.Okuno and M.Nakaoka "Electromagnetic Induction Eddy Current-based Fluid-Heating Boiler using New High-Frequency Inverters for Super Heated Steamer", Proceedings of International Symposium on Marine Engineering-Tokyo, pp183-188, October, 2000

[5] H.Tanaka, M.Kaneda, S.Chandhaket, M.Aubdallha AL, M.Nakaoka "Eddy Current Dual Packs Heater based Continuous Pipeline Fluid Heating using Soft Switching PWM High Frequency Inverter", Proceedings of International Symposium on Industrial Electronics-Mexico, pp306-311, December, 2000

[6] H.Terai, I.Hirota, T.Miyauchi, H.Omori, Koki Ogura, Y.Hirota, M.Nakaoka "Comparative Performance Evaluations of IGBTs and MCT in Single-Ended Quasi-Resonant Zero Voltage Soft Switching Inverter", Proceedings of Power Electronics Specialists Conference-Vancouver, pp2178-2182, June, 2001

[7] H.Terai, T.Miyauchi, I.Hirota, H.Omori, Mamun A. Al, M.Nakaoka "A Novel Time Ratio Controlled High Frequency Soft Switching Inverter using 4th Generation IGBTs", Proceedings of Power Electronics Specialists Conference-Vancouver, pp1868-1873, June, 2001

3 Phases-3 Devices AC Voltage Regulator With Quasi-Zero Switching

Qianzhi Zhou* (SM IEEE), Wenhua Hu** and Bin Wu*** (SM IEEE)

*Anhui University of Technology/School of Electric Engineering & Information, Maansan, China
**East China Jiaotong University/School of Electrical & Electronic Engineering, Nanchang, China
***Ryerson University/ Department of Electrical & Computer Engineering, Toronto, Canada

Abstract — This paper presents a novel voltage regulator with sine waves, which only consists of 3 switch devices in 3 phases- configuration. Simplest chopper control is adopted to operate the system in quasi-zero switch state (quasi-ZCS switching on and quasi-ZVS switching off) by means of reasonable parameter-match without Zero-switching branch circuit assisted. Analyses and experiment results are presented to prove the high efficiency and its appliance prospect.

Keywords-voltage regulator; chopper control; quasi-zero switching; parameter-match

I. INTRODUCTION

Bulky 3-phases of autotransformers are still widely used in industry and laboratories due to low cost and convenient operation, regardless of a lot of copper and iron materials wasted, and their high weight and large volume. Also most of soft-starters for motor start or break are still adopted SCR devices to form their power converter by phase-shift control, which makes the electrical network in industrial area undergo serious harmonics pollution during frequently motor start or break procedures. So, multi-kinds of harmonic compensation have to be adopted for qualifying the network. However, most economic ways should improve the natural performances of the starter products themselves. It is obviously that high frequency ac chopper formed by power switch devices should be a good choice to replace the phase-shift controlled SCR main circuit in present starters. However since the cost of switch device is much higher than that of SCR, also the total device numbers in 3 phases switch system are more than that in SCR system, the significant replacement have not yet realized. In fact, ac chopper technology is early developed rapidly since 80's years at last century, which is applied in other voltage regulating products for high performance, e.g. adjustable supplies, ac Line Conditioner, etc. Then, if make some progresses in ac chopper topology with less devices and simple control strategy, getting good waveforms with high efficiency and low cost, why not realize the creation in scheme and techniques for above products as soon as possible? This paper just introduces a novel ac chopper to satisfy the requirement for replacing old scheme in previous or more products, so as to reduce the product weight and volume, save medal materials or improve performances, be benefit to clear electrical environment, and reduce the energy consumption.

II. TOPOLOGIES AND SCHEME

Reference [1] and [2] show a kind of voltage regulator with ac chopper scheme consisting of least number of switch devices, where only one common switch device plus 6 fast diodes are adopted for positive and negative freewheeling paths, meeting the need to release the energy stored inside the inductor during phase-switch device at off state. Also for simplest chopper control strategy, only a single pulse from generator need to be controlled by variable Duty-Cycle for adjusting voltages, as long as 3 phase-devices are in step with it. Thus, an enough dead time must be set between 3 control pulses of phase-devices and the control pulse in opposite logic level of the common freewheeling device. Apparently, the switch devices must be shocked by high voltage spike caused by the dead time when all the devices are actually in off-state. Then, a related measure must be adopted to protect the devices. These problems are quite similar in all conventional ac choppers adopting simplest Duty-Cycle control.

Fig.1 (a) and (b) show the topologies of 3 phases-3lines ac chopper and 3phases-4lines ac chopper respectively, which all consist of switch unit and filter unit, where the total numbers of switch devices all less than that of conventional ac choppers. Comparing with [1],[2], it can be seen that the positive and negative freewheeling paths are replaced by a LC unit with π-configuration, so as to reduce a switch device, which makes the freewheeling procedure naturally. So no dead time problem need to be considered like that in the conventional topology among [1] and [2]. However in this situation, reasonable parameter-match is necessary. If excellent match is achieved, quasi-zero switching can be realized by such a simple control strategy and such less switch devices without additional any soft switch branch circuitry. Then, High frequency, high efficiency conversion with bi-directional power flowing can be easy implement. Also the problems in weight, volume, harmonics due to SCR, EMI due to high frequency hard switching, cost, etc. can be solved synthetically. Here the parameter-match is greatly emphasized.

1-4244-0448-7/06/$25.00 ©2006 IEEE

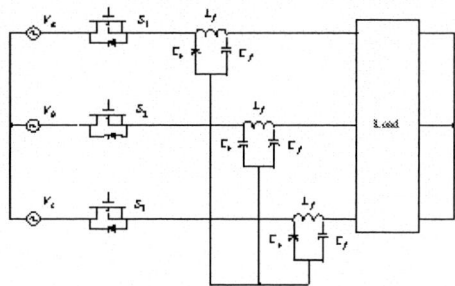

(a) 3 phases-3lines ac chopper

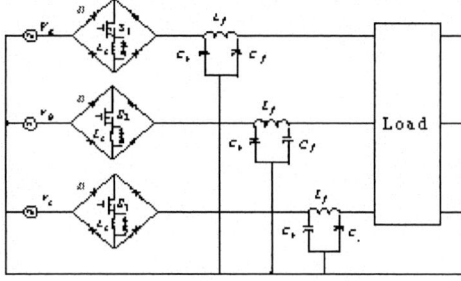

(b) 3 phases-4lines ac chopper

Figure 1.   Voltage regulator adopting new ac chopper

Take phase A from 3 phases-4 lines of system in Fig.1 (b) to show the operation scheme.

Suppose that the load is resistor in Y-connection, setting the Filter Capacitor $C_f \gg C_r$, take one switch-cycle in positive sine half-period to analyze. While in switch-on period of S1, the current flows via two paths, one is, i_f in main path via the fast diode in first bridge leg, S1, L_s, the fast diode in second bridge leg, L_f, load and Neutral back to supply. Another is, i_r via the fast diode in first bridge leg, S1, L_s, the fast diode in second bridge leg, C_r, and Neutral back to supply. At the switch-on transient, i_f is greatly limited by L_f approaching to zero, while setting the root square value of $C_r \cdot L_s$ small enough compared with the switch cycle, then the initial value of current i_r is also small up to zero, The total current i_s flowing via the switch device is the sum of i_f and i_r, also zero. Thus, ZCS is just realized.

While in switch-off period, the current will freewheel resonantly via the filter inductor L_f, capacitor C_f and resonant capacitor C_r. At the switch-off transient, the voltage on phase switch will rise with low dv/dt due to the capacitor across it and the supply, because low frequency supply can be seen as shorted for switching transient. Then, the soft-switch should be accepted as an approximate ZVS.

III. PARAMETER SELECTION & EXPERIMENTAL ANALYSES

A 1kVA of prototype in new topologies with 600W of load was operated at 5kHz switch-frequency, which does not cause much loss in electrical capacitors used as filters. The trade-off parameters are determined as following.

L_f: 1mH, L_s: 1 μ H, C_f: 10 μ F, C_r: 0.22 μ F. An approximate engineering estimation shows the prototype efficiency as about 94%. The test result is about 98% relative to mid Duty Cycle. The data in Table I can be used for calculating above practical efficiency.

Fig.2 and Fig.3 are actual experiment waveforms measured from a prototype, showing the component and total current waveforms through the phase switch respectively, as well as transient current and voltage current demonstrating quasi-zero switching.

TABLE I.
ORIGINAL DATA FOR CALCULATING EFFICIENCY

	Element1	Element2	Element3	Σ A	Σ B
Urms[V]	82.82	83.08	80.40	82.10	---------
Umn [V]	72.71	71.81	69.80	71.44	---------
Udc [V]	0.05	-0.30	-0.16	-0.14	---------
Uac [V]	82.82	83.08	80.40	82.10	---------
Irms[A]	1.405	1.412	1.429	1.416	---------
Imn [A]	0.844	0.851	0.869	0.855	---------
Idc [A]	0.002	0.005	0.002	0.003	---------
Iac [A]	1.405	1.412	1.429	1.416	---------
P [W]	0.0261k	0.0243k	0.0255k	0.0759k	---------
S [VA]	0.1164k	0.1173k	0.1149k	0.3486k	---------
Q [var]	-0.1134k	0.1148k	-0.1120k	-0.1107k	---------
λ [ ]	0.2240	0.2072	0.2223	0.2178	---------
φ [°]	ID 77.06	G 78.04	D 77.16	77.42	---------
fU [Hz]	2.5241k	2.7419k	2.5299k	---------	---------
fI [Hz]	3.5367k	3.7106k	3.7368k	---------	---------
U+pk[V]	371.17	364.99	376.91	---------	---------
U-pk[V]	-365.71	-372.18	-364.49	---------	---------
I+pk[A]	5.178	5.315	5.345	---------	---------
I-pk[A]	-5.148	-5.307	-5.322	---------	---------
CfU [ ]	4.48	4.48	4.69	---------	---------
Cf I [ ]	3.68	3.76	3.74	---------	---------
FfU [ ]	1.27	1.29	1.28	---------	---------
Ff I [ ]	1.85	1.84	1.83	---------	---------
Z [Ω]	58.923	58.850	56.250	173.98	---------
Rs [Ω]	13.198	12.195	12.505	37.891	---------

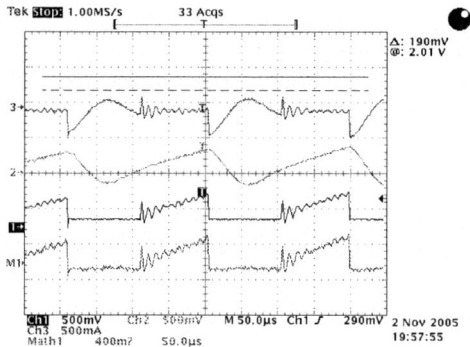

Trace 1: i_r, Trace 2: i_f, Trace 3: i_s, Trace 4: $I_r + i_f$

Figure 2.   Switch current waveforms

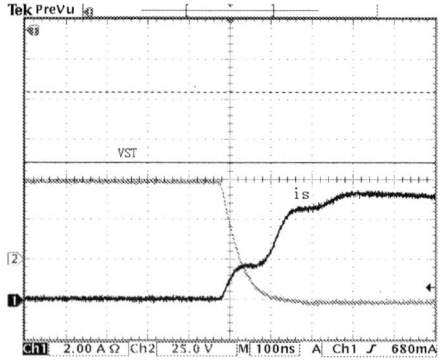

(a) Quasi- ZCS at on-transient

398

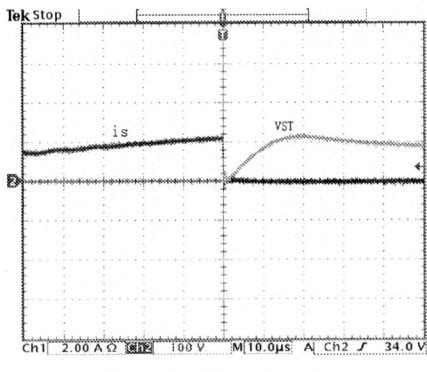

(b) Quasi- ZVS at off- transient

Figure 3. Switch voltage and current show ZCS & ZVS

It is not difficult to analyze the similar situation in negative sine half-period. The difference is only that the current directions in switch-on state or off state are just opposite to that in positive sine half-period. Fig.4 shows that the switch device must undergo twice half-period current from the supply. Also, It is unnecessary to worry about the switch frequencies' interferes shown in Fig.4 polluting the electrical network, because the fixed frequency ripples produced from chopping are easy to be filtered by small capacitors.

The scheme of 3 phases-3 lines topology shown in Fig.1 (a) is similar to that of 3 phases-4 lines topology described above. L_s is, indeed, small enough, which can be replaced by the action of discrete line inductance. However, the switch device only undergoes positive current while the anti-paralleled diode inside the switch device undergoes negative current. So the half- period current only pass through the switch device once time. Although the value is line peak current higher than that of the phase current via the switch twice in 3 phases-4 lines topology, the rms. value of current will be still lower than it. So the efficiency should be higher than that of 3 phases-4 lines topology. However sometimes the Neutral can be removed from the topology of 3 phases-4 lines, getting a 3 phases-3 lines, so that one set can be applied to two kinds of different system alternatively by easy exchanging the configuration. In addition, of course, better performance of the new topologies should be obtained from various kinds of close- loop control strategies discussed in many literatures including [1]. Another novel category of ac regulator by means of SLH (Switch-Line Hybrid Power Conversion) actually got very good THD and load robustness in open-loop operation, which just adopts the topologies in this paper to structure a switch-filter supply feeding a linear unit [3], [4] for synthetic optimal performance.

It is obviously that in the novel ac chopper, creating soft-switch with high frequency and high efficiency is actually along with a way just opposite to popular idea. In order to save devices, it does not increase branch circuits to realize zero-switch, avoiding complex control strategy, normal decoupling measures are replaced by coupling

measures. Several functions are often focused on one element. For instance, the filter inductor, L_f should filter the current ripples and act as a snubber to limit di/dt for creating soft-switch condition simultaneously. The filter capacitor, C_f must act to smooth the output voltage ripples, at the same time, offer a freewheeling path for resonant current, i_f during switch –off period. Capacitor, C_r is both an element to determine small resonant parameters with L_s creating the zero current condition as well as a snubber at the switch-off transient, and a freewheeling current path with resonant elements, L_f and C_f. In the case of strong parameter-coupling with Duty-Cycle modulation, it is difficult to design a perfect system to meet the needs of different load by mathematic model. Then simulation with the combination of experimental adjustment can be a good choice for selecting reasonable parameters. The trade-off parameters should be based on mid Duty-Cycle with half a load, also satisfy the condition $(L_f C_f)^{1/2} \gg T_s/2\pi \gg (L_s C_r)^{1/2}$, (where T_s is the switch cycle), so that the ripple-filtering is not too bad in smaller Duty-Cycle condition (i.e. low voltage), or too bad in lighter load condition. Therefore normally several resonant peak value of i_r will superpose on i_f, which makes the sum of them, total current, i_s flowing through the switch, get the waveform shape as in Fig.5.

It can be seen that only the first resonant peak value is highest. While the Duty-Cycle is adjusted from large to small the numbers of the peak will be changed from less to more. However this situation does not affect the device to deliver the energy normally during the relationship $(L_f C_f)^{1/2} \gg T_s/2\pi \gg (L_s C_r)^{1/2}$ has been satisfied.

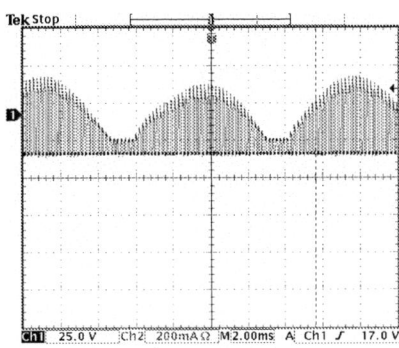

Figure 4. Switch current via S1

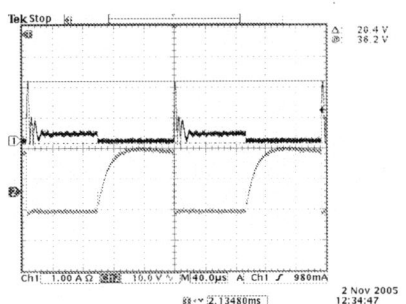

(a) Switch current i_s over on-state

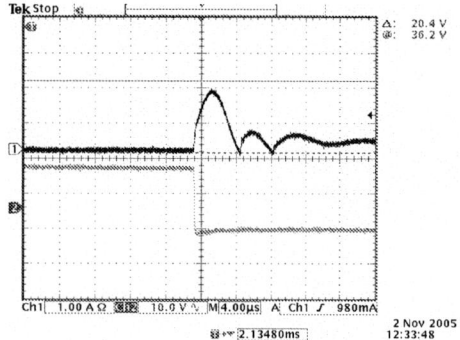

(b) The large scale of switch-on transient

Upper trace: switch current, i_s

Bottom trace: switch voltage, v_{si}

Figure 5. Multi-peak current i_s over on-state

IV. CONCLUSION

1． New topologies of voltage regulator with least switches in 3 phases system based on ac chopper is proposed.

2． Reasonable parameters' match can realize quasi-zero switch in strong coupling by simple circuitry and control strategy.

3． New topologies can be applied for developing high frequency high efficiency autotransformers, line conditioners, motor starters and other 3 phases voltage variable supplies with low cost and environment protection.

ACKNOWLEDGMENT

The authors thank to Dr. Congwei Liu and Mr. Mouzhi Dong for their assistances in recent experiment with great enthusiasm.

REFERENCES

[1] Donato Vincenti, Hua Jin, Phoivos Ziogas, "Design and Implementation of a 25-kVA Three-Phase PWM AC Line Conditioner," *IEEE Transactions on Power Electronics*, vol. 9, no. 4, pp.384-389, 1994.

[2] Haibin Xu, Qianzhi Zhou, " 3 Phase AC Chopper with Low THD," China patent, No.00221159.9, 2001.

[3] Qianzhi Zhou, Wenhua Hu, Bin Wu and Mouzhi Dong, "Switch-Linear Power Conversion (I)– The topology based on Source Follower," *IEEE Conference on Industrial Electronics, 1thed., Singapore*, May, 2006.

[4] Wenhua Hu, Bin Wu, Mouzhi Dong and Qianzhi Zhou "Switch-Linear Power Conversion (II) – Typical analyses in the topology's high efficiency," *IEEE Conference on Industrial Electronics, 1thed., Singapore*, May, 2006.

2006 5th International Power Electronics and Motion Control Conference

Study on Power Decoupling Control of Three Phase Voltage Source PWM Rectifiers

Wang Jiuhe[*], Yin Hongren[*], Zhang Jinlong[*], and Li Huade[**]

*Dept. of Information and communication Engineering, Beijing Information Technology Institute, Beijing, China
** Information Engineering School, University of Science and Technology Beijing, Beijing, China
wjhyhrwm@163.com

Abstract—This paper proposes a new power control strategy of three phase voltage source PWM rectifiers to solve the coupling between instantaneous active power and instantaneous reactive power. This strategy is obtained based on the power control math model of three phase voltage source PWM rectifiers in synchronous *dq* coordinates. A new power control switching table is set up according to new dividing of input voltage space and resultant voltage space vectors for improving the properties of system and saving complicated PWM modulation module. The switching functions is selected in new power control switching table according to the position of source voltage space vector and input voltage space vector of rectifier required. This paper gives a design method of system controller. The control strategy is proved feasible by simulink simulation with different loads.

Keywords—*PWM rectifier; unity power factor; power decoupling; sector; voltage space vector*

I. INTRODUCTION

PWM rectifier has many advantages, such as sinusoidal current control and unity power factor (UPF) on the AC side, and DC output voltage control on the DC side, and reversible power flow between both sides etc.. The power control strategy of PWM rectifier has simpler structure and algorithm, better dynamics, UPF etc, so that many researchers are interested in it. In order to improve the properties of PWM rectifier power control system, difference (or varies) control strategies are introduced in [1]. The virtual-flux-based power control strategy has lower sampling frequency, a simpler voltage and power algorithm, low total harmonic distortion (THD) current in[2]-[5]. The power control strategy of PWM rectifiers with the dead zone of sector borders eliminates the mistakes of selecting input voltage space vector of rectifier on sector borders and improves the waveforms of instantaneous active power and DC voltage in[6]. The power control strategy of PWM rectifiers via output regulation subspaces takes instantaneous active power and reactive power as output, and input voltage space vectors are chosen to control instantaneous active power and reactive power based on their derivatives, and good

control is obtained in[7]. Double switching tables power control strategy is presented based on using of instantaneous active power and reactive power switching tables in turn, and good control properties are achieved by the strategy in[8]. Meanwhile, DC dynamic voltage drop during load current disturbance is decreased and stable voltage drop is eliminated based on load current feedback. Despite of that above control strategies are used, instantaneous active power and reactive power are couple because the three phase voltage source PWM rectifier is hybrid nonlinear system, and influences the properties of the system. In this paper, power control model of voltage source PWM rectifier is obtained from its Math model in synchronous rotating *dq* coordinate, and a new power control strategy is proposed based on power decoupling, and a new power control switching table is set up for implementing the strategy therefore. The control strategy is proved feasible by simulink simulation under different loads.

The paper is organized as follow. In Section II Math model in synchronous rotating *dq* coordinate and power control model of voltage source PWM rectifier are set up. Section III presents power decoupling control strategy of voltage source PWM rectifier and gives a design method of power decoupling controller, and the strategy is implemented based on a new power control switching table. Section IV gives simulink simulation results of power decoupling control system. Finally, Section V states our conclusions and points to further improvements.

II. MATH MODEL OF VOLTAGE SOURCE PWM RECTIFIER

A. Math Model of Voltage Source PWM Rectifier in Synchronous Rotating dq Coordinate

The power circuit of three phase voltage source PWM rectifier is shown in Fig.1. In order to set up Math model, it is assumed that the AC voltage is a balanced three phase supply, the filter reactor is linear, IGBT is ideal switch and lossless. u_a, u_b and u_c are the phase voltages of three phase balanced voltage source and i_a, i_b, and i_c are phase current in Fig.1. S_a, S_b, and S_c are the switching function of PWM rectifier. S_j ($j = a$、b、c) is defined as

---

Project Supported by Science and Technology Development Project of Beijing Municipal Commission Of Education（KM200510772001）

1-4244-0448-7/06/$25.00 ©2006 IEEE   401

$$S_j = \begin{cases} 1, & S_j \text{ closed} \\ 0, & \overline{S}_j \text{ closed} \end{cases} \quad (1)$$

U_{dc} is the DC output voltage, R and L mean resistance and inductance of filter reactor, respectively, C is the DC side capacitance, R_L is the DC side load, u_{ra}, u_{rb} and u_{rc} are the input voltages of rectifier, and i_L is load current.

The math model of three phase voltage source PWM rectifier can be written as:

$$L\frac{d}{dt}\begin{bmatrix} i_a \\ i_b \\ i_c \end{bmatrix} = \begin{bmatrix} u_a \\ u_b \\ u_c \end{bmatrix} - R\begin{bmatrix} i_a \\ i_b \\ i_c \end{bmatrix} - \begin{bmatrix} u_{ra} \\ u_{rb} \\ u_{rc} \end{bmatrix} \quad (2)$$

$$L\frac{d\boldsymbol{i}}{dt} = \boldsymbol{u} - R\boldsymbol{i} - \boldsymbol{u}_r \quad (3)$$

$$C\frac{dU_{dc}}{dt} = i_{dc} - \frac{U_{dc}}{R_L} \quad (4)$$

where $u_{ra} = S_a U_{dc} + U_{ON}, u_{rb} = S_b U_{dc} + U_{ON}$, $u_{rc} = S_c U_{dc} + U_{ON}$, U_{ON} is the voltage between point O and point N, $\boldsymbol{i}$, $\boldsymbol{u}$, and $\boldsymbol{u}_r$ are current vector, voltage source vector, input voltage vector of rectifier, respectively, $i_{dc} = S_a i_a + S_b i_b + S_c i_c$.

We can transform (2) and (4) into synchronous rotating dq coordinate based on power conservation, and accordingly the math model of voltage-source PWM rectifier can be written as

$$L\frac{di_d}{dt} = u_d - Ri_d + \omega Li_q - u_{rd} \quad (5)$$

$$L\frac{di_q}{dt} = u_q - Ri_q - \omega Li_d - u_{rq} \quad (6)$$

$$C\frac{dU_{dc}}{dt} = i_{dc} - i_L = (i_d S_d + i_q S_q) - \frac{U_{dc}}{R_L} \quad (7)$$

where $u_{rd} = S_d U_{dc}, u_{rq} = S_q U_{dc}$, u_{rd}, u_{rq} and S_d, S_q are input voltage of rectifier, switch function in synchronous rotating dq coordinate, respectively. u_d, u_q and i_d, i_q are voltage source, current in synchronous rotating dq coordinate, respectively. ω is angular frequency.

B. the Power Control Math Model of Voltage Source PWM Rectifier

The instantaneous active and reactive power of PWM rectifier is defined by the product of the voltage vector and the current vector. The active power p is the scalar product of the voltage vector and the current vector,

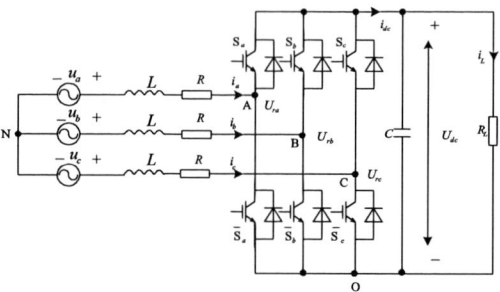

Fig.1 the power circuit of three phase voltage source PWM rectifiers

whereas the reactive power q is calculated as a vector product of them. In the case of a balanced three-phase product of them. In the case of a balanced three-phase supply, $u_d = \sqrt{3/2}U_m$ (U_m 为 is amplitude of the phase voltage), $u_q = 0$, the calculation of p and q can be Simplified as

$$p = \sqrt{\frac{3}{2}}U_m i_d \quad (8)$$

$$q = -\sqrt{\frac{3}{2}}U_m i_q \quad (9)$$

According to math model(5)（6）and power calculation(8)（9）, the math model with p and q variables is given by

$$L\frac{dp}{dt} = 1.5U_m^2 - Rp - \omega Lq - p_{rd} \quad (10)$$

$$L\frac{dq}{dt} = -Rq + \omega Lp + q_{rq} \quad (11)$$

Where $p_{rd} = \sqrt{3/2}U_m u_{rd}$ and $q_{rq} = \sqrt{3/2}U_m u_{rq}$ are defined as power control input of PWM rectifier.

On condition that resistance loss and switching loss are neglected,（7）can be written as

$$p = U_{dc}i_{dc} = CU_{dc}\frac{dU_{dc}}{dt} + \frac{U_{dc}^2}{R_L} \quad (12)$$

Let $U_d = U_{dc}^2$,（12）can be expressed as

$$p = \frac{1}{2}C\frac{dU_d}{dt} + \frac{U_d}{R_L} \quad (13)$$

p and q are couple by (10)(11), which causes trouble of power control. DC voltage is depended on p from (12) or (13).

III. POWER DECOUPLING CONTROL OF THREE PHASE VOLTAGE SOURCE PWM RECTIFIER

A. Structure of Power Decoupling Control

In order to implement power decoupling control, （10）（11）can be written as

$$L\frac{dp}{dt} = 1.5U_m^2 - Rp - \omega Lq - p_{rd} + p_{co1} + p_{co2} \quad (14)$$

$$L\frac{dq}{dt} = -Rq + \omega Lp + q_{rq} + q_{co} \quad (15)$$

where p_{co1}, p_{co2} and q_{co} are compensated terms.

Let $p_{co1} = \omega Lq$, $p_{co2} = -1.5U_m^2$ and $q_{co} = -\omega Lp$, power decoupling control is implemented.

Based on (14)（15）, structural diagram of power decoupling of three phase voltage source PWM rectifier is shown in Fig.2. Decoupled power control structure is shown in Fig.3.

B. Implement of power decoupling control

In order to save complicated PWM modulation module, p_{rd} and q_{rq} of satisfying control requirement are transformed into p_α and q_β by $T_{dq/\alpha\beta}$ (transform synchronous rotating dq coordinate variables into a stationary $\alpha\beta$ coordinate). According to p_α and q_β, the sectors of u_r is located, i.e. $\theta_p = arctgq_\beta / p_\alpha$ .Input voltage space is divided according to Fig.4 i.e.

$$(2n-3)\frac{\pi}{12} \le \theta_n \le (2n-1)\frac{\pi}{12} \quad n=1,2,\cdots,12 \quad (16)$$

New resultant voltage space vectors can be synthesized with voltage space vectors in adjacent voltage space vectors, such as U_{12} can be synthesized with U_1 and U_2, in Fig.4. In Fig.4, a sector has a non-zero voltage space vector. When θ_p is in a sector, a non-zero voltage space vector is chosen, and new power control switching table is obtained as shown in table Ⅰ.

Power decoupling control system of three phase voltage source PWM rectifier is shown in Fig.5. Control system is composed by DC voltage outer loop and power inner loop. Instantaneous active power p and instantaneous reactive power q are calculated by u_a, u_b and u_c from voltage transducers and i_a, i_b and i_c from current transducers. θ_n is calculated by u_α and u_β from u_a, u_b and u_c. Based on u_α and u_β, θ_n is calculated by sector locating device. p, q, p_{ref}, q_{ref} and U_m are inputed to power decoupling device, and p_{rd} and q_{rq} are obtained. p_{rd} and q_{rq} are transformed into p_α and q_β by $T_{dq/\alpha\beta}$.

θ_p is calculated by p_α and q_β. S_a, S_b and S_c are selected in switching table according to θ_n and θ_p . p_{ref} equals the product of DC voltage and output of PI regulator. q_{ref} equals zero in order to implement UPF. R is neglected in Fig.5.

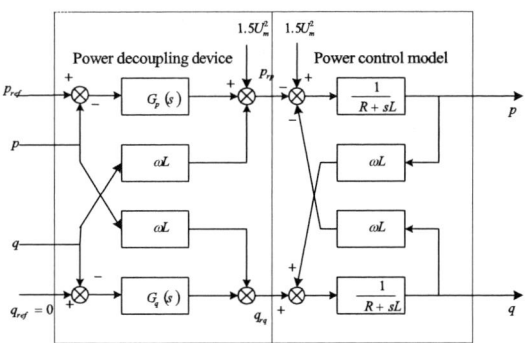

Fig.2 Structural diagram of power decoupling

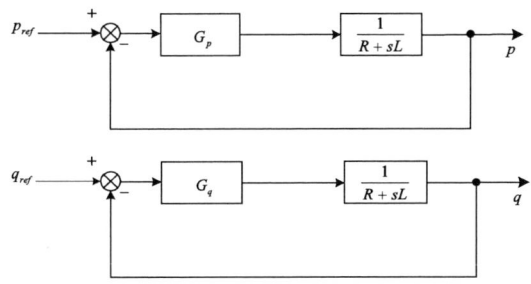

Fig.3 Power control structural diagram after decoupling

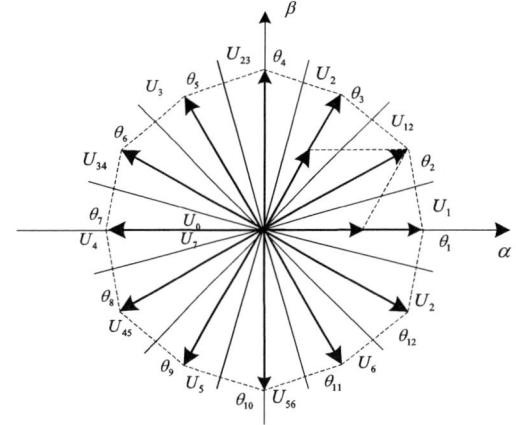

Fig.4 New resultant voltage space vectors and dividing of input voltage space

C. Design of Power Decoupling Controller

a Design of Power Loop

Based on Fig.3 and Fig.5, power decoupling control system structure of rectifier is shown in Fig.6. In Fig.6, the system is transformed into a linear system, and system is designed based on the structure.

From Fig.3, active power control structure is the same as reactive one, i.e. $G_p(s) = G_q(s)$. PI regulator is used in power inner loop, its transfer function is

$$G_P(s) = K_p \frac{\tau_p s + 1}{\tau_p s} = K_{pp} + \frac{K_{pl}}{s} \quad (17)$$

403

TABLE. I

POWER DECOUPLING CONTROL SWITCHING TABLE

S_a S_b S_c												
θ_p	θ_1	θ_2	θ_3	θ_4	θ_5	θ_6	θ_7	θ_8	θ_9	θ_{10}	θ_{11}	θ_{12}
U_1	U_{12}	U_2	U_{23}	U_3	U_{34}	U_4	U_{45}	U_5	U_{56}	U_6	U_{61}	
(110)		(110)		(010)		(011)		(001)		(101)		

Closed transfer function of power inner loop can be expressed as

$$G_{pc}(s) = \frac{\dfrac{K_{pp}}{L}s + \dfrac{K_{pI}}{L}}{s^2 + \dfrac{K_{pp}+R}{L}s + \dfrac{K_{pI}}{L}} \qquad (18)$$

For pole-zero cancellation, the parameters of the PI regulator should be chosen as

$$K_{pp} = K_{pI}\frac{L}{R} = K_{pI}T_L \qquad (19)$$

Substituting (21) into (20), $G_{pc}(s)$ is obtained as

$$G_{pc}(s) = \frac{p(s)}{p_{ref}(s)} = \frac{K_{pp}}{K_{pp}+Ls} = \frac{1}{1+T_p s} \qquad (20)$$

Where $T_p = \dfrac{L}{K_{pp}}$ is power time constant.

b   Design of   Voltage Loop

The transfer function of voltage regulator is

$$G_v(s) = K_v\frac{\tau_v s + 1}{\tau_v s} = K_{pv} + \frac{K_{iv}}{s} \qquad (21)$$

By Fig.6, the open transfer function of system can be expressed as

$$G_{Ov}(s) = \frac{K_v K_s U_{dcr}(\tau_v s + 1)}{\tau_v s(T_c s + 1)(T_p s + 1)(R_L C s + 1)} \qquad (22)$$

According to given parameters, (22) is transformed into canonical II system, that is

$$G_{Ov}(s) = \frac{K(\tau s + 1)}{s^2(Ts+1)} \qquad (23)$$

K_v and τ_v can be selected based on canonical II system.

IV.   SIMILATION OF POWER DECOUPLING CONTROL SYSTEM OF VOLTAGE SOURCE PWM RECTIFIER

System parameters followed as:

L=5mH, R=0.1Ω, R_L=10Ω, C=2200μF, U_{dcr}=200V, K_{pp}=0.4, K_{pi}=10, K_{iv}= 0.00072, K_{pv}=0.00026.

Simulink model of power decoupling control system of voltage source PWM rectifier is set up based Fig.5 and above parameters. Simulation results with different load are given as follows:

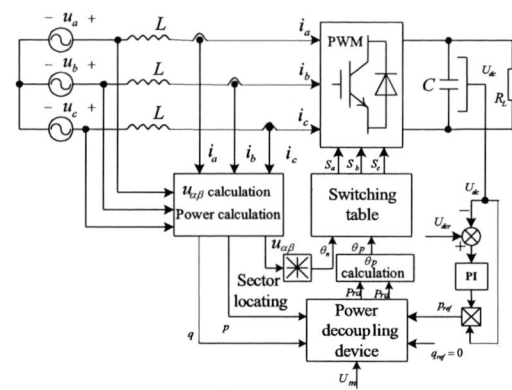

Fig.5 Power decoupling control system of three phase voltage source PWM rectifier

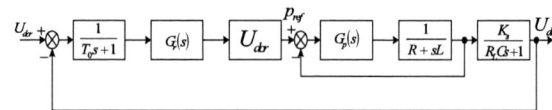

Fig.6 Power decoupling control structural diagram of three phase voltage source PWM rectifier

A .Nominal Load（R_L=10Ω）

Simulation results of nominal load are shown in Fig.7. The system has good static performances from Fig.7.

B. Overload（R_L=5Ω）

Simulation results of overload are shown in Fig.8.

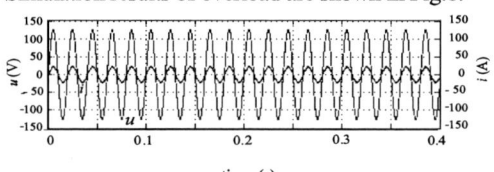

(a) phase voltage u, phase current i

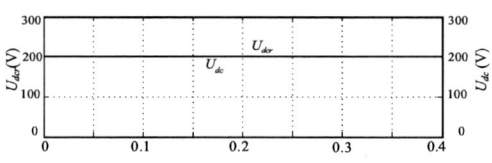

（b）DC voltage U_{dc}, DC reference voltage U_{dcr}

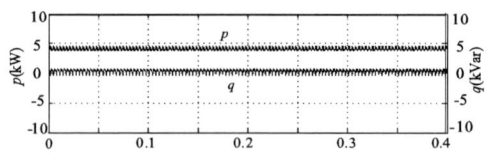

(c) active power p , reactive power q

Fig.7 Simulation results of power decoupling control system with R_L=10Ω

404

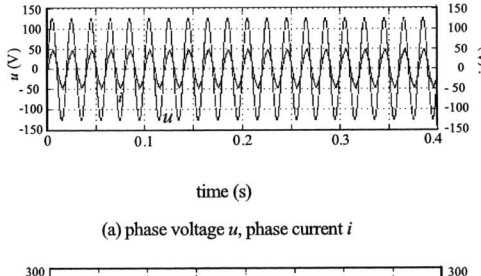

time (s)

(a) phase voltage u, phase current i

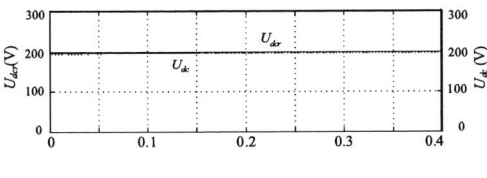

time (s)

（b）DC voltage U_{dc}, DC reference voltage U_{dcr}

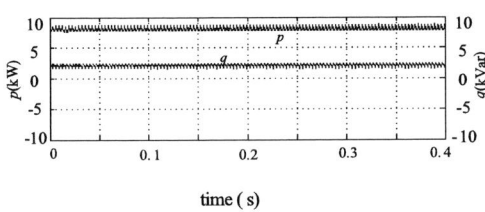

time (s)

(c) active power p , reactive power q

Fig.8 Simulation results of power decoupling control system with R_L=5Ω

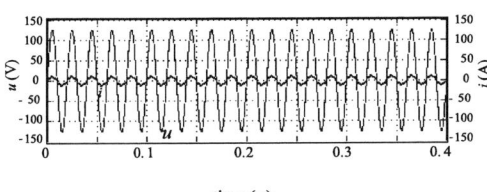

time (s)

(a) phase voltage u, phase current i

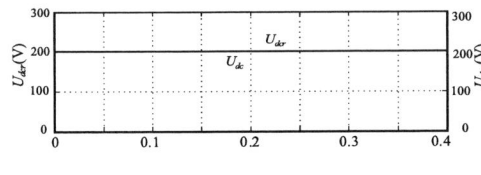

time (s)

(b)DC voltage U_{dc}, DC reference voltage U_{dcr}

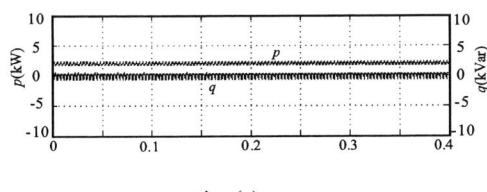

time (s)

(c) active power p , reactive power q

Fig.9 Simulation results of power decoupling control system with R_L=20Ω

C. Underload（R_L=20Ω）

Simulation results of underload are shown in Fig.9.

V. CONCLUSIONS

A power control model of voltage source PWM rectifier is obtained from its Math model in synchronous rotating dq coordinate. Based on the power control model, this paper proposes a new power control strategy of three phase voltage source PWM rectifiers based on power decoupling. The independent control of instantaneous active and reactive power is obtained by using this strategy. A new power control switching table is set up according to new dividing of input voltage space and resultant voltage space vectors to save complicated PWM modulation module. The control strategy is proved feasible by simulation with different loads. The power decoupling control system of voltage source PWM rectifier has many advantages, such as simple structure, higher DC voltage tracking precision, higher UPF etc.

From Fig.8(c), reactive power is not equal to zero.(Why?), and needs to be studied further.

REFERENCES

[1] Wang Jiuhe, Li Huade, Li Zhengxi, "Direct powercontrol technology of three-phase boost type PWM rectifiers," *Advanced technology of electrical engineering and energy*, 2004,Vol.23,No.3, pp64-67

[2] Toshihiko Noguchi, Hiroaki Tomiki, Seiji Kondo, Isao Takahashi, "Direct Power Control of PWM Converter Without Power Source Voltage Sensors,"*IEEE Trans on Industry Applications*, 1998, Vol.34, No.63, pp473-479.

[3] M. Sc.Mariusz Malinnowski. "Sensorless Control Strategies for Three-Phase PWM Rectifier," Warsaw University of Technology. Ph.D. Thesis, 2001.

[4] Mariusz Malinowski, Marian P.Kazmierkowski, Steffan Hansen, Frede Blaabjerg, and G.D.Marques, "Virtual-Flux_Based Direct Power Control of Three-Phase PWM Rectifiers," *IEEE Trans Industry Applications.*,2001, Vol.37, No.4, pp1019-1027.

[5] He Zhiyuan, Wei Wei, "Study on direct power control of PWM rectifier based on virtual flux," *Journal of Zhejiang University(Engineering Science)*, 2004, Vol.38, No.12, pp1619-1622.

[6] Wang Jiuhe, Li Huade, Yang Liyong, "Direct power control of three-phase boost-type PWM rectifyiers with the dead zone of sector borders," Journal of University of Science and Technology Beijing, 2005, Vol.27, No.3, pp380-384.

[7] Gerardo Escobar, AleksandarM.Stankovic, Juan.M.Carraso, Eduardo Galvan, RomeoOrtega, "Analysis and Design of Direct Power control (DPC)for a three Phase Synchronous Rectifier via Output Regulation Subspaces," *IEEE Trans on Power Electronics*, 2003, Vol.18, No.1, pp823-830.

[8] Wang Jiuhe, Li Huade, "A new direct power control strategy of three phase boost type PWM rectifiers," *Proceedings of the CSEE*, 2005, Vol.25, No.16, pp47-52

2006 5th International Power Electronics and Motion Control Conference

A Fully Digital Controlled 3KW, Single-Stage Power Factor Correction Converter Based on Full-Bridge Topology

HANG Li-jun, YANG Yue-feng, SU Bin, LU Zheng-yu, QIAN Zhao-ming

(College of Electrical Engineering, Zhejiang University, Hangzhou 310027, China)

E-mail Address: leejean_hang@hotmail.com

Telephone and fax numbers: +86-571-87951950

Abstract-This paper deals with a 3KW power factor correction (PFC) circuit based on the isolated full-bridge (FB) converter. The operation principle of the very converter is simply analyzed. Digital design consideration based on the small-signal dynamic modeling of the isolated PFC converter is specified in this paper. Implementation strategy and cost effectiveness of digital control for the single-stage PFC converter are studied. A transformer is used to provide galvanic isolation enables a wide range of output voltages, either lower or higher than the input voltage, allowing multi-output solutions. The adopted digital control strategy features fast dynamic response and excellent steady regulation characteristic without multiplying the input feedforward. High power factor and high efficiency are both achieved by combining FB main circuit and digital control. That flexibility is greatly increased can further contribute to system integration research on distributed power system[1]. Without multiplying the input feedforward, fast dynamic can be achieved due to the flexibility of digital control strategy. The above excellent performances are verified by an experimental prototype, whose output power is 3KW and PFC operating frequency is 110K.

Keywords-Single-stage; Full-bridge(FB); power factor correction (PFC); digital control; DSP

I. INTRODUCTION

Nowadays, a switch mode power supply (SMPS) is required to realize high power factor and low input current harmonics complied with harmonic standards such as IEC61000-3-2 together with a small size. A number of single-stage and two stage PFC circuits have been developed and reported in the literature [2]-[5]. Although the two-stage PFC circuits offer excellent performance in terms of input power factor, holdup time capability, no low-frequency ripple in the output voltage, and high efficiency, they have the disadvantages of low-power density and high cost. In order to provide a cost-effective and high-density solution for PFC ac/dc power supplies, particularly at low-power levels, a number of single-stage PFC circuits have been recently developed and reported in the literature. Increasing the switching frequency can make the size of SMPS even smaller.

In the single-stage PFC topology, performance of both PFC and DC/DC voltage regulation is implemented by one set of switches and controller. Provided that the PFC level offers constant output power, i.e. the output end obtain constant power, however, the input power is a periodical value, therefore, the instantaneous input does not equal the instantaneous output power and a capacitor is required for storing energy. Once the load is greatly

Project supported by National Natural Science Foundation of China (50237030ZD)

reduced, the duty-ratio can't be regulated in time, therefore the energy which charges to the capacitor is more than that which discharges from the capacitor, which results in the voltage across the capacitor greatly increases. For maintaining constant output voltage, voltage feedback loop works, i.e. the duty-ratio is reduced and further, the input energy is reduced. Generally, for the single-stage PFC topology, the dynamic performance can't be satisfied by analogy control. Nowadays, the prominent characteristic of digital control strategy is presented in switching mode power supply. Digital control processor dedicated to PWM converter, which possesses the merits of low loss, good immunity to the variation of parameters of the analogy circuits and powerful control ability, becomes more and more popular. Due to its powerful calculation ability, implementation of perfect and complex control technique can be easily realized and therefore digital PWM control system that is based on quantization of duty cycle is broadly employed. Generally, excellent dynamic performance can be achieved by adopting digital control. Furthermore, the front-end PFC regulator is required in every DPS (distributed power supply) system. Increasing the flexibility and intelligent characteristic of DPS system is the developing trend of power electronic system integration. Therefore, it is absolutely necessary to study the digital control strategy for isolated PFC.

This paper deals with a 3KW power factor correction (PFC) circuit based on the isolated full-bridge (FB) converter which is controlled by TMS320LF2407. The operation principle of the very converter is analyzed. Digital design consideration based on low frequency small-signal dynamic modeling of the PFC converter is specified. Implementation strategy and cost effectiveness of digital control for the converter are studied. A transformer is used to provide galvanic isolation enables a wide range of output voltages, either lower or higher than the input voltage, allowing multi-output solutions. The adopted digital control strategy features fast dynamic response and excellent steady regulation characteristic. High power factor and high efficiency are both achieved by combining FB isolated PFC main circuit and digital control strategy. The above excellent performances are verified by an experimental prototype, whose output power is 3KW and switching frequency is 110K.

II. ANALYSIS OF OPERATION PRINCIPLE AND MAIN TOPOLOGY EQUIVALENCE

Generally, topologies of full-bridge and push-pull are adopted in the medium-power level converter with output capacitance of several kilowatts [6]. However, the

1-4244-0448-7/06/$25.00 ©2006 IEEE

voltage stress of the later topology is twice of output, therefore it is difficult to choose the appropriate switches to make compromise of conduction dissipation and drain-source breakdown voltage. Considering IGBT, the

the characteristic of lower current/high voltage. Therefore, the application occasion of the converter with PFC input function is greatly extended by adopting single-stage FB PFC.

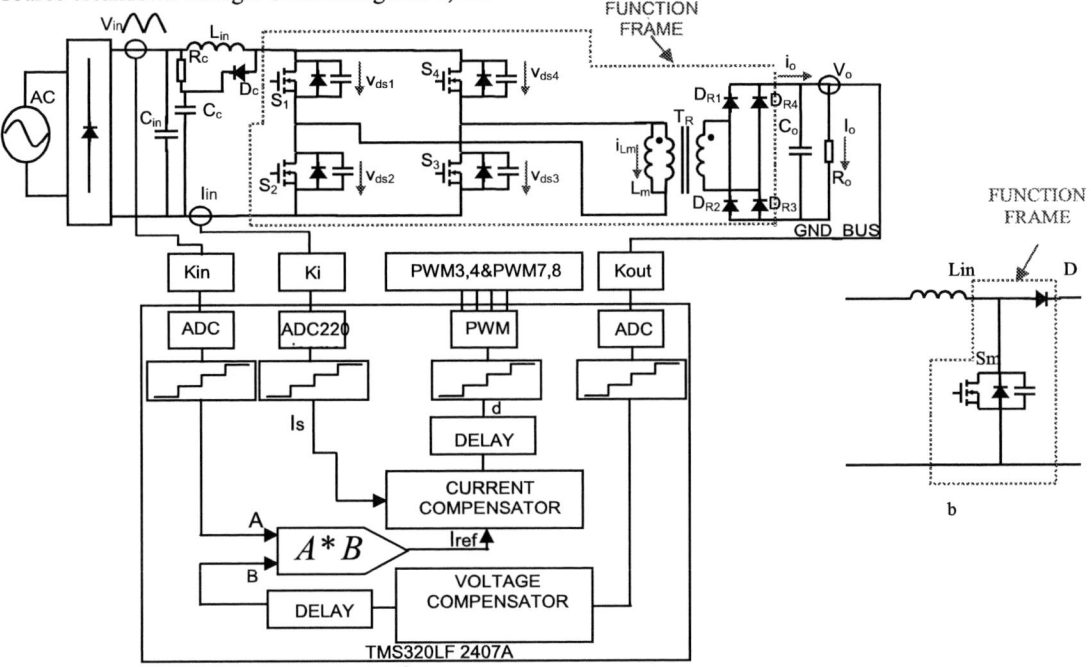

Fig.1 System Configuration of Fully Digital Controlled SS-FB PFC

switching frequency must be reduced, which limits the reduction of both transformer size and power density. Therefore, the former topology is a preferable one. The fully digital controlled single-stage PFC system is shown in Fig.1 (a), wherein, the circuit included in the dashed frame shows the main topology that adopted in the converter. The topology implements the function of both PFC and output regulation to acquire high power factor and steady output with high precision. By adding a transformer for galvanic isolation offers several interesting possibilities, such as: improved safety; simpler converters (which, consequently, means more efficient) can be used for additional output voltage regulation; multi-output configurations are easily achievable; moreover, in certain cases, using only the PFC converter will be enough to provide power of the desired quality. The full-bridge configuration, presented in Fig.1 (a), is better for high input currents and high input voltages. When intended for use with a lower input voltage (e.g. 115V mains) and/or lower output power, a full-bridge form can be replaced by half-bridge version, simplifying the circuit, but with the expense of doubled maximum voltage across the switches. Output voltage can be set to a value lower than the amplitude of the mains voltage depending on the addition of transformer. Along with galvanic isolation, it provides additional safety. Multiple outputs are easily achievable when desired for specific applications, for example, usually, the main output is intended for higher current while the auxiliary output has

In the following text, the operation principle is specified. First, it is assumed that the turn-ratio of the transformer is 1:1 and the circuit included in the dashed frame only implements the function of PFC, therefore the function frame, as shown in Fig.1(a), can be regarded as the equivalent of conventional Boost PFC. The following analysis is based on the function equivalence. The main operation waveforms based on the topology which is shown in Fig.1 (a) are shown in Fig.2. As shown in Fig.2,Vgs1, Vgs2, Vgs3 and Vgs4 offer the driving signals of main switches S1~S4. Wherein, S1 and S2 are respectively identical with S3 and S4. V_p indicates the primary voltage waveform of transformer, n is the turn-ratio between primary side and secondary side of the transformer. V_o is the output voltage of secondary side. I_L is the current waveform of inductor. T, which can be defined as PFC period, is the operating period of the inductor according to its charging and discharging process. T_{sw} is the switching frequency of the main switches, i.e. operating period of the transformer. Waveforms presented in Fig.2 show the pattern for driving the switches, together with the associated choke current. The input choke operates with doubled switching frequency, similar to other full-bridge converters. Soft-switching technique enables a significant increase in the operating frequency and that correspondingly further reduces dimensions of the input choke as compared to standard boost converters, therefore size of the converter can be greatly reduced [7][8].

407

Next, compared with the conventional Boost PFC, the definition of duty-ratio in the isolated FB-PFC is given and accordingly the operating modes of the SS-FB-PFC are specified. For conventional Boost type PFC, whose function frame is shown in Fig.1(b), once main active switch S_m is turned on, the input inductor L_{in} is charged and accordingly the current increases, therefore, energy is stored; similarly, all main switches involved in the SS-FB-PFC are turned on, i.e. the interval that all of the driving signals are overlapped with each other, the voltage after the rectifier charges the inductor and the energy is also stored, therefore, the fully overlapped interval is defined as duty-ratio. As shown in Fig.1 (b), after S_m is turned off, the input inductor charges to the output capacitor, the inductor discharges and consequently the inductor current decreases, the energy is transferred to the output end. Similarly, for the SS-FB-PFC, either S1 and S3 or S2 and S4 are turned on, i.e. the intervals of t1 or t2 as shown in Fig.2, the energy is transferred to the output end from the input inductor. Operating process of the SS-FB-PFC can be further understood by comparison.

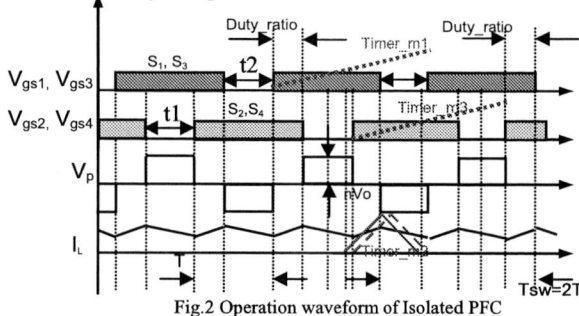

Fig.2 Operation waveform of Isolated PFC

According to the above analysis, the operation process can be divided into two modes, which are respective charging process and discharging process of input inductor, furthermore, the later mode can be regarded as the conventional FB operation mode that transfers the energy from the inductor to the output end by the transformer. By adding the transformer, the converter offers galvanic isolation. Furthermore, the output voltage range can be extended and multi-output can be realized. The basic equations defining this circuit are derived by following standard and simple procedures. The most important equation defines the output-to-input voltage ratio as shown in expression (1):

$$\frac{V_o}{V_i} = \frac{1}{n(1-\delta)} \qquad (1)$$

Where, V_o- output, V_i- input, n - transformer turns ratio, δ - duty-cycle.

III. SMALL MODELING AND DIGITAL CONTROL TRATEGY DESIGN OF THE SS-FB-PFC

Usually, the PFC controller is double-loop which includes the voltage and current loop [9]. According to operation principle analysis, the function frame of SS-FB-PFC can be substituted by the conventional Boost PFC. The equivalent small-signal model, as shown in

Fig.3, can be obtained. Accordingly, transfer function of control to input current can be derived as expression (2):

$$G_{id}(s) = \frac{\hat{i}(s)}{\hat{d}(s)} = \frac{V_o R C S + 2 V_o}{R L C S^2 + L S + R D'^2}$$

$$= \frac{2 V_o}{R(1-D)^2} \frac{1 + \frac{R C S}{2}}{1 + \frac{S L}{R(1-D)^2} + \frac{L C S^2}{(1-D)^2}} \qquad (2)$$

Wherein $\qquad V_o = I_g D' R = I_g (1-D) R$ (3)

Once operation frequency greatly exceeds the turn point frequency, the transfer function can be approximately regarded as: $G_{id}(S) = \dfrac{\hat{i}}{\hat{d}} = \dfrac{V_o}{LS}$ (4)

Based on the small-signal model of control to input current, the digital control compensator algorithm of the current loop can be deduced. The power stage can be digitized using the zero-order-hold method with sampling frequency of PFC operating frequency, as shown in Fig.4. The switching frequency f_{sw} is 55K and the PFC operating frequency f is 110K. Therefore, the sampling frequency f_s is 110K. From the dashed frame part of Fig.4, discrete duty to current transfer function is derived:

$$G_{idz}(z) = \frac{V_{out}}{L} * \frac{T_s}{(z-1)} \qquad (5)$$

Where $T_s = (1/110)*10^{-3}$ second. Generally, for analogy special IC controlled application, an analog low-pass filter with cutoff frequency equal to half of the switching frequency are inserted in both current and voltage feedback loops in order to reduce the aliasing effect. As a result, only the frequency range below half of the switching frequency is of concern in digital controller design and the transfer function of the low pass filter is ignored. When we are designing the current compensator, the influence of slow voltage loop can be ignored. In addition, the computation delay must be compensated. The current loop with digital compensator is illustrated in Fig. 4. The design target is similar to that of the analog compensator. For robustness, the phase margin is set to 45 degrees. It is defined that computation delay is T_{delay}, which is generally supposed to be equal to T_s. The gain of current loop which is illustrated in Fig.4 is:

$$T_c(z) = G_{idz}(z) * G_c(z) * K_i * z^{-\frac{T_{delay}}{T_s}} \qquad (6)$$

The compensated function $G_c(z)$ is supposed to be PI algorithm that can be expressed as follows:

$$G_c(z) = K_p + \frac{K_I}{1 - Z^{-1}} \qquad (7)$$

Fig.3 Current loop compensator

408

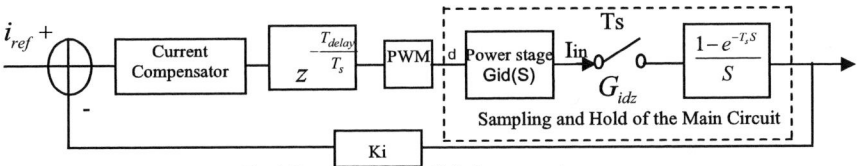

Fig.4 Current loop with digital compensator

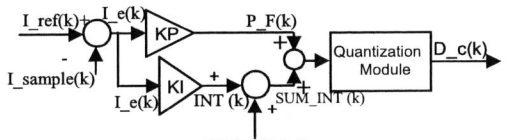

Fig.5 Current loop compensator

Furthermore, (6) can be specified as expression (8):

$$T_c(z) = \frac{V_{out} * T_s}{L(z-1)} * \frac{(K_P + K_I) * z - K_P}{z-1} * K_i * z^{-\frac{T_{delay}}{T_s}} \quad (8)$$

According to the design targets, which are described as expression (8):

$$\begin{cases} |T_c(e^{j\omega_c T_s})| = 1 \\ \angle T_c(e^{j\omega_c T_s}) = -180° + 45° \end{cases} \quad (9)$$

Crossover frequency ω_c and phase margin are used to determine the unknown parameters, gain K_P and integral coefficient K_I. Put (8) into equation (9), the following equation (10) can be obtained.

$$\begin{cases} \frac{V_{out} * T_s}{L|e^{j\omega_c T_s} - 1|} * (K_I + K_P) * \frac{|e^{j\omega_c T_s} - [K_P/(K_I + K_P)]|}{|e^{j\omega_c T_s} - 1|} * K_i = 1 \\ \angle e^{j\omega_c T_s} - [K_P/(K_I + K_P)]\} - 2\angle(e^{j\omega_c T_s} - 1) - \omega_c * T_{delay} * 180°/\pi = -135° \end{cases}$$

$$(10)$$

By solving the equation (10), K_P and K_I can be determined. The simplest way to realize the digital compensator that is expressed in Z domain is to convert it directly into discrete differential equation which can be easily realized in DSP. Transfer function of Z domain and its corresponding discrete differential equation of the general PID algorithm can be respectively expressed as (11) and (12):

$$U(z) = K_P E(z) + K_I \frac{E(z)}{1-z^{-1}} + K_D[E(z) - z^{-1}E(z)] \quad (11)$$

$$u(k) = K_P e(k) + K_I \sum_{j=0}^{k} e(j) + K_D[e(k) - e(k-1)] \quad (12)$$

IV. EXPERIMENTAL RESULS

For verifying the correctness of main topology principle and flexibility of control strategy, an experimental prototype adopting the topology as shown in Fig.1(a) with fully digital control is built, corresponding experiment results are given. For SS-FB-PFC system configuration, as shown in Fig.1(a), specification and main parameter are as follows:

Out put power: 3KW

Input voltage V_{in}: AC 220V±20% （after full-bridge rectifier circuit）

Output voltage V_{o1}: 373V

Operating frequency of PFC:110K

Main switches S_1, S_2, S_3, S_4: APT8014L2FLL

Rectifier diodes of main output D_{R1}, D_{R2}, D_{R3}, D_{R4}: APT75DQ60B

Input inductor: Lin=500uH

Transformer T: n=24:24, L_m=13.7mH, Leak inductor: 0.79uH, equivalent series resistor:70mΩ

Output capacitor: 9400uF

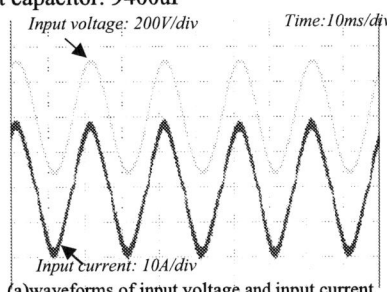

(a)waveforms of input voltage and input current

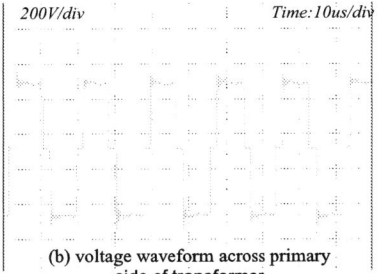

(b) voltage waveform across primary side of transformer

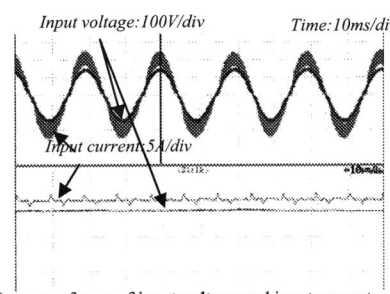

(c) waveforms of input voltage and input current

Fig. 6 experimental waveforms for SS-FB-PFC

Fig.6 shows the experimental waveforms based on the experimental prototype which specified as above. Fig.6 (a) shows waveforms of the input voltage and input current. It is obviously that high power factor is achieved, actually, PF value reaches 0.994 when the output power is about 2600w and the input voltage is 220V. Fig.6 (b) shows the voltage waveform across the

Table I Conversion efficiency comparison of various load for 110K PFC inductor operating frequency

Vin(V)	Vout(V)	Pout(W)	Pin(W)	η%
220	373	854	960	88.96
220	373	1546	1710	90.4
220	373	2227	2417.7	92.1
220	373	2366	2558	92.5
220	373	2642	2861	92.3

primary side of the transformer when the input voltage is at the peak point of sinusoidal, whereas, input current ripple is shown in (c). It is obviously illustrated that operation frequency of the transformer is half of PFC frequency. Waveform of Fig.6 (c) is just used to illustrate the ripple frequency of input inductor. Table.1 shows the conversion efficiency of SS-FB-PFC. It is indicated that the highest efficiency is 92.5%. Fi.g.8 (a) and (b) shows the waveforms of the prototype when the output power is about 3KW and input voltage is about 220V. In addition, it should be emphasized that Fig.6(c) shows the waveforms of prototype when the output power is further less than the rating value. Fig.7 shows the flowchart of current compensator, voltage compensator which is similar to current loop, is neglected. Without multiplying the input feedforward, fast dynamic can be achieved due to the flexibility of digital control strategy. Recover time is about 80ms. The resource of DSP can be greatly spared and the calculation speed can be greatly increased.

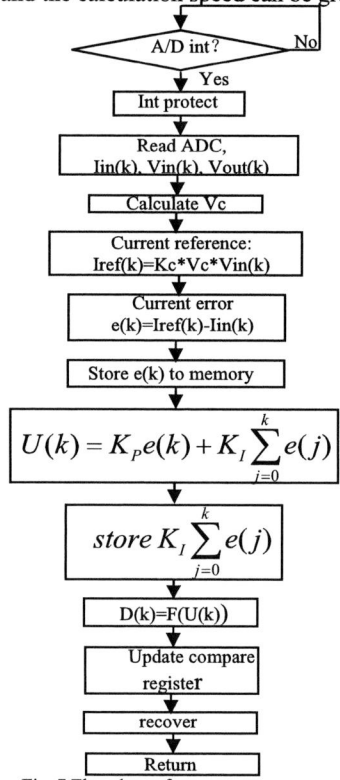

Fig. 7 Flowchart of current compensator

V. CONCLUSION

This paper deals with a 3KW power factor correction (PFC) converter based on the isolated full-bridge (FB) topology with fully digital control. The operation principle and compensator algorithm in Z domain are respectively analyzed in detail. Implementation strategy and cost effectiveness of digital control for the converter are studied. A transformer is used to provide galvanic isolation and furthermore offers extension of output range which can be either lower or higher than the input; in addition, multi-output solutions consideration can be adopted. The digital control strategy features fast dynamic response and excellent steady regulation characteristic and no DC part appears on the transformer. High power factor and high efficiency are both achieved by combining SS FB PFC main circuit and digital control strategy. Due to the excellent features of digital control strategy both in dynamic and steady process, it can greatly contributes to system integration research of DPS by adopting digital control in SS-PFC which is required in DPS.

ACKNOWLEDGMENT

The authors would like to thank National Natural Science Foundation of China (50237030ZD).

REFERENCES

[1] W.F. Ray and R.M. Davis, "The definition and importance of power factor for power electronics converters," Proc. European conference on Power Electronics and Applications (EPE), 1988, pp. 799-805

[2] F. Canales, D. Abud, J. Arau, and G. Jimenez, "Design of a two stage, 1 kW battery charger with power factor correction," in *Proc. IEE Power Electronics Variable-Speed Drives Conf.*, 1994, pp. 626–631.

[3] E. X. Yang, Y. Jiang, G. Hua, and F. C. Lee, "Isolated boost circuit for power factor correction," in *Proc. IEEE APEC*, 1993, pp. 196–203.

[4] J. Zhang, M. M. Jovanovic´, and F. C. Lee, "Comparison between CCM single-stage and two-stage boost PFC converters," in *Proc. IEEE APEC*,1999, pp. 335–341.

[5] C. Qiao and K. M. Smedley, "A topology survey of single-stage power factor corrector with a boost type input-current-shaper," *IEEE Trans.Power Electron.*, vol. 16, no. 3, pp. 360–368, May 2001..

[6] M. Qiu, G. Moschopoulos, H. Pinheiro, and P. Jain, "A PWMfull-bridge converter with natural input power factor correction," in *Proc. IEEE PESC*, 1998, pp. 1605–1612.

[7] A. K. S. Bhat and R. Venkatraman, "A soft-switched full-bridge singlestage AC-to-DC converter with low input current harmonic distortion," in *Proc. IEEE PESC*, 2000, pp. 799–804.

[8] C. A. Gallo, J. A. C. Pinto, L. C. de Freitas, V. J. Farias, E. A. A. Coelho, and J. B. Vieira Jr, "An unity high power factor power supply rectifier using a PWM ac/dc full bridge soft-switching," in *Proc. IEEE APEC*, 2002, pp. 1190–1194.

[9] A.Busse, and J.Holtz, "Multiloop control of a unity power factor fast switching ac to dc converter," Record of IEEE PESC 1982, pp.171-179.

2006 5th International Power Electronics and Motion Control Conference

A New ZVT Power Factor Corrected Three-Phase AC-AC Converter with Single-Phase HF Link

T. H. Abdelhamid, and A. Sabzali

College of Technological Studies, Department of Electrical Engineering, Kuwait

Abstract- This paper presents a new zero-voltage transition, power factor corrected three-phase ac-ac converter with single-phase high frequency link. It is a two-stage converter; the first stage is a boost-integrated bridge converter operated at fixed frequency and operates in two modes at ZVT for all switches and establishes a 1-ph square wave HF link. The second stage is a bi-directional PWM 3-ph bridge that converts the 1-ph HF link to a 3-ph voltage using a novel switching strategy. The converter modes of operation, and key equations are outlined. Simulation of the overall system is carried out using Simulink. The switching strategy and its corresponding control circuit are clearly described. Experimental verification of the simulation is carried out for a prototype of 500W at 10kHz link frequency.

Keywords; zero-voltage transition; power factor correction; ac-ac converter

I. INTRODUCTION

Power factor correction (PFC) circuits have become standard input stages for almost every medium and high power switching power supply operating at the mains voltage. Today's most often used configuration includes a boost converter as an input stage, stabilizing the input to the second stage at a level 10-20% over the maximum amplitude of mains voltage. The second stage is usually a half- or full-bridge converter which provides all necessary output voltages [1]. Efficiency of the boost PFC is high (90-95%), but the second stage converter has a maximum efficiency of about 85% (which is significantly reduced if high output currents are produced. Overall maximum efficiency is, therefore, about 80% [2].

Recently, soft-switching techniques eliminated variable frequency operation, further reducing component stress and improving efficiency. Zero-voltage and zero-current transition (ZVT and ZCT) modifications of almost every basic converter topology have been widely accepted, and the semiconductor industry has followed that trend by developing several fully integrated controllers. The full-bridge, phase-shift ZVS topology is the most frequently used [3].

A non-sinusoidal voltage clamper circuit [4] has been used to achieve soft switching with an inductive or current source load. However, excessive voltage stress due to the leakage inductance (typically about 4 times the input line voltage) has been reported.

In order to achieve a passive ZC turn-on and ZV turn-off, a small inductor and a capacitor must be added to the circuit. The inductor provides ZC turn-on of the active switches and limits the reverse recovery of the diodes while the capacitor provides ZV turn-off of the active switches. Typically, an inductor and capacitor have been placed in series and parallel with each active switch.

However, many other locations are possible and may yield lower component count, simplify the circuit, and improve performance.

Numerous ZVT converter topologies have been previously proposed [5], [6], however these drawbacks have been reported:

1. The auxiliary circuit consists of several components, and the auxiliary switch requires a floating gate drive.
2. The auxiliary circuit components have high voltage and current stresses and therefore, increased conduction losses.
3. Reduced overall efficiency due to the hard turn-off of the auxiliary switch.

To overcome these drawbacks, several kinds of soft switching PWM converters have been presented in the literature [7], [8]. They achieve high efficiency, minimum switching stresses, switching losses, and facilitate PWM control with constant frequency operation. All soft switching PWM methods utilize resonant techniques to soften the switching transition with ZVS and/or ZCS, and they can be generally classified into either passive or active ones. Passive methods use only resonant inductors, capacitors, and diodes to achieve soft switching turn-on and/or turn-off of the switches [9]. Active methods, as in [4], use resonant inductors, capacitors, diodes, and auxiliary active switches to reduce switching loss due to main power switches. The concepts of fundamental soft switching cells were employed to generate many families of soft switching PWM converters [7]-[11].

The ZVT-PWM soft-switching converters [10], [11] solve the existing problems of high switching losses of conventional PWM converters and high voltage and current stresses of resonant converters. By taking advantages of PWM and tank resonance, the soft-switching boost power converter with constant frequency operation is considered in this paper. These kinds of converters are especially useful in the application of high efficiency power converter systems.

This paper presents a new ZVT, power factor corrected three-phase ac-ac converter with single-phase HF link. The proposed system configuration enjoys the following advantages; ZVT, input PFC, constant frequency operation, and single-phase HF link that will make it possible to use a single-phase isolation transformer if isolation is needed in the system requirements. The principle of operation, different modes of operation, switching strategy, and control circuit are presented. Results obtained from Simulink simulation of a 500W, 100V laboratory prototype switching at 10kHz, verify the performance of the proposed configuration, highlight its advantages, and mention its drawbacks.

1-4244-0448-7/06/$25.00 ©2006 IEEE

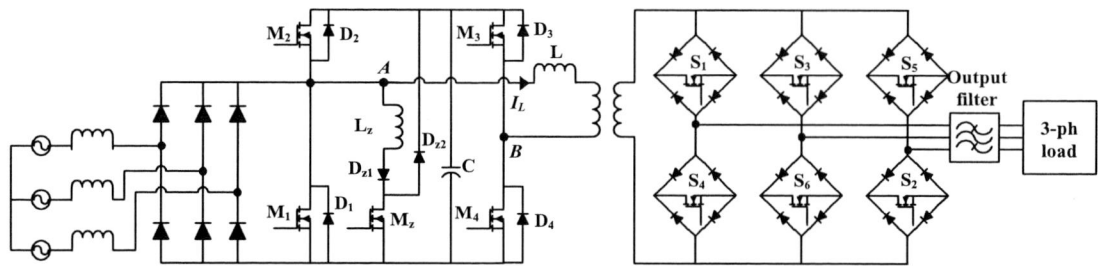

Fig. 1. The proposed ZVT power factor corrected 3-ph ac-ac converter with 1-ph HF link

II. CIRCUIT DESCRIPTION AND OPERATION

The detailed circuit diagram of the proposed 3-ph ac-ac converter employing input PFC and ZVT is shown in Fig.1. The proposed system consists mainly of two converters; the first converter is a combination of a 3-ph boost converter and a bridge converter, its function is to establish the constant amplitude-constant frequency 1-ph HF link voltage at ZVT for the bridge converter switches, while the boost converter incorporates the input PFC of the 3-ph supply currents. The second converter is a 3-ph PWM bi-directional bridge, its function is to reconstruct a balanced 3-ph low frequency output voltage from the constant amplitude-constant frequency 1-ph HF link voltage. A 3-ph filter is located at the 3-ph bridge's terminals in order to suppress the associated switching harmonics. A HF isolation transformer may be included in the circuit for galvanic isolation of the dc side from the load circuit.

A. The Boost Integrated ZVT Converter

This converter can be operated in continuous current mode (CCM) or in discontinuous current mode (DCM) depending on the loading conditions. In CCM, the gating signals of switches $M_1 \sim M_4$ have a 50% complementary duty cycle, and a square wave HF link is established. At light loads, the converter operates in DCM where pulse width is decreased with a dead gap and the HF link voltage is symmetrically cut by an angle α from both ends. The current waveform in the tank inductor is divided into several intervals for each mode of operation. These intervals depend on the polarity of voltage and direction of current in the tank circuit.

In CCM, the modes of operation can be summarized as follows: the cycle starts when diodes D_1 and D_3 are conducting, inductor current decreases from its initial value and before the end of this interval switch M_1 is turned-on at ZVT, then current in diode D_1 transfers to switch M_1; when switch M_3 conducts at ZVT, the current in diode D_3 transfers to the switch M_3;

$$i_L(t) = i_L(t_o) - \frac{(V_c - V_o)}{L}(t - t_o) \tag{1}$$

$$V_{AB} = V_L - V_o \tag{2}$$

$$i_{M1}(t) = i_{in} - i_L(t) \tag{3}$$

$$i_{M3}(t) = -i_L(t) \tag{4}$$

Switches M_1 and M_3 are turned-off when diodes D_2 and D_4 become forward biased,

$$i_L(t) = \frac{(V_c - V_o)}{L}(t - t_2) + i_L(t_2) \tag{5}$$

$$V_{AB} = V_c \tag{6}$$

$$i_{M2}(t) = i_L(t) - i_{in}(t) \tag{7}$$

$$i_{M4}(t) = i_L(t) \tag{8}$$

When the current in diode D_4 transfers to M_4 at ZVT, inductor current becomes negative and increases linearly; when switch M_2 turned-on ZVT, diodes D_2 and D_4 become reverse biased, and the link voltage equals the capacitor voltage. If overlap of triggering signals exists, the following interval will be added to the previous intervals; when diode D_3 becomes forward biased, and switch M_4 turns-off, inductor current decreases linearly from its initial condition; the inductor current keeps decreasing as long as both switch M_2 and diode D_3 are conducting, and the link voltage becomes zero.

$$i_L(t) = i_L(t_4) - (V_o / L)(t - t_4) \tag{9}$$

$$V_{AB} = 0 \tag{10}$$

$$i_{M2}(t) = i_L(t) - i_{in}(t) \tag{11}$$

$$i_{M3}(t) = -i_L(t) \tag{12}$$

In DCM, the modes of operation can be summarized as follows: the cycle starts when diode D_1 is conducting, the current in D_1 is a resonating sinusoid with a maximum value of $V_c / \sqrt{L_t / C_t}$, that provides the ZVT for switch M_1, in addition to D_3, switch M_3 and the auxiliary switch M_z are conducting, and the HF link voltage equals to the negative of the capacitor voltage; when switch M_1 turns-on at ZVT, the current in M_1 increases until the end of this interval while the inductor current decreases linearly, where the HF link voltage still equals the negative of the capacitor voltage,

$$i_L(t) = -\frac{(V_c - V_o)}{L}(t - t_1) \tag{13}$$

$$V_{AB} = -V_c \tag{14}$$

$$i_{M1}(t) = i_{in} - i_L(t) \tag{15}$$

$$i_{M3}(t) = -i_L(t) \tag{16}$$

The following three intervals are the same as those of CCM except that at the end of the last interval, currents in M_2, D_3, and inductor are all equal to zero, and the HF link voltage becomes zero due to the overlap between triggering signals of M$_2$ and M$_3$.

B. The 1-Ph to 3-Ph AC-AC Converter

Since the output of the bridge converter is of a constant amplitude, the modulation process should be incorporated in the output stage. Therefore, the output ac-ac converter should be controlled by a certain PWM technique in order to achieve the modulation process of the overall system.

The complete switching strategy of the 1-ph to 3-ph ac-ac converter is illustrated in Fig.2. The PWM control strategy is carried out by switching signal generation circuit where a carrier triangular waveform having a frequency much greater than that of the output voltage is compared with 3-ph reference sine-wave signals synchronized with the mains supply. The ratio of the frequencies of the triangle wave and the reference sine waves is defined as the frequency modulation ratio M_f. This ratio is typically in order of 200-400 for HF link applications. However, this ratio is taken as 9 in Fig. 2 in order to provide clear illustrating waveforms. The ratio of the amplitude of the sine and carrier waveforms is known as the amplitude modulation index, M_a, as it provides the modulation process of the amplitude of the output voltage. For linear modulation region, this value is restricted from 0 to 1 and it is taken as 0.8 in Fig. 2. The control strategy of the 1-ph to 3-ph ac-ac converter can be summarized as follows; a signal is taken from one of the phase voltages of the mains supply, where the other two signals of the following two phases are obtained via two cascaded phase shifting circuits such that control signals F_1, F_2, and F_3 are obtained. The output of the three comparators are denoted V_A, V_B, and V_C respectively. These signals are used to derive the 3-ph line voltages through subtracting circuits such that V_{AB}, V_{BC}, and V_{CA} are obtained. The 1-ph HF link signal, V_d, is a 50% duty ratio square-wave synchronized with the carrier waveform. The switching patterns of the 1-ph to 3-ph ac-ac converter G$_1$~G$_6$ are synthesized in the following novel way: The individual pulses of the line voltage V_{AB} are alternately carried by the switch pairs S$_1$, S$_6$ and S$_3$, S$_4$, while the pulses of V_{BC} are alternately carried by S$_5$, S$_6$ and S$_3$, S$_2$, and the pulses of V_{CA} are alternately carried by S$_5$, S$_4$ and S$_2$, S$_1$. The pulses carried by switch pairs S$_1$, S$_6$ and S$_3$, S$_4$ for the line voltage V_{AB} can be obtained by multiplying the line voltage V_{AB} by the normalized square-wave of the 1-ph HF link signal V$_d$, such that the switching signal G$_{ab}$ is obtained. Another two signals G$_{bc}$, and G$_{ca}$ are also obtained by multiplying V$_{BC}$, and V$_{CA}$ by the V$_d$. This step identifies the switching pulses of the each pair of switches S$_1$ and S$_6$ (G$_{16}$: the positive pulses of G$_{ab}$), and the switching pulses of the pair of switches S$_3$ and S$_4$ (G$_{34}$: the negative pulses of G$_{ab}$). This step is repeated for the line voltages V$_{BC}$ and V$_{CA}$, such that another four sets of gating signals are obtained, namely G$_{12}$, G$_{32}$, G$_{56}$, and G$_{54}$. The total switching pattern of each switch is the aggregate of the pulses carried out by the switch to supply different line voltage. This can be obtained by *ORing* the switching patterns accomplished by the same switch for different line voltages (for example; G$_1$=G$_{16}$ *OR* G$_{12}$). The final switching patterns of the 1-ph to 3-ph ac-ac converter G$_1$~G$_6$ shows that they possess a half-wave symmetry where the switches in the same arm don't conduct at the same time to avoid short-circuiting the transformer terminals. Also, there are three switches which are turned-

on at the same time (one from each arm). This novel switching strategy can be applied for any value of M$_f$ (odd or even) and also, not necessarily to be a multiple of three as required by the conventional switching strategies.

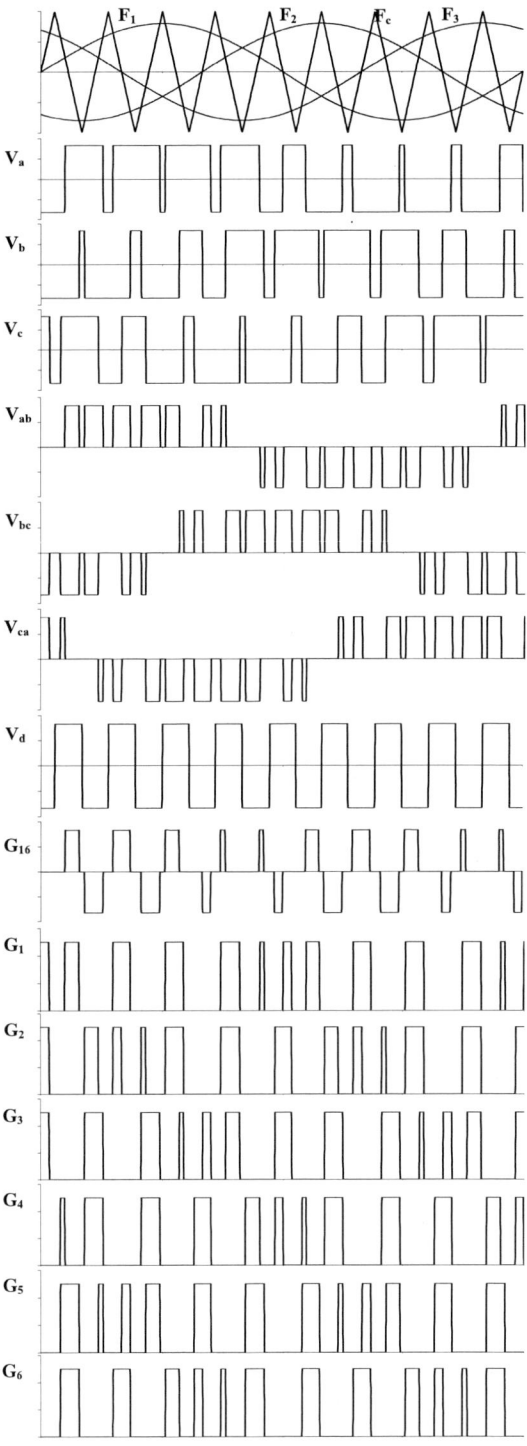

Fig. 2. Switching strategy of the 1-ph to 3-ph ac-ac converter

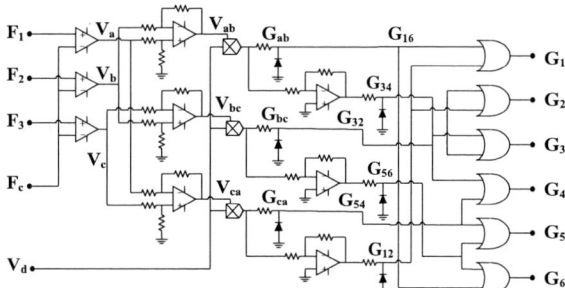

Fig. 3. Control circuit of the PWM 1-ph to 3-ph ac-ac converter

The control circuit used to verify this novel switching strategy is shown in Fig. 3. It consists of 3 comparators to obtain V_a, V_b, and V_c; 3 subtracting circuits to get the 3 line voltages V_{AB}, V_{BC}, and V_{CA}; 3 multipliers to get the 3 functions G_{ab}, G_{bc}, and G_{ca}; 3 negative clippers to get G_{16}, G_{32}, and G_{54}; and 3 inverting amplifiers followed by 3 negative clippers to get G_{34}, G_{56}, and G_{12}. Then 6 OR gates to obtain the final switching patterns G_1~G_6. Each signal is applied to its corresponding switch through appropriate buffer and driver circuits. This control circuit is very simple, reliable and has low cost. Closed loop control can be easily achieved by making the amplitudes of F_1, F_2, and F_3 controllable via a simple feed-back circuit comprises the output voltage and one of the reference signals such that the modulation index can be controlled to achieve output voltage regulation.

III. SIMULINK SIMULATION

In order to verify the validity of the proposed system, the overall system is simulated using Simulink. A design example at 10kHz switching frequency is chosen in order to save the disc space and shorten the computation time. The selection of the values of circuit elements is based on solving the steady-state relations of the described modes of operation at full-load. The full load rating of the design example is chosen at 500W, and 100V HF link.

Typical simulation waveforms at full-load for 50% duty cycle and CCM are shown in Fig. 4. The waveforms are presented in the following order: the switching patterns of switches M_1~M_4; HF link voltage, inductor current; and current in M_3 and M_4. Another set of simulation waveforms at light load in DCM are shown in Fig. 5, where an overlap between switching signals occurs, and a dead interval in the HF link voltage is observed. It can be shown that the ZVT is achieved during the entire range of load variation.

The un-rectified 3-ph output phase and line voltages are shown in Figs. 6 and 7 respectively, where the validity of the switching strategy of the 1-ph to 3-ph converter is verified. The harmonic spectra of the output phase and line voltages are shown in Fig. 8, where the associated harmonics appear at the link frequency and its multiples. The 3-phase input voltages and currents are shown in Fig. 9, where high quality input current waveforms are obtained with nearly unity power factor. This reflects the effectiveness of the PFC boost-integrated ZVT converter.

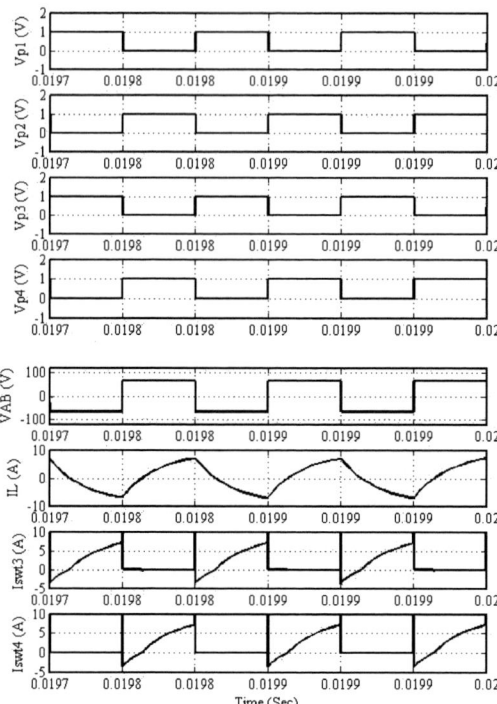

Fig. 4. Simulink simulation waveforms of the CCM of the boost-integrated ZVT converter

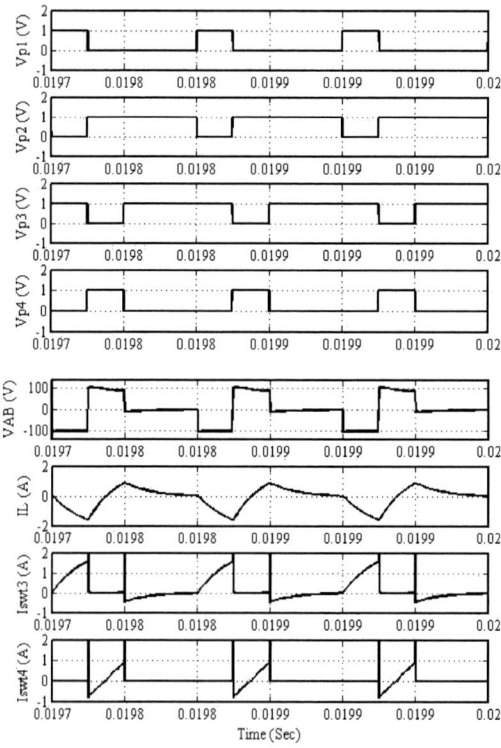

Fig. 5. Simulink simulation waveforms of the DCM of the boost-integrated ZVT converter

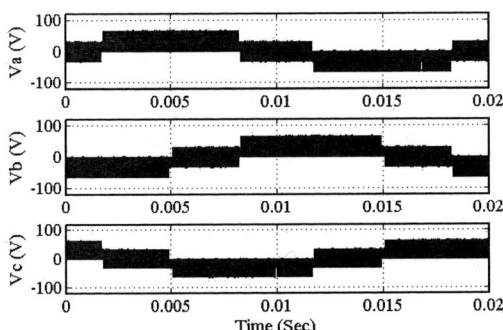

Fig. 6. The un-filtered 3-phase output phase voltages

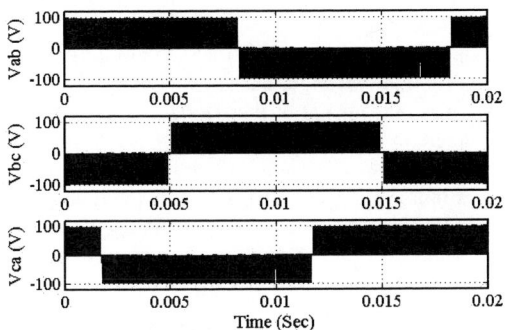

Fig. 7. The un-filtered 3-phase output line voltages

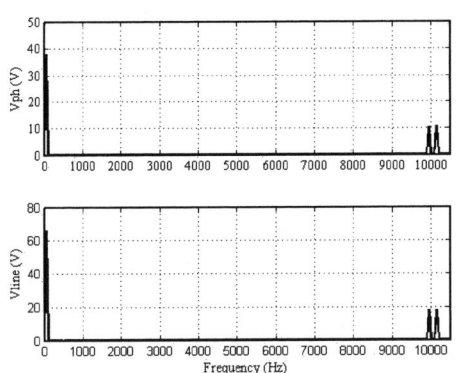

Fig. 8. Harmonic spectra of output phase and line voltages

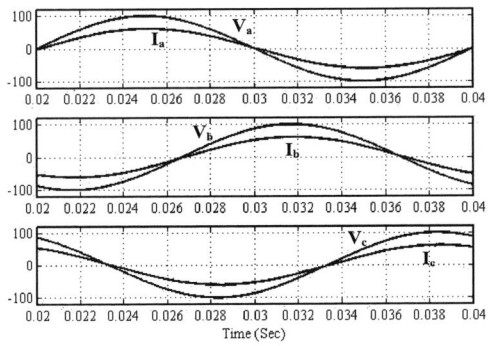

Fig. 9. The 3-phase input phase voltages and currents

IV. EXPERIMENTAL VERIFICATION

In order to verify the feasibility of the proposed converter, a 500W, 100V prototype is designed and built for illustration purpose at 10kHz link frequency. The available MOSFET switches IRF840A are used as the main controlled switches, and fast recovery diodes MUR860 are used in the whole system. Experimental results are obtained from a digital oscilloscope. The ZVT of switch M_3 of the boost converter in the DCM is illustrated in Fig. 10. The gating signals of switches S_1~S_6 are shown in Figs. 11, and 12. The unfiltered 3-ph output voltages are shown in Fig. 13, note that a scale factor of 10 is included in the probes used. Its harmonic spectrum is shown in Fig.14, where the output voltage has a low-frequency profile modulated at the link frequency, which can be easily filtered out using small size output filter. The input voltage and current of one of the input phases are shown in Fig. 15, where the input current is sinusoidal and in-phase with the input voltage, and the measured power factor ranges around 0.98 during the entire range of load variation. However, the overall system efficiency is about 88% due to the high on-resistance of the MOSFETs used, and the PWM nature of the 1-ph to 3-ph ac-ac converter. Extended work will be dedicated to apply soft switching to the output stage ac-ac converter in order to further improve the system efficiency.

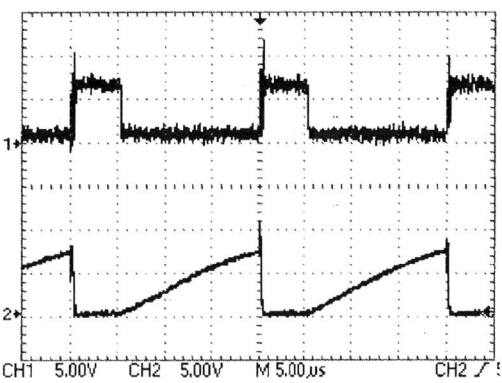

Fig. 10. The ZVT switch of switch M_3 in DCM

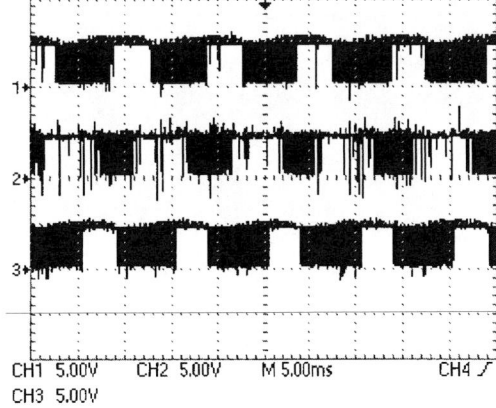

Fig. 11. The triggering signals of switches S_1, S_2, and S_3

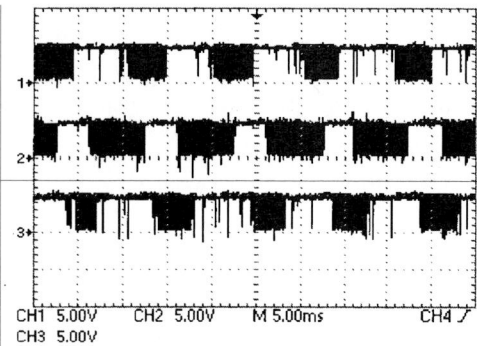

Fig. 12. The triggering signals of switches S_4, S_5, and S_6

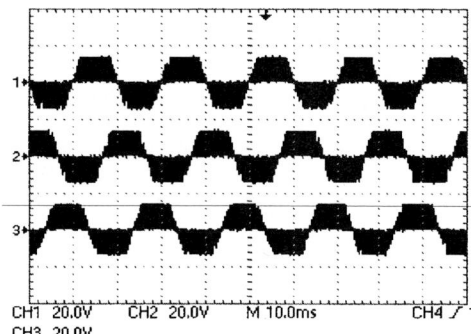

Fig. 13. The un-filtered 3-phase output voltages

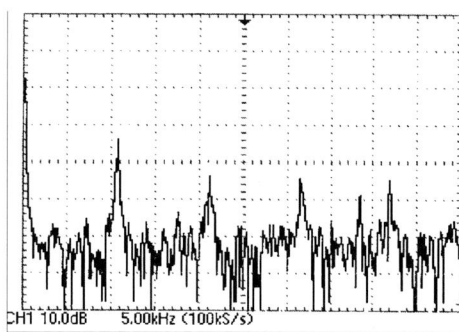

Fig. 14. Harmonic spectrum of the output line voltage

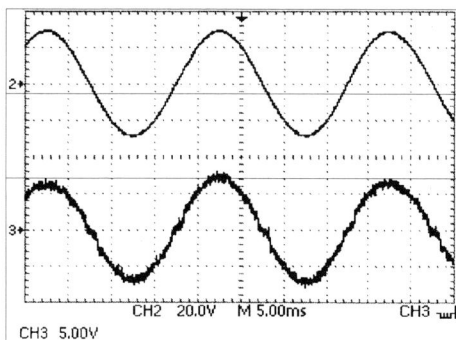

Fig. 15. The input phase voltage and current

V. CONCLUSIONS

A new configuration of 3-phase ac-ac converter has been proposed. The proposed system is based on a single-stage boost integrated ZVS converter, a 1-ph HF link, and a 1-ph to 3-ph ac-ac converter to reconstruct the low frequency output voltage. The overall system operates at a fixed frequency PWM control scheme, where synchronization between input and output voltages is taken into account through control circuit. Key equations, simulation and experimental results have been presented. ZVS is ensured, for all switches of the input stage, for the two modes of operation during the entire range of load variation. The novel switching strategy of the 1-ph to 3-ph ac-ac converter is verified experimentally, where the output harmonics appears at the link frequency. However, the system has the disadvantage that the harmonic content of the HF link is quit high due to its square wave nature, this will affect the design of the HF transformer. This can be prevailed by making the HF link has a sinusoidal nature, which, in turn, will affect the ZVS converter configuration. Also, the PWM nature of the output ac-ac converter lowers the overall system efficiency. Applying soft switching techniques to this converter will improve the overall efficiency.

ACKNOWLEDGMENT

The authors would like to thank the Public Authority for Applied Education and Training (PAAET) for supporting this work through project no: TS-04-004.

REFERENCES

[1] D. van der Berg, and J. A. Ferreira, "A family of low EMI unity power factor converters," *IEEE Trans. Power Electron.*, vol. 13, pp. 547-555, May 1998.

[2] J. M. Garcia, J. A. Cobos, R. Prieto, P. Alou, and J. Uceda, "Power factor correction: A survey," in *Proc. IEEE PESC'01*, 2001, pp. 8-13.

[3] H. Yu, B. M. Song, and J. S. Lai, "Design of a novel ZVT soft-switching chopper," *IEEE Trans. Power Electron.*, vol. 17, pp. 101-108, Jan. 2002.

[4] C. M. Duarte, and Barbi, "A family of ZVS-PWM active clamping dc-to-dc converters: synthesis, analysis, design, and experimentation," *IEEE Trans. Circuits Syst. 1*, vol. 44, pp. 698-704, Aug. 1997.

[5] T. W. Kim, H. S. Kim, and H. W. Ahn, "An improved ZVT PWM boost converter," in *Proc. IEEE PESC'00*, 2000, pp. 615-619.

[6] T. Wu, S. Liang, and Y. Chen, "A structural approach to synthesizing soft switching PWM converters," *IEEE Trans. Power Electron.*, vol. 18, no.1, pp. 38-43, Jan. 2003.

[7] G. Hua, and F. C. Lee, "Soft switching techniques in PWM converters," *IEEE Trans. Ind. Electron.*, vol. 42, pp. 595-603, Dec. 1995.

[8] A. Elasser, and D. A. Torry, "Soft switching active snubbers for dc/dc converters," *IEEE Trans. Power Electron.*, vol. 11, pp. 710-722, Sept. 1996.

[9] C. J. Tseng, and C. L. Chen, "Passive lossless snubbers for dc/dc converters," in *Proc. IEEE APEC'98*, 1998, pp. 1049-1054.

[10] J. G. Cho, J. W. Baek, G. H. Rim, and I. Kang, "Novel zero-voltage-transition PWM multiphase converters," *IEEE Trans. Power Electron.*, vol. 13, pp. 152-159, Jan. 1998.

[11] W. Guo, P. K. Jain, "A low frequency ac to high frequency ac inverter with build-in power factor correction and soft switching," *IEEE Trans. Power Electron.*, vol. 19, no. 2, pp. 430-442, March, 2004.

2006 5th International Power Electronics and Motion Control Conference

Simple Bridge-Type AC/DC Converters with Natural Input-Current-Shaper

Hsing-Fu Liu*, Chih-Yu Wu*, Chin Sun*and Lon-Kou Chang **

* Macroblock, Inc., Hsingchu, Taiwan, China
** Electrical and Control Engineering Department
National Chiao Tung University, Hsingchu, Taiwan, China
Lauren.ece88g@nctu.edu.tw

Abstract—**This paper presents a new design for bridge-type AC/DC converters with natural input current shaper. They have a fast output voltage regulation in full range input (85V~265V/AC). An additional winding in main transformer can naturally modify a line current. Further, the additional winding reduces an input inductance, so the weight and size of magnetic material are significantly reduced. The total harmonic distortion (THD) of the line current complies with IEC61000-3-2 class D. Furthermore, the proposed converters use current sensorless control and they can employ a conventional voltage-fed controller to simply the control circuit.**

Keywords- Switching Mode Power Supplies, Power Factor Correction, IEC 61000-3-2.

I. INTRODUCTION

In recent years, AC/DC power converters with power factor corrector (PFC) function are desirable and necessary in electronic products due to the requirement of the international standard IEC61000-3-2 [1]. Therefore, many papers were presented regarding the AC/DC converters with single-stage Input Current Shaper (ICS) in past several years [2]~[3]. These single-stage ICSs give a trade-off between high power factor, low cost and small size.

A prior proposed ICS has simplified a power stage to single-stage and line current comply with IEC61000-3-2, but the voltage across bulk capacitor will be probable over 450v in full range input [4]. A popular commercial voltage rating of aluminum electrolytic capacitor is under 450v. Basically, these prior proposed circuits of single-stage PFC have been a good compromises in main considerations but they although need a special signals to driver four main switching components (MOSFETs). Further, the special driver signals always lead to a complex control circuit [2]~[7].

New single-stage bridge-type AC/DC converters with natural ICS topology are presented in this paper. The new converters will naturally modify the line current via an additional winding in the transformer and also has a compromise in electrical performances, size and cost, the structure is shown in Fig. 1. Furthermore, the proposed

topologies can employ a conventional PWM controller to implement the control circuit.

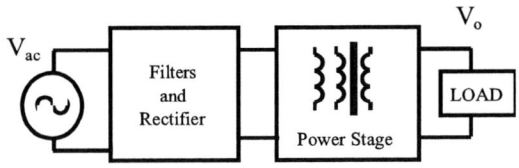

Figure 1. the structure of proposed AC/DC converters

II. PROPOSED CIRCUITS

The bridge-type AC/DC converters, with the functions of harmonic current elimination and fast output transient response, are proposed as shown in Fig. 2. In the full-bridge converter of Fig. 2, it is a single-stage AC/DC converter consisting of 4 switches S_1-S_4, an input filter C_1, a bulk capacitor C_2, an inductor L_1 and a transformer with two primary windings N_1&N_2, where N_1 plays a role of magnetic feedback winding. The winding N_1, inductor $L_1$and diode D_5&D_6 form an input current shaper. The winding N_1, inductor L_1, diode D_6&D_3 (or D_5&D_1), switch S_2 (or S_4) and bulk capacitor C_2 forms a boost circuit. The winding N_2&N_3, a bulk capacitor C_2, switches S_1-S_4, diodes D_7, D_8, inductor L_2 and output capacitor C_3 forms a full-bridge converter. The switches S_1~S_4 are always MOSFETs so D_1~D_4 are the body diode in switches S_1~S_4. The other two converters of Fig. 2 have the similar structure as full-bridge converter.

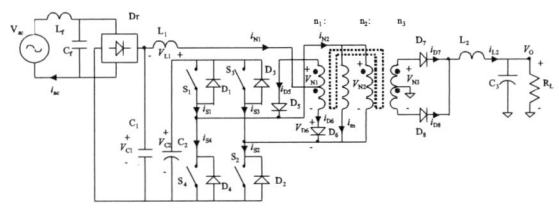

(a)

1-4244-0448-7/06/$25.00 ©2006 IEEE   417

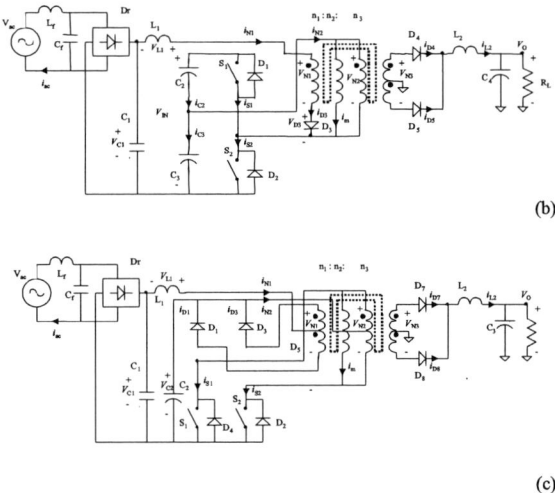

(b)

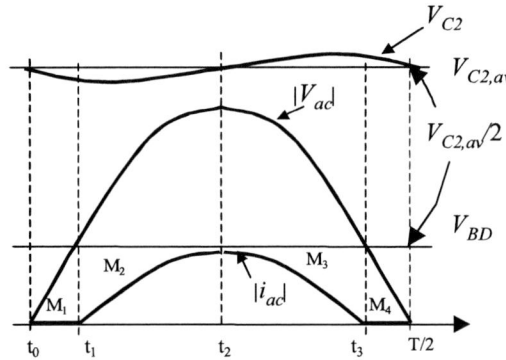

(c)

Figure 2. The proposed bridge-type AC/DC converters, (a) full-bridge converter (b) asymmetric half-bridge converter (c) push-pull converter

III. OPERATION THEOREIS

The operation principle of the proposed converter is somewhat similar to the boost-based forward AC/DC single-stage isolated power-factor-corrected power supply in the [8]. Magnetic energy is stored in inductor L_1, used as an energy-flow switch, when switch S_1&S_2 are on. Electric energy will be delivered to bulk capacitor C_2 through L_1 when switch S_1&S_2 turn off. Windings N_2 and N_3 provide the energy storing and transferring components of the full-bridge stage. Winding N_1 provides a path to charge L_1 and also transfers the line energy to output loads in the duty on duration. In the duty off duration while L_1 still conducts, winding N_1 also induce current $i_{N1} \times (n_1/n_3)$ to secondary side. However, when the line voltage is greater then the winding voltage $V_{N1}/2$ in the duty on duration, the power line can more strongly charges C_2 through L_1 and winding N_1. In the charging duration of C_2, the line current $|i_{ac}|$ is greater than zero and grows fast as the waveform shown in time duration t_1-t_3 of Fig. 3.

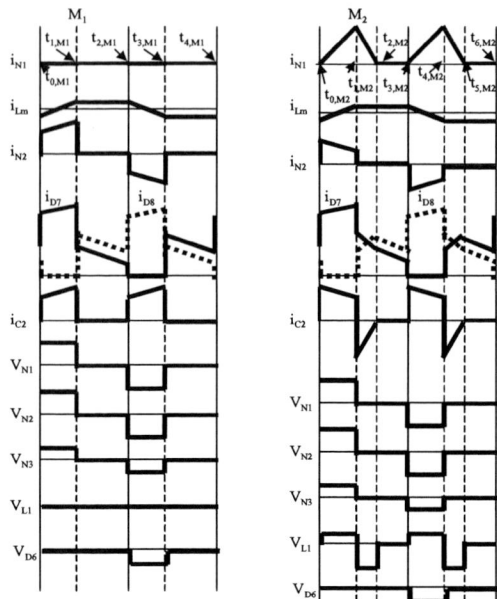

Figure 4. Voltage and current waveforms in a switching cycle in the two modes

Thus, the slope of V_{C2} is positive during t_1-t_3 and negative in other duration in each half line cycle. The resulting waveform of V_{C2} is sketched in Fig. 3. In this circuit the capacitor C_2 is arranged with a capacitance similar to that used in conventional AC/DC full-bridge converters. Since C_2 is large so that V_{C2} can approximate to its average voltage, $V_{C2,av}$. The proposed circuit has two operation modes. Fig. 3 shows these two operation modes that appear mirror-symmetrically in each quarter of a line cycle. Fig. 4 shows the relative voltage and current waveforms in one switching cycle in two operation modes.

A. Operation Modes M₁ or M₄ (during t₀-t₁ or t₃-Tl /2)

Within this mode, the line current $|i_{ac}|$ and i_{N1} are zeros. The operation principle of the converter is the same as that working in the conventional full-bridge converter. Since C_2 is large enough, V_{C2} can approximate to a constant value during a line cycle. The output inductor L_2 and capacitor C_3 provide a good low pass. Thus the output voltage can be regarded to a constant value and can be obtained as

$$V_o = V_{C2} \cdot \frac{n_3}{2n_2} \cdot D, \qquad (1)$$

where D is defined as $(t_{1,M1}-t_{0,M1})/(t_{2,M1}-t_{0,M1})$ in mode M_1 or $(t_{1,M2}-t_{0,M2})/(t_{3,M2}-t_{0,M2})$ in mode M_2. Since the capacitance C_2 is assigned large, V_{C2} is assumed constant. Thus, for fixed load the duty ratio D can be assumed approximate to a constant in mode M_1 and M_2.

Figure 3. Operation modes in one half of line cycle

418

Since M_2 starts at the time when V_{CI} reach to V_{NI}. Thus, the time bound of mode M_1 can be obtained by

$$V_{BD} = V_{C2} \cdot \frac{n_1}{2n_2} = V_o \cdot \frac{n_1}{D \cdot n_3} \qquad (2)$$

or

$$\omega t_1 = \sin^{-1}\left(\frac{V_o}{V_m} \cdot \frac{n_1}{D \cdot n_3}\right) \qquad (3)$$

Fig. 5 shows the current loop while the converter operating in mode M_1 or M_4.

B. Operation Modes M_2 or M_3 (during t_1-t_2 or t_2-t_3)

The operation mode initially starts at the condition, $V_{ac} = V_{BD}$. When S_1 and S_2 turn on, D_6 is forced to turn on and current i_{N1} flows through the winding N_1, L_1, D_6, and S_2. Consequently, i_{N1} linearly increases. The capacitor C_2 supplies current i_{N2} flowing through S_1, winding N_2, and S_2. Simultaneously D_7 starts to turn on and the power is delivered to the load. When S_1 and S_2 turn off, the current i_{N1} starts to charge capacitor C_2 through L_1, winding N_1, D_5, D_6, D_1, and D_3 and linearly decreases to zero at time $t_{2,M2}$. Simultaneously D_7 and D_8 will continuously turn on and the power is delivered to the load. Fig. 6 shows current loops in six operation stages in mode M_2/M_3.

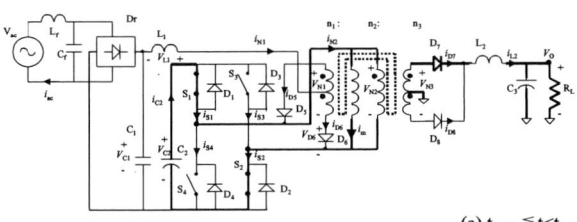

(a) $t_{0,MI} \leqq t < t_{1,MI}$

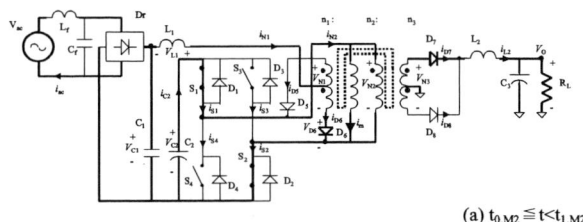

(a) $t_{0,M2} \leqq t < t_{1,M2}$

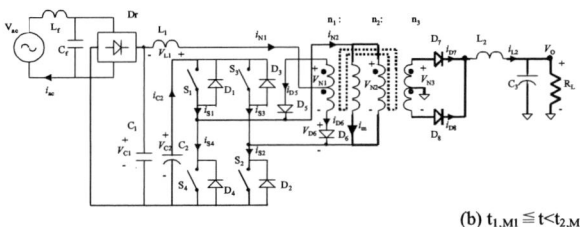

(b) $t_{1,MI} \leqq t < t_{2,MI}$

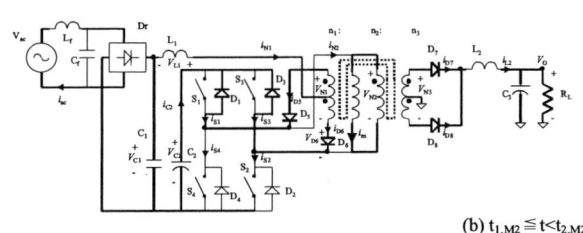

(b) $t_{1,M2} \leqq t < t_{2,M2}$

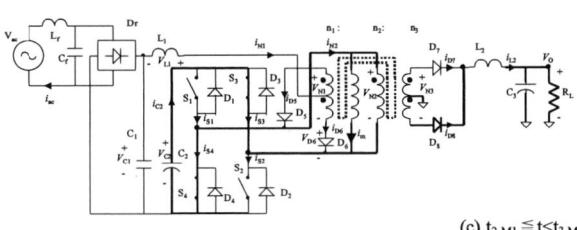

(c) $t_{2,MI} \leqq t < t_{3,MI}$

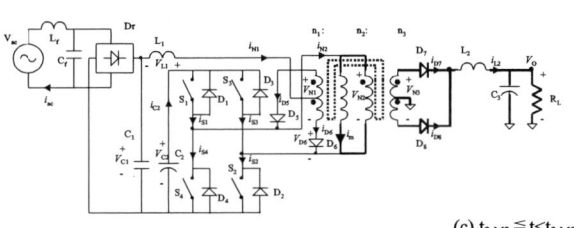

(c) $t_{2,M2} \leqq t < t_{3,M2}$

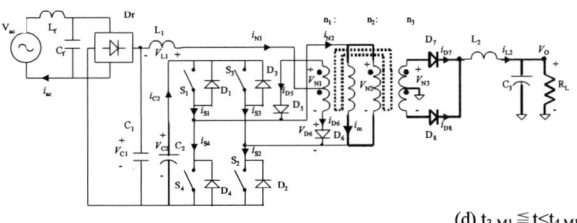

(d) $t_{3,MI} \leqq t < t_{4,MI}$

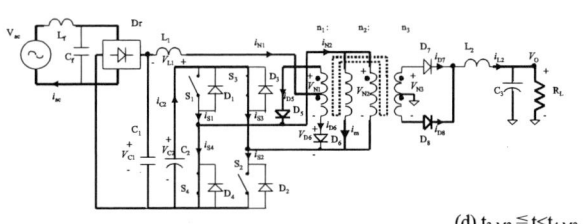

(d) $t_{3,M2} \leqq t < t_{4,M2}$

Figure 5. Current loops while (a) S_1&S_2 turn on, (b) S_1&S_2 turn off, (c) S_3&S_4 turn on, and (d) S_3&S_4 turn off in M_1/M_4

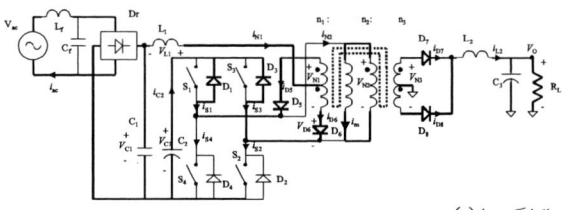

(e) $t_{4,M2} \leqq t < t_{5,M2}$

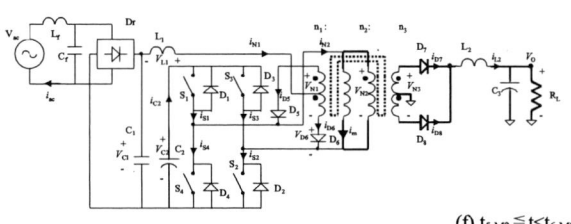

(f) $t_{5,M2} \leqq t < t_{6,M2}$

Figure 6. Current loops while S_1&S_2 (a) turns on, (b)~(c) turns off in mode M_2/M_3, and S_3&S_4 (d) turns on, (e)~(f) turns off in mode M_2/M_3.

IV. ANALYSIS OF CONVERTER OPERATION

A. Line current and Duty ratio

The line current i_{ac} mainly contains the line frequency part that is achieved by i_{N1} through low pass filter L_r-C_r. The line current $|i_{ac}|$ is approximate to the average current of i_{N1} in a switching cycle. Fig. 4 shows that i_{N1} is a linear function of V_{in} (or $|V_{ac}|$) during $t_{1,M2}$-$t_{0,M2}$, in mode M_2 and also in mode M_3 due to the mirror-like symmetry relation between these two modes.

Since the capacitance C_2 is assigned large, V_{C2} is assumed constant. Thus, for fixed load the duty ratio D can be assumed approximate to a constant. Under the assumptions the relation of V_{C2} and duty ratio D can be found through employing (1)

Additionally, in the CCM operation regarding i_{L2}, the duty ratio D will be also kept approximately a constant in different load current since the inductor L_2 is large and operates in a continuous current mode. The relation between V_{C2} and duty ratio is shown in (1) and Fig. 7.

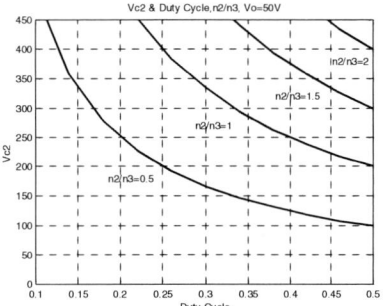

Figure 7. V_{C2} & Duty cycle, n_2/n_3 at V_o=50V

B. Corner angle of line current

A corner angle (CA) of line current is defined as $\omega(t_1 - t_0)$. Equation (3) shows the relation between CA and parameters. Equation (3) and Fig. 8 show that CA decreases as V_o/V_m or n_1/n_3 decrease. A smaller CA will result in a higher power factor and lower THD.

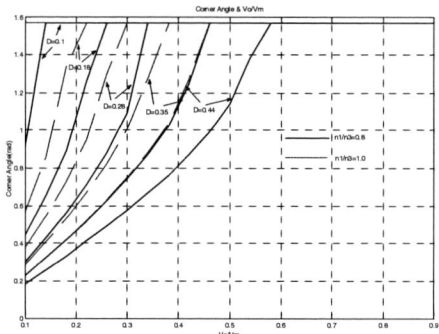

Figure 8. Corner angle and V_o/V_m

C. Voltage of bulk capacitor

V_{C2} is the voltage across bulk capacitor C_2. There is an easy way to estimate V_{C2}. That is to find it at the time when the corner angle (CA) is reached. The inductance current i_{L1} is near zero at CA time and V_{C2} can be obtained as

$$V_{C2} = (V_m \sin \omega t_1) \cdot \frac{n_2}{n_1} \qquad (4)$$

The relation between the voltage V_{C2}, n_2/n_1, CA, and input line voltage amplitude V_m, is shown in Fig. 9. In practical application, V_{C2} should be smaller than 450v/dc in full range input. Therefore, two parameters, n_2/n_1 and CA, have to meet the requirement in a given output V_o.

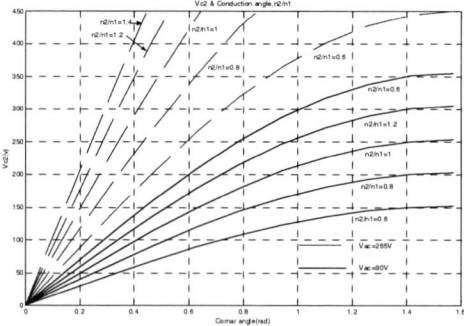

Figure 9. V_{C2}, Conduction angle and n_2/n_1

420

V. SIMULATION RESULTS

The proposed structure has been simulated in the specifications of 85v~265v/ac input voltage, 50v/dc output voltage and 500w output power. The turn-ratio of $n_1{:}n_2{:}n_3$ is 0.2:0.7:1 and the ratio of L_1/L_{N1} (L_{N1}=54μH) is 0.31. Fig. 10 shows the line current in a full line cycle. The experiments have shown that the harmonic distribution complies with a standard of IEC 61000-3-2. Table.1 shows that the detail harmonic distribution of the prototype and the harmonic contents meet the requirements of class D. Fig. 11 shows dynamic response from a half-load to full-load in 230V/AC input voltage. The output voltage of simulated circuit has a fast response and stable regulation. Fig. 12 shows the voltages across on bulk capacitor in different input voltage at full load. The bulk capacitor's voltage will dependent on V_{ac}, duty ratio D, and turn-ratio n_2/n_3 but it is almost independent of load current. The maximum voltage can be held under 450v/dc, which is a commercially available voltage ratio for electrolytic capacitor, by adjusting turn-ratio n_1/n_3.

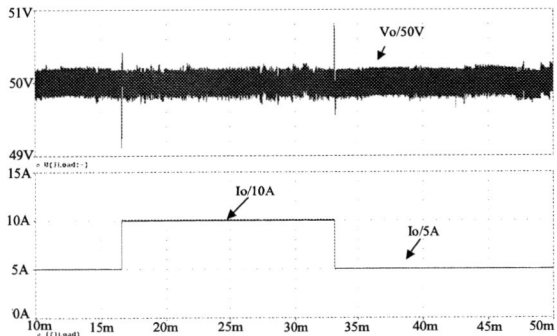

Figure 11. Dynamic response waveforms for V_{ac}, i_{ac}, V_o

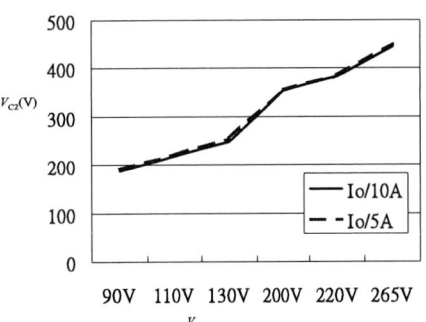

Figure 12. Voltage rating of bulk capacitor and line voltage

When the application doesn't need to deliver so much power and want to save MOSFET, user can employ half-bridge converter in his application. The proposed half-bridge converter is shown as Fig. 2(b). The simulation results are shown in Fig. 13 and Table 2.

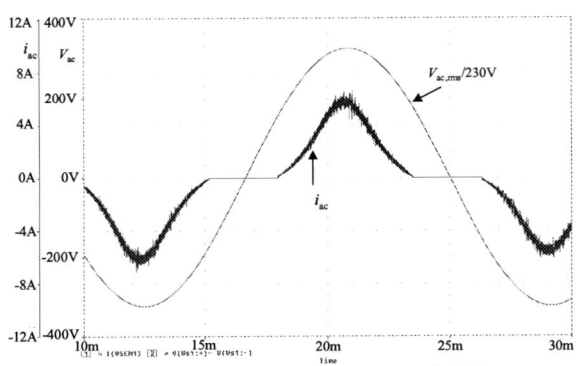

Figure 10 i_{ac} & V_{ac} waveform at V_{ac}=230v ,I_o=10A

Table. 1 The major harmonic components of the line current, P_o=50V ˙ 10A.

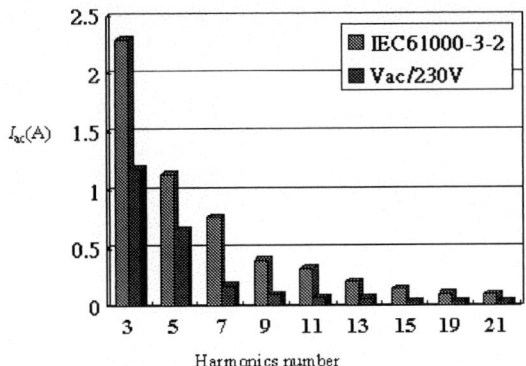

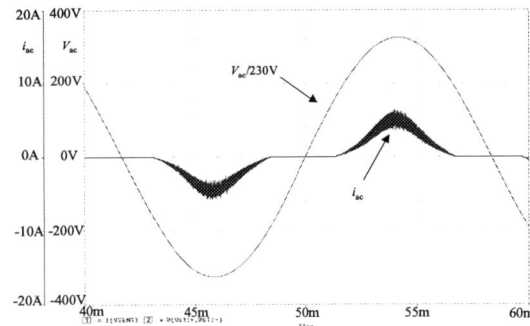

Figure 13. Simulation waveforms in half-bridge converter

Table 2. The major harmonic components of the line current, P_o=50V · 10A.

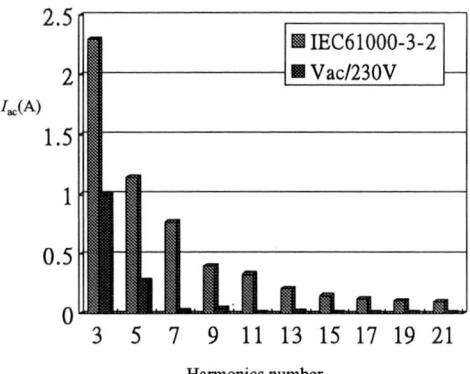

A push-pull converter could be another choice. The two switch S_1 and S_2 are low side switches so the driver method is simple and easy comparing to half-bridge or full-bridge converters. The proposed half-bridge converter is shown as Fig. 2(c). The simulation results are shown in Fig. 14 and Table 3.

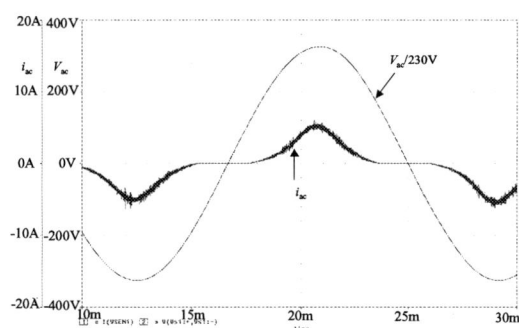

Figure 14. Simulation waveforms in push-pull converter

Table 3. The major harmonic components of the line current, P_o=50V · 10A

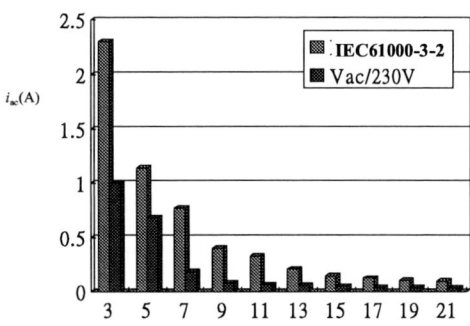

VI. CONCLUSION

The new structure of the bridge-type AC/DC converters is introduced in this paper. The proposed converters have line current shaper, fast dynamic response and tight voltage regulation. It is implemented by adopting a single-stage ICS and simple voltage-fed control loop. Therefore, the structure is simple and cost-effective. In the prior single-stage AC/DC PFC converters embed bulk inductor in boost cell. However, in the proposed design, an additional primary winding is added to replace the function of the bulk inductor of boost cell. Therefore, magnetic material's volume and weight are significantly saved. The simulation results have shown that the proposed converter's line current complies with standard IEC 61000-3-2 class D and the voltage regulation is tight under load change. The voltage across bulk capacitor can be kept under 450V by adjusting turn-ratio n_1/n_3 in full range operation.

References

[1] Electromagnetic Compatibility (EMC), Part 3, International Standard IEC61000-3-2, 2001.

[2] Chongming Qiao and Keyue M. Smedley, "A Topology Survey of Singlestage Power Factor Corrector with Boost Type Input-Current-Shaper," Proc. IEEE -APEC, 2000, pp. 460-467.

[3] O. Garcia, LA. Cobos, R. Prieto, P.Alou, and J.Uceda, "Power Factor Correction: A Survey," Proc. IEEE-PESC, 2001, pp. 8-13.

[4] Mei Qiu, Gerry Moschopoulos, Humberto Pinheiro, and Praveen Jain, "Analysis and Design of A Single Stage Power Factor Corrected Full-Bridge Converter," Proc. IEEE-APEC, 1999, pp. 119-125.

[5] Gerry Moschopoulos, Mei Qiu, Humberto Pinheiro, and Praveen Jain, "PWM Full-Bridge Converter with Natural Input Power Factor Correction," IEEE Trans. on Aerospace and Electronic System, Vol. 39, No. 2, April 2003, pp. 660-674.

[7] Gerry Moschopoulos, "A Simple AC-DC PWM Full-Bridge Converter with Integrated Power Factor Correction," Proc. IEEE-INTELEC, 2001, pp. 376-383.

[8] Lon-Kou Chang and Hsing-Fu Liu, "A Flexible and Cost-Effective Family for AC/DC Converters with Input-Current-Shaper and Fast Output-Voltage-Regulation," Proc. IEEE-PESC, 2004, pp. 3113-3119.

2006 5th International Power Electronics and Motion Control Conference

Rough Controlling TSC for Reactive Current Compensation in Traction Substations

Hongsheng Su and Qunzhan Li

School of Electrical Engineering, Southwest Jiaotong University, Chengdu 610031, P.R. China

shsen@163.com

Abstract—In process of reactive current compensation in electrified railway, due to applying fixed parallel compensation while not adapt to dynamic change of traction loads, redundant compensation or deficient compensation therefore frequently happen. Actually, while returning reactive power to power supply but suffering a positive measure, average power factor is lower than the primary one, thus, penalty is inflicted for poor power factor. To rescue the situation, in the paper we propose a novel capacitor switching strategy based on rough functions concept. The method firstly applies artificial neural networks to on-line monitor reactive current in traction load arms, then according to rough function, yields the discrete series space of reactive current for the intensity and speed of capacitor grouping switching, and generates decision-making rules for capacitor grouping switching. Meanwhile, considering the existence of three-time harmonics in traction loads, filter is designed in each grouping capacitor to filter a part of harmonics. The investigation indicates that the proposed method can overcome the deficiencies of traditional methods and duly trace dynamic change of loads, and power factor is also dramatically improved. In the end, an application example shows that the method is an effective dynamic compensation strategy for reactive power in traction substation.

Keywords-Traction substation; Reactive power compensation; Rough control; Parallel capacitor switching

I. INTRODUCTION

With the improvement of transportation abilities and the growth of traction loads year after year in electrified railway. Electric locomotive, a considerable reactive load, has an average power factor represented by 0.8, which is lower than 0.9 supported by nation. To deal with the problem, parallel reactive compensation installations are required to restrain reactive current and improve power factor in traction substations (see, e.g.,[1],[2]).

In the past, fixed parallel reactive power capacitor compensation is installed in each traction substation, compensation capacity is designed only according to average current of traction loads while not be able to automatically adapt dynamic change of loads, as a result, redundant compensation and deficient compensation frequently happen, in particular, while reactive power is sent back to power supply but suffering a positive measure

means implemented by power supply departments, in fact, power factor is lower than the primary one, compensation efficiency therefore become extremely low.

In recent years, Thyristor Switching Capacitor (TSC) compensation starts to receive welcoming due to overcoming the flaws of fixed parallel compensation. But, the compensation efforts seriously rely on compensation strategy and control algorithm. The average current and power factor of traction supply arms are selected as control variables in [3],[4],[5], since the linear relations between control variables and parallel capacitor can't be established well, control algorithm therefore becomes very complicated, compensation precision is very low, therefore, redundant compensation problem can't be tackled thoroughly. Moreover, because traction load current is a random variable related to the number of locomotives, traction loads, railway status and climate condition etc, precise math model for that is difficult to be established. A fuzzy controlling capacitor switching approach is presented in [6], the method better resolves the bottleneck problems in knowledge acquisition, but it is extremely difficult to precisely define fuzzy membership functions and some parameters related to it, and these definitions possess strong subjectiveness. By monitoring the dynamic fundamental current using artificial neural networks (ANN) in [7], supported by rough set theory, the paper selects reactive current as well as its increasing or decreasing speed close to electrical source side as control variables, a more reasonable control scheme for parallel capacitor grouping switching is proposed. The results of practical compensation indicate that the method can effectively improve power factor and avoid redundant or deficient compensation, and is a very ideal compensation means for reactive current in electrified railway.

II. ROUGH SET THEORY

A. Information System and Approximation Space

In rough sets, knowledge denotation system may be defined by

$$S=<U,A,V,F>. \tag{1}$$

where U is universe and expresses a set with finite objects, A is attribute set composed of condition attribute

C and decision attribute D, $A=C\cup D$, $C\cap D=\varnothing$, $a\in A$, $V=V_a$, V_a is range of a, $f:U\times A\rightarrow V$ is a information function, it specifies attribute values of every object in U.

Information system based on rough sets definition can be denoted using the table format, where columns express attributes and rows represent objects, each row describes information of an object. The table therefore is called decision table, which can generalize the relationships among data and educe the classification rules of the concepts. In rough sets, binary indivisible relationship *ind* (R) determined by $R\subseteq A$ can be expressed by

$$ind(R)=\{(x,y)\in U\times U|\ \forall\ a\in A, f(x,a)=f(y,a)\}. \quad (2)$$

It is very clear that if $(x,y)\in ind(R)$, then x and y can not be differentiated according to the existing information, they are an equivalent relation in U.

If R is a group of system parameters or only is a parameter of system, R can be applied to define or describe the correct level of the investigated object or abilities of classifying objects and expressing systems, it can be scaled using system parameter importance in [8].

Definition 1: Set X is characteristic vector set of input objects, x_i represents the ith characteristic in X, for output pattern classification Y, the importance factor of characteristic x_i is defined as

$$\alpha_{xi}(Y)=card(pos_X(Y)-pos_{X-xi}(Y))/card\ (U). \quad (3)$$

Obviously, for output pattern Y, if character variable x_i has larger importance factor $\alpha_{xi}(Y)$, this shows that its classification role is large, conversely, and small. If $\alpha_{xi}(Y)=0$, the variable x_i can be ignored, reduction of conditional attributes therefore is realized.

B. Rough function Theory[9]

Definition 2: Set R is a real number set, (a,b) is a open set in R, if real number sequence $S=\{x_0,x_1,...,x_n\}$ in (a,b) satisfies $a=x_0<x_1<...<x_n=b$, then $A=(R,S)$ is defined as the approaching space created by S, S is called discrete sequence. Each such S may define a partition in (a, b), which is expressed as $\pi(S)=\{(x_0),(x_0,x_1),(x_1),(x_1,x_2),(x_2),...,(x_{n-1},x_n),(x)\}$, namely, an equivalent relationship, then, for arbitrary $x\in(a,b)$, its upper approaching and lower approaching in S are respectively denoted as

$$\overline{apr_S}(x)=\inf\{x_i\in S:x_i\geq x, i=0,1,...,n\}. \quad (4)$$

$$\underline{apr_S}(x)=\sup\{x_i\in S:x_i\leq x, i=0,1,...,n\}. \quad (5)$$

Obviously, while $x=x_i$, then $\overline{apr_S}(x)=\underline{apr_S}(x)=x_i$; while $x_i<x<x_{i+1}$, then we have $\underline{apr_S}(x)=x_i$, $\overline{apr_S}(x)=x_{i+1}$, they are respectively defined as the former and the latter end of the equivalent classes (x_i,x_{i+1}).

Definition 3: Set real function $f:X\rightarrow Y$ as well as discrete sequences $S=\{x_1,x_2,...,x_n\}$ in X and $P=\{y_1,y_2,...,y_n\}$ in Y, the P_* and P^* approximation of f can be respectively denoted as $f_*:X\rightarrow Y$ and $f^*:X\rightarrow Y$, that is, for any $x\in X$, we have $f_*(x)=\underline{apr_P}\ (f(x))$, $f^*(x)=\overline{apr_P}\ (f(x))$

III. ROUGH CONTROLLING CAPCITOR GROUPING SWITCHING

A. Capacitor Grouping Switching

In range of normal voltage, due to the regulation abilities of electrical locomotive itself, the impact of bus voltage on parallel reactive power compensation may be neglected, meanwhile, the interaction between the two traction arms can be also considered very small, fetching current from the two supply arms therefore is considered to be single phase and independent, reactive compensation of the two arms may be designed, respectively. Figure 1 is design principle of parallel reactive power compensation with three-time filter circuit installed.

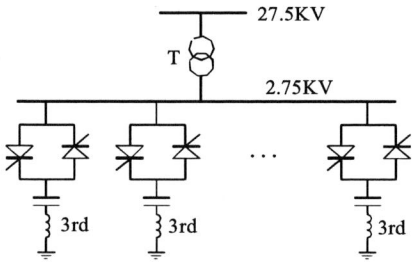

Figure 1 TSC Principle Diagram

Detail design process on reactive power compensator can be seen in [10], the followings mostly present its switching control principle. Figure 2 is switching control principle and connection diagram of TSC.

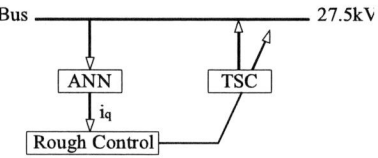

Figure 2. Rough Control Principle for TSC Switching

ANN can supply forceful technique support for TSC due to its self-adaptive detecting active current i_p and reactive current i_q close to power supply end shown in Figure 2, under reactive load condition, reactive current i_q lags the active current i_p by 90°, the aim of reactive power compensation is to improve power factor $\cos\varnothing$ from 0.8 to 0.9 upwards. In practice, if $\cos\varnothing$ can be guaranteed not lower than one value, switching frequency of capacitor groups may be greatly reduced. Therefore, after weighing compensation effects and parallel capacitor cost, $\cos\varnothing=0.92$ is confirmed to be the best, that is $\varnothing=23.07°$.

If the measured active current is $i_p \angle 0°$, then reactive current i_q' related to $\cos\varnothing=0.92$ is calculated by

$$i_q' = i_p \times \tan(-\varnothing) = -0.425\,i_p \tag{6}$$

Then the supplied reactive current value of parallel compensation should be

$$\Delta i_q = i_q - i_q' \tag{7}$$

Thus, if $\Delta i_q < 0$, then compensation capacity is considered scant; if $\Delta i_q = 0$, and just right. On the other hand, if $i_q > 0$, that is, i_q leads active current i_p, then compensation is considered redundant.

In addition to current, the increasing/decreasing speed of the current is also fully considered, and denoted as di_q. Considering the supplied reactive current of capacitor is right direct proportion to its capacity, linear relationship therefore can be established between current i_q and capacitance C.

Based on the above analysis, according to the operation characteristic of traction loads, fixed parallel capacitor compensation is divided into six groups, whose capacity is equal and circulation switching is implemented. The total capacitance C_{max} is calculated in accordance with reactive current $\Delta i_{max} = i_{qmax} - i_{qmax}'$, where i_{qmax} denotes reactive current of the most heavy load, and i_{qmax}' is the corresponding aim value. According to Δi_{max}, the 3rd times filter may be easily designed in [11].

B. Capacitor Grouping Switching Method Based on Rough Set theory

In accordance with the former analysis, $\Delta i_q < 0$ denotes scant compensation, $\Delta i_q = 0$ is suitable compensation, $i_q > 0$ indicates redundant compensation, $i_q = 0$ indicates right compensation. According to Δi_q, the overall capacitor is averagely divided into six groups called as C1,C2,C3,C4,C5 and C6, respectively related to $\Delta i_{q1}, \Delta i_{q2}, \Delta i_{q3}, \Delta i_{q4}, \Delta i_{q5}$ and Δi_{q6}. Set $S=\{\Delta i_{q1}, \Delta i_{q2}, \Delta i_{q3}, \Delta i_{q4}, \Delta i_{q5}, \Delta i_{q6}\}$, according to definition 2, $\pi(S)=\{\Delta i_{q1}, (\Delta i_{q1}, \Delta i_{q2}), \Delta i_{q2}, (\Delta i_{q2},\Delta i_{q3}), \Delta i_{q3}, (\Delta i_{q3}, \Delta i_{q4}), \Delta i_{q4}, (\Delta i_{q4},\Delta i_{q5}), \Delta i_{q5}, (\Delta i_{q5},\Delta i_{q6}),\Delta i_{q6}\}$, clearly, $\pi(S)$ has eleven equivalent classes. For arbitrary Δi_q, we select its lower approximation regarding S as control variable, then

$$apr_S\,(\Delta i_q)=\sup\{\Delta i_{qi}\in S:\ \Delta i_{qi}\le\Delta i_q,\ i=0,1,\ldots,6\}.$$

Further, if we use number i to discretize Δi_{qi}, then we have $S=\{1,2,3,4,5,6\}$, $\overline{apr}_S\,(\Delta i_q)\in S=\{1,2,3,4,5,6\}$. Likewise, we may dispose i_q in the same way, finally, we have $apr_T\,(i_q)\in T=\{1,2,3,4,5,6\}$,where T means discrete sequence of i_q. For the increasing/decreasing speed di_q, we use H to denote rapidness, and L means slowness, and also M is normal speed, and disposal method is as above. Because the locomotive itself can adjust its voltage, switching rules of capacitor groups therefore dose't consider impact of voltage, only consider Δi_q, i_q and di_q.

According to reactive power control rules of traction substations, Δi_q, i_q and di_q are select as input variables and C_i acts as output variable, then the formed decision table is shown in Table I.

TABLE I
THE DECISION TABLE FOR CAPACITOR SWITCHING

i_q	Δi_q	di_q	c	v
i	*	H	$i\downarrow$	H
i	*	M	$i\downarrow$	M
i	*	L	$i\downarrow$	L
*	i	H	$i\uparrow$	H
*	i	M	$i\uparrow$	M
*	i	L	$i\uparrow$	L

i: discretized number 1-6; V=H: high speed; V=M: normal speed; V=L: low speed; $\uparrow$: Capacitor plunge; $\downarrow$: Capacitor removal.

Seen from Table I, i_q and Δi_q are incompatible one another. For the output pattern $C\downarrow$, according to formula (3), $\alpha_{\square iq}\,(C\downarrow)=0$. Likewise, for pattern $C\uparrow$, $\alpha_{iq}\,(C\uparrow)=0$. Moreover, switching speed is only determined by di_q, namely, $\alpha_{iq}\,(V)=\alpha_{\square iq}\,(V)=0$. Thus the simplified decision rules can be gained as shown in Table II.

TABLE II
THE DECISION TABLE FOR CAPACITOR SWITCHING

i_q	Δi_q	di_q	c	v
i	-	H	$i\downarrow$	H
i	-	M	$i\downarrow$	M
i	-	L	$i\downarrow$	L
-	i	H	$i\uparrow$	H
-	i	M	$i\uparrow$	M
-	i	L	$i\uparrow$	L

Table II shows capacitor removal only relies on i_q, and plunge relies on Δi_q, switching speed is only relies on di_q, therefore, decision rules are greatly simplified. For instance, if $i_q=2$ and di_q=H, then $C_2\downarrow$ and V=H.

C. Capacitor Grouping Compensation Characteristics

Evaluation indices on capacitor compensation effect include improvement level of power factor, and the size of reactive current to power supply and compensator cost etc, and is a multi–attribute comprehensive evaluation system. Figure 3 shows the relationship between reactive current i_q close to power supply side and i_{Lq} close to loads.

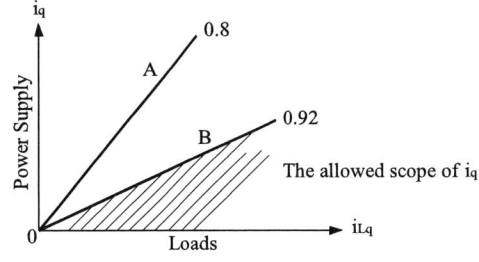

Fig.ure3. Evaluation Standard for TSC Compensation

In Fig. 3, i_{Lq} expresses reactive current close to traction load end, and i_q denotes reactive current close to power supply side, before no compensation, $i_{Lq} = i_q$, $\cos\varnothing=0.8$, as line A shown in Fig. 3. After compensation, i_q close to power supply end should reduce, but not be able to lower than zero, otherwise, redundant compensation may happen. Therefore, the allowed range of i_q is represented by hatching region between line B and horizontal axis i_{Lq} as shown in Fig. 3, where line B denotes $\cos\varnothing=0.92$. In addition, we can see from Fig. 3, with the increase of i_{Lq}, the allowed scope of i_q expands too, the required reactive power compensation capacity would become lower, meantime, the switching frequencies of capacitor groups are also reduced, under heavy loads, reactive compensation exhibits more economy for that than light loads.

Compensation characteristic of rough control is seen in Figure 4, thick fold line denotes capacitor groups switching process, after compensation, reactive current is constrained within the allowed range. Currently, we assume that one locomotive set to put in supply arm and produces reactive current i_L, before compensation its discrete value is 4, according to decision Table II, four groups of capacitors must quickly plunge into operation. After compensation, reactive current close to power supply end is denoted by point P in Figure 4, clearly, it lies inside allowed range.

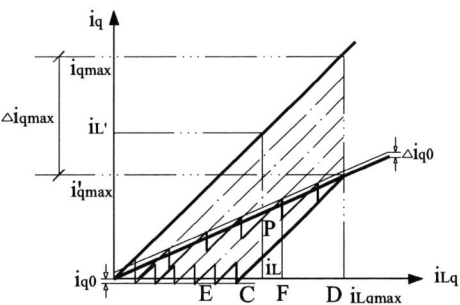

Figure 4 Compensation Characteristic of Rough Control

Since switching operation processes of capacitor groups can't be absolutely exact, therefore, error is unavoidable, like i_{q0} and Δi_{q0} shown in Fig. 4. Also, we can see from Fig. 4, after plunging into all six groups of capacitors, even if reactive current of locomotive varies in range of C to D, power factor is still kept between 0.92 and 1, and capacitor groups do not yield switching action, therefore, switching frequency is greatly reduced. Similarly, E to F is a variable scope when plunging into four groups of capacitors, obviously, the scope is smaller than one between C and D, and so switching frequency under heavy loads is lower than one under light loads.

IV. EXAMPLE ANALYSIS

The connection shape of the transformers of one traction substation is YN, d11, and whose capacities are

25MVA, under normal case one transformer is in operation, and another one is spare. The voltage level of input connection is 110kV and 27.5kV in output connection. The total short circuit capacity is represented by 500MVA. The reactive compensation installation in substation is fixed on the side of 110kV power supply, and positive measurement means is adopted to punish the situation that traction supply system adversely sends reactive current to power supply. Compensation capacity 1200kvar is fixed in A phase and 1200kvar in B phase, and filter can filter the 3rd times harmonic as seen in Fig. 1, thus a single-tuned filter id constituted. But the installed parallel compensation installations in A and B phase hasn't been grouped, fixed parallel compensation is implemented through the one-off overall switching by human resource.

In order to test the effects of the fixed parallel capacitor compensation, 50000 data or so of the fundamental harmonic current in two power loads arms are measured, the simulation results indicate that before compensation the average power factor is 0.805, if not consider reactive power into power supply, power factor may attain 0.96 upwards, but while consider it, then power factor is represented by a poor 0.74. Clearly, after the fixed compensation is applied, if positive measure means is applied for inversely sending reactive power into power supply, power factor is not improved, conversely, but becomes very low. To deal with the situation, we change the primary fixed parallel compensation scheme into grouping switching using the proposed method in the paper. Thus 6 groups of capacitors in all are produced and each group capacity is represented by 200kvar. The simulation results show that after compensation, average power factor is 0.934 under positive measure means for inversely sending reactive to power supply. This result is higher than one of fixed compensation, and attains national standard represented by 0.92. Compared with normal switching means, switching frequency and compensation cost are also decreased.

V. CONCLUSIONS

The proposed method in this paper overcomes the flaws of conventional fixed parallel capacitor compensation installations, effectively reduces capacitor groups switching frequency and compensation cost, and realizes dynamic reactive power compensation and improves power supply quality in traction supply system. Therefore, it possesses the excellent application prosperity and economic benefits. In addition, due to applying rough function controlling strategy, compared with fuzzy control scheme, it applies the approaching operators to replace fuzzy membership function, thus, the subjectiveness in process of defining fuzzy membership function as well as relevant parameters is avoided to some degree. Theoretically, it provides important instruction significant to those indeterminate discrete dynamic control systems, and is a newer research domain.

REFERENCES

[1] J. Y. Cao, *Tractive Supply System in Electrified Railway*, China Railway Press, Beijing, 1983.

[2] Q. Z. Li, *Parallel Integration Compensation and Its Application in Electrified Railway*, China Railway Press, Beijing, 1993.

[3] Z. Q. Huang, J. M. He and Q. Z. Li, "Supervisory system design and study of reactive power compensation" (in Chinese), *Journal of the China Railway Society*. vol..20, no.6, pp. 43-46, 1998

[4] Z. Q. Huang, Q. Z. Li and J. M. He, "Reliability design of voltage reactive power compensation controller"(in Chinese), *Journal of the China Railway Society*. vol..19, no.4, pp. 34-39, 1997

[5] Q. Z. Li, " A comprehensive analysis on parallel compensation used in A. C. traction substation," *Proceedings of International "SINGLES and SYSTEMS" Conference*, vol. 5, Chengdu, China, Sept. 1998,pp.167-178.

[6] R.Y. Wei, Z. Y. Li and Q. Z. Li, "Reactive current compensation of fuzzy controlling parallel capacitor switching," *Journal of the China Railway Society*, vol. 23, no.1, pp. 42-45,   2001

[7] R. Y. Wei and Z. Y. Li, "Approach of dynamic detecting harmonics and fundamental reactive current," *ICEE Symposium*, Hong Kong, 1999.

[8] H. Zeng, *Intelligence computation*, Chongqing University Press, Chongqing, China, 2004.

[9] J. J. Huang and S. Y. Li, "Generalized rough set models and their application," *PR & AI*, vol.17,no.2,,pp.184-1890,2004

[10] L. Zhang and Q. Z. Li, "Application of TSC in reactive power compensation in traction substations"(in Chinese), *Journal of the China Railway Society*. Vol.22, no.1, pp. 20-23, 2000

[11] J. C. Wu and S. S. Song, *Power Systems Harmonics*, China Power Press, Beijing,1985

2006 5th International Power Electronics and Motion Control Conference

A Digitally Controlled 4-kW Single-Phase Bridgeless PFC Circuit for Air Conditioner Motor Drive Applications

Yong Li and Toshio Takahashi

Energy Saving Products Division
International Rectifier
El Segundo, CA, U.S.A.   Email: yli3@irf.com

Abstract— Today's air conditioner applications demand a one-chip solution that simultaneously controls power–factor-correction (PFC) frond-end and two permanent magnetic (PM) motor sensorless drives, by using a single digital control IC. However, digital control has fundamental limitations that make most of conventional digital PFC control approaches unpractical. Implementation and integration of PFC control into digital ICs have been a quite challenge. International Rectifier recently introduced a new digital control IC that is based on the hardware computation engine called Motion Control Engine (MCE). Unique advanced digital PFC control algorithms have been developed, and it becomes practical to integrate digital PFC with 2 PM motor sensorless control into a single digital IC. Based the new control IC, a 4-kW air conditioner outdoor unit which uses a bridgeless PFC frond-end has been designed, built and tested.

This paper analyzes the fundamental limitations of conventional digital PFC control, presents the design and implementation of the 4-kW unit, and experimentally demonstrates the superior performance of the proposed digital PFC control.

I. Today's Desirable Air Conditioner Control Solution and Demand for Digital PFC Control

Environmental regulations are driving the development of energy-efficient motor drives for appliance such as air conditioners. Air conditioner control requires one of the most energy efficient controls among appliance electronics [1]. Typical home air conditioner units consist of two split units - indoor unit and outdoor unit. There are two motor loads in the outdoor unit - compressor and fan. Permanent magnet (PM) motors have been primary motors used for outdoor compressor and fan in order to increase the efficiency. The electrical power for the compressor is normally above 2kW, and can be as high as 4kW for single-phase AC input (220V/230V, 50Hz/60Hz). In order to comply with harmonic regulations, namely IEC/EN 61000-3-2 Class A and China CCC, active power-factor-correction (PFC) control is a must for the frond-end circuit.

Traditional solutions of the outdoor unit control often require at least three separate control ICs: An analog IC to control the PFC, a digital IC (typically 32-bit RISC microcontroller) to control the compressor PM motor without a position sensor, and another digital IC to control the fan PM motor also without a position sensor. These three ICs, together with their peripheral circuits, greatly increase the cost and complexity. It is highly desirable to have a one-chip solution that can simultaneously control PFC and two

PM motor sensorless drives, by using a single digital control IC. Digital control, however, has fundamental limitations that make most of conventional digital PFC control approaches unpractical. As a consequence, it has been difficult to implement and integrate PFC control into the same digital IC, mainly due to the limited computational power and lack of advanced PFC control algorithm development. To the best knowledge of the authors, up to now, for appliance there is no commercially available and practical digital control solution that can integrate PFC with sensorless PM motor control into a single digital IC.

II. A 4-kW Air-Conditioner Outdoor Unit Based on the iMOTION™ Design Platform

International Rectifier recently introduced a new digital controller, namely IRMCF312, that is based on the hardware computation engine called Motion Control Engine (MCE™) [2]. With the MCE, unique advanced digital PFC control algorithms have been developed, and it becomes possible and practical to integrate digital PFC with 2 PM motor sensorless control into one control IC. The MCE consists of a collection of control elements, motion peripherals, a dedicated motion control sequencer and dual port RAM to map internal signal nodes. Control programming is achieved using a dedicated graphical compiler integrated into the MATLAB/Simulink™ development environment. The user can design custom control loops (closed-loop current control, voltage control, speed control, etc.) based on application requirements. The compiler analyses the graphic design and automatically translates it into a sequence of MCE-specific machine code for integration control with the IRMCF312. Operating on the IRMCF312, the MCE code essentially customizes the device for the user's specific application requirements, such as PFC and sensorless PM motor control

A 4-kW air-conditioner outdoor unit drive system has been developed based on the MCE platform with IRMCF312. Figure 1 shows its circuit diagram and photo. The input stage uses a bridgeless PFC front-end circuit. The output power stage consists of two inverters, one for the compressor motor and the other for the outdoor unit fan motor. Three pieces of integrated power modules (IPMs) are used for the PFC (IRAMX25WF60B), compressor inverter (IRAMX-G42) and fan inverter (IRAMT02HP05B), respectively. This makes the 4-kW power stage very simple and compact. All the control tasks for the PFC and two inverters are accomplished by one chip IC (IRMCF312), with very simple peripherals including analog circuits.

1-4244-0448-7/06/$25.00 ©2006 IEEE

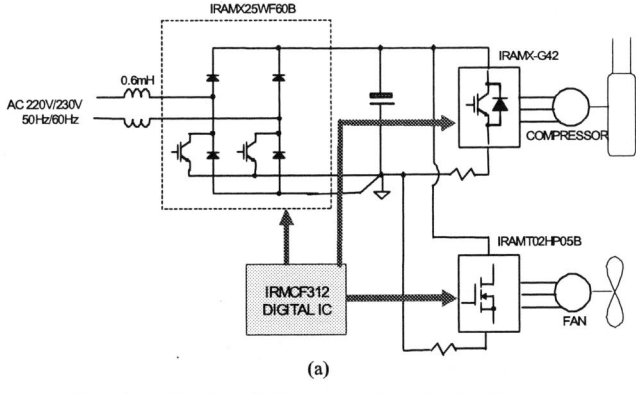

(a)

(b)

Figure 1. The 4-kW iMOTION™ air conditioner outdoor unit platform:
(a) Circuit diagram; and (b) Photo.

Different digital PFC control algorithms have been developed and implemented for the 4-kW bridgeless PFC circuit in the air conditioner outdoor unit. The reason to choose the bridgeless PFC topology is that it reduces the total forward voltage drop and conduction loss compared to conventional boost PFC topology, and can improve the efficiency. Meanwhile, the power loss and thermal dissipation are shared by 2 IGBTs alternatively within the line cycles, so the thermal stress applied on a single IGBT is reduced by 50%. These features make the bridgeless PFC more useful in high power applications. However, in terms of control principle and implementation, the bridgeless PFC is equivalent and transparent to the conventional boost PFC (In fact, this IRMCF312 control IC is designed to control both the bridgeless PFC and conventional PFC.) Therefore, for convenience, in the following discussion, the boost PFC circuit is used to illustrate the control design and implementation.

III. FUNDAMENTAL LIMITATIONS OF CONVENTIONAL DIGITAL PFC CONTROL

Figure 2 shows a control diagram of single-phase digital PFC, with conventional approach. It is similar to most of today's *analog* average-current-mode PFC control ICs, and it is a common practice and adopted by a number of digital PFC control research and designs. It mainly consists of a voltage regulator, a multiplier, a current regulator and a

PFC_PWM block. From the analog-to-digital conversion (ADC), the DC bus voltage VdcFdb, AC input voltage V_IN and AC input current I_IN are obtained. The voltage regulator consists of a proportional-integration (PI) regulator, and produces an output VAOut, which provides the amplitude information of the current control. The VAOut is sent to the multiplier and is multiplied with the sinusoidal V_IN signal and generates a sinusoidal output IREF_PFC, which serves as the reference for the current regulator. The current regulator consists of another PI regulator, and produces an output CAOut, which is the duty cycle command for the PWM generation. The CAOut signal is sent to the PWM modulator and generates the PWM signals.

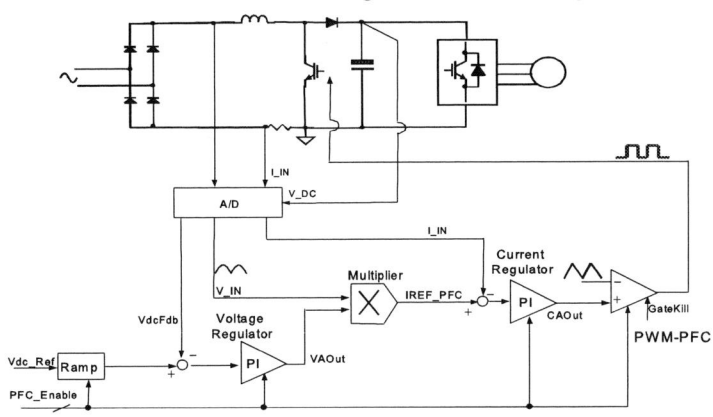

Figure 2. Conventional digital PFC control diagram, similar to conventional *analog* average-current-mode PFC control ICs.

The conventional digital PFC control was first implemented in the MCE platform. Unfortunately, as the test waveforms shown in Figure 3, it produced very poor performance: There is significant distortion and oscillation in the current waveforms. Measurement showed that this result actually failed to pass the EN/IEC 61000-3-2 Class A harmonic regulation.

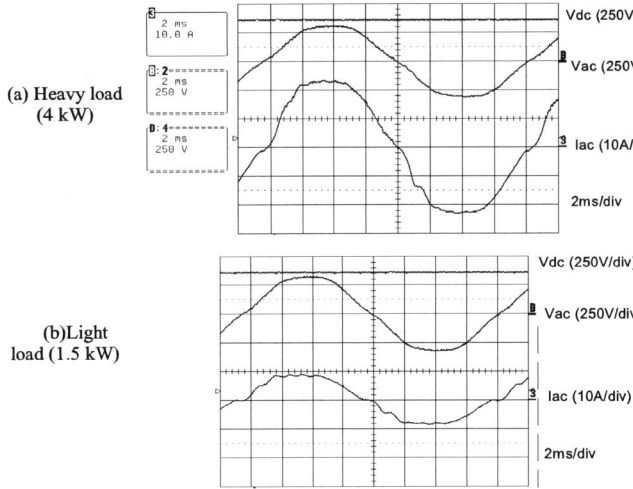

Figure 3. Test waveforms of the 4-kW bridgeless PFC circuit, with the conventional digital control (Vdc: DC Bus Voltage, Vac: AC Input Voltage, Iac: AC Input Current). ADC sampling rate 20kHz.

429

As a matter of fact, the conventional digital PFC control approach has fundamental limitations, mostly in two aspects. First, the key of PFC control relies on the current loop, which forces the inductor current waveform to track the rectified half-wave sinusoidal reference IREF_PFC, and the control regulation is solely done by a PI regulator. However, the PI regulator, by control theory, is good to regulate a signal that has a steady state with constant DC value, or steady-state DC operational point, but cannot regulate a time-varying signal, such as the sinusoidal current, unless the control bandwidth is extremely high. But for PFC control, the inductor current has to be always changing in the sinusoidal waveform, and indeed there is no steady state for the PI regulation. Second, with digital control, there are sampling-hold (S/H) and computation delays due to the ADC and digital process, which introduces additional phase shift to the current control loop. This phase shift reduces the phase margin of the closed-loop gain and therefore decreases the system stability. The slower the A/D sampling rate, the larger the phase shift. This means that in order to maintain the stability of closed loop control, the gain and bandwidth of the PI regulator has to be reduced. However, once the gain and bandwidth are reduced, the inductor current will have even poorer tracking to the sinusoidal reference, resulting in higher distortion and poorer power factor. In addition, it is well known that the PFC current-mode control itself has a sample-data effect that is similar to the S/H effect of digital control. Consequently, even with analog control, if the PWM carrier frequency is too low, the current control performance can still be suffer.

IV. ADVANCED DIGITAL PFC CONTROL ALGORITHM DEVELOPMENT AND EXPERIMENTAL DEMONSTRATION

One possible approach to solve the fundamental limitations of the conventional digital PFC control is to use higher ADC sampling rate, such as in the range of 50kHz-100kHz or even higher, which actually is a common practice in most of today's digital PFC control design. This may achieve desirable current waveform control, but will significant increase the cost of the digital IC: both high-speed A/D converter and high computation power of the digital IC are required. Moreover, as shown in Figure 1, PFC is just one of the three control tasks of the digital controller (the other two tasks are to control two PM sensorless motors simultaneously), so it is not affordable that the PFC control occupies too much computation resource. In addition, for the *4-kW* air conditioner application, the 50kHz-100kHz PWM switching is undesirable due to the high power loss and noises. Therefore, development of advanced digital PFC control algorithms is a must.

A) Improved Digital PFC Control with Unique Feed-Forward Loop (FFD)

We have developed an improved digital PFC control with a unique FFD implementation. Figure 4 shows a control diagram. Again, the control diagram is shown using the conventional boost PFC circuit, but it is also applicable to the bridgeless PFC circuit topology. In this control

scheme, the duty cycle command PWM_PFC is no longer provided solely by the PI regulator; instead, it is provided by the sum of the PI output added by a feed-forward loop, in the following way

$$PWM\_PFC = CAOut + FFD\_OUT, \qquad Eq(1)$$

where CAOut is the output of the PI regulator, and FFD_OUT is the output of the feed-forward loop. The FFD_OUT is generated based on the instantaneous AC input voltage and DC bus voltage information, and it basically provides the sinusoidal shape information to control the inductor current, or "quasi" steady-state operational point for the PI regulator. The PI feedback control loop still maintains. However, instead of handling large dynamic, the PI only regulates a small dynamic around the "quasi" steady-state operational point, and its output CAOut has a much smaller range of change. In this way, the inductor current can be controlled to have very good tracking of the sinusoidal reference, but without requiring the control loop to have high bandwidth and loop gain.

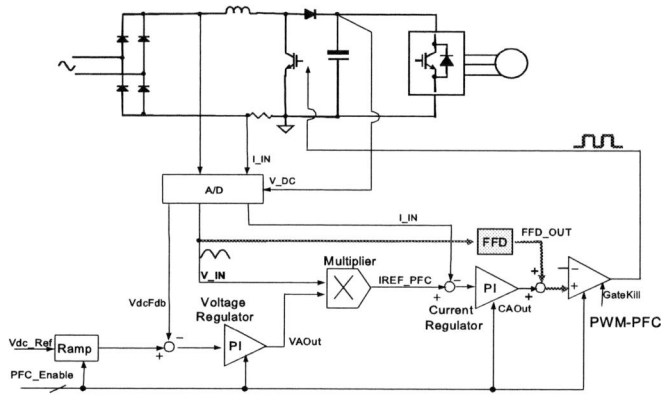

Figure 4. Improved digital PFC control with the feed-forward loop (FFD).

The improved digital PFC control with FFD has been implemented in the MCE proprietarily. Figure 5 shows one of the MATLAB/Simulink-based graphic design, and Figure 6 shows test waveforms of the 4-kW bridgeless PFC circuit. The test conditions and ADC sampling rate (20kHz) are the same as those used for generating the waveforms in Figure 3, but the current waveforms become much smoother and clearer, and there is no noticeable distortion or oscillation.

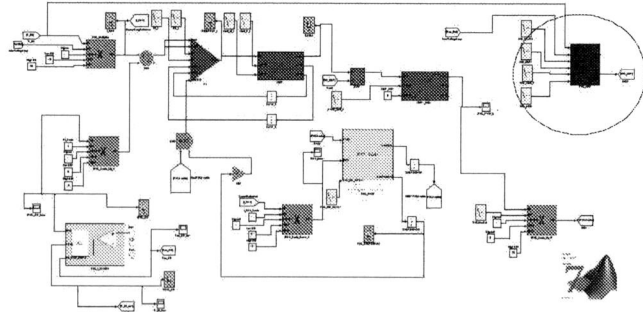

Figure 5. MATLAB/Simulink-based MCE implementation of the improved PFC control with FFD.

430

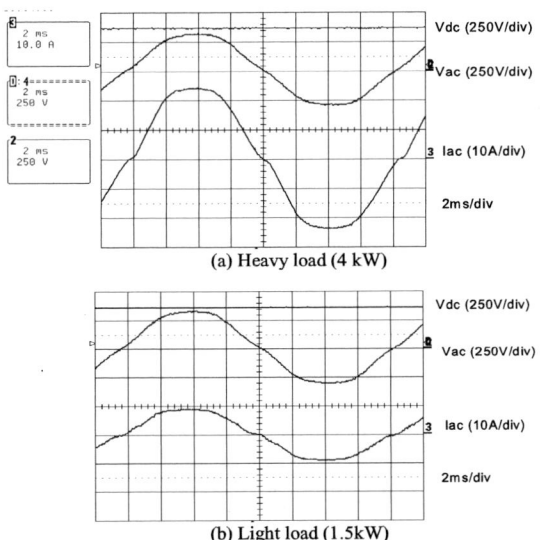

(a) Heavy load (4 kW)

(b) Light load (1.5kW)

Figure 6. Test waveforms of the 4-kW bridgeless PFC circuit, under the improve control with FFD. ADC sampling rate 20kHz.

B) Further Improved Digital PFC Control with FFD+VectorRotator

The improved digital PFC with FFD control presented in the preceding section solves the fundamental limitation of digital control to the current loop, but it does not change the way of generating the sinusoidal reference to the current loop. The reference to the current loop is still generated directly from the AC line voltage, via the ADC. One can see from the waveforms shown in Figure 6 that the current actually is not in a complete sinusoidal waveform; in particular that it has a flat top. This flat top is corresponding and caused by the flat top in the AC input voltage Vac, which is due to the heavy line impedance in the experimental set up (a 5-kVA VARIAC was used as the AC power supply which has quite heavy line impedance). This flat top increases low-order harmonic current, mainly fifth harmonic.

In order to produce a "clean" sinusoidal reference that is in phase with the AC input voltage regardless the influence of source/line impedance, the digital PFC control is further improved by using zero-crossing detection and VectorRotator control blocks, as shown in Figure 7. The zero-crossing detection block provides phase angle θ information based on the ADC result of AC voltage feedback, from which the VectorRotator produces the instantaneous sinusoidal value. It is well known that the VectorRotator block (Figure 8) is originally developed and mainly used for three-phase motor drive applications, to transfer variables between stationary frame and rotation frame in AC motor vector controls. This further improved digital PFC control fully utilizes the VectorRotator, but for the PFC application, without requiring extra hardware or firmware source. Otherwise, with conventional approaches, look-up tables would have to be used to store the instant sinusoidal value, or real-time computation has to be used, which increases the IC's memory and computation burden.

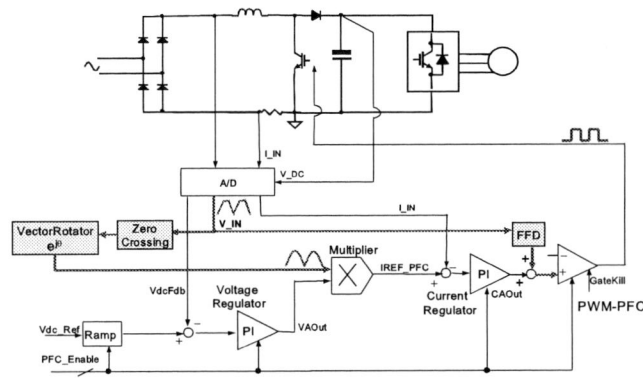

Figure 7. The further improved digital PFC control with FFD+VectorRotator.

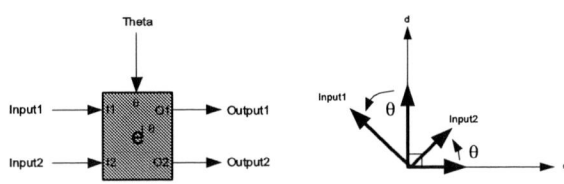

Figure 8. The VectorRotator.

The further improved digital PFC control with FFD+VectorRotator is also implemented in MCE. Figure 9 shows test waveforms under 20kHz ADC sampling rate. The test conditions and ADC sampling rate are the same as used for generating the waveforms in Figures 3 and 6. Clearly the current now is almost in a complete sinusoidal waveform without the flat top, and it is even more sinusoidal than the AC input voltage that has a flat top due to the heavy line impedance in the test set up. This result, if not impossible, will be difficult to achieve with conventional analog-based approaches.

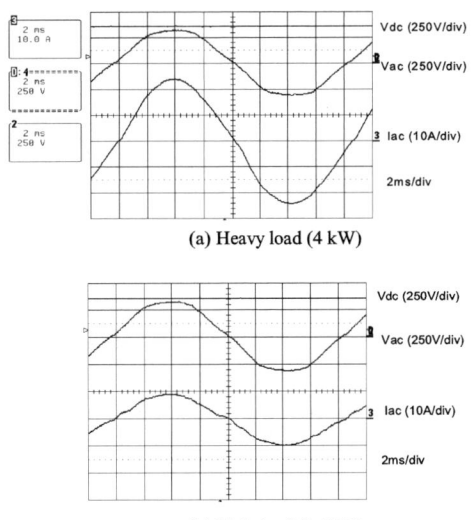

(a) Heavy load (4 kW)

(b) Light load (1.5kW)

Figure 9. Test waveforms of the 4-kW bridgeless PFC circuit, under the further improve control with FFD+VectorRotator. ADC rate 20kHz.

C) Performance Comparison

Compared to the conventional control approach, both the two improved digital PFC controls significantly increases the power factor and reduces the total harmonic distortion (THD) within the whole power range of 4kW (Figure 10). Under the maximum specified input current condition (16A) of EN61000-3-2 Class A, the measured harmonic currents under both improved digital PFC controls are well below the limits (Figure 11). The main difference is found on the 5th harmonic. As mentioned earlier the AC power supply itself used in the test has significant 5th harmonic, which reflects to the current spectrum if the VectorRotator is not used. It should be emphasized that the proposed FFD control implementation is the key to solve the fundamental problems with digital control, and the use of VectorRotator further reduces the low-order harmonic by providing the clean sinusoidal reference.

The choice of control algorithms is application-dependent. In most cases, the improved control with FFD is sufficient to comply the harmonic regulation. The further improved control with FFD+VectorRotator can be useful in applications where a complete sinusoidal current is required regardless the actual shape of AC input voltage or source/line impedance.

V. AIR CONDITIONER PLATFORM SYSTEM INTEGRATION

Thanks to the advanced digital PFC control development, together with other innovative control techniques, the 4-kW air conditioner outdoor unit has achieved simultaneous control of bridgeless PFC and two PM sinusoidal sensorless motor drives (compressor and fan), by using single digital IC (IRMCF312). Measurement shows that the total power conversion system, including the PFC and inverter stages cascaded together, achieves over 95% efficiency (Figure 12).

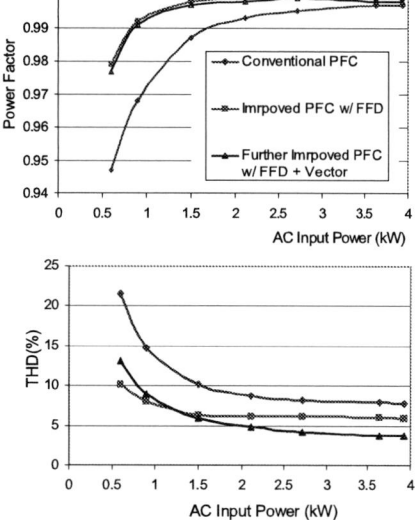

Figure 10. Measured power factor and THD.

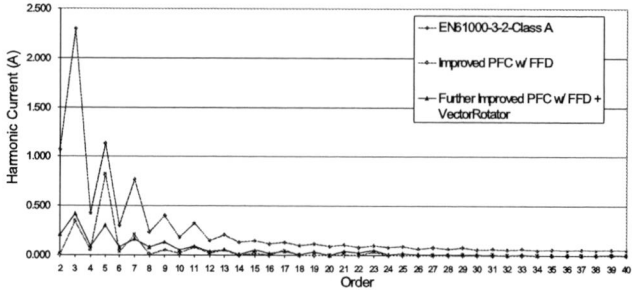

Figure 11. The measured harmonic currents under AC input current 16Arms, power 3.6kW, 230Vac, 380Vdc.

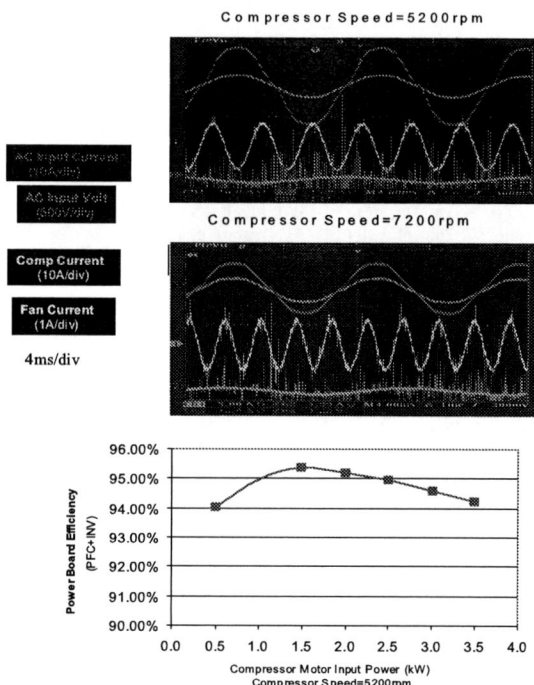

Figure 12. Test waveforms and measured power conversion system efficiency of the iMOTION™ air conditioner design platform.

VI. CONCLUSION

Conventional digital PFC control has fundamental limitations to be implemented and integrated into digital ICs. Unique advanced digital PFC control algorithms have been developed, and its superior performance is experimentally demonstrated by the 4-kW air conditioner platform. It has become practical to integrate digital PFC with two PM motor sensorless control into one digital IC. In fact, the IRMCF312 is the only single device in their class capable of simultaneously controlling all three systems. This will contribute to more energy-efficient, cost-effective solutions.

REFERENCES

[1] Toshio Takahashi, "New iMOTION appliance-motor control mitigates growing energy crisis in China, " in PCIM China 2006, pp.103-108.

[2] Guozhu Liang, Yong Li and Ming Li, "High efficiency and integrated digital PFC based air conditioer system by new PM sinusoidal sensorless control," in PCIM China 2006, pp. 152-154.

Optimized Electrical Design for Single Phase PFC Active IPEM

Qiaoliang CHEN, Xu YANG, and Zhao-an WANG
School of Electrical Engineering
Xi'an Jiaotong University
Xi'an, Shaanxi, CHINA, 710049
Email: QLCHEN@ieee.org

Abstract— **A 2kW single phase PFC active IPEM (Integrated Power Electronic Module), consisting of full bridge rectifier diodes, current sensing resistor and boost converter, has been developed employing CoolMOS and SiC diodes in this paper. The electromagnetic interference mechanism in the module is demonstrated. In order to reduce the voltage spike of power devices, three different design patterns are compared for reducing the parasitic self-inductance of critical loop. The approaches for reduction in inductive interference between power critical high di/dt loop and driver testing loop are investigated, including flux cancellation pattern for power high di/dt loop, reduction in electromagnetic interference source and inserting a copper shielding layer between them. Finally, the electrical design considerations are verified by the simulation and experimental results.**

Keywords-PFC; integrated power electronics modules; SiC; mutual inductance;

I. INTRODUCTION

As the front-end converter of distributed power system (DPS), high factor and low input-current harmonics are becoming mandatory design criteria in accordance with the EN61000-3-2 and JEIDA MITI standards, in addition to a tight output voltage regulation [1]. The active PFC solution is appreciated due to its better performance in the price of more complex circuit structure and more components compared with the passive solution.

The common rectifier diodes die, current sensor, MOSFET die and output diode die are soldered on one side of DBC (Direct Bonded Copper). Wire bonding technology is employed to interconnect the power dice and desired copper pattern of DBC. The driver and control board can be installed over the DBC layer to constitute a power module named as active IPEM, illustrated in Fig.1 and Fig.2. The use of active IPEM and passive components will allow systems assembly for customized applications with relative ease instead of having to design and build the systems from the component level [2, 3]. That will reduce the cost and size of power electronic converters as well as to improve electrical performance.

On the other hand, the higher switching frequency also leads to the smaller passive components size. However, the converter operating at higher switching frequency will generate more switching loss. The reduction in switching transition time will be helpful to reduce switching loss of MOSFET. However, it will test the limit of structural inductances associated with packaging and extend the EMI spectrum. At the same time, since the PFC circuit works at continuous current mode, there is a severe reverse recovery problem on the boost diode, which is the major source of switching loss and EMI.

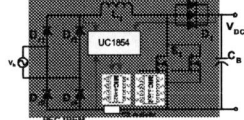

Fig.1. Circuit schematic for PFC IPEM

Fig.2. Structure demonstration for PFC IPEM

In this paper, a 2kW PFC active IPEM with switching frequency of 100 kHz, employing CoolMOS and SiC Schottky diode, is developed. CoolMOS has a much faster switching speed comparing with the conventional MOSFET, therefore the turn off loss of the MOSFETs is largely reduced. At the same time, the SiC Schottky diode doesn't have the reverse recovery current, and it gives the simple solution for the reverse recovery problem [4]. In order to reduce the parasitic inductance of high di/dt loop, different DBC patterns are compared. Furthermore, due to little distance between driver board and DBC power layer, the mutual inductive interference will be investigated. The solutions and analysis are verified by simulation and experimental results.

II. ELECTROMAGNETIC INTERFERE MECHANISM

The semiconductor power switches turn on or turn off accompanied by generated high di/dt and dv/dt which are regarded as the source of EMI. That will lead to voltage spike on semiconductor devices and electromagnetic interfere in control and driver circuit. Fig.3 shows the spectrum of MOSFET current for different rise & fall time. The period of rectangular current waveform is 10us and amplitude 10A with duty cycle of 0.5. The results indicate that slower switching transition time leads to lower amplitude in high frequency range. Therefore, there will have a tradeoff between less switching power loss and wider EMI spectrum range.

The electromagnetic field can be obtained by solving the Maxwell's equation under specific boundary conditions. It will be extremely complicated and lack of physical-origin understanding guiding the electrical design. However, if the electromagnetic field can be treated as the magnetoquasistatic field, the time dependent field can be described by the circuit parameters derived from static field, such as inductance and capacitance. It will be easy to handle and have clear physical origin. That will be true when the smallest electromagnetic wave length is much larger than the size of the phsical structure to be considered. The variation of electromagnetic wavelength with the frequency in air medium and FR4 is shown in Fig.4. The dimension of IPEM is 57.9mm×55.9mm×20mm. From the following results, the facts that the smallest electromagnetic wave length is much larger than the size of the structure are true within the interested frequency range.

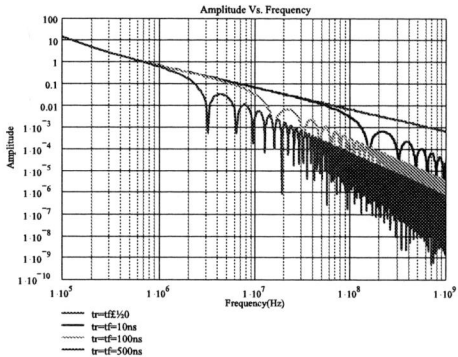

Fig.3 Spectrum of current waveform with different switching transition time

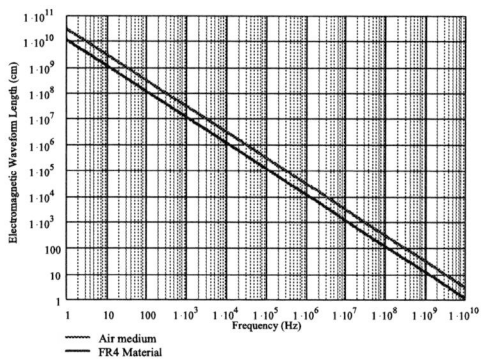

Fig.4 The variation of electromagnetic wave length with the frequency in different medium

According to the circuit theory, the inductive interference voltage within the IPEM is determined by the interference source di/dt and the mutual inductance, as shown in (1).

$$V_{jk} = M_{jk} \frac{di_k}{dt} \tag{1}$$

Where, M_{jk} is the mutual inductance between loop k and j. V_{jk} is inductive interference voltage of loop k on loop j through mutual inductance M_{jk}. If k and j represent the same current loop, the M_{jk} is self-inductance of the single current loop. In order to reduce the inductive interference voltage, the di/dt and mutual (self) inductance should be

reduced. In the following parts, the approaches to reduce them will be investigated.

For the power electronic circuit, the topology will change with the switches turn-on and turn-off. It is difficult to correctly estimate the current loops, therefore the concept of partial inductance is developed, which is defined on wire segments [5]. That is true when the dimensions are small with respect to the electromagnetic wavelength. Fig.5 shows the boost circuit topology labeling the number of branch segments, where the other rectangle loop represents the assumed driver board testing loop above power DBC layer. According to mesh circuit analysis and partial element equivalent circuit theory, the inductive interference voltage, V_{dp}, of power circuit on driver circuit loop can be expressed as (2).

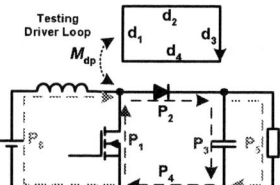

Fig.5 Demonstration of mutual inductance between power and driver loop

$$V_{dp} = \sum_{j=1}^{4} \sum_{k=1}^{6} \left(M_{d_j p_k} \frac{di_{p_k}}{dt} \right) \tag{2}$$

Where i_{p_k} is the branch current of power circuit and $M_{d_j p_k}$ is the partial mutual inductance between certain power circuit segment and driver loop segment, as illustrated in Fig.5.

According to the actual operation waveform of boost converter, the di/dt of some branches, such as $P_1 \sim P_4$, is much higher than that of others, i.e. P_5 and P_6, during the switching transition interval. Based on Kirchhoff's current laws, the $P_1 \sim P_4$ branches have the approximately same high frequency current and di/dt, denoted by di_C/dt. In the meanwhile, most parts of P_5 and P_6 branches are not packaged in the PFC IPEM, which indicates that mutual inductance between these two branches and driver loop will be reduced dramatically. Therefore, equation (2) can be simplified to (3). The first term in (3) represents the mutual inductance between driver testing loop and power critical high di/dt loop embraced by $P_1 \sim P_4$ branches.

$$V_{dp} = \left(\sum_{j=1}^{4} \sum_{k=1}^{4} M_{d_j p_k} \right) \frac{di_C}{dt} \tag{3}$$

In the same way, the sum of inductive interference voltage, V_S, on $P_1 \sim P_4$ segments of power circuit can be expressed in (4). The inductive interference voltage will lead to the voltage spike on semiconductor devices, such as power MOSFET and output diode.

$$V_S = \sum_{j=1}^{4} \sum_{k=1}^{6} \left(M_{p_j p_k} \frac{di_{p_k}}{dt} \right) \tag{4}$$

Where, $M_{p_j p_k}$ is the partial mutual inductance between

434

two power circuit segments. When j equals k, $M_{p_j p_k}$ represents the self-inductance of single segment.

$$V_S = \left(\sum_{j=1}^{4} \sum_{k=1}^{4} M_{p_j p_k} \right) \frac{di_C}{dt} \qquad (5)$$

According to the same reason demonstrated above, equation (4) can be simplified to (5). The first term represents the self-inductance of power critical high di/dt loop.

III. ELECTRICAL DESIGN CONSIDERATIONS FOR REDUCTION IN THE INDUCTIVE INTERFERENCE

From the analysis results, the effective methods to reduce the inductive interference include reduction in mutual inductance and inductive interference source, di_C/dt. Some practical considerations will be investigated.

A. Parasitic Self-Inductance of Critical High di/dt Loop

In order to optimize the layout design, three different designed layouts are analyzed and compared, as illustrated in Fig.6 and Fig.7. The parasitic inductance of critical loop mentioned above is calculated by Ansoft Maxwell 3D. From the results shown in Table I, the parasitic inductance of layout 1 is maximum, and critical loop parasitic inductance of layout 2 is only 1.5nH less than that of layout 3. The parasitic inductance calculation for layout 3 is carried out with eddy current effect and without it, respectively. That indicates that the eddy current effect of copper plate dramatically decreases the loop inductance. On the other hand, mutual cancellation of the fluxes by twisting the critical loop will reduce mutual effect on upper control and drive board, as demonstrated in next section. Therefore, the layout 3 was selected as the final version. The final fabricated module is shown in Fig.7 (b).

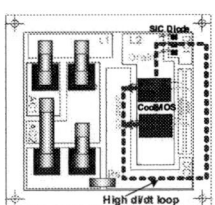 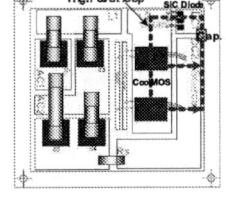

| (a) Layout 1 | (b) Layout 2 |

Fig.6. Designed DBC electrical layout

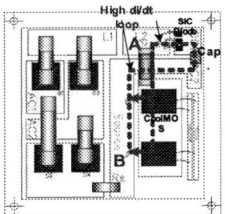 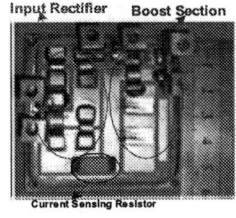

| (a) Layout 3 | (b) Final fabricated module |

Fig.7. The final selected DBC Layout and fabricated module

TABLE I
THE CRITICAL LOOP INDUCTANCE OF DIFFERENT LAYOUT CONSIDERATIONS

Layout consideration	Parasitic inductance of critical loop (nH)	
	With Eddy Effect	Without Eddy Effect
Layout 1	21.938	
Layout 2	10.126	
Layout 3	11.678	16.964

B. Mutual Inductance Reduction between Critical Loop and Driver Board

In the PFC module, the driver board is fabricated on top of DBC layer with the distance of less than 8mm. The power critical high di/dt loop will produce the driver voltage spike which may exceed the highest allowable gate-source voltage of MOSFET. On the other hand, the inductive interference will increase the switching transition time, that lead to higher switching loss. In order to reduce the mutual inductance, the less area of critical high di/dt loop is necessary. In the meanwhile, the structure producing the flux cancellation effect is proposed, shown in Fig.8 and Fig.9. The flux and radiating areas is reduced by twisted current paths similar to the twisted-pair wire [6]. As depicted in Fig.9, the driver layer loop magnetic flux generated by the critical loop current is cancelled dramatically through the twisted paths generating opposite flux in the driver layer loop. The mutual inductance between equivalent driver loop and different critical loop patterns illustrated in section A is calculated by ANSOFT Maxwell 3D, as shown in Fig.10. The eddy current effect by the copper plate isn't considered in the calculation. From the results listed in table □, the mutual inductance is reduced dramatically in the case of twisted critical high di/dt paths.

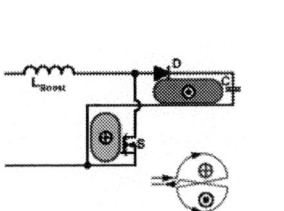 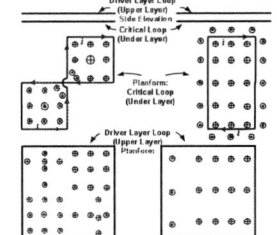

Fig.8. Flux and radiating areas reduction by twisted current paths   Fig.9. Flux cancellation in driver loop by twisting high di/dt loop

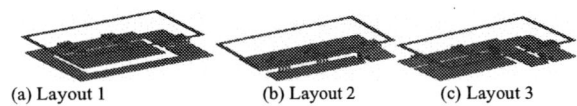

| (a) Layout 1 | (b) Layout 2 | (c) Layout 3 |

Fig.10. Simulation model for mutual inductance between equivalent driver loop and different critical loop considerations

TABLE II
THE MUTUAL INDUCTANCE BETWEEN DRIVER AND CRITICAL LOOP

Layout considerations	Mutual Inductance (nH)
Layout 1	15.232
Layout 2	6.4387
Layout 3	0.04

The experimental verification of the advantage of flux cancellation pattern of high di/dt is carried out on the following PCB test bed, whose assembly configuration is illustrated in Fig.11. The electrical patterns of power PCB and testing driver PCB is given in Fig.12. Two patterns of high di/dt loop, as illustrated in Fig.12 (b) and (c), are implemented respectively, where Boost topology is employed. The discrete power MOSFET is IRFP460 and the output diode MUR1560T (Ultra Fast Rectifier). The output diode current is measured by Tektronix TCP202.

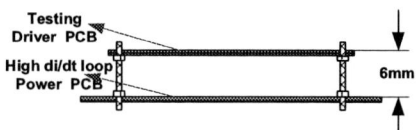

Fig.11. the testbed assembly configuration

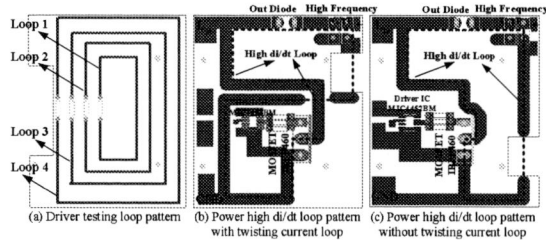

Fig.12. the electrical patterns of driver testing loop and power high di/dt loop

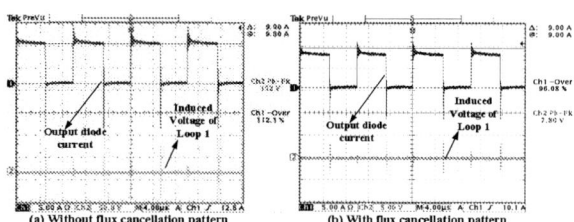

Fig.13 the waveform of output diode current and induced voltage of loop 1 for two different patterns with MUR1560

The waveforms of output diode current and the induced voltage of driver testing loop 1 for two power high di/dt loop patterns are given in Fig.13. The output MUR diode has very serious reverse recovery problem, featuring high di/dt, which causes high interference voltage on driver testing loop 1. From the results in Fig.13, with the flux cancellation pattern the peak-peak inductive interference voltage is reduced dramatically to 7.8V from 152V of without flux cancellation pattern.

C. Reduction in Interference Source

As mentioned in section B, the reverse recovey problem of MUR diode is the very serious interference source. The emerging SiC Schottky diode doesn't have the reverse recovery current with high reverse blocking voltage. The application of SiC diode in active PFC gives the simple solution for the reverse recovery problem. Fig.14 shows the compared output diode current waveform and induced voltage on loop 1 without flux cancellation pattern. The

spectrum of current waveform for MUR diode and SiC diode is given in Fig.15. The results indicate that perfect turn-off characteristic of SiC diode results in much less high frequency EMI components, and therefore less interference voltage. Fig.16 shows the inductive interference voltage waveforms of loop 1 with SiC Schottky diode, SDT12S60, in two cases, without and with flux cancellation pattern. In the latter case, the peak-peak voltage is dramatically reduced to 3.59V caused by the MOSFET turn-off.

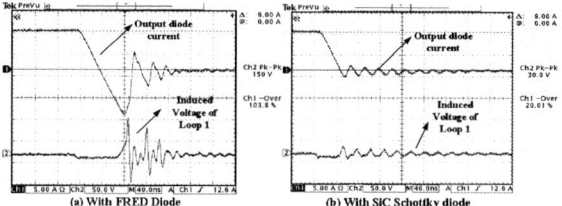

Fig.14. the current waveform for MUR and SiC Schottky diode

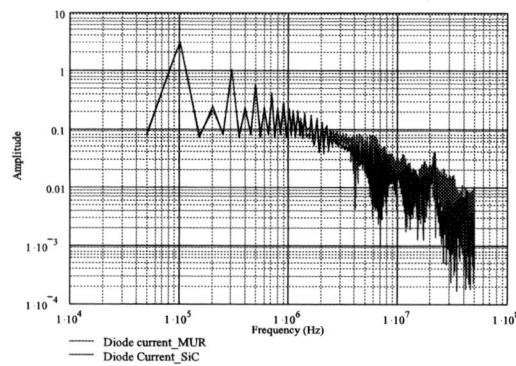

Fig.15. Spectrum of output diode current for MUR and SiC Schottky diode

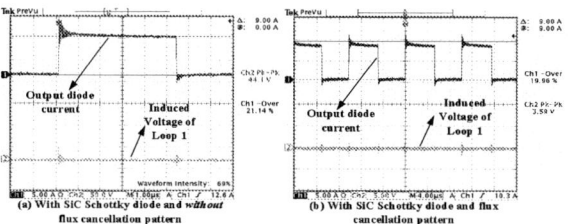

Fig.16. the waveform of output diode current and induced voltage with SiC Schottky diode for two different electrical patterns

D. Shielding for Reducing Inductive Coupling

According to Maxwell's equations, as the time-variable electromagnetic field passes certain medium, the field will decay. The penetration depth, δ, is a distance at which field falls off by a factor of 1/e.

$$\delta = \sqrt{\frac{1}{\pi \mu \gamma f}} \qquad (6)$$

Where, μ and γ are the magnetic permeability constant and conductivity of the medium respectively. The higher electromagnetic field frequency, f, is, the more difficultly the field passes the medium. Usually, some electromagnetic sensitive objectives are surrounded by the conductive material for shielding extern electromagnetic field. This is called as electromagnetic shielding. In PFC module, a conductive layer is inserted between power DBC and driver board for eliminating the magnetic interference by power high di/dt loop. The mutual inductance between them is expected to reduce by shielding approach, especially at high frequency range. Fig.17 shows the variation of mutual inductance between driver testing loop 1 and power high di/dt loop in Fig.12(c) with the frequency. The thickness of the shielding copper layer is 35um. The results indicate that the mutual inductance at high frequency range is reduced dramatically.

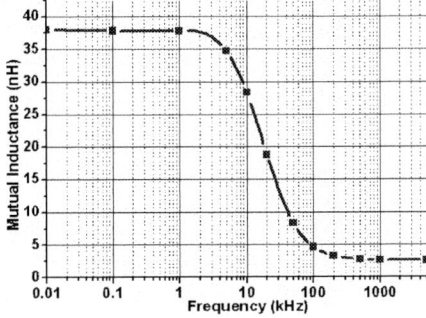

Fig.17: The variation of mutual inductance with the frequency

For DBC-based IPEM, the DBC-backside eddy current is also highly helpful to reduce the mutual inductance between power high di/dt loop on DBC and driver board [7].

IV. EXPERIMENTAL RESULTS FOR PFC IPEM

The 2kW IPEM-based PFC converter is prototyped under the practical electrical considerations demonstrated above. The CoolMOS and SiC Schottky diode are employed in the PFC IPEM including input rectifier diodes section. Fig.18 presents the drain-source voltage waveform of CoolMOS, gate-source driver voltage and input current waveform. The voltage overshoot of MOSFET drain-source voltage is 50V. The inductive interference of power high di/dt on driver waveform is slight.

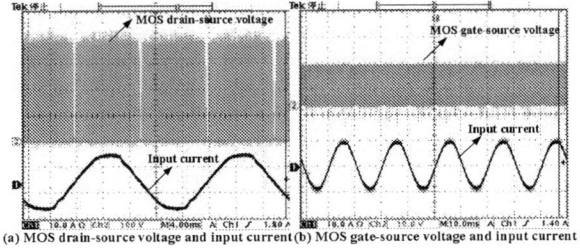

(a) MOS drain-source voltage and input current (b) MOS gate-source voltage and input current

Fig.18. Experimental waveforms on IPEM-based single PFC converter

V. CONCLUSIONS

This paper has presented the development of a 2kW IPEM-based PFC converter employing CoolMOS and SiC diode. Due to the more compact structure, the inductive interference of power high di/dt loop on driver board must be the considered carefully.

The electromagnetic interference mechanism in the module is analyzed. The electromagnetic field in the IPEM can be regarded as magnetoquasistatic field within the interested frequency range. Therefore, the mutual inductance derived from static field can be employed to describe the inductive interference.

Three different design layouts are analyzed and compared for reducing the parasitic self-inductance of critical loop. The reduction approach in inductive interference between power critical high di/dt loop and driver testing loop is investigated. The mutual inductance between equivalent driver loop and critical high di/di loop can be reduced with the flux cancellation pattern and by inserting a conductive shielding layer. The perfect turn-off characteristic of SiC Schottky diode results in much less high frequency EMI components, and therefore less interference voltage. Finally, the electrical design considerations are verified by the experimental results.

ACKNOWLEDGEMENT

This work is supported by Key Project of National Natural Science Foundation of China under award number 50237030. The author would like to acknowledge Infineon Technologies providing the power CoolMOS and SiC diode dice samples.

REFERENCES

[1] IEC 1000/3/2 International standard, "limits for harmonic current emissions (Equipment Input current < 16A per phase)", 1995.

[2] J. D. van Wyk, F. C. Lee, Z. X. Liang, R. G. Chen, S. Wang, B. Lu, "Integrating Active, Passive and EMI-Filter Functions in Power Electronics Systems: A Case Study of Some Technologies", *IEEE Trans. Power electronics*, vol. 20, no. 3, pp. 523-536, May. 2005.

[3] J. D. van Wyk, F. C. Lee, and D. Boroyevich, "Power Electronics Technology: Present Trends and Future Developments", *IEEE Proceedings*, vol. 89, no. 6, pp. 799-802, June 2001.

[4] L. Lorenz, G. Deboy and I. Zverev, "Matched pair of CoolMOS™ transistor with SiC-Schottky diode-advantages in applications," *IEEE IAS Conf. Rec.*, 2000, pp. 376-383.

[5] Frank B.J. Leferink, "Inductance calculations: methods and equations", *IEEE International Symposium on Electromagnetic Compatibility*, Atlanta, Georgia, USA, vol01.pp 16-22, 1995.

[6] L. Rossetto, S. Buso, G, Spiazzi, "Conducted EMI Issues in a 600-W Single-Phase Boost PFC Design", *IEEE Trans. Industrial Application*, vol. 36, no. 2, pp. 578-585, March. 2000.

[7] B.Gutsmann, P.Mourick, D.Silber,"Exact Inductive parasitic extraction for analysis of IGBT pararrel switching inclduing DCB-backside Eddy Currents", *IEEE 31st Annual Power Electronics Specialists Conference*, vol. 3, pp. 1291-1295, 2000.

2006 5th International Power Electronics and Motion Control Conference

A Novel Topology of APFC with On-Line Half-Bridge UPS Controlled by DSP

Xuejun Ma, Xuezhi Hu ,Hongxia Wu,XuWu Chen

Department of Electrical Engineering, Huangshi Institute of Technology Huangshi,435003,P.R.China

Email: hsmxj@tom.com

Abstract—In this paper, a novel topology of APFC with on-line half-bridge UPS controlled by DSP is presented.The operation principle is analyzed. In order to improve the dynamic response feature of APFC and make the design of LC filter easy ,improved current hysteretic control is proposed.A digital control scheme based on DSP is presented. The control model of APFC is simulated with Matlab. In order to verify the analysis, a prototype with 50Hz/220V/2.2kW is developed. The simulation and experimental results show the PF of the proposed topology is approximate to 1.

Keywords- APFC; improved current hysteretic control; digital control ;simulation

I. INTRODUCTION

In the inverter of middle and small power the topology of half-bridge is widely applied for its' low cost and good characteristic of anti-unbalance ability.

For an online inverter, AC- DC-AC can be realized. In order to decrease the effect of harmonic current in input power line and increase the power factor (PF) ,the technology of active power factor correction(APFC) should be considered. And, the topology of APFC with Boost converter is extensively applied for its' advantage.

In [1] an APFC topology with diode which is used for high frequency rectification plus Boost converter is presented. But this topology is only suitable for low ration rectification and not suitable for high frequency UPS.In [2] a Boost PFC for high power applications is proposed based on one cycle control.. It is simple and reliable but only suitable for full-bridge inverter.

In this paper ,for single-phase high frequency inverter, a novel APFC topology with double Boost converter will be presented. According to this topology ,a high PF and low harmonic current distortion can be gotten. At the same time, for positive bus and negative bus, a strict symmetrical voltage can be realized.

II. ANALYSIS OF TOPOLOGY AND CONTROL METHOD

A. A novel topology of APFC in half-bridge inverter

There are two main functions for APFC topology with double Boost converter. One of them is that current feedback is taken so that the current is in input power line can trace dynamically the AC input sine voltage v_s.So almost there is a same phase between i_s and v_s, the

PF will be approximately to 1.On the other hand, Boost converter can increase the output voltage in order to maintain the RMS of AC output voltage. Voltage feedback will be used to keep the voltage of positive and negative bus stable.So for the topology of APFC, double close loop i.e. voltage loop and current loop should be applied.The topology of APFC presented is shown as Fig.1.

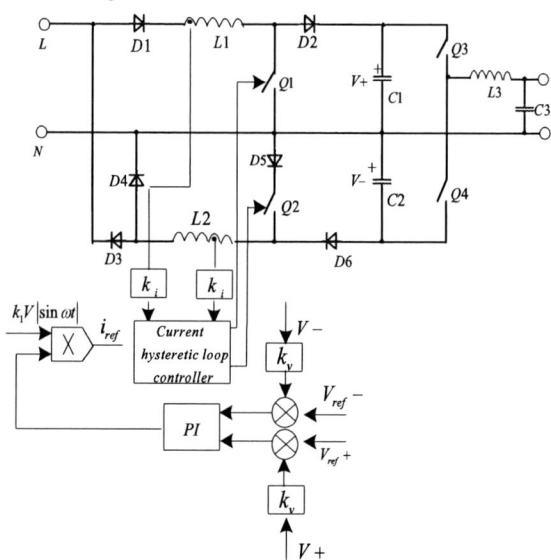

Fig.1 The topology of half-bridge inverter with APFC

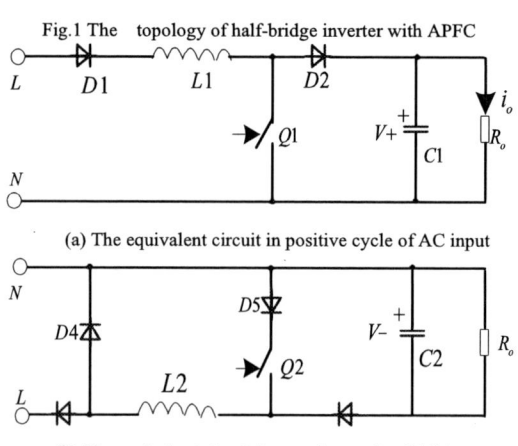

(a) The equivalent circuit in positive cycle of AC input

(b) The equivalent circuit in negative cycle of AC input

Fig.2 The equivalent circuit of APFC in one cycle of AC input

1-4244-0448-7/06/$25.00 ©2006 IEEE   438

From Fig.2. we can see that the proposed topology of APFC can be divided into two stages.In Fig.2. (a) in positive half cycle of AC input , D1、L1、Q1、D2、C1 consist of the topology of APFC with Boost converter. In Fig.2. (b) in negative half cycle of AC input , D3、D4、D5、D6、L2 、Q2、C2 consist of the topology of APFC with Buck-Boost converter..Each stage can supply DC 400V and supply energy for half-bridge inverter.

B.The control principle of current loop

In[3] a fast controller for a unity-powerf-factor PWM rectifier.For the presented topology of APFC, double closed-loop including external voltage loop and internal current loop is used in order to realize the output DC voltage stable and PF high enough. There are three control methods for current loop,i.e.. peak current control, current hysteretic loop control and average current control. In this paper, current hysteretic loop control is applied .And PI control is taken in voltage loop.

In fact, the topology of APFC can be regarded as a CCVS. The design of internal current loop is very important. For the method of current hysteretic loop control, there is a dynamic comparison between the current reference signal which is kept as a same phase with AC input voltage and the current in input inductance L1,L2. The switching device is turned off and the output current decrease when output current is higher than the reference current The switching device is turned on and the output current increase. In a switching cycle ,the duty ratio is dynamically adjusted so that the input current can trace the input voltage as sine waveform. For this method, the dynamic response is rapid than other control methods because of the current feedback.. On the other hand, the disturbance in internal current loop can be restricted effectively.

In current loop, current reference which is the product of the output of external voltage loop and unity sine waveform is a sine waveform with the same phase with AC input voltage.The sensed current of L1,L2 is represented with i_L .The width of hysteretic loop is represented as H.In a switching cycle,Q1,Q2 will turn off and i_L will decrease when $i_L \geq i_{ref} + H$ .On the other hand, Q1,Q2 will turn on and i_L will decrease when $i_L \leq i_{ref} - H$ .Obviously, i_L will follow i_{ref} with a high frequency of sawtooth waveform.The current waveforms of input inductance L1,L2 in hysteretic current loop is shown as Fig.3.

When the switching device is turned on, i_L increases linearly,

$$\Delta i_{L+} = \frac{V_{in}}{L} \cdot T_{on} = H \tag{1}$$

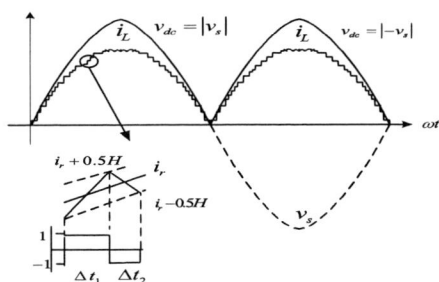

Fig.3 Current waveforms of input inductance L1,L2

When the switching device is turned off, i_L decreases linearly,

$$\Delta i_{L-} = \frac{V_O - V_{in}}{L} \cdot T_{off} = H \tag{2}$$

So the switching period of the switch device is

$$T = T_{on} + T_{off} = HL \ \frac{V_O}{V_{in}(V_O - V_{in})} \tag{3}$$

From above equations we can see that the switching frequency is decided by H, L, V_i, and V_o. Obviously the switching frequency is not constant. It is difficult for us to choose devices and filter with L,C.

In order to solve this problem, an improved hysteretic current loop with internal feedback can be taken.The contol block diagram is showm as Fig.4

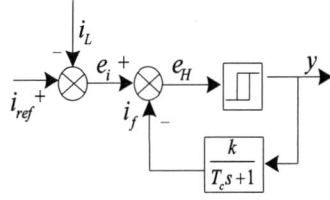

Fig.4 Improved current hysteretic loop controller

The transfer function of the proposed close loop can be expressed as:

$$G_d = \frac{y}{e_h} = \frac{1}{K}(1 + sT_C) \tag{4}$$

For a hysteretic current loop with negative feedback , the current error is controlled by PI controller.The switching frequency can be expressed as

$$f \approx \frac{K}{2T_c H} \tag{5}$$

From the equation (5) we can see that f will be approximate to constant with improved hysteretic current loop.

From the viewpoint of control this method is same

as a forward correction. So the system stability will be improved and a rapid response can be realized.

C. The control principle of voltage loop

In[4] a control method based on MCU is proposed . In this paper, the voltage control loop is realized based on DSP TMS320F240.In generally ,AC input current is regarded as to follow the reference completely in the design of this control loop. Current reference is equivalent to a AC current source whose frequency and phase are same as AC input voltage and amplitude is controlled by DC output voltage.On the other hand ,the reference sine waveform of current loop is produced by DSP with SPWM.

In Fig.1 ,a PI controller is taken in external voltage loop. The bus voltage is sampled and converted in A/D.The fundamental wave of SPWM waveforms is kept as the same phase with AC input voltage by digital phase locked loop(DPLL).SPWM waveforms is sent to a LPF in order to get the fundamental wave. The amplitude of fundamental wave is proportional to the output of PI controller and sent to the negative input of comparator. The current of L1,L2 is sensed and amplified, then is sent to the positive input of comparator after a precise rectifier. The output of hysteretic comparator controls the switching frequency and duty ratio of switching devices in Boost circuit. So the bus voltage can be kept as stable and high PF can be realized.The control block diagram is shown as Fig.5.

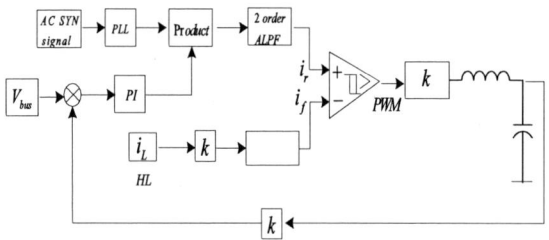

Fig.5 The APFC control block diagram based on DSP

III. THE PARAMETERS CALCULATION OF APFC

A.The filter capacitor

For Boost circuit, its output voltage Vo will almost keep constant for the close loop control. Co is discharged and supplies the energy for the load during the switching device is turned on. Co is charged,L1,L2 stores energy and load gets the energy from the source during the switching device is turned off.

Supposed that Vo falls down from Vo to Vomin, the output power is Po and the period of AC input voltage is Td.

$$C_O(V_o^2 - V_{o\min}^2)/2 \ge P_o T_d \qquad (6)$$

According the requirements of output power Po and ripple voltage Co can be calculated by equation (6).

B. Boost inductor

In APFC circuit,L1,L2 operate with CCM. Supposed that the maximum peak current is I_{Lp}, the maximum average current is I_{Lm}, the ripple current is Δi_L , the rectification average voltage is Ui. If we demand $\Delta i_L < 0.2 I_{Lm}$, then

$$L = \frac{U_i DT}{1.2 I_{Lm}} \qquad (7)$$

In Boost circuit ,

$$U_o = \frac{U_i}{1 - D} \qquad (8)$$

If we neglect the power loss,then

$$U_O I_O = U_i I_{Lm} \qquad (9)$$

From the equation (7),(8),(9),we can get

$$L = \frac{U_i^2 (U_o - U_i)}{1.2 f U_o^2 I_o} \qquad (10)$$

Obviously if Ui, Uo,f, Io is set up then L1,L2 can be calculated according to the equation (10).

IV. SIMULATION AND EXPERIMENTAL RESULTS

According to the operation principle of APFC with the control of presented hysteretic current loop and Fig.1,a simulation model is constructed by simlink of Matlab.The simulation results is shown as Fig.6.

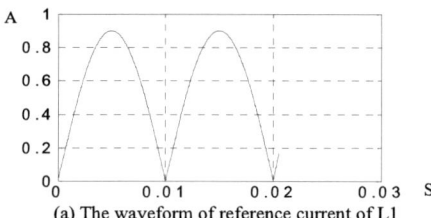

(a) The waveform of reference current of L1

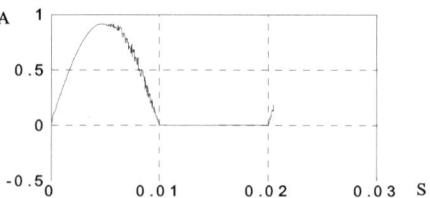

(b) The sensed current waveform of inductance L1

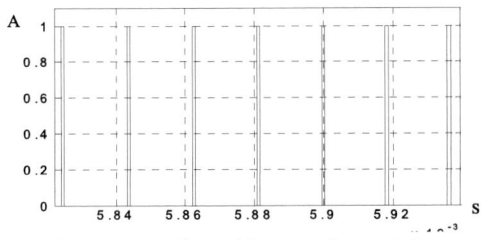

(c) Output waveform of the current hysteretic loop

Fig.6 The simulation waveforms

In order to verify the analysis a prototype of half-bridge inverter with APFC is developed with DC 400V bus voltage and 220V AC/50Hz/2kVA.In the circuit C1,C2 is 1000uF/450V,L1,L2 is 0.9mH.The reference and actual current waveforms under full load are shown as Fig.7. And the PF is 0.97.

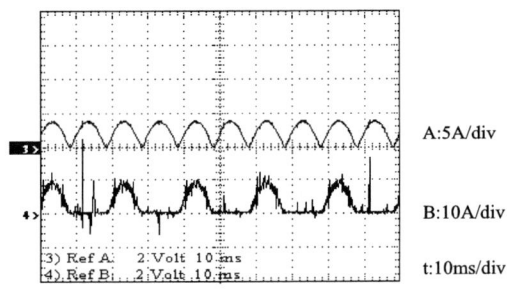

A:5A/div

B:10A/div

t:10ms/div

A : Reference current waveform

B: L1 current waveform

Fig.7   The experimental waveforms under full load

V. CONCLUSIONS

The novel CCM APFC with proposed current hysteretic loop based on DSP is introduced in this paper. The performance of APFC is satisfactory according to the simulation and experimental. Furthermore, the switching frequency varies in a small range so that the design of LC is easy. The control based on DSP enhances the system reliability and flexibility.

REFERENCES

[1] Sebastian , "An overview of power factor correction in single-phase power supply systems", IECON'94 ,Vol.3,pp.1688 - 1693

[2] S.Wall. and R.Jackson. "Fast controller design for single-phase power factor correction systems." IEEE. Trans. 1nd . Electr., Vol.44.No.5.Oct.1997.pp.654-660.

[3] M.O.Eissa.S.B.Leeb.G.C.Verghese.andA.M.Stankovie."Fast con -troller for a unity-power factor PWM rectifier." IEEE Trans.Power Electr.,Vol.11.No.1,Jan.1996,pp.1-6

[4] S.Buso.P.Mattavelli.L.Rosetto.and G.Spiazzi. "Simple digital control improving dynamic performance of power factor preregulators." IEEE Trans. Power Electr. Vol. 13. No.5. Sept. 1998. pp.814-823

2006 5th International Power Electronics and Motion Control Conference

Nonlinear Current Control of Single-Phase PFC Suitable for Mixed-Signal IC Implementation

Min Chen, Anu Mathew, and Jian Sun

Department of Electrical, Computer, and Systems Engineering
Rensselaer Polytechnic Institute, Troy, NY 12180-3590, USA
Telephone: (518) 276-8297; Fax: (518) 276-6226; E-mail: jsun@rpi.edu

Abstract–New nonlinear current control methods are developed for single-phase power factor corrected (PFC) ac-dc converters. The control methods combine input current feedforward with partial feedback control based on the switch current to achieve sinusoidal input current and unity input power factor. The nonlinear control methods don't suffer from the limitations of conventional linear (average) current control in terms of control bandwidth and sensitivity to noise. The resulting current control performance is superior to all existing single-phase PFC control methods, and can meet the requirements of various applications, including high-frequency (360-800 Hz) airborne power systems. The control methods are also ideally suited for integrated mixed-signal implementation, requiring an analog current controller that is very simple, insensitive to noise, and independent of converter and control design parameters, and a digital controller that operates with a sampling frequency much lower than the switching frequency. A prototype design is presented to validate the analysis and to demonstrate the performance of the control methods.

I. INTRODUCTION

Various current control methods have been reported in the literature for continuous-conduction mode (CCM) boost single-phase power factor corrected (PFC) rectifiers [1-11]. Among these, linear average current control [3] is most widely used due to its good control performance (compared to peak current control [2]) and the availability of integrated analog control chips developed by Unitrode in the 1990's. The method, however, has several limitations. First, its control performance becomes inadequate when the ratio of switching frequency to the line frequency is low, such as in the case of airborne systems where the line frequency can be as high as 800 Hz [12-14], or in high power applications where the switching frequency is low [15]. This is due to the fact that the current loop crossover frequency has to be limited to about one tenth of the switching frequency in order to ensure stability, while tracking a rectified sinusoidal reference current requires a crossover frequency of 50-100 times higher than the line frequency. The current loop under average current control is also very sensitive to noise due to the use of high-gain op-amp, making layout design very critical for the implementation of the method. Furthermore, average current control is not suitable for digital implementation due to the need for high-speed ADCs and processors.

For analog implementation, the use of multiplier in average current control is also a disadvantage as it significantly complicates the control IC design. Several nonlinear control methods were developed in the later 1990's [5-11] to overcome this problem. Most of these control methods also

eliminate the need for sensing the input voltage, leading to further simplification of control implementation. However, this is also a disadvantage of these control methods in terms of control performance, because input voltage feedforward, which is an effectively means to improve line regulation of the output voltage, becomes impossible without a measurement of the input voltage. The one-cycle control [5-7] is effectively a peak-current control method, hence has similar performance limitations as conventional peak-current control. Nonlinear-carrier control [9, 10] in effect controls the average of the input current, and its performance is even better than that under linear average current control. A disadvantage of the method is the dependency of the nonlinear carrier signal on the switching frequency, which complicates IC implementation.

This paper presents a set of new nonlinear current control methods for single-phase PFC converters which overcome the problems of existing linear and nonlinear control methods. They are also ideally suited for integrated mixed-signal implementation. The paper is organized as follows: The next section first introduces the basic principle of nonlinear current control for boost single-phase PFC based on cycle-by-cycle integration of the switch current. A primitive nonlinear control method is then presented and shown to be inherently unstable. Existing nonlinear control methods are then reviewed as variations to this primitive control, and are shown to have better stability characteristics. The basic form of the new nonlinear control methods is then introduced in Section III. Stability of the control method is analyzed by using a small-signal sampled-data model of the current loop, and improvement of the current loop stability by introducing additional low-frequency signals to the integrator as well as the pulse-width modulator is presented. Section IV discusses mixed-signal implementation of the proposed control methods. A prototype boost single-phase PFC converter using the proposed mixed-signal control structure is presented in Section V along with experimental measurements of its operation and performance.

II. NONLINEAR CURRENT CONTROL

Fig. 1 shows the power stage of a boost single-phase PFC converter. Unity-power-factor operation requires that the average of the boost inductor current follows a reference that is proportional to the rectified input voltage:

$$\bar{i}_L = i_{ref} = g_e|v_{in}| \tag{1}$$

1-4244-0448-7/06/$25.00 ©2006 IEEE

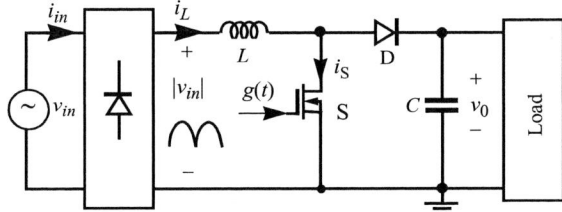

Fig. 1. Boost single-phase PFC converter power stage.

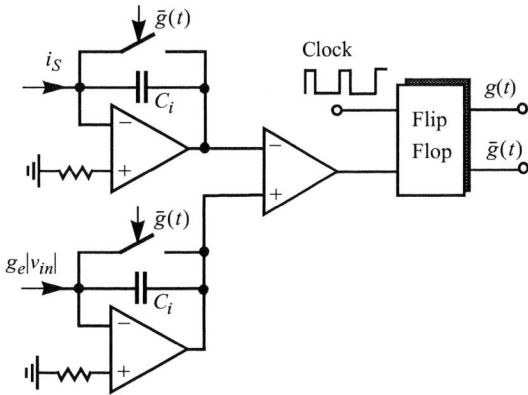

Fig. 2. A nonlinear current control for boost single-phase PFC based on (3) that is always unstable.

where $|v_{in}|$ is the rectified input voltage, and g_e is a constant determined by the output voltage controller. Since the switch current (i_s) is equal to the boost inductor current (i_L) in the on-time period of the switch, the average of the inductor current is related to the average of the switch current over each switching cycle by

$$d \cdot \bar{i}_L = \bar{i}_S = \frac{1}{T_s} \int_0^{dT_s} i_s dt, \qquad (2)$$

where d is the on-time duty ratio of the switch. Combining (1) with (2) gives

$$\frac{1}{T_s} \int_0^{dT_s} i_s dt = d \cdot g_e |v_{in}|. \qquad (3)$$

Equation (3) defines the duty ratio of the switch with which the average of the input current would follow the reference.

A nonlinear current control method can be formulated based on (3), as illustrated in Fig. 2. The control would work as follows: Simultaneously integrate the switch current and the reference current (note that $i_{ref} = g_e|v_{in}|$) from the beginning of each switching cycle, and turn off the switch when the two integrator outputs become equal. As one can see, the resulting duty ratio of the switch under this control would satisfy (3), hence the inductor current would follow the reference defined by (1). Also note that this control is nonlinear and uses no compensation components.

A. Stability Analysis

Unfortunately, the control method outlined above is inherently unstable. This can be seen from the following small-signal analysis of the current loop: Since the input voltage and the output voltage both vary slowly with time, the inductor current response over a switching cycle can be modeled by a first-order sampled-data model:

$$i_L[k+1] = i_L[k] + \frac{|v_{in}[k]|T_s}{L} - \frac{v_0[k](1-d[k])T_s}{L} \qquad (4)$$

where $i_L[k]$, $|v_{in}[k]|$, $d[k]$ and $v_0[k]$ represent the initial value of inductor current, the input voltage, the duty ratio, and the output voltage at the beginning of the kth switching cycle, respectively. To determine the duty ratio, $d[k]$, we note first that the current of the switch during its on-time in the kth switching cycle can be written as

$$i_s(t) = i_L[k] + \frac{|v_{in}[k]|}{L} t. \qquad (5)$$

Substituting this into (3) gives:

$$i_L[k]d[k] + \frac{|v_{in}[k]|}{2L} \cdot d[k]^2 T_s = g_e|v_{in}[k]|d[k]$$

from which the duty ratio, $d[k]$ can be determined:

$$d[k] = \frac{2L}{T_s} \cdot \frac{g_e|v_{in}[k]| - i_L[k]}{|v_{in}[k]|} \qquad (6)$$

This expression can now be substituted into (4), resulting in the following sampled-data model of the inductor current:

$$i_L[k+1] = \left\{1 - \frac{2v_0[k]}{|v_{in}[k]|}\right\} \cdot i_L[k] + $$
$$+ \frac{|v_{in}[k]|T_s}{L} - \left(\frac{T_s}{L} - 2g_e\right)v_0[k]$$

The above model indicates an unstable current loop, since the coefficient of $i_L[k]$ is always larger than unity.

B. Existing Nonlinear Current Control Methods

Although the nonlinear control method discussed in the previous subsection is not practically useful, the idea behind it can be found in most existing nonlinear control methods for boost single-phase PFC. Nonlinear carrier control [9, 10], for example, can be developed by replacing the rectified input voltage, $|v_{in}|$, in (3) by $v_0(1-d)$, which represents the steady-state voltage conversion ratio of the boost converter. The resulting control law is described by

$$\frac{1}{T_s} \int_0^{dT_s} i_s dt = g_e v_0 (1-d) d. \qquad (7)$$

Implementation of the method is similar to that for the nonlinear control discussed in the previous subsection, but requires two integrators to generator the non-inverted input to the comparator (nonlinear carrier), defined by

$$\frac{1}{T_s} \int_0^t g_e v_0 \left(1 - \frac{2\tau}{T_s}\right) d\tau. \qquad (8)$$

As (8) indicates, the nonlinear carrier signal generation circuit will require two integrators and is also dependent of the switching frequency.

If, instead of the average, the peak of the input current is required to follow the reference, (3) can be re-derived to be

$$i_s(dT_s) = g_e|v_{in}|.$$

Substituting $|v_{in}|$ by $v_0(1-d)$, as in the development of nonlinear carrier control, gives

$$i_s(dT_s) = g_e v_0(1-d),$$

which can be rearranged as follows:

$$\frac{1}{T_s}\int_0^{dT_s} g_e v_0 dt = g_e v_0 - i_s(dT_s) \qquad (9)$$

This can be translated into a nonlinear control method in a way similar to that for (3) and (7), and is the basic form of one-cycle control [5-7]. The method is clearly simpler than the nonlinear carrier control, since no integrator is required to generate the non-inverted input to the comparator. However, the input current will contain more harmonic distortion, particularly under high line and/or light load conditions when the ripple current is large compared to the average. On the other hand, both nonlinear carrier control and one-cycle control avoid the use of input voltage as well as a multiplier, which are required in the conventional average current control for generating the reference current. But this is achieved at the expense of not being able to implement input voltage feedforward [3], which is critical for improving line regulation of the output voltage.

III. NEW NONLINEAR CURRENT CONTROL METHODS

The new nonlinear control methods to be developed in this section are also derived from the basic control requirement for the switch as defined by (3). To avoid the instability problem discussed in the previous section, we replace the duty ratio, d, on the right-hand side of (3) by a "feedforward" duty ratio signal [14-17].

Two different feedforward duty ratio signals have been used in the literature for boost single-phase PFC converters:

1) Partial Feedforward [16]:

$$d_{ff} = 1 - \frac{|v_{in}|}{v_0} \qquad (10)$$

2) Full Feedforward [15]:

$$d_{ff} = 1 - \frac{|v_{in}| - Li'_{ref}}{v_0} \qquad (11)$$

These signals define the duty ratio of the switch under ideal operation conditions. The partial feedforward signal (10) is derived by ignoring the voltage across the boost inductor, hence is less accurate compared to the full feedforward signal. Use of both signals will be considered in this paper.

A. Basic Nonlinear Control Method

By replacing d on the right-hand side with either feedforward signal d_{ff} defined above, (3) becomes

$$\frac{1}{T_s}\int_0^{dT_s} i_s dt = d_{ff} \cdot g_e|v_{in}|. \qquad (12)$$

A new nonlinear current control can now be formulated based on (12). With reference to Fig. 1, the control method would work as follows:

1) Turn on the switch S at the beginning of each switching cycle (by a clock signal);
2) Start to integrate the switch current when the switch is turned on;
3) Compare the integrator output with a signal defined by the right-hand side of (12), and turn off the switch S when the two signals intersect.

This control method can also be implemented by using the circuit shown in Fig. 2. However, since the right-hand side of (12) doesn't change over a switching cycle, the second integrator used in Fig. 2 to generate the non-inverted input to the comparator is not needed. The overall control system including the generation of the non-inverted input to the modulator will be discussed in the next section.

To further illustrate the operation of this nonlinear current control method, a single-phase boost PFC converter employing this control was implemented and simulated in SABER. Fig. 3 shows details of the switch gate signal, the two inputs to the comparator, and the resulting switch current. The simulation was conducted with an open voltage loop, that is, with a constant g_e and output voltage.

B. Stability Analysis

To determine the stability of the new nonlinear current control method, the same sampled-data modeling method as used in the previous section is applied. Since the response of the inductor current is still described by (4), only a new duty ratio constraint needs to be developed for this new control method. This is obtained by substituting (5) into (12):

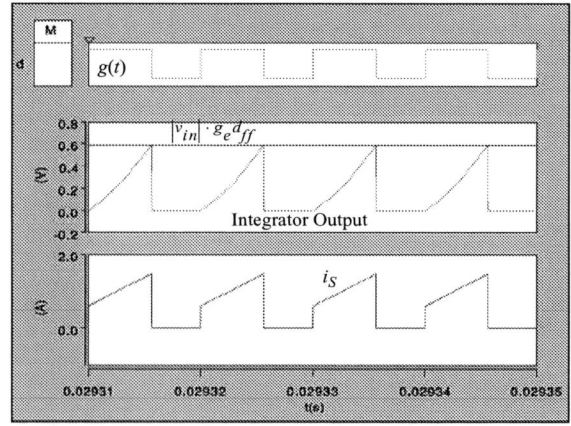

Fig. 3. Simulation waveforms of boost single-phase PFC converter with the a nonlinear current control based on (12).

$$i_L[k]d[k] + \frac{|v_{in}[k]|}{2L}d[k]^2 T_s = g_e|v_{in}[k]| \cdot d_{ff}[k] \quad (13)$$

The above equation together with (4) defines a complete sample-date model for the boost inductor current. The combined model is nonlinear, but can be linearized to give the following small-signal linear model for stability analysis:

$$\hat{i}_L[k+1] = \frac{i_L[k] + \frac{|v_{in}[k]| - v_0[k]}{L} \cdot d[k]T_s}{i_L[k] + \frac{|v_{in}[k]|}{L} \cdot d[k]T_s} \cdot \hat{i}_L[k] \quad (14)$$

From (14), the current loop under the new nonlinear control method is stable if

$$\left| \frac{i_L[k]L + (|v_{in}[k]| - v_0[k])d[k]T_s}{i_L[k]L + |v_{in}[k]|d[k]T_s} \right| < 1. \quad (15)$$

Further analysis indicated that (15) is satisfied in the region where the instantaneous input voltage is above certain value; more specifically, when

$$|v_{in}| > \frac{v_0 T_s}{2g_e L + 3T_s}. \quad (16)$$

This condition may not be satisfied around the zero crossing of the input voltage, which would lead to instability and, consequently, harmonic current distortion. These can be seen from the simulated input current waveform shown in Fig. 4.

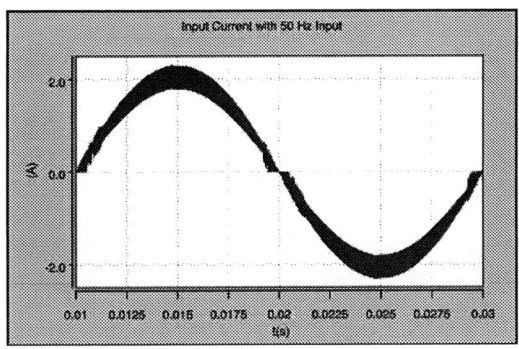

Fig. 4. Simulated input current response of boost single-phase PFC converter under a nonlinear current control defined by (12).

C. Stability Improvement

To improve the stability of the control around zero-crossing points of the line voltage, we will reformulate (3) by first splitting its right-hand side into two terms as

$$\frac{1}{T_s}\int_0^t i_s dt = (1+\alpha)g_e|v_{in}| \cdot d - \alpha g_e|v_{in}| \cdot d, \quad (17)$$

where $0 \le \alpha \le 1$, and then moving the second term to the right-hand side, as shown below:

$$\frac{1}{T_s}\int_0^{dT_s} [i_s + \alpha g_e|v_{in}|]dt = (1+\alpha)g_e|v_{in}| \cdot d \quad (18)$$

With a given parameter α, and by replacing d on the right-hand side by d_{ff} defined by either (10) or (11), the above equation can be used in the same way as (12) was to define different nonlinear current control methods. The additional term on each side of the equation requires some changes to the implementation circuit from that for (12):

1) The integrator input is now the combination of the switch current and a signal that is proportional to the reference current ($\alpha g_e|v_{in}| = \alpha i_{ref}$).

2) The non-inverted input to the comparator is now scaled up by a factor $1 + \alpha$.

To see how the additional terms improve the stability, we apply again small-signal analysis to the current loop. It was found that the current loop is stable if

$$\left| \frac{i_L[k] + \alpha g_e|v_{in}[k]| + \frac{|v_{in}[k]| - v_0[k]}{L} \cdot d[k]T_s}{i_L[k] + \alpha g_e|v_{in}[k]| + \frac{|v_{in}[k]|}{L} \cdot d[k]T_s} \right| < 1. \quad (19)$$

Compared to (15), both the numerator and denominator on the left-hand side of (19) contain an extra term $\alpha g_e|v_{in}[k]|$. Further analysis indicates that (19) holds if

$$\alpha g_e|v_{in}|L > d\left(\frac{v_0}{2} - |v_{in}|\right)T_s - i_L L. \quad (20)$$

Considering that in steady state,

$$v_0 \approx \frac{|v_{in}|}{1 - d}, \quad (21)$$

and by assuming that the inductor current follows exactly the reference, that is, $i_L = i_{ref} = g_e|v_{in}|$, we can rearrange (20) to give the following requirement for the parameter α under which stability of the current loop can be guaranteed:

$$\alpha > \frac{dT_s}{2g_e(1-d)L} - \frac{dT_s}{g_e L} - 1 \quad (22)$$

Different α can be selected to improve current loop stability. In addition to being a constant, α can also change over each line cycle, such as in the following cases:

1) Select $\alpha = d_{ff}(1 - d_{ff})^{-1}$, which would guarantee current loop stability if the converter and its operation is such that $(2g_e L)/T_s > 1$. The resulting control equation can be written as follows:

$$\frac{1}{T_s}\int_0^{dT_s} [i_s(t) + g_e v_0 d_{ff}]dt = g_e v_0 d_{ff} \quad (23)$$

2) Select $\alpha = (1 - d_{ff})^{-1}$, which would guarantee current loop stability if the converter and its operation is such that $(2g_e L)/T_s > d_{ff}$. The resulting control equation can be written as follows:

$$\frac{1}{T_s}\int_0^{dT_s} [i_s(t) + g_e v_0]dt = g_e(|v_{in}| + v_0)d_{ff} \quad (24)$$

Fig. 5 depicts a general implementation of the proposed nonlinear current control methods for a boost single-phase PFC converter. The signals $v_c(t)$ and $v_r(t)$ are defined as follows:

$$v_c(t) = v_r(t) = g_e v_0 d_{ff} \qquad (25)$$

for implementation based on (23), and

$$v_c(t) = g_e(|v_{in}| + v_0)d_{ff} \qquad (26)$$

$$v_r(t) = g_e v_0 \qquad (27)$$

for implementation based on (24). The digital control and current sensing circuitry will be discussed in the next section.

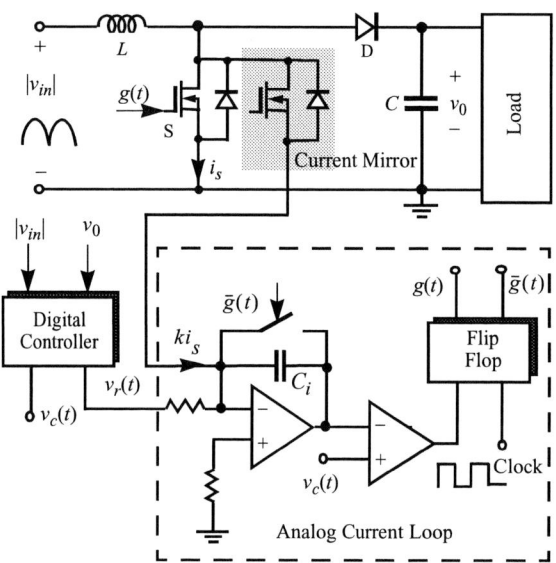

Fig. 5. A possible mixed-signal implementation of proposed nonlinear current control for a boost single-phase PFC converter.

Note that the stability analysis presented in this and the previous subsection assumed continuous conduction mode (CCM) of the boost inductor current. Characteristics and performance analysis of the proposed nonlinear current control methods under discontinuous conduction mode (DCM) will be presented in a future work.

IV. MIXED-SIGNAL IMPLEMENTATION

The proposed nonlinear current control methods are best suited for integrated mixed-signal implementation. A possible mixed-signal implementation scheme for a boost single-phase PFC converter is depicted in Fig. 5. The control circuitry consists of a digital controller that regulates the output voltage and provides necessary reference signals $v_c(t)$ and $v_r(t)$ as defined by (25)-(27), and an analog controller that controls the input current based on the nonlinear control method defined by (23) or (24) [or other variations from (18)]. The third part of the control circuitry is the sensing of the switch current. Each of these three parts will be discussed in more detail in the following.

A. Digital Voltage Controller

The digital controller samples the rectified input voltage, $|v_{in}|$, and the output voltage, $v_0(t)$, of the converter. Within the controller, the sampled output voltage is compared to a reference voltage, and the difference is processed by digital compensator, typically a PI controller, to generate the control parameter g_e, from which $v_r[k]$ and $v_c[k]$ as defined by (25)-(27) are generated. A zero-order hold can be used to interface each of these digital outputs with the analog current loop discussed in the next subsection. The feedforward duty ratio signal, d_{ff}, used in (25)-(27) can be calculated from either (10) or (11). In case (11) is used, the reference current, i_{ref}, in the expression will be replaced by $g_e|v_{in}|$.

It is apparent that all signals processed by the digital controller vary at the line frequency. Hence the necessary sampling frequency is very low and is independent of the switching frequency. For a 60 Hz line, a sampling frequency of 10 kHz will suffice under all conditions regardless of the operation frequency of the switch. Even for high-frequency applications, such as 360-800 Hz airborne systems, a sampling frequency of 50 kHz is enough. Moreover, the voltage loop of single-phase PFC converter is usually kept slow to avoid distortion in the input current, such that the digital voltage compensation function doesn't need to be executed more than a few times in every fundamental cycle. The combination of low sampling frequency and low computation requirements allows the digital controller to be implemented with a low-speed, low-cost digital device, thereby solving a major problem of full digital control.

B. Analog Current Loop

As shown in Fig. 5, the analog control loop consists of an integrator, a comparator, and a flip-flop. The integrator has two inputs: the sensed switch current, ki_s, and a compensation signal $v_r(t)$, which represents the second integral term in (23) or (24), and is generated by the digital controller. The integrator is reset at the beginning of every switching cycle by a signal complementary to the power stage switch (S) gate signal. The non-inverted input to the comparator represents the right-hand side of (23) or (24), and is generated by the digital controller. The clock signal, which initiates the turn-on the power stage switch, can be provided by the digital controller as well.

As is apparent from Fig. 5, there are only two design parameters in this analog control circuitry: the current sensing gain, k, and the integrator capacitor, C_i. The integrator output is proportional to the ratio of these two parameters, that is, k/C_i. Whatever this ratio is, the digital controller outputs can be adjusted proportionally such that the actual input current of the converter is independent of the two parameters. This allows the same analog control circuit to be used in different converter designs without having, making it possible to integrate the analog control circuitry with the digital controller to form a low-cost mixed-signal PFC control IC. This is a major advantage of the proposed nonlinear current control methods compared to existing current control methods.

C. Switch Current Sensing

In Fig. 5, the switch current is shown to be sensed by using a current mirror, which is essentially a small MOSFET in parallel with the main MOSFET. The current mirror diverts a small portion of the inductor current to the integrator in the current control circuit when the switch is on [18]. The current mirror is usually integrated with the main switch in a single silicon die to minimize the variation of current sensing gain with device layout design and operation temperature. Other current sensing methods can also be used and will be discussed in a future work.

V. EXPERIMENTAL RESULTS

A prototype boost single-phase PFC converter with the proposed nonlinear control methods was built and tested. The parameters of the converter power stage are: $L = 1$ mH, $V_0 = 385$ V, $V_{in,rms} = 115$ V, switching frequency $f_s = 100$ kHz. The main switch was implemented using IRC840, a power MOSFET with an integrated current mirror made by IR [19]. The digital controller was implemented using a digital signal processor ADMC401 from Analog Devices [20]. Although not necessary, the input voltage and the output voltage were sampled at the same frequency. The sampling frequency was set at 10 kHz for 60 Hz line and 50 kHz for 400 Hz line.

We implemented and tested all three control methods defined by (12), (23), and (24), respectively. The implementation based on (12) may be unstable around zero crossing of the input voltage under 50 or 60 Hz input, particularly when the load is light. The modified control laws (23) and (24) solved this problem and achieved similar control performance. Due to space limits, only results from the implementation based on (23) are presented below.

Fig. 6a) demonstrates the basic operation of the current control loop under the proposed nonlinear control over a few switching cycles. Note that, despite the large noise in the current sensing signal, the two inputs to the comparator are largely free of noise, indicating excellent noise immunity.

Fig. 7 shows the measured input current and voltage waveforms of the prototype with 60 Hz and 400 Hz line input, respectively. Operation under the 400 Hz line was intended to test the performance for airborne applications [14]. As can be seen from the measurements, the input current is almost purely sinusoidal and in phase with the input voltage under even under 400 Hz input. Unlike the conventional average current control where the harmonic distortion increases significantly with the line frequency [12], there is not much degradation in current control performance under the proposed nonlinear current. This makes the method also very attractive for airborne applications. Fig. 8 gives the spectrum of the measured current waveform under 400 Hz line and compares it with the harmonic current limits defined in DO-160D [13] for airborne equipment. As can be seen, the control performance meets DO-160D requirements.

As can be seen from Fig. 5, accurate sensing of the switch current requires that the integrator has infinite gain and its non-inverted input is at the same potential as the source of the main switch [18]. However, the two points in our current experimental setup are on two different circuit boards, which causes error in the current sensing signal. The effect of this sensing error is most pronounced around the zero crossing points where the useful signal level is the lowest, leading to control error in the input current, as observed in Fig. 7b). We expect that a better layout design will solve this problem.

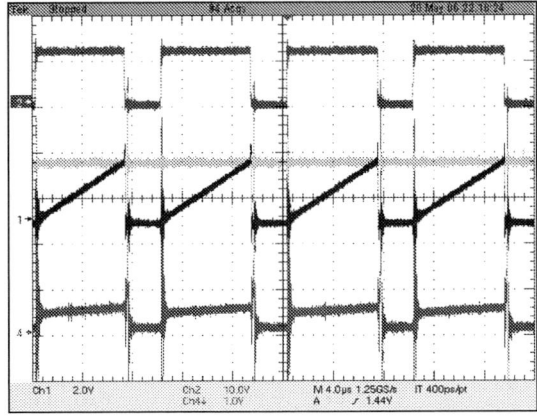

Fig. 6. Measured waveform of the current control loop under proposed nonlinear control. From top down: 1) main switch gate signal, 2) comparator non-inverted input, 3) comparator inverted input (integrator output), and 4) sensed switch current.

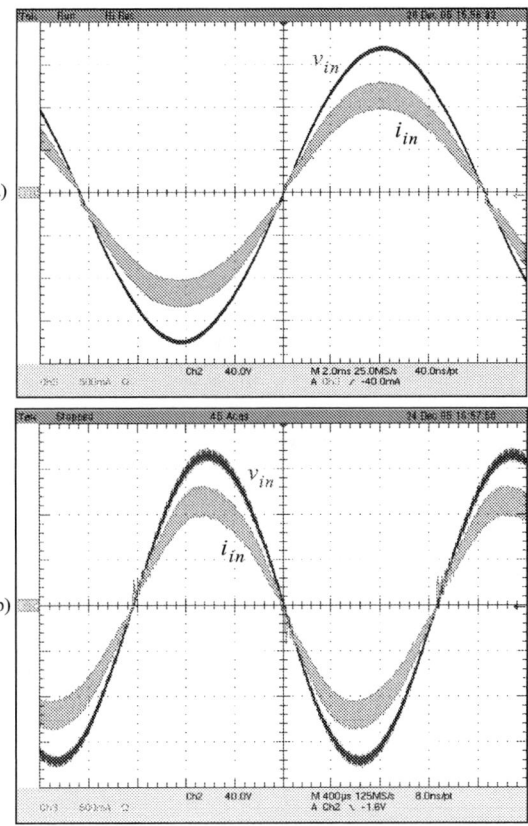

Fig. 7. Measured input current and voltage waveforms of the prototype boost single-phase PFC converter. a) 60 Hz line; b) 400 Hz line.

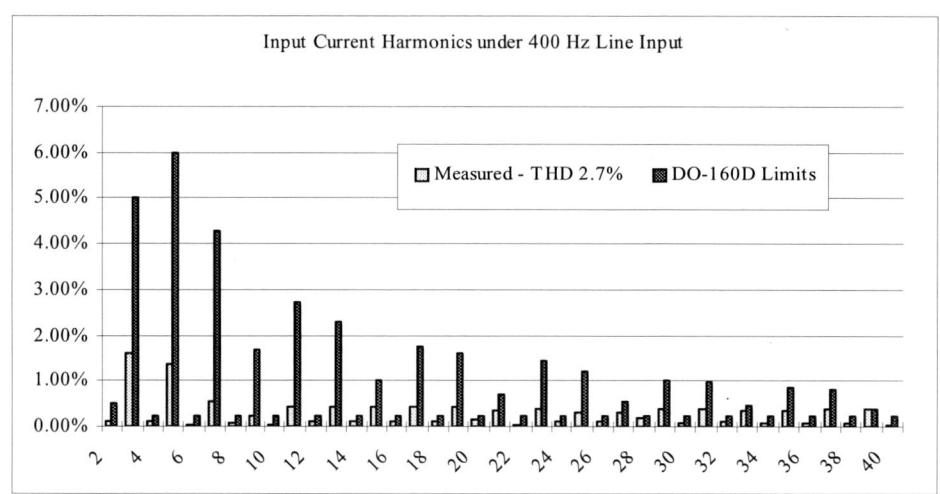

Fig. 8. Harmonic contents of the input current shown in Fig. 7b) in comparison with the limits defined in DO-160D for airborne equipment.

A comparison between the partial and the full feedforward signal defined by (10) and (11), respectively, was also conducted experimentally as well as by numerical simulation. It was found that the partial feedforward signal, although simpler to implement than the full feedforward signal, results in increased harmonics in the input current. This can be explained as follows: The partial feedforward signal differs from the full feedforward signal by an amount that is proportional to the derivative of the reference current. Since the reference current is proportional to the rectified input voltage, this derivative term is positive in the first half of each half line cycle and negative in the second half of each half line cycle. This leads to a quarter-wave asymmetry in the partial feedforward signal, which in turn causes harmonics in the input current. For this reason, the full feedforward signal defined by (11) should be used in high-performance applications.

VI. SUMMARY

This paper presented a new nonlinear current control method and several variations of it with enhanced stability around zero crossing points of the input voltage for boost single-phase PFC converters. The proposed methods are simple, robust, easy to design, and achieve input current control performance that can meet even the stringiest harmonic current limits for airborne applications. The methods are most suitable for integrated mixed-signal implementation, requiring a very simple analog control circuitry and a low-cost, low-speed digital device.

REFERENCES

[1] J. Sun, W. C. Wu, and R. Bass, "Large-signal characterization of single-phase PFC circuits with different types of current control," in *Proceedings of IEEE Applied Power Electronics Conference* (APEC'98), pp. 655-661, 1998.

[2] C. Zhou and M. M. Jovanovic, "Design trade-offs in continuous current-mode controlled boost power-factor correction circuits," *Conference Proceedings of HFPC'92*, pp. 209-220, 1992.

[3] L. H. Dixon, "Average current-mode control of switching power supplies," *Unitrode Power Supply Design Handbook*, 1990.

[4] J. B. Williams, "Design of feedback loop in unity power factor ac to dc converter," *Records of PESC'89*, pp. 959-967.

[5] K. Smedley and S. Cuk, "One-cycle control of switching converters," in *Record of IEEE PESC'91*, pp. 814-820, 1991.

[6] Z. Lai and K. M. Smedley, "A family of continuous-conduction-mode power-factor-correction controllers based on the general pulse-width modulator," *IEEE Transactions on Power Electronics*, vol. 13, no. 3, pp. 501-510, May 1998.

[7] R. Brown and M. Soldano, "One cycle control IC simplifies PFC designs," in *Proceedings of IEEE Applied Power Electronics Conference*, pp. 825-829, 2005.

[8] J. Hwang, A. Chee, and W.-H. Ki, "New universal control methods for power factor correction and dc to dc converter applications," in *Proceedings of APEC'97*, pp. 59-65, 1997.

[9] D. Maksimovic, Y. Jang, and R. W. Erickson, "Nonlinear-carrier control for high-power-factor boost rectifiers," *IEEE Transactions on Power Electronics*, vol. 11, no. 4, pp. 578-584, July 1996.

[10] R. Zane and D. Maksimovic, "A mixed-signal ASIC power-factor-correction (PFC) controller for high frequency switching rectifiers," in *PESC'99*, pp. 117-122, July 1999.

[11] J. Rajagopalan, et al., "A general technique for derivation of average current mode control laws for single-phase power-factor-correction circuits without input voltage sensing," *IEEE Transactions on Power Electronics*, vol. 14, no. 4, pp. 663-672, 1999.

[12] J. Sun, "On the zero-crossing distortion in single-phase PFC converters," *IEEE Transactions on Power Electronics*, vol. 19, no. 3, pp. 685-692, May 2004.

[13] Environmental Conditions and Test Procedures for Airborne Equipment, RTCA DO-160D, Section 16, June 2001.

[14] J. Sun, "Analysis and design of single-phase PFC converters for aerospace systems," in *Proceedings of 2003 IEEE Industrial Electronics Society Annual Meeting*, pp. 1101-1109, 2003.

[15] M. Chen and J. Sun, "Feedforward current control of boost single-phase PFC converters," in *Proceedings of IEEE Applied Power Electronics Conference*, pp. 1187-1193, Feb. 2004.

[16] S. Wall and R. Jackson, "Fast controller design for single-phase power-factor-correction systems," *IEEE Transactions on Industrial Electronics*, vol. 44, no. 5, pp. 654-660, 1997.

[17] P. T. Prathapan, M. Chen, and J. Sun, "Feedforward current control for boost-derived single-phase PFC converters," in *Proceedings of IEEE Applied Power Electronics Conference*, pp. 1716-1722, 2005.

[18] A. Sinkar, R. J. Gutmann, T. Paul Chow, and J. Sun, "Comparison of integrated current sensing circuits and implementation for smart control of synchronous rectifiers," in *Proceedings of 2005 CPES Annual Seminar*, pp. 301-304, 2005.

[19] International Rectifiers, IRC840 Datasheet (http://www.irf.com)

[20] Analog Devices Inc., *ADMC401 DSP Motor Controller Developer's Reference Manual*, March 2000.

2006 5th International Power Electronics and Motion Control Conference

A Novel Detection Method for Three-Phase Reactive Current

Zong Ming, Wang Fengxiang ,Hua Funian, Sun Yidan,

School of Electrical Engineering Shenyang University of Technology, Shenyang 110023, China

Abstract — Analogue multiplier has been widely used in the detection circuit of the reactive power. But an analogue multiplier is quite expensive so it is limited in some application case. And many analogue multipliers used in a circuit will reduce the detection accuracy. In this paper, a novel detection method for three-phase reactive current with analog multiplexer is proposed. The operation principle and simulation results are also presented. An available circuit is illustrated. The simulation result shows that it is feasible.

Keywords- detection reactive current; phase sensitive detector; analog multiplexer; analogue multiplier

I. INTRODUCTION

In the power system, many equipments, such as asynchronous motors, power transformers etc, consume considerable reactive power, which will affect the safe and stable operation of power system if reactive power can't be compensated in time. The compensation result of a compensation device to reactive power depends on the detection method for reactive power. So the suitable detection method for reactive power is very important to reactive power compensation.

Normally, in a reactive power compensation device, there are three methods, i.e. measuring power factor, measuring reactive current and measuring reactive power, to be used for obtaining compensation signal. Study shows that measuring reactive current as the compensation control signal is better than measuring power factor. In fact, if the voltage of power system is a constant, measuring reactive current and measuring reactive power is same.

How to detect reactive power is a key problem in reactive power compensation.

As we known, if voltage retains constant in power system, the reactive power magnitude changes with product of current and $\sin\varphi$, φ is power factor angle. So the analogue multiplier or Hall element is used for detecting reactive power magnitude. But it is high price or temperature character that their application area is limited.

A new method for detecting the reactive current magnitude using an analog multiplexer is studied in the paper. It is cheaper than analogue multiplier and has better temperature character than Hall element.

II. PRINCIPLE OF PHASE SENSITIVE DETECTION OF REACTIVE CURRENT

Phase sensitive detection technique is required for detecting reactive current. Phase sensitive detection means that detecting result should distinguish the sign of power factor angle. The block diagram of phase sensitive detector for detecting reactive current is shown in Figure 1.

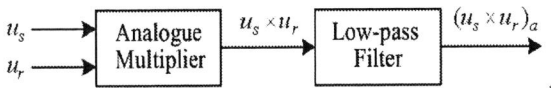

Figure 1. Block diagram of phase sensitive detector

In Figure.1, u_s is the input and u_r is the reference of analogue multiplier. $u_s \times u_r$ is the output of analogue multiplier or the input of low-pass filter and $(u_s \times u_r)_a$ is the output of low-pass filter or phase sensitive detector.

If A-phase current i_a is chosen as the input and voltage between B-phase and C-phase u_{BC} is the reference of analogue multiplier, then u_s and u_r will be

$$u_s = i_a = I \sin(\omega t - \varphi) \qquad (1)$$

$$u_r = u_{BC} = U \sin(\omega t - 90^o) = -U \cos \omega t \qquad (2)$$

Where φ is power factor angle.

The output of analogue multiplier is

$$u_s \times u_r = -UI \cos \omega t \times \sin(\omega t - \varphi)$$
$$= -UI \cos \omega t \times \sin \omega t \times \cos \varphi + UI \cos^2 \omega t \times \sin \varphi \qquad (3)$$

The average value of $u_s \times u_r$ is given by

$$(u_s \times u_r)_a = \frac{1}{2\pi} \int_0^{2\pi} (u_s \times u_r) d(\omega t)$$
$$= \frac{1}{2\pi} \int_0^{2\pi} -UI \cos \omega t \times \sin \omega t \times \cos \varphi \, d(\omega t)$$
$$+ \frac{1}{2\pi} \int_0^{2\pi} UI \cos^2 \omega t \times \sin \varphi \, d(\omega t) \qquad (4)$$

First part of the equation (4) must equal zero because that it's the integration of two orthogonal functions. The average value of $u_s \times u_r$ is given by second part of the equation (4) only.

$$
\begin{aligned}
(u_s \times u_r)_a &= \frac{1}{2\pi} \int_0^{2\pi} UI \cos^2 \omega t \times \sin \varphi \, d(\omega t) \\
&= \frac{UI}{2\pi} \sin \varphi \int_0^{2\pi} \frac{1 + \cos(2\omega t)}{2} d(\omega t) \\
&= \frac{1}{2} UI \sin \varphi = KI \sin \varphi
\end{aligned} \tag{5}
$$

Where $K = \frac{1}{2}U$.

Well-known the reactive power of power system is given by

$$
Q = UI \sin \varphi \tag{6}
$$

Comparing equation (5) with equation (6), it is can be seen that the output of phase sensitive detector is direct proportion with actual reactive power in power system. If the voltage of power system is a constant, it is correct that considering the output of phase sensitive detector is direct proportion with actual reactive current in power system.

III. PRINCIPLE OF NEW REACTIVE CURRENT DETECTION METHOD

A square-wave pulse is chosen as the reference u_r, which is shown in Figure 2.

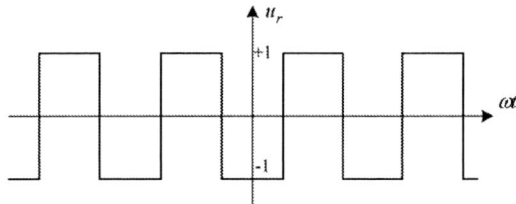

Figure 2. Square wave pulse

The reference u_r can be expressed by a Fourier series of the form

$$
\begin{aligned}
u_r &= -\frac{4}{\pi} \cos \omega t + \frac{4}{3\pi} \cos 3\omega t - \frac{4}{5\pi} \cos 5\omega t + \cdots \\
&= \frac{4}{\pi} \sum_{n=1}^{\infty} \frac{(-1)^n}{2n-1} \cos(2n-1)\omega t
\end{aligned} \tag{7}
$$

A-phase current i_a is also chosen as the input, seen equation (1). The product of u_r and u_s is given by

$$
\begin{aligned}
u_s \times u_r &= \frac{4}{\pi} I \sum_{n=1}^{\infty} \frac{(-1)^n}{2n-1} \cos(2n-1)\omega t \times \sin(\omega t - \varphi) \\
&= \frac{2}{\pi} I \sum_{n=1}^{\infty} \frac{(-1)^n}{2n-1} \{\sin(2n\omega t - \varphi) - \sin[(2n-2)\omega t + \varphi]\}
\end{aligned} \tag{8}
$$

To equation (8), if $n \geq 2$, the product $u_s \times u_r$ are all the AC portion and only if $n = 1$, the product contains DC portion. The DC component is

$$
(u_s \times u_r)_d = \frac{2}{\pi} I \sin \varphi = KI \sin \varphi \tag{9}
$$

Where $K = \frac{2}{\pi}$.

The minimum angular frequency of AC component is 2ω. So DC component can be obtained by means of the low-pass filter. Equation (9) shows that DC portion of the product $u_s \times u_r$ is direct proportion with actual reactive current in power system. If the voltage is a constant, it is also direct proportion with actual reactive power.

The simulation results of new reactive current phase sensitive detector are shown in Figure 3. 1 ampere is assigned for the magnitude of a phase current anda fourth order Butterworth low-pass filter with 20Hz passband edge frequency is chosen in the simulation.

The curve 1 is the reference u_r, the curve 2 is the product $u_s \times u_r$ and the curve 3 is average $(u_s \times u_r)_d$.

The simulation results indicate that the average $(u_s \times u_r)_d$ varies with power factor angle. Both inductive character and capacitive character reactive current can be detected.

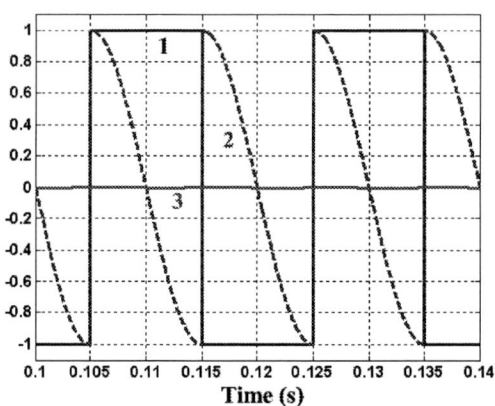

a). Simulation result for $\varphi = 0$

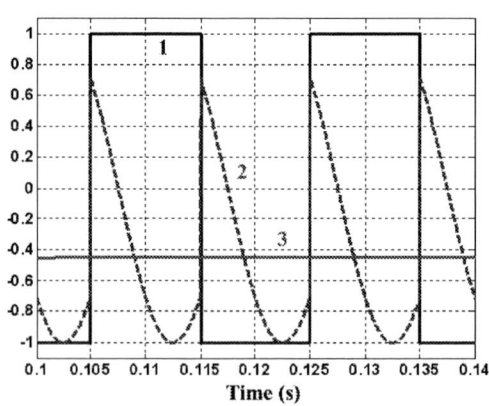

b). Simulation result for $\varphi = \pi/4$

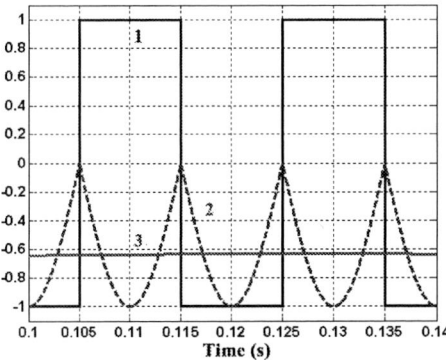

c). Simulation result for $\varphi = \pi/2$

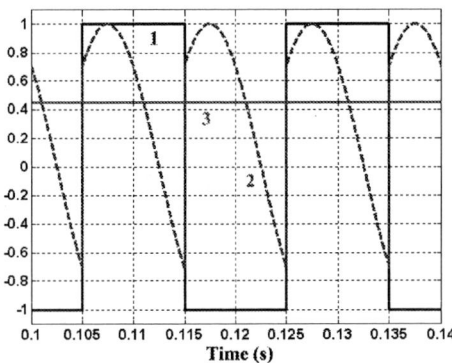

b). Simulation result for $\varphi = -\pi/4$

Figure 3. Simulation results of new reactive current phase sensitive detector

IV. DETECTION CIRCUIT OF REACTIVE CURRENT

It can be seen from equation (8) that multiplication is also needed in the method proposed in the paper. Do the analogue multipliers be required yet? The answer is not through analyzing the operation result carefully. Because u_r is an impulse function with ±1 values, the product result of $u_s \times u_r$ is that u_s changes its sign only at the different time which is decide by u_r. As long as u_s sign can be change with u_r, the result of equation (8) can be realized. The analog multiplexer is used to change u_s sign according to u_r in the paper. Refer to Figure 4 for schematic.

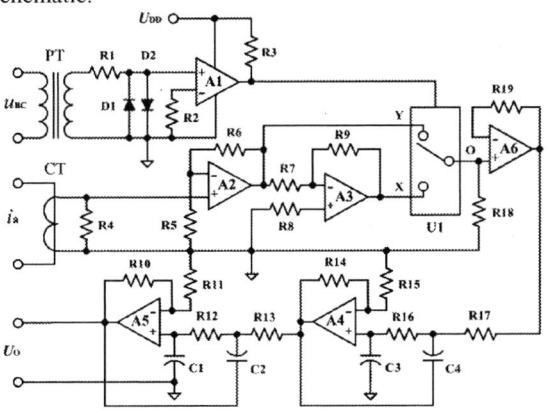

Figure 4. Schematic diagram of reactive current detection

Line voltage u_{BC} is reduced by step-down measuring transformer PT and the voltage comparator A1 is used to transform sinusoid voltage into square-wave voltage with $0V$ and $7V$ only. See Figure 5. The output of A1 is transmitted to the digital control input of the 2-channel analog multiplexer.

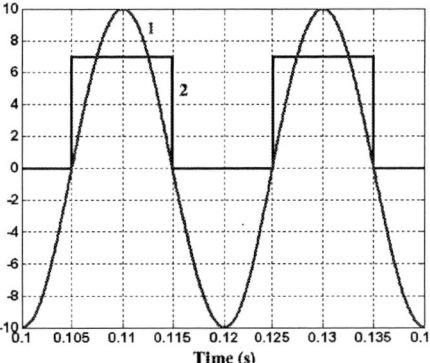

Figure 5. Schematic diagram of reactive current detection
Curve 1 is u_{BC}; Curve 2 is A1 output

A-phase current i_a is measured by the current transformer CT. The signal measured is enlarged and inverted by A2 and A3, which A2 is noninverting amplifier and A3 is inverter, then transmitted to the inputs, X, Y, of the 2-channel analog multiplexer. The gain of A2 should be adjusted according to current i_a and maximum input voltage of the analog multiplexer.

U1 is 2-channel analog multiplexer. O is output and X, Y are inputs. If $V_{DD} = 7V$, $V_{SS} = 0V$ and $V_{EE} = -7V$, analog signals from $-7V$ to $+7V$ can be controlled by digital control input of $0 \sim 7V$. When output of A1 equals $0V$, output O is connected to input X which is inverse of i_a. When A1 output is equal to $7V$, output O is connected to input Y which is in-phase signal of i_a. i_a changes its sign with u_{BC} is realized at the output of 2-channel analog multiplexer. A6 is voltage follower which is used for impedance conversion. The output and input signals of analog multiplexer see Figure 6.

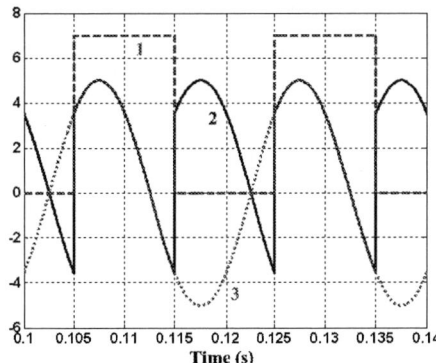

Figure 6. Output and input of analog multiplexer

In Figure 6, curve 1 is for digital control input; curve 2 is for output; curve 3 is for input Y. Power factor angle φ is selected $-\pi/4$.

The fourth-order Butterworth low-pass filter is composed of A4, A5 and peripheral circuit. In fact, it's made up of two second-order Butterworth low-pass filters connection in series. A6 output voltage which can be expressed by equation (8) is the input of Butterworth low-pass filter. Equation (8) indicates that the minimum frequency of A6 output signal is 628.3 rads/sec except DC portion. In order to choose the passband edge frequency ω_n rationally, simulation for analyzing filter effect have been down under the different ω_n. Simulation result is shown in Figure 7. Simulation condition is same as Figure 6.

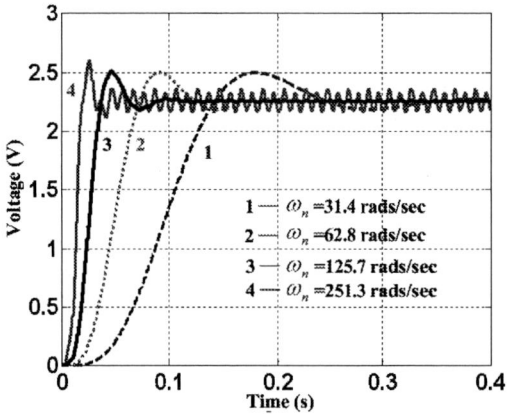

Figure 7. Simulation result for different ω_n

Simulation result indicates, the greater ω_n, the lesser delay and the worse filter effect. In the design, passband edge frequency ω_n =125.7 rads/sec, i.e. $f_n = 20Hz$, is selected. See curve 3 of Figure 7.

Because two second-order Butterworth low-pass filters have the same parameters, the transfer function of either one is given by

$$G(s) = \frac{1.579 \times 10^4}{s^2 + 177.715s + 1.579 \times 10^4} \quad (10)$$

According to equation (10), two second-order Butterworth low-pass filters can be designed.

The simulation result of reactive current detection circuit (see Figure 4) is illustrated in Figure 8.

In Figure 8, curve 1 is output of A1; curve 2 is output of A6; curve 3 is output of A5. Power factor angle φ is selected $-\pi/4$.

V. CONCLUSION

Form theoretical calculation and simulation analysis, the following conclusion can be drawn.

1. Reactive current detection circuit proposed in the paper can be used to measure the reactive current and reactive power if the voltage of power system is constant.

2. Detection result with this circuit is same as with analogue multiplier.

3. Both inductive character and capacitive character reactive current can be detected.

4. The reactive current detection result is not affected by fluctuation of system voltage.

5. Because analogue multiplier is instead of analog multiplexer, detection circuit is simplified.

Reactive current detection circuit mentioned in this paper has been used in an automatic compensation device of reactive power. The operational aspect is better up to now.

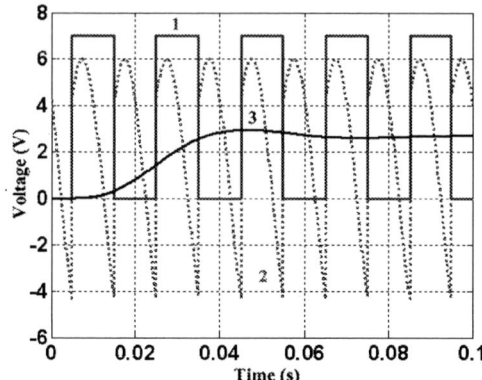

Figure 8. Simulation result of reactive current detection circuit

ACKNOWLEDGMENT

The work reported in this paper was supported by the National Natural Science Foundation of China(50437010).

REFERENCES

[1] Math Works. Using MATLAB, 2004
[2] Math Works. Simulink user's manual, 2004
[3] J. H. Wait, L. P. Huelsman, G. A. Korn, Introduction to Operational Amplifier Theory and Application, McGraw-Hill Book Company, 1975
[4] Wu Xiaojie, Dynamic Reactive Power Compensation Study Present State and Perspective, Colliery Automation, 5. 2000. pp19~22 (in chinese)
[5] Dong Yun-long, WU Jie, Wang Nian-chun, Summary of reactive power compensation technique, (in Chinese). Energy Conservation. 9 .2003. pp. 13-15,19.

2006 5th International Power Electronics and Motion Control Conference

Selective Harmonic Controlling for Three-Level High Power Active Front End Converter with Low Switching Frequency

Hui Zhang[*], hzhang@whu.edu.cn, Kaipei Liu [*], kpliu@whu.edu.cn[*]

M. Braun [**], braun@eti.uni-karlsruhe.de, C.C. Chan[*], chan@uni-poly.hk

[*] School of Electrical Engineering, Wuhan University, Wuhan, China
[**] Institute of Power Electronics and Electrical Drives, University of Karlsruhe, Karlsruhe, Germany

Abstract—Controlling of a specific harmonic by a switching converter with switching frequency lower than 1 kHz was originally implemented by two-level Optimal Pulse pattern with Unsymmetrical Switching angles [10], namely by OPUS-technique. That means, given a desired fundamental and one or more specific harmonics, the problem is to find the switching times (angles) that produce the fundamental while generating specifically all chosen harmonics. Based on the OPUS-technique a three-level optimal pulse pattern is proposed for high power level applications with low switching frequency. The transcendental mathematical equations are developed. Using the prior results of all complete solutions of selective harmonic elimination (SHE) the solutions of the new equations can be found by Newton iteration method. In particularly, the modulation performance of OPUS-technique for three-level Optimal Pulse pattern with Unsymmetrical Switching angles is demonstrated by the trajectory of the converter voltage space vector integral. Finally, the control range of a specific harmonic is given by OPUS-technique.

Keywords: partly unsymmetrical optimal pulse pattern; pulse-width modulation; voltage source inverter; harmonics control range

I. INTRODUCTION

With the development of the technique of Pulse Width Modulation (PWM), the control technique of four-quadrant operated inverters, the field-orientated control of synchronous machines and induction machines in low and middle power level on the one side and the rapid development of semiconductor devices, for example the appearing of GTO, IGBT and IGCT in Mega Watt power level on the other side, nowadays the application range of technology of power electronics is growing in the field of electrical drives, power propulsion systems for ships and utility apparatus in power systems towards the Multi-Mega watt power level, for example, with Flexible Alternating Current Transmission Systems (FACTS) devices.

Currently control of harmonics can be generated according to amplitude and phase angle reference by space vector modulation with switching frequencies of some kilohertz [4]. Unfortunately in the application of power range of tens of Megawatts, for example FACTS, the switching frequency of semiconductor is limited to a few hundred Hertz up to 1kHz [3]. According to these constraints, the switching frequency, pulse patterns and operating losses are of great concern for the application in converters of high power level.

For the purpose of control of harmonics a two-level Optimized Pulse pattern with Unsymmetrical Switching angles (OPUS) has been developed recently [10] (Fig. 1). With this two level pulse pattern specific harmonics are controlled in amplitude and phase angle even with switching frequencies lower than 1k Hz [12]. Besides the two-level converter, the three-level converter has been used for high power applications for a reasonable modulation performance; furthermore it has more attractive technical advantages for the application in the high power level from today's view.

Therefore, based on the idea of OPUS-technique, a three-level Optimal Pulse pattern with Unsymmetrical Switching angles is proposed for a three-level converter circuit in this work. The proposed pulse pattern can use the voltage level zero inherently and therefore has a better modulation performance than that of the two-level unsymmetrical pulse pattern. This proposed three-level pulse pattern is now considered by the case of six and ten independent switching angles separately. In particular the harmonic voltage control problem, represented by a set of transcendent equations, must be solved to determine the switching angles for turning the switches on and off in a three-level full bridge inverter, so as to produce a desired fundamental amplitude while generating selectively, for example, the 5th and 7th harmonics. Furthermore the demonstration of the modulation performance of the three-level OPUS-pulse pattern for three-phase PWM inverter is given by the trajectory of the converter voltage space vector integral. Finally, the applicable control range for specific harmonics is investigated.

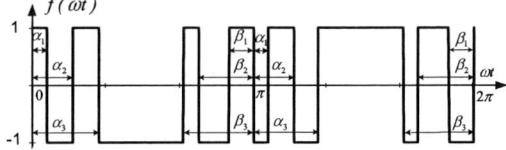

Figure 1. Two-level pulse pattern based on OPUS

1-4244-0448-7/06/$25.00 ©2006 IEEE   453

II. THREE-LEVEL PULSE PATTERN BASED ON OPUS-TECHNIQUE

A. Proposed unsymmetrical pulse pattern for three-level three-phase inverter

In Fig. 2 the principle of a three-level three-phase Voltage Source Inverter (VSI) is ideally modeled, while each phase-leg pair is shown by an ideal switch with three states. For a mathematical description with a switching function three switching variables are used, where each switching state is assigned to value +1 for switching state "up", to value 0 for switching state "zero" and to value –1 for switching state "down" separately.

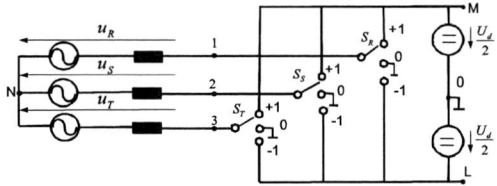

Figure 2. Simplified schematic diagram for a three-level three-phase inverter

In Fig. 3 a basic principle of a three-level Voltage Source Inverter (VSI) is shown by one phase-leg circuit. The one phase-leg circuit comprises 4 active valves. The switching rule is that only two valves, which are directly connected in series, can be switched on at a time, e.g. S_1 and S_2 or S_2 and S_3 or S_3 and S_4. Switching on S_1 and S_2 connects the output to the plus terminal, S_2 and S_3 connects the output via the clamping diodes to the mid-point (0), S_3 and S_4 connect the output to the minus terminal of the DC circuit. The switching rules are shown in TABLE I. The state of switching "1" means that the switch is turned on, the state of switching "0" means that the switch is turned off.

Thus, a switching function with unsymmetrical switching angles $\alpha_1, \alpha_2, \alpha_3, \beta_1, \beta_2, \beta_3$ is designed for a three-level pulse pattern based on OPUS-technique (Fig. 4). This optimal pulse pattern owns only inversely half-wave period symmetry. The resulting waveform of output line-neutral voltage is demonstrated in Fig. 4, too.

B. Mathematical procedure for OPUS-technique

1) Definition of phase angle of harmonic

In order to simplify the description, the OPUS technique is described by the help of space vectors with the complex coefficient $\underline{a} = e^{j2\pi/3}$. The output voltage space vector of the converter is defined by:

$$\underline{u} = 2(u_R + \underline{a} \cdot u_S + \underline{a}^2 \cdot u_T)/3$$
$$= u_\alpha + ju_\beta = \mathrm{Re}(\underline{u}) + j\,\mathrm{Im}(\underline{u}) \quad (1)$$

With the help of space vectors the v-th harmonic voltage can be described in a complex plane (Fig. 5).

The amplitude and phase angle can be shown more clearly in a rotating coordinate d-q system with $v\omega$ angular velocity (Fig. 6). The sinusoidal term and cosinusoidal term of the v-th harmonic are represented by a_v and b_v. The d-axis is shown as reference phase angle.

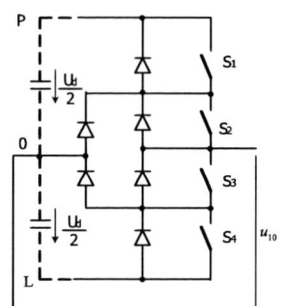

Figure 3. Basic principle of a 3 level VSI

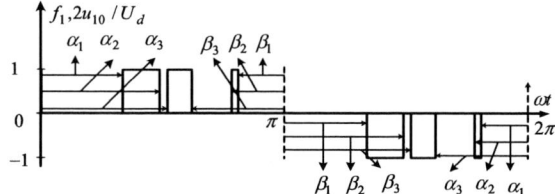

Figure 4. Waveform of a switching function and corresponding line to neutral output voltage of three level three-phase PWM converter based on OPUS-Technique with 6 independent switching angles

TABLE I. SWITCHING RULES FOR A THREE LEVEL VSI

State of switching	Output voltage of VSI
$S_1=1$, $S_2=1$, $S_3=0$, $S_4=0$	$U_d/2$
$S_2=1$, $S_3=1$, $S_1=0$, $S_4=0$	0
$S_3=1$, $S_4=1$, $S_1=0$, $S_2=0$	$-U_d/2$

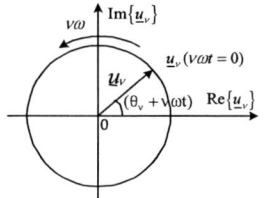

Figure 5. Space vector of v-th harmonic voltage in a complex plane

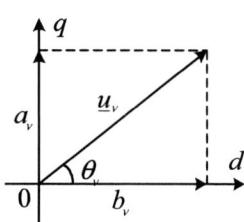

Figure 6. Space vector of v-th harmonic in a rotating coordinate system

2) Definition of modulation ratio of harmonics

The definition of the modulation ratio of the 1st, v-th harmonic is given as follows (2):

$$M_v = \hat{U}_v / (2U_d/\pi) \cdot \quad (2)$$

454

Where $\hat{U}_v$ is the voltage amplitude of the v-th order harmonic, $2U_d/\pi$ is the maximal fundamental component by square modulation technique.

3) Mathematical equations

Usually the bridge-leg pair circuit is so controlled that its output voltage owns half-period symmetry separately and in one-third fundamental period and two-thirds fundamental period are postponed with the other two bridge-leg circuits separately. With this condition the even harmonics and the consideration of all triple harmonics are cancelled. In other words, the odd and non-triple harmonics should be considered. Corresponding to Fig. 4 there are six degrees of freedom or independent switching angles. Thus three low order harmonics, usually fundamental, 5th and 7th harmonic can be controlled. The mathematical equations are developed by Fourier analysis (3).

$$\begin{cases}
\cos(\alpha_1) - \cos(\alpha_2) + \cos(\alpha_3) + \cos(\beta_1) - \cos(\beta_2) + \cos(\beta_3) = M_1 \cdot \cos(\theta_1) \cdot 2 \\
\sin(\alpha_1) - \sin(\alpha_2) + \sin(\alpha_3) - \sin(\beta_1) + \sin(\beta_2) - \sin(\beta_3) = M_1 \cdot \sin(\theta_1) \cdot 2 \\
\cos(5\alpha_1) - \cos(5\alpha_2) + \cos(5\alpha_3) + \cos(5\beta_1) - \cos(5\beta_2) + \cos(5\beta_3) = 5 \cdot M_5 \cdot \cos(\theta_5) \cdot 2 \\
\sin(5\alpha_1) - \sin(5\alpha_2) + \sin(5\alpha_3) - \sin(5\beta_1) + \sin(5\beta_2) - \sin(5\beta_3) = 5 \cdot M_5 \cdot \sin(\theta_5) \cdot 2 \\
\cos(7\alpha_1) - \cos(7\alpha_2) + \cos(7\alpha_3) + \cos(7\beta_1) - \cos(7\beta_2) + \cos(7\beta_3) = 7 \cdot M_7 \cdot \cos(\theta_7) \cdot 2 \\
\sin(7\alpha_1) - \sin(7\alpha_2) + \sin(7\alpha_3) - \sin(7\beta_1) + \sin(7\beta_2) - \sin(7\beta_3) = 7 \cdot M_7 \cdot \sin(\theta_7) \cdot 2
\end{cases} \tag{3}$$

The v-th harmonic is described by the corresponding modulation ratio M_v and phase angle θ_v. The sinusoidal term and cosinusoidal term of the v-th harmonic are shown on the right side of equations. With the help of Fourier analysis the Fourier coefficient of the output voltage is described by a predefined per unit value shown as a function of the switching angles in the left side of equations.

In order to guarantee for the physical meaning of these switching angles, the boundary conditions are developed by the following (4):

$$\begin{cases}
\alpha_1 \geq 0, \alpha_2 > \alpha_1, \alpha_3 > \alpha_2 \\
\beta_1 \geq 0, \beta_2 > \beta_1, \beta_3 > \beta_2, (\alpha_3 + \beta_3) < \pi
\end{cases} \tag{4}$$

For the case with more independent switching angle variables the mathematical equations can be created in a similar manner. As we know, the successful application of Newton's method to solve these non-linear equations needs a group of good starting values that can be obtained from prior results from the method of selective harmonic elimination [1].

C. Definition of the control range of harmonics based on OPUS-technique

The control range of a specific harmonic vs. fundament component can be studied, while the solution of the equation system (3) is searched for a specific phase angle θ_v ranging from 0 to 2π with increasing amplitude $\hat{u}_v$. The minimum value of $\hat{u}_v$ found by this way will be the maximum reference value for the respective harmonic. Then the control range for constant amplitude $\hat{u}_v$ independent from the phase angle θ_v is defined as the obtained specific harmonic control range [11].

D. Evaluation of modulation performance based on OPUS-technique

Evaluating the performance of OPUS-technique three factors are considered. The first one is the obtained fundamental control range, the second one is that the specific harmonic control range vs. fundamental component is as large as possible too, and the other not controlled higher order harmonic components are as low as possible; the third one is that the switching frequency is as low as possible for the lowest possible switching losses.

For a symmetrical three-phase voltage system the output voltages are defined as follows:

$$\begin{cases}
u_R = f_1(\omega t) \cdot U_d / 2 - u_{NO} \\
u_S = f_1(\omega t - 2\pi/3) \cdot U_d / 2 - u_{NO} \\
u_T = f_1(\omega t - 4\pi/3) \cdot U_d / 2 - u_{NO}
\end{cases} \tag{5}$$

$$u_{NO} = U_d (f_1(\omega t) + f_1(\omega t - 2\pi/3) + f_1(\omega t - 4\pi/3))/6$$

From (5) it is difficult to evaluate the modulation performance based on OPUS-technique. The description of voltage space vector (1) brings no helpful graphic diagram, therefore it is better to introduce a trajectory of a converter voltage space vector integral corresponding to (6) to show the modulation performance by a trajectory like the flux trajectory of an electrical machine.

The corresponding flux of an output voltage $\underline{u}$ is derived as follows:

$$\underline{\psi} = \int (u_\alpha + j u_\beta) dt = \int \underline{u} dt \tag{6}$$

Then, for the description of the modulation performance, the flux is defined as a per unit value by the help of U_d / ω:

$$\underline{\varphi} = \underline{\psi}/(U_d/\omega) = \underline{\psi}/(U_d \cdot T)/2\pi \qquad (7)$$
$$= 2\pi \cdot \underline{\psi}/(U_d \cdot T)$$

Corresponding to (6) the obtained relative flux trajectory $\underline{\varphi}$ is shown in (Fig. 8a). The zero voltage states are marked by the symbol of a small circle at those positions, where the space vector of the flux stands still, till the next active switching state with $|\underline{u}| \neq 0$ is switched on. The deviation of the converter voltage space vector integral is shown with a help circle line. It is possible to estimate roughly the expected harmonic currents by the deviation of the trajectory from the help circle line.

E. Relationship between the switching frequency and the number of switching angles

From the third column in TABLE II, it is shown that for 10 independent switching angles the harmonic free area ranges from the fundamental harmonic to 17th harmonic, meanwhile one or more specific harmonics can be generated according to the OPUS-technique. Particularly there is only 500Hz mean switching frequency necessary for each active converter switch.

TABLE II.   SWITCHING FREQUENCY NEEDED BY OPUS-TECHNIQUE

Number of independent switching angles within half period N	6	10
Order of controllable harmonic	1,5,7	1,5,7,11,13
Order of the lowest rest harmonic	11	17
Necessary mean switching frequency of a converter in 50Hz power system	300Hz	500Hz

III. RESULTS BASED ON THE THREE-LEVEL PULSE PATTERN

Amplitude and phase reference have been varied for a couple of specific harmonics. In every case solutions have been found. Table III gives the relation between the amplitudes and phase angles of the reference values and the corresponding figures. Fig.11 demonstrates the wide control range for the amplitude of the 5th, 7th, 11th and 13th harmonic for the full range of their control angles from 0 to 360 degrees with 6 and 10 switching angles, respectively. An example for 6 switching angles is given by a three dimensional diagram with a case of M_7=0.1 and parameter variables for M_1 from 0.2 to 0.8 and θ_7 from 0 to 360 degrees in Fig. 12. In Table III the symbol "-" means the corresponding harmonic can not be considered by OPUS-technique.

IV. CONCLUSION

A new modulation technique for the control of the fundamental and low order harmonics of the output voltage for the three-level inverter with low switching frequency is presented. The method to implement this technique is introduced. The good effect of the modulation performance shows that a new possibility of harmonic compensation for FACTS converters can be achieved by three-level optimal pulse patterns with unsymmetrical switching angles (OPUS-technique).

TABLE III.   OVERVIEW OF THE CALCULATION CASES

N	Calculation cases for given values with the number N of switching angles separately									Result
	M_1	θ_1	M_5	θ_5	M_7	θ_7	M_{11}	θ_{11}	M_{13}, θ_{13}	
6	0.8	0	0.2	$\pi/4$	0	0	-	-	-	Fig.4, Fig.8a Fig.9
6	[0,...,0.9]	0	[0,...,M_5]	[0,...,2π]	0	0	-	-	-	Fig.11a
6	[0,...,0.9]	0	0	0	[0,...,M_7]	[0,...,2π]	-	-	-	Fig.11b
6	[0.2,...,0.8]	0	0	0	0.1	[0,...,2π]	-	-	-	Fig. 12
10	0.8	0	[0,...,M_5]	$\pi/4$	0	0	0	0	0	Fig.7, Fig. 8b Fig.10
10	[0,...,0.9]	0	[0,...,M_5]	[0,...,2π]	0	0	0	0	0	Fig.11c
10	[0,...,0.9]	0	0	0	0	0	[0,...,M_{11}]	[0,...,2π]		Fig.11d

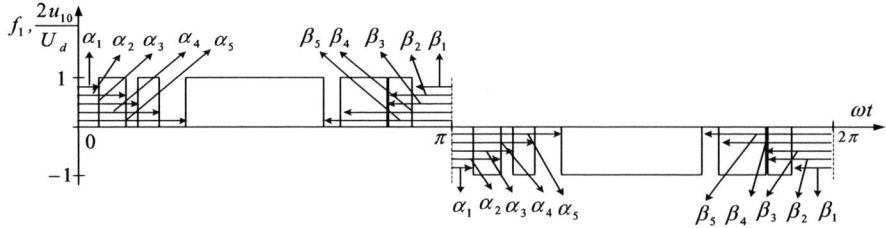

Figure 7. Switching function and corresponding line-neutral output voltage with 10 independent switching angles

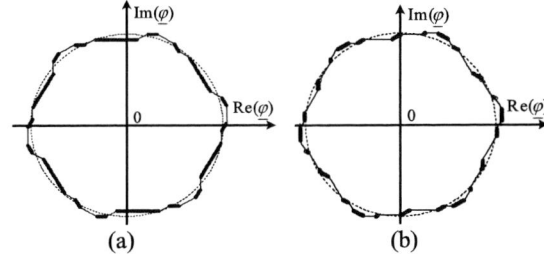

Figure 8. Trajectory of the converter voltage space vector integral
(a) corresponding to Fig. 4   (b) corresponding to Fig.7

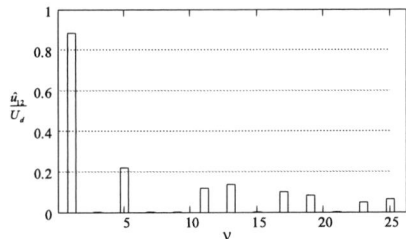

Figure 9. Influence on amplitude spectrum corresponding to Fig. 4

Figure 10. Influence of amplitude spectrum corresponding to Fig. 7

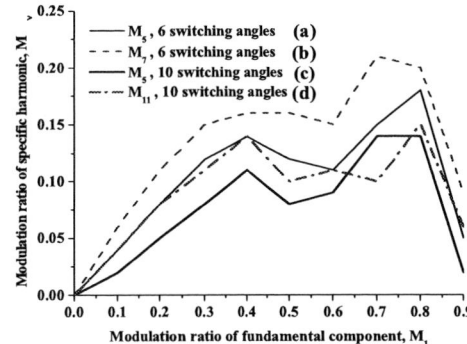

Figure 11. Control range of v-th harmonic vs. the fundamental component
with 6 and 10 independent switching angles respectively

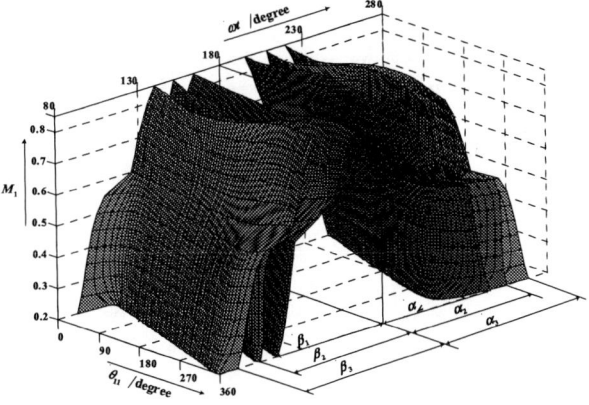

Fig. 12 Switching angles vs. M_1 and θ_7 (N=6)

REFERENCES

[1] J. N. Chiasson, L. M. Tolbert, K. J. McKenzie, Du Zhong, "A complete solution to the harmonic elimination problem Power Electronics, " IEEE Transl. on Ind. Elec. vol. 19, Issue 2, March 2004 pp. 491 – 499.

[2] R., Dugane, "Electrical Power Systems Quality," McGraw-Hill, pp. 175, 1996.

[3] N. Hingorani, L. Gyugyi, "Understanding FACTS, "concepts and technology of flexible AC transmission systems," IEEE Press, 2000.

[4] J. Holz , "Pulsewidth Modulation - A Survey," IEEE Trans. Ind. Elec. Vol. 39, No. 5 pp. 410-420, December 1992.

[5] H. S. Patel and R. G. Hoft , "Generalized techniques of harmonic elimination and voltage control in thyristor inverter: Part I-Harmonic elimination," IEEE Tra[s. Ind. Applicat. vol. IA-9, No. 3, pp. 310-317, May/June 1973.

[6] W. Stephan, "Leistungselektronik interaktiv – Aufgaben unter Simplorer und Mathcad, " Fachbuchverlag Leipzig, 2001.

[7] J Sun and H. Grotstoollen, "Solving nonlinear equations for selective harmonic eliminated PWM using predicted initial values, " in Conf. Proc. IECON'92, November pp. 259-264, 1992.

[8] I. Takahashi, "A New Control of PWM Inverter Waveform for Minimum Loss Operation of an Induction Motor Drive, " IEEE Trans. Ind. Elec. vol.IA-21, No.4, May/June 1985, pp. 580-587.

[9] F. Zach, H. Ertl, "Efficiency Optimal Control for AC Drives with PWM Inverters, " IEEE Trans. Ind. Elec. vol. IA-21, No. 4, pp. 987-1000 July/August 1985.

[10] H. Zhang, M. Braun, "A new partly unsymmetrical PWM technique for   harmonic compensation," 10th European Conference on Power Electronics and Applications (EPE) September Toulouse 2003.

[11] H. Zhang, M. Braun, "Application of a new partly unsymmetrical PWM technique for harmonic compensators," 11th European Conference on Power Electronics and Applications (EPE) September Dresden 2005.

[12] H. Zhang, "Untersuchung eines neuen Modulationsverfahrens zur Verringerung der Oberschwingungsbelastung elektrischer Netze, Shaker Press (in German) 2005.

2006 5th International Power Electronics and Motion Control Conference

A Unity Power Factor Three-Phase Buck Type SVPWM Rectifier Based on Direct Phase Control Scheme

LI Yabin, LI Heming and PENG Yonglong

Department of Electrical Engineering, North China Electric Power University, Baoding 071003, China

Abstract—This paper describes a direct phase control scheme for unity power factor three-phase Buck type rectifier modulated with space vector PWM method (SVPWM). The advantages of the developed system are low harmonic distortion in ac supply currents, nearly unity power factor over wide operating scope, easily realization. The theory analysis and modeling method are introduced first. Second, the control scheme and implementation based on FPGA is put forward. Finally, the characteristics of the developed rectifier are tested from experiments. It is proved from simulation and experiments that the developed rectifier, using direct phase control scheme, has good dynamic and static performance though the PWM frequency is rather low.

Keywords-Three-Phase Buck Type Rectifier; SVPWM; Direct Phase Control; Unity Power Factor; FPGA

I. INTRODUCTION

With developing applications of power electronic devices in industrial fields, increasing emphasis has been put on the power quality. The conventional rectifier using uncontrolled diode bridge or phase-controlled SCR bridge lead to poor power factor and serious harmonic distortion in ac power supply. In order to meet harmonic control standards such as IEEE 519, passive and/or active harmonic compensation techniques have to be employed. In recent years, researchers are working toward to propose and develop inherently clean new power converter topologies, which operate at nearly unity power factor and inject very low harmonic content into the power supply with relatively high converter efficiency.

Many unity power factor rectifier topologies such as the Buck, Boost and Buck-Boost derived topologies have been proposed and analyzed in the literatures [1-5]. Among these, the unity power factor Boost type PWM rectifier is one of the most common topologies for its relatively high efficiency and simple structure. However, the output dc voltage of Boost type rectifier is higher than the peak line-line voltage of power supply, which needs a second part to translate the dc voltage to a low level in most industrial applications. On the contrary, Buck type PWM rectifier offers a good solution for these

applications. It directly converts the ac power supply to dc part over a wide scope voltage below peak line-line value of power supply at high power densities to meet the strict power factor penalty limits by electricity authorities.

There are many different PWM modulation techniques, such as sinusoidal PWM (SPWM), space vector PWM (SVPWM), delta modulation techniques. It has been analyzed theoretically [6] and proved in experiments [4-5] that the SVPWM technique is maybe the best modulation solution on the whole. Since instantaneous space vector representation is employed to treat three-phase quantity as a whole, the implementation is more complex than SPWM. In order to achieve unity power factor and low harmonic distortion, the usual SVPWM method is realized through d/q transformation, which make the control system complicated and is difficult to design the control system parameters. The indirect current control such as phase-amplitude control scheme has a simple structure and easy to design control system [7-8]. But, the dynamic response and the control accuracy of the presented system are not very satisfied.

This paper presents a direct phase control scheme, which is a modified phase-amplitude control, and the design and implementation is described as well. The dynamic and static response performance in a wide output range is held through simulations and experiments.

II. SYSTEM DESCRIPTION

The circuit diagram of present three-phase unity power factor Buck type rectifier is shown in Fig.1. The block diagram of control system, and dc voltage and dc current feedback loops are shown in the same figure. Each power semiconductor switch consists of an IGBT connected in series with an ultra-fast recovery diode, resulting in reverse voltage blocking capability and unidirectional current flow. A low pass LC filter is connected to the input side of the converter to filter out the switching frequency harmonic components in the line current of power supply. The inductor (L_d) that feeds the dc bus acts as a stiff current source. If the Buck type rectifier acts as a voltage source for the load (R_L), a parallel connection dc capacitor (C_d) is needed across the dc bus. It is noted that the additional freewheeling diode (D_w) is not indispensable in the system. However, on the aspect of

1-4244-0448-7/06/$25.00 ©2006 IEEE

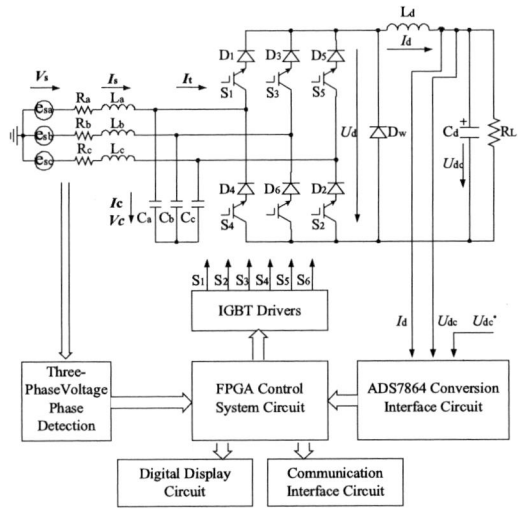

Fig.1 The circuit diagram of the unity power factor three-phase Buck type PWM rectifier

safety and reducing switching losses of IGBTs, the freewheel diode is suggested to be used to avoid device damage caused by error trigger and IGBT broken.

In most applications, the control circuit is accomplished with DSP or other microcontroller, which need some additional IC working together to achieve display and communication purposes. All these make the control circuit complicated and poor in reliability. In this work, the control part is realized by one chip FPGA (Xilinx Spartant Ⅱ 2s200), which fulfill the whole control functions including the control strategy calculation, switching signal generation, A/D transformation control, digital display function and communication with other devices. The A/D chip is ADS7864 that can sample and transform 6 analog input signals to digital signals in parallel.

Compared with DSP and other microcontroller, FPGA is more suitable for digital control system due to its flexible configuration and parallel operation structure, which enable the control system to operate at very high speed and with minimal time delay.

III. DIRECT PHASE CONTROL BASED ON SVPWM

A. Principle of Current Space Vector of Buck Type Rectifier

For a three-phase balanced power system, the three-phase variables can be transformed into instantaneous space vectors V_s (source voltage), I_s (source current), V_c (capacitor voltage) and I_t (converter input current) in $\alpha - \beta$ stationary coordinate plane [1-2]. If the real axis α is oriented as the axis of phase as shown in Fig. 2, the current space vector of the converter is then given by,

$$I_t = \frac{2}{3}(i_{ta} + i_{tb}e^{j\frac{2}{3}\pi} + i_{tc}e^{-j\frac{2}{3}\pi}) \qquad (1)$$

It can be simply expressed as the following equation,

$$I_t = |I_t| e^{j(\omega t + \delta)} \qquad (2)$$

where $|I_t|$ is the magnitude of I_t, ω is the constant angular speed and δ is the phase angle between I_t and V_s.

The switching function of a switch semiconductor shown in Fig.1 is defined as,

$\sigma_k = 1$,   upper switch on and lower switch off.

$\sigma_k = 0$,   both switch on or off.

$\sigma_k = -1$,   upper switch off and lower switch on. (3)

where $k = a, b, c$ denotes the three phase switch bridge.

For a current source rectifier, the following equation must be satisfied at any time,

$$\sum_{k=a,b,c} \sigma_k = 0 \qquad (4)$$

So, there are nine switch states for the three phase bridges a, b and c defined as I_i (i=1~9),

$$I_i = \frac{2}{\sqrt{3}} I_d e^{j(\frac{i\pi}{3} - \frac{\pi}{6})} \qquad \text{for i =1~6}$$
$$I_i = 0 \qquad \text{for i =7,8,9} \qquad (5)$$

where I_d is the dc current of dc bus.

In order to improve the device safety and minimized the switching loss of active semiconductors, the zero vectors are achieved through freewheeling diode instead of the switches all on in the same bridge. All the current vectors and axis are shown in Fig.2.

SVPWM modulation method is to use these nine currents to combine the needed current vector. There are many different algorithms. The common one, single triangle mode, is shown in Fig. 2 for instance.

B. Direct Phase Control Scheme

On the aspect of space vector, the equivalent circuit and the space vectors in steady state is shown in Fig.3 (a, b, c), respectively.

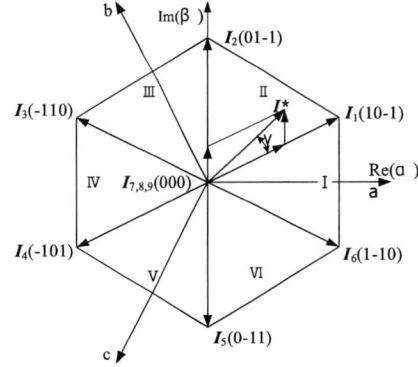

Fig.2 Diagram of current vectors of buck type rectifier

459

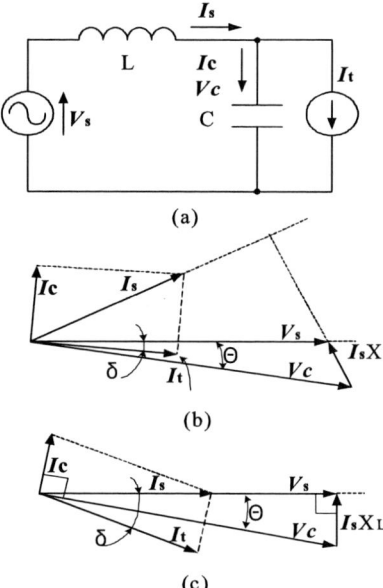

(a)

(b)

(c)

Fig.3 (a) Equivalent circuit using space vectors, (b) Non-unity power factor operation (leading power factor), (c) Unity power factor operation.

The equations describing the circuit can be represented by,

$$V_s = L\frac{d}{dt}I_s + V_c \tag{6}$$

$$C\frac{d}{dt}V_c = I_s - I_t \tag{7}$$

The popular method to get the reference current vector I^* and then to get the vector I_t introduced in literatures [1-3] is complicated, which needs sample the source currents and coordinates transformation calculation. The control parameters of the control system are also hard to design. The simple methods described in literatures [6-8] are not suitable for high accuracy system.

If the fundamental components of the variables are considered only, the mathematical model of Buck type rectifier can be expressed by,

$$\begin{cases} i_{1ta} = \sigma_{1a}I_d \\ i_{1tb} = \sigma_{1b}I_d \\ i_{1tc} = \sigma_{1c}I_d \\ U_d = v_{ca}\sigma_{1a} + v_{cb}\sigma_{1b} + v_{cc}\sigma_{1c} \end{cases} \tag{8}$$

$$\sigma_{1a,b,c} = m\begin{bmatrix} \cos(\omega t + \delta) \\ \cos(\omega t + \delta - 120^0) \\ \cos(\omega t + \delta + 120^0) \end{bmatrix} \tag{9}$$

And,

$$v_{sa,b,c} = \begin{bmatrix} \cos(\omega t) \\ \cos(\omega t - 120^0) \\ \cos(\omega t + 120^0) \end{bmatrix} \tag{10}$$

where $i_{1ta,b,c}$ is the three phase input current of converter respectively, U_d is dc output voltage, $\sigma_{1a,b,c}$ is the fundamental component of switching function defined in (4), m is modulation index ($0 \le m \le 1$), δ is the phase angle between v_s and switching function σ .

From Equations (6-10), the mean dc output voltage can be expressed as,

$$\begin{aligned} U_{dc} &= \frac{3}{2} \times \frac{m\cos\delta}{1 - \omega^2 LC}|V_s| \\ &\approx \frac{3}{2} \times m\cos\delta|V_s| \end{aligned} \tag{11}$$

And, from steady vector relations shown in Fig.3 (c), the following equation can be obtained,

$$\begin{aligned} \cos\theta &= \frac{|V_s|}{|V_c|} \\ \frac{\sin(90^0 - \theta)}{|I_t|} &= \frac{\sin\delta}{|I_c|} \\ |I_c| &= \omega C|V_c| \\ |I_t| &= mI_d \end{aligned} \tag{12}$$

Then from Equation (12) we can get,

$$\sin\delta = \frac{\omega C|V_s|}{mI_d} \tag{13}$$

where the $|\cdot|$ is the amplitude of vectors.

For a three phase Buck type rectifier operating at unity power factor, Equation (13) indicates that the phase angle of converter input current can be determined by modulation index and the current value of dc bus, and has no relation with filter inductance L, which make the control system robust to parasitical parameters of the line. Equation (11) indicates that the output dc voltage can be modified through modulation index m and phase angle δ .

In fact, the output dc voltage is more sensitive to the modulation index m than to the phase angle δ in a wide range. Then, the direct phase control presented in this paper is to modify the output dc voltage through modifying modulation index m rather than modifying the phase angle δ , and the phase angle δ can be obtained directly through Equation (13). The block diagram of control system is shown in Fig. 4(a), and the simplified transfer function diagram to design the control system is also shown in Fig.4 (b).

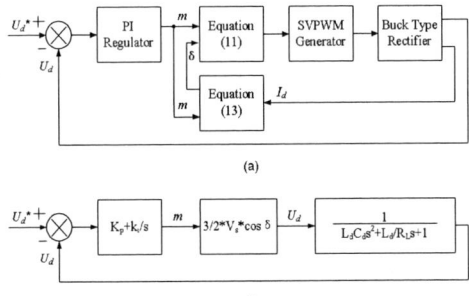

(a)

(b)

Fig.4 (a) Block diagram of direct phase control scheme, (b) transfer function of buck type rectifier system.

IV. SIMULATIONS AND EXPERIMENTS

Based on the analysis above, the simulation model of the whole system is built based on Matlab/Simulink. The switching frequency is selected 3.6 kHz. The dynamic and steady simulations are shown in Fig.5, the drive pulses of semiconductors S1, S3 and S5 are shown in Fig.6, and the detailed input current of the power source and its harmonic spectra is shown in Fig.7. From Fig.6, one can find that the pulses are not similar to the conventional ones. The reason is that the zero vectors in this paper are achieved by freewheel diode rather than the active switches. So, when the zero vectors are active, the

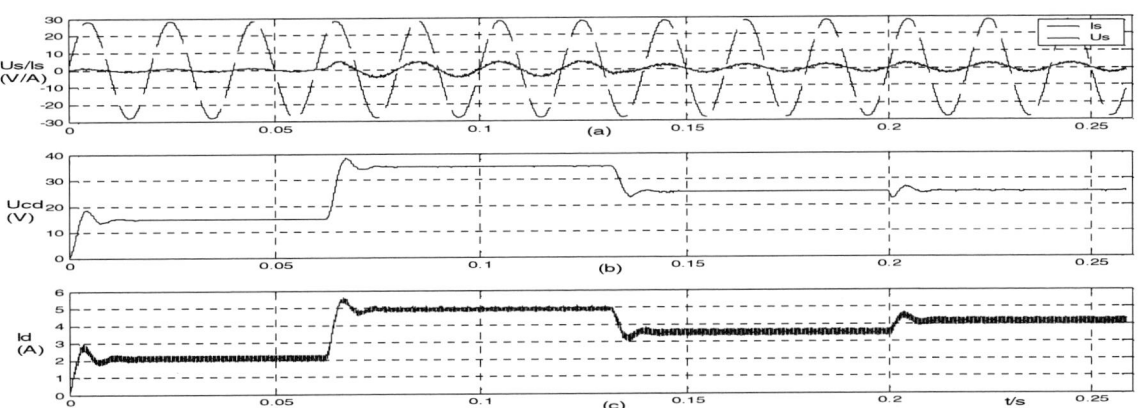

Fig.5 Dynamic and steady simulation of buck type rectifier based on direct phase control
(a) Source voltage and current, (b) dc voltage, (c) dc current.
(10V/div, 10A/div)

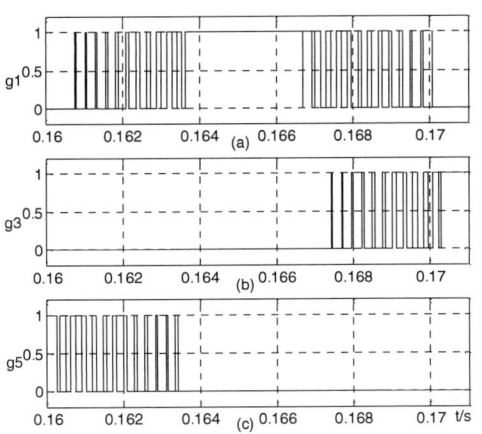

Fig.6 Drive pulses of switches S1, S3 and S5

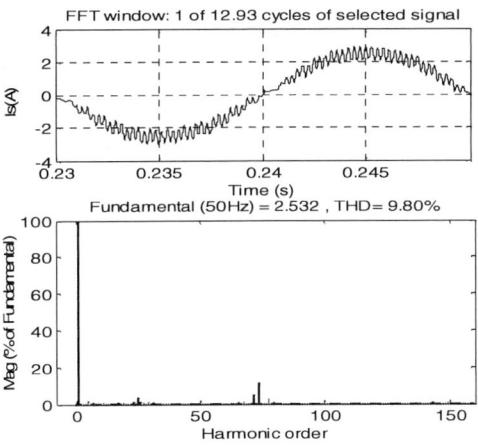

Fig.7 Power source current and harmonic spectra at dc voltage 25 V. top, current, Bottom harmonic spectra.

main switches can keep its former states. This technique can reduce the power losses of switches by 30%. The performances of the whole system are satisfied. And, Fig. 8, 9, 10, 11 is the experimental results, respectively.

The experimental parameters in experiments are set as, L=0.48mH,C=30uF,v_s=60V,R_L=10ohm,L_d=8.4mH,C_d=100uF and switching frequency is 3.6kHz.

V. CONCLUTION

A simple unity power factor Buck type rectifier is introduced in this paper. By using the direct phase control scheme, the structure of the system and the control circuit is simplified. Simulations and experiments prove that the performance of the whole system is satisfied. The power factor can be kept at unity in a wide range and the

Harmonics THD is low although the switching frequency is not high. By using FPGA, performing very high speed and very short time delay, the rectifier has good stability, fast dynamic response, and the surge of voltage and current during the transient duration is small.

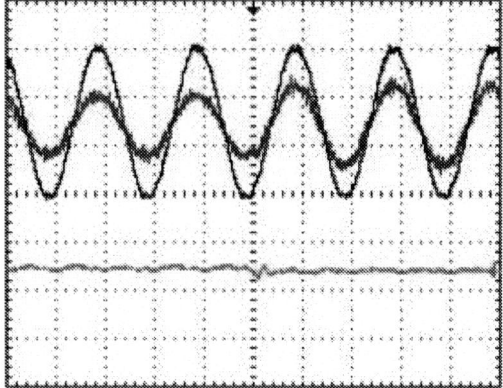

Fig.11 Step change in load from 0.85 to 1 pu. (50V/div, 10A/div).
From top to bottom V$_s$ and I$_s$, U$_d$

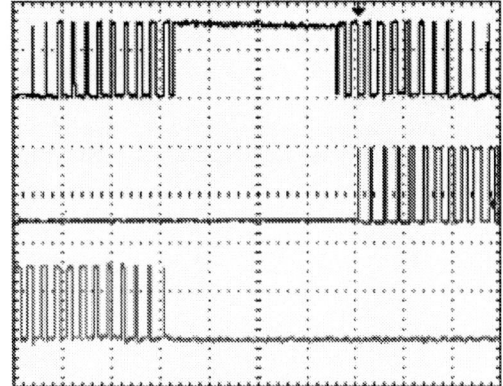

Fig.8 Drive pulses of switch, from top to bottom S1, S2 and S3.
(10V/div)

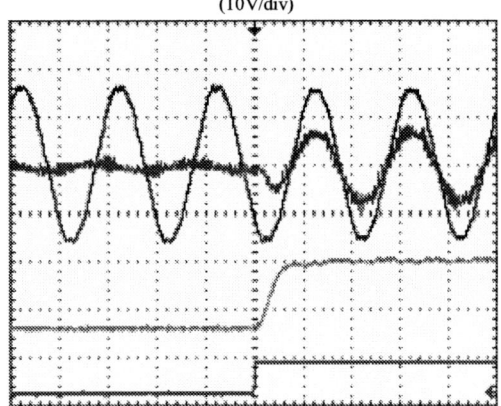

Fig. 9 Step change in dc voltage from 0 to 65 V (50V/div,10A/div).
From top to bottom V$_s$ and I$_s$, U$_d$, step signal.

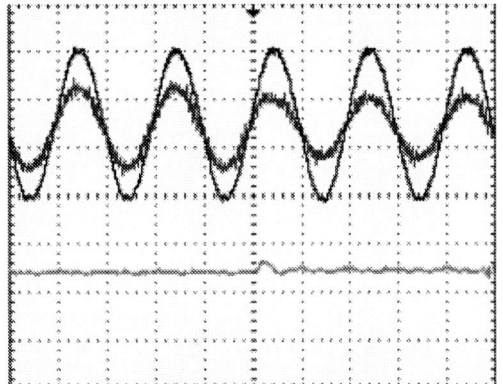

Fig.10 Step change in load from 1.0 to 0.85 pu. (50V/div, 10A/div).
From top to bottom V$_s$ and I$_s$, U$_d$

REFERENCES

[1] Soo-Bin Han, Nam-Sep Choi, and Gyu-Hyeong, "Modeling and analysis of static and dynamic characteristics for buck-type three-phase PWm rectifier by circuit DQ transformation," *IEEE Transactions on Power electronics*, vol. 13, no. 2, pp. 323–336, March 1998.

[2] Yang Ye, Mehrdad Kazerani and Victor H. Quintana, "Modeling,control and implementation of three-phase PWM converters," *IEEE Transactions on Power electronics*, vol. 18, no. 3, pp. 857–864, May 2003.

[3] Eduardo P. Wiechmann, Rolando P. Burgos and Joachim Holtz, " Active front-end converter for medium-voltage current-source drives using sequential-sampling synchronous space-vector modulation," *IEEE Transactions on Industrial Electronics*, vol. 50, no. 6, pp. 1275–1289, December 2003.

[4] Ian Wallace, Ashish Bendre, Jonathan P. Nord and Giri Venkataramanan, "A unity-power-factor three-phase PWM SCR rectifier for high-power application in the metal industry," *IEEE Transactions on Industry Applications*, vol. 38, no. 4, pp. 898–908, July/August 2002.

[5] Poh Chiang Loh and Donald Grahame Holmes, "A variable band universal flux/charge modulator for VSI and CSI modulation," *IEEE Transactions on Industry Applications*, vol. 38, no. 3, pp. 695–705, May/June 2002.

[6] Shoji Fukuda and Yoshitaka Iwaji, "Introduction of the harmonic distortion determining factor and its application to evaluating real time PWM inverters," *IEEE Transactions on Industry Applications*, vol. 31, no. 1, pp. 149–154, January/February 1995.

[7] ZHANG Chun-jiang, Gu He-rong, WANG Bao-cheng, ZHU Yan-ping and LIU yan-min, "Mathematical model of three-phase PWM rectifier based on a novel phase and amplitude control," Proceeding of the CSEE, vol. 23, no. 7, pp.28-31, July 2003.

[8] Wang Ying, ZHANG Chun-jiang and CHEN Hui-ming, "A new phase and amplitude control strategy and mathmatics model of three-phase voltage rectifier," Proceeding of the CSEE, vol. 23, no. 11, pp.22-27, November 2003.

2006 5th International Power Electronics and Motion Control Conference

3-Phase Current-Source SMES-UPS Based on TFSC and its Control Strategies

WANG Fu-sheng, LI Hong-mei

Hefei University of Technology, Department of Electrical Engineering, Hefei, China

wfs@ipp.ac.cn

Abstract—3-Phase Current-Source SMES-UPS based on toroidal field superconducting coils (TFSC) is introduced in this paper, which serves for the exciting power supply (EPS) for TFSC when the power grid is in order, as well as for uninterrupted power supply (UPS) for the cryogenic and vacuum system once the deep voltage sag or other grid faults are detected. In the case of former, by controlling active and reactive current of ac side of converter, not only the dc current in TFSC slopes up to a steady value according to the referenced rate, but also the ac side of converter can achieves unit power factor and low harmonic distortion; in the case of latter, by controlling the ac current of converter with the feedback of protected loads voltage, not only the amplitude and frequency of protected loads voltage is steady, but also its phase is invariable. A 1.0KVA prototype has been established and the experiment when it serves for EPS for TFSC, as well as UPS for protected loads are carried out, the results show that the proposed control strategies may fulfill the goals of SMES-UPS based on TFPS perfectly on either of above condition.

Keywords-Current-Source; SMES-UPS; TFSC; EPS; UPS; Control Strategies

I. INTRODUCTION

The first full superconducting tokamak fusion experimental device in the world–EAST is being assembled at Institute of Plasma Physics, Chinese Academy of Sciences, and is arranged to run in Year 2006 [1-2]. In every experimental cycle, cryostable condition of superconducting coils (SC) would be kept for 2-3 months, large voltage sags or short interruptions caused by thunderbolts or other grid faults will cause circulating compressors and turbo-expanders of the helium liquefier/refrigerator immediately trip, and result in the quench of SC. Restoring the cryostable condition of SC would expense much [3]. To avoid these accidents, a SMES for uninterruptible power supply (SMES-UPS) application is designed in this paper, which is shown as fig.1.

The difference between the SMES-UPS presented in this paper and others is that it reuses the toroidal field superconducting coils (TFSC) of EAST as the superconducting magnetic, which saves much money to

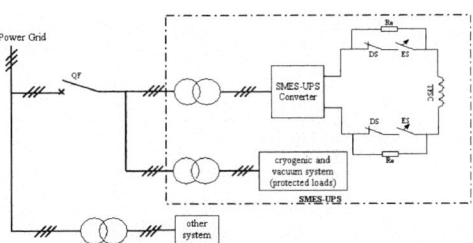

Figure 1.   Diagram of SMES-UPS

rebuilt one [4-6]. When the power grid is in order, SMES-UPS serves for exciting power supply (EPS) for TFSC to provide 3.5T toroidal magnetic field at the plasma center of tokamak restricting the action of plasma; once any power gird fault is detected, the breaker QF connecting the power gird and SMES-UPS is opened first, and SMES-UPS serves for UPS taking out the energy stored in TFSC for the protected loads (compressors and turbo-expanders); and after a short time when the power grid restores, QF is closed and SMES-UPS serves for EPS again. Since TFSC' inductance are 2.95H,and work in a steady state of 14.3KA current under the normal experimental condition, the storage energy in TFSC is approximately 300MJ; On the other hand, TFSC can endure 200V average dc voltage, and the total active power of protected loads is about 1MW, so before the current in TFSC decreases to 5KA, enough power can be provided for protected loads, and the maintainable time is about 260s, the feasibility of constructing a SMES-UPS based on TFSC is obvious.

II. MAIN CIRCUIT OF SMES-UPS AND ITS CONTROL STRATEGIES

A.   Main Circuit of SMES-UPS

To simplify SMES-UPS，3-Phase Current-Source-Converter (3PCSC) topology is adopted as the main circuit for SMES-UPS, which is parallel connection with protected loads [7-8]. The main circuit of SMES-UPS is shown as fig.2.

This work is sponsored by the National Science of China and Foundation (50577013)，

1-4244-0448-7/06/$25.00 ©2006 IEEE

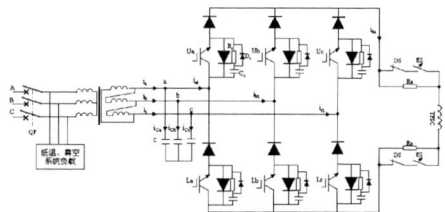

Figure.2   Main Circuit of SMES-UPS

B. *Control strategy when SMES-UPS serves for EPS*

When the power line is in order and SMES-UPS serves for EPS, the control aims are that the dc current in TFSC slopes up to a given steady value according to a referenced rate, and the ac side of the converter achieves unit power factor and low harmonic distortion. The AC side of 3PCSC based on dynamic trilogic pwm can be considered as balanced 3-phase controlled current source [9-11], the equivalent circuit of each phase is shown as fig.3, L, R is leakage inductance and resistor of transformer.

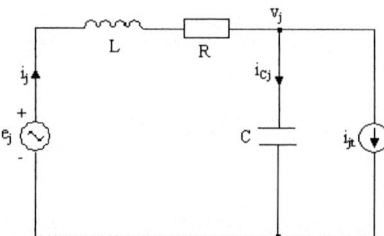

Figure 3.   Equivalent Circuit of 3PCSC

According to Ref.9 to Ref.11, the current of the ac side of converter i_{jt} (j=a, b, c) can be decomposed as two parts: one is quadrature phase with power line voltage - reactive current i_{jtd}, and the other is in phase with power line voltage –active current i_{jtq}. The rms of i_{jtd} and i_{jtq} is shown as (1):

$$\begin{cases} I_{jtd} = \dfrac{\sqrt{3}}{2\sqrt{2}} i_{dc} M \sin\varphi \\ I_{jtq} = \dfrac{\sqrt{3}}{2\sqrt{2}} i_{dc} M \cos\varphi \end{cases} \quad (1)$$

φ in (1) is the phasic difference between modulate signal with power line voltage. To insure the ac side of converter achieving unit power factor, the rms of reactive current i_{jtq} should meet (2):

$$I_{jtd} = -I_{jc} \approx -\omega CE . \quad (2)$$

E in (2) is the rms of power line voltage. The rms of active current is determined by the active power the converter absorbing from power grid, and is given by the

close loop controller of dc side. Control strategy when SMES-UPS serves for EPS is shown as fig.4, φ_j ($j = a,b,c$) in fig.4 is the preliminary phase of 3-phase power line voltage.

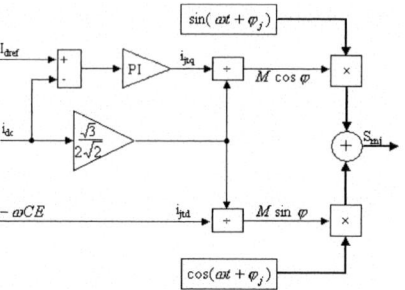

Figure 4.   Control Strategy when SMES-UPS Serves for EPS

C. *Control strategy when SMES-UPS serves for UPS*

When some power gird fault is detected by the control system [12], the breaker QF is opened and SMES-UPS serves for UPS, the control aims are to provide 3-phase balanced rated voltage for protected loads, whose frequency and phase is steady, which can be fulfilled by controlling the 3-phase current injected into the protected loads of ac side. So the correlativity between voltage and current of protected loads should be found first.

Assuming the protected loads are 3-phase balanced, the 3-phase state reference frame – (a, b, c) of ac side can be changed into 2-phase rotate reference frame – (d, q). The transform matrix for space vector between (a, b, c) and (d, q) is shown as (3) and (4) if the preliminary phase of 'd' axis is in phase with 'a' axis.

$$\begin{bmatrix} VI_q \\ VI_d \end{bmatrix} = \frac{2}{3}\begin{bmatrix} -\sin\omega t & -\sin(\omega t -120) & -\sin(\omega t +120) \\ \cos\omega t & \cos(\omega t -120) & \cos(\omega t +120) \end{bmatrix}\begin{bmatrix} VI_a \\ VI_b \\ VI_c \end{bmatrix} \quad (3)$$

$$\begin{bmatrix} VI_a \\ VI_b \\ VI_c \end{bmatrix} = \begin{bmatrix} -\sin\omega t & \cos\omega t \\ -\sin(\omega t -120°) & \cos(\omega t -120°) \\ -\sin(\omega t +120°) & \cos(\omega t +120°) \end{bmatrix}\begin{bmatrix} VI_q \\ VI_d \end{bmatrix} \quad (4)$$

The voltage of protected loads in (a, b, c) provided by 3PCSC is shown as (5)

$$V_j = V\sin(\omega t + \varphi + \varphi_j) \quad (j = a,b,c$$
$$、\; \varphi_a = 0, \varphi_b = -2\pi/3, \varphi_c = 2\pi/3) \quad (5)$$

Substitution (3) with (5), the projection of space vector voltage on q and d axis is shown as (6)

464

$$\begin{cases} V_q = -V\cos\varphi \\ V_d = V\sin\varphi \end{cases} \tag{6}$$

Assuming the amplitude of protected loads is I and its phase is lag δ with voltage. Similarly, the projection of space vector current on q and d axis is shown as (7)

$$\begin{cases} I_q = -I\cos(\varphi-\delta) \\ I_d = I\sin(\varphi-\delta) \end{cases} \tag{7}$$

For 3-phase balanced protected loads, assuming its impedance is Z, the relation between V and I is shown as (8)

$$V = IZ \tag{8}$$

Spreading (7) according to trigonometric function and considering (6) and (8), the relation between voltage and current in (d, q) is calculated as (9)

$$\begin{bmatrix} I_q \\ I_d \end{bmatrix} = \frac{1}{Z}\begin{bmatrix} \cos\delta & -\sin\delta \\ \sin\delta & \cos\delta \end{bmatrix}\begin{bmatrix} V_q \\ V_d \end{bmatrix} \tag{9}$$

The power factor angle of protected loads $-\delta$ can be calculated according to (10)

$$\begin{cases} \cos\delta = \dfrac{V_q I_q + V_d I_d}{\sqrt{(V_q^2 + V_d^2)(I_q^2 + I_d^2)}} \\ \sin\delta = \dfrac{V_q I_d - V_d I_q}{\sqrt{(V_q^2 + V_d^2)(I_q^2 + I_d^2)}} \end{cases} \tag{10}$$

By the analysis above, control strategy when SMES-UPS serves for UPS is shown as fig.5. If the protected loads are pure resistor – that is $\delta = 0$, the part of broken line in fig.5 can be omitted. The control strategy shown in fig.5 not only ensures the voltage amplitude and frequency steady, but also controls the voltage phase, and the phase can be unchanged if the referenced voltage preliminary phase is the same as that of protected loads voltage once the power grid fault is detected.

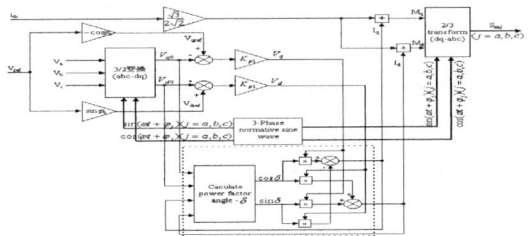

Figure 5. Control Strategy when SMES-UPS Serves for UPS

III. EXPERIMENTAL RESULTS OF SMES-UPS PROTOTYPE

To validate the feasibility, effectively of proposed control strategies, a 1KVA experimental prototype is established [13], whose main circuit is shown in Fig.2 and parameters are shown in Table 1.

TABLE I.
MAIN CIRCUIT PARAMETERS OF EXPERIMENTAL PROTOTYPE

Phase voltage of power grid	220V (rms)
Filter Capacitance	193uF/phase
Commute switch	IGBT/50A, 600V
Switch frequency	5K
Inductance of TFSC	0.1H
Capability of transformer	1KVA
Leakage impedance of transformer	4.7% (p.u.)
Winding connection	Y-Δ 11
Rated voltage of first/second winding	380V/47V

The experimental results of SMES-UPS when it serves for EPS are shown in Fig.6. Ch1, Ch2, Ch3 and Ch4 in Fig.6 are phase A voltage of power grid, phase A line current of power grid, dc current in TFSC and phase A modularized signal respectively. Fig.6-a, b are the experimental results of transition when TFSC is charged up on the condition of final referenced current Idref=15A, and the rate of dc current ascending di/dt=15A/s, di/dt=30A/s respectively. It is shown in Ch4 of Fig.6-a, b that dc current in TFSC can slope up to final referenced current according to the referenced rate of dc current ascending on either of conditions. Fig.6-c is the experimental results when SMES-UPS is steady state, and it is seen that the line current of phase A is 1.59ms (28.6^0) before its voltage, which shows that ac current of converter is in phase with voltage and the power factor of converter is unit.

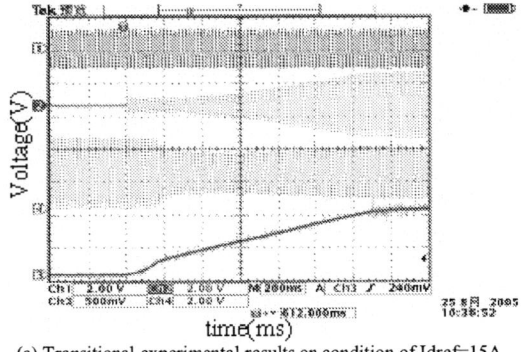

(a) Transitional experimental results on condition of Idref=15A, di/dt=15A/s

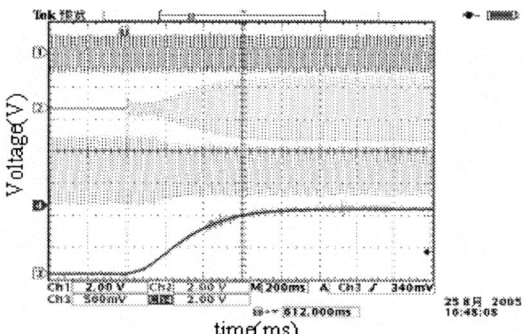

(b) Transitional experimental results on condition of Idref=15A,
di/dt=30A/s

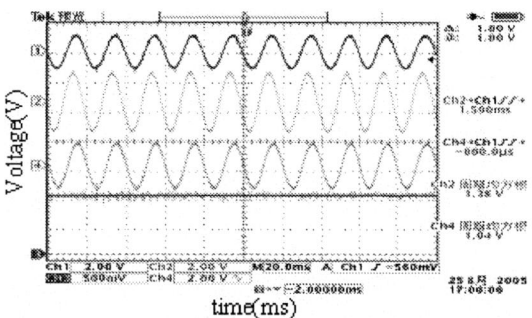

(c) Steady experimental results and measure of current and voltage
on condition of Idref=15A, di/dt=15A/s

Figure 6.   Experimental results when SMES-UPS serves for EPS

The experimental results of SMES-UPS when it serves for UPS on condition of zero-loads, and 3-phase balanced 145W-loads are shown in Fig.7-a, b respectively. Ch1, Ch2, Ch3 and Ch4 in Fig.7 are phase A modularized signal, phase A voltage of power grid, phase A voltage of protected loads and dc current in TFSC respectively. It is seen in Fig.7 that SMES-UPS is fully disconnected with the power grid after about one period transition once any power grid fault is detected and the instruction of opening breaker is executed; the voltage of protected loads stands, as before power grid fault happened, its amplitude and phase are changeless. The results show that the storage energy in TFSC is taken out properly by SMES-UPS converter for protected loads to operate uninterruptedly, which fulfilled the goal of SMES-UPS based on TFSC when power grid is fault.

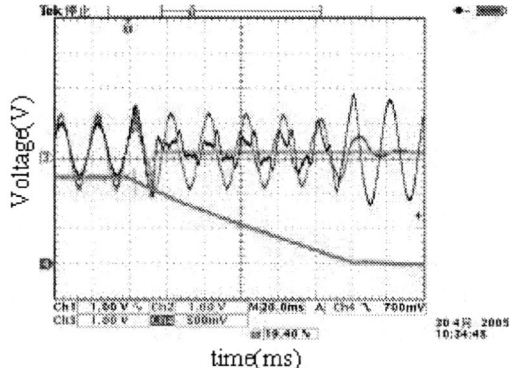

(a) Experimental results of SMES-UPS when it serves for UPS
on condition of zero-loads

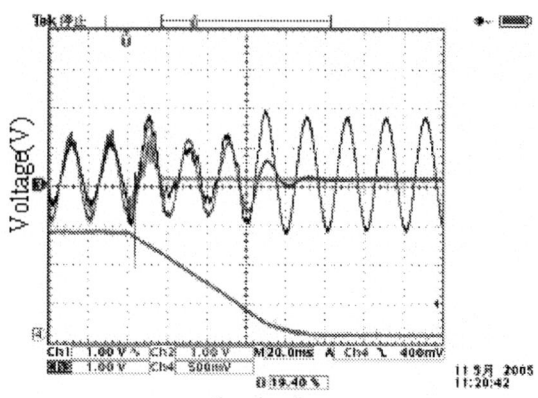

(b) Experimental results of SMES-UPS when it serves for UPS with
3-phase balanced 145W-loads

Figure 7.   Experimental results when SMES-UPS serves for UPS

IV.   CONCLUSIPONS

3-Phase Current-Source SMES-UPS based on TFSC is introduced in this paper. It can serve for EPS of TFSC when the power system is in order, as well as for uninterrupted power supply (UPS) to provide steady 3-phase voltage for protected loads once the deep voltage sag or other grid faults are detected. In the case of former, not only the dc current in TFSC can slope up to a steady value according to the referenced rate, but also the ac side of converter can achieve unit power factor and low harmonic distortion; in the case of latter, not only the voltage amplitude and frequency of protected loads are steady, but also its phase can be invariable, by the control strategies proposed in this paper. The experimental results of 1KVA prototype show the strategies fulfill the dual function of SMES-UPS serves for EPS and UPS perfectly, which saves the expense of constructing SMES-UPS and improves the security of tokamak during its long-term running.

REFERENCES

[1]   Wu Songtao, Weng Peide, "The HT-7U project and its preliminary engineering design", Fusion Engineering, 1997 (1), pp.249-252

[2]   Weng Peide, "Progress of HT-7U superconducting tokamak", Fusion Engineering, 2002 (1), pp.282-284

[3]   Liu Xiao-ning, Wang Fu-sheng, "Design and control strategies of converter for SMES-UPS based on TF superconducting coils of tokamak, Proceedings of the CSEE, 2004,24(11), pp.172-176.

刘小宁，王付胜，"基于托卡马克超导纵场线圈的 SMES-UPS 变流器设计及其控制策略"，中国电机工程学报，2004, 24(11), pp.172-176.

[4]   Buckles. W. E, Hassenzahl. W. V., "Superconducting magnetic energy storage", IEEE Power Engineering Review，2000, 20(5), pp.16-20

[5]   H. Salbert, D. Krischel, M. Schillo, et al., "2 MJ SMES for uninterruptible power supply", IEEE Transactions on Applied Superconductivity, 2000,10(1), pp. 777-779

[6] Xu Dehong, Eisuke Masada, "GTO PWM current converter array for superconducting magnetic energy storage", Proceedings of the CSEE, 1998,18(2), pp.124-129.

徐德鸿, 正田英介, "超导储能装置用 GTO PWM 电流型变流器模块方阵", 中国电机工程学报, 1998,18(2), pp.124-129.

[7] Hou Yong, Jiang Xiaohua, Jiang Jianguo, "Superconduction magnetic energy storage based parallel processing uninterruptible power system and its control strategy", Power System Technology, 2004, 28(16), pp.1-6.

侯勇, 蒋晓华, 姜建国, "基于超导储能的并联处理不间断供电系统及其控制策略的研究", 电网技术, 2004,28(16), pp.1-6.

[8] LI Jun, XU Dehong, K. W. E. Cheng, et al., "Carrier-Swapping method to equalize current in a multimodular current source converter for SMES", Proceedings of the CSEE, 2004, 24(7), pp.106-111.

李君, 徐德鸿, 郑家伟, 等., "超导储能系统用多模块电流型变流器载波轮换均流方法", 中国电机工程学报, 2004, 24(7), pp.106-111.

[9] Xiao Wang, Boon-Teck Ooi, "Unity PF Current-Source rectifier based on dynamic trilogic PWM", IEEE Transactions on Power Electrionics, 1993, 8(3), pp. 288-294

[10] Xiao Wang, Boon-Teck Ooi., "Real-Time Multi-DSP control of Three-Phase Current-Source unity power factor PWM rectifier" , IEEE Transactions on Power Electronics, 1993, 8(3), pp.295-300.

[11] Wang Fu-sheng, Liu Xiao-ning, Pan Sheng-ming, "3-Phase Current-Source rectifier based on DSP and CPLD", Advanced Technology of Electrical Engineering and Energy, 2005,24(4), pp.26-29.

王付胜, 刘小宁, 潘胜明, "基于 DSP 与 CPLD 的三相电流源型变流器", 电工电能新技术, 2005, 24(4), pp.26-29.

[12] Hirotaka Chikaraishi, Kazuo Hayashi, Toshiyuki Mito, Kagao Okumura, and Ryo Abe, "Line voltage detector for SMES system designed to protect from momentary voltage drop", IEEE Transactions on Applied Superconductivity, 2004, 14(2), pp.754-757

[13] Iglesias. I. J, Bautista. A, Visiers. M, "Experimental and simulated results of a SMES fed by a current source inverter", IEEE Transactions on Applied Superconductivity, 1997, 7(2), pp.861-864

A novel control scheme of 230kA DC power source using thyristor, Phase-shifting rectifier transformer and On-load tap changer

Qiao Shutong[*], Jiang Jianguo[*] *Member, IEEE*, Zuo Dongsheng[*] and Wu Xiaojie[**]

*Department of Electrical Engineering, Shanghai Jiao Tong University, Shanghai, 200240, China
**China University of Mining and Technology, Xuzhou, Jiangsu, 221008, China
Email:qiao_shu_tong@sjtu.edu.cn

Abstract—This 230kA DC Power source is composed of four set separated DC power in parallel operation. There are +3.75°,-3.75°, +11.25°, -11.25° phase shifting between each rectifier transformer and original source. Each set device mainly contains on-load tap changer, phase-shifting rectifier transformer and con-phase counter parallel rectifier. In this paper, a current control scheme is realized based on SIEMENS corp. product. According to the required function, the hardware is chose and described, and the software is designed and realized. The design control system is organized by current control level, logic control level and human level. The relation between these levels is determined. The system is validated by electrolytic aluminum plant. From the measured data, the current control precision can reach 0.1% in dynamic process. The rectifier efficiency is more than 97.5%. The whole control device's realized zero current changing on-load tap changer. This developed DC power source can utilize for electrochemical, electro winning, plasma torches etc.

Key words — power source; phase-shifting rectifier transformer; Control system; Parallel operation

I. INTRODUCTION

High power direct current source with power semiconductors have been applied in several important processes such as electrochemical, electro winning, dc arc furnaces, plasma torches, etc[1]. Semiconductor rectifiers have more efficient, healthful and precision than traditional mercury rectifier. In electrolytic aluminum industry, the current efficient and electro-bath temperature are associated with the direct current precision, in addition to the high-precision current can reduce anode effect and lighted labor intension [2].

There are mainly three type combinations in the high-power DC sources for electrolytic aluminum plants. Firstly, On-Load Tap Changer (OLTC) transformer and diode rectifier are combined as DC source equipment. This combination use ampere-hour automotive current stability method. The steady error is 0.1-0.25% during

This work is supported by 973 Project (2005CB221505) and SRFDP (20050248058) in China.

several hours. The dynamic error is 5-10% during anode effect. Secondly, OLTC transformer, phase shifting rectifier transformer, diode rectifier and saturation reactance are combined. This combination use proportional or proportional- integral current control method. The steady error is less than 0.1%. The dynamic error is 0.1-0.25%[2]. Thirdly, OLTC transformer, phase shifting rectifier transformer and thyristor rectifier are combined. The steady error approaches to zero error. The dynamic error is less than 0.1%. Now, most of all aluminum plant adopt the third combination in china.

In addition to, the current control unit's reliability is a critical issue to be considered in the operation of these high power dc sources units due to its impact on the productivity.

This paper describes the main equipments of the high power dc source and their characteristics. The requirements and specifications of control units are established. The following sections include the control system hardware choose and software design description. The system is validated by electrolytic aluminum plant. The experimental results and system issues are given and analyzed. In the last section, conclusions and future work are summarized.

II. DIRECT CURRENT POWER SOURCE CONFIGURATION

The large direct current power source system utilized for the electrolytic aluminum process consists of four sets of phase difference +3.75°,-3.75°, +11.25°,-11.25° subsystem compared with the main alternating current power phase. Each rectifier system is composed of oil-filled on-load tap changer transformer, oil-filled phase shifting rectifier transformer, water-cooled co-phase counter parallel connection rectifiers, and associated equipments such as switchgear, cooling water equipment, dc disconnect switch, heat changers, control equipment, alarm, passive power filters, etc. The main equipment is shown in Fig.1a). The diagram of co-phase counter parallel connection rectifier is shown in Fig.1b). The phasor diagram of phase- shifting rectifier transformer is shown in Fig.1 c).

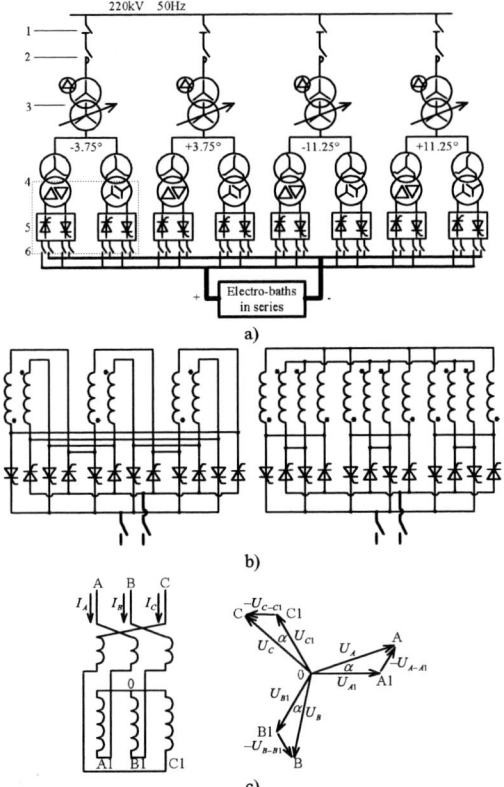

Figure 1. a) DC power source configuration 1-isolating switch, 2-contact switch, 3-on-load tap-changing transformer,4-phase shifting rectifier transformer,5-co-phase counter parallel connection rectifier, 6-dc disconnect switch; b) the detail of the dashed line inside in a), the diagram of co-phase counter parallel connection rectifer; c) the phasor diagram of phase-shifting rectifier transformer, A-A1,B-B1,C-C1 phase shift winding, A1-0,B1-0,C1-0 basic winding, α -shifting phase angle [3].

III. REQUIREMENTS AND SPECIFICATIONS OF CONTROL SYSTEM

Form the load side, the control equipment has the capability to deliver a controlled dc current in a range of 0%-100% of rated value. In general, no additional requirements are applied to the quality of current ripple, which is acceptable for most processes.

A. Requirements of the rectifier system

One is high efficiency (usually >97%) due to the large power and need to dissipate out the losses.

High reliability is required to avoid unexpected downtimes. This is one of the most important specifications which is satisfied by over sizing the power semiconductors in current and voltage. Usual engineering practice is the requirement of $n-1$ redundancy for thyristors. This means that the rectifier must continue its normal operation at rated current even when one thyristor fails.

The refrigeration of transformers and semiconductors is a key issue in the operation of large rectifiers. For this reason, the use of redundant cooling system is mandatory.

B. Specifications of the control unit

Utilize 32-bit digital processor and high precision current sensor to enhance calculation precision and measurement precision. Utilize redundant control to enhance reliability. And system module design for ease replace and high reliability. Utilize human-machine interface, field-bus (Profibus-DP and RS485) to enhance the communication and automation performance [5,6].

The configuration of control subsystem is shown in Fig.2. The system must have the following functions.

a) A closed-loop control system for constant current operation, configured with current measure and synchronous voltage with Power source. The rectifier system drives electric-baths in series, is fed by thyristors, and the SIMADYN D controls the output trigger phase shift with the natural commute point.

b) An open-loop control system for debugging and installation. This operation is used for adjusting phase sequence according with the main source and determining the original shifting phase angle and initialization of trigger phase.

c) The united control system operation. Each set subsystem in parallel outputs the same current which equal to a quarter of total current value. And three sets output the same current which equal to one third total current value while one of them is downtime.

The open-loop configuration can be obtained from the close-loop configuration by removing the current feedback. On the other hand, operation c) results if the operation a) and b) control system is bypassed.

d) Urgent operation is applied for such conditions as many fuse failure and damage accident.

IV. HARDWARE DESCRIPTION AND SIGNAL DISTRIBUTION

The main diagram is shown in Fig.1 a). Each set control unit of subsystem is shown in Fig.2. The control unit is composed of SIMADYN D double Processors controller, PLC Controller and Panel. The SIMADYN D double Processors is the current control processor, which consists in sub-rack which accommodate SIMADYN D modules, processor module PM5, converter gating module ITDC, coupling buffer memory module MM3, interface module SE20.2 which implements mechanical conversion and the electrical isolation of the standard SITOR interface from the ITDC to the current sensor and the rectifiers, communication support module CS7, communication module SS52 and SS4 [7]. The PLC Controller is the logic control processor, which is composed of rail, power supply module, CPU module, digital input-output signal module, communication module, PROFIBUS- DP bus cable [9,10,11]. The panel is a touch Human- Machine Interface which can communication with PLC by RS485 standard Multiple Point Interface profile cable [12].

The topology of control system configuration is show in Fig.3. The information between these controllers is listed in Table I.

V. SOFTWARE DESCRIPTION

The SIMADYN D controllers use the graphic-based programming method [8]. The PLC controllers use the ladder diagram programming method [10]. The HMI panels use Protool software for function design [12].

The SIMADYN D double processors can make two individual current closed-loops simultaneously. The current control flowchart is shown in Fig.4. The run sequence of these algorithm blocks is firing angle actual generation, set point, current actual value sensing, current set point calculation, current pre-control in the discontinuous range, current controller, firing angle controller.

Traditionally, for prolonging the lifetime of OLTC switch, the current control error must be enlarged to reduce the action times of OLTC switch. And the output voltage of OLTC cannot meet the requirement of current control precision. However, using appropriate control algorithm can make OLTC switch action at zero current. This can make the lifetime of OLTC longer. The zero current action algorithms are shown in Fig.5.

Four set subsystems can be single operation or united operation. The conversion between two operations is automated by PLC software control. The conversion algorithm is show in Fig.6 and Fig.7 respectively. The 5# PLC is the united control processor in Fig.3 and Fig.6.

During the emergent operation, the control algorithm is applied to protect the system safety. For example, the multiple thyristors' fault is an emergent mode. The action sequence is shown in Fig.8.

VI. RESULTS

The whole control system was installed and debugged at electrolytic aluminum plant. The control equipment is shown in Fig.9 a). The OLTC and phase shifting rectifier transformer is shown in Fig.9 b). The co-phase counter thyristor rectifiers are shown in Fig.9 c). The performance of DC power source system is measured.

At the beginning, the efficient of system is calculated according to the measured data. Equation (1) is used to calculate the operation efficiency of AC/DC rectifier equipment [4]

$$\eta = \frac{P_O}{P_I} = \frac{U_d I_d}{p_1 + p_2 + p_3 + p_4} \times 100\% \qquad (1)$$

η the operation efficiency of rectifier; P_O the DC side output power, which is equal to output voltage multiplied by output current; P_I the total input active power at AC side; U_d, I_d the output voltage and current at DC side; p_i, $i=1,2,3,4$ the input active power of each set rectifier. In this experiment, there are 84 electro-baths are connected in series. In the steady state, the output current and voltage are 230kA and 353.5V respectively. The measuted data are listed in Table II.

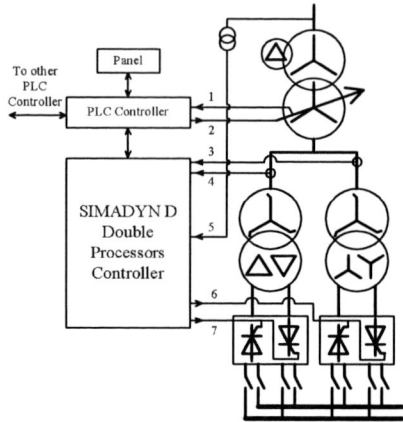

Figure 2. Electrical diagram of experimental system,1-OLTC level signal; 2- OLTC level changing action; 3,4-current signals from two alternating current sensors respectively; 5-synchronous signal from AC main source; 6,7-trigger pulses for two rectifiers, respectively.

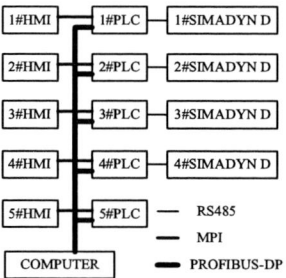

Figure 3. Topology of control system

TABLE I.
INFORMATION TRANSFER BETWEEN THESE CONTROL COMPONENTS

Transfer Direction (i=1,2,3,4)	Information Content
i#HMI → i#PLC	Operation mode choice Given current value Synchronous shifting angle Given triggering angle Controller parameters OLTC level up/down
i#PLC → i#HMI	Operation mode Actual current value Triggering angle OLTC level
i#PLC → i#SIMADYN D	Given current value Synchronous shifting angle Given triggering angle Controller parameter
i#SIMADYN D → i#PLC	Actual current value Actual Triggering angle Over change
5#HMI → 5#PLC	Choose operation mode Given total current value Stop choice
5#PLC → 5#HMI	Operation mode Actual total current value Each rectifier system current value
i#PLC → 5#PLC	i # PLC operation mode Actual current value Triggering angle OLTC level
5#PLC → COMPUTER	System operation mode Actual current value i # Triggering angle i # OLTC level

Figure 4. The close-loop current control flowchart

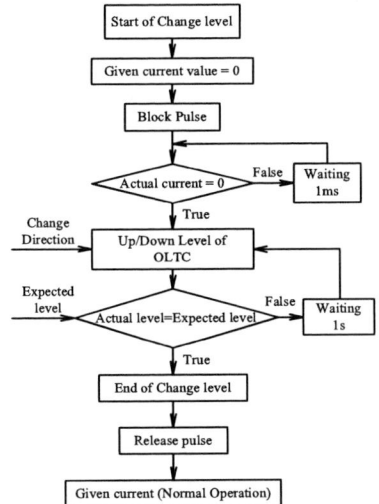

Figure 5. Flowchart of OLTC level adjust at zero current

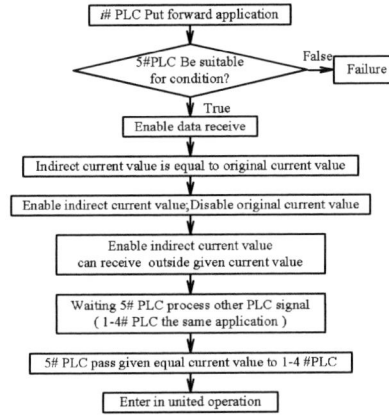

Figure 6. Flowchart of single to united software

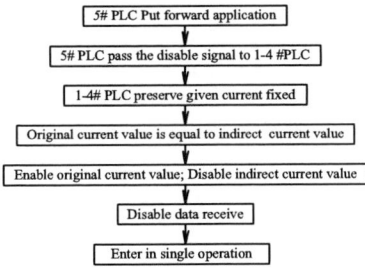

Figure 7. Flowchart of united to single software

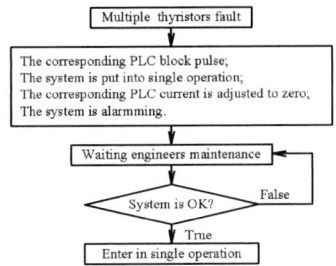

Figure 8. Emergent protection software (thyristors are failure)

The power factor of input side is measured. The active power and reactive power are calculated from the voltage sensor and current sensor. The thyristor rectifiers have low power factor. Four set power filters are connected with the corresponding compensative winding for enhancing power factor. The comparison is listed in Table III. Each set subsystem has the same measure device and the same size passive power filters. According to the data of 1-subsystem with power filters and 2-subsystem without power filters in Table3, the power factor is enhanced by passive power filters from 0.80 to 0.92.

The output voltage waveforms of each 6-pulses rectifier at trigger angle 30° and 10° are shown in Fig.10. The 48-pulses output voltage waveforms at trigger angle 30° and 10° are shown in Fig.11. The comparison between Fig.10 and Fig.11 is shown that the 48-pulses output voltage is smoother than the single 6-pulses output voltage.

At anode effects, the performance of closed-loop operation is measured. The output voltage and output current of system is shown in Fig.12. Because the closed-loop control algorithm is applied, the output current is fixed and the output voltage varies from load change. The output voltage is boosted 30-40V at that time than general output voltage.

a)

b)   c)

Figure 9. a) The designed control system is installed at plant; b) The OLTC and phase shifting rectifier transformer; c) The thyristor rectifiers.

TABLE II.
EXPERIMENTAL DATA OF RECTIFIERS EQUIPMENTS

Operation mode	p_1(MW)	p_2(MW)	p_3(MW)	p_4(MW)	P_o(MW)	η
Four set rectifiers In parallel	21.36	20.29	20.90	20.75	81.31	0.9761
Three set rectifiers In parallel	28.38	0	27.47	27.47	81.31	0.9759

TABLE III.
EXPERIMENTAL DATA OF RECTIFIERS EQUIPMENTS

Operation mode	Subsystem Number	1	2	3	4
Four set rectifiers In parallel	Active power (MW)	21.36	20.29	20.90	20.75
	Reactive power (MVar)	8.81	15.11	8.85	8.85
	Power Factor	0.92	0.80	0.92	0.92
Three set rectifiers In parallel	Active power (MW)	28.38	0	27.47	27.47
	Reactive power (MVar)	13.12	0	12.82	12.82
	Power Factor	0.91	✕	0.91	0.91

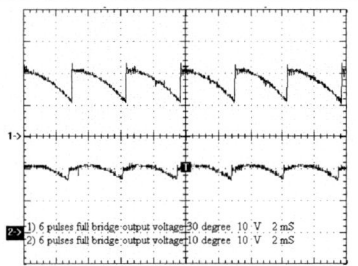

Figure 10. The 6 pulses output voltage waveforms at 30° and 10°

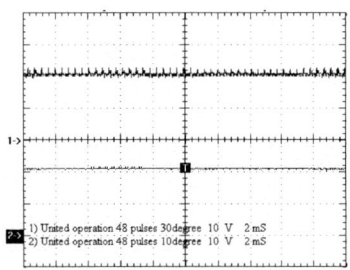

Figure 11. The 48-pulses output voltage waveforms at 30° and 10°

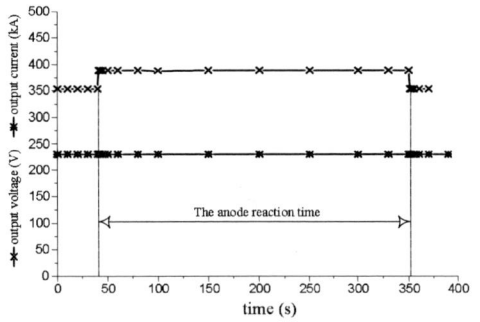

Figure 12. The output voltage and output current of closed-loop system including a period of anode effect time

VII. CONCLUSIONS

The novel control system is designed and installed at industry plant. The control system reliability and stability is validated by the plant. From the results, the conversion efficiency of DC power source exceeds 97.5% and the input side power factor exceeds 0.90 by utilizing passive power filters. The control system can response the load variety and keep the output current constant. Furthermore, the control system can make OLTC switch action at zero current; the lifetime of OLTC switch is extended and the degree of safe action is high.

The control system can communicate with PLC, HMI or PC by Profibus-DP standard field-bus protocol. So the control system is open and can be connected with other related devices to enhance industrial automation and information level. The control system adopted HMI substituted for single button and indicator light. This makes the system design flexible. The HMI make the local display function change without addition hardware even label.

The effective performance of DC power source system is proved at electrolytic aluminum plant. The control system can be adopted by other high power DC thyristor rectifier source. In addition to described functions, the digital control system has overload protection, rectifiers and control equipments monitoring. The digital control system proved to be a versatile function and efficient control tool in industrial power converter application.

REFERENCES

[1] Jose R. Rodriguez, Jorge Pontt, Cesar Silva etl. "Large Current Rectifiers: State of the Art and Future Trends" *IEEE Transaction ON Industrial Electronics,* vol.52,no.3, pp.738-746.

[2] Shen Yang Lv Mei She Ji Yan Jiu Yuan Dian Li shi,*"Gui Zheng Liu Suo Dian Li She Ji".* BeiJing: Metallurgical Industry Press. 1983. (In Chinese).

[3] XIN Cheng-shan. "Further analysis of Winding current in Phase Shifting Rectifier Transformers". *TRANSFORMER,* vol.39, no.12.pp.5-10.

[4] *Measuring and calculation method on efficiency of rectifiers and their consumers energy-conversion efficiency in electro-chemical industry.* GB7228-87.UDC 621.3.017.8,(In Chinese).

[5] Jiang Jianguo, Dai Peng, Zuo Dongsheng etl. "Development of full digital control system of TCR dynamic reactive power compensation based on Simadyn-D". *IEEE ICIT '02.* vol.1, pp.125 -128.

[6] Wu Xiaojie, Jiang Jianguo, Dai Peng, Zuo Dongsheng. "Full digital control and application of high power synchronous motor drive with dual stator winding fed by Cyclo-converter", *Proceedings of PEDS 2003.*vol.2, pp.1194 -1199.

[7] *SIEMENS SIMADYN D Hardware Manual,* SIEMENS AG Corporation, 2002

[8] *SIEMENS SIMADYN D System and Communication Configuring D7-SYS Manual,* SIEMENS AG Corporation, 2003.

[9] *SIEMENS SIMTIC S7-300 Programmable Controller Hardware and Installation Manual,* SIEMENS AG Corporation, 2002

[10] *SIEMENS SIMTIC S7-300 Programmable Controllers CPU Specifications, CPU 31xC and CPU 31x Reference Manual,* SIEMENS AG Corporation, 2002

[11] *SIEMENS SIMTIC S7-300 Programmable Controllers Module Specifications Reference Manual,* SIEMENS AG Corporation, 2002

[12] SIEMENS SIMATIC HMI Protool Configuration Figure Display Manual, SIEMENS AG Corporation, 1999

2006 5th International Power Electronics and Motion Control Conference

Research on Control Method of Double-Mode Inverter with Grid-Connection and Stand-Alone

Herong Gu, Zilong Yang, Deyu Wang, and Weiyang Wu
Yanshan University, Qinhuangdao, P. R. China
Email: ydghr@ysu.edu.cn

Abstract—Inverter system, as the interface device between the renewable energy system and grid, plays an important role in the distributed generation system. A double-mode single-phase inverter system, which can be operated in grid-connected mode or stand-alone mode with seamless transitions control, is proposed in this paper. The different control strategies applied on the system for the different modes are designed. In the grid-connected mode, the grid governs the load voltage, the inverter operates as a current source, and the grid current is directly controlled by a three-level hysteresis controller for faster dynamic responsibility and lower current ripple. In the stand-alone mode, the inverter is voltage-controlled, fuzzy controller with parameter self-adapting on-line is introduced to improve the system adaptability to the varieties of load and system parameters, so good quality of output sine-voltage can be obtained under various load conditions. In order to make the distributed generation utility-interactive, an algorithm, which controls the inverter to transfer seamlessly between grid-connected mode and stand-alone mode, is present. The DSP (TMS320LF2407A) based inverter prototype is developed to verify the analysis.

Keywords-inverter; grid-connected; stand-alone; fuzzy control; hysteresis control; mode-transition

I. INTRODUCTION

The developments of renewable energy and distributed power generation have attracted more and more attentions because of the lack of energy sources. Distributed generation systems using renewable sources like PV or wind power offer many advantages such as compensation for power system, which can reduce tension of power supply, and standby generation which provides power to sensitive and mission-critical industrial loads during system outages until service can be restored. Both of the above advantages can be effectively utilized if the distributed generation system is utility-interactive [1].

Inverter system plays an important role as the interface device between the distributed generation system and grid. In order to implement the utility-interactive, it is necessary that the inverter system can be operated in grid-connected mode or stand-alone mode and has the ability of seamless transitions between the two modes. The PWM inverter is operated in current-control mode when it is connected to

the grid, the current injected into grid should be regulated to fellow the reference. In the stand-alone mode, the PWM inverter is operated as a voltage source. The inverter is operated to regulate the voltage across the load, and steady sine-voltage is expected. In addition, the PWM inverter has to be capable of shifting between current-control mode and voltage-controlled mode, and maintain the voltage across the load in the presence of faults on grid.

The different control strategies applied on the system for the different modes are designed. When the inverter is grid-connected, it is operated in current-controlled mode with a novel three-level hysteresis control method that has double-frequency function. Compared with other controllers, it has not only the general advantages of hysteresis controller such as automatic peak-current limitation, simple implementation, a rapid dynamic response, load parameter independence and unconditional stability [2] [3], but also low switch frequency, small current ripple, low impact on DC line. When the inverter is stand-alone, it is operated in voltage-controlled mode with a self-adaptive fuzzy control scheme. An assistant fuzzy controller is induced to modify the scaling factor of primary fuzzy controller, so a fuzzy controller with parameter self-adapting on-line is used to improve the system adaptability for the varieties of load or parameters, so high quality sine-voltage can be obtained under different load conditions [4] [5].

Refer to the transition between these two modes, there may be voltage spikes or inrush currents drawn by the load due to the sudden change in the load voltage and it is harmful to the grid and inverter. In case of sensitive and mission-critical industrial load, maintaining a continuous, uninterrupted AC power is of utmost importance. It is necessary that the utility-interactive system has the ability that load is unaffected by transition from one mode to the other. The transition algorithm takes care of all the above-mentioned requirements associated with the transfer of a PWM inverter between grid-tied mode and off-grid mode. Moreover, the algorithm is completely independent of the actual realization of the current and voltage controllers.

II. CONFIGURATION OF THE INVERTER SYSTEM

A configuration of proposed grid-connecting inverter system that can be operated in stand-alone mode or grid-connected mode is shown in Fig. 1, and the circuit

1-4244-0448-7/06/$25.00 ©2006 IEEE

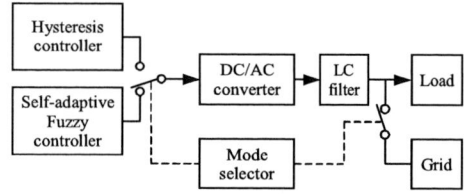

Figure 1. Configuration of the inverter system operated in stand-alone mode or grid-connection mode

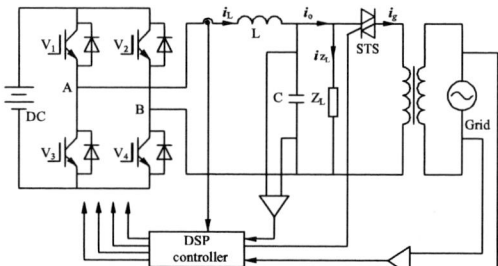

Figure 2. Single-phase grid-connection inverter system

diagram of inverter system is shown in Fig. 2. Operation of the inverter system is divided into two modes, one is grid-connected mode, and the other is stand-alone mode. The load is connected across the output of the PWM inverter. In order to disconnect from the grid in the least and exact time, a triac is used as a static transfer switch. The traic would ensure that the grid could be disconnected from load within half of a fundamental frequency cycle, in the event of a grid fault.

III. INVERTER CONTROL FOR GRID-CONNECTION

A. Equivalent Model of Grid-connection System

In many examples of grid connected inverter, the inverter is operated as a voltage source or a current source [6] [7]. When the inverter is operated as a voltage source, the grid-connection system can be equivalent to the parallel connection of two voltage sources. In this mode the inverter is controlled to generate a sine waveform voltage, the output current injected into grid depends on the grid voltage quality. If the grid voltage is distorted, the exported output current is distorted. So this scheme is not a good choice for grid-connection system.

Current-controlled mode has been chosen for this work to operate the inverter as a current source, because within the control frequency range higher output-impedance is observed from the point of view of the grid voltage. This minimizes the effect of voltage harmonics on the output current and improves the power quality. The equivalent model of the grid-connected inverter system operated in current-controlled mode is shown in Fig. 3. The grid is equivalent to a voltage source u_g in series with an equivalent resistance Z_g. The inverter is equivalent to a current source i_o. i_g is the current injected into grid, i_o is the output current of inverter, Z_L is the load impedance.

The grid is assumed to be relatively stiff and maintains the voltage across the load, so the impedance of grid is very small, that is $Z_g=0$. While the load is regarded as a part of the grid, $u_g=u_o$. So i_o is regards the current injected into grid. The Predigested mode of grid-connection system is shown as Fig. 4. Controlled by (1), the current injected into grid can match the grid voltage on frequency and phase, so the power can be injected into grid with PF=1.

$$i_o = k\dot{u}_g \qquad (1)$$

Fig. 5 shows the proposed control scheme for grid-connection system. Phase locked logic (PLL) control is employed to match the frequency and phase of grid voltage. The reference current is generated by the sine generator based on the voltage phase and the given amplitude of output current. The reference current i_{ref} subtracts the inductance current i_L for the error current i_e, which goes through the three-level hysteresis controller to obtain signals for switching full-bridge inverter.

B. Three-level Hysteresis Current Controller Design

Compared with two-level hysteresis control, three-level hysteresis control has lower inductor current ripple under the same average switching frequency. Moreover, much lower frequency and higher efficiency can be obtained with a double-frequency strategy presented in this paper.

Fig. 6 shows the sequential state machine of the three-level hysteresis current controllers. The state machine has three inputs: Q identifies the last zero state, B_1 is the logical output of hysteresis comparator #1, B_2 is the logical output of hysteresis comparator #2. A and B are the phase-leg states of the inverter. When the inverter output current is positive, u_{AB}, the output voltage of full-bridge inverter, has 1 and 0 stage. When the inverter

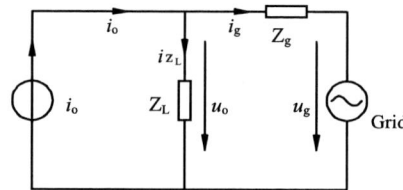

Figure 3. Equivalent model of grid-connected inverter system operated in current-controlled mode

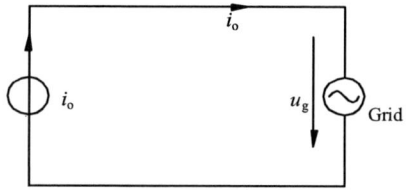

Figure 4. Predigested model of grid-connection inverter system

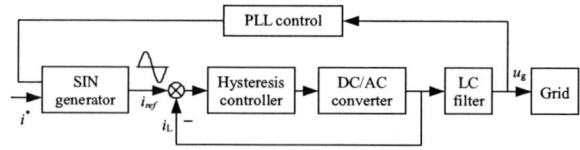

Figure 5. Control scheme for grid-connected inverter

output current is negative, u_{AB} has -1 and 0 stages. Thus, the output voltage of the inverter will be unipolar square pulse in period of half-sinusoid, under the three-level hysteresis control. The state machines can be easily implemented by DSP.

C. Experimetal Results

An experimental system has been constructed to verify the proposed current controller. The system parameters are: L=5mH, C=9.4uF. A TMS320LF2407 DSP is used to implement the control algorithms and generates PWM signals to drive the inverter. Fig. 7 shows the waveforms of drive signals of V_1 and V_3, the inductor current i_L and u_{AB}. The frequency of current ripple is twice as switches'. The output current of inverter i_o and grid voltage u_g are shown in Fig. 8.

IV. INVERTER CONTROL FOR STAND-ALONE

When the inverter is stand-alone, it is operated as a voltage source. A functional block diagram of the inverter system with double-loop control is shown in Fig. 9. The current loop acts as the current reference compensated by P-regulator. For the voltage loop, a self-adaptive fuzzy control (SAFC) scheme with scaling factor self-adjusting is designed to improve the adaptive ability of inverter to the varities of inverter's parameters and different load

conditions.

The structure of fuzzy control system is shown in Fig. 10. The assistant fuzzy controller is introduced to identify the system states, and then can generate a modifying signal to correct parameters of the primary fuzzy controller. In the primary fuzzy controller, the scaling factor α has relativity with the error E and the change of error EC, it can be described by (2).

$$\alpha(K) = f[E(K), EC(K)] \qquad (2)$$

f is a no-linear function of E and EC. The α is decided by the instantaneous states, and has no relation with the model of system. In fact this self-adaptive controller is a no-linear controller independent of system model.

The fuzzy rules of α, output of assistant fuzzy controller, are shown in Table I. The basic rules of assistant fuzzy controller are designed to make sure that the output response of the inverter system has small overshoot and short rise-time. When error E is large and has the inverse sign with the change of error EC, it is indicated that the system response is making for the set value, so the scaling factor should be reduced for avoiding output overshoot; on the contrary, when error E is large and has the same sign with the change of error EC, it is indicated that the system is deviating from the set value, so the scaling factor should be increased for a shorting rise-time. On the other hand, when the error E is small, the scaling factor should have a big variation range.

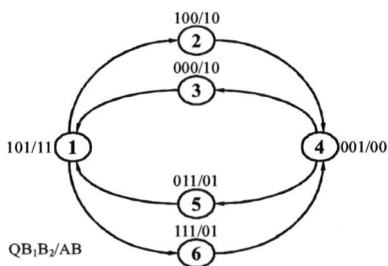

Figure 6. State-transition diagram for three-level hysteresis control

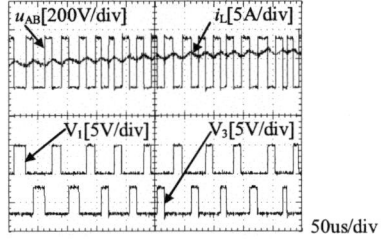

Figure 7. Experimental waveforms with three-level hysteresis control

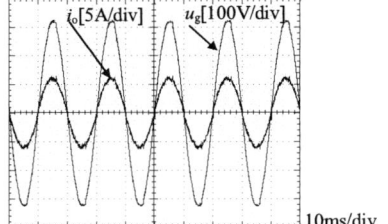

Figure 8. The output current of inverter and grid voltage

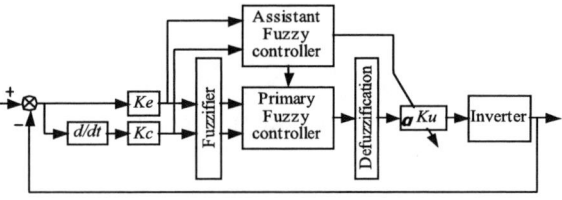

Figure 9. Block diagram of inverter system operated in stand-alone mode

Figure 10. Block diagram of SAFC controller

TABLE I.
ASSISTANT FUZZY CONTROL RULES

EC \ E	NB	NM	NS	ZE	PS	PM	PB
NB	VB	VB	VB	B	M	SM	VS
NM	VB	VB	B	B	BM	SM	S
NS	VB	BM	NM	B	VB	S	S
ZE	SM	M	BM	VS	M	BM	SM
PS	S	SM	S	VB	B	BM	VB
PM	S	SM	BM	B	B	VB	VB
PB	VS	SM	M	B	VB	VB	VB

The experiment results in the cases of inductor load and rectifier load are given out in Fig. 11 and Fig. 12, which indicated that the inverter system with SAFC controller is relatively robust to variations of system component values and load conditions. The rationality and validity of the proposed control method are verified.

V. SEAMLESS TRANSITION BETWEEN THE GRID-CONNECTION AND STAND-ALONE MODES

The control system of grid-connection inverter can be divided into two parts with the current-controlled module and the voltage-controlled module. A mode-transition algorithm, which is completely independent of the current controller and voltage controller, is designed to control the transition between two modes.

A. Stand-alone mode to grid-connected mode

Assume that initially there is a fault on the grid and the inverter is operating in the voltage-controlled mode with the static transfer switch open. When the fault on the grid is cleared and the grid voltage comes back, the phase and amplitude of the load voltage (maintained by the PWM inverter) may not match those of grid voltage. It is necessary that both the magnitude and phases of the inverter's output voltage should be adjusted for a matching state; before the triac can be turned on to reconnect the inverter to the utility.

Thus, the steps to be performed in this phase of the algorithm can be summarized as follows:

1. Detect that the grid is normally operating.

2. Adjust the inverter's output voltage (or load voltage) to match the magnitude and phase of the grid voltage.

3. Once the load voltage is equal to the grid voltage, turn on the triac and switch inverter from voltage-controlled mode to current-controlled mode, with the reference current being equal to the load current.

4. Change the reference current slowly to the desired current (both magnitude and phase).

B. Grid-connection mode to stand-alone mode

Assume that initially the inverter is operating in the grid-connection mode. The PWM inverter is current-controlled and the grid governs the voltage at the PCC. When there is a fault on the grid, the voltage at the PCC would drop. The fault detection circuitry turns off the triac when the grid voltage goes below a pre-set minimum value. So, the inverter is disconnected from the grid at the first zero crossing of the current. Instantly, the PWM inverter can be shifted from current controlled mode to the voltage controlled mode.

Thus, the steps to be performed in the algorithm can be summarized as follows:

1. Detect a fault on the grid and give a turn off signal to the triac.

2. Monitor the magnitude and phase of the load voltage.

3. When the triac current goes to zero, transit the inverter to a voltage-controlled mode, with the voltage reference being derived from the load voltage.

4. Ramp up the magnitude of the load voltage from its initial value to the rated value.

C. Experimetal Results

The mode-transition algorithm is tested on a hardware prototype. The details of the hardware setup are given below:

Load voltage (rms.)	: 220V
Load	: 50Ω
Grid current (rms.)	: 5A

A solid-state relay (SSR) is used to the static transfer switch. Fig. 13 shows the progression of the phase-locking. The phase difference at start is about $\pi/2$ rad. As time progresses, the phase difference decreases to zero, as expected. Fig. 14 shows the load voltage and grid current during the progresses of shifting from voltage-controlled mode to current-controlled mode. Fig. 15 shows the progress of shifting form current-controlled mode to voltage-controlled mode. The load voltage is found to be fairly smooth during the transition.

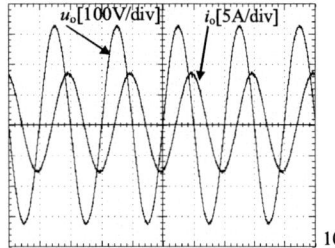

Figure 11. Waveforms of output current i_o and voltage u_o of inverter under inductor load

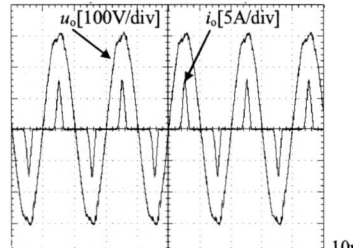

Figure 12. Waveforms of output current i_o and voltage u_o of inverter under rectifier load

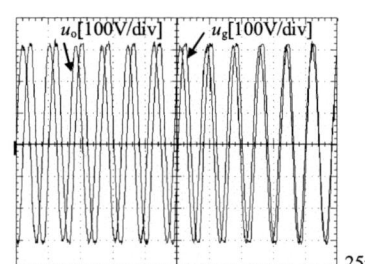

Figure 13. Load voltage and grid voltage during the phase-matching

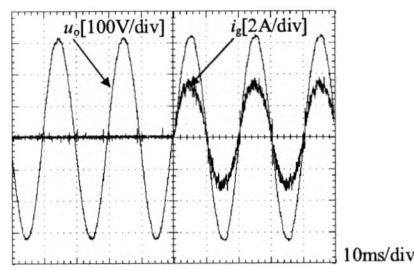

Figure 14. Load voltage and grid current during stand-alone to grid-connection

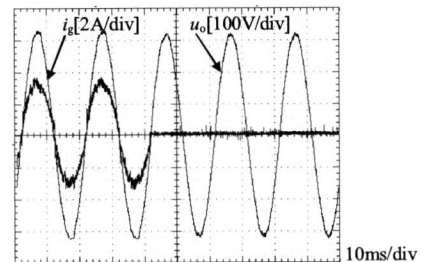

Figure 15. Load voltage and grid current during grid-connected to stand-alone

VI. CONCLUSION

The proposed inverter system not only controls the energy from renewable generations into grid, but also realizes a utility-interactive system. A continuous, uninterrupted AC power is supplied to load. The characteristics of this system are shown as follows.

The current injected into the grid is directly controlled to follow the grid voltage, with good stability performance, fast responsibility and low ripple, employing the proposed novel three-level hysteresis control when the inverter is operated in grid-connected mode.

The inverter system can obtained relatively robust to variation of system component values and load conditions, employing the self-adaptive fuzzy control (SAFC) strategy.

The proposed algorithm which can be implemented easily using digital signal processors (DSP) for the mode-transition seamlessly between grid-connected mode and stand-alone mode, so there are no inrush currents drawn by the load due to the sudden change in the load voltage.

ACKNOWLEDGMENT

This work was supported by the National Natural Science Foundation of China, NO. 50237020.

REFERENCES

[1] T. Rihit, M. Ned, and H. Chris, "Seamless transfer of grid-connected PWM inverters between utility-interactive and stand-alone modes," *IEEE APEC2002.* Vol. 2, pp. 1081-1086, March 2002.

[2] G. H. Bode and D. G. Holmes, "Implementation of three level hysteresis current control for a single phase voltage source inverter," *IEEE PESC 2000*, Vol. 1, pp. 18-23, June 2000.

[3] P.A. Dahono, "New current controllers for single-phase full-bridge inverters", *2004 International Conference on Power System Technology*, Vol. 2, pp.1757-1762, Nov. 2004.

[4] Ling. Luo, Y. Zhou, J. Xu, S, Wan, "Parameters self-adjusting fuzzy PI control with repetitive control algorithms for 50 Hz on-line UPS controlled by DSP Industrial Electronics Society", *IEEE IECON 2004*. Vol. 2, pp.1487-1491, Nov. 2004.

[5] G. Chen, W. Cai, T. Cai, H. Liu, "A hybrid fuzzy current regulator for three-phase voltage source PWM-inverter Machine Learning and Cybernetics", *Proceedings of 2002 International Conference on Machine Learning and Cybernetics*, Vol. 2, pp. 919-923, Nov.2002.

[6] T. Komiyama, K. Aoki, E. Shimada, T. Yokoyama, "Current control method using voltage deadbeat control for single phase utility interactive inverter with FPGA based hardware controller", *IEEE IECON 2004*,Vol. 2, pp.1594 – 1599, Nov. 2004

[7] M. Prodanovic, T.C. Green, "Control and filter design of three-phase inverters for high power quality grid connection", *IEEE Transactions on Power Electronics*, Vol. 18, pp.373-380, Jan. 2003

2006 5th International Power Electronics and Motion Control Conference

Power and Energy Management of a Dual-Energy Source Electric Vehicle - Policy Implementation Issues

P.C.K. Luk, L.C. Rosario

Cranfield University/Department of Aerospace Power and Sensors, Swindon, United Kingdom
email: p.c.k.luk@cranfield.ac.uk

Abstract — **This paper offers a new design-oriented approach to address implementation issues associated with a power and energy management policy for an Electric Vehicle (EV) powered by a dual-sourced system. A special sequential decision process, which determines the power split ratio and reference energy levels of the dual battery-ultracapacitor sources, has been formulated as a state transition mechanism. Experimental results from a test vehicle were used to validate simulation models of tractive power demands and battery behavior.**

Keywords - Electric vehicle, ultracapacitor, battery, power management policy

I. INTRODUCTION

Continuous and pulsed power requirements are common for electric vehicle applications. These requirements, when met by a single sourced system, are closely related with one another. To meet both usually results in an over-sized power storage system. With a dual-source, it is possible to decouple the requirements leading to both performance and weight advantages. The success of implanting a dual-source system, however, depends on how the power and energy flow between the two sources is managed during the drive cycle. Here, we use a battery-ultracapacitor system to create an energy storage system with the high energy density attributes of batteries and the high power density of ultracapacitors. In essence power and energy management policies for such dual energy storage systems in Electric Vehicles (EV) involve determining the proportional power split between two sources as well as the optimal State of Charge (SoC) throughout the drive cycle. In effect, it represents a sequential decision process over the mission profile.

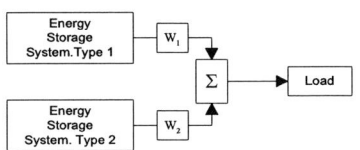

Figure 1: Power split between two energy sources in an EV

The power split of different types of energy storage systems within an EV can be concisely described as follows. Considering the block diagram of Figure 1, the contribution of power to meet a particular load requirement is split between two energy storage types. W_1 and W_2 represent the weighing factors corresponding to the proportion of energy extracted from the two storage units. Due to the difference in Power to Energy ratios (P/E) of Type 1 and Type 2 systems, a strategy and infrastructure to coordinate power flow by dynamically varying the weighting factors is required. For successful operation of the vehicle, the power availability must always meet the power requirement. This has to be done with further consideration to the system constraints, for example the depletion level of the energy storage units.

This vexing issue of controlling power flow of two or more sources has been addressed through various approaches. Jalil *et al.* [5] suggested a rule-based framework for power split between a battery pack and ICE. The proposed strategy ensured that both power sources operate at maximum efficiency whenever possible and demonstrated an increase in energy efficiency. Steinmauer and del Re [11] however stated that techniques that use a fixed controller structure and then searches for optimal parameters to minimise a cost function yields only a solution that is a consequence of the selected structure. They proposed to tackle the dual source power split problem in terms of optimal control using statistical data of vehicle power demands for known drive cycles. According to Langari and Won [6], optimal control methods, due to its dependency on the drive cycles used to generate the control actions may not yield optimal power split for misclassified or arbitrary drive cycles. As an alternative, they proposed a concept of a fuzzy logic (FL) based energy management to capture driving situational awareness. Similarly, Hellgren and Jonasson [3] studied the comparison of a fuzzy logic approach and an analytical formula for a hybrid powertrain. Their findings showed that the FL method proved more flexible but required three times as many design variables. West *et al.* [12] introduced a Model Predictive Control (MPC) method to coordinate the power flow from two sources in a pure electric vehicle. Employing a constrained MPC with zone control, they demonstrated that the net energy expenditure of a battery bank in a battery-ultracapacitor system was significantly less compared to a DC-Link voltage control method. Leading more towards the

1-4244-0448-7/06/$25.00 ©2006 IEEE

practical implementation of power split strategies, which require instantaneous management of power flow, Paganelli et al. [9] introduced a general supervisory control policy for charge-sustaining HEVs.

Baisden and Emadi [1] demonstrated a control strategy based on selecting operating modes of a DC-DC converter to determine the power split between an ultracapacitor bank and battery pack. Through simulations of predefined drive cycles, Baisden showed that this hybridisation of battery and ultracapacitors allowed the battery pack to be downsized to 70%. Although ultracapacitors were added to the energy storage system, the significant reduction in battery mass plus the increase in battery life justified the addition of 35 ultracapacitor cells.

As with practical scenarios, the complete mission profile is not known *a priori*. To this end, recent research efforts [2,8,10] tend to focus on more implementable methods. The problem of integrating multiple energy storage systems then requires the appropriate power electronics infrastructure and a causal controller to determine the optimised power split. Non-causal methods using previous and future information of the power demand trajectories would yield the optimum instantaneous power split and maximum end SoC. However, for practical applications, suboptimal but implementable methods are required.

In our progressive research, we investigate EV power and energy management policy implementations that do not have load demand profile for the complete drive cycle as a priori information. In order to develop the policies, we define and represent the operating modes of the dual energy storage systems as having a finite number of states. Transitions between states and the power split are determined based on the vehicle speed, target SoC and DC link variation.

II. EXPERIMENTAL ELECTRIC VEHICLE

In order to obtain more realistic data of power demand dynamics and identify the practical limitations when designing power management systems, a test vehicle was developed for this work. With successive tests, beginning from a battery - alone configuration, we then progressively develop power and energy management policies.

The experimental vehicle as illustrated in Figure 2 consists of a modified go-kart chassis propelled by a 10kW Lynch motor, driven by a 4-quadrant DC Motor controller. The power-pack units comprises of deep discharge SLA batteries and self-balancing ultracapacitor modules. Instrumentation and control is designed around the National Instruments Compact Field Point (CFP) architecture. This system was chosen due to its robust operating condition tolerance and flexibility of

developing real-time control process treads. To facilitate online monitoring of the vehicle during test runs, a wireless link was added to transmit system parameters to a Lab View visualisation environment running on a remote PC.

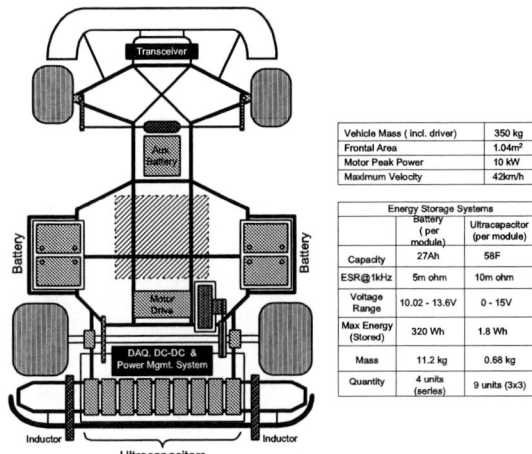

Vehicle Mass (incl. driver)	350 kg
Frontal Area	1.04m²
Motor Peak Power	10 kW
Maximum Velocity	42km/h

Energy Storage Systems		
	Battery (per module)	Ultracapacitor (per module)
Capacity	27Ah	58F
ESR@1kHz	5m ohm	10m ohm
Voltage Range	10.02 - 13.6V	0 - 15V
Max Energy (Stored)	320 Wh	1.8 Wh
Mass	11.2 kg	0.68 kg
Quantity	4 units (series)	9 units (3x3)

Figure 2: Experimental vehicle setup

A. System architecture

To implement a control intervention of battery and ultracapacitor power flow, a dual source power electronics converter architecture based on single stage buck-boost topology was designed. During an unloading (discharging) process, the converter on either side is in boost mode whilst during loading (charging), the converter operates in buck mode. With reference to Figure 3, the Vehicle Power and Energy Management (VPEM) block functions as the decision-making mechanism that provides reference power levels to a power stage transfer function block. Reference duty cycles generated are then fed to the DCDC converter. It should be noted that the power stage block performs the necessary voltage and current regulation.

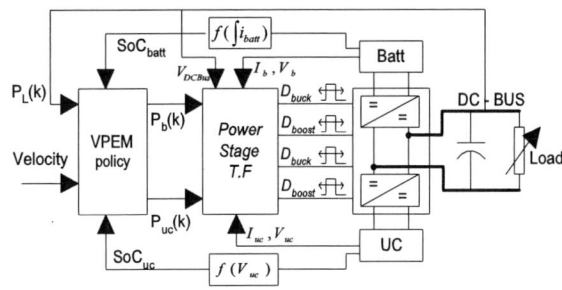

Figure 3: Overall system architecture

B. Dual source Power and Energy level boundaries

The characteristic constraints of ultracapacitors and batteries lead to the definition of a 'usable energy' level for the ultracapacitors and a 'usable power' level for the batteries. The battery performance under increasing power levels will deteriorate and influence the energy capacity [7]. The effect can be seen in the Figure 4 as a reduction in energy for increasing power levels particularly if the power level is higher than the specified Ah rating. Global efficiency of the battery may be defined as the ratio of the energy extracted form the battery to the energy required to recharge the battery on the basis of a full charge/discharge cycle [2].

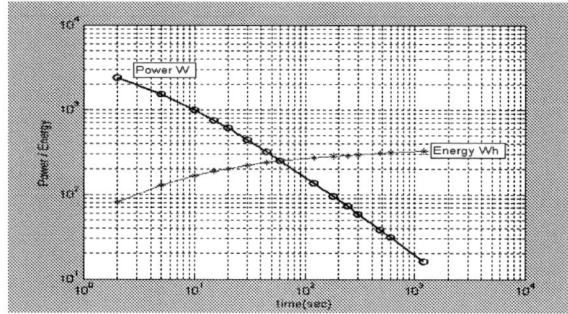

Figure 4: Plot indicating lower available energy as power increase
(Empirical Data from Odyssey®SLI)

The VPEM policy is non-charge sustaining, which means that the initial SoC of the battery is not constrained to equal the end SoC after a mission profile duration N. That is:

$$E_{batt}(t_0) \neq E_{batt}(t_N) \tag{1}$$

The objective rather is to extend the vehicle run time by operating the battery at a high global efficiency. Limiting both the maximum battery power and the step change in power has the positive effect of extending the run time and the long-term life cycle of the battery [7].

The lead acid battery pack consists of four Odyssey® SLI Drycell™ batteries. The batteries are specified to operate until a minimum terminal voltage of 10.02V. The ultracapacitor pack consists of nine Maxwell® BMOD0350-15 modules with non-dissipative module-to-module series charge balancing circuitry. The ultracapacitor modules are configured as a 3x3 matrix, effectively creating a 58F / 45V pack with a maximum energy storage capacity of 16.3Wh (58.7kJ).

Since the available energy bandwidth of the ultracapacitor bank is bounded by the maximum (V_{max}) to minimum (V_{min}) terminal voltage swing, the usable ultracapacitor energy E_{UC} reduces according to

$$E_{UC} = \left(1 - \frac{V^2_{min}}{V^2_{max}}\right)E_{max} \tag{2}$$

where E_{max} is the maximum storable energy of the ultracapacitor pack. Using Equation (2), with a minimum ultracapacitor terminal voltage of 20V, the usable ultracapacitor energy of the system becomes 13.1Wh (47.1kJ). This would normally translate to the capability of delivering maximum traction power of 10kW for approximately 4.7seconds. In practice, this would only be valid if the DC-DC converter has the corresponding current handling capacity, in this case 400A.

C. *Vehicle and energy system model verification*

A theoretical model of the experimental vehicle was developed to test VPEM policy concepts before final implementation. This simulation model is based on fundamental vehicle kinetics [4] to capture the tractive effort and tractive power while considering only the vehicle longitudinal dynamics. As depicted in Figure 5,

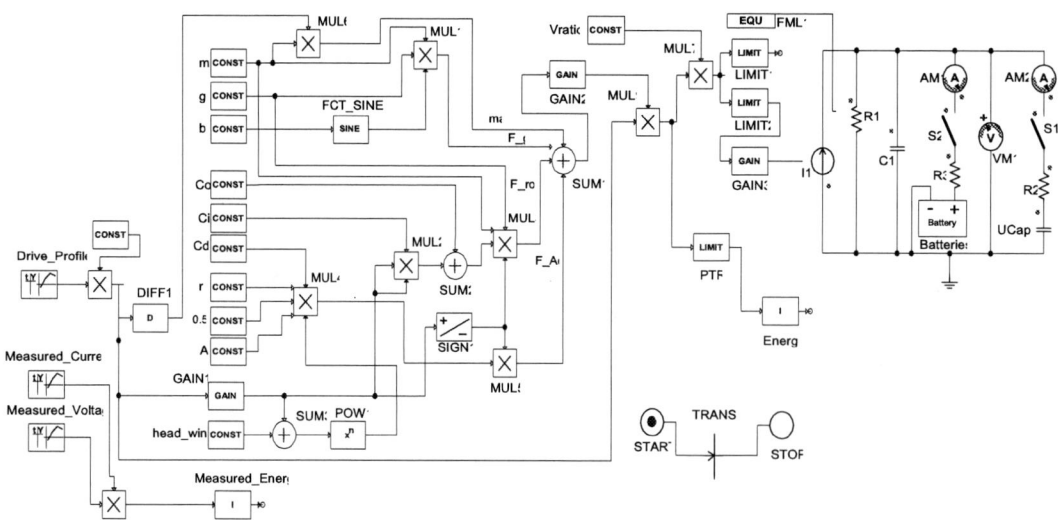

Figure 5: Vehicle and energy system simulation model

the tractive power requirement, calculated from tractive force and velocity is modelled as a varying load resistor and a current source on the DC BUS. The tractive force is calculated as:

$$F_{TR} = F_{la} + F_{gxT} + F_{roll} + F_{AD} \qquad (3)$$

$$F_{TR} = ma + mg \sin \beta + \text{sgn}[\, v_{xT}\,]mg\,(C_0 + C_1 v_{xT}{}^2) \\ + \text{sgn}[\, v_{xT}\,]\{0.5\rho C_D A_F (v_{xT} + v_0)^2\} \qquad (4)$$

where F_{la} is the linear acceleration force, F_{gxT} is the gravitational force, F_{roll} is the rolling resistance force, and F_{AD} is the aerodynamic drag force. m is the total vehicle mass, g is the acceleration due to gravity (9.81m/s²). C_0 and C_1 are rolling resistance coefficients with reference values from [2]. β is the grade angle which is set to zero in this work. ρ is the air density, C_D is the aerodynamic drag coefficient, A_F is the vehicle equivalent frontal area and v_0 is the head wind velocity. v_{xT} represents the vehicle tangential velocity.

With the tractive force F_{TR}, the instantaneous tractive power is

$$P_{TR}(t) = F_{TR}(t) \cdot v(t) \qquad (5)$$

Considering only the propulsion load, the energy storage system is required to meet this instantaneous load demand throughout the drive cycle. The next section reports experimental results to validate the vehicle and battery model.

III. EXPERIMENTAL RESULTS

A 600 second test cycle was carried out to obtain actual propulsion power profiles. This baseline test was conducted with only the battery pack servicing the load. Vehicle velocity, battery current and battery voltage was acquired and compared to a simulation model of the vehicle. Through parameter extraction, measured data was used to tune the simulator. Results indicate a good agreement between simulated and measured battery current and voltage as shown in Figure 6. The experimental data also provides useful indication of the vehicle power demand dynamics and energy consumption.

IV. POWER AND ENERGY MANAGEMENT POLICY

The power and energy management policy is built upon on a kinetic to electric energy balance correlation. To be concise, the 'power' portion of the policy is designed to ensure that the load power demands are always met and the 'energy' portion regulates the SoC of the ultracapacitor bank as a function of the vehicle velocity. For a typical power demand profile as in Figure 7, the operating modes and power proportioning of the dual energy systems depends on load power level and rate of change. The VPEM policy automates the state selection

and power proportioning for a given discretised power demand.

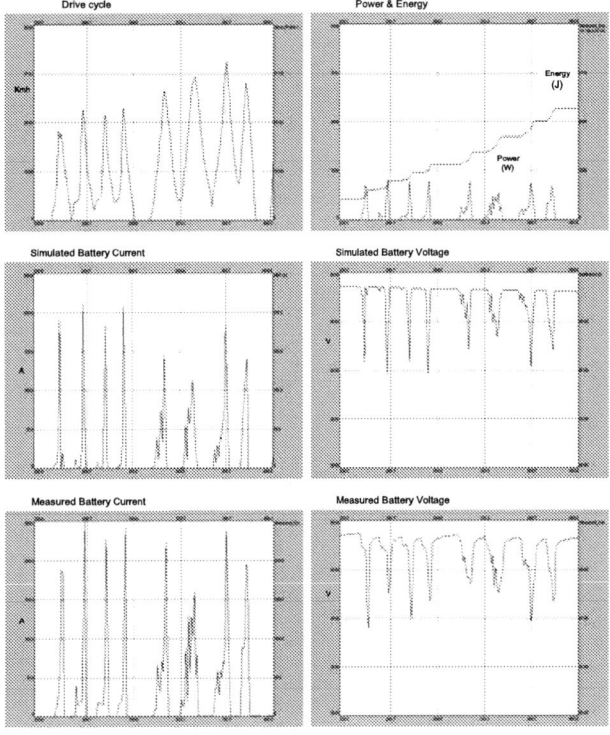

Figure 6: Experimental and simulated battery current and voltage
(t=230sec to t=400sec of a 600sec test)

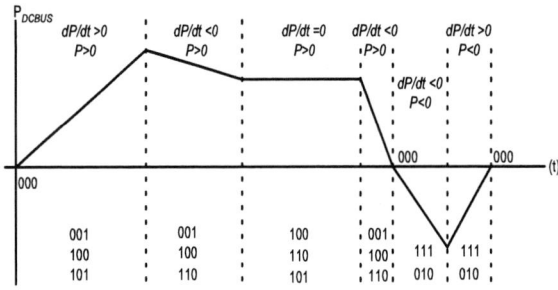

Figure 7: Segments of a typical EV power demand profile

The decision mechanism behind the policy implementation is shown in Figure 8 as state transitions with state description given in Table 1. It is imperative to note that the DC bus capacitor is sized sufficiency large to service any load request during the discrete 'sample and decide' time window.

481

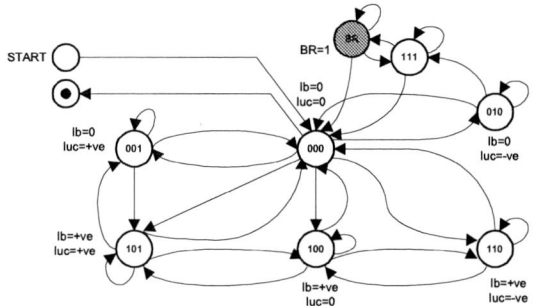

Figure 8: Power and Energy management policy states

State	Batteries	Ultracaps	Operating condition
000	Quiescent / Transition	Quiescent / Transition	Zero load condition or zero battery/ ultracapacitor current crossing to next state
001	Quiescent	Discharging	Conditions require only the ultracapacitors to service the requested load demand.
100	Discharging	Quiescent	Batteries are discharging within operating constraints and supercapacitor is at its target SOC
101	Discharging	Discharging	Both batteries and ultracapacitors are discharging within specified maximum discharge rate and power level
110	Discharging	Charging	Batteries are servicing all load demands and charging the ultracapacitors
010	Quiescent	Charging	Ultracapacitors are charging via regenerative DC Bus power
111	Charging	Quiescent	Ultracapacitors are fully charged and surplus regenerative power is diverted to batteries at limited charging rate
BR	*Quiescent*	*Quiescent*	*Atypical condition requiring activation of dissipative brake resistor for failsafe operation*

Table 1 Description of operating states

The operating states of the batteries and ultracapacitor can only transit between seven normal operating states. The additional state (BR) only occurs when the ultracapacitors are fully charged and the batteries are unreceptive to regenerative power. In such an event, the DC bus voltage rise is limited by the activation of the dynamic brake resistor.

The policy is designed to satisfy the discretised power balance equation

$$P_L(k) = P_b(k) + P_{uc}(k) \quad \forall k \tag{6}$$

where P_L is the load power demand, P_b is the battery power and P_{uc} is the ultracapacitor power.

To generate the reference ultracapacitor SoC, we assume a linear relation between the vehicle speed and the ultracapacitor target SoC according to the stored electrical energy in the ultracapacitor and the kinetic energy of the vehicle. This is expressed as

$$E_{uc} + E_{kin} = K \tag{7}$$

Thus,

$$(0.5CV_{uc}{}^2) + (0.5mv_T{}^2) = K \tag{8}$$

where C is the capacitance, V_{uc} is the ultracapacitor voltage, m is the vehicle mass and v_T is the velocity. Since the state of charge of the ultracapacitor is directly proportional to the square of its voltage, it is more useful to express the reference state of charge in terms of a reference ultracapacitor voltage. Thus,

$$V_{uc}ref = \lambda + \sqrt{\frac{2K - mv_T{}^2}{C}} \tag{9}$$

The λ term is added as an offset to shift the maximum $V_{uc}ref$ to the maximum design voltage of the ultracapacitor bank. From experimental results, the maximum achievable vehicle velocity of 11.11m/s was verified. With this, the computed design values are $K = 21605$ and $\lambda = 18$. The discrete time intervals used in Equation (6) is intentionally excluded in Equation (9) because the SoC (Energy) balance loop does not run at the same iteration rate as the fast power demand loop.

Whilst the detailed descriptions of all states are very involved, the mechanism behind one state (101) is illustrated as a flowchart in Figure 9. In the diagram, $P_b max$ is a function of P_L and the battery SoC and implicitly performs as a varying threshold and step change limiter on the battery power-proportioning ratio. Thus,

$$P_b max = f(P_L(k), SoC_b(k)) \tag{10}$$

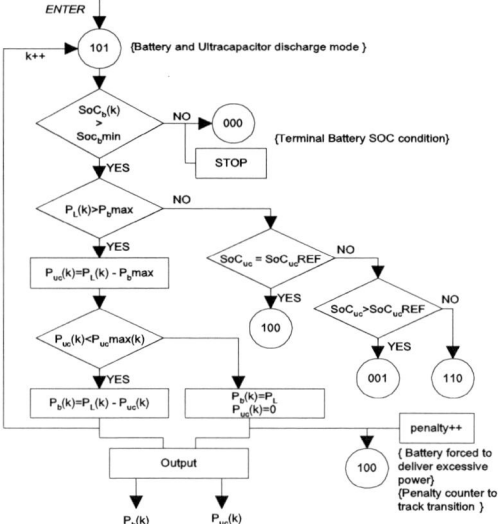

Figure 9: Flowchart of state (101)

In states (101) and (100) conditions exist to overwrite the battery power limits and deliver all the required power if necessary. As this is an unfavourable condition, a penalty counter is included to track these occurrences. The penalty accumulator serves as a quick policy performance quantifier. In Equation (11), $P_{uc}max$ is the product of the

482

maximum ultracapacitor current I_{uc}max and the measured ultracapacitor voltage V_{uc}. By design, I_{uc}max is limited to the DC-DC converter inductor rating of 250A.

$$P_{uc}\max(k) = \underbrace{I_{uc}\max}_{fixed} \cdot \underbrace{V_{uc}(k)}_{measured} \qquad (11)$$

The output of each active state generates corresponding values of P_b and P_{uc} at each loop iteration, which is then transformed to duty cycle values.

V. CONCLUSION

An implement-able power and energy management policy based on experimentally verified parameters has been outlined. Segmenting the operating modes of the battery and ultracapacitor leads to the development of a state transition mechanism that services the power demands while attempting to manage the energy balance. The framework is realisable for online/onboard computation of the reference power ratios. Future work focuses on hardware in loop testing and policy verification.

REFERENCES

[1] A.C. Baisden, A. Emadi, "ADVISOR-based model of a battery and an ultra-capacitor energy source for hybrid electric vehicles",*IEEE Transactions on Vehicular Technology*, Vol. 53, Issue: .pp. 199 – 205. (2004)

[2] L. Guzzella, A.Sciarretta. "Vehicle Propulsion Systems", *Springer*. ISBN-10 3-540-25195-2. (2005)

[3] J. Hellgren, K. Jonasson, "Comparison of two algorithms for Energy management of hybrid Powertrains", Lund University, (http://www.elkraft.ntnu.no/norpie/10956873/Final%20Papers/060%20-20 Jonasson_Hellgren_060_2.pdf (Nov.2004)

[4] I. Husain. "Electric and Hybrid Vehicles", *CRC Press.* ISBN 0-8493-1466-4 (2003)

[5] N. Jalin, N. A. Kheir, M. Salman, "A Rule-Based Energy Management Strategy for a Series Hybrid Vehicle", *Proceedings of the American Control Conference,*. Vol 1, pp. 689-693. (1997)

[6] R. Langari, J. S.Won, "Integrated Drive Cycle Analysis for Fuzzy Logic Based Energy Management in Hybrid Vehicles", *IEEE International Conference on Fuzzy Systems*, Vol. 1, pp. 290 – 295.(2003)

[7] J. Larminie, J. Lowry. "Electric Vehicle Technology", Wiley. ISBN 0-470-85163-5 (2003)

[8] J. Moreno, J.Dixon, M.Ortúzar. "Energy management system for an Electric Vehicle using ultracapacitors and neural networks", *IEEE Vehicle Power and Propulsion.* (2004)

[9] G. Paganelli, G. Ercole, A. Brahma, Y. Guezennec, G. Rizzoni. "General supervisory control policy for the energy optimization of charge-sustaining hybrid electric vehicles", *ELSEVIE RJSAE 22*,pp.511-518.(2001).

[10] P.Pisu, G. Rizzoni. "A supervisory control strategy for series hybrid electric vehicles with two energy storage systems", *IEEE Vehicle Power and Propulsion*, pp. 65-72. (2005).

[11] G.Steinmauer, L. del Re, "Optimal control of dual power sources", *Proceedings of the 2001 IEEE International Conference on Control Applications*, pp. 422 – 427(2001)

[12] M.J . West, C. M. Bingham, N. Schofield, "Predictive control for energy management in all/more electric vehicles with multiple energy storage units", *IEEE International Electric Machines and Drives Conference*, Vol. 1, pp. 222 – 228.(2003)

2006 5th International Power Electronics and Motion Control Conference

Study on Non Contact Automatic on-Load Voltage Regulating Distributing Transformer Based on Solid State Relay

Zhao-Yulin*, Dong-Shoutian*, Li-Jiahui**, Yao-Xin*, Zheng-Na*and Liu-Xueli***

* Northeast Agricultural University, Harbin, China
** Harbin University of Science and Technology, Harbin, China
***Shanghai Jiao Tong University,Shanghai,China
e-mail: zyl5631@163.com

Abstract—**Automatic on-load voltage regulation of transformer is an effective method to stabilize load voltage. Mechanic contact tap changing switch is not adequate for distributing transformer because of its high cost and low capability. The paper presents the structure and automatic on-load voltage regulating principle of distributing transformer, which employs solid state relay as non contact automatic on-load voltage regulating tap changing switch. The generation mechanism and limiting measure of circular current that is occurred in the process of switching tap joint and the harmonic suppressing method are also introduced. By theoretical analysis and experimental verification, it is concluded that it has different occurring process of circular current when voltage is regulated to higher or lower value. And it is proved that the application of power electronic component in on-load voltage regulation can not generate harmonic. The sample has run for 3 months in power network. It achieves good result. And it can meet the demand for system running.**

Keywords-distributing transformer; solid state relay; on-load voltage regulation; automatic control; tap joint; non contact

I. INTRODUCTION

Automatic on-load voltage regulation of transformer is an effective measure to stabilize the power network voltage. At present, automatic mechanic contact tap changing switch is widely used in power system. It can not operate frequently. In addition, arc will occur when the switch changes, so it must be installed in independent oil tank. Cooke G H solved the problem that arc would occur when the switch changes by using on-load voltage regulating tap changing switch in which thyristor as its auxiliary switch[1]. But the device also has mechanic part and can't operate frequently so the drawback still remains. The author of ref. [2] series connects the primary side of auxiliary transformer with the main

transformer. And he uses thyristor to change the secondary tap joint of auxiliary transformer to regulate voltage. It can operate quickly. And arc won't occur. But this method needs complex auxiliary transformer and large capacity thyristor. It adds costs of the transformer and limits the switch's application field. Therefore it is not fit for distributing transformer for its cost and performance.

The power electronic component can be easily controlled and it can switch quickly. In addition, arc will not occur in the process of switching. If it is used as on-load tap changing switch of transformer, the switch will has low cost and long lifetime and it can be regulated frequently. The paper presents a distributing transformer, which employs solid state relay as non contact automatic on-load voltage regulating tap changing switch. And the solid state relay is based on bidirectional thyristor.

II. MAIN CIRCUIT AND OPERATION PROCESS OF NON CONTACT AUTOMATIC ON-LOAD VOLTAGE REGULATING DISTRIBUTING TRANSFORMER

A. Main circuit of Non Contact Automatic On-Load Voltage Regulating Distributing Transformer

The device comprised of transformer, non contact tap changing switch, circular current limiting circuit and control unit. Its wiring diagram is showed in Fig. 1. w_{10} is the primary operating winding of transformer. It works during energization of power network. w_{11} and w_{12} are primary regulating winding. w_2 is secondary winding of distributing transformer. $SSR_i(i=1,2,3,4)$ is zero crossing type solid state relay. Under the action of controlling voltage, it can be on when AC voltage crosses zero and be off when AC current crosses zero. R_X is circular current limiting resistance. It is only connected to the circuit in the process of switching working tap joint. The function of single chip monitoring system is to monitor the output voltage of distributing transformer and to send controlling signal according to the prearranged program. CB is

1-4244-0448-7/06/$25.00 ©2006 IEEE

measuring transformer. It offers power and measuring signal for monitoring system.

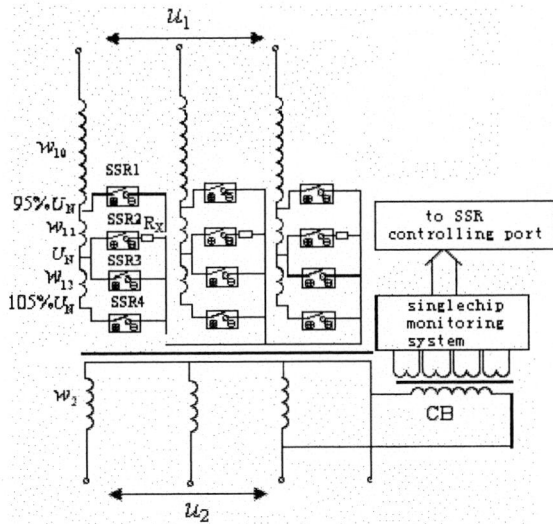

Figure 1   Main circuit of non contact automatic on-load voltage regulating distributing transformer

B. Automatic Voltage Regulating Process of Automatic On-Load Voltage Regulating Distributing Transformer

The storage battery supply power for monitoring system before the distributing transformer is powered on to make SSR3 in on state. When the transformer is switched on, the output voltage of transformer is detected by singlechip monitoring system automatically. If the voltage exceeds permitted range of fluctuation, the automatic on-load voltage regulating program will be operated. The flow chart of automatic on-load voltage regulation is shown in Fig. 2.

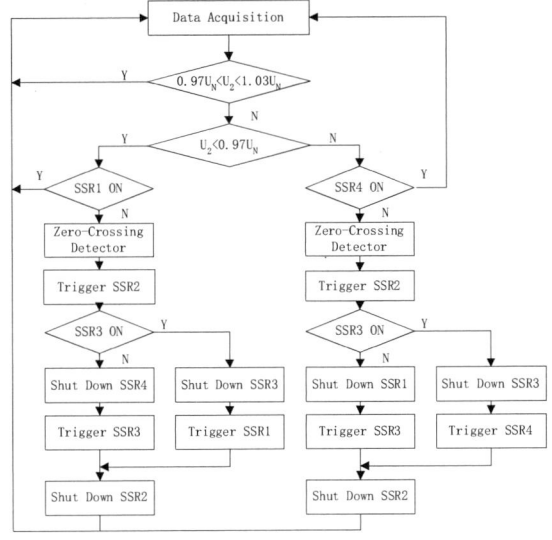

Figure. 2   Flow chart of automatic on-load voltage regulation

It can be seen from Fig.2 that it connects the current limiting resistance R_x to the tap joint switching circuit firstly, and then switches the tap joint. It can limit the value of circular current.

III.   THEORETICAL ANALYSIS OF REGULATING WINDING'S CONDITION WHEN TAP JOINT IS CHANGED

It can be seen from the operating sequence of each solid state relay in the process of switching tap joint that the regulating winding and solid state relay will make up of closed loop in this process. And the circular current will occur in the loop. The solid state relay will be damaged if the value of circular current excess permitted value. The value of circular current is related to the impedance of the loop and the power factor $\cos\varphi_1$ of transformer's primary side. $\cos\varphi_1$ is smaller as the value of circular current is bigger. $\cos\varphi_{1\min} = 0.2, namely \varphi_{\max} = 78.5°$.

The sample is S_9—50kVA/10kV energy saving distributing transformer. There are 3 tap joints $95\%U_N, U_N$ and $105\%U_N$ at primary side. They are near neural point. When the line voltage of transformer's primary side is rated voltage 10kV, current is rated current 2.75A, SSR3 is on, other SSRs is off and w_{10} and w_{11} are connected in series as working winding, then line voltage of transformer's secondary side is rated voltage 400V and it work under rated condition. The transformer's short circuit voltage percentage $U_k\% = 4$, short circuit loss $\Delta P_k = 1150$ W, no load current $I_0\% = 2.5$, no load loss $\Delta P_0 = 190$ W. It can be calculated that the equivalent resistance which is converted to the transformer's primary side $R_T = 23\Omega$, equivalent impedance $X_T = 32.7\Omega$.

It is assumed that the transformer works at tap joint U_N and current limiting resistance $R_X=0$. Then we analyze the mechanism of the circular current which is occurred in the process of switching tap joint.

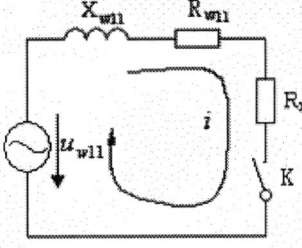

Figure. 3   Equivalent circuit in the process of switching tap joint

If the output voltage of transformer is lower than permitted value, the working tap joint will switched from $U_N to 95\%U_N$. During the period from turning on the SSR1 to shutting off the SSR2, the closed loop is composed by regulating winding w_{11}, SSR1 and SSR2. And circular current i is occurred. i can be calculated by the equivalent circuit which is shown in Fig. 3. The circuit is also adequate for the condition that tap joint is switched from $105\%U_N$ to U_N. X_{w11} and R_{w11} is equivalent impedance and resistance of regulating winding w_{11} at transformer's primary side respectively. K is equivalent switch of SSR1. And u_{w11} is equivalent power source voltage of the circuit. It is assumed that the winding is uniform. Then according to the relation that winding's resistance is proportional to its turns and impedance is proportional to the square value of its turns, and w_{10} and w_{11} is in series connection, it can be calculated that $X_{W11} = 0.82\Omega$ and $R_{W11} = 1.15\Omega$ [4]. u_{w11} is approximate to 5% of primary side phase voltage $u_{1\varphi}$ $(=8164\sin\omega t)$, namely $u_{w11} \approx 408\sin\omega t$. The voltage equation of the loop in Fig. 3 is[5]:

$$iR_{W11} + L_{W11}\frac{di}{dt} = 408\sin\omega t \qquad (1)$$

The solution of (1) is:

$$i = i' + i'' \qquad (2)$$

Where i' is stable component of loop current. If $R_X = 0$ then

$$i' = \frac{U_m}{\sqrt{R_{W11}^2 + X_{W11}^2}}\sin(\omega t - tg^{-1}\frac{X_{W1}}{R_{W11}}) \qquad (3)$$
$$= 289\sin(\omega t - 35.49°)$$

i'' is transient component of loop current:

$$i'' = Ae^{-\frac{t}{\tau}} \qquad (4)$$
$$= Ae^{-440.37}$$

So

$$i = i' + i''$$
$$= 289\sin(\omega t - 35.49°) + Ae^{-440.37t} \qquad (5)$$

Where A is relatively to the power factor angle φ and current value of primary side which are the value that tap joint is not switched. If $\varphi = \varphi_{max} = 78.5°$, $i_1(0^-) = -3.81A$ when $t = 0^-$. The direction of i is contrary to that of $i_1(0^-)$ in Fig. 3.

So at the time of $t = 0$

$$3.81 = -289\sin35.49° + A$$
$$A = 171.59$$

So

$$i = i' + i''$$
$$= 289\sin(\omega t - 35.49°) + 171.59e^{-440.37t} \qquad (6)$$

For observing the variation of current that is through SSR2 when SSR1 is on and off conveniently, it draws the

current $-i_1$, which is through SSR2 by broken line. The direction of the current is the same as that in Fig. 3. The circular current wave according to (6) is shown in Fig. 4.

It can be seen from the figure that after SSR1 is on at the time of $t = 0$, value of the current which is through

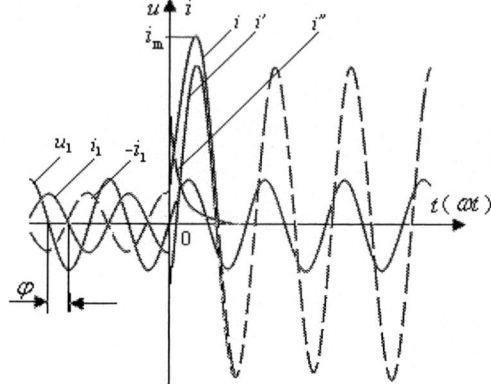

Figure. 4   Circular current waveform in the process of

regulating voltage to higher value

SSR2 (its direction is the same as that in Fig. 3) is increasing from 3.81A, and it be smaller gradually after it reaches i_m. And SSR2 is off at the time of $i = 0$. The overcurrent will occur before SSR2 is shut off in the process of voltage is regulated to a higher value.

If the value of output voltage is higher than permitted value, the working tap joint should be regulated to $105\%U_N$. The equivalent circuit in Fig. 3 can be used to calculated circular current i too. K is equivalent switch of SSR4. The circuit parameters adopt parameters of w_{12}. If $R_X = 0$, the value of circuit parameters are not changed because $w_{11}=w_{12}$. Thus the equation of circular current is the same to (5). And $i_1(0^-) = -3.81A$. Under this condition, the current direction of SSR2 is not changed whether SSR4 is on or off. Therefore at the time of $t = 0$

$$-3.81 = -289\sin35.49° + A$$
$$A = 164.97$$
$$i = i' + i''$$
$$= 289\sin(\omega t - 35.49°) + 164.97e^{-440.37t}$$

The waveform is shown in Fig. 5.

It can be seen from Fig. 5 that circular current i (the current through SSR2) changed from −3.81A to zero crossing point quickly after SSR4 is on at the time of $t = 0$. It is different from the process of regulating tap joint for higher the voltage value. SSR2 is off at the time of $i = 0$. The circular current i doesn't reach its maximum value before SSR2 is off. There are not big circular current occur in the switching process.

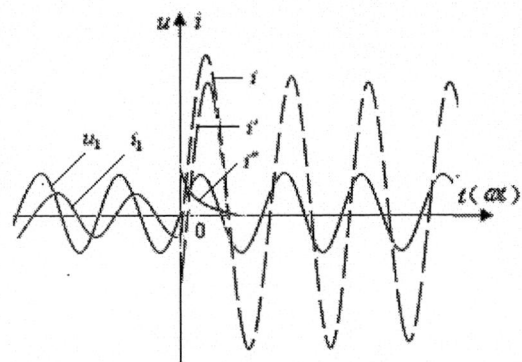

Figure. 5 Circular current waveform in the process of regulating voltage to lower value

It can be concluded that the occurring mechanism of circular current in the process of regulating output voltage to higher value is different from that in the process of regulating output voltage to lower value. In the former, circular current is occurred by overcurrent in the process of switching tap joint. And the overcurrent value is in relation to impedance of the circuit. In the latter the overcurrent won't occur in the process of switching tap joint whatever the value of circuit impedance is. And the conclusion is verified in laboratory.

AC voltage regulator applies voltage to sample transformer with several voltage levels. And each voltage level is 1kV. It begins with 0V. The output voltage range of the voltage regulator is 0~11kV. During the period of switching tap joint to regulate the transformer's output voltage to higher value, when the applied voltage of primary side is 3kV, overcurrent will occur and fuse will blow. And during the period of switching tap joint to regulate the transformer's output voltage to lower value, overcurrent is not occur even primary side voltage reaches 11kV.

Because overcurrent will occur and its value is in relation to circuit resistance, current limiting resistance must be connected to the circuit during the process of switching tap joint. That is R_X can't be zero.

The connection of R_X reduces amplitude of circular current's stable component and time constant of the circuit. It quicken the attenuation of transient component i'' of circular current i and diminish the influence of i'' on overcurrent amplitude i_m.

If $R_X=6.8\,\Omega$ then $\tau = 0.227\text{ms}$. It indicates that the transient component is attenuate to zero after 1.6ms. The magnitude of $|i_m|$ is determined by stable component. Then

$$i' = \frac{408}{\sqrt{7.95^2 + 0.82^2}}\sin(\omega t - tg^{-1}\frac{0.82}{7.95})$$
$$= 51.05\sin(\omega t - 5.89°)$$
$$i_m = 51.05 \text{ A}$$

The sample transformer adopts 40A solid state relay. Its permitted maximum current is 200A in half cycle.

And its safety factor is approximate to 4. It is proved that the choice can meet the demand of safety by experiment and sample operation.

IV. HARMONIC AND ITS SOLUTION

The solid state relay has non-linear characteristics if its applied voltage is less than 2V. The solid state relay is not on near zero point if controlling pulse doesn't trigger synchronously at the time of AC voltage crosses zero. It causes AC voltage is not continuity thus harmonic is generated.

The harmonic can be generated only at the range of $\pm 0.014°$ near AC voltage zero crossing point. Its value is very small and can be ignored. Solid state relay which is on when voltage crosses zero can be on continuously if the controlling voltage is always applied, and the harmonic can not generate.

V. EXPERIMENT AND OPERATION RESULT

A. Experiment

The input voltage of sample is regulated by 0~11kV AC voltage regulator. If the input voltage is changed in the range of 9kV~11kV, with the tap joint is regulated on-load automatically by singlechip monitoring system, the range of output voltage is 379V~419V. If there are no on-load voltage regulation, the range of output voltage is 360V~440V. Thus the device is effective to stable voltage.

The output voltage waveform of sample during the period when it operates with load is shown in Fig. 6. The waveform is recorded by DS5202CA type digital oscillograph. It can be seen from the figure that the non contact automatic on-load voltage regulating distributing

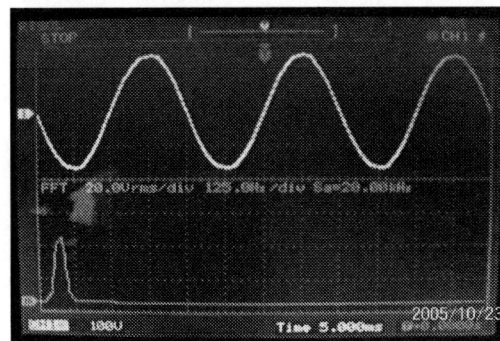

Figure. 6 Output voltage waveform of automatic on-load voltage regulating distributing transformer

transformer can not generate harmonic pollution.

The primary side voltage of transformer is regulated at the range of 9kV~11kV, and SSR is repeating switched, SSR is not damaged because of overcurrent. After simulated operation in laboratory for 2 months, SSR is not damaged. Thus it can be concluded that adopting $6.8\,\Omega$ current limiting resistance can meet the demand for limiting current. And the reliability of SSR can meet the demand for transformer.

B. Operation Result

The sample begins to switch tap joint if its output voltage exceed the range of 380~412V. It began to work in power system at Binxian, Heilongjiang province, China on 16:00, Oct. 24 2005. The output voltage variation of the sample on Oct. 24 and Oct. 26 is shown in Fig. 7.

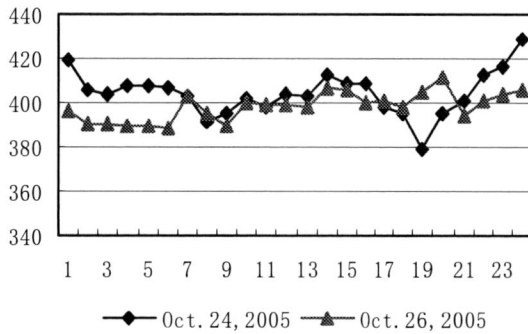

Figure. 7 Comparison of output voltage variation on Oct. 24 and Oct. 26

It can be seen from the figure that the output voltage of transformer is higher than permitted value because of power network voltage is increasing and load is lessening. The tap joint switched to $105\%U_N$ to lower the output voltage and stabilize it. In no-load voltage regulation, the range of output voltage variation of transformer is 428.6~376.6V. And it is 412~388V in on-load voltage regulation. The voltage quality is enhanced after adopting on-load voltage regulation. It is an effective method to stabilize voltage in the area in which voltage fluctuates heavily if adopting on-load voltage regulation in distributing transformer.

VI. CONCLUSION

It is proved that the automatic on-load voltage regulation is an effective method to stabilize load voltage by theoretical analysis and experiment. Adopting SSR as on-load tap changing switch can not generate harmonic. The variation rule of circular current is different between the process of regulating voltage to higher and lower value. In the former process, the circular current crosses zero after its amplitude reaches the peak value so the current limiting resistance must be series connected to the circuit. While in the latter process, the circular current crosses zero before its amplitude reaches the peak value.

Therefore it needn't connect with current limiting resistance. The value of current limiting resistance can be determined according to the criterion that suppresses the amplitude of circular current stable component near rated value of SSR.

ACKNOWLEDGMENT

The project is sponsored by Heilongjiang science and technology office (the contact number is GC05A312) and Harbin science and technology office (the contact number is 0011211098). Heilongjiang Binxian Power Company provided help for sample operation. And Harbin zhenhua electrical joint-stock limited company provided help for transformer body design. The authors would like to express their sincere thanks to them their imbrues.

REFERENCES

[1] Cooke G. H. et al. "New Thyristor Assisted Diverter Switch for On Load Transformer Tap Changers," *IEE Proceedings.* Vol. 139. No. 6. pp. 507-511. September 1992

[2] Guorong Zhu. Minzu Li, Xiaodong Liu and Tingting Wang "Voltage Regulation Method in Series Thyristor of Distribution transformer transformer," *Transformer.* Vol. 22. No. 5. pp. 1001-8425. May 2002

[3] Jiansheng chang "Detect and Transform Technique," *Mechanics and Industry Publishing Company Beijing,* pp64, 2004

[4] Heyangzan, and Wenzengyin, "Power System Analysis," *Huazhong University of Science and Technology Publishing Company,* pp.16-18, June 1984

[5] Yudaguang. "Fundamentals of Electric and Electronic Engineering." *People Education Publishing Company,* vol. □, pp. 108~114, 1965

BIOGRAPHY

Zhao-Yulin, male, born in 1956, Professor, main areas: power system automation

Li-Jiahui, male, born in 1978, Ph.D., main areas: electrical engineering

2006 5th International Power Electronics and Motion Control Conference

The Principle of a Novel Arc-suppression Coil and its Implementation

Cheng Lu, Chen Qiaofu, Zhang Yu, Zhang Changzheng
Dept. of Electric Machinery
College of Electrical and Electronic Engineering
Huazhong University of Science and Technology
Wuhan, Hubei, 430074, P.R.China
e-mail: chemlu_1@163.com

Abstract—In this paper, a novel arc-suppression coil based on transformer theory is proposed. According to this theory, the reactance of the arc-suppression coil can be controlled exactly by controlling the current of secondary winding of the reactor, which is implemented by a single phase inverter. This fire-new arc-suppression coil has some excellent characteristics such as linear regulation capacity and few harmonics pollution. Another feature is that the capacitance of power network can be detected real time, which makes automatic tuning of the arc-suppression coil available. These features make this type arc-suppression coil competitive comparing with others in industrial applications. An apparatus is built up in terms of the theory and its structure will be introduced in the paper. In the last section, the experimental results are presented to prove the feasible of the theory and the validity of control system.

Keywords-arc-suppression coil; transformer; current control; inverter; multiple secondary windings

I. INTRODUCTION

Arc-suppression coil is a widely used apparatus in 10-35KV level power network, which is employed to suppress the fault current between fault point and ground when one phase of the power utility grid connects to the ground. Arc-suppression coil is an adjustable reactor essentially. The reactance can be regulated automatically or manually when fault current varies. There have been developed many kinds of arc-suppression coil based on different principles[1][2], some typical methods are: (1)change winding turns of the reactor; (2)change the air-gap between winding and iron-core; (3)change the magnetizing current injected into the winding; (4)thyristor controlled reactor. There are corresponding problems in different kinds of arc-suppression coils. e.g. The reactance can not be adjusted continually in type(1); the response speed is not fast enough and the accuracy is not satisfactory in type(2); the reactance can not be regulated in a wide region in type(3); the power harmonics pollution is serious in type(4)[3][4]. These disadvantages limit the performance of arc-suppression coil, therefore other methods are chosen to solve the

problem in some countries when single phase short circuit fault occurs.

This paper has brought forward a new type arc-suppression coil based on transformer theory. It's novel either in principle or in implementation comparing with aforementioned ones. According to the theory, the equivalent reactance of the transformer can be regulated linearly by controlling the current injected into the transformer's secondary winding in specific proportion to the current in primary winding, which can be implemented by a voltage source inverter(VSI) connected to the secondary winding of the transformer. This type of arc-suppression coil has excellent characteristics such as the reactance can be regulated in a wide region linearly without generating power harmonics and fast response speed, which make it competitive in industrial applications.

II. ANALYSIS OF TRANSFORMER WITH CONTROLLABLE CURRENT SOURCE LOAD

Fig.1(a) shows a simplified circuit of a transformer with current-source load. Fig.1(b) is the equivalent T-type circuit. The turns of primary and secondary winding of the transformer are N_1 and N_2, respectively; the turn ratio is represented by $k = N_1/N_2$. According to the transformer theory, the potential equations can be established as follow:

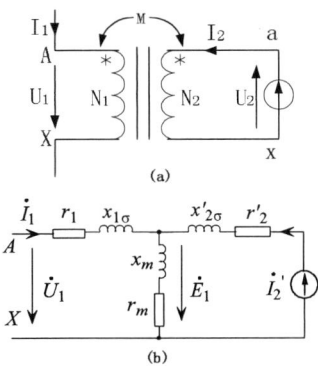

Figure 1. Transformer circuit with current source load(a) and its equivalent T-type circuit (b)

1-4244-0448-7/06/$25.00 ©2006 IEEE   489

$$\dot{U}_1 = (r_1 + jwL_{11})\dot{I}_1 + jwM\dot{I}_2 \quad (1)$$

$$-\dot{U}_2 = (r_2 + jwL_{22})\dot{I}_2 + jwM\dot{I}_1 \quad (2)$$

Here, r_1, L_{11}, r_2, L_{22} represent the resistance and the inductance of primary winding and secondary winding of the transformer respectively, M represents the mutual inductance between primary and secondary winding. Equation (1) and (2) can be rewritten as:

$$\dot{U}_1 = [r_1 + jw(L_{11} - kM)]\dot{I}_1 + jwkM(\dot{I}_1 + \dot{I}_2/k) \quad (3)$$

$$-\dot{U}_2 = [r_2 + jw(L_{22} - kM)]\dot{I}_2 + jwkM(\dot{I}_2 + \dot{I}_1/k) \quad (4)$$

If the secondary current $\dot{I}_2$ satisfies the condition below:

$$\dot{I}_2 = -k\alpha\dot{I}_1 \quad (5)$$

Here α is a real, denotes the proportion of the secondary current, its value region is [0,1].

Then (3) can be rewritten as:

$$\dot{U}_1 = [r_1 + jw(L_{11} - kM)]\dot{I}_1 + jwkM(\dot{I}_1 - \alpha\dot{I}_1)$$

$$= (r_1 + jwL_{1\sigma})\dot{I}_1 + jwL_m(1-\alpha)\dot{I}_1 \quad (6)$$

Here $L_{1\sigma} = L_{11} - kM$ is the leakage inductance of transformer's primary winding, $L_m = kM$ is the magnetizing inductance of the transformer. Then from terminals AX, the equivalent impedance of the transformer is:

$$Z_{AX} = \dot{U}_1/\dot{I}_1 = r_1 + jwL_{1\sigma} + (1-\alpha)jwL_m$$

$$= Z_1 + (1-\alpha)Z_m \quad (7)$$

In above equation, $Z_1 = r_1 + jwL_{1\sigma}$ is the leakage impedance of transformer's primary winding; $Z_m = jwL_m$ is the magnetizing impedance of the transformer.

It is worth noting that the equivalent impedance of the transformer (7) is a variable resting with coefficient α, when α changes from 0 to 1, the impedance will change from Z_1 to $Z_1 + Z_m$.

From above analysis, when the coefficient α varies, the transformer will exhibit an adjustable reactance. This feature makes it available in arc-suppression coil application. The current control of transformer's secondary winding can be implemented by a voltage source single phase inverter. The inverter works as a controllable load of the transformer essentially, variable input impedance can be gotten from terminals AX by controlling the current source load.

III. SYSTEM CONFIGURATION

Fig.2 shows the structure of the novel arc-suppression coil based on proposed principle in section II. One terminal of transformer's primary winding connects to the neutral point of power network. Another connects to the ground. The current of transformer's primary winding is detected and tracked by an inverter circuit, which makes the current in the transformer's secondary winding satisfying condition (5).IGBT is chosen to be the switching element, for its superiority in high switching

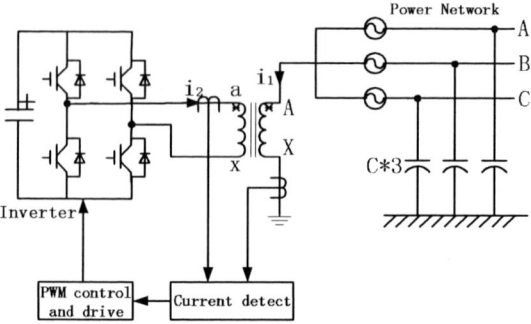

Figure 2.   Topology of the system

frequency, which makes the output current of the inverter containing fewer harmonics. However, the capacity of IGBT is not large enough comparing with thyristor. In order to solve this problem, a multiple secondary windings scheme is adopted. Its structure is shown in fig.3. All secondary windings are designed to have same shape and turns, so the parameters are nearly the same in different windings. Assume that the mutual inductance between the primary winding and each secondary winding are $M_1, M_2 ... M_N$. N represents the number of secondary windings. It can be considered that the mutual inductance is equal approximately:

$$M_1 \doteq M_2 \doteq ... \doteq M_N = M \quad (8)$$

If the currents injected into each secondary winding are $\dot{I}_{21}, \dot{I}_{22} ..., \dot{I}_{2n}$ respectively, then the potential equation is:

$$\dot{U}_1 = (r_1 + jwL_{11})\dot{I}_1 + jwM\dot{I}_{21} + jwM\dot{I}_{22} + \cdots + jwM\dot{I}_{2n}$$

$$= [r_1 + jw(L_{11} - kM)]\dot{I}_1 + jwkM(\dot{I}_1 + \frac{\dot{I}_{21}}{k} + \cdots + \frac{\dot{I}_{2n}}{k}) \quad (9)$$

If controlling:

$$\dot{I}_{21} = \dot{I}_{22} = \cdots = \dot{I}_{2n} = -\frac{k\alpha}{N}\dot{I}_1 \quad (10)$$

Combining (9) and (10), then：

$$\dot{U}_1 = (r_1 + jwL_{1\sigma})\dot{I}_1 + jwL_m(1-\alpha)\dot{I}_1$$

$$= [Z_1 + (1-\alpha)Z_m]\dot{I}_1 \quad (11)$$

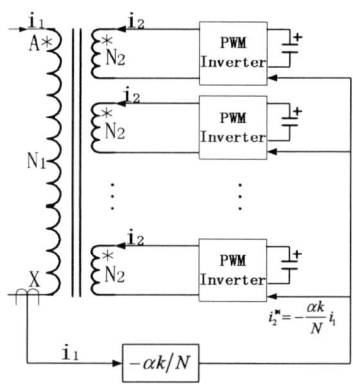

Figure 3.   Structure of multiple secondary windings arc-suppression coil

Comparing (7) and (11), it can be concluded that the equivalent impedance of the multiple secondary windings transformer is the same as single secondary winding transformer aforementioned. By adopting the multiple secondary windings topology, the current level of each inverter can be reduced, which is proved by comparing (10) with (5). The current of each inverter is in inverse proportion to the number of secondary windings. Accordingly, low capacity switching elements can be adopted to realize high capacity apparatus.

Another key point of arc-suppression coil is that the reactance should match the capacitance of power network, and then the arc-suppression coil can sharply reduce the fault current flowing through fault point to ground. So it's significant to detect the capacitance of power network accurately, which is denoted as C in fig.2. By adopting the topology as Fig.2 shown, the capacitance can be detected easily and exactly without adding additional apparatus. The principle is that: when there is no short circuit fault occurs in power network, the inverter circuit which connects to the secondary winding of the transformer is used to inject a sinusoidal current. The frequency of the injected current changes step by step and circularly in a region. Fig.4(a) shows the equivalent circuit of the system. I_{inj} represents the injected current, $L_{1\sigma}, L'_{2\sigma}$ represent the leakage inductance of primary winding and secondary winding of the transformer respectively, here the prime denotes referred quantities of secondary winding to primary winding. L_m is the magnetizing inductance, C, R are equivalent capacitance and resistance between neutral point and ground in power network. For I_{inj} is a current source, $L'_{2\sigma}$ can be ignored. The resistance R is very small so it also can be neglected. Fig.4(b) shows the simplified circuit. Assume the frequency of the injected current is f_{res} when parallel resonant occurs, according to the resonant condition, the capacitance of power network can be calculated as follow:

$$C = \frac{1}{(2\pi f_{res})^2 (L_m + L_{1\sigma})} \qquad (12)$$

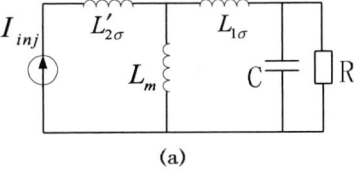

(a)

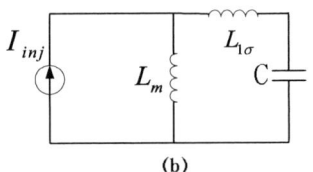

(b)

Figure 4. Equivalent capacitance detect circuit(a) and its simplified circuit(b)

Once the capacitance value is obtained by (12), the reactance value desired to compensate the fault current can be calculated as follow:

$$X_L = \frac{2\pi f_{res}^2 (L_\sigma + L_m)}{f_1 (1-\gamma)} \qquad (13)$$

Here, f_1 is the frequency of power system, γ is defined as: $\gamma = \frac{I_c - I_L}{I_c} = 1 - \frac{I_L}{I_c} = 1 - \frac{X_c}{X_L}$, which is an important criterion in industrial application. γ should be below $\pm 10\%$ after compensation. Then the coefficient α defined in (5) can be calculated by combining (7) and (13):

$$\alpha = \frac{(L_m + L_{1\sigma})[f_1^2 (1-\gamma) - f_{res}^2]}{f_1^2 (1-\gamma) L_m} \qquad (14)$$

Here the resistance r_1 in (7) is neglected.

IV. CONTROL SYSTEM

According to the system configuration introduced above, the crucial part of control system is current control, either in capacitance detection status or in compensation status. Linear current control scheme is adopted for the sake of simplicity and fixed switching frequency. Fig.5 shows a block diagram of linear current PWM controller[5]. The current detective and PWM generation is implemented by a digital control system based on T1 2407DSP, which is very convenient to get pairs of PWM pulse with programmable dead-beat time. A single polarity double frequency technology is adopted to double the frequency of generated pulse, in the other words, the actual switching frequency of power electronic elements can be reduced.

For the multiple secondary windings scheme is adopted to improve the capacity of the apparatus, a communication module is needed to control and synchronize the work of all subsystems. CAN module in 2407DSP is used to communicate between subsystems and main DSP board, the SCI module is used to communicate between main board and computer. Fig.6 shows the integrated communication diagram. The main DSP board scouts the neutral point's potential of power network, when the value is normal, the main board transmits a "capacitance detection" instruction to one of the subsystems, and then the subsystem controls one inverter injecting frequency-changing current to detect the capacitance of power network periodically, and then

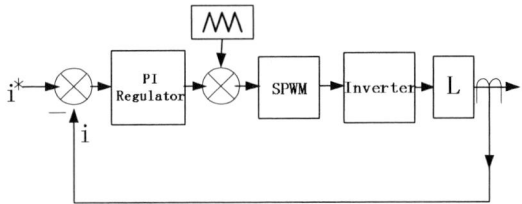

Figure 5. Block diagram of linear current controller

491

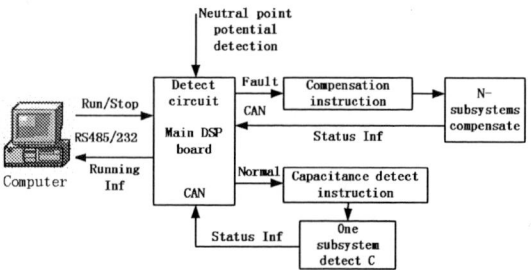

Figure 6.   Integrated communication diagram

the subsystem transmits the detected capacitance value to the main board. The main DSP board calculates the compensate coefficient α by (14). Once the potential of the neutral point exceeds a nominal value previously set, a short circuit fault is considered to be occurred, then it transmit a "compensation" instruction and the coefficient "α" to all subsystems by CAN module immediately, all inverters work in compensation status synchronously. All running information is transmitted to a computer by RS232/485 communication protocol, so the working status can be observed real time.

V. EXPERIMENTAL RESULTS

An apparatus is manufactured according to the system configuration in Fig.3. Some capacitors connect in parallel between terminals A and X to simulate the distributed capacitor in power network. A booster is used to regulate the voltage between AX in the purpose of simulating the potential variation of the neutral point. The arc-suppression coil has $N = 8$ secondary windings, the turn ratio is $k = 20:1$. The magnetizing inductance and primary leakage inductance are $L_m = 4.5H$ and $L_{1\sigma} = 0.139H$ respectively. When the rated value of the capacitor is $8.1\mu f$ and the frequency of the injected current changes from $23Hz$ to $43Hz$, the resonant frequency gotten from detection is $f_{res} = 25.8Hz$, the capacitor calculated by DSP is $8.2\mu f$ according to (12), so the result is accurate. When the voltage between AX is regulated over a setting value(e.g. $1000V$), the apparatus work in compensation status immediately. Table I collects the running data when compensate coefficient α is 0.9. The calculated reactance is 185.0Ω by (7), so the actual value match the calculated value, the maxim relative error bellows 3%. Fig.7 is the voltage-ampere curve described by computer according to the data collected in table I. It is worth noting that the voltage-ampere curve is nearly linear when α is fixed. That's mean the reactor's voltage-ampere characteristic is excellent. Table II collects the running data when α changes from 0.75 to 0.9. Fig.8 is the $Z_{AX} = f(\alpha)$ curve drawn by computer according to the data in table II. It can be seen that the impedance is reduced almost linearly along with the increasing of α, which matches the theory's prediction. A group of typical waveforms were recorded by digital oscilloscope. Fig.9 shows the voltage and current waveform of reactor's primary winding. It can be seen that the waveform is nearly sinusoidal, that's

mean the reactor takes few harmonic pollution to power network. Fig.10 shows the current waveforms of primary and secondary windings of the reactor, which indicates the current tracing control system working effectively.

TABLE I.   COMPENSATION INFORMATION WHEN $\alpha=0.9$

α	U_1 (V)	I_1 (A)	$Z_{AX}=\dfrac{U_1}{I_1}$ (Ω)
0.9	2800	15.0	186.7
0.9	3200	17.5	182.9
0.9	3600	19.8	181.8
0.9	4000	22.3	179.4

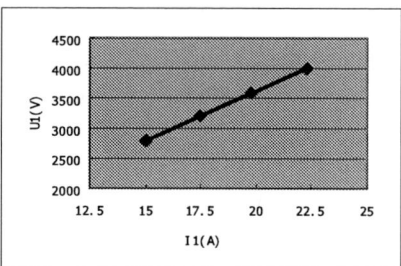

Figure 7.   The voltage-ampere curve

TABLE II.   COMPENSATION INFORMATION WHEN α CHANGES FROM 0.75 TO 0.9

α	U_1 (V)	I_1 (A)	$Z_{AX}=\dfrac{U_1}{I_1}$ (Ω)
0.75	5200	12.7	409.4
0.8	5200	15.5	335.5
0.85	5200	19.7	264.0
0.9	4000	22.3	179.4

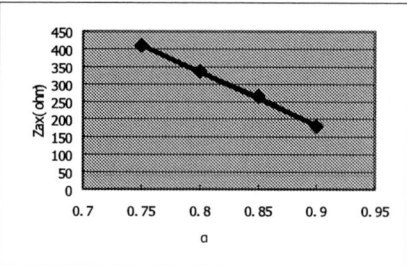

Figure 8.   The $Z_{AX}-\alpha$ curve

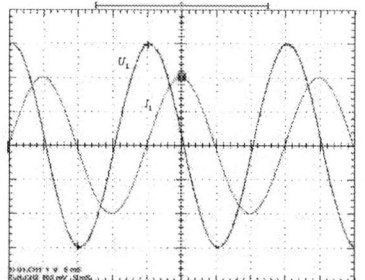

Figure 9.   Voltage and current waveform of reactor's primary winding

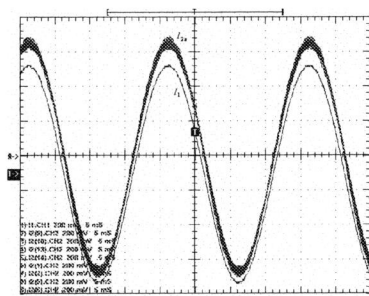

Figure 10. Current waveform of primary winding and secondary windings

VI. CONCLUSION

In the paper, the principle of a novel arc-suppression coil is proposed. When the current of transformer's secondary winding satisfies the condition (5), the equivalent impedance from the primary winding of transformer exhibits an adjustable reactance in relation with the coefficient α. Accordingly the reactance desired for compensating the capacitance could be adjusted by regulating α. Comparing with other types arc-suppression coil, it has the superiority of linear regulation capacity and by adopting high switching frequency power electronics elements, harmonics pollution is limited effectively.

The injecting-current capacitance detection method is verified accurately, which can be implemented by the inverter circuit easily without adding any additional apparatus.

A multiple secondary windings scheme is employed to realize high capacity. Its validity is proved by theory and experiments.

REFERENCES

[1] C. L. Wang, R. Liang, J. H. liu, et al. "Analysis on Principle of operation of Arc-suppression Coil Based on Thyristor Controlled Reactor," Proceedings of Electrical Machines and Systems Conference, ICEMC 2005, Vol.2, pp.1305-1308

[2] Y. C. Liu, T. C. Lu, C. Fu, "Automatic Arc Suppression Equipment with Fast Response Based on TSC Control Technology," Proceedings of Industrial Electronics Conference 2004, Vol.1, pp.520-523

[3] Leonard J. Bohmann, Robert H. Lasseter, "Harmonic Interactions in Thyristor Controlled Reactor Circuits," IEEE Trans. on Power Delivery, Vol.4, No.3, pp.1919-1926,1989

[4] Z. Z. Liu, H. Y. Chen, Z. H. Yu, "Harmonic Analysis of Arc-Suppression Coil Based on Transformers with High Short Circuit Impedance," Proceedings of Power System Technology Conference 2002, Vol.1, pp.130-133

[5] Buso S, Malesani L, Mattavelli P, "Comparison of current control techniques for active filter applications," IEEE Trans. on Industrial Electronics, Vol.45, No.5, pp.722-729,1998

2006 5th International Power Electronics and Motion Control Conference

Grid Connection to Stand Alone Transitions of Slip Ring Induction Generator During Grid Faults

G. Iwanski, W. Koczara

Warsaw University of Technology
Institute of Control and Industrial Electronics
75, Koszykowa str., 00-662 Warsaw, Poland
iwanskig@isep.pw.edu.pl , koczara@isep.pw.edu.pl

Abstract—A grid connected power generation systems based on the superior controllers of an active and reactive power are useless during a grid failures like grid short-circuit or line braking. Therefore the change of operation mode from grid connection to stand alone allows for uninterruptible supply of a selected part of grid connected load. However, in the stand alone operation mode the superior controllers should provide fixed amplitude and frequency of the generated voltage in spite of the load nature. Moreover, a soft transition from grid connection mode to stand alone operation requires that, the mains outage detection method must be applied. A grid voltage recovery requires change of the generator operational mode from stand alone to grid connection. However, the protection of a load from rapid change of the supply voltage phase is necessary. This may be achieved by synchronization of the generated and grid voltages and controllable soft connection of the generator to the grid.

The paper presents the transients of controllable soft connection and disconnection to the grid of the variable speed doubly fed induction generator (DFIG) power system. A description of the mains outage detection methods for the DFIG is based on the grid voltage amplitude and frequency measurement and comparison with a standard values. Also an angle controller, between generated and grid voltages, for synchronization process is described. The short description of the sensorless Direct Voltage Control of the autonomous doubly fed induction generator (ADFIG) is presented.

All the presented methods are proved based on PSIM simulation software and in a laboratorial conditions and the oscillograms with a test results are presented in the paper. A 2.2kW slip-ring induction machine was applied as a generator and 3.5kW DC motor was used as a primary mover to speed adjusting. A switching and sampling frequencies are equal to 8kHz. For filtering the switching frequency distortions in the output voltage external capacitances equal to 21uF per phase are connected to the stator. The control algorithm is implemented in a DSP controller build on a floating point ADSP-21061 with an Altera/FPGA support.

Keywords - distributed power generation; variable speed generation; stand-alone power systems, mains outage, DFIG.

I. INTRODUCTION

The stand alone operation mode of the power generation system can be used in island power plants for isolated load supplying, as well as in a grid connected systems during a mains outage for supplying a part of grid connected load. The variable speed generators, used more and more frequently, especially in a renewable energy sources like wind or water turbines are equipped with a power electronics converters. Using power electronics converter the stand alone operation can be easily achieved and the load independent, high quality generated voltage can be obtained.

Using doubly fed induction machine the grid connected power generation systems are mainly described in a publications [1][2][3]. However, active and reactive power controllers, used there control methods, developed for this kind of operation, do not allows for correct operation during grid failures. Thus the power generation system is useless during the grid voltage failures caused e.g. by grid line breaking or other generator disconnecting.

Stand alone power DFIG generation system producing high quality voltage is presented on the Figure 1. System requires the stator connected capacitance for filtering a switching frequency distortions on the stator voltage produced by rotor connected power electronics converter PCo1 and front-end converter PCo2. The output low pass LC filter, consists of stator capacitances and an equivalent rotor/stator leakage inductance and then does not requires any additional inductances connected to the rotor neither the stator.

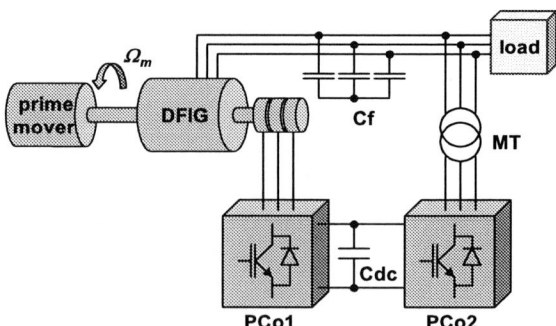

Figure 1. Stand alone power generation system with a doubly fed induction generator.

1-4244-0448-7/06/$25.00 ©2006 IEEE

The stand alone operation is described in [4], [5]. [6]. However, the grid soft connection and disconnection problems are not considered there and the possibility of stand alone operation in grid connected systems is not discussed.

Moreover the presented methods have few disadvantages, like used speed sensor [4] or the rotor position encoder [5], or additional rotor connected large inductances (10mH [5] and 30mH!!! [6]). Also the presented results are obtained with pure resistive load, that is a significant simplification in stand alone mode operation.

A power generation system presented on Figure 2 can operate in stand alone mode and in the grid connection mode as well. The change of operation mode is possible by use a grid connection switch GCS controlled by superior control block. This block includes an algorithms of the methods of mains outage detection and procedure of generated and grid voltages synchronization after the grid voltage reappearing.

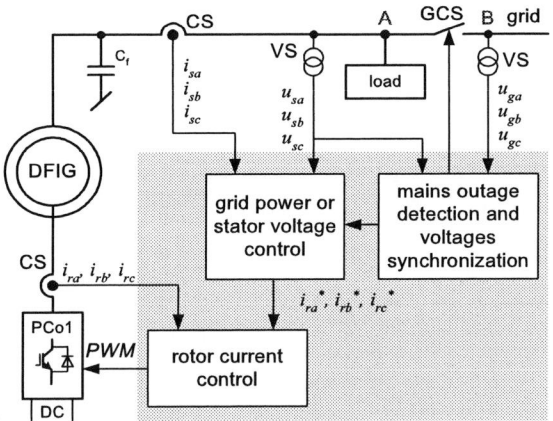

Figure 2. Grid connected and/or stand alone power generation system with a doubly fed induction generator

The GCS opening results in the stand-alone mode, whereas the GCS closing is related to the grid connection mode. During the stand alone operation the isolated load is supplied from the DFIG system. Controllable grid connection and disconnection of the system allows for uninterruptible load supply.

During grid connection mode a grid voltage u_g and a stator voltage u_s are the same, because the GCS switch is closed. Also during the mains outage transients the grid voltage u_g in node B and stator voltage u_s are the same because to GCS opening while the voltage in B point is maintained by the doubly fed induction generator.

During grid connection mode an active and reactive power controllers are used, but this applied control method is out of scope of the paper and will not be presented in details. Also other method for grid connection mode can be considered [1][2][3].

II. A STAND ALONE OPERATION

A. Sensorless Control Method of the Output Voltage

The stand alone operation of the power generation system requires the fixed output voltage and frequency. The proposed control method, based only on the stator voltage and rotor current sensors, is presented on Figure 3. There are no any rotor position encoders neither the speed sensors or estimators [7][8][9].

A transformation of the three phase stator voltages u_{sa}, u_{sb}, u_{sc} from $abc_{(s)}$ stator coordinates to rotated with synchronous speed dq coordinates system, and next from dq to polar coordinates $A\phi$, gives the stator voltage vector u_s defined by its amplitude $|u_s|$ and an angle α_{us} related to the d axis. Both amplitude and angle of the vector u_s are obtained from d and q vector axis components:

$$|u_s| = \sqrt{u_{sdp}^2 + u_{sqp}^2} \qquad (1)$$

$$\alpha_{us} = \mathrm{atan}\!\left(\frac{u_{sqp}}{u_{sdp}}\right) \qquad (2)$$

The reference angular speed $\Omega_{ir}{}^*$ of the rotor current vector i_r is achieved from α_{us} controller Rα. This way the actual u_s and reference $u_s{}^*$ vectors of the stator voltage are synchronized and the actual stator voltage vector has fixed angle related to d axis in the steady state. It means the frequency of the actual stator voltage u_s is fixed and corresponds to the reference angular speed $\Omega_s{}^*$ of the rotated dq coordinates. Obtained angular speed $\Omega_{ir}{}^*$ of the rotor current i_r vector from Rα controller is integrated (3) to achieve a reference angle $\phi_{ir}{}^*$ of the rotor current in an $abc_{(r)}$ coordinates connected with a rotor.

$$\phi_{ir}^*(t) = \int \Omega_{ir}^*(\tau)\,d\tau \qquad (3)$$

Simultaneously the amplitude of the stator voltage vector is controlled using an RU regulator, where output signal is responsible for reference amplitude $|i_r|^*$ of the rotor current vector. It is justified because the system is stationary and here is only one electromotive force (PCo1), that means the amplitude of the output voltage is proportional to the amplitude of the rotor current. Using polar coordinates system the decoupled control of the voltage amplitude and frequency is obtained.

A transformation of the rotor current vector, from polar $A\phi$ coordinates to $abc_{(r)}$ connected with a rotor, produces a reference signals for the three phase currents $i_{ra}{}^*$, $i_{rb}{}^*$, $i_{rc}{}^*$.

$$\begin{bmatrix} i_{ra}^* \\ i_{rb}^* \\ i_{rc}^* \end{bmatrix} = |i_r|^* \, cos \begin{bmatrix} \phi_{ir}^* \\ \phi_{ir}^* - 2\pi/3 \\ \phi_{ir}^* + 2\pi/3 \end{bmatrix} \qquad (4)$$

Next the proportional rotor current controller RI, based on the reference and actual rotor current signals, is applied. A proportional type regulator introduces a steady-state error of the rotor current, but the RI is an inner controller, that doesn't influence negatively on the controlled stator voltage, because the superior voltage controllers are PI type.

495

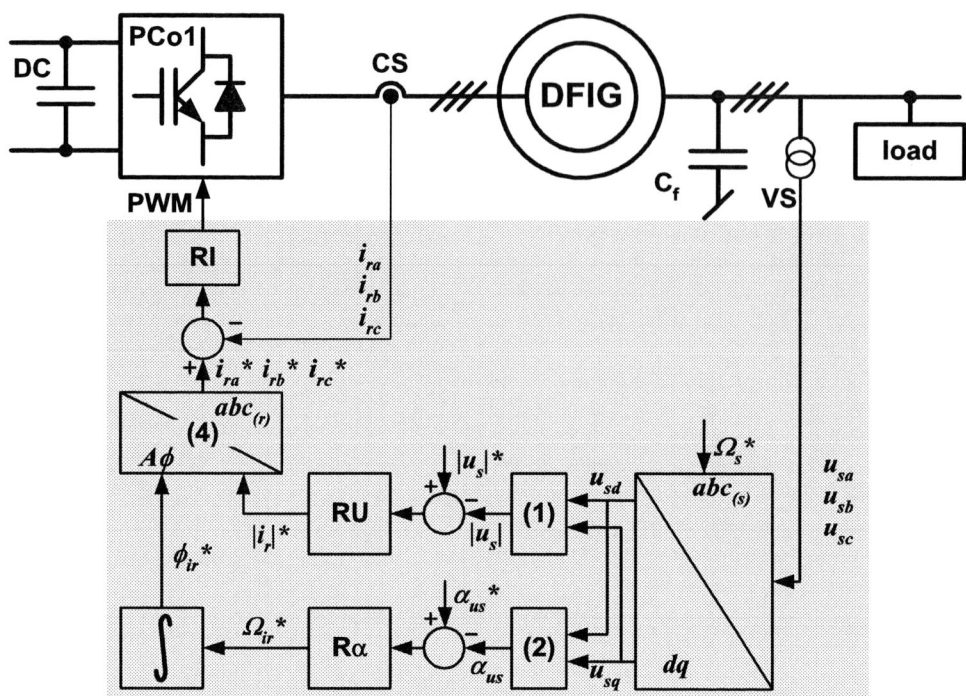

Figure 3. Scheme of the sensorless control method of the output voltage of slip-ring induction generator

B. Experimental Results of Stand Alone Operation.

The experimental setup with 2.2kW, 220/380V, 6 poles slip ring machine was build and tested in a laboratory (Fig. 4). Other data of the slip-ring machine are following: stator current 9.8/5.7A, $\cos\phi = 0.71$, rated speed 950rpm, rotor voltage 108V, rotor current 13A.

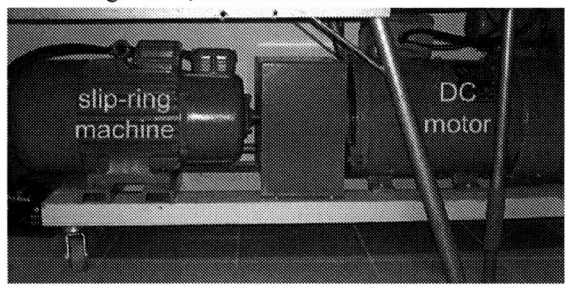

Figure 4. A machine set used in an experimental investigations build on slip-ring generator and a DC motor as a primary mover.

As the DLL file, used in simulation, was created using C code, the same C source file is used directly for the DSP controller programming [10]. The control system was build on the floating point processor ADSP-21061 with an FPGA/ALTERA support. The sampling and switching frequency is equal to 8kHz. A stator connected filtering capacitance is equal to 21uF in each phase, that significantly compensates the generator's reactive power.

The stand alone operation requires that, the initial power, for excitation, should be provided to the DC link. This is achieved by using an additional capacitances connected to the stator, that overcompensates the slip-ring machine and then the self excitation is obtained. [7].

During the stand alone operation, it should be possible to supply different type of load. Proposed method of output voltage control allows for strongly nonlinear load supply, that can be observed on Figures 5 and 6. A nonlinear load of 50% of rated load is supplied from the generator, and the voltage amplitude and frequency are fixed in spite of the variable rotor speed (Fig. 5b). This is obtained thanks to the rotor current frequency control.

Moreover, the voltage amplitude is not sensitive on the load change. In both cases –step-loading (Fig. 6a) and step-unloading (Fig. 6b) – the transients are short and control method allows for high quality voltage for nonlinear load supply as well as during no load operation.

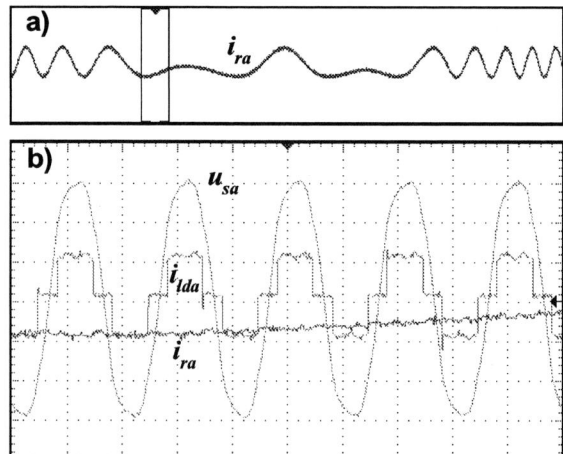

Figure 5. Oscillogram presenting a speed change transient; a) rotor current i_{ra}, b) stator voltage u_{sa} (100V/div), rotor current i_{ra} (20A/div), load current i_{lda} (2A/div)

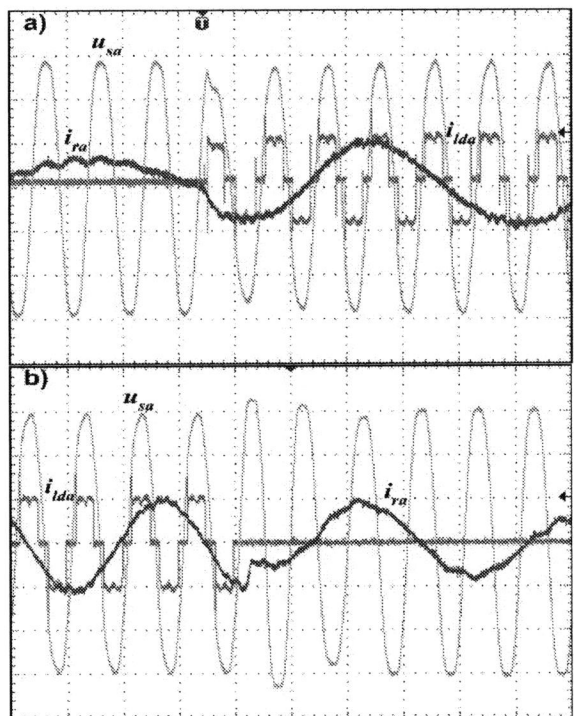

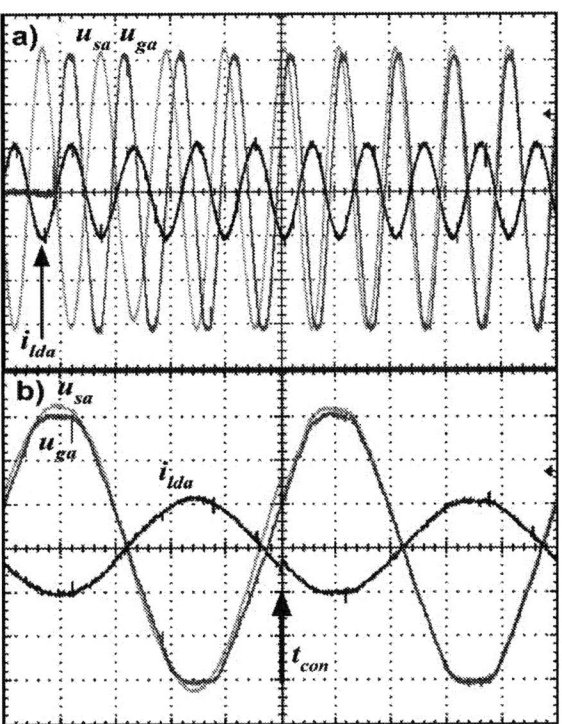

Figure 6. Oscillograms presenting a load change transients; stator voltage u_{sa} (100V/div), rotor current i_{ra} (20A/div), load current i_{lda} (2A/div) during step-on load (a), and step-off load (b).

Figure 7. Oscillograms presenting a synchronization of the stator and grid voltages (a), and grid connection instant (b); stator voltage u_{sa} (100V/div), grid voltage u_{ga} (100V/div), load current i_{lda} (2A/div)

III.   GRID CONNECTION AND DISCONNECTION TRANSIENTS

A.   Voltages synchronization and grid connection

The stand alone operation can be used in an island power generation systems for independent load supply as well as in the grid connected systems during grid failures. The soft transition of the change of the operation mode from stand alone to the grid connection after grid voltage recovery is obtained by a stator u_{sa} and grid u_{ga} voltages synchronization. A voltages synchronization is obtained by elimination of an angle β_u between the vectors u_s, u_g of grid voltage and stator voltage. As the amplitudes $|u_s|$, $|u_g|$ of the voltage vectors are different from zero, the function $sin\beta_u$ obtained from a cross product of the vectors (5) can be used for voltages synchronization. This function is equal to zero when the vectors are parallel, that means the voltages are synchronized and the stand alone generator is ready to grid connection.

$$sin\,\beta_u = \frac{u_s \otimes u_g}{|u_s||u_g|} = \frac{u_{sd}u_{gq} - u_{sq}u_{gd}}{|u_s||u_g|} \qquad (5)$$

The synchronization process is presented on Figure 7a, whereas a grid soft connection as the final part of synchronization is presented on Figure 7b. The synchronization time can be much longer that voltage period, that causes the load is protected from the voltage phase rapid change. In the case of unload generator the synchronisation process is shorter and equal to few periods of the generated voltage.

B.   Mains outage detection methods

A controlled grid disconnection of the generator during mains outage and soft change of operational mode to stand alone with saved amplitude, frequency and phase of the stator voltage, allows for uninterruptible load supply with protection of the voltage phase rapid change. However some methods of the mains outage detection are necessary.

In the moment of the supply grid voltage is cut-off the voltage in node B (Fig.3) is maintained by the induction generator. Transient states are shown in Figures 8a,b. The switch GCS cut-off the failed system and the generator supplies the independent load.

As the load power during grid connection is clearly higher or lower than the stator power, the grid current in point B is ~~clearly~~ different to zero. In this case, during line breaking transient, the amplitude of the grid voltage, maintained by the generator, is much higher or lower than standard value and allows for easy detection of the mains outage caused by line breaking. While the load is more less equal to the stator power, the grid current in point B is close to zero and the grid voltage has its rated value after the line breaking, therefore it is impossible to mains outage detection based on the voltage amplitude analysis.

Hence, other method of the mains outage detection is needed. As the control method for grid connection mode do not allows for frequency control, during stand alone mode, the second method of the mains outage detection bases on the voltage frequency is proposed. The voltage frequency in the line breaking transient is not equal to the reference 50Hz, and the measurement of it allows for detection of line failure.

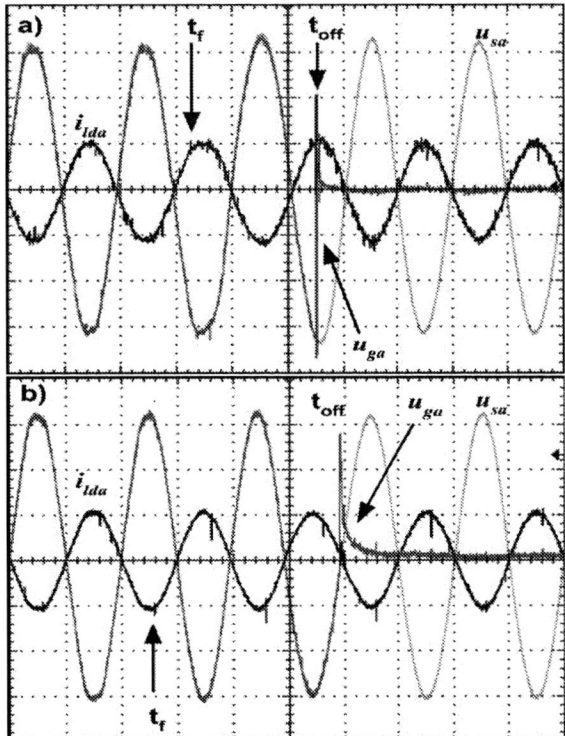

Figure 8. Oscillograms presenting a generator disconnection from grid using methods based on the voltage amplitude (a) and frequency (b); stator voltage u_{sa} (100V/div), grid voltage u_{ga} (100V/div), load current i_{lda} (2A/div), t_f – fault instant, t_{off} – grid disconnection instant

The mains outage is caused not only by line breaking but also by other grid disturbances e.g. grid short circuit. The grid short-circuit causes the grid voltage fails and its value (amplitude) is much less than the rated level. A voltage sag caused by short circuit which is shorter than one period of the generated voltage is not dangerous for majority part of load and there is not necessary a generator disconnection. Also during the grid overload a disconnection of the generator is not obligated in each case. Using a delay of the reaction time of the control system, the grid connection mode is maintained during grid overload (Fig. 9).

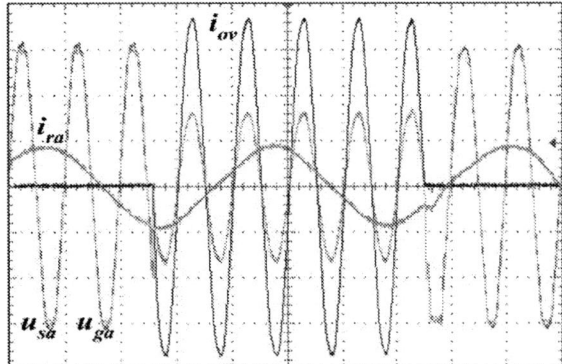

Figure 9. Oscillogram presenting a grid voltage sag during overload; stator voltage u_{sa} (100V/div), grid voltage u_{ga} (100V/div), rotor current i_{ra} (20A/div), short circuit current i_{sc} (50A/div),

CONCLUSION

Adjustable/variable speed generation system with slip-ring induction machine is presented. The generation system control method does not need speed signal. In case of cut-off grid the control system provides stand alone operation applying direct voltage control method. Resynchronization is fully controlled. To provide safe transients during grid failures methods of the grid state detections are developed. Computer calculation were used to tune system controllers. Laboratory tests confirms the system high quality operation in spite of grid disturbances. The quality of the system operation is referred to produced voltage.

REFERENCES

[1] R. Datta and V.T. Ranganathan, „Direct power control of grid connected wound rotor induction machine without rotor position sensors", *IEEE Transactions on Power Electronics*, vol. 16, No. 3, May 2001.

[2] Z. Krzeminski, "Sensorless multiscalar control of double fed machine for wind power generators" *Proc. of Power Conversion Conference – PCC'02*, Osaka, Japan.

[3] A. Petersson, L. Harnefors and T. Thiringer, "Comparison between stator-flux and grid-flux-oriented rotor current control of doubly-fed induction generators", *Proceedings of the 35th Annual IEEE Power Electronics Specialists Conference – PESC'04*. Volume 1, 20-25 June 2004 pp. 482 - 486

[4] Y. Kawabata, Y. Morine, T. Oka, E. Ejiogu, and T. Kawabata, "Variable speed constant frequency power generating system by the use of rotor excitation of induction machine" *Proceedings of Power Conversion Conference - PCC 02*, Osaka, Japan

[5] D. Forchetti, G. Garcia and M.I. Valla, "Vector control strategy for a doubly-fed stand-alone induction generator". *Proceedings of Industrial Electronics Conference - IECON 02* Volume 2, 5-8 Nov. 2002 pp:991 - 995

[6] R. Cardenas, R. Pena, J. Proboste, G. Asher and J. Clare, "Sensorless Control of a Doubly-Fed Induction Generator for Stand Alone Operation" in *Proceedings the 35th IEEE Power Electronics Specialist Conference – PESC'04*, Aachen, Germany

[7] G. Iwanski and W. Koczara, "Control system of the variable speed autonomous doubly fed induction generator" *Proceedings of International Power Electronics and Motion Control Conference – EPE-PEMC'04*, Riga, Latvia

[8] G. Iwanski and W. Koczara, "Sensorless stand alone variable speed system for distributed generation", *Proceedings of the 35th Annual IEEE Power Electronics Specialist Conference – PESC'04*, Aachen, Germany

[9] G. Iwanski and W. Koczara, "Sensorless Direct Voltage Control Method for Stand-Alone Slip-Ring Induction Generator" *Proceedings of 11th European Conference on Power Electronics and Applications – EPE'05*, Dresden, Germany

[10] B. Kaminski, K. Wejrzanowski, W. Koczara "An application of PSIM simulation software for rapid prototyping of DSP based power electronics control systems" *Proceedings of the 35th Annual IEEE Power Electronics Specialist Conference – PESC'04*, Aachen, Germany

[11] PSIM Software, www.powersys.fr

2006 5th International Power Electronics and Motion Control Conference

System Control of Power Electronics Interfaced Distribution Generation Units

D. Feng   Z. Chen

The Institute of Energy Engineering, Aalborg University, Denmark
Email: zch@iet.aau.dk

Abstract— Unlike conventional power system, the system voltage and frequency of some renewable energy resources based stand-alone distribution generation (DG) system maybe controlled by power electronics converters. This paper presents the control of parallel operation multi-converters in a DG system. The investigated stand-alone DG system is constructed by two local DG subsystems; in each of them, two power electronic converters operation in parallel. A hybrid control scheme is developed to control the system voltage and frequency, hence achieves the load sharing between two DG subsystems and between the converters in each DG subsystem. The LCL filters are designed to filter out the high frequency components from the converters. The simulation results show that it is possible to use power electronics interfaced DG system to supply quality power.

Keywords- DG system; parallel operation; power sharing; power quality; harmonics; PLL system; LCL filter

I. INTRODUCTION

In the past thirty years, combine the increasing of global population and growing of world economy, the global electricity consumption has been fast increasing. In energy resources, fossil fuels (oil, coal, natural gas) have the primary contribution. However, it is known that the fossil fuels resources are limited resource and cause greenhouse emission. In order to limit the usage of fossil fuel and reduce the greenhouse emissions, sustainable energy technologies are being developed quickly.

The last decade has shown an increasing interest in developing alternative and cleaner forms of energy production, such like wind energy, solar energy, biomass and fuel cell etc. The features of renewable energy systems make it possible that to invest a distribution generation (DG) system in a local area and to form a new ownership of generation *[1]*. Many of these renewable resources based DG units have very different features in comparison with the power system using conventional fossil fuel; these units need power electronic based interfaces, which could be DC to AC inverters or AC-DC-AC converters.

The converters present very different characteristic with the conventional electrical generators. The operation and control of a DG system present great challenge to the

power system operator. Particularly the connection of large amount of power electronic interfaced DG units to a weak grid in rural areas has to be carefully designed with respect to voltage and frequency fluctuations. Also the control of electrical generator and interface power electronic converter will form an important part of the control system. Unlike conventional power system, in stand-alone DG systems, there are not grid voltage and frequency to reference. It makes the control and operation of converter becomes a great concern in DG system design.

This work will investigate the characteristics to operate a power system with power electronic interfaced distributed generation units; develop a hybrid control method of parallel operating converters to achieve good load sharing and optimize the energy production.

II. SYSTEM STRUCTURE

A. System structure

Many renewable resources, like wind and solar, are not controllable sources. An energy storage system and a secondary generator must be employed in the DG system to supply reliable power.

Currently, the power quality and the stability of a DG system are the main concern. For a stand-alone DG system, especially a weak system, it is quite difficult to produce very good quality power. It is valuable to investigate a group of DG systems with multiple converters and to develop a control system to make the system working reliably. The structure of this investigated system is shown in figure 1. Based on this system structure, this paper studies a DG system with two DG subsystems. In each of them, there are two voltage source converters and load.

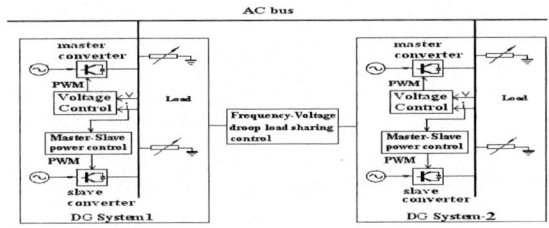

Fig.1 DG systems with two parallel operation converters and control schemes

1-4244-0448-7/06/$25.00 ©2006 IEEE   499

By controlling the converters, the system voltage and frequency are controlled at the required level. Hence, to achieve the power balance and load sharing when the system has a load variation.

B. LCL filter Design

The voltage source converter can be controlled in both the AC voltage and the power factor [2]. But the high power devices switching frequency (generally from 2 *kHz* to 15 *kHz [3]*) causes high order harmonics, that can disturb the sensitive load or equipments. In order to reduce the current harmonics injection, a high value inductance filter should be selected. But for applications about several hundred kW DG system, it becomes so costly [4].

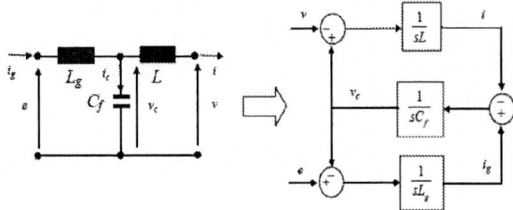

Figure 2 Equivalent circuit of LCL filter and control model

Papers [3], [4], [5] present a method to design LCL filter. The structure of the LCL filter is shown in figure 2 [5].

The transfer function of the LCL filter can be expressed as,

$$G(s) = \frac{i_g(s)}{v(s)} = \frac{s^2 + \dfrac{1}{L_g \cdot C_f}}{L \cdot s \cdot (s^2 + \omega_{res}^2)} \qquad (1)$$

The open loop properties of the designed LCL-filter could be indicated in a Bode diagram (figure 3). The Bode plot shows that around the resonance frequency, the filter has a very high attenuation.

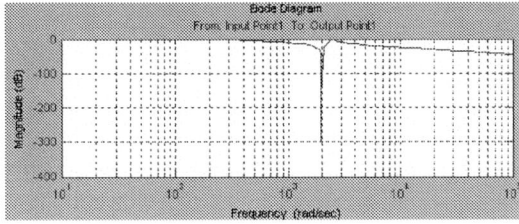

Figure 3 Bode diagram of open loop transfer function of LCL filter

III.   DESIGN OF CONTROLLERS

An overall of the system control structure is shown in figure 1. There are three main control modes:

- Voltage control mode: the system voltage is controlled by a voltage controller

- Master-slave control mode: Using the voltage control mode to control the master converter. The current of common line is controlled by a slave converter; hence achieve the load sharing in a DG subsystem.

- Real power-frequency and Reactive power-voltage characteristics: The load sharing of the DG system is programmed by frequency and voltage droop characteristics.

A. Double loops current-regulated voltage control scheme

As known, the system voltage and current depend on the active power and reactive power produced by generation units. In the upper part of figure 5, the inner loop is current regulation loop and the outer loop is voltage control loop. The rms line-to-neutral voltage V_{abc} and the rms line current i_{abc} are measured and used to generate the reference voltage V_{ref}. The sum of the V_{ref} and V_{abc} is the voltage error signal (e_v). The controller G_v is used to compensate the e_v and the compensated voltage error signal is to be reference current command i_{ref}. The reference current i_{ref} subtracts the line current i_{abc} to find out the current error signal e_i, which goes though the controller G_i to obtain the PWM signal for driving the full bridge converter.

The reference voltage V_{ref} is generated by three main blocks, which are list following,

- Power calculation block is used to calculate the active and reactive power of three-phase circuit. The mathematical equations [6] are,

$$P_{3\phi} = 3 V_{LN} I_L \cos(\phi) \qquad (2)$$

$$Q_{3\phi} = 3 V_{LN} I_L \sin(\phi) \qquad (3)$$

Where, V_{LN} is the rms line-to-neutral voltage; I_L is the rms line current; ϕ is the phasor angle between line-to-neutral voltage and line current.

- Frequency and voltage characteristic block: this block functions the active power–frequency (P-ω) characteristic and reactive power–voltage (Q-v) characteristic.

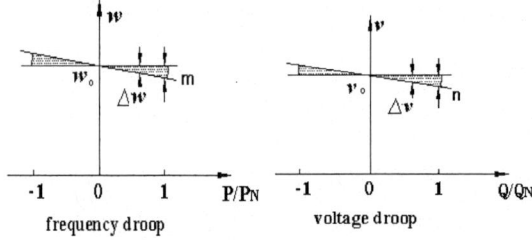

Figure 4   Schematic diagram of P-ω characteristic and Q-v characteristic

The P-ω characteristic and Q-v characteristics could be determined by following equations,

$$\omega = \omega_0 - m \cdot P \quad (4)$$
$$v = v_0 - n \cdot Q \quad (5)$$

Where, ω_0 is electrical angular speed at no loads; v_0 is voltage amplitude at no load; m, n are the droop coefficients.

- Reference voltage calculation block: is to generate the three-phase reference voltage using the voltage magnitude and frequency calculated by the frequency and voltage characteristic block. The three-phase reference voltage can be expressed as,

$$v^*_{ref,a} = v \cdot \cos(\omega \cdot t) \quad (6)$$

$$v^*_{ref,b} = v \cdot \cos(\omega \cdot t - 120^0) \quad (7)$$

$$v^*_{ref,c} = v \cdot \cos(\omega \cdot t + 120^0) \quad (8)$$

B. Master-slave control scheme

In this control scheme, there are two parts of controllers. One is voltage control mode; another is current control mode. See figure 5.

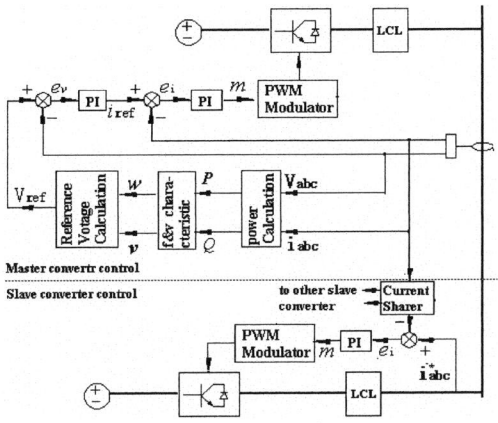

Figure 5 Master-slave control strategies

The system AC bus voltage is held by the master converter control. The control method in this study is the double loop current-regulated voltage control mode. All the slave converters (in figure 7, only one slave converter is shown) following the voltage and frequency of master converter. The AC bus current is shared by all slave converters. Each shared current signal i_{ref} is to be the reference signal and compared with the output current of slave converter to generate a current error signal. This current error signal goes through a controller to produce a modulation index to drive the slave converter, further to achieve the load sharing and power balance.

IV. CONTROLLERS DESIGN AND MODULATION

A. Controllers design of master converters

The controller of the master converters is a double-loop current regulated voltage controller. In this control

strategy, the inner loop is a current regulation loop. The block diagram of this current loop is shown in figure 8.

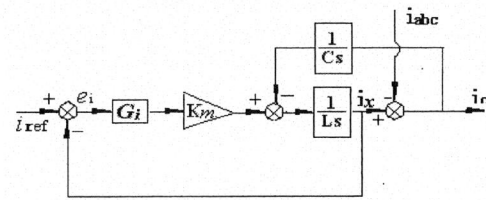

Figure 6 Current regulation loop diagram

In this control loop, i_{ref} is the outputs of voltage compensator (to be present in later section). The i_{ref} acts the current reference signal and be fed in a current compensator G_i. The inner inductor current compensator is included here mainly to stabilize the system and to improve the system dynamic response by rapidly compensating for variations in the load voltages, whose rate of change is indirectly sensed by measuring the inductance current. The closed-loop transfer function of the inner current regulation loop can be expressed as:

$$i_c = \frac{k_m \cdot G_i \cdot Cs}{LCs^2 + k_m \cdot G_i \cdot C \cdot s + 1} \cdot i_{ref} - \frac{LCs^2}{LCs^2 + k_m \cdot G_i \cdot C \cdot s + 1} \cdot i_{abc} \quad (9)$$

The voltage control loop is similar to the current loop. The control logical diagram is shown in figure 7.

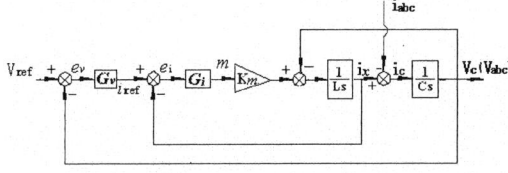

Figure 7 Control logical diagram of double loop current regulate voltage controller

The output signal i_c of inner current loop going through the filter capacitor produces a capacitor voltage signal V_c, which is fed back into the voltage compensator G_v. The voltage compensator commonly introduced by a PI compensator to avoid phase errors. The transfer function of outer voltage control loop is,

$$V_c = \frac{k_m \cdot G_i \cdot G_v}{LCs^2 + k_m \cdot G_i \cdot G_v + 1} \cdot v_{ref} - \frac{Ls}{LCs^2 + k_m \cdot G_i \cdot G_v + 1} \cdot i_{abc} \quad (10)$$

B. Selection of the voltage-frequency drop coefficients m and n

Paper [7] presents a method to determine the power–frequency (P- ω) characteristic and reactive power–voltage (Q-v) characteristic, those should coordinated to make each DG system supplying active and reactive power in proportion its power rating. The equations are:

$$m = \frac{\omega_0 - \omega_{min}}{p_0 - p_{max}} \quad \text{and} \quad n = \frac{v_0 - v_{min}}{Q_0 - Q_{max}} \quad (11)$$

Where, ω_0 is the operation frequency when the system is no load (Po =0). p_{max}, ω_{min} are the maximum active power output of DG system and the minimum operation frequency. v_0 is the initial value of voltage magnitude, it is set as 318.2 V (225 V phase voltage). Q_0 is the initial reactive power, it is set as zero. Q_{max}, v_{min} are the maximum reactive power output of DG system and the minimum voltage magnitude.

C. Design of current controllers for slave converters

This study selects the PI-dq control method to design the controller, which can force the slave converter following the voltage and frequency of its master. The design of current controller is shown in the figure 8.

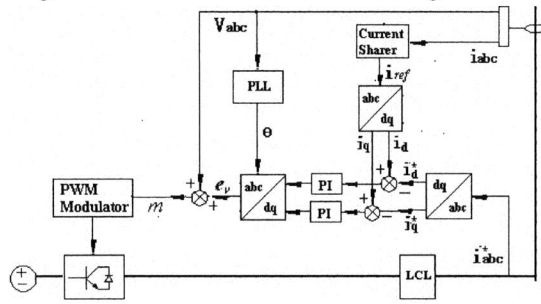

Figure 8 Block diagram of PI-dq current controller

The investigated PI-dq current controller measures the ac line current per unit value as the reference and feeds it into a current sharer, which is designed with a predetermined share factor for the slave controller. The shared current i_{abc-pu} and the output current of converter filter i^*_{abc-pu} are transformed into dq-axis quantities, i_d , i_q and i^*_d i^*_q . By calculating the error between i_d and i^*_d ; i_q and i^*_q . The errors compensated by two PI controllers and then transform them back to three phase signals e_v.

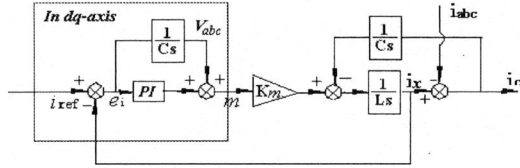

Figure 9 Block diagram of current control

The line-to-phase voltage signal V_{abc} is a feed-forward signal to force the slave converter follow the voltage of master converter. This voltage is also a phase-angle reference signal, through a phase-locked loop (PLL), generates a phase angle and send it into the dq-to-abc block to force the slave converter produce same phasor voltage to the master converter. The sum of the voltage feed-forward signal and the error signal goes through a PWM modulator block to produce the modulation index to drive the slave converter.

The closed-loop current control block diagram is shown in figure 9. The transfer function can be expressed as:

$$i_c = \frac{k_m \cdot (G_d + \frac{1}{Cs})}{LCs^2 + k_m \cdot (G_d + \frac{1}{Cs}) \cdot C \cdot s + 1} \cdot i_{ref} - \frac{LCs^2}{LCs^2 + k_m \cdot (G_d + \frac{1}{Cs}) \cdot C \cdot s + 1} \cdot i^*_{abc} \quad (12)$$

The control of three-phase sinusoidal current in dq-axis provides much better performance than the control in stationary reference frame. Because, for dc signal, the PI controllers can compensate the steady-state error much better than in ac system.

V. SIMULATION RESULTS AND ANALYSIS

The simulation studies have been performed. The simulation parameters are given in table 1.

Table 1: simulation parameters

components	Parameters	symbols	Quantity (units)
DG system	Nominal Frequency	f	$50 H_z$
	Initial Frequency	f_0	$50.1 H_z$
	Minimum Frequency	f_{min}	$49.82 H_z$
	Nominal rms phase-voltage	V	220V
	Initial rms voltage	V_0	225V
	minimum rms voltage	V_{min}	214V
Converter LC filter	inductance	L	2mH
	capacitance	C	100 μF
AC Bus	resistance	R	0.01 Ω
	inductance	L_g	2mH
Master Controller	frequency-droop characteristics	m	0.0000094
	voltage-droop characteristics	n	0.000177
	PI controller G_i	K_p	4
		K_i	0.2
	PI controller G_v	K_{p-v}	20
		K_{i-v}	0.2
Slave Controller	Current sharer factor		0.5
	d-axis controller	K_{p-d}	500
		K_{i-d}	0.01
	d-axis controller	K_{p-q}	500
		K_{i-q}	0.01
Slave controller	Low-pass filter Cut-off frequency		$200 H_z$
PLL system	Minimum frequency		$45 H_z$
	Initial frequency		$50 H_z$
	Damping factor	ζ	0.707
	Regulator gains	K_p^*	222
		K_i^*	49333
Simulation setup	Sampling time		50e-6 s
	Load disturbance time	In DG1	0.30s
		In DG2	0.75s
		common	1.00s
	Simulation time		1.2s

502

By implementing the simulation model, some results are presented and analyzed in this section.

1. System frequency

In figure 10, the frequency of two DG subsystems is shown.

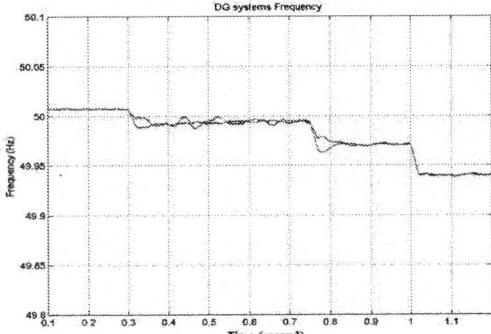

Figure 10 Frequency of two DG subsystems

The frequency feature corresponds to the load changing in different periods, in each of them; system has some extra load switching in. That causes the frequency drop. For each load varying, the two systems need a short time to regulate the generated power frequency. In figure 12, from time t =0.3s to t = 1.2 s, there are three period with different frequency features.

- Period t = 0.3~0.75 s: in this period, two DG subsystems have different active power load, that causes the two systems have different frequency drop. This frequency difference causes the generation sets in lower frequency side generate more power than the higher frequency side (this feature will be represented with the load sharing in next section). The controllers of two DG systems will control the generated power until the subsystems achieve a steady active power sharing according to the predetermined droop characteristics.

- Period t = 0.75~1.0 s: some active power load is added in the low-load side to make the two subsystems have the same load. It obviously shows in figure10, the systems reach the same frequency much fast than the system has unequally load.

- From time t = 1.0 to 1.2 s, a disturbance load is added in the common load. This time, the frequency drops of two DG subsystems are completely synchronous.

2. System voltage

This simulation work measures the voltages of the common load, the outputs of the DG subsystems and the output of converters. It is found that these voltages have different profiles as shown in figure 11.

Similar to the analysis of system frequency, the analysis of the voltages is also divided into three periods.

- Period t = 0.3~0.75 s: in this period, the varying active power flow and frequency causes the power generation variation. The DG subsystem-1 is closer to the increased load, that makes the system has a relative lower voltage. Contrarily, the shared by DG subsystem-2 increases, and the system voltage increase a little bit to send the increased power to the load. This feature could be seen in figure 14.

- Period t = 0.75~1.0 s: the two subsystems have balanced load. The subsystems reach the same frequency. The voltages of two subsystems back to the same level again.

- From time t = 1.0 to 1.2 s, some reactive power load is also added in the common load, that makes a voltage drop of the DG system.

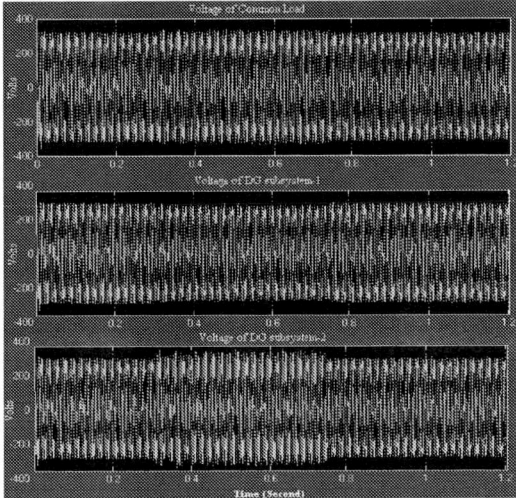

Figure 11 System voltages

3. Load sharing

The feature of the active power load sharing between two DG subsystems is quite similar to the frequency. In figure 12, the curves like mirror images. That indicates that the system frequency is varying corresponds to the load variation and frequency droop characteristics.

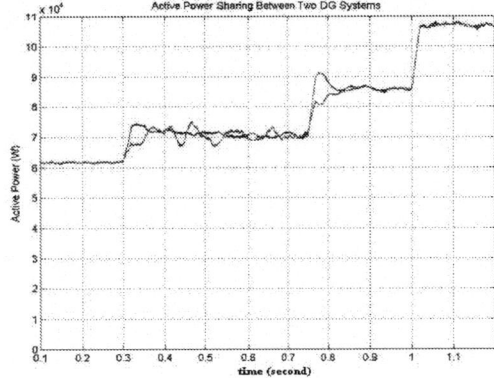

Figure 12 Real power sharing between two DG subsystems

The active power sharing in DG subsystems is also analyzed. Figure 13 shows the active power sharing between two converters in DG subsystem-1(upper) and in DG subsystem-2 (lower).

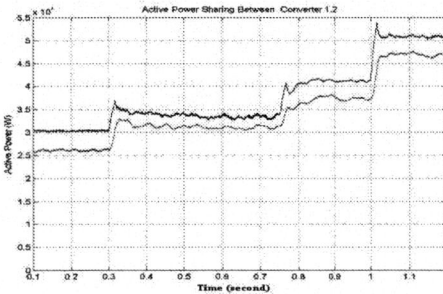

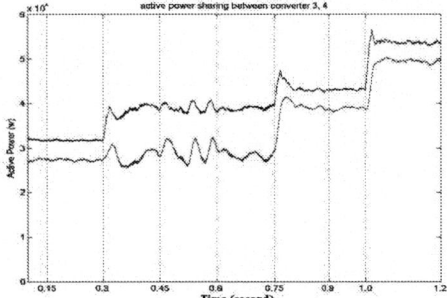

Figure 13 Active powers sharing in DG Subsystems

The reactive power sharing is shown in figure 14. Like the analysis of system frequency, the curve could be divided into four periods.

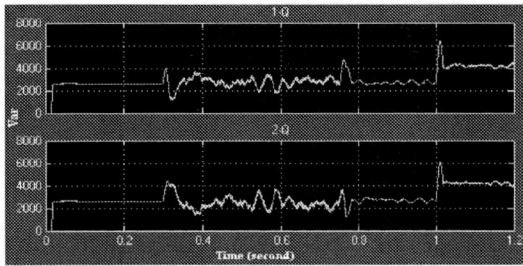

Figure 14 Reactive power sharing of two DG systems

- Period 1 (t = 0.1~0.3 s): in this period the reactive power load is equally shared by two DG subsystems.

- Period 2 (t = 0.3~0.75 s): in this period, the frequencies of two DG subsystems are varying. This frequency varying causes the reactive power generated in two systems are different. This reactive power difference will exist until the frequencies of two DG subsystem are regulated to a same value.

- Period 3 (t = 0.75~1.0 s): an overshoot appears in the curves and then goes much more 'flat' than the second period, means the variation of reactive power is smaller.

- From time t = 1.0 to 1.2 s, some reactive power load is added in the common load, the curves in figure 14

show that the reactive power is almost equally shared by two systems.

By analyzing the simulation results, it could be concluded that

- It is feasible that by controlling the converters of a stand-alone DG system to achieve the load sharing and generates good quality power.

- The designed controllers are suitable to the proposed DG system.

- The unequally loads of the DG subsystems will cause frequency variation and voltage variation. The system needs a short time to reach the steady frequency.

- The disturbance load may cause different response from generation units at different locations.

VI. CONCLUSION

The proposed control method shows that it is possible to form a DG system with several parallel operation subsystems without any communication between them. And it is feasible that by controlling the power electronics interfaced converters to regulate the system voltage and frequency. These features make it possible to construct a new generation network by using renewable energy resources based generators.

REFERENCES

[1] N. Jenkins, R. Allan, P.Crossley, D. Kirschen and G. Strbac *Embedded Generation*. IEE Power and Energy Series, London, 2000 ISBN: 0 85296774 8.

[2] R. Wu; S. B. Sewan; G.R. Slemon, *Analysis of an ac-to-dc voltage source converter using PWM with phase and amplitude control*. IEEE Trans. On Industry Application. Volumn 27, No.3, March/April 1991. page(s): 355-364.

[3] Liserre, M; *Blaabjerg*, F; Hansen, S. *Design and control of an LCL-filter based three phase active rectifier*. Industry Application Conference, 2001. Thirty-Sixth IAS Annual Meeting.

[4] Liserre, M; Teodorescu, R; Blaabjerg, F. *Stability of grid-connected PV inverters with large grid impedance variation*. Power Electronics Specialists Conference, 2004. IEEE 35th Annual Volume 6, 20-25 June 2004 Page(s): 4773-4779

[5] Teodorescu, R; Blaabjerg, F; Liserre, M; Dell'Aquila, A. *A stable three-phase LCL-filter based active rectifier without damping*. Industry Applications Conference, 2003. Thirty-Eighth IAS Annual Meeting PESC 04; Conference Record of the Volume 3, 12-16 Oct. 2003 Page(s): 1552-1557

[6] Hadi Saadat, *Power System Analysis*, McGraw-Hill, 1999 ISBN: 0-07-012235-0

[7] Mukul C. Chandorkar; Deepakraj M. Divan; Rambabu Adapa. *Control of Parallel Connected Inverters in Standalone ac Supply Systems*. IEEE Trans. On Industry Application. Volumn 29, No.1, Jan./Feb. 1993. page(s): 136-143.;.

Author Index

A

Abbasian, M.A. 1043
Abdelhamid, T. H. 411
Abedini, A. 224
Abjadi, N. R. 1917
Abo-Khalil, Ahmed G. 1477
Abramovitz, A. 1412
Agarwal, Anant K. 157
Agarwal, Vivek 281
Ahmed, Nabil A. 242
Ahn, C. H. 1198
Aide, Xu 1162
Ai-Juan, Jin 1421, 1426
Ait-Amirat, Y. 1882
Ajjarapu, Venkataramana 505
Akagi, Hirofumi 23
Akimasa, Koji 1613
Andersen, Henrik Rosendal 1032
Ando, Tatsuo 947
Arpilliere, M. 249
Ashida, M. 2000
Askari, J. 1917

B

Badica, M. 1751
Bai, Haijun 219, 826
Bai, Zhifeng 1581
Baihua, ... 161
Balda, J.C. 1353
Banaei, M. R. 759, 764
Bao, G.Q. 813
Baocheng, Wang 569, 1991
Baoming, Ge 918
Barsoum, ... 1148
Bendjedia, M. 1882
Berthon, A. 515, 1882, 2005
Bhattacharya, Subhashish 1450
Bin, Su 406
bin, Wu 1368
Binder, A. 842
Bing, Chen 1401
Bisogno, F.E. 1117
Blaaberg, F. 46, 1107, 2029
Bo, Chen 122
Böcker, Joachim 1112
Bodson, M. 1912
Bojoi, R. 1651
Boroyevich, D. 249
Boroyevich, Dushan 1836
Bréhaut, Stéphane 92
Brouji, H. El 1663

C

C, Sreekumar 281
Cailin, Wang 1167

Calderon-Lopez, G. 1328
Calverley, S.D. 977
Camara, M.B. 515
Câmpeanu, A. 1751
Cao, Binggang. 1581
Cao, R. X. 510
Cao, Yanjie 2015
Carazo, A. V. 1117
Cartes, David 774
Cen, Yuwan 1986
Chan, C.C. 57
Chang, Chung-Hsing 291
Chang, Duan Qi 1551
Chang, H.-H. 1343
Chang, Jie (Jay) 102
Chang, Lon-Kou 417
Chang, Yuan 1722
Changhong, Wang 1793
Changzheng, Zhang 489, 739
Chau, K. T. 1788
Chen, Bin 1450
Chen, C.-C. 117
Chen, Cheng-Hu 913
Chen, Chern-Lin 332
Chen, Chien-An 967
Chen, Guiyou 1202
Chen, Guocheng 1560
Chen, Guozhu 794
Chen, H. 933
Chen, H. G. 1129
Chen, J. 194
Chen, Jian 1218
Chen, Jiann-Fuh 361, 1178
Chen, Jiaxin 346, 831
Chen, Jie 113
Chen, Jun-Ning 199, 286, 1283, 1392
Chen, Min 442
Chen, Qiaoliang 433, 642
Chen, Rui 1171
Chen, Ruijuan 1253, 1877
Chen, Tso-Min 332
Chen, Wei 171, 1081
Chen, Xi 1507
Chen, Xiangjun 607
Chen, XuWu 438
Chen, Y.-M. 108, 117
Chen, Yao 1454
Chen, Yen-Ming 1763
Chen, Yuan-rui 1397
Chen, Yunpeng 236
Chen, Z. 49, 499, 1773, 2029
Chen, Zongxiang 142
Chenchen 386
Cheng, Chun-An 1178
Cheng, Ming 1746, 1815
Cheng, Ming-Yang 913

A-1

Author Index

Cheng-ning, Zhang 1027
Chengsheng, Wang 589
Cherifi, A. 574
Chi, Song 890, 1825
Chiang, Huann-Keng 967
Chiasson, J. N. 1703, 1912
Chiu, Huang-Jen 291
Cho, Yun-hyun 1238, 1784
Choi, E.S. 1382
Chongjian, Li 589
Chun, Dong 1623
Chun, YonDo 1784
Chung, Jung Kee 1736
Chunjiang, Zhang 554, 559, 1473, 1618
Corzine, Keith A. 637
Costa, François 92
Crausaz, A. 2005
Cui, Bo 798
Cui, Jiefan 657
Cui, Junwei 1436
Cvetkovski, G. 254

D

Dai, Ke 789
Dai, Renchang 1122
Dai, Yue-Hua 199
Da-ming, Liu 1674
Danhe, Li 1991
De Doncker, R. W. 31
Deng, Jianming 923
Deng, Yan 1931
Dianguo, Xu 301, 1713
Ding, Xiaoyu 1560
Ding, Ye 1223
Divan, D. 16
Divan, Deepak M. 2010
Doi, Toshimitsu 356, 1302, 1307
Dong, Jiang 880
Donghua, Luo 1634
Dongsheng, Zuo 468
Dong-Shoutian, ... 484
DongYu, ... 1623
Dou, Sen 537
Du, Guiping 316
Du, Zhong 1450
Duan, Baoxing 70
Duan, Huijuan 798
Duan, Shan Xu 1522
Duan, Shanxu 1218
Duarte, Jorge L. 784

E

Ebrahimi, Yousef 779
Eiuo, Bin 1358
El Din, Ashraf Salah El Din Zein 847
Elbanhawy, Alan 342. 1967

Endo, Tsunehiro 947
Ertugrul, Nesimi 147, 962

F

Fa, Naiguang 1253, 1877
Fang, Liang 817
Fang, Xin 789
Fang, Xu-Peng 166
Fang, Yu 1406
Fang, Zhuo 122, 1542
Fathy, Khairy 356, 1302, 1307, 1358, 1363
Fei, Wanmin 1138
Fei-peng, Xu 647
Feng, D. 499
Feng, Huang 1401
Feng, L. 842
Feng, Zhao 622, 679
Feng, Zheng 585
Fengxiang, Wang 449, 903
Feyzi, M. Reza 204, 1228
Forrest, S. J. 977
Forsyth, A. J. 1323, 1328
Francis, Jerry 1836
Friedrichs, Peter 132
Fröhleke, Norbert 1112
Fuchs, F.-W. 325
Fujita, Kouetsu 1971
Fukuda, S. 1468
Fukushima, Kentaro 1333
Funian, Hua 449
Furuya, Atsushi 1598
Fu-sheng, Wang 463
Futami, M. 1468

G

Gallay, R. 2005
Gang, Ma 378
Gao, F. 1107
Gao, Yan 157
Gao, Yang 113, 1159
Gao, Yong 1198
Gao, Z. Y. 1071
Garinto, Dodi 306
Ge, L.S. 1458
Ge, Lu-sheng 1368, 1576
Ge, Qiongxuan 1171
Geng, Pan 789
Goharrizi, A. Yazdanpanah 1697
Gong, Yu 1223, 1788
Grabner, C. 999
Graczkowski, J. J. 1096
Grantham, Colin 1207, 1858
Gruenberger, Hans Pert 219, 826
Grundmann, Frank 870, 1442
Gu, G. 175
Gu, Herong 473, 1585

Author Index

Gu, Yilei 171, 276
Gualous, H. 515, 2005
Guan, Xiaohan 688
Guang, Zeng 734, 1669
Guangzheng, NI 1091
Guenther, D. 842
Guilan, Chen 554, 1006, 1049
Gui-xin, Shao 1027
Guiyou, Chen 1630, 1634
Guo, Hongche 612, 853
Guo, Qingding 612, 853, 896
Guo, Wei 952
Guo, Xin 337
Guo, Youguang 346, 831
Guobiao, Gu 808
Guocheng, San 1473
Guojun, Lu 802
Guoxin, Zhu 1802
Gustin, F. 515

H

Habetler, Thomas G. 836
Haibing, Hu 937
Haijie, Xu 937
Haiping, Xu 684, 1298
Haitao, Zhang 161
Halász, S. 693
Han, B.D. 1143
Han, Chong 1450
Han, Chong Zhao 652
Han, F. T. 1071
Han, S.K. 1382
Hang-Tian, Li 1421, 1426
Harley, Ronald G. 836, 2010
Hartavi, A.E. 2018
Hashimoto, Takayoshi 1333
He, Guofeng 657
He, Junping 1081
He, Shijie 1463
He, Xiangning 83, 1931
He, Zhongyi 1537
Hemin, Wang 1849
Heming, Li 458
Hendrix, Marcel A. M. 784
Herong, Gu 559, 1618
Hirao, Mitsuhiro 1768
Ho, Chien-Yeh 1527, 1995
Ho, S. L. 1901
Hong, Peng 899
Hong, Shen 559, 1273, 1618
Hong-mei, LI 463
Hongren, Yin 401
Hori, Yoichi 1797
Hosseini, S. H. 759, 764, 1697
Hosseini, Seyyed Hossein 753, 779, 1679
Howe, D. 908. 928, 1841

Hsu, Kai-Sheng 967
Hsu, Ken-Chuan 1957
Hsu, W.P. 718
Hu, D.Q. 1143
Hu, Haibing 127, 1183
Hu, Jiangang 703
Hu, Qing 1806
Hu, Qingbo 526
Hu, Songqin 351
Hu, Weihao 321, 585
Hu, Wenhua 397
Hu, Xuezhi 438
Hu, Y. 2029
Hu, Z. L. 1708
Hu, Zongbo 316
Hua, Li 729, 1571
Hua, Wei 1746, 1758
Huade, Li 401
Huang, Zhenyue 1213
Huang, Alex Q. 113, 157, 1159, 1450
Huang, Chien-Lan 748, 1278
Huang, Congsheng 1288
Huang, Jin 1017
Huang, Xuwen 1288
Huang, Yafeng 542
Huang, Yi 1076
Huang, Yuehui 580, 1512
Hui, Li 729
Hui, Wu 862
Hui, Zhang 1032, 1492
Hui-jie, Xiang 1942

I

Ichinose, M. 1468
Inoue, Kaoru 1233, 1613
Inoue, Shigenori 23
Iov, F. 46
Iwanski, G. 494

J

Jang, Jeong-Ik 1482
Jangwanitlert, A. 1353
Járdán, R.K. 1338
Jeon, K. S. 1198
Jewell, G. W. 977
Ji, Yanchao 627, 1507
Jia, C. 1323
Jia, Y. 2000
Jia, Y.P. 1143
Jiag, Maoh-Chin 1527
Jian, Chen 74, 769
Jian, Cui 1431
Jian, Liu 272
Jian, Wu 1713
Jiang , Chang 1213
Jiang-Hui, Chen 1401

A-3

Author Index

Jiang, J. Z. 1788
Jiang, J.G. 1458, 1952
Jiang, J.J. 152
Jiang, J.Z. 813
Jiang, Jianguo 1081
Jiang, Jianzhong 1223
Jiang, Xianglong 1608
Jiang, Xiaochun 1896
Jianguo, Jiang 468
Jianlin, Zhu 1557
Jianru, Wan 1273
Jian-Ru, Wan 1431
Jian-wen, Zhang 1657
Jianze, Wang 1693
Jiarong, Kan 1532
JiaYi, Yuan 808
Jie, Shuo 1806
Jie, Wang 569
Jiefan, Cui 862, 1849
Jin, Jianxun 831
Jin, Mengjia 885, 1872
Jin, Shun 617
Jin, Tianjun 1183
Jin, Wenxi 127
Jin, Xin Min 1454
Jing, Liu 88
Jin-gang, Li 549
Jing-Gang, Zhang 1669
Jinjun, Liu 1061, 1492, 1722
Jinlong, Zhang 401
Jinupun, P. 1887
Jiqiang, Wang 903
Jiuhe, Wang 401
Jo, WonYoung 1784
Johal, H. 16
Johnson, C. M. 977
Joseph, Alan 1076
Jou, H.L 718
Jun, Liu 1012
Jun, Wang 899
Jung, Kun-seok 1238
Junjuan, Sun Xiaofeng Wu 569
Junmin, Zhang 1684
Junzhu, Wan 1849
Jwo, Ko-Wen 1590

K

Kaijie, Feng 1741
Kaipei, Liu 209, 1684
Kang, B.W. 1143
Kang, Ju-Sung 1358
Kang, Y. 194
Kang, Yong 97, 564, 789, 1218, 1522, 1981
Karimi, E. 1697
Kato, Tomohiko 1971
Kato, Toshiji 1233, 1613

Ke, Dao-Ming 199
Ke, Fu-Jing 1154
Ke, Yi-Jing 1392
Kerkman, Russel J. 1054
Kesong, Ye 699, 1832
Khaehintung, Noppadol 137, 368
Khajee, M. Darkalee 759
Khan, Mahamnad Mansoor 386
Kim, E. D. 1198
Kim, Jang-Hwan 662
Kim, Joo Han 1736
Kim, Young-Sin 1482
Kimura, Noriyuki 1768
Kiranon, Wiwat 137
Kita, H. 1468
Koczara, W. 494
Konghirun, M. 972
Koo, DaeHyun 1784
Kou, X. 1096
Krishnaswami, Sumi 157
Ku, Chung-Ping 1590
Kumar, Pavan 537
Kun, Li 1447
Kunakorn, Anantawat 368
Kuo, J. -S. 1343

L

L., M. 1148
Lai, Ching-Ming 1590
Lai, Stephen L. 1502
Lai, Y. M. 296
Lang, Yongqiang 708
Lee, Chi-Yang 1192
Lee, Dong-Choon 1477, 1482
Lee, Fred C. 1
Lee, Hyun Woo 392, 1302, 1307, 1358, 1363, 1372, 1377
Lee, Se-Hyun 1477
Lemberg, Nicholas 989
Li, Chongjian 995
Li, Dong 1202
Li, Dongsheng 947
Li, F. 152
Li, H. 1773
Li, Han 1006, 1049
Li, Hongtao 688, 1248
Li, M. 1703, 1912
Li, Ma 88, 209
Li, Mingzhu 1537
Li, Min-zu 97
Li, Qi 79
Li, Qunzhan 423
Li, Rongyuan 1112
Li, Shijie 1171
Li, Tianbo 1947
Li, Wen 1797

Author Index

Li, Wenguang 1815
Li, Xia 1674
Li, Y.W. 1101
Li, Yaohua 995
Li, Yong 428
Li, Yongbin 942
Li, Yongdong 1892
Li, Zhanlong 1416
Li, Zhaoji 70, 79
Li, Zheng-Ping 286
Li, Zhou 1630, 1634
Liang, L. 1129
Liang, Tsorng-Juu 361, 1178
Liang, Zhonghua 607
Liao, Changming 102
Li-Jiahui, ... 484
Lijie, Chen 1595
Li-jun, Hang 406
Lijun, Zhao 862, 1849
Lili, Jiang 1849
Liming, Liu 74
Lin, Bor-Ren 748, 967, 1278
Lin, Chang-Hua 1957
Lin, Fei 184, 1976
Lin, Liangrui 1243
Lin, Li-Wei 291
Lin, Ray-Lee 361
Lin, Ruan 808
Lin, Ruiguang 885, 1872
Lin, W.-C. 108
Lin, Yang-Sheng 1178
Lin, Ying-De 1154
Lin, Yu-Tzung 1192
Ling, Xia 899
Lingjie, Meng 674
Lipo, T.A. 989
Liqiang, Yuan 161
Liu, Cheng-Tsung 1763
Liu, Ching-Hsiung 361
Liu, Guiqiu 896
Liu, Hongwei 798
Liu, Hsing-Fu 417
Liu, Jian 1267
Liu, Jianqiang 184
Liu, Jingbo 703
Liu, Jinjun 713, 1726
Liu, K. 1071
Liu, Kaipei 453
Liu, Shu-Lin 1267
Liu, Tien-Shuo 1957
Liu, Wei-Shih 361, 1178
Liu, Wenhua 542
Liu, Wenji 1248
Liu, Xiang 229
Liu, Xiaodong 351
Liu, Xinhua 1223

Liu, Yuanchao 236, 1248
Liu, Zhengang 688
Liu-Xueli, ... 484
Liwei, Zhang 1012
Loh, P. C. 1107
Lorenz, L. 39
Lou, Z. L. 373
Lu, Bin 836
Lu, Bing 1
Lu, Cheng 489, 739
Lu, Haihui 1054
Lu, P.-C. 117
Lu, Shuai 637
Lu, Xiaodong 83
Lu, Zhengyu 127, 171, 276, 526, 1183
Luk, P.C.K 478, 1872, 1887
Luo, Fang 789, 1522
Lyons, James 1122

M

Ma, Hao 1312, 1637
Ma, Hongfei 708
Ma, Wenchuan 627
Ma, Xiangfei 1836
Ma, Xuejun 438, 1288
Maeda, Toshihiro 1971
Manmek, Thip 1207
Mansouri, O. 574
Mao, Hong 1267
Mathew, Anu 442
Matsumoto, Shuji 1598
Matsuse, Kouki 1598
Mayor, J. Rhett 2010
Meghriche, K. 574
Member, Student 1825
Meng, Zheng 236
Mi, Chris 942
Miao, Guan 744
Miao, Zhao 549
Miller, Nicholas 1122
Ming, Cheng 1758
Ming, Zhou 1431
Ming, Zong 449, 903
Ming-fu, Zhao 1623
Mingli, Ding 1793
Min-qian, Ke 734
Miyatake, Masafumi 242
Moghbelli, H. 597
Mohr, M. 325
Molinas, Marta 63
Moon, G.W. 1382
Morimoto, Keiki 356, 1302, 1307
Morizane, Toshimitsu 1768
Mou, Shann-Chyi 291
Mu, Gang 542
Mudannayake, Chathura P. 1207

Author Index

N

N., N. 1148
Na, He 1713
Nagy, I. 1338
Naidu, S. R. 1731
Nakaoka, Mutsuo 356, 392, ... 1302,
1307, 1358, 1363, 1372, 1377
Nakayama, Y. 1468
Nan, C. H. 1708
Nan, Liu 214, 1942
Nan, Zhao 918
Nasiri, A. 224
Neff, K. L. 1096
Niasar, A. Halvaei 597
Ning, Gaidi 1463
Ninomiya, Tamotsu 1333
Nishimae, Kazuya 1233
Nittayarumphong, S. 1117
Niu, Shuangxia 1788
Nolle, Eugen 219, 826
Nondahl, Thomas A. 1054
Notohara, Yasuo 947
Nozawa, Yusuke 1598
Nuttall, D. R. 1328

O

Ogiwara, Hiroyuki 1307, 1358
Ohara, S. 1468
Oka, Kazuo 1598
Okude, Takaaki 1363
Oleschuk, V. 1651
Omata, Ryuji 1598
Omori, Hideki 392, 1358, 1363, 1372, 1377
Ou, Chung-Lun 1957
Ouyang, Wen 989

P

Pan, Junmin 142, 267, 1348
Pan, Ming-Ho 1590
Pan, Sanbo 1348
Pang, Da-Chen 1763
Paponpen, K. 972
Park, J. D. 1198
Paweletz, A. 842
Payam, A. Farrokh 1906
Pedersen, John K. 1773
Pei, Yunqing 321
Peng, Fang Z. 1076
Pengcheng, Zhu 74
Petchjatuporn, Panom 137
Petkovska, L. 254
Piwko, Richard 1122
Poon, N. K. 1502
Poure, P. 1663
Prado, R. N. do 1117

P (right column continued)

Pratt, Annabelle 537
Profumo, F. 1651

Q

Qi, Feng 1637
Qi, Wang 1793
Qian, Lewei 774
Qian, Zhaoming 127, 171, 276, 1076, 1183
Qiang, Li 1853
Qiang, Mei 1926
Qiao, Ermin 337
Qiao, Wei 836
Qiaofu, Chen 489, 739
Qi-gang, Fu 734
Qing, Sun 862
Qingding, Guo 1802, 1846
Qingdong, Zhou 1793
Qingfan, Zhang 1634
Qinglin, Zhao 532, 1387, 1517
Qingyu, Yang 699, 1832
Qinmu, Wu 1820
Qiu, Dongyuan 316, 1293
Qiu, Jianqi 885, 1872
Qiu, Zhiling 794
Qizhi, Zhan 1542

R

Radecker, M. 1117
Rafik, F. 2005
Ragon, S. 249
Rahman, M. F. 983, 1646, 1867, 2023
Rahman, M. Faz 1858
Rahnavard, Reza 779
Rajagopalan, Satish 2010
Ren, Hai Peng 652
Ren, Shi 699, 1832
Rentschler, A. 842
Rhyu, Se Hyun 1736
Rosado, Sebastian 1836
Rosario, L.C. 478
Ruan, L. 175
Ruan, Xinbo 1936
Ruixia, Wang 1853
Ruliang, Zhang 1167
Ruxi, Wang 1061

S

Saadate, S. 1663
Sabahi, Mehran 753, 1228, 1679
Sabzali, A. 411
Saha, Bishwajit 392, 1372
Sahinkaya, M.N. 2018
Sakamoto, Kiyoshi 947
Sanchez-Gasca, Juan 1122
Scozzie, Charles 157
Segawa, Takeshi 1333

Author Index

Shancheng, Xing 1023
Shanxu, Duan 769
Shao, Changhong 607
Shao-De, Zhang 1447
Shaojun, Xie 1532
Shao-Long, Li 1421, 1426
Sharifian, M. B. B. 204
Shen, Guoqiao 1566
Shen, Hong 1603
Shen, J.X. 908, 928
Shen, Miaosen 1076
Shen, W. 249
Sheng, K. 1188
Sheng, Weihui 995
Shergin, V. V. 1133
Shi, Cenwei 885, 1872
Shi, Y. F. 1841
Shiang, J. -Z. 1086
Shibata, R. 2000
Shi-feng, Zhang 1447
Shiri, A. 821
Shoulaie, A. 821
Shu, Mantang 1560
Shu, Zhibing 1811
Shun, Jin ....
Shutong, Qiao 468
Shyu, Kuo-Kai 1590
Sibo, Ge 699, 1832
Sirisuk, Phaophak 137, 368
Skorokhod, Y. Y. 1133
Sneineh, Anees Abu 1318
Soltani, J. 1038, 1043, 1906, 1917
Song, Wenchao 1450
Song, Wenxiang 1560
Songboonkaew, J. 1353
Songhua, Shen 744
Soong, Wen Liang 962
Souza, E. V. N. 1731
Stankovic, A.M. 1651
Stefanovic, V. 249
Su, Hongsheng 423
Sugimoto, Hidehiko 1258
Sugimura, Hisayuki 392, 1372, 1377
Sul, Seung-Ki 662
Sun, Chin 417
Sun, Jia-E 199
Sun, Jian 442
Sun, Sizhou 351
Sun, Wei 866
Sun, Xiaofeng 674
Sun, Yuxin 179
Sunat, Khamron 137
Sung, Ha Kyeong 1736
Suul, Jon Are 63
Suzuki, Takahiro 947

T

Tai, Wei-Chih 913
Takahashi, Toshio 428
Tan, Guang-Hui 627, 1507
Tan, Ruimin 1032
Tan, Siew-Chong 296
Tanaka, Chikara 947
Tang, Yan 866
Tang, Yupeng 1416
Tang, Yu-peng 669
Taniguchi, Katsunori 1768
Tao, Haimin 784
Tao, Liu 122
Tenconi, A. 1651
Teodorescu, R. 46
Tezcan, Ibrahim 1546
Thammasiriroj, W. 1353
Tian, Kai 1318
Toba, Akio 1971
Tolbert, L. M. 1703, 1912
Tongjing, Sun 1630
Tsai, C.-T. 108
Tsai, Ming-Fa 1154
Tsay, Shuh-Chuan 1278
Tse, Chi K. 296, 580, 1512
Tseng, S. -Y. 1086, 1343
Tseng, S.-H. 1086
Tuncay, R.N. 2018
Tzou, Ying-Yu 1192

U

Undeland, Tore 63

V

Vahedi, A. 597, 821
van der Broeck, Heinz 1546
Vansencc, Flalph 875
Varjasi, I. 693
Venkataramanan, Giri 259
Volskiy, S. I. 1133

W

Walther, B. 1882
Wan, Deyu 1585
Wan, Shuyun 1608
Wang, Bin 674
Wang, Bingsen 259
Wang, Changkun 1258
Wang, Chengxue 2015
Wang, Chien-Ming 1527, 1995
Wang, Deyu 473
Wang, F. 249
Wang, Fred 1836
Wang, Gang 1986
Wang, Hua 1507

Author Index

Wang, J. 977
Wang, J.K. 813
Wang, Jianhui 1436
Wang, Jian-quan 229
Wang, Juan 798
Wang, Jui-Kum 1154
Wang, Li-Li 1283
Wang, Linbing 311
Wang, Liqiao 632, 724
Wang, Ming-Yan 1318
Wang, Pei-zhen 1576
Wang, Qingyi 1608
Wang, Qun-jing 857
Wang, Shuo 1
Wang, Xiaofeng 1931
Wang, Xiaoyu 713, 1726
Wang, Y.F. 1458
Wang, Yaonan 866
Wang, Yue 1463
Wang, Yunfei 1896
Wang, Z. A. 1708
Wang, Z. S. 373, 1901
Wang, Zhaoan 321, 433, 642, 713, 1463, 1726
Wang, Zhixin 520, 1487
Watkins, S. J. 1551
Wei, Dong 189
Wei, Guo 880
Wei, Liu 1473
Wei, Shi 1061
Wei, Wen 684, 1298
Wei, Xueliang 789, 1522
Weibin, Cheng 602
Weiguo, Liu 679
Wei-ping, Zhou 1674
Weiyang, Wu 189, 532, 554, 559, 569,
1273, 1387, 1473, 1517, 1618, 1991
Wei-Yang, Wu 1926
Wen , Z. 175
Wen, H.-T. 1343
Wen, Xuhui 337, 622
Wenjuan, Dong 1542
Wenlang, Deng 1557
Wenlong, Qu 378
Wenqing, Shi 684, 1298
Wenxi, Yao 937
Wetzel, Hermann 1112
Wiseman, J. 1101
Wu, B. 1101
Wu, Bin 397
Wu, C.-Y. 117
Wu, Chih-Yu 417
Wu, Hongxia 438, 1288
Wu, J.C 718
Wu, Jiaju 1258
Wu, Jiande 83
Wu, Li 1487

Wu, Li 520
Wu, Q. P. 1071
Wu, Shanshan 1892
Wu, T.-F. 108, 117
Wu, Tao 1936
Wu, Wei-yang 1603
Wu, Weiyang 473, 632, 674, 724, 1585
Wu, Wilson 537
Wu, Yong 1608

X

Xi, Zhai 122
Xia, Kun 857
Xiangjun, Zhang 301
Xiangrong, Li 301
Xiangyun, Fu 1693
Xianmin, Ma 1922
Xianmin, Mu 1693
Xiao, D. 983, 1867, 2023
Xiao, G. C. 1708
Xiao, Lei 209
Xiao, Wenxun 1293
Xiao, Zheng 1447
Xiaobo, Yang 1273
Xiaofeng, Sun 189, 1991
Xiaofeng, Zhang 958, 1626
Xiaohuan, Wang 1273
Xiaojie, Wu 468
Xiao-ping, Yang 1718
Xiaoqiang, Guo 532
Xiaotan, Zhao 589
Xiaoxia, Wei 1693
Xiaoyi, Jin 189
Xiaoyu, Wang 1722
Xie, Jian 870, 1442
Xie, S.S. 1143
Xie, Yong 1406
Ximei, Zhao 1802, 1846
Xindong, Tian 808
Xing, Yan 1406, 1537
Xinming, Huang 1492
Xinxin, Wang 1162
Xu, Cai 1657
Xu, D. 1101
Xu, Dehong 1566
Xu, Dianguo 708
Xu, Jianping 1066
Xu, Jia-peng 669
Xu, Jinbang 1608
Xu, Longya 703, 890, 1779
Xu, Ming 1
Xu, Wancai 627
Xu, Y. N. 194
Xu, Yanping 1863
Xu, Longya 1825
Xuan-fang, Yang 1674

A-8

Author Index

Xue, H. 1952
Xue, Shan 622
Xuhui, Wen 684, 1006, 1012, 1049, 1298
Xun, Li 769

Y

Yabin, LI 458
Yamamura, N. 2000
Yan, Caizhong 1811
Yan, Chen 1623
Yan, Gangui 542
Yan, Wang 1718
Yanchao, Ji 1693
Yanfeng, Wu 729
Yang, Bo 83, 311
Yang, Chun-Sheng 1278
Yang, Geng 1896
Yang, Hui 1863
Yang, J.J. 718
Yang, Jia-qiang 1017
Yang, Junyou 657, 1253, 1877
Yang, R. 152
Yang, S. Y. 510
Yang, Sheng 505
Yang, X.J. 1458
Yang, Xiao-bo 1603
Yang, Xi-jun 229
Yang, Xing-hua 229
Yang, Xu 433, 642
Yang, Zhaoning 1450
Yang, Zhongping 184
Yang, Zilong 473
Yanhui, He 1061
Yanliang, Xu 1741
Yan-min, Su 734, 1669
Yan-ru, Zhong 382, 549, 602, 1718
Yansong, Hou 1571
Yao, Duan 880
Yao, Lei 1463
Yao, Ruoping 989
Yao, Tianjun 127
Yao, Yue-feng 1397
Yaogang, ...386
Yaohua, Li 589
Yao-Xin, ...484
Yazdanpanah, R. 1038
Ye, Min 1581
Ye, Pengsheng 142
Ye, Peng-sheng 229
Yeic, Zhuliang 875
Yesong, Li 1820
Yeung, Heidi H.T. 1502
Yi, Qin 1820
Yi, Wen 1387
Yi, Zhang 862
Yidan, Sun 449, 903

YII, ...1148
Yin, Qiang 1054
Ying, Jiang 1942
Ying, Li 147
Yinhai, Fan 1162
Yong, Gao 88, 1167, 1595
Yong, Kang 74, 769
Yong, Wang 744
Yongchang, Zhang 161
Yonglong, Peng 458
Yoon, H.K. 1382
You, Keping 1646
You, Xiaojie 1976
Youbin, Zhao 739
Yougui, Guo 1557
Youn, M.J. 1382
Yu, Dongmei 1806
Yu, Hongxiang 627
Yu, L.C. 1188
Yu, Y. H. 1129
Yu, Zhang 489
Yuan, Chang 713, 1726
Yuan, Xiaoming 593, 1122, 1566
Yuan, Yang 1595
Yuan, Zhou 647
Yuanbin, Wang 272
Yuanfang, Wen 802
Yuanyuan, Liu 378, 1162
Yuda, Chen 739
Yue, Wang 1061
Yue-feng, Yang 406
Yu-fan, Xi 1669
Yu-gang, Yang 1942
Yugang, Yang 214
Yun-qing, Pei 585
Yun-Xiang, Xie 1401

Z

Zargari, N. 1101
Zeng, Fanpeng 1507
Zeng, Guohong 1689
Zhai, Xiaohua 866
Zhang , B. 152
Zhang, Bo 70, 316, 505, 1293
Zhang, C. L. 1198
Zhang, C. W. 510
Zhang, D. 813
Zhang, Dong 1788
Zhang, Dongyan 236, 1243
Zhang, Fengge 219, 826
Zhang, Hairong 1811
Zhang, Handong 1986
Zhang, Hongyan 794
Zhang, Hui 453
Zhang, Jia 1689
Zhang, Jianzhong 1746

A-9

Author Index

Zhang, Jun 1858
Zhang, Kai 564, 1981
Zhang, L. 1551
Zhang, Luan-guo 229
Zhang, Qian 1976
Zhang, Qiang 774
Zhang, Shifu 219, 826
Zhang, Tengchao 179
Zhang, Weiping 236, 688, 1243, 1248
Zhang, X. 510
Zhang, Xi 267
Zhang, Xianmiao 1183
Zhang, Xiaoqiang 1248
Zhang, Xueguang 708
Zhang, Yanli 1138
Zhang, Yingchao 952
Zhang, Yingqi 593
Zhang, Yonggao 564, 1981
Zhang, Yongping 316
Zhang, Yu 1218
Zhang, Yuan 1779
Zhao, Wenxiang 1746, 1815
Zhao, Xiaotan 995
Zhao, Xusen 688, 1243
Zhao, Y. 194
Zhao, Zhengming 952
Zhaoan, Wang 122, 1061, 1542, 1722
Zhaomin, Fanyinhai 1962
Zhao-ming, Qian 406
Zhao-yong, Zhou 647
Zhao-Yulin, ... 484
Zhe, Chen 802, 1387
Zhe, Zhang 559, 1618
Zheng- guo, Wu 1674
Zheng, Shi-cheng 1576
Zheng, Trillion Q. 184, 1012, 1976
Zheng, Zedong 1892
Zhengfeng, Ming 382, 1091
Zhengguo, Wu 729, 1023
Zhengming, Zhao 161, 880
Zheng-Na, ... 484
Zhengyu, Lu 406, 937, 958, 1626
Zhen-lin, Xu 1926
Zhi, Na 1198
Zhili, Tan 769
Zhi-Qiang, Wei 1431
Zhi-yuan, Zhang 1623
Zhong, Yanru 1863
Zhong, Yan-ru 617
Zhongmin, Wang 1630
Zhongnan, Guo 554
Zhongying, Chen 1517
Zhou, Qian-zhi 1368
Zhou, Qianzhi 397
Zhou, Tao 1066
Zhou, Wenqi 1312, 1637

Zhou, Y. M. 1129
Zhou, Yang 923, 1947
Zhou, Yu-Fei 286, 1283, 1392
Zhou, Yunbin 564, 1981
Zhu, Guo-rong 97
Zhu, Huangqiu 179, 817, 923, 1213, 1947
Zhu, Jianguo 346, 831
Zhu, Jingwei 962
Zhu, Xiaoyong 1746, 1815
Zhu, Yuran 798
Zhu, Yu-wu 1238
Zhu, Z.Q. 908, 928, 1841
Zou, X. D. 194

A-10

9781424404483